BOTANY

Plant Form and Function

BOTANY

Plant Form and Function

Randy Moore
University of Akron

W. Dennis Clark
Arizona State University

Wm. C. Brown Publishers

Dubuque, IA   Bogota   Boston   Buenos Aires   Caracas   Chicago
Guilford, CT   London   Madrid   Mexico City   Sydney   Toronto

Book Team

Editor *Margaret J. Kemp*
Production Editor *Kay J. Brimeyer*
Designer *Christopher E. Reese*
Design Assistant *Kathleen F. Theis*
Art Editor *Rachael Imsland*
Photo Editor *Carrie Burger*
Permissions Coordinator *Karen L. Storlie*

Wm. C. Brown Publishers
A Division of Wm. C. Brown Communications, Inc.

Vice President and General Manager *Beverly Kolz*
Vice President, Publisher *Kevin Kane*
Vice President, Director of Sales and Marketing *Virginia S. Moffat*
Vice President, Director of Production *Colleen A. Yonda*
National Sales Manager *Douglas J. DiNardo*
Marketing Manager *Craig Marty*
Advertising Manager *Janelle Keeffer*
Production Editorial Manager *Renée Menne*
Publishing Services Manager *Karen J. Slaght*
Permisions/Records Manager *Connie Allendorf*

Wm. C. Brown Communications, Inc.

President and Chief Executive Officer *G. Franklin Lewis*
Senior Vice President, Operations *James H. Higby*
Corporate Senior Vice President, President of WCB Manufacturing *Roger Meyer*
Corporate Senior Vice President and Chief Financial Officer *Robert Chesterman*

Copyedited by *Nick Murray*

Photo Research by *Kathy Husemann*

Cover Photo © Craig Tuttle/The Stock Market

The credits section for this book begins on page 541 and is considered an extension of the copyright page.

Copyright © 1995 by Wm. C. Brown Communications, Inc. All rights reserved.

A Times Mirror Company

Library of Congress Catalog Card Number: 94–70837

ISBN 0–697–16656–2

No part of this publication may be reproduced, stored in a retrieval system, or transmitted, in any form or by any means, electronic, mechanical, photocopying, recording, or otherwise, without the prior written permission of the publisher.

Printed in the United States of America by Wm. C. Brown Communications, Inc., 2460 Kerper Boulevard, Dubuque, IA 52001

10 9 8 7 6 5 4

"To Kris—with all my love and respect—for unfailing love, patience, and humor."
R. M.

"To Lisa and our children, Elizabeth and William, whose love and support are the greatest."
W. D. C.

A personal library is a lifelong source of enrichment and distinction. Consider this book an investment in your future and add it to your personal library.

Brief Table of Contents

CHAPTER 1
An Introduction to Botany 2

UNIT ONE
Cells as the Fundamental Units of Life . . . 19

CHAPTER 2
Atoms and Molecules: The Building Blocks of Life 20

CHAPTER 3
Structure and Function of Plant Cells 44

CHAPTER 4
Membranes and Membrane Transport 72

UNIT TWO
Plants and Energy . . . 91

CHAPTER 5
Energy and Its Use by Plants 92

CHAPTER 6
Respiration 110

CHAPTER 7
Photosynthesis 130

UNIT THREE
Genetics . . . 163

CHAPTER 8
Patterns of Inheritance 164

CHAPTER 9
The Cell Cycle 186

CHAPTER 10
Meiosis, Chromosomes, and the Mechanism of Heredity 210

CHAPTER 11
Molecular Genetics and Gene Technology 230

UNIT FOUR
The Form and Function of Plants . . . 259

CHAPTER 12
Plant Growth and Development 260

CHAPTER 13
Primary Growth: Cells and Tissues 280

CHAPTER 14
Primary Growth: Stems and Leaves 304

CHAPTER 15
Primary Growth: Roots 332

CHAPTER 16
Secondary Growth 354

CHAPTER 17
Reproductive Morphology 378

UNIT FIVE
Regulating Growth and Development . . . 409

CHAPTER 18
Plant Hormones 410

CHAPTER 19
How Plants Respond to Environmental Stimuli 436

UNIT SIX
Nutrition and Transport . . . 459

CHAPTER 20
Soils and Plant Nutrition 460

CHAPTER 21
Movement of Water and Solutes 488

Table of Contents

foreword xv
list of tables xvi
list of boxed readings xvii
to the student xix
from the editor xxv
acknowledgments xxvii
about the authors xxx

CHAPTER 1

An Introduction to Botany 2

PLANTS AND LIFE 4
A BOTANIST'S VIEW OF LIFE 10
   The Scientific Method 10
   Using the Scientific Method 11
   We All Think Like Scientists At Times 14
THE UNIFYING THEMES OF BOTANY 14
Chapter Summary 18
Questions for Further Thought and Study 18

UNIT ONE

Cells as the Fundamental Units of Life . . . 19

CHAPTER 2

Atoms and Molecules: The Building Blocks of Life 20

ELEMENTS OF LIFE 22
LARGE MOLECULES: POLYMERS AND THEIR MONOMERS 22
   Carbohydrates 22
   Proteins 27
   Nucleic Acids 31
   Lipids 33
SECONDARY METABOLITES 37
   Functions of Secondary Metabolites 37
   Alkaloids 38
   Terpenoids 38
   Phenolics 39
   Minor Classes of Secondary Metabolites 40
Chapter Summary 41
Questions for Further Thought and Study 41

CHAPTER 3

Structure and Function of Plant Cells 44

SCALES OF MICROSCOPIC OBSERVATION 46
METHODS OF CYTOLOGY 48
   Electron Microscopy 50
   Freeze-Fracturing and Electron Microscopy 52
   Cell Fractionation 52
   Biochemical Cytology 52
FUNCTIONAL ORGANIZATION OF CELLS 53
   Why Are Cells So Small? 53
   Membranes and Cell Compartments 54
   The Cytoskeleton 54
   How Does the Cytoskeleton Work? 55
THE CELL WALL 56
   How Cell Walls Grow 56
   Connections Between Cells 57
THE NUCLEUS 58
RIBOSOMES 59
THE MEMBRANE SYSTEM 59
   The Plasma Membrane 60
   The Endoplasmic Reticulum 60
   Dictyosomes 61
   Vacuoles 62
   Microbodies 63
ORGANELLES FOR ENERGY CONVERSION 64
   Chloroplasts 64
   Mitochondria 65
CELL MOVEMENTS 65
   Internal Movements 65
   Cells That Swim 66
CELL THEORY: THE DOGMA AND AN ALTERNATIVE 69
   The Cell Theory: Postulates and Problems 69
   An Alternative: The Organismal Theory 70
Chapter Summary 71
Questions for Further Thought and Study 71

CHAPTER 4
Membranes and Membrane Transport 72

OVERVIEW OF MEMBRANE STRUCTURE AND FUNCTION 74
- Structure of Membranes 74
- Functions of Membranes 76

THE MOVEMENT OF WATER AND OTHER MOLECULES THROUGH MEMBRANES 78
- Movement of Solutes 78
- Water Potential 78
- Osmosis 79
- Turgor 80
- Inducing Osmosis: The Control of Turgor in Plants 81

DIFFERENTIAL PERMEABILITY OF MEMBRANES 82
- Facilitated Diffusion 82
- Active Transport 82
- Bypassing Membrane Transport 83

MOVEMENT OF IONS ACROSS MEMBRANES 85
- Ion Pumps 85
- ATP Synthesis 86

CELLULAR COMMUNICATION 86
- Hormone Receptors 87
- Membrane Interactions with Other Organisms 87
- Lectins and Glycoproteins 88

Chapter Summary 89

Questions for Further Thought and Study 89

UNIT TWO
Plants and Energy . . . 91

CHAPTER 5
Energy and Its Use by Plants 92

WHAT IS ENERGY? 94
MEASURING ENERGY 94
ENERGY CONVERSIONS 96
THE LAWS OF THERMODYNAMICS 96
- The First Law of Thermodynamics 97
- The Second Law of Thermodynamics 97

METABOLISM: ENERGY FOR LIFE'S WORK 99
FREE ENERGY 100
- Free Energy and Chemical Equilibrium 101
- Oxidation, Reduction, and Energy Content 101

ATP: THE ENERGY CURRENCY OF CELLS 102
- Coupled Reactions 103
- Other Compounds Involved in Energy Metabolism 103

ENZYMES AND ENERGY 105
- Regulating Metabolism 106

THE MAJOR ENERGY TRANSFORMATIONS IN PLANTS: PHOTOSYNTHESIS AND RESPIRATION 107
- The Flow of Energy 107

Chapter Summary 108

Questions for Further Thought and Study 109

CHAPTER 6
Respiration 110

RETRIEVING GLUCOSE FROM OTHER MOLECULES 112
- Retrieval from Sucrose 112
- Retrieval from Starch 114

HARVESTING ENERGY FROM GLUCOSE: AN OVERVIEW 114
GLYCOLYSIS 115
- Potential Energy of Glucose 115
- Breakdown of Glucose to Pyruvic Acid 115

THE KREBS CYCLE 117
- Steps in the Krebs Cycle 117

ELECTRON TRANSPORT AND OXIDATIVE PHOSPHORYLATION 118
- Electron Transport 119
- Chemiosmosis and ATP Synthesis 123

OTHER TYPES OF RESPIRATION 125
- Anaerobic Respiration 125
- Respiration of Pentose Sugars 126
- Respiration of Lipids 126
- Cyanide-Resistant Respiration 126
- Photorespiration 127

Chapter Summary 128

Questions for Further Thought and Study 129

CHAPTER 7
Photosynthesis 130

HOW WE LEARNED ABOUT PHOTOSYNTHESIS 132
THE NATURE OF LIGHT 136
PIGMENTS 138
- Pigments in Plants 138
- Accessory Pigments 139
- Making and Destroying Pigments 142

CHLOROPLASTS 143
- Complexes of Pigments in Chloroplasts 143
- What Happens When Pigments Absorb Light? 145
- Photophosphorylation: Chemiosmosis in Chloroplasts 145

THE PHOTOCHEMICAL AND BIOCHEMICAL REACTIONS OF PHOTOSYNTHESIS 147
- The Photochemical Reactions of Photosynthesis 147
- The Biochemical Reactions of Photosynthesis 150
- The Efficiency of Photosynthesis 153

PHOTOSYNTHESIS IS NOT PERFECT: PHOTORESPIRATION 153
C_4 PHOTOSYNTHESIS 154
- Why Don't C_4 Plants Dominate the Landscape? 157
- C_3-C_4 Intermediates 158

CRASSULACEAN ACID METABOLISM (CAM) 158
CONTROL OF PHOTOSYNTHESIS 159
THE FATE OF PHOTOSYNTHATE 159

Chapter Summary 160

Questions for Further Thought and Study 161

Unit Three
Genetics... 163

Chapter 8
Patterns of Inheritance 164
- THE THEORY OF INHERITANCE: GREGOR MENDEL'S DISCOVERY 166
  - Setting the Scene: Plant Hybridization Before Mendel 166
  - How Did Mendel Begin? 167
  - Mendel's Results: A Theory of Inheritance 168
  - Genotypes and Phenotypes 171
  - Experiments Using Multiple Traits 171
  - A Note about Meiosis and Chromosomes 172
- THE SEARCH FOR THE HEREDITARY MATERIAL 173
  - The Importance of the Nucleus in Heredity 173
  - Choosing Between Protein and DNA 173
  - Genes 175
- COMPLEX INHERITANCE 176
  - Types of Dominance 176
  - Multiple Alleles of the Same Gene 176
  - Multiple Genes 177
  - Serial Gene Systems 177
  - Polygenic Inheritance 177
  - Pleiotropy 178
- NON-MENDELIAN INHERITANCE 180
  - Linkage 180
  - Cytoplasmic Inheritance 181
  - Mutations 181
  - Transposable Elements 181
- *Chapter Summary* 183
- *Questions for Further Thought and Study* 184

Chapter 9
The Cell Cycle 186
  - The Cell Cycle in Plants 188
- OVERVIEW OF THE CELL CYCLE 189
- INTERPHASE 190
  - The G_1 Phase 191
  - Beginning a New Era: The Theory of DNA Structure and Duplication 191
  - The Structure of Chromosomes During Interphase 194
  - Making Chromatin: The Major Task of the S Phase 194
  - Evidence for Semiconservative Replication 196
  - Enzymes of Replication: Keys to Understanding the S Phase 198
  - Replicons 200
  - The G_2 Phase 203
- MITOSIS AND CYTOKINESIS 204
  - Before Prophase 204
  - Chromosome Condensation: Prophase 204
  - Chromosome Alignment: Metaphase 205
  - Chromosome Separation: Anaphase 206
  - Formation of Daughter Nuclei: Telophase 206
  - Division Into Cells: Cytokinesis 206
- MYSTERIES OF THE MITOTIC SPINDLE: HOW ARE CHROMOSOMES MOVED? 207
  - The Push of Polar Microtubules 207
  - The Pull of Kinetochore Microtubules 207
- *Chapter Summary* 208
- *Questions for Further Thought and Study* 208

Chapter 10
Meiosis, Chromosomes, and the Mechanism of Heredity 210
- SEXUAL REPRODUCTION IN PLANTS 212
  - Spores and Gametes in Flowering Plants 212
  - Pollination and Fertilization 212
- MEIOSIS 215
  - Prophase I 215
  - The Remainder of the First Meiotic Division 217
  - The Second Meiotic Division 218
- SYNAPTONEMAL COMPLEX 218
  - Recombination Nodules 219
  - Chiasmata and Chromosome Segregation 220
- GENETIC RECOMBINATION 220
  - Recombination by Chromosomal Separation 220
  - Evidence for Chromosomal Exchange During Crossing-Over 222
  - Gene Conversion 223
  - Unequal Crossing-Over and Gene Duplication 225
- WHY SEX? 226
  - DNA Repair Hypothesis 227
  - Transposon Hypothesis 228
- *Chapter Summary* 228
- *Questions for Further Thought and Study* 229

Chapter 11
Molecular Genetics and Gene Technology 230
- HOW GENES WORK 232
  - RNA Synthesis: Transcription 233
  - Different Kinds of RNA 235
  - Polypeptide Synthesis: Translation 240
  - Making the Finished Product: Polypeptide to Protein 241
- THE GENETIC CODE 242
  - The Structure of Eukaryotic Genes 243
  - Roles of Gene Interruptions 244
  - The Exon-Shuffling Hypothesis 244
- GENETIC ENGINEERING 245
  - DNA Cloning 245
  - Finding the Gene of Interest 248
  - Industrial Uses of Genetically Engineered Bacteria 249
  - Making Transgenic Plants: Problems and Solutions 251
  - Monocots vs. Dicots in Genetic Engineering 252
  - Transgenic Crops 253
- THE MOLECULAR GENETIC BASIS OF ROUND VS. WRINKLED PEAS 255
- *Chapter Summary* 257
- *Questions for Further Thought and Study* 257

Unit Four
The Form and Function of Plants... 259

Chapter 12
Plant Growth and Development   260

A Perspective on Plant Form and Function   262
The Lives of Plants   263
Meristems   263
- Functions of Apical Meristems   263
- Derivatives of Meristems   265
- How Apical Meristems Are Organized   267

Plant Growth and Development   268
- Growth   268
- Development   271

Signals that Regulate Plant Growth and Development   274
- Electrical Currents   274
- Hormones   274
- Positional Controls   276
- Biophysical Controls   277
- Genetic Controls   277
- Competence of Cells to Respond to Developmental Signals   277

Modular Plant Growth   278
Chapter Summary   278
Questions for Further Thought and Study   279

Chapter 13
Primary Growth: Cells and Tissues   280

Ground Tissue   282
- Parenchyma   282
- Collenchyma   285
- Sclerenchyma   285

Dermal Tissue: The Epidermis   288
- Cuticle   288
- Epidermal Cells and Gas Exchange   289
- Trichomes   291
- Epidermal Cells and Cellular Recognition   295
- The Fate of the Epidermis   295

Vascular Tissues: Xylem and Phloem   295
- Xylem   295
- Phloem   296

Secretory Structures   299
- External Secretory Structures   299
- Internal Secretory Structures   299

Plant Tissues and Life on Land   301
Chapter Summary   302
Questions for Further Thought and Study   303

Chapter 14
Primary Growth: Stems and Leaves   304

Stems   306
- Stems and Their Functions   306
- Control of Stem Growth   306
- The Structure of Stems   307
- Axillary Buds and Branching   309
- Modified Stems   309
- The Economic Importance of Stems   311

Leaves   311
- How Leaves Form   311
- Phyllotaxis   312
- The Structure of Leaves   315
- Environmental Control of Leaf Variation   320
- Leaf Movements   323
- Modified Leaves   323
- How Leaves Defend Themselves   325
- Leaf Abscission   327
- The Economic Importance of Leaves   327

Chapter Summary   329
Questions for Further Thought and Study   330

Chapter 15
Primary Growth: Roots   332

Kinds of Root Systems   334
- Taproot System   334
- Fibrous Root System   335
- Adventitious Roots   335

Functions and Structure of Roots   336
- Root Tip   336
- Subapical Region   338
- Mature Region   339
- Transition Region Between the Root and the Shoot   344

The Root-Soil Interface   344
Factors Controlling the Growth and Distribution of Roots   345
The Growth of Roots vs. Shoots   347
Modified Roots   348
- Nutrition   348

The Economic Importance of Roots   351
Chapter Summary   351
Questions for Further Thought and Study   352

CHAPTER 16
Secondary Growth   354

THE VASCULAR CAMBIUM   356
- Where and How the Vascular Cambium Forms   358
- What Controls the Activity of the Vascular Cambium?   359
- Extent of Secondary Growth   362

SECONDARY XYLEM: WOOD   362
- Kinds of Wood   362
- Characteristics of Wood   364
- Reaction Wood   367

BARK: SECONDARY PHLOEM AND PERIDERM   368
- Secondary Phloem   368
- Periderm   368

UNUSUAL SECONDARY GROWTH   370
- Dicots   370
- Monocots   371

USES OF SECONDARY XYLEM AND PHLOEM   374
- Commemorating Trees   375

Chapter Summary   375

Questions for Further Thought and Study   376

CHAPTER 17
Reproductive Morphology   378

FLOWERS   380
- Stamens   380
- Carpels   382
- Petals   383
- Sepals   384
- The Nature of Flower Parts   384

REPRODUCTIVE MORPHOLOGY AND PLANT DIVERSITY   386
- Flower Variation   386
- Inflorescences   392
- Breeding Systems   392
- Vegetative Reproduction   394
- Pollination Mechanisms   395

FRUITS   398
- Types of Fruit   398
- Fruit Development   399

SEEDS   400
- Seed Structure   400
- Seed Germination   401
- Seed Banks   401
- Seedling Development   403

DISPERSAL OF FRUITS AND SEEDS   404
- Dispersal by Wind and Water   404
- Dispersal by Animals   405
- Self-Dispersal   406

Chapter Summary   407

Questions for Further Thought and Study   407

UNIT FIVE
Regulating Growth and Development . . .   409

CHAPTER 18
Plant Hormones   410

AUXIN   412
- Discovery   412
- Synthesis   417
- Synthetic Auxins   417
- Transport   418
- Effects of Auxin   418
- Auxin and Calcium   422

GIBBERELLINS   422
- Discovery   422
- Synthesis and Transport   422
- Effects of Gibberellins   423

CYTOKININS   425
- Discovery   425
- Synthesis and Transport   426
- Effects of Cytokinins   426
- Cytokinins and Calcium   427

ETHYLENE   427
- Discovery   427
- Synthesis and Transport   427
- Effects of Ethylene   428
- Ethylene and Auxin   431

ABSCISIC ACID   431
- Discovery   431
- Synthesis and Transport   431
- Effects of Abscisic Acid   431

OLIGOSACCHARINS AND OTHER PLANT HORMONES   432

CONTROLLING THE AMOUNTS OF PLANT HORMONES: HORMONAL INTERACTIONS   432

DO PLANT HORMONES REALLY EXIST?   432

Chapter Summary   433

Questions for Further Thought and Study   434

Chapter 19

How Plants Respond to Environmental Stimuli 436

TROPISMS 438
- Phototropism 438
- Gravitropism 441
- Hydrotropism 443
- Thigmotropism 444

NASTIC MOVEMENTS 445
- Seismonasty 445
- Nyctinasty 447

THIGMOMORPHOGENESIS 448

SEASONAL RESPONSES OF PLANTS TO THE ENVIRONMENT 448
- Flowering 448
- Senescence 455
- Dormancy 455

CIRCADIAN RHYTHMS 455
- What Controls Circadian Rhythms in Plants? 456
- Biological Clocks and Entrainment 457

Chapter Summary 457

Questions for Further Thought and Study 458

UNIT SIX
Nutrition and Transport... 459

Chapter 20

Soils and Plant Nutrition 460

SOILS 462
- How Soils Form 462
- Components of Soil 463

PLANT NUTRITION 469
- Essential Elements 469
- Obtaining Essential Elements 474

Chapter Summary 486

Questions for Further Thought and Study 486

Chapter 21

Movement of Water and Solutes 488

MOVING WATER AND MINERALS IN THE XYLEM 490
- Leaf Architecture and Transpiration 490
- Structure of the Conducting Cells 491
- Water Potential: The Force Responsible for Water Movement 492
- How Does Water Move in Plants? 494
- Factors Affecting Transpiration 496
- A Model to Regulate the Photosynthesis-Transpiration Compromise 502
- The Adaptive Value of Transpiration 503
- Coping with Extremes: Too Much or Too Little Water 503

TRANSPORTING ORGANIC SOLUTES IN THE PHLOEM 504
- Structure of Conducting Cells 505
- How Substances Move in the Phloem 506
- Influence of the Environment on Phloem Transport 509
- What Moves in the Phloem? 509

EXCHANGE BETWEEN THE PHLOEM AND XYLEM 510

Chapter Summary 511

Questions for Further Thought and Study 511

Appendices A-1
Glossary G-1
Credits C-1
Index I-1

Foreword

Botany is in the midst of a renaissance. New research methods, powerful instruments, and fresh ideas are transforming how biologists study plants at all levels, from the molecular mechanisms of cells to the ecological dynamics of the biosphere. Developmental biologists are using DNA technology to unravel how genes program a mass of undifferentiated cells to become a flower. Taxonomists are reassessing plant classification by applying new methods for measuring evolutionary relationships. Plant ecologists are analyzing field data by coupling computers to remote environmental sensing instruments placed in tropical forests and other ecosystems. Biotechnologists are engineering plants to improve agricultural productivity. Plant biologists are reinventing every field of botany.

If anything about botany is constant, it is its relevance. The activities of plants are vital to the welfare of nearly all organisms that share Earth, including humans. Renewed interest in environmental issues has brought the importance of plants back into sharp focus at the same time that new approaches are catalyzing botanical research. This is a wonderful time to study plants—a great time to take a botany course!

Botany is a new textbook that captures the excitement of the botanical revolution and reflects how the objectives of botany courses are changing. Randy Moore, Dennis Clark, and Kingsley Stern are award-winning teachers, distinguished researchers, and gifted writers. More than these talents, it is the authors' shared commitment to emphasize concepts over facts and to engage students in the process of science that sets this book apart. Scan the table of contents and you'll find the requisite topics, including plant structure and function, life cycles, metabolic pathways, and plant diversity. Read any chapter and you'll discover that it is *presentation*, not content, that defines this book. In crafting each chapter, the authors have constructed a strong framework of concepts that places botanical facts and terms in context and discourages rote memorization. Moore, Clark, and Stern build the process of science into every chapter by explaining not only *what* we know about plants, but *how* we know it. They also highlight what we *don't* know, and encourage students to frame questions and pose testable hypotheses of their own.

A good textbook must be clear, current, and correct; a great one also has a point of view. The authors of *Botany* see scientific literacy not as mastery of a vocabulary list, but as development of skills for lifelong enjoyment of nature and for thinking more critically about all subjects. The wonder and significance of plants are not lost along the way. Moore, Clark, and Stern love writing about plants, and their enthusiasm is infectious. Reading one chapter made me realize, more than ever why it is so much fun to learn about plants. I gave a chapter to a few students to read, and they came back asking for more.

The publication of a new botany textbook is a rare event. This one is worth the wait.

Neil A. Campbell
Visiting Scholar
Department of Botany and Plant Sciences
University of California, Riverside

List of Tables

TABLE 2.1 The Most Common Elements in Plants   23

TABLE 2.2 Examples of Well-Known Secondary Metabolites   37

TABLE BOX 6.1 Potential Production of ATP by Respiration of One Molecule of Glucose   120

TABLE 7.1 Radiant Energies of Different Wavelengths of Light   137

TABLE 7.2 Photosynthetic Characteristics of C_3, C_4, and CAM Plants   156

TABLE 7.3 Maximum Photosynthetic Rates of Major Types of Plants in Natural Conditions   156

TABLE 8.1 Mendel's Experimental Results for Seven Characteristics of Garden Pea (These Data Represent the F_2 Generation)   169

TABLE 8.2 Mendel's Results from a Dihybrid Cross   172

TABLE 9.1 Postulates of the Theory of DNA Structure and Duplication   194

TABLE 11.1 Goals of Recombinant DNA Technology in Agriculture   254

TABLE 12.1 Functions of Plant Hormones   275

TABLE 17.1 Main Differences between Monocots and Dicots   392

TABLE 17.2 Features of Outcrossing vs. Inbreeding Plants   394

TABLE 17.3 Floral Features Associated with Pollination by Different Kinds of Insects   397

TABLE 17.4 Dichotomous Key to Major Types of Fruit   399

TABLE 20.1 Physical and Chemical Properties of Soil as Affected by Soil Texture   464

TABLE 20.2 Some Functions and Deficiency Symptoms of Essential Elements   472

List of Boxed Readings

BOXED READING 1.1 Coffee: Even If You Don't Know Beans 6
BOXED READING 1.2 Breakfast at the Sanitarium: The Story of Breakfast Cereals 13
BOXED READING 2.1 Reading Chemical Structures 24
BOXED READING 2.2 What's Wrong With Tropical Oils? 35
BOXED READING 3.1 How to Prepare Cells for Microscopy 51
BOXED READING 3.2 Cellular Invasion: Origin of Chloroplasts and Mitochondria 66
BOXED READING 4.1 Membrane Transport and Making Beer 84
BOXED READING 5.1 Earth's Energy 95
BOXED READING 6.1 How Efficient is Respiration? 120
BOXED READING 6.2 Why Skunk Cabbage Gets So Hot 128
BOXED READING 7.1 The Evolution of Photosynthesis: Why Aren't Plants Black? 141
BOXED READING 8.1 Artificial Hybridization in Plants 168
BOXED READING 8.2 Examining Proteins by Gel Electrophoresis 178
BOXED READING 8.3 Cytoplasmic Male Sterility in Corn 182
BOXED READING 9.1 Mutations and DNA Repair: Do Mistakes Have to Be Corrected? 200
BOXED READING 9.2 Determining the Size and Composition of Genomes 202
BOXED READING 10.1 Polyploidy in Plants 219
BOXED READING 10.2 Multigene Families 227
BOXED READING 11.1 Ribozymes and the Origin of Life 236

BOXED READING 11.2 The Discovery of Interrupted Genes 238
BOXED READING 11.3 Genetic Engineering with a Gene Gun 253
BOXED READING 12.1 Making Apples and Oranges 266
BOXED READING 13.1 Feeling their Way 292
BOXED READING 14.1 Leonardo the Blockhead 314
BOXED READING 14.2 Clogging the Waterways 321
BOXED READING 15.1 Finding a Host 350
BOXED READING 16.1 The Bats of Summer: Botany and our National Pastime 366
BOXED READING 16.2 Cork 371
BOXED READING 16.3 The Lore of Trees 373
BOXED READING 17.1 Orchids that Look Like Wasps 388
BOXED READING 17.2 The Dodo Bird and the Tambalacoque Tree 402
BOXED READING 18.1 Measuring Plant Hormones 416
BOXED READING 18.2 Plant Hormones in Plant Pathology 428
BOXED READING 19.1 Tracking the Sun 440
BOXED READING 19.2 Is There a Flowering Hormone? 449
BOXED READING 19.3 Vernalization 452
BOXED READING 20.1 Putting Things Back 465
BOXED READING 20.2 Losing the Soil 466
BOXED READING 21.1 The Risks of Having Vessels 492
BOXED READING 21.2 Living in Salty Soil 495
BOXED READING 21.3 Tapping Wood for Maple Syrup 500

To The Student

Why should you study botany? Besides the obvious—a course requirement for your college degree—consider the botanical basis of the following items that appeared recently in newspapers across the country:

The people of France consume more wine and more fat-laden food than the people of most other industrialized countries, yet the French have a lower than expected rate of heart disease caused by high-fat diets. Based on population surveys of these kinds of data, some medical researchers have concluded that wine somehow inhibits heart disease.

The people of France consume more fruits and vegetables with their otherwise high-fat diets than the people of most other industrialized countries. Based on population surveys of these kinds of data, other medical researchers have concluded that fruits and vegetables, not wine, somehow inhibit heart disease.

In a 20-year study of older men in the Netherlands, those who consumed fruits, vegetables, and teas containing the highest amounts of vitamin C, vitamin E, and flavonoids (a group of chemicals produced by plants) had the lowest rate of heart disease. Researchers who did this study concluded that flavonoids somehow inhibit heart disease.

Which of the above are you going to believe: all of the conclusions, some of the conclusions, or none of the conclusions? You are bombarded with these kinds of conclusions in the popular press every day. Whether you accept some and not others may cause you to make decisions that affect your life. In a broad context, therefore, the quality of your life would improve if you could make thoughtful decisions based on how we acquire knowledge in science. Or, more specifically, based on how we acquire knowledge about plants. The basic skill that you need to make these decisions is how to think critically in a scientific context. The main purpose of this textbook is to help you develop that skill.

The abilities to think critically and communicate effectively are the most important skills that a student can develop during his or her formal education. Consequently, we have written this book to help you develop those skills as you learn about plants: what plants are, how they function, how they interact with each other and the environment, where they came from, and how we use them. As is the nature of all textbooks, ours contains an abundance of interesting "facts," but the real emphasis of this book is *how* we know. We have, as much as possible, de-emphasized the details of some of our knowledge and reduced the overwhelming number of new terms that usually appear in a text. In their place, we have substituted more of the process of science.

Our emphasis on the scientific process involves explaining botany as botany is done. Specifically, we describe the competing hypotheses that botanists have devised to answer questions about botanical phenomena, the experiments done by botanists to test these hypotheses, interpretations of data, and the many unanswered questions and unresolved conflicts that remain. This approach differs significantly from that of merely presenting definitions and the conclusions of experiments (i.e., the "facts" of botany).

We also pose many questions, some of which we don't answer and others of which we answer by showing you how botanists have addressed those questions. We do this to mirror the nature of botany: botanists generate knowledge by posing and testing hypotheses. Nevertheless, there are many things that we still don't know about plants. We hope that these questions, both answered and unanswered, will trigger your curiosity and help you sharpen your critical thinking skills.

In addition to learning about how we know, throughout this book you'll also learn about *what* we know. We present this information in the context of various principles (e.g., the

relation of structure and function) and integrate it into an informational framework. In *Botany*, that framework is created by the following themes:

The importance of plants. Plants affect virtually everything that we do. If you understand plants, you can better appreciate their influence on your life.

The process of science. We stress the process of science throughout the textbook. For example, you'll not just read about the products of evolution (i.e., the diversity of plants). Rather, you'll see the dynamics of evolutionary thought, as well as how botanists are applying various techniques to test competing hypotheses that explain what evolution is and how it operates.

Interactive learning. We designed this book to be a private tutor that will not only help you learn, but will also make you a more effective learner. Consequently, we have included a variety of learning aids. For example, chapter overviews introduce the topics of each chapter, artwork is integrated with and complements the text, and in-text concepts summarize themes. We hope that these features will encourage you to learn by making you an active participant in your learning.

We've used these features to present information with authority, breadth, and quality. Although we've treated some topics in depth, we've not written a botanical encyclopedia. That is, this text is not a compendium, but rather a stimulus. Because the first step to knowing is the curiosity of a question, we hope you'll generate your own questions to study. Indeed, we'll help you develop those questions in every chapter.

Unit Openers, Chapter Outlines, and **Chapter Openers** will stimulate your curiosity and prepare you for what you'll learn. These features will also motivate, orient, and tell you what you'll learn, and why.

UNIT FIVE

Regulating Growth and Development . . .

Plants seem to keep track of time because they do certain things at specific times of the day or year. For example, morning glories open their flowers in the morning, but flowers of the Queen-of-the-night cactus open at night. Mulberries and aspens drop their leaves in the fall and grow new leaves in the spring; the timing is never the other way around. Plants also sense gravity, light, and touch. All of these activities show that plants sense their environment and respond to many different stimuli. Environmental stimuli vary, as do the responses of plants to these stimuli.

How do plants perceive a stimulus, communicate the stimulus to appropriate cells in the plant, and change their growth or development accordingly? This complex question can be addressed by learning about the roles of hormones (also called *plant growth-regulating substances*) and other chemical means of communication in plants and about the kinds of changes plants undergo in response to different stimuli. These topics are the main subject of Unit 5, Regulating Growth and Development.

As you read the chapters in this unit, keep in mind that the search for compounds that regulate plant growth and development is ongoing. New compounds that may affect plant growth and development are discovered every year. Research on the established hormones continues because there are many unanswered questions about how they work.

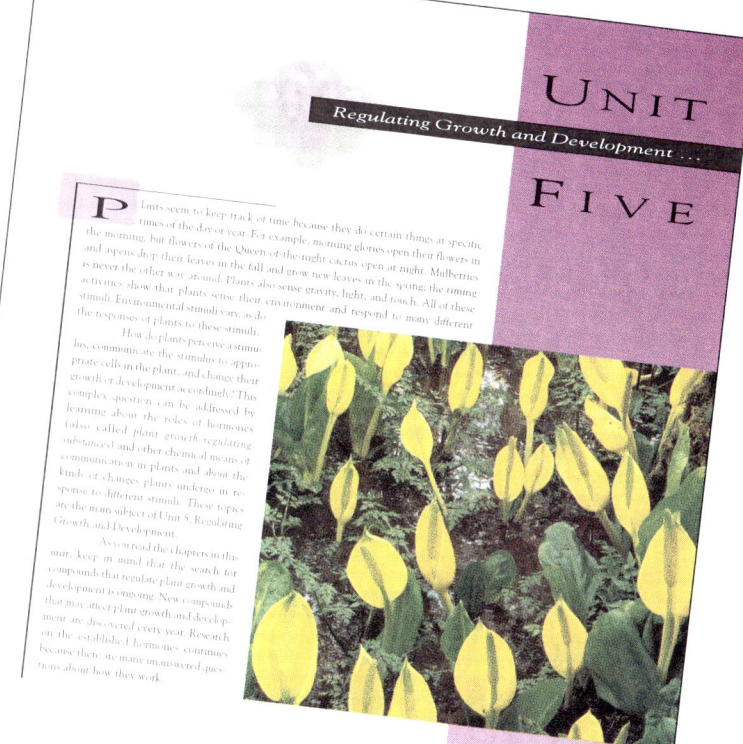

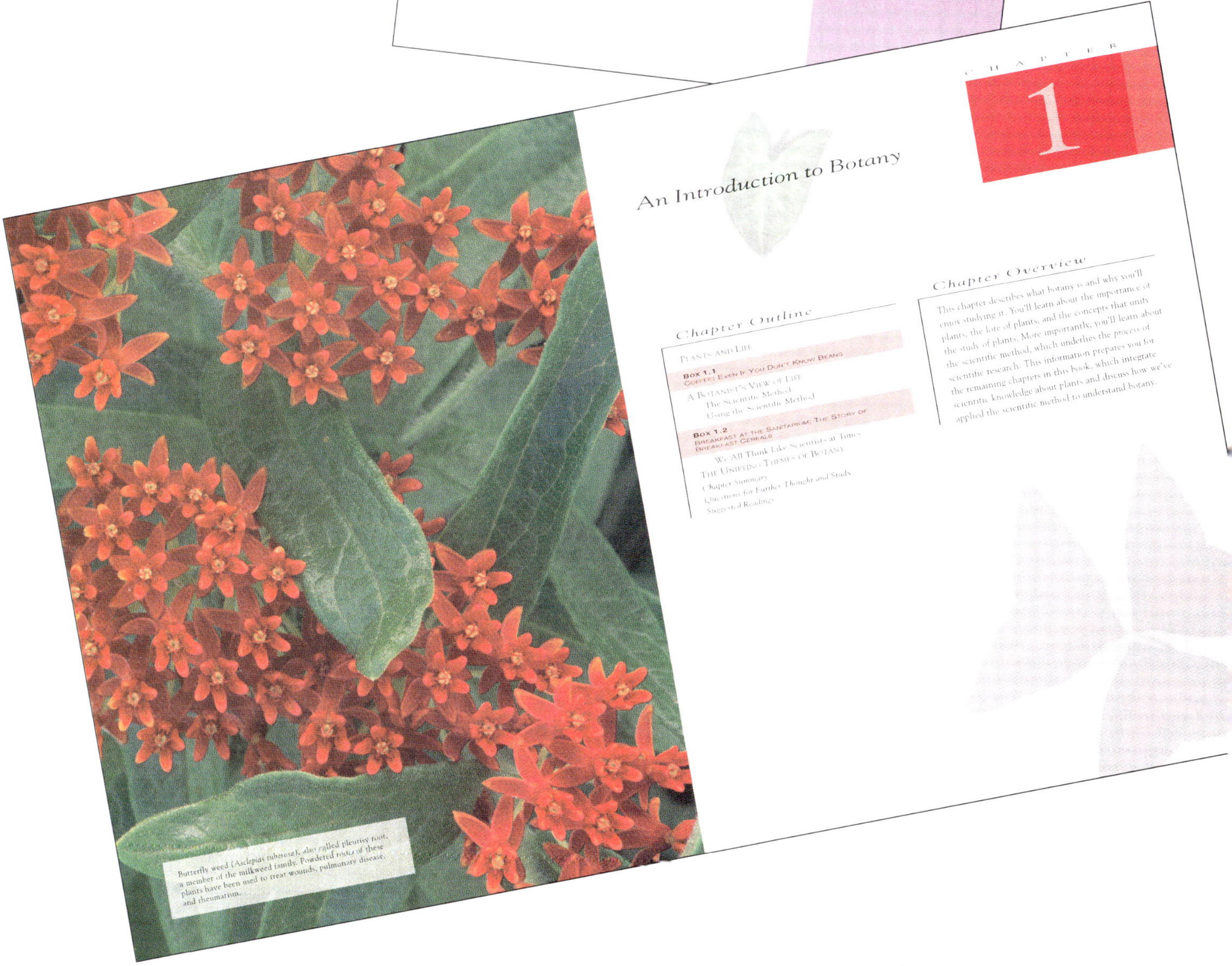

CHAPTER 1

An Introduction to Botany

Chapter Outline

PLANTS AND LIFE

BOX 1.1 COFFEE: EVEN IF YOU DON'T KNOW BEANS

A BOTANIST'S VIEW OF LIFE

The Scientific Method

Using the Scientific Method

BOX 1.2 BREAKFAST AT THE SANITARIUM: THE STORY OF BREAKFAST CEREALS

We All Think Like Scientists at Times

THE UNIFYING THEMES OF BOTANY

Chapter Summary

Questions for Further Thought and Study

Suggested Readings

Chapter Overview

This chapter describes what botany is and why you'll enjoy studying it. You'll learn about the importance of plants, the lore of plants, and the concepts that unify the study of plants. More importantly, you'll learn about the scientific method, which underlies the process of scientific research. This information prepares you for the remaining chapters in this book, which integrate scientific knowledge about plants and discuss how we've applied the scientific method to understand botany.

Butterfly weed (*Asclepias tuberosa*), also called pleurisy root, a member of the milkweed family. Powdered roots of these plants have been used to treat wounds, pulmonary disease, and rheumatism.

Throughout the text you'll also encounter four features that will help you extend your learning beyond the classroom and textbook:

The Lore of Plants

will tell you about the fascinating—and even bizarre—history of plants and their uses by humans.

What are Botanists Doing?

will challenge you to learn the newest ideas about various botanical subjects.

Boxed Readings will expand the scope of botany to include people, controversies, curiosities, and new developments.

Chapter Summaries will help you integrate what you've learned.

Questions for Further Thought and Study will help you apply what you've learned. Because these questions emphasize critical thinking rather than the mere recall of isolated "facts," you'll find no true-false or multiple-choice questions at the end of chapters. Nor will you have to fill in the blanks of someone else's thinking. Instead, you'll be asked to put *your* ideas together.

Suggested Readings provide sources of additional information about topics that you've read about.

Doing Botany Yourself

will challenge you to design experiments to test hypotheses and, in doing so, help you develop an experimental approach to botany.

Writing to Learn Botany

will encourage you to use writing as a tool to learn botany. This feature will help you develop your communications skills, which are crucial to success in virtually any profession.

We have tried to make this textbook entertaining and accessible; we believe these features are critical to making students better learners, and instructors better teachers. To accomplish this, we've rejected the awkward, stoic writing style typical of "scientific writing." In its place, you'll find an engaging text that stresses communication and understanding.

We hope that *Botany* helps you learn about and appreciate plants. If you have questions, suggestions, or comments about what you read, let us know.

Randy Moore
Akron, Ohio

Dennis Clark
Tempe, Arizona

April, 1994

Acknowledgments

I cannot adequately thank my many teachers, friends, colleagues, reviewers, and students; their suggestions, teachings, and encouragement have improved this book immensely. I also thank Marge Kemp, Carol Mills, Kathy Husemann, Kay Brimeyer, and Kevin Kane for their continued help and guidance all writers should be so lucky to have editors and colleagues like these people.

I thank Jill Hammann and Flo Fiehn for their ongoing (and usually undeserved) patience, Marc Low and David Jamison for their encouragement, and Doyle and Tillie for their helpful advice. I also thank Stella, who has never said anything bad about me.

R. M.

List of Reviewers

Many instructors have contributed to the making of this text. We gratefully acknowledge the invaluable assistance of the many reviewers, whose constructive criticism was integral to the development of this text.

Bonnie Amos	Angelo State University
Robert Antibus	Clarkson University
Michael Barbour	University of California-Davis
Linda Barham	Meridian Community College
Jerry Baskin	University of Kentucky
William Beasley, Jr.	Paducah Community College
Rolf Benseler	California State University-Hayward
David Bilderback	University of Montana
Allan Bornstein	Southeast Missouri State University
Jack Bostrack	University of Wisconsin-River Falls
David Buckalew	Xavier University of Louisiana
Linda Chalker-Scott	SUNY College at Buffalo
Lee Anne Chaney	Whitworth College
Ronald Cheetham	Worcester Polytechnic Institute
William Chrouser	Warner Southern College
Tommy Cole	Phillips College
Bill Cook	Midwestern State University
Jonathan Cumming	University of Vermont
Ronald Darey	SUNY-Cortland
Archie Dickey	Yavapai College
Rebecca McBride DiLiddo	Suffolk University
David Domozych	Skidmore College
Max Dunford	New Mexico State University
Roland Dute	Auburn University
G. F. Estabrook	University of Michigan-Herbarium
Lance Evans	Manhattan College
Gerald Farr	Southwest Texas State University
Mary Fields	Ursinus College
Rosemary Ford	Washington College
Ronald Frank	University of Missouri-Rolla
Janice Glime	Michigan Technological University
Sue Harley	Weber State University
Marcia Harrison	Marshall University
Jon Huston	Murray State College
Julius Ikenga	Mississippi Valley State University
Susanne James	Berry College

George Johnson	*Arkansas Tech. University*	J. Kenneth Shull, Jr.	*Appalachian State University*
Ken Kilborn	*Shasta College*	Linda Spessard-Schueth	*University of Nebraska-Kearney*
Mark Knauss	*Shorter College*	Stephen Stocking	*San Joaquin Delta Community College*
Craig Landgren	*Middlebury College*		
Peter Lindeman	*Madisonville Community College*	Richard Storey	*Colorado College*
	University of Kentucky Community College System	William Suder	*Brevard College*
		Lawrence Swails	*Francis Marion University*
William Lipke	*Northern Arizona University*	Herbert Tepper	*SUNY College of Environmental Science & Forestry*
Karen Lustig	*Harper College*		
M. A. Madore	*University of California-Riverside*	Stephen Timme	*Pittsburg State University-Kansas*
George Manaker	*Temple University*	Dwain Vance	*University of North Texas*
Lawrence Matten	*Southern Illinois University*	Terry Vassey	*Horry-Georgetown College*
James Matthews	*University of North Carolina at Charlotte*	Judith Verbeke	*University of Arizona*
		Dennis Walker	*Humboldt State University*
Jack McDonald	*Monroe County Community College*	Leon Walker	*University of Findlay*
Conley McMullen	*West Liberty State College*	James Wallace	*Western Carolina University*
L. Maynard Moe	*California State University-Bakersfield*	Frank D. Watson	*St. Andrews Presbyterian College*
		Cherie Wetzel	*City College of San Francisco*
Lytton Musselman	*Old Dominion University*	Erleen Whitney	*Clark College*
Craig Nessler	*Texas A&M University*	Charles Wimpee	*University of Wisconsin-Milwaukee*
Rebecca Nystrom	*Jamestown Community College*	George H. Wittler	*Ripon College*
Wayne Owen	*Boise State University & U.S. Forest Service*	Dan Wujek	*Central Michigan University*
		Steve Adams	*Lake City Community College*
Lois Peck	*Philadelphia College of Pharmacy & Science*	Charla Rae Armitage	*Lees-McRae College*
		Marilyn Barker	*University of Alaska Anchorage*
R. Gordon Perry	*Fairleigh Dickinson University*	Norlyn Bodkin	*James Madison University*
Jerry Pickering	*Indiana University of Pennsylvania*	Cecile Boehmer	*Alice Lloyd College*
William Pietraface	*SUNY-Oneonta*	Frank Bowers	*University of Wisconsin-Stevens Point*
Paul Redfearn	*Southwest Missouri State University*	Joe Bruner	*Columbus College of Art & Design*
Christa Schwintzer	*University of Maine*	Richard Churchill	*Southern Maine Technical College*
Fred Searcy, Jr.	*Broward Community College*	Opal Dakin	*Hinds Community College*
Stephen Shimmel	*Chipola Junior College*		

Jill Deikman	*Pennsylvania State University*	James Rasmussen	*Southern Arkansas University*
Roger del Moral	*University of Washington*	Maralyn Renner	*College of the Redwoods*
C. Ron Dillon	*Clemson University*	Herbert Robbins	*Dallas Baptist University*
Ronald Doney	*SUNY at Cortland*	Joseph Rohrer	*University of Wisconsin-Eau Claire*
Robert Dorrance	*Herkimer County Community College*	Robert Ross	*Linn-Benton Community College*
Jim Ebersole	*Colorado College*	Robert Rupp	*Ohio State University/Agricultural Technical Institute*
Daniel Flisser	*Cazenovia College*		
Royce Granberry	*Texarkana College*	Barbara Schumacher	*San Jacinto College*
Danny Ingold	*Muskingum College*	John Sehloff	*Bethany Lutheran College*
Jody Jellison	*University of Maine*	Steven Selva	*University of Maine-Fort Kent*
J. Morris Johnson	*Western Oregon State College*	Richard Sims	*Jones County Junior College*
William Kinnison	*Central Arizona College*	Ken Smith	*Montcalm Community College*
M. Joseph Klingensmith	*Rochester Institute of Technology*	Richard Stalter	*St. John's University*
Penelope Koines	*University of Maryland*	Richard Storey	*Colorado College*
Duane Kolterman	*University of Puerto Rico-Mayaguez Campus*	Jon Stucky	*North Carolina State University*
		Nicholas Sturm	*Youngstown State University*
Vaughnda Larson	*West Virginia Northern Community College*	Jerry Tammen	*Rochester Community College*
		Rahmona Thompson	*East Central University*
Robert Lebowitz	*Mankato State University*	Michael Timko	*University of Virginia*
Jerri Dawn Martin	*Sue Bennet College*	Thomas Vierheller	*Prestonsburg Community College*
William Mathena	*Kaskaskia College*	Joyce Weary	*Oakland Community College*
Richard Alan Mayes	*Greensboro College*	Edgar Webber	*Keuka College*
W. D. McBryde	*Central Texas College*	David Williams	*Ann Arundel Community College*
Sheila McCormick	*Plant Gene Expression Center*	Donald Williams	*Sterling College*
Robert McGuire	*University of Montevallo*	Kathryn Wilson	*Indiana University-Purdue University at Indianapolis*
Jerry Melaragno	*Rhode Island College*		
Robert Mellor	*University of Arizona*	Kenneth Wilson	*Miami University*
Jeanette Oliver	*Flathead Valley Community College*	Robert Winget	*Brigham Young University-Hawaii*
P. C. Pendse	*Cal Poly University*	James Winsor	*Penn State University*
Ed Perry	*James Faulkner State Community College*	Todd Christian Yetter	*Cumberland College*

About the Authors

Dr. Randy Moore is currently Dean of Arts and Sciences and professor of biology at The University of Akron. In addition to writing numerous articles and textbooks, he is Editor-in-Chief of the *American Biology Teacher*. Recently, Dr. Moore was awarded the 1993 Teacher Exemplar Award by the Society for College Science Teachers. Other awards and honors include many research grants (e.g., NSF, Nasa) a Fulbright Scholarship, a listing in the *Who's Who in Science and Engineering*, 1986 Most Outstanding Professor from Baylor University, and 1993 Most Outstanding Professor at Wright State University. He received his Ph.D. in plant development from UCLA.

Dr. W. Dennis Clark is currently a professor at Arizona State University in Tempe, where he has taught for the past 17 years. A prolific researcher, Dr. Clark has authored numerous scientific papers, while at the same time being active in many professional societies. He received his Ph.D. in botany from the University of Texas at Austin.

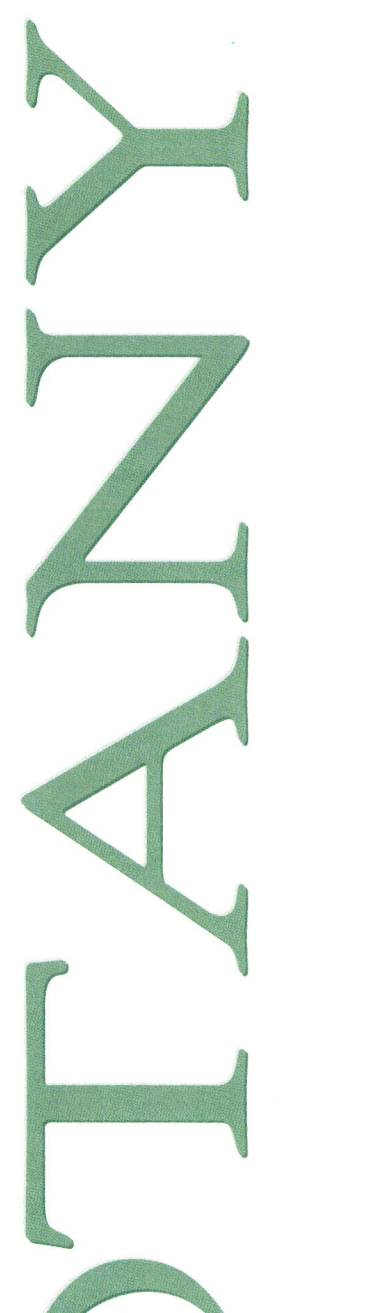

BOTANY

Plant Form and Function

Butterfly weed (*Asclepias tuberosa*), also called pleurisy root, a member of the milkweed family. Powdered roots of these plants have been used to treat wounds, pulmonary disease, and rheumatism.

CHAPTER 1

An Introduction to Botany

Chapter Outline

PLANTS AND LIFE

BOX 1.1
COFFEE: EVEN IF YOU DON'T KNOW BEANS

A BOTANIST'S VIEW OF LIFE
    The Scientific Method
    Using the Scientific Method

BOX 1.2
BREAKFAST AT THE SANITARIUM: THE STORY OF BREAKFAST CEREALS

    We All Think Like Scientists at Times

THE UNIFYING THEMES OF BOTANY

Chapter Summary
Questions for Further Thought and Study
Suggested Readings

Chapter Overview

This chapter describes what botany is and why you'll enjoy studying it. You'll learn about the importance of plants, the lore of plants, and the concepts that unify the study of plants. More importantly, you'll learn about the scientific method, which underlies the *process* of scientific research. This information prepares you for the remaining chapters in this book, which integrate scientific knowledge about plants and discuss how we've applied the scientific method to understand botany.

PLANTS AND LIFE

For many centuries, humans have relied on plants for survival and pleasure. Asian civilizations were based on rice, Middle Eastern civilizations were based on wheat and barley, and American civilizations were based on corn. Our dependence on plants persists: almost everything we do is influenced, either directly or indirectly, by plants. We've even fought wars over plants. For example, the First Opium War (1839–1842) was fought over opium, an extract of the opium poppy, *Papaver somniferum* (Britain won that war and, as part of the settlement, claimed Hong Kong). Today we use plants to make clothes, string, rope, resin, lumber, musical instruments, sports equipment, furniture, fabrics, cardboard, medicines, and explosives. Even these many uses of plants, however, are only a small part of their importance to us (see fig. 1.1). Consider the following examples.

- Green plants and algae generate the oxygen and sugars that sustain life on earth.

- Plants supply our food and many of our drinks. For example, about 95% of our food comes from only twenty species of plants; 80% of our food comes from six of those species. We've made tequila from the century plant (*Agave*) for hundreds of years, and Egyptians made beer from barley as early as 500 B.C. Tea and coffee, the world's two most popular beverages, are also made from plants (see Box 1.1 "Coffee: Even If You Don't Know Beans").

- We use extracts of plants and plantlike organisms to make paint, plastics, soap, oils, adhesives, natural rubber, waxes, dyes, spices, and drugs such as morphine, cocaine, aspirin, caffeine, codeine, digitoxin, quinine, vinblastine, and most antibiotics. We use flowers for decorations, perfumes, and to express our feelings.

- About 33 million people in the United States have home gardens. Thirty percent of these people use their gardens as a source of fresh fruits and vegetables, 25% as a source of better-tasting vegetables, 15% as a source of income, and a lighthearted 22%—the croquet and lawn tennis group—use them as a source of fun. These gardens provide $16 billion worth of food; that's about 23 kilograms (50 pounds) of food per gardener, down from 57 kilograms (126 pounds) per gardener in 1910. This decrease in production over the last eighty years is probably due to our increasing urbanization. Today, the people who don't have gardens blame a lack of space (35%) or lack of time (28%), followed closely by the hammock-lovers who say that gardening is too much work (27%).

- On the negative side, plants clog our rivers, damage our crops, cause allergies, and poison us. Socrates was poisoned with poison hemlock, and Claudius, the father of Nero, was poisoned with mushrooms and monkshood juice. Many children are poisoned each year from eating poisonous plants.

Today, plants dominate our lives and economy, just as they have in all civilizations.

Plants affect virtually everything that we do.

Our need-driven uses of plants, along with our curiosity about our world, have spawned **botany,** the scientific study of plants. Today, botany is a thriving, exciting science that, directly or indirectly, deals with the largest part of our national and world economy.

But what exactly is a plant? Although you probably already have some notion as to what a plant is—a quiet, green organism that we eat, mow, and use for decoration and shade—it's difficult to come up with a complete definition for the word *plant*. For example, not all plants are green and some plants consume animals. Indeed, some plants do not look or act like plants at all. Nevertheless, if you're like most people, you've probably not given plants much thought. You might have assumed that they're all basically alike and, because they don't move around, that they're uninteresting. To dispel this notion, consider the giant sequoia (*Sequoiadendron giganteum*) and its taller but slimmer relative, the coastal redwood (*Sequoia sempervirens*) (fig. 1.2).

- These cloud-piercing trees are the world's tallest plants: they would tower above our nation's Capitol, dwarf the Saturn V rockets that flew astronauts to the moon, and reach from home plate to the outfield bleachers of any baseball park in the country. A twenty-year-old sapling, a mere sprig of a thing, is often more than 15 meters (50 feet) high.

- The General Sherman Tree in California weighs 1,400 tons—that's as much as 13 space shuttles, 20 million boxes of toothpicks, 200 elephants, or 10 blue whales, the animal kingdom's largest representative. Interestingly, these giant trees are supported by roots that seldom go deeper than 1.8 meters (6 feet) into the soil.

Foxglove (*Digitalis purpurea*) contains cardiac glycosides used to treat congestive heart failure. These plants, along with *Digitalis lanata*, have saved the lives of many heart-attack victims and are helping millions of people with heart problems to lead normal lives.

Chocolate is made from the seeds of cocoa (*Theobroma cacao*). Because these seeds were believed to be of divine origin, botanist Carolus Linnaeus named the plant *Theobroma*, meaning "food of the gods." Most famous as the main ingredient of the drink given to Cortés by Montezuma in 1519, cocoa beans were also used as currency by the Aztecs, who paid their taxes with them until 1887. Chocolate has long been considered an aphrodisiac, and Casanova is said to have preferred chocolate to champagne as an inducement to romance. Today, chocolate remains a popular gift for romantics.

Piper nigrum is the source of black pepper, the most commonly used spice. The lure of pepper, the only spice that could make decaying or heavily salted meat edible, drew Columbus and medieval merchants to "discover" the rainforested areas of earth. For Europeans, North America was a "by-product" of the maritime search for pepper.

Flax (*Linum usitatissimum*), with its wiry stems and distinctive flowers, has been used for centuries for its fibers (from stems, used to make linen) and linseed oil (crushed from its seeds). Flax fibers are about three times stronger than cotton fibers. The luster of linen results from the fibers' capacity to reflect light.

Each year more than 2 million tons of tea are produced from about 25 countries. North Americans drink about 40 billion cups of tea per year. Tea is made from leaves of *Camellia sinensis*, a small evergreen shrub related to the garden camellia.

Money, one of the most important things we make from plants. Paper money is made from fibers of flax.

FIGURE 1.1

Plants affect virtually all aspects of our lives. Many plants are economically important; our paper money is even made from plants.

BOXED READING 1.1
COFFEE: EVEN IF YOU DON'T KNOW BEANS

Throughout this book you'll see essays such as this that describe the excitement, lore, and recent discoveries of botany. Here, in the first of them, is "the rest of the story" about a plant extract that millions of people enjoy every day: coffee.

Coffee originated in the Middle East more than one thousand years ago. Since then, it's been used as food, medicine, wine, and even as an aphrodisiac. Ever since the Boston Tea Party in 1773, coffee has been America's most popular beverage. Today, many people treasure quality coffees and pay up to $100 per pound for gourmet brands. Such value isn't a recent phenomenon—coffee beans are used as currency in some isolated parts of Africa, and the failure of a Turkish husband to provide his wife with coffee was once considered as "grounds" for divorce.

Coffee (*Coffea arabica*) is grown from seeds. After two years, the seedlings are about 0.5 m tall and are transplanted to large plantations. Two to three years later, the plants produce white, jasmine-scented flowers that form fruit. When ripe, the fruits are red and sweet. Each fruit contains two green coffee-beans that are harvested, dried, and sold.

Different kinds of coffee plants produce different kinds of beans. Most plants produce robusta beans, which are used to make grocery-store varieties of coffee. Each of these trees produces about 5,000 fruits—1.4 kilograms (3 pounds) of coffee—per year.

Gourmet coffees are made from arabica beans, which are more susceptible to disease and more costly to harvest. These plants grow at elevations exceeding 1,800 meters (6,000 feet) and produce only about 2,500 fruits—0.7 kilograms (1.5 pounds) of coffee—per plant per year.

The coffee industry generates $12 billion per year, and the more than 10 billion pounds of coffee produced annually in fifty countries provide more than 20 million jobs. Brazil produces more than one-third of the world's coffee, helping to make coffee the world's second-largest commodity—second only to oil—in international trade.

BOX FIGURE 1.1
Coffee berries mature in 7 to 9 months, turning from dark green to yellow then red.

- The oldest giant sequoias are about 3,200 years old, meaning that the trees that grow today were already 200 years old when King David ruled the Israelites.[1] In its seventies—the sunset years of humans—a sequoia is still a teenager and bears its first seeds. Each of these seeds weighs less than 0.01 gram (i.e., about eight-thousandths of an ounce), a mere one-hundred-billionth of the weight of a full-size tree.

- The average sequoia needs more than 1,100 liters of water per day. The energy required to lift this water to the tree's leaves each day is enough to launch a can of Pepsi into a low orbit of earth.

Clearly, these giants belie the notion that plants are too static or too quiet to be as thrilling as stampeding elephants or tail-pounding dinosaurs. However, the stories of fascinating and useful plants don't stop with sequoias and redwoods. Indeed, plants have a fascinating lore:

- Although George Washington never chopped down his father's cherry tree, others have used cherry trees for many purposes. Cherry bark has been used to make a tea to ease the pains of childbirth and to cure coughs, colds, and dysentery. Queen Victoria and others since her time have enjoyed cherry brandy. Daniel Boone carved several coffins out of cherry and always kept one under his bed. He gave away all but the last one.

- The Republican Party was founded under a white oak in Jackson, Michigan in 1856. Oaks have also played prominent roles in some of Hollywood's greatest movies. For example, Twelve Oaks, the name of Ashley Wilkes' estate in *Gone With The Wind*, was named for its dozen kinds of live oaks. Incidentally, the Spanish moss that covered those oaks is not a moss, but a bizarre relative of pineapple (fig. 1.3). Similarly, poison oak is not an oak at all, but rather is a cousin of poison ivy, mango, and cashew.

- Most wood connoisseurs don't like elm. The wood is hard to split, is too wet to burn, and usually smells bad—hence the name *piss elm*. Nevertheless, elms have

---

1. Even the huge redwoods seem like youngsters when compared to bristlecone pines (see fig. 16.8b)—gnarled, low-lying conifers that seldom grow higher than about 9 meters (30 feet). Many of these pines are more than 5,000 years old, meaning that they had been growing for 500 years before the pyramids were built.

FIGURE 1.2

The sequoias are among the most impressive members of the plant kingdom. For example, the volume of the sequoias' trunk is almost twice that of a typical two-story house.

FIGURE 1.3

Spanish moss (*Tillandsia usneoides*) is an epiphyte that often grows on oaks (*Quercus*). You are more likely to see *Tillandsia* on bald cypress in swamps or on telephone wires.

witnessed many important events of history. For example, physicists at the University of Chicago met beneath an elm in 1943 to discuss the experiments that led to the atomic bomb. Today, most elms in the United States are being killed by Dutch elm disease, a disease caused by the fungus *Ceratocystis ulmi*[2] (fig. 1.4).

- Despite warnings to the contrary, Americans eat only about 8.1 kilograms (18 pounds) of apples per person per year—far less than the one per day to keep the

---

2. To try to help save these majestic trees, botanists have recruited an unlikely soldier: the bacterium *Pseudomonas syringae*. Some strains of this bacterium produce a compound that kills *C. ulmi*. In 1987 Gary Strobel and his colleagues at Montana State University discovered a genetically altered strain of *P. syringae* that made large amounts of the bacterial toxin. When preliminary experiments with greenhouse-grown elms worked well—injecting the bacterium into the trees protected the trees from later infections—Strobel and his colleagues injected fourteen fifteen-year-old elms on their campus with the genetically altered fungus. However, Strobel did this without the approval of the Environmental Protection Agency, whose mission includes preventing scientists from releasing unauthorized organisms into the environment. Strobel's release of *P. syringae* was the first deliberate release of genetically altered organisms without the government's approval. The furor that resulted from Strobel's experiment caused the project to be put on hold. On 3 September 1987, Strobel ended the experiment by cutting down all of the experimental elms.

FIGURE 1.4

Dead American elm trees ravaged by Dutch elm disease.

doctor away. Italians do better—25 kilograms (56 pounds) per year. Only the Dutch, who each eat an average of 45 kilograms (100 pounds) per year, keep the doctor away.

The Lore of Plants

Sweet potatoes were once considered strong aphrodisiacs. When Shakespeare's Falstaff shouted, "Let the sky rain potatoes!" in *The Merry Wives of Windsor*, he was hoping for sweet potatoes, not the spuds we use to make French fries.

- Chlorophyll, the photosynthetic pigment that makes plants green, was wildly popular as a deodorizer in the 1950s. Pepsodent marketed a toothpaste named Chlorodent in 1950, and soon thereafter started selling a chlorophyll-based dog food. Other chlorophyll-based products included gum, mouthwash, deodorant, diapers, popcorn, and cologne, and there were plans for chlorophyll-based salami and beer. Although the chlorophyll industry was damaged by tests done by the Food & Drug Administration (FDA) showing that chlorophyll was a poor deodorizer, the folly of the craze was summarized by a magazine article which pointed out that although goats eat almost nothing but green grass, they nevertheless stink. Soon thereafter, chlorophyll returned to its everyday job.
- We use spruce to make the 23,000 tons of newsprint needed each day to produce the United States's 65 million newspapers. A bumper issue of *The New York Times* clears about 400 hectares (990 acres) of trees, and the average tree produces enough newsprint to produce about 400 copies of a 40-page tabloid. Each person in the United States uses an average of about 290 kilograms (640 pounds) of paper each year—that's roughly equivalent to 0.8 cubic meters (25 cubic feet) of wood. Much of that paper produces the dreaded "paperwork" of many people's jobs. Indeed, 2 trillion pieces of paper flow through offices each year; the 120 billion sheets that are filed pack about 5 million filing cabinets.
- Not all products of spruce are as inexpensive as newsprint. One of the most valuable is the spruce wood that's part of a 300-year-old violin made by Antonio Stradivari (fig. 1.5). If you're lucky enough to find one of those violins in your attic, be sure to play it often; vibration keeps the wood elastic, while neglect stiffens it. Don't let your violin end up like Paganini's: his was kept locked away in a vault for fifty years, after which it was useless.

A.

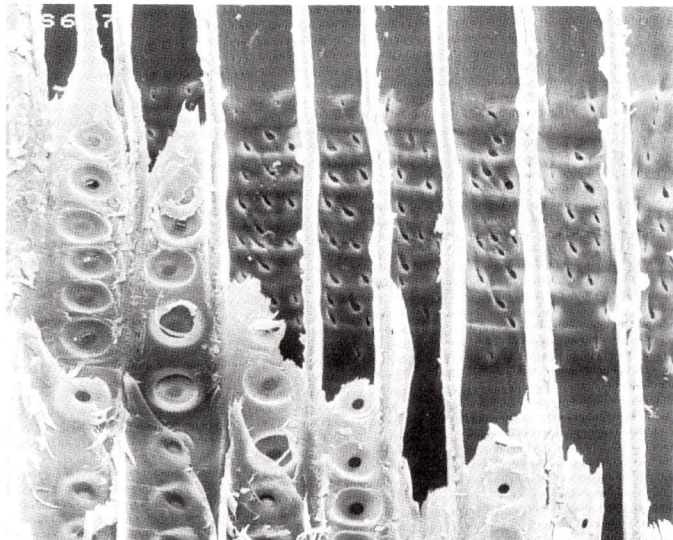

B.

FIGURE 1.5

(a) Violins such as this Stradivarius are made of the wood of spruce (*Picea*). (b) The lower photo is a scanning electron micrograph of the wood of spruce, ×500. This wood was taken from the Betts Stradivarius violin.

- Although cartoons and nagging parents try to convince you otherwise, spinach contains little that is nutritious except vitamin A.[3] Spinach might have given Popeye the Sailor an edge as a night pilot, but it probably did him little good in his dockside brawls.
- Some of Isaac Newton's great ideas were triggered when he saw an apple fall from a tree. Other important ideas have been stimulated by plants; for example, Velcro was invented by George deMestral after he examined the structure of a cocklebur clinging to his clothes.

---

3. Although spinach contains much iron, humans do not absorb it during digestion.

FIGURE 1.6

Charles and Anne Lindbergh, whose son was kidnapped. Botanical evidence helped convict Bruno Hauptmann of the crime.

FIGURE 1.7

This is one of the seedless navel orange trees that gave rise to the seedless navel orange industry. It grows at the head of Magnolia Street in Riverside, California.

- Plants have been the star witnesses in several criminal trials, most notably in the trial of Bruno Hauptmann. Botanical evidence helped convict him of kidnapping the baby boy of Charles and Anne Lindbergh (fig. 1.6). Similar evidence has been used to convict rapists, murderers, and thieves.
- The preserving effects of peat moss have shown us the gruesome details of ancient religious ceremonies and have provided us with ancient human DNA (see "Secrets of the Bog," in chapter 27).
- Cox's orange pippin apples can be traced to a neglected apple tree that Richard Cox discovered while walking across an abandoned farm. Similarly, the seedless navel orange industry can be traced to an orange tree growing near the corner of Magnolia and Arlington Avenues in Riverside, California (fig. 1.7).
- The Union army used onion juice to cleanse gunshot wounds. General Ulysses Grant, deprived of onions in one campaign, sent a terse memo to the War Department saying, "I will not move my troops without onions." The War Department sent him three cartloads.
- People have used fibers for more than ten thousand years. We use hemp to make rope, sisal to make brooms and brushes, ramie to make textiles, and flax to make linen, wrappings for Egyptian mummies, cigarette paper, and money (fig. 1.1).
- People have shown their fascination with plants by sculpting them into various shapes and naming their children after them. We've also done some rather bizarre things with plants. For example, people in Charlevoix, Michigan—presumably taking a break from their diets—baked a 7-ton cherry pie, and a fellow in Tulsa claimed to have four rooms of his house filled with turnips. However, nothing beats the story of Jay Gwaltney of Chicago who, in 1980, ate a 3.4 meter (11-foot) birch tree. It took him just over 89 hours, but he ate the *whole plant*. There was no word on what made him do it.

CONCEPT

Botany is the scientific study of plants. Plants are the basis for our national and world economy and have a fascinating history and lore.

A Botanist's View of Life

Botanists, other scientists, and nonscientists often share an almost insatiable curiosity about life and its diverse but related forms. We watch "nature shows" on television, visit zoos and botanical gardens, buy pets, and tend our gardens. Scientists, however, also often watch life in more exotic places—for example, while perched in a treetop of a tropical rain forest or in a greenhouse surrounded by a variety of plants bred specially for experiments. Such intense observation has generated many questions about plants, most of which can be grouped as follows:

- How are plants constructed?
- How do plants work?
- How did plants get here?
- Why are plants important?

Botanists have answered these and other questions by using an experiment-based process called the **scientific method**, a systematic way to describe and explain the universe based on observing, comparing, reasoning, predicting, testing, concluding, and interpreting. All in all, it's not very different from the way that many nonscientists think about this fascinating thing we call life.

The Scientific Method

Superficially, botany is a collection of "facts" that describe and explain the workings of plants and the living world. Although these facts are interesting—even fascinating—they are not the essence of botany. Rather, the excitement of botany lies in the intriguing observations and the carefully crafted experiments that botanists and others have devised to help us learn about plants. The important thing here is the *process* of botany—not just knowing the facts, but appreciating *how* botanists discovered (and continue to discover) those facts. This process is the distinguishing feature of science that you'll read about throughout this book.

The scientific method begins with things that we are all familiar with: observation and curiosity (fig. 1.8). Such observations can happen just about anywhere—in a research laboratory, garden, or while you're watching a television program. For example, the observation that leaves of many plants can follow the movement of the sun catches our attention and piques our curiosity. This may prompt us to ask, "What enables the leaves of some plants to follow the movement of the sun?" Similarly,

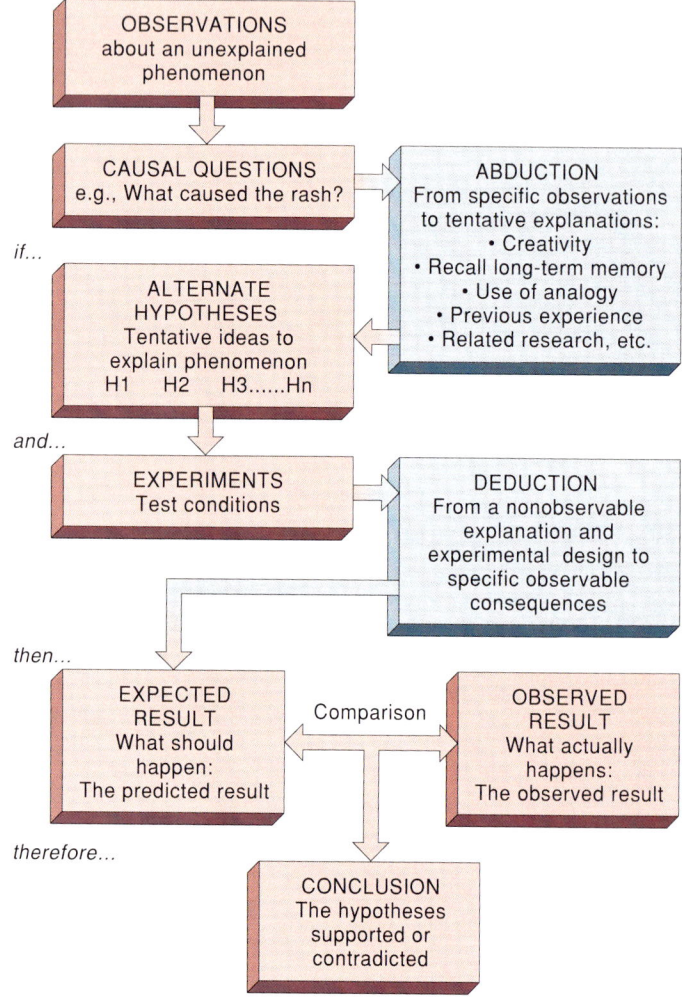

FIGURE 1.8

The process of the scientific method.

the observation that some plants in our infested garden are not attacked by Polish whiteflies while others are infested may lead us to ask, "What enables resistance to Polish whiteflies in some plants but not others?" These are the kinds of **causal questions** that are at the heart of the scientific method. Science is fundamentally about finding answers to such questions. To find these answers, botanists use past experiences, ideas, and observations to propose **hypotheses**—a fancy name for guesses—that may produce predictions. To determine whether predictions are accurate, botanists do experiments. If experimental results match the predictions of a hypothesis, the hypothesis is accepted; if results do not match the predictions, the hypothesis is rejected. In this way some hypotheses are accepted and some are not. Our understanding improves by eliminating some of the explanations (rejecting hypotheses). However, hypotheses are accepted or rejected provisionally, for experiments can neither prove nor disprove a hypothesis. The net effect of all this is to make scientific progress by revealing answers piece by piece.

Posing hypotheses is perhaps the most creative step in the scientific method. It requires a type of reasoning called **abduction,** which is the process of devising explanations for observations. Abduction involves sensing ways that a new situation is somehow similar (analogous) to other known situations, and using this similarity to make hypotheses about the new situation.

Let's see how the scientific method works in real life. Suppose that you wake up one night with a bothersome rash (observation). You want to know what caused this rash (causal question). You first try to figure out how this situation is similar to other known situations. Perhaps a classmate had a similar rash that was caused by an allergy to the grass at the neighborhood baseball park. You realize that your rash has appeared after all of the baseball games that you've played at that park. The similarity (analogy) between your friend's allergy and your rash leads to the explanation (hypothesis) that your rash is caused by an allergy to the grass, too.

If your hypothesis is correct, you would not expect to get a rash from other kinds of grass (prediction). When you play baseball at several other parks, all with a different kind of grass than your neighborhood park (experiment), you do not get a rash (results). Your results match the prediction of the grass-allergy hypothesis, from which you decide that your rash is caused by the grass at your neighborhood park (conclusion). However, if this was your only hypothesis, you have not actually identified the cause. Perhaps your rash was caused by the paint used in the dugout. Or perhaps it was caused by fertilizers, pesticides, bacteria, molds or other agents associated with the grass or soil at one baseball park and not another. By testing a single hypothesis, you have not ruled out any of these other possibilities. To do so, you would have to devise alternative hypotheses, make predictions from them, and obtain experimental results to compare with the predictions. By this process, you may be able to reject some but not all of the alternatives, or you may be able to reject all hypotheses, even the grass-allergy hypothesis. Either way, you make progress by testing several hypotheses, not just one.[4] If you can reject all but the grass-allergy hypothesis, you can more confidently conclude that the grass caused your rash. Furthermore, if the grass is in the outfield, but not in the infield, you may decide to become an infielder so that you can avoid the irritating plant (use of new knowledge).

---

4. The danger of trying to do science armed with just one hypothesis is that the scientific method rapidly degenerates into a hunt only for supporting evidence. In the example above, if the rash happened to be caused by a mold that infects only the kind of grass at your neighborhood park, then predicting that you would not get a rash at a park with different grass is correct even though the grass-allergy hypothesis is false. You would come to the wrong conclusion because you did not examine other possible explanations (alternative hypotheses).

If all of this seems like common sense, it is. As Thomas Henry Huxley said, "Science is nothing but trained and organized common sense." Although a conclusion marks an end to the scientific method for a particular experiment, it seldom ends the process of scientific inquiry. Any conclusion must be placed into perspective with existing knowledge. Moreover, an [...] that the se[...] cting, testing, [...] deas spawned [...] conclusion is [...] more to study,

[...] open mind—
[...] hod requires
[...] ts—and not
[...] ation of re-
[...] n nature to
[...] es not "fit"
[...] le believed
[...] dies—espe-
[...] he air and
[...] that grew on it—"proved" that life spontaneously arises from the nonliving, a conclusion that brilliant scientists such as Newton, Harvey, and Linnaeus never questioned. However, spontaneous generation was finally rejected by another set of different, more critical experiments done by Louis Pasteur. The conclusion from those experiments—namely, that life does not arise from decaying meat—surprised many who believed that mice were created by mud, that flies came from rotting beef, and that beetles sprouted from cow dung. Similarly, biologists initially rejected the idea that DNA is the genetic material, thinking that protein was a more likely candidate.

Perhaps no one knew the conflicting feelings of joy and frustration that accompany a scientist's discovery of a quirk of nature better than the late Barbara McClintock (fig. 1.9). In the 1940s, McClintock studied the inheritance of kernel color in corn—by, as she put it, "Asking the maize (corn) plant to solve specific problems and then watching it respond." While watching her carefully bred plants respond, McClintock noticed that some kernels had a peculiar pattern of spotting. McClintock concluded that the units of inheritance (genes) moved around in corn cells. This seemed preposterous at the time, for genes were thought to be immovable parts of larger structures called chromosomes. Despite McClintock's evidence, the scientific community refused to accept her conclusion. However, after a decade of additional discoveries of mobile genes in other organisms, Nobel Prize winner Barbara McClintock was finally recognized as the brilliant botanist that she had always been. Her observation of "jumping genes" in corn plants almost five decades ago may ultimately help explain how roving genes cause certain cancers in humans.

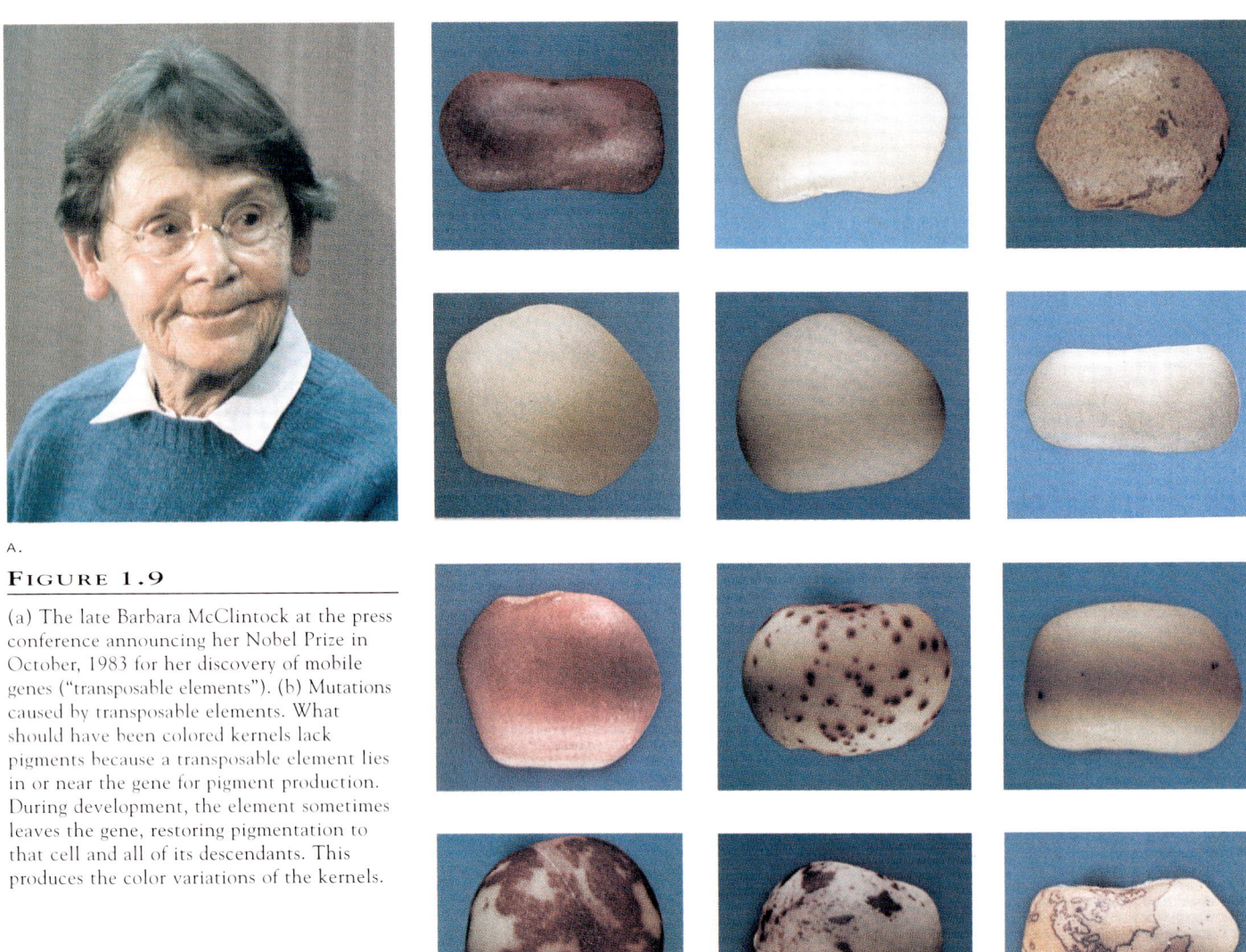

FIGURE 1.9

(a) The late Barbara McClintock at the press conference announcing her Nobel Prize in October, 1983 for her discovery of mobile genes ("transposable elements"). (b) Mutations caused by transposable elements. What should have been colored kernels lack pigments because a transposable element lies in or near the gene for pigment production. During development, the element sometimes leaves the gene, restoring pigmentation to that cell and all of its descendants. This produces the color variations of the kernels.

CONCEPT

The scientific method is a systematic way to describe and reveal the universe based on observing, comparing, reasoning, predicting, testing, and interpreting. Everyone uses the scientific method at times.

Is the Scientific Method Foolproof?

Although the scientific method is a powerful tool for answering some kinds of questions, it is not foolproof. For example, several studies have shown that animals that are fed large doses of vitamin E live longer than those receiving only average amounts of the vitamin. From those studies, many concluded that vitamin E retards aging, thereby prompting all sorts of claims by skin-cream manufacturers. However, the animals fed large amounts of vitamin E also lost weight, another factor that correlates positively with increased life expectancy. The original experiment did not distinguish these possible interpretations. As is usually the case, more experiments are necessary to distinguish the other possible interpretations.

Many experiments that biologists do are difficult to validate. For example, consider the experiments of Stanley Miller, who sought to re-create the chemical reactions that might, on the primitive earth, have formed the compounds found in living organisms today. Miller's experiments on the origin of life are difficult to validate. Although he provided interesting results and inspired much more research, we can't really know whether he successfully re-created the conditions and events of the planet's early history—unless, of course, we invent a time machine to go back and see.

BOXED READING 1.2

BREAKFAST AT THE SANITARIUM: THE STORY OF BREAKFAST CEREALS

Cornflakes and several other breakfast cereals can be traced to Seventh-Day Adventists, an American religious sect. In the 1860s, the Adventists organized the Western Health Reform Institute in Battle Creek, Michigan. This institute was later renamed the Battle Creek Sanitarium and, due to the sect's religious beliefs, served only vegetarian meals. Each morning, John Kellogg (a physician) and his brother Will made breakfast for the sanitarium's patients by running a potful of cooked wheat through some rollers to form sheets of wheat. These sheets, when toasted and ground up, became a sort of cereal.

One night in 1895, the Kellogg brothers were called away on an emergency and left some wheat soaking for more hours than usual. When they tried to make their wheat-sheets the next day, each wheat grain formed its own flake. Patients at the sanitarium had a new breakfast food: wheat flakes. Four years later, the Kellogg brothers tried their new approach with corn, added some malt flavoring, and produced the first cornflakes. Patients at the sanitarium liked the flakes, especially ones made of corn, and thus was born Kellogg's Corn Flakes of the Kelloggs' cereal empire, which still exists in Battle Creek.* Although the Kellogg brothers later had a falling out (Will won use of the family name, which explains why his signature appears on the boxes of cereal), their product remained immensely popular. Indeed, within fifteen years, more than forty cereal companies were centered in or near Battle Creek, some of them operating out of tents.

Interestingly, one of the patients (a suspender salesman) at the sanitarium saw how the Kelloggs made their cornflakes. He later began a competing business that produced breakfast cereals, naming his first cornflakes "Elijah's Manna." However, the American clergy considered that name blasphemous, prompting the former patient in 1908 to change the name of the flakes. That patient was C. W. Post, who named his flakes Post Toasties.

*Interestingly, wheat flakes (e.g., Wheaties) weren't marketed until twenty-six years after cornflakes were sold. The process used by the Kelloggs to make flakes was soon modified to produce grains, shreds, and puffs. The puffs are made by heating the grains, causing pressure to build. When this pressure is released, the water vapor in the grain explodes the grain into a puff.

Writing to Learn Botany

What are the strengths and weaknesses of the scientific method?

Creativity in Science

Science is not always as planned and organized as the scientific method suggests. Sometimes discoveries involve luck and creativity, as with Barbara McClintock. The creative side of science involves making the mental connections to take advantage of accidental observations. For example, consider the case of Alexander Fleming. In 1928 Fleming left a culture dish of disease-causing *Staphylococcus* bacteria uncovered in his lab. Other organisms quickly contaminated the culture. Just before throwing out the contaminated plate, Fleming noticed several clear areas on the plate where the bacteria were not growing. This keen observation suggested to Fleming that some contaminant from the air had stopped the growth of the bacteria. Fleming tried to isolate and characterize the substance responsible for this inhibition of growth, but failed because he was a poor chemist. However, other scientists later identified and purified penicillin, a product of the fungus whose effects Fleming had accidentally discovered. Fleming's observation, along with the follow-up work, saved millions of lives and eased much pain.

Was Fleming lucky? Perhaps—but he would not have discovered penicillin if he had not been observant and able to recognize the potential importance of what he saw. His discovery, like so many others, exemplifies a famous saying: "Luck is when preparation meets opportunity." Most others would not even have recognized the clear areas on the plate, much less thought that those areas could be significant. Such observations are critical to science.

The Role of Technology in Scientific Discovery: A Case Study of the Cell Theory[5]

Discoveries are determined by our ability to observe, which often depends upon technology. To appreciate this, consider the development of the cell theory, the set of postulates that holds the cell to be the fundamental unit of life. Before the

---

5. A theory is a collection of assumptions (postulates) that, taken together, try to explain a set of related phenomena.

middle of the seventeenth century, no one even knew that cells existed. Because most cells are too small to be seen with the naked eye, their discovery awaited the invention of microscopes. In 1665, the English scientist Robert Hooke used a crude, hand-built microscope to study a thin slice of cork. To his surprise, Hooke saw a honeycomb of compartments that he described and named *cells,* meaning "little rooms." Although Hooke's discovery was a landmark event in the history of biology, Hooke thought that these cells were unique to cork, and therefore never realized the full significance of his discovery. Several years later, Dutch naturalist Anton van Leeuwenhoek built several tiny microscopes, each of which would fit easily in the palm of your hand. With these microscopes, van Leeuwenhoek studied pond water and became the first to describe microscopic organisms (he also described other organisms and discovered red blood cells). However, like Hooke, Leeuwenhoek did not fully appreciate the importance of his own work. Neither did other biologists, for the discoveries of Hooke and van Leeuwenhoek were ignored for almost two centuries.

By the early 1800s, technology had produced stronger and easier-to-use microscopes. In 1838, German botanist Matthias Jakob Schleiden used one of these microscopes to study a variety of plants. He concluded that all plants "are aggregates of fully individualized, independent, separate beings, namely the cells themselves." This was the first postulate of what we now call the **cell theory,** (see Chapter 3) and was a brilliant example of inductive reasoning—making a generalization based on concurring observations. The next year, German zoologist Theodor Schwann reported that animals are made of cells and proposed a cellular basis of life. Later, in 1855, the great German pathologist Rudolf Virchow completed the modern cell theory when he induced that "the animal arises only from an animal and the plant only from a plant." This strong statement refuted the idea of spontaneous generation—that life could arise from nonliving matter—that had been suggested by Aristotle and had persisted for many centuries. Today, the cell theory is tremendously important because it emphasizes the similarity of all living systems and thereby unites the many studies of different kinds of organisms. The cell theory guides the work of many biologists and physicians as they study disease and food production. However, the development of the cell theory was limited by technology (the availability of microscopes), and depended upon keen observation and inductive reasoning.

Today, much of science is driven by new technology. Just as microscopes revolutionized our understanding of cells, so too will new technology produce remarkable discoveries and new insights. That technology will overturn many of the "facts" that you read about in this or any other science book. This is why *Botany* concentrates so strongly on the *process* of science; the facts may change, but the process used to discover them—that is, the process that enables you to think critically, solve problems, and even become a professional scientist—remains the same.

We All Think Like Scientists At Times

The cycle of questioning and answering that defines the scientific method is second nature to a practicing scientist; it's more a philosophy than a set of rules. However, the use of the scientific method is not restricted to scientists: we all observe, compare, guess, plan, do experiments, and interpret and use the results of experiments. For example, our ancestors used experimental trial-and-error to determine which plants were edible. In a specific experiment in 1820, a New Jersey colonel named Robert Johnson bravely ascended the steps of a county courthouse and ate a basketful of tomatoes, much to the amazement of the 2,000 townspeople who thought that tomatoes were poisonous. His bold experiment and its results (his survival) showed the townspeople that tomatoes are not poisonous. Parents do a similar experiment when they introduce food one item at a time to their children to check for allergic reactions. They also experiment—by tasting the food in question—to check its safety or taste.

Our use of the scientific method has greatly advanced knowledge and improved our quality of life. For example, how would Gregor Mendel, a monk who described the principles of heredity more than a century ago, feel if he knew that his observations of pea plants would one day help to explain how a perfectly healthy couple could have a child afflicted with a crippling disease?[6] Similarly, it took keen observation, experimentation, and reasoning to discover that the mild and supposedly harmless tranquilizer called thalidomide could cause 10,000 European children to be born without arms and legs, and that a tiny virus could destroy the immune system and cause AIDS.

THE UNIFYING THEMES OF BOTANY

Our studies of plants have shown us that plants, like all organisms, share several important features. Just as you appreciate an opera or symphony if you know its themes, so too will you better understand plants if you understand their repeating themes (fig. 1.10). These themes are integrated throughout this book.

Plants consist of organized parts.

Just as builders use the same kinds of bricks to make large or small houses, plants use the same building blocks to make all of their parts. Building all of these parts from the same building blocks is analogous to making thousands of words from the twenty-six letters of the alphabet—the secret is in the arrangement of the building blocks. The differences we see in organs such as leaves and roots result primarily from differing arrangements of similar tissues rather than to the presence of unique tissues.

---

6. Mendel did his famous experiments in the monastery garden at Brünn, Austria, in the mid-1800s. He also taught natural science at Brünn high school, but never managed to pass the exams necessary to get a teaching certificate.

FIGURE 1.10

The unifying themes of botany. (a) Plants consist of organized parts. The cells in the center of this cross section of a buttercup (*Ranunculus*) root conduct fluids, while those surrounding the center of the root store food. (b) Plants exchange energy with their environment. Virtually all energy that powers life is captured from sunlight. (c) Plants respond and adapt to their environment. The growth of these stems toward light increases the chances of the leaves being able to absorb light for photosynthesis. (d) Plants reproduce. Flowering plants such as this Easter lily (*Lilium longiflorum*) reproduce sexually by producing flowers. (e) Plants share parts of a common ancestry and have coevolved with other organisms. Here, an Anna's hummingbird visits a flower. Both of these organisms evolved with compatible anatomy.

Plants exchange energy with their environment.

Although organization is a requirement for life, it alone does not make something alive. Plants, like all organisms, absorb energy from their surroundings and, in turn, impact their environmemt as they use that energy for their activities. This conversion of energy from one form to another occurs via a set of chemical reactions collectively called *metabolism*.

Plants respond and adapt to their environment.

Such adaptations are inherited characteristics that help to ensure survival. These responses include the ability of plants to detect and respond to stimuli such as gravity; to convert light energy to chemical energy; to gather nutrients; and to lure animals into helping pollinate, defend, and disperse the plants. All of these adaptations and responses are important because they increase a plant's chances of survival and reproduction. Today's more than 260,000 species of plants show that there are at least 260,000 ways that plants adapt to their environment.

Plants reproduce.

Their reproduction can be as "simple" as splitting a cell, or as complex as orchestrating flower production, mobilizing energy reserves, or luring insects, birds, and other helpers. Although most plants can reproduce sexually or asexually, the greatest variation results from sexual reproduction, in which pieces of DNA called *genes* move from one generation to the next. This explains why plants resemble their parents, just as children resemble theirs. Genes, combined with the influence of the environment, produce the characteristic features of plants. The diversity of plants results from their diverse environments and differences in their DNA, and is a product of evolution. Understanding the language of DNA—that is, how cells read DNA and translate its instructions into metabolism and growth—is a major goal of many biologists and the basis of an exciting business: biotechnology.

Plants share parts of a common ancestry.

This ancestry was best explained in the most important science book ever written: Charles Darwin's *The Origin of Species*, published in 1859. This book, "one long argument" for the diversity of life, used evidence, logic, and analogy to explain the central role of variation in the evolution of life. Although Darwin's ideas were based largely on the ideas of other scientists, no one had ever produced such a convincing argument. Darwin's argument forced scientists to accept his theory of natural selection despite all opposing arguments, primarily those of religious dogma. Darwin's book shook the foundations of biology and remains one of the most important books ever written.

The products of evolution surround us, and we've used geological and biochemical evidence to infer how and when life's diverse forms appeared on earth. Life on earth probably began about 3.6 billion years ago as a simple cell. Since then, natural selection has produced a variety of plants and plantlike organisms. For example, the algae are about a billion years old, and ferns appeared on earth at least 350 million years ago. Cone-bearing trees such as pines may have appeared 300 million years ago, and the flowering plants, which are relative newcomers, appeared a mere 200 million years ago. These flowering plants are the most diverse group and include monocots (plants such as grasses, lilies, and orchids, whose leaves have parallel veins) and dicots (plants such as apples, oaks, and sunflowers, whose leaves have netted veins). Today, we use and study all of these kinds of plants.

CONCEPT

Plants consist of organized parts, exchange energy with their environment, respond and adapt to their environment, reproduce, and share parts of a common ancestry.

Botany discusses these themes in several contexts, including the relation of structure and function, the evolution and diversity of plants, and the lore, importance, and uses of plants. Throughout this book you'll also learn about a new, exciting, multimillion dollar business: plant biotechnology. Biotechnology is a way of using organisms to make commercial products. Although we've used biotechnology for several decades, our ability to directly manipulate a plant's genome began in 1983 when botanists transferred a gene from a bacterium into a plant. Today, biotechnology and molecular biology are revolutionizing biology and biology-based industries. Here are a few of the ways that we use biotechnology:

- To make vitamins and drugs. For example, just as pigs are now being used to make human hemoglobin, plants are being transformed into factories that make drugs and oils. A promising new vaccine for hepatitis B is made with baker's yeast, and plants are now used to make serum albumen, which is used for fluid replacement in burn victims.

FIGURE 1.11

Using genetic engineering to breed resistance to insects. Both of these tomato plants were exposed to destructive caterpillars. The nonengineered plant (left) was eaten; the engineered plant (right) was not damaged.

- To produce plants that resist drought, disease, and insects. Botanists at Monsanto have used genetic engineering to produce disease-resistant plants. Others have transferred a gene from *Bacillus thuringiensis*, a common soil bacterium, into tomato and tobacco plants. This gene produces a protein which, when eaten by pests such as the tomato hornworm, induces paralysis and death (fig. 1.11). The protein has no effect on other organisms, making it environmentally safe. Although most *B. thuringiensis* is used to fight caterpillars, other strains of the bacterium are being developed to fight beetles and mosquitoes.[7] Botanists are searching for ways to help plants resist diseases that lay waste, on average, up to 12% of crops worldwide.

- To produce plants resistant to herbicides. Genetically engineered plants can now tolerate glyphosate, the active ingredient in Roundup, a nonspecific herbicide that degrades quickly in the environment. Consequently, farmers can spray fields with Roundup and kill all the plants except the crop plant. This could increase yields significantly, because weeds reduce agricultural productivity by more than 10%.

- To make foods and beverages such as yogurt, cheese, bread, and beer. Botanists at Quaker Oats are using biotechnology to increase the protein content of their oats, and botanists at Kraft are using similar techniques to decrease the amount of saturated fat in the soybean oil used in their products. Enzymes from engineered tobacco plants are used to make bread and low-calorie beer, and

---

[7]. Our use of *B. thuringiensis* as a soldier to fight pests has been significant. For example, 90% of the pesticide used today to control spruce budworm in Canada is *B. thuringiensis*, up from 10% in 1979. This amounts to an annual release of about 2.3 million kilograms of *B. thuringiensis*—that's about 4.5×10^{20} bacteria, a number roughly equivalent to the estimated number of stars in the universe.

to clarify wine and juices. Genetically engineered potatoes containing increased amounts of starch are easier and cheaper to process for fries and chips.

- Other uses of biotechnology include making better-tasting food, cleaning the environment, recycling wastes, preventing tooth decay, and producing antibiotics, biodegradable plastics, and fragrances. Microbiologists currently use about a dozen genera of fungi and bacteria to make about 4,500 antibiotics.

CONCEPT

Plant biotechnology is a multibillion-dollar industry that is changing our lives and society.

What are Botanists Doing?

Botanists have used biotechnology to produce a variety of genetically engineered plants. Many of these plants could greatly benefit society—for example, by capturing heavy metals from groundwater and locking them in their tissues.

Go to a library and read two articles that describe uses of genetically engineered plants. How could these plants improve the quality of our lives? What restrictions, if any, should be placed on botanists who try to engineer such plants? Explain your reasoning.

Throughout this book you'll learn about the biotechnological revolution, as well as its more unusual applications. For example, genetically engineered tobacco plants now produce human antibodies useful for diagnosing and treating disease. Similarly, biologists have cloned and transferred a gene for an unusual protein—the one that keeps winter flounder from freezing—into tomato and tobacco plants. This "antifreeze gene" could help extend the growing range of these plants.

Another important goal of *Botany* is to show you the process of science and the doings of scientists. You'll see the scientific method in action—scientists formulating hypotheses, doing experiments, gathering data, and drawing conclusions. You'll also see that scientists, like other people, are often passionate about their work. For example, Rachel Carson wrote *Silent Spring* in response to a letter that she received in January of 1958 from Olga Huckins describing how a small part of the world had been made lifeless by pesticides. Carson's response was to say, "There would be no peace for me if I kept silent."[8] Such passion fuels the work of many successful botanists.

---

8. Chemical companies and several manufacturers spent thousands of dollars to block the publication of *Silent Spring*, and Carson was maligned as "ignorant" and "hysterical." Nevertheless, *Silent Spring* became a best-seller that prompted President Kennedy to create a special panel of the Science Advisory Committee to study the use of pesticides. That panel's report vindicated Carson and stimulated the government to act against pollution. *Silent Spring* launched the environmental movement.

A final goal of *Botany* is to help you appreciate living organisms. This is perhaps the most important goal of any biology book and of any biologist's teaching, for such an appreciation of life is the first step toward respecting and conserving life. Appreciation and respect for life are critical, especially in light of our neglect of the environment for many decades—a neglect that is now costing us money and killing earth's organisms. For example:

- In 1992 businesses spent more than $70 billion to try to clean up their messes in the environment. All of these costs are passed to consumers as higher prices. Pollutants such as lead, mercury, and DDT kill many organisms.

- Producing 1,000 tons of low-grade paper from virgin pulp produces more than 40 tons of pollutants and requires 1,700 trees, 25 million gallons of water, and 20 billion joules of energy. That amount of energy is equivalent to the energy released during the explosion of five tons of TNT.

- In 1980 we cleared about 11.3 million hectares of rain forest—an area equivalent to that of Pennsylvania. By clearing these forests, we drive animals and plants to extinction and therefore lose many potentially valuable drugs—drugs that could save lives. Botanists estimate that as many as 50,000 species of tropical plants may be extinct within the next few decades because of the destruction of rain forests.

- More than 200 species of plants are thought to have become extinct during the past two hundred years. About 900 species are endangered (that is, in danger of becoming extinct in part of or throughout the world), and more than 1,400 are threatened (that is, likely to become endangered). About 5% of the 80,000 species of plants native to temperate regions are near extinction. This has been caused by habitat destruction, overgrazing by domestic animals, the introduction of foreign plants, and the destruction of pollinators.

- Red spruce on Mt. Mitchell, North Carolina, are dying because of acid rain. Similarly, almost 40% of the trees in Germany are affected by acid rain, and many ponds in the Adirondack Mountains of upstate New York are nearly lifeless because of low pH caused by acid rain. Acid rain has dropped the pH of these ponds to about 4.1—far too low for most organisms. These trees and ponds are like the canaries carried into mines to warn of natural gas: unless we heed their message, we'll continue to pay an ever-increasing price.

Although people today are more environmentally aware than those of previous generations, our environmental problems don't seem to be improving. For example, in 1980 most environmentalists talked about the "dangerous" rate of deforestation. By 1990, that rate had nearly doubled. This year, we'll clear rain forests having an area equal to that of Florida *plus*

Maine. Similarly, our land, air, and water are becoming increasingly polluted. Food production is high, yet millions of people are starving. What are we to do?

Such problems will not be solved by governmental panels, presidential commissions, blue-ribbon committees, or expert testimony before Congress. Such groups have been convened for decades, during which time our problems have only worsened. These problems will be solved only by a scientifically literate public. This textbook uses botany to teach such literacy; we hope it will enable you to help solve the many problems that we face.

Doing Botany Yourself

Botanists, like all other scientists, often face ethical questions when doing their work. To appreciate this, consider taxol, a drug that shrinks tumors in many women whose ovarian cancer resists other therapies. Taxol is present in bark of Pacific yew, a relatively rare tree that grows in forests of the Pacific northwest. Because the harvesting of taxol kills the tree, many botanists fear that a large-scale harvest of taxol could endanger the Pacific yew and disrupt the forest ecosystem (it takes 12,000 trees to make only 1.1 kilograms of taxol).

Should we harvest taxol so that we can treat everyone who needs the drug, even if this means endangering the yew and upsetting the ecological balance of the forests? As a scientist and concerned citizen, what ethical principles would guide your work when faced with such a dilemma?

Botany will help you appreciate what botanists know, what they don't know, and what they do. You'll learn not only the "facts," but the *process* of doing botany. Your observations and ideas concerning life will be more meaningful if you have more experience and learning on which to build. This book and your course will give you some of this valuable background, which should enable you to apply the scientific method to your own observations of the living world. There's no reason why you can't be the next person to make a major scientific discovery.

Chapter Summary

Botany is the scientific study of plants. Plants have a fascinating history and lore, and our national and world economies are based on plants. Virtually everything we do is influenced by plants.

Botanists learn about plants and the world by using the scientific method, a systematic way to describe and explain the universe, based on observing, comparing, reasoning, predicting, testing, and interpreting. Everyone uses the scientific method at times.

Plants consist of organized parts, exchange energy with their surroundings, respond and adapt to the environment, reproduce, and share a common ancestry. Plant biotechnology is a multibillion-dollar industry that is changing our lives and society.

Questions for Further Thought and Study

1. Why study plants?
2. Interferon is a protein that animals make to fight viral infections. Botanists recently inserted the gene for interferon into turnips, hoping the protein would make the plants more resistant to viral diseases. The experiment failed; engineered plants were no more resistant than were untreated plants. However, the interferon from the turnips appears to be active in animals, including humans. What does this tell you about the process of science?
3. How do you use the scientific method in your daily life?
4. Is the scientific method foolproof? Explain your answer.
5. How has our understanding of life been affected by technology?
6. Can the scientific method be used to make any judgments? Explain your answer.
7. How do plants influence your life?

Suggested Readings

ARTICLES

Darley, W. M. 1990. The essence of "plantness." *The American Biology Teacher* 52:354–357.

Davis, Jerry D., Ann Korschgen, and Barbara W. Saigo. 1989. Employment prospects in biotechnology. *The American Biology Teacher* 51:346–348.

Gibbons, Ann. 1990. Biotechnology takes root in the third world. *Science* 248:962–963.

Gibbs, A., and A. E. Lawson. 1992. The nature of scientific thinking as reflected by the work of biologists and by biology textbooks. *The American Biology Teacher* 54:137–152.

Moffat, A. S. 1992. Plant biotechnology: High-tech plants promise a bumper crop of new products. *Science* 256:770–771.

Roberts, Leslie. 1988. Extinction imminent for native plants. *Science* 242:1508.

Stern, W. L. 1988. Wood in the courtroom. *World of Wood* 41:6–9.

BOOKS

Morton, A. G. 1981. *History of Botanical Science. An Account of the Development of Botany from Ancient Times to the Present Day.* New York: Academic Press.

Overfield, R. A. 1993. *Science with Practice: Charles E. Bessey and the Maturing of American Botany.* Ames: Iowa State University Press.

Simpson, B. B., and M. Conner-Ogorzaly. 1986. *Economic Botany: Plants in Our World.* New York: McGraw-Hill.

UNIT ONE

Cells as the Fundamental Units of Life...

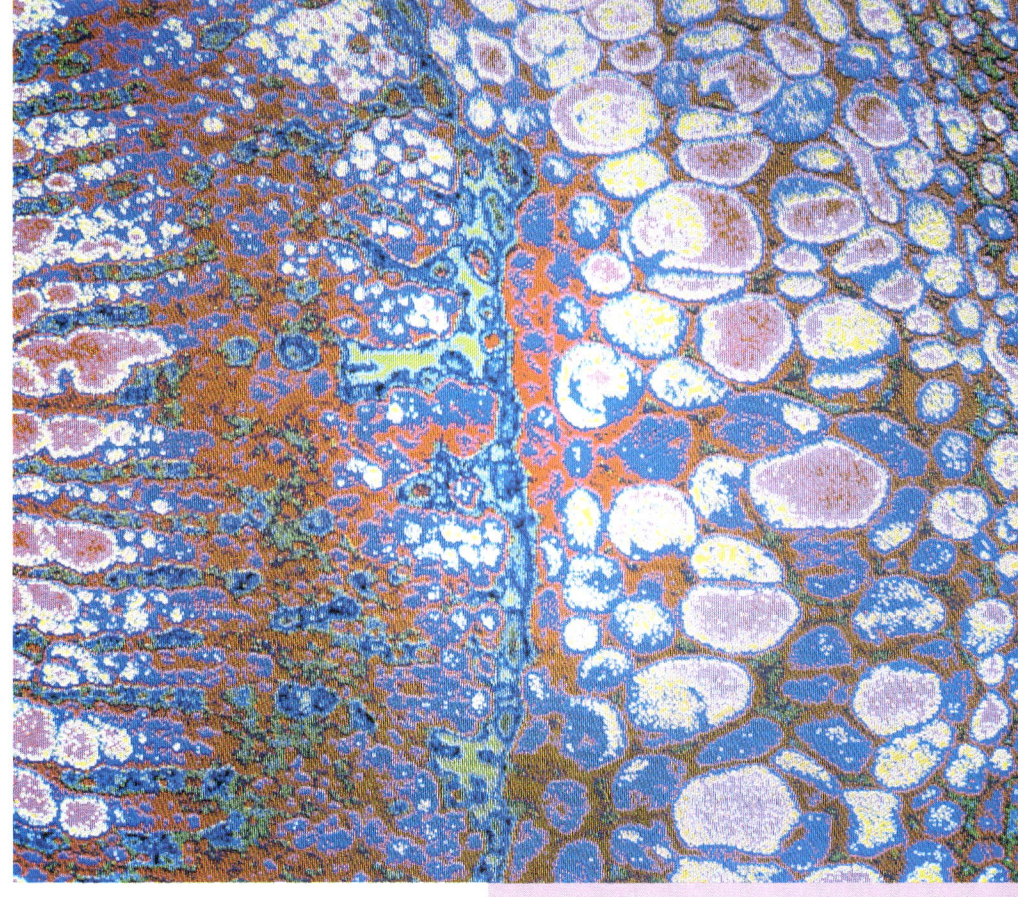

One of the unifying themes of botany, as mentioned in Chapter 1, is that plants consist of organized parts. This construction includes several levels of organization, from atoms and molecules to cells, tissues, organs, and whole plants. Specialties within botany often follow these organizational lines. Biochemistry involves molecules, physiology covers biochemical pathways, anatomy seeks information about structure, and morphology compares variation among whole plants and their different organs.

The most common approach to learning botany begins at the molecular level and moves up the organizational ladder to whole plants, as we do in this textbook. We begin in Unit 1 by discussing the molecular aspects of plant design (Chapter 2); then we describe the structure of cells (Chapter 3) and the main features of membranes and how they work (Chapter 4). Unit 1 should therefore give you a good foundation for understanding the key cellular processes of plants that are presented in Unit 2 ("Energetics") and Unit 3 ("Genetics"). Unit 1 will also help you understand how plants grow and reproduce, which are the main subjects of Unit 4 ("The Form and Function of Plants"), Unit 5 ("Regulating Growth and Development"), and Unit 6 ("Nutrition and Transport").

We could just as easily have saved Unit 1 for the end of the text and instead presented botany from the top; that is, we could have begun with whole plants and gone down the organizational ladder. This descending order of presentation is justifiable because the organization of plants may be controlled at the organism level, not at the cellular or molecular level. Evidence for this point of view is discussed in Chapter 3. This issue is significant, because a "molecules-to-cells-to-tissues" approach implies that this is the only way plants can be organized. However, remember that there is an alternative hypothesis. We present botany in ascending order more as a matter of convenience and because of the historical tradition of our own professional training.

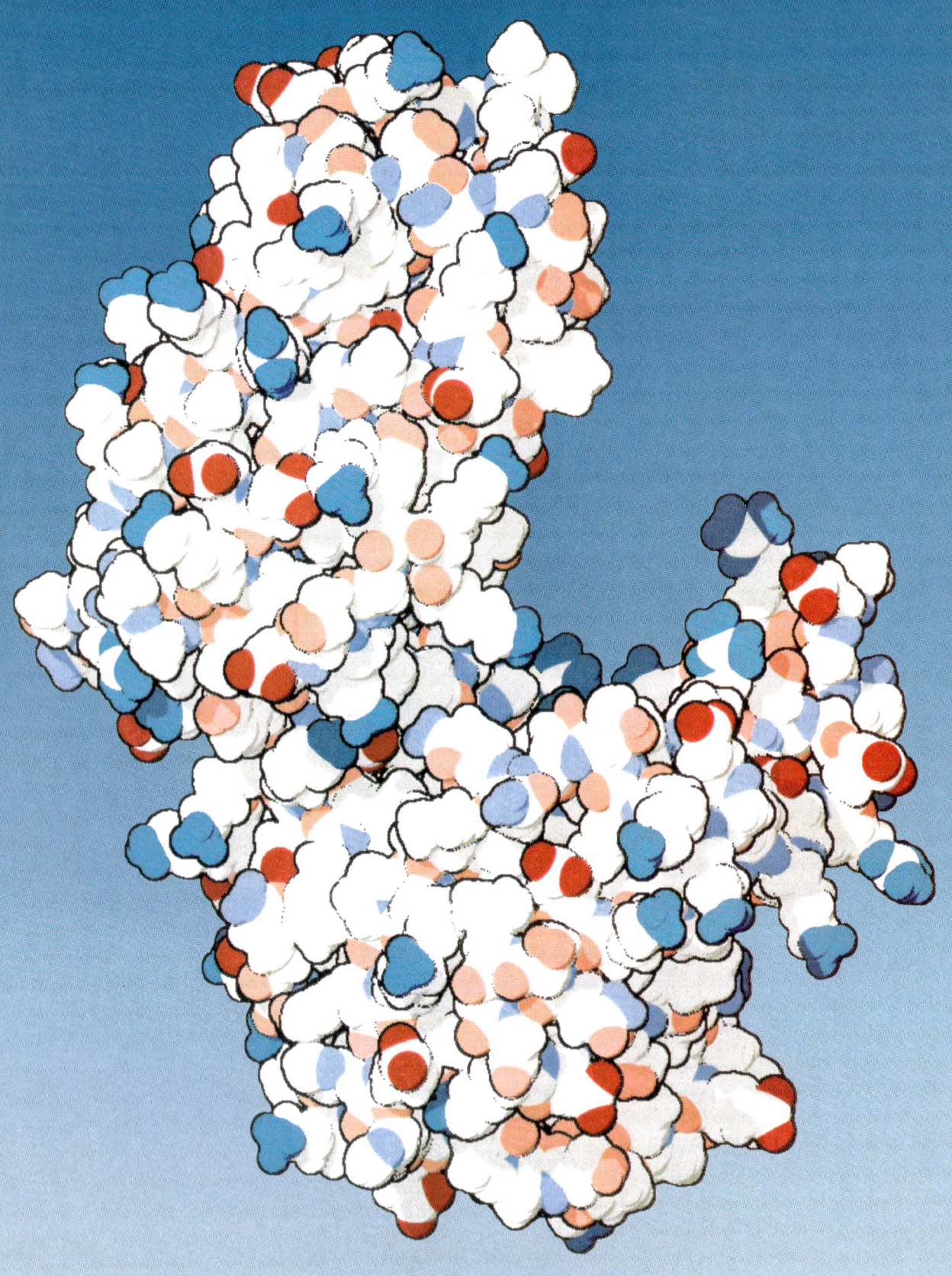

Computer graphic models of the enzyme phosphoglycerate kinase. This enzyme transfers a phosphate group from 3-phosphoglycerol phosphate to adenosine diphosphate, which becomes adenosine triphosphate (ATP). The diagram here shows what phosphoglycerate kinase would look like at about 8 million times its normal size.

Atoms and Molecules: The Building Blocks of Life

CHAPTER 2

Chapter Outline

INTRODUCTION
ELEMENTS OF LIFE
LARGE MOLECULES: POLYMERS AND THEIR MONOMERS
    Carbohydrates

BOX 2.1
READING CHEMICAL STRUCTURES

    Proteins
    Nucleic Acids
    Lipids

BOX 2.2
WHAT'S WRONG WITH TROPICAL OILS?

SECONDARY METABOLITES
    Functions of Secondary Metabolites
    Alkaloids
    Terpenoids
    Phenolics
    Minor Classes of Secondary Metabolites
Chapter Summary
Questions for Further Thought and Study
Suggested Readings

Chapter Overview

Organisms are a succession of increasingly complex levels of organization, from molecules to individuals and communities. The traits that distinguish a leaf from a root, or a giant redwood from a dandelion, emerge from the type and arrangement of their molecules. The tree you see out the window consists of thousands of different kinds of molecules, some of which are shared by all life, and some of which are unique to that kind of tree. These molecules are joined to form cellular structures such as membranes, cell walls, and other cell organs. Different types of cells are grouped into tissues that form the leaves, roots, and branches of the tree. The properties of each level of complexity emerge from its components.

In this chapter we present the major kinds of molecules that form living systems. We emphasize the kinds of molecules common to all life, the kinds of molecules common to plants, and the kinds of molecules that distinguish one plant from another. These molecules include proteins, carbohydrates, nucleic acids, fats, and secondary metabolites. The study of these molecules and how they are made provide the first insight into how plants live.

Introduction

The physical and chemical behavior of matter can be anticipated, at least partly, by knowing its behavior at the atomic level. Similarly, cellular processes can be understood by knowing how biological molecules interact. For instance, our knowledge of chlorophyll helps us understand how plants harvest and use the energy of light to make food during photosynthesis. Similarly, we get basic information about the mechanisms of inheritance from studies of DNA. We understand how plants control the uptake of foreign substances because we know something about the chemistry of cell membranes. Such **reductionism**—studying simpler components to understand the functions of complex systems—is the most dominant strategy in biological research.

In this chapter we use the reductionist strategy to explain the properties of the main ingredients in living cells. However, this perspective cannot explain what life is, because a cell is more than a bag of chemicals. Nevertheless, the study of cellular components is a good place to start our study of living organisms. Note that a basic background in chemistry is necessary before reading this chapter. An overview of some of the most important background information is provided in Appendix A, "Fundamentals of Chemistry."

Elements of Life

Only about sixteen of the ninety-two naturally occurring elements are essential to most plants. Four of these elements—hydrogen, carbon, oxygen, and nitrogen—make up more than 99% of the mass of most plant tissues (see table 2.1).

Most molecules that contain carbon are referred to as **organic** compounds. Almost all organic compounds also include hydrogen, and most include oxygen. However, pure carbon (graphite, diamond, buckminsterfullerene) and carbon compounds lacking hydrogen (carbonates, carbon dioxide, carbon monoxide, cyanide salts) are classified with noncarbon compounds as **inorganic** chemicals. **Biochemicals** are organic and inorganic molecules that occur in living organisms.

The major elements of living cells occur in thousands of combinations, each combination forming a different kind of molecule. Except for water, the mass of a plant or other organism consists primarily of four kinds of molecules: carbohydrates, proteins, nucleic acids, and lipids. However, plants also make many other kinds of compounds, usually in lesser amounts, including chemicals having such exotic names as phenolics, alkaloids, terpenoids, sterols, and flavonoids.

Large Molecules: Polymers and Their Monomers

Many biochemicals exist individually as small, single-unit molecules called **monomers;** however, most plant tissues consist of **polymers,** which are compounds made of many identical or similar monomers assembled into large molecules. Examples of polymers include cellulose, starches, enzymes, DNA, waxes, lignins, and tannins. Glucose is the only monomer in cellulose and starch. In contrast, enzymes and other proteins consist of twenty different amino acid monomers; there are probably more than 50,000 unique polymers made from amino acids in every plant. Only four nucleotides make a polymer of DNA, but every organism has a different set of DNA molecules. Variations in these and other classes of polymers account for the main differences we see among plants and animals that exist now or have existed in the past. In addition to the major classes of large molecules common to all organisms—carbohydrates, proteins, nucleic acids, and lipids—plants may also contain such polymers as lignins, tannins, polyterpenes, and polyacetylenes.

This section presents the main features of the major classes of biological polymers. Polymers unique to plants are discussed with other kinds of plant chemicals later in the chapter.

CONCEPT

Most plant tissues are made of large molecules that are polymers of simpler molecules. The major biological polymers are divided into four classes: carbohydrates, proteins, nucleic acids, and lipids.

Carbohydrates

Glucose and other sugars, as well as their polymers, are called **carbohydrates.** Carbohydrates generally contain one oxygen and two hydrogens for every carbon. For example, glucose and fructose consist of six carbons, twelve hydrogens, and six oxygens, and have the formula $C_6H_{12}O_6$. Galactose, mannose,

Table 2.1

The Most Common Elements in Plants

Element	Symbol	Atomic Mass	Average Percent of Plant Tissue	Example of Role in Plants
Hydrogen	H	1	47.4	One of the two components of water; also occurs in all organic molecules
Carbon	C	12	27.6	Backbone of all organic molecules
Oxygen	O	16	23.7	The other of the two components of water; also occurs in nearly all organic molecules; molecular oxygen is used in aerobic respiration
Nitrogen	N	14	0.8	Component of proteins, nucleic acids, chlorophylls, and alkaloids
Potassium	K	39.1	0.2	Prevalent ion in plants; regulates water uptake; activates certain enzymes
Calcium	Ca	40.1	0.1	Important in synthesizing pectin in cell walls; activates enzymes involved in chemical communication in cells
Magnesium	Mg	24.3	0.06	Part of chlorophyll and of some enzymes; helps to stabilize ribosomes
Phosphorous	P	31	0.05	Part of phosphate in energy-transfer molecules, nucleic acids, coenzymes, and phospholipids
Sulfur	S	32	0.02	Ingredient of most proteins and of some enzyme cofactors
Chlorine	Cl	35.4	0.002	Possible role in some reactions of photosynthesis
Iron	Fe	55.8	0.0016	Chlorophyll synthesis; part of active site of many important oxidation-reduction enzymes
Boron	B	10.8	0.0016	Unknown; may work in translocation of sugars
Manganese	Mn	54.9	0.0008	Prevalent enzyme-activating metal in plants
Zinc	Zn	65.4	0.0002	Activates many enzymes; involved in synthesis of the plant hormone indoleacetic acid
Copper	Cu	63.5	0.00008	Activates many enzymes; occurs in plastocyanin, an electron carrier of photosynthesis
Molybdenum	Mo	95.9	0.0000007	Important in nitrate reduction

Note: Several other elements, such as cobalt, either occur in smaller amounts or they are important only in some plants. This latter group includes, for example, sodium, selenium, and silicon. Still other elements occur sporadically in plants because they happen to be in the soil. These include metal ions (gold, aluminum, lead, mercury), fluorine, and even uranium. Such elements are often toxic to the plants. We'll discuss minerals and plant nutrition in more detail in Chapter 20.

and many other monomers have this same formula, differing only in the arrangement of the elements (see box 2.1, "Reading Chemical Structures"). This means they are **isomers,** which are different compounds with the same formula. Structural differences between isomers can impart significantly different chemical properties to the molecules. In the example of glucose and fructose,[1] one has little flavor (glucose), while the other (fructose) is the sweetest sugar known. Common carbohydrate monomers having different chemical formulas include ribose, xylose, and arabinose ($C_5H_{10}O_5$); deoxyribose ($C_5H_{10}O_4$); glucuronic acid and galacturonic acid ($C_6H_{10}O_7$); and rhamnose ($C_6H_{12}O_5$).

Doing Botany Yourself

Imagine that you are given several grams of a previously unknown carbohydrate monomer from the leaves of a Brazilian tree. Describe an experiment for determining the sweetness of this chemical.

---

1. Glucose is also called *grape sugar*, *blood sugar*, or *dextrose*. Fructose is often referred to as *fruit sugar* because of its abundance in ripening fruits.

Carbohydrate monomers rarely occur as free sugars in plants; rather, they are usually bound to other kinds of molecules or linked together into larger carbohydrates. To distinguish different sizes of carbohydrates, monomers and **dimers** (molecules with two monomers) are called **monosaccharides** and **disaccharides,** respectively; polymers are called **polysaccharides.** The most common disaccharide in plants is sucrose, which is a glucose and a fructose linked into a dimer (fig. 2.1). Sucrose is common table sugar, also called *cane sugar* or *beet sugar*. This disaccharide is the major form of carbohydrate that moves in plants. People exploit such transport when they collect sap from sugar maples (*Acer saccharum*) to make maple syrup (see "Tapping Wood for Maple Syrup" in Chapter 21). People also use maltose, a disaccharide of glucose, to make beer (see boxed reading 4.1, p. 84).

You are now prepared for an interesting and important exercise in carbohydrate chemistry: reading food labels for their sugar content. The different forms of sugar listed on food labels include: sugar, sucrose, glucose, dextrose, fructose or high-fructose corn syrup. Sugar is an ingredient of many foods in one or more of these forms. Labels on the sweetest foods (e.g., candy, cookies, breakfast cereals) include several of these names. Even ketchup, salad dressings, bread, and other nonsweet foods often include sweeteners.

BOXED READING 2.1
READING CHEMICAL STRUCTURES

At first encounter, chemical structures are often a pesky problem for students, because different chemicals are represented in variable and seemingly inconsistent ways. The common goal of all structural representations, however, is to portray chemicals simply but informatively. Such simplification is important because complete structures are cluttered and usually too tedious to read or write. For example, examine the complete structures of some common laboratory chemicals—glucose, hexane, and toluene—shown on the left in the accompanying diagrams. These structures include a C for each carbon, an H for each hydrogen, and an O for each oxygen. Each chemical bond is shown as a single line. Contrast these three structures with the streamlined versions of each compound shown on the right. These are the "bare bones" versions that eliminate the need to write some of the carbons, hydrogens, and chemical bonds. The unwritten atoms and bonds are still there, but by general agreement they are not shown; it is up to the reader to mentally fill in the appropriate atoms or bonds.

BOX FIGURE 2.1

Diagrams of three complete chemical structures (left) and their abbreviations (right).

There are no rules that encompass all structural simplifications. Generally, however, the streamlining involves only carbon and hydrogen. For example, unwritten carbons are the corners of rings or the joints of chains. Hydrogens attached to carbons are either not written, as in hexane and toluene, or they are written without bonds to the right of the carbon they belong to, as in the CH_2 of glucose. Some inconsistencies do occur, however, such as between carbohydrates and other organic chemicals. The hydrogens on the carbons of the glucose ring are abbreviated as lines (bonds) without H's, but those of hexane are not shown in any way. Moreover, the bond sticking out of toluene represents a methyl group ($—CH_3$) and not a hydrogen as it does in glucose.

Perhaps the best advice on reading simplified chemical structures is to satisfy carbon—that is, to make sure that each carbon bonds to four other atoms. In hexane, for example, where each carbon is shown to have two bonds, the two unwritten bonds must be to hydrogens. As a brief exercise, apply this principle to fill in the unwritten hydrogens and carbons for strychnine (fig. 2.13). (Hint: The formula for strychnine is $C_{21}H_{22}N_2O_2$.) The structure of this compound would be very messy if all the carbons and hydrogens were written.

Structural Polysaccharides

Polysaccharides that hold cells and organisms together are called **structural polysaccharides.** Cellulose is the most abundant structural polysaccharide in plants and is the most abundant polymer on earth; between 40% and 60% of a cell wall is cellulose. Cellulose also occurs in almost pure form in some plants, for example, in the fibers surrounding cotton seeds (*Gossypium hirsutum*).

The strength of cellulose comes from its organization (see fig. 2.2). **Beta-glucose** units are linked by oxygen between the number 1 carbon of one glucose and the number 4 carbon of the next glucose. (*Beta* refers to a monomer whose hydroxyl at the number 1 carbon points up. When the hydroxyl points down, it is the *alpha* configuration, as in **alpha-glucose**.) This beta-1,4 linkage is repeated to make a linear molecule consisting of 100–15,000 glucose units (fig. 2.2). A higher level of organization occurs in cellulose when a thousand or more polymers twist together into **microfibrils.** Microfibrils are like strong, tiny cables. The tension that microfibrils can withstand before breaking—that is, that which is called *tensile strength*—is as high as 110 kilograms per square millimeter. This means that if a microfibril could be made with a diameter about the size of this letter *o*, it could suspend your instructor. This tensile strength is about 2.5 times that of the strongest steel.

Several microfibrils intertwine to make cellulose fibrils. Layers of fibrils are cemented together into strong three-dimensional grids by other kinds of structural polysaccharides (fig. 2.3). The gluey polysaccharides that hold cellulose fibrils together are called **pectins** and **hemicelluloses.** Cell-wall

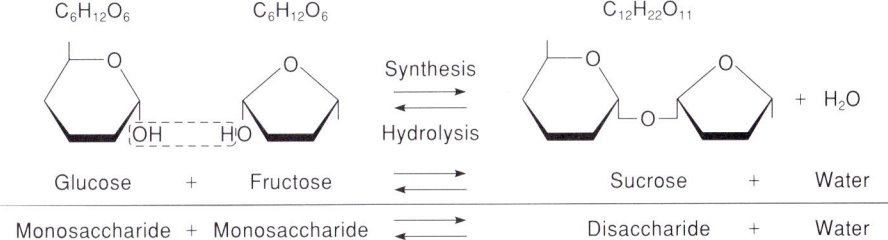

FIGURE 2.1

Diagram showing the synthesis and hydrolysis of sucrose. During synthesis, a bond forms between glucose and fructose, and a molecule of water is removed. Hydrolysis occurs as a molecule of water is added and the bond between glucose and fructose is broken. The enzyme that drives hydrolysis is sucrase; the synthesis of sucrose is actually a multistep reaction that involves several enzymes.

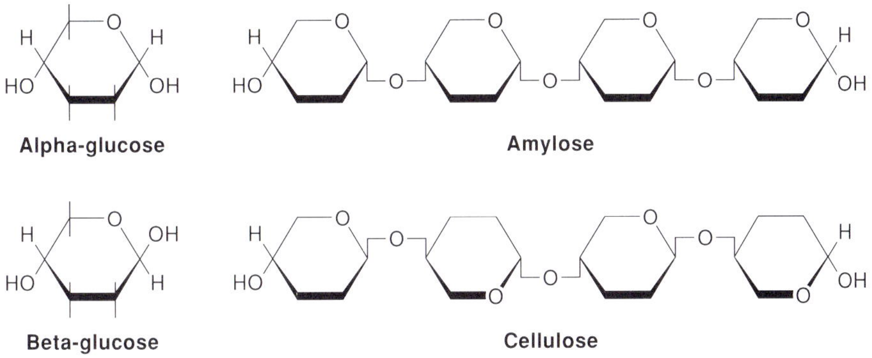

FIGURE 2.2

Line diagrams of alpha-glucose and beta-glucose, plus primary structures of amylose and cellulose. The main difference is that amylose is made of alpha-glucose, and cellulose is made of beta-glucose.

pectins are mostly polymers of galacturonic acid with alpha-1,4 linkages. Hemicelluloses are not related to cellulose; they are made of different kinds of monomers. Linkages in hemicelluloses are variable and complicated, but the composition of these polysaccharides can be studied by hydrolyzing them in acid and determining their component monomers. For example, hemicelluloses in grasses consist mostly of xylose, with lesser amounts of arabinose, galactose, and uronic acids. In contrast, hemicelluloses in legumes (the pea family) are high in uronic acids, galactose, and arabinose, with little xylose. Some of the most interesting hemicelluloses may not be structural polysaccharides at all. These are usually exuded from stems, roots, leaves, or fruits in a sticky mixture called **gum.** Gums are complex, branched polysaccharides consisting of several kinds of monomers. For example, a gum called **gum arabic** (from *Acacia senegal*) consists of arabinose, galactose, glucose, and rhamnose. Gum arabic is almost everywhere in our daily lives: it is used to stabilize postage-stamp glue, beer suds, hand lotions, and liquid soaps.

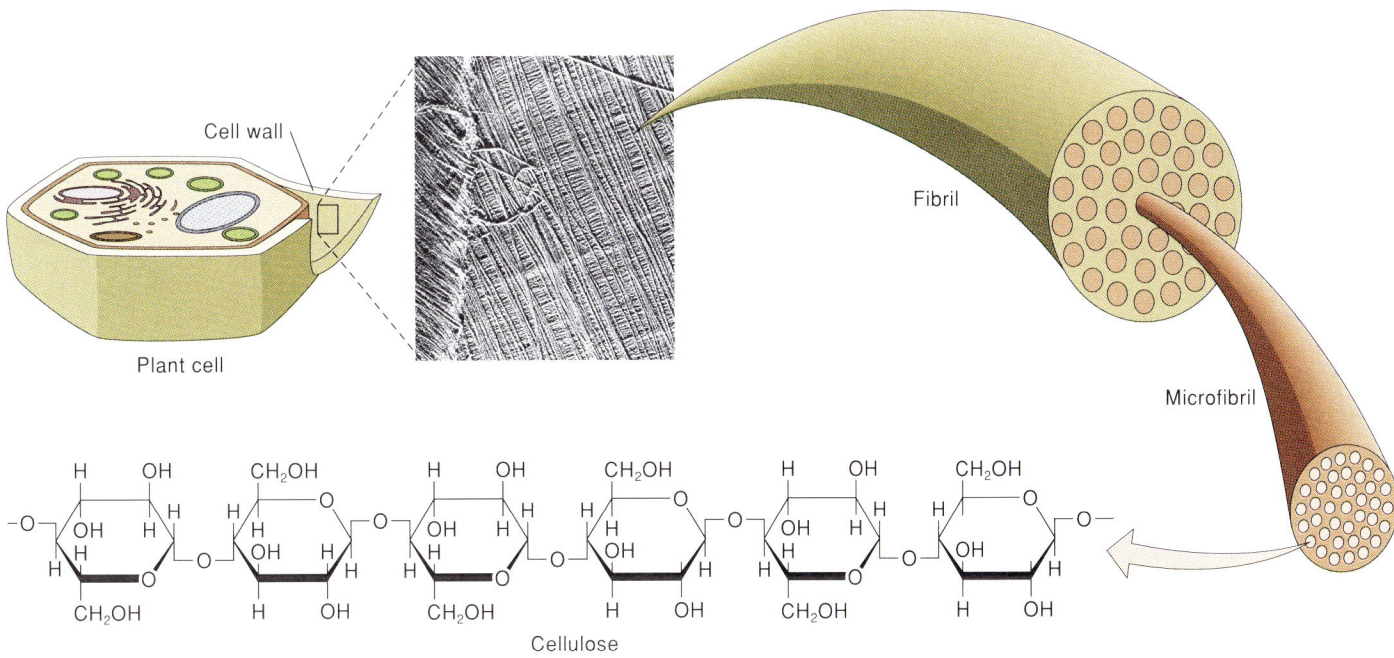

FIGURE 2.3

Model for the arrangement of fibrils, microfibrils, and cellulose in cell walls. The scanning electron micrograph shows the fibrils in a cell wall of the green alga *Chaetomorpha*, ×30,000.

CHAPTER TWO *Atoms and Molecules: The Building Blocks of Life*

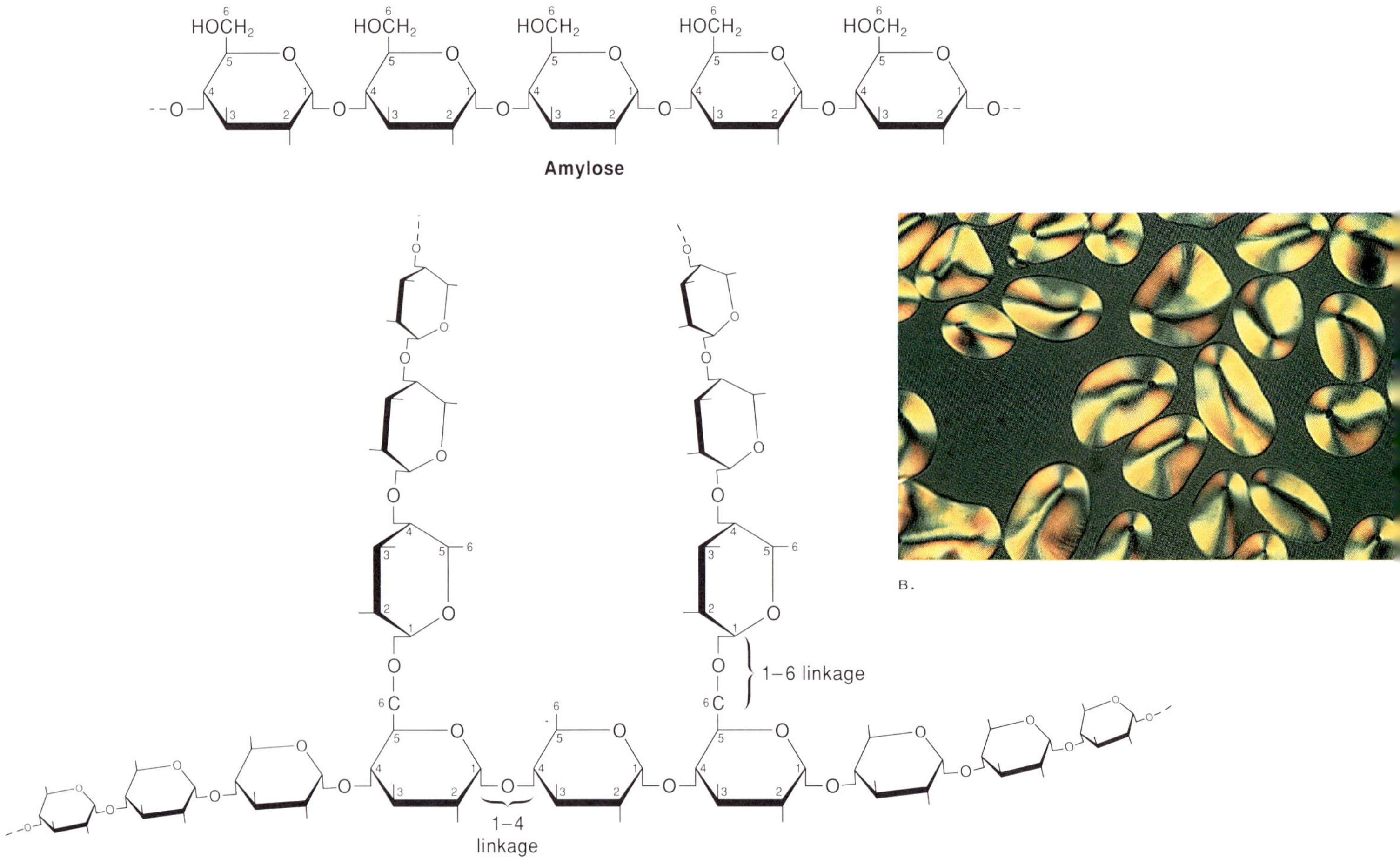

FIGURE 2.4

Forms of glucose-containing storage polymers in plants. (a) Amylopectin is made of short, helical chains that consist of branched amylose molecules. (b) Molecules of amylopectin aggregate into starch grains, as shown here from potato tubers, ×83.

Two commercially important polysaccharides of algae are **agar** and **carrageenan.** These polymers are the slimy substances that surround the cellulosic cell walls of certain red algae. Each polymer is a different mix of alpha-galactose sulfates having 1,4 and 1,6 linkages. Agar, which is harvested mostly from *Gelidium robustum* in the United States, is used to make drug capsules, cosmetics, gelatin desserts, and temporary preservatives; it is also used in scientific research as a medium for growing microorganisms and plant tissue cultures. Carrageenan, which comes mostly from species of the genus *Chondrus*, is used primarily as a stabilizer in paints and cosmetics and in foods such as salad dressings and dairy products.

Storage Polysaccharides

Two of the most common food reserves in plants are **amylose** and **amylopectin** (fig. 2.4a); they are the polymers of alpha-glucose that make up starch. Amylose is the smaller and simpler of the two: it consists of chains of a hundred to several thousand monomers with 1,4 linkages. In contrast, amylopectin is a highly branched polymer of up to 50,000 monomers. This polymer consists of short chains with 1,4 linkages, which are cross-linked to other chains. Most starch grains are about 20% amylose and 80% amylopectin, but this depends on the plant. For example, starches in some types of corn (*Zea mays*) and rice (*Oryza sativa*) are almost entirely amylopectin. At the other extreme, a variety of wrinkled pea (*Pisum sativum* var. Steadfast) makes starch that is 80% amylose.

Starch grains have many shapes and structures (fig. 2.4b). These shapes depend not only on the linkages of amylose and amylopectin but also on how these molecules are folded into secondary structures. In general, amylose twists into coils, groups of which are surrounded by the larger amylopectin molecules.

Although starches are the most common storage polysaccharides in plants, they are not the only ones. Many plants make **inulin** instead of or in addition to starch. Inulin is a polymer of fructose having beta-2,1 linkages (the number 2 carbon of fructose reacts like the number 1 carbon of glucose). Inulin is a storage polysaccharide in dahlias (*Dahlia* species),

Jerusalem artichoke (*Helianthus tuberosum*), globe artichoke (*Cynara scolymus*), chicory (*Cichorium intybus*), and sweet corn. The storage organs of these plants (e.g., tubers of Jerusalem artichoke, kernels of sweet corn) taste sweet because inulin releases fructose.

Amylose vs. Cellulose: Feast or Famine

As you learned from the preceding discussion, there is only a slight difference between alpha-glucose and beta-glucose (fig. 2.2); however, this difference gives the molecule significantly different properties. Cellulose is a strong structural polysaccharide because it forms microfibrils and fibrils. More importantly, cellulose is impervious to the enzymes that degrade starch. One such enzyme, called **alpha-amylase,** digests only alpha-glucose linkages in amylose. Thus, the calories stored in cellulose are unavailable for plant growth or for herbivorous animals. This is why plants cannot use cellulose as a storage reserve and why we get little energy from eating a stalk of celery or a spoonful of bran. Conversely, plants produce alpha-amylase in seeds, roots, and tubers when they harvest the energy from stored starch. Because human saliva contains alpha-amylase, the digestion of starch in our food begins even before it is swallowed.

> **Writing to Learn Botany**
>
> Enzymes that are beta-glucosidases hydrolyze bonds between beta-glucose molecules. What nutritional benefits would animals get from having this kind of enzyme? What nutritional problems would humans have if we had this kind of enzyme?

Some organisms can digest cellulose because they make enzymes called **cellulases.** Examples include some fungi and bacteria, earthworms, and certain insects. Cows, sheep, termites, and other herbivores digest cellulose indirectly by housing cellulase-producing microorganisms in their digestive systems. The microorganisms digest the cellulose, and the animals digest the glucose released from it.

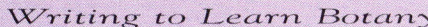

Carbohydrates include monomers called *monosaccharides* and polymers called *polysaccharides*. Polysaccharides differ by the kind of monomer they contain and by the linkages between monomers. The two most widespread polysaccharides in plants are cellulose and starch. Cellulose is a structural polysaccharide, and starch is a storage polysaccharide.

Proteins

After cellulose, proteins make up most of the remaining biomass of living plant cells. A protein consists of one or more **polypeptides** and may also include sugars or other kinds of

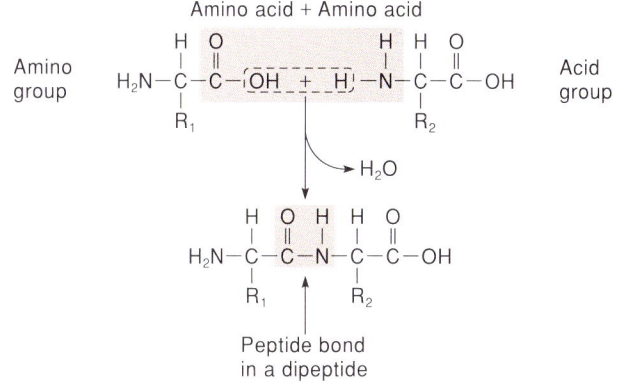

FIGURE 2.5

Peptide synthesis occurs as a bond forms between the amine-nitrogen of one amino acid and the carboxy-carbon of another, with the removal of a molecule of water.

small molecules. A polypeptide is a chain of amino acids linked together by carbon-nitrogen bonds called **peptide bonds** (fig. 2.5). The smallest polypeptides have fewer than 100 amino acids, and the largest have several thousand. There are thousands of different kinds of proteins.

Like polysaccharides, proteins are important in cell structure and as storage reserves. Many proteins are also enzymes that catalyze biochemical reactions. Unlike polysaccharides, proteins do not occur as chains of the same monomer with different linkages. Instead, each protein contains different amino acids linked by the same peptide bonds. The diversity of proteins results from the different sequences of the amino acids. Thus, each amino acid is like a letter in an alphabet, and each polypeptide is like a long word that usually contains the entire alphabet, with some letters repeated many times.

Each of the twenty amino acids in plants has two main parts. The first part, which all amino acids share, consists of a carbon with both a carboxylic acid group (–COOH) and an amino group (–NH$_2$ or –NH–) attached to it. The second part, which is attached to the same carbon as the amino and carboxyl groups, is the side chain. The side chain, or *R group*, differs for each amino acid. The R group can simply be a hydrogen atom, as in glycine, or a complex ring structure, as in tryptophan (fig. 2.6).

Protein Structure

The sequence of amino acids in a protein is the protein's **primary structure** (fig. 2.7). However, proteins are not simply long molecules that wave randomly in the cell; rather, they fold and twist into three-dimensional shapes. The shape of a protein is determined by bonds and other electronic interactions between amino acids at different positions in the primary structure. Hydrogen bonds occur at regular intervals and hold specific parts of a polypeptide together in a **secondary structure.** The secondary structure is shaped either like a coil or like a pleated sheet (fig. 2.7).

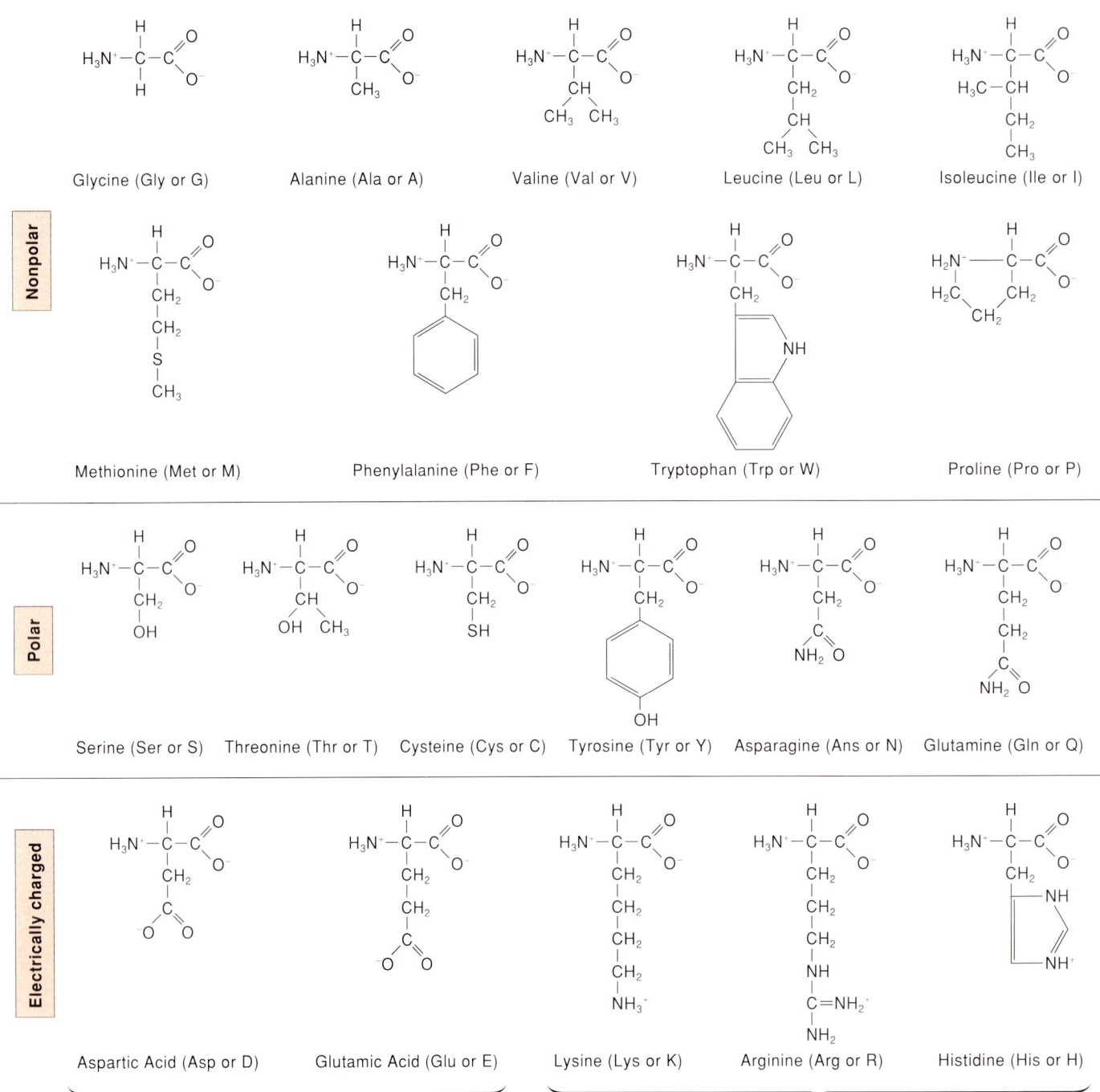

Figure 2.6

Structures of the twenty amino acids that occur most commonly in proteins. Each amino acid has one carbon that is bonded to both an amine group (H$_2$N—) and a carboxyl group (—COOH). At neutral pH (pH 7), the prevailing form of the amine is H$_3$N$^+$ and of the carboxyl is COO$^-$. The amino acids are arranged according to the main properties of their side chains (in pink shade).

Disulfide bonds, which are covalent bonds between the sulfurs of two cysteines, maintain the rigid **tertiary structure** of a protein. The tertiary structure is also maintained by weak bonds between ionized carboxyl and amino groups, and by adjacent nonpolar amino acids that aggregate by excluding water from their vicinity. The result of all of these interactions is a protein with a unique three-dimensional structure.

Many proteins consist of two or more polypeptide chains called **subunits.** For example, the plant enzyme ribulose-1,5-bisphosphate carboxylase/oxygenase (known as *rubisco*) has

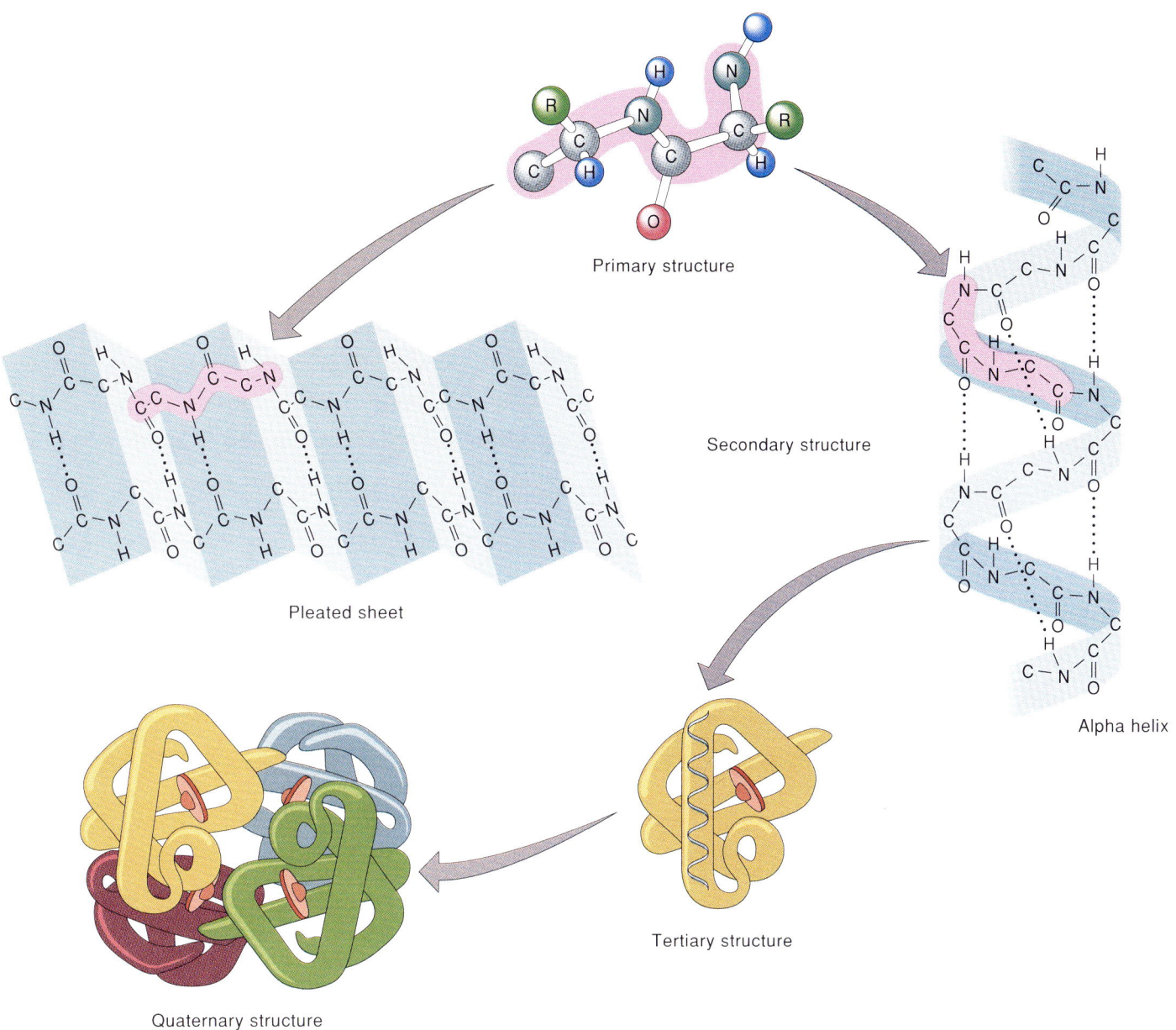

FIGURE 2.7

Models of the primary, secondary, tertiary, and quaternary structures of a protein. The secondary structure may consist of coils (alpha helixes) or folds (pleated sheets), depending on which amino acids form hydrogen bonds with each other.

sixteen subunits. In this enzyme, one polypeptide makes up eight large subunits, and another polypeptide makes up eight small subunits. The **quaternary structure** of this protein refers to the way that the different subunits attach to each other. Only proteins having more than one subunit have a quaternary structure.

In solution, the three-dimensional structure of a protein may be somewhat flexible. Proteins may also be **denatured** in laboratory experiments by adding urea or other chemicals. Urea inhibits hydrogen bonds that contribute to secondary structure, thereby causing a protein to lose its shape and function. However, denaturation may be reversible, since proteins often regain their original shapes when the urea is removed from the solution. Proteins may also be irreversibly denatured by heat or harsher chemicals. For example, a cooked egg, which is mostly denatured protein, cannot be "uncooked" (i.e., **renatured**).

Proteins are difficult to classify because of their great diversity. The most commonly used schemes of protein classification are based primarily on the protein's solubilities in water or detergent solutions and on their functions. For simplicity, we will discuss proteins relative to their three major functions: structural proteins, storage proteins, and enzymes. Most of the proteins mentioned in this book fit into one of these categories.

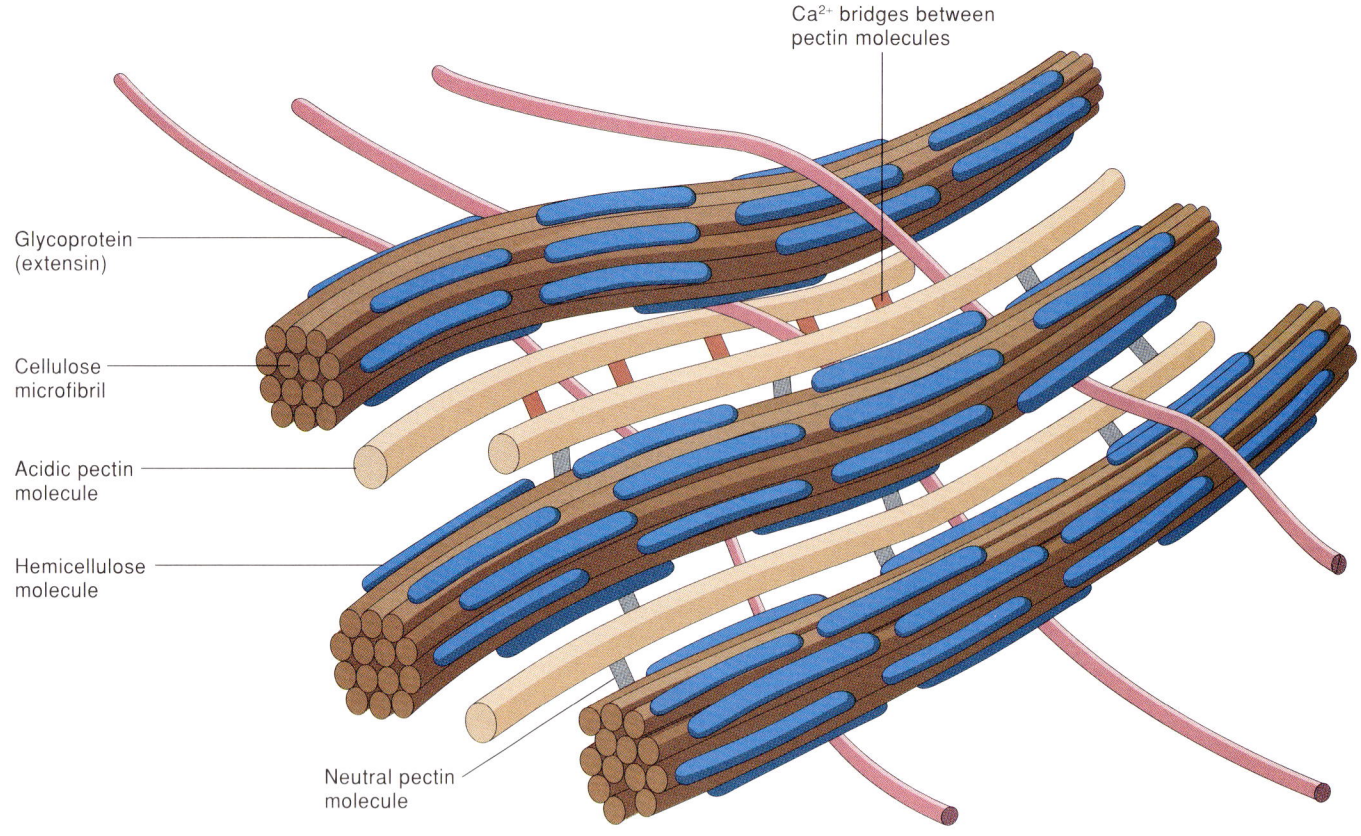

FIGURE 2.8

Model of the interconnections among major components of primary cell walls. Hemicellulose molecules bind to the surface of cellulose microfibrils by hydrogen bonds and to acidic pectin molecules by cross-links with neutral pectin molecules. Calcium (Ca^{2+}) holds acidic pectin molecules to one another. Glycoproteins (extensins) are tightly woven into the cell-wall matrix.

Structural Proteins in Cell Walls and Membranes

In addition to containing carbohydrates, cell walls also include from 2% to 10% protein. Although some of this protein consists of enzymes that help build cell walls, most of it is structural. Because this structural protein was originally thought to play a role in the expansion of cell walls, it was named **extensin.** Extensin was more recently found to be a family of related proteins. Moreover, the synthesis of extensin is induced when cells are damaged by wounding, infection, or freezing. This is taken as indirect evidence that extensins somehow help protect or repair damaged cells.

Extensins are rich in hydroxyproline (a derivative of proline), serine, threonine, and aspartic acid. Extensins also consist of up to 30% carbohydrate, which makes them **glycoproteins,** that is, "sugar-protein." The carbohydrates of extensins include mostly arabinose and galactose. Although the exact structure of the extensin-polysaccharide complex in cell walls is unknown, the arabinose or galactose of extensins probably attaches to one or more kinds of cell-wall polysaccharides (fig. 2.8).

Cells contain several kinds of membranes, each having different protein composition. For example, the internal membranes of mitochondria and chloroplasts are about 75% protein, whereas the membrane that surrounds the cell is usually about 50% protein. Lipids comprise most of the nonprotein portion of membranes. The composition, structure, and functions of different membranes are discussed in more detail in Chapters 3–4 and 6–7.

Storage Proteins

Storage proteins are stored mostly in seeds and are used as a source of nutrition for the early development of seedlings. The composition of seed proteins depends on the plant species. For example, corn produces a storage protein called **zein,** which consists of nearly thirty polypeptides that occur in two major subunits and one minor subunit. In comparison, wheat produces a storage protein called **gliadin,** which has four major subunits that are made up of at least forty-six polypeptides. The most complex storage proteins may be the glutenins of wheat, which consist of up to fifteen different proteins, ranging in size from 1,100 to more than 13,300 **kilodaltons** (1 kilodalton is equal to a molecular mass of 1,000).

Seed storage proteins in soybean and cereal grains (oats, rice, wheat, corn, barley) are especially important because they are a major source of nutrition for humans and cattle. However, corn and barley are low in the amino acids lysine, threonine, and tryptophan. Soybean and other legume seeds are deficient in the sulfur amino acids, cysteine and methionine, but their lysine content is adequate.

Seeds of some plants also contain proteins that have undesirable nutritional effects on the animals that eat them. For example, up to 10% of the protein in many cereal grains inhibits certain digestive enzymes of animals. These proteins are called **protease inhibitors** because they inhibit proteases, the enzymes that digest proteins. The seed proteins of a few plants are toxic. These include glycoproteins such as **ricin D** from the castor bean (*Ricinus communis*) and **abrin** from the rosary pea (*Abrus precatorius*).

Enzymes

Most proteins in a living cell are **enzymes,** which are the catalysts for biochemical reactions. This means that enzymes speed up reactions without being consumed in the process.

Enzymes usually have flexible, globular shapes. A specific place on each enzyme, called an **active site,** binds to one or more substrates (reactants). Enzymes are often named by adding the ending *-ase* to the name of a reactant or product of the reaction that the enzyme catalyzes. One example is alpha-amylase, the enzyme that digests amylose. Enzymes also catalyze reactions that make or digest cell walls, membranes, storage polymers, proteins, DNA, RNA, pollen walls, seed coats, chlorophyll, amino acids, and other plant metabolites. Examples of enzymes and their roles in the lives of plants are described in almost every chapter in this textbook.

Enzymes often remain active even after they are removed from the cell. Alpha-amylase, for example, digests amylose in a test tube if it is provided with the appropriate temperature, pH, and cofactors. Pure enzymes that maintain their activity are commercially important. These enzymes include the proteases **papain** and **chymopapain** from papaya (*Carica papaya*). Papain digests the muscle tissue of animals, which is why it is a major ingredient in meat tenderizers. Chymopapain is a drug used to treat the slippage of a disk in the spinal column. Chymopapain is injected directly into the problem area, where it dissolves the proteinaceous cartilage of which the disk is made. You will learn more about how enzymes work in Chapter 5.

CONCEPT

Proteins consist of chains of up to twenty different amino acids bonded into polypeptides. The primary structure of a polypeptide is the amino acid sequence. Secondary and tertiary structures are the three-dimensional shapes created by various nonpeptide bonds and other interactions among amino acids within the polypeptide. Quaternary structure is the arrangement of more than one polypeptide in a protein. Like carbohydrates, proteins have roles in cell structure and in storage. Unlike carbohydrates, proteins may also be enzymes.

The Lore of Plants

There are no pineapple-flavored gelatin desserts, because pineapples contain a protease (bromelain) that hydrolyzes gelatin. Hydrolyzed gelatin cannot form a gel, so gelatin desserts containing pineapple juice would be too runny to appeal to the typical consumer.

Nucleic Acids

The most complex biological polymers are **nucleic acids.** Their complexity emerges from a primary structure that contains mostly four different monomers, each of which is called a **nucleotide.** Each nucleotide consists of a phosphate group, a simple sugar, and a nitrogen-containing base (fig. 2.9). The simple sugar is either ribose, which occurs in **ribonucleic acid (RNA),** or deoxyribose, which is the sugar in **deoxyribonucleic acid (DNA).** There are two classes of nitrogenous bases: pyrimidines, which have one ring, and purines, which have two rings. The pyrimidines in DNA are cytosine and thymine; the purines are guanine and adenine. The base composition of RNA is the same as that of DNA, except that RNA includes uracil instead of thymine. Furthermore, RNA occurs as a single strand, whereas DNA occurs as a two-stranded spiral called a **double helix.**

Nucleic acid molecules vary in size from about thirty nucleotides to several million nucleotides. The smaller nucleic acids are primarily RNA. Even the smallest nucleic acids, however, have a seemingly infinite number of nucleotide sequences. For example, four different nucleotides can be arranged in a ten-monomer sequence in 4^{10} (1,048,576) different ways. This is analogous to using an alphabet with four letters to write more than a million ten-letter words.

Although DNA consists of the same four nucleotides in all organisms, the amounts of different nucleotides are variable. These amounts are generally measured as the percentage of G (guanine) + C (cytosine) since guanine and cytosine always complement each other, and adenine and thymine

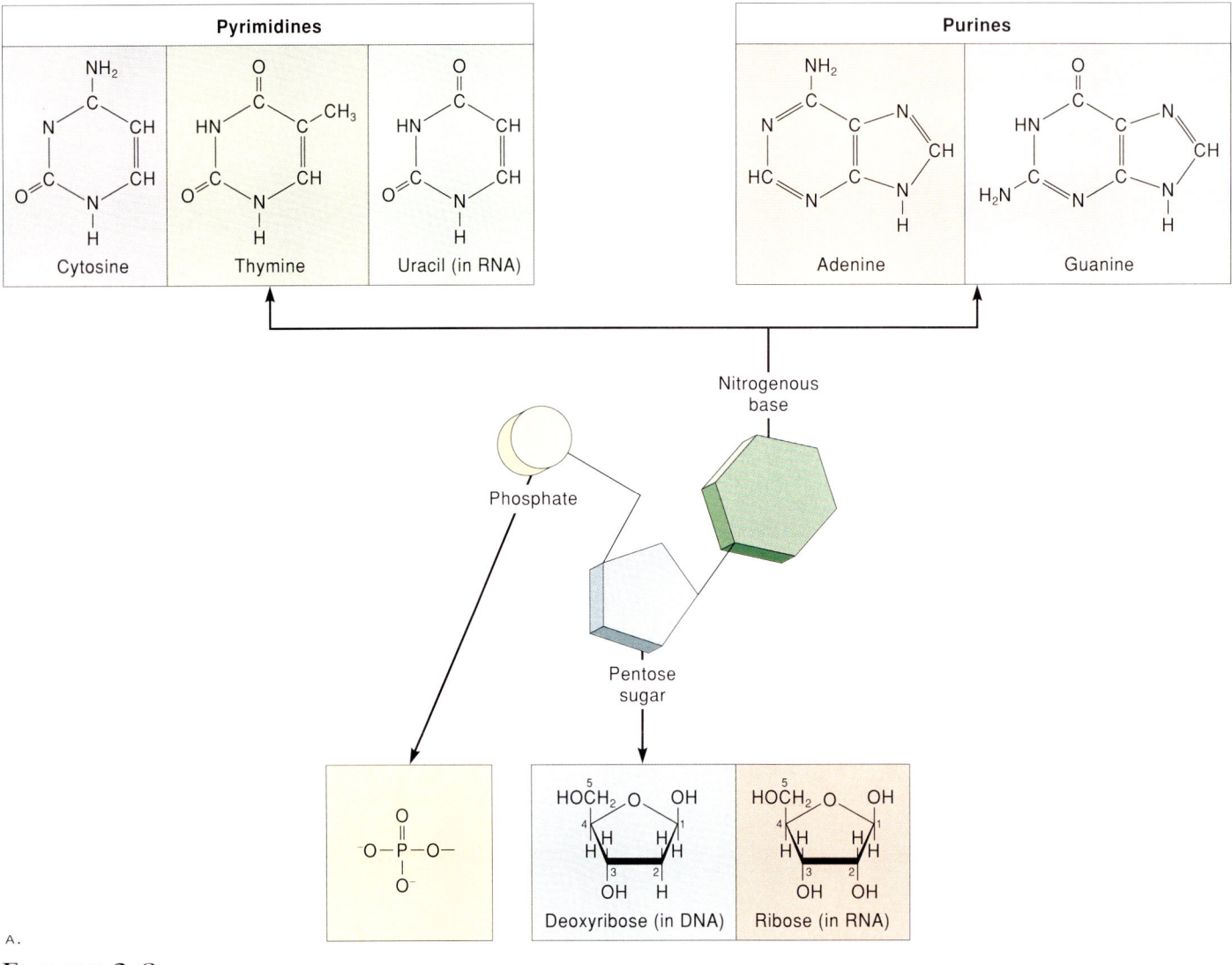

FIGURE 2.9

The structure of DNA. (a) Each nucleotide monomer consists of three smaller building blocks: a nitrogenous base, a pentose sugar, and a phosphate group. (b) Nucleotide monomers are bonded to each other by covalent bonds between the phosphate of one nucleotide and the sugar of the next nucleotide. (c) DNA is usually a double strand held together by hydrogen bonds between nitrogenous bases; A pairs only with T, and C pairs only with G. The double strand is twisted spirally into a double helix. (d) A space-filling model shows the close stacking between nitrogenous bases. Color code for atoms: yellow = P, dark blue = C, red = O, turquoise = N, and white = H.

always complement each other (the reason that G/C and A/T are complementary is explained in Chapter 9). For example, the GC content of plants is usually between 36% and 43%, but in the grass family (*Poaceae*) it exceeds 48%. In contrast, the bacterium that causes botulism (*Clostridium botulinum*) has 30% GC, and the bacterium that makes the antibiotic neomycin (*Streptomyces fradiae*) has 80% GC.

Nucleic acids are unique because they can replicate themselves. Furthermore, DNA can make RNA, which guides the assembly of proteins. Nucleic acids form the molecular foundation for every living organism. The roles of nucleic acids are so central to life that all of Unit 3 in this text ("Genetics") is devoted to their biology, chemistry, and importance.

Other Nucleotides

Nucleotides other than adenine, guanine, cytosine, and thymine occasionally occur in DNA in trace amounts. One that is abundant, however, is 5-methylcytosine, an otherwise normal cytosine that has a methyl group ($-CH_3$) added to it. The seemingly minor change from cytosine to 5-methylcytosine inactivates the DNA.

As much as 30% of the cytosine in plant DNA is replaced with 5-methylcytosine. Since cytosine is methylated and demethylated at different times during the life of a cell, these reactions probably regulate the activity of genes.

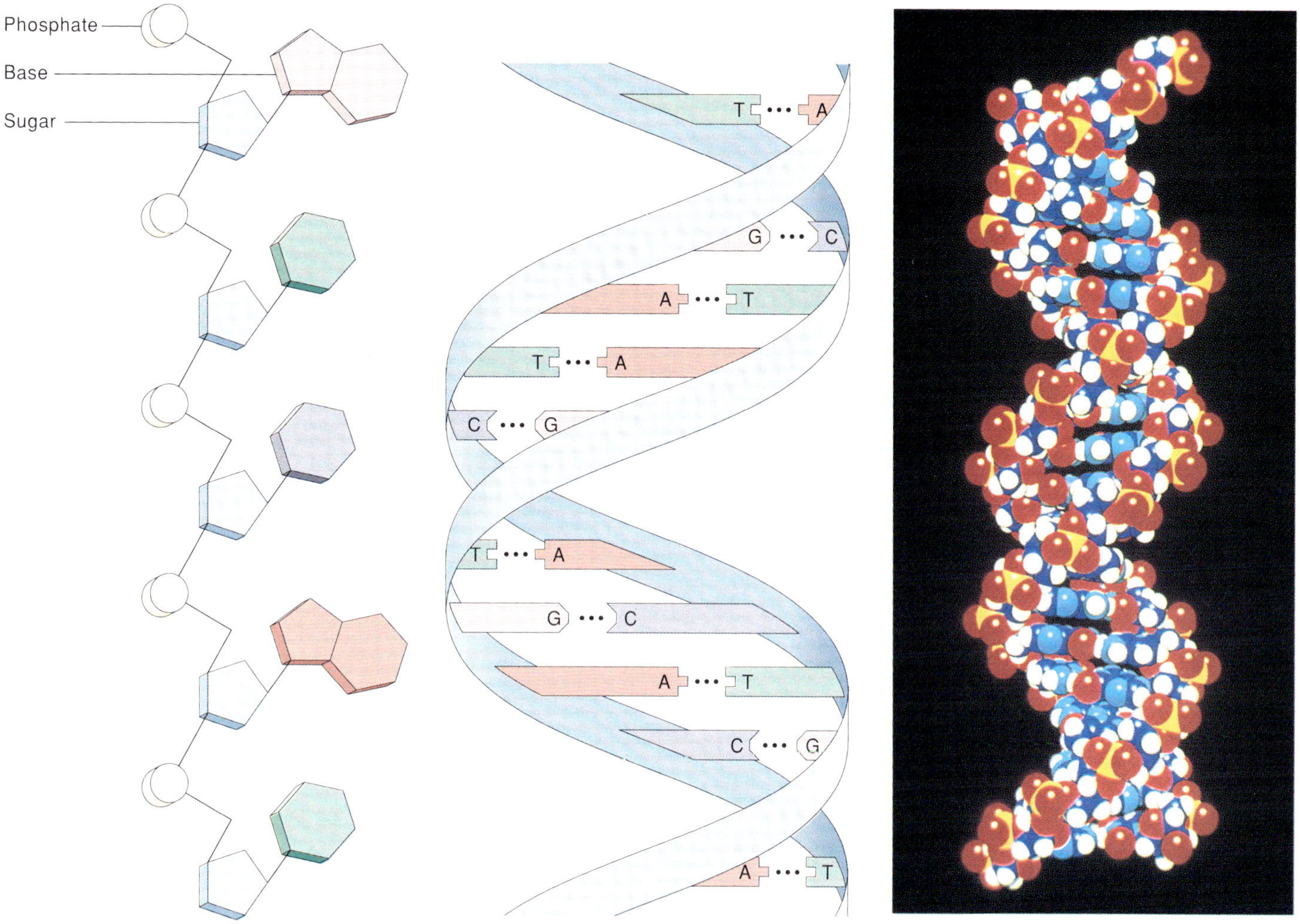

B.   C.   D.

FIGURE 2.9 *continued*

CONCEPT

Nucleic acids are chains of nucleotides. Different nucleic acids vary in their nucleotide composition and sequence. The two main types of nucleic acids, RNA and DNA, also contain different sugars: RNA contains ribose, and DNA contains deoxyribose.

Lipids

Unlike other biological polymers, lipids are not defined by specific, repeating monomeric subunits. Rather, they are defined by their water-repellent property (*lipid* means "fat"). The only common structural theme shared by all lipids is a large proportion of nonpolar hydrocarbon groups ($-CH_3$, $-CH_2$, or $-CH$). These hydrocarbon groups are often made from polymers of a two-carbon compound called *acetate*. The major plant lipids with acetate-derived hydrocarbon chains include oils, phospholipids, and waxes.

Oils

Oils are fats that are liquid at room temperature. A fat is a combination of a molecule of glycerol with three long-chain organic acids, called **fatty acids,** linked to it (fig. 2.10). The linkage between a fatty acid and glycerol is called an *acylglyceride linkage*. Thus, a fat is a **triacylglyceride.**

Fatty acids of triacylglycerides may differ in length and in the number and placement of double bonds in the chain. A fatty acid having no carbon-carbon double bonds is **saturated.** Oils are liquid because their fatty acids are **unsaturated;** that is, they have double bonds between one or more pairs of carbon atoms. Double bonds provide molecular rigidity at sharp angles, which prevents the molecules from packing tightly into a solid. This is why unsaturated lipids such as corn oil and peanut oil are liquids at room temperature.

Although triacylglycerides occur in all parts of a plant, they are most abundant in seeds. Like carbohydrates, seed oils are a form of chemical energy that is harvested when the seed germinates. Some seeds contain enough oil to be commercially

CHAPTER TWO *Atoms and Molecules: The Building Blocks of Life*   33

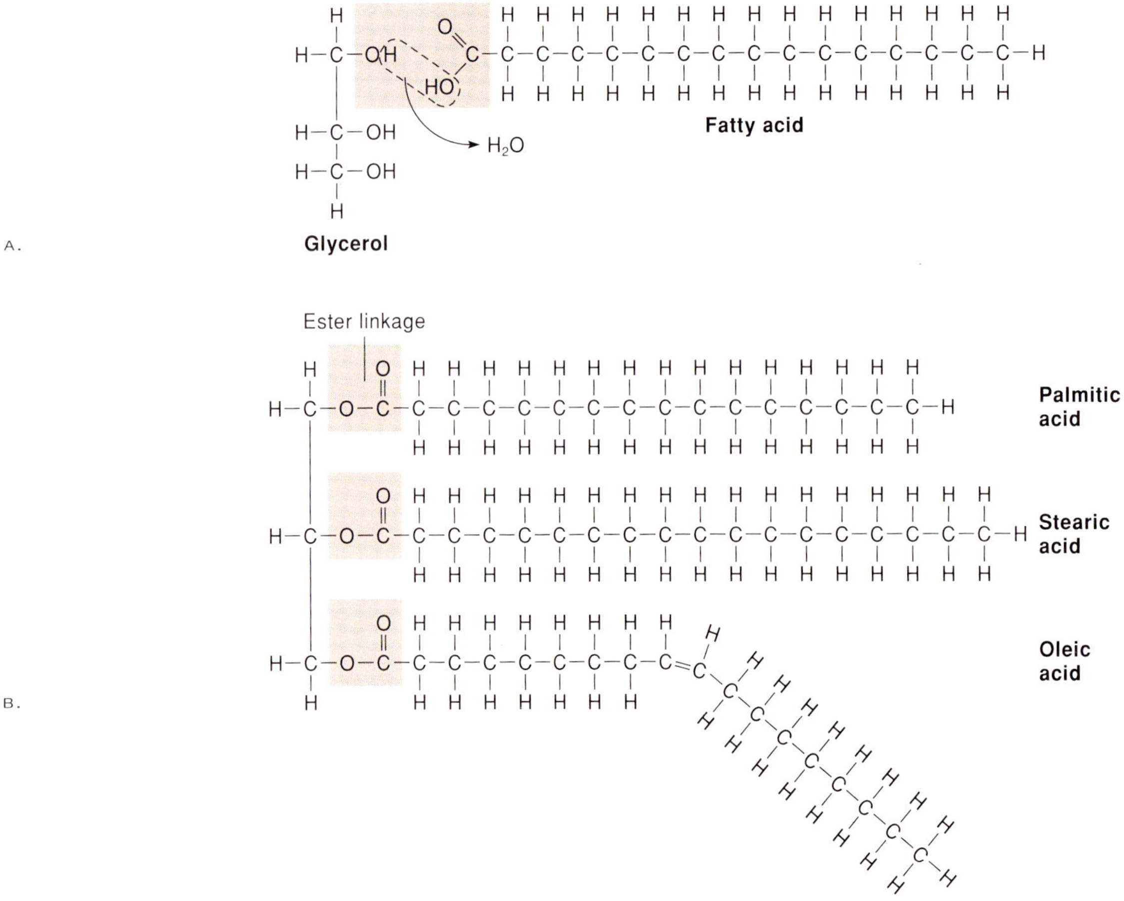

FIGURE 2.10

The structure of a fat. (a) An acylglyceride bond forms when the carboxyl group of a fatty acid links to the hydroxyl group of a glycerol, with the removal of a water molecule. (b) Fats are triacylglycerides whose fatty acids vary in length and also in the presence and location of carbon-carbon double bonds.

valuable. The best known of these are cotton, sesame, safflower, sunflower, olive, coconut, peanut, corn, castor bean, and soybean. You can buy most of these oils at grocery stores.

Fatty acids in plant oils are usually either monounsaturated or polyunsaturated with two or more double bonds. The most common fatty acids are oleic acid (one double bond), linoleic acid (two double bonds), and linolenic acid (three double bonds). However, seed oils from palms, coconuts, and other tropical plants contain mostly palmitic acid, which is saturated. Box 2.2, "What's Wrong with Tropical Oils?" explains how the composition of different plant oils affects human health.

Phospholipids

Membranes contain lipids in which a phosphate group is substituted for one of the fatty acids (fig. 2.11). Such lipids are called **phospholipids.** The phosphate group gives the compound a polar end that dissolves in water or forms a covalent bond with a membrane protein. This property causes phospholipids to have a dual solubility, since the phosphate end is water-soluble (hydrophilic), and the fatty end is water-repellent (hydrophobic). Consequently, phospholipids interact with the polar groups of proteins at one end, and form an oily matrix at the other. This versatility enables membranes to control the passage of polar and nonpolar substances through them. Chapter 4, "Membranes and Membrane Transport," presents more details about the role of phospholipids in membrane structure and function.

Waxes and Waxlike Substances

Waxes are complex mixtures of fatty acids linked to long-chain alcohols. This mixture also contains free fatty acids, fatty acids with hydroxyl groups (e.g., –OH), and long-chain hydrocarbons. Waxes that comprise the outermost layer (*cuticle*) of leaves, fruits, and herbaceous stems are called **epicuticular wax.** Waxes

BOXED READING 2.2

WHAT'S WRONG WITH TROPICAL OILS?

In most people, saturated fats raise blood-cholesterol levels more than do monounsaturated or polyunsaturated fats. In fact, blood-cholesterol levels are raised more by saturated fats than by cholesterol in the diet. High blood cholesterol has been linked with an increased risk of heart disease. Thus, it is important to minimize saturated fats in our diets.

How does this often-repeated advice relate to tropical oils? Tropical oils are the most common "hidden" source of saturated fats. They have long been used to make a variety of processed foods, including breakfast cereals, crackers, cookies, coffee creamers, baked goods, and microwave popcorn. Palm kernel oil and coconut oil are also used in cocoa mixes, diet desserts, instant soup, and chocolate.

It seems like a simple task to avoid tropical oils, because all you have to do is read the ingredients list on a food label. Some manufacturers are already making this easier by printing "NO TROPICAL OILS" in bold, colorful letters on the fronts of packages. However, many such products still contain saturated fats in the form of "hydrogenated vegetable oil." When hydrogen is added to the double bonds of a polyunsaturated fat, the fat becomes saturated. This means that linolenic acid, with three double bonds, becomes palmitic acid when it is hydrogenated. Since palmitic acid from linolenic acid does not come from palms or coconuts, the "no tropical oils" claim is legitimate; however, the saturated fat content may still be high.

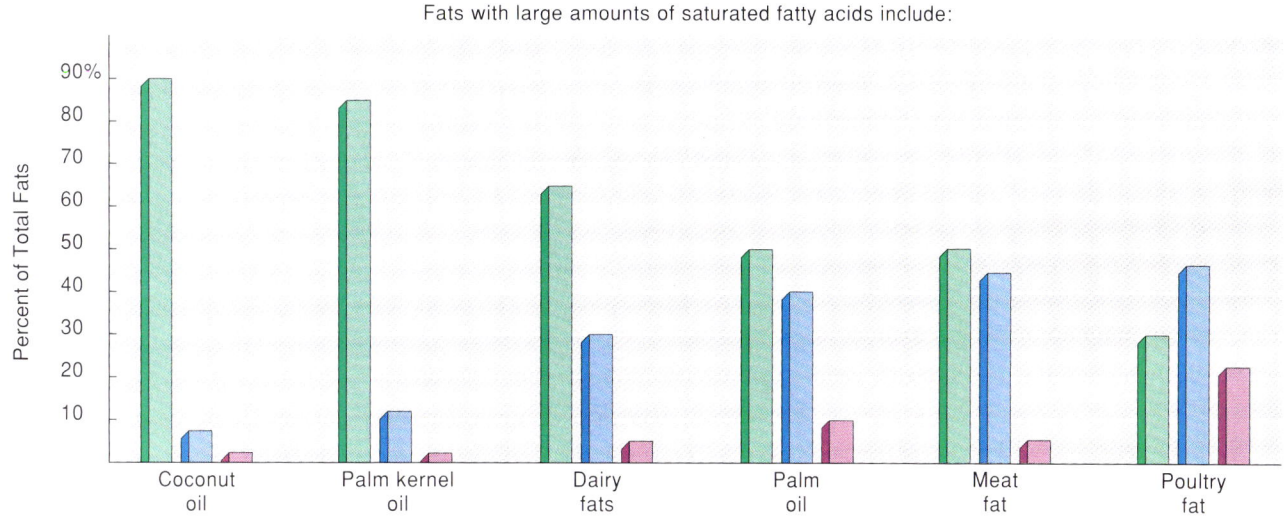

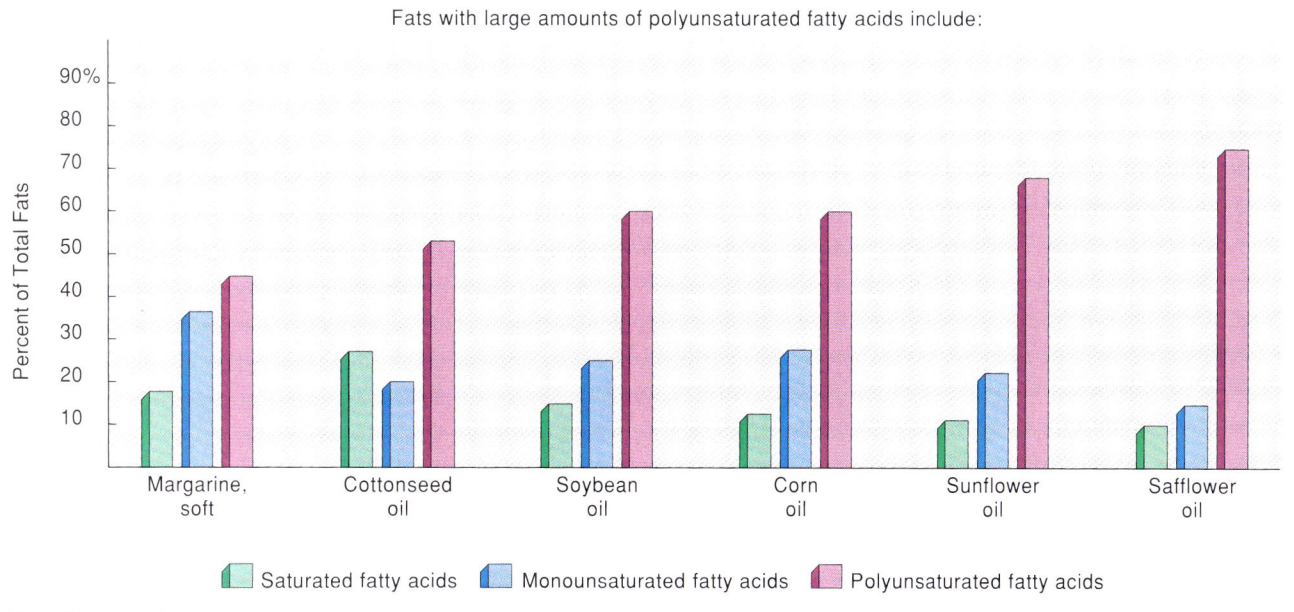

BOX FIGURE 2.2

Dietary fat and fatty acid proportions in different foods.

CHAPTER TWO *Atoms and Molecules: The Building Blocks of Life*

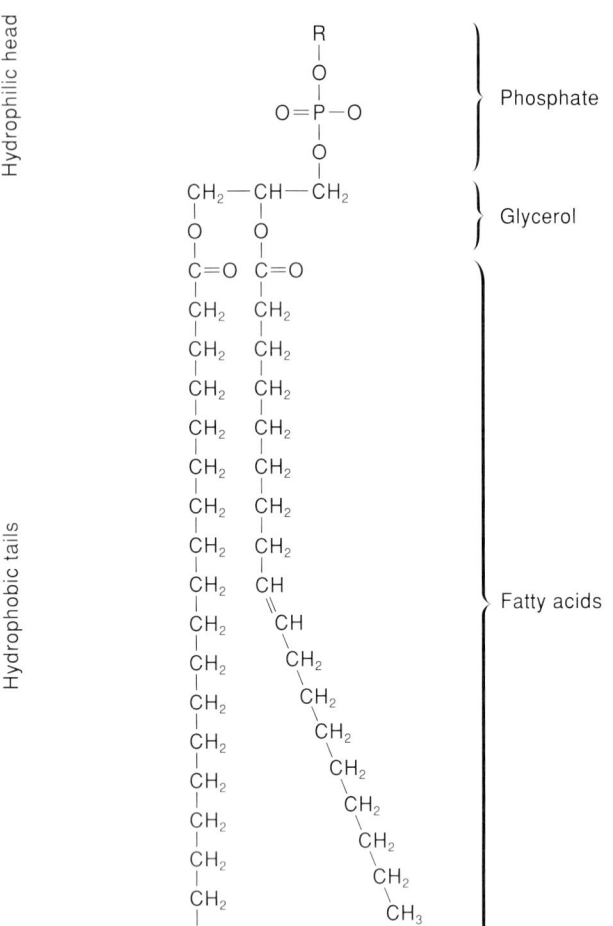

FIGURE 2.11

The structure of a phospholipid. A phospholipid consists of glycerol that is bonded to two fatty acids, which are hydrophobic, and to one phosphate group, which is hydrophilic. Phospholipids vary by their fatty acids and by side chains (R groups) that are attached to the phosphate. The R groups include glycerol, sugars, and amine-containing carbon chains.

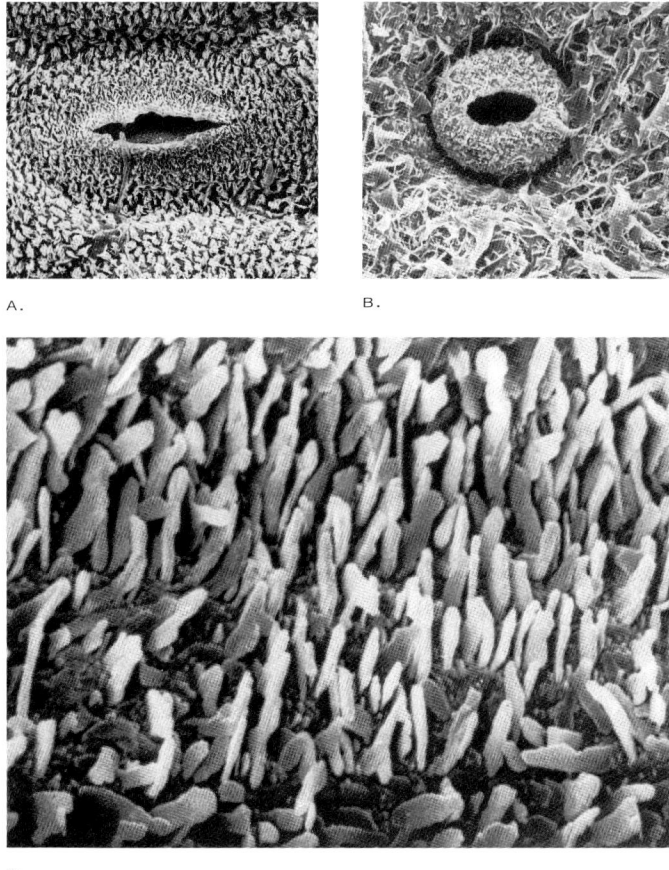

FIGURE 2.12

Scanning electron micrographs of leaf surfaces, showing different structures of epicuticular waxes. These are all from different orchids. (a) *Cleistes rosea*, ×700. (b) *Stelis gemma*, ×550. (c) *Eulophia paivaeana*, ×1400.

embedded in the cuticle are called **cuticular wax.** The structures of different waxes vary depending on the plants that produce them (fig. 2.12).

Waxes are usually harder and more water-repellent than other fats, which is why you don't "fat" your car or kitchen floor. Wax from plants such as the carnauba palm (*Copernicia cerifera*, the source of carnauba wax) are especially hard. Other commercial waxes come from the bayberry (*Myrica pensylvanica*), which is used to make novelty candles, and from candellila (*Euphorbia antisyphilitica*). Candellila wax, which is often substituted for carnauba wax, was once the main component of the hard "melts in your mouth, not in your hands" coating of M&M chocolate candies. (The *antisyphilitica* part of the scientific name implies another use for this plant, but there is no evidence that it can be used against sexually transmitted diseases.) Unfortunately, candellila is near extinction because the plants have been overcollected for their wax. Collecting this plant is forbidden in the United States, but it is still heavily collected in Mexico.

Although most waxes are hard, one significant exception is the wax from jojoba seed (*Simmondsia chinensis*). About half of a jojoba seed is liquid wax. This wax is sometimes called an oil because it is liquid at room temperature, but it has the chemical makeup of a wax. Jojoba "oil" is used in cosmetics because it penetrates the outer layer of human skin. It is also used as a lubricant. Jojoba oil is similar to the oil of sperm whales, which are an endangered species. Botanists and chemists are trying to modify jojoba oil to substitute for sperm whale oil as a lubricant for high-speed industrial machinery.

Cuticular wax is also associated with **cutin,** another waxy substance, which makes up most of the cuticle. Cutin consists of a variety of hydroxylated fatty acids linked together. A similar substance, called **suberin,** occurs in cork cells in bark and in the same cells of underground plant parts. Cutin and suberin differ mainly in the kinds of fatty acids they contain, but both function as barriers to water loss.

TABLE 2.2
Examples of Well-Known Secondary Metabolites

Compound	Example Source	Comment
Alkaloids		
coniine	poison hemlock	nerve toxin; killer of Socrates
strychnine	strychnine tree	potent nerve stimulant and convulsant
tubocurarine	curare tree	component of arrow poisons; used as muscle relaxant during surgery
tomatine	tomato leaves	inhibits feeding by Colorado potato beetles, which are also pests of tomatoes
morphine	opium poppy	main painkiller used worldwide
codeine	opium poppy	cough suppressant
atropine	belladonna	used in eye exams to dilate pupils and as an antidote to nerve gas
vincristine	Madagascar periwinkle	main treatment for certain kinds of leukemia
quinine	quinine tree	bitter flavor of tonic in gin and tonic; used to prevent malaria
Other Nitrogen-Containing Compounds		
canavanine	jack beans	toxic nonprotein amino acid
betalin	roots of garden beet	red pigment
sinigrin	horseradish	"burning" spice of mustards
amygdalin	apricot seeds	releases cyanide; may be active ingredient of unapproved cancer drug called Laetrile
Terpenoids		
menthol	mints and eucalyptus tree	strong aroma; used in cough medicines
camphor	camphor tree	component of disinfectants and plasticizers
nepetalactone	catnip	very attractive to cats
smilagenin	sarsaparilla	steroidal glycoside
digitalin	purple foxglove	cardiotonic used to stimulate heart action
oleandrin	oleander	heart poison (similar to digitalin)
limonin	grapefruit	bitter flavor
lycopene	tomatoes	red/orange pigment
rubber	rubber tree	component of rubber tires
taxol	Pacific yew	drug that inhibits cancerous tumors, especially of ovarian cancer
Phenolics		
salicin	willow tree	folk medicine against headaches and fever; precursor of aspirin*
gallic acid	walnut husks	main component of some tannins
myristicin	nutmeg	main flavor of the spice
rutin	buckwheat	common "bioflavonoid" sold in nutrition stores
cyanidin glucoside	chrysanthemums	deep red pigment

Opium poppy (*Papaver somniferum*)

Garden beet (*Beta vulgaris*)

Pacific yew (*Taxus brevifolia*)

Nutmeg (*Myristica fragrans*)

*The medicinal reputation of willow (*Salix alba*) stems from salicin, a chemical first isolated from willow bark in 1827. Although salicin was medically useless because of its severe side effects, related compounds were later isolated from other plants (e.g., salicylic acid from meadowsweet, *Spiraea ulmaria*). Finally, in 1899, Felix Hoffman produced acetylsalicylic acid to help his father deal with arthritis. Acetylsalicylic acid, which produced no bad side effects, was later mass-produced by Friedrich Bayer & Co. Bayer named the compound "aspirin"—*a* for acetyl and *spirin* for *Spiraea*—which, although catchy, is misleading, because *Salix* was there first. Today, aspirin is the world's most frequently taken medicine; Americans swallow almost forty tons of it every day.

CONCEPT

Lipids are water-repellent compounds made mostly of carbon and hydrogen. The major lipids of plants are triacylglycerides, phospholipids, and waxes, which all contain fatty acids. Triacylglycerides function mainly as storage in oilseeds. Phospholipids are important components of membranes. Waxes and waxlike substances prevent water loss from plants.

SECONDARY METABOLITES

Plants make a variety of less widely distributed compounds such as morphine, caffeine, nicotine, menthol, and rubber. These compounds are the products of **secondary metabolism,** which is the metabolism of chemicals that occur irregularly or rarely among plants and that have no known general metabolic role in cells.

Most plants have not yet been examined for their secondary products, and new compounds are discovered almost daily. Most secondary products can be grouped into classes based on structural similarities, biosynthetic pathways, or the kinds of plants that make them. The largest such classes are the alkaloids, the terpenoids, and the phenolics. Examples of several kinds of secondary metabolites are presented in table 2.2.

Secondary metabolites often occur in plants in combination with one or more sugars. Such combination molecules are called **glycosides.** The most common sugars in glycosidic secondary products are glucose, galactose, and rhamnose. Several other sugars also occur in glycosides, some of which are rare. For example, digitoxose is known only from the genus of the purple foxglove (*Digitalis purpurea*; fig. 2.14c), and apiose is unique to parsley (*Petroselinum*) and its relatives.

Functions of Secondary Metabolites

The most common roles suggested for secondary products are ecological roles that govern the interactions between plants and other organisms. For example, many secondary products are brightly colored pigments that attract insects and other animals for pollination or for dispersal of fruits and seeds. Conversely, nicotine and other toxic chemicals may protect plants against attacks by hungry insects or invasion by pathogenic microbes. This defensive role seems to be true in some cases but not in others. For example, phenolics in soybeans provide resistance to fungal infection, and nicotine from tobacco is toxic to many insects. Conversely, tobacco hornworms have little difficulty eating tobacco leaves that contain up to 30% nicotine, and monarch butterflies are unaffected by the cardiac glycosides of milkweeds. Instead of protecting the plants from being eaten by butterfly caterpillars, these toxins remain in the adult butterflies and protect them from insect-eating birds. Although there are many examples of the ecological functions of these kinds of chemicals, most secondary metabolites have not been examined for their potential ecological roles in plants.

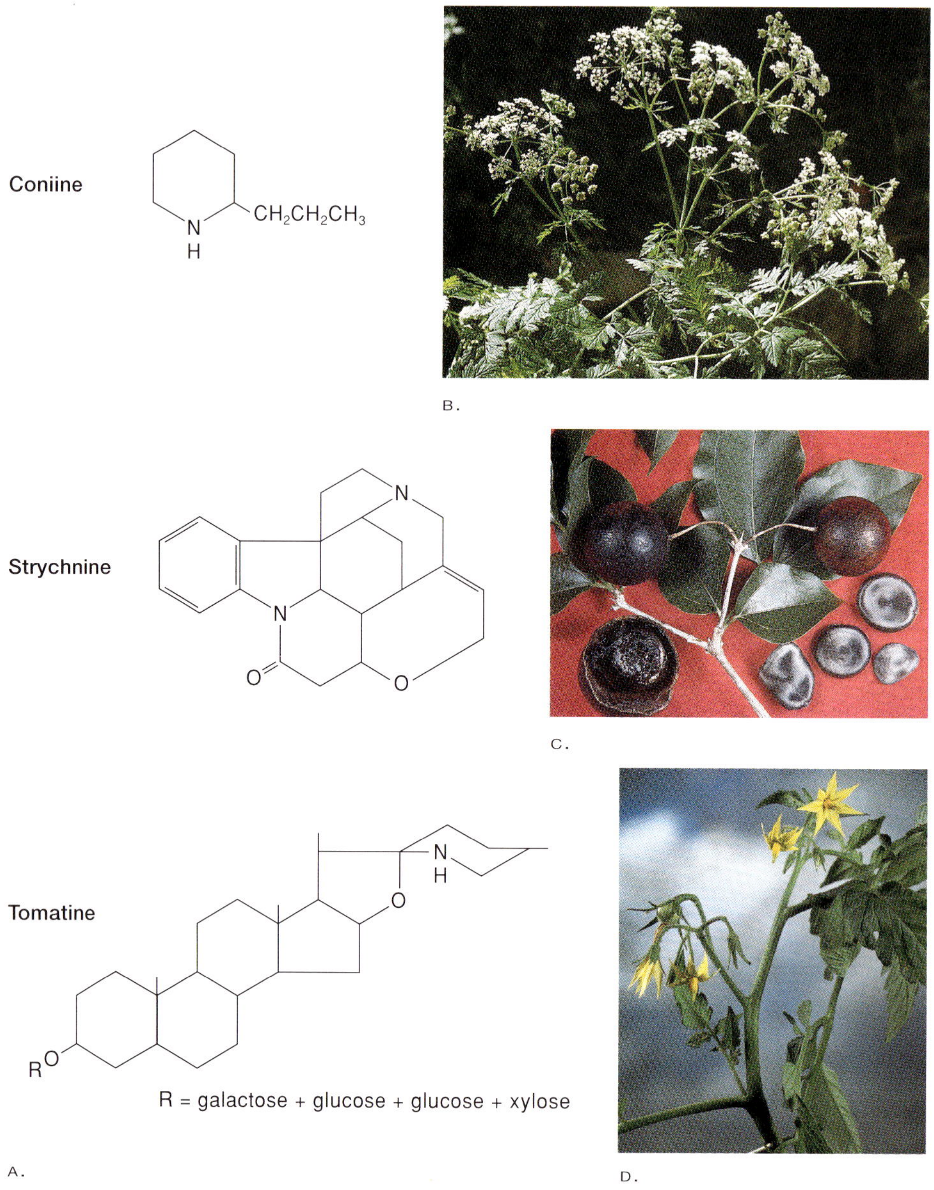

Alkaloids

Alkaloids generally include alkaline substances that contain nitrogen as part of a ring structure. The alkaloids, of which more than 6,500 are known, comprise the largest class of secondary metabolites. They occur in several plant families, especially the pea family, the sunflower family, the poppy family, the citrus family, and the potato family. Alkaloids are unknown in mosses, ferns, conifers, and most families of flowering plants.

Alkaloids are a diverse group of secondary products, ranging from simple compounds like coniine to complex compounds like strychnine and tomatine (fig. 2.13). They often produce dramatic physiological effects in humans and other animals. For example, coniine, strychnine, and tubocurarine are infamous toxins, while morphine, codeine, atropine, and vincristine are important theraputic drugs. Alkaloids are often bitter; one of the most bitter substances known is the alkaloid quinine.

Terpenoids

Terpenoids are dimers and polymers of five-carbon precursors called **isoprene** units (fig. 2.14). The smallest terpenoids, the **monoterpenes,** have two isoprene units. Monoterpenes such as geraniol and menthol, from the leaves of mints and eucalyptus, are volatile and usually have strong odors. Pines and other resinous plants also synthesize terpenoids made from four isoprene units (**diterpenes**) or from six isoprene units (**triterpenes**).

One subclass of triterpenes is the **sterols,** which are chemically similar to the steroidal hormones of animals. Sterols may be combined with nitrogen to form alkaloids, as in tomatine (fig. 2.13), or with sugars in steroidal glycosides like digitalin (fig. 2.14). Terpenoids having eight isoprene units form a class

FIGURE 2.13

(a) Structures of three alkaloids. (b) Poison hemlock (*Conium maculatum*) produces coniine in its leaves. (c) Strychnine plant (*Strychnos nux-vomica*) produces strychnine in seed coats. (d) Tomato leaves (*Lycopersicon esculentum*) are a source of tomatine.

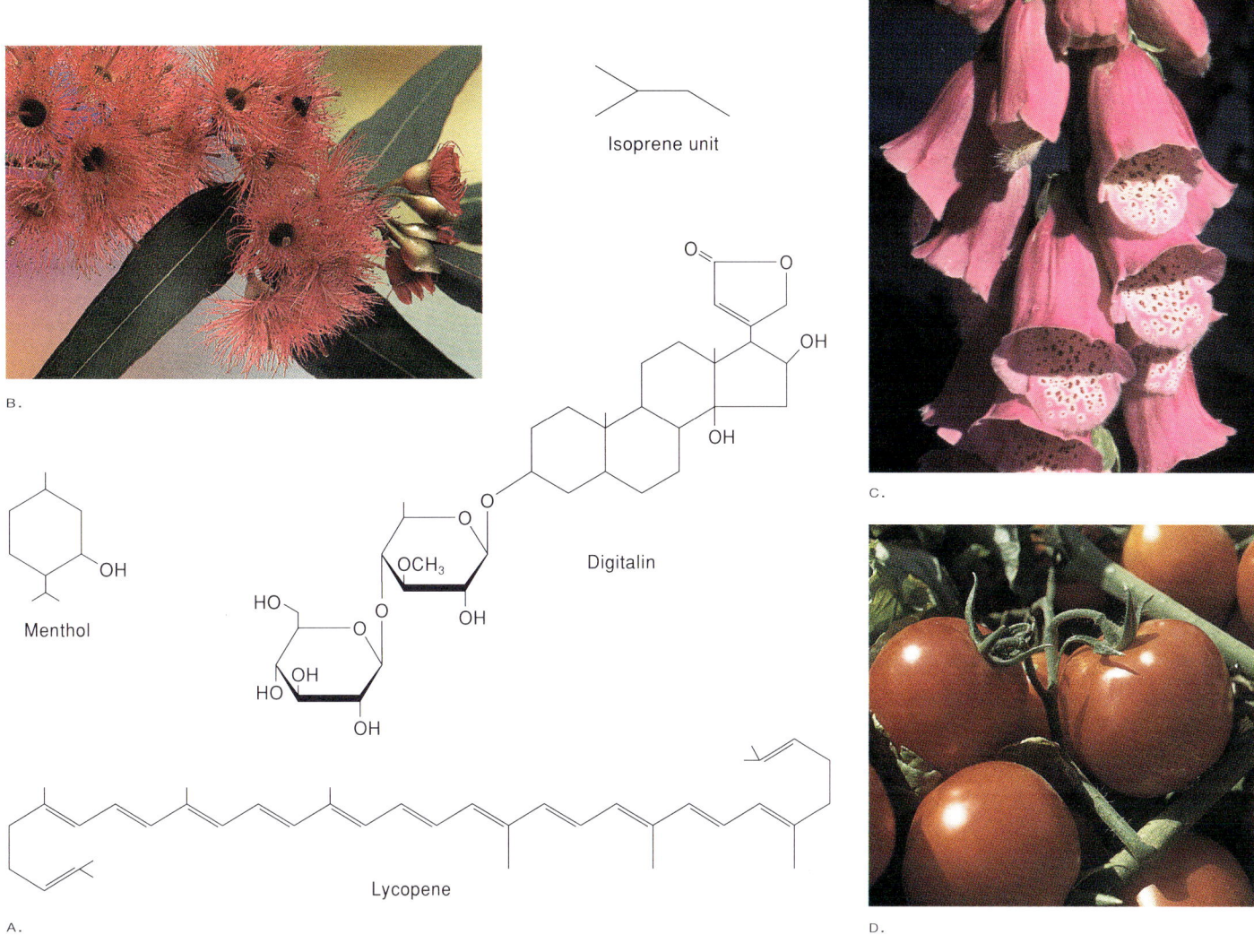

FIGURE 2.14

(a) Structures of three terpenoids. (b) *Eucalyptus* species produce menthol. (c) The purple foxglove (*Digitalis purpurea*) is a source of digitalin. (d) Lycopene is the main red pigment of tomatoes.

of yellow to red pigments called **carotenoids** (see Chapter 7). The largest terpenoids, which can have more than 6,000 isoprene units in a single molecule, are known as **rubber.** Rubber is made by about 2,000 plant species, but most of them make this substance in amounts too small for commercial use.

Although terpenoids are considered secondary metabolites, they include some compounds that have clear roles in plants. The most prominent terpenoids are **abscisic acid,** which is made from three isoprene units, the **gibberellins,** which are diterpenes, and the carotenoid **beta-carotene.** Abscisic acid and the gibberellins are important plant hormones that regulate plant growth and development; Chapter 18 discusses these and other plant hormones. Beta-carotene is a universally occurring accessory pigment in photosynthesis. Abscisic acid, the gibberellins, and beta-carotene are examples of basic metabolites that are derived from secondary metabolic pathways.

Phenolics

Compounds that contain a fully unsaturated, six-carbon ring linked to an oxygen are called **phenolics.** Simple phenolics, such as salicylic acid, are single-ring chemicals adorned by simple side groups (fig. 2.15). Complex phenolics that have a three-carbon side chain are called **phenylpropanoids** (*phenyl* refers to the ring; *prop* refers to the three-carbon side chain); they are derived from tyrosine and phenylalanine, which are phenylpropanoid amino acids. Myristicin, the main flavor of nutmeg, is an example of a familiar phenylpropanoid (fig.2.15).

Phenylpropanoids commonly occur as parts of more complex molecules. For example, an entire subclass of secondary metabolites, called **flavonoids,** includes phenylpropanoids that are condensed into complex three-ringed structures. Certain flavonoids, called **anthocyanins,** are the red and blue pigments of many flowers; most others are colorless except in ultraviolet light. Flavonoids are sold in nutrition centers and health food

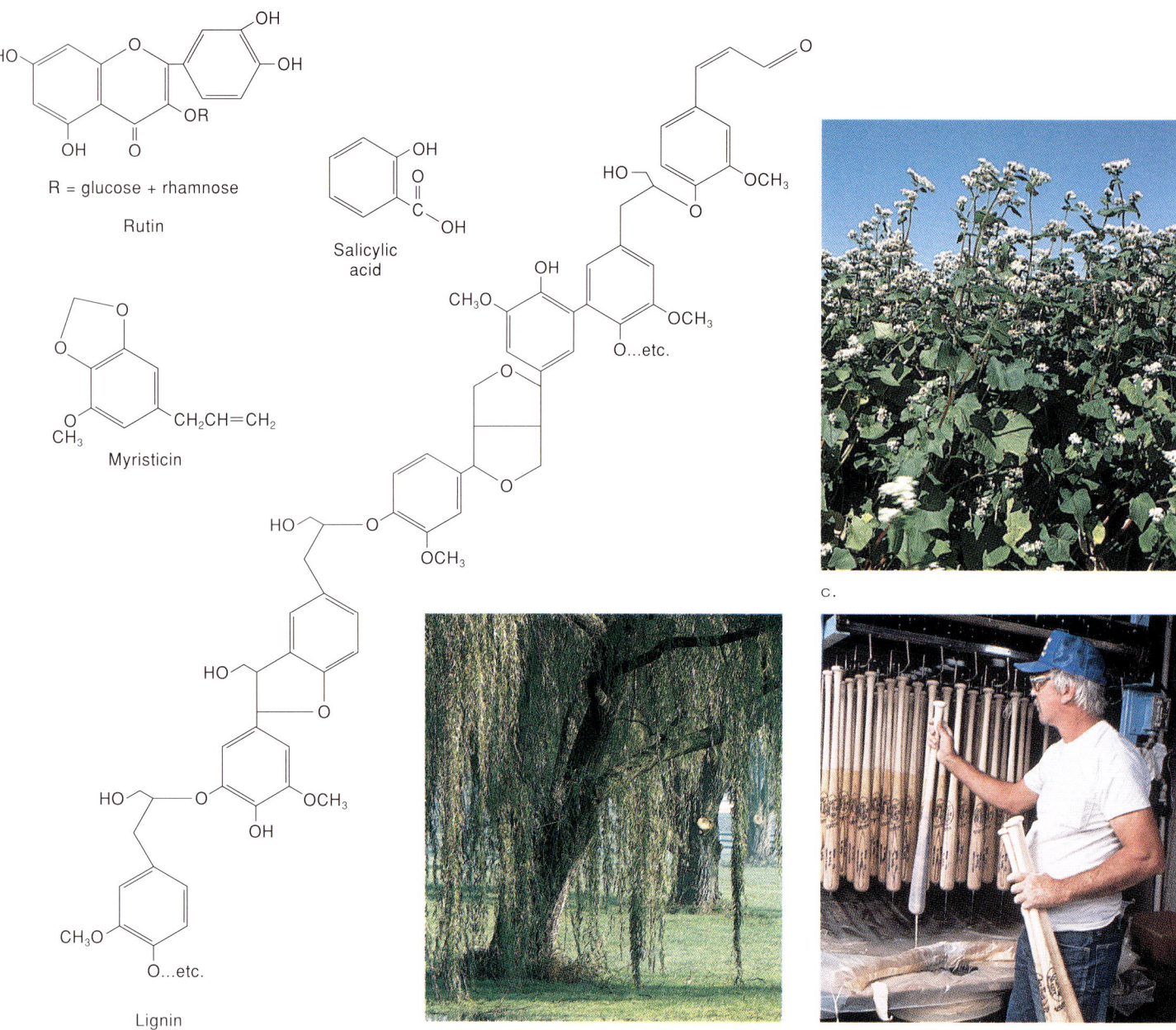

FIGURE 2.15

(a) Structures of four phenolics. (b) Willows (*Salix* species) accumulate salicylic acid in their bark. (c) Rutin is produced abundantly in buckwheat (*Fagopyrum esculentum*) and many other plants. (d) Lignin is the strengthening polymer that makes wood valuable commercially. These baseball bats are made of the wood of white ash (*Fraxinus americana*). To learn how these bats are made, see the boxed reading entitled "The Bats of Summer: Botany and our National Pastime" on page 366 in Chapter 16.

stores, usually as supplements with vitamin C. The most commonly available flavonoid is rutin, which is easily obtained from eucalyptus or buckwheat leaves (fig. 2.15c).

What are Botanists Doing?

Use the computerized database or the journal abstracts (e.g., Biological Abstracts) in your library to find out what kinds of studies have been done on flavonoids during the past twelve months.

Like the terpenoids, the phenolics also form polymers. Flavonoid polymers, called **tannins,** can tan leather and are astringent to the taste—they are the compounds that impart the "dryness" (astringency) to dry wines. Perhaps the most significant phenolic polymer is **lignin,** a major structural component of wood. Lignin consists of polymeric phenylpropanoids whose combined structure is unknown.

Minor Classes of Secondary Metabolites

Besides the "big three" (alkaloids, terpenoids, phenolics), plants also make several minor classes of secondary metabolites. For example, **mustard oil glycosides** are nitrogen-sulfur compounds that occur in cabbage, broccoli, horseradish, watercress, and other members of the mustard family. These compounds produce the main flavor and odor of plants in this family.

Members of the pea family produce amino acids that are not incorporated into proteins. About four hundred of these so-called **nonprotein amino acids** have been discovered.

Other minor classes of secondary products, such as **polyacetylenes** and **cyanogenic glycosides,** occur in several families of plants. Polyacetylenes are long-chain derivatives of fatty acids that contain carbon-carbon triple bonds. Most known polyacetylenes are from the sunflower family or from the magnolia family. Cyanogenic glycosides are sugar-containing compounds that release cyanide gas when they are hydrolyzed. They occur in many plant families, but they are especially common in the pea and rose families. Cyanide from such plants is not usually fatal, but two cups of well-chewed apple seeds would probably kill an adult human. The seeds would have to be well-chewed because cyanogenic glycosides are hydrolyzed only when the cells are damaged.

There are many other minor classes of secondary products. Many are familiar to us as the unique flavors and odors of common edible plants or the colors of garden flowers. New secondary products from these and other plants are reported in scientific journals almost daily. The tremendous diversity of secondary metabolites shows that plants are amazing biochemical factories.

CONCEPT

Plants have common metabolic processes that involve carbohydrates, proteins, nucleic acids, and lipids. They also have a highly diverse secondary metabolism that makes many rare chemicals. Most secondary products belong to one of three main groups: terpenoids, phenolics, or alkaloids. The functions of secondary metabolites are generally unknown for most plants.

Chapter Summary

Organisms are organized into a complex hierarchy of atoms, molecules, and cells. Beginning with the structure of atoms, the successively higher levels of complexity depend on the properties of their simpler components. Elements comprise the simplest level that cannot be broken down by ordinary chemical means. The main elements in organisms are carbon, hydrogen, oxygen, and nitrogen. These and other elements bond in a variety of combinations to make carbohydrates, proteins, nucleic acids, lipids, and many other classes of plant chemicals.

Carbohydrates are monomers and polymers of sugars. The most abundant carbohydrate polymers are cellulose and starch, which exhibit the two main functions of carbohydrates: structure (cellulose) and storage (starch). Cellulose is a major component of cell walls, and starch is a common energy reserve. The structures of carbohydrate polymers depend on linkages between monomers and on hydrogen bonding between different chains of the polymer.

Proteins, like carbohydrates, are also important in structure and storage. Many proteins are enzymes that catalyze the chemical reactions of cells. Structural proteins occur mostly in membranes, but certain glycoproteins, such as the extensins, occur in cell walls. Storage proteins are common in seeds but not in other parts of plants.

The primary structure of a protein is determined by its sequence of amino acids. Secondary and tertiary structures depend on bonding and interactions between amino acids in the protein. Proteins consisting of more than one polypeptide have quaternary structure, which is the configuration of the subunits relative to each other.

Nucleic acids are the largest and most complex polymers in cells. The two kinds of nucleic acids, DNA and RNA, differ in the type of sugar they contain and in one of their four nitrogen-containing bases. The basic monomer of each type of nucleic acid includes a sugar, a nitrogenous base, and a phosphate group.

Lipids are water-repellent compounds such as fats. Fats consist of hydrocarbon chains of fatty acids that are attached three at a time to glycerol to make triacylglycerides. Phospholipids consist of two fatty acids plus a phosphate group attached to glycerol. Phospholipids are important parts of membranes, whereas triacylglycerides are primarily storage reserves in seeds. Waxes are combinations of fatty acids and long-chain alcohols mixed with other kinds of long-chain hydrocarbons.

Plants also make many other classes of chemicals. The distribution of the chemicals in these other classes is irregular. Most of the chemicals in them have no obvious or universal function in plant metabolism. Thus, they are called secondary metabolites to distinguish them from the widespread, basic classes of plant compounds (i.e., carbohydrates, proteins, nucleic acids, lipids). Some products of secondary metabolism, however, are widespread and important in basic plant metabolism.

Questions for Further Thought and Study

1. Biochemical reactions form bonds between fatty acids and glycerol (ester bonds), between amino acids (peptide bonds), and between carbohydrates (glycosidic bonds). What do the reactions that make all of these kinds of bonds have in common?

2. How do the chemical and physical properties of sunflower oil, jojoba oil, and petroleum oil differ? How are they alike?

3. One of the earliest explanations for the function of secondary metabolites was that they were waste products of plant metabolism. Suggest reasons why this might be a good explanation; also suggest reasons why it might not be a good one. How could you test each idea?
4. How do the properties of amino acids contribute to the three-dimensional structures of proteins?

Suggested Readings

ARTICLES

Duchesne, L. C., and D. W. Larson. 1989. Cellulose and the evolution of plant life. *BioScience* 39:238–241.

Patton, A. 1991. Tropical oils. *Diabetes Forecast* 44, no. 4:56–59.

Richards, F. M. 1991. The protein folding problem. *Scientific American* 264(1):54–65.

BOOKS

Atkins, P. W. 1987. *Molecules.* New York: Scientific American Library.

Harborne, J. B. 1988. *Introduction to ecological biochemistry.* 3d ed. New York: Academic Press.

Lehninger, A. L., D. L. Nelson, and M. M. Cox. 1992. *Principles of Biochemistry.* 2d ed. New York: Worth.

Sackheim, G. 1991. *Introduction to Chemistry for Biology Students.* 4th ed. Redwood City, CA: Benjamin/Cummings.

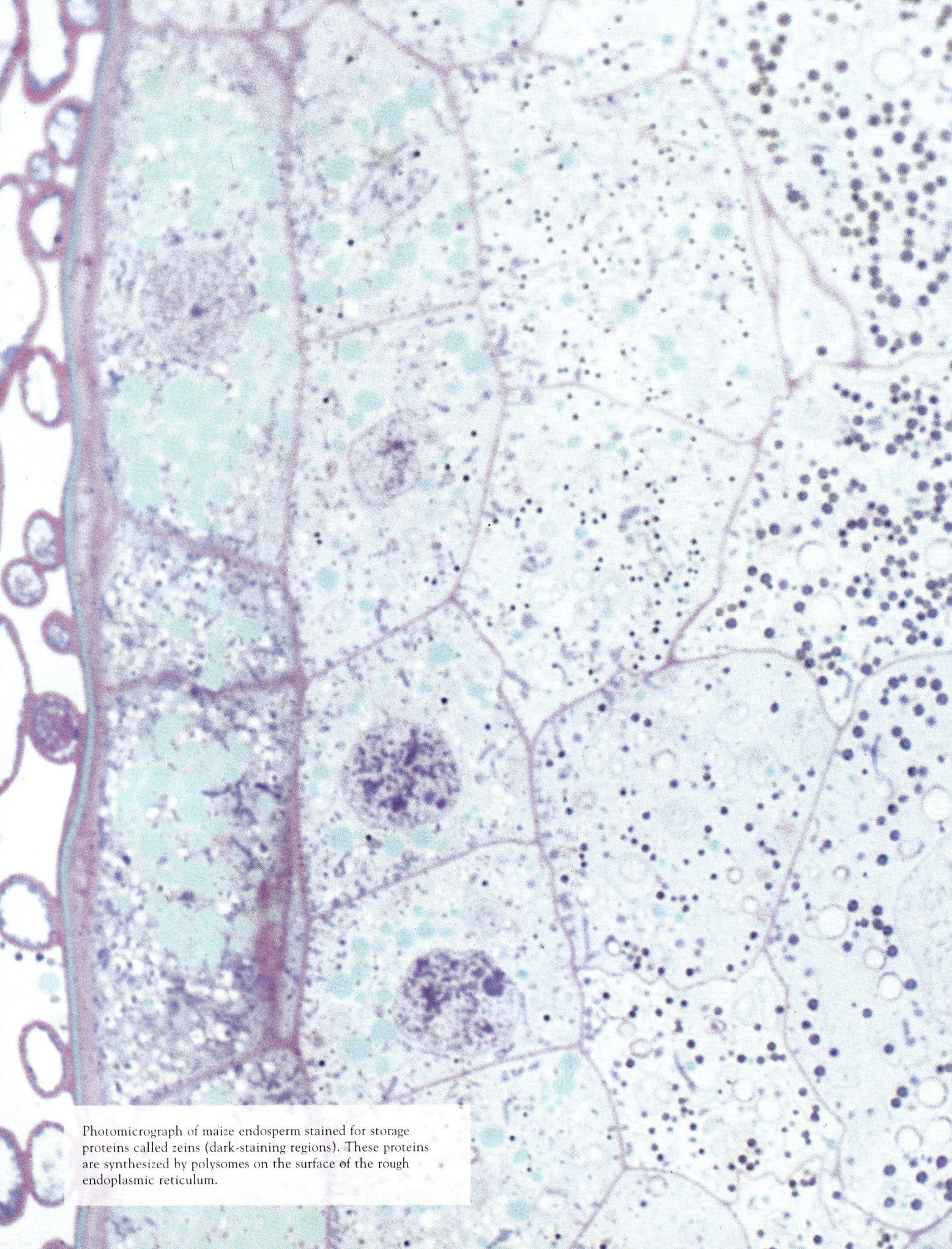

Photomicrograph of maize endosperm stained for storage proteins called zeins (dark-staining regions). These proteins are synthesized by polysomes on the surface of the rough endoplasmic reticulum.

CHAPTER 3

Structure and Function of Plant Cells

Chapter Outline

INTRODUCTION
SCALES OF MICROSCOPIC OBSERVATION
METHODS OF CYTOLOGY
    Electron Microscopy

BOX 3.1
HOW TO PREPARE CELLS FOR MICROSCOPY
    Freeze-Fracturing and Electron Microscopy
    Cell Fractionation
    Biochemical Cytology
FUNCTIONAL ORGANIZATION OF CELLS
    Why Are Cells So Small?
    Membranes and Cell Compartments
    The Cytoskeleton
    How Does the Cytoskeleton Work?
THE CELL WALL
    How Cell Walls Grow
    Connections Between Cells
THE NUCLEUS
RIBOSOMES
THE MEMBRANE SYSTEM
    The Plasma Membrane
    The Endoplasmic Reticulum
    Dictyosomes
    Vacuoles
    Microbodies
ORGANELLES FOR ENERGY CONVERSION
    Chloroplasts
    Mitochondria
CELL MOVEMENTS
    Internal Movements

BOX 3.2
CELLULAR INVASION: ORIGIN OF CHLOROPLASTS AND MITOCHONDRIA
    Cells That Swim
CELL THEORY: THE DOGMA AND AN ALTERNATIVE
    The Cell Theory: Postulates and Problems
    An Alternative: The Organismal Theory
Chapter Summary
Questions for Further Thought and Study
Suggested Readings

Chapter Overview

After atoms and molecules, the next higher level of complexity in living organisms includes cells and their components. All plants are made of cells. Some cell components occur in all living cells, while others occur only in the cells of leaves, roots, or other parts of plants. Depending on their components, cells can divide, grow, transport sugar or water, photosynthesize, secrete nectar or resin, or harvest energy from organic molecules. Most types of cells also contain genetic material that controls the activities of the cell and is inherited by new cells after cell division.

In this chapter we introduce the main components of cells and the primary functions of these components in plant structure and metabolism. We emphasize the common components of cells, and also present examples to show that these components vary among different cell types and among different plants. For example, components such as the plasma membrane occur in all living cells, while chloroplasts occur only in the cells of green tissue. These and other examples discussed in this chapter show how tissues can differ at the cellular level from one part of a plant to another.

INTRODUCTION

All plants consist of cells, which are the simplest units of a plant that can live independently. The smallest organisms are single cells, but plants are made of billions of cells. Plant cells have many shapes and sizes, the smallest of which are the dividing cells at the tips of roots and stems. These cells are usually about 12 μm in diameter; it would take about one hundred of these cells to equal the thickness of a dime. Conversely, the largest cells are long, thin fiber cells. Fibers of jute (*Corchorus capsularis*), which are used to make burlap and rope, can be more than 2.3 meters long (fig. 3.1c). In comparison, your height is probably between 1.5 and 2.0 meters.

Each cell in a plant consists of a cell wall that surrounds a **plasma membrane,** which encloses many smaller parts called **organelles.** Because each organelle has its own set of functions, the job of each cell in a plant is determined by how many and which organelles it contains, and what the organelles do. For example, leaf cells contain chloroplasts, nectar-secreting cells contain many dictyosomes, and the storage cells of oil-containing seeds have many glyoxysomes. These organelles have roles in photosynthesis, secretion, and oil metabolism, respectively. They and the other organelles of plant cells are discussed further in later sections of this chapter.

Like plant chemistry, plant cell biology can be reduced to the study of the structure and function of smaller components. By analyzing the anatomy of a cell, we can find clues to how the cell works. In this chapter we present an overview of plant cell structure and how it controls cellular processes. Many of these processes are discussed in greater detail in later chapters.

SCALES OF MICROSCOPIC OBSERVATION

Before reading about cells, you should have a good perspective on the sizes of cells and organelles relative to what you can see with your unaided eyes. Sizes are described either in metric units or in **magnifications** (see below; see also Appendix B). Figure 3.2 shows metric units based on the meter, which is slightly longer than a yard.

The power of the naked eye is 1, or 1×. An ordinary magnifying lens can give a 5× magnification, meaning it can make an object five times larger than the object appears to the naked eye (fig. 3.3). Magnifying power can be increased further by combining two or more lenses, as is done in a **light microscope,** which consists of a series of lenses. A light microscope that has two 5× lenses has a magnifying power of 25× (5× · 5× = 25×; fig. 3.4). Modern light microscopes generally offer an optional combination of lenses that magnify from 100× to 1000×.

Another factor important in microscopy is **resolving power,** which is the minimum distance necessary to distinguish two points from one another. For example, the resolving power of the human eye is about 100 μm, meaning that two dots less than 100 μm apart appear to us as one dot. The best light microscopes can distinguish objects that are about 0.2 micrometers apart (0.2 μm); this is about five hundred times more resolving power than the human eye, and is sufficient to see plant cells and some of their parts. The resolving power of a microscope also determines the sharpness of the magnified image.

In light microscopy, resolving power is limited by the wavelengths of visible light used to create the image. The shortest wavelength of light visible to the human eye is around 400 nanometers (nm), which is 0.4 μm. However, greater resolution is obtained by using beams of electrons, which have shorter wavelengths. A microscope that focuses an electron beam instead of light is called an **electron microscope** (fig. 3.5). The wavelengths of electron beams depend on the amount of voltage used to generate them. For example, with 60,000 volts, an electron microscope can generate a beam of electrons having a wavelength of about 0.005 nm. This wavelength can achieve a resolution of up to 0.4 nm, which is about one thousand times more resolution than that of a light microscope. At this level of resolution, sharp images may be obtained at magnifications exceeding 100,000×.

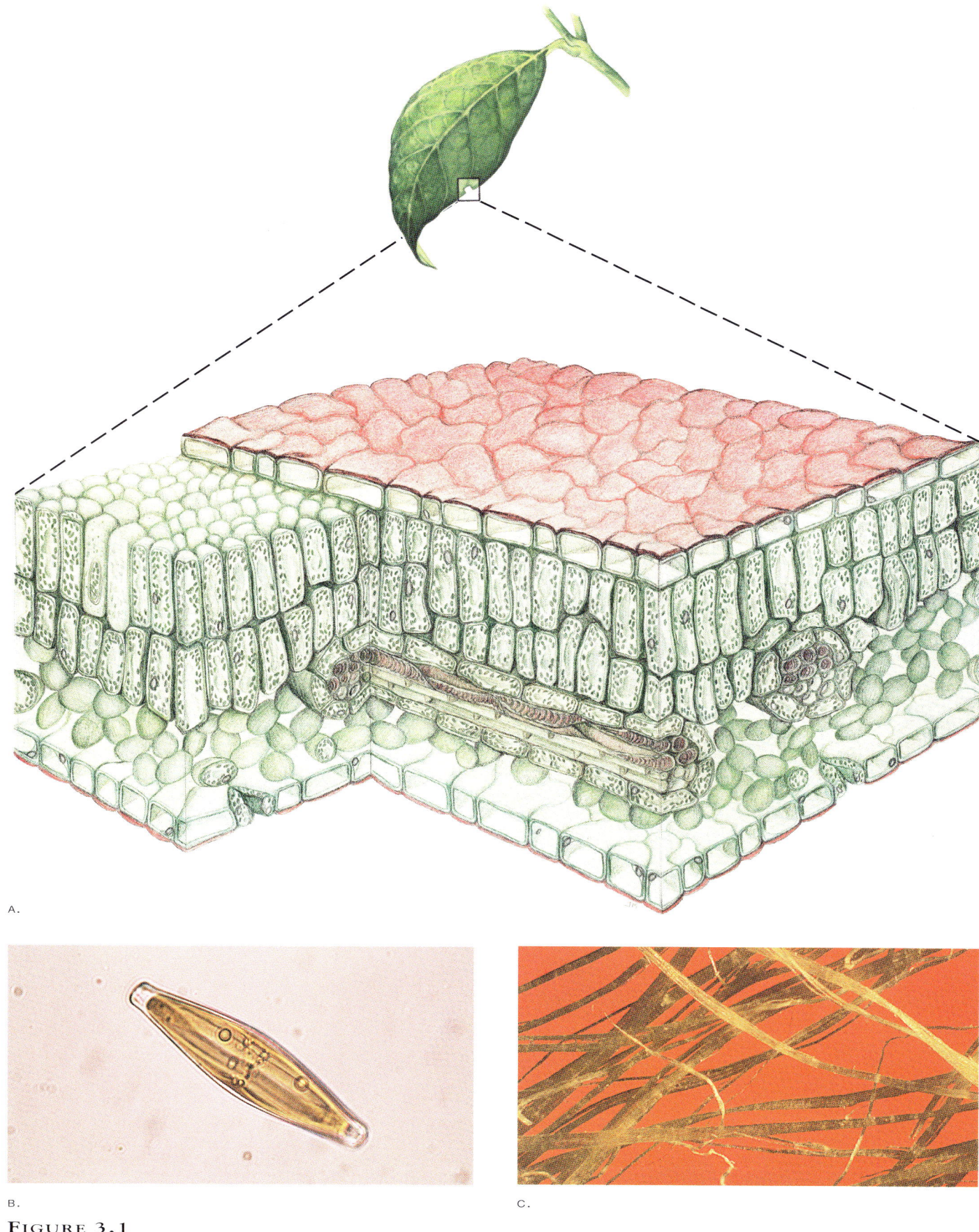

Figure 3.1

Cells of plants. (a) Diagram of a leaf with small portion enlarged and sectioned to show cellularity. (b) Light micrograph of a green alga (*Desmidiaceae* family); the entire organism is a single cell, ×450. (c) Photo of jute fibers, ×20. These fibers are used to make burlap and rope.

Chapter Three *Structure and Function of Plant Cells*

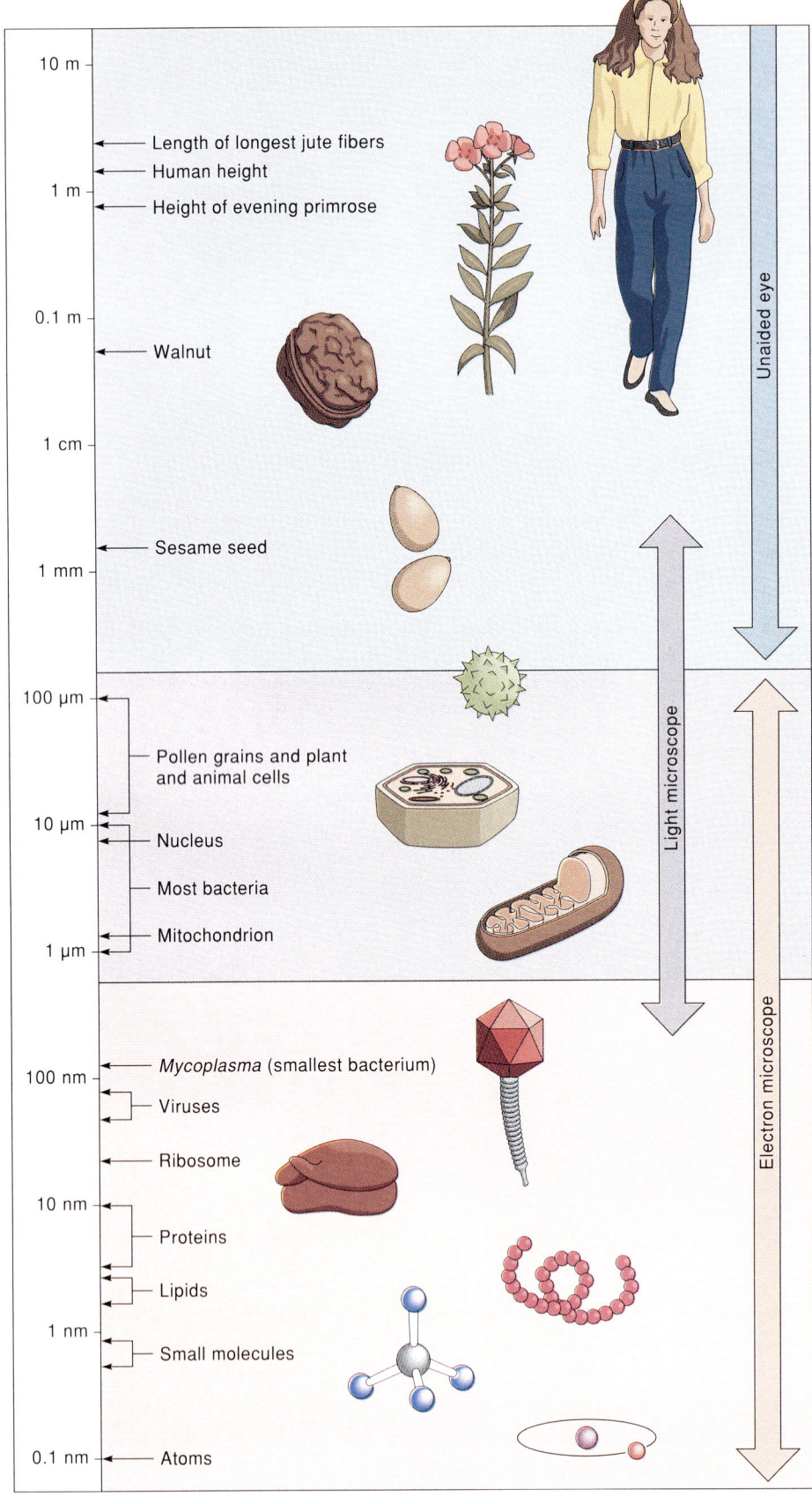

FIGURE 3.2

The relative size of cells and cellular parts, in metric units.

This chapter contains many photographs of plant cells and their parts. It is important that you keep a clear perspective on the scales of measurement in these photographs so that you can better understand how these structures fit together to make a cell. As an aid to your perspective of scales, figure 3.2 shows the size range of familiar objects relative to molecules and to cells and some of their components.

CONCEPT

Most cells cannot be seen with the unaided eye. However, cells and their components can be seen with light and electron microscopes. Light microscopes magnify objects up to about 1000×; electron microscopy magnifies them to more than 100,000×. The quality of the magnified image is determined by resolving power, which is limited by the wavelength of the light or electron beam that is used for observation.

METHODS OF CYTOLOGY

Cytology is the study of cell structure and function. Cytologists use light microcopy, electron microscopy, and cell chemistry to obtain information about cells. Several kinds of light microscopy can enhance images of cells; for example, differential-interference microscopy exaggerates differences in density within a specimen. Figure 3.6 shows images obtained by this and other kinds of light microscopy. Light microscopy is also aided by chemicals that stain specific parts of the cell; for example, acetocarmine stains the nucleus deep red or purple (fig. 3.7). The next few paragraphs describe how electron microscopy and cell chemistry also provide information about cell structure and function. Before you read about these methods, see box 3.1, "How to Prepare Cells for Microscopy."

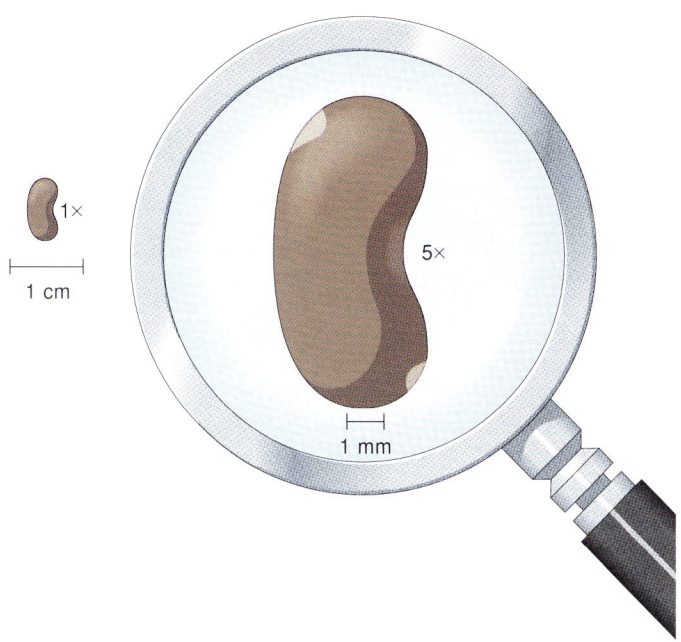

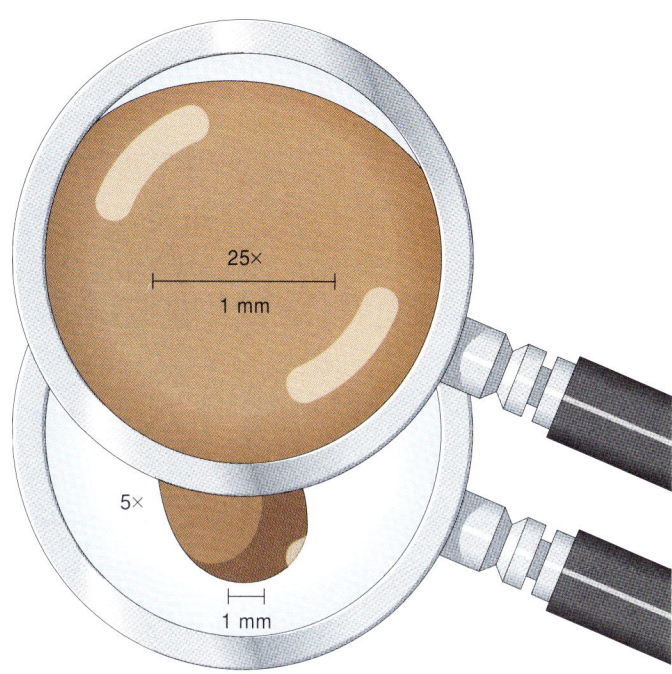

Figure 3.3

A 5× magnifying glass makes this kidney bean look five times larger than its actual size.

Figure 3.4

The kidney bean looks 25 times larger than normal under two 5× magnifying lenses (5× · 5× = 25×).

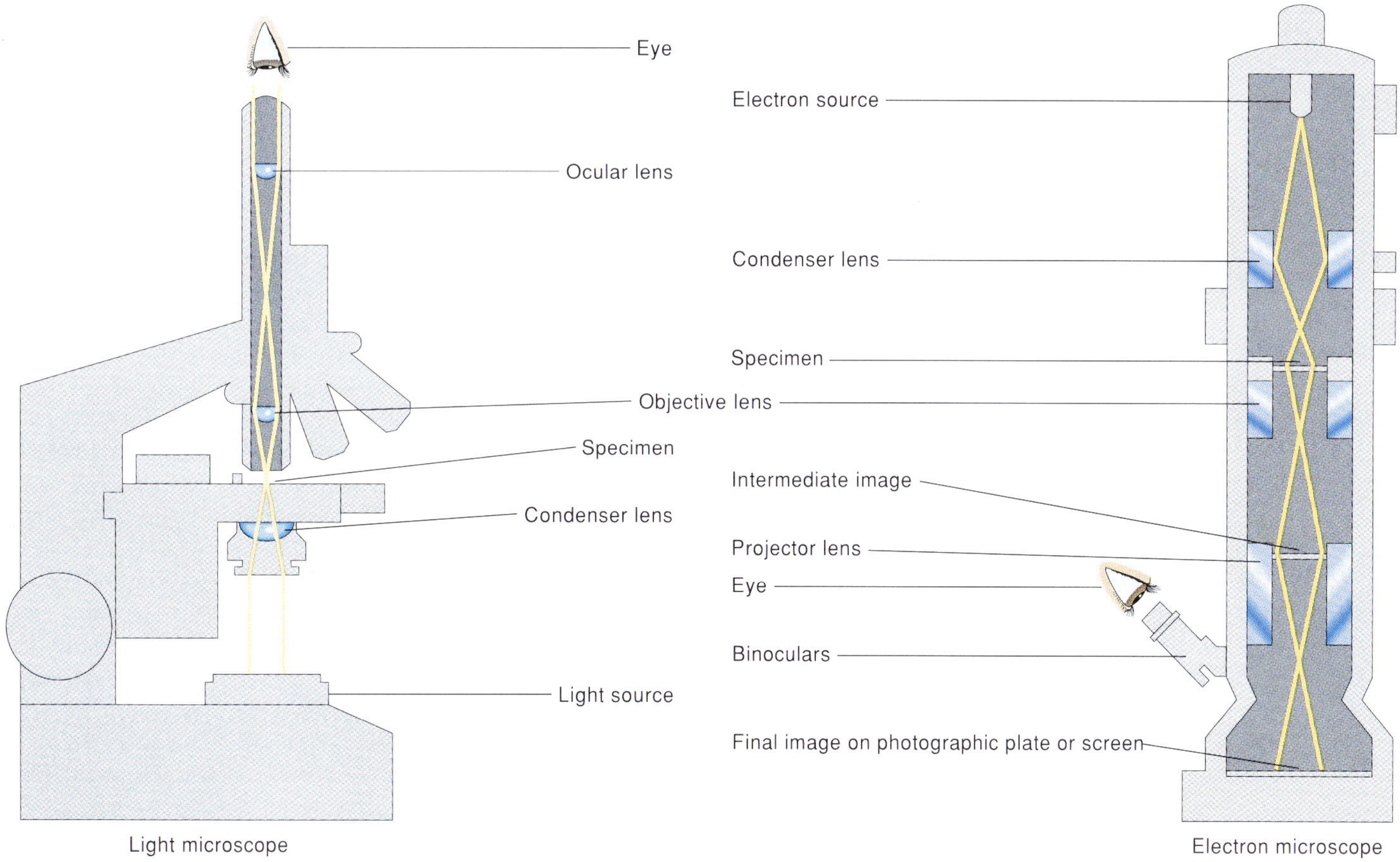

Figure 3.5

Comparison of light microscopes and transmission electron microscopes. In standard light microscopy, light is focused on the specimen by a condenser lens. The image is subsequently magnified by an objective lens and an ocular lens. All lenses are made of glass. In contrast, electron beams are focused by electromagnetic lenses in transmission electron microscopy. Electrons cannot pass through glass lenses.

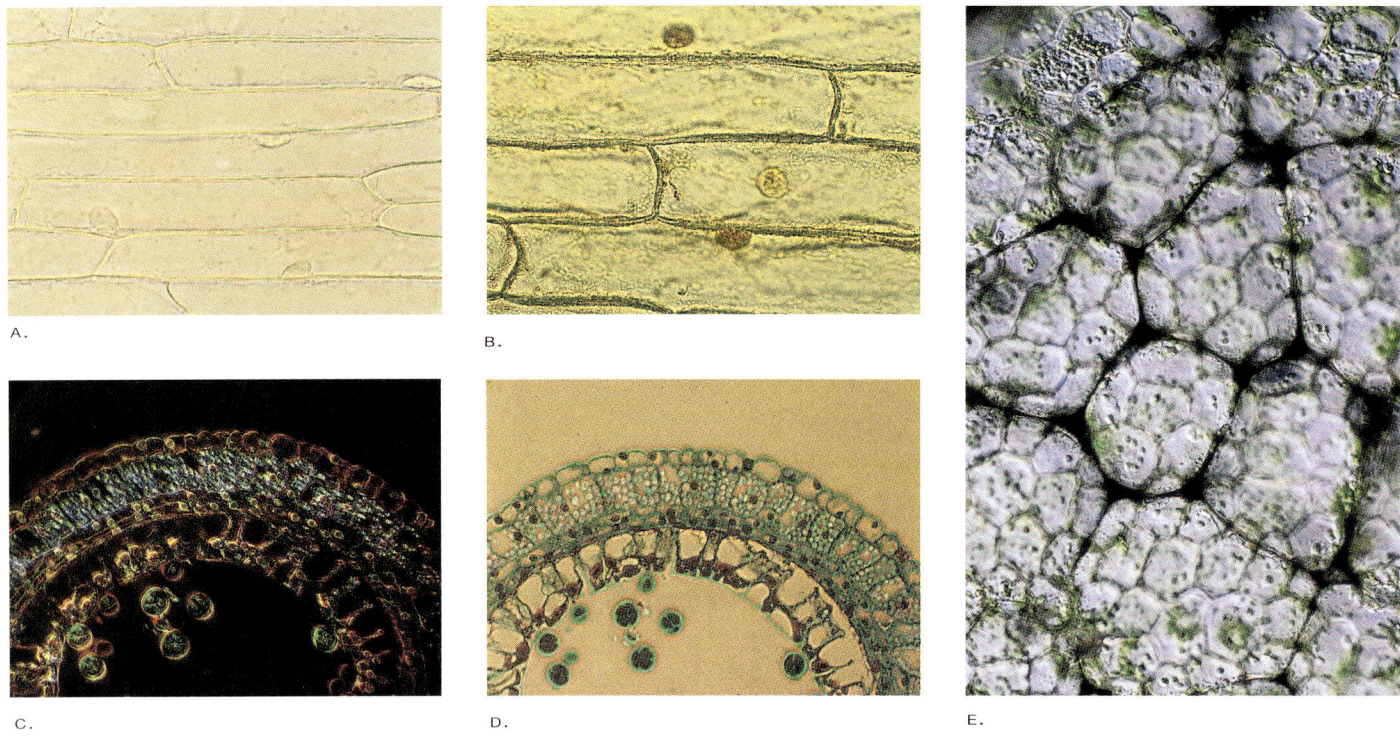

FIGURE 3.6

Comparison of different types of light microscopy. (a) Standard (brightfield), with fresh or living specimen. Light passes directly through specimen, but details are washed out where there is no natural pigment. (b) Standard, with specimen that has been fixed and stained. Staining with different dyes enhances contrast. (c) Darkfield. Light passes through specimen at an angle, so only scattered light is seen. (d) Phase-contrast. Amplifies different densities within the specimen; especially useful for unpigmented structures in living cells. (e) Differential-interference. Intensifies differences in density within the specimen.

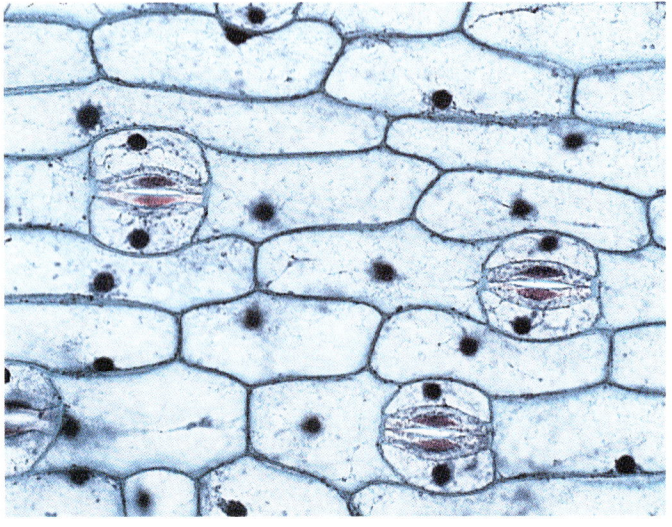

FIGURE 3.7

Light micrograph of leaf cells. Nuclei (dark red) were stained by acetocarmine.

Electron Microscopy

The pattern of contrast in an electron micrograph reveals the ultrastructure of the cell—that is, the structure that can be seen only with an electron microscope. For example, a membrane 10 nm thick can be seen with an electron microscope but not with a light microscope (fig. 3.8a).

Electron microscopy that transmits electrons through a specimen is called **transmission electron microscopy (TEM).** Observations using this type of microscopy are limited to nonliving specimens that are sliced into thin sections and treated with stains containing elements that deflect the beam of electrons (fig. 3.8a). Images from TEM come from the differences between areas of the specimen that deflect the electrons (dark areas) and areas that do not (light areas). By contrast, in **scanning electron microscopy (SEM),** bright areas (high spots) and dark areas (low spots) show the actual surface structure of the specimen. Thus, SEM specimens appear as three-dimensional images (figs. 2.3, 2.12, 3.9). Unlike TEM, SEM does not require that tissues be sliced into thin sections and stained.

BOXED READING 3.1
HOW TO PREPARE CELLS FOR MICROSCOPY

Although living cells can be studied by light microscopy, much greater detail about cellular structure can be obtained from cells that are preserved and stained. Tissue preserved in alcohol, for example, is generally soaked in a dye that stains a specific cellular component. One of the most widely used dyes is leucofuchsin, which stains the nucleus because the dye is specific for DNA. Other common dyes are periodic acid for carbohydrates and Sudan red for lipids. Specimens for the electron microscope must be stained with materials that deflect electrons. The most frequently used stain for electron microscopy is a heavy metal such as osmium tetroxide, which creates dark areas wherever the electron-dense metal binds in the cell.

For best results, cells are generally cut into thin sections before staining. Before being sliced, they must first be embedded in wax (light microscopy) or plastic (electron microscopy) so they will be rigid and not collapse when they are cut. Sections are cut with a **microtome,** which resembles a small-scale meat saw. Steel knives are used to make sections for light microscopy. For electron microscopy, sharp glass or diamond knives are used to cut thin sections through cells embedded in hard plastic.

A remarkable array of subcellular structures can be seen after cells are preserved, embedded, sectioned, and stained. However, two kinds of difficulties arise from preparing cells for microscopy. One is that cellular structure may change during preparation of the fixed cells; for example, proteins that seem to connect cytoskeletal filaments may be artifacts caused by cell preservation. The other kind of difficulty is that thin sections give a static, two-dimensional view of cellular components that were formerly living, three-dimensional structures. A mitochondrion, for example, may appear round or cylindrical, depending on the angle of the section through it. This second problem can be partially overcome by the use of computers with graphic imaging programs that add images together from many thin sections. In this way a simulated three-dimensional structure can be reconstructed from two-dimensional photographs.

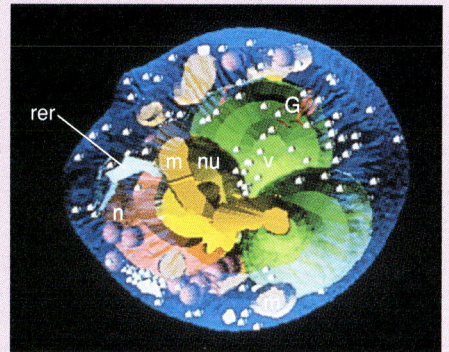 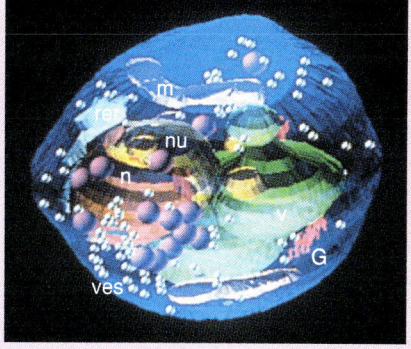

BOX FIGURE 3.1

Computerized three-dimensional reconstruction of a yeast cell. (a) Top view. (b) Side view. (G = dictyosome; m = mitochondrion; n = nucleus; nu = nucleolus; rer = rough endoplasmic reticulum; v = vacuole; ves = membrane vesicle)

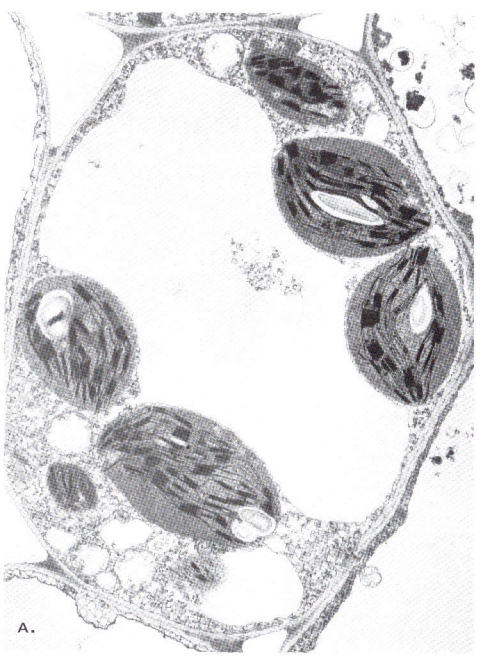 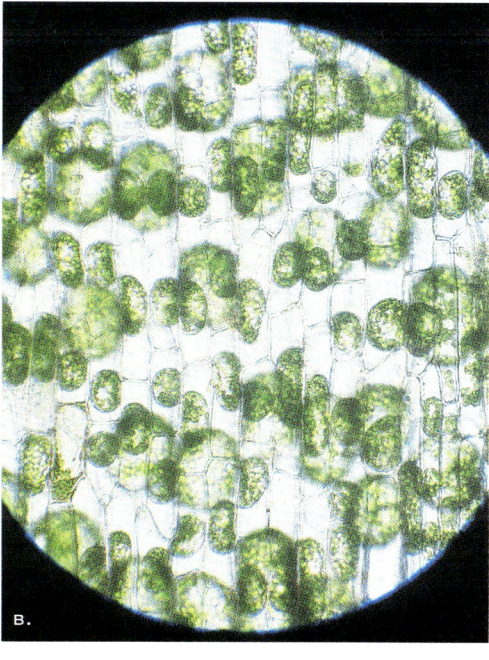

FIGURE 3.8

(a) Transmission electron micrograph of a plant cell, showing organelles and membranes, ×16,000. (b) Light micrograph of plant cells, showing organelles whose membranes cannot be seen at this magnification and resolution, ×225.

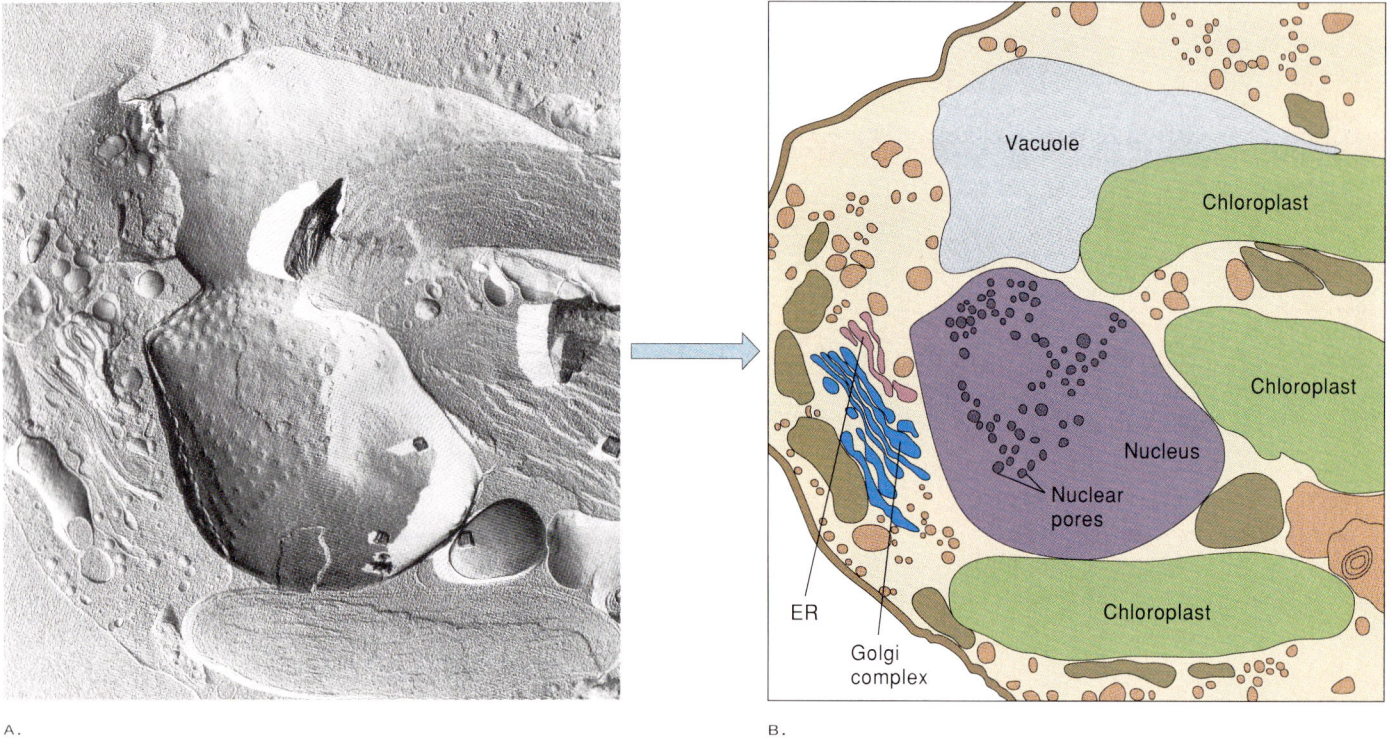

FIGURE 3.9

(a) Scanning electron micrograph of freeze-fractured cell from an onion root tip, showing organelles and the surface of a membrane, ×20,200; this is not an image of the original specimen but a carbon-platinum replica of it. (b) Diagram showing interpretation of structures in the micrograph in (a).

Freeze-Fracturing and Electron Microscopy

While SEM is useful for studying surfaces, cells must be broken open before internal structures can be studied with scanning electron microscopy. This is done by quick-freezing tissues in liquid nitrogen and then fracturing the frozen material. The plane of fracture generally follows lines of weakness, where there was little water in the cell that could form strong ice. Because such regions are common in membranes, freeze-fractures often reveal the surfaces of membranes.

Photographs of freeze-fractured cells are made from replicas of the original surface. After all of the ice is vaporized in a cold vacuum, the specimen is coated with a film of carbon and platinum. The carbon-platinum replica of the fractured surface is then examined with SEM (fig. 3.9).

Cell Fractionation

The functions of organelles, which are the membrane-bound components of a cell, are not usually obvious from their structure. Therefore, botanists often purify and concentrate each type of organelle so that its function can be determined and correlated with its structure. This is the purpose of **cell fractionation**: to obtain large quantities of organelles so their functions can be determined more easily.

In preparation for cell fractionation, cells are first homogenized into small fragments, and the homogenate is laid atop a sucrose solution in a test tube. The test tube is then spun at high speed in an instrument called an **ultracentrifuge.** An ultracentrifuge can spin at more than 100,000 revolutions per minute, which is more than fifty times faster than an idling car engine. During spinning, different organelles separate on the basis of their buoyancy in the sucrose solution. Nuclei, which are the densest organelles, sink the farthest. Chloroplasts are usually intermediate in density and therefore form a bright green band near the middle of the test tube.

 *Writing to Learn Botany*

What might be the advantages and disadvantages of studying isolated organelles?

Biochemical Cytology

Specific substances in cells can be stained with chemicals that bind to them but not to other parts of the cell. Such selective staining is useful for finding specific kinds of molecules in cells prepared for microscopy (see fig. 3.7). This method is one of the tools used in a subdiscipline of cytology called **biochemical cytology,** or **cytochemistry.** Biochemical cytology uses the biochemical properties of cell components in conjunction with techniques of microscopy to unravel the details of cellular structure and function.

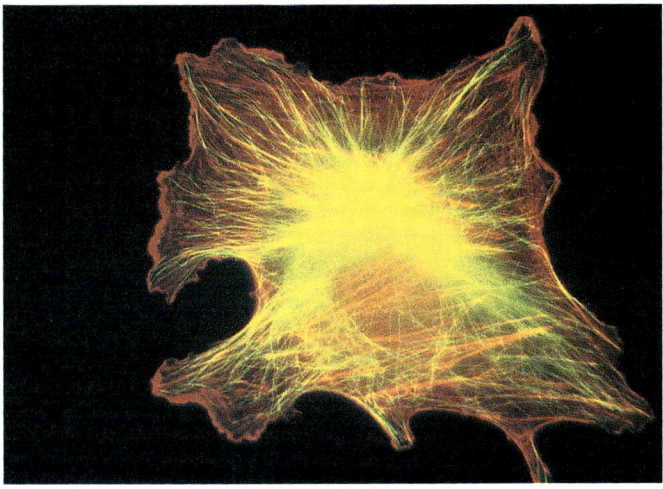

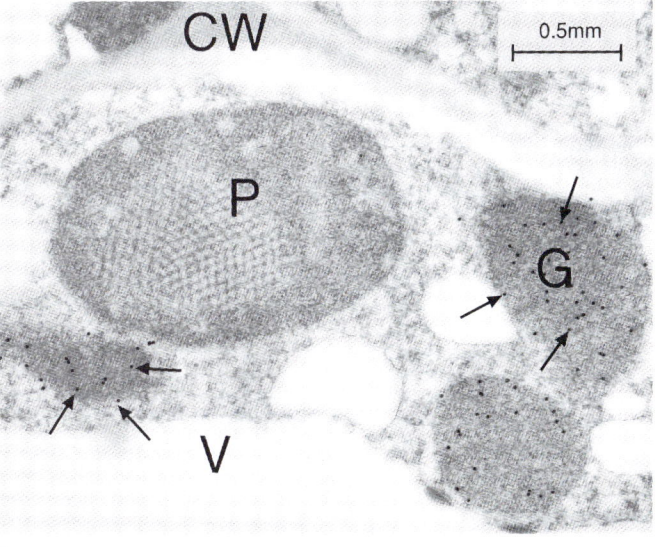

FIGURE 3.10

(a) Light micrograph of a cell that has been immunofluorescent-stained twice—once for the proteins of microtubules (green) and once for the proteins of actin filaments (red) (see fig. 3.12).
(b) Electron micrograph of section through a cotton (*Gossypium*) cotyledon, showing staining by antibody-gold complex. Colloidal gold particles (arrows) are bound to malate synthase in glyoxysomes (G). Lack of gold particles in the cell wall (CW), plastid (P), and vacuole (V) indicates an absence of the enzyme in these parts of the cell.

One of the most sophisticated and informative techniques of biochemical cytology uses the immune systems of animals. Large molecules from plant cells, especially proteins, become **antigens** (antibody inducers) when they are injected into an animal, usually a rabbit or a mouse. The animal makes **antibodies** (protein complexes that bind to antigens), and the antibodies induced by the plant protein are then used to locate the antigen in the plant cell. The antigen-antibody complex is visible because it has been linked to another complex that includes a dye, such as fluorescein. The location of the antigen shows as a fluorescent area when it is exposed to ultraviolet light. An example of this method is shown in figure 3.10a; in this example, protein from microtubules was used as the antigen. Another example is shown in the micrograph that opens this chapter (p. 44). In that example, the technique was used to localize zein (a storage protein) in seeds of corn.

Immunological cytochemistry is also useful in electron microscopy, although antibody complexes used in electron microscopy must be linked to a heavy metal that deflects electrons. For example, an antibody-gold complex linked to malate synthase, the enzyme that converts glyoxylic acid to malic acid, shows where this protein occurs in cells of cotton seedlings (fig. 3.10b).

C O N C E P T

Cytology uses microscopy and cell chemistry to study cellular structure and function. Cell chemistry is useful in staining specific parts of the cell to enhance their visibility by microscopy. Organelles or other parts of cells can be separated by cell fractionation and studied as isolated components.

FUNCTIONAL ORGANIZATION OF CELLS

Why Are Cells So Small?

For convenience, most of the contents of a cell are referred to as its **cytoplasm;** the only component of the cell that is not part of the cytoplasm is the nucleus. The plasma membrane, which surrounds the cytoplasm, is a barrier that protects the cell from harmful substances. It has the consistency of salad oil, and must allow the passage of gases and nutrients into and out of the cell. However, the surface area of the membrane can service only so much cellular volume—that is, the surface-to-volume ratio must have a lower limit. Multicellular organisms avoid this limit by making more but smaller cells for a given volume. Consider the example of a cube-shaped cell that is 250 μm on each side; it has a volume of 1.56×10^7 μm^3 (250 μm × 250 μm × 250 μm) surrounded by 3.75×10^5 μm^2 (250 μm × 250 μm × 6 sides) of plasma membrane (fig. 3.11). The surface-to-volume ratio is 0.024, which is unacceptably small for most cells. This means that such a cell would be too big for its plasma membrane. However, if this single large cell were to be divided into cells 25 μm on a side, the total volume would be the same, but the surface area of all plasma membranes would be 3.75×10^6 μm^2 (25 × 25 μm per side × 6 sides per cell × 1,000 cells). The surface-to-volume ratio would be 0.24, which is ten times more membrane for the same volume as the single large cell. A ratio of 0.24 is about what many plant cells have for the relative area of plasma membrane surface to cell volume.

The need for a sufficient amount of plasma membrane for the volume of the cell only partially explains why most cells are so small. Other important factors include the limits on

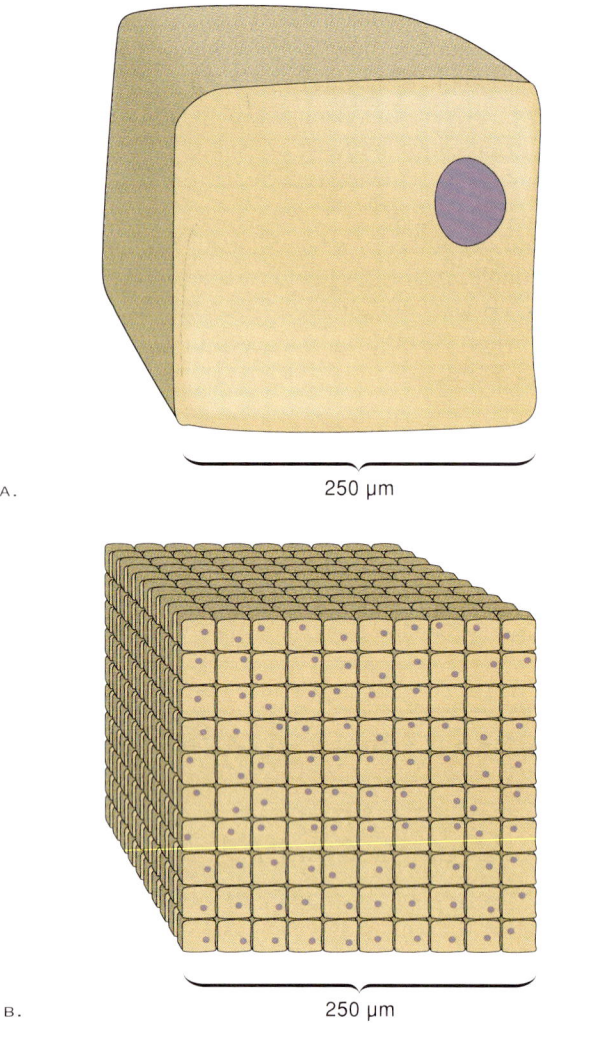

FIGURE 3.11

Diagrams showing (a) a single large cell compared with (b), a cube ten cells across having the same total volume as the single cell.

rates of synthesis and transport of molecules within the cell and the requirement that a single nucleus manage the metabolism of the entire cell.

Cells are small because cell volume is limited by the amount of plasma membrane that can service it. A group of smaller cells has a greater ratio of plasma membrane area to cell volume than a single large cell.

Membranes and Cell Compartments

Many metabolic functions in a cell occur in or on membranes. The plasma membrane alone is inadequate for all of these processes, regardless of cell size. Additional internal membranes, either attached to the plasma membrane or included in organelles, compensate for the insufficiency of the plasma membrane. These other membranes also form compartments that maintain different environments within the cell. For example, incompatible reactions, such as the hydrolysis and dehydration of carbohydrates, are isolated from each other in different compartments. These compartments allow many different kinds of reactions to occur in the cell simultaneously.

Biological membranes are usually 7–10 nm thick and consist mostly of phospholipids and proteins. In addition to the plasma membrane, there are membranes that surround nuclei, mitochondria, chloroplasts, vacuoles, and microbodies. Membranes also pervade the cell between organelles in a complex system that complements the functions of the organellar membranes. Further characteristics of organelles and the internal membrane systems of cells are presented in later sections of this chapter.

The plasma membrane is supplemented by internal membranes that provide additional areas for enzymatic reactions and that divide the cell into compartments.

The Lore of Plants

Cooks are the plant cell biologists of the kitchen. Every cookbook describes how to prevent browning of fruits and vegetables before they are canned or frozen. Browning is caused by the enzyme-catalyzed polymerization of phenolics in damaged cells. The enzymes (phenolic oxidases) and the phenolics are kept in separate compartments in intact cells. The polymerization reaction can be stopped by covering the freshly cut tissue with ascorbic acid (vitamin C), as crystals or in lemon juice, or with vinegar, which contains acetic acid. These organic acids inhibit the reactions that are driven by phenolic oxidases.

The Cytoskeleton

The **cytoskeleton** is a network of filaments that forms a mechanical support system in the cell. These filaments also help maintain organelle position, direct cell expansion, and control the movement of chromosomes during nuclear division. Groups of filaments may also form channels for transporting large molecules within the cell.

There are at least three kinds of filaments that comprise the cytoskeletons of cells in plants and animals (fig. 3.12). The largest filaments are **microtubules,** which are hollow tubes about 18–25 nm in diameter. Microtubules are made of two types of globular proteins, **alpha tubulin** and **beta tubulin.** These proteins are bound to each other as dimers, which aggregate into microtubules.

The smallest filaments in the cytoskeleton are **actin filaments,** which consist of two intertwined strands of the globular protein **actin.** Actin filaments are 4–7 nm in diameter. The

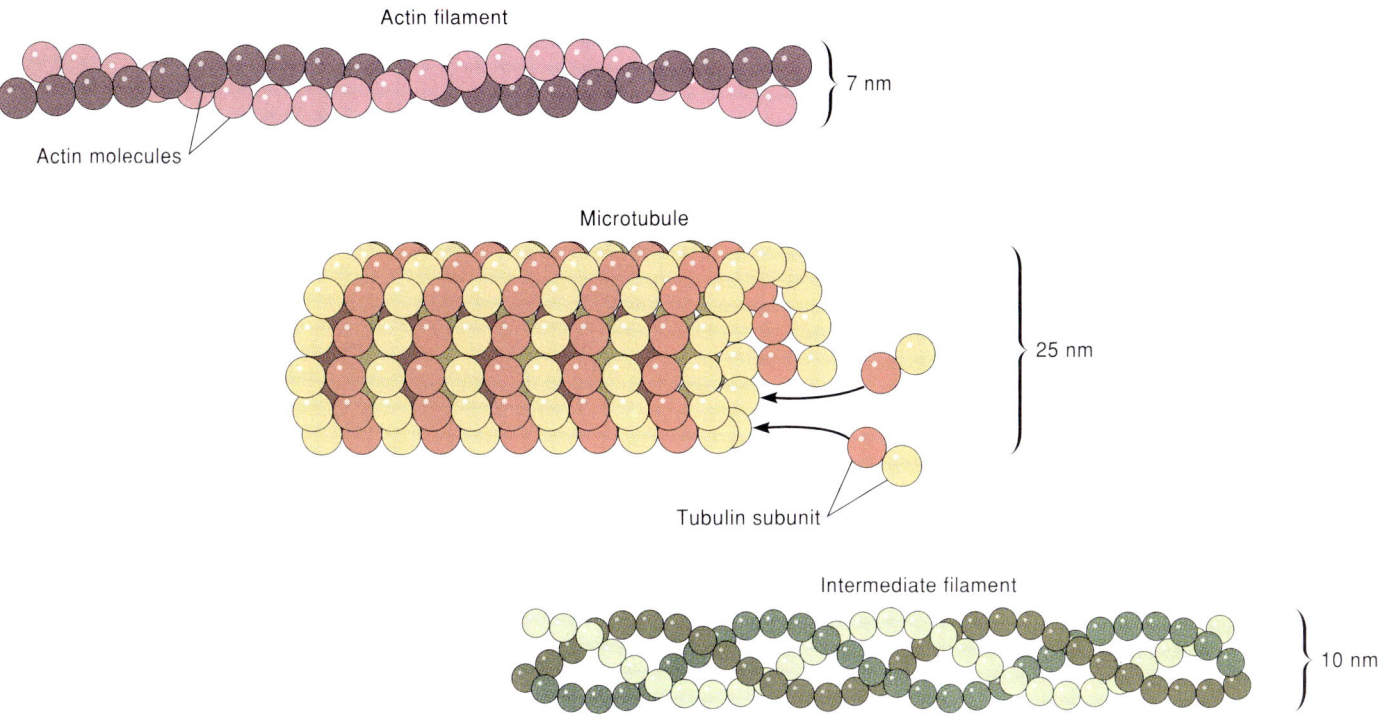

FIGURE 3.12

Structural models of the three different kinds of cytoskeletal filaments.

proteins of microtubules and actin filaments are the same in plants and animals. This similarity suggests that these filaments work the same way in both kingdoms, even though the tissues that contain them, such as muscle tissue and leaf tissue, are different.

The third kind of cytoskeletal filaments are intermediate in size between microtubules and actin filaments; for this reason, they are called **intermediate filaments.** These filaments are made of fibrous proteins wound into coils having diameters of 8–12 nm. Several kinds of proteins in intermediate filaments occur in animals, but little is known about them in plants. Unlike the proteins of microtubules and actin filaments, the proteins of intermediate filaments in animals are fibrous and more numerous. Moreover, different proteins occur in different types of cells.

The proteins of intermediate filaments are probably similar in both plants and animals. The evidence for this suggestion includes cytochemical studies showing that antibodies made against intermediate filaments from animals bind to intermediate filaments from plants. At least three kinds of proteins of intermediate filaments have been identified in plants by this method.

How Does the Cytoskeleton Work?

It is easy to imagine the cytoskeleton as a stationary lattice of filaments that holds cells in specific shapes and keeps cell components in specific places; the name *cytoskeleton* implies such a static function. However, cells often change shape, and organelles and other cellular components are in almost constant motion, which means that the cytoskeleton cannot be a rigid structure. This assertion is confirmed by observations that expansion of and internal movements in cells occur in conjunction with the growth and breakdown of microtubules and actin filaments. Such observations are indirect evidence that the cytoskeleton has many dynamic functions in the cell. Experimental evidence comes from treating cells with colchicine, an alkaloid from the meadow saffron (*Colchicum autumnale*), which disrupts the assembly of microtubules. The formation of new cell walls during cell division is inhibited by colchicine, which suggests that cell division requires microtubule synthesis.

What are Botanists Doing?

Use the reference section of your library to find articles that have been published in the past twelve months that involve the use of colchicine in plant cell biology. Answer the question, "What is the prevalent use of colchicine in cell biology today?"

The least understood parts of the cytoskeleton are the intermediate filaments. Because their proteins are fibrous, intermediate filaments do not have globular subunits that are easily assembled and disassembled like those of microtubules and actin filaments. Thus, intermediate filaments may be relatively static, tension-bearing structures that are not as dynamic as their larger and smaller counterparts in the cytoskeleton.

Although the various filaments of the cytoskeleton must somehow cooperate to control the overall organization of the cell, we do not yet know how their functions are integrated.

The cytoskeleton consists of microtubules, actin filaments, and intermediate filaments. Each type of filament is made of different kinds of proteins. The cytoskeleton functions not only as a structural support for the cell but also as a dynamic network that controls certain aspects of cell growth and internal movement.

THE CELL WALL

The most easily observed part of a plant cell is the cell wall. In some cells, such as the cork cells in bark, the cell wall is the only remnant of a formerly living cell. Cell walls are dynamic parts of cells that can grow and change their shape and composition. Their composition varies in different cell types and from one species to another. Up to 60% of a cell wall may be cellulose; other components include hemicelluloses, pectins, lignins, and proteins. Almost all plant cells have cellulose-containing cell walls.

Young cells and cells in actively growing areas have **primary cell walls** that are relatively thin and flexible. Examples of such cells include the dividing cells at tips of roots and shoots. The primary wall is usually less than 25% cellulose, the remainder being hemicelluloses, pectins, and glycoproteins. Some primary cell walls also contain small amounts of lignin. However, many cells having only primary cell walls can change shape, divide, or differentiate into other kinds of cells.

Certain kinds of cells stop growing when they reach maturity. When this occurs, these cells form a **secondary cell wall** inside the primary cell wall (fig. 3.13). The secondary cell wall is more rigid than the primary cell wall and therefore functions as a strong support structure. Although cellulose is one of their main components, the secondary walls of cells in wood are up to 25% lignin, which adds hardness and resists decay. Because of its lignin content, wood is one of the strongest materials known.

Some cell walls, such as those of cork cells, also contain suberin. Suberized tissues inhibit water loss through bark, which is why cork from the cork oak (*Quercus suber*) is useful in making stoppers for wine bottles (see Chapter 16, box 16.2, entitled "Corks"). Unlike primary cell walls, secondary cell walls are rigid and lack glycoproteins. Most types of cells that have secondary cell walls die when they reach maturity.

Cells that adjoin one another are probably held together by pectins. The pectic layer between cells is called the **middle lamella.** Some tissues, such as the flesh of apples, are so rich in pectins that these polysaccharides are extracted for use as thickening agents in making jams and jellies.

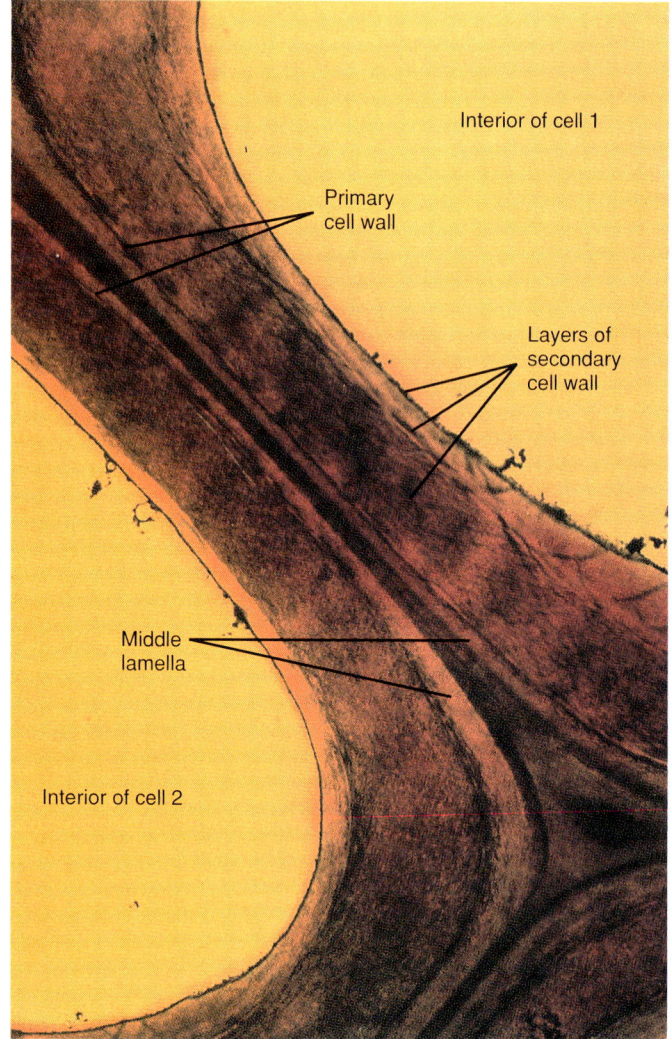

FIGURE 3.13

Transmission electron micrograph of cell walls. The primary wall is constructed when the cell is young. Thicker secondary walls are added later when the cell stops growing, ×3,000.

How Cell Walls Grow

All materials necessary for making cell walls come from inside the cell. Dictyosomes play a role in this process by transporting pectins, hemicelluloses, and glycoproteins to the plasma membrane. These substances pass through the membrane in the region of cell-wall synthesis, where they are assembled inside the existing wall. This is how cell walls thicken.

Cellular expansion requires the cell wall to stretch or deform as the cell absorbs more water. As the wall enlarges, the existing cellulose microfibrils must separate because they cannot stretch. During expansion, new cellulose microfibrils are deposited inside the loosened cell wall in different patterns, depending on the cell type. In stem cells microfibrils are oriented mostly perpendicular to the direction of cell expansion. Thus, cells may grow twenty times their original length, with

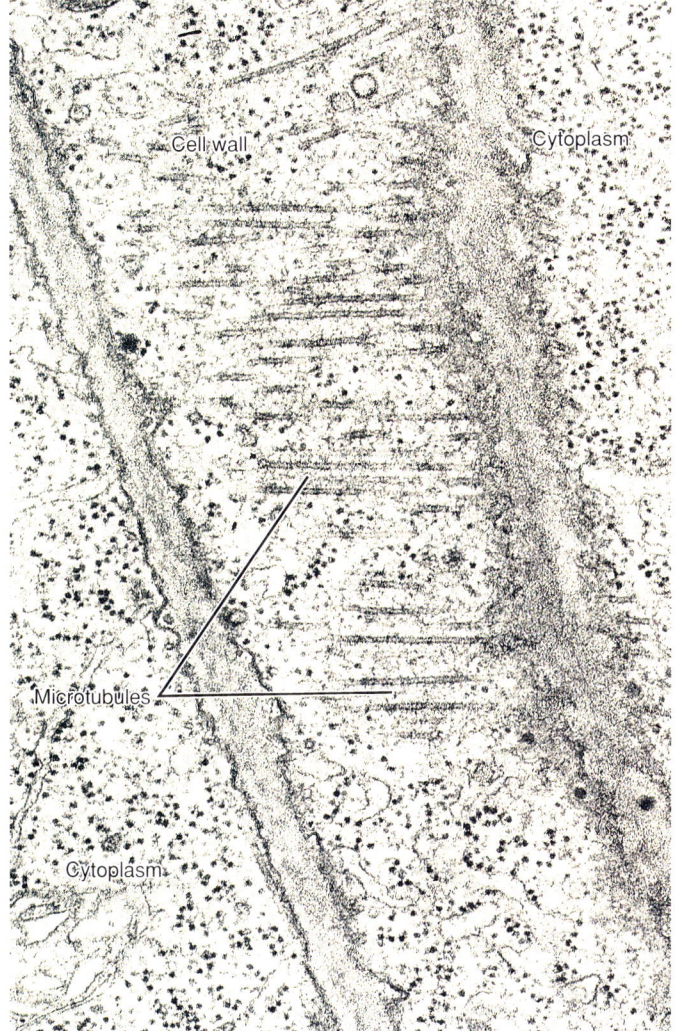

FIGURE 3.14

Transmission electron micrograph of a grazing section of a root tip cell, showing orientation of microtubules.

little growth in width. In comparison, microfibrils are deposited in random arrays in cells of storage tissues and tissue cultures. This pattern enables growth in such cells to be more or less uniform in all directions.

As cell walls grow, the sites and orientation of new microfibrils are controlled by arrays of microtubules that occur immediately inside the plasma membrane. These microtubules are arranged in the same orientation as the microfibrils (fig. 3.14). Evidence for the role of microtubules in this process is based on the growth of cells that are treated with colchicine. For example, in water-conducting cells of stems, colchicine changes the normal development of regular wall thickenings to a disorganized smear of wall material. Despite this effect, colchicine apparently does not affect the synthesis of new microfibrils,

and in some cases colchicine has little or no effect on the orientation of new microfibrils that are deposited in a preexisting pattern. This means that the role of microtubules in cell-wall growth is not as straightforward as might have been expected. Although results from experiments with colchicine suggest that cell-wall components outside the cell connect directly or indirectly to microtubules across the plasma membrane, there is no direct evidence for such connections, and their molecular basis is unknown.

Primary and secondary cell walls are made mostly of a variety of carbohydrates, but the most rigid walls also contain lignin. Secondary cell walls are deposited inside primary cell walls. Microtubules have a role in the pattern of cell-wall deposition.

Connections Between Cells

Primary cell walls have thin areas where many tiny connections, called **plasmodesmata** (singular, **plasmodesma**), occur between adjacent cells. Plasmodesmata are lined by the plasma membrane, thereby forming an uninterrupted channel for the movement of materials from one cell to another. This means that all cells in a plant are interconnected and have the potential to exchange substances through plasmodesmata.

Plasmodesmata often occur in clusters where primary cell walls are particularly thin. These regions are called **primary pit-fields** (fig. 3.15). Primary pit-fields and plasmodesmata are abundant in conducting cells and secretory cells, such as those in nectar glands or oil glands. Plasmodesmata are usually about 40 nm in diameter, but they may enlarge to 1,000 nm (1 μm) in some cells.

The structure of plasmodesmata and the frequency of their occurrence in conducting and glandular cells suggest that these connections function in transport between cells. Direct evidence for this function comes from experiments with dyes and electric currents. When a dye that does not easily cross the plasma membrane is injected into one cell, it quickly passes into neighboring cells. Similarly, the high electrical resistance of the plasma membrane can be bypassed by electric currents that pass through plasmodesmata. Despite such evidence, cells probably do not exchange all materials freely: neighboring cells can differentiate into different cell types and maintain different internal concentrations of various chemicals.

Water-conducting tissue is an important exception to the general occurrence of plasmodesmata. Cells of this tissue die as they mature, so they have no living material to share between them. Instead, they function as inanimate "straws" formed by many cells. In flowering plants, water moves through perforations in the primary cell wall, around which there is no secondary cell wall (fig. 3.16). The movement of water through plants is discussed in more detail in Chapter 21.

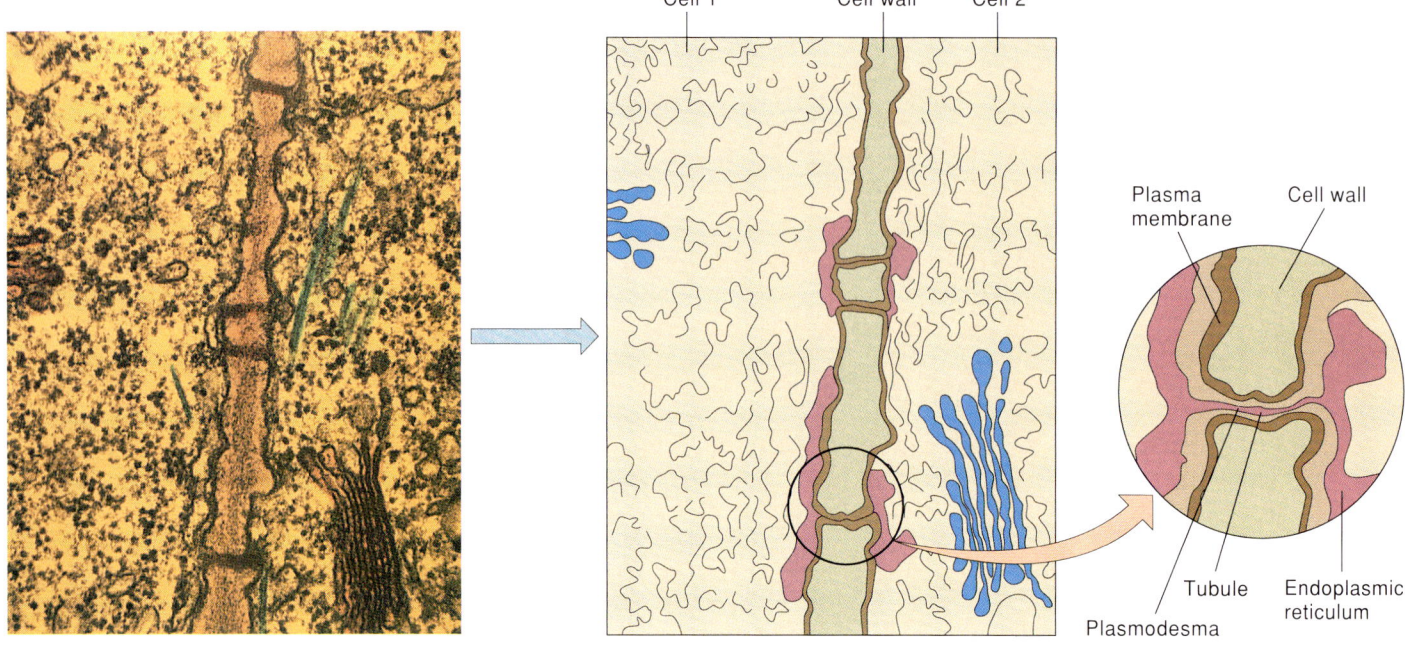

FIGURE 3.15

The transmission electron micrograph, ×17,000, and accompanying drawing of primary cell wall show the plasmodesmata (which contain plasma membrane), and a tubule that connects the endoplasmic reticulum between adjacent cells.

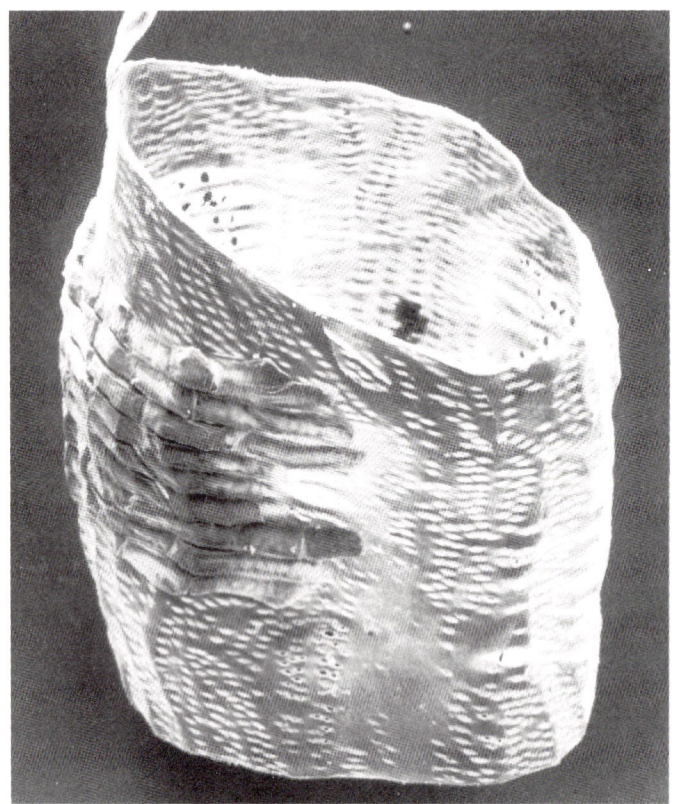

FIGURE 3.16

Scanning electron micrograph of a water-conducting vessel element, showing perforations and pitting, ×325.

CONCEPT

All living cells in a plant are interconnected by plasmodesmata, which are lined with a continuous plasma membrane. Some substances can move from cell to cell through plasmodesmata. In spite of these intercellular connections, neighboring cells can maintain different chemical composition.

THE NUCLEUS

The **nucleus** is usually the most conspicuous organelle in a cell; when stained, it can be seen easily with a light microscope (fig. 3.7). The nucleus contains most of the cell's DNA, which occurs with proteins in threadlike chromosomes (see Chapter 10). The nucleus is surrounded by two membranes, together called the **nuclear envelope** (fig. 3.17). The outer membrane is continuous with the membrane of the endoplasmic reticulum, which is discussed later in this chapter. The inner and outer nuclear membranes are separated by a space of 20–40 nm, except where they fuse to form pores in the envelope. These nuclear pores are small circular openings, 30–100 nm in diameter, bordered by proteins that probably influence the passage of molecules between the nucleus and the rest of the cell. For example, certain proteins move into the nucleus, where they join with ribosomal RNA to make the subunits of ribosomes (see next section). In turn, ribosomal

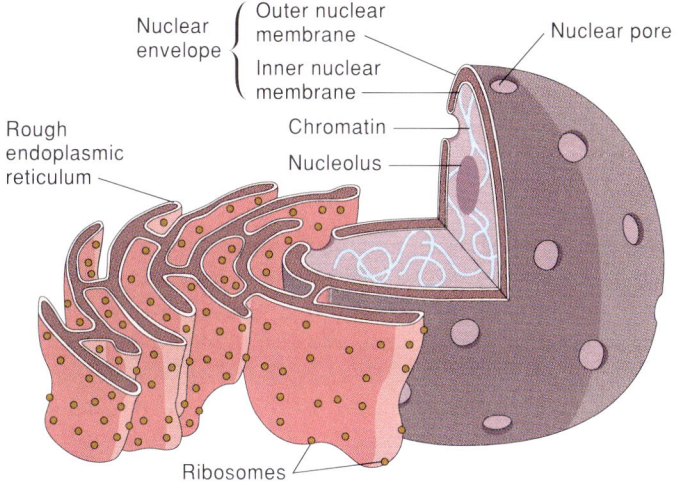

FIGURE 3.17

The nuclear envelope has pores that link the cytoplasm with the inside of the nucleus. The outer nuclear membrane is continuous with the endoplasmic reticulum.

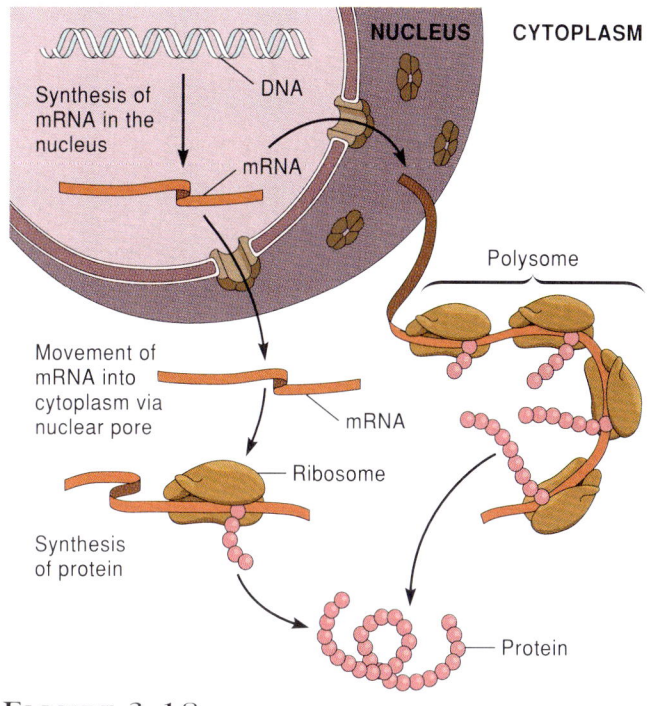

FIGURE 3.18

Ribosomes play a central role in protein synthesis. Unlike other organelles, ribosomes do not have membranes.

subunits and other RNA-containing molecules that are made in the nucleus move out into the cytosol through the nuclear pores. The **cytosol** is the semifluid matrix between organelles.

The complexity and importance of the nucleus and its activities are the focus of modern research in genetics. An entire unit of this text (Unit 3: "Genetics") is devoted to this branch of biology.

RIBOSOMES

Ribosomes are where proteins are made in the cell. Ribosomes are about 20 nm in diameter and consist of approximately equal amounts of protein and ribosomal RNA (**rRNA**). Each ribosome is assembled from two subunits that are produced in the nucleus and exported to the cytosol. The two subunits are joined when they attach to a molecule of messenger RNA (**mRNA**). Ribosomes usually occur in clusters, called **polysomes,** on a single molecule of mRNA (fig. 3.18). Unlike the nucleus and other organelles, ribosomes are not surrounded by membranes.

Ribosomes are either attached to membranes or move freely in the cytosol. Proteins made by cytosolic ribosomes are usually also cytosolic; that is, they are not associated with membranes. These proteins include enzymes that degrade sucrose and glucose in the cytosol. Conversely, membrane-bound ribosomes usually make proteins that will be attached to or embedded in membranes. Examples of such proteins are those that control the transport of ions and other substances through the plasma membrane.

The number of ribosomes varies among cell types and in different stages of cell development: A growing cell can make about 10,000 ribosomes per minute; ribosomes are especially abundant in dividing cells because these cells make large amounts of protein. The mechanisms of protein synthesis by ribosomes and the nucleus are discussed in detail in Chapter 11, "Molecular Genetics and Biotechnology."

THE MEMBRANE SYSTEM

All of the membranes in a cell are connected either by direct contact with each other or by the exchange of membrane segments. These interconnected membranes function together as a **membrane system** that includes the plasma membrane and the various organellar membranes.

Many biochemical processes occur in or on membranes. For example, the enzymes involved in photosynthesis and in ATP synthesis are embedded in membranes. Membranes also provide a framework for making more membranes.

All membranes have a similar structure, which consists of a double layer of phospholipids that is impregnated with proteins (see Chapter 4, fig. 4.4). However, not all membranes have exactly the same structure and function; for example, the permeability of the plasma membrane differs from that of the nuclear envelope. Each membrane controls how much and what kinds of materials pass through it; that is, each membrane has a different **membrane selectivity,** which depends on its composition. Moreover, the composition and physical properties of a membrane can change during cell development.

In principle, internal membranes increase the ratio of membrane surface to cell volume. However, internal membranes do more than this. Indeed, they are physically and chemically distinct from the plasma membrane and from each other. Furthermore, they divide the cell into functionally distinct compartments (i.e., organelles) that are bounded by their own differentially permeable membranes. Thus, the cell is separated into a set of individual reaction vessels. Each reaction vessel has its own specialized functions, which are performed by a unique set of enzymes bound to or held within the membrane.

The Plasma Membrane

The plasma membrane is a remarkably changeable, multipurpose membrane. As already mentioned, it acts as a differentially permeable barrier to substances that enter and leave the cell; some substances can pass through it, and some cannot.

Partly because of its outer position, the plasma membrane receives and translates chemical and environmental signals from outside the cell. Signals translated by the plasma membrane change cellular metabolism. For example, hormones received by the plasma membrane can initiate a series of enzymatic reactions that cause the cell to enlarge. The plasma membrane also accepts packets of raw materials from other membranes inside the cell and directs the assembly of these materials into cell-wall microfibrils.

The Endoplasmic Reticulum

Most of the membrane surface area inside cells occurs in the **endoplasmic reticulum (ER),** which is an extensive network of sheetlike membranes distributed throughout the cytosol. When viewed with an electron microscope, the ER appears to meander throughout the cell (fig. 3.19a). Seen in three dimensions, however, the ER is a system of flattened tubes and sacs (fig. 3.19b)

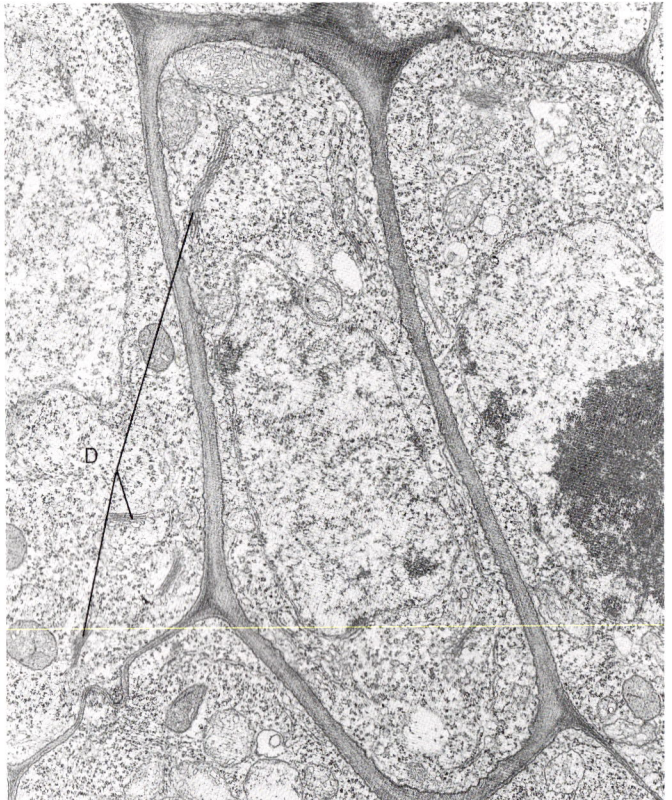

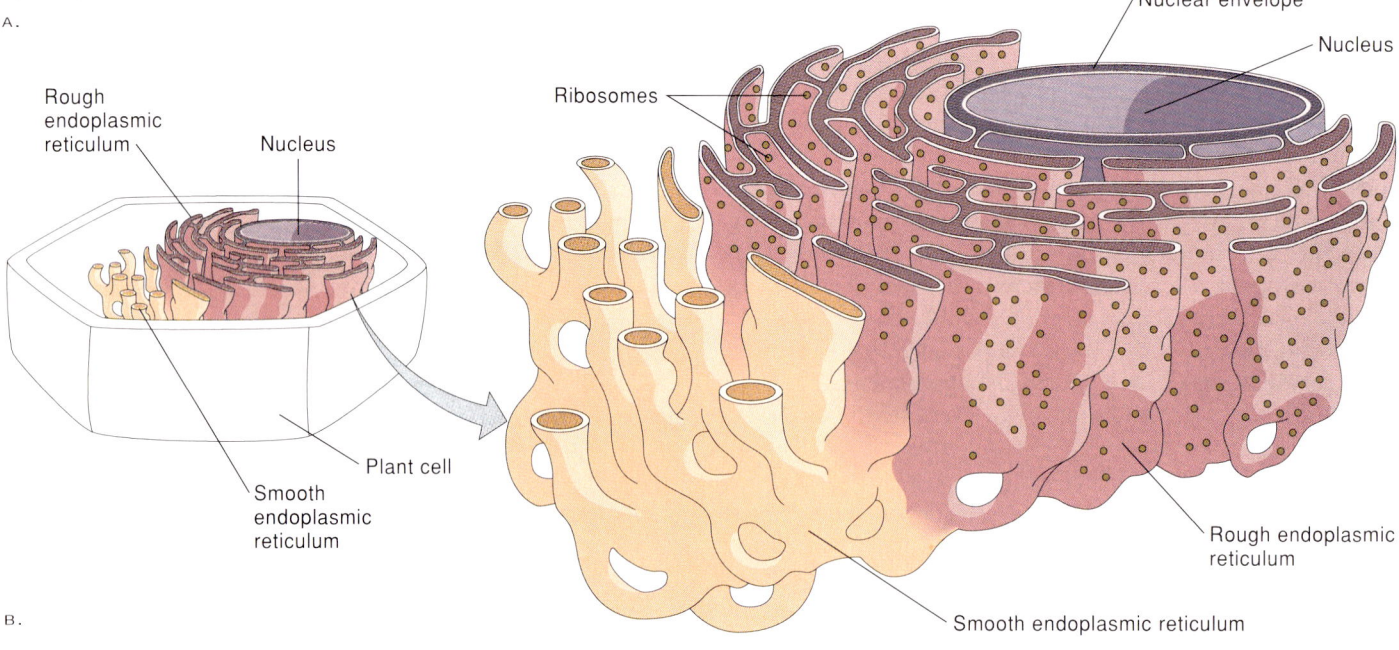

Figure 3.19

(a) Transmission electron micrograph of a differentiating cell in a root tip of bean (*Phaseolus vulgaris*) showing dictyosomes (D), ×22,000; the endoplasmic reticulum criss-crosses the cell. (b) Three-dimensionality of the ER, as shown here, cannot be seen in the electron micrograph.

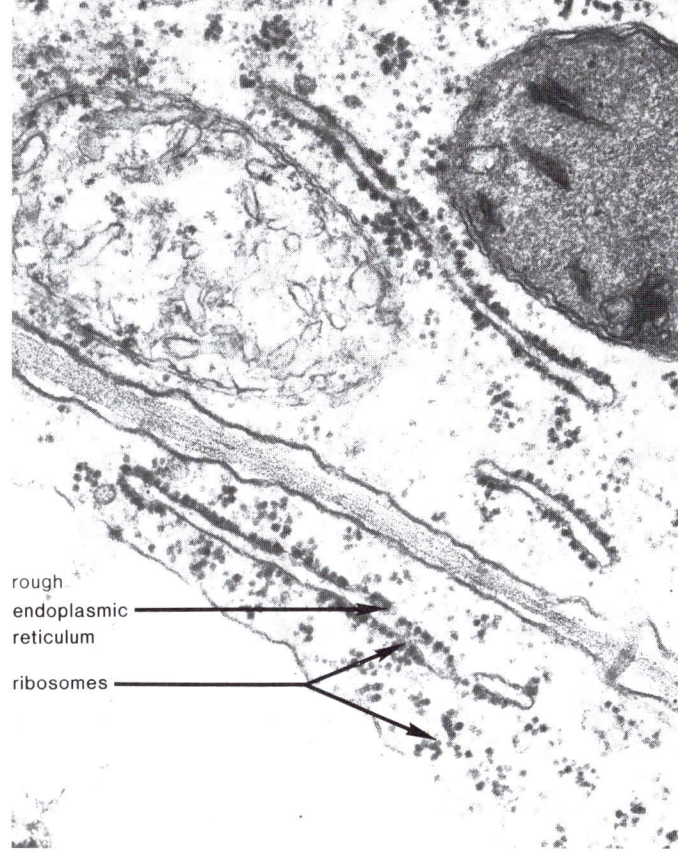

FIGURE 3.20

Transmission electron micrograph from a young leaf cell of maize. Areas of the ER with numerous ribosomes attached are the rough endoplasmic reticulum.

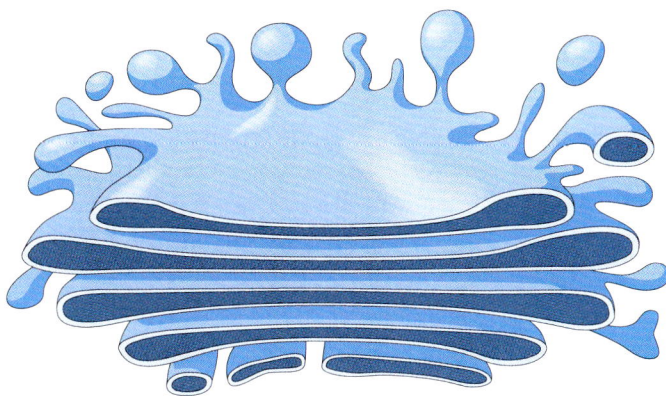

FIGURE 3.21

Three-dimensional representation of a dictyosome (Golgi body). Vesicles near the stacked membrane are considered to be part of the dictyosome.

C O N C E P T

The endoplasmic reticulum is a complex internal membrane that has many functions. It contains enzymes that catalyze phospholipid synthesis, the construction of other membranes, and the synthesis of cell-wall precursors. The rough endoplasmic reticulum also bears ribosomes for protein synthesis.

Dictyosomes

Stacks of flattened, membranous vesicles are called **dictyosomes,** or sometimes **Golgi bodies** (fig. 3.21). Dictyosomes are usually two-sided, with one side facing the smooth ER and one side facing the plasma membrane. Membrane-bound vesicles occur near the edges of the dictyosomes and are considered to be part of them.

Dictyosomes receive material from the smooth ER, either through direct connections or in vesicles released by the ER. Transport vesicles from the ER fuse with the inner face of the dictyosome and release their contents into its interior. These vesicles contain proteins, lipids, and other substances, which are often chemically modified in the dictyosome and then sorted into separate packets. For example, sugars in dictyosomes become linked to proteins, converting them into glycoproteins. Packets of glycoproteins or other materials eventually move to the edge of the dictyosome near the outer face, where the dictyosome membrane is pinched off into another vesicle. This new vesicle moves to the plasma membrane or to other sites in the cell.

Vesicles that move to the plasma membrane are secretory vesicles, because they fuse with the plasma membrane and secrete their contents to the exterior of the cell. This type of secretion is called **exocytosis.** Polysaccharides secreted by exocytosis become part of the cell wall. Other substances are secreted from specialized cells with the aid of dictyosomes. These substances include nectar that flowers secrete as a reward for

that is continuous between the plasma membrane and the outer membrane of the nuclear envelope. Internal compartments of the ER are isolated from the rest of the cytoplasm but in contact with the space between the two membranes of the nuclear envelope. Plasmodesmata also contain portions of ER, which form a continuous internal membrane between cells.

Two regions of ER can be distinguished in electron micrographs. One region is called the *rough* ER because the many ribosomes attached to it give it a rough appearance (fig. 3.19b, fig. 3.20). In contrast, the other region is called the *smooth* ER because it has no ribosomes attached to it (fig. 3.19b).

Biochemical cytology and studies of isolated ER membranes have shown that the rough ER is the major region of protein synthesis in the cell. The smooth ER is involved with the synthesis of phospholipids and the assembly of new membranes. Both types of ER form vesicles that break away and fuse with other membranes. When one of these vesicles fuses with the plasma membrane, its contents are secreted from the cell; cell-wall components, waxes, lipids, and mucilages are secreted by such vesicles. Other vesicles can fuse with organellar membranes and become part of them.

hummingbirds or other pollinators, and the oils or similar resinous chemicals secreted from the glands on leaves and stems of mints and many other highly fragrant plants.

In dividing cells, dictyosomes help build new primary walls after the nuclei have divided. Wall construction occurs when dictyosomes are guided by microtubules to a region between nuclei. Vesicles containing cell-wall precursors fuse in this region to form a disk-shaped structure called a **cell plate.** The cell plate grows by the addition of more vesicles until it fuses with the parent cell wall, which completes the partition between the two new cells.

A logistical problem arises in some cells as a consequence of building secondary walls. As secretory vesicles continually fuse with the plasma membrane, large amounts of new membrane are added to it (estimates are that active cells can double the amount of plasma membrane every twenty minutes), but plasma membranes have little room to grow in cells that are forming rigid secondary walls. Cells must therefore have a system for recycling excess plasma membrane. In animal cells, certain regions of the plasma membrane become coated with bristlelike structures. These regions, called **coated pits,** form vesicles that pinch off into the cell, thereby removing a portion of the plasma membrane. Coated pits have also been discovered in plant cells, which suggests that the mechanism for recycling plasma membrane may be the same in plants as it is in animals.

C O N C E P T

Dictyosomes are membranous organelles that modify chemicals from the endoplasmic reticulum, build primary cell walls between newly divided nuclei, secrete substances through the plasma membrane, and probably help to recycle excess plasma membrane.

Vacuoles

Vesicles from the ER and dictyosomes often fuse to form larger sacs called **vacuoles.** Immature cells of plants and animals may contain several small vacuoles, but in most plant cells these small vacuoles fuse into larger ones as the cell matures. A mature plant cell typically has one large vacuole that can occupy up to 95% of the cell volume (fig. 3.22a). The membrane of the central vacuole has its own name, the **tonoplast.**

As plant cells grow, most of their enlargement results from the absorption of water by vacuoles (fig. 3.22b), which expand and push the rest of the cell's contents into a thin layer against the cell wall. Vacuoles that are filled with water create pressure, called **turgor pressure,** on the cell walls, which contributes to structural rigidity of the cell. When a plant receives too little water, turgor pressure decreases, and the plant wilts. We can see the effects of turgor pressure by letting carrots or celery dry out and become flaccid; we can make them crisp

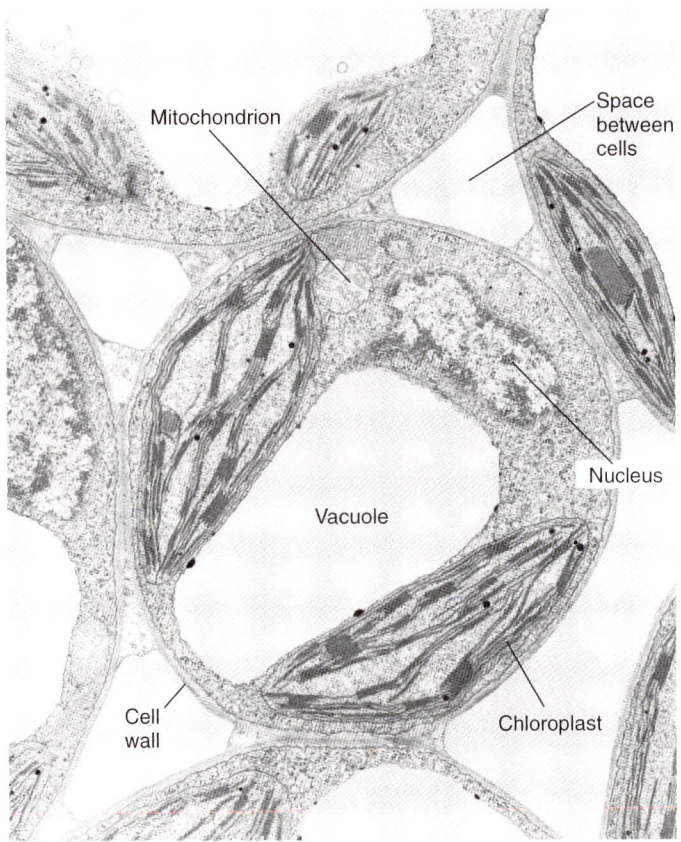

A.

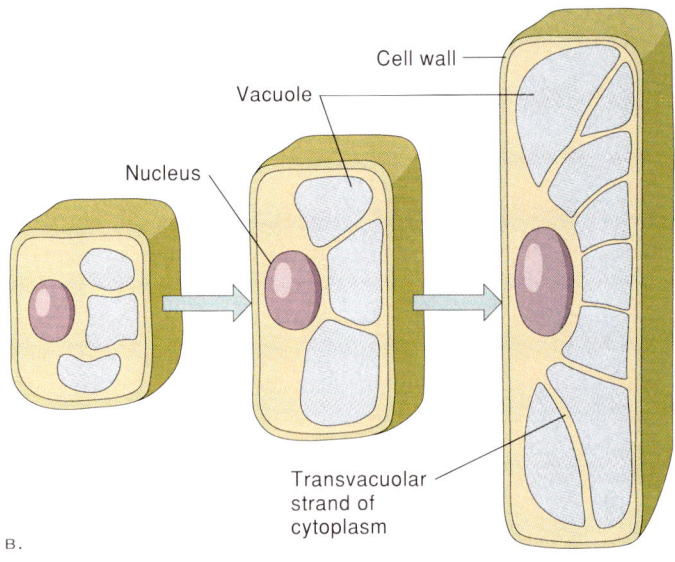

B.

FIGURE 3.22

Vacuoles. (a) Transmission electron micrograph from a coleus leaf (*Coleus blumei*). The expanding central vacuole compresses the cytoplasm into a small space against the cell wall, ×20,000. (b) Schematic diagram of cell growth, showing how an increase in cellular volume can occur without a large increase in the amount of cytoplasm. The peripheral layer of cytoplasm may be interconnected through the vacuole by cytoplasmic strands that radiate from the region of the nucleus.

again by putting them in water. This reacquired crispness (turgor) is caused by vacuoles that have refilled with water.

Vacuoles are versatile organelles, as indicated by the diversity of substances that occur in them. In addition to water, vacuoles contain enzymes and other proteins, water-soluble pigments, growth hormones, and ions. Vacuolar enzymes digest storage materials and components from other organelles for recycling into the cytosol. Pigments in vacuoles, especially red and blue anthocyanins, impart bright colors to flowers, fruits, and other plant parts. Some plants may harbor toxic alkaloids or other secondary products in their vacuoles. These alkaloids, which are water-soluble at the acidic pH of vacuoles, may deter insects and other animals from eating the plants that contain them.

Ions such as potassium and chloride are stored in vacuoles for easy retrieval to the cytosol when needed for cellular metabolism. In plants specialized for high-salt habitats, such as those along coastlines and near marine estuaries, vacuoles can accumulate chloride salts to concentrations several thousand times greater than in the cytosol. The cytoplasm of these plant cells is protected from salt toxicity, enabling the plants to thrive in their harsh environment. In other plants, such as rhubarb (*Rheum rhabarbarum*), oxalic acid accumulates in the vacuoles and forms crystals of calcium oxalate. The tartness of these plants comes from these oxalates.

CONCEPT

Most plant cells have a large central vacuole that arises from the fusion of vesicles and many smaller vacuoles. The central vacuole contains mostly water, but it may also contain enzymes, salts, pigments, alkaloids, and other kinds of chemicals.

Microbodies

The smallest membrane-bound organelles in a cell are called **microbodies.** Microbodies, which are bound by a single membrane, are usually spherical and 0.5 to 1.5 μm in diameter (fig. 3.23). These tiny organelles are often associated with membranes of the ER, but they may also be closely associated with chloroplasts and mitochondria. Different types of microbodies have specific enzymes for certain metabolic pathways. Two of the most important kinds of microbodies are **peroxisomes,** which occur primarily in leaves, and **glyoxysomes,** which are common in germinating oil-bearing seeds and the young seedlings that grow from them.

Peroxisomes are so named because they metabolize hydrogen peroxide (H_2O_2). The enzyme that catalyzes this breakdown is **catalase.** However, peroxisomes also contain **oxidases** that catalyze the production of H_2O_2. The importance of having these oxidases and catalase in the same organelle is a mystery. It might seem logical to suggest that peroxisomes protect cells by destroying peroxide, but this hypothesis does not explain why peroxisomes make peroxide, too. Furthermore, the

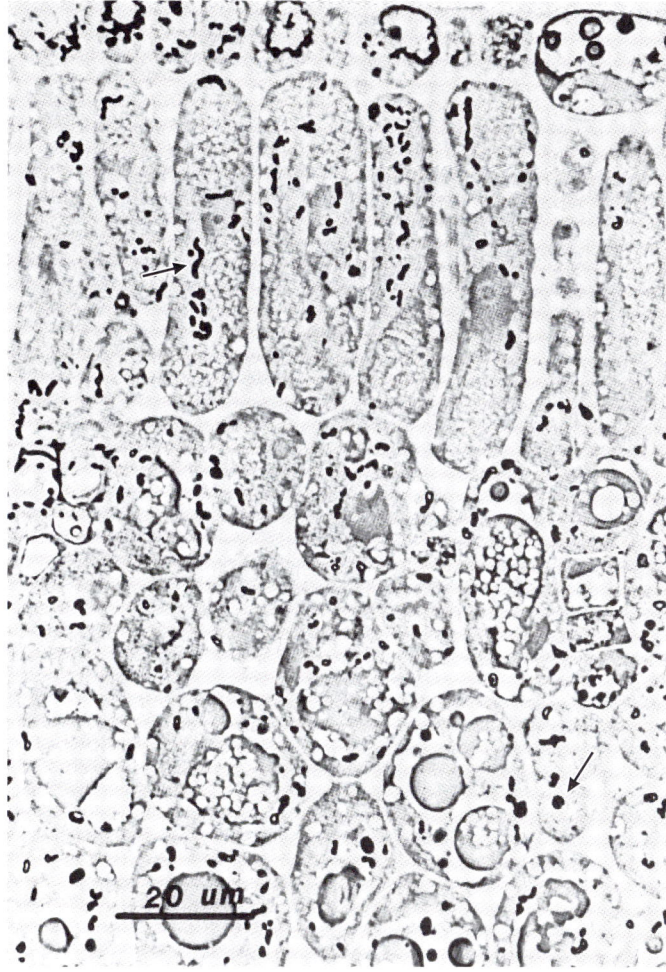

FIGURE 3.23

Phase-contrast light micrograph of section through a cotyledon of cotton (*Gossypium*). Cells were stained to localize the enzyme catalase (arrows), a microbody enzyme that degrades hydrogen peroxide; the localization of catalase shows where microbodies occur. The microbodies in these cells are peroxisomes.

cytosol and mitochondria also contain H_2O_2-producing oxidases, which enable these cell compartments to make peroxide that is not metabolized in peroxisomes. Moreover, many cells that make H_2O_2 lack peroxisomes. Thus, it seems that peroxisomes are not just peroxide-detoxifying organelles; their importance may be in their use of H_2O_2 to oxidize other cell toxins, such as ethanol and nitrites.

Glyoxysomes also contain enzymes that catalyze the breakdown of fatty acids into acetyl-CoA. This acetyl-CoA is used to make organic acids that can be exported from the glyoxysome and used in other metabolic pathways, such as respiration and sucrose synthesis. The pathway for acetyl-CoA metabolism in glyoxysomes is the **glyoxylic acid cycle** (fig. 3.24). Unlike peroxisomes, glyoxysomes rarely occur in animals. Consequently, plants can convert lipids to carbohydrates, but animals generally cannot.

CHAPTER THREE *Structure and Function of Plant Cells*

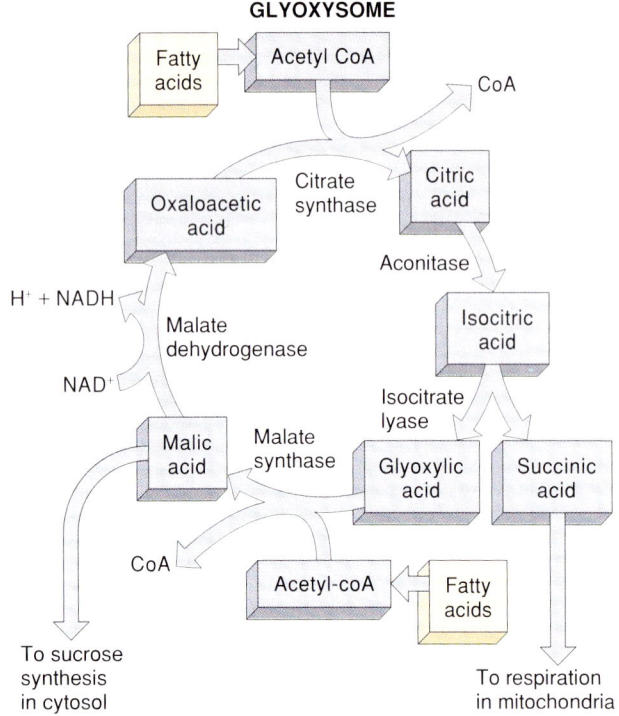

FIGURE 3.24

The glyoxylic acid cycle. Acetyl-CoA from fatty acid degradation enables the synthesis of succinic acid, which may be transported to mitochondria, and the synthesis of malic acid, which may be used to make sucrose in the cytosol.

CONCEPT

Microbodies include peroxisomes and glyoxysomes. Peroxisomes break down some of the toxic products of cell metabolism by oxidizing them with hydrogen peroxide. Glyoxysomes house the glyoxylic acid cycle, which links the breakdown of fatty acids to the synthesis of sucrose.

ORGANELLES FOR ENERGY CONVERSION

Cells thrive on the energy of ATP. Two kinds of organelles, **chloroplasts** and **mitochondria,** produce most of the ATP needed for cellular metabolism. These organelles are similar in several respects. For example, both are bounded by two membranes, and much of their internal membranes is folded and stacked to form complex compartments. Their internal membranes contain the enzyme ATPase, which uses the electrochemical energy of protons to phosphorylate ADP into ATP. Chloroplasts and mitochondria also contain DNA that controls the synthesis of many of the enzymes necessary for their respective metabolic pathways (nuclear DNA provides instructions to make additional enzymes). Finally, chloroplasts and mitochondria are semiautonomous: they grow and divide in the cell on their own (see boxed reading 3.2, p. 66).

The differences between chloroplasts and mitochondria include their respective sources of energy for making ATP, their appearance, and their composition. Chloroplasts use the energy of light, while mitochondria use the energy of chemical bonds. Chloroplasts contain chlorophyll, which makes them green, while mitochondria are colorless. Photosynthesis occurs in chloroplasts, and most of respiration occurs in mitochondria. Each process requires a different set of enzymes. Chloroplasts have many shapes and sizes, but they are generally larger than mitochondria, which are often cigar-shaped. A brief overview of these two organelles follows; details of their functions are presented in Chapters 6 and 7.

Chloroplasts

The fluid inside chloroplasts is called the **stroma** (fig. 3.25a). Membranes occurring throughout the stroma are called **thylakoids.** Thylakoids are either aggregated into stacks, called **grana,** or they form connections between stacks (fig. 3.25b). Because these connections are so common, all thylakoids in a chloroplast are probably formed by the same, continuous membrane system.

Thylakoid membranes contain a rich diversity of enzymes and pigments that characterize chloroplasts. Because these enzymes catalyze the reactions of photosynthesis, they are distinct from the enzymes of mitochondria. Pigments in chloroplasts include the chlorophylls, which create the green color of leaves and other green organs, and the carotenoids, which are the yellow, orange, or red colors of autumn leaves, tomatoes, and carrots. In addition to enzymes and pigments, chloroplasts often contain starch or oil.

The greenest cells of a leaf may each contain more than fifty chloroplasts. However, chloroplasts are just one type of **plastid.** Other plastids are classified according to the kinds of pigments or storage products they contain; for example, amyloplasts store starch, and elaioplasts store oils. Colored, nongreen plastids are chromoplasts, and are usually red, orange, or yellow.

All plastids develop from proplastids in young, unspecialized cells, and each type of plastid inherits the same genetic material from its proplastid precursors. Depending on environmental conditions, such as light versus dark, plastids may change. When green tissues are kept in the dark, chloroplasts become colorless, but upon reexposure to light, they become green again.

Doing Botany Yourself

Examine the living leaf cells of two or three plants whose chloroplasts can be seen easily with a light microscope. Determine whether the orientation of the chloroplasts (end vs. side view) is random or ordered (nonrandom). Make at least two hypotheses to explain the pattern you found.

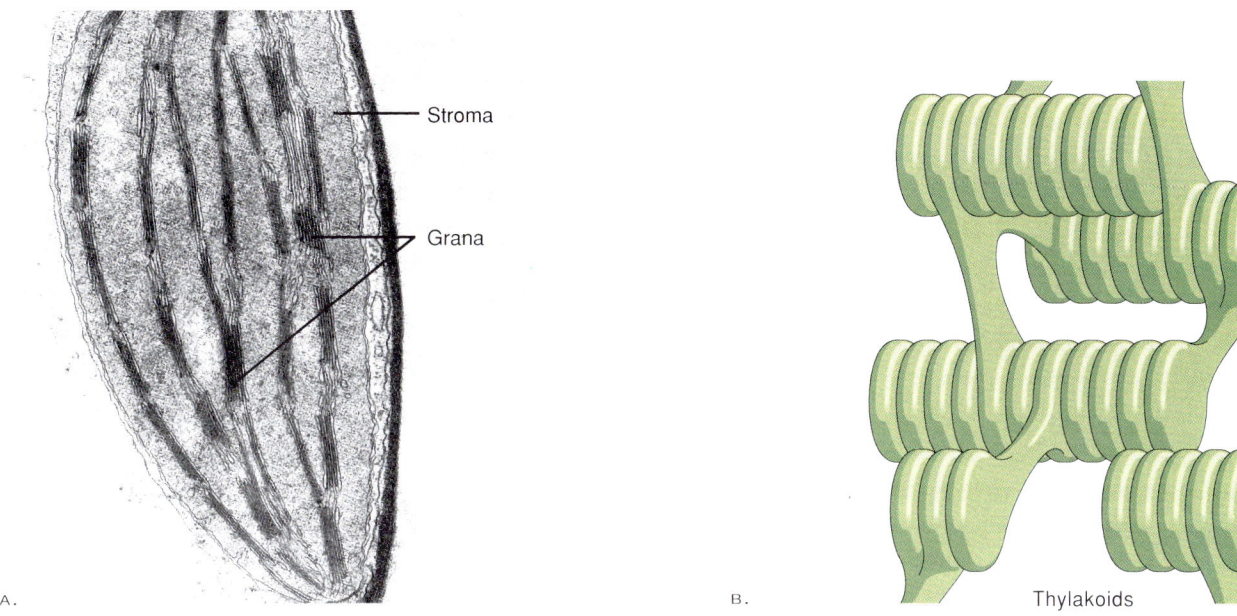

FIGURE 3.25

Chloroplasts. (a) Transmission electron micrograph of a chloroplast, showing stroma, thylakoids, and thylakoid stacks (grana), ×35,000. Also see the chloroplasts in figure 3.8a. (b) Three-dimensional model of chloroplast membranes, showing vesicle-like thylakoids in grana and continuous thylakoid connections between grana.

Mitochondria

A mitochondrion consists of a smooth outer membrane, and an inner membrane that is folded into tubular or vesicle-shaped **cristae** (fig. 3.26). The internal membrane system of mitochondria arises from the inner membrane of the mitochondrial envelope, whereas the outer membrane of the envelope is smooth. This arrangement of membranes creates two compartments within the mitochondrion: one is the space between the two membranes, and the other is enclosed by the inner membrane.

Many of the reactions of aerobic respiration are catalyzed by enzymes bound to mitochondrial membranes. Other reactions occur either in the space between the inner and outer membranes or in the matrix that is enclosed by the inner membrane (fig. 3.26).

A cell may contain several hundred mitochondria; the actual number of mitochondria in a cell is usually proportional to its requirements for ATP. Dividing cells and cells that are metabolically active need large amounts of ATP and usually have the largest numbers of mitochondria. You will learn more about the structure and function of mitochondria in Chapter 6 ("Respiration").

CELL MOVEMENTS

Cells prepared for study with microscopes are usually arrested in static positions that conceal their dynamic nature. However, the contents of cells are mobile; internal movements can be seen directly in living cells with a light microscope, or they can be tracked by time-lapse photography. Neither of these techniques is possible with electron microscopy because specimens must be killed and fixed before they can be studied with the electron microscope. By means of three-dimensional reconstructions from numerous serial sections, however, electron microscopy can also reveal internal cell movements. Unlike observations with light microscopy, serial reconstructions from electron micrographs reveal the changeability of small organelles and membranes.

In addition to internal movements, some entire cells are motile; that is, they can swim. In plants, only sperm cells swim. Only seedless plants and a few seed plants have swimming sperm cells; the sperm cells of flowering plants cannot swim.

Internal Movements

When we observe living plant cells by light microscopy, we see that their cytoplasm moves constantly. Organelles and other particles usually move in a circle around the central vacuole. This streaming movement is called **cyclosis** because of its circular path.

Chloroplasts and mitochondria tumble along definite paths that are associated with actin filaments and microtubules of the cytoskeleton. The outermost region of the cytoplasm is more anchored and relatively immobile, whereas the innermost region is more fluid. Rows of actin filaments occur between the two regions and induce streaming. Cyclosis enhances the exchange of materials among organelles, between membranes and organelles, and even between cells.

BOXED READING 3.2

CELLULAR INVASION: ORIGIN OF CHLOROPLASTS AND MITOCHONDRIA

Chloroplasts and mitochondria are similar to bacteria: they are about the same size, they contain DNA arranged in circular form, they have ribosomes that are smaller than those made in nuclei, and they have a common gene structure that differs from that of nuclear genes. Furthermore, the inner membranes of chloroplasts and mitochondria are similar to the plasma membranes of bacteria. Protein synthesis in chloroplasts and mitochondria is also inhibited by antibiotics that have similar effects on ribosomes in bacteria but not those made in nuclei. Such similarities between these organelles and bacteria are evidence for the **endosymbiotic hypothesis** of the origin of chloroplasts and mitochondria. According to this hypothesis, small bacteria invaded or were engulfed by larger bacteria. The smaller bacteria, now represented by chloroplasts and mitochondria, developed symbiotic relationships with their larger hosts. The mitochondrial precursors may have provided aerobic (oxygen-using) metabolism for a host that was probably unable to use oxygen. The photosynthetic precursors of chloroplasts provided the ability to make carbohydrates.

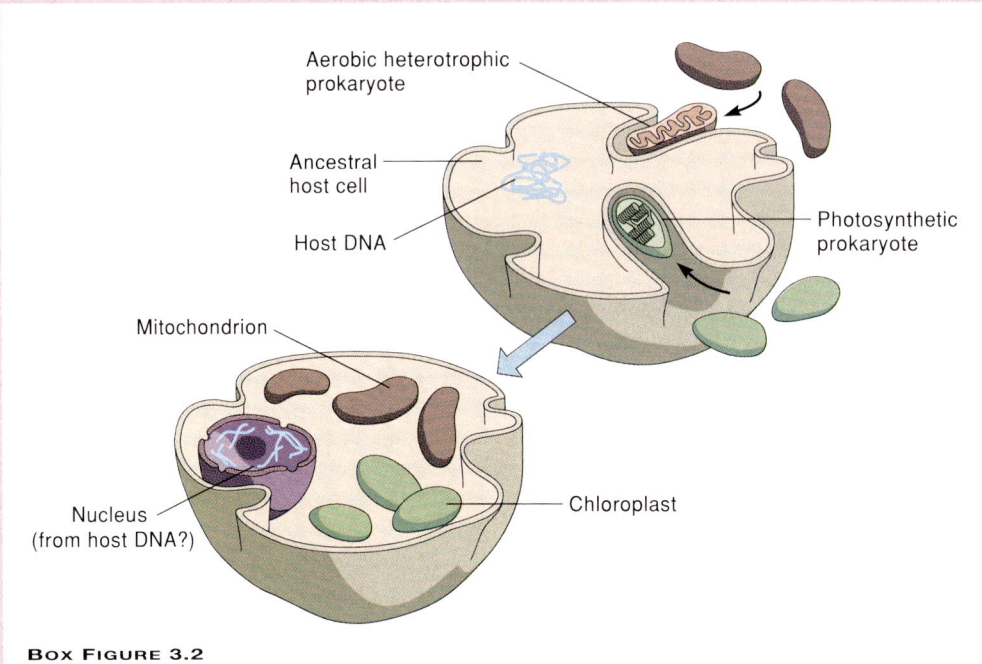

BOX FIGURE 3.2
Model of cellular invasion to illustrate the endosymbiotic hypothesis of the origin of chloroplasts and mitochondria. Origin of the nucleus is not explained by this hypothesis.

Bacterialike organisms probably originated more than 3.5 billion years ago. The first endosymbiotic fossils are about 1.5 billion years old. Nevertheless, modern organisms provide good clues about the bacterial relatives of chloroplasts and mitochondria. Physiological and genetic similarities point to the largest group of bacteria as a source of ancestors for these organelles. For example, one extant member of this group, *Prochlorothrix*, contains chlorophyll b, which is otherwise known only in plants and some algae. Other photosynthetic bacteria lack this pigment. However, certain other bacteria have pigments similar to those in chloroplasts of the red algae. This may indicate that chloroplasts originated at least twice, from two different kinds of bacteria. Nevertheless, botanists do not generally agree on how many times chloroplasts arose or which kinds of bacteria gave rise to them.

Cells That Swim

Cells that swim have hairlike locomotor organelles that protrude into the medium surrounding the cell. In plant cells these hairs can be up to 150 µm long and are called **flagella** (singular, **flagellum**). The only such swimming cells in plants are sperm cells, which may have two flagella, as in mosses, or several thousand flagella, as in cycads (fig. 3.27). Other plants that have flagellated sperm cells are ferns and other seedless vascular plants and the maidenhair tree (*Ginkgo biloba*), which is a seed plant. Some algae, water molds, and animals also have flagellated sperm cells. Certain water molds and algae have other kinds of cells that swim by flagella as well.

Regardless of their occurrence in diverse organisms, all flagella in plants, animals, fungi, and protists' have the same internal structure and the same mechanism of action. Each flagellum consists of a membrane that surrounds ten pairs of microtubules. One pair occupies the center of the flagellum, and nine pairs occur in a ring around the central pair (fig. 3.28a). Each outer pair of microtubules is connected to its neighboring pairs by sidearms that are evenly spaced along the length of the flagellum. Similarly, spokelike extensions from the outer microtubules connect to the inner microtubules (fig. 3.28b). The sidearms and spokes are made of the protein **dynein**.

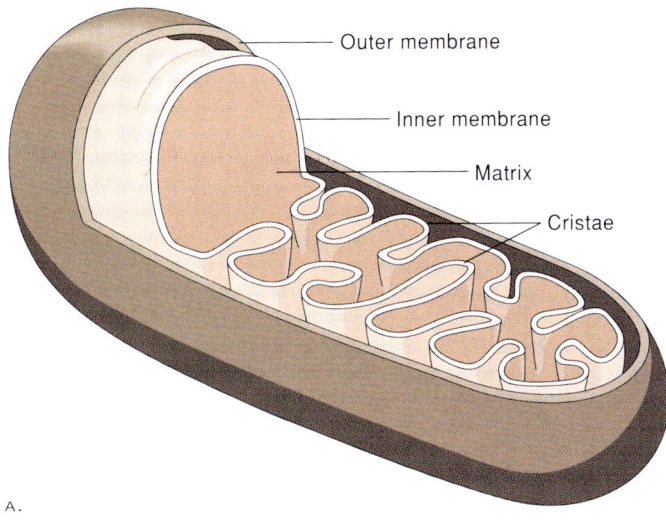

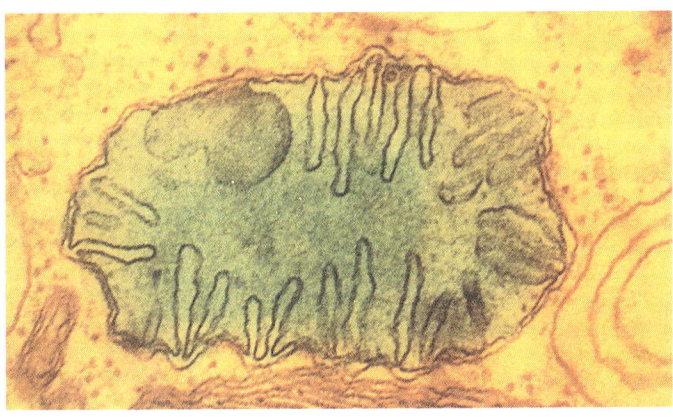

FIGURE 3.26

Mitochondria. (a) Three-dimensional model of a mitochondrion. (b) Transmission electron micrograph of a mitochondrion, ×20,000.

Dynein arms use ATP as a source of energy for their movement, which entails bending and sliding between pairs of microtubules. This pattern of dynein movement means that as the arms on one pair of microtubules slide or bend and the arms on the adjacent pair remain motionless, the first pair of microtubules becomes shorter; this causes the flagellum to contract in the region of the shortened pair of microtubules. This mechanism of flagellar motion is coordinated among nine pairs of microtubules, which enables the flagellum to whip in a powerful beating motion.

CONCEPT

The nucleus and cytoplasm of most plant cells are in constant motion. Organelles and other particles move by cyclosis, which is associated with microtubules and actin filaments in the cytoskeleton. Another kind of cellular movement occurs in the sperm cells of some plants, which have flagella that enable the sperms to swim.

FIGURE 3.27

(a) Light micrograph of sperm cell from a cycad (*Zamia*), showing numerous flagella, ×250. (b) Photo of a cycad that produces flagellated sperm shown in figure (a).

CHAPTER THREE *Structure and Function of Plant Cells* 67

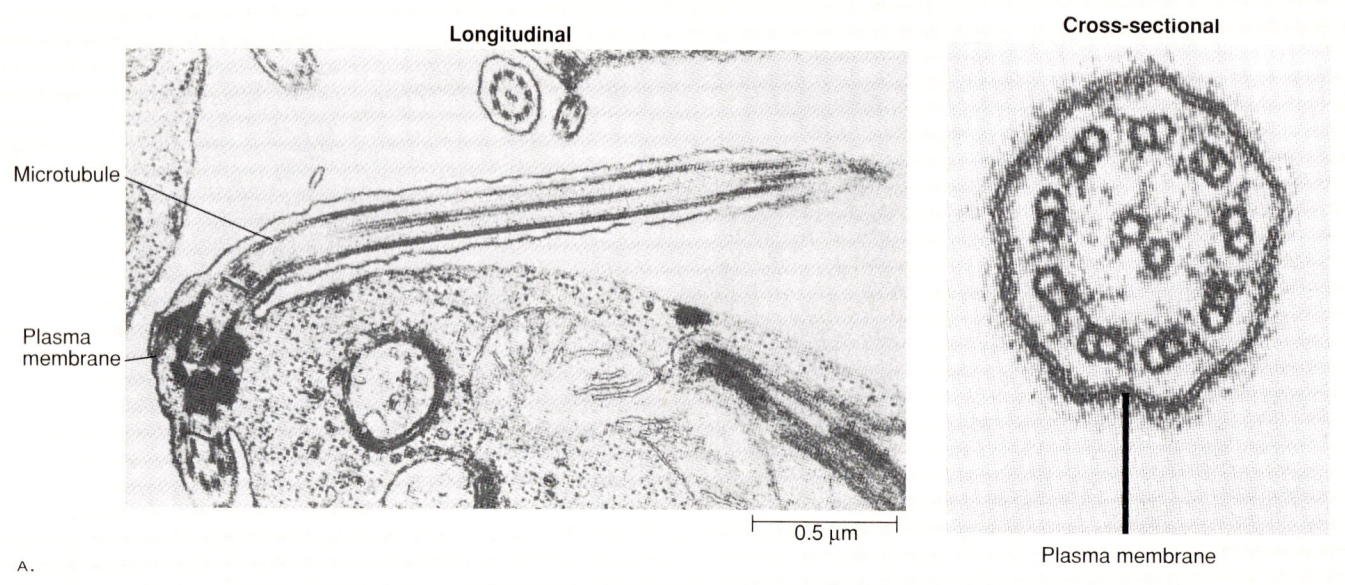

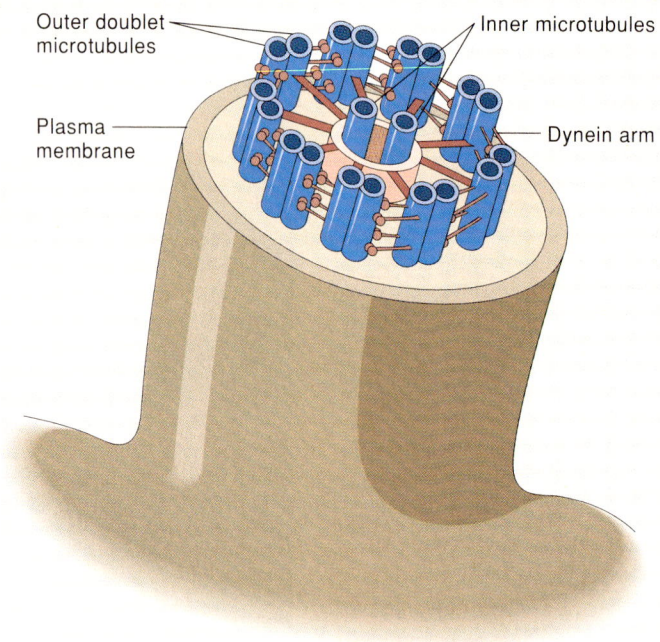

FIGURE 3.28

Flagella. (a) Transmission electron micrographs of a flagellum in the reproductive cell of the green alga *Ulvaria*. Longitudinal section shows that the membrane surrounding the flagellum is continuous with the plasma membrane. Cross section shows ten pairs of microtubules, one central pair surrounded by nine peripheral pairs in a ring. (b) Model of flagellum, showing three-dimensional arrangement of internal components. Microtubules are attached to each other by dynein arms.

CELL THEORY: THE DOGMA AND AN ALTERNATIVE

The history of cell science, like that of scientific development in general, parallels the invention of instruments that expanded our observations beyond their previous limits. For example, the discovery of cells was made possible by the invention of the light microscope. In 1665, Robert Hooke used a hand-made microscope to see cells in the dead outer bark of a cork oak. He made thin slices of cork tissue and observed the cell walls. Cork tissue was ideal for Hooke because it is rigid enough to slice without collapsing and because it has a simple structure due to the absence of living matter. These advantages do not exist for most other tissues, which may help explain why it was 150 years after Hooke's discovery before any significant advances were made in the study of cells.

The most significant advance in cytology was based on the idea that cells are the fundamental units of life, not just an oddity of cork. This idea is a basic postulate of the **cell theory,** which was proposed in 1838 by Matthias Schleiden and Theodor Schwann and became generally accepted by the 1860s (see Chapter 1). The main evidence for the cell theory was the observation that organisms are made of cells. However, there are many exceptions to the cell theory and its consequences for biological organization.

The Cell Theory: Postulates and Problems

As the cell theory was developed, several principal postulates were proposed for explaining the organization of living organisms:

1. All living substance is concentrated in cells.
2. Cells in an organism are all individuals of the same organizational rank.
3. The cell is the basic unit of structure and function.
4. An organism is an aggregate of cells, which are its building blocks.
5. The action of an organism is the sum of many actions of different kinds of collaborating cells.
6. The appearance of an organism depends on multicellularity.

In sum, these postulates explain that an organism is analogous to a "cell republic": each cell in an organism is at the same level of organization as each cell in another organism. The equivalency of cells in different organisms, according to the cell theory, is shown in figure 3.29.

One organism whose organization is consistent with all of the postulates of the cell theory is the slime mold, *Dictyostelium discoideum*, which exists as an aggregation of independent, amoebalike cells for most of its life history. When it begins to reproduce, the individual cells come together to form a spore-

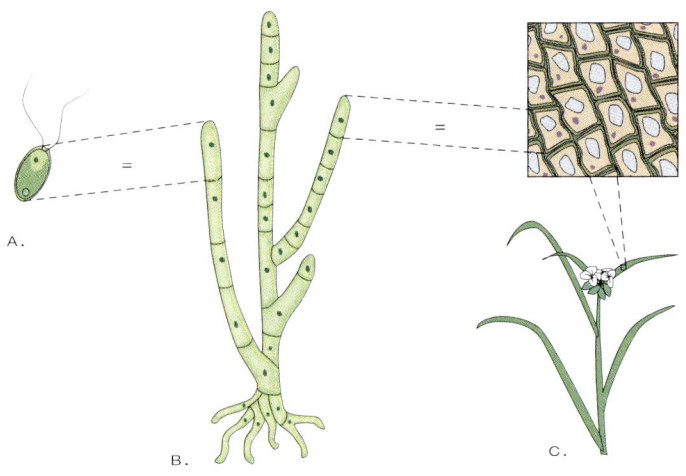

FIGURE 3.29

Equivalency of plant organization according to the cell theory. (a) Vegetative cell of the single-celled green alga *Chlamydomonas*. (b) Branched filament of the green alga *Spongomorpha*. (c) Leaf cells from the leaf of spiderwort (*Tradescantia*).

containing organ. Similar examples occur in certain green algae. Such organisms provide support for the cell theory. However, many biologists question the cell theory because this kind of organization is rare and because it does not occur in multicellular plants or in animals. Flowers and leaves, for example, are not aggregations of formerly dissociated cells.

Further doubt arises when we compare a wider variety of organisms than is shown in figure 3.29. Such a comparison is presented in figure 3.30, which shows that the equivalent level of organization among green algae and flowering plants may be the whole organism, not the cell. Moreover, the noncellular body of *Derbesia* in figure 3.30 shows that appearance does not depend on multicellularity, at least in green algae. Similar noncellularity also occurs in certain developmental stages of other plants. For example, the nutritional tissue of many seeds is multinucleate cytoplasm before it becomes cellular; coconut milk is a common example of such noncellular tissue. The embryos of pines and other cone-bearing trees are also noncellular in their early development. Thus, their development is independent of multicellularity. Similar examples of noncellularity also occur in fungi, protozoans, and animals.

The cell theory was derived primarily from animal development. However, the multicellular construction of plants is distinctly different from that of animals. Unlike animals, plants are characterized by incomplete cleavage between dividing cells, by continuous cytoplasmic connections between cells, and by the absence of cellular motility during development. The fixed position and interconnectedness of all cells in a plant provide evidence for the concept of the **symplast.** This term describes the entire living mass of the plant as a continuous unit, contrary to the idea that cells are separate individuals.

FIGURE 3.30

Equivalency of plant organization according to the organismal theory. (a) *Chlamydomonas*. (b) Branched filament of the noncellular green alga *Derbesia*. (c) Branched filament of the multinucleate but cellular green alga *Cladophora*. (d) *Spongomorpha*. (e) Diagram of *Vicia faba* (broad bean).

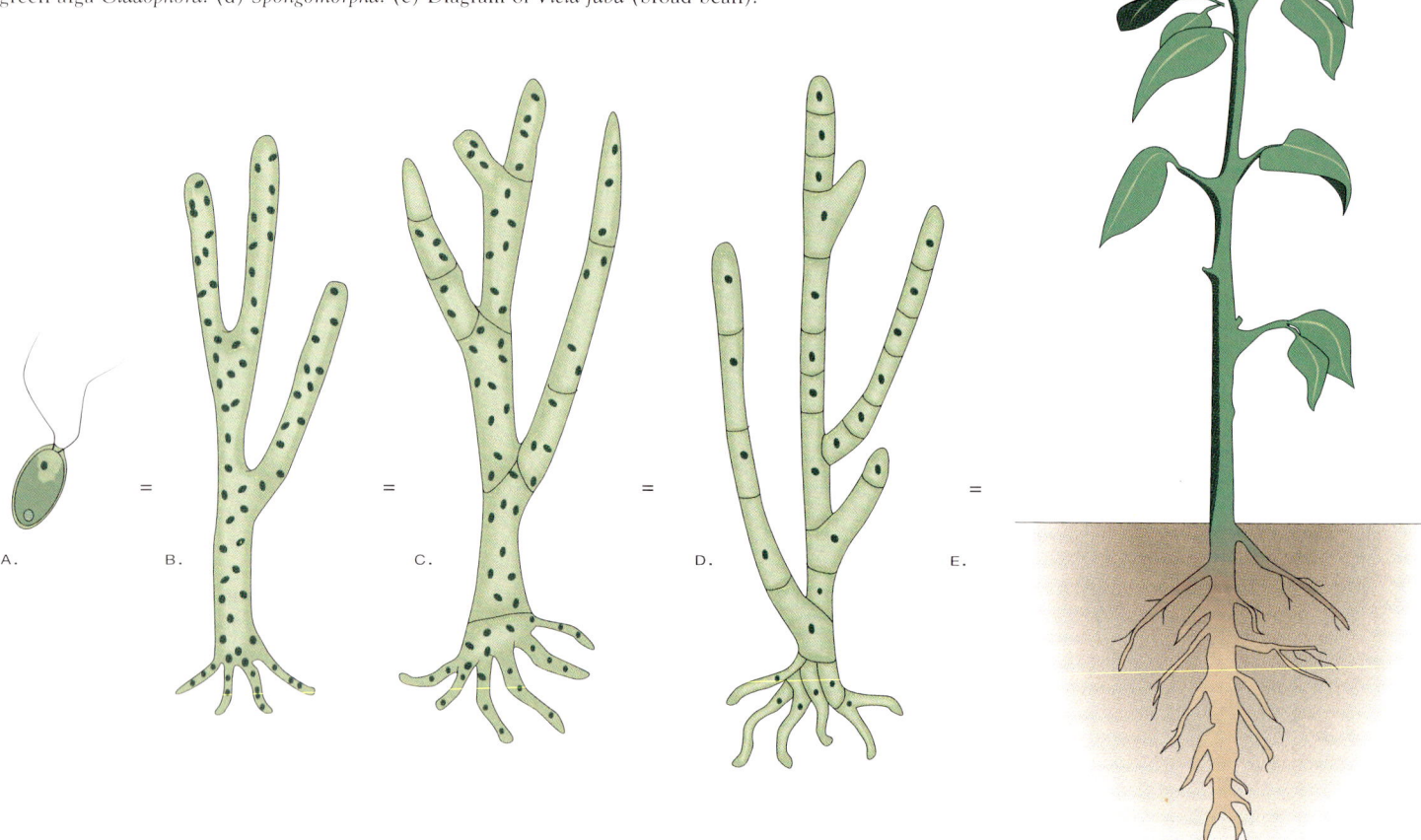

An Alternative: The Organismal Theory

If cells are not the basic units of biological organization, then what is the alternative? Perhaps the best alternative comes from the **organismal theory,** which holds that the whole organism is the basic unit of organization. This means that an entire plant is like a large cell that is compartmentalized into many parts. According to this theory, a whole plant constitutes the same level of organization as a single-celled green alga (fig. 3.30). One of the main postulates of this theory is that the appearance of an organism is independent of multicellularity; that is, development is not coupled to cell division. *Derbesia* presents observational evidence for this suggestion (fig. 3.30).

Botanists draw additional support for the organismal theory from the lemon tree, in which they have observed that the cuticle is first formed around the zygote, which is the first cell of the tree. The cuticle expands and maintains its coverage of the embryo, the seedling, and the tree as the plant grows. This observation suggests that the cuticle is a characteristic of the whole organism rather than of a particular cell type. This means that the plant differentiates by compartmentation of the zygote and all other cells within the same cuticle, and not by the aggregation of separate cells, each with its own cuticle.

Support for the organismal theory also comes from experimental evidence that shows a lack of a cause-and-effect relation between cell division and development. The experiment involved wheat grains irradiated with gamma rays, which suppressed cell division in the seedlings. In spite of this treatment, the wheat grew to normal size but more slowly and with fewer cells than untreated wheat (see Chapter 12). This experiment provides dramatic evidence that cell division does not cause development, a result that supports the organismal theory. Although the focus here is on plants, support for the organismal theory has also been found in animals.

CONCEPT

The cell theory states that the cell is the basic unit of organization of living organisms. In contrast, the organismal theory states that the whole organism is the basic unit of living organisms. The cell theory is best represented in slime molds and certain algae, but observations and experiments supporting the organismal theory come from plants and animals.

Chapter Summary

Cells are the simplest units of a plant that can live independently. Each plant cell consists of a cell wall that surrounds a plasma membrane, which encloses the contents of the cell. The contents of a plant cell usually include a nucleus and the cytoplasm. The cytoplasm includes all organelles except the nucleus, plus all internal membranes and the cytosol.

The study of cells is aided by instruments such as the light microscope and the electron microscope. The quality of microscopic observation depends both on magnification and on resolving power. Resolving power is limited by the wavelength of the light or the electron beam that is used for observation. Our study of plant cells has been aided by examining the functions of organelles that have been isolated by cell fractionation and by selectively staining different parts of cells prior to microscopy.

Membranes divide cells into many interconnected compartments. These compartments are connected by membranes that either move between or are attached to the nucleus and other organelles. All membranes consist of a double layer of phospholipids that has enzymes attached to or embedded in it. The composition of membranes varies from one organelle to another.

The distribution and movement of membranes and other cell compartments are affected by the cytoskeleton, a network of filaments that includes microtubules, actin filaments, and intermediate filaments. All three types of cytoskeletal filaments are made of different kinds of proteins.

Young cells and actively growing cells have flexible primary cell walls. As cells mature, however, they often form rigid secondary cell walls just inside the primary walls. Cell-wall synthesis is controlled by arrays of microtubules just inside the plasma membrane.

Plant cells are connected to one another by plasmodesmata. These connections contain cytoplasm and a plasma membrane that is continuous between cells.

The nucleus, chloroplasts, and mitochondria are organelles that are surrounded by double membranes. Microbodies have single membranes, and ribosomes have no membranes. Dictyosomes and the endoplasmic reticulum are composed mostly of membranes that enclose relatively little internal fluid. Many of the chemical reactions in a cell are catalyzed by enzymes that are in or on membranes, including the plasma membrane.

Sperm cells in plants and animals, as well as other kinds of cells in algae and water molds, have flagella that enable them to swim. The structure and function of the flagella are identical in these organisms.

The cell theory has been the dominant theory of how living things are organized. It explains how cells are the basic units of organization. An alternative to this idea is the organismal theory, which holds that the basic unit of organization is the whole organism. Some evidence supports both theories, but observational and experimental evidence supports the organismal theory better in plants.

Questions for Further Thought and Study

1. Discuss the importance of having both proteins and lipids as components of membranes.
2. How does the function of rough ER differ from that of smooth ER?
3. Where are microtubules in cells? What are their roles?
4. How does microscopy provide evidence for the functions of cellular organelles?
5. What evidence supports the organismal theory?
6. The cell theory is an excellent example of inductive reasoning. How do you use inductive reasoning in your life?

Suggested Readings

ARTICLES

Albersheim, P. 1975. The walls of growing plant cells. *Scientific American* 232 (April):81–95.

de Duve, C. 1983. Microbodies in the living cell. *Scientific American* 248 (May):74–84.

Kaplan, D. R., and W. Hagemann. 1991. The relationship of cell and organism in vascular plants. *BioScience* 41:693–703.

Lane, M. A., et al. 1990. Forensic botany. *BioScience* 40:34–39.

Niklas, K. J. 1989. The cellular mechanisms of plants. *American Scientist* 77:344–349.

Rothman, J. E. 1985. The compartmental organization of the Golgi apparatus. *Scientific American* 253 (September):74–89.

Storey, R. D. 1990. Textbook errors and misconceptions in biology: Cell structure. *American Biology Teacher* 52:213–218.

Symmons, M., et al. 1989. The shifting scaffolds of the cell. *New Scientist* (18 February):44–47.

Vogel, S. 1987. Mythology in introductory biology. *BioScience* 37:611–614.

BOOKS

Becker, W. M., and D. W. Deamer. 1991. *The World of the Cell.* 2d ed. Redwood City, CA: Benjamin/Cummings.

de Duve, C. 1984. *A Guided Tour of the Living Cell.* Scientific American Library. New York: W. H. Freeman.

Gunning, B. E. S., and M. W. Steer. 1986. *Plant Cell Biology: An Ultrastructural Approach.* Copyright M. W. Steer.

This *Coleus* shows that plants wilt when they lose water (right) but stiffen again when given water. In both cases, water passes freely across cellular membranes.

Membranes and Membrane Transport

CHAPTER 4

Chapter Outline

INTRODUCTION
OVERVIEW OF MEMBRANE STRUCTURE AND FUNCTION
  Structure of Membranes
  Functions of Membranes
THE MOVEMENT OF WATER AND OTHER MOLECULES THROUGH MEMBRANES
  Movement of Solutes
  Water Potential
  Osmosis
  Turgor
  Inducing Osmosis: The Control of Turgor in Plants
DIFFERENTIAL PERMEABILITY OF MEMBRANES
  Facilitated Diffusion
  Active Transport
  Bypassing Membrane Transport

BOX 4.1
MEMBRANE TRANSPORT AND MAKING BEER

MOVEMENT OF IONS ACROSS MEMBRANES
  Ion Pumps
  ATP Synthesis
CELLULAR COMMUNICATION
  Hormone Receptors
  Membrane Interactions with Other Organisms
  Lectins and Glycoproteins
Chapter Summary
Questions for Further Thought and Study
Suggested Readings

Chapter Overview

In Chapter 3 you learned that all living cells have membranes. Membranes surround cells, connect cells to one another, form complex internal networks, and divide cells into distinct compartments. The abundance of membranes in cells underscores their importance, but it does not begin to show the diversity of their functions. Membranes are active and changeable participants in cellular metabolism: everything that happens in a cell is directly or indirectly associated with a membrane. In this chapter we explain more about membrane-dependent metabolism and how membranes control the life of a cell. You will also see how, in spite of their wide range of capabilities, all membranes have the same basic structure. Moreover, you will learn how plasma membranes enable organisms to grow in salty soil or water, how organellar membranes harvest energy from ion movement, and how membranes help cells communicate. All of these functions of membranes are essential to life. In fact, one of the criteria for describing life is the presence of a dynamic membrane system.

Introduction

In the preceding chapter you learned that membranes have various functions in cellular metabolism. Most of the important activities of cells are associated with membranes. For example, proteins destined for secretion or for insertion into cell membranes are made by ribosomes that are attached to a membrane system called the endoplasmic reticulum. The separation of cytosol and organellar contents from one another is maintained by membranes.

Suggestions about the structure of membranes appeared in the 1920s, long before membranes were seen with the electron microscope. The first ideas about membrane structure were based on the soaplike properties of phospholipids in artificial membranes. Phospholipids, like soaps, have a dual solubility. Their long hydrocarbon "tails" are nonpolar and hydrophobic. *Hydrophobic* literally means "water-fearing" and refers to chemicals that do not dissolve in water. In contrast, the ionic phosphate "head" of a phospholipid is polar and hydrophilic. *Hydrophilic* means "water-loving" and refers to chemicals that can dissolve in water. In water, phospholipids aggregate spontaneously into a **bilayer,** which is a double membrane with an interior of hydrophobic hydrocarbons and an exterior of hydrophilic phosphates (fig. 4.1). As you will learn in this chapter, this artificial membrane resembles the phospholipid backbone of biological membranes.

However, the great diversity of membrane functions cannot be explained by such a simple structural model. The properties of membranes also depend on proteins, which are their other main ingredient.

Many functions of membranes relate to the general properties of phospholipids and proteins. However, different membranes are made of different proteins and lipids. In this chapter, we first discuss the basic functions of membranes in terms of the general characteristics of proteins and lipids. Later, we examine different kinds of membranes and their functions.

Overview of Membrane Structure and Function

Early ideas about the structure of phospholipid bilayers explained how some molecules, including nonpolar molecules of gases (N_2 and O_2) and small polar molecules such as water could flow across membranes (fig. 4.2). However, lipid bilayers could not account for the ready passage of larger polar molecules such as monosaccharides and amino acids. The passage of these molecules through membranes was first explained in the 1930s by H. Davson and J. F. Danielli, who suggested that the lipid bilayer was coated on both sides with hydrophilic proteins that were attached to the polar heads of phospholipids. According to this model, the hydrophilic proteins absorbed polar molecules and somehow eased their passage through the nonpolar layer of the membrane. In the 1950s, the first electron micrographs of membranes seemed to confirm this model (fig. 4.3). These pictures showed membranes to have an electron-transparent inner region sandwiched between two electron-dense outer layers. The outer layers were assumed to be made of proteins and phospholipid heads, and the inner region was believed to be made of the hydrocarbons in phospholipid tails.

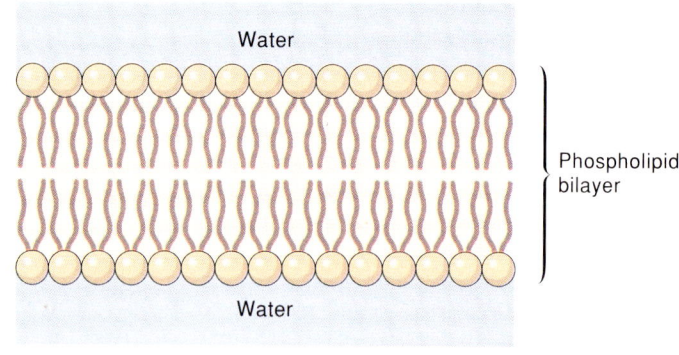

Figure 4.1

Model of an artificial membrane consisting of phospholipids. Phospholipids aggregate spontaneously into a bilayer.

Structure of Membranes

Despite the apparent support from electron microscopy, flaws in the Davson-Danielli model began to accumulate. For example, cell biologists found that this model could not explain structural and biochemical variations among different kinds of membranes. Mitochondrial membranes, for instance, are

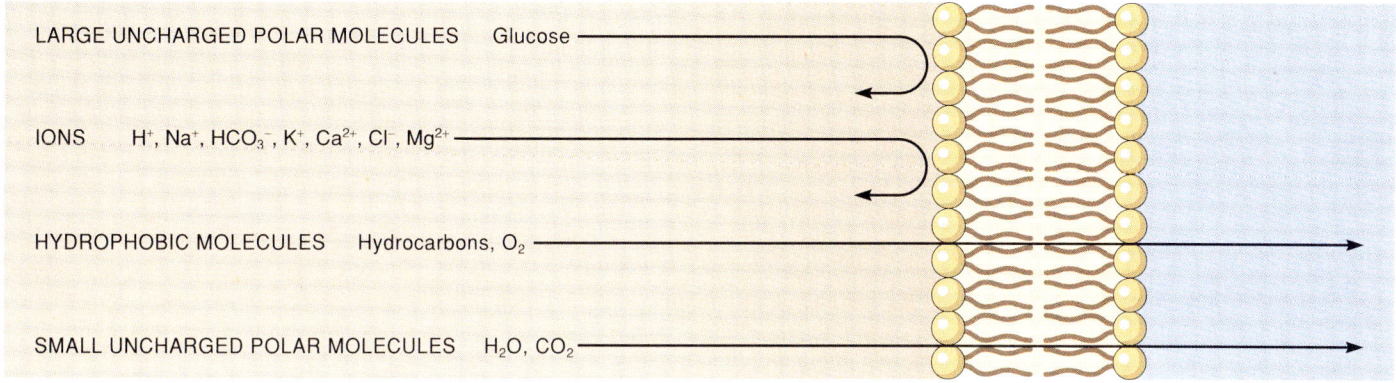

FIGURE 4.2

Selective permeability of a phospholipid bilayer. The bilayer is much more permeable to hydrophobic molecules and small uncharged polar molecules than it is to ions and large polar molecules.

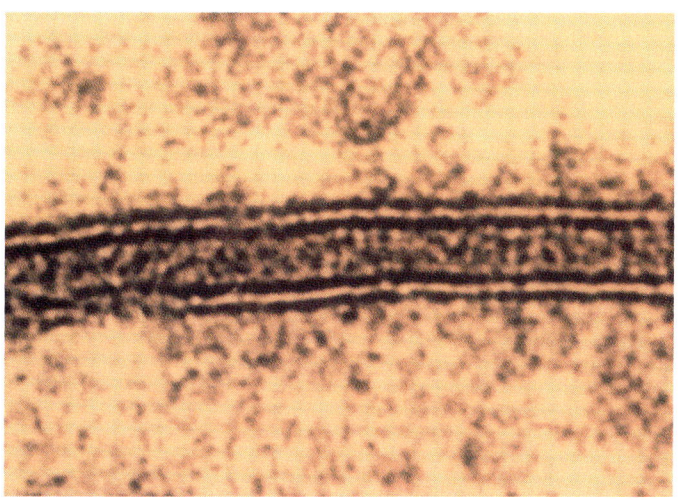

FIGURE 4.3

Transmission electron micrograph of two membranes. Each membrane appears as two dark lines that are separated by a lighter region. At first, micrographs of this sort seemed to confirm the Davson-Danielli model of membrane structure, but this interpretation was later found to be incorrect.

thinner than plasma membranes and contain a much higher proportion of protein. This and other findings contradicted the Davson-Danielli model, which held that membrane proteins must be hydrophilic and that all membranes have the general structure of a protein-lipid-protein "sandwich."

The current view of membrane structure entails several modifications of the Davson-Danielli model. This newer version, proposed in 1972, is called the **fluid mosaic model** because it holds that proteins occur as a mosaic in a fluid bilayer of phospholipids (fig. 4.4). Visual evidence for the fluid mosaic model comes from scanning electron microscopy of freeze-fractured membranes. Recall from Chapter 3 that freeze-fracturing often splits membranes between the two layers of phospholipids. The interior of the membrane is then seen as a dotted plain: the plain is a sea of lipids, and the dots are proteins inserted into them (fig. 4.5). The proteins are partly hydrophobic and partly hydrophilic, and are anchored at the protein-lipid interface by hydrophobic amino acids. Hydrophilic amino acids occur inside the proteins and also where the proteins protrude from the membrane into the aqueous environment on either side of the membrane (fig. 4.6).

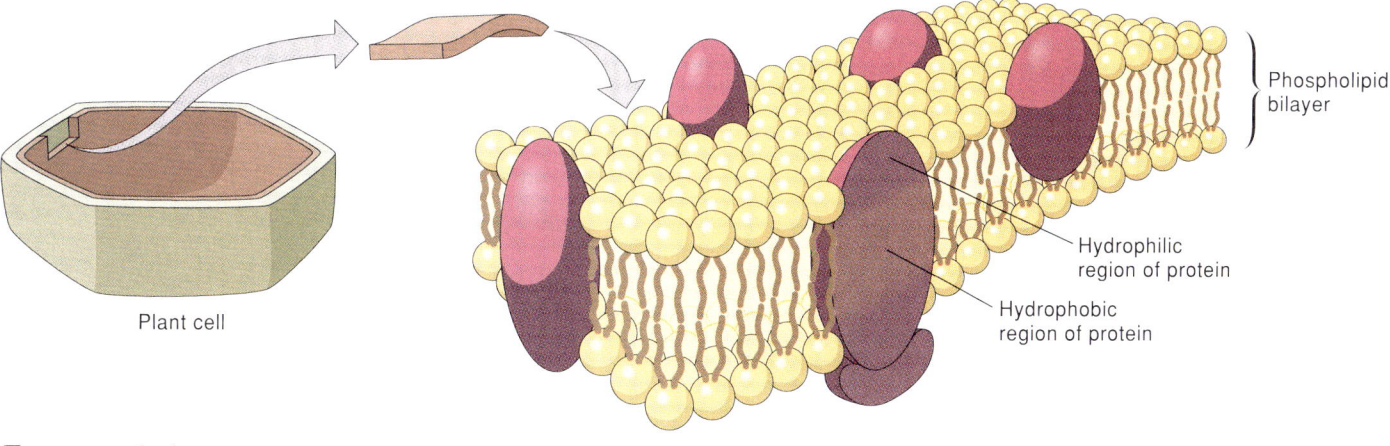

FIGURE 4.4

The fluid mosaic model of membrane structure. Proteins are dispersed in the phospholipid bilayer according to this model.

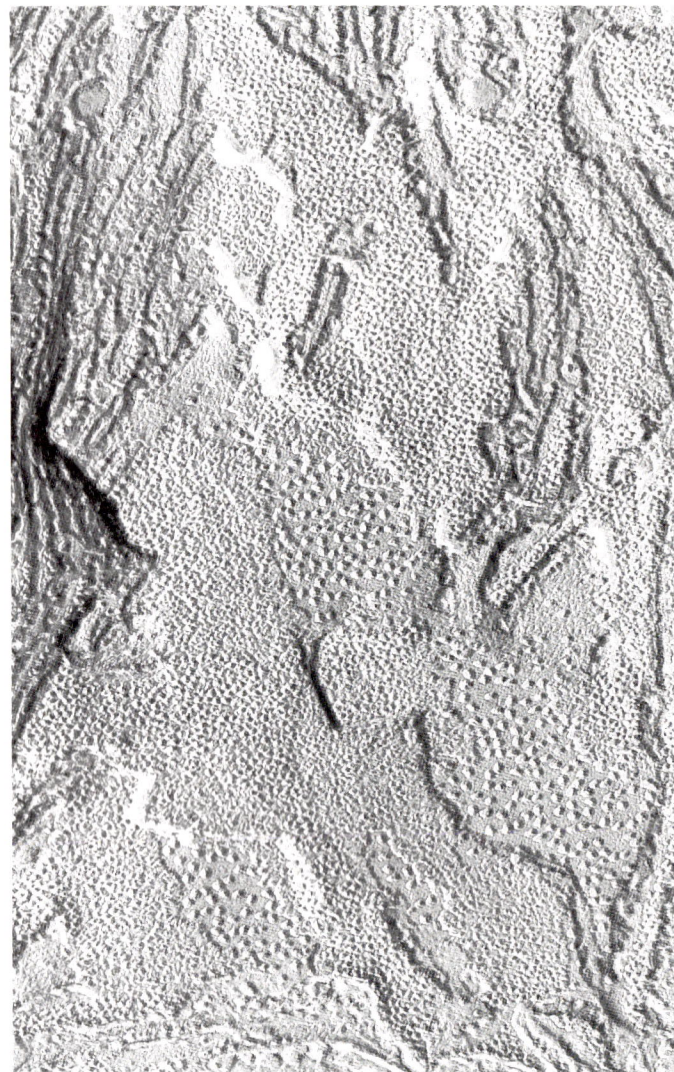

FIGURE 4.5

Scanning electron micrograph of freeze-fractured chloroplast membrane. Proteins appear as dots embedded in the lipid matrix.

The name *fluid mosaic model* implies that the membrane contains liquid. In fact, the lipid bilayer is oily; its fluidity is due to the loose packing of the fatty acid tails of the phospholipids. The mosaic of proteins actually floats through the lipids. The mobility of proteins is a significant feature of the fluid mosaic model; it accounts for the movement and intermingling of proteins that must touch each other to function.

Another important feature of membrane structure, one that is not immediately obvious from the fluid mosaic model, is its asymmetry; that is, one side of a membrane is different from the other. This difference comes mostly from the carbohydrates that are attached to proteins on the outside surface of the membrane. Proteins with carbohydrates attached to them are called **glycoproteins,** and they do not usually occur on the inner surfaces of membranes. Examples of the roles of glycoproteins are discussed later in this chapter in the section on "Cellular Communication."

C O N C E P T

Membranes are a mosaic of proteins that move around in a fluid double layer of phospholipids. This is called the *fluid mosaic model* of membrane structure. Proteins that protrude from the outer surface of membranes are often glycoproteins.

Functions of Membranes

Proteins control most of the functions of membranes. There may be fifty or more different kinds of proteins in a plasma membrane, and perhaps as many in the tonoplast or other organellar membranes. This diversity of proteins is reflected in the enormous range of activities associated with membranes. Some of the more important of these activities are discussed in the next few sections of this chapter and are summarized below.

Movement of water and solutes. The plasma membrane generally allows the unrestricted passage of water and certain dissolved substances into or out of the cell. Water balance is crucial for maintaining turgor pressure, which makes the cell rigid and drives cellular expansion during growth.

Differential permeability. Membranes control or block the passage of some kinds of molecules; such membranes are referred to as *differentially permeable* membranes. Differential permeability is different for different membranes.

Ion pumps. Certain ions, such as K^+ and H^+, are pumped through membranes. Ion pumps in the plasma membrane use energy from ATP to move ions from the cell, while ion pumps in mitochondrial and chloroplast membranes are important for making ATP.

Enzyme activity. Enzymes that cooperate in multistep processes, such as ATP synthesis or the absorption of light energy, often occur together in a particular spot on a membrane.

Cellular communication. The plasma membrane contains proteins that bind molecules released from other cells. Once bound to an external molecule, these proteins activate other proteins in the membrane that cause metabolic changes in the cell.

C O N C E P T

Membranes have many roles in a cell. Besides blocking or facilitating the entry of substances into the cell, membranes are also the sites of enzyme activity, ATP synthesis, and cellular communication.

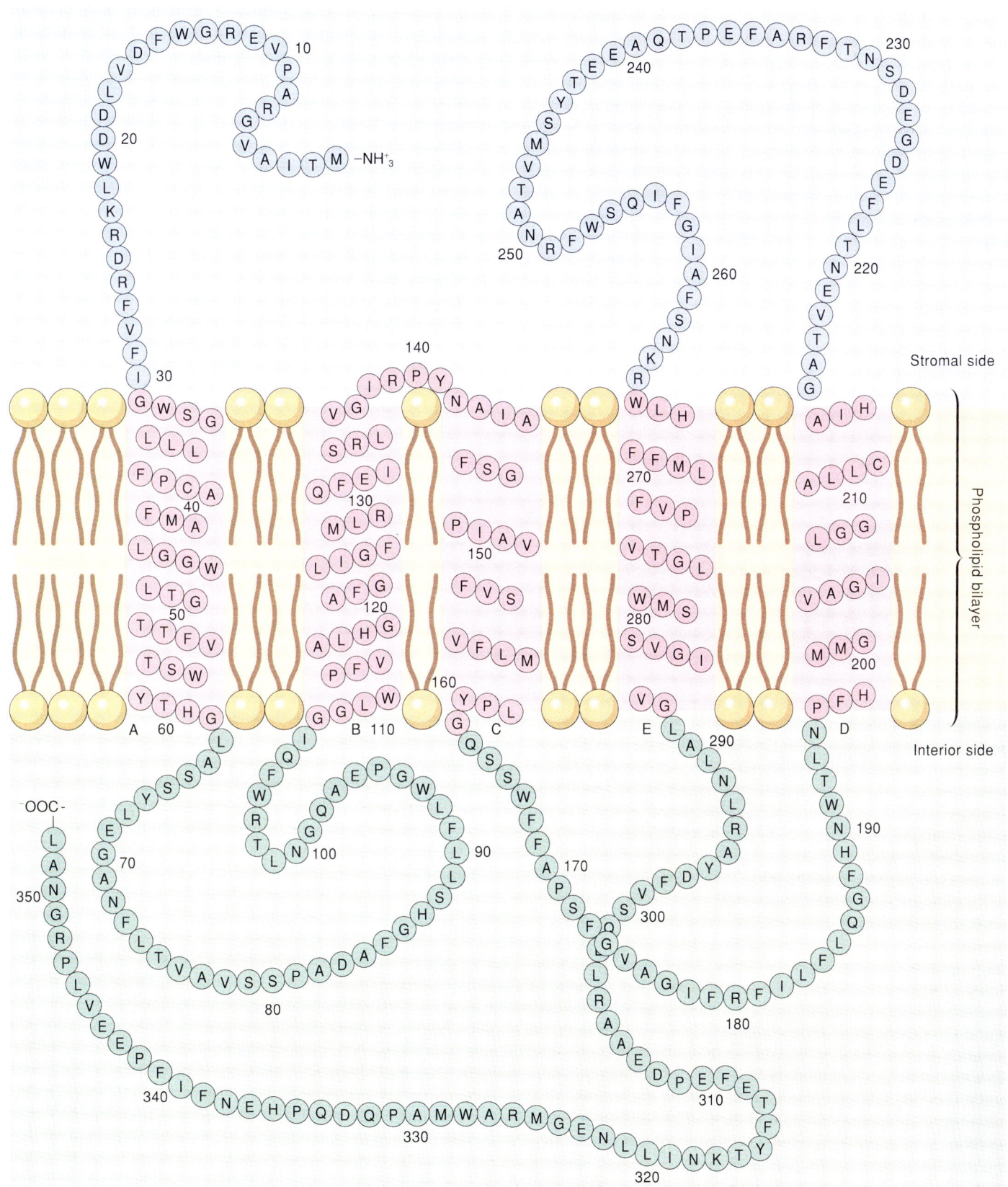

FIGURE 4.6

Primary structure of a thylakoid protein. Five regions of the protein occur in the membrane (boxes A–E); the remaining regions protrude either into the stroma or into the interior of the thylakoid. This membrane protein consists of 352 amino acids, each designated by a one-letter abbreviation (see Chapter 2, fig. 2.6).

CHAPTER FOUR *Membranes and Membrane Transport*

The Movement of Water and Other Molecules through Membranes

The most common kind of molecule in cells is water. Ions and other polar molecules are dissolved in this water; that is, they are **solutes,** and water is the **solvent** that dissolves them. Solutes include protons (H⁺), mineral ions such as potassium (K⁺) and magnesium (Mg⁺⁺), and organic compounds such as sugars and amino acids. The passage of these substances through membranes is determined both by the phospholipid bilayer and by the proteins embedded in it. Nonpolar molecules, such as hydrocarbons and oxygen, pass easily through a membrane's lipid core. Small and uncharged polar molecules, such as water and carbon dioxide, also pass through membrane lipids without hindrance. Ions, however, are almost entirely prevented from diffusing through the phospholipid bilayer. The concentrations of different solutes in cells and organelles are closely regulated, despite the free passage of water and certain solutes through the plasma membranes. This inhibits substances from leaking freely into or out of the cell.

Some terms involving water and solute movement are introduced in this chapter because they relate to membrane function. A more complete discussion of these terms is presented in Chapter 21, "Movement of Water and Solutes."

Movement of Solutes

All molecules display random thermal motion, or kinetic energy; that is, a solute molecule has a tendency to move around in a solution. One result of this random movement is that molecules diffuse outward from regions of high concentration to regions of lower concentration. **Diffusion** by random movement continues until the distribution of molecules becomes homogeneous throughout the solution. For example, when a crystal of dye is placed in a container of pure water, the dye diffuses as the crystal dissolves. The rate of diffusion depends on the size of the dye molecules (larger molecules move slower) and the temperature of the solution (higher temperatures cause faster movement). Regardless of this rate, the dye concentration will eventually become uniform throughout the solution. This phenomenon is easily illustrated by placing a drop of a water-soluble dye into a glass of water (fig. 4.7).

Because diffusion is based on random movements, each molecule has an equal chance of moving toward or away from the region of high solute concentration. However, there are more solute molecules near the crystal, which means there is a greater chance for **net movement** away from the source. Molecular movement continues after homogeneity is reached, even though there is no longer a net change in concentration from one region of the solution to another. This means that, after such a balance is reached, for every molecule that vacates a spot, another molecule takes its place just by chance.

Figure 4.7

Beakers of water before and after diffusion of the dye bromthymol blue. Random movements of water and dye molecules cause diffusion, eventually resulting in a uniform concentration of the dye.

Diffusion is the net movement of solutes from an area of greater concentration to an area of lesser concentration—that is, down a **concentration gradient.** We can think of solutes as marbles and the gradient as a hill; marbles rolling down a hill would be analogous to solutes diffusing down a concentration gradient. Furthermore, a steeper hill would be analogous to the steeper gradient caused by a higher concentration of solutes. Thus, a steeper gradient causes a higher net rate of solute movement. Just as marbles can move up the hill if there is a force to push them, so too can solutes move from lower to higher concentrations if there is energy to push them. As you will see later in this chapter, cellular energy is often used for moving solutes up their concentration gradients.

Anything that can fall, change, or flow from a higher level to a lower one has the potential to do work, which is called **potential energy.** When the solute is moving, it has energy of movement, which is called **kinetic energy.** We discuss these forms of energy in this chapter in relation to membrane function, but a broader discussion of energy and its use by plants is presented in Chapter 5.

Water Potential

Like solutes, water also has potential energy to flow to where it is less concentrated. The potential energy of water has a special name: **water potential.** Water tends to move down a water-potential gradient, that is, from a region of high water potential to a region of low water potential. Also, like solutes, water requires energy to move up a water-potential gradient.

By general agreement, the water potential of pure water is zero. This means that the water potential of a solution has a negative value because the water is less concentrated than pure water. Also by general agreement, water potential is expressed

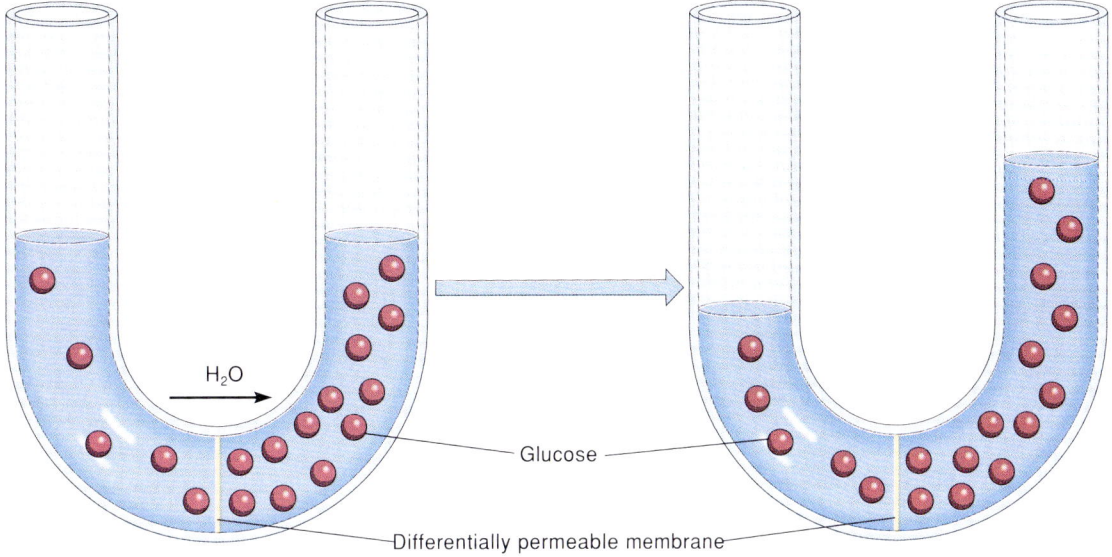

FIGURE 4.8

Osmosis is demonstrated by the movement of water through a differentially permeable membrane in a U-tube. Glucose, which cannot pass through the membrane, is more concentrated in the right-hand part of the tube than in the left-hand part. Net movement of water into the more concentrated glucose solution causes the volume of the right-hand solution to increase.

in units of pressure instead of units of energy, because pressure is simpler to measure. Thus, water potential may be expressed in **bars** or in **megapascals (MPa)**. One bar equals atmospheric pressure at sea level and room temperature, and 0.1 MPa is 1 bar. The potential of sea water is about −25 bars, the water potential in the cells of freshly watered herbs is more than −1 bar (meaning closer to zero), and the water potential in dry seeds is about −200 bars. For comparison, the recommended air pressure for most car tires is about 2 bars. We'll discuss water potential in more detail in Chapter 21.

The random thermal motion of all molecules causes them to move about in a solution. Because of this motion, molecules move from a region of higher concentration to a region of lower concentration. This net movement of molecules down a concentration gradient is called *diffusion*. The potential for solutes to move down a gradient is called *potential energy*. The potential energy of water to move down a gradient has a special name, *water potential*.

Osmosis

Many substances, including water, move through biological membranes as freely as they move through an aqueous solution. Such unrestricted movement of a substance through a biological membrane is called **passive transport.** The energy for passive transport is the kinetic energy that is inherent in all molecules. That is, passive transport does not require energy from cellular metabolism.

The passive transport of water influences many activities of the cell, including cell growth, structural rigidity, and photosynthesis. Because of its roles in so many processes, the diffusion of water through a selectively permeable membrane has a special name: **osmosis.**

Osmosis is influenced by different water potentials on either side of a membrane, as in the following example. Consider a membrane that is permeable to water but impermeable to glucose. When such a membrane separates two halves of a container, each having a different concentration of glucose, water diffuses by osmosis into the side having the higher glucose concentration (i.e., lower water potential) (fig. 4.8). The net movement of water will stop either when both sides of the container have the same concentration of glucose, or when the force of gravity equals the force of water movement, whichever occurs first. Note that the side that began with a higher concentration of glucose increases in volume.

The foregoing example of osmosis also hints at another feature of this process. Immediately before osmosis begins, water on the **hypotonic** side (low solute concentration) of the membrane has a higher water potential, which means a greater potential to move. Its movement can be prevented, however, if a piston is placed on the **hypertonic** side (high solute concentration), with just enough downward pressure to keep the volume constant. A pressure gauge on this piston measures the force required to maintain a constant volume (fig. 4.9).

The potential of pure water to move into a solution on the other side of a membrane is called **osmotic pressure** (fig. 4.10). Conversely, the potential of the solution to cause osmotic pressure is called **osmotic potential.** Solutions with high concentrations of solutes (i.e., more negative osmotic potential) cause

CHAPTER FOUR *Membranes and Membrane Transport*

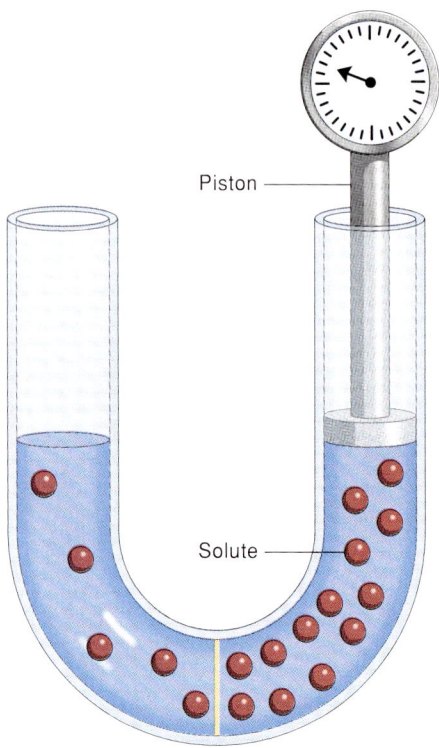

FIGURE 4.9

Higher water potential in the left-hand solution causes pressure for water to move into the right-hand solution. The amount of pressure that is necessary to maintain constant volume on the right equals the force of water movement across the membrane.

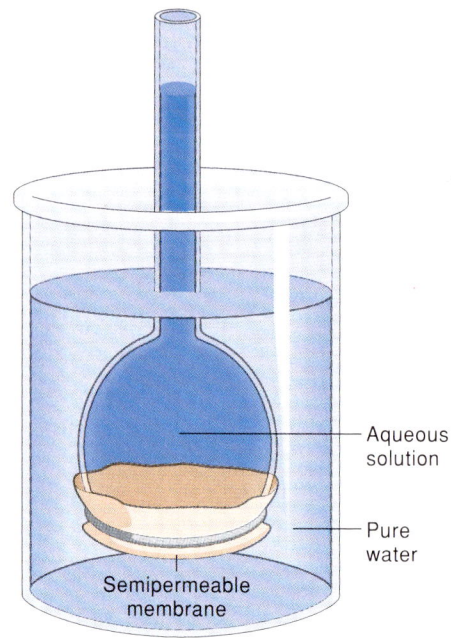

FIGURE 4.10

Osmotic pressure is the pressure of pure water to move into a solution on the other side of a membrane. The osmotic potential of the solution is the potential to cause osmotic pressure.

high osmotic pressure. Osmotic potential and water potential are expressed in the same units of measurement and may be easily confused. However, although osmotic potential occurs only across a membrane, water potential has no such limitation.

Writing to Learn Botany

Suppose you wanted to measure the osmotic pressure of a living cell. How would you do it? Explain the rationale for the techniques you've decided to use. What would be the strengths and shortcomings of the technique you used?

CONCEPT

Osmosis is the diffusion of water across a differentially permeable membrane. The potential for water movement into a cell is continuous because the concentrations of solution are generally higher in the cytoplasm than in the matrix outside the cell. This potential causes pressure, called *osmotic pressure*, from outside the cell. Inside the cell, the counterpart of osmotic pressure is called *osmotic potential*.

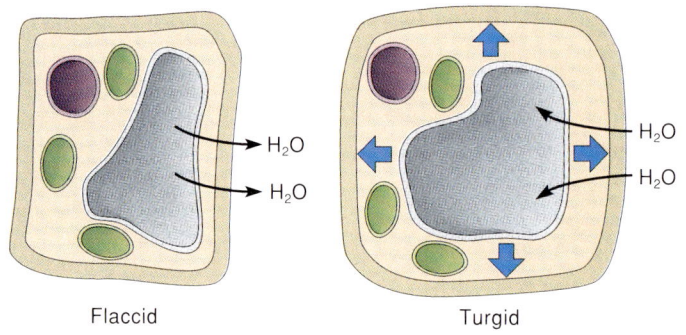

FIGURE 4.11

Flaccid and turgid cells. The plumpness of the turgid cell is maintained by turgor pressure.

Turgor

Most plant cells are surrounded by a hypotonic environment. As a result, cells absorb as much water as they can hold. The outward pressure of the plasma membrane against the cell walls is called **turgor pressure** because it keeps the cells turgid (fig. 4.11).

Turgor pressure is vital to plants in many ways. During growth, cell expansion is caused by turgor pressure on cell walls that have become relaxed. Turgor pressure also keeps herbaceous (nonwoody) plants upright and supports the fleshy stalks and leaves of trees and shrubs, and it keeps supermarket vegetables crisp when they are sprayed with water. Changes in turgor also cause movements in plants, such as the opening and closing of stomata and the curling of grass leaves (fig. 4.12).

FIGURE 4.12

Scanning electron micrograph of an open stoma, ×640. Stomata are open when guard cells are turgid; they are closed when guard cells lose turgidity (see Chapter 21).

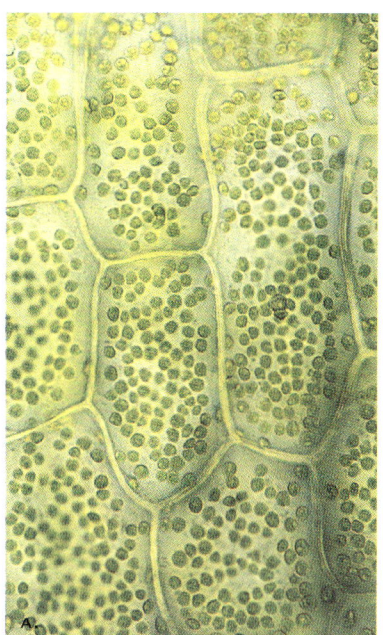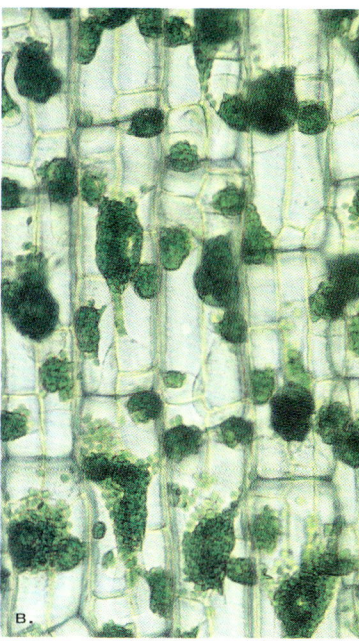

FIGURE 4.13

Light micrographs of (a) turgid cells, ×400, and (b) plasmolyzed cells, ×100, showing the effects of plasmolysis.

Some movements caused by changes in turgor are dramatic, such as leaf movement in the sensitive plant (see Chapter 19).

Cells lose turgor when they are placed in a hypertonic solution. The continued loss of turgor causes the cytoplasm to shrink away from the cell wall (fig. 4.13). Osmotically induced shrinkage of the cytoplasm is called **plasmolysis.** This phenomenon occurs in crop and garden plants when salt accumulates in the soil from extensive use of hard (i.e., mineral-rich) water. It also occurs when people apply too much fertilizer. The loss of turgor in these plants causes their leaves and stems to wilt.

Doing Botany Yourself

Design an experiment to determine which concentration of sucrose causes half of the cells in a plant tissue to plasmolyze. Use this experiment to compare the plasmolytic responses of an aquatic plant, a desert plant, and a plant containing high amounts of sucrose (e.g., sugar beet, sugarcane, sweet fruits). Explain the results of your experiment by suggesting two or three alternate hypotheses.

Inducing Osmosis: The Control of Turgor in Plants

Although water moves freely across biological membranes, plants can control osmosis by regulating the concentrations of solutes in their cells. Loss of turgor causes the uptake of potassium ions (K^+), which are **osmotically active** because they change the cell's osmotic potential. As the concentration of K^+ increases, the osmotic potential increases, and more water enters the cell. Conversely, cells reduce their osmotic potential by secreting K^+. This causes water to leave the cell.

The uptake of K^+ occurs against its concentration gradient; that is, K^+ moves from a region of low concentration outside the cell to a region of high concentration inside the cell. This process, which is called **active transport,** requires metabolic energy (see following page).

Plants that live in high-salt environments such as salt flats, near ocean bays, or inland where oceans once were, accumulate large amounts of osmotically active solutes such as the amino acid proline and the carbohydrate mannitol (fig. 4.14). These organic solutes help the plant absorb water (via osmosis) from dry, salty soil.

CONCEPT

Osmotic potential causes water pressure to push the plasma membrane against the cell wall. This pressure, called *turgor pressure*, keeps the cell turgid. Decreased osmotic potential causes a loss of water and, consequently, a loss of turgor. When turgor is lost, the plasma membrane shrinks from the cell wall in a process called *plasmolysis*. Although membranes allow the free passage of water, turgor pressure is controlled by the active transport of ions or the synthesis of other osmotically active solutes. These processes require energy from cellular metabolism.

FIGURE 4.14

Photograph of the salt bush (*Atriplex occidentalis*) growing in salt flats along the Pacific coast. These plants produce large amounts of osmotically active compounds that enable them to absorb water from the dry, salty soil.

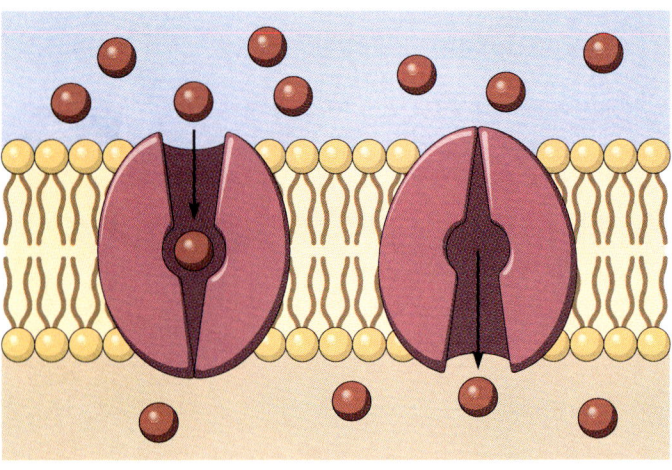

FIGURE 4.15

A possible mechanism for facilitated diffusion. The transport protein (purple) accepts solute molecules (red spheres) on one side of the membrane and releases them on the other side. The transport protein alternates between two forms, depending whether it is "open" to one side of the membrane or the other.

DIFFERENTIAL PERMEABILITY OF MEMBRANES

As already mentioned in this chapter, biological membranes are differentially permeable. This is one of their most important properties, because it keeps metabolically important substances inside the cell or organelle and prevents inappropriate or toxic substances from entering. These substances include ions and larger polar molecules, such as sugars, which only pass through specific membrane proteins called **transport proteins.**

There are two main types of transport proteins. One type includes proteins that work by active transport, which requires energy from ATP to move solutes up a concentration gradient; the other includes passive transport proteins, which do not require metabolic energy. In passive transport, proteins merely act as channels for the diffusion of certain molecules down their concentration gradients. Each passive transport channel is specific for one or two solutes. In contrast to simple diffusion through the phospholipid bilayer, passive transport through protein channels is called **facilitated diffusion.** As in simple diffusion, potential energy is also released during facilitated diffusion.

Facilitated Diffusion

Like simple diffusion, facilitated diffusion is driven by a concentration gradient. Solutes move through transport proteins from the hypertonic side of the membrane to the hypotonic side of the membrane (fig. 4.15). Each transport protein forms a continuous, hydrophilic pathway for polar molecules. Some proteins allow only one solute to diffuse at a time, whereas others only work when two solutes move at the same time, by cotransport.

Little is known about how transport proteins work. The best guess is that they alternate between two forms. One form of the protein accepts a solute molecule on one side of the membrane, which changes the protein to the other form. That second form of the transport protein releases the solute on the other side of the membrane. According to this model, we must also assume that "empty" proteins flip-flop randomly between the two forms. Because of their greater number, molecules on the hypertonic side of the membrane would have more frequent contact with transport proteins than those on the hypotonic side of the membrane; thus, solutes would move down their concentration gradient.

Sugars typically move by facilitated diffusion that involves cotransport with another solute. For example, sucrose moves into conducting cells of leaf veins by hitching a ride with hydrogen ions (fig. 4.16): the energy for sucrose transport comes from the force of H^+ diffusion. The force of H^+ diffusion can be powerful enough to move sucrose against its own concentration gradient.

Active Transport

Many substances move into or out of cells and organelles against a concentration gradient without the aid of facilitated diffusion by cotransport. The uptake of potassium, mentioned earlier in this chapter, is one example. Likewise, marine algae secrete sodium, even though the sea water surrounding them is much saltier than their cytoplasm. In both cases, the transport of solutes requires energy from the cell to overcome the energy of thermal motion that drives passive transport. This energy-requiring process is called *active transport*.

The energy required for active transport usually comes from the hydrolysis of ATP. This reaction is catalyzed by membrane-bound enzymes called *ATP phosphohydrolases (ATPases)*, which are transport proteins that use the energy of ATP hydrolysis (fig. 4.17). Many ATPases actively transport ions against the ion's concentration gradient, thereby creating potential energy for the passive cotransport of other solutes back across the

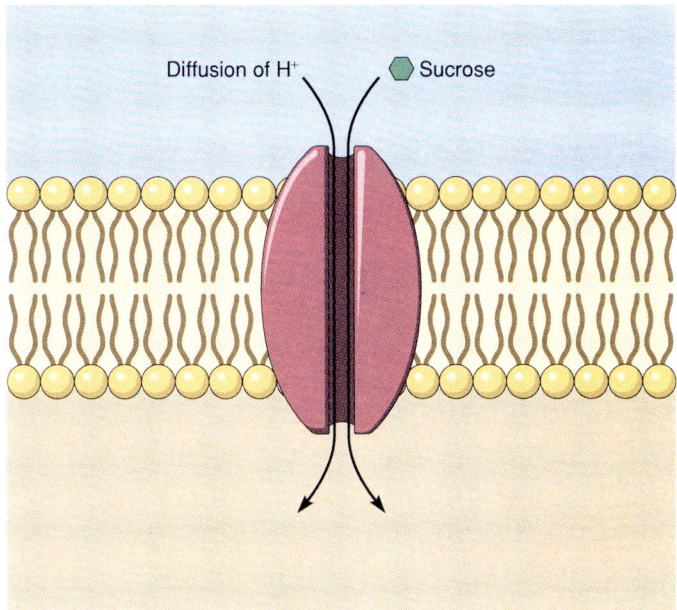

FIGURE 4.16

Cotransport across membranes uses energy from the force of diffusion of one solute (H^+) to move another solute (sucrose) against its concentration gradient.

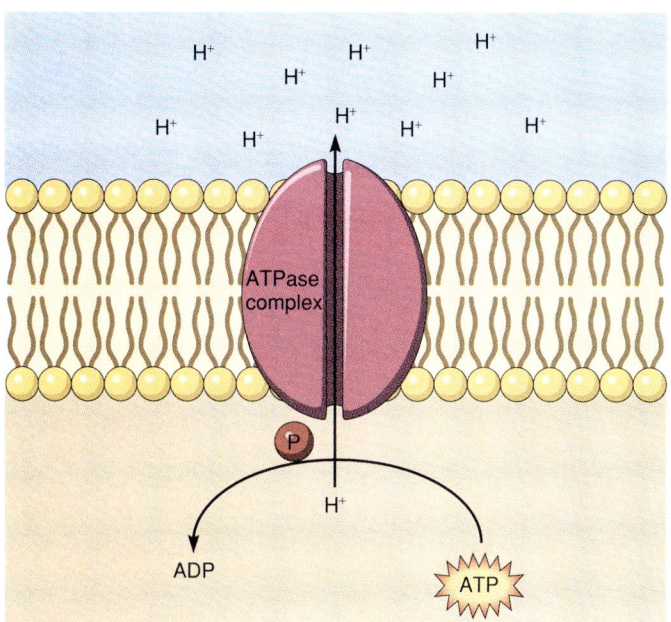

FIGURE 4.17

Active transport uses energy released by the hydrolysis of ATP by ATPases in the membrane. This energy is spent in transporting ions (H^+) against their concentration gradient. Phosphate (P) from ATP binds to ATPases during hydrolysis.

membranes. The cotransport of sucrose with H^+, described, depends on a higher concentration of H^+ outside the cell. This gradient is maintained by the active transport of H^+ across the plasma membrane. This is an example of a **coupled cotransport system,** so-called because it uses energy from active transport to create a gradient that drives the passive cotransport of two solutes.

In addition to a concentration gradient, H^+ and other ions also have an electrical gradient because they are charged particles. Thus, ion transport is also influenced by an electrical gradient. The combination of the concentration gradient and the electrical gradient of ions is called an **electrochemical gradient.** The effects of electrochemical gradients on ion transport are described more fully later in this chapter.

C O N C E P T

Ions and large polar molecules are prevented from diffusing through the lipid bilayer of membranes. Instead, membranes selectively allow these kinds of solutes to pass through transport proteins. The movement of solutes through transport proteins, down a concentration gradient, is called *facilitated diffusion*. When solutes move against a concentration gradient, metabolic energy from ATP is required for *active transport*.

Bypassing Membrane Transport

Simple diffusion, facilitated diffusion, and active transport all entail direct movement through the phospholipid bilayer or

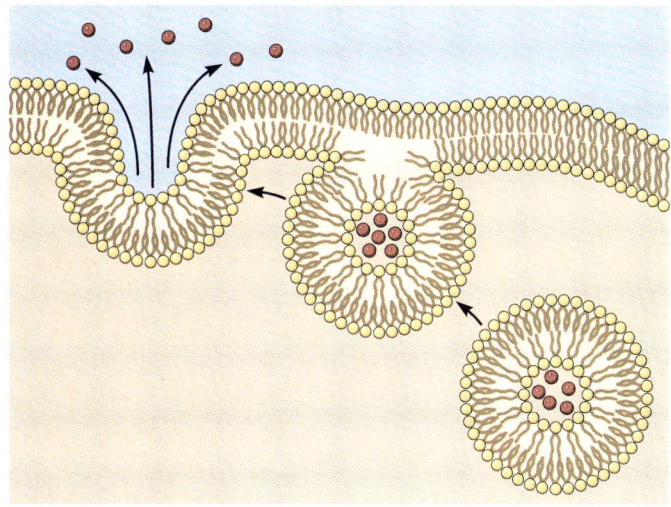

FIGURE 4.18

Exocytosis transports large molecules out of cells. This kind of transport involves membrane-bound vesicles that fuse with the plasma membrane.

through proteins embedded in it. Nevertheless, membrane transport is often bypassed by **exocytosis** (fig. 4.18).

Plant cells secrete polysaccharides and proteins across the plasma membrane for assembly into cell walls. Moreover, cells of root tips secrete a slimy polysaccharide that lubricates their passage through soil as they grow, and cells covering leaves exude waxy substances onto their surfaces to inhibit water loss. Leaves of the Venus's-flytrap and other insectivorous plants

BOXED READING 4.1
MEMBRANE TRANSPORT AND MAKING BEER

Malting is the process of soaking grains of barley (*Hordeum vulgare*) in water to induce seed germination and starch hydrolysis. Malting is of interest to plant biologists because of its importance in brewing beer and because it is a model system for studying membrane transport. During malting, different solutes move by a variety of transport mechanisms. The result is that glucose is provided to the rapidly growing embryo. The accompanying photograph and diagram show where the principal steps of malting occur in a barley grain.

The primary enzyme that digests starch is alpha-amylase. Synthesis of this enzyme begins when the embryo sends a chemical signal to the **aleurone layer** of the grain. This chemical signal is a gibberellin, a plant hormone (see Chapter 18). Thus, the role of gibberellin in seed germination is to "tell" aleurone cells to make alpha-amylase. The enzyme is then transported to the **endosperm** where starch is stored. Finally, starch there is hydrolyzed by alpha-amylase into maltose units (dimers of glucose) and then by maltase into glucose units, which are carried to the embryo.

Membrane transport moves gibberellin from the embryo to the aleurone layer, alpha-amylase from aleurone cells to the endosperm, and glucose from the endosperm to the embryo. The movement of alpha-amylase and glucose requires energy, but little is known about the movement of gibberellin. Studies of barley-seed germination generally begin when gibberellin is applied to grains whose embryos have already been removed.

A.

BOX FIGURE 4.1

(a) A germinating barley grain. (b) Alpha-amylase is synthesized in the aleurone layer and then transported to the endosperm, where it hydrolyzes starch.

When gibberellin induces the synthesis of alpha-amylase, aleurone cells show an increase in the amount of endoplasmic reticulum. Vesicles from the ER move the newly synthesized enzyme to the plasma membrane, where it is secreted by exocytosis. The enzyme must then be transported into cells of the endosperm and into amyloplasts (i.e., where starch is stored). When starch is completely hydrolyzed, the osmotic potential of the amyloplasts increases because, unlike starch, glucose is an osmotically active solute. Glucose probably moves out of the amyloplasts and into the cytoplasm by facilitated diffusion, but its transport from the endosperm to the embryo is hastened by a cotransport system that uses energy from H^+-ATPase.

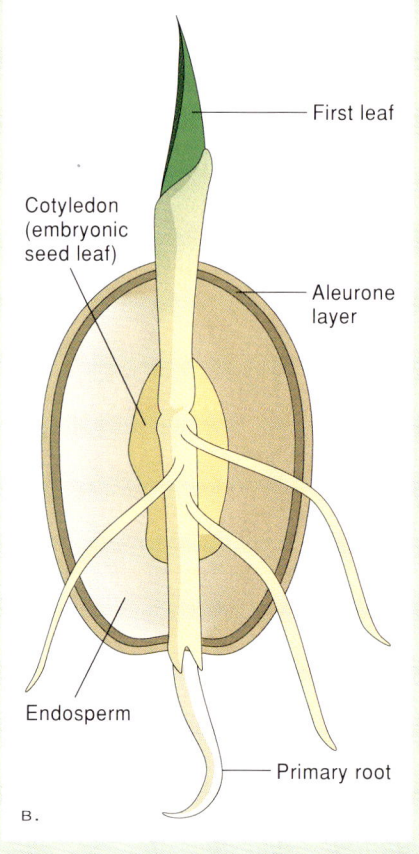

B.

During malting, the temporary buildup of glucose in endosperm cells has a regulatory effect on starch hydrolysis, because glucose inhibits the secretion of alpha-amylase from aleurone cells. This inhibition leads to incomplete malting, which affects the flavor of beer adversely. Commercial breweries add gibberellin during malting to stimulate the production of alpha-amylase, which drives the malting to completion and, ultimately, makes a better tasting beer.

---

secrete enzymes that digest insects. However, unlike the exocytosis of cell-wall materials, which occurs via dictyosome vesicles, the secretion of digestive enzymes relies on vesicles derived from the endoplasmic reticulum.

The movement of starch-digesting enzymes in cereal grains has probably been studied more thoroughly than any other example of exocytosis in plants. The physiology of cereal grains is important to brewers of beer and other beverages. Box 4.1, "Membrane Transport and Making Beer," presents a brief discussion of membrane transport and starch digestion in cereal grains.

Substances can also bypass membrane transport into the cytoplasm by pinching off small coated pits in the plasma membrane. This process, called **endocytosis,** is common in animal cells, but it is not readily observed in plants. Plant cells do have coated pits (fig. 4.19), however, which is indirect evidence for endocytosis.

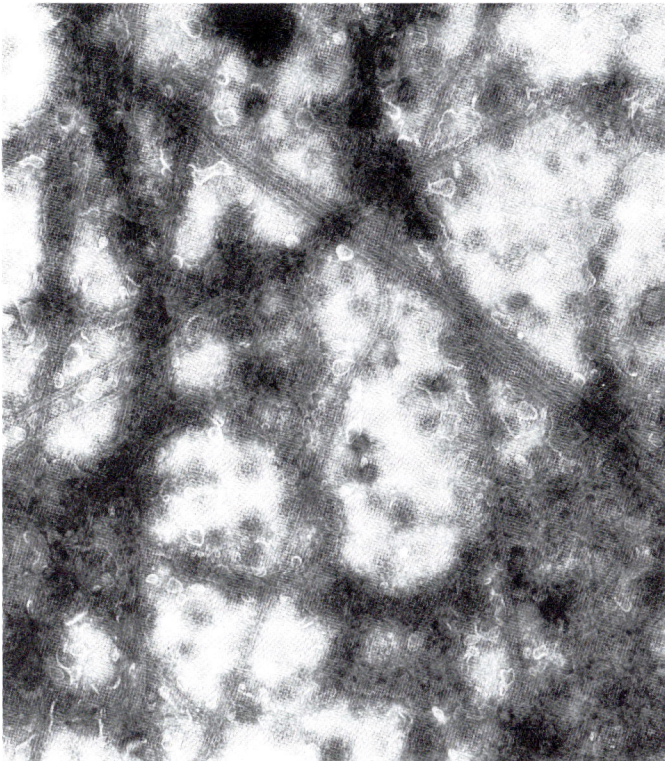

FIGURE 4.19

Transmission electron micrograph of the plasma membrane from an unspecialized tobacco cell in culture, showing coated pits, ×67,000. The occurrence of coated pits is indirect evidence for endocytosis in plant cells.

Endocytosis is apparently more difficult in plants than in animals, because the plasma membrane of plant cells is usually pressed against the cell wall by turgor pressure. This turgor pressure hinders the plasma membrane from invaginating into the cytoplasm.

CONCEPT

The fusion of membrane-bound vesicles to the plasma membrane, and the expulsion of their contents, is called *exocytosis*. Exocytosis bypasses the diffusion and transport mechanisms of membranes. The reverse process, where the plasma membrane invaginates to release vesicles into the cytoplasm, is called *endocytosis*.

MOVEMENT OF IONS ACROSS MEMBRANES

Plasma and organellar membranes have unequal concentrations of negatively charged ions (*anions*) and positively charged ions (*cations*) on one side versus the other. For example, the cytoplasm has a higher concentration of anions and a lower con-

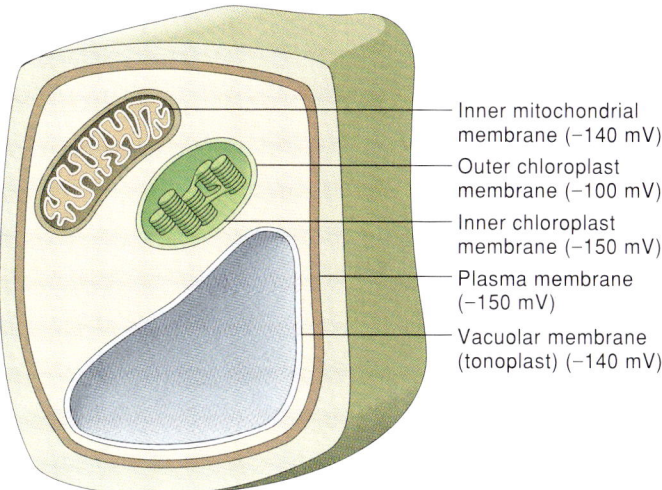

FIGURE 4.20

Membrane potentials of different membranes in a leaf cell. Potentials are relative to the extracellular matrix at zero millivolts (mV; 1mV = 10^{-3} volt).

centration of cations than does the matrix of the cell wall. This unequal distribution of ions creates an electrical gradient that is analogous to a concentration gradient. However, because an electrical gradient is based on electrical charge, the diffusion force of a charge is an electrical potential instead of a chemical potential. Because membranes selectively control the passage of ions, this electrical potential is called the **membrane potential.** Like any other electrical potential, membrane potential is measured in volts. Figure 4.20 shows some typical membrane potentials for a leaf cell. Note that membrane potentials are negative to indicate that the cytoplasm or organelle has more anions than does the matrix outside the cell or organelle.

What are Botanists Doing?

Look through two or three general biology textbooks for their discussions of membrane potential. Then use reference materials at your library to get a general idea of why botanists are interested in membrane potential. Compare the textbook coverage with the information you obtained from the library.

Ion Pumps

Membrane potentials are maintained by proteins that actively transport ions. A membrane protein that pumps ions is called an **electrogenic pump** because it generates voltage across a membrane. Different ions are pumped by different proteins, but the main electrogenic pumps of plants are proton pumps, the H^+-ATPases (fig. 4.17). One function of proton pumps, discussed earlier in this chapter, is to provide energy for the coupled cotransport of uncharged solutes such as sucrose. Another function is to regulate pH. Chemical reactions in a cell often

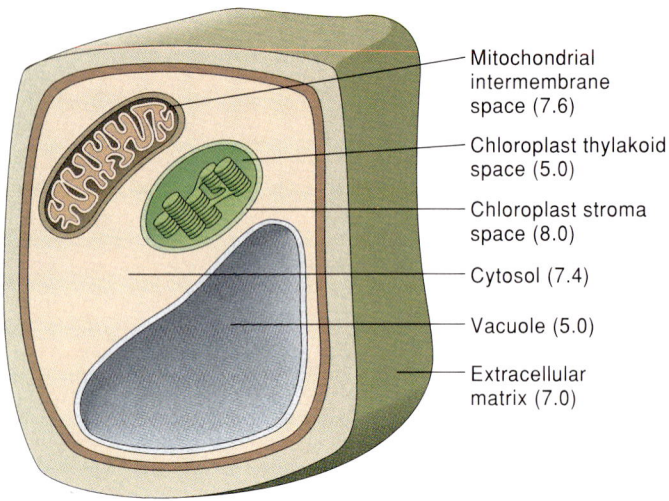

Figure 4.21

The pH of different compartments in a leaf cell. Proton pumps use energy from the hydrolysis of ATP to move protons out of the cell (raises pH of the cytosol) or into the vacuole (lowers pH).

incorporate or release ions that affect the pH of cells; the uptake of ions from soil also affects pH. Pumping protons out of the cell keeps the cytoplasm at a constant pH of about 7.4. Similarly, pumping protons into vacuoles keeps the pH there at about 5.0. This low pH is ideal for enzymes that break down organic compounds that are dumped into vacuoles for disposal. Proton pumps also regulate the pH of other organelles (fig. 4.21).

Proton pumps also influence cellular elongation. When H^+-ATPases in the plasma membrane are stimulated, the outward transport of hydrogen ions decreases the pH in the surrounding cell wall; this causes certain enzymes in the cell wall, which are activated at a lower pH, to begin to degrade cellulose microfibrils. This degradation loosens the cell wall, thereby allowing the cell to expand because of turgor pressure. Loosening of the cell wall can also be induced by applying **auxin,** a plant hormone. Physiologists suspect that auxin stimulates cellular elongation by stimulating the proton pump. Auxin and other hormones are discussed in more detail in Chapter 18, "Plant Hormones."

ATP Synthesis

There are two types of proton pumps. One type uses ATP and occurs mainly in the plasma membrane and in the tonoplast; the other produces ATP and occurs in the membranes of mitochondria and chloroplasts. In these ATP-synthesizing organelles, ATP is made from ADP and a phosphate group when the diffusion of H^+ down its electrochemical gradient releases energy (fig. 4.22). This is the opposite of what happens in a proton pump that is driven by ATP. However, ATP is made only when a gradient of H^+ already exists; therefore, energy must be used to maintain this gradient. In chloroplasts, the energy for such a gradient comes from light energy during photosynthesis. In mitochondria, the energy comes from the rearrangement of chemical bonds during respiration. Both of these processes are explained more fully in Unit 2, "Plants and Energy."

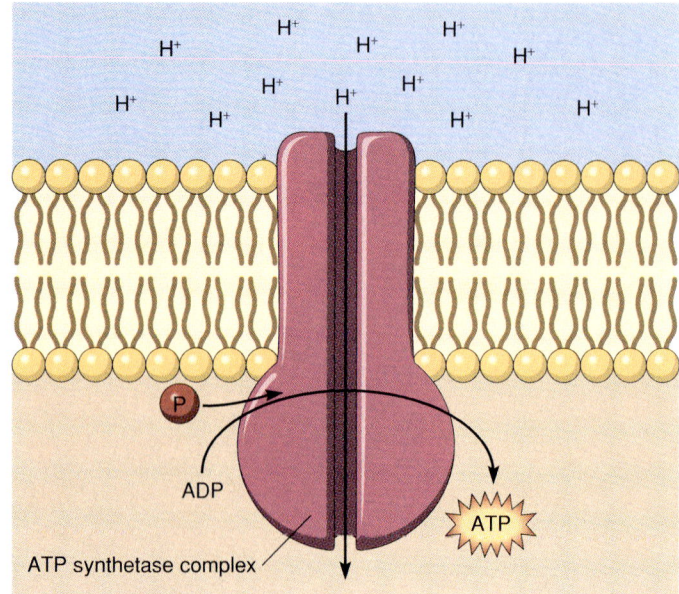

Figure 4.22

Membranes of chloroplasts and mitochondria use energy stored in a proton gradient to make ATP from ADP and phosphate (P). Energy is harvested from protons as they diffuse down their concentration gradient, through ATP synthase complexes in the membranes.

CONCEPT

The concentrations of ions are different on either side of a membrane. This difference produces an electrical potential, called *membrane potential*, which is measured in volts. Membrane potential is maintained by proteins in the membrane that actively transport ions against their electrical gradient. Protons are pumped by ATPases that use the energy of ATP in the plasma membrane and tonoplast. In contrast, proton pumps in mitochondrial and chloroplast membranes produce ATP.

CELLULAR COMMUNICATION

Cells in a complex organism interact with their environment (e.g., gravity—see Chapter 19), with one another, and with the cells of other organisms. Cell-to-cell interactions occur when chemical or electrical signals released from one cell are received by another, where they change some aspect of metabolism. Auxin is an example of an internal chemical signal—that is, a signal that moves from cell to cell in the same plant. External signals are those that pass between different organisms, for example, between plants and bacteria or fungi (fig. 4.23).

The reception of chemical signals and the transmission of their messages are important functions of proteins in

FIGURE 4.23

Auxin is an example of an internal signal between cells. Microbial polysaccharides may function as a signal from bacteria to roots.

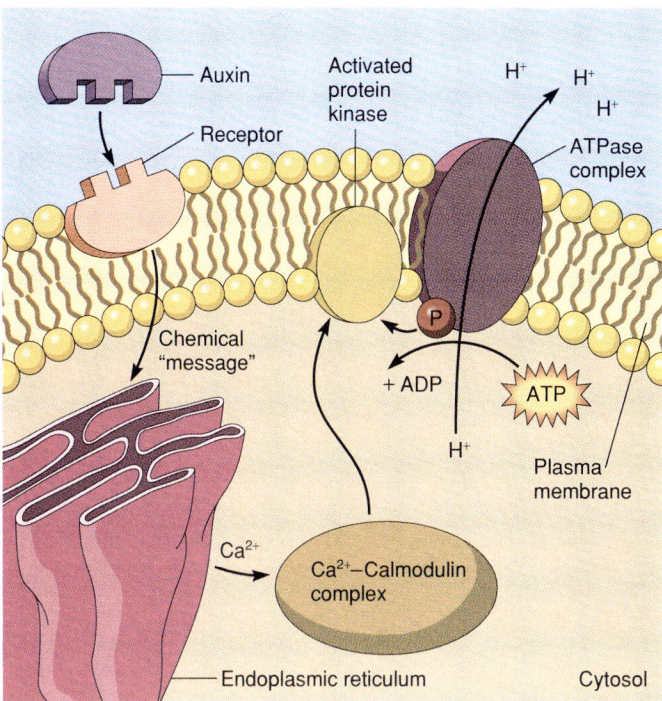

FIGURE 4.24

Auxin can induce and amplify proton pumping. Possible role of auxin in amplifying proton pumping: one molecule of auxin starts a series of signals that results in the activation of protein kinase. One activated molecule of protein kinase removes several phosphates from ATPases, thereby enabling hydrolysis of more ATP molecules.

membranes. Studies of signal transduction in plants have focused on the role of calcium ions (Ca^{2+}) and **calmodulin,** a protein that is activated when it binds to calcium. In its active form, the Ca^{2+}-calmodulin complex activates enzymes in membranes, essentially telling them to get to work. As much as 2% of the plasma membrane may be calmodulin.

An example of how calmodulin works is its role in transmitting a chemical "message" from auxin. The message causes the endoplasmic reticulum to release Ca^{2+}, which activates calmodulin. Calmodulin, in turn, activates protein kinase, which is an enzyme that activates or inactivates still other enzymes by transferring phosphates to and from them. Protein kinase removes phosphate groups from, for example, ATPases. The ATPases can then accept more phosphates and hydrolyze more molecules of ATP to ADP. In this case, the result is that energy is harvested to fuel the proton pump. This series of steps in signal transduction may seem unreasonably complicated, but it is effective in amplifying proton pumping. In this example, a single molecule of auxin can, by causing the activation of a series of enzymes, influence many protons to be transported across the membrane (fig. 4.24).

Hormone Receptors

Signal transduction in plant cells begins when a hormone binds to a receptor protein on the plasma membrane. Plants make several different hormones, each of which must be recognized by a different receptor. Studies of plant hormone receptors have concentrated on auxin receptors because auxin has so many effects on plant growth and development (see Chapter 18). Each auxin receptor causes different metabolic changes, depending on where it occurs in a plant. Furthermore, the amount of binding varies from one tissue to the next; for example, auxin receptors in leaf stalks bind more than one hundred times more auxin than do receptors in fruits. The multiple characteristics of hormone binding mean that there are probably many different receptors for auxin, as well as for each of the other plant hormones. Little is known about how they work or what they have in common, but the subject of hormone reception and signal transduction is an active area of botanical research.

Membrane Interactions with Other Organisms

Each plant is surrounded by other organisms, including animals, bacteria, fungi, and other plants, and interactions are common among plants and many of the organisms in their environment. Most of our knowledge about these kinds of interactions concerns the effects of microbes that cause plant diseases.

Two kinds of molecules are commonly secreted by microbes that cause plant diseases: small polypeptides and small polysaccharides. For example, a toxic polypeptide called syringomycin is secreted by *Pseudomonas syringae*, a species of bacteria that infects corn, beans, stone fruits (cherries, peaches, etc.) and many other plants. Syringomycin alters the membrane potential of the plasma membrane, probably by inducing protein kinase to phosphorylate the ATPase. This saturates ATPase with phosphates and inactivates the proton pump. Inactivated proton pumps allow protons to diffuse back into the cytoplasm. These and other effects of syringomycin are poorly understood, but they ultimately cause the membrane to leak and disintegrate. Enzymes from the bacteria digest the cell as it disintegrates.

To fight infections, the plasma membranes of many plant cells can send signals into their own cytoplasm that induce a defensive response. For example, polysaccharides secreted by the fungus *Colletotrichum lindemuthianum* cause cells in kidney bean leaves to make several different kinds of chemicals that inhibit the growth and spread of the fungus. Among these chemicals are phenolics that inhibit fungal growth. Cells at the edge of the infection also make lignin, which is indigestible to the fungus and limits its spread. Although the mechanisms for this response are not well understood, the process probably starts when the plasma membrane recognizes the fungal polysaccharides and sends a signal to the nucleus. This signal activates genes that control the synthesis of lignin, phenolics, and many other products that help to defend the plant. The result is that the leaf inhibits the growth of the fungus in the infected cells and blocks fungal growth at the edge of the infection.

Lectins and Glycoproteins

Proteins that bind to carbohydrates on cell surfaces are called **lectins.** Many lectins are glycoproteins. Lectins occur in all parts of the cell and in different kinds of tissues, and are usually associated with the endoplasmic reticulum and other membranes, including the plasma membrane. A lectin recognizes a specific carbohydrate on the basis of which monomers the carbohydrate contains and the linkages between them. Because of this specificity, plant lectins can recognize the cell-wall carbohydrates of specific bacteria and fungi.

The Lore of Plants

Plants can be used to determine your blood type. Lectins from jack beans, lima beans, and lotus bind to glycoproteins on the plasma membranes of red blood cells. Because the cells of different blood types have different glycoproteins, cells of each blood type bind to a specific lectin. This is one example of the many clinical or research applications of plant lectins in human medicine.

FIGURE 4.25

Photograph of roots of pea plants with bacteria-containing nodules. The infection of roots by bacteria is enabled by root-hair lectins that bind to bacterial cell walls.

Cellular recognition is perhaps best understood among plants in the legume family and bacterial species of the genus *Rhizobium*. For example, lectins in the root hairs of white clover (*Trifolium repens*) bind only to the cell walls of *Rhizobium trifolii*. Once bound to the root hairs, these bacteria infect the roots and cause them to swell into nodules (fig. 4.25). The interaction between root-nodule bacteria and plants is mutually beneficial because the organisms trade metabolites: the plants supply carbohydrates for bacterial respiration, and the bacteria convert nitrogen from the air into ammonia, a metabolically useful form of nitrogen for plants.

Proteins, most often glycoproteins, also have important roles in sexual reproduction. For example, they enable the reproductive cells of some algae to recognize their appropriate

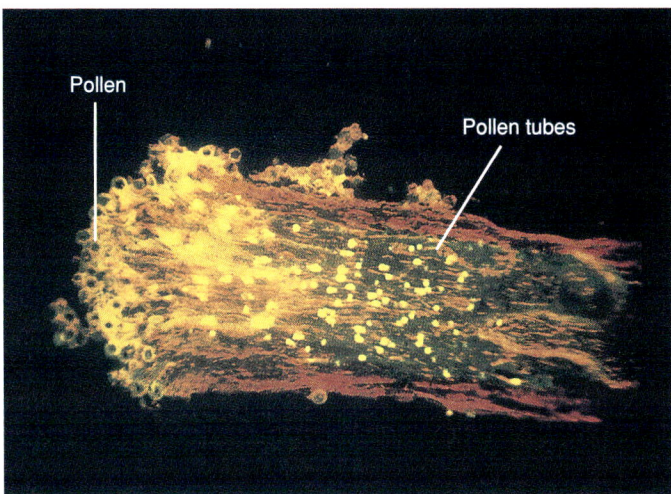

FIGURE 4.26

Light micrograph of fluorescent-stained pollen tubes growing through the stigma, ×20.

mates for fertilization. In flowering plants, reproductive compatibility is controlled by interactions between the glycoproteins of pollen, which carry the sperm, and the glycoproteins of stigmas, which are borne on the seed-producing organ. If their respective glycoproteins fit together appropriately, pollen tubes grow normally, and fertilization can occur; if they do not fit together, the pollen tubes grow irregularly and incompletely, and fertilization does not occur. In some cases, glycoproteins of the pollen and stigma within the same flower do not fit together, which makes the plant self-incompatible (fig. 4.26).

C O N C E P T

Cells communicate with other cells by chemical signals. Receptor proteins on plasma membranes are activated by binding to hormones and other chemicals. The activated proteins induce enzymes and other proteins to change the metabolism of the cell. Membrane surfaces also recognize specific secretions or cell-wall components from other organisms. Glycoproteins and other proteins are probably the most common receptors for recognizing other cells by their carbohydrates. Glycoproteins also have a role in controlling sexual reproduction.

Chapter Summary

Membranes consist of two main components: phospholipids and proteins. The structure of membranes is best described by the fluid mosaic model. The phospholipids form a fluid bilayer, with the hydrophobic tails of fatty acids at its core and the hydrophilic heads of phospholipids on both sides. Proteins move in the fluid. The lipid layer allows the unrestricted diffusion of small molecules across the plasma membrane, but membrane lipids block the diffusion of ions and large polar molecules. This property of membranes is called *differential permeability*. The diffusion of hydrophilic solutes is facilitated by proteins embedded in the membranes.

The diffusion of water through a differentially permeable membrane is called *osmosis*. Plants regulate osmosis by controlling the uptake of ions or by making osmotically active solutes. In so doing, cells maintain turgor pressure against the cell walls. Because turgor pressure prevents further uptake of water, the water outside the cell exerts pressure, called *osmotic pressure*. Its counterpart inside the cell is the osmotic potential of the cell, which is expressed as a negative number.

Osmotic potential results from the potential of water to diffuse down its concentration gradient. This concentration gradient of water is controlled by the uptake or synthesis of osmotically active solutes, which requires energy from the hydrolysis of ATP. The energy-requiring transport of solutes is called *active transport*. The most common actively transported solutes are ions, especially hydrogen ions. Transport proteins for hydrogen ions are also called *ATPases* because they use energy from the hydrolysis of ATP. The lipid bilayer and membrane proteins are bypassed by exocytosis, the process by which proteins and other large molecules are secreted in vesicles that fuse with the plasma membrane.

Ion transport is affected by the concentration gradient of ions and by the electrical gradient of their charges. Together, these two gradients create an electrochemical gradient across a membrane. The electrical component of this gradient is measured in volts and is called the *membrane potential*. Ion transport proteins, called *electrogenic pumps*, maintain the membrane potential. The most common electrogenic pumps are proton pumps. In the plasma membrane and tonoplast, proton pumps use the energy of ATP to move protons against an electrochemical gradient. In the membranes of mitochondria and chloroplasts, proton pumps are reversed—that is, they produce ATP. The ATP required to run these pumps comes from respiration (in mitochondria) or photosynthesis (in chloroplasts).

Membrane proteins are also receptors for hormones and other chemicals from other cells or organisms. As receptors, these proteins signal other parts of the cell to change their metabolism. These changes contribute to cell growth, defense against disease, recognition between mating cells, and many other aspects of plant life.

Questions for Further Thought and Study

1. How does the outside surface of the plasma membrane differ from the inside surface? In what ways does this difference reflect the functions of the plasma membrane?

2. How is the Davson-Danielli model of membrane structure deficient in explaining what we now know about membranes?
3. What features of phospholipids account for the following characteristics of a biological membrane: (a) its double layer; (b) its fluidity; (c) its hydrophobic and hydrophilic properties; (d) its selective permeability?
4. Can you explain why glucose is osmotically active and starch is not?
5. Certain brands of water purifiers for the home clean water by "reverse osmosis." How do you think this process might relate to osmosis?
6. Why do you think ATP synthetase complexes are often referred to as ATPases?
7. Some botanists think that the response of kidney beans to infection by *Colletotrichum* is evidence that plants have an immune system. How can you support or refute that idea?

Suggested Readings

ARTICLES

Bretscher, M. S. 1985. The molecules of the cell membrane. *Scientific American* 253 (October):100–108.

Lavenda, B. H. 1985. Brownian motion. *Scientific American* 252(2):70–84.

Longenecker, N. E., and E. T. Hibbs. 1986. Active transport. *The American Biology Teacher* 48:304–306.

Maloney, P. C., and T. H. Wilson. 1985. The evolution of ion pumps. *BioScience* 35:43–48.

Nelson, N., and L. Taiz. 1989. The evolution of H^+-ATPases. *Trends in Biological Sciences* 14:113–116.

Satir, B. 1975. The final stage in secretion. *Scientific American* 233 (October):28–37.

Slayman, C. L. 1985. Proton chemistry and the ubiquity of proton pumps. *BioScience* 35:16–17.

___. 1985. Plasma membrane proton pumps in plants and fungi. *BioScience* 35:34–37.

Unwin, N., and R. Henderson. 1984. The structure of proteins in biological membranes. *Scientific American* 250 (February):78–94.

BOOKS

Gennis, R. 1989. *Biomembranes: Molecular Structure and Function*. New York: Springer-Verlag.

Robinson, D. G. 1985. *Plant Membranes: Endo- and Plasma Membranes of Plant Cells*. New York: John Wiley & Sons, Inc.

Tosteson, D. 1989. *Membrane Transport: People and Ideas*. London: Oxford University Press.

UNIT TWO

Plants and Energy ...

Corn is the most widely planted crop in the United States. To people who have driven across the midwestern United States (also called the Corn Belt) during the summer, this is no surprise, for in many areas there is little except mile after monotonous mile of corn. These crops cover a combined area about the size of Arizona and produce about 8 billion bushels of corn per year. According to one mathematically minded urbanite, that's enough corn to bury all of Manhattan Island about 5 meters deep.

Corn is the most efficient of all major grain crops; that is, corn is the champion at converting sunlight to sugars that the plant stores as starch in grain. We use these grains for a variety of purposes. For example, we feed about half of our corn to livestock; another 25% of the crop is exported. About 10% of the crop—optimally within 5 minutes of picking—appears on our dinner tables. The rest of the crop passes into an enormous number of peripheral products, including corn syrup, corn meal, corn oil, cardboard, crayons, firecrackers, aspirin, wallpaper, gasohol, pancake mix, shoe polish, ketchup, chewing gum, marshmallows, soap, and corn whisky.

Nearly a fourth of the Calories in people's diets can be traced to corn. Where do those Calories come from? How does corn convert this energy to sugars? How do corn and other organisms extract this energy from the sugars?

In this unit you'll learn about energy and its use by plants. After an introduction to the laws governing energy conversions, you'll learn about the most important set of chemical reactions on earth: photosynthesis. These are the reactions that convert sunlight to chemical energy, thus providing fuel for virtually all other organisms. You'll also learn how organisms—plants included—extract the energy trapped in sugars and other compounds for work. In the process, you'll come to appreciate one of the themes of life: that all organisms, in addition to requiring energy to stay alive, exchange energy with their environment.

Like all organisms, plants convert energy from one form to another. This corn (*Zea mays*) plant uses photosynthesis to convert sunlight to chemical energy. This chemical energy is then used to fuel the plant's growth and development.

CHAPTER 5

Energy and Its Use by Plants

Chapter Outline

INTRODUCTION
WHAT IS ENERGY?
MEASURING ENERGY

BOX 5.1
EARTH'S ENERGY

ENERGY CONVERSIONS
THE LAWS OF THERMODYNAMICS
    The First Law of Thermodynamics
    The Second Law of Thermodynamics
METABOLISM: ENERGY FOR LIFE'S WORK
FREE ENERGY
    Free Energy and Chemical Equilibrium
    Oxidation, Reduction, and Energy Content
ATP: THE ENERGY CURRENCY OF CELLS
    Coupled Reactions
    Other Compounds Involved in Energy Metabolism
ENZYMES AND ENERGY
    Regulating Metabolism
THE MAJOR ENERGY TRANSFORMATIONS IN PLANTS:
    PHOTOSYNTHESIS AND RESPIRATION
    The Flow of Energy
Chapter Summary
Questions for Further Thought and Study
Suggested Readings

Chapter Overview

Although the word *energy* is common today, it was not coined until about two hundred years ago when the Industrial Revolution shifted our concept of energy from horses (i.e., horsepower) and falling water to combustion engines. Thereafter, a quantitative understanding of the concepts of work and energy became not only practical, but essential. Biologists then realized that these concepts could help them understand how organisms function. Applying the concepts of energy to living organisms began the study of **bioenergetics,** a fascinating discipline that helps us understand life.

Today, few things are more important than energy: countries fight over oil supplies, and accidents at nuclear power plants endanger lives. Everything involves energy, including music, games, tides, seasons, and the orbits of planets. Plants and animals are no exceptions: all aspects of their lives, ranging from maintenance and repair to exotic tasks such as producing beautiful flowers, require energy. Consequently, we can't understand or appreciate plants unless we know something about energy, including what it is and how it is used.

In this chapter, you'll learn how organisms use energy for their various activities. These principles will help you understand the topics of the following two chapters, respiration and photosynthesis.

INTRODUCTION

Metabolism is a bit like a capitalist economy: labor is hired and paid for, products are made from raw materials that have to be bought, and goods are packaged and shipped to different locations by a service industry that requires compensation. However, the gold standard of metabolism is energy: plants harvest energy from the sun, convert it to a metabolically useful form, and move it in different forms ("currencies") within cells according to an energy budget. Energy is needed for active transport across membranes, for biosynthesis of large molecules from smaller ones, and for moving molecules through the cytosol and into and out of organelles and cells.

In this chapter we look at what energy is and at the different forms it takes in plants. Some of the discussion is general, and some applies only to living organisms. The latter is referred to as *bioenergetics* because it involves the compounds of energy metabolism.

WHAT IS ENERGY?

Energy is the ability to do work—that is, to bring about change or move matter against an opposing force such as gravity or friction. Because energy is an *ability* to do work, it is not always as obvious to us as matter, which has mass and occupies space. We describe energy according to how it affects matter. Humans use energy for conspicuous activities such as dancing, mowing the lawn, playing baseball, and studying plants. However, plants expend most of their energy in subtle, nearly unrecognizable ways. For example, consider the philodendron (*Philodendron*) plants shown in figure 5.1. Their large leaves slowly gather the energy available in sunlight and use it to fuel their metabolism and growth. However, the calm existence of these plants changes when they reproduce. Their large flowers open for only a couple of days. At night, when air temperatures often hover near freezing, their flowers can reach temperatures exceeding 46° C/115° F (by comparison, butter melts at 30° C/88° F). These furnacelike flowers maintain their high temperature for many hours in the cold night air. Plants such as skunk cabbage and voodoo lily also generate large amounts of heat. This heat may be used to help the plants reproduce. The heat produced by voodoo lily, for example, helps to disperse compounds that smell like dung or rotting flesh. The insects attracted to the plants by these odors assist in pollination. Thus, understanding the bioenergetics of the plant helps us understand how they live.

MEASURING ENERGY

Since energy exists in many forms, it's not surprising that energy is described with many units. Most scientists measure energy in calories (cal) or joules (J). A **calorie** (**cal;** note the small *c*) is the amount of energy required to raise the temperature of 1 gram of water by 1° C. The most common unit for

FIGURE 5.1

The clublike spadix in the center of this *Philodendron* flower heats up to temperatures as high as 46° C.

BOXED READING 5.1
EARTH'S ENERGY

The energy available on earth can be traced to the sun (e.g., fossil fuels, biomass, wind, and incoming radiation), cosmic evolution preceding the origin of our solar system (nuclear power), lunar motion (tidal power), or the earth's core (geothermal power). Of these, sunlight provides the most energy for organisms. Sunlight originates inside the sun, where temperatures of 10,000,000° C fuse hydrogen to form helium and release gamma rays, which then produce electrons and photons:

H + H → He + gamma rays + energy
gamma rays → electrons + photons + energy

Life depends on transforming the energy contained in photons of sunlight into chemical energy before it is transformed to heat. The sun's total energy output is about 3.8 sextillion (3,800,000,000,000,000,000,000) megawatts of electricity. Earth intercepts only about 1/2,000,000,000 of this energy; this amounts to the energy-equivalent of about 2×10^{14} tons of coal per year. Of this,

- Thirty percent is reflected back to space. This is what astronauts see while orbiting the earth.
- Fifty percent is absorbed, converted to heat, and reradiated.

Nineteen percent powers the hydrologic cycle, creates winds, and drives photosynthesis. Of this 19%, only 0.05 to 1.5% is incorporated into plant material. However, this relatively small amount of energy sustains life. Without photosynthesis, almost all life would quickly disappear from the earth.

The amount of sunlight energy at the earth's surface is about 2 million trillion calories per second, an amount of energy equivalent to that of 400,000,000 of the atomic bombs dropped during World War II.* Stated another way, the annual amount of energy that strikes the earth as sunlight is 15,000 times greater than the world's present supply of energy. This is more than 20 billion times greater than our present rate of energy consumption.

Despite this vast resource of energy, we've learned relatively little about how to use renewable sources of energy such as sunlight. These so-called "alternate sources" of energy differ from oil, gas, and nuclear power because they are plentiful and free of pollution. Today, renewable forms of energy (e.g., hydropower and biomass) supply only 18% of the world's energy needs, while nuclear power supplies 4%. The rest of our energy demands are met by increasingly scarce, nonrenewable sources such as fossil fuel. These demands are huge. For example, to exert the power equivalent to the amount of electricity used by U.S. manufacturers in one week, one worker, doing average manual labor for fifty 40-hour weeks per year, would have to work for 145,193,740 years.

*Every day, the earth receives about 1.5 billion times more energy than is contained in all of the electricity used in the United States each year.

---

measuring the energy content of food and the heat output of organisms is the **Calorie** (**Cal;** note the large C), which is the energy required to raise the temperature of 1 liter of water by 1° C.[1] A **joule** (**J**) is the amount of energy needed to move 1 kilogram through 1 meter with an acceleration of 1 meter per second per second (1 m/sec^2; for comparison purposes, 1 cal = 4.12 J).[2] To help you put these units into better perspective, consider that a slice of apple pie provides enough energy (1.5×10^6 J, or 365 Cal) for a woman to run for an hour or for a typist to enter about 15 million characters on a manual typewriter (almost 11,000 typewritten pages) (fig. 5.2).

---

1. The Calorie (written with a capital C) used to measure the energy content of food is equivalent to 1,000 calories (written with a lowercase c), or 1 kcal.

2. Units to measure energy and power throughout the world (but not necessarily in the United States) include ergs, British thermal units (Btu), watts (W), kilowatt hours (kW h), and horsepower. An erg is the energy needed to move 1 g of mass 1 cm with an acceleration of 1 cm sec^{-2}. A Btu is the energy needed to raise the temperature of 1 lb of water 1° F. A watt is the energy provided when 1 J is used for 1 sec. A kilowatt hour is the energy used when 1 kW is available for 1 h. A horsepower is the energy needed to raise 550 lb 1 ft in 1 sec. Although these different units often bewilder consumers, they are interconvertible. Here are the conversion factors for these units:

1 erg = 10^{-7} J
1 Btu = 1055 J
1 cal = 4.1 J = 0.001 kcal = 0.001 Cal
1 hp = 746 W
1 W = 0.001341 hp

FIGURE 5.2

The energy for all of our activities comes either directly or indirectly from plants.

Here's the energy involved in some common (and a few not-so-common) events:

Severe earthquake (Richter 8)	10^{18} J
Atomic bomb dropped on Hiroshima	8.4×10^{13} J
Man running for 1 h	2.5×10^{6} J
Woman running for 1 h	1.8×10^{6} J
Lethal dose of X rays	7×10^{2} J
Depressing a key of a manual typewriter	10^{-1} J
Chirrup of a cricket	9×10^{-4} J
Wingbeat of a honeybee	8×10^{-4} J
Moonlight on a person's face for 1 sec	8×10^{-5} J
Energy released by splitting one uranium atom	4×10^{-11} J
Energy of a photon within the visible range	2.5–5.1×10^{-19} J

Organisms must continually obtain energy to stay alive—heterotrophs such as humans must eat food, and plants must absorb light for photosynthesis. These transformations of food or light into usable energy occur via energy conversions.

ENERGY CONVERSIONS

All activities—ranging from cellular division and heat production by flowers to home runs and nuclear explosions—involve converting energy from one form to another. For example, we convert the energy contained in oil to electricity, and then convert the electricity into light energy to illuminate our homes, streets, and parks. Similarly, plants convert sunlight into chemical energy that they use to reproduce, repair their DNA, and build new parts. This conversion of light energy to chemical energy is **photosynthesis,** which sustains almost all life on earth. Animals stay alive by eating other animals and/or plants and their stored energy. All aspects of the lives of organisms center on energy and energy conversions.

There are two types of energy: *potential energy* and *kinetic energy* (fig. 5.3). **Potential energy** is stored energy—that is, energy available to do work. Examples of potential energy include a teaspoon of sugar, an unlit firecracker, and a rock atop a hill. Potential energy is determined by the position (e.g., water held at an altitude behind a dam) or arrangement (e.g., the type of chemical bonds) of matter. In organisms, potential (i.e., latent) energy is stored subtly in chemical bonds such as those in sugars, starch, and fats.

Kinetic energy is energy being used to do work. Examples of kinetic energy include burning sugar, an exploding firecracker, a rock rolling down a hill, light emitted by a firefly, and a root forcing its way through soil. Kinetic energy affects matter by transferring motion to other matter, much as a moving billiard ball transfers its kinetic energy to other billiard balls. Similarly, flowing water can be used to turn a turbine, and a growing root can break a concrete sidewalk. Kinetic energy moves objects, whether they be mountains, molehills, or molecules.

Heat is kinetic energy because it involves the movement of molecules. Again study the example shown in figure 5.3.

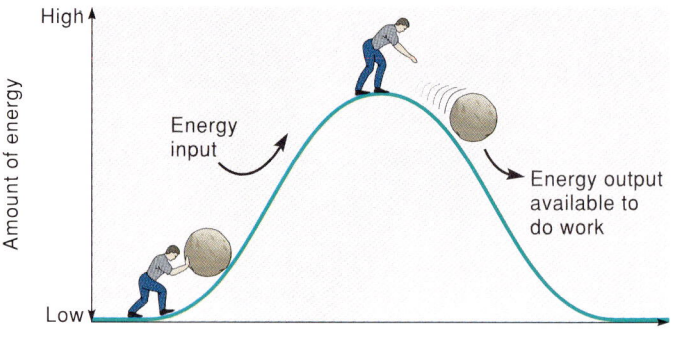

FIGURE 5.3

Pushing a boulder to the top of a hill requires an input of energy because it increases the potential energy of the boulder. The boulder atop the hill has potential energy because it can do work as it rolls down the hill. As the boulder rolls down the hill, potential energy is converted to kinetic energy.

From Postlethwait and Hopson, Nature of Life, 2d ed. Copyright © 1992 McGraw-Hill, Inc., New York, NY. Reproduced with permission of McGraw-Hill, Inc.

The rock atop the hill contains much potential energy—the *capacity* to do work—because of its position, but since it is at rest, it has no kinetic energy. If the rock were given a nudge, it would spontaneously roll down the hill, transforming its potential energy into kinetic energy, which could be used to do work.

Just as the mainspring of a tightly wound watch can propel the hands around the dial, potential energy can be converted to kinetic energy to do work. Likewise, winding a watch transforms kinetic energy from the person winding the watch into potential energy stored in the watch's mainspring. Thus, under most conditions, potential and kinetic energy are freely interconvertible, though not at 100% efficiency.

C O N C E P T

Energy is the ability to do work. Potential energy is energy available to do work, and kinetic energy is energy being used to do work. Under most conditions, kinetic energy and potential energy are freely interconvertible, though not at 100% efficiency.

THE LAWS OF THERMODYNAMICS

Life depends on energy transformations. For example, our bodies transform the chemical energy in food to mechanical energy that enables us to study, play, and dance; and our appliances convert electrical energy to light for reading and to heat for warming our homes. Combustion engines convert the chemical energy in gasoline to mechanical energy that runs our automobiles and mows our lawns. Plants convert sunlight to chemical energy that sustains life on the planet.

Energy transformations are regulated by laws of thermodynamics. These laws involve a system and its surroundings. The collection of matter being studied is called the *system,* and

Some more good stuff. Better grades. Easier class.

Life is good.

1 Conserve your energy.

[**Student Study Art Notebook**
ISBN 0-697-24309-5]

Why work harder than you have to? If you spend more class time sketching pictures from the overhead than you do paying attention, or if you have trouble seeing the writing on the overhead—then the *Student Study Art Notebook* is the thing to make class go by a little easier. We've copied down the overhead pictures into this notepad and left room for your notes, to bring out the scholar—not the artist—in you.

3 Is your writing as good as it could—and should—be?

[**Writing to Learn Botany**
ISBN 0-697-17455-1]

You can't get around it: In today's world, you must write well—no matter what your major. And you can *always* be better. In *Writing to Learn Botany*, author Randy Moore— an acclaimed writer himself—offers friendly, clear suggestions about writing research papers and lab reports. This book's direct references to your text *Botany* will help you during this course, and its broader applications will serve you long after this course is over.

**Contact your bookstore now.
Or call Wm. C. Brown Publishers Customer Service: 800-338-5578.**

2 OK. Here's what you should study.

[**Student Study Guide**
ISBN 0-697-03776-2]

If you always know exactly what to study, never mind. But if you're like most of us, you could use a little extra help. Inside the *Student Study Guide* are learning objectives, concept maps that tie ideas together so they make sense, and fill-in-the-blank and multiple-choice questions—so you can check to make sure you do indeed know it all. It also contains an answer key—just in case you're not entirely sure of the solutions—as well as suggested experiments for you adventurous types.

4 Conquer your science courses.

[**How to Study Science**
ISBN 0-697-14474-7]

Science courses aren't always easy. But they never have to be torture. *How to Study Science*, by Fred Drewes, gives "how to succeed" tips on just about every aspect of science courses—from taking notes and interpreting text figures, to knowing where to sit in class and overcoming "science anxiety."

 Such a deal—
cut your study
time and improve
your grades.

[**Computerized Study Guide,**
by Medi-Sim
ISBN 2X117101]

Focus your study time on only those areas where you need help. This easy-to-use software helps you identify and target those areas. Simply test yourself on each text chapter. The program not only gives you the correct answers and a short rationale for the question, but creates *your own personalized study plan* based on your incorrect answers—with page references for each chapter.

This great program is available for MS-DOS and Macintosh computers.

To order the Computerized Study Guide, call MEDI-SIM at 800-748-7734.

 m. C. Brown Publishers **has** *all the*  **tools you need to**

nail down  **botany and other science courses—materials**

designed to *make your life easier*  **and not cost you a**

small  *fortune.* **To order, contact your** *local bookstore.*

Or call our Customer Service Department  **at** **800-338-5578.**

the rest of the universe is referred to as the *surroundings*. A closed system, such as that approximated by a thermos bottle, is isolated from (i.e., does not exchange energy with) its surroundings; conversely, an open system exchanges energy with its surroundings. Don't let all of this intimidate you: the laws of thermodynamics are rather simple and are based on common sense. More importantly, they are unbreakable laws that apply to *all* energy transformations, whether they be the combustion of gasoline in an engine, the breakdown of glucose in a cell, the closing of a Venus's-flytrap, or the generation of heat by philodendron flowers. The laws of thermodynamics are important because they govern the existence of all organisms.

The First Law of Thermodynamics

The **first law of thermodynamics** is the law of conservation of energy. This law states that energy cannot be created or destroyed, but only converted to other forms. It can also be stated in other ways:

- In any process, the total amount of energy in a system and its surroundings remains constant.
- The total amount of energy in any isolated (i.e., closed) system is constant.
- The amount of energy in the universe is constant.
- You can't get something for nothing.

For example, the energy used to wind a watch comes from the person winding the watch. Similarly, a power plant does not create energy—it merely transforms energy from one form (e.g., fossil fuel) to another (e.g., electricity). Likewise, green plants are not energy producers—they merely trap and convert the energy in sunlight into chemical bonds. The first law asserts that the energy in sunlight that warms our planet and drives photosynthesis must come from somewhere else in the system. In this case, it comes from the sun.

Energy conversions often generate much heat. For example, as you read this book, your body generates the heat of a 100 W lightbulb (a class of ten students would produce the heat equivalent of a kilowatt heater). However, your body temperature is neither increasing nor decreasing because the heat generated by your body is radiated into your surroundings. That is, there is no change in the total amount of energy in the system—the energy radiated from your body that heats the room you're in can be traced to the energy contained in the food that you ate. According to the first law of thermodynamics, the amount of energy released by your body cannot exceed the amount of energy contained in the food that you eat. If you stop eating, you'll eventually run out of energy, die, and stop releasing heat.

The first law of thermodynamics has tremendous implications for everyday life. For example, it explains why an automobile can go only a limited distance, regardless of its fuel efficiency or mileage rating: the energy used to move the car cannot exceed that contained in the chemical bonds of its fuel. When the car runs out of gas, it can go no farther until more energy (i.e., gasoline) is added to the system. This addition of energy to the car corresponds to a loss of energy from somewhere else—the storage tank at the gas station.

The first law of thermodynamics also dictates that the energy trapped by leaves during photosynthesis cannot exceed the energy of the absorbed light. For example, if 100 units of light energy strike a leaf, no more than 100 units of energy can be trapped in carbohydrates produced by photosynthesis. No matter how hard you try, you can't get more energy out of a system than you put in. There are no exceptions to this law.

The Second Law of Thermodynamics

The **second law of thermodynamics** is the law of **entropy,** or disorder. This law states that all energy transformations are inefficient; that is, that the amount of concentrated, useful energy decreases in all energy transformations. The following statements express the second law in other ways:

- Systems tend toward increasing entropy (disorder).
- Any system tends spontaneously to become disorganized.
- In all energy transformations, some usable energy is lost as heat.
- Any spontaneous change decreases the amount of usable energy.
- A perpetual motion machine cannot exist; that is, no real process can be 100% efficient.
- You can't break even.

These restatements of the second law are based on our world being overwhelmingly irreversible. This irreversibility is all around us, including our aging, the futility of trying to unscramble an egg, and the impracticality of coasting downhill in a car to refuel its tank. Irreversibility results from the message of the second law of thermodynamics, namely, the loss of *usable* energy as heat during an energy transformation.

To better understand this, consider a person throwing a ball or heating a cup of tea. Both of these processes require energy. According to the first law, no energy has been created or destroyed in moving the ball or heating the tea: the energy used to heat the tea probably came from breaking the chemical bonds of natural gas, while that used to throw the ball came from energy in food used to contract muscles. However, the energetics of the moving ball and of heating the tea are drastically different. The moving ball heats the air and any object that it strikes, and is in *coherent* motion: all of its parts move together in an orderly way. In contrast, energy in the heated tea is contained in the random, *incoherent* motion of its molecules. There's no order to it; the heat energy results only from random molecular motion. This randomness distinguishes heat from all other forms of energy: any other form of energy can be converted completely to heat, but heat cannot be completely converted to other forms of energy. This again goes back to our everyday experience: the thrown ball heats the air and the glove that it hits, but applying the same amount of heat to the air doesn't move the ball. All energy is ultimately converted to heat, and heat is not usable energy.

Consider again the example of a combustion engine in a car. Gasoline is a concentrated, orderly source of energy—its energy resides in the covalent bonds of octane. However, when these bonds break and release energy in the car's engine, less than one-fourth of the energy is used to move the automobile (i.e., combustion engines are less than 25% efficient). According to the first law of thermodynamics, no energy was lost when the car moved: the amount of energy used to move the car, heat the engine block (and the air around it), power the radio, and heat the tires equals that originally contained in the gasoline. However, applying heat to a car does not move the car. That is, the energy contained in the heated tires, pavement, air, and engine-block cannot be recycled to run the car, because heat energy resides in randomly moving molecules and is therefore not in a concentrated, useful form. This heat represents the inefficiency inherent in any energy transformation and is the basis for the second law of thermodynamics.[3] Because all energy transformations produce heat (i.e., an unusable form of energy), all things naturally become more disorganized.

The consequences of the second law of thermodynamics are important and familiar to everyone. For example, rocks tumble downhill rather than uphill, and pieces of a jigsaw puzzle never spontaneously fall into place when they're poured from the box. In short, the second law of thermodynamics explains why disorder in the universe is increasing continually. All naturally occurring processes tend to reach maximum entropy (i.e., maximum disorder). Life exists at the expense of the universe, which is constantly running down.[4]

Cells derive energy from sugars and fats for growth, repair, and reproduction. The chemical reactions that free this energy are inefficient and release much heat; indeed, the cells of most organisms extract only about half of their fuel's energy for useful work (e.g., the energy used to power your brain while you sleep is equivalent to that of a 60 W bulb). Thus, although organisms can channel the transformation of energy from one form to another—they can hoard the energy in reserves (e.g., fat or starch) or use it for repair, movement, or reproduction—these diversions are only temporary. Eventually, all energy is transformed to heat.

Writing to Learn Botany

Creationists frequently claim that the evolution of increasingly complex organisms during the history of life contradicts the second law of thermodynamics, which is an unbreakable rule. Therefore, they claim, biological evolution is invalid. How would you respond to this argument?

The loss of useful energy as heat during energy transformations increases the entropy in a system, and only processes that decrease the amount of useful energy occur spontaneously. Therefore, there is a natural tendency for things to become less organized. Although the entropy of one system, such as an organism or cell, may decrease (i.e., the system may become more organized), the entropy of the universe is always increasing. On a more local level, our apartments and desks quickly become messy unless we periodically straighten them, and parents often point out that children are "entropy's little helpers."

Because organisms are highly ordered, it might seem that they are exceptions to the laws of thermodynamics. However, the second law applies only to closed (i.e., isolated) systems. Organisms remain organized because they are *not* closed systems: they use inputs of matter and energy such as food and sunlight to reduce randomness (i.e., decrease entropy) and stay alive. The energy that keeps organisms alive comes ultimately from the sun—that is, plants transform sunlight into the chemical bonds of carbohydrates, which humans and other organisms use as an energy source. Life is possible only because organisms temporarily store and later use some of the energy flowing through the system. Plants and plantlike organisms (such as algae) are the first and most important part of the scheme (see fig. 5.4).

Doing Botany Yourself

Design an experiment to measure how efficiently a leaf converts light energy to chemical energy. What would be the possible sources of error in the experiment? What precautions would you take to improve the accuracy of your measurements?

We're most familiar with energy transformations such as explosions, home runs, moving vehicles, and electricity, which release large amounts of energy at once. However, energy transformations in cells each involve small amounts of energy. It is the cumulative effect of these many small transformations that we see as growth and development.

The two primary energy transformations in plants are photosynthesis and cellular respiration (fig. 5.5). Photosynthesis uses light energy to convert CO_2 and H_2O—both of which

---

3. Heat can be used to do work only when there is a steep temperature gradient, as in a steam engine. These gradients do not exist in living organisms, and nature has never evolved a steam engine. Indeed, such gradients would quickly denature enzymes and kill cells. In the uniform conditions characteristic of living organisms, heat is useless except for warming.

4. The laws of thermodynamics determine the energy relationships of cells and organisms. They also predict the fate of our universe. For example, the first law of thermodynamics indicates that the energy present when the universe formed 14 billion years ago is all that can ever exist. Similarly, the second law means that our sun (and all other suns) will eventually burn out (i.e., all of its energy will be transformed to heat), thus leaving the earth without an energy source. When this occurs, all of the energy originally present when the universe formed will remain present, but there will be no usable energy for work. Everything will be at the same temperature, and there will be no more energy transformations; only small, randomly moving molecules will remain in the universe. All usable energy will have been frittered away into heat, and all life in the universe will end, a scenario described by biochemist science-fiction writer Isaac Asimov in *The Last Question*. In short, the lights will be out and the party will be over. But don't be overly concerned—even the most pessimistic folks don't expect this to happen for several billion years.

FIGURE 5.4

Plants are producers; that is, they transform energy in sunlight into chemical energy. This energy then flows through and is used by other organisms. Some energy is converted to heat during each transformation.

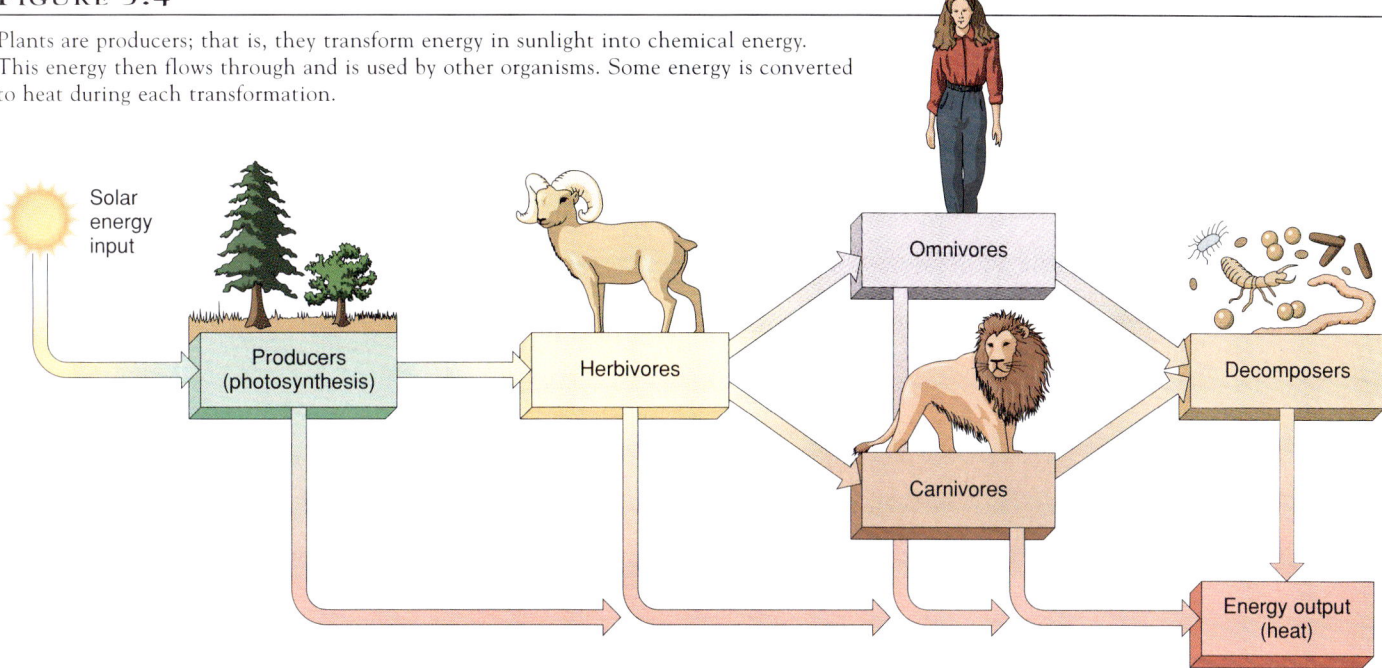

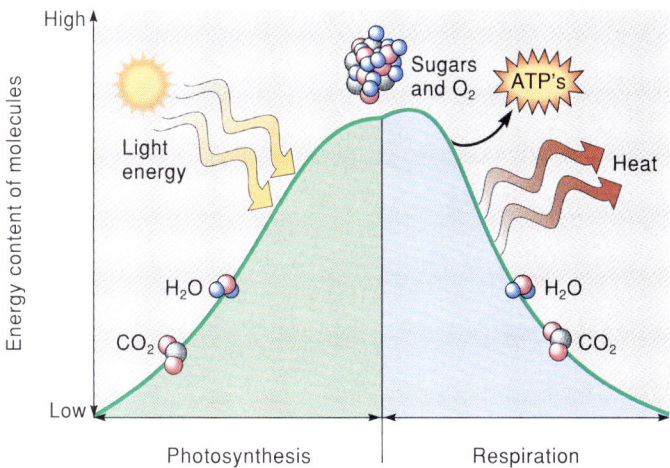

FIGURE 5.5

Plants use photosynthesis, shown here as an uphill reaction, to make energy-rich sugars (and O_2) from energy-poor CO_2 and water. Plants and other organisms use respiration, shown here as a downhill reaction, to convert sugars to CO_2 and water. Some of the energy released during respiration is conserved in chemical bonds of other molecules, especially ATP.

are energy-poor—into sugars. In the process, oxygen gas (O_2) is released. Cells extract energy from sugars via cellular respiration. Some of this energy is stored in molecules such as ATP (to be discussed later in this chapter). If all of the energy in the chemical bonds of sugars were released at once (as in an explosion), the energy would be converted mostly to heat and produce lethally high temperatures. To avoid these problems, cells extract energy from glucose and other molecules by slowly oxidizing the molecule in a *series* of chemical reactions. During each reaction there is a drop in the potential energy of the molecule. Some of this energy is lost as heat; however, much of it is trapped in the chemical bonds of other molecules in the cell. The chemical reactions that transform energy in cells are collectively called *metabolism*.

CONCEPT

All of life's activities involve changing energy from one form to another. These energy transformations are governed by the laws of thermodynamics. The first law states that energy cannot be created or destroyed, but only converted to other forms. The second law states that all energy transformations are inefficient because they involve the loss of usable energy as heat. The quantitative measure of disorder created by any spontaneous reaction is entropy. All spontaneous reactions increase entropy.

METABOLISM: ENERGY FOR LIFE'S WORK

Metabolism (from the Greek word meaning "change") is a fundamental property of life arising from energy transformations in cells; it is the sum of the vast array of chemical reactions that occur in an organism. These reactions do not occur randomly; rather, they occur in step-by-step sequences called **metabolic pathways,** in which the product of one reaction becomes the substrate (i.e., the starting point) for another. The various metabolic pathways in a cell are much like the roads on a map (fig. 5.6).

CHAPTER FIVE *Energy and Its Use by Plants*

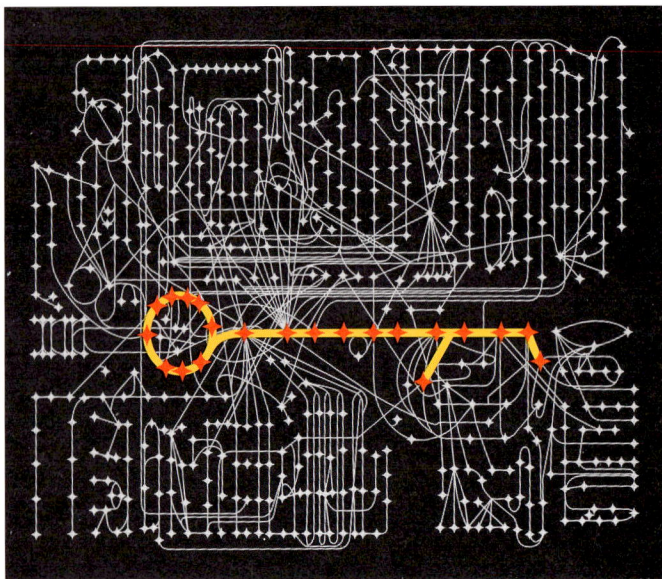

FIGURE 5.6

Cellular metabolism. This diagram traces some of the metabolic reactions that occur in a cell. Dots represent molecules, and lines represent reactions—each catalyzed by a specific enzyme—that change them. The stepwise sequences of reactions are called *metabolic pathways*. The pathway shown in yellow is central to most other pathways.

Each reaction in a metabolic pathway rearranges atoms into new compounds, and each one either absorbs or releases energy. The amount of energy required to break a particular bond is equivalent to the amount of energy required for its formation. This amount of energy is called the **bond energy**. For example, consider the energies of the following bonds:[5]

Bond	Energy (kcal mol^{-1})
C–C	83
C–O	84
C–H	99
C=O	174
O=O	118
O–H	111

Thus, C=O bonds are much stronger than C–C bonds. This is important because metabolism continually breaks and reforms these bonds to obtain energy for growth, movement, and repair.

During chemical reactions, the net release or uptake of energy equals the difference of energy released and energy consumed. For example, burning a mole (16 g) of methane releases 160 kcal of energy:

$$CH_4 + 2O_2 \rightarrow CO_2 + 2H_2O$$

This heat of reaction is the heat that you feel from the stove (i.e., the net amount of energy released by the reaction) and is represented by ΔH (delta H). It is derived from the total potential energy of the molecules, a measure called **enthalpy**. Therefore, we say that the heat released into the surroundings comes from the enthalpy of the reacting molecules. In the case of burning methane, the products (CO_2 and H_2O) have 160 kcal less enthalpy than the reactant (CH_4). Such heat-releasing reactions are called **exothermic** reactions and change the molecules so that their energy content decreases.

CONCEPT

Metabolism is the sum of the chemical and energy transformations in cells. Metabolism occurs in stepwise sequences called *metabolic pathways*. Each reaction of a metabolic pathway rearranges atoms into new compounds.

Although most exothermic reactions are spontaneous, some are not. To accurately predict whether a reaction will occur spontaneously, we must introduce a new concept: free energy.

FREE ENERGY

The potential energy of a compound is contained in its chemical bonds. When these bonds break, the energy that is released can be used to do work, such as form other bonds. The amount of energy available to form other bonds is the **free energy** of the molecule and is represented by G (for its discoverer, Yale physicist Josiah Gibbs).

Chemical reactions change the amount of free energy available for work. This change in free energy is called ΔG (delta G) and is the most fundamental property of a chemical reaction: it is equivalent to the change in the heat content minus the change in entropy. These relationships are represented by the following formula:

$$\Delta G = \Delta H - T\Delta S$$

where:

ΔG is the change in free energy of the reaction and is the part of the potential energy that can do work. The remaining energy is not available for work because of entropy.

ΔH is the change in enthalpy (heat content), which is the energy in chemical bonds.

T is the temperature measured on a scale of °C above absolute zero.[6]

ΔS is the change in entropy, or disorder.

---

[5]. These bond energies vary somewhat because they are affected by neighboring bonds of the molecule.

[6]. Temperature in the equation for calculating the change in free energy is expressed in degrees Kelvin (K), which are units above absolute heatlessness (i.e., where molecular motion stops). Since temperature measures the intensity of molecular motion, all Kelvin temperatures exceed zero. A kelvin (the unit) is the same as a degree centigrade, and to get from centigrade to Kelvin add 273 (i.e., K = °C + 273). For example, water freezes at 273 K, room temperature is ~ 293 K, and body temperature is ~ 310 K. Note that there is no degree sign used with Kelvin measurements.

Entropy is amplified at higher temperatures because temperature measures random molecular motion (i.e., the intensity or potential of heat), which increases disorder. Therefore, higher temperatures speed reactions and increase disorder. This is also why water evaporates faster at higher temperatures.

Spontaneous reactions change bond energies and release heat.[7] These reactions usually increase entropy and are called **exergonic** (energy outward) reactions. They have a ΔG less than zero and therefore form products with less free energy than their reactants; this energy is potentially available for cellular work. All spontaneous reactions decrease the amount of free energy because some energy is dissipated, thereby increasing entropy. Because this energy can do no work, life is a constant struggle against entropy.

Not all reactions are spontaneous. For example, consider the formation of sucrose (table sugar) and water from glucose and fructose:

$$\text{glucose} + \text{fructose} \rightarrow \text{sucrose} + H_2O$$

This reaction has a ΔG of +5.5 kcal, meaning that its products have 5.5 kcal mol^{-1} *more* energy than its reactants. This reaction absorbs energy from the surroundings and is not spontaneous. Such reactions are called **endergonic** (energy inward) reactions and will not occur without a net input of energy.

The free energy of a particular reaction determines many of a reaction's properties. Most important, the ΔG of a reaction dictates how much work a reaction can perform. For example, consider the oxidation of a mole of glucose to carbon dioxide and water:

$$C_6H_{12}O_6 + 6O_2 \rightarrow 6CO_2 + 6H_2O + \text{energy}$$
$$\Delta G = -686 \text{ kcal mol}^{-1}$$

Since ΔG for this reaction is less than zero, this reaction is exergonic and spontaneous. The carbon dioxide and water it produces store 686 fewer kcal than does glucose.

Free Energy and Chemical Equilibrium

When a chemical reaction reaches equilibrium, ΔG equals zero. Similarly, ΔG increases as one moves away from equilibrium. Because all of life's processes require work, cells must remain far from equilibrium to stay alive. They accomplish this by continually preventing the accumulation of any of the reactants of metabolic pathways. For example, the huge difference in free energy between glucose and its oxidation products (carbon dioxide and water) pulls cellular metabolism strongly in one direction; as soon as reactants form, they are quickly converted to new compounds by other reactions.

---

7. *Spontaneous* indicates only that the reaction will occur, not its rate.

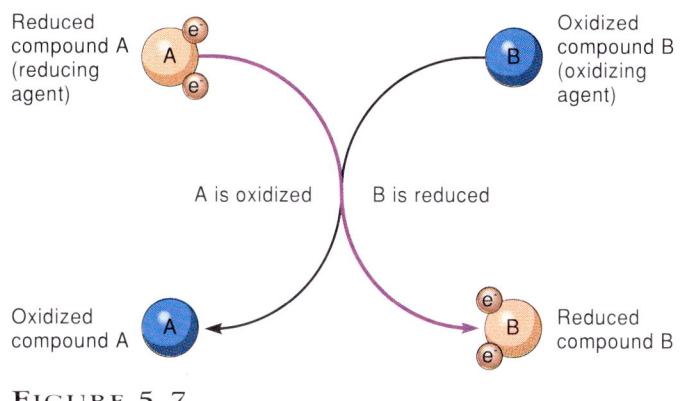

FIGURE 5.7

Oxidation and reduction occur simultaneously; as compound A is oxidized (i.e., donates electrons), compound B is reduced (i.e., accepts electrons).

Chemical reactions change the amount of free energy available for work. This change in free energy is the amount of energy available to form other bonds.

Oxidation, Reduction, and Energy Content

Most energy transformations in organisms involve chemical reactions called *oxidations* and *reductions* (fig. 5.7). **Oxidation** is the loss of electrons, either alone or with hydrogen, from a molecule. This is equivalent to adding oxygen because oxygen is strongly electronegative and therefore attracts electrons from the original atom. Oxidative reactions such as the breakdown of glucose to carbon dioxide and water degrade molecules into simpler products and are therefore examples of **catabolism.** They are "downhill" reactions that release energy.

Reduction is the addition of electrons, either alone or with hydrogen, to a molecule. Reduction changes the chemical properties of a molecule, not necessarily its size. Electrons removed from a molecule during oxidation reduce another molecule. Reduction reactions such as the formation of lipids usually involve synthesis of more complex molecules and are therefore examples of **anabolism.** They are "uphill" reactions that require a net input of energy. Oxidation and reduction reactions always occur simultaneously; if something is reduced, something else must be oxidized.

Many energy transformations in living systems involve oxidation and reduction of carbon. Reduced carbon contains more energy than oxidized carbon. This explains why reduced molecules such as methane (CH_4) are explosive, while oxidized molecules such as carbon dioxide (CO_2) are not. The same

FIGURE 5.8

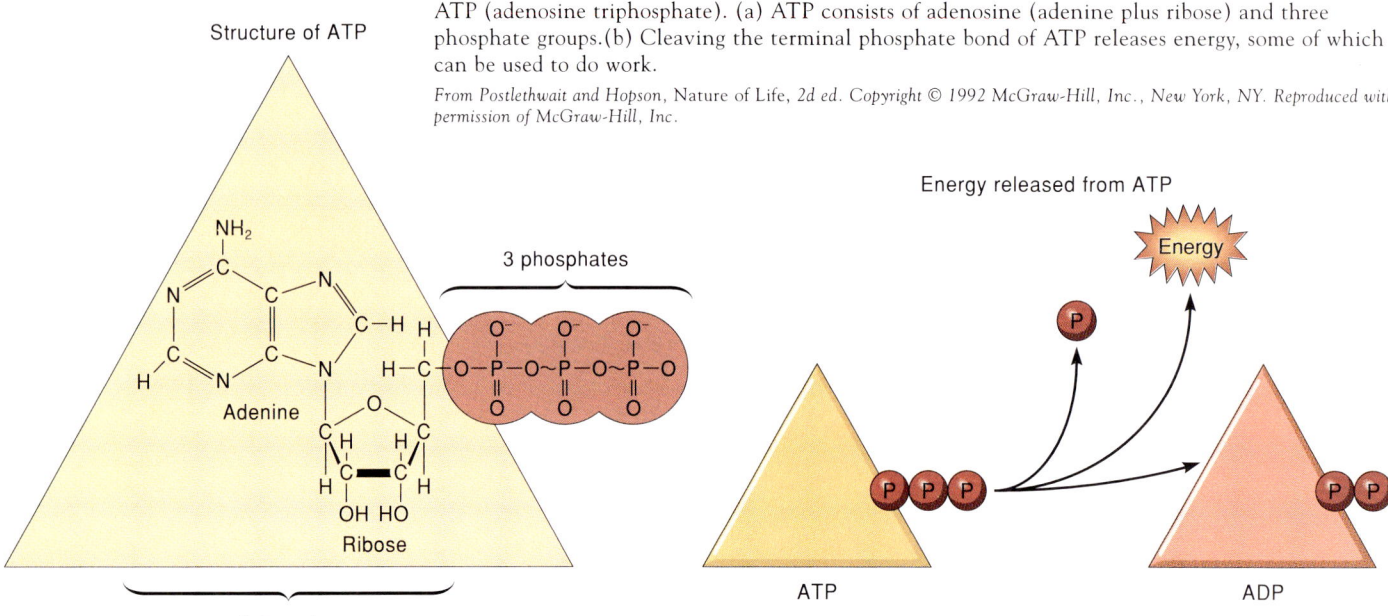

ATP (adenosine triphosphate). (a) ATP consists of adenosine (adenine plus ribose) and three phosphate groups. (b) Cleaving the terminal phosphate bond of ATP releases energy, some of which can be used to do work.

From Postlethwait and Hopson, Nature of Life, 2d ed. Copyright © 1992 McGraw-Hill, Inc., New York, NY. Reproduced with permission of McGraw-Hill, Inc.

principle applies to other compounds: the more reduced they are, the more energy they contain. For example,

Substance	Energy Content (J kg^{-1})	
Hydrogen	122,000,000	
Natural gas	55,000,000	
Gasoline	42,000,000	
Wood (dry)	17,000,000	
Starch	17,000,000	
Sucrose	17,000,000	
Salad oil (or any fat)	37,000,000	
Wheat flour	15,000,000	(mostly starch)
Lima beans, raw	5,000,000	(67% water)
Potatoes, raw	3,000,000	(80% water)
Beef, T-bone steak	17,000,000	(37% fat)
Chicken (skinless)	4,000,000	(2% fat)

This example should help you better appreciate how the different energy content of foods affects our lives. Most college students require about 3,000 Calories per day—that's equivalent to the energy in 6 pounds of TNT or 1 pound of butter. If you had to obtain all of that energy from lettuce or celery, you'd have to eat about 60 pounds of lettuce or about 450 celery sticks every day. Although your friends might start mistaking you for a rabbit, an "Eat All the Lettuce You Want and Still Lose Weight" diet would be effective.

CONCEPT

Many energy transformations in organisms involve chemical reactions called *oxidations* and *reductions*. Oxidation is the loss of electrons from a molecule, and reduction is the addition of electrons to a molecule. Oxidation and reduction reactions always occur simultaneously.

Organisms extract energy from energy-rich compounds such as sugars and fats via catabolic reactions collectively called **cellular respiration.** They use this energy to drive anabolic reactions that do work—for example, moving flagella, building cell walls, and replicating genetic information. An important hub through which energy passes during cellular metabolism is adenosine triphosphate, a compound more commonly known as *ATP*.

ATP: THE ENERGY CURRENCY OF CELLS

When cells need energy, they hydrolyze **adenosine triphosphate,** or ATP. ATP is a nucleoside triphosphate made of adenine (a nitrogen-containing base), ribose (a five-carbon sugar), and three phosphate (HPO_4^{2-}) groups (fig. 5.8). ATP molecules contain much energy that is released when the terminal phosphate group (represented by a squiggly line, ~) is cleaved from the molecule.[8] Because the breakdown of ATP links energy exchanges in cells, ATP is the energy currency of the cell: when cells need energy to do something, they "spend" ATP by converting it to adenosine diphosphate (or ADP), inorganic phosphate (Pi), and energy (fig. 5.8).

$$ATP + H_2O \rightarrow ADP + Pi + energy$$
$$\Delta G = -7.3 \text{ kcal mol}^{-1}$$

---

[8]. The terminal phosphate bond is often called a *high-energy* bond. This is incorrect. The energy contained in ATP is not simply in this phosphate bond; it resides in the complex interaction of atoms in the entire molecule.

Several properties of ATP make it ideally suited as the energy currency of cells:

- The amount of energy released by converting ATP to ADP + Pi (7.3 kcal mol^{-1}) is about twice as much as is needed to drive most cellular reactions. The rest of the energy is dissipated as heat.
- Much of its energy is immediately available to cells. Although fats and starch also store large amounts of energy, their energy must first be converted to ATP before it can be used. ATP represents the readily available energy "cash" of a cell; fats and starch are analogous to energy stocks and bonds.
- Unlike the covalent bonds linking carbon and hydrogen in molecules such as methane and glucose, the terminal phosphate bond of ATP is unstable—that's why it breaks so easily.

ATP is a common energy currency—all cells of all organisms use ATP for energy transformations. Like all of the different appliances that can plug into an electrical outlet and do different things, so too can different chemical reactions use a cell's ATP to do different kinds of work. Organisms use ATP for virtually all of their work, including making new cells and macromolecules, pumping materials, and moving materials through cells and throughout the organism. Accomplishing all of this work requires huge amounts of ATP. For example, a typical adult uses the equivalent of about 200 kg (440 lb) of ATP per day, but has only a few grams of ATP on hand at any one time. Even the hungriest of us doesn't eat 200 kg of food in a week, much less in a day. Nevertheless, organisms that run out of ATP—that is, those that go "bankrupt"—immediately die. Where does all this ATP come from? Obviously, we do not make 200 kg of ATP from scratch each day. Rather, we recycle our ATP at a furious pace, turning over the entire supply every minute or so:

$$ADP + Pi + energy \rightarrow ATP$$
$$\Delta G = +7.3 \text{ kcal mol}^{-1}$$

Since the ΔG for this reaction is positive, each reaction requires an input of energy, which comes from molecules that are broken down in other reactions that are coupled to the synthesis of ATP.

Coupled Reactions

Just as cells couple the breakdown of food to the production of ATP, so too do they couple the breakdown of ATP to other reactions that occur at the same time and place in the cell. These **coupled reactions** drive other reactions that do work or make other molecules (fig. 5.9). For example, consider again the formation of sucrose (table sugar) and water from glucose and fructose:

$$glucose + fructose \rightarrow sucrose + H_2O$$
$$\Delta G = +5.5 \text{ kcal mol}^{-1}$$

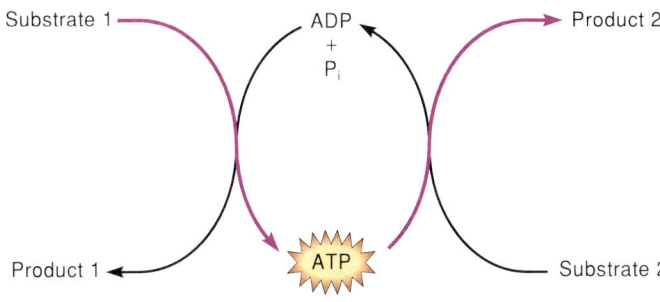

FIGURE 5.9

Coupled reactions. Energy released by an exergonic reaction can be used to make ATP from ADP and Pi. This ATP can then be used to drive an endergonic reaction.

Because the ΔG for this reaction exceeds zero, the reaction is not spontaneous and will not occur without a net input of energy. This input of energy is provided by two moles of ATP, each of which provide 7.3 kcal mol^{-1} of energy:

$$ATP \rightarrow ADP + Pi + energy$$
$$\Delta G = -7.3 \text{ kcal mol}^{-1}$$

This changes the equation for sucrose synthesis to

$$glucose + fructose + 2ATP$$
$$\downarrow$$
$$sucrose + H_2O + 2ADP + 2Pi + energy$$

and makes the $\Delta G = 5.5 - 14.6 = -9.1$ kcal mol^{-1}. Thanks to the expenditure of energy by the cell (i.e., as ATP), the reaction proceeds because its overall ΔG is negative. In this reaction, the breakdown of ATP is coupled to the formation of sucrose.

ATP accomplishes much of its work by transferring its phosphate group to another molecule in a process called **phosphorylation.** Phosphorylations energize the molecules receiving the phosphate group, so that they can be used in later reactions. The original "cost" of this phosphorylation is returned in subsequent reactions.

Adenosine triphosphate, or ATP, is the energy currency of cells. That is, cells use ATP to do work. ATP releases about twice as much energy as is needed to drive most cellular reactions. ATP is used by all cells for energy transformations. Many of these transformations involve coupled reactions in which the energy released by ATP drives other reactions.

Other Compounds Involved in Energy Metabolism

Several other compounds besides ATP affect energy transformations in plant cells. Nonprotein helpers such as Mn^{2+} and Na$^+$ aid cellular metabolism and are called **cofactors.** Cofactors are often ions; for example, Mg^{2+} is a cofactor required to transfer phosphate groups between molecules. Organic cofactors are called **coenzymes** and

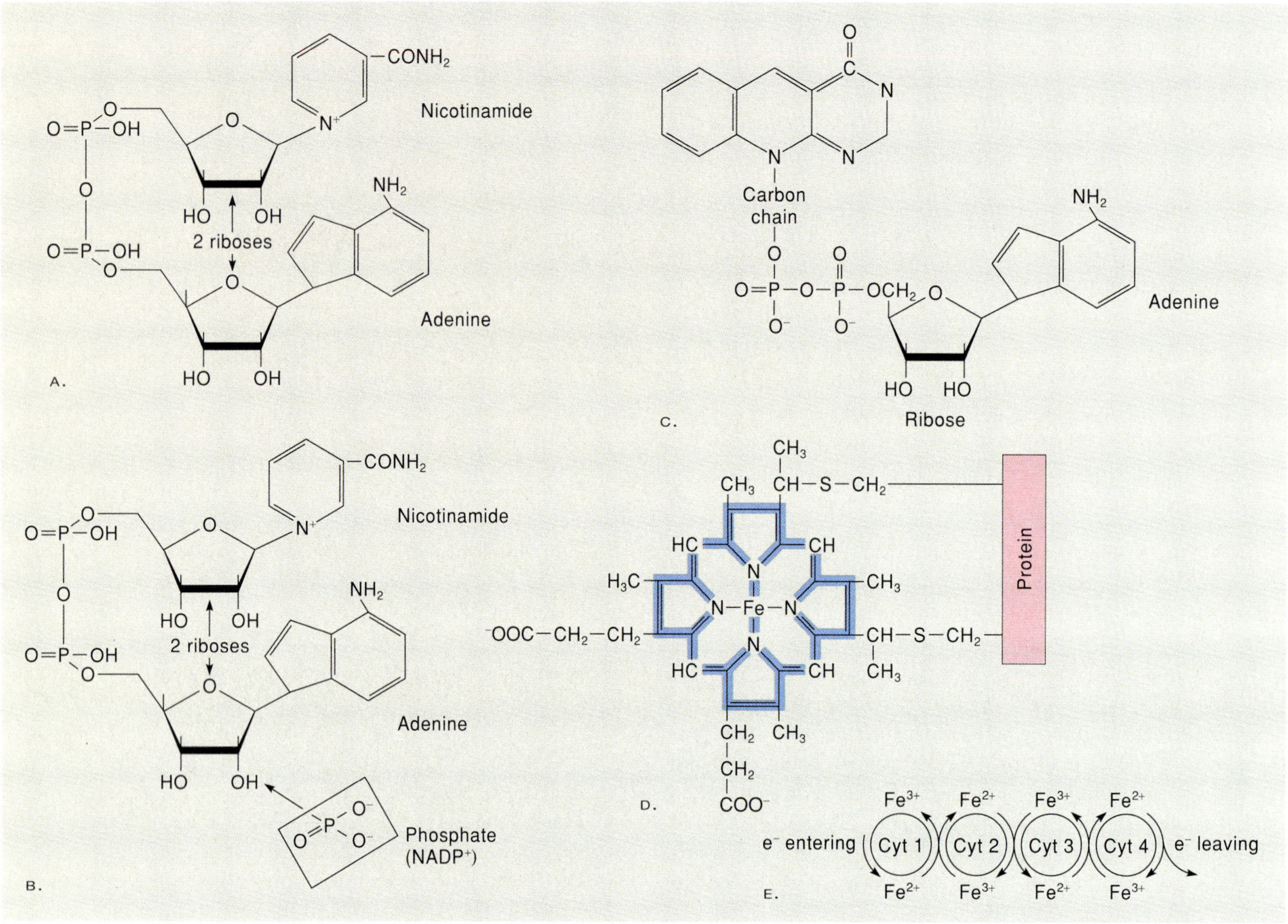

FIGURE 5.10

Structures of electron-carriers in plant cells. (a) NAD⁺ (nicotinamide adenine dinucleotide). The nicotinic acid ring is the part of NAD⁺ that is oxidized or reduced during energy transformations. (b) NADP⁺ (nicotinamide adenine dinucleotide phosphate). NADP⁺ differs from NAD⁺ only in that NADP⁺ has a phosphate group. (c) FAD (flavin adenine dinucleotide). The active group of FAD is isoalloxazine. (d) Cytochrome c, one of several kinds of cytochromes in cells. Cytochromes contain a heme group (outlined in blue), which is a ring of nitrogens with an iron atom at its center. (e) The iron atoms of cytochromes alternate between a reduced (Fe^{2+}) and an oxidized (Fe^{3+}) state during electron transport. At each step there is a drop in free energy; some of the energy that is released is used to make ATP.

---

usually carry protons or electrons. These are often nucleotides; unlike ATP, their energy content depends on their oxidation state, not on the presence or absence of a particular phosphate bond. Coenzymes are vitamins that occur in all cells. Humans and other animals must obtain vitamins from food; plants produce their own vitamins.

NAD⁺: Nicotinamide Adenine Dinucleotide

NAD⁺ is similar to ATP in that it is made of adenine, ribose, and phosphate groups (fig. 5.10a). However, the active part of NAD⁺ is a nitrogen-containing ring, called *nicotinamide,* which is a derivative of nicotinic acid (niacin, or vitamin B_3, one of the compounds added to products such as cornflakes to make them "vitamin fortified"). NAD⁺ is reduced when it accepts two electrons and a proton from the active site of an enzyme or from a substrate. NAD⁺ removes two protons and two electrons from a substrate, and both electrons and one proton are added to the NAD⁺:

$$NAD^+ + 2H^+ + 2e^- \rightarrow NADH + H^+$$
$$\Delta G = -52.6 \text{ kcal mol}^{-1}$$

NADH + H⁺ is fully reduced and is therefore energy rich. It is used to make ATP and to reduce other compounds in cells. You'll learn in Chapter 6 how other pathways "cash in" the energy contained in NADH + H⁺ for ATP.

NADP⁺: Nicotinamide Adenine Dinucleotide Phosphate

NADP⁺ has a structure similar to NAD⁺ with an added phosphate group (fig. 5.10b). NADPH supplies the hydrogen that reduces CO_2 to carbohydrate during photosynthesis. NADPH also supplies the hydrogen used to reduce nitrate to ammonia.

FAD: Flavin Adenine Dinucleotide

FAD is one of the coenzyme forms of riboflavin (vitamin B_2; fig. 5.10c). FAD, like NAD, carries two electrons; however, FAD accepts both protons to become $FADH_2$. FAD functions in cellular respiration.

Other Nucleoside Triphosphates

Several other nucleoside triphosphates function in cellular metabolism. For example, uridine triphosphate (UTP) is involved in making cell walls, guanosine triphosphate (GTP) is involved in protein synthesis, and cytidine triphosphate (CTP) is involved in membrane production.

Cytochromes

Like chlorophyll and hemoglobin, cytochromes are a group of metal-containing molecules that participate in metabolism by transferring electrons (fig. 5.10 d and e). When oxidized, the iron in cytochromes exists as Fe^{3+}. When this iron accepts an electron, it is reduced to Fe^{2+}. There are several types of cytochromes, all of which carry electrons in cells. These molecules participate in virtually all energy transformations in living organisms, suggesting that the machinery for energy transformation in all organisms has a common heritage.

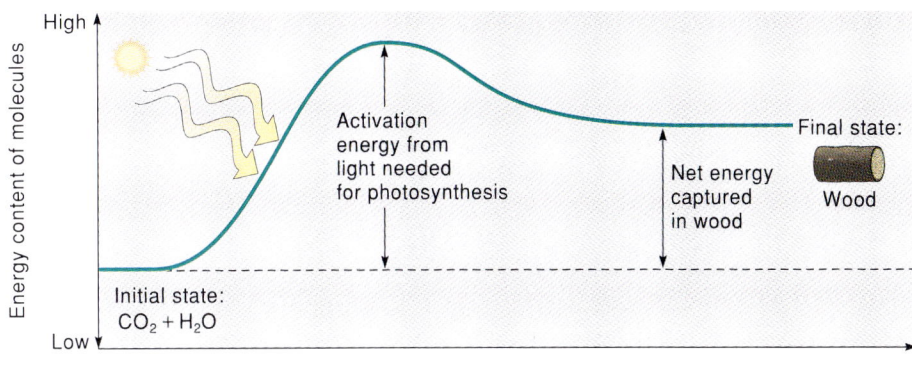

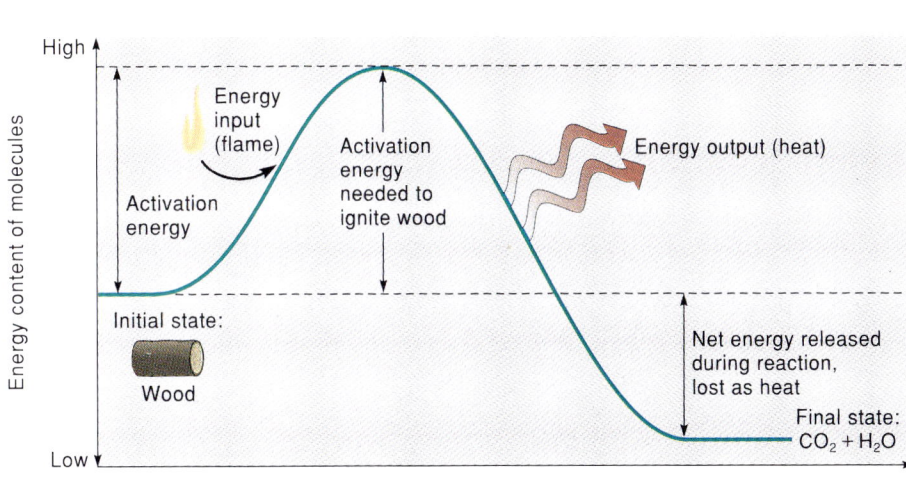

FIGURE 5.11

Energy of activation. (a) Plants use photosynthesis to make sugars that are, in turn, used to make products such as wood. The energy to do this—that is, the activation energy for photosynthesis—comes from sunlight. (b) The energy captured in wood is released when wood burns. To start this reaction, activation energy must be provided to ignite the wood.

ENZYMES AND ENERGY

You learned earlier in this chapter that exergonic reactions (e.g., the combustion of methane) are spontaneous. If this is true, then why haven't all of these reactions already occurred? The answer is that these reactions require an energy input to start. To understand this, consider again the example shown in figure 5.3. The rock atop the hill contains potential energy, but is at rest. If given a nudge, the rock will roll down the hill and release its kinetic energy. The nudge is the energy input required to start the reaction. Once started, the rock rolls to the bottom of the hill spontaneously (i.e., on its own).

What are Botanists Doing?

Go to the library and browse through a botany journal such as *Plant Physiology*, *American Journal of Botany*, or *Planta*. Read an article that describes a study of metabolism or bioenergetics. What did the authors do? What did they conclude? What experiments would you do if you wanted to extend their work?

Many chemical reactions are similar to the rock atop the hill: they proceed only when activated by an energy input, called the **energy of activation** (E_{act}). In photosynthesis, the energy of activation is provided by sunlight and the net gain of energy is trapped in products such as wood (fig. 5.11a). Although this wood contains much energy, the wood is too stable to burn spontaneously. That is, although it contains much potential energy, it will not burn unless it is first heated (fig. 5.11b). The heat necessary to ignite the wood is the energy of activation. Once ignited, the wood will continue to burn and release energy.

Although heat is generally an effective catalyst, it is usually not effective for cells. Indeed, the heat needed to activate most metabolic reactions would quickly kill most cells. Thus, cells rely on biocatalysts called *enzymes* to start their reactions.

Enzymes are globular proteins that speed reactions by decreasing the energy of activation of a reaction (fig. 5.12). They do this by binding the substrate so that the reaction can occur. For example, the energy of activation to degrade casein (a protein in milk) is 20,600 kcal mol^{-1} without the reaction's enzyme, but is only 12,600 kcal mol^{-1} in the presence of the

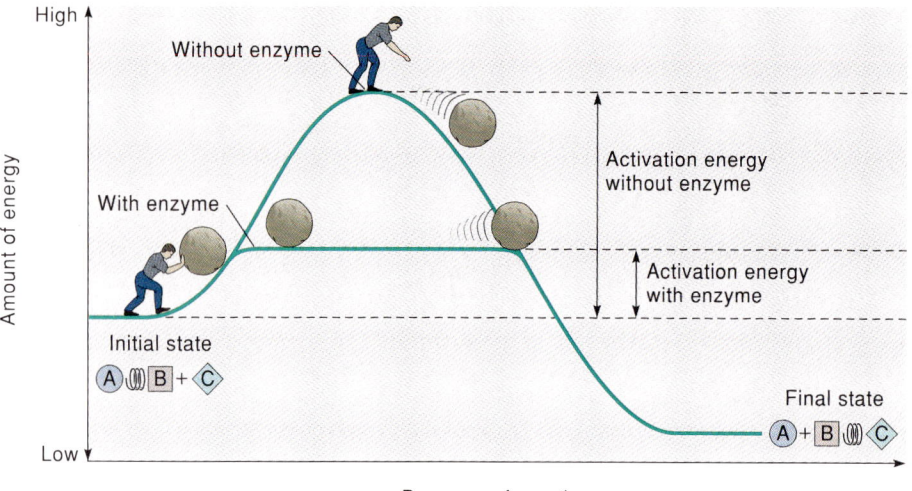

FIGURE 5.12

Enzymes speed the rate of chemical reactions by lowering the energy of activation of the reaction.

From Postlethwait and Hopson, Nature of Life, 2d ed. Copyright © 1992 McGraw-Hill, Inc., New York, NY. Reproduced with permission of McGraw-Hill, Inc.

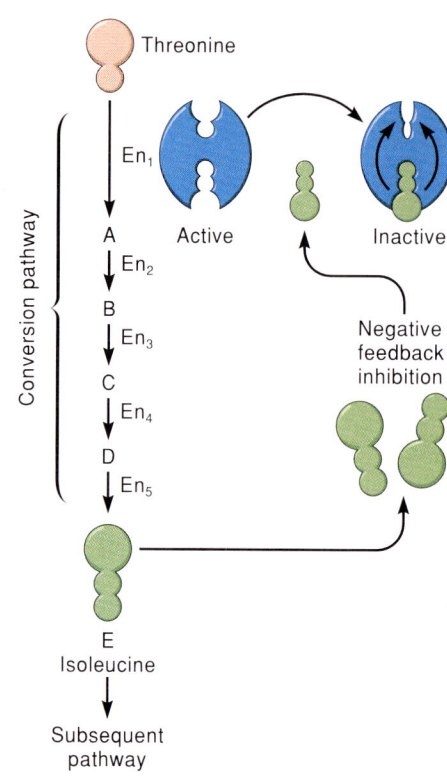

FIGURE 5.13

Feedback inhibition. The metabolic conversion of threonine to isoleucine consists of five steps, each controlled by an enzyme (En_1, En_2, En_3, …). An accumulation of isoleucine inhibits the first enzyme of the pathway (En_1), thereby preventing an unnecessary buildup of isoleucine.

enzyme. Decreasing the energy of activation greatly speeds the reaction. For example, consider the formation of H_2CO_3 from CO_2 and H_2O, a reaction involved in gas exchange and catalyzed by carbonic anhydrase:

$$CO_2 + H_2O \xrightarrow{\text{carbonic anhydrase}} H_2CO_3$$

Without carbonic anhydrase, only about 1 molecule of H_2CO_3 forms per second. This is far too slow to be useful to organisms. However, in the presence of carbonic anhydrase, H_2CO_3 forms at a rate of about 600,000 molecules per second, an increase of more than 10^7. This emphasizes the importance of enzymes; they are critical to life because they speed spontaneous reactions to a biologically useful rate.

Higher temperatures typically increase enzymatic activity. Indeed, up to a point enzymatic activity typically doubles for every increase of 10° C. However, beyond about 60° C, entropy wins out as the protein is denatured and the reaction stops. Although a few organisms can tolerate high temperatures (e.g., bacteria in hot springs of Yellowstone National Park thrive at 70° C), most enzymes work best at much lower temperatures. For example, most enzymes in our bodies function best near body temperature (37° C/98.6° F).

Regulating Metabolism

Enzymes regulate energy transformations by controlling metabolic reactions. If metabolism is likened to a series of interconnected roads (fig. 5.6), then enzymes would function as traffic lights that control the flow of energy in cells. How do enzymes do this?

HYPOTHESIS #1:

Enzymes decrease the ΔG of a reaction. *No.*
Enzymes do not change the ΔG of a chemical reaction.

HYPOTHESIS #2:

Enzymes heat cells to increase reaction rates. *No.*
Enzymes do not heat cells. The heat required to speed most metabolic reactions to useful rates would kill the cells.

How, then, do enzymes regulate metabolism? The products of a pathway often affect the activity of the pathway. For example, plants use a five-step pathway to make isoleucine from threonine (fig. 5.13). When isoleucine accumulates, it inhibits the first enzyme of the pathway, thereby decreasing production of isoleucine until the current supply is used. This means of slowing a pathway when its products aren't needed is called **feedback inhibition** and is common in plants and animals. Feedback inhibition balances supply and demand in cells, thereby averting unnecessary excesses and deficiencies.

One means of feedback inhibition involves **allosteric regulation**, in which the product binds weakly to a receptor site on the enzyme that differs from the active site. Allosteric regulation is common in enzymes having more than one subunit. The binding of a molecule to an allosteric site (usually located where subunits join) changes the enzyme's activity, thereby affecting the cell's metabolism. For example, isoleucine allosterically inhibits the first enzyme of its synthetic pathway and thus prevents the unnecessary buildup of isoleucine.

Many enzymes are also inhibited by other molecules that compete for the enzyme's active site. These so-called **competitive inhibitors** mimic the substrate and can be overcome by increasing the concentration of substrate (i.e., diluting the concentration of the inhibitor). Drugs such as sulfanilamide competitively inhibit enzymes.

Other compounds inactivate enzymes by binding to parts of the enzyme that are different from the active site, thereby preventing the enzyme from binding the substrate at its active site. Compounds that do this, such as lead and nerve gas, are called **noncompetitive inhibitors.** Although many noncompetitive inhibitors bind reversibly to an enzyme, many do not. For example, penicillin is a noncompetitive inhibitor that binds irreversibly to an enzyme that makes cell walls in bacteria. This blockage of cell-wall synthesis ultimately kills the bacteria, thus accounting for the antibiotic effect of penicillin.

CONCEPT

Enzymes catalyze biological reactions by lowering the energy of activation. Enzymes are critical to life because they speed spontaneous reactions to a biologically useful rate. Enzymes, and therefore metabolism, are controlled by feedback inhibition, competitive inhibition, and noncompetitive inhibition.

THE MAJOR ENERGY TRANSFORMATIONS IN PLANTS: PHOTOSYNTHESIS AND RESPIRATION

The energy transformations that sustain life occur similarly in all organisms. The most important energy transformations in plants are photosynthesis and respiration, the topics of the next two chapters. To appreciate these reactions, again examine figure 5.5. The energy-requiring uphill stage of the process is photosynthesis. During this process, light energy absorbed by chloroplasts is used to release oxygen and reduce carbon dioxide (a low-energy compound) to carbohydrate (a high-energy compound):

carbon dioxide + water + light
↓
carbohydrate + oxygen

Carbohydrate fuels the activities of plants and all other organisms.

The energy-releasing (i.e., exergonic), downhill stage of the process is cellular respiration. During this process, energy-rich molecules such as sugars are oxidized to carbon dioxide and water:

carbohydrate + oxygen
↓
carbon dioxide + water + energy

Respiration drives the cellular economy of most organisms.

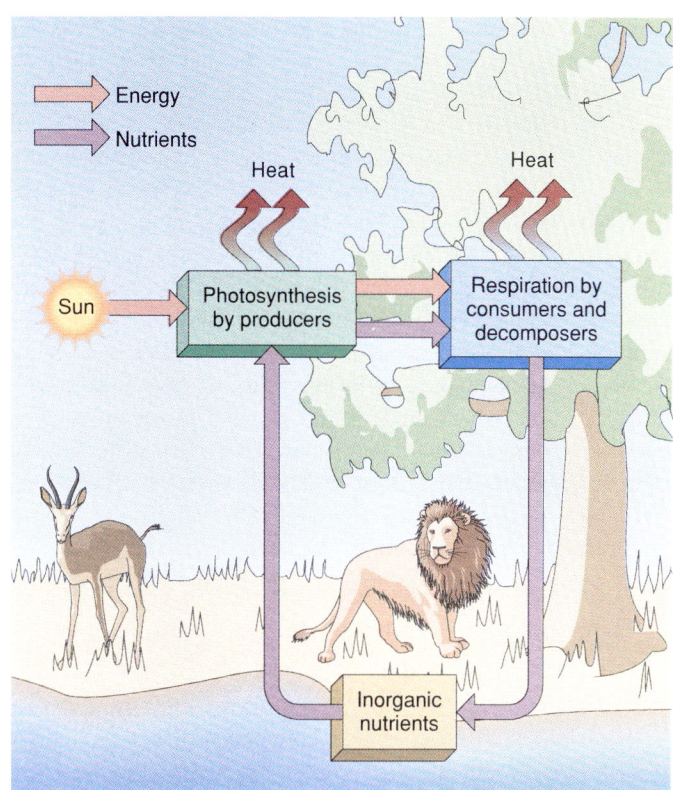

FIGURE 5.14

The major transformations of energy by organisms: photosynthesis and respiration. Photosynthesis provides food for virtually all organisms on earth. Organisms use cellular respiration to extract energy from food. Although nutrients cycle in an ecosystem, energy flows through an ecosystem. All energy is eventually converted to heat.

The Flow of Energy

Examine again the equations for photosynthesis and respiration. Each process involves oxidation or reduction of carbon by the addition or removal of hydrogen. Photosynthesis removes hydrogen from water and adds it to carbon; thus, carbon is reduced during photosynthesis. Cellular respiration removes hydrogen from carbon and adds it to oxygen to form water; thus carbon is oxidized during respiration. Although one equation is the reverse of the other, they proceed by different mechanisms.

Now examine figure 5.14, which shows how photosynthesis and respiration are integrated in nature. This diagram illustrates an important concept about energy: *Energy flows through a system and is ultimately converted to heat.* For example, sunlight is the energy input that sustains life on earth. This is possible only because plants convert sunlight into chemical energy (e.g., sugars). Animals stay alive by eating plants (or by eating other animals that ate plants) and using their stored energy to power their activities. These animals are then eaten by other animals to fuel their activities. Ultimately, the organisms die and are decomposed by bacteria and fungi. In the process, the energy stored temporarily by organisms in this so-called "food chain" is released as heat. Organisms may temporarily store various amounts of the energy, but the net effect is that it flows through the system and is ultimately transformed to heat.

According to the second law of thermodynamics, all of the energy transformations at every step in the food chain are less than 100% efficient. Indeed, there is a 90% loss of usable energy at each stage. This has tremendous implications. For example, consider the crop of corn shown at the beginning of this unit. Most of the energy striking the plants' leaves is reflected or converted to heat. Only a small amount of the energy is trapped by photosynthesis in sugars. When these sugars are converted to starch, more of the energy is lost. When this starch is fed to cattle, only about 10% of the usable energy is stored by the cow; the rest is lost as heat. By the time we eat a steak made from the cow, another 90% of the usable energy has been lost. Thus, the amount of usable energy in the steak is only about 1% of that contained in the starch of the corn.

| 100 units of energy in corn | → | 10 units of energy in an herbivore | → | 1 unit of energy in a carnivore |

The inefficiency of these energy transformations, as predicted by the second law of thermodynamics, makes this an inefficient process because much of the energy is converted to heat. By inserting an extra energy-transformation between the corn plant and ourselves (i.e., the cow), we lose much of the usable energy.

Look again at figure 5.14. Notice that although energy moves through the system in a one-way flow (i.e., toward heat), nutrients are cycled. That is, photosynthesis produces sugars and oxygen from carbon dioxide, light, and water, and the sugars and oxygen are recycled to carbon dioxide and water during cellular respiration. Thus, *nutrients cycle in an ecosystem.* You'll learn more about these nutrients in Chapter 20.

CONCEPT

Photosynthesis and respiration represent the major energy transformations in organisms. Photosynthesis uses light energy to reduce carbon dioxide and water to carbohydrate, while respiration oxidizes carbohydrates, producing ATP in the process. Together, photosynthesis and respiration compose a system by which energy flows through an ecosystem and is ultimately converted to heat. Unlike energy, nutrients cycle in an ecosystem.

The Lore of Plants

On average, vegetables and fruits make up about 10% of our daily intake of calories, 7% of our daily intake of protein, and an unimpressive 1% of our daily intake of fat.

Chapter Summary

Energy is the ability to do work. It is measured with a variety of units, the most common of which are calories and joules. Potential energy is energy available to do work, while kinetic energy is energy being used to do work.

All of life's activities require that energy be transformed from one form to another. These energy transformations are governed by the laws of thermodynamics. The first law of thermodynamics is the law of conservation of energy; it states that in all chemical and physical changes, matter and energy cannot be created or destroyed, only converted to other forms. The second law of thermodynamics states that all energy transformations are less than 100% efficient. This inefficiency results from the loss of usable energy in the form of heat. The quantitative measure of the disorder created by any spontaneous reaction is entropy. All spontaneous reactions increase entropy.

Metabolism is the sum of the vast array of energy and matter transformations in cells, which occur in step-by-step sequences called *metabolic pathways*. Each reaction of a metabolic pathway rearranges atoms into new compounds. Chemical reactions change the amount of free energy available for work. This change in free energy is the amount of energy available to form other bonds.

Many energy transformations in organisms involve chemical reactions called *oxidations* and *reductions*. Oxidation is the loss of electrons from a molecule; reduction is the addition of electrons to a molecule. Oxidation and reduction reactions always occur simultaneously.

Adenosine triphosphate, or ATP, is the energy currency of cells, which use it to do work. ATP is suited for this function because (1) most of its energy is available immediately to cells, and (2) it releases about twice as much energy as is needed to drive most cellular reactions. ATP is used by all cells for energy transformations. Many of these transformations involve coupled reactions in which the energy released by ATP is used to drive other reactions. Other compounds involved in cellular metabolism include NAD, NADP, FAD, and cytochromes.

Enzymes catalyze biological reactions by lowering the energy of activation of the reaction. Enzymes are critical to life because they speed spontaneous reactions to a biologically useful rate. Enzymes, and therefore metabolism, are controlled by feedback inhibition, competitive inhibition, and noncompetitive inhibition.

Photosynthesis and respiration are the major energy transformations in organisms. Photosynthesis uses light energy to reduce carbon dioxide to carbohydrate, while respiration converts carbohydrate to ATP. Together, photosynthesis and respiration comprise a system by which energy flows through an ecosystem and is ultimately converted to heat. Unlike energy, nutrients cycle in an ecosystem.

Questions for Further Thought and Study

1. Why are the laws of thermodynamics important?
2. Does gravity affect the potential energy of an object? If so, how?
3. What's the difference between potential and kinetic energy?
4. What is entropy? How does the second law of thermodynamics affect plants?
5. Why isn't heat considered usable energy?
6. Why haven't all spontaneous reactions already occurred?
7. Which contains more energy, ATP or reduced NAD? How do you know?
8. How are ATP and NADP similar? How are they different?
9. What are coenzymes, and why are they important?
10. What is energy, and why is it important?
11. What are coupled reactions? Why are they important?
12. Several catabolic enzymes involved in cellular respiration have allosteric sites that fit ATP and ADP. How could this regulate cellular respiration?
13. What does ΔG tell us about a chemical reaction?
14. How could a cell minimize the amount of enzyme needed to produce or concentrate a metabolic product?
15. As you read this book, your body uses about 25 cal sec^{-1}. Your brain alone consumes the equivalent of 250 M&M's worth of sugar per day. Where does this energy ultimately come from?

Suggested Readings

ARTICLES
Davis, Ged R. 1990. Energy for planet earth. *Scientific American* 263:53–62.

Storey, Richard D. 1992. Textbook errors and misconceptions in biology: Cell energetics. *The American Biology Teacher* 54:161–166.

BOOKS
Becker, W. M. 1986. *The World of the Cell.* Menlo Park, CA: Benjamin/Cummings.

Harold, F. M. 1986. *The Vital Force: A Study of Bioenergetics.* New York: W. H. Freeman.

Stryer, L. 1989. *Biochemistry.* 3d ed. San Francisco: W. H. Freeman.

The starch-rich endosperm of this corn kernel has been stained blue-black with iodine. Starch is fuel for the emergent seedling.

Respiration

CHAPTER 6

Chapter Outline

INTRODUCTION
RETRIEVING GLUCOSE FROM OTHER MOLECULES
  Retrieval from Sucrose
  Retrieval from Starch
HARVESTING ENERGY FROM GLUCOSE: AN OVERVIEW
GLYCOLYSIS
  Potential Energy of Glucose
  Breakdown of Glucose to Pyruvic Acid
THE KREBS CYCLE
  Steps in the Krebs Cycle
ELECTRON TRANSPORT AND OXIDATIVE PHOSPHORYLATION
  Electron Transport

BOX 6.1
HOW EFFICIENT IS RESPIRATION?

  Chemiosmosis and ATP Synthesis
OTHER TYPES OF RESPIRATION
  Anaerobic Respiration
  Respiration of Pentose Sugars
  Respiration of Lipids
  Cyanide-Resistant Respiration
  Photorespiration

BOX 6.2
WHY SKUNK CABBAGE GETS SO HOT

Chapter Summary
Questions for Further Thought and Study
Suggested Readings

Chapter Overview

All organisms, including plants, harvest energy from stored chemicals in much the same way. The metabolic pathways by which organisms liberate stored energy are referred to as *cellular respiration*. This process breaks down complex molecules into simpler chemicals and harvests the energy stored in them.

The products of the complete oxidation of glucose are the same whether the glucose is burned in the laboratory or respired in a cell: CO_2, H_2O, and energy. When it is burned, a mole of glucose (180 grams) makes a small fire that lasts for several minutes. In a cell, respiration controls this process, so the heat does not destroy the tissue. Instead, some of the energy from glucose is trapped in ATP. The heat lost during respiration never causes smoke or fire because it is released slowly by many chemical reactions.

There are several kinds of respiration, but the most common ones require oxygen and produce ATP. Other kinds of respiration make little or no ATP; rather, they lose energy as heat or transfer energy into the synthesis of organic molecules other than ATP.

Introduction

In the previous chapter you learned that all organisms convert energy from one form to another. Plants convert light energy to chemical energy by photosynthesis, a process that you will learn more about in the next chapter. Plants and animals convert stored chemical energy to ATP by respiration. The ATP harvested from energy stored in carbohydrates and other organic molecules fuels the energy-requiring activities of the cell, including the movement of organelles and small particles within a cell, active transport across membranes, and the transport of organic compounds from one part of a plant to another. The carbon from storage chemicals goes into carbohydrates, proteins, fats, nucleic acids, and other kinds of organic molecules.

Beginning with glucose, respiration harvests energy in three stages. The first stage is **glycolysis,** in which glucose is split into smaller molecules in the cytosol. The second stage, called the **Krebs cycle,** completely metabolizes the products of glycolysis (fig. 6.1). In the third stage several electron carriers, together called the **electron transport chain,** harvest energy from the movement of electrons (fig. 6.1). The Krebs cycle and the electron transport chain occur in mitochondria.

All three stages of respiration produce ATP. In this chapter you will learn how respiration accomplishes this task. Although the process is complicated, remember the theme of the chapter: Cells make ATP from energy stored in organic molecules.

CONCEPT

Energy stored in glucose is retrieved in the three stages of respiration: glycolysis, the Krebs cycle, and electron transport. Part of this energy is converted into ATP, which can be used for cellular work.

RETRIEVING GLUCOSE FROM OTHER MOLECULES

Most discussions of respiration focus on what happens to glucose, probably because glucose metabolism is so similar in all organisms. However, glucose is usually not abundant in cells; thus, respiration begins by converting storage compounds to glucose, which can then be shunted into glycolysis. Such storage compounds vary from one organism to another or, in plants, from one organ to another. For example, humans and other vertebrates store the polysaccharide glycogen; potatoes and bananas store starch; and Jerusalem artichokes and onions store polymers of fructose (fig. 6.2). Furthermore, sucrose is the common starter molecule for respiration in leaves, but in the same plant respiration may begin with starch in nongreen storage organs such as roots and stems.

In other plants respiration may begin with different carbohydrate polymers, with different sugars, with fats, with organic acids, or with proteins. Nevertheless, the most well-known and probably most common sources for glucose in plants are sucrose and starch.

Retrieval from Sucrose

Sucrose, abundant in sugarcane stems and sugar beet roots, is common in plants and is commercially important because humans consume so much of it as table sugar. Besides being a common energy source in photosynthetic cells, sucrose is the carbohydrate that moves most readily through conducting cells to growing tissues.

In slow-growing and mature cells, the degradation of sucrose is mostly an irreversible hydrolysis to its component monomers, glucose and fructose:

$$\text{sucrose} + H_2O \rightarrow \text{glucose} + \text{fructose}$$

Enzymes that catalyze this reaction are called **sucrases;** they occur in the cytosol, in the central vacuole, and sometimes in cell walls. Hydrolysis by sucrases frees glucose and fructose for glycolysis.

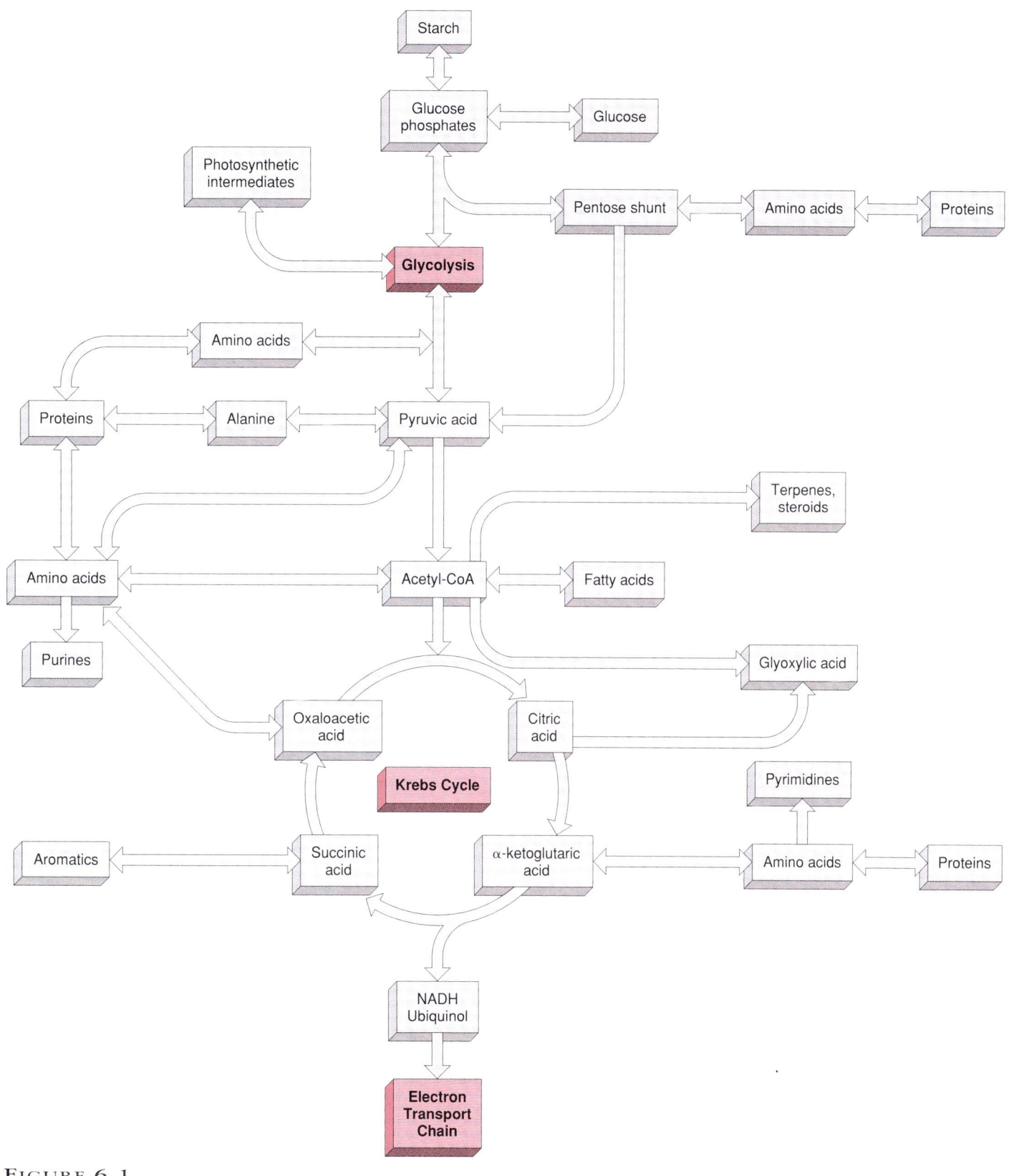

FIGURE 6.1

Respiration and its major metabolic relationships. The main pathway of respiration begins with the breakdown of carbohydrates, continues through the Krebs cycle, and ends with the electron transport chain. Many kinds of organic compounds are made from metabolic spinoffs of respiration. (The metabolic relationships shown here represent many of the hundreds of reactions that are depicted in fig. 5.6.)

CHAPTER SIX *Respiration*

FIGURE 6.2

Different organisms store polysaccharides in different forms.

In starch-storing organs and rapidly growing cells, sucrose from starch degradation is usually hydrolyzed by the enzyme **sucrose synthase** in a reversible reaction that releases fructose but binds glucose to a carrier molecule, **uridine diphosphate (UDP):**

$$\text{sucrose} + \text{UDP} + H_2O \rightleftharpoons \text{fructose} + \text{UDP-glucose}$$

Fructose from this reaction is available for glycolysis, but glucose is either used in cell-wall synthesis or released from the UDP-glucose complex for further respiration.

Retrieval from Starch

Starch is a complex branched polymer made of glucose monomers (Chapter 2). Because of the complexity of starch, its breakdown requires several enzymes for the complete retrieval of glucose. These enzymes attach directly to starch grains in chloroplasts and amyloplasts. Three general kinds of enzymes degrade starch. One kind is the **amylases,** which hydrolyze alpha-1,4 glucose linkages. Another kind of enzyme is **starch phosphorylase,** which cleaves glucose at one end of a glucose polymer by adding a phosphate to it:

$$\text{starch} + H_2PO_4 \rightleftharpoons \text{glucose-1-phosphate}$$

Although this reaction is reversible, starch synthesis from it is negligible. Also, glucose-1-phosphate from this reaction bypasses the need for an ATP in the first step of glycolysis, as discussed later in this chapter.

Neither the amylases nor starch phosphorylase can break the alpha-1,6 glucose linkages at the branch points in starch. Only a third group of enzymes, called **debranching enzymes,** can hydrolyze starch at branch points.

Although glucose is retrieved from starch in chloroplasts and amyloplasts, it must be exported to the cytosol for glycolysis. However, intact glucose does not move out of plastids readily. Therefore, glucose in plastids must be modified before it is exported to the cytosol. This usually entails breaking the glucose into two smaller molecules, exporting them, and then either reassembling them into glucose in the cytosol or shunting them directly into glycolysis.

HARVESTING ENERGY FROM GLUCOSE: AN OVERVIEW

The overall equation for the respiration of glucose is as follows:

glucose + oxygen → carbon dioxide + water + energy

$$C_6H_{12}O_6 + O_2 \longrightarrow CO_2 + H_2O + \text{energy}$$

This reaction links the solar energy obtained during photosynthesis, which is in glucose in this equation, to the energy that is needed by cells in all organisms. Some of the energy on the right side of this equation ends up in ATP; the rest either resides in CO_2 and H_2O or is lost as heat. The efficiency of the overall process is the topic of boxed reading 6.1 on pp. 120–121.

The Lore of Plants

The amount of CO_2 released by running a car engine is about 100 kilograms per 40-liter tank of gasoline. If one mole of glucose (180 grams) yields six moles of CO_2 (264 grams) during aerobic respiration, then burning a tank of gasoline is equivalent to the respiration of about 68 kilograms of glucose, or the mass of the average adult human. This is roughly the amount of glucose respired by 10,000 young sunflower plants during a warm summer night.

Synthesis of ATP occurs in two ways. The first involves the transfer of phosphate from organic compounds to ADP. Because a substrate provides the phosphate for ATP directly, this reaction is called **substrate-level phosphorylation** (fig. 6.3). The transfer of phosphate is catalyzed by an enzyme that binds the substrate and ADP.

The energy of substrate-level phosphorylation comes from the phosphate-containing substrate. Regardless of which substrate is used, the removal of a phosphate from it must release more energy than necessary for making ATP from ADP. For example, removal of a phosphate from phosphoenolpyruvic acid (PEP) releases 14.8 kcal mol^{-1} (fig. 6.3). This is more than enough energy to phosphorylate ADP, which requires 7.3 kcal mol^{-1}. The remaining energy (7.8 kcal mol^{-1}) is lost as heat.

The second way that ATP is made involves coupling energy from an electron donor to an electrochemical gradient that spans the inner mitochondrial membrane. The electron donor is usually NADH, a molecule that you learned about in

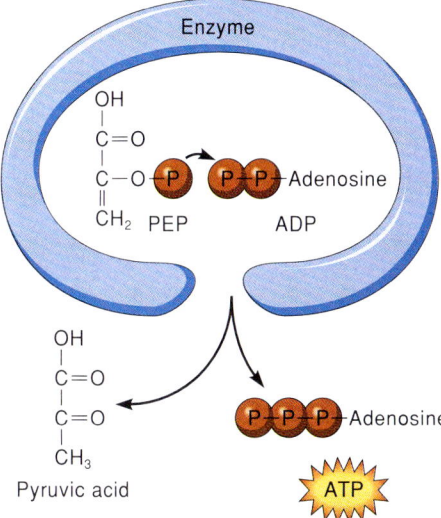

FIGURE 6.3

Substrate-level phosphorylation. Phosphate is enzymatically transferred from a substrate, in this case phosphoenolpyruvic acid (PEP), to ADP. The reaction proceeds because the phosphate bonds of ATP are more stable than the phosphate bond of PEP.

Chapter 5. Electrons from NADH are passed through a chain of electron carriers by a series of oxidation-reduction reactions. The energy from this movement of electrons is used by proton pumps to generate a proton electrochemical gradient that fuels the phosphorylation of ADP to ATP. The overall process is called **oxidative phosphorylation.**

Membranes play an important role in coupling energy between the electron transport chain and the phosphorylation of ADP. This energy-coupling is usually referred to as **chemiosmosis** because it involves chemical reactions and transport across membranes. Chemiosmosis is also used to make ATP in chloroplasts during photosynthesis (see Chapter 7). The mechanism of chemiosmosis in mitochondria is described later in this chapter.

Of the two ways that ATP is made during respiration, substrate-level phosphorylation is the simpler, but, it accounts for only a small percentage of ATP synthesis. Most ATP is made by oxidative phosphorylation.

In glycolysis and the Krebs cycle, chemical bond energy from different substrates is used to bond phosphate to ADP to make ATP. This is called *substrate-level phosphorylation*. In electron transport, energy for phosphorylating ADP comes from a series of oxidation-reduction reactions. This is called *oxidative phosphorylation*, or phosphorylation by *chemiosmosis*.

GLYCOLYSIS

Potential Energy of Glucose

As you learned in Chapter 5, glucose contains 686 kcal mol^{-1}. This means that, with the addition of some heat (i.e., energy of activation, E_{act}) to get it going, one mole of glucose will yield 686,000 calories when it is completely oxidized to CO_2 and H_2O. However, cells recover only a small amount of this energy during glycolysis; the rest is lost as heat. Because the energy is released in small increments, the excess heat does not damage the cell.

Breakdown of Glucose to Pyruvic Acid

During glycolysis, glucose is split into two three-carbon compounds. The entire process requires ten steps, all of which occur in the cytosol (fig. 6.4). The first step phosphorylates glucose by using one ATP. This phosphorylation activates glucose so that the appropriate enzyme can do the next step. Phosphorylated glucose is rearranged and phosphorylated again by a second ATP before it is split in half. After it is split, each of the three-carbon products has one phosphate. One of the products is glyceraldehyde-3-phosphate, which is further metabolized in glycolysis. The other product is dihydroxyacetone phosphate, which is not metabolized directly but is instead converted to glyceraldehyde-3-phosphate. These products represent the halfway point of glycolysis (see steps 1–5 in fig. 6.4). By this point, energy has been used but not yet harvested.

The first energy-harvesting step of glycolysis occurs in the second half of the pathway. This step is the reduction of NAD$^+$ to NADH by the oxidation of glyceraldehyde-3-phosphate. After this step, some of the energy from glucose is stored in the energy-rich electrons of NADH. This oxidation also releases enough energy to add a second phosphate group to glyceraldehyde-3-phosphate, converting it to 1,3-bisphosphoglyceric acid. At this point, glycolysis is finally ready to make ATP. Substrate-level phosphorylation occurs when one of the phosphates of 1,3-bisphosphoglyceric acid is transferred to ADP. The three-carbon molecule that remains is then rearranged to form phosphoenolpyruvic acid, which becomes pyruvic acid when it gives up its phosphate to a second ADP. In this way, each glyceraldehyde-3-phosphate from the first half of glycolysis is used to make two molecules of ATP and one molecule of pyruvic acid.

Since two molecules of glyceraldehyde-3-phosphate can be made from one molecule of glucose, each glucose yields four ATPs and two pyruvic acids from glycolysis. However, two ATPs are used in the first half of glycolysis, so the net gain is two ATPs for each glucose.

By the end of glycolysis, a small amount of the chemical energy that started out in glucose ends up in ATP and NADH.

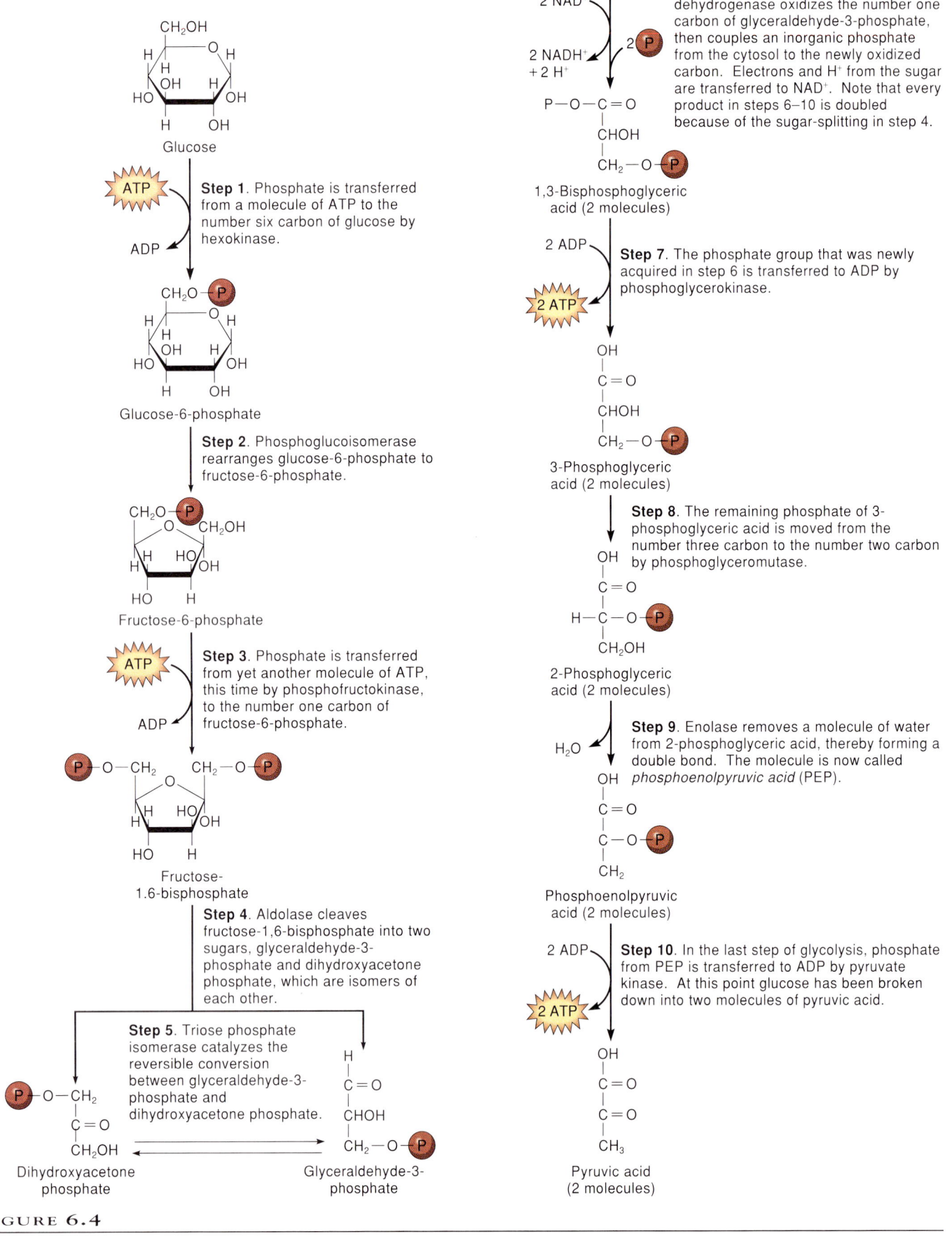

FIGURE 6.4

The ten steps of glycolysis. Each step requires a specific enzyme. All enzymes of glycolysis occur in the cytosol, outside the mitochondria.

Most of the energy of glucose remains in pyruvic acid. The energy stored in pyruvic acid is used to make more ATP in the mitochondrion. The synthesis of ATP from the energy of pyruvic acid occurs in the second phase of respiration, the Krebs cycle.

CONCEPT

During the first half of glycolysis, glucose is phosphorylated twice and split into two three-carbon compounds. This first half of glycolysis uses ATP. The three-carbon compounds are oxidized to make pyruvic acid during the second half of glycolysis, which also makes ATP and NADH.

THE KREBS CYCLE

Pyruvic acid that is transported into the mitochondrion is not used in the Krebs cycle directly. Instead, pyruvic acid first loses a molecule of carbon dioxide. The remaining acetyl group is attached to a coenzyme to form **acetyl coenzyme A (acetyl-CoA).** Figure 6.5 outlines the conversion of pyruvic acid to acetyl-CoA in the mitochondrion. Note that when carbon dioxide is released from pyruvic acid, NAD⁺ is reduced to NADH.

Glycolysis and the Krebs cycle are linked by the conversion of pyruvic acid to acetyl-CoA. Pyruvic acid is the final product of glycolysis, and acetyl-CoA is the compound that enters the Krebs cycle.

Steps in the Krebs Cycle

The Krebs cycle, named in honor of Sir Hans Krebs, is a cycle because the last step regenerates the starter chemicals for the first step (fig. 6.6). In all, there are eight steps, seven of which occur in the mitochondrial matrix. Step 6 occurs in the mitochondrial membrane.

In the first step, coenzyme A is cleaved from acetyl-CoA, and the acetyl group is attached to **oxaloacetic acid,** which has four carbons. The resulting six-carbon compound is **citric acid.** (The Krebs cycle is also known as the **citric acid cycle** because it begins with citric acid, or the **tricarboxylic acid cycle** because citric acid has three carboxyl groups.) Citric acid is then rearranged to **isocitric acid** (step 2), which is the beginning substrate for two oxidative steps that entail the removal of two molecules of carbon dioxide. These steps also reduce two molecules of NAD⁺ to NADH.

In the first oxidative step from isocitric acid (step 3), the removal of carbon dioxide yields **alpha-ketoglutaric acid.** In the second oxidative step (step 4), carbon dioxide is removed from alpha-ketoglutaric acid, and the product is bound to coenzyme A to make **succinyl-CoA.** Each step also reduces a molecule of NAD⁺ to NADH. Coenzyme A is then cleaved from succinyl-CoA, thereby producing **succinic acid** and providing energy for the substrate-level phosphorylation of ADP to ATP (step 5).

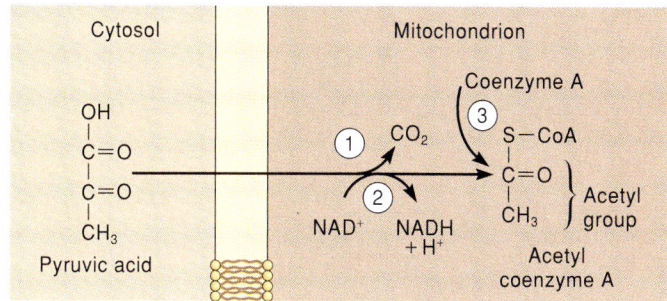

FIGURE 6.5

Conversion of pyruvic acid to acetyl-CoA. (1) CO_2 is removed from pyruvic acid that has been imported into the mitochondrion. (2) The remaining two-carbon fragment is oxidized to an acetyl group, while NAD⁺ is reduced to NADH. (3) The acetyl group is attached to the sulfur atom of coenzyme A (CoA).

Following the formation of succinic acid, three more oxidative steps occur (steps 6–8): succinic acid is oxidized to **fumaric acid,** fumaric acid to **malic acid,** and malic acid to oxaloacetic acid. The oxidation of succinic acid also reduces the electron carrier **ubiquinone** to **ubiquinol** (fig. 6.7)[1]; the oxidation of malic acid to oxaloacetic acid reduces NAD⁺ to NADH. Thus, for each acetyl-CoA that enters the Krebs cycle, only one ATP is made by substrate-level phosphorylation. Most of the energy derived from the oxidative steps of the Krebs cycle is stored in the high-energy electrons of NADH and ubiquinol. The energy in these molecules is harvested in the third phase of respiration, oxidative phosphorylation.

CONCEPT

Pyruvic acid from glycolysis is moved into mitochondria and converted into acetyl-CoA, which enters the Krebs cycle. Each turn of the Krebs cycle uses one molecule of acetyl-CoA and produces one molecule of ATP, one molecule of ubiquinol, and three molecules of NADH. The acetyl carbons are released as carbon dioxide.

What are Botanists Doing?

Plant scientists are finding that an atmosphere of artificially enriched CO_2 affects plant respiration. Using the reference resources at your library, find out how plant respiration is affected by a high-CO_2 environment. What hypotheses, if any, have plant scientists proposed to explain these effects?

---

1. Most textbooks show that the oxidation of succinic acid reduces FAD to $FADH_2$. (Recall that $FADH_2$ is similar to NADH; see Chapter 5.) Although this reaction does occur, FAD and $FADH_2$ are highly transient, enzyme-bound intermediates during the reduction of ubiquinone to ubiquinol; therefore, ubiquinol is the mobile product from this step in the Krebs cycle, not $FADH_2$.

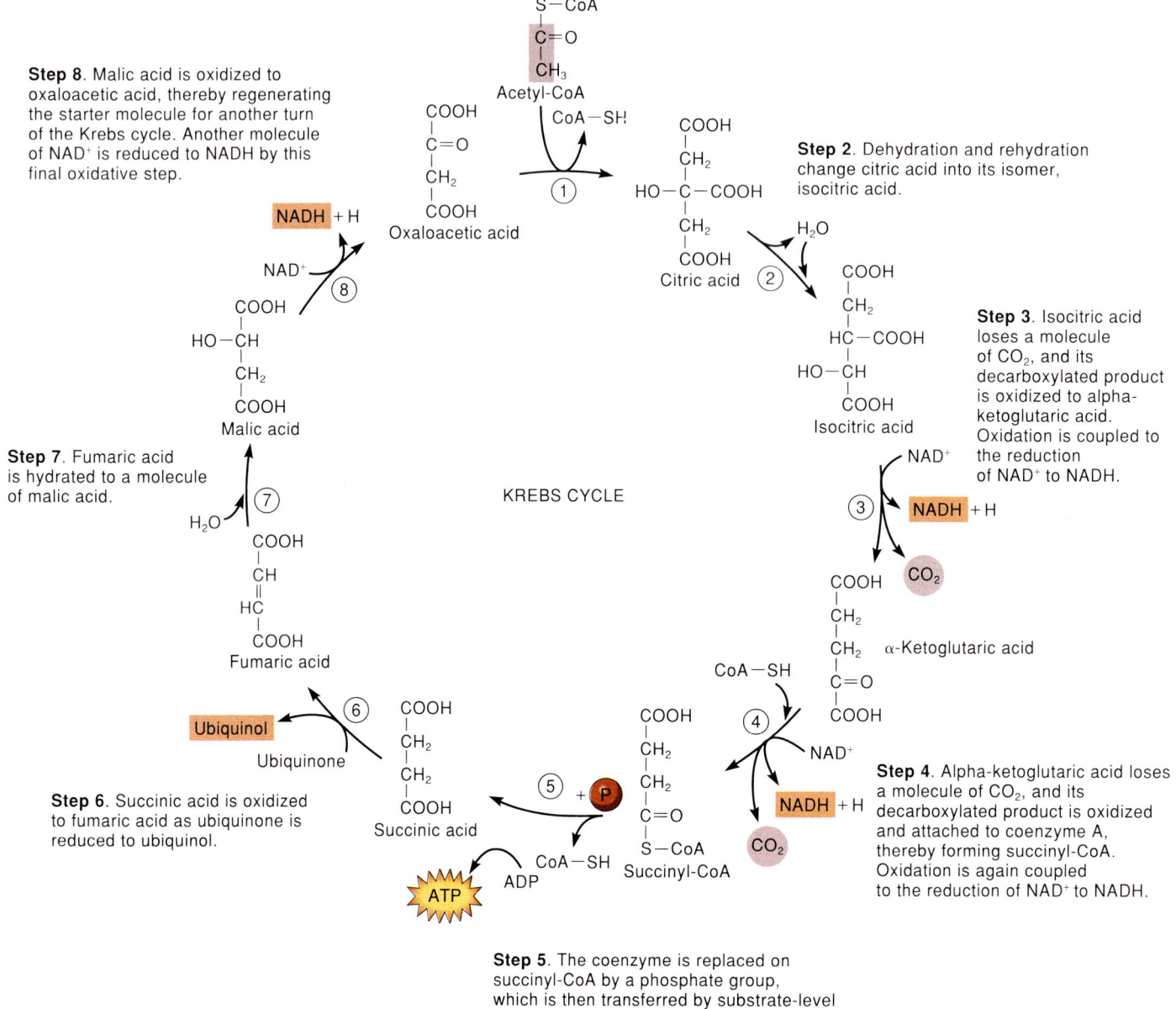

FIGURE 6.6

The Krebs cycle. Purple carbons are either added from acetyl-CoA or removed as CO_2. Red boxes contain reduction products whose formation depends on oxidative steps in the cycle.

ELECTRON TRANSPORT AND OXIDATIVE PHOSPHORYLATION

Relative to glycolysis and the Krebs cycle, oxidative phosphorylation is a bonanza for ATP synthesis. Energy for oxidative phosphorylation comes from NADH and ubiquinol, which are produced by the first two phases of respiration. The high-energy electrons of NADH and ubiquinol are not used directly for ATP synthesis; instead, they start a series of oxidation-reduction reactions that move electrons through several carriers. The sequence of electron carriers is known, appropriately enough, as the **electron transport chain.** Energy from electron flow through these carriers maintains a proton gradient across the inner mitochondrial membrane, which drives the synthesis of ATP. *Oxidative phosphorylation* refers to the combination of

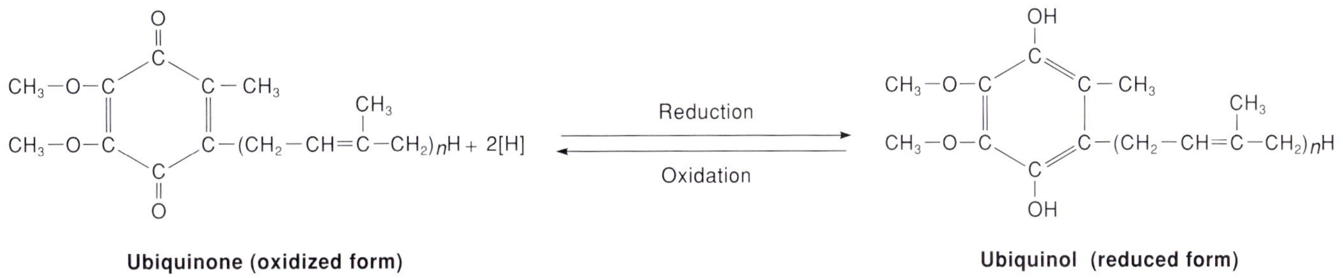

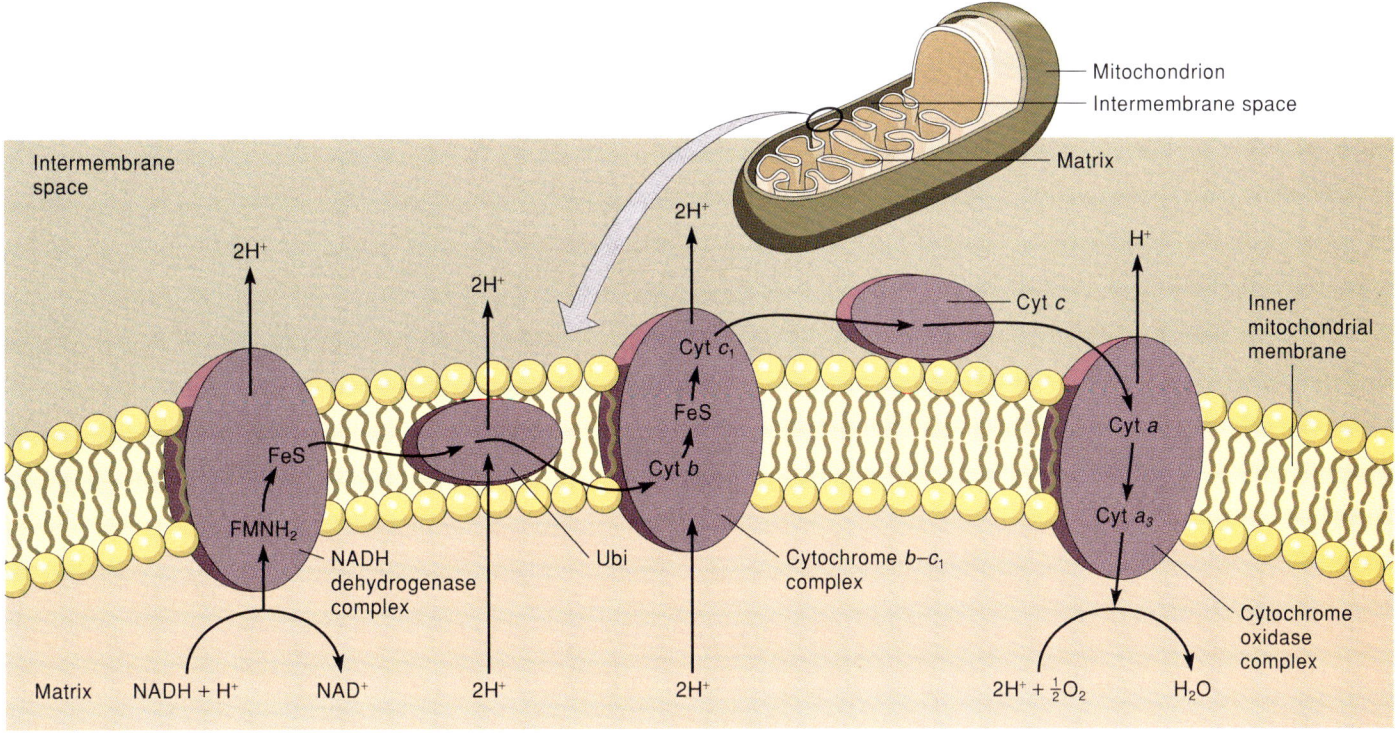

FIGURE 6.7

Reversible conversions between ubiquinone and ubiquinol. The n in each formula stands for the number of times the 5-carbon side chain is repeated. Naturally occurring ubiquinones have 6–10 of these 5-carbon units (30–50 carbons in the side chain).

FIGURE 6.8

Model for the sequence of steps in the electron transport chain. Two mobile electron carriers, ubiquinone (Ubi) and cytochrome c (Cyt c), transfer electrons between three electron-carrier complexes that are embedded in the membrane: the NADH dehydrogenase complex, the cytochrome b-c_1 complex, and the cytochrome oxidase complex. Electron flow along the chain causes proton pumping from the mitochondrial matrix into the intermembrane space. This model shows that one NADH fuels the transport of six protons, although the number of protons is variable and can be as high as eleven.

the oxidation-reduction reactions of electron transport that enable a cell to use the energy in NADH and ubiquinol to phosphorylate ADP to ATP.

Electron Transport

The electron transport chain is like a series of tiny, successively stronger magnets; that is, each has a higher potential than its predecessor. This means that each component of the chain pulls electrons away from a weaker neighbor and gives them up to a stronger one. Thus, the strongest acceptor in the chain is the terminal acceptor, oxygen, which has a potential of 0.82 volts.

There are at least nine electron carriers in the series, most of which are proteins. Some of these carriers also accept protons and release them into the intermembrane space. Although scientists are uncertain how these components work together for electron and proton transport, figure 6.8 shows a widely accepted model for the sequence of steps in the chain.

BOXED READING 6.1
HOW EFFICIENT IS RESPIRATION?

An age-old question in biology and botany classes is, "How many ATPs can a cell make from a molecule of glucose?" This seems like a simple but important question that can be answered by adding up the number of times ADP is phosphorylated between glucose and the end of the electron transport chain. If we knew this number, we could calculate respiratory efficiency by comparing the number of calories in glucose to the number of calories in ATP made by respiration. However, such calculations require some simple and probably incorrect assumptions.

The Dogma of Calculating Respiratory Efficiency

The usual calculation shows that a maximum of thirty-six molecules of ATP can be made from one molecule of glucose (see Table Box 6.1). This number is obtained by adding the presumed maximum net number of ATPs from glycolysis, the Krebs cycle, and oxidative phosphorylation. For example, substrate-level phosphorylation yields two ATPs from glycolysis and two ATPs from the Krebs cycle (one ATP each from two turns of the cycle). But these are the only steps where ATP can be counted directly. Most ATP from respiration is made by the electron transport chain via oxidative phosphorylation. To estimate the number of ATPs made by the electron transport chain, we assume that one ATP is produced for each pair of actively transported protons. This means that, based on the model in figure 6.8, three ATPs would be produced from each NADH, and two ATPs from each ubiquinol. Thus, we calculate ATP production from the number of NADHs and ubiquinols that enter the chain. Hence, one molecule of glucose yields two NADHs from glycolysis, two NADHs from decarboxylating two molecules of pyruvic acid to acetyl-CoA, and six NADHs and two ubiquinols from two turns of the Krebs cycle. The ten NADHs would yield thirty ATPs, and the two ubiquinols would yield four more ATPs. Added to the four ATPs from substrate-level phosphorylation, the total is thirty-eight ATPs. However, NADH made from glycolysis must be actively transported into mitochondria, perhaps using as many as two ATPs. This would reduce the net production of ATPs from a molecule of glucose to thirty-six. Glucose contains 686 kcal mol^{-1} of potential energy, and ATP contains 7.3 kcal mol^{-1} of potential energy. Based on the ratio of thirty-six ATPs to one glucose, the maximum efficiency of respiration is $(36 \times 7.3) / 686 = .383$, or about 38%.

An Alternative View

The ledger for ATP synthesis is simple for substrate-level phosphorylation. However, such ATP-counting is much more complex for oxidative phosphorylation than our ledger shows. The reason is that NADH production does not translate directly into a specific number of ATPs by chemiosmosis.

Some variability in the energy harvest comes from the use of NADH from glycolysis. This variability occurs because NADH made in the cytosol cannot cross the inner mitochondrial membrane. Instead, NADH from the cytosol may trigger a series of oxidation-reduction reactions that span the membrane. The final reactant produces a new molecule of NADH in the mitochondrial matrix (see Box Figure 6.1). Because some energy is lost in this process, each original NADH cannot power the complete reduction of another NAD$^+$. The amount of lost energy probably varies among different cell types and at different stages of cell growth, so the amount of energy from each NADH from glycolysis that actually gets to the electron transport chain cannot be calculated accurately.

Another difficulty in ATP-counting is that the amount of proton transport caused by each NADH is probably variable. For example, laboratory measurements show that each NADH can transport up to eleven protons

TABLE BOX 6.1

Potential Production of ATP by Respiration of One Molecule of Glucose

			ATPs
I. Glycolysis			
Substrate-level phosphorylation:	2 ATP	→	2 ATP
Reduction of NAD$^+$:	2 NADH*		
	(to mitochondrion)		
II. Pyruvic acid→Acetyl-CoA (×2)			
Reduction of NAD$^+$:	2 NADH*		
III. Krebs cycle (×2)			
Substrate-level phosphorylation:	2 ATP	→	2 ATP
Reduction of NAD$^+$:	6 NADH*		
Reduction of ubiquinone:	2 ubiquinols**		
IV. Electron transport			
*Oxidation of 10 NADH:			30 ATP
**Oxidation of 2 ubiquinols:			4 ATP
			38 ATP
Minus energy for active transport of NADH from glycolysis into mitochondrion			−2 ATP
		TOTAL:	36 ATP

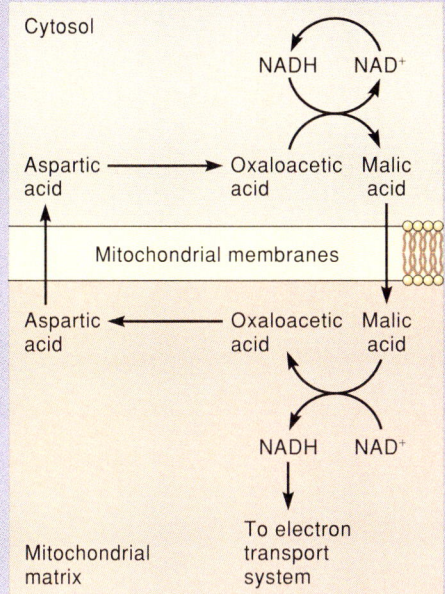

BOX FIGURE 6.1

One of the pathways for energy transfer from cytosolic NADH to mitochondrial NADH. In this pathway, cytosolic NADH reduces oxaloacetic acid to malic acid, which is imported into the mitochondrion. Oxidation of malic acid to oxaloacetic acid regenerates NADH in the mitochondrion by reducing NAD^+. The transfer cycle begins again when oxaloacetic acid is converted to the amino acid aspartic acid, which is exported to the cytosol and converted back to oxaloacetic acid. Energy is lost during this transfer cycle, so more than one cytosolic NADH is required to regenerate each mitochondrial NADH.

from the mitochondrial matrix to the intermembrane space, not six as predicted by the model in figure 6.8. If a pair of protons could fuel the production of one ATP, then one NADH could produce 5.5 ATPs instead of three, as previously assumed. Conversely, recent measurements of chemiosmosis show that the energy of at least four protons is needed to make each ATP. Thus, the maximum number of ATPs from NADH would be 2.75—that is, eleven protons divided by four protons per ATP. Ignore for the moment that three-quarters of a phosphate bond makes little sense. The main problem with this kind of counting is that it implicitly assumes that *all* protons diffusing back through the inner mitochondrial membrane go through ATP synthases. This is incorrect; the proton gradient is used for other things besides making ATP. For example, inorganic phosphate and other metabolites are actively transported into the mitochondrial matrix by the proton gradient. The proton gradient is also used to export ATP from the mitochondrial matrix (and simultaneously import ADP), so that it can be used for metabolism elsewhere in the cell. Finally, some protons leak through the membrane, and their energy is not harvested at all.

We cannot calculate the number of ATPs made by the proton gradient because energy from the gradient is used for so many other processes at the same time. The energy demand by these processes varies among cell types and at different stages of cell development; therefore, the amount of ATP synthesis also varies under the same circumstances.

Regardless of the complexity of respiration, trying to balance the books on respiratory efficiency is still one of the favorite activities of biochemists. Current speculation is that a minimum of twenty-one ATPs are produced by one glucose, and that the maximum of thirty-six ATPs is probably unreachable. Thus, average respiratory efficiency is probably between 22%–38% (21–36 ATPs per glucose). For comparison, the lower limit of respiratory efficiency is probably equivalent to the efficiency of your car in harvesting energy from gasoline—about 20%–25%.

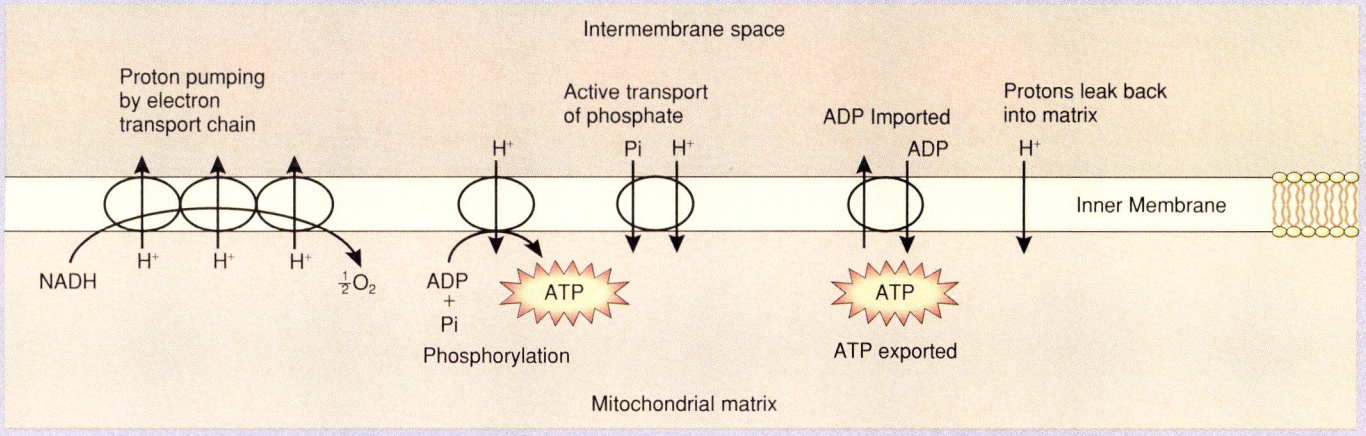

BOX FIGURE 6.2

Uses of the proton gradient generated by electron transport energy. The amount of energy consumed by phosphorylation and active transport or lost by ATP export and proton leakage is unknown.

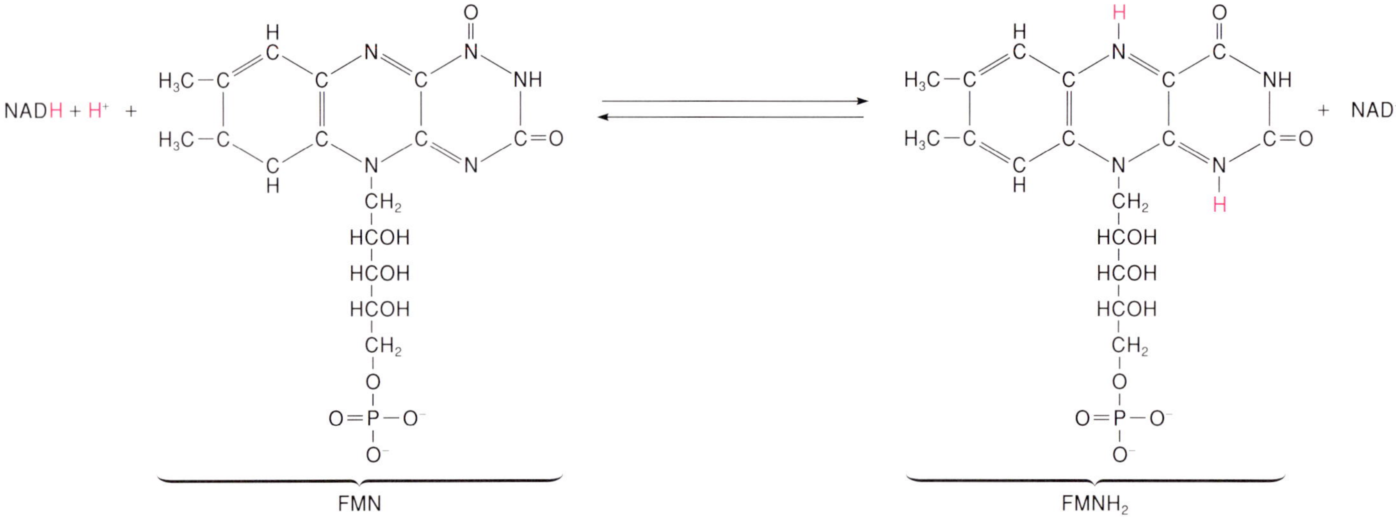

FIGURE 6.9

In the first step of the electron transport chain, flavin mononucleotide (FMN) is reduced in a reversible reaction to $FMNH_2$ by NADH and H^+ from the mitochondrial matrix. (FMN is also called *riboflavin* or *vitamin B_2*.)

According to the model, in the first step of electron transport, **flavin mononucleotide (FMN)** takes electrons from NADH in the mitochondrial matrix. FMN also takes two protons (H^+), one from NADH and one from the aqueous matrix (fig. 6.8). In this step, NADH is oxidized to NAD^+, and FMN is reduced to $FMNH_2$ (fig. 6.9). In the second step an iron- and sulfur-containing protein is reduced as $FMNH_2$ is oxidized back to FMN (i.e., the iron-sulfur protein pulls electrons from $FMNH_2$). Meanwhile, the protons from $FMNH_2$ are released into the intermembrane space. The net result of these first two steps is that energy from oxidation-reduction reactions drives the transport of protons from one side of the inner mitochondrial membrane to the other. The carriers responsible for these steps are called the **NADH dehydrogenase complex,** named after the enzyme that removes hydrogens from NADH.

In the third step of electron transport, electrons are pulled from the NADH dehydrogenase complex by ubiquinone (also called **coenzyme Q**), which shuttles the electrons to a complex of two **cytochromes** and another iron-sulfur protein. This complex is called the **cytochrome b-c_1 complex.** Protons are again transported from the matrix side of the inner membrane to the intermembrane space, using the energy of electron flow from ubiquinone through the cytochrome b-c_1 complex. Thus, ubiquinone acts as a second proton pump in the electron transport system, and the cytochrome b-c_1 complex acts as a third proton pump. Note that ubiquinone can also accept electrons directly from ubiquinol in the matrix.

Finally, electrons from the cytochrome b-c_1 complex are pulled away by a mobile cytochrome, cytochrome c, and shuttled to the terminal carrier complex. The terminal complex consists of two more cytochromes, cytochrome a and cytochrome a_3, which form the **cytochrome oxidase complex.** Cytochrome a_3 donates its electrons to oxygen, which is the terminal electron acceptor in the electron transport chain. The reduced oxygen then combines with protons in the mitochondrial matrix, producing water.

One of the key features of the electron transport chain is that it involves iron. This element occurs not only in the iron-sulfur proteins but also in the cytochromes. Thus, the oxidation and reduction of these electron carriers entails interchanges between the reduced (Fe^{2+}) and the oxidized (Fe^{3+}) forms of iron. In cytochromes, this change occurs in a complex ring group, called a **heme,** that is attached to the protein (fig. 6.10). The heme of cytochromes is similar to the heme of hemoglobin in the blood of animals.

Note that the transport of protons from one side of the inner mitochondrial membrane to the other requires energy, which is provided by the movement of electrons through the electron transport chain. Electron movement yields energy because each reduced electron-carrier has less free energy than does the donor immediately before it in the series. The lowest level of free energy in the chain occurs in the water that forms when oxygen is reduced in the last step.

Between the NADH at the start of electron transport and the oxygen at the end, the total change in free energy is -53 kcal mol^{-1}. The energy change between ubiquinol and oxygen is about one-third less because ubiquinol enters the chain at a lower energy level (fig. 6.11).

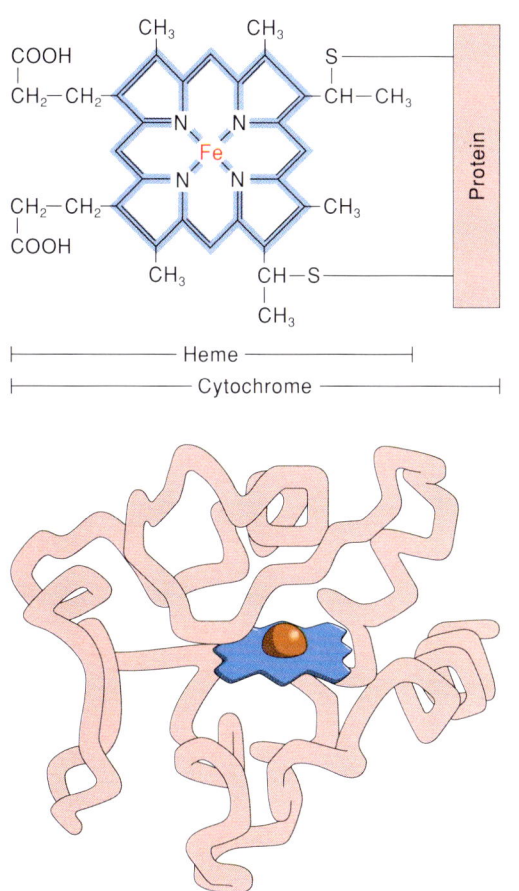

FIGURE 6.10

Structural models of cytochrome c, showing the iron-containing heme group. The heme group is blue and the iron is red. During electron transport, each iron accepts one electron and is reduced from Fe^{3+} to Fe^{2+}.

Why not have NADH and ubiquinol simply donate electrons directly to oxygen? If this happened, the total free-energy change would be the same. However, there are at least two reasons why this does not occur. The first is that such a single-step reaction would probably release too much heat for the cell to absorb without damage. The second is that a single step from NADH or ubiquinol to oxygen would skip the proton pumping that is necessary to make ATP.

CONCEPT

Electrons from NADH and ubiquinol reduce electron carriers that pass electrons to other carriers. The drop in free energy from one carrier to the next provides energy for transporting protons from the mitochondrial matrix into the intermembrane space. Thus, mitochondrial proton pumps are fueled by the energy of electron transport.

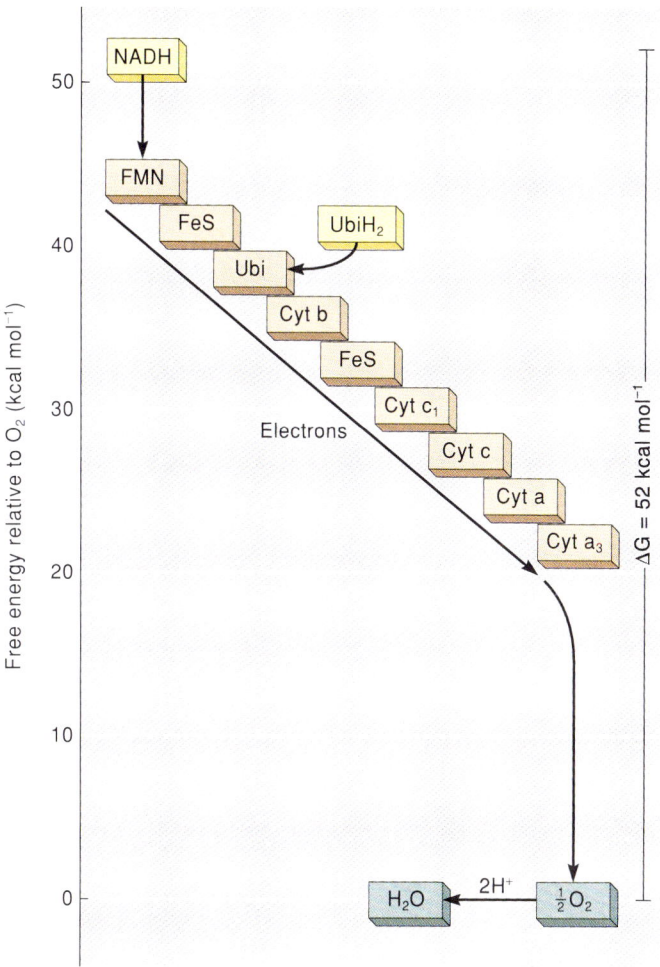

FIGURE 6.11

Free-energy scale of the electron transport chain. The energy drop for electrons from NADH to oxygen is 53 kcal mol^{-1}. The energy drop from ubiquinol is about one-third less.

Chemiosmosis and ATP Synthesis

How do the protons that are pumped by the electron transport chain provide energy for ATP synthesis? While many scientists searched for high-energy compounds formed directly by the electron transport chain, British scientist Peter Mitchell suspected that he could understand *how* ATP was made if he knew *where* it was made. Mitchell and others knew that all of the carriers of the electron transport chain were part of the inner mitochondrial membrane. Consequently, Mitchell suspected that the membrane must have a role in ATP synthesis. To test this idea, he first isolated the inner membranes of mitochondria and then made vesicles of the membranes. When he added oxygen and NADH to the medium containing the vesicles, the electron transport chain functioned as it does in intact cells. However, something else also happened: the pH of the medium surrounding the vesicles decreased (recall that a lower pH indicates a greater concentration of protons). Mitchell concluded that the protons were pumped out of the vesicles during electron transport. Scientists then used

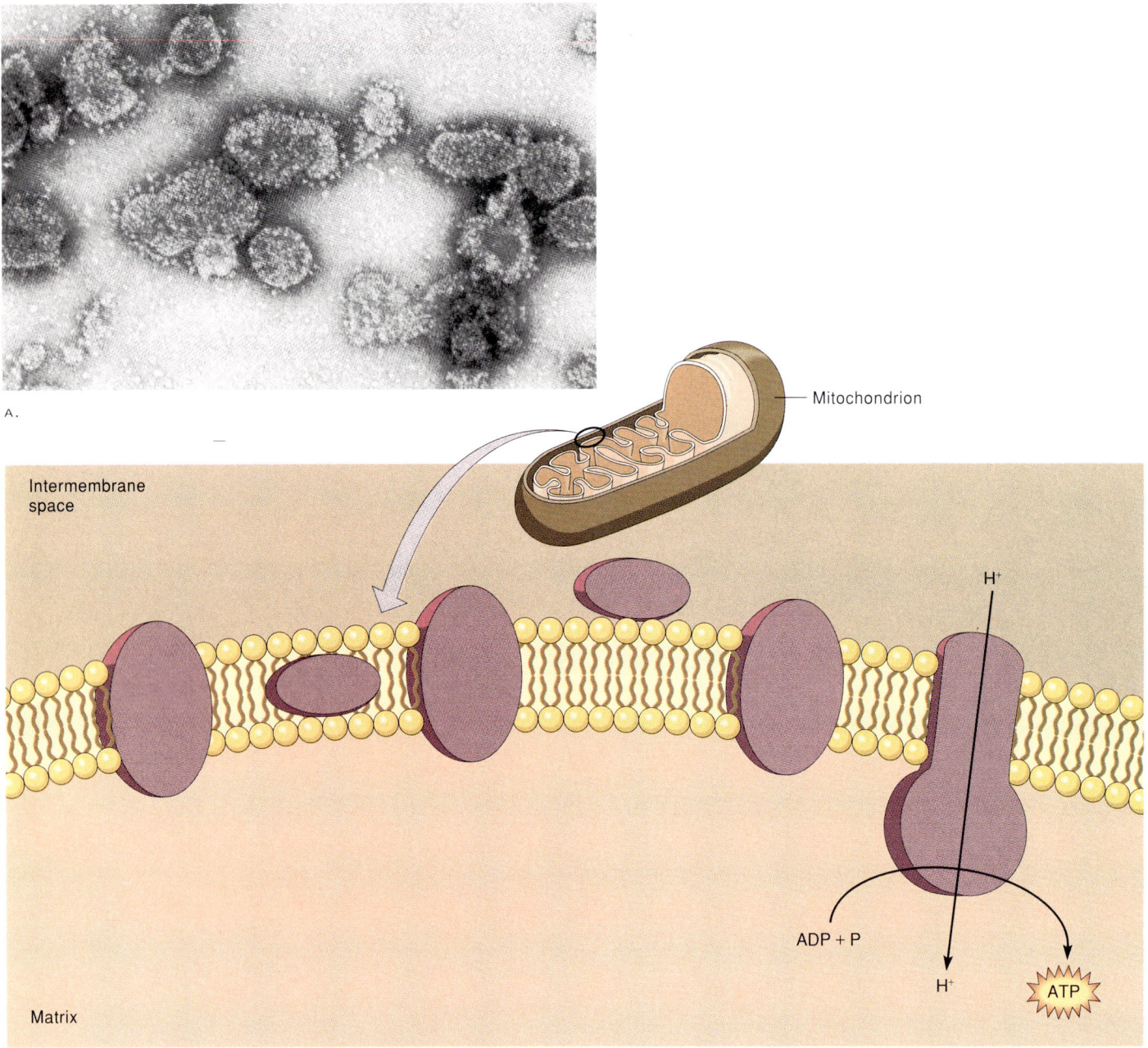

FIGURE 6.12

(a) Transmission electron micrograph of vesicles made of inner membrane fragments from mitochondria. Bulblike protrusions are portions of ATP-synthase complexes on the matrix side of the membrane, ×145,000. (b) Model of ATP-synthase complex in the inner mitochondrial membrane. Energy from electron transport pumps protons out of the matrix. Energy from the proton gradient is used for phosphorylating ADP when protons diffuse back into the matrix through ATP-synthase complexes.

electron microscopy to show that **ATP synthases,** the enzymes that make ATP, protruded from one side of the vesicles (fig. 6.12).

Mitchell used all this information to formulate his chemiosmotic theory of ATP synthesis:[2]

- Carriers of the electron transport chain are embedded in the inner mitochondrial membrane. The flow of electrons through this series of carriers is used to pump protons across the membrane. Because the membrane is not very permeable to protons, relatively few protons leak back across the membrane.

- Continued electron transport pumps more and more protons into the space between the inner and outer membranes of the mitochondria. This creates a gradient of protons across the membrane.

---

2. Mitchell won a Nobel Prize in 1978 for his study of chemiosmosis.

- The proton gradient serves as a battery to do work, including making ATP.

According to Mitchell's model, which is supported by much experimental evidence, the interaction between electron transport and ATP synthesis is indirect: electron transport produces a proton gradient, and that gradient drives the synthesis of ATP.

The inner membrane of a mitochondrion is folded many times, so that much surface area is available for membrane-dependent metabolism. Thousands of copies of the ATP-synthase complex are embedded throughout the membrane, making the inner mitochondrial membrane a powerhouse for making ATP.

CONCEPT

The proton gradient that is maintained by electron transport provides the energy necessary for making ATP. Protons diffuse down their electrochemical gradient through enzyme complexes that harness the energy of proton flow for the phosphorylation of ADP.

Doing Botany Yourself

Consider the question, "What effect does temperature have on aerobic respiration?" Without actually doing the laboratory work to get the answer, design an experiment and describe how you could do it. In your description, include any assumptions you must make, equipment and plant materials you might need, and methods of measurement you would have to learn.

OTHER TYPES OF RESPIRATION

The foregoing discussion of respiration focuses on the main type of respiration in most organisms, which is called **aerobic respiration** because it requires oxygen as the terminal electron acceptor. When oxygen is not available, other electron acceptors can be used in a process called **anaerobic respiration.** This is one of several alternatives to aerobic respiration that are discussed in the next few paragraphs.

Anaerobic Respiration

Although oxygen is required only at the end of electron transport during aerobic respiration, electron transport and the Krebs cycle are both inhibited when oxygen is not available. Glycolysis works normally in an oxygen-free environment, but energy stored in the electrons of the NADH from glycolysis can become a problem if the NADH cannot be used fast enough elsewhere in the cell, or if the supply of NAD^+ is depleted. In both cases, there would be no NAD^+ available to oxidize glyceraldehyde-3-phosphate; consequently, glycolysis would stop. In animals, certain fungi, and bacteria, pyruvic acid is reduced

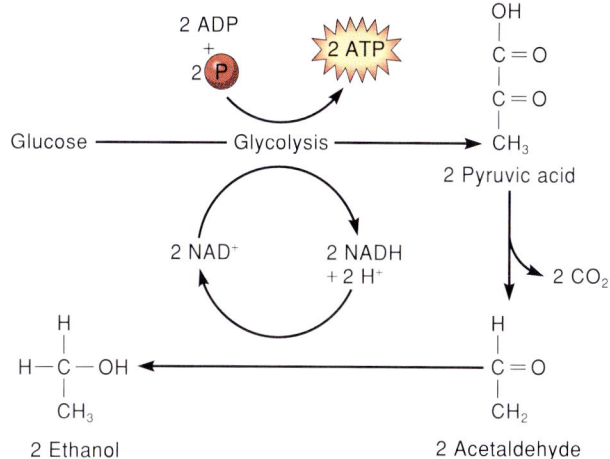

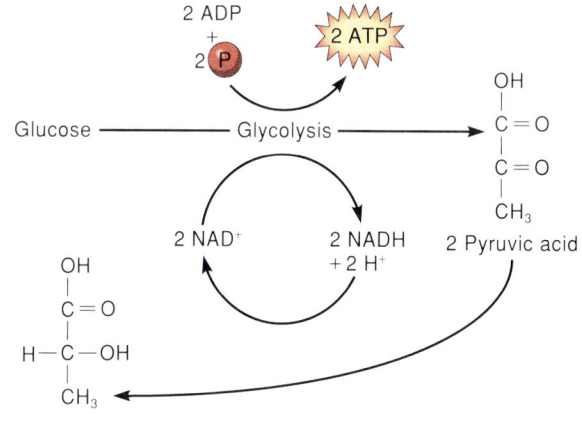

FIGURE 6.13

Anaerobic respiration usually produces ethanol in plants and certain fungi (a); in other fungi and in bacteria and animals it produces lactic acid (b).

to lactic acid by excess NADH. In other fungi and in plants, the abundance of NADH is relieved when pyruvic acid is converted to acetaldehyde, which is then reduced by NADH to ethanol (fig. 6.13). Together these anaerobic reactions are called **fermentation.** Unless they have adaptations (such as aerenchyma) to facilitate diffusion of oxygen to their roots, many plants ferment when they grow in mud or in oxygen-poor water; this is why most houseplants die when they are overwatered. Anaerobic respiration is inefficient because it produces little or no ATP.

Anaerobic respiration is economically important. For example, fermentation by bacteria and fungi is important for flavoring cheese and yogurt. Similarly, fermentation by yeast is important for making alcoholic beverages and bread; wine, beer, and bread are made by different strains of brewer's yeast (*Saccharomyces cerevisiae*). The carbon dioxide produced during fermentation makes bread rise, but the alcohol from fermentation evaporates when the bread is baked. Wine is usually made by yeast that grows on grapes, although fermented honey (mead), apples (cider), and other carbohydrate-rich substrates are also used to make wine or winelike beverages. Likewise, although beer is usually made from barley, rice beer is common in the Orient, and corn beer is made by the Tarahumara tribe of northern Mexico.

Respiration of Pentose Sugars

Glucose-6-phosphate from glycolysis is often redirected to a series of reactions that involve several five-carbon sugars (pentoses) in what is usually called the **pentose phosphate pathway** (fig. 6.14). Like glycolysis, the pentose phosphate pathway releases carbon dioxide and reduces an electron carrier—in this case NADP$^+$ instead of NAD$^+$. The NADPH is not used in electron transport, and no other useful energy comes from this pathway. However, NADPH is used in certain reactions that do not use NADH, such as the reduction of nitrate to ammonia. Nitrogen in ammonia can then be incorporated into the amino groups of amino acids. Of equal importance are the intermediates of pentose metabolism that are precursors of other kinds of molecules. These include ribose, which is used to make DNA and RNA, and erythrose, which is a precursor of many phenolic compounds, including lignin.

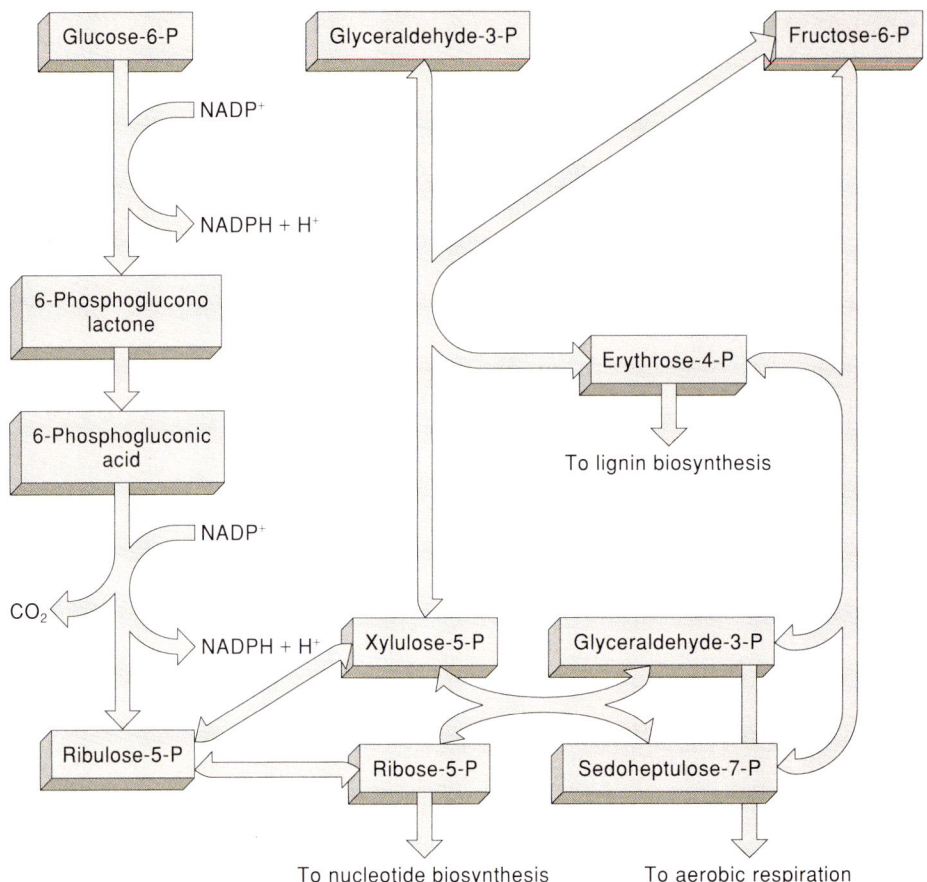

FIGURE 6.14

The major intermediates of the pentose phosphate pathway. Single-headed arrows indicate reversible conversions whose direction depends on metabolic demand for nucleotide synthesis, lignin synthesis, and aerobic respiration.

Writing to Learn Botany

What are the advantages and disadvantages of aerobic versus anaerobic respiration?

Respiration of Lipids

Lipids that include fatty acids are important storage compounds in oil-containing seeds. When such seeds germinate, their fatty acids are metabolized in glyoxysomes to recover the energy stored in them. Fatty acids are first removed from glycerol in storage triacylglycerides, then snipped into two-carbon pieces that are released as acetyl-CoA (fig. 6.15). This reaction, called **beta-oxidation,** is repeated for every pair of carbons until all of the fatty acid is converted into acetyl-CoA molecules. Thus, a molecule of stearic acid, which has eighteen carbons, yields nine molecules of acetyl-CoA. Beta-oxidation occurs either in the cytosol or in glyoxysomes. The release of acetyl-CoA drives the reduction of NAD$^+$ and FAD to NADH and FADH$_2$, respectively.

Acetyl-CoA from beta-oxidation can be used in the Krebs cycle for further recovery of the energy stored in fatty acids, or it can be routed to other metabolic pathways (see fig. 6.1). NADH and FADH$_2$ may be transported to mitochondria, where their electrons can be used in the electron transport chain. (FADH$_2$ donates electrons to an iron-sulfur protein that donates to ubiquinone.)

Cyanide-Resistant Respiration

Aerobic respiration may be inhibited when the terminal electron carrier, cytochrome oxidase, combines with cyanide (CN$^-$), azide (N$_3^-$), or certain other negatively charged ions. These ions

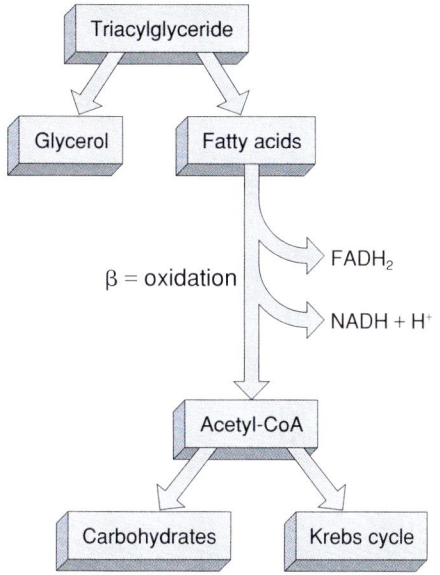

FIGURE 6.15

Respiration of lipids. Triacylglycerides are hydrolyzed into glycerol and fatty acids. Fatty acids are degraded by beta-oxidation into acetyl groups that are attached to coenzyme A, which can be used in respiration or for carbohydrate synthesis. FAD and NAD^+ are also reduced during beta-oxidation.

complex with iron in the oxidase, thereby poisoning the enzyme and halting electron transport. However, in many plant tissues respiration continues despite such respiratory poisons. Such respiration is called **cyanide-resistant respiration.** Certain plants, fungi, and bacteria have cyanide-resistant respiration, but it is rare in animals.

The reason respiration does not stop when cytochrome oxidase is poisoned is that mitochondria have an alternative short chain of electron carriers that branch off at ubiquinone. Because this chain also ends in an oxidase, it is aerobic—that is, oxygen is the terminal electron acceptor (fig. 6.16). Little or no oxidative phosphorylation is coupled to this chain, however, so the energy from the oxidation of NADH in this pathway produces heat instead of ATP.

Cyanide-resistant respiration is most active in sugar-rich cells, where glycolysis and the Krebs cycle occur unusually fast. This observation is indirect evidence that the main electron transport chain becomes saturated, and the cyanide-resistant pathway takes up the overflow of electrons. Thus, the main benefit of this pathway for plants is probably that it dissipates excess energy from respiration.

In some plants, such as the aroid lilies, the amount of heat generated from cyanide-resistant respiration is remarkable. For example, in the eastern skunk cabbage (*Symplocarpus*

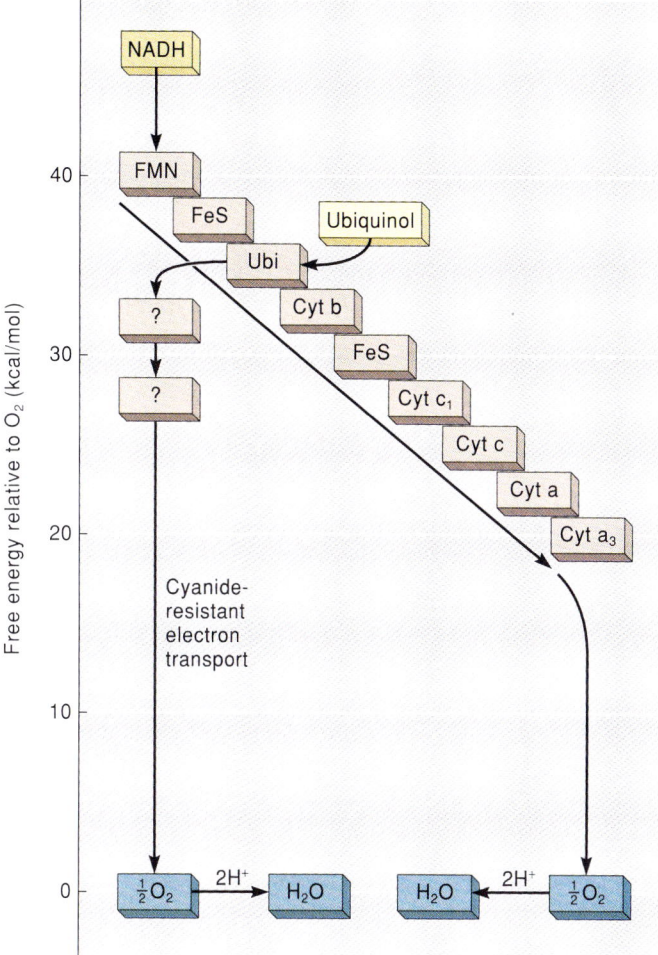

FIGURE 6.16

Model of electron transport chain for cyanide-resistant respiration. Question marks represent unknown electron carriers. As in cyanide-susceptible electron transport (fig. 6.8), the final carrier of cyanide-resistant electron transport is an oxidase, because the terminal electron acceptor is oxygen.

foetidus), this heat can raise temperatures in some parts of the plant more than 20° C above the air temperature. Such extreme temperatures are important in the ecology of these plants (see box 6.2; also see Chapter 5, p. 94).

Photorespiration

Some plants increase their use of oxygen when available CO_2 diminishes. This light-dependent process, called **photorespiration,** is discussed in more detail in Chapter 7. Photorespiration, which interferes with photosynthesis, causes large losses in the yield of many crops.

BOXED READING 6.2
WHY SKUNK CABBAGE GETS SO HOT

Skunk cabbage spends the spring and summer packing starch into its fleshy rhizome. The root then acts as a furnace during late winter by providing the energy needed to heat the flowering shoot and melt the snow around it as it grows up through the snow (see the accompanying photo). Skunk cabbage burns this fuel and uses oxygen at a rate comparable to that of a hummingbird. Heat is conserved by a thick, spongy structure that acts as an insulator around the flowering shoot. The overall effect is that the flowers develop in a near-tropical climate even though the plant is buried in snow.

Respiration that generates so much heat is costly to the plant; moreover, it makes little or no ATP. There is also no evidence that getting such an early jump on spring gives skunk cabbage an advantage over other plants. Why, then, does skunk cabbage get so hot? One possible explanation is an evolutionary one—that is, heat-generating respiration was passed down unchanged from the ancestors of skunk cabbage.

To study this hypothesis, botanists have compared skunk cabbage with its tropical relatives, which include many well-known houseplants, such as caladiums, philodendrons, voodoo lilies, and dumbcanes. These plants, as well as skunk cabbage, belong to the Arum family (Araceae), most of whose members still live in the tropics. When some aroids heat up, they vaporize foul odors that diffuse into the air around the flowering shoot. These odors include such chemicals as putrescine, cadaverine, and skatole. They smell like dung or decaying flesh, which attracts flies and other insects seeking suitable places for laying their eggs. Instead of finding places to lay their eggs, however, the flies and insects are often trapped temporarily by the plant and become covered with pollen. After they are released, they may get trapped again and leave pollen in another flowering shoot. In this way they transfer pollen from one flower to another.

Unlike its tropical relatives, skunk cabbage does not seem to benefit from its vaporized fragrance by attracting pollinators. For one thing, few insects are active when skunk cabbage is hot. Also, although skunk cabbage smells a lot like its tropical relatives, it has no mechanism for trapping insects.

The heat-generating ability of skunk cabbage and some of its relatives has intrigued botanists for decades. Is the heat-generating process an adaptation for surviving cold weather or is it merely an evolutionary remnant of a feature of some tropical plants? What kinds of information and experiments would help you answer this question?

Chapter Summary

Respiration harvests energy from organic molecules. There are several kinds of respiration, but the most common kind is aerobic respiration, which requires oxygen and produces ATP. Respiration occurs in three stages. The first stage, glycolysis, entails the breakdown of glucose into smaller organic compounds; it occurs exclusively in the cytosol and converts some of the energy stored in glucose to ATP and NADH. During glycolysis, ATP is made by substrate-level phosphorylation. The remainder of the energy from glucose is either lost as heat or remains in pyruvic acid, which is shunted into the mitochondria. In the mitochondria, pyruvic acid is converted into acetyl-CoA in a reaction that also reduces NAD^+ to NADH.

Acetyl-CoA enters the second stage of respiration, the Krebs cycle, which is a series of oxidation-reduction reactions that produce ATP, NADH, ubiquinol, and CO_2. As in glycolysis, ATP is made in the Krebs cycle by substrate-level phosphorylation.

Energy-rich electrons from NADH and ubiquinol fuel the electron transport chain, which is the third stage of respiration. Electrons move through a series of carriers that release energy at each step. The terminal electron acceptor is oxygen, which is reduced to form water. The energy of electron transport is used to pump protons from the mitochondrial matrix into the intermembrane space. As protons diffuse by chemiosmosis back into the matrix through ATP-synthase complexes, their energy is coupled to the phosphorylation of ADP to ATP. This synthesis of ATP from the energy of electron transport is called *oxidative phosphorylation*.

There are several kinds of respiration. Anaerobic respiration occurs in the absence of oxygen. The pentose phosphate pathway uses products from the breakdown of glucose to synthesize other organic molecules. Lipids are respired by beta-oxidation to yield acetyl units that enter the Krebs cycle. Cyanide-resistant respiration uses the energy of NADH to generate heat. Photorespiration occurs when photosynthesis cannot proceed due to the depletion of CO_2 in leaves.

Questions for Further Thought and Study

1. If the proton gradient in mitochondria worked only to make ATP, how many ATPs could be made from each NADH? Explain why such maximum ATP production cannot be attained.
2. Suggest reasons why the absence of oxygen inhibits both electron transport and the Krebs cycle but not glycolysis.
3. Without distillation, wine and other fermentation products do not exceed about 12% alcohol. Suggest reasons why anaerobic respiration does not continue after this concentration of alcohol is reached.
4. How does aerobic respiration differ from cyanide-resistant respiration?
5. What might the similarities and differences be between the inhibition of respiration by carbon monoxide versus carbon dioxide?
6. Where could an organism live if it could only respire anaerobically?
7. Too much oxygen is toxic to aerobic organisms. Explain how too much oxygen might affect respiration.

Suggested Readings

ARTICLES

Amthor, J. S. 1991. Respiration in a future, higher-CO_2 world: Opinion. *Plant, Cell and Environment* 14:13–20.

Babcock, G. T., and M. Wikstrom. 1992. Oxygen activation and the conservation of energy in cell respiration. *Nature* 356:301–308.

Cammack, R. 1987. $FADH_2$ as a "product" of the citric acid cycle. *Trends in Biochemical Sciences* 12:377.

Davies, D. D. 1987. Introduction: A history of the biochemistry of plant respiration. In D. D. Davies, ed., *The Biochemistry of Plants*. Vol. 11. New York: Academic Press, 1–38.

McCarty, R. E. 1985. H^+-ATPases in oxidative and photosynthetic phosphorylation. *BioScience* 35:27–33.

Meeuse, B. J. D., and I. Raskin. 1988. Sexual reproduction in the arum lily family, with emphasis on thermogenicity. *Sexual Plant Reproduction* 1:3–15.

Prince, R. C. 1988. The proton pump of cytochrome oxidase. *Trends in Biochemical Sciences* 13:159–160.

Ryan, M. 1991. Effects of climate change on plant respiration. *Ecological Applications* 1:157–167.

Storey, R. D. 1991. Textbook errors & misconceptions in biology: Cell metabolism. *The American Biology Teacher* 53:339–343.

Wivagg, D. 1987. Research reviews: How many ATPs per glucose molecule? *The American Biology Teacher* 49:113–114.

BOOKS

Amthor, J. S. 1989. *Respiration and Crop Productivity*. New York: Springer-Verlag.

Douce, R., and D. A. Day, eds. 1985. Higher Plant Cell Respiration. *Encyclopedia of Plant Physiology*. Vol. 18. New York: Springer-Verlag.

Egginston, S., and H. F. Ross, eds. 1992. *Oxygen Transport in Biological Systems: Modelling of Pathways from Environment to Cell*. New York: Cambridge University Press.

Kramer, S. P. 1986. *Getting Oxygen: What Do You Do If You're Cell Twenty-Two?* New York: Crowell.

Salisbury, F. B., and C. W. Ross. 1992. *Plant Physiology*. 4th ed. Belmont, CA: Wadsworth.

Plants use photosynthesis to convert sunlight to chemical energy.

CHAPTER 7

Photosynthesis

Chapter Outline

INTRODUCTION
HOW WE LEARNED ABOUT PHOTOSYNTHESIS
THE NATURE OF LIGHT
PIGMENTS
    Pigments in Plants
    Accessory Pigments

BOX 7.1
THE EVOLUTION OF PHOTOSYNTHESIS: WHY AREN'T PLANTS BLACK?

    Making and Destroying Pigments
CHLOROPLASTS
    Complexes of Pigments in Chloroplasts
    What Happens When Pigments Absorb Light?
    Photophosphorylation: Chemiosmosis in Chloroplasts
THE PHOTOCHEMICAL AND BIOCHEMICAL REACTIONS OF PHOTOSYNTHESIS
    The Photochemical Reactions of Photosynthesis
    The Biochemical Reactions of Photosynthesis
    The Efficiency of Photosynthesis
PHOTOSYNTHESIS IS NOT PERFECT: PHOTORESPIRATION
C_4 PHOTOSYNTHESIS
    Why Don't C_4 Plants Dominate the Landscape?
    C_3-C_4 Intermediates
CRASSULACEAN ACID METABOLISM (CAM)
CONTROL OF PHOTOSYNTHESIS
THE FATE OF PHOTOSYNTHATE
Chapter Summary
Questions for Further Thought and Study
Suggested Readings

Chapter Overview

Photosynthesis is a light-driven reaction that converts energy-poor compounds such as carbon dioxide and water to energy-rich carbohydrates. In plants, photosynthesis oxidizes water and releases oxygen. Over time, the oxygen released by photosynthesis dramatically changed the earth's atmosphere, and enabled the evolution of aerobic respiration. Today, virtually all of life depends on photosynthesis.

Introduction

In Chapter 5 you learned that cells are open systems—systems that can absorb, but not create, energy. Before the evolution of photosynthesis, virtually all organisms used organic compounds in the planet's "primeval broth" as sources of energy; they were *heterotrophs*—that is, organisms that make organic compounds from other energy-rich organic compounds that they absorb. As these organisms oxidized the carbon compounds from the primeval broth, they released carbon dioxide into the environment. Although this life-style was adequate in the short term, there were long-term problems: the primeval broth was nonrenewable, thereby creating a constant competition for the limited and ever-dwindling supply of carbon compounds. Since earth's earliest organisms couldn't fix carbon dioxide, they were doomed to extinction as soon as organic carbon was gone from the broth.

The evolution of photosynthesis about 3 billion years ago gave many organisms a new source of energy. Photosynthetic organisms, rather than relying on the ever-dwindling amount of primeval broth for energy, began to use sunlight as an energy source. In doing so, they became solar-powered and could exploit a reliable and abundant supply of energy. These organisms were the earth's first *photosynthetic autotrophs*—organisms that use light energy to make organic compounds from inorganic compounds such as water and carbon dioxide. Their ability to package light-energy into chemical energy soon supported almost all other forms of life on the planet. Moreover, these autotrophs radically changed the planet and its remaining organisms by decreasing the atmospheric concentration of carbon dioxide, diminishing the greenhouse effect, and filling the atmosphere with a waste product that some of the other organisms ultimately found essential for life: oxygen. All of the oxygen in air that we breathe has been cycled through plants via photosynthesis, "life's grand device." This oxygen allowed the evolution of aerobic respiration and higher life-forms.

Today, only about 500,000 of the earth's 3–5 million species of organisms are photosynthetic. However, the importance of these organisms cannot be overstated: Without photosynthesis, humans and virtually all other animals would become extinct.

Photosynthesis packages light into chemical bonds. Indeed, each year photosynthesis produces about 1.4×10^{14} kg (3.1×10^{14} lb) of carbohydrates—enough sugar to fill a string of boxcars reaching to the moon and back fifty times. Here's another way of appreciating the annual contribution of photosynthesis: If all of the sugars made by photosynthesis were converted to an equivalent amount of coal loaded into standard railroad cars (each holding about 50 tons of coal), then the earth's photosynthesis would fill more than 100 cars per second with coal. Photosynthesis is the reaction of life.

How We Learned about Photosynthesis

Considering how much we know about biology (witness the size of this and other biology books), it's easy to forget that for most of recorded history, scientists had no idea that the sun supplies the earth with virtually all of its energy, or that green plants trap energy and produce the invisible gases that we breathe. Ancient Greeks, noting that fertilizing the soil increases plant growth and that the lives of animals depend on the food that they eat, reasoned that plant growth must result from "food" that the plants "eat" from the soil. This concept of plants as soil-eaters went unchallenged until 1648, when the Dutch physician named Jan-Baptista van Helmont reported a simple but elegant experiment (fig. 7.1).

> That all vegetable matter immediately and materially arises from the element of water alone I learned from this experiment. I took an earthenware pot, placed in it 200 lb of earth dried in an oven, soaked this with water, and placed in it a willow shoot weighing 5 lb. After five years had passed the tree growth therefrom weighed 169 lb and about 3 oz. But the earthenware

pot was constantly wet only with rain or (when necessary) distilled water; and it was ample in size and imbedded in the ground; and, to prevent dust flying around from mixing with the earth, the top of the pot was kept covered with an iron plate coated with tin and pierced with many holes. I did not compute the weight of the deciduous leaves of the four autumns. Finally, I again dried the earth of the pot, and it was found to be the same 200 lb minus about 2 oz. Therefore, 164 lb of wood, bark, and root had arisen from the water alone.

Although van Helmont's conclusion would later be proven wrong, he did show that plants are not soil-eaters. His careful measurements also set the stage for similar studies of how plants grow. Near the same time, other scientists were studying combustion, a topic that interested not only medieval alchemists but also their successors, who laid the foundation of modern chemistry. One of the people interested in the changes made in air by combustion was an English chemist and radical, nonconformist Unitarian minister named Joseph Priestley (fig. 7.2). Priestley's studies showed that combustion somehow "injured"

A.

FIGURE 7.1

(a) J. B. van Helmont (1578–1644). (b) Van Helmont's willow experiments. Van Helmont concluded from these experiments that 164 lb of wood, bark, and root had arisen from H_2O alone.

B.

CHAPTER SEVEN *Photosynthesis*

the air: If a candle were burned in a closed container, it would go out. If a mouse were then put in the container, the mouse would die (fig. 7.2). In 1771 Priestley described his findings:

> I flatter myself that I have accidentally hit upon a method of restoring air which has been injured by the burning of candles, and that I have discovered at least one of the restoratives which nature employs for this purpose. It is vegetation. In what manner this process in nature operates, to produce so remarkable an effect, I do not pretend to have discovered; but a number of facts declare in favour of this hypothesis. I shall introduce my account of them, by reciting some of the observations which I made on the growing of plants in confined air, which led to this discovery. One might have imagined that, since common air is necessary to vegetable, as well as to animal life, both plants and animals had affected it in the same manner, and I own I had that expectation, when I first put a sprig of mint into a glass-jar, standing inverted in a vessel of water; but when it had continued growing there for some months, I found that the air would neither extinguish a candle, nor was it at all inconvenient to a mouse, which I put into it.
>
> Finding that candles burn very well in air in which plants had grown a long time, and having had some reason to think, that there was something attending vegetation, which restored air that had been injured by respiration, I thought it was possible that the same process might also restore the air that had been injured by the burning of candles.
>
> Accordingly, on the 17th of August, 1771, I put a sprig of mint into a quantity of air, in which a wax candle had burned out, and found that, on the 27th of the same month, another candle burned perfectly well in it. This experiment I repeated, without the least variation in the event, not less than eight or ten times in the remainder of the summer. Several times I divided the quantity of air in which the candle had burned out, into two parts, and putting the plant into one of them, left the other in the same exposure, contained, also, in a glass vessel immersed in water, but without any plant; and never failed to find, that a candle would burn in the former, but not in the latter. I generally found that five or six days were sufficient to restore this air, when the plant was in its vigour; whereas I have kept this kind of air in glass vessels immersed in water many months without being able to perceive the least alteration had been made in it.

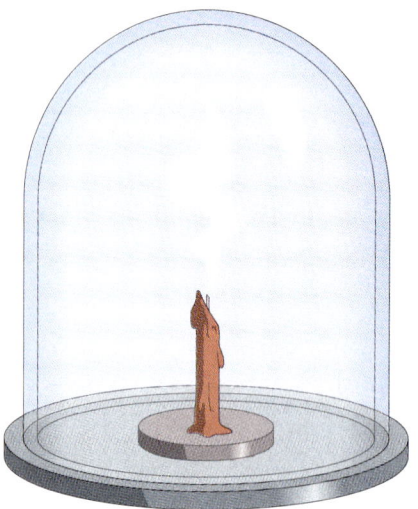
A.

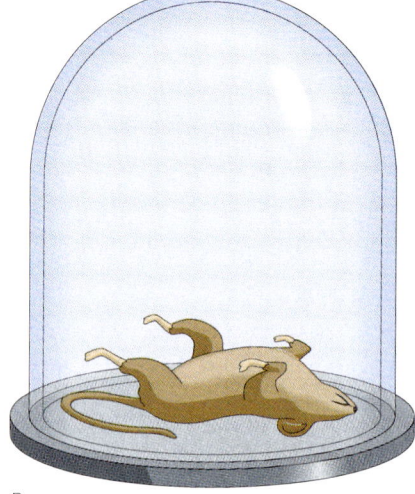
B.

C.

Figure 7.2

Joseph Priestley (1733–1804). Priestley's experiments showed that (a) a candle burning in an airtight jar went out, (b) a mouse died if kept in an airtight jar, and (c) the mouse lived if a plant were placed in the airtight jar with the mouse. Today, we explain Priestley's results by saying that plants use CO_2 produced by combustion or exhaled by animals, and that animals inhale and use the O_2 released by plants.

Today we explain Priestley's results by saying that plants use CO_2 produced by combustion or exhaled by animals, and that animals inhale and use the O_2 released by plants.

Priestley's experiments offered the first logical explanation of how air remained "pure" and able to support a mouse despite the burning of candles and the breathing of animals. Although he did not realize it, Priestley's experiments were the first demonstration that plants produce oxygen. Nor did Priestley realize that light was essential for photosynthesis. Nevertheless, he received a medal and a citation that read, in part, as follows: "For these discoveries we are assured that no vegetable grows in vain … but cleanses and purifies our atmosphere." Priestley implicated oxygen (*dephlogisticated air*, as he called it) in photosynthesis by showing that green plants could renew air made bad by the breathing of animals. However, a serious problem appeared: Others could not repeat Priestley's work. Indeed, even Priestley could not get the same results when he repeated his experiments (he had probably moved the experimental apparatus to a dark part of his lab). Seven years later, however, the Dutch physician Jan Ingenhousz confirmed Priestley's work and made an important addition by showing that light is necessary for plants to release oxygen (although, like Priestley, he knew nothing about oxygen at the time and explained it in other terms). Ingenhousz reported that plants in the dark "contaminate the air" and make it "harmful to animals." He also made a bold and accurate statement: "The sun by itself has no power to mend air without the concurrence of plants." Ingenhousz published a wonderfully titled book, *Experiments upon Vegetables, Discovering Their Great Power of Purifying the Common Air in the Sun-Shine, and of Injuring It in the Shade and at Night*, in which he also reported that only the green parts of a plant could photosynthesize and that plants, too, "injure" air when kept in darkness. Ingenhousz must have been rather alarmed by his findings, because he recommended that plants be removed from houses at night to avoid poisoning the occupants.

In 1782 a Swiss preacher and part-time scientist named Jean Senebier showed that photosynthesis depended on a particular kind of gas, which he called *fixed air* (and we call CO_2). Senebier also claimed that this air (CO_2) produced by animals and plants in darkness stimulated production of purified air (O_2) by plants in light. Thus, by the late 1700s biologists knew that at least two gases participate in photosynthesis. Work done by Lavoisier and others (especially P. S. Laplace, a French mathematician) showed that these gases were CO_2 and O_2.[1] Ingenhousz adapted the ideas of Lavoisier and suggested that plants do not just exchange "good air" for "bad air." Rather, he suggested that plants absorb carbon from carbon dioxide, "throwing out at that time the oxygen alone, and keeping the carbon to itself as nourishment." Ingenhousz's work was extended in 1804 by the Swiss botanist and physician Nicholas de Saussure, who noted that approximately equal volumes of CO_2 and O_2 are exchanged during photosynthesis. He added the final component of the overall photosynthetic reaction when he showed that photosynthesis requires water:

carbon dioxide + water + light

↓

organic material + oxygen

By this time, many botanists were studying photosynthesis. Among the most clever was T. W. Engelmann, who in 1883 designed an elegant experiment that simultaneously studied the light requirements and biochemistry of photosynthesis (fig. 7.3). Engelmann studied photosynthesis in *Spirogyra*, a

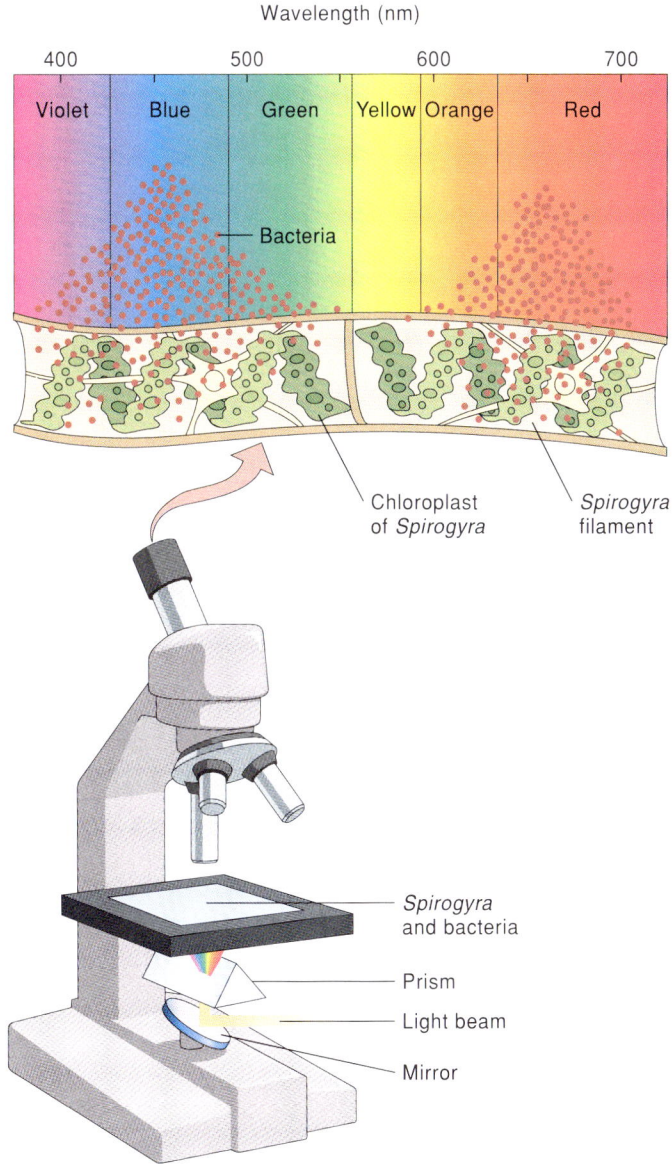

FIGURE 7.3

Engelmann's experiment. Engelmann illuminated a filament of *Spirogyra* with light dispersed into a spectrum by a prism inserted into a beam of light. Oxygen-motile bacteria gathered in the blue and red light, suggesting that these colors of light are most effective for photosynthesis.

---

1. Lavoisier was a French chemist who named oxygen and hydrogen and wrote the first modern textbook of modern chemistry. He was also a tax farmer—a term used to designate a tax collector empowered to make a profit. Although Lavoisier used his profits to fund his research, such activities were not popular with the general public. Soon after the French Revolution, Lavoisier's career ended at the guillotine.

green alga that has a long, spiral chloroplast in each of its cells. He placed the alga in a drop of water containing an oxygen-requiring bacterium and used a prism to illuminate different parts of the chloroplast with different colors of light. When Engelmann viewed the *Spirogyra* through a microscope, he saw that the bacteria clustered around parts of the chloroplast illuminated by red and blue light. He concluded that red and blue light are most effective for producing oxygen during photosynthesis.

Other biologists began to follow up Ingenhousz's finding that light was required for the release of O_2. Julius Sachs reported that chlorophyll, the photosynthetic pigment, occurs in chloroplasts and that photosynthesis forms carbohydrates only in light. Thus, Sachs revised the overall reaction of photosynthesis:

$$nCO_2 + nH_2O + light \xrightarrow{chlorophyll} (CH_2O)_n + nO_2$$

In this reaction CH_2O is an abbreviation for starch or other carbohydrates. Further research showed that Sachs's conclusion was correct: there is no known exception to his linking of chlorophyll with oxygen production.

By the turn of the twentieth century, most biologists accepted Ingenhousz's suggestion that the oxygen released during photosynthesis came from carbon dioxide. Others, however, questioned this assumption. In the 1920s, a Stanford University graduate student named C. B. van Niel began a study that would resolve this question and become a milestone in biological research. Van Niel studied a photosynthetic bacterium that uses H_2S as an electron source and deposits sulfur as a by-product. Photosynthesis in these bacteria occurred as follows:

$$CO_2 + 2H_2S \xrightarrow{light} CH_2O + H_2O + 2S$$

Van Niel's work didn't attract much attention until he made a bold extrapolation: he asserted that the oxygen released during photosynthesis came from water, not carbon dioxide. His reasoning was based on analogies between the roles of H_2S and H_2O, and of O_2 and sulfur.

Sulfur bacteria:
$$CO_2 + 2H_2S \xrightarrow{light} CH_2O + H_2O + 2S$$

General equation:
$$CO_2 + 2H_2X \xrightarrow{light} CH_2O + H_2O + 2X$$

Green plants:
$$CO_2 + 2H_2O \xrightarrow{light} CH_2O + H_2O + O_2$$

Van Niel's conclusion that O_2 released by plants comes from water rather than CO_2 was tested in 1941 by Samuel Ruben and Martin Kamen, who exposed cultures of *Chlorella* (a green alga) to H_2O labeled with ^{18}O, a nonradioactive isotope of oxygen that they could detect with a mass spectrometer. Ruben and Kamen reasoned that if oxygen released during photosynthesis came from water, then the oxygen would be tagged with ^{18}O. Conversely, if the oxygen were derived from carbon dioxide, it would not be labeled with ^{18}O. Their results were striking: the oxygen, not the carbohydrates, were labeled with $^{18}O_2$:

$$CO_2 + 2H_2^{18}O \xrightarrow{light} CH_2O + H_2O + {}^{18}O_2$$

These results confirmed van Niel's claim that oxygen released during photosynthesis comes from water, not carbon dioxide.

Further evidence for this conclusion was provided by Robin Hill and his coworkers, who discovered that isolated chloroplasts could release O_2 in the absence of CO_2 if given a suitable electron acceptor for the electrons removed from water. This light-driven splitting of water in the absence of CO_2 became known as the **Hill reaction** and showed that (1) whole cells are unnecessary for some of the reactions of photosynthesis, and (2) the light-driven release of O_2 during photosynthesis is not linked directly to the fixation of CO_2. In 1951 botanists discovered that the electron acceptor in chloroplasts is $NADP^+$, a coenzyme that can accept electrons. Later studies showed that $NADP^+$ reduced during photosynthesis was used to reduce CO_2 during photosynthesis. Similarly, the reduction of $NADP^+$ is driven by light.

CONCEPT

Botanists have studied photosynthesis for hundreds of years. During that time, they have discovered that (1) photosynthesis uses carbon dioxide; (2) photosynthesis requires light and water and releases oxygen; (3) light for photosynthesis in plants is absorbed by chlorophyll; and (4) oxygen released during photosynthesis comes from water.

THE NATURE OF LIGHT

The sun is a giant thermonuclear reactor—each minute more than 120 million metric tons of solar matter are converted to radiant energy. Much of this energy is released as radiation that moves in rhythmic waves similar to those created by pebbles dropped in a pond (however, light moves as disturbances of electric and magnetic fields, not as disturbances of the surface of water; see fig. 7.4a). Eight minutes later, about two-billionths (5×10^{24} kcal; 1.73×10^{17} W) of this energy has traveled 160 million km and hit the earth's upper atmosphere. About one-third of the light hitting the atmosphere is reflected back to space (fig. 7.4b), and only about 1% of the light is used for photosynthesis. Clearly, we must know something about light to understand photosynthesis.

Light is the part of the electromagnetic spectrum having wavelengths visible to the human eye (about 390 to 760 nm). Virtually all life depends on this light. Our understanding of light began about three hundred years ago, when Isaac Newton showed that white light that passed through a prism, droplet of water, or soap bubble would separate into a band of colors that, if passed through another prism, could be recombined to form white

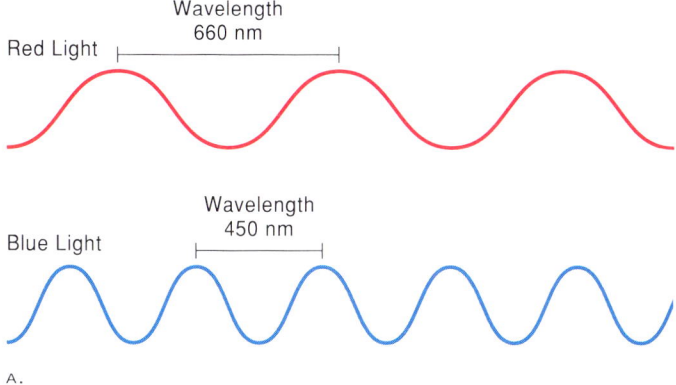

TABLE 7.1
Radiant Energies of Different Wavelengths of Light

Wavelength (nm)	Color	Energy (joules μmol^{-1})
300		0.399
400	violet	0.299
450	blue	0.277
500	green	0.239
600	orange	0.199
700	red	0.171

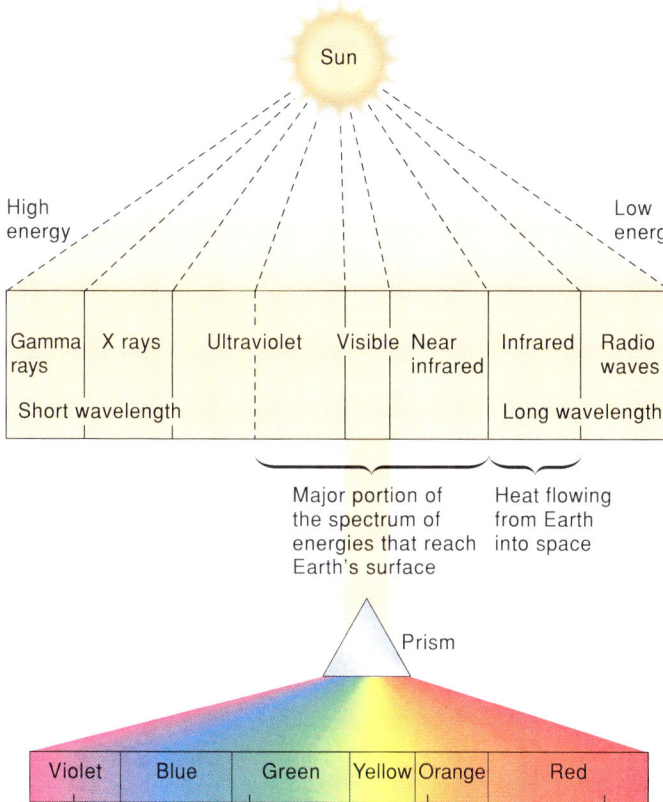

FIGURE 7.4

Properties of light. (a) The wavelike nature of light. (b) Visible light is only a small part of the electromagnetic spectrum. Visible light consists of a rainbow of colors ranging from violet to red.

In 1905, Einstein linked the ideas of Newton and Maxwell by proposing that light consists of packets of energy called **photons.** The intensity (i.e., brightness) of light depends on the number of photons (i.e., the amount of energy) absorbed per unit of time. Each photon carries a fixed amount of energy that is determined by how the photon vibrates: the slower the vibration, the less energy carried by the photon (table 7.1). The distance moved by the photon during a complete vibration is referred to as the photon's **wavelength** (λ) (fig. 7.4a). Stated another way, the wavelength is the distance between vibrational crests of a photon. The wavelengths of visible light are measured in nanometers (nm), or billionths (10^{-9}) of a meter. We perceive the different wavelengths as different colors. For example, violet light has a wavelength of about 400 nm, which is about one-fortieth the thickness of this page. The energy of a photon is a **quantum** and is inversely proportional to the wavelength of the light: the longer the wavelength (i.e., the longer the distance traveled during a vibration), the less energy per photon (table 7.1).

Sunlight consists of about 4% ultraviolet radiation, 52% infrared radiation, and 44% visible light. Each of these kinds of light has different energy and affects organisms differently.

Ultraviolet radiation (UV) contains too much energy for most biological systems. Indeed, its high-energy photons often drive electrons from molecules, thus explaining why UV is also called **ionizing radiation.** UV breaks weak bonds and causes sunburn. It is absorbed by O_2, ozone (O_3), and glass. You won't get a tan by sitting in front of a glass window.

Infrared radiation (IR) doesn't contain enough energy per photon to be useful to living systems. Cells absorb IR radiation, but this energy is insufficient to excite electrons. Consequently, most of the energy of IR is converted immediately to heat. IR is absorbed by water and carbon dioxide, but goes through glass. Thus, although you won't tan in front of a window, you will warm up.

Visible light contains just the right amount of energy for biological reactions such as photosynthesis. To have an effect, however, light must first be absorbed. Light is absorbed by pigments.

light. Based on that experiment, Newton proposed that white light is actually a spectrum of colors ranging from violet to red (fig. 7.4b). Two hundred years later, in the 1860s, a Scottish mathematician named James Maxwell showed that the visible light that Newton separated into a spectrum of colors is only a small part of a much larger spectrum of radiation (fig. 7.4b).

CONCEPT

Virtually all of life depends on light, which powers photosynthesis. Light moves in waves, and its energy is contained in packets called *photons*. The energy of a photon is inversely proportional to the wavelength of the light: the longer the wavelength, the less energy per photon. Sunlight consists of a spectrum of colors of light. Red light and blue light are the most effective for photosynthesis.

PIGMENTS

Light striking an object is either reflected, transmitted, or absorbed. Only light that is absorbed can have an effect. Light is absorbed by molecules called **pigments,** which are colored because they transmit particular colors of light. Black pigments absorb all wavelengths of light, while white pigments absorb no wavelengths of light.

Pigments in Plants

Evolution repeats its inventions and, in the process, adapts existing mechanisms to new and different purposes: many kinds of pigments are similar, but have different functions. For example, certain pigments are derived from tetrapyrroles, a group of compounds present in all organisms. Tetrapyrroles are large rings made of four smaller rings, each of which is called a *pyrrole ring* and consists of four carbons and one nitrogen (fig. 7.5). The four pyrrole rings of the tetrapyrrole are linked by one-carbon bridges that sequester a metal atom. If the metal is iron, the pigment is a heme, such as in hemoglobin or cytochrome (iron in hemoglobin is always reduced as Fe^{2+}, while in cytochromes the Fe shuttles between Fe^{2+} and Fe^{3+}; see fig. 5.10e). If the metal is copper, the pigment is turacin, the pigment in many feathers. If the metal is magnesium, the pigment is chlorophyll (or the more reduced bacteriochlorophyll of bacteria), the primary pigment of photosynthesis (fig. 7.5).

Chlorophylls are fat-soluble pigments that occur in plants, algae, and all but one primitive group of photosynthetic bacteria. Unlike the color of hemoglobin, which is insignificant for the molecule's function as an oxygen carrier, the color of chlorophyll *is* significant. Chlorophyll absorbs maximally at wavelengths of 400–500 nm (violet-blue) and 600–700 nm (orange-red) (fig. 7.6). The synthesis of chlorophyll and several other pigments in plants is stimulated by light, which explains why plants grown in the dark (i.e., etiolated plants) contain no chlorophyll. Similarly, the light-stimulated synthesis of anthocyanin, another plant pigment, explains why apples are always redder on the sunny side of an apple tree.

There are several types of chlorophyll, the most important of which is **chlorophyll *a*,** the primary photosynthetic pigment. Chlorophyll *a* ($C_{55}H_{72}O_5N_4Mg$) is a grass-green

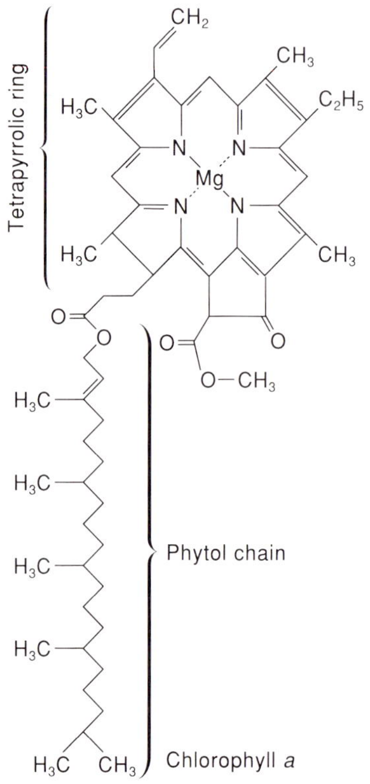

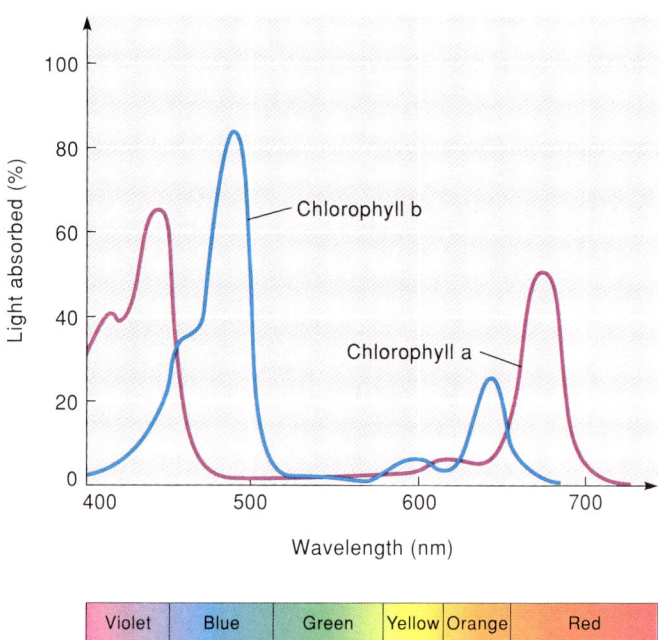

FIGURE 7.5

Chlorophyll *a*, the primary pigment of photosynthesis.

FIGURE 7.6

The absorption of light by chlorophylls *a* and *b*. Chlorophylls absorb maximally at wavelengths of 400–500 nm (violet-blue) and 600–700 nm (orange-red).

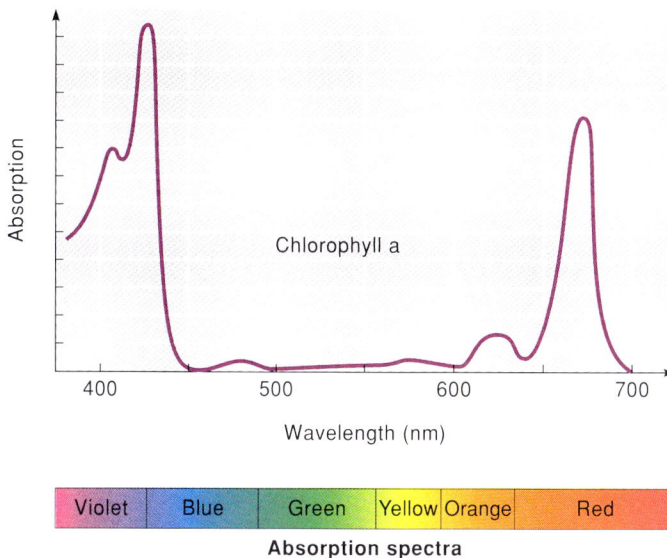

FIGURE 7.7

The absorption spectrum of chlorophyll *a* vs. wavelength. Chlorophyll *a* absorbs maximally at 430 nm and 662 nm and is the primary photosynthetic pigment.

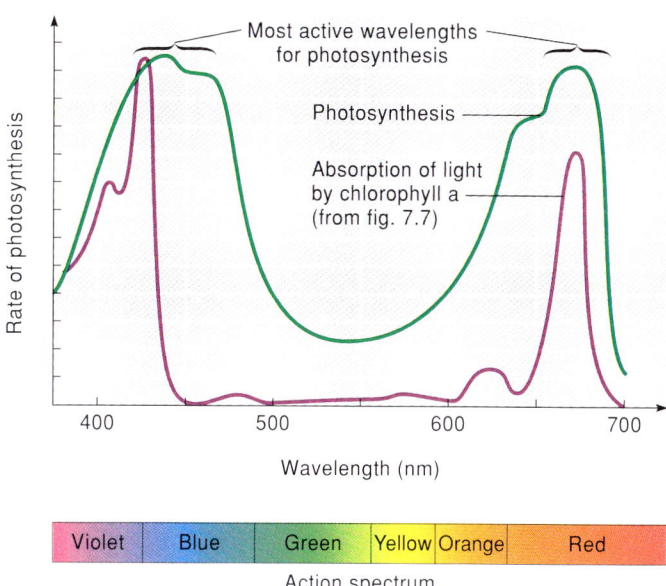

FIGURE 7.8

The rate of photosynthesis in different wavelengths of light is similar to the ability of chlorophyll *a* to absorb those wavelengths. The similarity of the action and absorption spectra of chlorophyll *a* suggests that light absorbed by chlorophyll *a* drives photosynthesis.

pigment whose structure includes an atom of magnesium (Mg) (fig. 7.5). It occurs in all photosynthetic organisms except photosynthetic bacteria and absorbs maximally at 430 and 662 nm (fig. 7.6).

The relationship of light absorption by chlorophyll *a* versus wavelength is shown in figure 7.7. This **absorption spectrum** shows that chlorophyll *a* strongly absorbs all visible light except green light; it reflects and transmits green light, and therefore appears green. This absorption spectrum for chlorophyll *a* closely matches the graph showing how photosynthesis varies with different wavelengths of light (fig. 7.8). This so-called **action spectrum** shows that the pattern of photosynthesis closely follows that of chlorophyll *a*, suggesting that chlorophyll *a* is the primary pigment in photosynthesis. However, the absorption spectrum of chlorophyll *a* does not perfectly match the action spectrum for photosynthesis. Therefore, we conclude that other pigments must be involved in photosynthesis. Those pigments are called *accessory pigments*.

Accessory Pigments

Plants also contain a rainbow of other pigments. Those pigments extend the range of light useful for photosynthesis by absorbing photons not absorbed by chlorophyll *a* and transmitting that energy to chlorophyll *a*. The most common of these **accessory pigments** in plants are chlorophyll *b* and carotenoids.

Chlorophyll *b* ($C_{55}H_{70}O_6N_4Mg$) is a bluish-green pigment that absorbs maximally at 453 and 642 nm. It occurs in all plants, green algae, and some prokaryotes. Plants usually contain about half as much chlorophyll *b* as chlorophyll *a*.

Carotenoids are accessory pigments that occur in all photosynthetic organisms. They contain forty carbons, are fat-soluble, and often contain no oxygen (fig. 7.9). Carotenoids absorb maximally at wavelengths between 460 and 550 nm; therefore, they are red, orange, and yellow. Carotenoids are chemically unrelated to chlorophylls and consist of carbon rings linked by long carbon chains having alternating single and double bonds.

Like other accessory pigments such as chlorophyll *b*, carotenoids extend the range of photosynthesis by absorbing light that is not absorbed by chlorophyll *a*. Carotenoids also protect plants against photo-oxidation, which occurs when excited chlorophyll transforms oxygen into high-energy radicals. These radicals can attract hydrogen from nearby molecules, thereby destroying the molecules and killing the cells. Mutants that lack carotenoids are susceptible to such radicals, which explains why they are bleach-white when grown in light, and soon die (fig. 7.10). Many herbicides kill plants by blocking the synthesis of carotenoids, thereby causing the plant to photo-oxidize itself.

The most common carotenoid is **beta-carotene,** a reddish-yellow pigment consisting of two six-carbon rings connected by an eighteen-carbon chain (fig. 7.9). Beta-carotene absorbs maximally at wavelengths between 400 and 500 nm. When split in half, beta-carotene becomes two molecules of vitamin A, which is a precursor of retinal, a pigment essential for human vision. Carotenoids occur throughout the plant kingdom and produce the colors of tomatoes, carrots, squash, bananas, avocados, and autumn's colorful leaves.

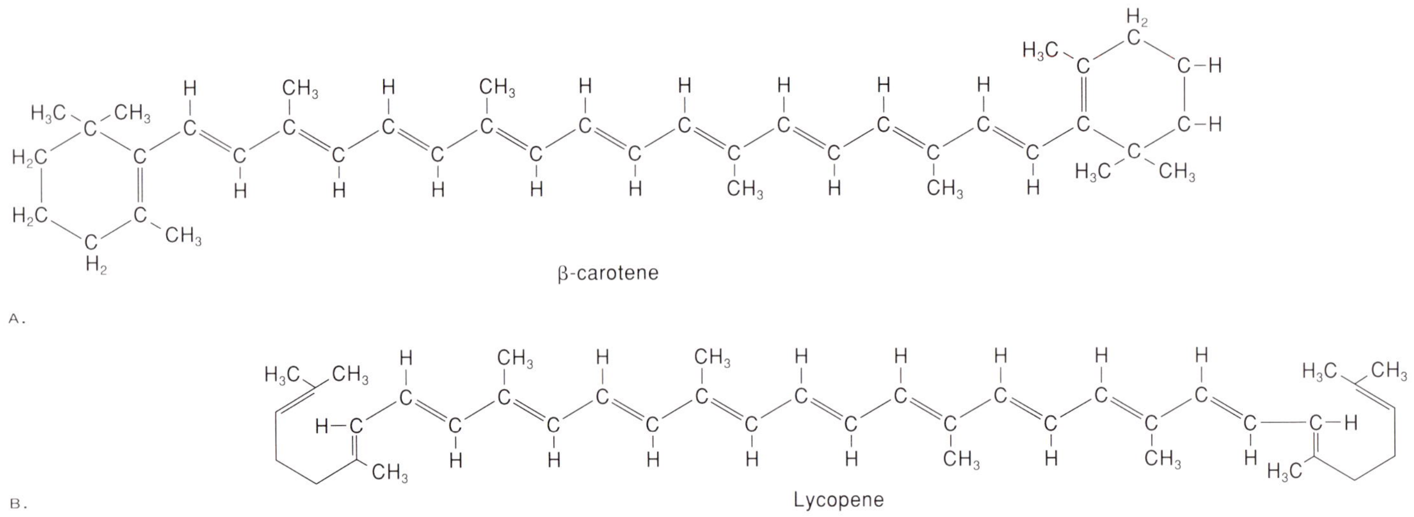

FIGURE 7.9

Carotenoids. (a) Beta-carotene, a reddish-yellow carotenoid. (b) Lycopene, a carotenoid that colors tomatoes. Beta-carotene and lycopene each have the empirical formula of $C_{40}H_{56}$.

FIGURE 7.10

Carotenoids protect against photo-oxidation. The plants on the left were treated with an herbicide that blocks synthesis of carotenoids. Consequently, the plants became photo-oxidized (bleached) and died when grown in light. Untreated plants (right) make carotenoids and therefore are not photo-oxidized by light.

BOXED READING 7.1

THE EVOLUTION OF PHOTOSYNTHESIS: WHY AREN'T PLANTS BLACK?

Photosynthetic autotrophs use light-energy to reduce carbon dioxide to carbohydrate. Because the amount of energy absorbed by an organism largely determines its rate of photosynthesis, why then aren't plants black? Stated another way, Why do plants reflect, and therefore waste, green light? Why don't they use all of the spectrum of visible light for photosynthesis? The answers to these questions lie in the evolution of photosynthesis.

The earliest photosynthetic organisms were aquatic bacteria, several of which are around today. Chief among these is *Halobacterium halobium,* a purple bacterium that grows only in extremely salty water (3–4 times saltier than seawater). No other organisms can tolerate this environment.

Since water absorbs most light outside the visible part of the spectrum, most ultraviolet and infrared light was unavailable to the first photosynthetic organisms. Natural selection therefore favored the evolution of pigments such as bacteriorhodopsin, the photosynthetic pigment in *Halobacterium.* Bacteriorhodopsin, a purple pigment that resembles rhodopsin (the light-sensing pigment of our eyes), absorbs broadly in the middle of the visible spectrum of light. When it absorbs light, it pumps protons out of the cell, thereby creating a proton gradient that the cell uses to make ATP. That ATP powers *Halobacterium.*

Organisms living on the surface of the sediment faced a different problem. The bacteriorhodopsin in the *Halobacterium* swimming above the sediments absorbed most of the green light. Consequently, the light that reached the bottom-dwelling organisms was poorly suited for another bacteriorhodopsin-based type of photosynthesis. Similarly, the surrounding water absorbed all of the infrared and ultraviolet light of sunlight. Natural selection thus favored the evolution of a pigment that could absorb the remaining wavelengths—red and blue. This pigment was chlorophyll, which—in addition to pumping protons out of the cell—also pumped electrons into the cell.

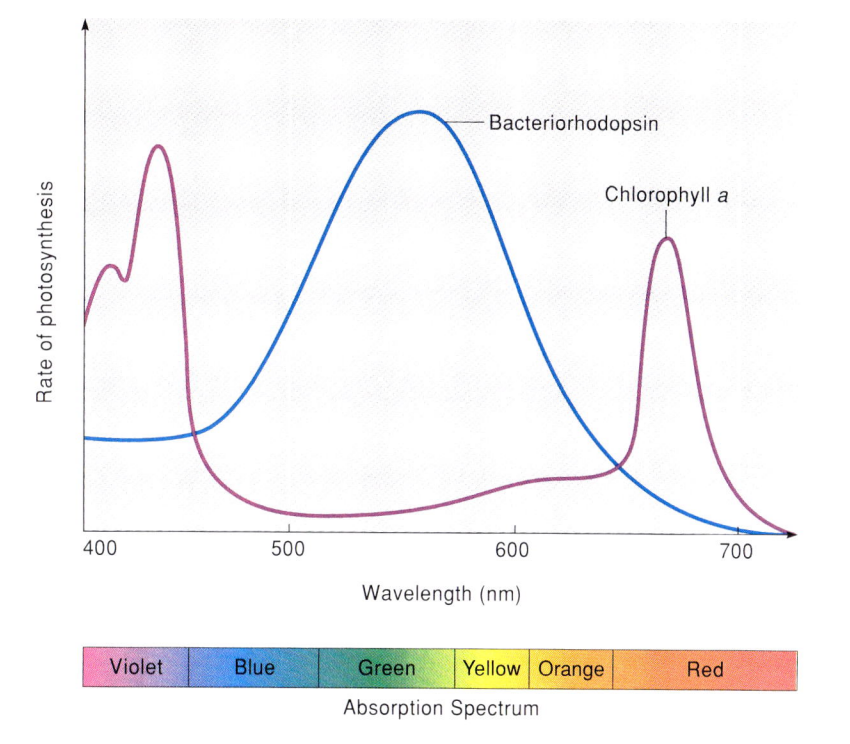

BOX FIGURE 7.1

The absorption spectrum of bacteriorhodopsin peaks in the green wavelengths, while that of chlorophyll *a* peaks on either side, in the red and blue wavelengths.

Their ability to pump electrons enabled these organisms to reduce carbon dioxide to sugars.

Plants and plantlike organisms have evolved other pigments to help compensate for their poor use of light. Accessory pigments absorb wavelengths near the center of the visible spectrum and pass this energy to chlorophyll. Cyanobacteria and red algae have phycocyanin and allophycocyanin as accessory pigments, which absorb orange light, but don't cover all of the wavelengths missed by chlorophyll. They also contain a red pigment called phycoerythrin, which absorbs green light and thus extends the range of photosynthesis beyond that of most plants. Indeed, despite their name, many red algae are dark-colored, approaching black, making them ideally suited for photosynthesis. Similarly, the dark colors of brown algae result from chlorophyll and fucoxanthin, an accessory pigment. Not surprisingly, many red and brown algae grow in deep water, where light is extremely dim. Some darkly colored red algae in the western Atlantic live 268 m (884 ft) down, where the light is less than 1% of full sunlight.

And what about land plants? Since light is abundant on land, there has been relatively little selection pressure for the evolution of a wider range of pigments. In other words, unlike red and brown algae, land plants grow in adequate light and do not have to extract as much light as possible. Consequently, land plants depend almost entirely on chlorophyll for photosynthesis. Carotenoids partly fill the gap, especially toward blue, but they do not pass this energy efficiently to chlorophyll. Since no pigment in plants absorbs significantly in the green part of the spectrum, this light is wasted. It also creates a paradox: the most advanced plants have the least advanced system for absorbing light. Nevertheless, this is not all bad—few of us would trade a green countryside for a black one in the name of photosynthetic efficiency.

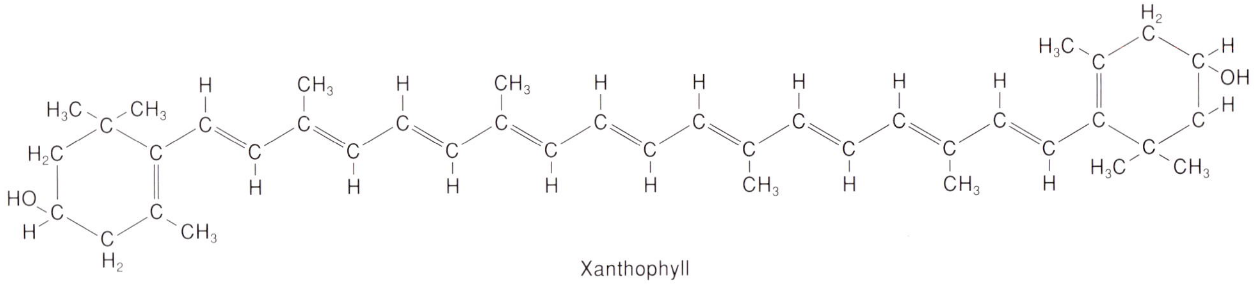

FIGURE 7.11

Xanthophyll. Xanthophylls are abundant in carrots, leaves, and many algae.

Unlike chlorophyll, carotenoids also occur in animals. Animals cannot make carotenoids, but they can metabolize and use carotenoids from plants. For example, carotenoids color egg-yolks, flamingo feathers, insect wings, the black ink released by squid, and the bodies of corals, fish, and amphibians. The dazzling array of colors of carotenoids includes blue, green, red, violet, gray, chocolate, and black; they result largely from proteins that are attached to the carotenoid. When heated, this protein breaks off and frees the carotenoid (this accounts for the red pigment released when we cook lobsters).

The Lore of Plants

Tomatoes are red because of the carotenoid lycopene, the same pigment that colors the flesh of pink grapefruits and watermelon. White tomatoes lack carotenoids, and yellow tomatoes make yellow carotenoids but no red carotenoids. There is even a variety of tomato that ripens but does not break down its chlorophyll. This variety, which stays confusingly green, is called Evergreen.

Oxidizing carotenes produces **xanthophylls,** which are red and yellow pigments in tomatoes, carrots, leaves, algae (e.g., fucoxanthin in brown algae), and photoautotrophic bacteria (fig. 7.11). Xanthophylls are less efficient at transferring energy during photosynthesis than is beta-carotene.

Other accessory pigments include chlorophylls *c* and *d*, phycoerythrin, a red pigment, and phycocyanin, a blue pigment (fig. 7.12). Different forms of phycoerythrin and phycocyanin occur in red algae and cyanobacteria.

CONCEPT

Light is absorbed by pigments. Chlorophyll *a* is the primary pigment for photosynthesis and occurs in all photosynthetic organisms except photosynthetic bacteria. Accessory pigments such as carotenoids and chlorophyll *b* absorb light that chlorophyll *a* cannot absorb, thereby extending the range of light useful for photosynthesis.

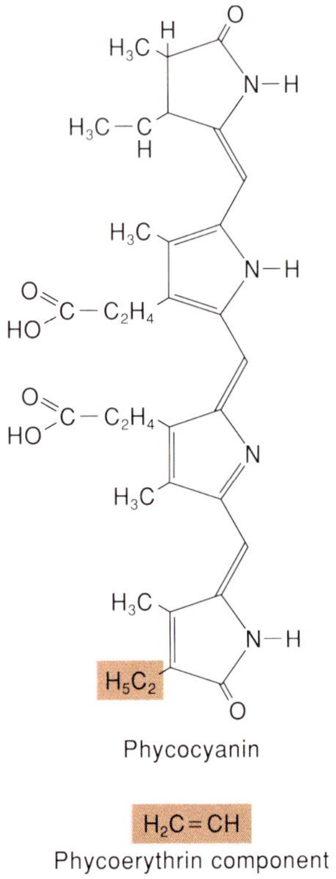

FIGURE 7.12

Phycocyanin and phycoerythrin are accessory pigments in red algae and cyanobacteria. Phycoerythrin is obtained by exchanging the highlighted groups.

Making and Destroying Pigments

Most plants regularly destroy and resynthesize their chlorophyll. Indeed, each year the 300 million tons of chlorophyll on earth are turned over an average of three times. When the rate of chlorophyll synthesis lags behind that of its breakdown, leaves begin a photochemical suicide. The decreasing amounts of chlorophyll no longer mask the other pigments, which results in the colors of autumn that are praised by poets and songwriters. In many trees, this "turning of the leaves" is spectacular. Tourists

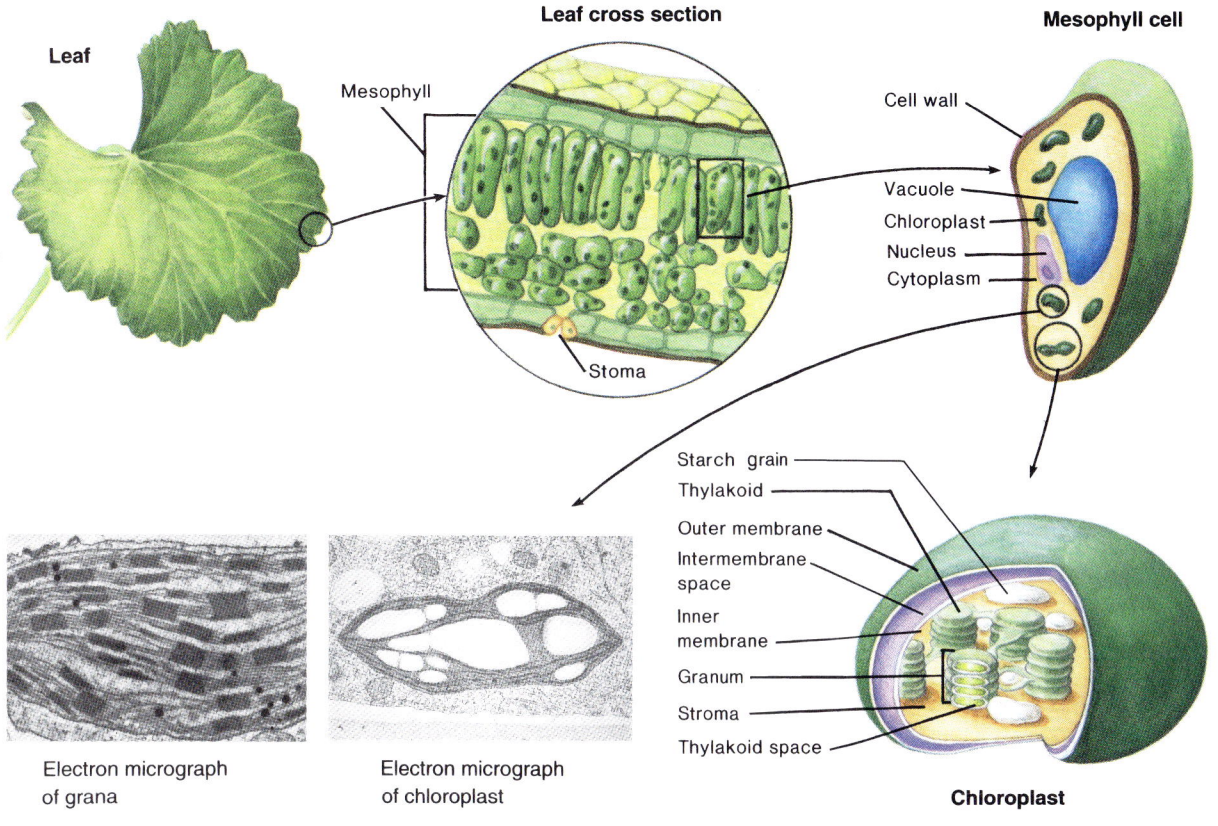

FIGURE 7.13

In plants, photosynthesis occurs in chloroplasts. Light is absorbed and converted to chemical energy in thylakoids and grana; this chemical energy is then used in the stroma to make sugars.

who visit New England to see the carotenoids of trees generate a billion-dollar-per-year industry, most of which happens in just three weeks. States such as New Hampshire even have a fall foliage hotline to keep people up to date on the breakdown of chlorophyll.

CHLOROPLASTS

The green color of leaves is due to solar chemical factories called **chloroplasts,** the site of photosynthesis in eukaryotes (fig. 7.13). Chloroplasts in plants are usually shaped like footballs with a diameter of 5–10 μm and a depth of 3–4 μm. Most photosynthetic cells have 40–200 chloroplasts, which amounts to about 500,000 chloroplasts per square millimeter of leaf area. To help you put the size of a chloroplast in perspective, consider that it would take about 2,000 chloroplasts to stretch across your thumbnail.

Each chloroplast is surrounded by two membranes that enclose a gelatinous matrix called the **stroma** (fig. 7.13). The stroma contains ribosomes, DNA, and means of making carbohydrates. The sugars involved in photosynthesis are also made in the stroma. Suspended in the stroma are neatly folded sacs of membranes called **thylakoids** (fig. 7.13). These membranes are unique to chloroplasts. In some parts of chloroplasts, 10–20 thylakoids are stacked into **grana** (fig. 7.13). Thylakoids and grana contain chlorophyll (chlorophyll constitutes about half of the lipid in thylakoids) and are located where light is absorbed during photosynthesis. Most of the proteins in chloroplasts are coded by nuclear genes, produced in the cytoplasm, and then shipped to the chloroplast.

Complexes of Pigments in Chloroplasts

In the early 1950s Robert Emerson at the University of Illinois made an unusual observation: he noted that red light having wavelengths exceeding 690 nm was ineffective for photosynthesis despite the fact that it was absorbed by chlorophyll. However, when this light was supplemented by light having a shorter wavelength, photosynthesis occurred faster than it did

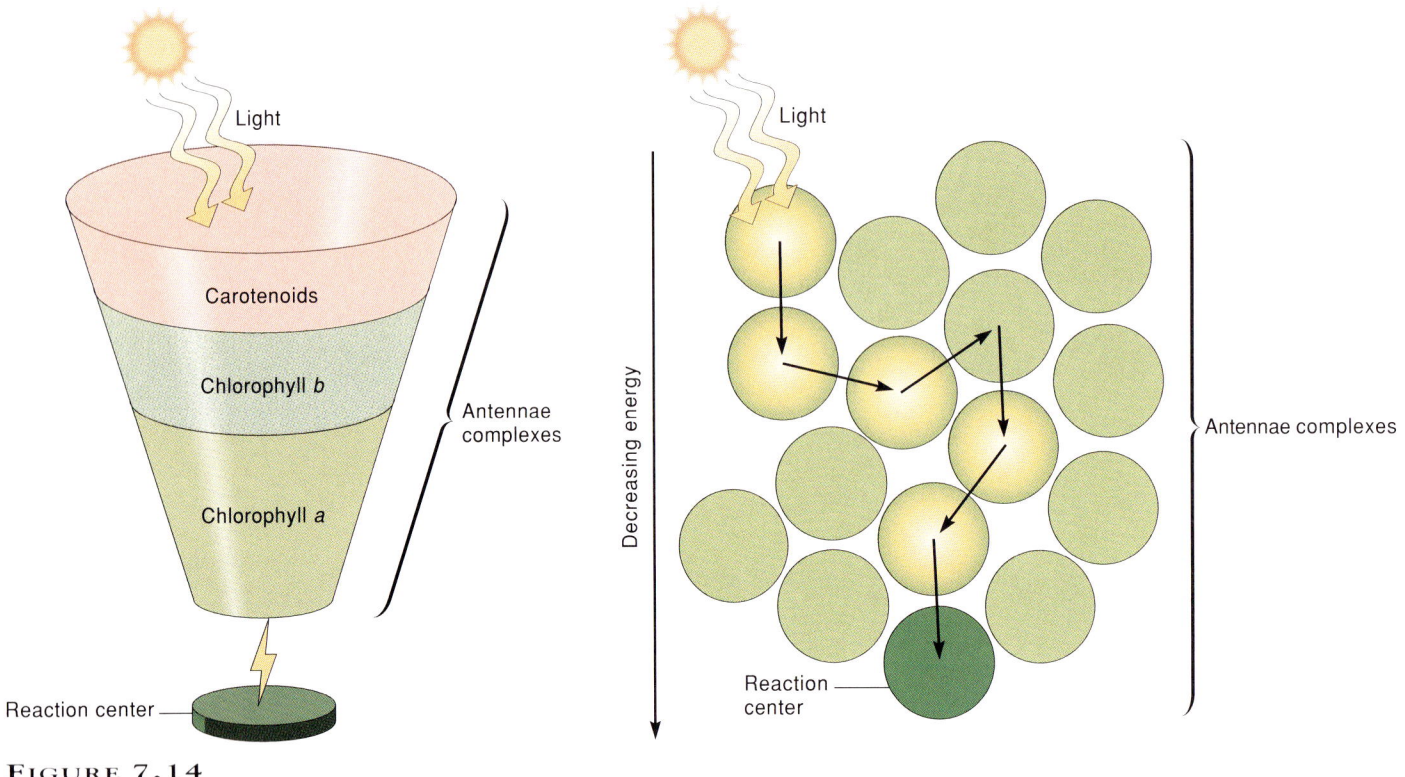

FIGURE 7.14

The light-harvesting system in chloroplasts functions like a funnel; it collects photons and passes their energy to the reaction center.

in either light alone. This effect became known as the *Emerson enhancement effect*, and it showed that plants contain two light-harvesting systems.

In all but the most primitive bacteria, light is captured by a network of chloroplast pigments arranged in aggregates on thylakoids. These aggregates, called **antennae complexes** (fig. 7.14), are anchored in a protein matrix and consist of proteins, about three hundred molecules of chlorophyll *a*, and about fifty molecules of carotenoids and other accessory pigments that gather light. Energy absorbed by antennae complexes flows energetically "downhill" to a special pair of energy-collecting molecules of chlorophyll *a* and associated proteins called a **reaction center** (fig. 7.14). Although reaction centers comprise less than 1% of the chlorophyll in plants, chlorophyll *a* in the reaction center is the electron acceptor that participates directly in photosynthesis; all other photosynthetic pigments function as antennae.

There are two kinds of reaction-center chlorophylls, each having light-harvesters in their antennae. One of these chlorophylls absorbs maximally at 700 nm and is called **P700** (for pigment 700) (fig. 7.16). The complex containing P700 is called *Photosystem I* and consists of eleven polypeptides, six of which are coded in the nucleus, and five of which are coded in the chloroplast. The core of Photosystem I consists of about forty molecules of chlorophyll *a*, several molecules of beta-carotene, lipids, four manganese, one iron, several calcium, several chlorine, two molecules of plastoquinone, and two molecules of pheophytin, a colorless form of chlorophyll *a*.

The chlorophyll *a* molecule in *Photosystem II* resonates from energy transmitted by about one hundred molecules of chlorophyll *a* and *b* (in a 4:1 ratio) bound to nuclear-coded proteins in the antennae. Three peripheral polypeptides bind calcium and chlorine, which explains why these nutrients are essential for photosynthesis. The chlorophyll *a* molecule in Photosystem II is identical to that of Photosystem I, but it is associated with different proteins; in Photosystem I, the reaction center is bound to a large protein (molecular weight 110,000), while that in Photosystem II is bound to a smaller protein (molecular weight 47,000). The associated protein in Photosystem II shifts the maximal absorption to about 680 nm. Consequently, the reaction center in Photosystem II is called **P680** (fig. 7.16). P680 resonates from energy transmitted by about 250 molecules of chlorophyll *a* and *b* (in equal numbers). Its core includes numerous xanthophylls (but no beta-carotenes) and integral proteins coded in the nucleus and made in the cytoplasm.

What Happens When Pigments Absorb Light?

Light must be absorbed by a pigment before it can have an effect. When light is absorbed, the energy of its photons is captured by the pigment and is used to boost the energy of electrons—that is, it changes the configuration of electrons by boosting them to higher-energy orbitals. These excited electrons can have several fates (fig. 7.15):

- The energy can be released as heat (i.e., molecular motion; fig. 7.15a). This is what is happening to the electrons in the pigment (i.e., the ink) that you're looking at now.
- The energy can be released as an afterglow of light via a process called **fluorescence.** Light released during fluorescence has a longer wavelength (and therefore, less energy) than the light that excited the pigment. Chlorophyll fluoresces deep red. Isolated chlorophyll fluoresces because no molecules are nearby to accept the energized electrons (fig. 7.15b).
- The energy can be passed to a neighboring molecule. Chlorophyll in thylakoids is surrounded by other molecules that trap the energy of the excited electrons (fig. 7.15c).

The light-driven reactions of photosynthesis, which pass the energy of photons to other molecules, occur on photosynthetic membranes. In plants and algae, these photosynthetic membranes are enclosed in chloroplasts, the evolutionary descendants of photosynthetic prokaryotes (which lack chloroplasts, but do have membranes similar to thylakoids; see boxed reading 3.2 on p. 66). In these organisms, chlorophyll is on membranes, in vesicles, or—as in cyanobacteria—in parallel stacks of flattened sacs.

Photophosphorylation: Chemiosmosis in Chloroplasts

Photons absorbed by pigments energize electrons. Plants pass that energy through a series of molecules called an *electron transport chain*. This transfer of electrons involves reduction and oxidation, or *redox*, reactions: the electron donor is oxidized as the electron acceptor is reduced. For electrons to flow "down" this chain, each receiver must attract the electrons more strongly than does the donor.

The potential energy of the electrons drops at each step of the electron transport chain. Plants couple this exergonic flow of electrons to an endergonic reaction that makes ATP. This light-driven production of ATP from electron transport is called **photophosphorylation** and depends upon a proton

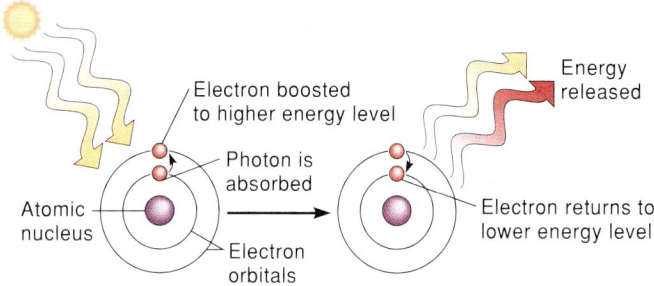

A. Release of energy as heat

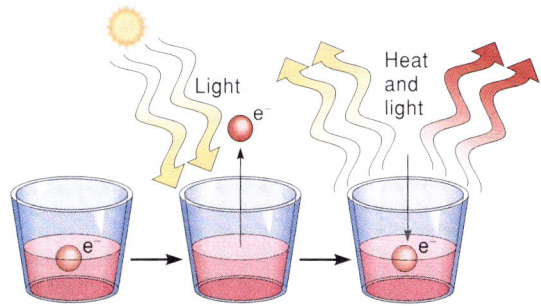

B. Isolated chlorophyll molecule: Release of energy as heat and light

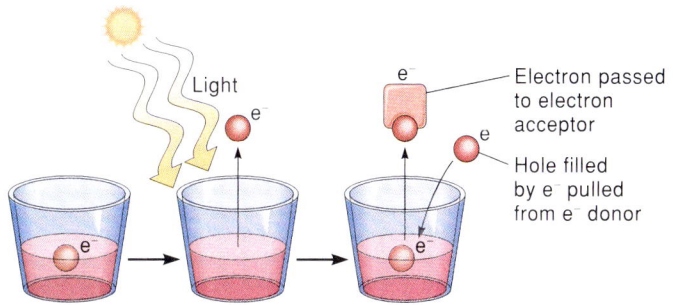

C. Chlorophyll in intact chloroplast: Passage of energy to other molecules

FIGURE 7.15

Fates of energy in energized electrons of chlorophyll. (a) Release of energy as heat. (b) Isolated chlorophyll: electrons that had been raised fall to original energy level; light energy absorbed is released as heat and light (fluorescence). (c) Intact chloroplast: energy passes to electron acceptors.

gradient (fig. 7.16). Photons captured by pigments excite electrons, which are shuttled along carriers embedded in the thylakoid membrane. When these electrons reach transmembrane H^+-pumps, their arrival induces transport of H^+ across the membrane, thereby creating the proton gradient that drives the synthesis of ATP.

As you've probably noticed, chemiosmosis in chloroplasts (photophosphorylation) resembles chemiosmosis (oxidative phosphorylation) in mitochondria (see Chapter 6). In both cases, (1) protons are pumped through a series of carriers that

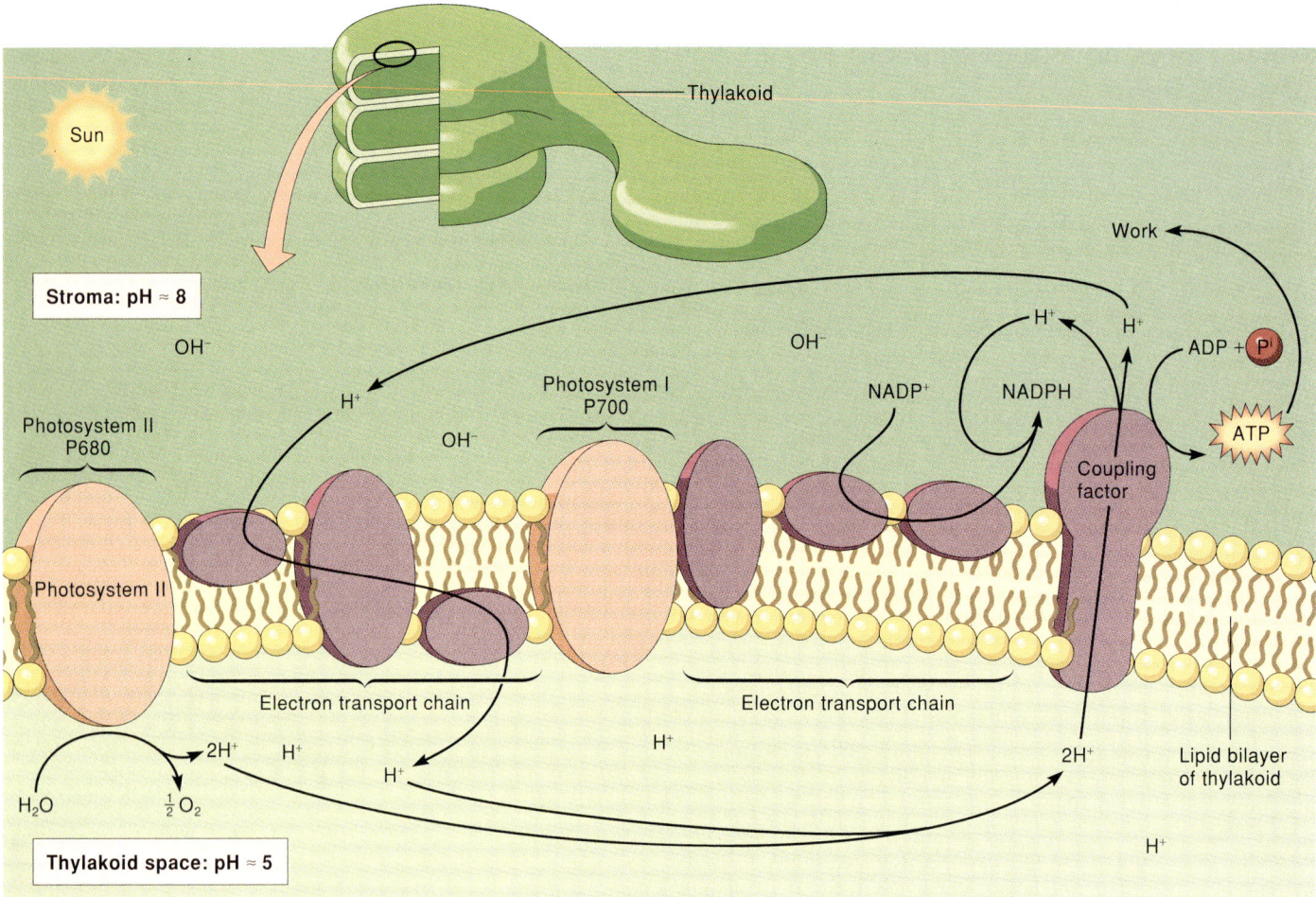

FIGURE 7.16

The light-driven transport of electrons during photosynthesis creates a proton gradient across the thylakoid membrane. This proton gradient drives the formation of ATP and reduced NADP⁺.

are progressively more electronegative, and (2) the free energy released by electron transport generates and maintains a proton gradient across a membrane. However, there are some differences. For example, in mitochondria, energized electrons moving through the electron transport chain are extracted by the oxidation of food; in chloroplasts, light energizes the electrons, and no food is necessary. In mitochondria, the inner membrane pumps protons from the matrix out to the intermembrane space; this is the reservoir of protons that powers the synthesis of ATP. In chloroplasts, electron carriers in the thylakoid pump protons from the stroma into the lumen (i.e., thylakoid space), which is the proton reservoir. This light-driven pumping of protons into the lumen decreases the pH there to about 5, while the pH of the stroma increases to 8 (i.e., a thousandfold difference in the concentration of protons). Light is required to generate this proton gradient; the difference in pH across the thylakoid membrane disappears quickly in the dark. The pH gradient is discharged during the synthesis of ATP as protons flow into the stroma across channels whose catalytic heads protrude like knobs into the stroma (fig. 7.16). Coupling-factor proteins in these channels include ATP synthase, an enzyme that harnesses the flow of protons to make ATP. The ATP produced by the ATP-synthase complexes is released into the stroma, where it is used to make carbohydrates.

CONCEPT

Photosynthesis in eukaryotes occurs in chloroplasts. Chloroplasts consist of a gelatinous matrix called the *stroma* and stacks of membranes (thylakoids) called *grana*. Aggregates of pigments in grana absorb light and funnel its energy to special pairs of chlorophyll and proteins called *reaction centers*. The energy of the light is used to energize electrons, which in turn are used to pump protons from the stroma into the thylakoid space. The resulting pH gradient is discharged during the synthesis of ATP as protons flow back into the stroma.

Doing Botany Yourself

How could you test whether a proton gradient is necessary for ATP synthesis in chloroplasts?

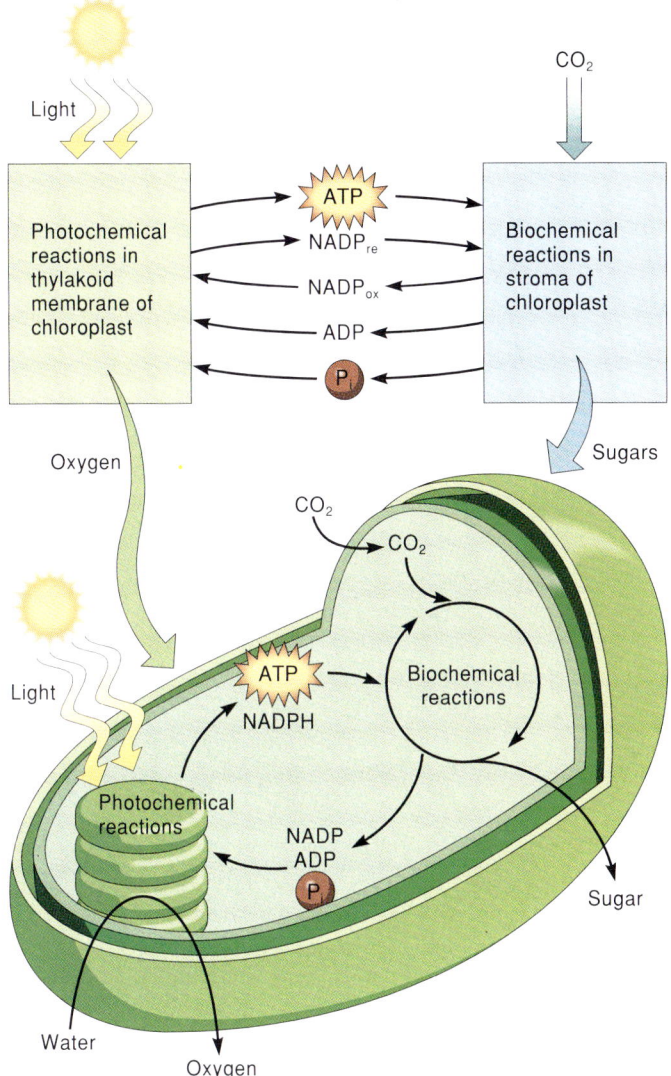

FIGURE 7.17

Photosynthesis consists of photochemical and biochemical reactions. The photochemical reactions convert light-energy to chemical energy: ATP and NADPH. The biochemical reactions use the ATP and NADPH produced by the photochemical reactions to reduce CO_2 to sugars. The photochemical reactions occur on thylakoid membranes, whereas the biochemical reactions occur in the stroma.

THE PHOTOCHEMICAL AND BIOCHEMICAL REACTIONS OF PHOTOSYNTHESIS

In 1905, F. F. Blackman reported an interesting set of experiments showing that photosynthesis was more complex than most botanists had realized. Blackman reported that in dim light, increasing the intensity of light increased the rate of photosynthesis, but increasing the temperature did not. Thus, these reactions were light-dependent and temperature-independent.

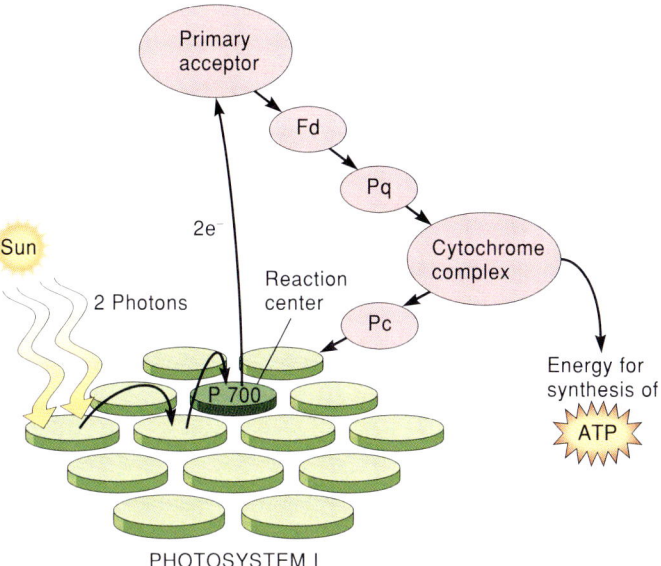

FIGURE 7.18

Cyclic flow of electrons. Electrons ejected from Photosystem I are passed through a series of electron carriers: ferredoxin (Fd, an iron-containing protein), plastoquinone (Pq, a mobile electron carrier similar to ubiquinone in the electron transport chain of mitochondria), a complex of two cytochromes, and plastocyanin (Pc, a copper-containing protein). The cycle is completed when plastocyanin returns the electron to P_{700}. The loss of potential energy at each step is used to create a proton gradient that, in turn, is used to generate ATP (see fig. 7.16).

However, in bright light, increasing the temperature increased the rate of photosynthesis, while further increases in light intensity did not. Thus, these reactions were temperature-dependent and light-independent.

Blackman's experiments provided the first evidence that photosynthesis is a two-stage process (fig. 7.17). The light-dependent reactions became known as the **photochemical (or light) reactions** of photosynthesis; these reactions are insensitive to changes in temperature. The temperature-dependent reactions became known as the **biochemical (or dark) reactions** of photosynthesis because of their insensitivity to light. However, these enzymatic reactions occur only in light—*dark* means only that light is not directly involved in the reactions. Further studies showed that the photochemical reactions oxidize water, release oxygen, and produce ATP and reduced $NADP^+$ that are used in the biochemical reactions, which reduce carbon dioxide to carbohydrate (fig. 7.17).

The Photochemical Reactions of Photosynthesis

Cyclic Electron Flow

Photosynthetic prokaryotes use only **cyclic photophosphorylation,** a process geared to energy production rather than to biosynthesis. Photophosphorylation is relatively simple and involves only Photosystem I (fig. 7.18). Electrons energized by light are recycled through the electron transport chain and return to the same reaction center from which they originated. Electron carriers move the electrons only in the company of hydrogen ions, which are transported across the membrane as the electrons move to the next member of the chain. In the process, the energy-releasing flow of electrons is coupled to the energy-requiring manufacture of ATP from ADP and Pi. The coupling mechanism is chemiosmosis, as in mitochondria. At each step of the electron transport chain, electrons lose potential energy, finally returning to their ground-state energy in P700. Absorption of more light excites the reaction center again, restarting the cycle. Electrons that return to P700 have only about half of the energy they had when they left; the remainder of the energy is the photosynthetic payoff.

Cyclic phosphorylation short-circuits production of reduced $NADP^+$ by shuttling the energized electrons from P700 to ferredoxin and then to plastoquinone. As the electrons again move down the electron transport chain, ATP forms. This process generates no oxygen or reduced $NADP^+$.

Noncyclic Electron Flow

Cyclic photophosphorylation alone occurred for more than a billion years and is geared to energy production, not biosynthesis. It suffices for prokaryotes that use molecules such as H_2S as an electron source. However, plants strip electrons from water. This requires a huge change in biochemistry, because the voltage from one light-energized reaction center of P700 doesn't have enough energy to strip electrons from water and give them to carbon dioxide. Plants overcame this problem by grafting onto the bacterial system a second, more powerful photosystem that enabled a linear, **noncyclic electron flow** (fig. 7.19). These two pumps are connected in series, much as we use two batteries to increase the voltage in a flashlight. The second pump is Photosystem II, whose unique arrangement of pigments allows it to harvest shorter wavelengths of light. This new photosystem acts first in what is now called the *Z scheme*, a model for noncyclic flow of electrons (and energy) during the photochemical reactions of photosynthesis (see fig. 7.19). Noncyclic photophosphorylation occurs in green plants, algae, and photosynthetic bacteria, and involves two photosystems that (1) make ATP in the electron transport chain linking photosystems I and II, and (2) reduce NADP.

A Summary of the Photochemical Reactions

The photochemical reactions of photosynthesis start at P680 and end at $NADP^+$. These reactions are linear rather than cyclic, and produce ATP and NADPH. Follow in figure 7.19 how they occur:

- A photon hits the end of the reaction center nearest the inner surface of the membrane, exciting an electron. This electron carries its energy to the other end of the reaction center (the end nearest the outer surface of the membrane).

- Three more photons cause other electrons to do the same thing. Thus, four photons strike P680 of Photosystem II, which functions like a tiny capacitor by creating a separation of charge. Although the reaction center captures 98%–99% of the photons that it absorbs, only about half of their energy is stored in the charge separation. Thus, Photosystem II is only about half as efficient as a battery.

- Like all energy-rich substances, the excited electrons are unstable and release much of their energy as heat. The electrons ejected from P680 leave "holes" that are filled by electrons from Z protein, a manganese-containing protein that gets its electrons from water. Thus, Photosystem II removes electrons from water, releases oxygen, and uses the electrons to replace those ejected by photons absorbed by the reaction center. All of this occurs in less than a billionth (10^{-9}) of a second.

- Splitting two molecules of water releases a molecule of O_2 and four electrons, all at once. However, only one electron at a time can be accepted by P680. The electrons stripped from water are managed by manganese, whose stable oxidation states, ranging from +2 to +7, suit it ideally for its function as a charge accumulator. At saturating intensities of light, one molecule of oxygen is released per about 2,500 molecules of chlorophyll.

- Within a few trillionths of a second, the electrons ejected from P680 of Photosystem II reduce pheophyton, a pigment that accepts electrons from chlorophyll. The resulting separation of charge widens as the electrons move across the thylakoid membrane and are accepted by plastoquinone (Pq), which is embedded on the outer, stromal side of the membrane.

- The electrons then descend an electron transport chain of molecules (such as cytochromes) in thylakoids linking Photosystem II and Photosystem I,[2] which are connected in series in thylakoid membranes. As the electrons move down this chain, they form a proton gradient that generates ATP. Herbicides such as monuron (CMU) and diuron (DCMU) kill plants by blocking this electron transport.

- The final electron acceptor in the electron transport chain is P700, the reaction center of Photosystem I. There, the electrons have a higher energy than when they left P680 a split second earlier. At P700, four more photons absorbed by antennae transfer their energy to the reaction center, where the energy is used to eject electrons from P700's electron donor, located on the thylakoid space side of the membrane. These energized

---

[2]. These photosystems were named in their order of discovery, not their organization in chloroplasts. This explains why electrons move from Photosystem II to Photosystem I, rather than from I to II.

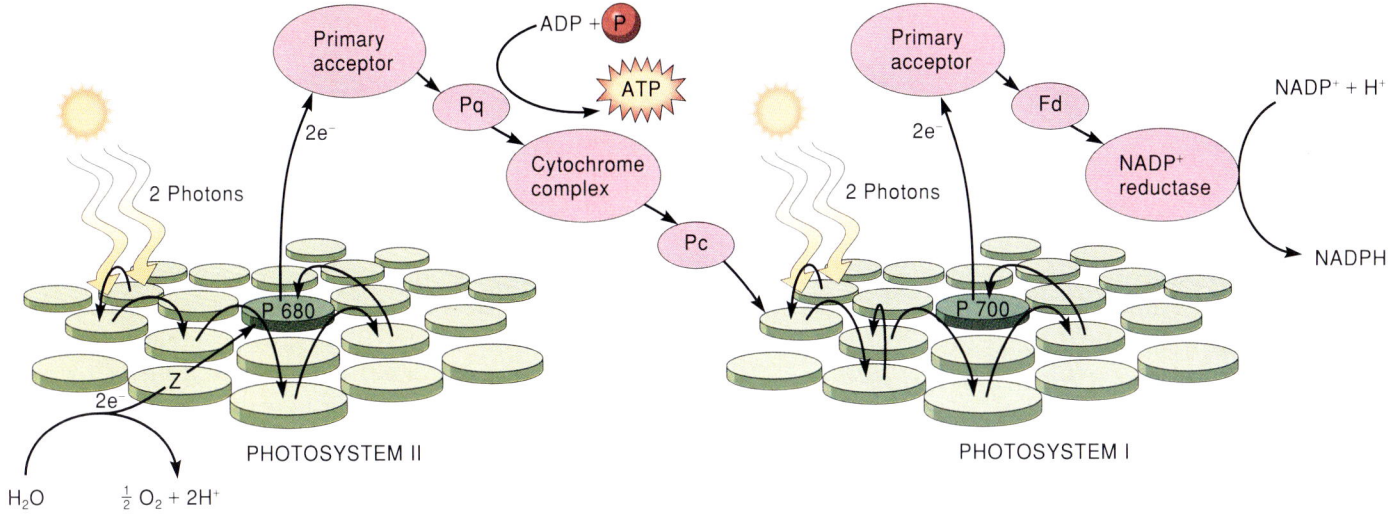

A.

B.

Figure 7.19

(a) Summary of noncyclic flow of electrons. When Photosystems I and II are illuminated, electrons flow from water to $NADP^+$. Electrons ejected from P680 are replaced by electrons from water, and a by-product of this process is O_2. The energized electrons from P680 flow through an electron transport chain to P700, generating an ATP in the process. Illumination of P700 energizes the electrons again; these electrons are used to reduce $NADP^+$. (b) Photosystems I and II are linked in series in the thylakoid membrane. Noncyclic flow of electrons produces ATP, reduced $NADP^+$, and O_2.

CHAPTER SEVEN *Photosynthesis* 149

electrons cross the membrane and reduce ferredoxin, a small, iron-containing protein that is the electron acceptor of Photosystem I. Ferredoxin then reduces NADP⁺ to NADPH. The herbicides diquat and paraquat kill plants by accepting electrons from Photosystem I before ferredoxin. They transfer these electrons to oxygen and make superoxide (O_2^-), a free radical that destroys unsaturated fatty acids and kills the cells.

- Because ferredoxin is on the outer, stromal side of the membrane, NADP⁺ forms in the stroma, where it is used in biochemical reactions to reduce carbon to carbohydrate, the basic nutrient of life.

- The eight photons[3] used in the photochemical reactions make three molecules of ATP and reduce two molecules of NADP⁺.

$$2H_2O + 2NADP^+ + 3ADP + 3Pi$$
$$\downarrow 8\ photons$$
$$O_2 + 2NADPH + 3ATP + 4e^- + 2H^+$$

- These ATPs and NADPHs are used to reduce CO_2 in the biochemical reactions of photosynthesis.

The evolution of Photosystem II forever changed life on earth. For example, it made photosynthetic organisms independent of H_2S from decaying organic compounds as sources of electrons. Moreover, it liberated oxygen into the atmosphere, thereby allowing the evolution of aerobic respiration.

CONCEPT

The photochemical reactions of photosynthesis convert light-energy into chemical energy. In plants, the photochemical reactions oxidize water, produce ATP, reduce NADP⁺, and involve two photosystems connected in series. The ATP and reduced NADP⁺ made in the photochemical reactions are used in the biochemical reactions of photosynthesis to reduce carbon dioxide to carbohydrate.

The Biochemical Reactions of Photosynthesis

The biochemical reactions of photosynthesis reduce carbon dioxide to carbohydrate, a storage depot of chemical energy. Straightforward as this may sound, the actual mechanisms of photosynthesis have been difficult to explain. The most important contribution to our understanding of how plants reduce carbon dioxide to carbohydrate was provided by Melvin Calvin and his colleagues at Berkeley in the early 1950s. These botanists studied photosynthesis by using a product of World War II: ¹⁴C, or radioactive carbon (¹⁴C has identical chemical properties to the common ¹²C, but is radioactive because it has more neutrons). Calvin first exposed a dense culture of *Chlorella* (a green alga) growing in a lollipoplike chamber to radioactively labeled CO_2 (fig. 7.20). Then, using chromatography (to separate the compounds) and autoradiography (to identify the separated compounds), Calvin traced the path of the ¹⁴C as it moved from CO_2 to carbohydrate. When Calvin exposed the alga to labeled CO_2 for 60 seconds, he found that many compounds in the alga were labeled with ¹⁴C. Calvin reasoned that he needed to modify the experiment and expose the alga to the ¹⁴CO_2 for less time. When he repeated the experiment with a 7-second exposure, most of the radioactive carbon appeared in 3-phosphoglyceric acid (3-PGA), a three-carbon molecule (fig. 7.20). This suggested to Calvin that the molecule that joined with CO_2 (i.e., the CO_2 acceptor) was a two-carbon compound, since combining such a two-carbon acceptor with CO_2 (a one-carbon molecule) would produce a three-carbon compound. Calvin searched for this two-carbon acceptor, but did not find it.

Faced with these results, Calvin began searching for another molecule that, if combined with CO_2, would produce a three-carbon product. He found the elusive acceptor in ribulose-1,5-bisphosphate (RuBP), a five-carbon sugar. The product of RuBP and CO_2 is an unstable six-carbon molecule that immediately breaks down into two molecules of 3-PGA:

$$CO_2 + RuBP \rightarrow 2\ 3\text{-PGA}$$

Calvin used a similar approach to figure out the rest of the biochemical reactions of photosynthesis. Follow in figure 7.21 how they occur:

- CO_2 diffuses into the chloroplast stroma and is fixed; that is, it is incorporated into an organic compound.[4] This occurs by the covalent bonding of CO_2 to RuBP. The enzyme that catalyzes this reaction is *RuBP carboxylase/oxygenase* (also called *rubisco*), a large (molecular weight 490,000) enzyme consisting of eight large subunits and eight small subunits. The large subunits are coded in the chloroplast; the mRNA for these subunits is part of the chloroplast RNA and is made on ribosomes in the chloroplast.[5] The eight small subunits are coded in the nucleus; their mRNA is associated with cytoplasmic ribosomes and includes a polypeptide leader that directs the mRNA into the chloroplast. The assembly of the large and small subunits into rubisco is controlled by light.

---

[3]. Theoretically, plants need only eight photons to reduce one molecule of carbon dioxide. Most plants need 15–20 photons to do this in field conditions. In ideal lab conditions, plants need 9–12 photons to reduce one CO_2.

[4]. CO_2 actually moves into cells as bicarbonate ion (HCO_3^-). CO_2 dissolves in water on the wet surface of mesophyll cells. This forms carbonic acid (H_2CO_3), which then breaks down into a proton and a bicarbonate ion:

$$CO_2 + H_2O \leftrightarrow H_2CO_3 \leftrightarrow H^+ + HCO_3^-$$

[5]. Each photosynthetic cell has several hundred to several thousand copies of the gene to make the large subunit of rubisco. This results not only from the 40–200 chloroplasts per cell, but also from the polyploid nature of chloroplast DNA; there are 50–1,000 copies of the genome per chloroplast.

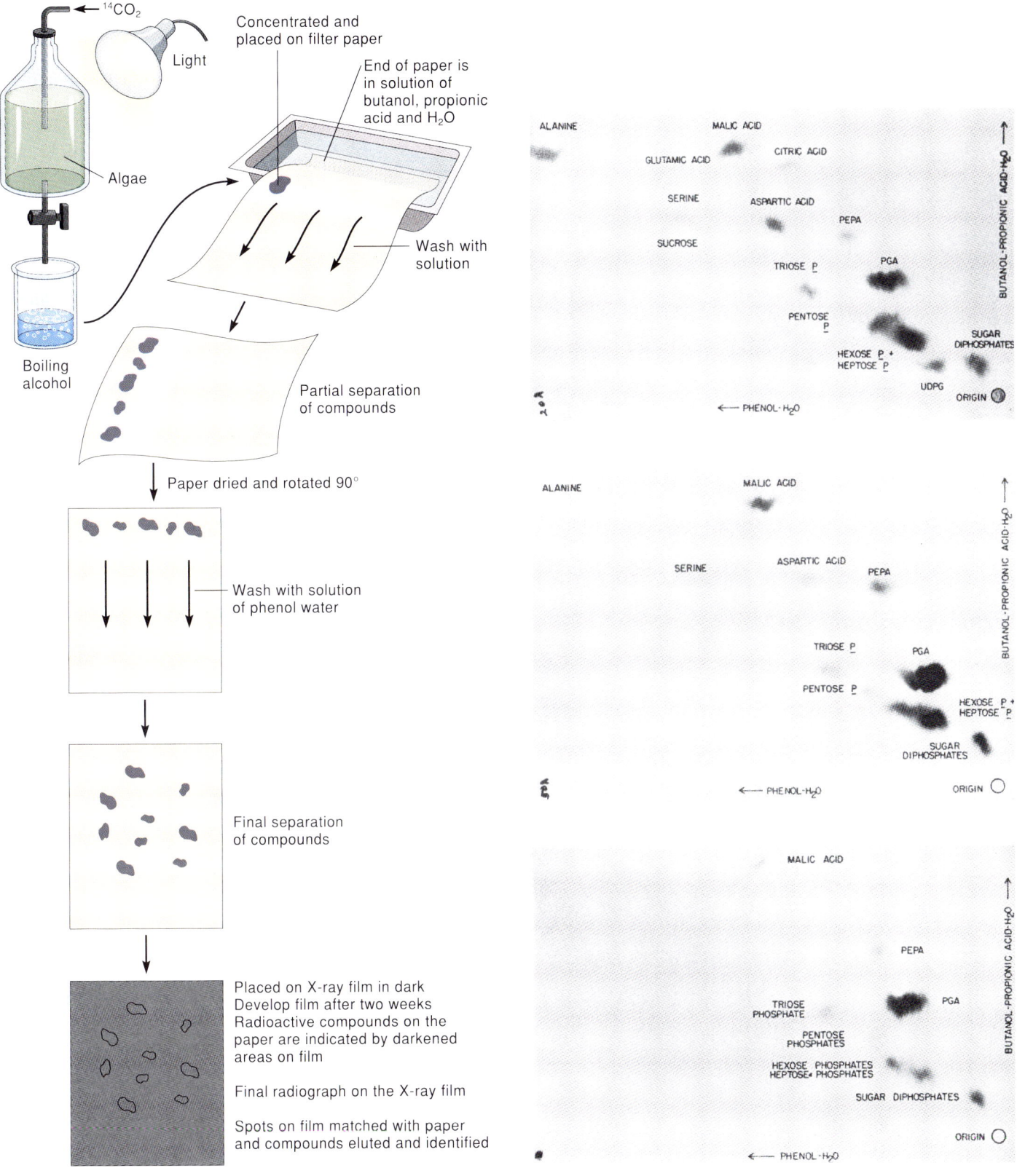

Figure 7.20

A summary of Calvin's experiments and results. Calvin exposed algae to radioactive CO_2 for various times (shown here are results for 2, 7, and 60 sec.) and then killed the algae at different times by placing them in boiling ethanol. Calvin then applied an extract of the algae onto pieces of paper and ran solvents through the paper from one direction. This process, called *chromatography*, separated the compounds in the extract: different compounds moved different distances in the solvent moving through the paper. Calvin then placed film against the pieces of chromatography paper. The radioactive compounds in the original extract—now at different points on the paper—exposed the film. Calvin then identified the compounds by repeating the process with known standards. This technique is known as *autoradiography*. Calvin used chromatography and autoradiography to decipher the biochemical reactions of photosynthesis.

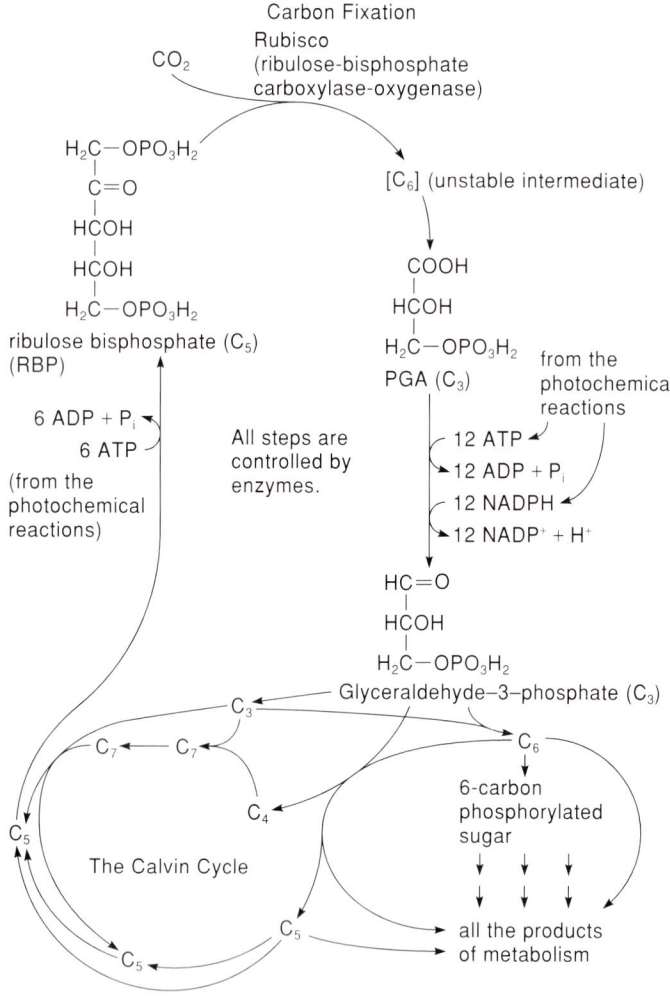

FIGURE 7.21

The Calvin (C_3) cycle.

- Rubisco occurs in all autotrophs except a few species of bacteria, and is the most important and abundant protein on earth. It makes up about 20%–25% of the soluble protein in leaves and is made at a rate of 4×10^{10} g yr^{-1}, which is the equivalent of about 10^6 g sec^{-1}. Every person on earth is supported by about 44 kg of rubisco.

- The reaction of CO_2 with RuBP produces an unstable six-carbon compound that immediately breaks into two molecules of 3-phosphoglyceric acid (3-PGA) (fig. 7.22). This three-carbon molecule is the first stable product of photosynthesis, which explains why botanists refer to this series of reactions as the **C_3 cycle** or Calvin cycle.[6] Similarly, plants that use only the Calvin cycle to fix carbon dioxide are called **C_3 plants**. About 85% of plant species are C_3 plants, including cereals (e.g., barley, oats, rice, wheat), peanuts, cotton, sugar beets, tobacco, spinach, soybeans, most trees, and lawn grasses such as rye, fescue, and Kentucky bluegrass. The features of C_3 plants are summarized in tables 7.2 and 7.3 later in this chapter.

---

6. Calvin's research earned him a Nobel Prize in 1961.

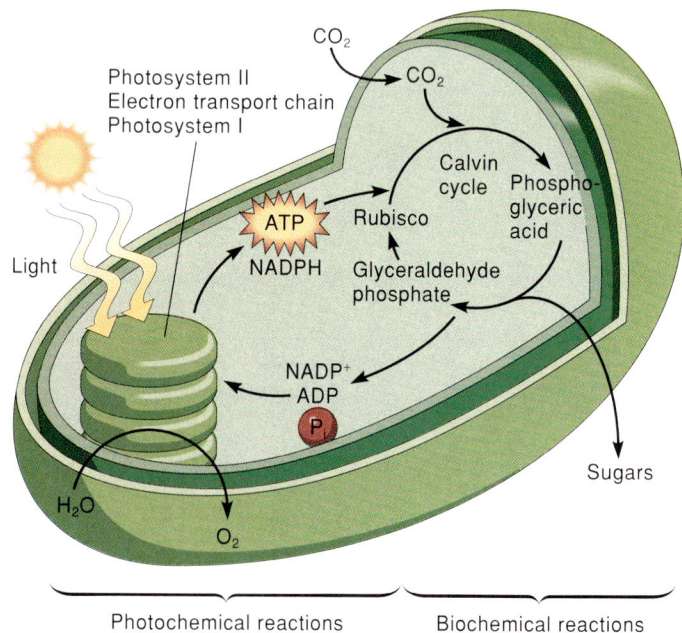

FIGURE 7.22

A summary of photosynthesis. The photochemical reactions, which occur in thylakoids and grana, produce O_2, ATP, and reduced NADP$^+$. The Calvin cycle (i.e., biochemical reactions), which occurs in the stroma, uses the ATP and reduced NADP$^+$ to reduce CO_2 to carbohydrate (three key intermediates are shown here). The by-products ADP, Pi, and NADP$^+$ are returned from the Calvin cycle to the photochemical reactions.

- A phosphate group is cleaved from each molecule of 3-PGA, and ATP and electrons from NADPH (produced by the photochemical reactions) are then used to reduce diphosphoglycerate (DPGA) to glyceraldehyde-3-phosphate (G-3-P), a three-carbon sugar. This sugar, not glucose, is the carbohydrate produced by the Calvin cycle and is the starting point for many other metabolic pathways in the plant.

- Some of the G-3-P is used to re-form RuBP. This can occur in the dark and requires ATP made in the photochemical reactions. The rest of the G-3-P moves through a series of chemical reactions to form fructose diphosphate, which is used to make glucose, sucrose, starch, and other compounds needed by the plant.

A Summary of the Biochemical Reactions

The first steps of the Calvin cycle make precursors for glucose and other carbohydrates, while the later steps regenerate RuBP. Since the Calvin cycle takes in only one carbon (as CO_2) at a time, it takes six turns of the cycle to produce a net gain of six carbons (i.e., two molecules of glyceraldehyde-3-phosphate, a three-carbon sugar). These six turns of the cycle require eighteen ATPs (three per carbon) and twelve NADPHs (two per carbon), all of which come from the photochemical reactions of photosynthesis (fig. 7.22):

$$6CO_2 + 18ATP + 12NADPH + 12H_2O$$
$$\downarrow$$
$$C_6H_{12}O_6 + 18ADP + 18Pi + 12NADP^+ + 6H_2O + 6O_2$$

Although this summary equation is convenient for understanding the basics of photosynthesis, it is imperfect; some of the intermediates of the Calvin cycle are siphoned off to make other compounds (e.g., amino acids).

CONCEPT

The biochemical reactions of photosynthesis use ATP and reduced NADP⁺ made in the photochemical reactions to reduce carbon dioxide to carbohydrate. These reactions are collectively referred to as the *Calvin*, or *C₃*, *cycle* and begin with the fixation of carbon dioxide by an enzyme called *RuBP carboxylase/oxygenase*. The product of the biochemical reactions is glyceraldehyde-3-phosphate, a sugar that is the starting point for several metabolic pathways. Fixing each molecule of carbon dioxide requires three ATPs and two reduced NADPs. Plants that use only the Calvin cycle to fix CO_2 are called C_3 plants.

Writing to Learn Botany

How does photosynthesis differ from respiration?

The Efficiency of Photosynthesis

How efficient is photosynthesis? To reduce two molecules of NADP⁺ requires that photosystems I and II each absorb four photons (i.e., a total of eight photons). Since 1 mole of photons having a wavelength of 600 nm carries 47.6 kcal of energy, 8 moles of such photons carry 381 kcal. It takes 114 kcal of energy to reduce 1 mole (44 g) of carbon dioxide to hexose. Therefore, the theoretical efficiency of photosynthesis is 114/381 = 30%. On cloudy days, field measurements of the photosynthetic efficiency of individual plants average about 0.1%, while those for intensively cultivated plants average about 3%. Most crops range from 1%–4% (as do algae), and values as high as 25% have been recorded in laboratory conditions. A photosynthetic superstar of field-growing plants is *Oenothera claviformis*, an annual winter-evening-primrose that grows in Death Valley: its value of 8% is the highest of all plants studied in natural conditions. Sugarcane is a close second at about 7%.

PHOTOSYNTHESIS IS NOT PERFECT: PHOTORESPIRATION

Photosynthesis took a long time to evolve, but it still has many shortcomings. For example, it wastes much light energy (see box 7.1) and often does a poor job of fixing carbon. The first biologist to show this imperfection of photosynthesis was Otto Warburg, who in 1920 showed that increased amounts of oxygen inhibit photosynthesis in C_3 plants. This effect became known as the Warburg effect. You may wonder how this is relevant to plants, since the amounts of CO_2 and O_2 in air are relatively constant. The answer is that it depends on what air you're talking about. Indeed, the composition of air outside a leaf often differs significantly from that inside a leaf where photosynthesis occurs.

During photosynthesis, plants fix CO_2 from the atmosphere. Although the concentration of CO_2 in this air is only about 0.034%, this represents an almost endless supply of CO_2 for photosynthesis—indeed, there are about 7×10^{11} tons of carbon (as carbon dioxide) in the atmosphere (for comparison, there are about 4.5×10^{11} tons of carbon in vegetation as carbohydrate).[7] Consequently, you wouldn't think that plants would have much trouble obtaining enough carbon dioxide for photosynthesis. However, all of this carbon dioxide is available to a plant only when the plant's stomata are open. As CO_2 diffuses into the leaf for photosynthesis, water diffuses out through the stomata. As long as the plant has plenty of water, this is not a serious problem, but on hot, dry days, plants close their stomata to conserve water. When this occurs, the continued fixation of carbon dioxide (via the Calvin cycle) decreases the concentration of carbon dioxide inside the leaf, which greatly increases the relative amount of O_2. This has a tremendous effect on the plant: when the concentration of carbon dioxide inside the leaf dwindles to about 50 ppm, rubisco stops fixing CO_2 and starts fixing oxygen. The products of this oxygenase reaction (remember that the enzyme is ribulose-1,5-bisphosphate carboxylase/*oxygenase*) are phosphoglycolic acid, a two-carbon compound, and PGA, a three-carbon compound:

$$O_2 + RuBP$$
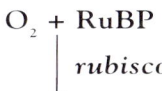
phosphoglycolic acid + PGA

The PGA stays in the Calvin cycle, but the phosphoglycolic acid leaves the cycle. In a complex series of reactions occurring in mitochondria and peroxisomes (fig. 7.23), the phosphoglycolic acid is hydrolyzed to glycine, a two-carbon amino acid. Two molecules of glycine then combine to form CO_2 and serine, a three-carbon amino acid. The serine is converted to 3-PGA (at the expense of ATP), which can enter the Calvin cycle. However, the CO_2 is often lost from the plant.

The reactions initiated by rubisco's fixation of oxygen rather than carbon dioxide are collectively called **photorespiration** because they occur only in light, consume oxygen, and release carbon dioxide. Unlike respiration, however, photorespiration produces no ATP. Rather, rubisco operates at only about 25% of its maximal rate as it squanders large amounts of carbon

---

7. About 80% of the earth's carbon is in rocks, and the remainder is in organic compounds such as cellulose and oil. The atmosphere contains about 1/1000 of 1% of the total, or 720 billion metric tons of carbon. About 560 billion metric tons of carbon are in living organisms.

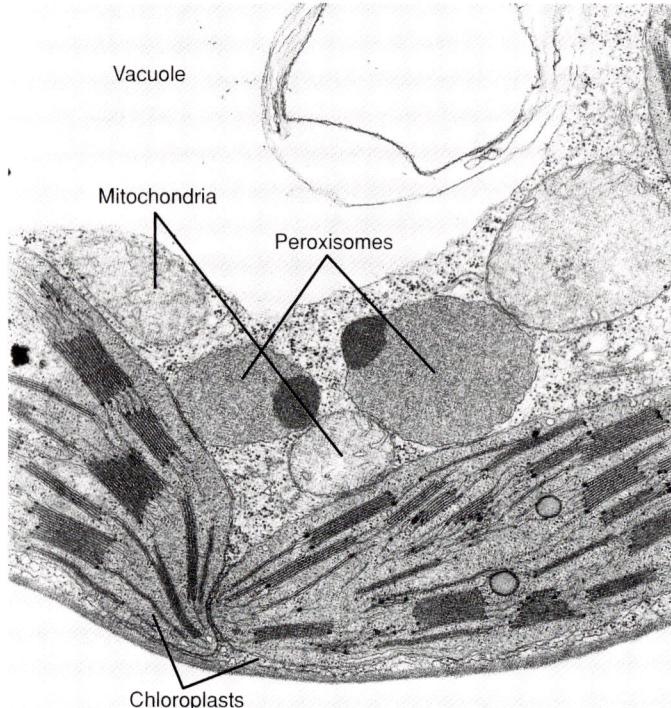

FIGURE 7.23

Electron micrograph showing the close association of chloroplasts, peroxisomes, and mitochondria, ×58,000. All of these organelles participate in photorespiration.

originally fixed at great expense by the Calvin cycle. Although some of the CO_2 released during photorespiration is recycled by the Calvin cycle, as much as half of the carbon fixed in photosynthesis by C_3 plants is reoxidized and lost as CO_2. Thus, photorespiration undoes photosynthesis.

How did plants evolve a wasteful process such as photorespiration? Some botanists suggest that the oxygenase reaction of rubisco is inevitable because of rubisco's structure: its active site makes it impossible for the enzyme to discriminate between CO_2 and O_2. Another suggestion is that photorespiration represents evolutionary baggage—a metabolic relic from when the atmosphere contained no free oxygen and competition for CO_2 didn't matter. The buildup of oxygen derived from photosynthesis led to photorespiration after millions of years.

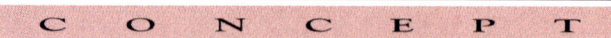

CONCEPT

Photorespiration occurs in C_3 plants when the internal concentration of carbon dioxide becomes low, such as when stomata close during drought. During photorespiration, rubisco fixes oxygen, and carbon dioxide is later released. Consequently, photorespiration undoes photosynthesis: as much as half of the carbon fixed in the Calvin cycle is released by photorespiration during hot, dry days.

Photorespiration peaks in hot, dry conditions (especially at temperatures above 28° C) when plants begin to close their stomata to conserve water. When this occurs, photorespiration causes plants to lose carbon. Clearly, plants that could avoid photorespiration would have a significant advantage. That advantage was provided by the evolution of C_4 photosynthesis.

C_4 PHOTOSYNTHESIS

Soon after Calvin reported his brilliant study, other biologists began to repeat Calvin's study with other plants. Most of these studies confirmed Calvin's findings, but one study of sugarcane in 1965 stood out. In this plant, a 1-second exposure to radioactive CO_2 resulted in 80% of the radioactivity being found not in a three-carbon compound such as PGA, but in a four-carbon acid called malic acid. Moreover, most of this malic acid appeared only in thin-walled mesophyll cells which, strangely enough, lack most of the enzymes of the Calvin cycle. Only after about 10 seconds did the label appear in PGA in bundle-sheath cells, which are thick-walled cells surrounding the vascular bundles (fig. 7.24). This kind of leaf anatomy—veins encased by thick-walled photosynthetic bundle-sheath cells that are in turn surrounded by thin-walled mesophyll cells—is characteristic of tropical grasses such as sugarcane and is referred to as **Kranz (halo** or **wreath) anatomy** (fig. 7.24). More interesting, at least initially, was the fact that CO_2 being fixed into malic acid was inconsistent with the idea that the Calvin cycle was the only means of fixing CO_2 in plants. What was going on here?

Careful studies of the structure and function of these leaves finally unlocked the puzzle—sugarcane *does* fix CO_2 via the Calvin cycle, but only *after* fixing it via another set of reactions that preface the Calvin cycle. Here's the explanation (fig. 7.25):

- Carbon dioxide diffuses into the leaf through stomata and is fixed in mesophyll cells. These cells lack rubisco; during greening, light blocks the transcription of the mRNA to make the large subunit of rubisco.

- Carbon dioxide in mesophyll cells combines with a three-carbon compound called *phosphoenolpyruvic acid (PEP)*, a respiratory intermediate. This reaction is catalyzed by PEP carboxylase and produces oxaloacetic acid (OAA), or oxaloacetate, a four-carbon acid (and an intermediate in the Krebs cycle):

$$CO_2 + PEP$$
$$\downarrow PEP\ carboxylase$$
$$\textbf{oxaloacetic acid (OAA)}$$

- The four-carbon acid is quickly converted to malic acid (malate) or aspartic acid (aspartate), which is then moved (at the expense of ATP) through plasmodesmata to the adjacent bundle-sheath cell. There, these acids are split into CO_2 and a three-carbon compound. Thus,

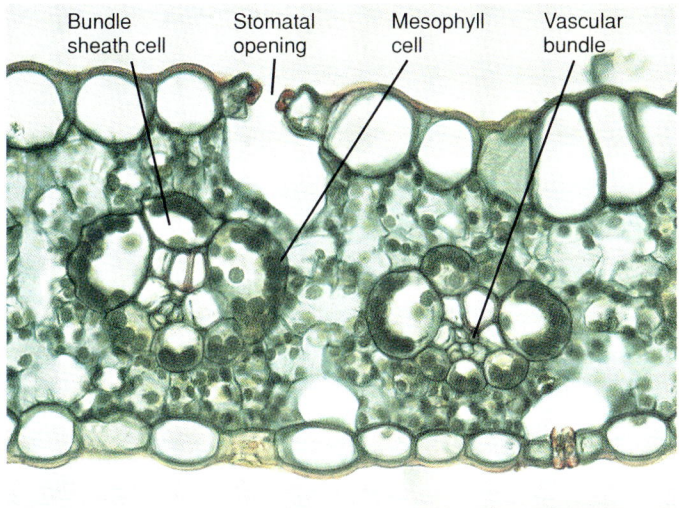
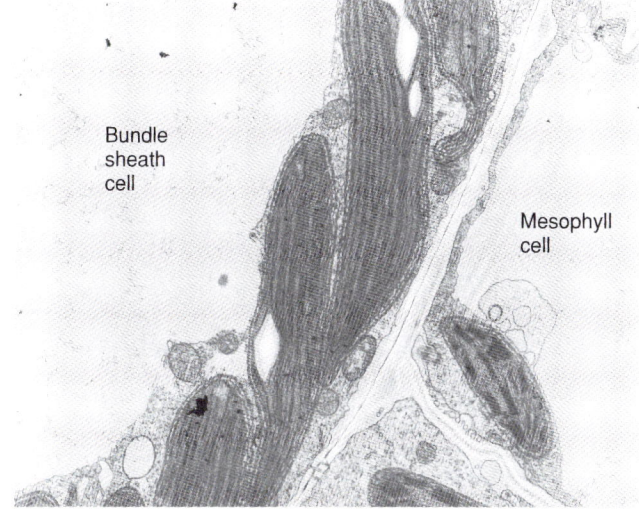

FIGURE 7.24

Structure of leaves of C_4 plants. (a) Leaves of C_4 plants have a characteristic Kranz anatomy: veins are tightly surrounded by photosynthetic bundle-sheath cells, which in turn are surrounded by photosynthetic mesophyll cells. This is a light micrograph of a cross section of a leaf of corn (*Zea mays*). (b) Electron micrograph of cross section of a leaf of *Echinochloa colonum*, a C_4 grass.

malic acid and aspartic acid function as short-lived reservoirs of CO_2 (they turn over in about 10 seconds).

- The pumping by C_4 plants of CO_2 into bundle-sheath cells keeps the internal concentration of CO_2 in bundle-sheath cells 20–120 times greater than normal. This high concentration of CO_2 allows rubisco to fix CO_2 at maximal rates (recall that in C_3 plants, rubisco operates at only about 25% of its maximum rate). This eliminates photorespiration, thereby enabling C_4 plants to fix CO_2 more efficiently than C_3 plants do.

- The CO_2 released in the bundle-sheath cells is fixed by the Calvin cycle. Meanwhile, the three-carbon compound is shuttled back to the mesophyll cell (again at the expense of ATP), where it is converted to PEP, the initial CO_2 acceptor in C_4 photosynthesis.

Plants such as sugarcane are called **C_4 plants** because a four-carbon acid is the first stable product of their photosynthesis. Although none of

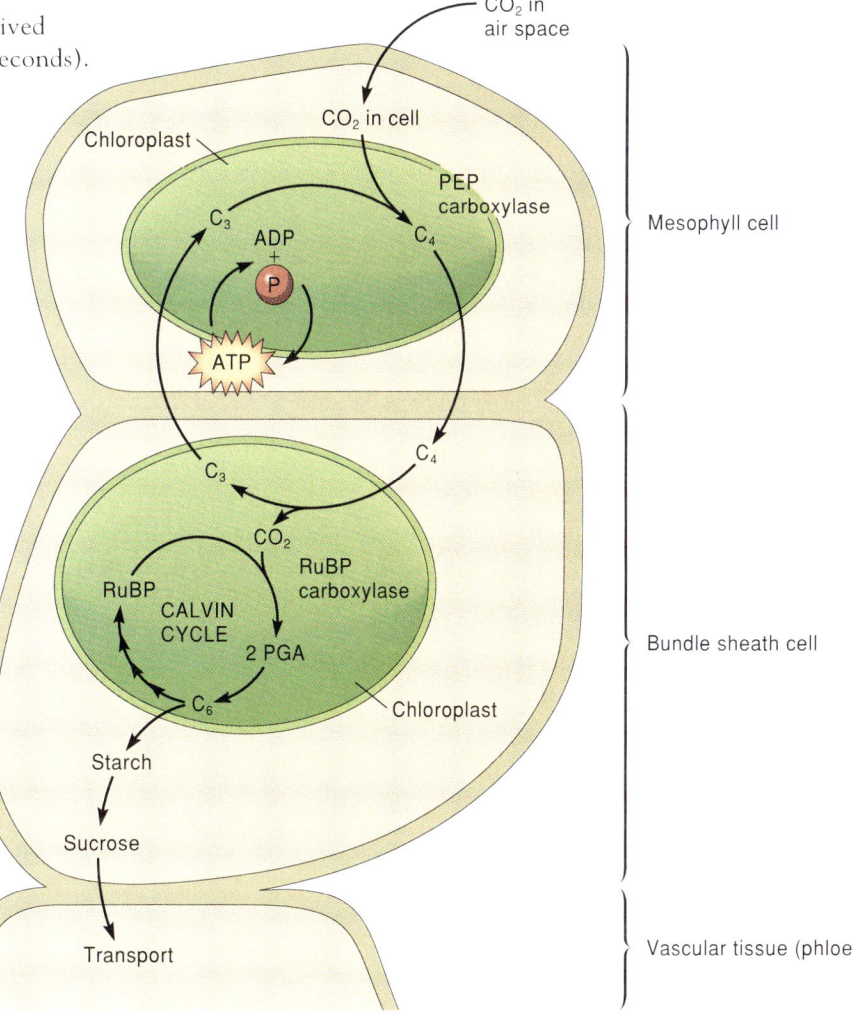

FIGURE 7.25

Summary of C_4 photosynthesis. Mesophyll cells fix CO_2 into four-carbon acids that move into bundle-sheath cells. There, the acids are decarboxylated; the CO_2 released by this decarboxylation is fixed via the Calvin cycle in the bundle-sheath cell. The three-carbon compound moves back to the mesophyll cell, where it is converted to PEP, the CO_2 acceptor.

CHAPTER SEVEN *Photosynthesis*

TABLE 7.2
Photosynthetic Characteristics of C_3, C_4, and CAM Plants

Characteristic	C_3	C_4	CAM
Leaf anatomy	Bundle-sheath cells without dense arrangement of chloroplasts	Bundle-sheath cells with dense arrangement of chloroplasts; no distinctive palisade cells	Large vacuoles
Primary carboxylating enzyme	Ribulose bisphosphate carboxylase/oxygenase	PEP carboxylase	PEP carboxylase at night
Grams of water required to produce 1 g of dry matter	450 to 950	250 to 350	50 to 55
Requires sodium as micronutrient	No	Yes	Probably
Minimum atmospheric CO_2 concentration required for photosynthesis	0.03%–0.07%	0%–0.01%	0%–0.005%
Photorespiration detectable	Yes	Only in isolated bundle-sheath cells	No
Optimum temperature for photosynthesis	15°–25° C	30°–40° C	About 35° C
Approximate tons of dry matter produced per hectare per year	20–25	35–40	Usually low and variable

TABLE 7.3
Maximum Photosynthetic Rates of Major Types of Plants in Natural Conditions

Type of Plant	Example	Maximum Photosynthesis mg CO_2 dm^{-2} h^{-1}
CAM	*Agave americana* (century plant)	1–4
C_3		
Tropical, subtropical, and Mediterranean evergreen trees and shrubs; temperate zone evergreen conifers	*Pinus sylvestris* (Scotch pine)	5–15
Temperate zone deciduous trees and shrubs	*Fagus sylvatica* (European beech)	5–20
Temperate zone herbs and C_3 pathway crop plants	*Glycine max* (soybean)	15–30
C_4		
Tropical grasses, dicots, and sedges with C_4 pathway	*Zea mays* (corn)	35–70

the photosynthetic reactions of C_4 plants are unique, they produce dramatic differences in plant growth. The advantage of the added set of reactions in C_4 photosynthesis is that PEP carboxylase scavenges CO_2 and does not react with oxygen. This, combined with the pumping of CO_2 by surrounding mesophyll cells into bundle-sheath cells, enables C_4 plants to fix CO_2 with rubisco (in bundle-sheath cells) that is insulated from high concentrations of oxygen (fig. 7.25). Consequently, C_4 plants exhibit no photorespiration; they can fix CO_2 until the internal concentration of CO_2 reaches zero. This allows them to continue to fix CO_2 even in hot, dry weather when stomata begin to close (fig. 7.26a). As a result, they need only about half as much water as C_3 plants for photosynthesis.

C_4 plants also use nitrogen more efficiently than do C_3 plants, largely because they saturate their rubisco (a nitrogen-rich protein) with CO_2, thereby maximizing its efficiency. This allows C_4 plants to make less rubisco than C_3 plants and offsets the expense of the additional reactions of the C_4 cycle. Moreover, it is almost impossible to light-saturate C_4 plants (fig. 7.26b). Consequently, they usually out-compete C_3 plants in hot, dry weather. However, if given enough water, C_3 plants such as peanut and sunflower fix carbon at rates comparable to those of C_4 plants. Other features of C_4 plants are summarized in tables 7.2 and 7.3.

C_4 photosynthesis evolved in the hot conditions of the tropics. The primary selecting mechanism for C_4 photosynthesis was probably the decrease in CO_2 concentrations during the last 50 to 60 million years. Indeed, hot and arid conditions have been common throughout history, but only during the last 50 to 60 million years have CO_2 concentrations dropped to levels that give C_4 plants a selective advantage over C_3 plants.

Today, C_4 plants are all angiosperms that are most common in hot, open ecosystems. They occur in at least seventeen families, none of which includes only C_4 plants. Because these families are diverse, distantly related, and have no common C_4 ancestors, C_4 photosynthesis probably evolved independently several times. Although most C_4 plants are monocots (especially grasses and sedges), more than three hundred are dicots (including some trees and shrubs). Examples of C_4 plants include members of the Poaceae (corn, sorghum, sugarcane, millet, crabgrass, and Bermuda grass), pigweed (*Amaranthus*), and *Atriplex*. There are no known C_4 gymnosperms, bryophytes, or algae.

CONCEPT

C_4 photosynthesis occurs in tropical grasses and fixes carbon dioxide into a four-carbon acid in mesophyll cells. This acid then moves to bundle-sheath cells, where it is broken down to carbon dioxide and a three-carbon molecule. The three-carbon molecule moves back to the mesophyll cell to accept another molecule of carbon dioxide. Meanwhile, the carbon dioxide is fixed by the Calvin cycle in bundle-sheath cells. This pumping of carbon dioxide into bundle-sheath cells eliminates photorespiration, which makes C_4 plants more efficient than C_3 plants during hot, dry conditions.

Why Don't C_4 Plants Dominate the Landscape?

If C_4 plants are more efficient than C_3 plants, why then don't they dominate our landscape? The answer is that C_4 plants are more efficient than C_3 plants *only in bright, hot conditions* (fig. 7.26). To best appreciate this, examine the summary equation for C_4 photosynthesis:

$$6CO_2 + 30ATP + 12NADPH + 12H_2O$$
$$\downarrow$$
$$C_6H_{12}O_6 + 30ADP + 30Pi + 12NADP^+ + 6H_2O + 6O_2$$

Compare this reaction with that of C_3 photosynthesis:

$$6CO_2 + 18ATP + 12NADPH + 12H_2O$$
$$\downarrow$$
$$C_6H_{12}O_6 + 18ADP + 18Pi + 12NADP^+ + 6H_2O + 6O_2$$

When it's not hot and dry, C_4 plants are *less* efficient than C_3 plants because of the extra two ATPs per carbon they use to run the C_4 part of their photosynthesis. However, in a hot, dry environment in which photorespiration would otherwise remove much of the carbon fixed in photosynthesis, the additional expense of C_4 photosynthesis is the best compromise available. Indeed, many of the plants that grow in hot climates are C_4 plants. However, this added expense is the reason they cannot compete effectively with C_3 plants in wet environments when temperatures are less than 25° C or in habitats such as dark forest floors.

Only about 0.4% of the 260,000 known species of plants are C_4 plants. So why all the fuss about C_4 photosynthesis? The answer is that several economically important plants such as corn and sorghum are C_4 plants.[8] Moreover, in hot, dry

---

8. Each year the United States produces 7 million bushels of corn from about 700 billion plants. These plants cover about 70 million acres, a cornfield the size of Arizona. On a smaller scale, the 10,000 plants that grow on one acre fix 2,500 kg (5,512 lbs) of carbon in one growing season. This requires about 10 metric tons (11 tons) of CO_2.

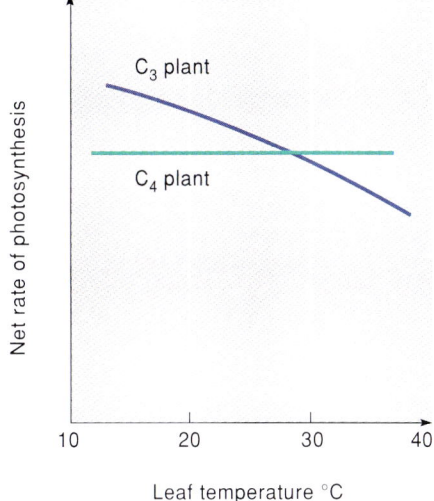

A.

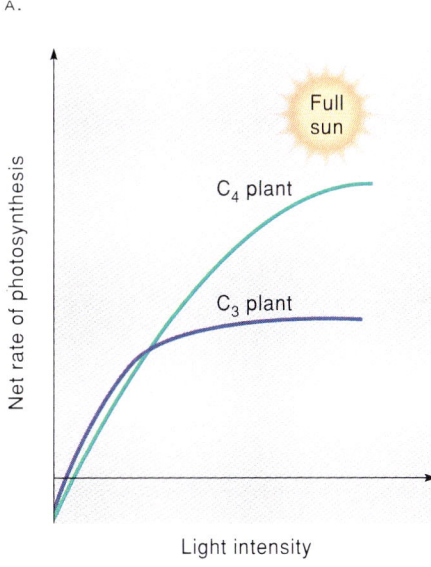

B.

FIGURE 7.26

C_4 plants are photosynthetically more efficient than C_3 plants only in hot, dry, bright conditions. Their increased efficiency in these conditions is due to the absence of photorespiration in C_4 plants.

conditions they produce more biomass than do C_3 plants. Thus, the C_4 machinery represents a potential way for botanists to increase crop yields.

What are Botanists Doing?

Use a reference source at the library to document the global distributions of C_3 and C_4 plants. What do these distributions tell you about the evolution of these plants?

C_3-C_4 Intermediates

Genera such as *Flaveria* (Asteraceae), *Panicum* (Poaceae), and *Alternanthera* (Amaranthaceae) contain C_3 and C_4 species as well as species intermediate between C_3 and C_4 photosynthesis. These so-called C_3-C_4 intermediates have the following characteristics:

- *Intermediate leaf anatomy.* Their leaves contain bundle-sheath cells, but they are indistinct and poorly developed.

- *Reduced photorespiration.* The CO_2 compensation point is the concentration of CO_2 at which net photosynthesis stops. In C_3 plants it ranges from 50 to 100 parts per million (ppm; 1 ppm = 1 µmol mol^{-1}), while in C_4 plants it is less than 5 ppm. The CO_2 compensation point for C_3-C_4 intermediates ranges from 7 to 28 ppm.

- Interestingly, closely related taxa often have different types of photosynthesis. For example, diploids of *Alloteropsis semialata* are C_3, while polyploids are often C_4.

CRASSULACEAN ACID METABOLISM (CAM)

The renewed interest in photosynthesis stimulated by Calvin in the 1950s prompted many botanists to begin studying the biochemistry of a variety of plants. Biologists familiar with the literature knew that their predecessors in the nineteenth century had shown that some members of the Crassulaceae family become acidic at night and progressively more basic during the day. The significance of these diurnal changes was not appreciated until 1958, when botanists reported that these plants open their stomata at night and fix CO_2 into malic acid that is stored overnight in vacuoles of the large, succulent photosynthetic cells (fig. 7.27); the concentration of malic acid in the vacuole can reach 0.3 M, dropping the pH to as low as 4. The next day, the stomata close to conserve water. Plants exhibiting this diurnal acidity and nocturnal fixation of CO_2 described above are called **Crassulacean acid metabolism (CAM) plants** because their unusual photosynthesis was discovered in a member of the Crassulaceae.

During the day the acids made the previous night are decarboxylated, and the liberated CO_2 is fixed via the Calvin cycle in chloroplasts (fig. 7.27). Unlike C_4 plants, in which the initial fixation of CO_2 occurs in a different cell than in the Calvin cycle, CAM plants perform all of their photosynthesis in the same cell. Therefore, while the reactions of C_4 plants are separated spatially (i.e., they occur in different cells), the reactions of the Crassulaceae are separated both spatially (i.e., in different parts of the cell) *and* temporally (i.e., at different times of the day) (see figs. 7.25 and 7.27). The PEP that reacts with CO_2 during the night comes from the breakdown of starch, which is resynthesized during the day (fig. 7.27).

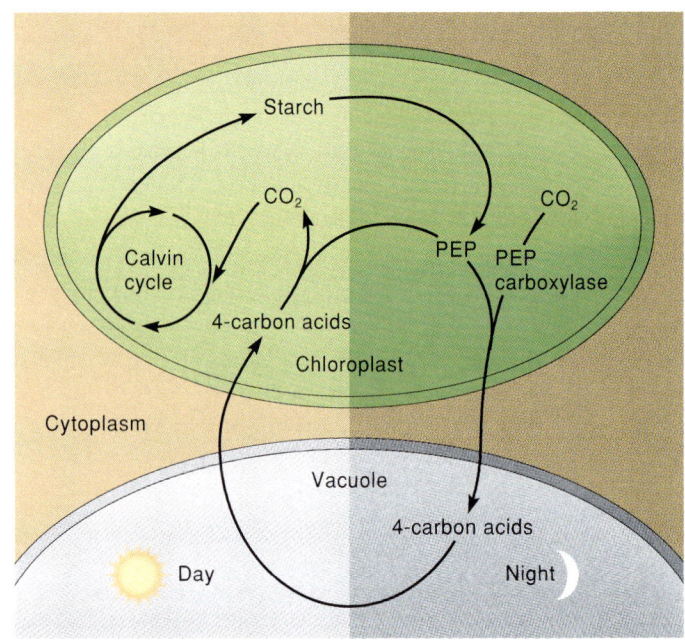

FIGURE 7.27

Photosynthesis in CAM plants. At night, CAM plants fix CO_2 into four-carbon acids, which are stored in the vacuole. During the following day, these acids are decarboxylated, and the CO_2 is fixed via the Calvin cycle. CAM photosynthesis conserves water, thus enabling CAM plants to grow in dry environments (e.g., deserts) where most other kinds of plants cannot grow.

CAM plants open their stomata at night when temperatures drop and relative humidity increases. This enables CAM plants to use water more efficiently than do C_3 plants. Indeed, although CAM plants grow slowly, they need much less water than do either C_3 or C_4 plants. Other features of CAM plants are summarized in tables 7.2 and 7.3.

CAM is more widespread than C_4 photosynthesis: it occurs in more than twenty families that include monocots, dicots, and primitive plants. Most CAM plants are succulents (fleshy plants having a low surface-to-volume ratio), but not all succulents are CAM plants. For example, pineapple is a CAM plant that is not succulent, and many halophytes (plants that grow in salty soil) are succulents but are not CAM. Examples of CAM plants include cacti, orchids, maternity plant (*Kalanchoë daigremontiana*), wax plant (*Hoya carnosa*), pineapple (*Ananas comosus*), Spanish moss (*Tillandsia usneoides*), and at least two genera of ferns (*Pyrrosia*). About 10% of plant species are CAM plants, but the only ones that are agriculturally significant are pineapple and an *Agave* species used to make tequila and as a source of fibers.

There are several variations of CAM. Some CAM plants revert to C_3 photosynthesis at the end of the day when their stored acids are gone and when given adequate water. Many CAM plants revert to C_3 photosynthesis when they're given large amounts of water, and many C_3 plants become CAM plants during drought. Plants such as *Peperomia camptotricha* lead double lives—the seedlings of these plants use C_3 photosynthesis, while mature plants use CAM.

CONCEPT

Crassulacean acid metabolism (CAM) occurs in plants living in extremely dry habitats. CAM plants fix carbon dioxide at night into a four-carbon acid that is stored in the vacuole until daylight. The acid is then decarboxylated, and the carbon dioxide is used in the Calvin cycle. All of these reactions occur in the same cells (in C_4 plants they occur in different cells). Fixation of carbon dioxide at night (i.e., when the relative humidity increases) helps CAM plants conserve water.

CONTROL OF PHOTOSYNTHESIS

The previous sections discussed one factor that controls photosynthesis: temperature. Other factors that control photosynthesis include light, the concentration of carbon dioxide, the availability of water, and metabolic sinks.

Light

Chlorophyll absorbs red and blue light strongly; therefore, these wavelengths stimulate photosynthesis more than does green light, which is reflected or transmitted. Maximum rates of photosynthesis occur near noon when light is brightest, and fix about eight times more carbon dioxide than is released during respiration. Averaged over a day, photosynthesis fixes about six times more carbon dioxide than is released by respiration.

The Calvin cycle includes twelve enzymes, at least five of which are rate-limiting to photosynthesis. These same five enzymes are all activated by light.

In darkness, there is no photosynthesis, and respiration produces carbon dioxide that is released from the plant. As light intensity increases, so too does photosynthesis, eventually reaching a point at which the rate of photosynthesis equals the rate of respiration. This **light-compensation point** varies in different plants. "Sun plants," such as corn and sugarcane that grow in bright sun, have a high light-compensation point and require much light to saturate photosynthesis. Conversely, plants that grow in low light (e.g., on forest floors) have a low light-compensation point and saturate at low light intensities. The rates of photosynthesis in these plants also differ significantly: shade plants fix about 5 mg C dm^{-2} h^{-1}, while those of sun plants fix about 20 mg C dm^{-2} h^{-1}. You'll learn more about sun and shade plants in Chapter 14.

Concentration of Carbon Dioxide

The concentration of carbon dioxide at which plants show no net fixation of carbon dioxide is the CO_2 **compensation point.** In C_3 plants such as wheat and rice, the CO_2 compensation point is about 50 ppm (at 25°C and 21% oxygen). For C_4 plants, the CO_2 compensation point is 0–5 ppm.

Increasing the concentration of carbon dioxide in the air to 0.10% doubles the rate of photosynthesis in C_3 plants by reducing photorespiration. As you'd expect, such "fertilizing" of the air with CO_2 does not significantly affect the rate of photosynthesis in C_4 plants—they saturate at CO_2 concentrations near 0.04%, just above normal. The benefit of increased concentrations of CO_2 is limited, however, because stomata close and photosynthesis stops at CO_2 concentrations exceeding about 0.15%.

Water Availability

The inability of plants to maintain enough water in their leaves causes stomata to close, thereby inhibiting photosynthesis. The lack of water is the most limiting factor controlling photosynthesis.

Metabolic Sinks

Metabolic sinks are parts of a plant where photosynthate is stored or used, such as roots and growing areas of the plant. Photosynthesis decreases when sinks are removed, and increases when sinks are created (e.g., by wounds or infections). In general, high rates of photosynthesis require high rates of translocation of the photosynthate to sinks.

CONCEPT

The primary factors that control photosynthesis are temperature, water availability, light, the concentration of carbon dioxide, and metabolic sinks.

THE FATE OF PHOTOSYNTHATE

The products of photosynthesis have many fates:

- About half of all photosynthate is used as fuel for cellular respiration and photorespiration.

- Some of the 3-PGA moves into cytoplasm, where it is used to make amino acids.

- The PGAL is used to make fructose-1,6-diphosphate, which, in turn, is used to make other sugars and starch. Fructose-6-phosphate is used to regenerate RuBP.

- Chloroplasts use ATP to reduce sulfate to sulfhydryl groups. Ferredoxin and NADPH in chloroplasts are used to reduce nitrite to ammonia for incorporation into amino acids (nitrate is reduced to nitrite in the cytoplasm).

- Some of the photosynthate is used to make glucose. Molecules of glucose are smaller and store more energy than does ATP, which is reactive, polar, and difficult to move across membranes.

- Some of the glucose (along with fructose) is used to make sucrose, a disaccharide that is shipped throughout the plant. Glucose is often stockpiled as dense, inert granules of starch, which is stored in roots and stems. Small amounts of starch are stored in chloroplasts. Unlike glucose starch has little effect on the osmotic properties of the cell, and serves as a starting point for making other compounds that the cell needs.

- Much of the photosynthate is used to make cellulose, the most abundant organic compound on earth. Cellulose, you'll recall, is a primary ingredient in cell walls (see Chapter 2, pp. 24–25).

- Plants use some of their photosynthate to make secondary metabolites such as latex. Some of these metabolites resemble crude oil; for example, the grape-size fruits of *Pittosporum undulatum* have been used for centuries as torches. Melvin Calvin and other botanists are studying how to use the hydrocarbons in the latex of some plants (e.g., members of the spurge family and relatives of poinsettia) as fuel. Some of the results are promising. For example, extracts from the Amazonian copa iba tree (*Copaifera langsdorfii*) are a diesel fuel that can be burned in automobile engines with no further refining. Similarly, one acre of gopher plants (*Euphorbia lathyris*) can produce enough latex to make 12 barrels of oil. Plantations of *Euphorbia* would be economically feasible when oil prices reach about $31 per barrel.

- We use photosynthate of corn to make almost 90% of the ethanol used in the United States, some of which is mixed with gasoline (in a 1:9 ratio) to make gasohol. Most cars made in Brazil run on 100% ethanol, and in the United States gasohol accounts for about 10% of gas sales. One bushel of corn will make about 3 gallons of ethanol.

- Every year we use photosynthesis to produce about 1.5 billion tons of grain, the staple of the world's diet.

Each year, photosynthesis fixes about 10% of the carbon in the atmosphere. Humans, either directly or indirectly, use about 40% of the net products of photosynthesis from terrestrial ecosystems. These uses affect all aspects of our lives: textiles, food (e.g., cereals such as oats and wheat provide half of the world's food), fuels such as coal and oil (formed millions of years ago), wood (the United States uses more than 3 billion cubic feet of wood per year to make paper), drugs (e.g., caffeine, quinine, cocaine, nicotine), waxes, oils (e.g., rose, jasmine), rubber, spices (e.g., pepper, cinnamon, peppermint), and even the perfumes that Cleopatra used to entice Marc Antony and Julius Caesar. This page, the oxygen you breathe, and all of the atoms in your body were once parts of plants.

No process can match the importance or magnitude of photosynthesis. It sustains virtually all life on earth. Without photosynthesis, all other biological reactions would be irrelevant.

Chapter Summary

Photosynthesis is the light-driven conversion of water and carbon dioxide to carbohydrates. Virtually all of life depends on light. Light moves in waves, and its energy is contained in packets called *photons*. The energy of a photon is inversely proportional to the wavelength of the light: the longer the wavelength, the less energy per photon. Light occurs in a spectrum of colors. Red and blue light are most effective for photosynthesis.

Only light that is absorbed can have an effect. Light is absorbed by pigments. Chlorophyll *a* is the primary pigment for photosynthesis and occurs in all photosynthetic organisms except photosynthetic bacteria. Accessory pigments such as carotenoids and chlorophyll *b* absorb light that chlorophyll cannot absorb, thereby extending the range of light used for photosynthesis.

Photosynthesis in eukaryotes occurs only in chloroplasts, which consist of a gelatinous matrix called *stroma* and stacks of membranes (thylakoids) called *grana*. Aggregates of pigments in grana absorb light and funnel its energy to special pairs of chlorophyll and proteins called *reaction centers*. The energy of the light energizes electrons, which are used to pump protons from the stroma into the thylakoid space. The resulting pH gradient is discharged during the synthesis of ATP as protons flow back into the stroma.

The photochemical reactions of photosynthesis convert light-energy into chemical energy. In plants, the photochemical reactions oxidize water, produce ATP, reduce $NADP^+$, and involve two photosystems connected in series. The ATP and reduced NADP made in the photochemical reactions are used in the biochemical reactions of photosynthesis to reduce carbon dioxide to carbohydrate. The biochemical reactions are collectively referred to as the *Calvin*, or C_3, *cycle* and begin with the fixation of carbon dioxide by an enzyme called *ribulose bisphosphate carboxylase/oxygenase* (*rubisco*). The product of the biochemical reactions is glyceraldehyde-3-phosphate, a sugar that is the starting point for several metabolic pathways. Fixing each molecule of carbon dioxide requires three ATPs and two reduced $NADP^+$s. Plants that use only the Calvin cycle are called C_3 *plants*.

Photorespiration occurs in C_3 plants when the internal concentration of carbon dioxide becomes low, such as when stomata close during drought. During photorespiration, rubisco fixes oxygen, and carbon dioxide is later released. Consequently, photorespiration undoes photosynthesis: as much as half of the carbon fixed in the Calvin cycle is released by photorespiration during hot, dry days.

C_4 photosynthesis occurs in tropical grasses and involves fixation of carbon dioxide into a four-carbon acid in mesophyll cells. The acid then moves to bundle-sheath cells, where it is broken down to carbon dioxide and a three-carbon molecule. The three-carbon molecule moves back to the mesophyll cell to accept another molecule of carbon dioxide. Meanwhile, the carbon dioxide is fixed by the Calvin cycle in bundle-sheath cells. This pumping of carbon dioxide concentrates carbon dioxide in bundle-sheath cells, thereby eliminating photorespiration. This, in turn, makes C_4 plants more efficient during hot conditions.

Crassulacean acid metabolism (CAM) occurs in many plants that live in dry habitats. CAM plants fix carbon dioxide at night into a four-carbon acid that is stored in the vacuole until daylight. The acid is then decarboxylated, and the carbon dioxide is used in the Calvin cycle. Fixation of carbon dioxide at night helps CAM plants conserve water. The primary factors that control photosynthesis are temperature, water availability, light, the concentration of carbon dioxide, and metabolic sinks.

Questions for Further Thought and Study

1. You learned in Chapters 5 and 6 that the complete combustion of one mole of glucose to carbon dioxide and water releases 686 kcal of energy (i.e., a mole of glucose stores about 686 kcal of energy in its chemical bonds). However, making glucose requires almost 2,000 kcal. What happens to the rest of the energy?
2. What is the significance of using NAD^+ in respiration and $NADP^+$ in photosynthesis?
3. Geldikkop is a rare disease in sheep in which chlorophyll that gets into the blood is energized by light and causes lesions. Why would chlorophyll in blood cause such a problem?
4. Zooplankton are transparent and eat phytoplankton. Most zooplankton avoid bright light by migrating down during the day and up during the evening and night. They also contain large amounts of carotenoids. What is the significance of this?
5. The chloroplast genome is the same in the mesophyll and bundle-sheath cells of C_4 plants. How, then, can these chloroplasts be so different?
6. Explain how the adaptations of CAM and C_4 plants enhance photosynthesis in hot, dry environments.
7. Life has been defined as the heat generated between photosynthesis and respiration. Explain the basis for this statement.
8. How does visible light differ from ultraviolet light? From infrared light?
9. How is an absorption spectrum different from an action spectrum?
10. How does cyclic electron flow differ from noncyclic electron flow? What is the significance of these differences?
11. What is photorespiration? Why is it significant?
12. Why don't C_4 plants dominate the landscape?
13. What are some of the things that we do not know about photosynthesis? What experiments would you do to learn more about these mysteries?
14. Use a reference source at the library to document the effect of temperature on photosynthesis. What does this information tell you about the process?

Suggested Readings

ARTICLES

Anderson, L., A. Ashton, A. Mohamed, and R. Scheibe. 1982. Light/dark modulation of enzyme activity in photosynthesis. *BioScience* 32:103–107.

Andrews, T. J., and G. H. Lorimer. 1987. Rubisco: Structure, mechanisms, and prospects for improvement. In *The Biochemistry of Plants*, vol. 10, M. D. Hatch and N. K. Boardman, eds. San Diego: Academic Press, 131–218.

Ehleringer, J. R., R. F. Sage, L. B. Flanagan, and R. W. Pearcy. 1991. Climate change and the evolution of C_4 photosynthesis. *Trends in Ecology and Evolution* 6:95–99.

Ellis, R. J. 1979. The most abundant protein in the world. *Trends in Biochemical Science* 4:241–244.

Govindjee, and William J. Coleman. 1990. How plants make oxygen. *Scientific American* (February):50–58.

Hendry, George. 1990. Making, breaking, and remaking chlorophyll. *Natural History* 90(5):36–41.

Keeley, J. 1988. Photosynthesis in quillworts, or why are some aquatic plants similar to cacti? *Plants Today* (July): 127–132.

O'Leary, M. 1988. Carbon isotopes in photosynthesis. *BioScience* 38:328–336.

Storey, Richard D. 1989. Textbook errors and misconceptions in biology: Photosynthesis. *The American Biology Teacher* 51:271–274.

Zelitch, Israel. 1992. Control of plant productivity by regulation of photorespiration. *BioScience* 42:510–516.

BOOKS

Barber, J., and N. R. Baker. 1985. *Photosynthetic Mechanisms and the Environment*. Topics in Photosynthesis, vol. 6. Amsterdam: Elsevier.

Foyer, C. H. 1984. *Photosynthesis*. New York: Wiley-Interscience.

Galston, A. W. 1994. *Life Processes of Plants*. New York: Scientific American.

Gregory, R. P. F. 1989. *Photosynthesis*. Glasgow: Blackie.

UNIT THREE

Genetics...

"Like begets like." This old saying has been around for thousands of years because people have long recognized that, for example, only roses can make more roses, and only camels can make more camels. This means that information passes from one generation to another, thereby defining what each organism can be. The inheritance of such information is the foundation for the discipline of genetics, the subject of this unit.

Plants manage inherited information at the molecular level. Such information management is intertwined with how plants obtain and use energy for cellular metabolism, which was discussed in the previous unit. Reproduction requires energy from photosynthesis and respiration, and the synthesis of molecules that guide metabolism relies on directions from information inherited from previous generations.

The most sophisticated technology in biological research is now used to understand the molecular basis of inheritance. New discoveries are probably being made faster in genetics than in any other discipline. These discoveries underscore the importance of genetics to science and society in the late twentieth century. Moreover, they also mean that some of the information presented in this text is probably changing as you read it. Some of the controversies in genetics that are discussed here are being resolved even as others are being raised.

Garden pea (*Pisum sativum*) with flowers and pods. Gregor Mendel used garden pea to establish the basic postulates of the theory of inheritance.

CHAPTER 8

Patterns of Inheritance

Chapter Outline

INTRODUCTION

THE THEORY OF INHERITANCE: GREGOR MENDEL'S DISCOVERY
  Setting the Scene: Plant Hybridization Before Mendel
  How Did Mendel Begin?

BOX 8.1
ARTIFICIAL HYBRIDIZATION IN PLANTS

  Mendel's Results: A Theory of Inheritance
  Genotypes and Phenotypes
  Experiments Using Multiple Traits
  A Note about Meiosis and Chromosomes
THE SEARCH FOR THE HEREDITARY MATERIAL
  The Importance of the Nucleus in Heredity
  Choosing Between Protein and DNA
  Genes
COMPLEX INHERITANCE
  Types of Dominance
  Multiple Alleles of the Same Gene
  Multiple Genes
  Serial Gene Systems
  Polygenic Inheritance

BOX 8.2
EXAMINING PROTEINS BY GEL ELECTROPHORESIS

  Pleiotropy
NON-MENDELIAN INHERITANCE
  Linkage
  Cytoplasmic Inheritance
  Mutations
  Transposable Elements

BOX 8.3
CYTOPLASMIC MALE STERILITY IN CORN

Chapter Summary
Questions for Further Thought and Study
Suggested Readings

Chapter Overview

Every organism is the expression of thousands of genes. Such genetic complexity would seem to prevent an understanding of how inheritance works. Nevertheless, the current theory of inheritance was proposed more than a century ago. The nineteenth-century assumptions of this theory are still correct, at least for characteristics that are under simple genetic control. However, many apparent exceptions and complex patterns of inheritance have been discovered since the beginning of the twentieth century. The search for explanations to complex inheritance is one of the primary subjects of modern genetics. However, unlike the earliest geneticists, we have a better understanding of what genes are and how they work.

In this chapter we present a historical perspective on the theory of inheritance and how early experiments laid the foundation for explaining heredity. We also discuss how different patterns of inheritance arise and how they relate to our current understanding of genetics. Much of this chapter is a springboard for more detailed discussions of cell reproduction, the mechanisms of heredity, and how genes work, which are presented in subsequent chapters.

Introduction

People have always been interested in heredity and its importance in their families, in crop production, and in animal husbandry. We have known for centuries that all organisms inherit traits from their parents. The earliest farmers, thousands of years ago, selected bigger, more desirable seeds for continued cultivation. By so doing, they selected heritable characteristics even though they had little or no understanding of inheritance.

Many theories on the nature of heredity, some based on religion or mythology and some based on factual observations, have held sway at one time or another during the past few centuries. The modern theory of inheritance was established in the mid-nineteenth century. The basic assumptions of this theory were first made by Gregor Mendel, an obscure amateur botanist working in a monastery garden in Austria (fig. 8.1). Several decades passed before biologists fully appreciated Mendel's work. Once they did, these botanists made many discoveries involving the chemical and biological nature of heredity. Among the more recent of these discoveries are that DNA is the hereditary material and that genes are the basic units of inheritance.

THE THEORY OF INHERITANCE: GREGOR MENDEL'S DISCOVERY

Gregor Mendel (1822–1884) studied a simple genetic system and recognized the importance of his results for understanding patterns of inheritance that are basic to all organisms. Remarkably, Mendel chose an appropriate plant species, designed powerful experiments, and interpreted his results without knowing about chromosomes or genes. By doing so, he derived a set of correct assumptions about heredity in the garden pea (*Pisum sativum*) that form the theory of inheritance. Subsequent sections of this chapter describe how Mendel achieved such a significant scientific advancement and how his work fits into the framework of modern genetics.

Setting the Scene: Plant Hybridization Before Mendel

The first plant breeders were probably ancient Babylonians and Assyrians who cultivated date palms (*Phoenix dactylifera*) in the region of modern-day Iraq more than 4,000 years ago. Like many plants, date palms are **dioecious,** meaning that each plant produces either pollen or fruits, but not both. Date growers discovered that pollen from just a few trees could be transferred by hand to several hundred fruit-producing trees, causing them to make dates. This practice was important economically because only a few fruitless (pollen) trees had to be maintained in each plantation. Unknown to the growers, pollen transfer was also important biologically because it promoted variation in the size, flavor, shape, and color of the dates. As a result, desirable variations were selected for replanting through many generations, producing more than four hundred different varieties of dates (fig. 8.2). Even though the patterns of inheritance

FIGURE 8.1

Gregor Mendel (1822–1884). Mendel's experiments with the garden pea established the basic postulates for the theory of inheritance.

FIGURE 8.2

Some of the more than four hundred different varieties of dates. Selection of dates for different traits probably began more than 4,000 years ago.

were exploited, how plants inherited traits from their parents remained unknown for many centuries.

By the end of the seventeenth century, the need for pollen in sexual reproduction had been shown experimentally in several species. During the eighteenth and nineteenth centuries, European botanists began to study heredity by artificial hybridization (see box 8.1: "Artificial Hybridization in Plants"). The most notable early experiments were done in the 1760s by Josef Kölreuter of Germany, who discovered that hybrid offspring had features that were either intermediate between their parents or from only one parent. He also showed that continued self-pollination of hybrids over several generations produced much variability, including the reappearance of features that had disappeared in earlier generations. Hence, Kölreuter discovered a crucial piece in the puzzle of inheritance: that parental traits could be absent in one generation of offspring but reappear in a subsequent generation.

The most extensive studies of plant hybridization were done by Karl Friedrich von Gaertner, who was from the same district in Germany as Kölreuter. Beginning in the 1820s, Gaertner did nearly 10,000 experiments involving artificial hybridization that established the role of pollen in transmitting traits of the pollen parent to the hybrid offspring. Also, Gaertner discovered that certain traits are expressed preferentially over others. For example, purple flowers are **dominant** over white flowers. A dominant trait is one that masks the alternative trait for a particular feature. The masked trait is said to be **recessive.** Gaertner determined the dominant and recessive forms of several traits, including flower color (purple over white), seed shape (round over wrinkled), and pod color (green over yellow) in the garden pea. By confirming the tendency of these dominant traits to be more commonly expressed among offspring, Gaertner further illuminated the theory of inheritance.

CONCEPT

Extensive genetic studies of plants were done a century before Mendel began his experiments with garden peas. These studies showed that pollen transmits traits to offspring. Dominance was known, and the reappearance of recessive traits in offspring had been observed. This knowledge provided important background information for Mendel's studies.

How Did Mendel Begin?

The voluminous work of Gaertner and his predecessors was a rich resource for Mendel. The challenge for Mendel was to find order in seemingly overwhelming complexity. The problem, as he perceived it, was that

> Whoever surveys the work in this field will come to the conviction that among the numerous experiments not one has been carried out to an extent or in a manner that would make it possible to determine the number of different forms in which hybrid progeny [offspring] appear, permit classification of these forms in each generation with certainty, and ascertain their numerical interrelationships.

Armed with a strong background in biology and mathematics, Mendel tried to measure the results of hybridization experiments—to "ascertain their numerical interrelationships." His work is an excellent example of how expertise in one discipline (mathematics) can provide insight into another (biology). Although many people had done the same experiments as Mendel, it was his use of mathematics that led to his great discoveries.

Mendel chose the garden pea as an experimental organism because it was easy to grow and manipulate in hybridization experiments. Before making hybrids, Mendel grew his pea plants for two years, through several generations, until he could select **true-breeding** (purebred) strains for each of several traits. He chose nine sets of traits based on the ease of distinguishing one trait from another. Three of these (flower color, seedling axil color, seed coat color) were so well correlated, however, that he regarded them as expressions of the same factor. Thus, he followed the inheritance of seven separate pairs of traits, shown in figure 8.3. By focusing on these simple and easily distinguished traits, Mendel eliminated the potential for complex variability.

In his initial experiments, Mendel crossed (hybridized) pairs of plants that differed in only one trait. For example, he crossed purple-flowered plants with white-flowered plants, and wrinkled-seeded plants with smooth-seeded plants. In so doing, he made reciprocal crosses between parental strains—that is, he transferred pollen from purple flowers to white flowers, and pollen from white flowers to purple flowers.[1] He then

---

1. Reciprocal crosses occur when one plant is the pollen parent and the other plant is the seed parent for the first cross, and the pollen parent and seed parent are reversed for the second cross.

CHAPTER EIGHT *Patterns of Inheritance*

BOXED READING 8.1

ARTIFICIAL HYBRIDIZATION IN PLANTS

The offspring of parents with different traits is called a **hybrid.** Reproductive structures of many kinds of flowers allow easy manipulation of the pollen-containing anthers, which means that plant hybrids can be made artificially—that is, by human hands. In hybridization experiments, anthers are typically removed from each flower to prevent self-pollination. This is necessary only when a plant is self-compatible—that is, when it can fertilize itself. Not all plants are self-compatible. Mendel took the precaution of removing anthers because the garden pea is a self-compatible species.

After the anthers are removed, pollen from another individual is brushed onto the stigma, which is the receptive surface of the seed-producing organ (carpel). The pollen can be from any other plant, but successful fertilization depends upon genetic compatibility between the pollen parent and the seed parent. It is impossible to fertilize a garden pea with pollen from an orchid, because peas and orchids are too dissimilar genetically. However, pollen from one variety of garden pea can fertilize another variety of garden pea. The limits of interfertility are inconsistent from one type of plant to another. For example, many species of pines cannot be hybridized artificially, but hybrids of orchids, mustards, carnations, and many other plants have been made.

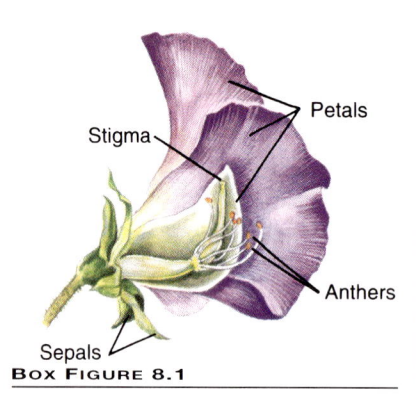

BOX FIGURE 8.1
A pea flower (longitudinal section); the petals enclose the anthers and stigma, thereby promoting self-pollination.

Plant breeders develop many new kinds of plants by selecting flowers having the most desirable traits (size, color, etc.) for making artificial hybrids. Most commercially available garden plants are the products of many generations of hybridization. These are referred to as *cultivars* to denote cultivated varieties that are not found in nature. As a commercial enterprise, the business of making new cultivars has produced thousands of particularly remarkable or unique forms. The most valuable of these are patented.

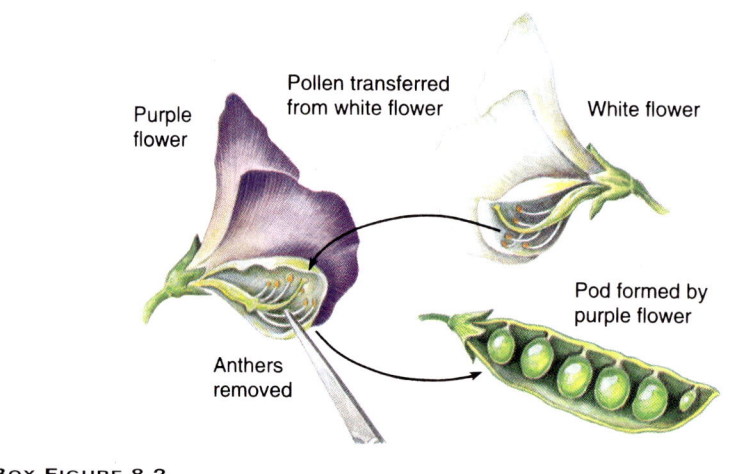

BOX FIGURE 8.2
Artificial hybridization in pea flowers.

---

allowed the offspring of the various crosses to self-pollinate for several generations and counted the plants with each trait in every generation.

Writing to Learn Botany

Hybridization experiments should ideally include reciprocal crosses. Why?

Mendel's Results: A Theory of Inheritance

The first generation of offspring is called the F_1 generation. (The *F* comes from **filial,** which is derived from the Latin word for daughter.) In every experiment, only the dominant trait was expressed in the F_1 generation. However, when F_1 plants were allowed to self-pollinate, the recessive trait reappeared in the F_2 generation. By counting the offspring in the F_2 generation, Mendel found that the ratios of dominant to recessive traits were the same for all seven pairs of traits, about 3:1 (table 8.1). Mendel concluded that "factors" from each parent somehow controlled the traits he observed. Moreover, he suggested that each offspring gets one of two factors from each parent. In the F_1 generation all of the offspring have both factors, but the factor for the dominant trait masks the factor for the recessive trait. Mendel also predicted that any plant having two different factors would donate the recessive factor to one half of the offspring and the dominant factor to the other half. Thus, the probability for F_2 plants to receive two recessive factors would

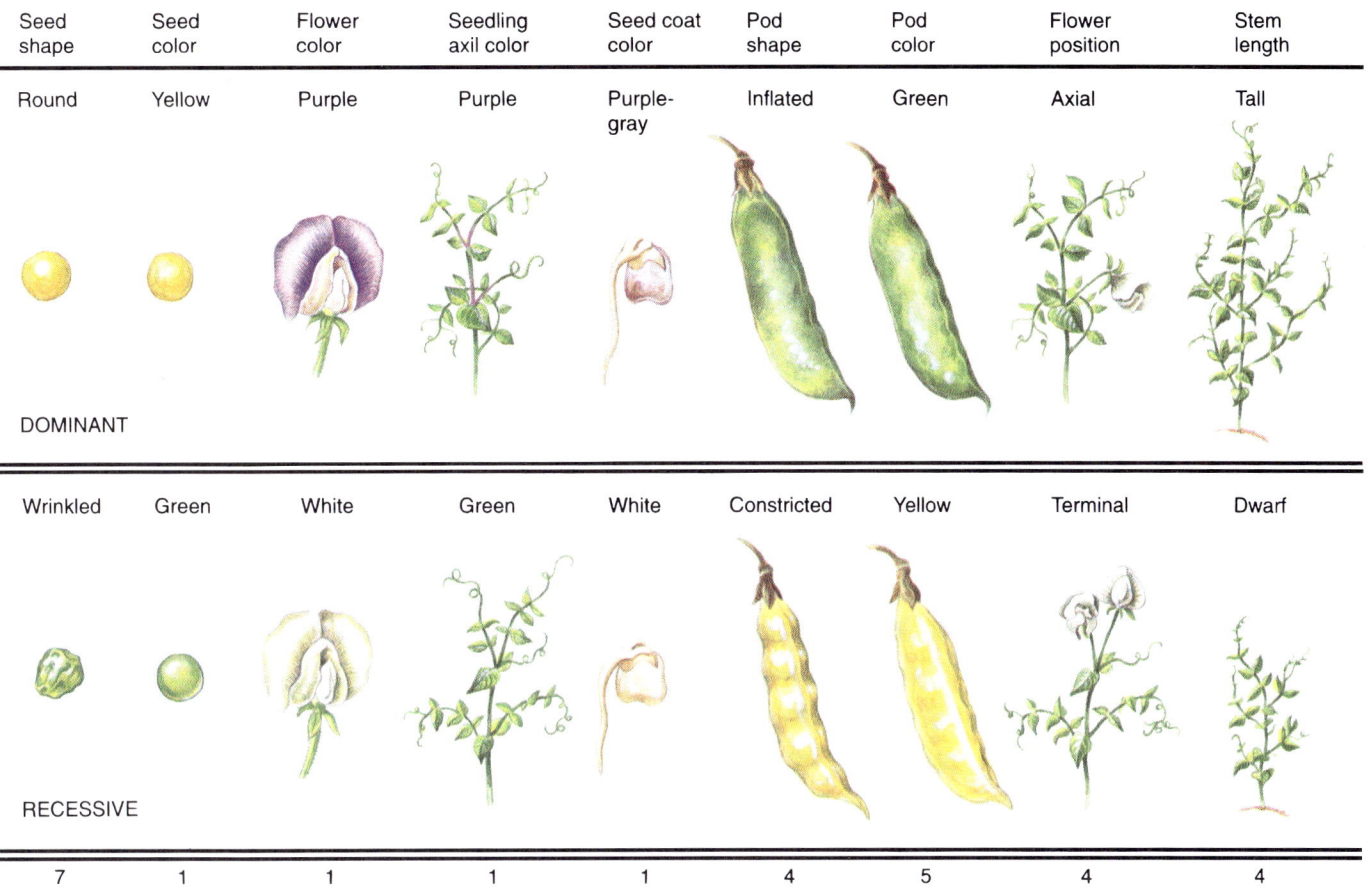

FIGURE 8.3

Characteristics of the garden pea studied by Mendel. Mendel regarded flower color, seedling axil color, and seed coat color as expressions of the same factor. The number below each characteristic is the number of the chromosome that bears the gene for it. This information was not discovered until the twentieth century.

TABLE 8.1

Mendel's Experimental Results for Seven Characteristics of Garden Pea (These Data Represent the F_2 Generation)

	Sample Size	Dominant Form	Recessive Form	Ratio
	7,324 seeds	5,474 round	1,850 wrinkled	2.96:1
	8,023 seeds	6,022 yellow	2,001 green	3.01:1
	929 plants	705 purple-flowered	224 white-flowered	3.15:1
	1,181 plants	882 with inflated pods	299 with constricted pods	2.95:1
	580 plants	428 with green pods	152 with yellow pods	2.82:1
	858 plants	651 with axial flowers	207 with terminal flowers	3.14:1
	1,064 plants	787 with long stems	277 with short stems	2.84:1
TOTAL:	19,959	14,949	5,010	2.98:1

Note: Two additional characters, seedling axil color and seed coat color, were also examined by Mendel. However, purple seedling axils and purple-gray seed coats were always associated with purple flowers, so Mendel regarded these three features as expressions of the same factor. He therefore recorded data for flower color and disregarded the other two.

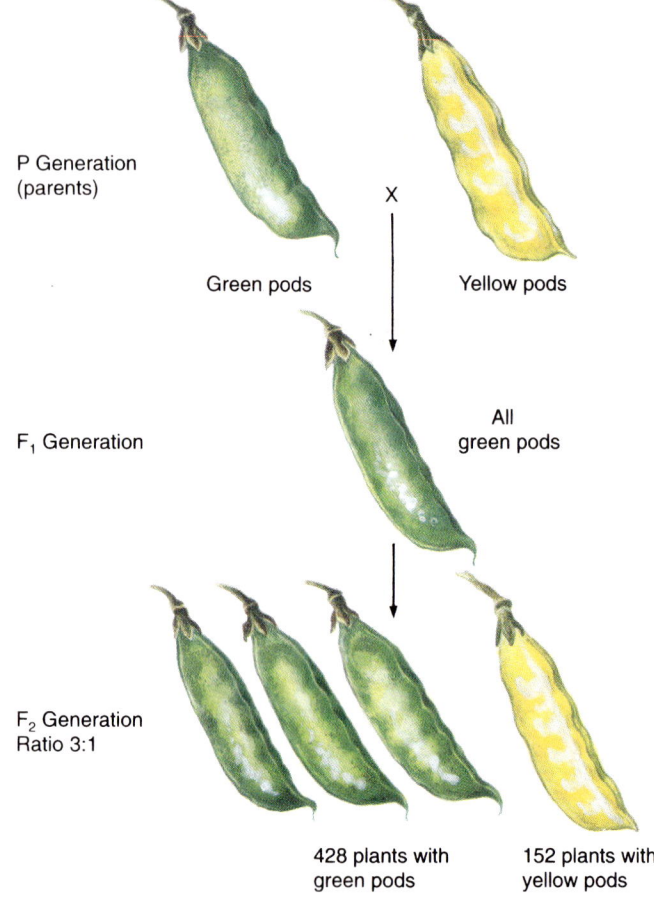

FIGURE 8.4

The inheritance pattern of one of Mendel's crosses (green vs. yellow pods) through two generations. All plants in the F_1 generation had green pods. About three-fourths of the F_2 generation had green pods and about one-fourth had yellow pods.

be one-half times one-half, or one-quarter. This prediction matches the observed ratio of three dominant traits (3/4) to one recessive trait (1/4) in the F_2 generation (fig. 8.4).

Mendel accounted for heritable factors and the probabilities of inheritance for each trait *only because he counted the offspring* of each cross. Several assumptions, or *postulates*, can be derived from his initial experiments. Taken together, these postulates are referred to as Mendel's **theory of inheritance.** Note, however, that the postulates of this theory, which are listed next, are modernized versions of Mendel's ideas; Mendel did not actually state them as postulates, nor did he propose a theory of inheritance. In fact, Mendel barely mentioned the ideas behind postulates 7 and 8.

1. Traits are controlled by heritable factors.
2. Factors are passed from parent to offspring in reproductive cells.
3. Each individual contains pairs of factors in every cell except reproductive cells.
4. Paired factors segregate during the formation of reproductive cells so that each reproductive cell gets one of the factors of a pair.
5. There is an equal chance that a reproductive cell will get one or the other factor of a pair.
6. Each factor from one parent has an equal chance of combining either with the identical factor or with the other factor from the other parent during fertilization.
7. Sometimes one factor dominates the other factor; in such cases, the dominant factor controls that feature of the plant.
8. When considering two or more pairs of traits, the factors for each pair of traits assort independently to the reproductive cells.

Postulates 4, 7, and 8 are often called Mendel's **laws of inheritance:**[2] the **law of segregation,** the **law of dominance,** and the **law of independent assortment,** respectively. Unfortunately, the exalted status of these postulates as laws came from twentieth-century geneticists who were a little overzealous in clarifying Mendel's ideas. Mendel proposed no such laws. On the contrary, Mendel did not regard the segregation of factors as a law because he could not observe it directly. He did not believe dominance to be a law because he knew of exceptions to this pattern, and he only did two sets of experiments that showed independent assortment, which he barely mentioned as if it were an afterthought. (See "Experiments Using Multiple Traits" later in this chapter.) Nevertheless, many modern biology textbooks still perpetuate these myths about Mendel's experiments.

In modern genetics, each of Mendel's paired factors is called a **gene.** Each factor of a pair is an **allele,** which is an alternative form of the gene. This means that in true-breeding parents, the genes controlling flower color or other features used by Mendel had two identical alleles for each gene. A plant that has the same alleles for a gene is said to be **homozygous** for that gene. In contrast, a plant that has different alleles for a gene is **heterozygous** for that gene. Thus, the parental generation in Mendel's experiments was homozygous for each gene, either dominant or recessive. The F_1 generation was therefore heterozygous for each gene, and the F_2 generation included both homozygous and heterozygous plants.

---

2. The term *law* is derived from a time when naturalists believed they were observing the universal truths of God's plan. Scientists no longer make this assumption. On the contrary, our modern view of science prohibits us from ever finding ultimate or universal truth (see the discussion in Chapter 1 on the scientific method). In our opinion, therefore, the term law is no longer useful. (Physicists would disagree however, since they still refer to many natural phenomena as laws, such as the laws of thermodynamics that were discussed in Chapter 5.)

Genotypes and Phenotypes

For clarity, we distinguish between an organism's genes, individually or collectively, which we call its **genotype,** and the observable characters they control, which we call its **phenotype.** For example, in the garden pea, phenotypes for flower color are purple and white. The genotype can be designated in several ways, but the most common way is by the first letter of the dominant trait. Accordingly, the gene for flower color is P for the dominant allele, and p for the recessive allele (fig. 8.5). Consequently, the genotype for flower color is PP (homozygous) or Pp (heterozygous) for purple, and pp (homozygous) for white.

Note that Mendel counted phenotypes and calculated phenotypic ratios. The phenotypic ratios for all seven features of garden pea in the F_2 generation were the same: 3 dominant to 1 recessive (fig. 8.4). However, what do we know about the genotypic ratios? We know that the genotype for white flowers (pp) occurs 25% of the time because of the probability of matching two recessive alleles from heterozygous parents (Pp). Similarly, the probability of matching two dominant alleles (PP) would be the same. The remaining 50% of the genotypes are therefore heterozygous. Thus, the 3:1 phenotypic ratio in the F_2 generation is based on a genotypic ratio of 1 PP to 2 Pp to 1 pp (1:2:1).

Genotypic and phenotypic ratios can be conveniently calculated by using a gametic grid called a **Punnett square,** named for R.C. Punnett. This is a checkerboardlike diagram that has the gametic genotype of one parent across the top and the gametic genotype of the other parent down one side (fig. 8.5). In this way, we show that the **gametes** (i.e., sperm and egg) from a cross between two heterozygotes for flower color ($Pp \times Pp$) would be P and p across the top and P and p down the side. When the gametes of such a cross fuse (fertilize), we add them into an expected ratio of 1:2:1 in the genotypes of the offspring. This ratio is perhaps as simple to obtain by direct inspection as it is by using a Punnett square. Nevertheless, the Punnett square is indispensable for calculating ratios from multiple traits, as discussed next.

Experiments Using Multiple Traits

Mendel also did experiments using two or three pairs of traits at a time. For example, he studied combined inheritance of seed shape (dominant round versus recessive wrinkled) and seed color (dominant yellow versus recessive green) in the same plants. One set of parents was homozygous for both dominant traits ($RRYY$), and another set of parents was homozygous for both recessive traits ($rryy$); this means he crossed plants that were true-breeding for round and yellow seeds with plants that were true-breeding for wrinkled and green seeds. All of the seeds produced from this cross were round and yellow. Moreover, since they received dominant alleles from one parent and recessive alleles from the other parent, these seeds (F_1) were heterozygous for both genes ($RrYy$). This pattern of inheritance

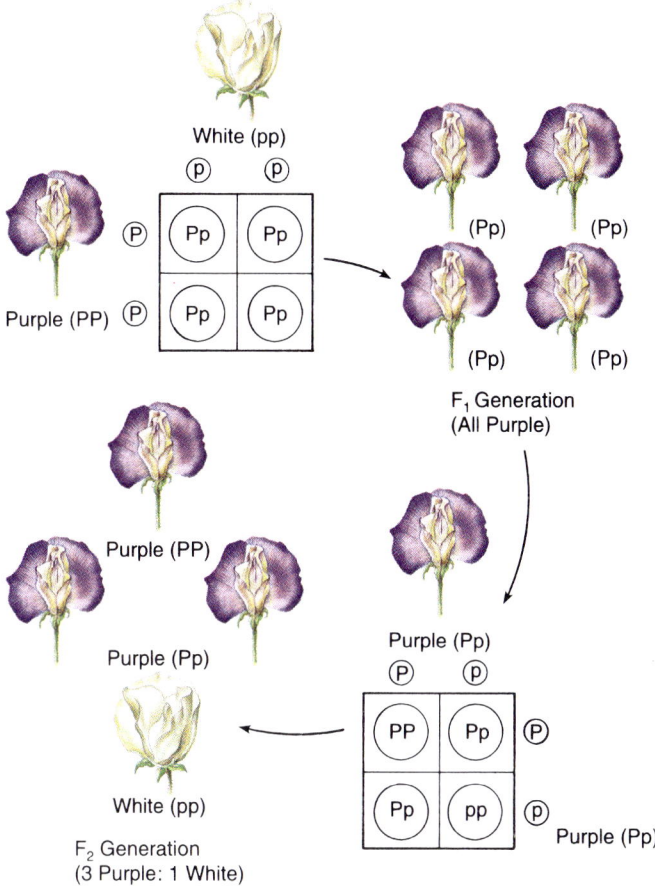

FIGURE 8.5

Use of the Punnett square to show the inheritance of phenotypes and underlying genotypes in Mendel's cross with flower color. P and p at top and side of the Punnett squares represent gametes that have dominant (P) or recessive (p) alleles of the gene for flower color. The genotypes of the offspring are predicted by combining alleles from different gametes into the boxes of the Punnett square.

conforms to predictions from experiments using one pair of traits at a time. (Note that a Punnett square of this cross would be simple: One parent has only gametes with the genotype RY, and the other parent has only gametes with the genotype ry. Thus, the sole genotype of their offspring is $RrYy$.)

When the hybrid seeds were grown into mature F_1 offspring and allowed to self-pollinate, they produced the F_2 generation of seeds, which included four phenotypes in the quantities listed in table 8.2. The approximate ratios among these phenotypes are 9 round yellow to 3 round green to 3 wrinkled yellow to 1 wrinkled green (9:3:3:1). This is the expected result of a **dihybrid cross,** which follows two genes that are both heterozygous (fig. 8.6). Using the pea seed example, a dihybrid cross is written $RrYy \times RrYy$. In a Punnett square, each parent has four gametic genotypes: RY, Ry, rY, and ry (fig. 8.6). This means that the Punnett square of a dihybrid cross will have 16 boxes (4×4). These 16 boxes will contain 9 different genotypes that underline the 4 phenotypes of the F_2 generation (fig. 8.6).

TABLE 8.2

Mendel's Results from a Dihybrid Cross

	Sample Size	Phenotypes	Ratio
Seed shape/Seed color	556 seeds	315 round/yellow	9.84
		108 round/green	3.38
		101 wrinkled/yellow	3.16
		32 wrinkled/green	1.00

Note: Ratios are calculated by using the wrinkled/green phenotype as the common denominator.

The phenotypic ratios from Mendel's dihybrid cross can be explained only if the segregation of one pair of traits (green vs. yellow seeds) is not influenced by the other pair of traits (round vs. wrinkled seeds)—that is, if different genes are inherited independently of each other. Although Mendel did not recognize the 9:3:3:1 ratio of F_2 phenotypes, he did note that different pairs of traits were inherited independently of one another. From this observation we derive postulate 8 (independent assortment) of Mendel's theory of inheritance.

Twentieth-century geneticists derived several postulates from Mendel's experiments on inheritance in garden peas. These postulates make up Mendel's theory of inheritance.

The Lore of Plants

Mendel was not only an amateur botanist but also an active member of the local beekeeper's society. Some historians have suggested that he was inspired to begin his genetic studies by a paper published on inheritance in bees. We can imagine, therefore, that he at least considered using bees in his experiments. If he had done so, generations of biology students would have been reading about Mendel's bees instead of Mendel's peas.

A Note about Meiosis and Chromosomes

Meiosis is a type of nuclear division that produces daughter nuclei that have half of the alleles of the parent nucleus. This feature of meiosis explains how **haploid** cells are produced in certain parts of flowers. Haploid cells have half the amount of nuclear genetic material that their parent cells have. Cells that have the full amount of nuclear genetic material are **diploid** cells. Such a simple description of meiosis accounts for the segregation of alleles into haploid cells that will become gamete-producing structures. Hence, Mendel's peas had two factors for each feature of the plant because they were diploid, and because of meiosis, each diploid parent passed on only one of the factors in each haploid gamete.

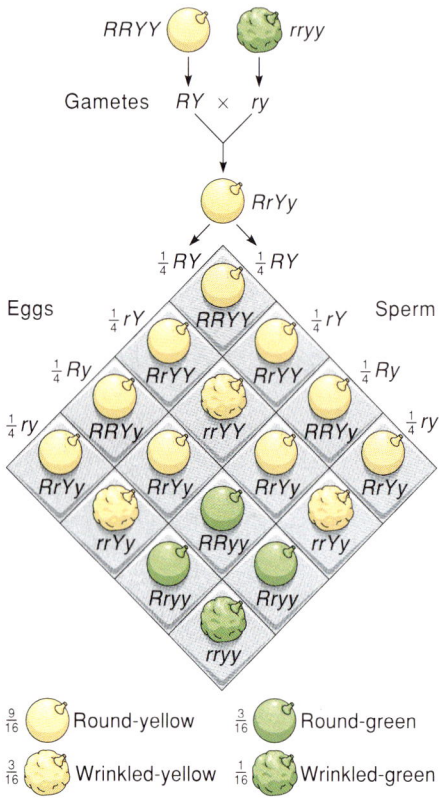

FIGURE 8.6

Inheritance in the garden pea. This Punnett square shows the expected pattern of inheritance in a dihybrid cross according to independent assortment of alleles for seed color and seed shape. It predicts four phenotypes in the F_2 generation in a ratio of 9:3:3:1.

Chromosomes were seen as threadlike structures in plant nuclei as early as the 1830s. By the start of the twentieth century, biologists knew that chromosomes existed in pairs and that the pairs separated during meiosis. The parallel separation between pairs of chromosomes and between pairs of alleles during sexual reproduction was the first indication that genes were associated with chromosomes. This was the main postulate of the **chromosomal theory of heredity.** This postulate accounts for the separation of two alleles for each gene on **homologous** chromosomes, which are chromosome pairs that have alleles for the same genes. Our current view on the meiotic separation of homologous chromosomes and their alleles is shown in figure 8.7 for one pair of traits from the garden pea. The figure shows that chromosomes bear genes, that alleles separate on homologous chromosomes during meiosis, and that homologous chromosomes reunite during fertilization.

Genes occur on chromosomes. Chromosomes and alleles segregate during meiosis. Haploid nuclei have one set of chromosomes and one allele for each gene.

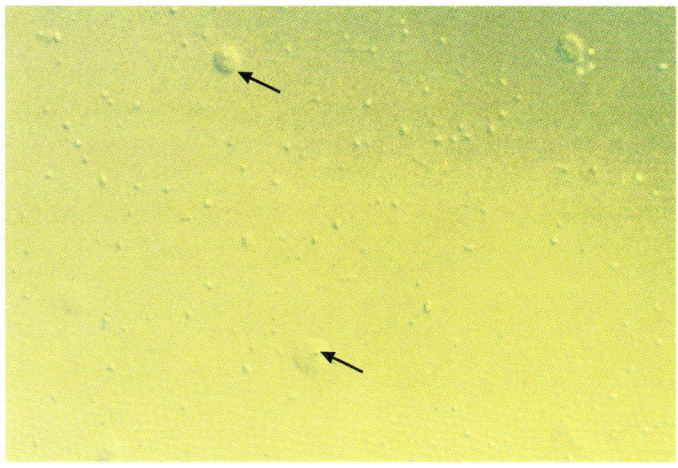

FIGURE 8.7

Sperm cells of corn (arrows) as seen by Nomarski interference light microscopy. Sperms have a nucleus and little or no cytoplasm.

Meiosis is necessary for sexual reproduction, which means that it has many important consequences for inheritance. A more complete description of the roles of meiosis and chromosomes in heredity is presented in Chapter 10, "Meiosis, Chromosomes, and the Mechanism of Heredity."

THE SEARCH FOR THE HEREDITARY MATERIAL

In the decades following Mendel's experiments, biologists made important discoveries about the cellular and chemical basis of heredity. New methods and improvements in light microscopy permitted detailed studies of the nucleus and of chromosomes, while advances in chemistry led to the discovery of DNA and nuclear proteins. Although the role of chromosomes in heredity was suspected, this role was not demonstrated until after 1900. The importance of DNA in inheritance was not confirmed until halfway into the twentieth century.

The Importance of the Nucleus in Heredity

The nucleus was the only subcellular body that could easily be seen with a light microscope in the 1800s. Because of this limitation, early studies of the substances of inheritance focused on the potential role of the nucleus. But it was not until the 1870s that scientists agreed that the nucleus was separate from the cytoplasm (fig. 8.8) and that fertilization involved the fusion of two nuclei.

Choosing Between Protein and DNA

By the 1880s, nuclei were known to consist mostly of proteins and DNA, together referred to as **chromatin** because this material can be stained with various dyes (fig. 8.9). The dual chemical nature of chromatin created a difficult puzzle: Which of the two components are genes made of? Without evidence one way or another, the logical choice seemed to be proteins. They consist of complex arrays of amino acids which form many different kinds of proteins. Complex genetic processes, it was thought, must be controlled by complex protein molecules. Conversely, since there were only four different nucleotides in DNA, chromosomal DNA was thought to be too simple a molecule to meet complex cellular demands.

Resolving the issue of DNA versus proteins was difficult because nuclear DNA is associated with proteins in chromosomes. In bacteria and viruses, however, DNA does not occur in chromosomes and is not associated with protein. Two experiments, one using bacteria and one using viruses, provided substantial support for DNA as the hereditary material.

By 1928, a British microbiologist, Frederick Griffith, had discovered that pathogenic (i.e., disease-causing) strains of the bacterium now called *Streptococcus pneumoniae* could transform nonpathogenic strains into infectious strains (fig. 8.10). Later, in 1944, a group of researchers led by Oswald Avery discovered the "transforming principle" of this bacterium to be DNA. They showed the activity of this principle by introducing DNA from killed, pathogenic strains into cultures of nonpathogenic strains. As a result, the nonpathogenic strains became infectious because of the DNA they absorbed. Although some scientists doubted the validity and significance of Avery's interpretations, it was the first demonstration that DNA is the hereditary substance. General acceptance of this notion, however, did not come until nearly a decade later, when further evidence for the role of DNA was obtained in experiments using viruses.

Viruses have even simpler genetic systems than do bacteria. By the 1950s, viruses were already characterized as being made of nucleic acids surrounded by a protein coat (see Chapter 25). Certain types of viruses, called **bacteriophages,** parasitize bacteria: they attach to bacterial cells, inject their genes into the host, and cause the host cell to make more viruses. In other words, viruses transform bacterial cells into miniature factories for making more viruses. In 1952, Alfred Hershey and Martha Chase grew a strain of virus, called bacteriophage T2, with radioactive isotopes of either sulfur (^{35}S) or phosphorus (^{32}P). The ^{35}S was incorporated into the protein coat of one culture, and the ^{32}P was incorporated into the DNA of the other culture. Each radiolabeled phage was then allowed to infect cells of *E. coli*, its host. However, before the bacterial cells were totally destroyed, they were agitated to remove the viral coats attached to their cell walls. After they were separated from the cells, the empty viral coats contained ^{35}S, which meant that the coats contained only protein (DNA does not contain sulfur). Conversely, the bacterial cytoplasm contained ^{32}P that had been incorporated into the viral DNA (proteins lack phosphorus). Thus, only genes made of DNA could have entered the bacteria and caused their transformation to produce more viruses (fig. 8.11).

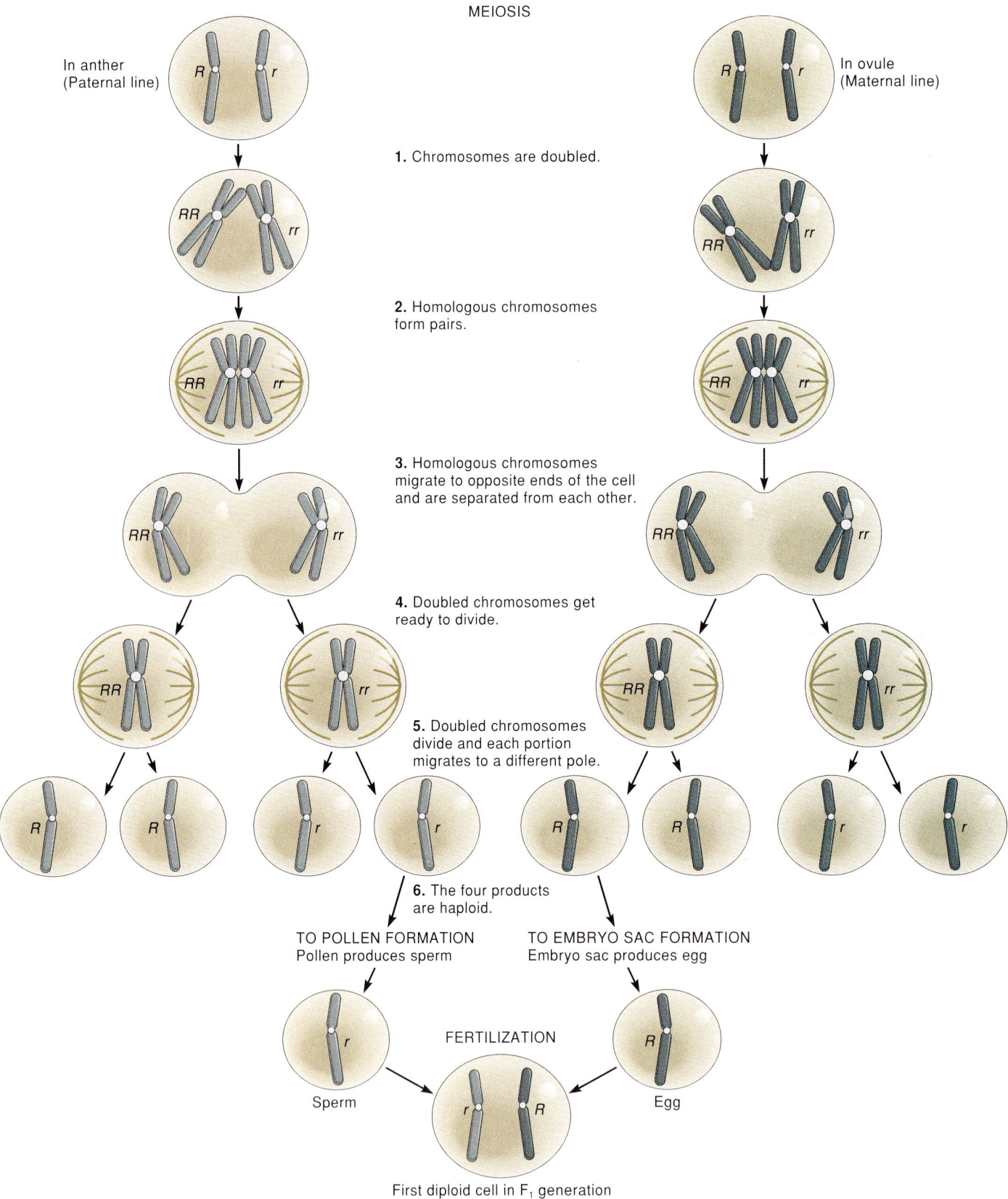

FIGURE 8.8

In anthers, alleles separate during meiosis and form the paternal genotype in the sperm. In ovules, alleles separate during meiosis and form the maternal genotype in the egg. Maternal and paternal alleles are reunited by fertilization. The parental genotype (Rr) is reproduced in the F_1 cell in this example, but different combinations of sperm and egg alleles will yield RR and rr genotypes in other offspring of the same parents.

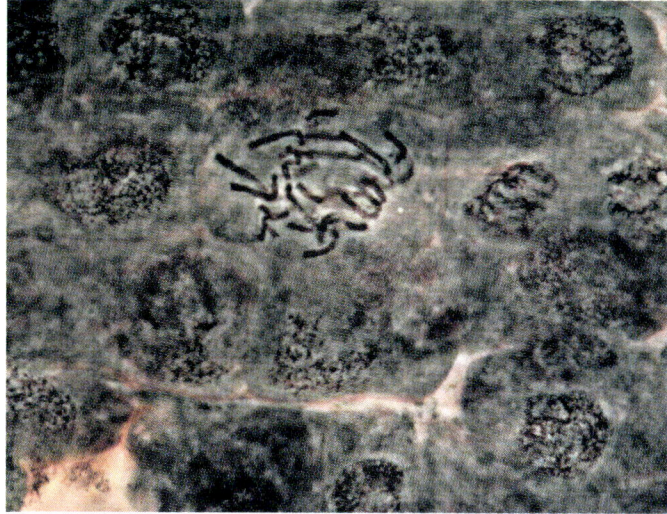

A. B.

FIGURE 8.9

(a) Flowers of daffodil (*Narcissus* sp.). (b) Root-tip cells of daffodil, showing stained chromatin in threadlike chromosomes. Chromatin consists of protein and DNA.

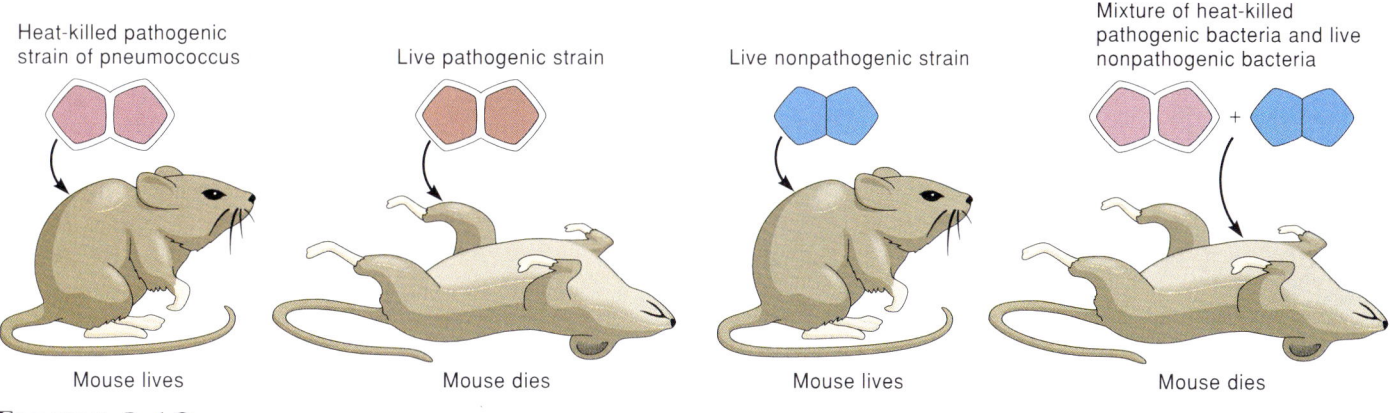

FIGURE 8.10

Evidence for bacterial transformation. Bacteria must be alive and have a polysaccharide coat to be pathogenic (*red*). Dead cells (*violet*) or strains without the polysaccharide coat (*blue*) are not pathogenic. However, a mixture of living, coat-free cells (harmless) and dead, coated cells (harmless) kills mice. Bacterial cells cultured from those mice have coats, which means that the living cells are "transformed" by the dead ones.

These bacteriophage experiments confirmed the role of DNA in heredity and added to our rapidly growing knowledge of this molecule.

Genes

Mendel's results can be readily appreciated in light of the accepted role of DNA as the hereditary substance and of the nature of genes as sequences of nucleotides. We now know that each of the seven pairs of traits that Mendel studied in garden peas is controlled by a unique sequence of nucleotides—that is, by a single gene. The position of a gene on a chromosome is referred to as that gene's **locus** (plural, **loci**). Thus, Mendel's experiments dealt with seven loci.

At the molecular level, genes are codes for making different molecules, including proteins or protein subunits and different kinds of RNA. The size of a locus ranges from a few dozen to several thousand bases for different genes. Details about how DNA works as a code for proteins and RNA are discussed in Chapter 11.

How does a gene produce a phenotype? Although many genes have been described both by their base sequences and their coded products, we do not understand how the sum of this molecular information becomes a complex organism. Our understanding is generally limited to biochemical reactions catalyzed by enzymes, such as when a single enzyme in flowers of garden pea makes a colored pigment from a colorless precursor. This reaction, controlled by the product of one

CHAPTER EIGHT *Patterns of Inheritance*

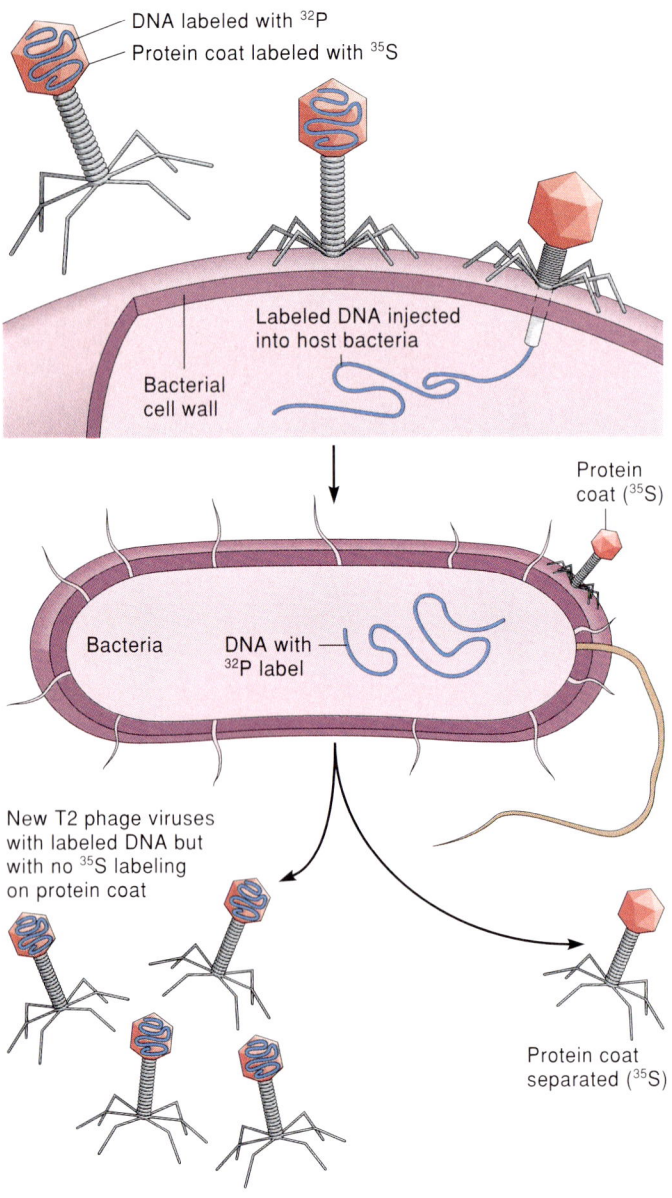

FIGURE 8.11

Evidence for the role of DNA in heredity. Bacteriophage T2 is radiolabeled with ^{35}S in its protein coat and ^{32}P in its DNA. The bacteriophage reproduces after infecting bacterial cells. New bacteriophages contain radioactive DNA but not radioactive protein, which means that the DNA was inherited, but the protein was not.

gene, explains how that gene influences flower color. However, we have no idea how other genes made the flower in the first place.

Genes are codes for making other molecules, which make up the phenotype. Exactly how they interact to make a complex organism is mostly unknown.

COMPLEX INHERITANCE

The traits Mendel used and the patterns of inheritance he discovered are often described as **Mendelian inheritance.** Many other traits in Mendel's peas, however, are inherited differently than the ones he studied. Moreover, the same kinds of traits that Mendel studied in peas may be inherited differently in other plants, and some hereditary patterns do not conform to Mendelian predictions at all. Such complex and non-Mendelian patterns of inheritance are discussed in the next few sections of this chapter. As you will see, Mendel was lucky to have avoided some pitfalls that would have greatly complicated his task.

Types of Dominance

The seven genes studied by Mendel all exhibit **complete dominance,** which is a relatively rare type of inheritance. Complete dominance occurs when one trait completely masks its recessive allele. More frequently, the phenotype for one allele is only partly masked by the other, a condition called **incomplete dominance.** Incomplete dominance occurs when hybrids have a phenotype intermediate between those of the two parents. For example, the allele for red flowers in camellia (*Camellia japonica*) is incompletely dominant over the allele for white flowers. As a result, the F_1 progeny from a cross between red-flowered and white-flowered camellias all have pink flowers. The phenotypic ratio in the F_2 offspring is 1:2:1 (25% red, 50% pink, 25% white) (fig. 8.12). Accordingly, in cases of incomplete dominance, the phenotypic and genotypic ratios are the same.

Codominance occurs when both alleles of a heterozygote are expressed equally, so there is really no dominance at all. Codominance is common for heterozygous genes that code for two equally functional enzymes. This means that there is more than one form of the same enzyme. The differing forms of enzymes made by different alleles of the same locus are called **allozymes.** Although allozymes catalyze the same reaction, they differ from each other by one or a few amino acids, which makes them slightly different from each other in size and overall electric charge. For example, in wild sunflower (*Helianthus debilis*), there are allozymes of phosphoglucomutase, which catalyzes one of the first reactions in glycolysis (Chapter 6). Heterozygotes produce both forms of the enzyme, but homozygotes produce only one or the other.

Multiple Alleles of the Same Gene

Diploid plants have only two alleles at a single locus. The two allozymes at a locus are usually referred to as *fast* or *slow* forms, depending on how far they move on an electrophoresis gel (see box 8.2 "Examining Proteins by Gel Electrophoresis"). Also, some populations of plants have more than two allozymes for some loci. For example, most wild sunflowers have fast or slow forms of the allozymes of the phosphoglucomutase gene,

FIGURE 8.12

Camellias (*Camellia* sp.) show incomplete dominance in flower color. Pink flowers are heterozygous for flower color (*Rr*); red and white are homozygous for flower color (*RR* and *rr*, respectively).

Pgm–3, but other individuals of this species have additional forms that replace either the fast or slow allozyme. This means that there are more than two allozymes for *Pgm–3*. This example shows that even though each diploid plant can have a maximum of two allozymes, a population of plants can have more than two allozymes.

Botanists are not sure why populations of plants have multiple forms of the same enzyme, but such variation may be adaptive. There are two kinds of indirect evidence for this explanation. One is that allozymes work at different optimum pHs, temperatures, or other conditions. We imply from this evidence that allozyme variation enables plants to thrive under a range of environmental conditions. The second kind of evidence is that certain allozymes occur more frequently in populations, for example, at higher elevations, in wetter soils, or within shadier forests. In this case, allozymes are thought to be adaptive because their occurrence is correlated with where certain populations live. Nevertheless, we have no strong direct evidence for the adaptiveness of allozymes.

Multiple Genes

Allelic variation of a single gene is often complicated by the presence of more than one gene for the same enzyme. The same enzymes from different genes are called **isozymes** to distinguish them from allozymes. Multiple isozymes are common in plants. For example, *Helianthus debilis* has two nuclear genes and two chloroplast genes for phosphoglucomutase (phosphoglucomutase and other glycolytic enzymes in chloroplasts are thought to be holdovers from the prokaryotic ancestors of these organelles). The four genes for phosphoglucomutase in this sunflower make a total of nine forms of the enzyme.

The most frequent application of isozyme/allozyme studies involves population genetics. Botanists routinely study the variability patterns of more than two dozen types of enzymes, many having isozymic and allozymic forms. By studying the distribution of these enzymes in plant populations, we can learn about the reproductive biology of plants.

For example, populations with relatively high numbers of heterozygous loci indicate a high level of cross-pollination (i.e., pollination between plants). Fewer heterozygous loci are associated with a high level of self-pollination (i.e., pollination within a single flower or between flowers of the same plant). Further, when all samples from a population have the same allozymes and isozymes, the "population" is probably a clone of the same genotype—that is, sexual reproduction is absent. In contrast, a relatively high level of allozyme/isozyme variability at the edge of a population may indicate hybridization with a nearby species.

Serial Gene Systems

Unlike the independently expressed genes of Mendel's peas, genes often act together to control one or more characteristics. This phenomenon can occur, for example, in the multistep biosynthesis of a complex molecule. Such a serial system of genes consists of two or more Mendelian genes acting together to complete a developmental sequence. This interaction of two or more genes is called **epistasis.**

In most plant species, flower color is controlled by the epistatic effects of several genes in a series. In snapdragon (*Antirrhinum majus*), seven different flower colors are controlled by a series of four genes (fig. 8.13). One of these, called the *nivea* gene (abbreviated *niv*), blocks the synthesis of a precursor molecule when both of the gene's alleles are recessive. When this occurs, there is no pigment, and the flowers are white. When at least one dominant allele occurs at the *niv* locus, flower color depends on the genotypes of one or more of the other three genes. Figure 8.14 shows the genotypes of flower colors in snapdragons.

Polygenic Inheritance

In his less well-known experiments with garden beans (*Phaseolus multiflorus*), Mendel noticed that flowers of F_2 plants exhibit a continuous series of colors ranging from purple to white. He suggested that two or three independent factors control a single color trait, but he provided no further evidence to support this hypothesis. We now know that heritable variation in many

BOXED READING 8.2

EXAMINING PROTEINS BY GEL ELECTROPHORESIS

Gel electrophoresis separates large, charged molecules by their different rates of movement through a gel in an electric field. A protein extract is put into a small well at one end of the gel, which is placed on a Plexiglas plate and immersed in an aqueous solution. Direct current is supplied to a negative electrode at the same end as the protein well and to a positive electrode at the other end of the gel. Proteins move toward the positive end because their overall charge is negative. A protein's rate of movement is determined by its size and charge: larger proteins are slowed down by the gel, while proteins having higher negative charges move faster. Complications between size and charge are generally eliminated by treating extracts with a detergent that gives all proteins nearly the same charge. Detergent-treated proteins are therefore separated on the basis of size alone.

A protein extract may contain thousands of proteins, none of which are visible on the gel without some kind of chemical modifications. To analyze allozymes, gels are treated with a substrate appropriate for the enzyme being studied. The enzyme reacts with the substrate to produce a colored band on the gel. Specific color-generating substrates are known for only about forty enzymes, which means that the inheritance of most enzymes cannot be studied by gel electrophoresis.

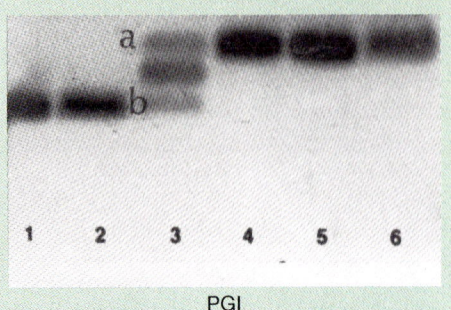

BOX FIGURE 8.3
Starch gel electrophoresis showing stained proteins.

Inheritance of enzymes typically involves codominant alleles. This means that, for simple enzymes, heterozygous and homozygous genotypes can usually be identified directly just by inspecting the gel. Homozygotes have one band for a gene and heterozygotes have two bands. Enzymes that consist of more than one polypeptide often show more complicated patterns.

Nucleic acids can also be analyzed by gel electrophoresis. Analysis of DNA by gel electrophoresis is discussed in Chapter 11.

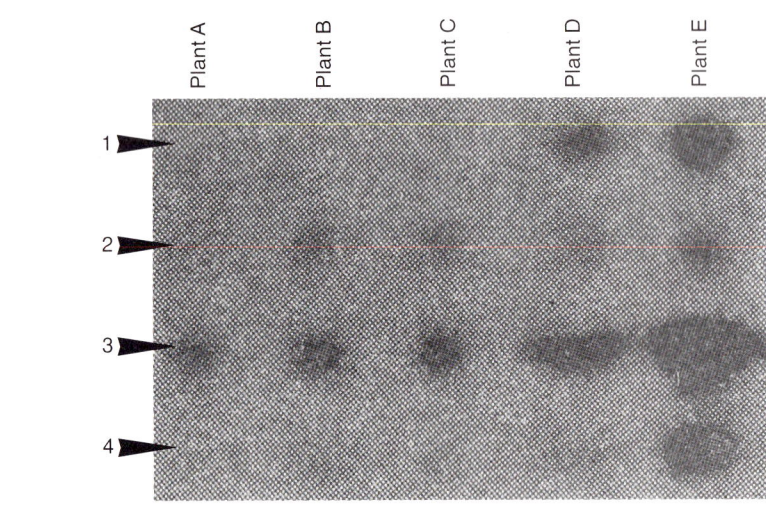

BOX FIGURE 8.4
Starch gel electrophoresis of different forms of the enzyme phosphoglucomutase from the wild sunflower, *Helianthus debilis*. Letters (A–E) are different lanes on the gel; each lane is from an extract of a different plant. Numbers (1–4) indicate each of the four forms of the enzyme. Note that each plant has only one or two allozymes but that four allozymes occur among the five plants.

---

characteristics appears to be continuous. Body weight, plant height, quantity of seed oil, and length of corn ears are examples of continuous characters.

Multiple genes can act additively, without dominance, to control a continuous trait. Such genes are referred to as **polygenes**. For example, polygenes control the inheritance of corolla length in *Nicotiana longiflora* (fig. 8.15a), a species of wild tobacco. When long-corolla plants were crossed with short-corolla plants, measurements of this character in the F_1 generation yielded the distribution of phenotypes shown in figure 8.15b. The F_2 generation showed a similar distribution of phenotypes. This pattern of inheritance is attributed to the polygenic control of corolla length.

Expression by polygenes, especially those involving size, can be modified by temperature, the availability of water and nutrients, or by other environmental factors. Because of the potential complexity of interactions between polygenes and the environment, the phenomenon of polygenic control is poorly understood. We can only assume that traits which vary continuously are at least partially controlled by polygenes.

Pleiotropy

Genes that affect more than one phenotypic characteristic are called **pleiotropic genes**. Mendel may have encountered a pleiotropic gene in the garden pea—the one that influences the color of flowers and seed coats (fig. 8.3). Purple or purplish coloration occurs in all of the

FIGURE 8.13

Snapdragons (*Antirrhinum majus*) have seven different flower colors, most of which are shown in this photograph. These colors are controlled by different combinations of alleles from four genes (see fig. 8.14).

Color	Genotype for four genes			
	Nivea	Sulfurea	Incolorata	Eosinea
white	*niv/niv*	NA	NA	NA
yellow	*niv+/—*	*sulf/sulf*	*inc/inc*	NA
ivory	*niv+/—*	*sulf+/—*	*inc/inc*	NA
bronze	*niv+/—*	*sulf/sulf*	*inc+/—*	*eos/eos*
pink	*niv+/—*	*sulf+/—*	*inc+/—*	*eos/eos*
crimson	*niv+/—*	*sulf/sulf*	*inc+/—*	*eos+/—*
magenta	*niv+/—*	*sulf+/—*	*inc+/—*	*eos+/—*

FIGURE 8.14

Control of flower color in snapdragons by four genes. Gene loci are designated by three-letter or four-letter abbreviations in italics. The "+" superscript indicates the dominant allele. A blank is used for the second allele when the dominant allele is present, because the second allele has no effect on the phenotype. NA means not active when the preceding gene in the table is homozygous recessive.

Source: Data from Hans Stubbe, 1966.

organs or in none of them, thereby behaving like a group trait controlled by a single gene. A clearer and more dramatic example of pleiotropy occurs in common tobacco (*Nicotiana tabacum*). In this species, the sizes and shapes of leaves, flowers, anthers, and fruits segregate into two sets of phenotypes controlled by the S gene. Plants with at least one dominant allele (SS or Ss) grow longer and narrower organs. Plants that are homozygous for the recessive allele (ss) have shorter and broader structures (fig. 8.16).

A pleiotropic gene also influences development in the European columbine (*Aquilegia vulgaris*), which has a dwarf form in which secondary cell walls thicken earlier than normal. Because of this early wall development, dwarfs are shorter than normal plants and have brittle stems, erect flower buds, smaller flower parts, and more branches. This set of traits is controlled by a single homozygous recessive gene.

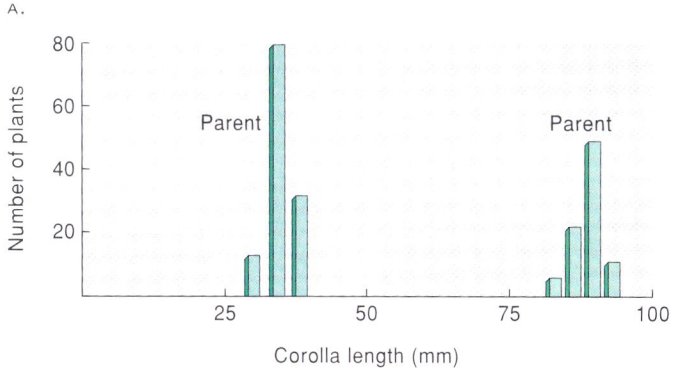

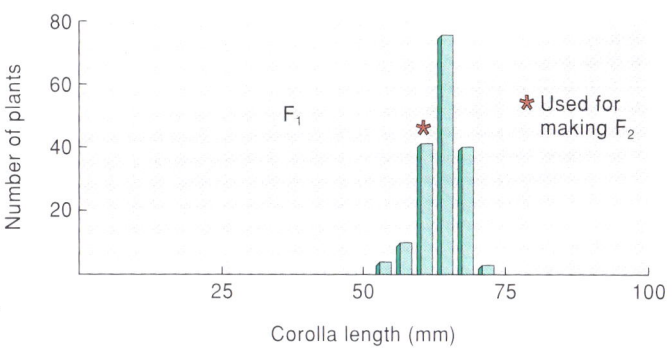

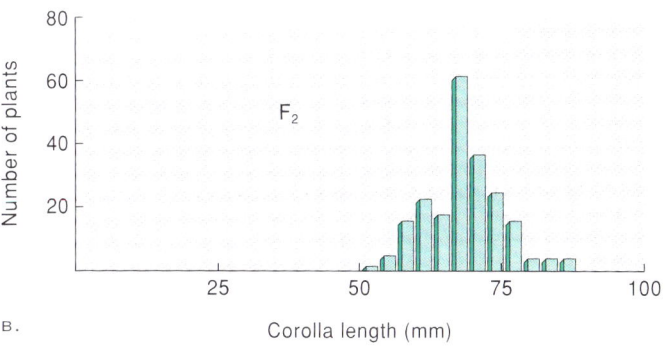

FIGURE 8.15

(a) Flower of *Nicotiana longiflora*. (b) Frequency distribution of corolla lengths in *Nicotiana longiflora*. Both the F_1 and F_2 generations show continuous variation of corolla length, which indicates polygenic inheritance.

CHAPTER EIGHT *Patterns of Inheritance*

FIGURE 8.16

The pleiotropic effects of two alleles of the S gene on various parts of the tobacco plant. The S allele is dominant; genotypes SS and Ss have longer and narrower organs, whereas organs of the ss genotype are shorter and broader.

Complex inheritance is revealed by various combinations of Mendelian genes. These include polygenic control of a single trait, multiple enzymic phenotypes from different loci, and single genes that control more than one phenotypic characteristic.

NON-MENDELIAN INHERITANCE

Aside from such complicating factors as polygenic control or pleiotropy, even the inheritance of many single-gene traits does not yield Mendelian ratios. The most common causes of non-Mendelian inheritance are linkage, cytoplasmic inheritance, mutations, and transposable elements.

Linkage

Many genes occur on each chromosome. When two or more genes occur on the same chromosome, it seems as if they should be inherited together, not independently. The simultaneous inheritance of genes on the same chromosome is called genetic **linkage.** Linkage violates the postulate of independent assortment of genes.

Linkage was first reported in the sweet pea (*Lathyrus odoratus*), a relative of the garden pea. In 1906, geneticists at Cambridge University discovered that genes for flower color and pollen shape did not assort into the expected 9:3:3:1 phenotypic ratio in the F_2 generation. Instead, the ratio was 7:1:1:7 (fig. 8.17). This means that almost 44% (7/16) of the F_2 plants had the dominant flower color (purple) and the dominant pollen shape (oblong), while an equal proportion had the recessive flower color (red) and the recessive pollen shape (spherical). This pattern indicates that the purple-oblong and red-spherical phenotypes are inherited together, which means that the genes for these traits are linked.

The main puzzle of linkage, however, was that some of the F_2 plants were red-oblong and some were purple-spherical, unlike the purebred parents or the hybrids of the F_1 generation. If linkage is perfect, it seems that these phenotypes should not occur. Imperfect linkage occurs when chromosomes exchange complementary fragments during meiosis. As a result, fragments that have different alleles for linked genes may be rearranged to produce nonparental combinations of alleles. This process is called **crossing-over** and is discussed more completely in Chapter 10, "Meiosis, Chromosomes, and the Mechanism of Heredity."

How did Mendel avoid results influenced by linkage? He noted that traits in a dihybrid combination and a trihybrid (three-character) combination showed independent inheritance. The simplest explanation for this pattern is that each of the genes used in Mendel's experiments occurs on a separate chromosome. This reasoning is correct for his dihybrid cross, because the gene for seed shape is on chromosome 7 and the gene for seed color is on chromosome 1 (fig. 8.3). However, this reasoning is incorrect for color of flowers, seedling axils, and seed coat because genes for these features are all located on chromosome 1. Nevertheless, we can explain independent assortment of linked genes on chromosome 1 by frequent crossing-over between them. Such frequent crossing-over causes genes to assort independently even though they are on the same chromosome.

Our assumption of frequent crossing-over can account for all but one of the pairs of linked genes in Mendel's peas: the genes that control pod shape and plant height. Modern geneticists have shown that they do not assort independently because they are too close to each other for frequent crossing-over to occur between them. Fortunately for Mendel, his experiments were not confounded by this result because he apparently did not make hybrids with this gene combination.

Each chromosome has many genes. Genes on the same chromosome may not assort independently because they are linked. Because of crossing-over, however, linked genes can be rearranged into nonparental combinations of alleles and mimic independent assortment.

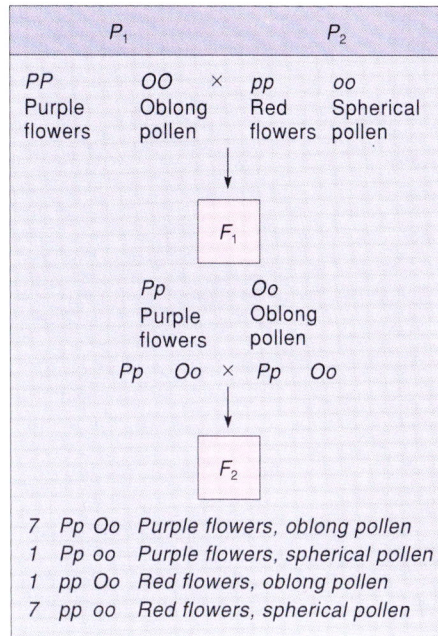

FIGURE 8.17

(a) Sweet pea (*Lathyrus odoratus*) in bloom. (b) A dihybrid experiment with sweet pea does not yield the expected 9:3:3:1 ratio of phenotypes in the F_2 generation because of linkage between the loci for flower color and pollen shape. Instead, most of the offspring are like the parents (P_1 generation). However, linkage is not perfect, since a small number of nonparental combinations (red/oblong and purple/spherical) appear in the F_2.

Cytoplasmic Inheritance

You learned in Chapter 2 that chloroplasts and mitochondria contain DNA. Genes in these organelles control certain aspects of photosynthesis and respiration, respectively. Inheritance of these genes is independent of sexual reproduction because they are transmitted to offspring with the cytoplasm, usually that of the maternal parent (see box 8.3, p. 182).

One example of cytoplasmic gene control occurs in certain forms of the cultivated four-o'clock (*Mirabilis jalapa*) that have yellowish-white leaves instead of green leaves. This difference in leaf color is caused by defective chloroplast genes. Phenotypic expression depends solely on the seed parent. Thus, when pollen from a white-leaved plant is transferred to a green-leaved plant, all the offspring have green leaves. In contrast, all of the offspring of the reciprocal cross have white leaves. This is an example of the **cytoplasmic inheritance** of non-nuclear genes.

The cooperation of organellar and nuclear genes is often necessary for normal metabolism. For example, the photosynthetic enzyme ribulose-1,5-bisphosphate carboxylase/oxygenase (rubisco; Chapter 7) has two subunits, one derived from a nuclear gene and one from a chloroplast gene. Similarly, some ATPases (Chapter 6) have a dual origin between the nucleus and mitochondria. In each case, the final product—that is, a complete and functional enzyme—depends on genes from two sources in the same cell.

Mutations

Dutch botanist Hugo de Vries was one of the rediscoverers of Mendel's work. De Vries studied patterns of inheritance in several kinds of evening primrose (*Oenothera* species). Most of his results conformed to Mendelian inheritance, but occasional characteristics appeared that were not present in either parent. De Vries called these spontaneous hereditary changes **mutations**, a term that remains in widespread use today. In modern genetics, the term *mutation* includes a variety of genetic changes, including chromosomal rearrangements and sequence changes in DNA. Details of both types of mutations are presented in Chapters 9 and 10.

Transposable Elements

One of the most important kinds of mutations involves sequences of DNA that seem to multiply and move spontaneously among an organism's chromosomes. These movable pieces of DNA are called **transposable elements,** and have been estimated to account for as many as half of the plant's spontaneous mutations. Such fragments of DNA may contain one or more genes that can function as inhibitors, modifiers, or mutators, thereby affecting the action and phenotypic expression of ordinary genes in many ways. Inserting transposable elements into or near functional genes produces unpredictable patterns of inheritance.

Transposable elements were first identified in the 1940s by Barbara McClintock (see fig. 1.9), who observed their

BOXED READING 8.3

CYTOPLASMIC MALE STERILITY IN CORN

The development of anthers and pollen in hybrids is often sensitive to interactions between nuclear and cytoplasmic genes. In many species, certain nuclear genotypes interact with specific mitochondrial genes to produce sterile pollen. The phenotype is then said to show **cytoplasmic male sterility (cms).** Crop-plant breeders can take advantage of male-sterile plants in controlled hybridizations. One crop that has been used in this manner is corn (*Zea mays*). Corn plants have two kinds of flowers, one for pollen and one for kernels, both occurring on the same plant. The pollen flowers occur in **tassels** at the tops of each plant, so that most of the pollen falls downward to fertilize the kernels on the same plant. To make hybrids, the tassels must be removed. This step becomes very expensive when whole fields must be cross-pollinated to produce large quantities of hybrid corn. Fortunately, certain varieties of corn have cytoplasmic male sterility (i.e., produce sterile pollen). Consequently, it is unnecessary to remove tasels before making hybrid kernels.

One particular variety of corn, which contains what is called Texas cytoplasm, or *cms-T,* was widely used for the commercial production of hybrid seed corn before about 1970. At its peak use, it was estimated that more than 85% of the corn grown in the United States contained *cms-T.* Unfortunately,

BOX FIGURE 8.5

Corn plant (*Zea mays*) in full bloom. Upper arrow: tassels of pollen flowers. Lower arrow: husks containing cobs of seed flowers. "Cornsilk" threads protruding from the husks are stigmas of the seed flowers.

BOX FIGURE 8.6

Corn husks infected with southern corn blight (*Helminthosporium maydis*). This corn has a strain of cytoplasmic male sterility called *cms-T*, which is susceptible to infection by the fungus.

heavy reliance on *cms-T* became disastrous for growers, beginning in 1970. At that time, all *cms-T* varieties became susceptible to southern corn blight, a disease caused by the fungus *Helminthosporium maydis*. This fungus destroyed *cms-T* varieties of crops in epidemic proportions. The susceptibility factor was subsequently found to be controlled by mitochondrial genes associated with the T-cytoplasm.

Male-sterile corn continues to be widely used for making hybrid corn. This means that there is a continuing need for studying the interactions between cytoplasmic and nuclear genes. With the advent of molecular biology, cytoplasmic male sterility has become one of the most intensely studied subjects in plant genetics.

---

effects on pigment patterns in corn kernels. Specifically, McClintock noticed that a fully pigmented grain was often flanked by a totally unpigmented grain on one side, and on the other side by a grain having pigment over only a part of its area (fig. 1.9b). If all of the cells in, for example, the endosperm of a given grain are derived from a single cell, why weren't they all alike? McClintock proposed that portions of kernels remained white, even though genes for pigment synthesis were present, because the pigment genes were disrupted by "controlling [transposable] elements." McClintock noted that reversion to the wild, nondisrupted genotype occurred when these elements moved away from the affected genes. Normal pigmentation appeared in cells in which reversion occurred. The pigment-inhibiting mutation was so unstable

that many groups of cells reverted to the wild-type pigmentation as each kernel developed. This produced a patchwork of colored spots and streaks, distributed in seemingly random patterns among colorless portions of the kernel (fig. 1.9). For McClintock, examining an ear's variously colored grains was like reading the history of the movement of transposable elements in cells.

Recent research has shown that coloration patterns in snapdragons and morning glories (*Ipomoea purpurea*) are also caused by transposable elements. Variegated flower colors result from complex interactions between mobile genes and genes for pigment synthesis (fig. 8.18). Similar color patterns are also probably caused by transposable elements in many other species of plants.

FIGURE 8.18

Photograph of a morning glory flower shows the effects of transposable elements on flower color. White sections of the flower occur where transposable elements disrupt the biosynthesis of pigments.

© Evelyne Cudel-Epperson, University of California, Riverside, 1992.

Transposable elements are now accepted as a general feature of many organisms, and are widely used to induce mutations. Scientists believe, however, that since not all genes are active at the same time, the potential effects of transposable elements on gene regulation vary. For example, transposable elements may help regulate cancer-causing genes, which are widespread among humans but are normally not activated. The importance of transposable elements in cancer research led to the 1983 Nobel Prize in the category "Physiology and Medicine" for Barbara McClintock.

What are Botanists Doing?

Compile a list of plants that are known to have transposable elements. What kind of observations do plant scientists use to indicate the presence of transposable elements? What kind of evidence do they use to confirm such a suggestion?

CONCEPT

Many kinds of non-Mendelian inheritance patterns cannot be explained by linkage. Changes in the base sequences of DNA, either by random mutation or by the insertion of transposable elements, alter inheritance patterns in unpredictable ways. Also, the inheritance of organellar genes does not conform to Mendelian predictions. In most plants, plastid and mitochondrial chromosomes are inherited with the cytoplasm of the female gamete.

Chapter Summary

The basic postulates for the theory of inheritance were derived from experiments by Gregor Mendel. However, we now know that the heredity of an organism is much more complex than it initially appeared. Exceptions to the basic postulates occur for several reasons, including linkage, cytoplasmic inheritance, mutations, and transposable elements. Moreover, complex patterns of inheritance arise from multiple genes, serial genes, or interactions between genes.

Chromosomes are made of DNA and proteins. Genes were originally thought to be made of proteins, but experiments with bacteria and viruses, which do not have DNA associated with protein, showed that the hereditary substance is DNA.

Different alleles control the production of alternative forms of an enzyme. These enzymes are called *allozymes* when they are made by the same gene locus, and *isozymes* when they are produced by different loci. Several allozymes often occur for a single locus, indicating that multiple alleles of the locus exist among the individuals of a population.

Interactions of many genes can contribute to the expression of a quantitative characteristic. The expression of such polygenic characters (such as plant height) can be plotted on a continuous curve. Polygenic inheritance is usually influenced by environmental factors.

Genes that occur on mitochondrial or chloroplast chromosomes control cytoplasmic inheritance. The expression of nuclear genes occasionally depends on these organellar genes. Examples include enzymes (such as rubisco) having at least two subunits, one derived from a nuclear gene and one from a mitochondrial or chloroplast gene.

Mutations are changes in the positions or composition of genes. One type of mutation that occurs frequently involves spontaneous movement of DNA segments called *transposable elements*. The expression of ordinary genes is often influenced by their proximity to these mobile elements.

Questions for Further Thought and Study

1. Suppose you have a purple-flowered garden pea. What would its genotype be if it were crossed with a white-flowered individual and the phenotypic ratio of F_1 plants was 1:1?

2. Assume that a red and a white allele exist for flower color and a blue and a yellow allele exist for pollen color. What phenotypic ratios would be predicted in a dihybrid cross in which both traits showed incomplete dominance?

3. What would the genotypic ratios be in the cross described in question 2?

4. Why was DNA discounted as the hereditary substance for so long?

5. Avery's results in 1944, which showed that genes are made of DNA and not protein, were not widely accepted until 1952. What objections to his experiments do you think might have caused this delay?

6. What happened in 1952 that confirmed Avery's findings?

7. Explain why a 9:7 phenotypic ratio in the F_2 generation indicates control by two genes.

8. Why is the inheritance of chloroplast and mitochondrial genes non-Mendelian?

9. How can one parent plant reproduce sexually?

10. Describe how the movement of transposable elements could produce the patchwork pigmentation of the corn kernels shown in figure 1.9. Account for fully pigmented kernels, unpigmented kernels, and kernels having varying degrees of pigmentation.

11. Barbara McClintock's work was either neglected or disbelieved for almost 40 years, after which microbiologists obtained independent evidence for the existence of transposable elements in bacteria. Well into her eighties, McClintock was awarded a Nobel Prize and saw her work embraced and extended by a new generation of biologists. That is, McClintock (who died in 1992 at the age of 90) lived to witness the results of the revolution she had wrought. What does her story tell you about science and scientists?

Suggested Readings

ARTICLES

Blixt, S. 1975. Why didn't Mendel find linkage? *Nature* 256:206.

Corcos, A., and F. Monaghan. 1985. Some myths about Mendel's experiments. *The American Biology Teacher* 47:233–236.

Corcos, A., and F. Monaghan. 1990. Mendel's work and its rediscovery: A new perspective. *Critical Reviews in Plant Sciences* 9:197–212.

Dahl, Hans-Henrik M. 1993. Things Mendel never dreamed of. *Medical Journal of Australia* 158:247–254.

Federoff, N. 1984. Transposable genetic elements in maize. *Scientific American* 249 (June):85–98.

Hartl, D. L. 1992. What did Mendel think he discovered? *Genetics* 131:245–254.

Huckabee, C. J. 1989. Influences on Mendel. *The American Biology Teacher* 51:84–88.

Janick, J. 1990. Gregor (Johann) Mendel (1822–1884). *HortScience* 25:1211–1213.

Laughnan, J. R., and S. Gabay-Laughnan. 1983. Cytoplasmic male sterility in maize. *Annual Review of Genetics* 17:27–48.

Rogers, J. 1991. Mechanisms Mendel never knew. *Mosaic* 22(3):2–7.

von Tschermak-Seysenegg, E. 1951. The rediscovery of Mendel's work. *Journal of Heredity* 42:163–71.

Books

Corcos, A. F., and F. V. Monaghan. 1993. *Gregor Mendel's Experiments on Plant Hybrids: A Guided Study.* New Brunswick, NJ: Rutgers University Press.

Fowler, C., and P. Mooney. 1990. *Shattering: Food, Politics, and the Loss of Genetic Diversity.* Tucson: University of Arizona Press.

Galston, A. W. 1994. *Life Processes of Plants.* New York: Scientific American Library.

Grant, V. 1975. *Genetics of Flowering Plants.* New York: Columbia University Press.

John, B., and G. Miklos. 1988. *The Eukaryote Genome in Development and Evolution.* Boston: Allen and Unwin.

Murphy, T. M., and W. F. Thompson. 1988. *Molecular Plant Development.* Englewood Cliffs, NJ: Prentice Hall.

Olby, R. C. 1985. *The Origins of Mendelism.* 2d ed. Chicago: University of Chicago Press.

Roberts, H. F. 1929. *Plant Hybridization Before Mendel.* Princeton, NJ: Princeton University Press.

Russell, P. J. 1990. *Genetics.* 2d ed. Boston: Scott, Foresman.

Watson, J. D. 1968. *The Double Helix.* New York: Atheneum.

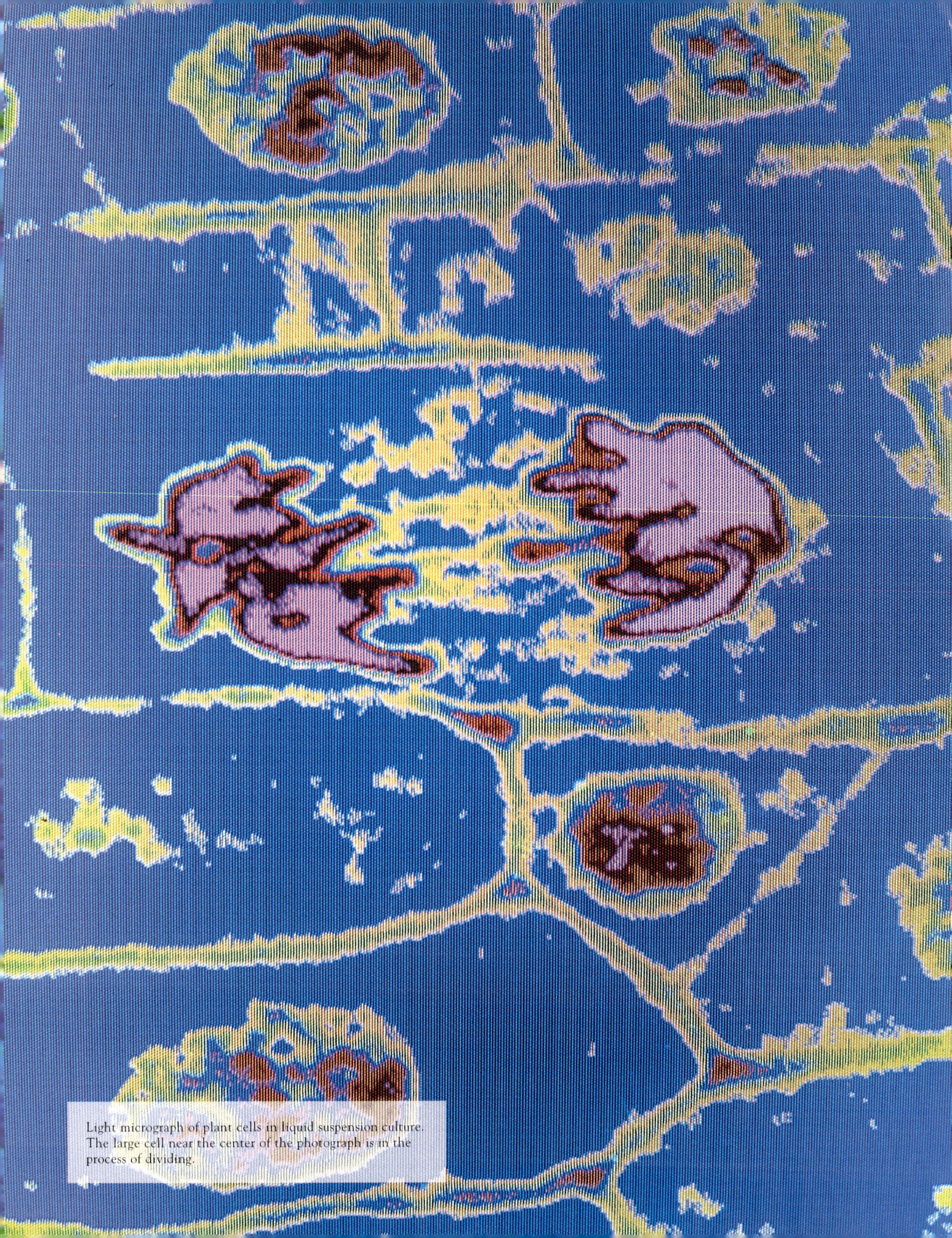
Light micrograph of plant cells in liquid suspension culture. The large cell near the center of the photograph is in the process of dividing.

CHAPTER 9

The Cell Cycle

Chapter Outline

INTRODUCTION
    The Cell Cycle in Plants
OVERVIEW OF THE CELL CYCLE
INTERPHASE
    The G_1 Phase
    Beginning a New Era: The Theory of DNA Structure and Duplication
    The Structure of Chromosomes During Interphase
    Making Chromatin: The Major Task of the S Phase
    Evidence for Semiconservative Replication
    Enzymes of Replication: Keys to Understanding the S Phase

BOX 9.1
MUTATIONS AND DNA REPAIR: DO MISTAKES HAVE TO BE CORRECTED?

    Replicons

BOX 9.2
DETERMINING THE SIZE AND COMPOSITION OF GENOMES

    The G_2 Phase
MITOSIS AND CYTOKINESIS
    Before Prophase
    Chromosome Condensation: Prophase
    Chromosome Alignment: Metaphase
    Chromosome Separation: Anaphase
    Formation of Daughter Nuclei: Telophase
    Division Into Cells: Cytokinesis
MYSTERIES OF THE MITOTIC SPINDLE: HOW ARE CHROMOSOMES MOVED?
    The Push of Polar Microtubules
    The Pull of Kinetochore Microtubules
Chapter Summary
Questions for Further Thought and Study
Suggested Readings

Chapter Overview

Every plant begins as a single cell and grows as cells repeatedly expand and divide. As the plant matures, however, only a few cells continue to divide. Most cells instead become specialized for a limited set of functions, such as photosynthesis, storage, and transport. Cells that continue to divide occur only in specific regions of the plant, such as stems, root tips, and buds; in these regions, cellular expansion and division can theoretically continue indefinitely. The uninterrupted repetition of cell expansion and division, together called the cell cycle, gives plants the potential for unlimited growth. Some organisms, such as giant redwoods, come closer to this potential than others.

Most of the attention devoted to the cell cycle involves the mechanisms of DNA synthesis and the behavior of chromosomes during mitosis. These subjects are the focus of this chapter.

INTRODUCTION

Virtually all living cells in a plant have identical sets of chromosomes in their nuclei. This observation leads to the prediction that each cell is **totipotent**—that is, has the same genes and therefore the same genetic potential to make all other cell types. The first confirmation of this prediction came from experiments in the 1950s by F. C. Steward and his colleagues at Cornell University (fig. 9.1). They began by growing small pieces of tissue from carrots in a nutrient broth. Cells that broke free from the fragments dedifferentiated, meaning that they reverted to unspecialized cells. As these unspecialized cells grew, however, they divided and redifferentiated back into specialized cell types. Eventually, cell division and redifferentiation produced new plants. Each unspecialized cell from the nutrient broth expressed its genetic potential to make all the other cell types in a plant.

Whole plants are now routinely regenerated from cell cultures in many plant species. This process is used, for example, to grow clones of orchids and other ornamental plants inexpensively and to make disease-free clones of potatoes and other crop plants. Cloning by cell culture is also important in genetic engineering in plants, which is discussed in Chapter 11.

Experiments that show totipotency led to one of the biggest puzzles in genetics: How do identical genomes dictate the differential expression of the same genotype in different cells? In other words, how can some genes be active while others are not? Solutions to such a complex puzzle lie in the answers to a multitude of questions about the functions of chromosomes and their DNA. Some questions have been answered, but many seemingly acceptable hypotheses still lack experimental support. Most processes remain largely unknown.

Much of our understanding of genetic mechanisms comes from studying the changing structure and behavior of chromosomes during cellular growth and division, especially during DNA synthesis.

The repeating processes of cellular growth and division are known collectively as the **cell cycle.** These processes occur in specific tissues during normal plant growth. In this chapter, we introduce the concept of the cell cycle briefly from a historical perspective and then present the major events of cellular growth and reproduction in plants.

The Cell Cycle in Plants

The cell cycle is complete only in cells that divide. The first dividing cell in the life history of a plant is the **zygote,** which forms by fertilization between a sperm and an egg. The zygote grows into a pre-embryo by cellular division and expansion. The pre-embryo becomes an embryo that has specialized regions, called **meristems,** that undergo cell division (fig. 9.2). As a seed germinates, cells at the growing tips of embryonic roots and shoots never stop dividing; the cell cycle in these cells is complete. In contrast, cells in leaves and certain tissues of stems and roots stop dividing and become specialized; the cell cycle in these cells is arrested.

Cells also divide and grow in several other sites in plants. For example, flowers, fruits, and branches arise by cell division in buds, which contain meristems that are active only during specific periods of plant development. Such buds usually occur in the axils of leaves. Branches in roots also arise from meristems, but lateral meristems in roots are internal and are not derived from axillary buds (Chapter 12). Furthermore, many kinds of nonmeristematic cells in stems and roots, and in the leaves of some plants, can be stimulated to divide when they are damaged. Cell division in damaged tissues continues until the wound is sealed with a layer of protective cork. Thus, the cell cycle is complete, when needed, for the formation of reproductive organs, stem and root branches, and wound-repair tissue.

CONCEPT

The cell cycle is a set of processes that foster cell growth and division. It is therefore complete only in dividing cells. Most cells do not divide after they mature. Because all living cells are totipotent, however, they may dedifferentiate and revert to being meristematic when induced to do so.

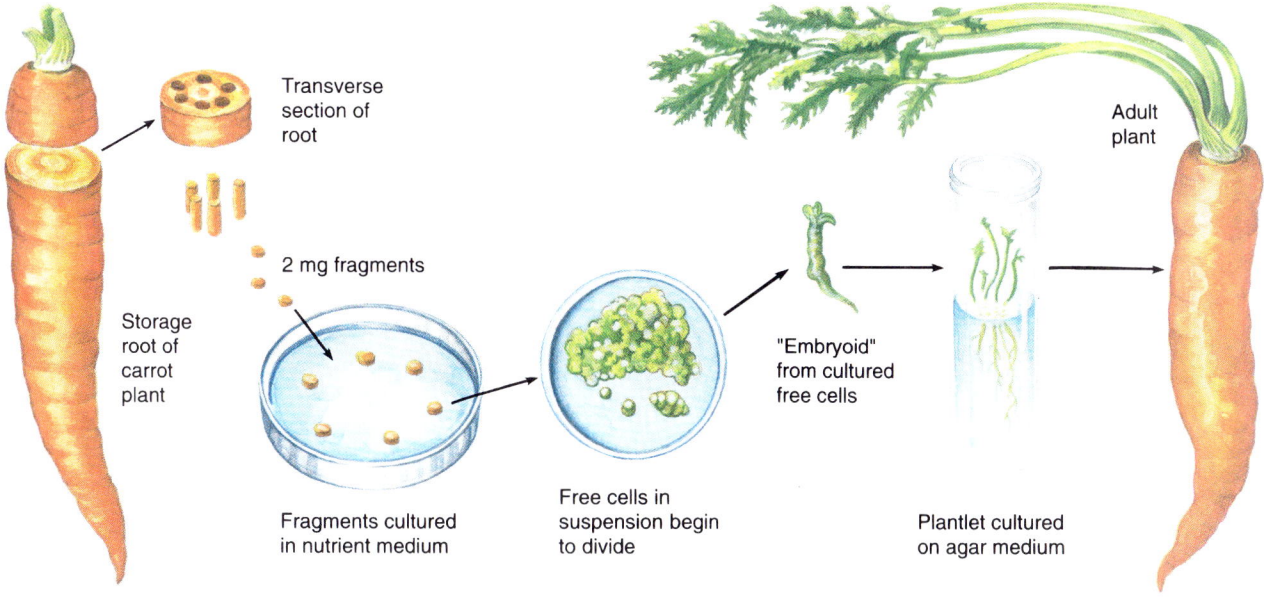

FIGURE 9.1

The main steps involved in culturing vegetative cells and regenerating a new plant from cell culture. Unspecialized cells in suspension divide and redifferentiate into all of the cell types needed to grow into another plant. The results of this experiment are evidence that each cell has the genetic potential to make all other cell types.

FIGURE 9.2

The cell cycle is complete in meristems, which divide throughout the life of the plant. Meristems usually occur at the tips of roots and shoots.

OVERVIEW OF THE CELL CYCLE

In 1858, nearly two hundred years after cells were first seen through a microscope, the German physiologist Rudolf Virchow proposed that all cells must come from preexisting cells. Since then, we have discovered that the formation of new cells includes both nuclear and cytoplasmic processes. During the past century, however, much more attention has been given to events that occur in the nucleus. Emphasis on the nucleus began because chromosomes were relatively easy to see with light microscopy. Nuclear processes still comprise the most intensely studied aspects of the cell cycle.

The cell cycle has two main parts: cell growth and cell division. The phases of cell growth are collectively called **interphase,** and the phases of cell division are collectively called **mitosis** and **cytokinesis.** Mitosis is the division of nuclei, and cytokinesis is the division of cytoplasm. Several shorter phases can be distinguished in each of the two stages of the cell cycle. Interphase includes three phases. The first is the **G_1 phase** (*first gap*), which occurs between the end of mitosis and the onset of DNA synthesis. This is followed by the **S phase,** which is when

CHAPTER NINE *The Cell Cycle*

DNA is synthesized. The third part of interphase is the **G₂ phase** (*second gap*), which begins at the end of the S phase (i.e., when DNA synthesis is complete) and lasts until the beginning of mitosis.

Mitosis is a continuous process that is divided for convenience into four phases: **prophase, metaphase, anaphase** and **telophase.** Cytokinesis usually follows telophase but may begin before telophase is completed. In certain tissues of plants and other organisms, cytokinesis is delayed or does not occur at all. Cells in these tissues are therefore multinucleate, either temporarily or permanently (see Chapter 3).

CONCEPT

The two main stages of the cell cycle are growth (interphase) and division (mitosis and cytokinesis). The cell cycle is a continuous process, but several phases can be recognized in each stage.

What are Botanists Doing?

Look through several journals related to cell biology, or conduct a computer search of the scientific literature in your library to estimate how many research articles have been published in the past 12 months on the cell cycle or topics related to it. What proportion of these articles is based on plants? How can you explain the difference in the number of botanical versus nonbotanical studies on the cell cycle?

INTERPHASE

Most meristematic cells spend about 90% of their time in interphase (fig. 9.3). However, throughout the first half of this century, little was known about what happens during interphase, since nothing but diffuse chromatin could be seen by light microscopy for this part of the cell cycle (fig. 9.4). It was instead believed, incorrectly, that everything important in the cell cycle happened immediately before and during mitosis. This included chromosome doubling and other mysterious processes, both nuclear and cytoplasmic, which prepare the cell for division.

As techniques of light microscopy, cell preparation, and chemical staining improved, biologists made discoveries that contradicted the initial assumptions about interphase nuclei. One of these discoveries came by using the now famous Feulgen stain, a chemical mixture that forms a purple complex with DNA. The absorbance intensity of this complex, which is proportional to the amount of DNA, was used by Hewson Swift in 1950 to estimate the relative amounts of DNA in meristematic cells of corn (*Zea mays*) and spiderwort (*Tradescantia* species). By making such determinations of many individual nuclei, Swift

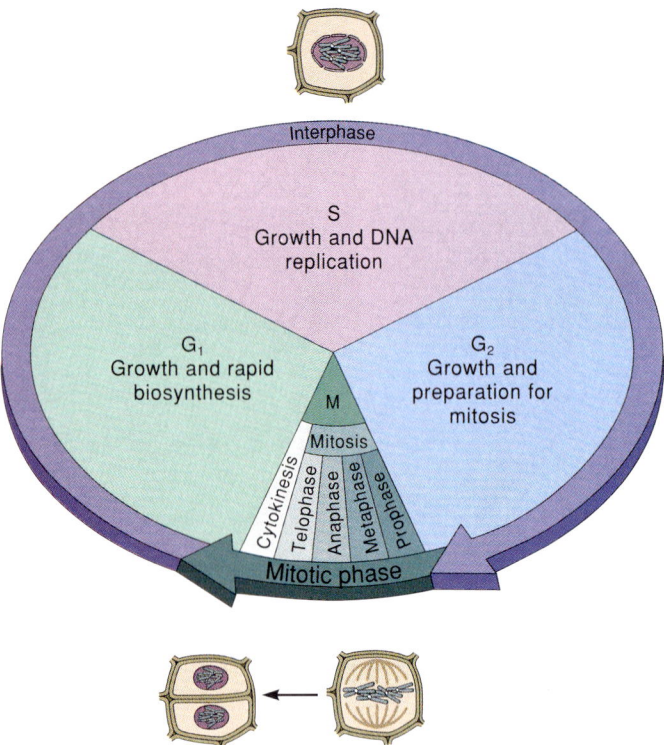

FIGURE 9.3

Periods of the cell cycle. Most meristematic cells spend about 90% of their time in interphase.

discovered that the amount of DNA increases during interphase, long before mitosis begins. This was the first indication that interphase nuclei are active and that they play a key role in preparing for nuclear division.

In 1953, Alma Howard and S. R. Pelc determined precisely when DNA is synthesized during interphase and how long it takes. They did this by measuring the incorporation of radioactive phosphorus (^{32}P) into the growing root tips of broad beans (*Vicia faba*). Since phosphate is a major ingredient of DNA, the uptake of ^{32}P indicates that DNA is being made. After extracting other phosphorous-containing compounds (ATP, inorganic phosphate, phosphorylated proteins, etc.), Howard and Pelc carefully measured changes in radioactivity caused by DNA synthesis. They found that DNA synthesis in broad beans (1) begins about twelve hours after the previous mitosis, (2) requires about six hours for completion, and (3) stops about eight hours before the next mitosis. Accordingly, about 87% of the thirty-hour cell cycle in broad beans is spent in interphase, and about 20% of the cell cycle is spent in DNA synthesis. Howard and Pelc were the first to designate the phases of interphase as G_1, S, and G_2. Their work is the foundation for our current model of the cell cycle.

CONCEPT

The longest period of the cell cycle is interphase. Most studies of interphase involve the S phase, when DNA is synthesized.

190   UNIT THREE   *Genetics*

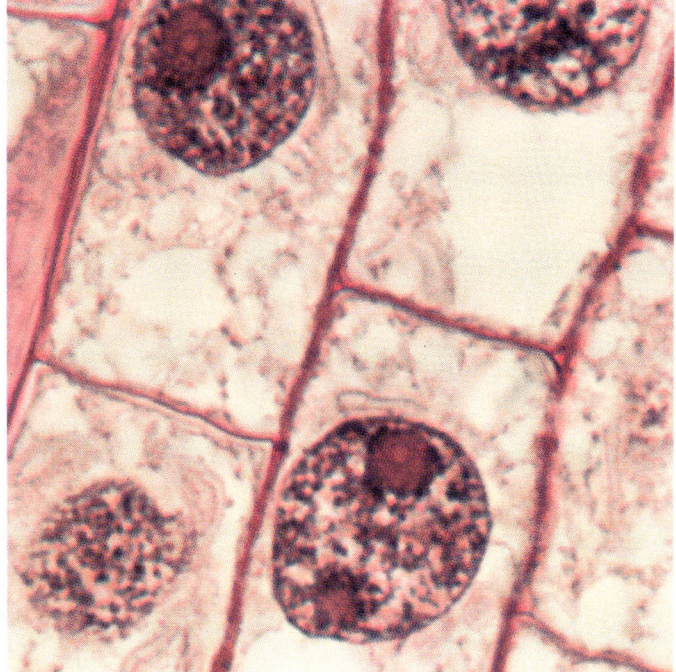

FIGURE 9.4

Interphase nucleus from a cell of onion root tip (*Allium cepa*), stained to show diffuse chromatin, ×500.

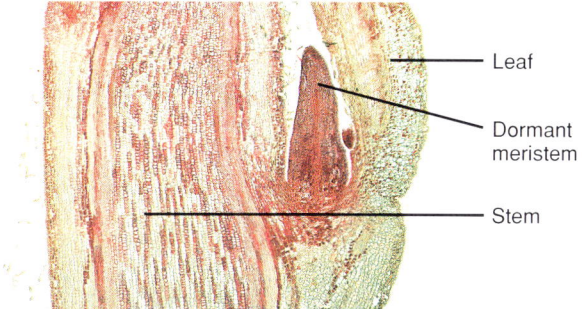

FIGURE 9.5

Dormant axillary bud in a longitudinal section of a pear stem (*Pyrus communis*), ×40. Cells in these meristems are usually arrested in the G_1 phase of the cell cycle.

FIGURE 9.6

"Sea urchin" galls on an oak leaf. Leaf cells are normally arrested in the G_1 phase of the cell cycle. Gall-forming insects induce the cells to continue the cycle so that the leaf will make new tissue to house the parasite.

The G_1 Phase

Cellular activities that are slowed or arrested during mitosis are reactivated during the G_1 phase. For example, cytoskeletal microtubules reassemble to support the growing cell. Also, the cell enlarges, organelles multiply, mitochondrial and plastid DNA increases manyfold, and many enzymes and structural proteins are made during the G_1 phase. One of the main tasks of G_1 is the synthesis of nucleotides that will be used to make DNA in the S phase to follow.

G_1 is the most variable phase in the cell cycle; it can be almost absent in rapidly dividing cells, or it can last for hours or days in slowly dividing cells. When cell cycles are suspended during winter dormancy, they are usually arrested in a G_1 phase that lasts for several months (fig. 9.5). The cell cycle in dormant cells speeds up when the cells resume activity in the spring. Conversely, the G_1 phase is permanent in cells that are differentiated. Such cells can continue the cell cycle only if they are induced to do so by such things as wounds, infections, or gall-forming organisms (fig. 9.6).

Beginning a New Era: The Theory of DNA Structure and Duplication

Progress in understanding the cell cycle, especially the S phase, was accelerated in the 1950s by a new theory of the structure of DNA. At that time, interest in the structure of DNA was at its peak. Recent evidence indicated that DNA was the substance of genes (see Chapter 8), and several different kinds of studies began to reveal the complexity of the DNA molecule. These studies were like the pieces of a puzzle—the structure of DNA being the puzzle. The relevance of various kinds of information to understanding the structure of DNA was first correctly recognized by James Watson and Francis Crick of Cambridge University. They are therefore credited with discovering the structure of DNA, the most significant discovery in biology of this century. Watson and Crick accomplished their feat by comparing the structure of proteins to a possible structure of DNA and by rationalizing all of the chemical and physical data on DNA into a unified structural proposal. The pieces for the puzzle were available, and Watson and Crick put them together.

Watson and Crick were aided in their quest for the structure of DNA by results from two other laboratories. One of these was at the California Institute of Technology, where Linus

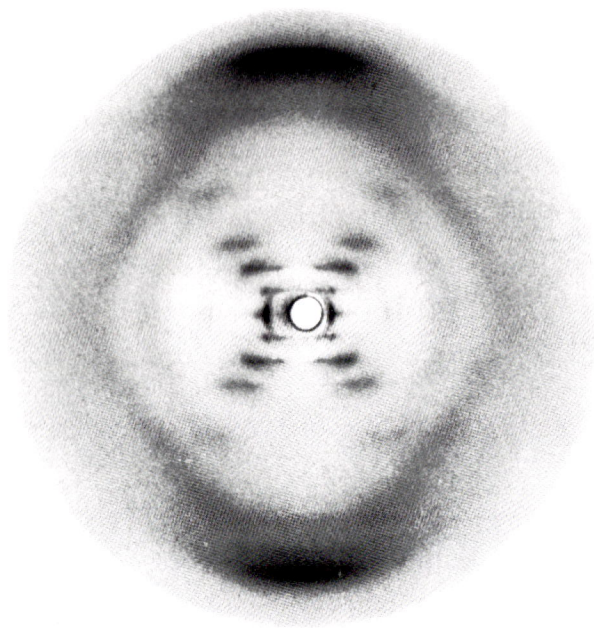

FIGURE 9.7

Photograph of the X-ray diffraction pattern of DNA. This photograph, which was made by Rosalind Franklin, shows the characteristic X-shaped pattern of helical structures.

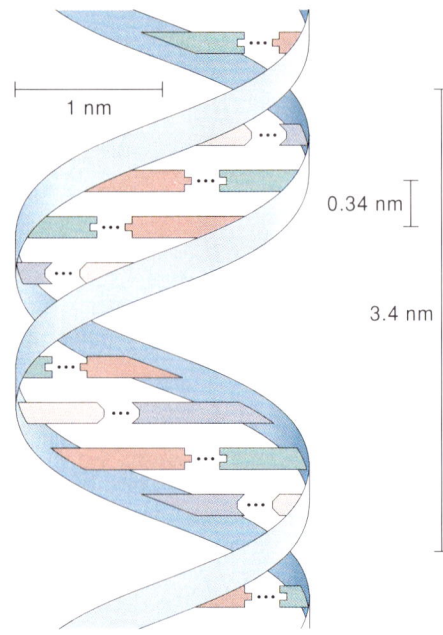

FIGURE 9.8

Diagrammatic representation of the double helix of DNA. Ribbons represent the deoxyribose-phosphate backbones of each strand. The bars between ribbons represent paired nitrogenous bases.

Pauling had found that parts of a protein form a **helix** (spiral) in three dimensions. Pauling's interpretation of protein structure was derived from X-ray diffraction patterns of a protein crystal (such patterns are obtained when X rays are shot through a crystal and bent according to the distances between different parts of the molecule). Pauling further explained that the twists of a protein helix are held together by hydrogen bonds between different parts of the molecule. These results led to the second significant piece in the DNA puzzle, which was provided in 1953 by Rosalind Franklin and Maurice Wilkins at King's College in London. They showed that X-ray diffraction patterns of DNA resemble those of proteins, thereby also indicating a helical structure for DNA (fig. 9.7).[1] Based on this information, several models of DNA structure were proposed, but only the one by Watson and Crick reconciled all of the available data. Their model was a **double helix**—that is, a two-stranded spiral (fig. 9.8). The use of the term *model* in this case is not abstract. Watson and Crick actually built a DNA molecule by putting together pieces of metal to represent the chemical components of DNA (fig. 9.9).

The double-helix model incorporated the property of macromolecules, such as Pauling's protein, to form hydrogen bonds at normal cytoplasmic pH. Specifically, Watson and Crick showed that hydrogen bonding could pair only certain nucleotides with each other (fig. 9.10). Furthermore, in summing all

---

[1]. Franklin and Wilkins were the first to publish the suggestion that DNA is a helix. Before publication, Wilkins gave the necessary X-ray data to Watson and Crick, without Franklin's knowledge. Franklin thought Watson and Crick had simply beat her to the punch on figuring out the structure of DNA.

FIGURE 9.9

Watson and Crick with their model of DNA in 1953.

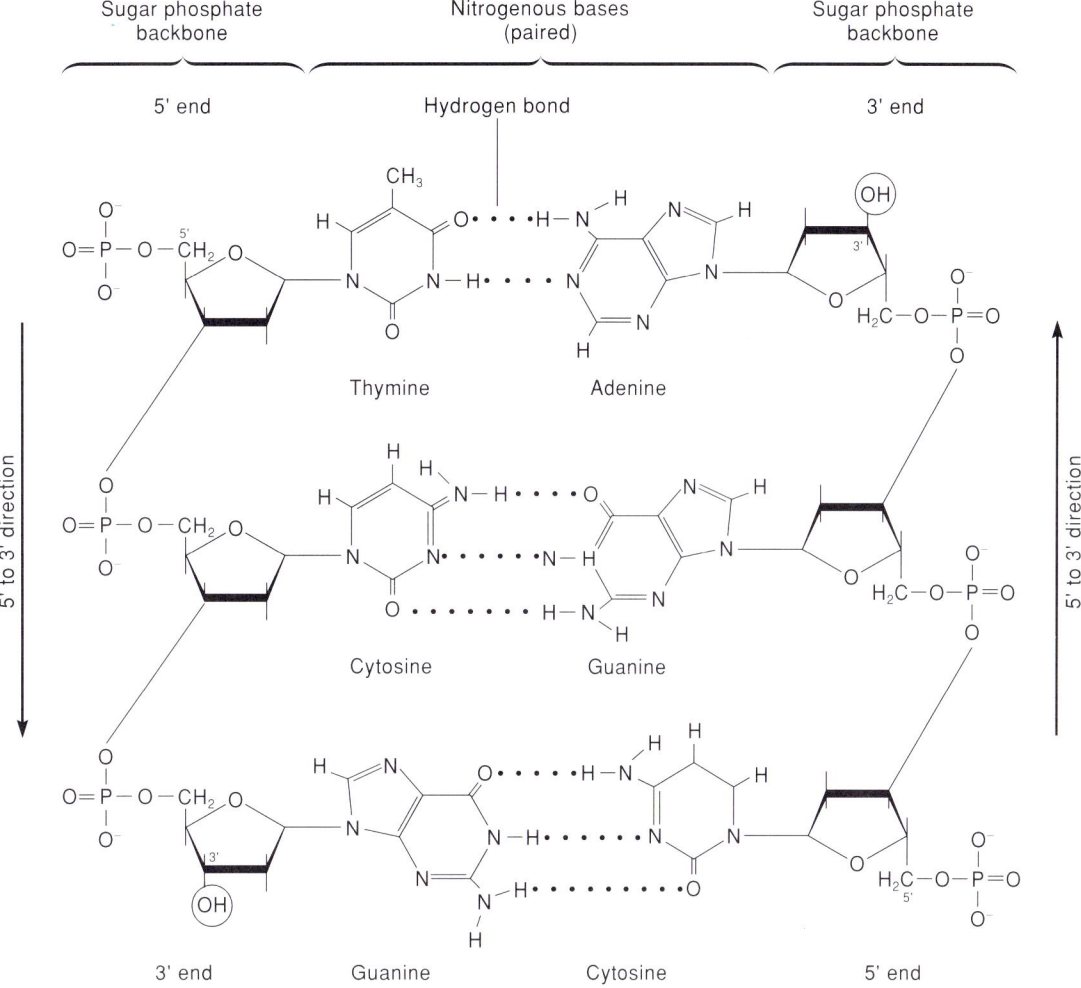

FIGURE 9.10

Components of a portion of a DNA molecule. The two strands are held together by hydrogen bonds between pairs of nucleotides, adenine pairing with thymine and cytosine pairing with guanine. The 3´-carbon of each sugar is bound to a phosphate group that is attached to the 5´-carbon of the next sugar.

of the appropriate electronic interactions and bond angles for each string of nucleotides (base + deoxyribose sugar + phosphate; see Chapter 2), Watson and Crick accounted for the twisting forces that form the helix itself. When assembled, the model immediately revealed that the two strands are oriented in opposite directions—that is, they are **antiparallel.** By convention, the directionality of a DNA strand derives from the orientation of its deoxyribose sugars. Each sugar has a phosphate at its 3´ (read "3-prime")-carbon and one at its 5´-carbon. Thus, one strand of the double helix is said to be in the 3´ to 5´ direction, and the other is in the 5´ to 3´ direction.

| C | O | N | C | E | P | T |

Discovery of the helical structure of DNA was aided by its similarity with a known protein structure; however, DNA is a double helix, unlike the single helix of the protein.

The impact of the Watson-Crick model was immediate. It explained how DNA could exist in many different sequences of nucleotide bases, thereby being sufficiently complex to produce genes. The model hinted that genes could exist as base sequences of defined length. It also accounted for the constant ratios between cytosine and guanine and between adenine and thymine that had been found in all organisms. Finally, the complementary nature of opposite strands suggested a potential mechanism for self-replication of the molecule. The postulates made by Watson and Crick make up the theory of DNA structure and duplication, which is summarized in table 9.1. Explanations of DNA structure and synthesis presented in this chapter are based on this theory.

In 1962, Watson, Crick, and Wilkins received the Nobel Prize in the category Physiology or Medicine, for their landmark studies. Franklin did not receive the prize because she died before the award was made (the Nobel Prize is not awarded posthumously).

TABLE 9.1

Postulates of the Theory of DNA Structure and Duplication

1. DNA consists of a double helix with the two strands of the molecule made of alternate molecules of deoxyribose and phosphate (fig. 9.8).
2. Pairs of bases form links by hydrogen bonding between opposite deoxyribose molecules in the two strands. The base-pairs are adenine-thymine (A-T) and cytosine-guanine (C-G) (fig. 9.10).
3. Base-pairs may be in any sequence along a double strand.
4. When DNA duplicates, the hydrogen bonds between base-pairs break, permitting the two strands to separate (fig. 9.13).
5. Complementary nucleotides are added in sequence to the separated strands (fig. 9.13).
6. As nucleotides are added together, the deoxyribose of each additional nucleotide is chemically bonded to the phosphate of the preceding nucleotide, thus forming the second strand of a new double helix (fig. 9.16).

The Watson-Crick model for DNA structure explains how DNA can be complex enough to accommodate a large number of genes in a multitude of nucleotide sequences. The pairing of adenine with thymine and cytosine with guanine also explains why base composition ratios are constant in each organism. The model also suggests how DNA can replicate itself.

The Structure of Chromosomes During Interphase

When the chromatin of an interphase nucleus is examined by electron microscopy, it appears as regularly spaced beads on a thin string (fig. 9.11e,i). The beads are called **nucleosomes** and are the basic structural unit of the chromosome. Nucleosomes are stable complexes of negatively charged DNA and positively charged proteins called **histones.** Histones are positively charged because they contain large amounts of arginine and lysine, which have positively charged amino groups in their side chains. DNA packs tightly around the histones in a form that does not react with most enzymes (fig. 9.11e).

A nucleosome usually contains about three hundred nucleotides, but the exact number can be different for different organisms. There are eight histones in each nucleosome, each composed of two copies of four different proteins. The arrangement of histones and DNA in the nucleosome is not known precisely, but DNA probably wraps around the histone (fig. 9.11e,i). A double-helical strand that is 50 nm long and 2 nm across is packaged into a nearly spherical nucleosome having a diameter of about 10 nm. Nucleosomes are separated by spacer DNA, usually 40–60 nucleotides long, which is associated with another histone (H1). This histone binds nucleosomes together in a regular, repeating array that is stabilized in a **chromatin fiber** about 30 nm in diameter (fig. 9.11d).

Chromatin fibers are further packed into folds called **looped domains** (fig. 9.11c), which may consist of 20,000 to 100,000 nucleotide pairs. The loops extend from the main axis of the chromosome, but the mechanisms of loop formation and maintenance are unknown. One suggestion is that DNA-binding proteins hold each end of a loop (fig. 9.12). Another suggestion is that the ends of loops attach to proteins in a chromosome axis. There is no direct evidence for either of these models, but regions near the ends of loops are enriched for certain enzymes.

Some looped domains coil and fold even further, until stained chromatin becomes visible by light microscopy during interphase. This visible chromatin is called **heterochromatin.** The more open form of chromatin, which cannot be seen easily by light microscopy, is called **euchromatin.** Heterochromatin is more condensed than euchromatin, which may mean that the genes in heterochromatin are packed too tightly to be functional. This is one explanation for how some genes can be inactive at different times during the cell cycle. Chromosomes condense even further during mitosis, which is discussed later in this chapter.

Chromatin contains DNA and proteins that are arranged in a succession of levels of organization. The basic level of organization is the chromatin fiber, which may be further packed into looped domains and heterochromatin.

Making Chromatin: The Major Task of the S Phase

Chromatin synthesis dictates that approximately equal amounts of DNA and histones be made during the S phase. This means that two different products must be made: DNA and protein. Thus, DNA must not only duplicate itself but also direct the synthesis of histones. Most studies of chromatin synthesis have been devoted to the DNA portion of chromosomes. The synthesis of DNA is called **self-replication,** because DNA makes exact copies of itself. The involvement of DNA in protein synthesis is a function of genes, which will be discussed in Chapter 11.

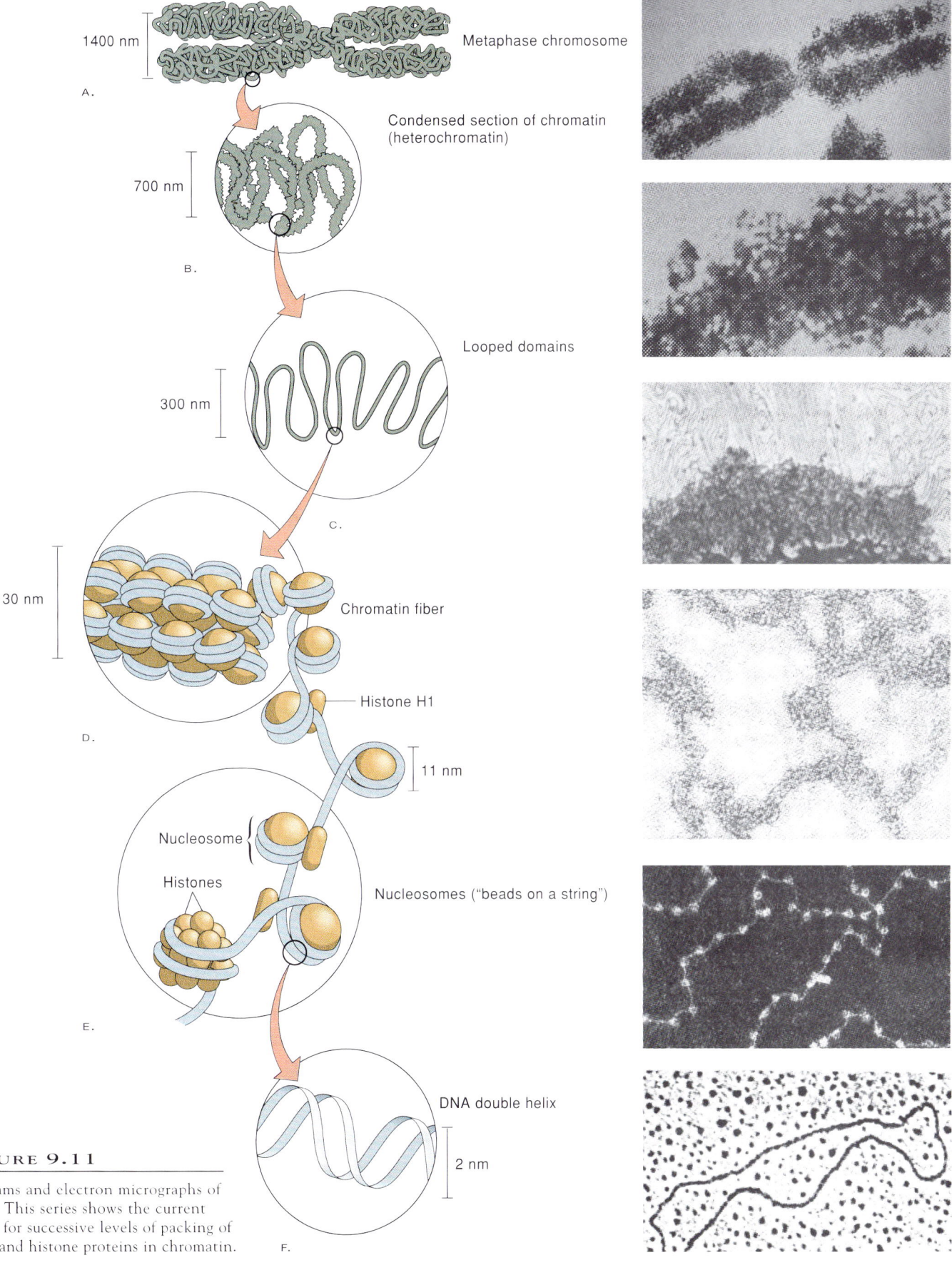

FIGURE 9.11

Diagrams and electron micrographs of DNA. This series shows the current model for successive levels of packing of DNA and histone proteins in chromatin.

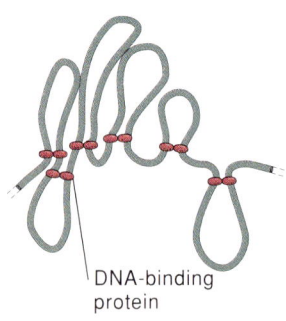
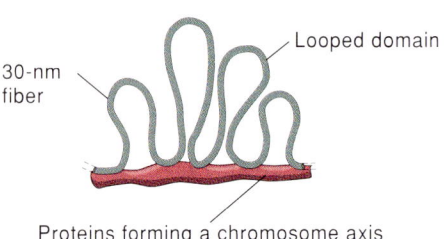

FIGURE 9.12

Two models for the maintenance of looped domains. *Top:* DNA-binding proteins hold each end of a loop together. *Bottom:* The ends of loops are all attached to a chromosome axis of proteins.

Source: Alberts, Molecular Biology of the Cell, 2d ed. Copyright © Garland Publishing Company, New York, NY.

The Lore of Plants

As often happens in science, "normal" processes, such as those of the cell cycle, have many exceptions. One common exception to the cell cycle is that, in many organisms, the S phase can occur many times before mitosis. The result is that the nuclei of some cells can have much more than the expected amount of DNA. For example, hair cells and glandular cells may have from 16 to more than 4,000 times the haploid amount of DNA. The record for plants is probably in certain cells of the seeds of the lords-and-ladies arum (*Arum maculatum*), which have more than 24,000-fold the haploid amount of DNA. But this number pales in comparison with the giant neurons of a certain species of mollusk, which have at least 75,000 times the haploid amount of DNA.

When Watson and Crick proposed the double-helix model for the structure of DNA, they immediately saw that this structure revealed a potential mechanism for self-replication. They envisioned that the two strands would "unzip" between base-pairs by breaking the hydrogen bonds between them. Each single strand could then act as a template for guiding the formation of a new chain onto the old one (fig. 9.13). According to this postulate, two double-stranded molecules would be produced, each made of one parent strand and one new strand. This is referred to as **semiconservative replication,** because half of each daughter molecule is conserved from one strand of the parent double helix. The semiconservative model seemed logical, but it lacked experimental support for about five years after it was proposed.

Evidence for Semiconservative Replication

In 1957, J. Herbert Taylor and others at Columbia University grew seedlings of the broad bean in a medium containing thymidine bound to tritium (^{3}H), a radioactive isotope of hydrogen. Because emissions from radioisotopes react with photographic film, Taylor could photograph radioactive chromosomes in root-tip cells. After one cell cycle, the DNA in each chromosome was radioactive throughout (fig. 9.14). After the next cell cycle, which occurred in the absence of the ^{3}H-thymidine, the parent DNA remained radioactive (conserved) but the newly made DNA was not radioactive. This was the first evidence that replication was semiconservative.

Further support for semiconservative replication of DNA was obtained in 1958 by Matthew Meselson and Franklin W. Stahl at the California Institute of Technology. They devised a simple but ingenious method involving the incorporation of heavy nitrogen (^{15}N) into cells of *Escherichia coli*, a common gut bacterium. After they grew the bacteria for many generations on a medium containing ^{15}N-ammonium chloride, the DNA of virtually all cells was heavier than ordinary. Meselson and Stahl showed that heavy DNA moves farther than normal DNA when centrifuged at high speed in a gradient of cesium chloride (fig. 9.15; see Chapter 3 for an explanation of cell fractionation). They then found that *E. coli* containing only heavy DNA would make DNA of intermediate density after growing for one cell cycle in a medium with normal ammonium chloride. This intermediate band consisted of hybrid DNA that contained equal amounts of ^{14}N and ^{15}N. Furthermore, after one more cycle on normal medium, two equal bands appeared, one at intermediate density and one at normal density. This means that each bout of DNA synthesis conserved one old strand for each new strand.

C O N C E P T

DNA synthesis is semiconservative, meaning that one strand of a new double helix is newly synthesized and the other strand is conserved from the double helix of the parent cell.

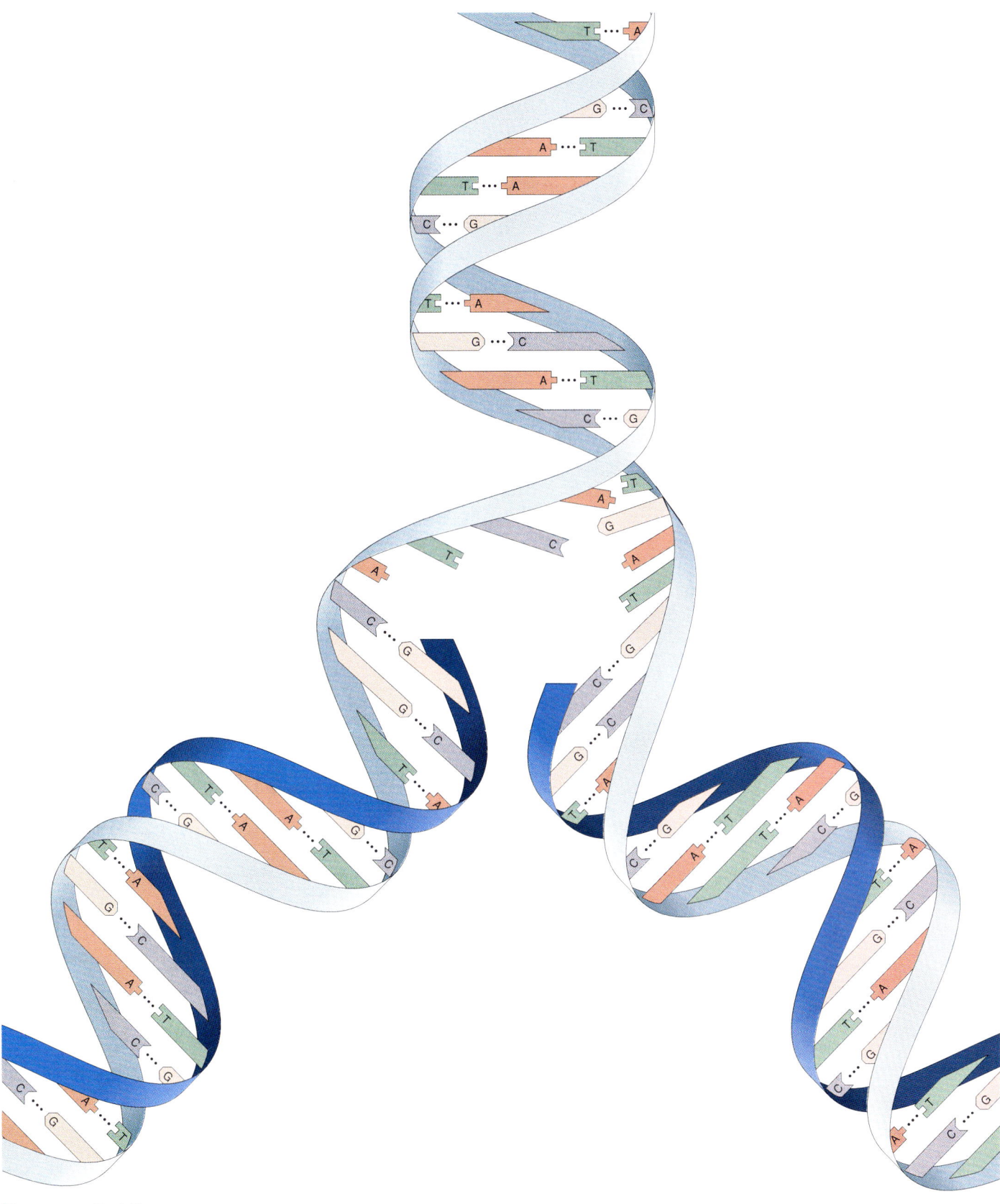

FIGURE 9.13

Semiconservative replication means that half of each daughter molecule is conserved from the parent molecule. After replication, the two daughter molecules are identical to the parental double helix.

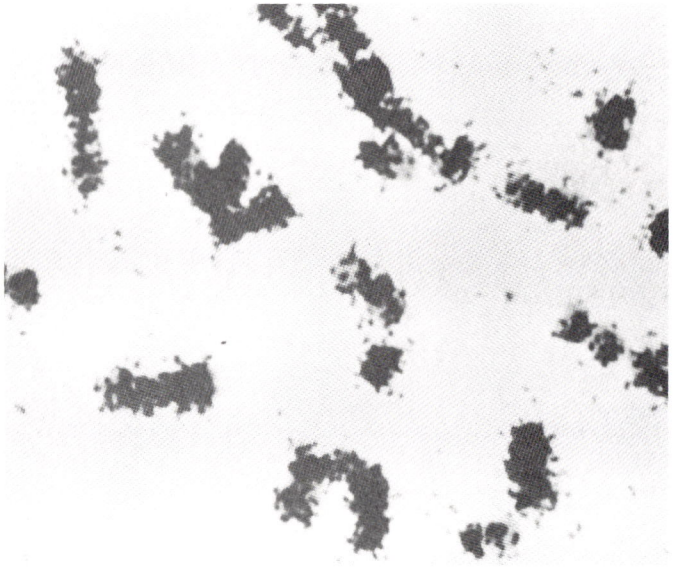

A.

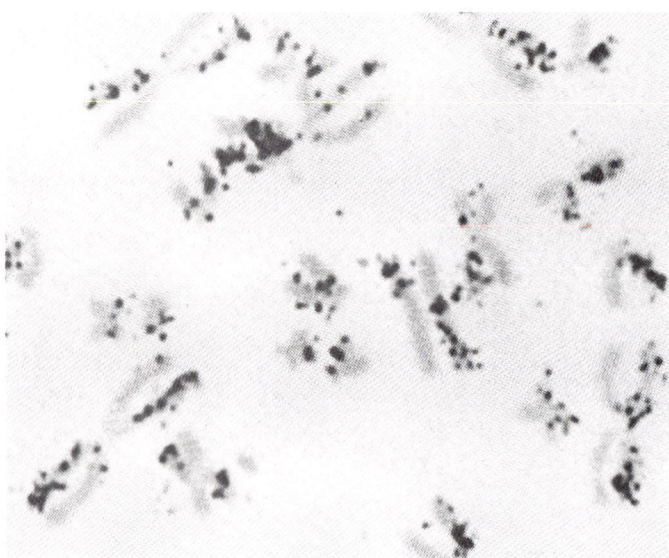

B.

FIGURE 9.14

Semiconservative replication as shown by radioactive labeling. (a) Metaphase chromosomes after DNA synthesis in the presence of tritium-labeled thymidine (^{3}H-thymidine), which is radioactive. Radioactivity is shown by dark areas. (b) One cell cycle later, after DNA synthesis in the absence of ^{3}H-thymidine. The radioactive parent DNA remains, and the newly synthesized DNA is not radioactive.

Enzymes of Replication: Keys to Understanding the S Phase

Although the concept of semiconservative replication is relatively simple, the process is a complex set of biochemical reactions. Each step is catalyzed by a separate enzyme that works with other enzymes to accomplish the overall task.

As figure 9.16 shows, several reactions occur more or less simultaneously. The DNA double helix must first be untwisted

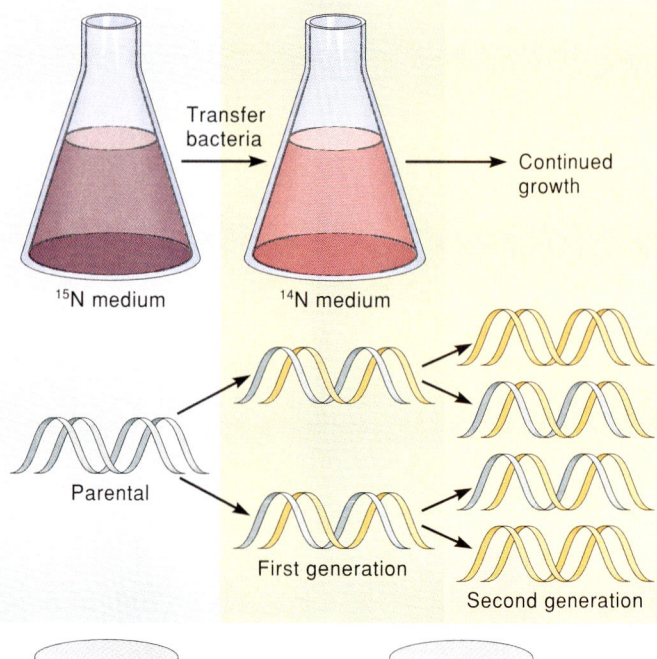

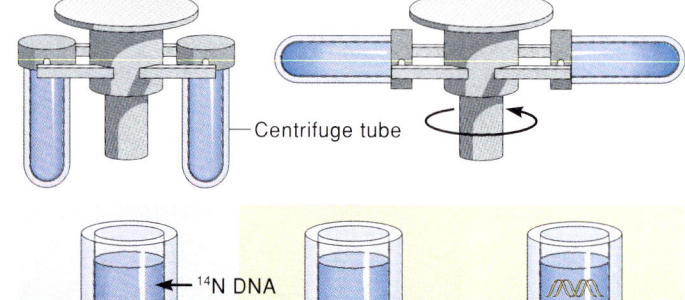

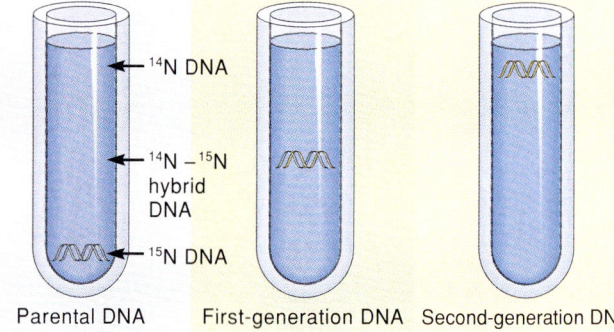

FIGURE 9.15

Incorporation of heavy nitrogen (^{15}N) into DNA makes the DNA heavier. Upon centrifugation, heavy parental DNA sinks farther into a gradient of CsCl. After one generation in a normal culture medium containing ^{14}N, DNA becomes more buoyant. One strand in each double helix still has ^{15}N because of semiconservative replication; the other strand of the "hybrid" DNA has ^{14}N. After the second generation in normal culture medium, half of the DNA in the cell is "hybrid" DNA, and half contains only ^{14}N (i.e., is most buoyant).

and split into two single strands. We do not know how this starts in plants, but in bacteria the initial untwisting of DNA is associated with special enzymes, called **helicases,** which open the helix by breaking hydrogen bonds between complementary base pairs. As the strands unwind and separate, their natural tendency to fuse back together and rewind is controlled by **single-strand binding proteins.** These proteins prevent refusion and keep the single strands from becoming tangled.

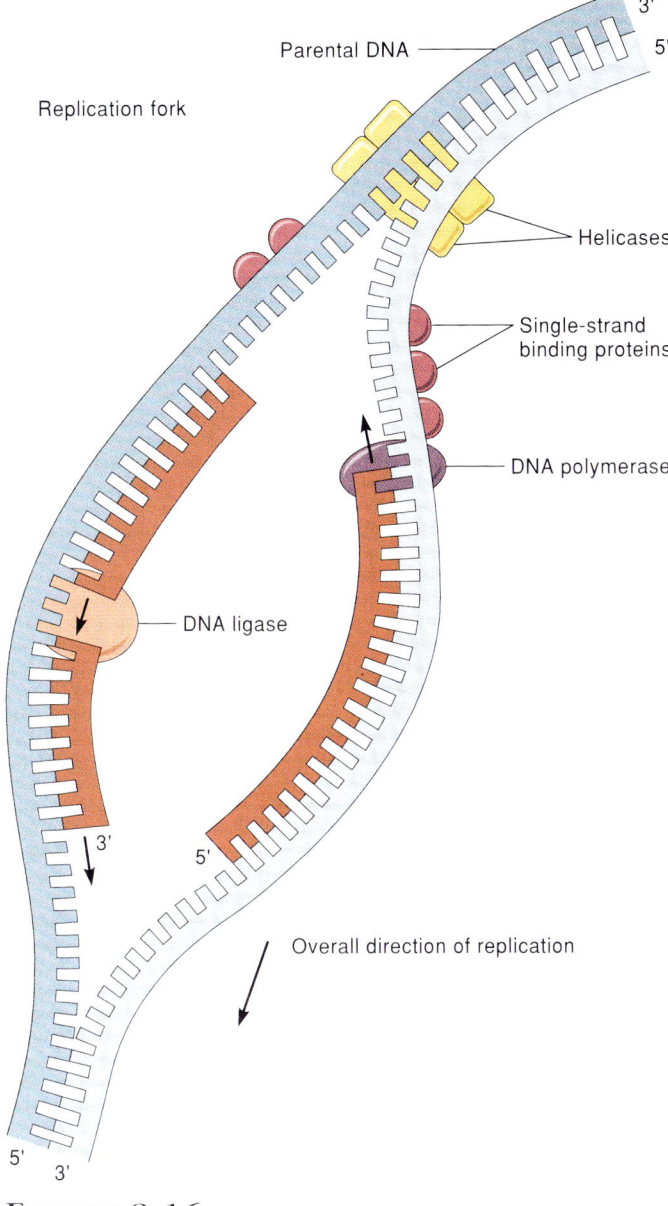

FIGURE 9.16

DNA replication requires helicases to unwind the double helix, single-strand binding proteins to stabilize the unzipped double strand, DNA polymerase to assemble nucleotides into a new strand, and DNA ligase to bind fragments together from discontinuous synthesis.

Synthesis occurs when **DNA polymerases** catalyze the covalent bonding of newly made nucleotides. The order of nucleotides is directed by the sequence of the single-stranded DNA template.[2] Thus, only bases complementary to the template can be added to the growing chain, which keeps adenines opposite thymines and cytosines opposite guanines in the developing double strand (but see box 9.1: "Mutations and DNA Repair: Do Mistakes Have to Be Corrected?"). Together, the processes of unwinding the helix and separating the strands create a moving **replication fork,** so named because of its forklike appearance in electron micrographs (fig. 9.17). Two replication forks move in opposite directions from a **replication origin,** which is the point of initiation for DNA synthesis. These moving forks form an expanding **replication bubble** between them.

When DNA polymerases were first studied, they were found to have a seemingly severe limitation: They could make chains of DNA in only one direction, from 5′ to 3′. This means that uninterrupted synthesis (**continuous synthesis**) can occur on only one of the two parent strands. It was later found, however, that synthesis on the other strand occurs when polymerase starts closer to the replication fork and then builds the new chain backward in the 5′ to 3′ direction. Pieces of this new chain are built in the opposite direction of the growing replication fork. Because it occurs in fragments, this interrupted synthesis is called **discontinuous synthesis.** Nevertheless, the net growth of a new chain by discontinuous synthesis occurs in the same direction as continuous synthesis.

More than a dozen other kinds of enzymes are involved in DNA replication. Discussion of the roles of all of these would be exceedingly complicated, but one additional type of enzyme has attracted considerable interest. This is the type called **topoisomerases,** so named because they influence the architecture (topology) of DNA. In the nucleus, DNA is often so severely twisted that it is like an overwound rubber band. Topoisomerases relieve the formation of kinks that would otherwise block the movement of replication forks. Topoisomerases can break one or both strands, allowing them to uncoil by swiveling around one another (fig. 9.18). After swiveling, DNA is then reconnected in its original form by the same enzymes. In this manner, topoisomerases loosen knots of DNA and allow access to the newly exposed double strands by replication enzymes. Such access is crucial for replication, since enzymes cannot react with segments of DNA that are tied in knots. Furthermore, by breaking and swiveling, DNA does not have to uncoil as replication occurs through a chromosome. If DNA did not break and swivel, then uncoiling would have to stay ahead of a replication rate of about 50 nucleotides per second, which converts to 5 revolutions per second (10 nucleotides per turn of the helix). This means that parts of chromosomes would be spinning at about 300 revolutions per minute, or about half the speed of a car engine at low idle. The havoc from such chromosome spinning would be hard to imagine.

Interestingly, topoisomerases can also *form* knots. This process hides DNA segments from biochemical reactions, perhaps in the form of heterochromatin. Furthermore, one type of topoisomerase is abundant at the ends of loops in looped domains, which may indicate a role for this enzyme in making loops. Because of their many roles in controlling chromatin structure, topoisomerases are studied intensely for their potential involvement in regulating gene expression. These studies underscore the importance of chromosome structure for controlling the functions of DNA.

---

2. DNA can also be made in a test tube by mixing isolated DNA with purified DNA polymerase. A short fragment (usually 10–30 nucleotides) of artificially made DNA is added as a primer to get the reaction started. This procedure, which is called the **polymerase chain reaction (PCR),** and its application to molecular biology are discussed in Chapter 11.

BOXED READING 9.1

MUTATIONS AND DNA REPAIR: DO MISTAKES HAVE TO BE CORRECTED?

Mutations are permanent alterations in chromosomal DNA. Many types of changes fit this definition. For example, in Chapter 8, we discussed mutations caused by inserting mobile sequences (transposable elements) into genes. Sequences may also be duplicated or deleted. Finally, mutations may involve the substitution of one nucleotide for another, or the gain or loss of one or more nucleotides.

An essential function of DNA polymerases is to add nucleotides in a specific direction to a growing chain of DNA during replication. Polymerases follow the lead of primers, which are short pieces of RNA complementary to the DNA template. Because the synthesis of RNA primers does not also depend upon primers, their sequences are relatively error prone. Errors in RNA primers cause about 1 in every 10,000 bases to be mismatched. Nevertheless, repairing these "mistakes" is possible because they can be recognized by DNA polymerases during replication. When a mismatch is encountered in the primer, polymerase backs up and other enzymes cut out the offending base. This means that DNA polymerases "proofread" replication. Imagine what havoc there would be if mistakes in RNA primers were not corrected. Even in the smallest of plant genomes, that of the common wall cress *Arabidopsis thaliana*, there could be as many as 280,000 mutations in every cell cycle! Nevertheless, we can find evidence that mutations have occurred and not been repaired in different genomes, although not at such an enormous rate. This evidence comes from comparing DNA sequences for the same genes among different organisms.

The advent of modern molecular biology has enabled biologists to determine nucleotide sequences of specific genes. DNA sequences vary in different species, yet remain functional in their respective plants. Such comparisons enable biologists to determine both the magnitude and the rate of mutations.

One of the more popular genes for study in this regard is *rbc*L, which occurs in the chloroplast genome and is responsible for the large protein subunit of rubisco (ribulose-1,5-bisphosphate carboxylase/oxygenase; see Chapter 7). This gene is studied intensively because it is important for photosynthesis, is relatively easy to extract from many plants, and because its inheritance is not confounded by sexual reproduction. Comparisons between the *rbc*L of maize (*Zea mays*) and tobacco (*Nicotiana tabacum*) show that more than 20% of the nucleotide positions differ in the *rbc*L of these plants. This comparison shows that even though DNA repair is reasonably dependable, it is not perfect. Some mutations escape the process and, if not lethal, are maintained without apparent harm to the organism.

CONCEPT

Clues to the mechanisms of DNA replication come from identifying the enzymes involved and determining their functions. Enzymes must stabilize and unwind the double helix, temporarily break hydrogen bonds between strands, and build a new chain along the parent DNA template.

Replicons

Replication starts at replication origins, and it stops where the forks of adjacent replication bubbles meet. The block of DNA between two replication origins is a **replicon** (fig. 9.19). Replicons vary in size from 30,000 to 300,000 pairs of nucleotides. This means that, based on an average replication rate of 50 nucleotides per second, the time required for completing one replicon ranges from 10 to 100 minutes. Replication of a whole genome, therefore, requires that different replicons be active at different times during the several hours of the S phase.

Different patterns of replicon activation may explain why genomes of vastly different sizes do not have S phases of vastly different lengths of time. For example, the haploid genome of common wall cress (*Arabidopsis thaliana*) is about 70 million nucleotide pairs, which is the smallest plant genome known. In contrast, onion (*Allium cepa*) has about 4.4 billion nucleotide pairs in its haploid genome, nearly sixty-three times greater than that of wall cress (see box 9.2: "Determining the Size and Composition of Genomes"), yet its S phase lasts only 3.9 times longer than that of wall cress; the S phase in wall cress lasts 2.8 hours and in onion, 10.9 hours.

Based on observations of a few organisms, including wall cress, replicons are apparently activated along a chromosome in clusters of 50–80 at a time. Many clusters operate on different chromosomes at the same time, with some finishing before others start. Furthermore, families of replicon clusters are active throughout the genome at specific times during the S phase. In wall cress, two replicon families accommodate all ten chromosomes; clusters in the first family activate 36 minutes before synthesis begins in the second. In comparison, DNA synthesis in onion probably depends on many more replicons per cluster and larger replicon families than it does in wall cress.

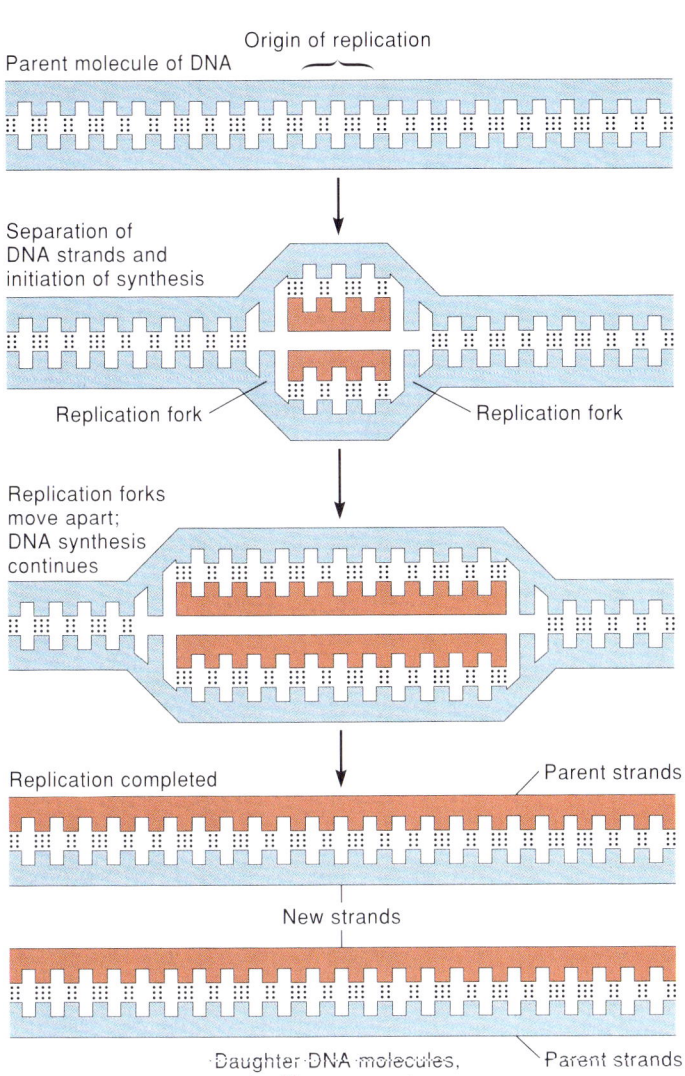

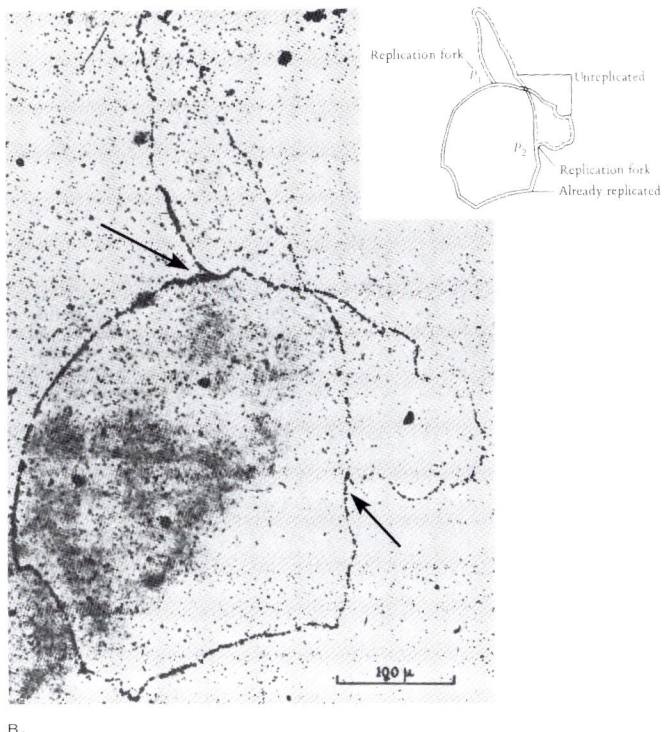

FIGURE 9.17

Origin of replication. (a) Model of the initiation and expansion of a replication bubble for DNA synthesis. (b) Electron micrograph of a replication bubble. Arrows point to replication forks at either end of the bubble.

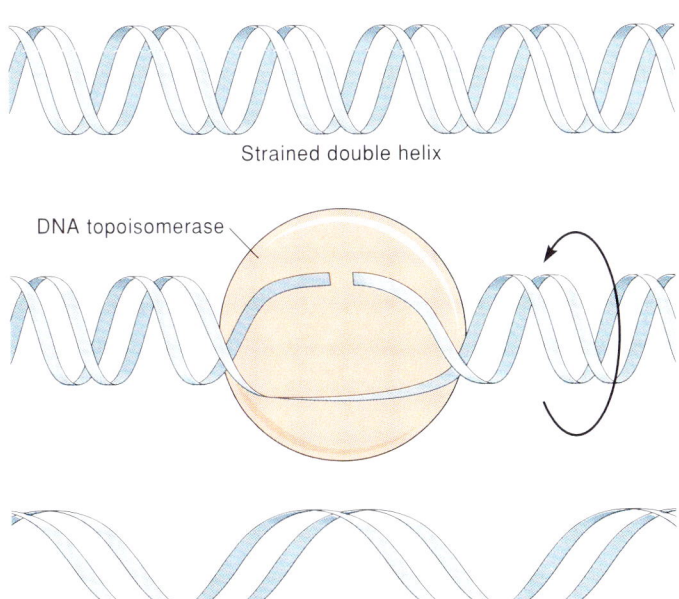

FIGURE 9.18

A strained double helix can form coils and kinks, but DNA topoisomerases can break one strand and allow the strained region to swivel. Once the strain is relieved, the topoisomerases also repair the broken strand.

CHAPTER NINE *The Cell Cycle*   201

BOXED READING 9.2
DETERMINING THE SIZE AND COMPOSITION OF GENOMES

The first measurements of genome size used *cytophotometry* to estimate the mass of DNA in a nucleus. This method entails staining the chromatin of cells in the G_1 phase and then recording its absorbance by a photometer through a light microscope. The density of the darkened nucleus is proportional to the size of the genome, which is usually expressed in picograms (pg; 1 pg = 10^{-12} gram) per nucleus. Cytophotometry is relatively simple and can be used to measure genome sizes of many plants in a short time. The genomes of several hundred species have been measured in this manner. The results reveal a tremendous variation in genome sizes among plants, ranging from about 0.52 pg in *Arabidopsis thaliana* to more than 680 pg in a species of lily (*Lilium longiflorum*). A picogram is equivalent to approximately 269 million pairs of nucleotides, so the genome of *Arabidopsis* is about 140 million base-pairs and that of the lily is more than 180 billion base-pairs (for comparison, the human genome is almost 24 pg, or about 6.5 billion base-pairs). The massive genomes of some species are due to polyploidy, wherein many copies of the same genome occur in a single nucleus. Most variation cannot be explained so easily, however. For example, the amount of DNA in diploid species of the bean genus *Vicia* varies sevenfold.

The phenomenon of different genome sizes has fostered much interest in the molecular biology, ecology, and evolution of genome size variation. For example, as discussed in this chapter, the mechanics of the

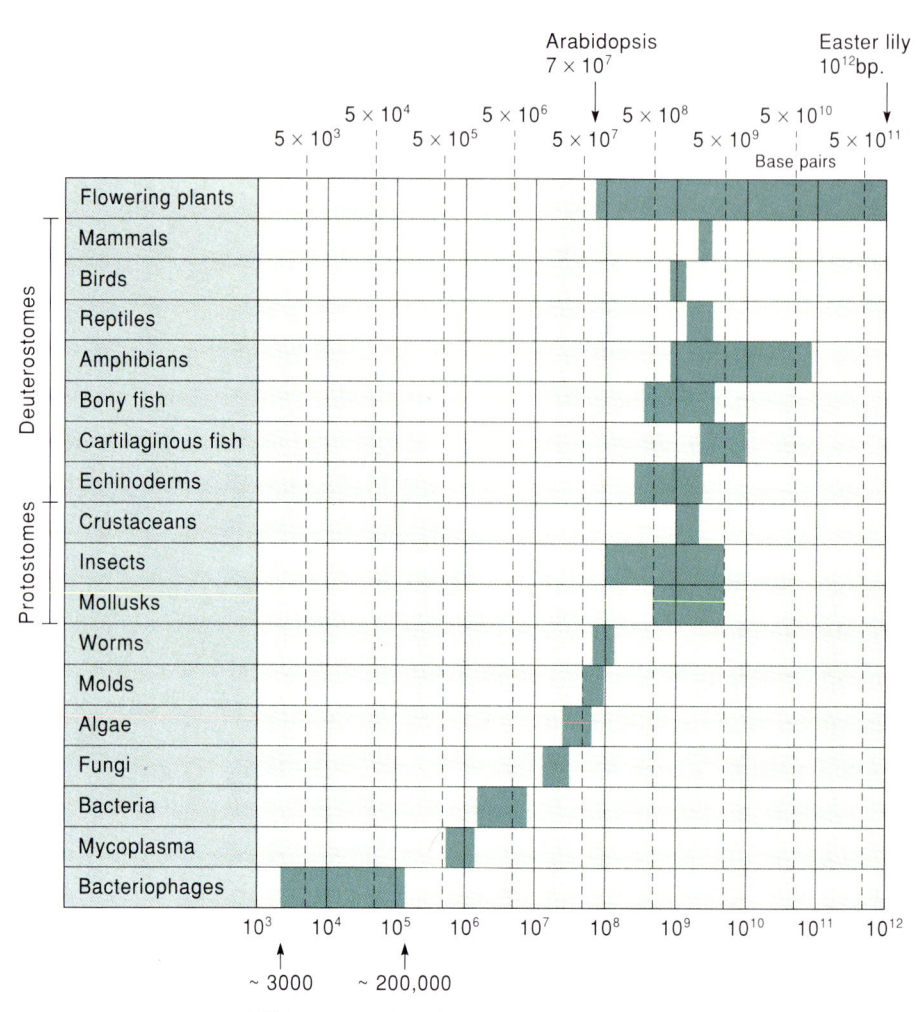

BOX FIGURE 9.1

Chart showing range of amounts of DNA among plants and other organisms.
Source: Fristrom and Clegg Principles of Genetics, *2d ed. Copyright © Chiron Press, Cambridge, MA.*

FIGURE 9.19

A replicon is the block of DNA between two replication origins.

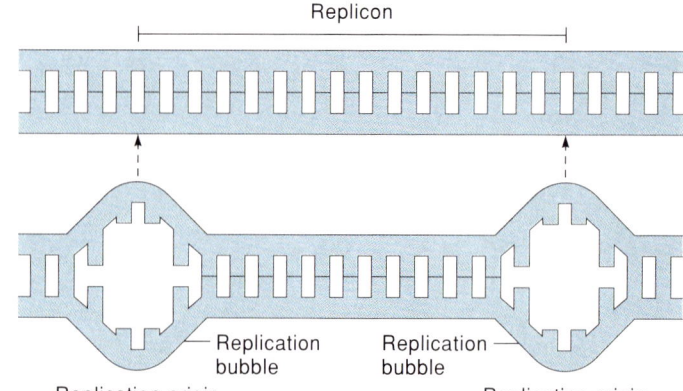

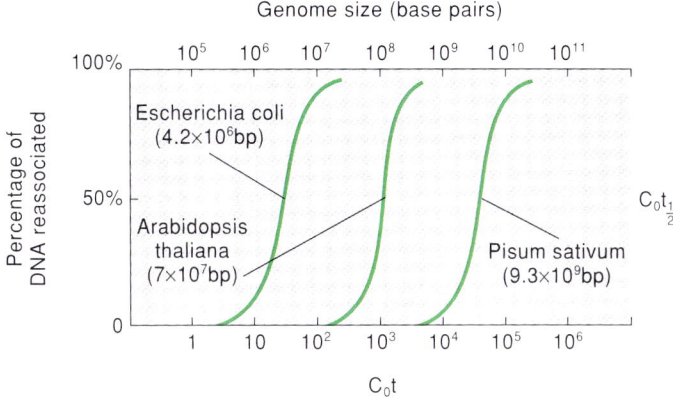

BOX FIGURE 9.2

Examples of C_0t curves. The point at which half of the DNA is reassociated ($C_0t_{1/2}$) is correlated with the amount of DNA in the genome. *Escherichia coli* is a bacterium; *Arabidopsis thaliana* and *Pisum sativum* are flowering plants (bp = base-pairs).

cell cycle must account for the amount of DNA made in the S phase. Accordingly, replicon size, number, and activity patterns have been studied intensively at the molecular level. At the organismic level, however, the reasons for excessive amounts of DNA in species with large genomes have been elusive.

One of the most informative methods for measuring the size of a genome exploits the property of DNA strands to reassociate after they are physically separated. Separation is induced by heating native DNA (i.e., double-stranded), usually to about 90° C, until hydrogen bonds between the two strands break, thus forming **denatured** (i.e., single-stranded) DNA. When cooled, the single strands reassociate into native DNA. In this renaturation process, DNA strands collide randomly until a small region on one strand bumps into a complementary region on another and forms hydrogen bonds with it. Then the two strands zip in both directions from that point until the native DNA completely reforms. The extent of strand reassociation depends upon the initial DNA concentration (C_0) and the incubation time (t) required to complete it, or $C_0 \times t$. When plotted on a logarithmic scale, the reassociation graph, called a **cot curve,** has a sigmoidal shape with an inflection point that is halfway to complete reassociation. This halfway point, referred to as the $C_0t_{1/2}$ value, is compared with a standard of known genome size, such as that of *E. coli*. The genome sizes of plants can be extrapolated from such comparisons.

Early measurements of genome size also revealed some unexpected surprises about genome composition. Much of the DNA in plants and other eukaryotes reassociates faster than expected, based on comparisons with prokaryotes and viruses. Since rates depend on concentration, some portions of a genome occur in higher concentrations than others. That is, some DNA sequences are repeated many times. The portion of DNA which reassociates faster than expected is the percentage of **repetitive DNA** in the genome. Species with larger genomes usually have higher proportions of repetitive DNA. For example, the small genome of *Arabidopsis* has almost no repetitive DNA, but the genomes of garden peas and mung beans (*Vigna radiata*) are about 75% repetitive. Repeated sequences can account for more than 90% of the genome of the lily, *Trillium erectum*. For comparison, the human genome is also comprised of at least 90% repetitive DNA.

Reassociation occurs a million times faster in some sequences than in others. This means that these DNA sequences are present in a million (10^6) copies per nucleus. Such highly repetitive DNA accounts for the massive genomes found in some plants.

The presence of huge amounts of DNA in some but not all organisms raises the question about the necessity of repetitive DNA. If many organisms have little repetitive DNA, why do other genomes maintain large amounts of it? There is as yet no convincing answer for this question.

CONCEPT

Genomes are duplicated in specific and highly organized patterns. The basic unit of replication is the replicon, which is the sequence between adjacent replication origins. Replicons are activated simultaneously in clusters. Replicon clusters operate on different chromosomes at specified times during the S phase of the cell cycle.

The G_2 Phase

The G_2 phase begins after chromatin synthesis is complete, when newly replicated chromatin gradually begins to coil and condense into a compact form. During G_2, the cell finishes preparing for mitosis. This preparation includes making tubulins for mitotic microtubules, making proteins for processing chromosomes, and breaking down the nuclear envelope.

The average length of a G_2 phase is 3–5 hours, and it continues until mitosis begins. The end of G_2 is the end of interphase.

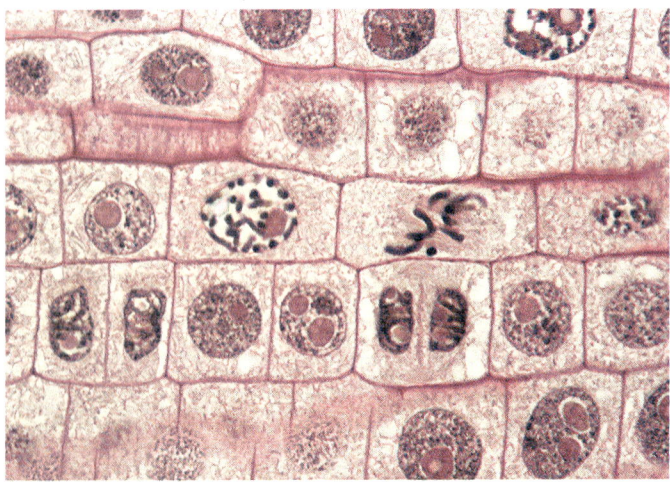

FIGURE 9.20

Mitosis can be seen easily among the nuclei of actively dividing cells, such as those of onion root tip (*Allium cepa*). At any one time, most nuclei are in interphase, ×280.

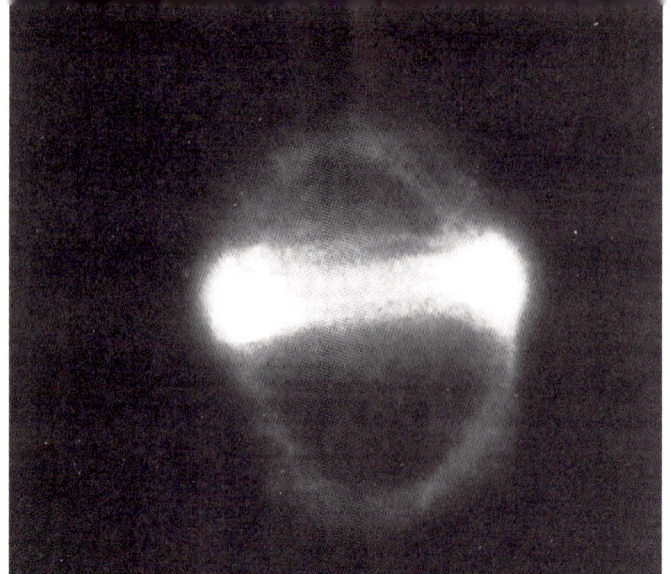

FIGURE 9.21

Photomicrograph of dividing cell stained by immunofluorescence to show microtubules in preprophase band and in cytoskeleton. The plane of the preprophase band is perpendicular to the direction that chromosomes will move in mitosis.

MITOSIS AND CYTOKINESIS

In one of the earliest descriptions of dividing nuclei, a German cytologist named Walther Flemming referred to the various stages of chromatin appearance as *mitosen* (threads), giving rise to the name **mitosis** for the process of nuclear division. Flemming's work in the 1870s inspired many studies of mitosis. By the start of the twentieth century, biologists knew that this process was the foundation for nuclear division in all plants and animals.

Mitosis refers to the separation of chromosomes and the formation of two genetically identical daughter nuclei. It is usually followed by cytokinesis, but there are many exceptions to this pattern. For example, in seed development, the nuclei of some tissues divide many times before they become separated by cell walls. Furthermore, in many species of algae cytokinesis occurs only during sexual reproduction. Vegetative stages of these algae are permanently multinucleate.

Mitosis can be studied in several types of meristematic cells, either in whole plants, in cell cultures, or in wound or tumor tissues. For example, cells undergoing mitosis can be readily observed in onion root tips, as illustrated in figure 9.20. At any one time, most nuclei are in interphase, and the remainder are in various stages of nuclear division.

Before Prophase

Before prophase, microtubules and actin filaments that are just inside the plasma membrane begin to form a narrow bundle, called the **preprophase band,** around the nucleus. The plane of the preprophase band is perpendicular to the direction of chromosome movement during mitosis (fig. 9.21). As mitosis begins, the microtubules disappear, but the actin filaments remain and provide a circular support network between the dividing nucleus and the plasma membrane. Toward the end of mitosis, this network guides the deposition of precursors for the new cell wall. This means that the radial strands of actin act as a "memory" of the preprophase band, thereby predetermining where cytokinesis will occur.

Chromosome Condensation: Prophase

Chromatin of the interphase nucleus condenses rapidly in early prophase. As condensation proceeds, chromatin appears as a mass of elongated threads. By late prophase, individual chromosomes are visible, appearing as two parallel threads attached at a constriction point called the **centromere** (fig. 9.22a). The centromere consists of a specific sequence of DNA that is required for segregation of chromosomes later in mitosis. In this double-thread form of the chromosome, each thread is called a **chromatid.** The double chromatid is still considered to be one chromosome because it has just one centromere.

The end of prophase is marked by the disappearance of nucleoli and the nuclear envelope. Without a nuclear boundary, chromosomes protrude into the cytoplasm, and their behavior appears to be chaotic and uncontrolled. As we shall see, this could not be further from the truth.

Chromosomes are at their shortest near the end of prophase and the beginning of metaphase. In the garden pea, the seven pairs of chromosomes have a combined length of about 335 µm at their shortest stage. This represents a reduction from a total DNA double-strand length of about 7 m for the 20.4 billion nucleotide pairs that are in the diploid nucleus after the S phase. This is more than a 20,000-fold shrinkage from the double-helical strands of DNA to the most highly coiled form of chromatin. Such shrinkage would be equivalent to reducing the length of a football field to about the height of this line of type.

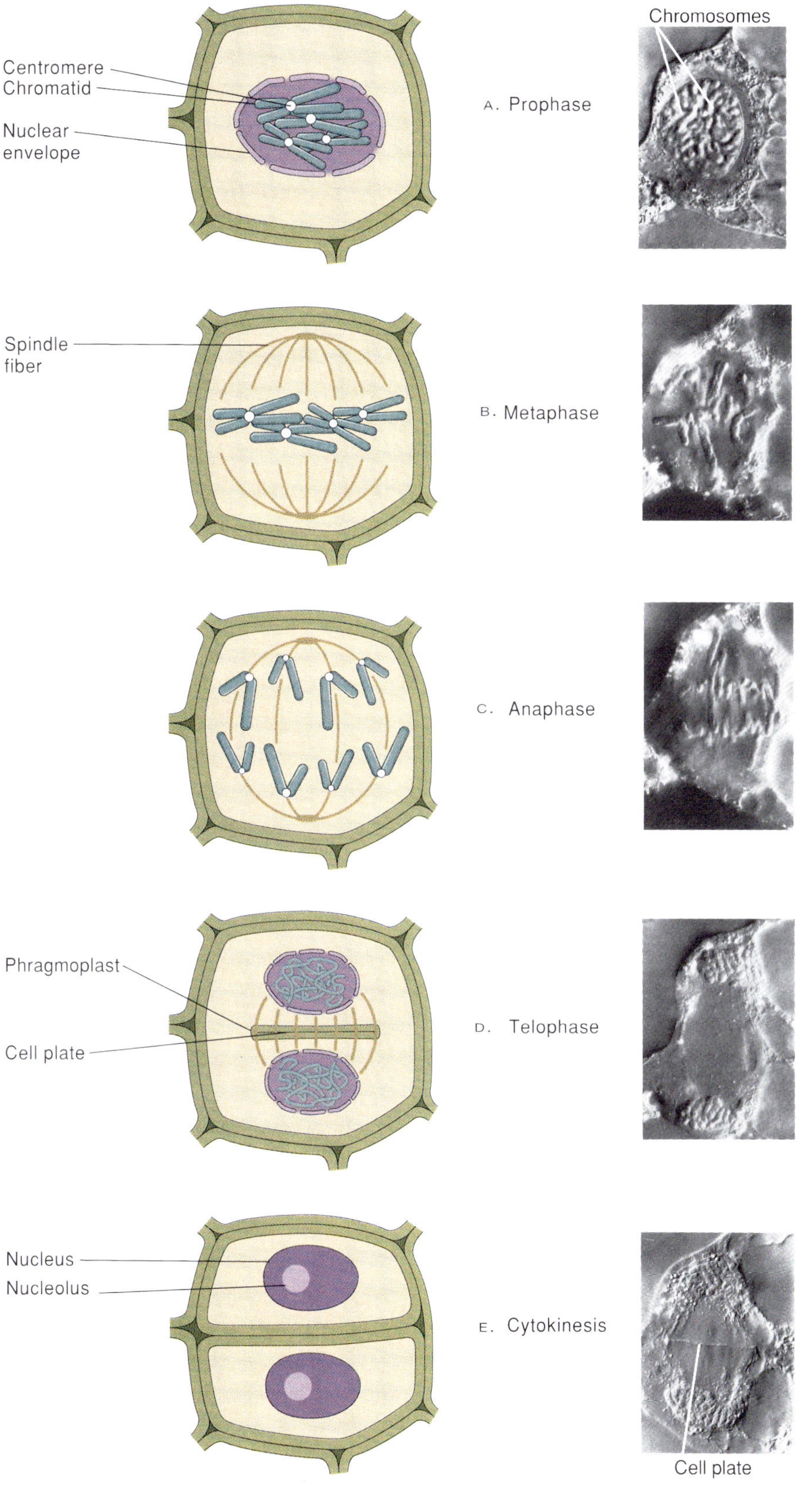

Figure 9.22

Mitosis and cytokinesis. (a) Chromosome condensation: prophase. (b) Chromosome alignment: metaphase. (c) Chromosome migration: anaphase. (d) Chromosomes decondense and new nuclei are formed: telophase. The cell plate forms at this point if telophase is to be followed by cytokinesis. (e) Cells divide: cytokinesis.

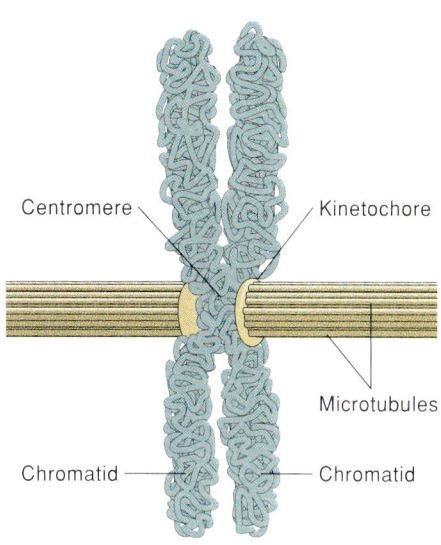

Figure 9.23

Diagram of a metaphase chromosome with spindle fibers from opposite ends of the cell attached to opposing kinetochores. A kinetochore is a complex of proteins that binds to the centromere.

Chromosome Alignment: Metaphase

After the nuclear envelope dissolves, several nearby parallel fibers, called **spindle fibers,** accumulate near the chromosomes in early metaphase. Spindle fibers, which are made of microtubules, are collectively called the **spindle apparatus** (*spindle* for short). The mature spindle apparatus is elongated, with its ends pointing to opposite poles of the cell (fig. 9.22b). Fibers from opposite poles of the spindle attach to each chromosome on either side of its centromere, to a disc-shaped structure called the **kinetochore** (fig. 9.23), a complex of proteins that binds to the centromere. Most fibers are not attached to kinetochores, because there are usually many more fibers than chromosomes. The unattached fibers are called *polar fibers*, to distinguish them from kinetochore-bound fibers.

As viewed by light microscopy, the metaphase chromosomes line up independently of one another in a circle around the circumference of the spindle apparatus. The plane of this circle, called the **metaphase plate,** is perpendicular to the axis of the spindle. The position of chromosomes along the metaphase plate is maintained by the balance of forces between two sets of kinetochore-bound fibers, one set from each pole of the spindle.

By the end of metaphase, the chromosomes have all been forced into a tight plane at a specific position in the cell. Disorder has become order. The chromosomes are now ready for the splitting of their centromeres and the separation of their chromatids.

Chromosome Separation: Anaphase

During anaphase, which is the shortest stage of mitosis, chromatids attached to the same centromere (i.e., *sister chromatids*) separate and move to opposite poles of the spindle (fig. 9.22c). Centromeres split before the chromatids separate and are pulled along the axis of the spindle. When the centromere is near the center of a chromosome, the two chromosome arms form a V as they are dragged through the cytoplasm. Upon separation, the term *chromatid* no longer applies; each structure is considered a separate (i.e., daughter) chromosome because it has its own centromere.

In most cells, the spindle elongates during anaphase, thereby increasing the distance between the poles. This movement further separates the two sets of chromosomes as they move along the spindle. By the end of anaphase, chromosomes have been separated into two genetically identical nuclei.

Formation of Daughter Nuclei: Telophase

During telophase, daughter nuclei seem to mimic prophase in reverse. The spindle apparatus disappears, a nuclear envelope forms around each of the two sets of chromosomes, and nucleoli appear in each new nucleus (fig. 9.22d). The chromosomes steadily elongate and decondense once again into diffuse chromatin.

CONCEPT

Mitosis is a continuous process that produces two genetically identical nuclei. For convenience, biologists recognize several stages of the process: chromosome condensation (prophase), alignment (metaphase), separation (anaphase), and formation into daughter nuclei (telophase).

Doing Botany Yourself

Explain how you can determine the proportion of the cell cycle that is spent in interphase versus in mitosis for meristematic cells by examining prepared microscope slides or photographs (e.g., fig. 9.20). Use this procedure to determine such a proportion in onion root tip cells and the meristem of another type of plant, if available. How do the two plants compare?

Division Into Cells: Cytokinesis

As early as late anaphase or telophase, the cell begins to prepare for cytokinesis. As part of this preparation, a cylindrical system of short microtubules starts to form between the daughter nuclei, parallel to the spindle. This system of microtubules, called a **phragmoplast,** is divided by the plane that was established by the preprophase band, which is also where the metaphase plate had been. This plane will be the division plane of the cytoplasm. As cytokinesis continues, dictyosome vesicles laden with cell-wall precursors become trapped by the phragmoplast. These vesicles fuse into a single platelike vesicle, in which the new cell walls of the daughter cells begin to form. Together, the phragmoplast, central vesicle, and new walls become the first structure of cytokinesis that is easily visible by light microscopy. This structure is called the **cell plate** (fig. 9.22d).

The cell plate grows as dictyosome vesicles are added to the outer edge of the central vesicle. The new walls in the central vesicle grow outward along their edges. The cell plate continues to grow in this manner until the central vesicle fuses with the plasma membrane of the parent cell, thereby becoming part of the plasma membranes of the daughter cells. At the same time, new cell walls in the central vesicle join the wall of the parent cell. Fusion of the cell walls is the last step in cytokinesis (fig. 9.22e). As cytokinesis nears completion, the wall that surrounds the parent cell loosens and stretches, thereby enabling subsequent growth of the daughter cells by cell expansion. Shortly after cytokinesis, the daughter cells deposit additional primary cell-wall materials on both sides of the newly formed cell wall and on all of the preexisting walls as well.

Two features of the cytokinesis described above characterize plants and a few algae. One is the formation of a cell plate that involves a phragmoplast. In most other organisms having rigid cell walls, cytokinesis occurs by constriction of the parent wall and plasma membrane. In animal cells, which lack cell walls, cells are pinched off by constriction of the parent plasma membrane. Neither of these nonplant types of cytokinesis involves a cell plate. In some algae, however, a cell plate forms from microtubules oriented perpendicular to the spindle apparatus, not parallel as in a phragmoplast. In such algae, the cell-plate forming structure is called a **phycoplast** (*phyco* refers to algae).

The second feature of cytokinesis that characterizes cell division in plants and some algae is the formation of plasmodesmata (see Chapter 3). As dictyosome vesicles accumulate in the growing cell plate, they trap parts of the endoplasmic reticulum. After cytokinesis, parts of ER that were caught between fused vesicles connect the cytoplasms of the daughter cells. Although the formation of plasmodesmata by this mechanism seems to be random, it is probably controlled. Indirect evidence for this suggestion is that the number of plasmodesmata varies among different types of cells (see Chapter 3).

CONCEPT

Cytokinesis splits the cytoplasm of a dividing cell into daughter cells. In plants and a few algae, cytokinesis involves a phragmoplast and a cell plate. The cell plate from a phragmoplast forms in a division plane that is perpendicular to the mitotic spindle. The new cell wall and plasma membranes of daughter cells are derived from the cell plate.

MYSTERIES OF THE MITOTIC SPINDLE: HOW ARE CHROMOSOMES MOVED?

Mitosis is dominated by the spindle apparatus. The mechanisms of spindle function have received much attention in studies of mitosis, but the precise way that forces are formed by the spindle is still not fully understood.

Recall from earlier in this chapter that two kinds of forces contribute to chromosome movement during anaphase (fig. 9.24). One force is the pull by kinetochore-bound microtubules on either side of each centromere, which pulls sister chromatids in opposite directions. The other force is the push by microtubules that converge at the poles of the spindle but are not attached to kinetochores. This pushing force moves the poles of the spindle apparatus in opposite directions from each other. How can some microtubules pull while others push?

The Push of Polar Microtubules

Polar spindle fibers from one pole overlap those of the other pole near the middle of the spindle apparatus. This suggests that opposing fibers somehow slide past one another, thereby pushing their respective poles in opposite directions. This *sliding-microtubules hypothesis* is supported by studies of isolated spindles of *Stephanopyxis turris,* a freshwater diatom (a type of alga). Unlike the spindles of plants, those isolated from *Stephanopyxis* remain active after being removed from the organism. In isolation, opposing spindle fibers separate until the overlap zone disappears. Once the overlap zone is gone, spindle movement stops, suggesting that movement depends upon contact between opposite microtubules. In contrast, the overlap zone is not exhausted in living cells, presumably because polar microtubules continue to be made.

The mechanism by which microtubules slide during mitosis is unknown, but one suggestion is that it resembles the movement of flagella. Spindle movement and flagellar movement are similar because they can both be blocked by inhibiting the hydrolysis of ATP. This means that ATP provides energy for microtubular sliding and for flagellar movement (see Chapter 3). Unlike spindles, however, flagella contain **dynein,** a large protein complex that converts energy from ATP into a sliding force between flagellar microtubules. Dynein has not

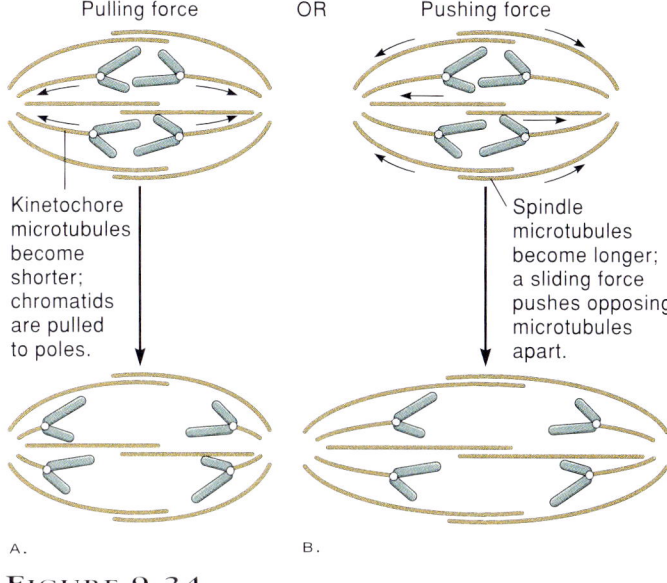

FIGURE 9.24

Forces that move chromosomes. (a) Pulling force on chromosomes is created by shortening of kinetochore-bound microtubules. (b) Pushing force against poles of the spindle apparatus comes from sliding action between opposite polar microtubules. Arrows indicate the location and direction of forces during anaphase.

yet been found in the mitotic spindle. Furthermore, spindle microtubules are oriented in directions opposite to their slide mates, while flagellar microtubules are oriented in the same direction.

The Pull of Kinetochore Microtubules

The pulling force of microtubules that are attached to a chromosome is probably caused by disassembly of microtubules. Partial evidence for this hypothesis is that the alkaloid taxol, which suppresses microtubule disassembly, inhibits chromosome movement. Further evidence is that the alkaloid colchicine, which inhibits assembly of microtubules, causes spindle fibers to shorten faster than normal; this accelerates the movement of chromosomes.

It is not known whether the disassembly of microtubules occurs only at the kinetochores or both at the kinetochores and at the poles (fig. 9.25). The model for microtubule disassembly is also incomplete, because the force generated by kinetochore-attached fibers is proportional to fiber length. This suggests that the pulling force is distributed along the entire length of the fiber, rather than only at the polar ends where disassembly occurs. Consequently, we still cannot fully explain how chromosomes are separated during mitosis.

CONCEPT

Microtubules create both pulling and pushing forces that separate chromosomes during mitosis.

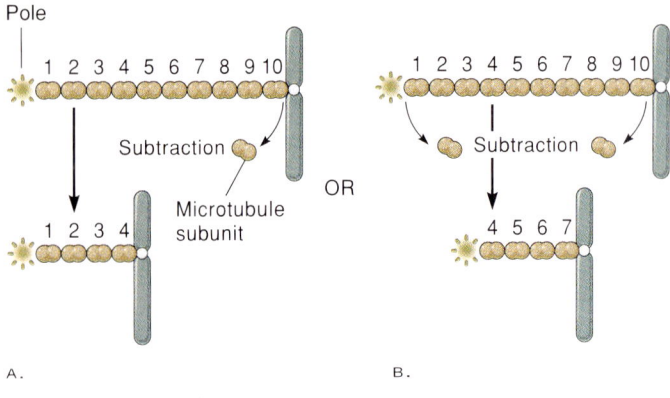

FIGURE 9.25

Possible patterns of disassembly of kinetochore-bound microtubules. (a) Disassembly at kinetochore end only. (b) Simultaneous disassembly at pole and at kinetochore.

Writing to Learn Botany

At different times during the cell cycle, microtubules seem to appear, disappear, and reappear at specific places in the cell. This pattern can be explained by the hypothesis that microtubules in all of these places are derived from the same tubulin subunits, which are disassembled in one place and reassembled where they are needed. How could you test this hypothesis?

Chapter Summary

The cell cycle is a set of repeated processes that foster cell growth and division. When cell growth and division stop in mature, nonmeristematic cells, the cell cycle also stops. Such mature cells remain in the phase that precedes DNA synthesis, unless they are induced to divide by wounding or some other external influence.

The two main stages of the cell cycle are interphase, which mostly entails cell expansion and DNA synthesis, and cell division. Interphase is further recognized to have three phases: (1) the gap phase between the end of the previous mitosis and DNA synthesis (G_1 phase), (2) the phase involving DNA synthesis (S phase), and (3) the gap phase between the end of DNA synthesis and the beginning of mitosis (G_2 phase). In comparison, cell division has several phases involving chromosome behavior, together called *mitosis*, and a phase called *cytokinesis* that divides cytoplasm into daughter cells.

Most of the interest in interphase is directed at the S phase. Our knowledge of the S phase relies heavily on understanding the structure of DNA and chromatin, which is the DNA-protein complex that makes up chromosomes. DNA synthesis occurs only after chromosomes are unknotted and their DNA helices are unwound and split into single strands. Such manipulations of chromosomes require several kinds of enzymes that work together to replicate DNA. These enzymes are referred to as *replication enzymes*.

DNA synthesis is referred to as *semiconservative replication* because each new double helix contains just one newly synthesized strand. The other strand remains from the previous double helix. Semiconservative replication uses a single strand of parent DNA as a template for guiding the synthesis of the new strand. The sequence of nucleotides in the new strand complements that of the parent strand.

The main phases of mitosis are easily observable by light microscopy. Chromosomes condense and become visible in prophase, align along a division plane in metaphase, split into daughter chromosomes and move to opposite poles of the cell in anaphase, and organize into separate daughter nuclei in telophase. Mitosis is usually followed by cytokinesis. When it occurs, cytokinesis begins as early as late anaphase by forming a microtubule-containing structure called a *phragmoplast*. The phragmoplast guides the accumulation of dictyosome vesicles into a large central vesicle that becomes the new cell walls and plasma membranes between daughter cells. Plasmodesmata are formed by pieces of the endoplasmic reticulum that are trapped between dictyosome vesicles. Cytokinesis that involves a phragmoplast, cell plate, and the formation of plasmodesmata is unique to plants and a few algae.

Chromosome movement in mitosis is caused by the behavior of microtubules in the spindle apparatus. Microtubules generate two kinds of forces: one pulls chromosomes to opposite poles of the spindle, and one pushes the poles of the spindle apart.

Questions for Further Thought and Study

1. Before 1950, why did scientists believe that chromosomes doubled immediately before prophase? How was chromatin synthesis discovered to occur several hours before the start of mitosis?

2. What function could centromeres have in nuclear division?

3. Mitosis is said to be "observable" with a light microscope. However, meristematic cells are killed before they are mounted onto a glass microscope slide for study. How, then, do we observe this living process by examining dead cells?

4. DNA polymerases catalyze the assembly of DNA chains in one direction only. Can you suggest reasons for this seemingly severe limitation?

5. Organellar and nuclear genomes must be passed on to daughter cells during cell reproduction. During which period(s) of the cell cycle are plastids and mitochondria most likely to be reproduced? Why?

6. Measurements of genome sizes have been obtained by measuring rates of reassociation of denatured DNA. How can reassociation rates be used to determine genome size?

7. By examining *cot* curves, we can learn about the composition of chromatin. The earliest comparisons of *cot* curves between prokaryotes and eukaryotes yielded some surprising results. What were these results, and why were they unexpected?

8. Onion and wall cress have a 63-fold difference in their respective genome sizes, but less than a 4-fold difference in S-phase duration in their root tip cells. How can we account for the comparatively much faster duplication of the onion genome?

9. What properties of histones do you think are especially important for the function of these proteins in maintaining chromatin structure? Why?

10. How could heterochromatin function in gene regulation?

11. Suggest how a topoisomerase could sometimes unkink chromatin fibers and at other times tie them in knots.

12. The genomes of organisms contain gigantic amounts of information. For example, each human parent contributes to an embryo a set of DNA consisting of almost 3.3 billion nucleotides arranged in 23 chromosomes. If such a DNA sequence were typed in a 10-point font on a continuous ribbon, the ribbon would stretch from San Francisco to Chicago, then to Baltimore, Houston, Los Angeles, and finally back to San Francisco. Even individual genes consist of huge amounts of information (recall that one looped domain consists of 20,000 to 100,000 nucleotide pairs). How would you organize that much information so that you could compare it with the findings of other biologists?

Suggested Readings

ARTICLES

Allen, R. D. 1987. The microtubule as an intracellular engine. *Scientific American* 256 (February):42–49.

Andrews, B. J., and S. W. Mason. 1993. Gene expression and the cell cycle: A family affair. *Science* 261:1543–1544.

Blow, J. J., and R. A. Laskey. 1988. A role for the nuclear envelope in controlling DNA replication within the cell cycle. *Nature* 332:546–548.

Chang, F., and P. Nurse. 1993. Finishing the cell cycle: Control of mitosis and cytokinesis in fission yeast. *Trends in Genetics* 9:333–335.

Grime, J. P., and M. A. Mowforth. 1981. Variation in genome size—An ecological interpretation. *Nature* 299:151–153.

Hartwell, L. H., and T. A. Weinert. 1989. Checkpoints: Controls that ensure the order of cell cycle events. *Science* 246:629–634.

Hollenbeck, P. J. 1985. Mitotic spindles in isolation. *Nature* 316:393–394.

Koshland, D. E., T. J. Mitchison, and M. W. Kirschner. 1988. Polewards chromosome movement driven by microtubule depolymerization *in vitro*. *Nature* 331:499–504.

Koshland, D. E., Jr. 1989. The cell cycle [editorial]. *Science* 261:545.

Murray, A. W. 1989. Dominoes and clocks: The union of two views of the cell cycle. *Science* 246:614–621.

Murray, A. W. 1993. Cell-cycle control: Turning on mitosis. *Current Biology* 3:291–293.

Murray, A., and M. Kirschner. 1991. What controls the cell cycle? *Scientific American* 264 (March):56–63.

North, G. 1985. Eukaryotic topoisomerases come into the limelight. *Nature* 316:394–395.

Pickett-Heaps, J. D., D. H. Tippit, and K. R. Porter. 1982. Rethinking mitosis. *Cell* 29:729–744.

Sloboda, R. D. 1980. The role of microtubules in cell structure and cell division. *American Scientist* 68:290–298.

Van't Hof, J., and C. A. Bjerknes. 1979. Chromosomal DNA replication in higher plants. *BioScience* 29:18–22.

Ward, G. 1988. How-to-do-it: A handy model for mitosis. *American Biology Teacher* 50:170–172.

BOOKS

Alberts, B., D. Bray, J. Lewis, M. Raff, K. Roberts, and J. D. Watson. 1989. *Molecular Biology of the Cell*. 2d ed. New York: Garland.

Anderson, J. W., and J. Beardall. 1991. *Molecular Activities of Plant Cells*. Oxford: Blackwell Scientific Publications.

Becker, W. M., and D. W. Deamer. 1991. *The World of the Cell*. 2d ed. Menlo Park CA: Benjamin/Cummings.

Bernstein, J. 1978. *Experiencing Science*. New York: Basic Books, Inc. (See Chapter 4, "A Sorrow and a Pity: Rosalind Franklin and *The Double Helix*")

Bryant, J. A., and D. Francis, eds. 1985. *The Cell Division Cycle in Plants*. New York: Cambridge University Press.

Lehninger, A. L., D. L. Nelson, and M. M. Cox. 1992. *Principles of Biochemistry*. 2d ed. New York: Worth.

Portugal, F. H., and J. S. Cohen. 1977. *A Century of DNA*. Cambridge, MA: MIT Press.

Rawn, J. D. 1990. *Biochemistry*. Burlington, NC: Neil Patterson Publishers.

Pollen grains contain cells that are haploid. These cells descended from spores that developed by meiosis in diploid parent cells.

Meiosis, Chromosomes, and the Mechanism of Heredity

Chapter Outline

INTRODUCTION
SEXUAL REPRODUCTION IN PLANTS
   Spores and Gametes in Flowering Plants
   Pollination and Fertilization
MEIOSIS
   Prophase I
   The Remainder of the First Meiotic Division
   The Second Meiotic Division
SYNAPTONEMAL COMPLEX

BOX 10.1
POLYPLOIDY IN PLANTS

   Recombination Nodules
   Chiasmata and Chromosome Segregation
GENETIC RECOMBINATION
   Recombination by Chromosomal Separation
   Evidence for Chromosomal Exchange During Crossing-Over
   Gene Conversion
   Unequal Crossing-Over and Gene Duplication
WHY SEX?

BOX 10.2
MULTIGENE FAMILIES

   DNA Repair Hypothesis
   Transposon Hypothesis
Chapter Summary
Questions for Further Thought and Study
Suggested Readings

Chapter Overview

All types of organisms, except bacteria, undergo sexual reproduction that involves a regular alternation between meiosis and fertilization. Meiosis produces haploid nuclei from diploid nuclei, and fertilization combines two haploid nuclei into a new diploid nucleus. Offspring that arise from new diploid nuclei usually differ genetically from each other and from their parents. We assume that sexual reproduction and genetic variation are advantageous, since most plants and animals have developed it. However, sex is not necessary for reproduction; many organisms do without it, either periodically or permanently. For example, plants may propagate vegetatively by sprouting from roots or stems. Aspens, dandelions, strawberries, and cholla cacti are all examples of species that reproduce asexually. Unlike organisms that reproduce sexually, those that reproduce asexually rely solely on mitosis for producing offspring. Asexually produced offspring are genetically identical clones of the parent organism.

Sexual reproduction is fundamentally different in plants and animals because plants have multicellular haploid and diploid phases of their life cycle. In contrast, gametes are the only haploid cells in an animal's life cycle. Nevertheless, all sexual organisms rely on meiosis as a major source of genetic variation.

This chapter focuses on the basic features of the meiotic cell cycle and how genetic variation arises from it. It expands on the classical genetics presented in Chapter 8 and shows how the cell cycle of sex cells differs from the asexual cell cycle discussed in Chapter 9. This chapter will also help you understand how genes work, which is the main subject of Chapter 11.

Introduction

The discovery in the 1880s that gametes of a particular worm are haploid led to the idea that some cells are produced by a special type of nuclear division, which was named **meiosis** (from *meion*, meaning "less"). It was not until about 1900 that meiosis was described in plants. At first, meiosis was thought to be a simple process whereby half the chromosomes in a diploid nucleus migrated to one side of the cell and half to the other, followed by cytokinesis to produce two haploid cells. The behavior of chromosomes during meiosis turned out to be much more complex than expected, even more so than during mitosis.

The major features of meiosis were described by the 1930s. By that time the mechanisms of meiosis had become a central theme in cytogenetic studies of plants. The role of meiosis in generating genetic variation was established as a result of such work.

In this chapter we first describe the role of meiosis for sexual reproduction in plants and then provide an overview of chromosome behavior during meiosis. These first sections give you a foundation for understanding later discussions of how meiosis produces genetic variation. By the end of this chapter you will be prepared to read Chapter 11, which discusses how genes work and how we can manipulate them by genetic engineering.

SEXUAL REPRODUCTION IN PLANTS

Meiosis is basically the same in all organisms, but the products of meiosis vary. In many organisms, including all plants, the haploid cells produced by meiosis are **spores,** which grow into multicellular structures that make gametes. This is unlike meiosis in animals, which produces gametes directly. Thus, in plants but not in animals, the life cycle includes a spore-producing generation, called the **sporophyte** (spore-plant), and a gamete-producing generation, called the **gametophyte** (gamete-plant).

Diploid plants that produce stems, roots, and leaves also produce haploid spores (fig. 10.1a). Hence, oak and pine trees, Bermuda grass, and lilies are examples of sporophytes. Conversely, pollen grains and unfertilized seeds contain haploid cells that make up the gametophytes (fig. 10.1b,c).

The following paragraphs present a brief overview of sexual reproduction in flowering plants, emphasizing the features of reproduction that they all share. The diversity of reproductive structures in this plant group is described in detail in Chapter 17.

Spores and Gametes in Flowering Plants

Meiosis produces two kinds of spores in flowering plants. Spores produced in the anthers of flowers are called **microspores** (fig. 10.2). Microspores are produced from specialized cells called **microspore mother cells.** Each microspore mother cell is diploid and divides meiotically to form four haploid microspores. Following meiosis, each microspore divides by mitosis and cytokinesis inside the spore wall and produces an immature male gametophyte. At this stage, the combined structure that includes a microspore wall and a male gametophyte is called a **pollen grain.** The male gametophyte matures when one of the cells later divides into two sperm cells, which are the male gametes. This means that the sexually mature, full-fledged male plant consists of three cells.

Spores that are produced in immature seeds (**ovules**) are called **megaspores.** In each ovule only one of the diploid cells, the **megaspore mother cell,** undergoes meiosis to form four haploid megaspores. Three of the four megaspores then disintegrate, leaving the ovule with only one megaspore, the **functional megaspore.** The nucleus of the functional megaspore usually divides by mitosis three times, which produces eight free nuclei. Seven of the eight nuclei then become cellular, one of which is the female gamete (egg). The seven-celled, eight-nucleate structure is the female gametophyte, also called the **embryo sac** (fig. 10.3). Hence, the sexually mature, full-fledged female plant consists of seven cells. The two nuclei that remain free are called the **polar nuclei,** because they come from opposite ends of the embryo sac.

Pollination and Fertilization

Sexual reproduction in flowering plants depends on **pollination,** which is the transfer of pollen grains from an anther to the receptive region of a **carpel.** Carpels are the ovule-bearing structures of flowers. Each carpel has three main parts: (1) the **stigma,** which is the receptive surface for pollen grains; (2) the **style,** which raises the stigma to a receptive position; and, (3) the **ovary,** where ovules form. In nature, pollen is usually transferred from anther to stigma either by animals or by wind.

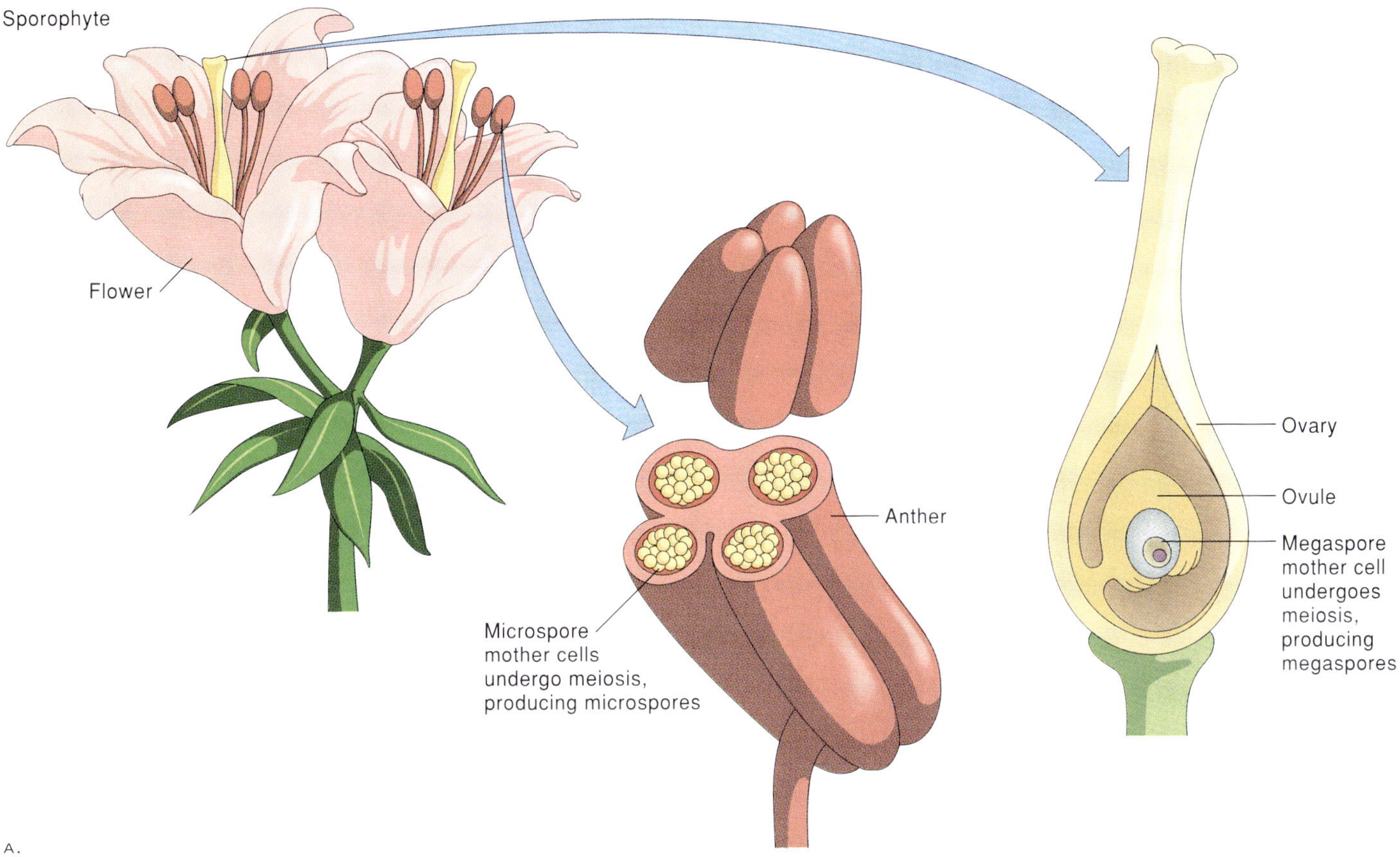

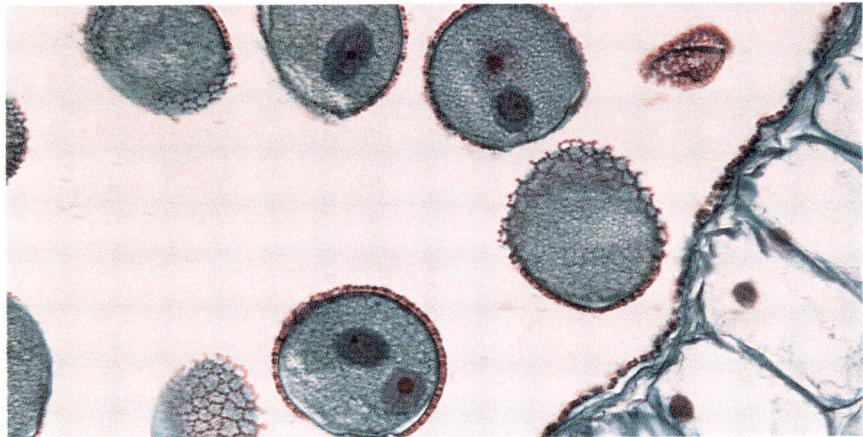

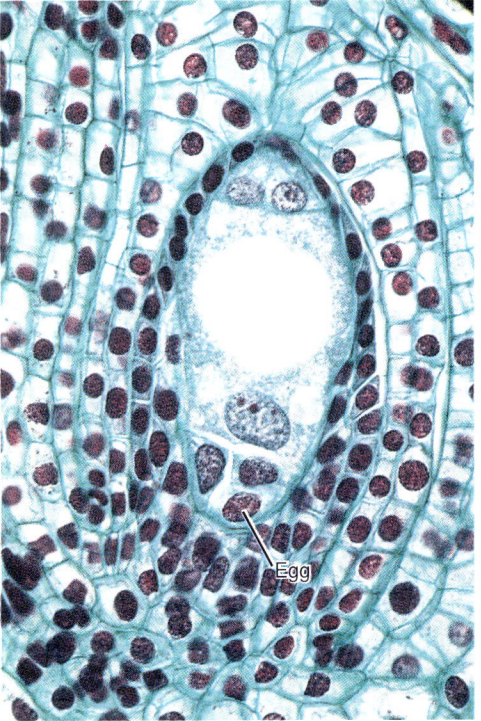

FIGURE 10.1

Reproductive parts of a flowering plant. (a) The sporophyte. Cells of the sporophyte are diploid; those that will undergo meiosis to produce haploid spores are in yellow. (b) Pollen grain of *Lilium*. Microspore mother cells produce microspores, which grow by one mitotic division into pollen grains. Each grain of lily pollen contains a two-celled male gametophyte, one cell (i.e., the generative cell) of which will divide by mitosis into two sperm cells. The other cell (i.e., the tube cell) will produce the pollen tube that delivers the sperm to the female gametophyte (see figs. 10.2, 10.3), ×100. (c) The megaspore mother cell of lily undergoes meiosis, followed by mitosis, forming a seven-celled female gametophyte, or embryo sac (not all cells are visible). One of the cells is the egg, ×100.

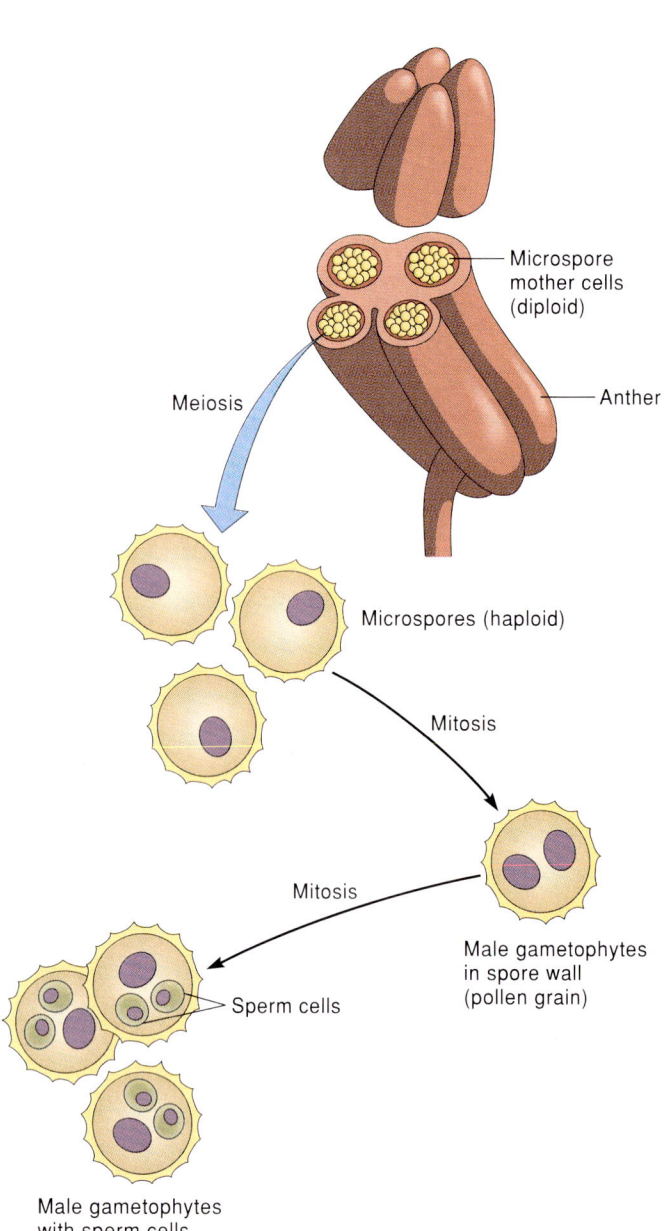

FIGURE 10.2

Development of the male gametophyte. Microspore mother cells in the anther undergo meiosis. Microspores grow into two-celled male gametophytes, each producing two sperm cells (see fig. 10.1b).

After a pollen grain reaches a stigma, it germinates, forming a **pollen tube.** The cell that forms the pollen tube is the **tube cell.** The second cell in the pollen grain is called the **generative cell** because it divides and generates two sperm cells. This cell divides as the pollen tube grows, and the two sperm cells move through the tube to a small opening in an ovule (fig. 10.4). The two sperm cells enter the embryo sac through one of the cells next to the egg. The first sperm then fertilizes the egg, and the second sperm moves to the polar nuclei and fuses with them. This process is called **double fertilization** because both sperm cells undergo fusion. The diploid cell from the fusion of

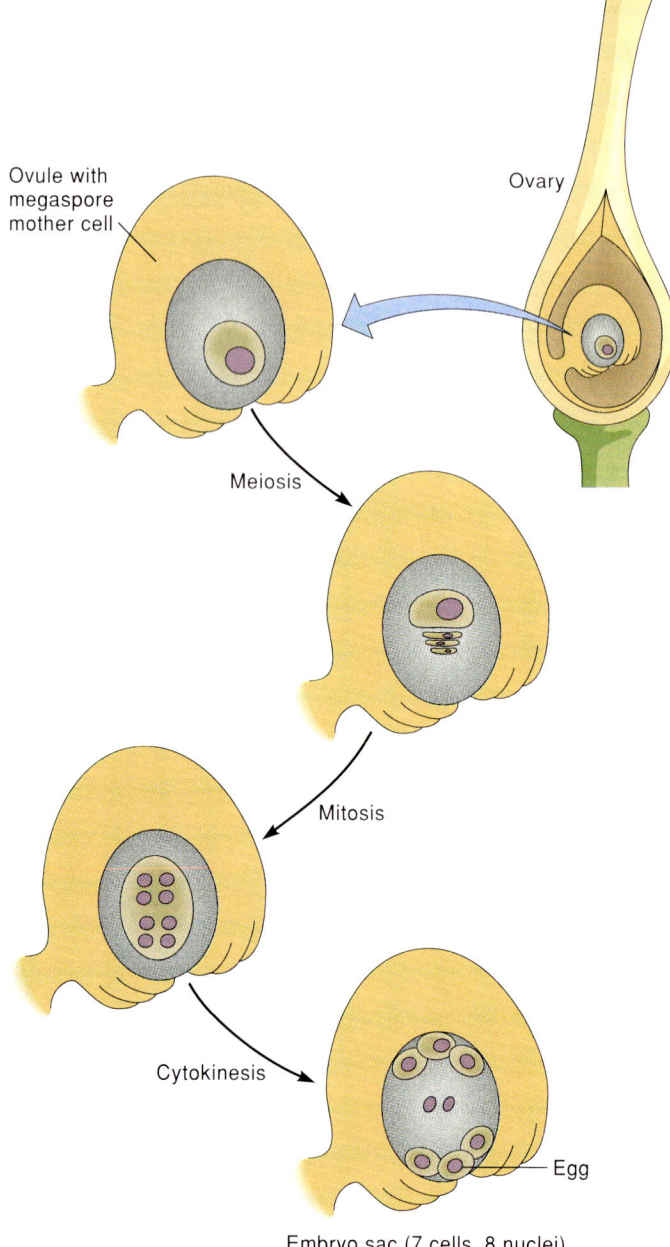

FIGURE 10.3

Development of the female gametophyte. A single megaspore mother cell in each ovule undergoes meiosis. In most plants, three of the megaspores disintegrate, leaving one functional megaspore. Three mitotic divisions in the functional megaspore produce the embryo sac, which consists of eight nuclei, which are walled off into seven cells. One of the cells is the egg (see fig. 10.1c).

sperm and egg is the **zygote,** which will become the embryo (fig. 10.5). The zygote is therefore the first cell of the new sporophyte generation. The triploid cell formed by the fusion of sperm and polar nuclei will usually form the **endosperm,** a nutritive tissue for the developing embryo. After fertilization, the ovule matures into a seed.

Some of the genetic variation among offspring of the new sporophyte generation comes from mixing the genomes of different parents. For example, carpels from several flowers of the

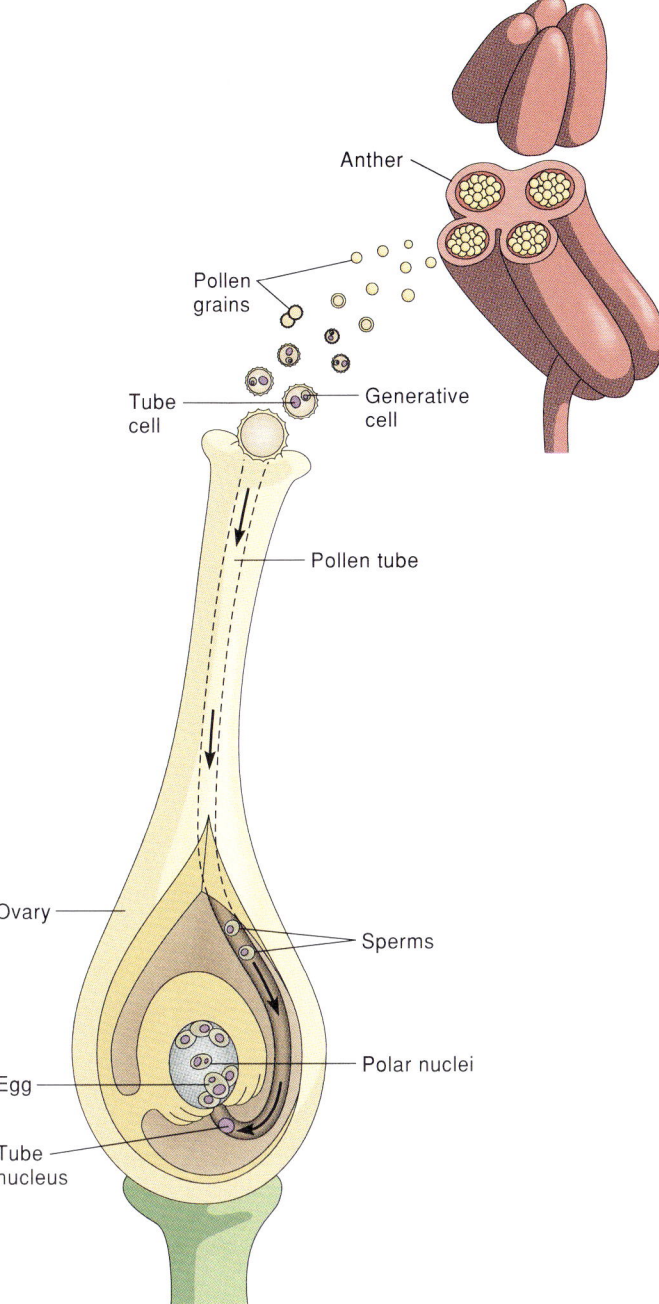

Figure 10.4

Pollination and fertilization. Pollen is transferred from the anther to the stigma of the carpel. Two sperms move from the pollen grain through the pollen tube to an opening in the ovule. One sperm fertilizes the egg, and the other fertilizes the polar nuclei.

same plant may receive nonidentical pollen from many other plants. Variation occurs because pollen from different plants contains different genomes. Thus, ten sources of pollen will produce embryos with at least ten different genotypes. However, ten pollen-producing plants will produce many more than ten different pollen genotypes. Such additional genetic variation develops as a result of meiosis. The remainder of this chapter describes how meiosis works and how it gives rise to genetic variation.

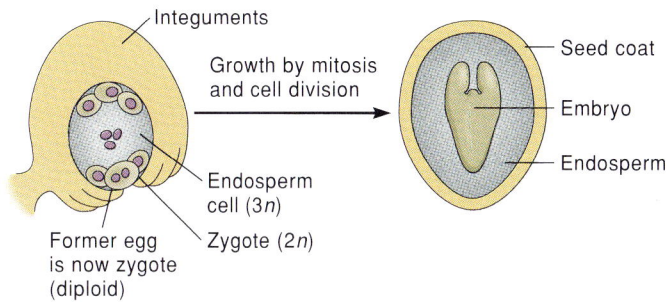

Figure 10.5

The zygote forms the embryo, which is the beginning of new sporophyte generation. The product of the second fertilization develops into a triploid (i.e., 3n) endosperm. Integuments harden and darken into a seed coat.

Sexual reproduction in plants involves regular alternation between diploid and haploid generations. The diploid generation gives rise to the haploid generation by meiosis. The haploid generation restores the diploid generation by fertilization.

Meiosis

Chromosomes are duplicated during the interphase portion of the meiotic cell cycle. After duplication, chromosomes enter the first of two successive nuclear divisions: **meiosis I** and **meiosis II**. The phases of meiosis are distinguished by a *I* for the first division and a *II* for the second. Accordingly, meiosis I consists of **prophase I, metaphase I, anaphase I,** and **telophase I**; in meiosis II nuclei go through **prophase II, metaphase II, anaphase II,** and **telophase II**. The names of the phases in meiosis are the same as those of mitosis because chromosomes behave similarly in both types of nuclear division. Nevertheless, starting with prophase I, meiosis exhibits several major differences from mitosis. The most obvious distinction is that **homologous** chromosomes form pairs in meiosis but not in mitosis. Chromosomes are homologous to one another when they have the same genes (but not necessarily the same alleles). Starting with chromosome pairing in prophase I, meiosis is a much more complicated and time-consuming process than mitosis.

Prophase I

Prophase I is the longest and most complex stage of meiosis; it typically lasts about 90% or more of the total time required for meiosis. Prophase I involves several stages, which include chromosome condensation, pairing **(synapsis),** and unpairing **(desynapsis).** The major events of prophase I are summarized in figure 10.6. Pairing between homologous chromosomes is achieved by a series of proteins, collectively called the **synaptonemal complex.** Details about how the synaptonemal complex works and its importance in fostering genetic variation appear later in this chapter.

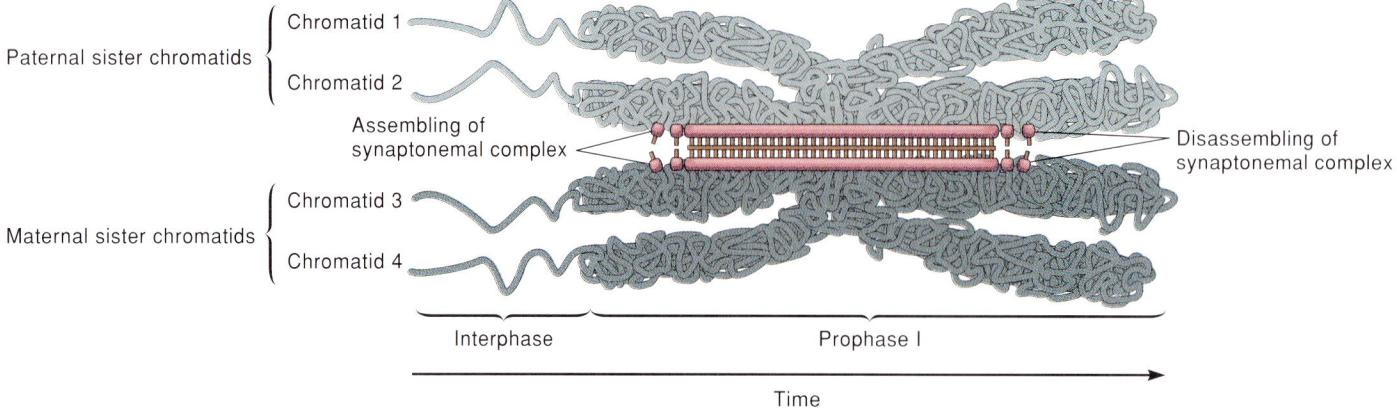

Figure 10.6

Prophase I. Homologous chromosomes form pairs that attach to a synaptonemal complex. The synaptonemal complex disintegrates during desynapsis in late prophase I.

As prophase I begins, the chromosomes condense into long threads. In contrast to mitotic prophase, chromosomes in prophase I of meiosis appear to consist of single chromatids (fig. 10.7). Evidence from cytophotometry, however (see Chapter 9), shows that they have a doubled amount of DNA. This means that each chromosome is actually a pair of sister chromatids, even though it looks like one chromatid. Paired chromosomes in prophase I are called **bivalents** (fig. 10.8a). Because each member of a bivalent has doubled during interphase, a bivalent consists of four chromatids and two centromeres (fig. 10.8b).

The Lore of Plants

One of the oddest modifications of meiosis in plants occurs in a group of North American evening primroses (*Oenothera* species). They are all diploid, with a diploid chromosome number of 14. Instead of forming seven bivalents in synapsis, however, they form a ring of up to fourteen chromosomes, seemingly arranged end-to-end. Botanists discovered this oddity in the early nineteenth century, and they are still trying to figure out the hows and whys of sexual reproduction in these plants.

Soon after chromosomes pair, nonsister chromatids may twist around each other and exchange genetic material by **crossing-over.** Crossing-over occurs when two intertwined, nonsister chromatids break, and each broken end re-fuses with its nonsister chromatid instead of with its original chromatid (fig. 10.9). Thus, fragments of chromosomes switch places between homologs. Synapsis and crossing-over can therefore change the genetic makeup of two chromatids when the alleles on one fragment differ from those on the other. Such chromosomal rearrangements during crossing-over are a type of **genetic recombination,** which is the general term for producing offspring that are genetically different from the parents. Several kinds of genetic recombination are derived from meiosis in addition to

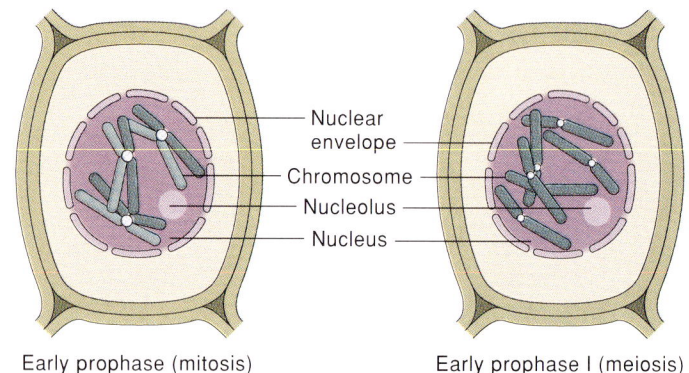

Figure 10.7

Comparison of early prophase in mitosis and meiosis. Sister chromatids are distinguishable from each other in early mitosis but not in early meiosis.

chromosomal rearrangements (see "Genetic Recombination" later in this chapter).

As prophase I continues, the synaptonemal complex dissolves and the nuclear envelope disintegrates. At this time the arms of homologous chromosomes seem to repel each other, but they are still attached at their centromeres at attachment points called **chiasmata** (singular, **chiasma**). Chiasmata occur where two of the four chromatids have crossed over. Near the end of prophase I, chiasmata move to the ends of the chromosomes; the chiasmata eventually all disappear, but homologs are still held together at their centromeres.

CONCEPT

Prophase I begins when already replicated, homologous chromosomes synapse along their entire lengths. While fused, chromosomes exchange genetic material by crossing-over. By the end of prophase I, homologs are paired only at their centromeres.

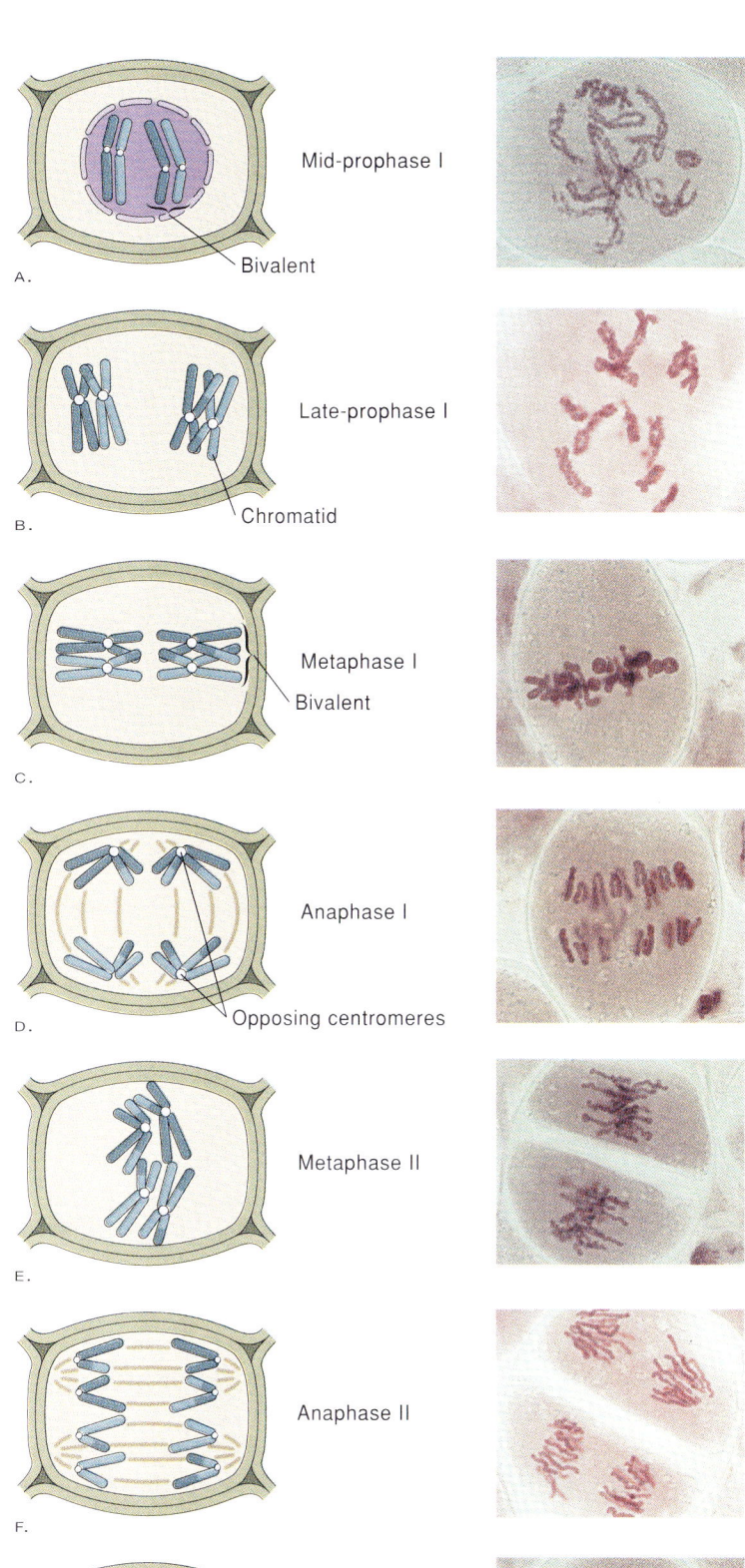

The Remainder of the First Meiotic Division

Metaphase I is characterized by the formation of a meiotic spindle apparatus, which moves the bivalents to a metaphase plate at the center of the cell (fig. 10.8c). Microtubules attach to the kinetochore on only one side of each centromere in a bivalent. Because of this pattern of spindle attachment, opposing centromeres are pulled apart during anaphase I (fig. 10.8d). This means that homologous chromosomes separate into different nuclei. The separation of centromeres in meiosis I differs from the splitting of centromeres in mitosis. By splitting mitotic centromeres, mitosis maintains diploidy in the daughter nuclei. However, because intact centromeres separate in meiosis I, the daughter nuclei are haploid. Thus, meiosis I is called the **reduction division** of meiosis.

Chromosomes reach their destinations and begin to decondense in telophase I. Nuclei do not enter another interphase, however, since each chromosome still contains a pair of chromatids; instead, they enter meiosis II. In some organisms, continued development at this stage is not direct, because their chromatin uncoils fully in telophase I and a new nuclear envelope appears around the daughter nuclei. In meiosis II, therefore, chromosomes must recondense in these organisms, and the newly made nuclear envelope must disintegrate. In other organisms meiosis is more direct, (because most or all of telophase I is bypassed), and a nuclear envelope does not form after meiosis I. When telophase I is skipped, the already condensed chromatin of meiosis I also skips prophase II and enters metaphase II directly. The shortcut from anaphase I to metaphase II is not accompanied by the formation and disintegration of a nuclear envelope.

FIGURE 10.8

Stages in meiosis. (a) Mid-prophase I. Sister chromatids are distinguishable. (b) Late-prophase I. Nonsister chromatids cross over between homologous chromosomes, making it possible to exchange genetic material between homologs. (c) Metaphase I. Chromosomes align along the metaphase plate. (d) Anaphase I. Homologs separate and move to opposite ends of the spindle. Chromosome fragments that were exchanged during prophase move with recombined chromosomes. (e) Metaphase II. Chromosomes align along metaphase plates in a plane that is perpendicular to the direction of the plate in metaphase I. (f) Anaphase II. Centromeres divide and former sister chromatids move to opposite poles of the spindle. (g) Late telophase II. Chromosomes decondense and nuclear envelopes form around each haploid daughter nucleus. At this stage the products of meiosis are called spores, which form spore walls and are later released from the parent cell. Light micrographs are of meiosis in the anthers of Easter lily (*Lilium longiflorum*).

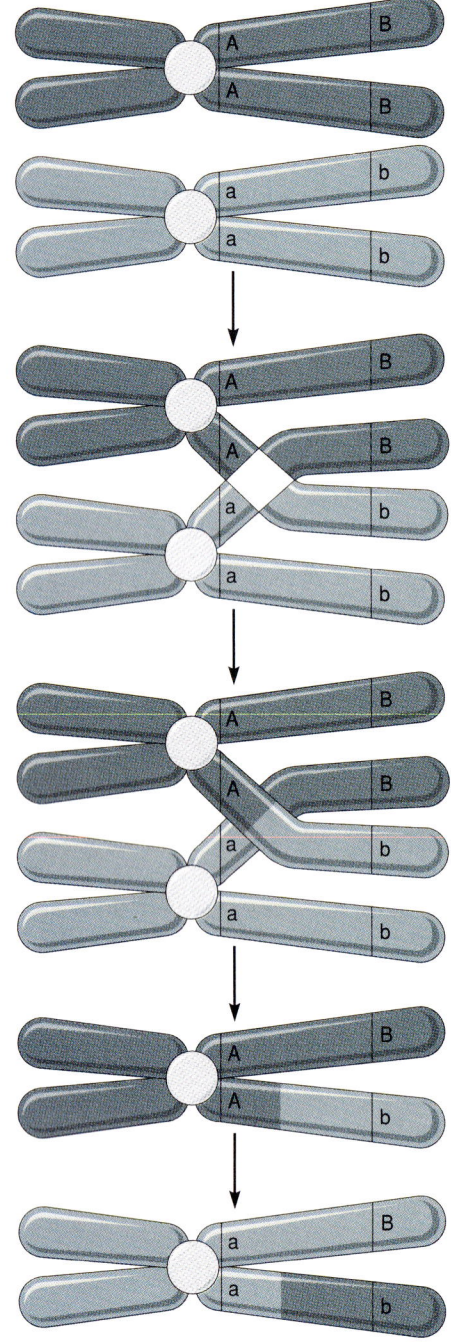

FIGURE 10.9

Homologous chromosomes align during prophase I, enabling nonsister chromatids to cross over. After breakage and refusion, fragments of chromosomes switch places. In this way, nonsister chromatids exchange genetic material, resulting in genetic recombination.

The Second Meiotic Division

The second meiotic division is the same as in mitosis. Chromosomes either condense in prophase II or, if telophase was bypassed, they go straight to metaphase II (fig. 10.8e), where the spindle apparatus forms and aligns chromosomes in a plane. Once they are aligned, chromosomes enter anaphase II, when centromeres divide and sister chromatids become individual chromosomes. The new centromeres are pulled apart during anaphase II, thereby separating duplicate chromosomes from each other (fig. 10.8f). In telophase II, chromosomes decondense and a nuclear envelope forms around each of the four daughter nuclei (fig. 10.8g).

By the end of meiosis II, two divisions have been completed. In sum, the diploid parent nucleus that entered meiosis has divided twice to form four haploid nuclei. After the second meiotic division in plants, cytokinesis divides the cytoplasm into four spores. Cytokinesis may occur after meiosis I and after meiosis II, or only after meiosis II.

Meiosis is normally an orderly process that produces four haploid nuclei from one diploid nucleus. However, mistakes in meiosis have apparently occurred many times during the evolution of plants. The evidence for such meiotic errors is that many plants are **polyploid,** which means that they have more than two sets of chromosomes. Polyploidy can be induced by treating cells in prophase I with colchicine, which disrupts the assembly of microtubules in the spindle apparatus; the products of meiosis are therefore diploid instead of haploid. This artificial treatment shows how meiosis may have failed in the past. As explained in box 10.1, "Polyploidy in Plants," meiotic mistakes have played an important role in plant evolution and, more recently, in the origin of cultivated plants.

C O N C E P T

After prophase I, homologous chromosomes align on a metaphase plate and then segregate into two haploid nuclei for the remainder of meiosis. If necessary after meiosis I, chromosomes condense once again as they enter meiosis II. They then align on a metaphase plate, their centromeres split, and the former sister chromatids move to opposite poles. The net result of the two meiotic divisions is four haploid nuclei from one diploid nucleus.

Writing to Learn Botany

Instead of undergoing meiosis, diploid nuclei could simply segregate their chromosomes into equal parts. This occurs in some fungi. How would the haploid nuclei of such a process differ from haploid nuclei produced by meiosis? After fertilization occurred between gametes from these "haploidized" cells, how would the new diploid generation differ from offspring produced by normal sexual reproduction?

SYNAPTONEMAL COMPLEX

The synaptonemal complex, which forms during synapsis, begins as a protein axis along each chromosome. The synaptonemal complex links homologous parts of chromosomes, allele for allele (fig. 10.10). Such alignment can be best seen with a light microscope when homologs differ in allele sequence. For example,

> ### BOXED READING 10.1
> ### POLYPLOIDY IN PLANTS
>
> About 35% of the species of flowering plants are polyploid. The most common level of polyploidy is tetraploidy, which means that plants have four sets of chromosomes instead of two. This also means that the gametes are diploid instead of haploid. Most of the easily observed effects of polyploidy are associated with increased size, and they are usually observed in plants that have been induced to become polyploid by treatment with colchicine. In nature, however, polyploid plants are often indistinguishable from diploid plants, at least upon initial inspection. Instead, for example, polyploids may differ less obviously by being better adapted to temperature or water stress than their diploid relatives. Thus, polyploids can survive in habitats that are not as suitable for their diploid counterparts.
>
> Many cultivated plants are polyploids. One of these is bread wheat (*Triticum aestivum*), the most commonly cultivated crop in the world. Bread wheat probably arose in the Middle East at least 8,000 years ago. It evolved by hybridization between a durum wheat, which is a tetraploid that has 28 chromosomes, and one of the goat grasses, which are wild diploid relatives of wheat that have 14 chromosomes. Thus, bread wheat has 42 chromosomes, which means that it is hexaploid. Durum wheat arose as the polyploid descendant of a diploid hybrid between two kinds of einkorn wheat and is still cultivated as the principal grain used in macaroni products.
>
> In a polyploid series such as that of wheat, distinct genomes can be distinguished by light microscopy. Thus, haploid genomes of the two einkorn wheats are designated *A* and *B*, respectively, and the haploid genome of goat grasses is designated *D*. The tetraploid genome of durum wheat is therefore *AABB*, and that of bread wheat is *AABBDD*.
>
>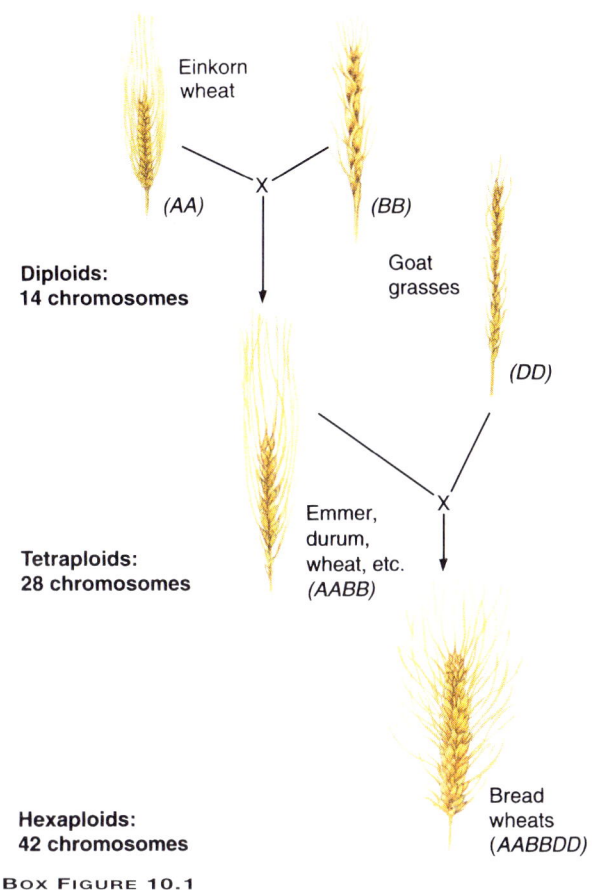
>
> **BOX FIGURE 10.1**
> Domesticated wheats evolved through hybridization and polyploidy. Different haploid genomes are designated *A*, *B*, and *D*.
> Adapted from "Wheat" by Paul C. Mangelsdorf. Copyright © 1953 by Scientific American, Inc. All rights reserved.

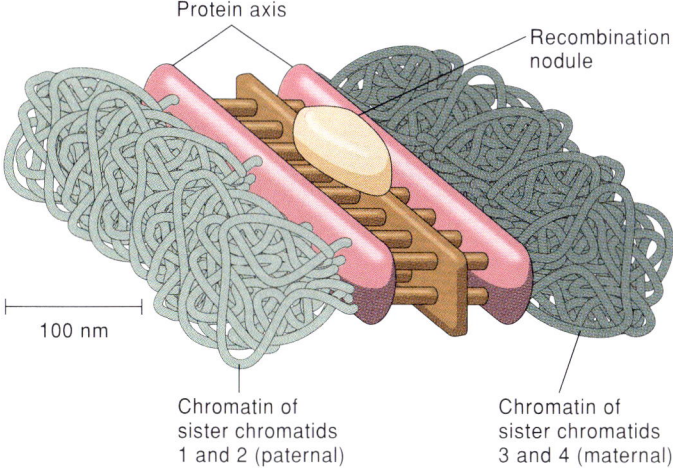

FIGURE 10.10

Model of a synaptonemal complex. It consists of a protein axis that binds homologous chromosomes together at alleles of the same gene locus.

when one chromosome has an inverted segment relative to the other, the bivalent makes a loop to ensure proper pairing of chromosome (fig. 10.11). Early evidence that allelic matching occurs between homologs came from studies of such inversion loops.

In contrast to the fidelity of allele matching, exact base-pairing is not required along the entire length of each interacting pair of chromosomes. Homologous chromatin strands, which are at least 100 nm apart, are too widely separated by the synaptonemal proteins for base-pair recognition to occur. Also, the synaptonemal complex can join regions of chromosomes that are different; this is crucial for allowing nonidentical alleles at the same locus to fuse and exchange different genetic material.

Recombination Nodules

Chromosome exchange occurs in specific regions of the synaptonemal complex that are called **recombination nodules;** these regions are so-named because of the large, protein-containing

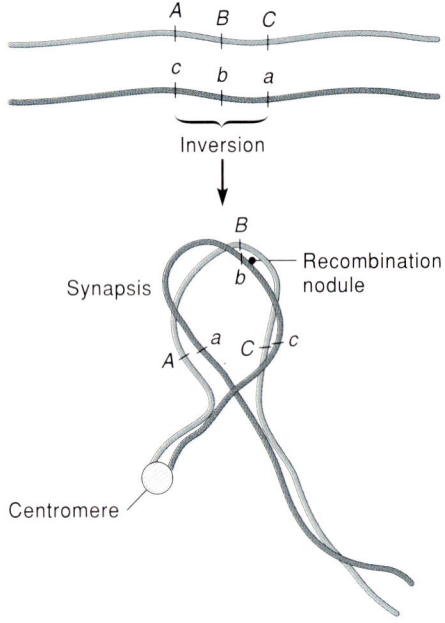

Figure 10.11

Drawing of synapsed chromosomes with an inversion loop. An inversion loop forms when the order of alleles in a segment of one homolog is inverted relative to the other homolog. The loop enables the appropriate matching of alleles during synapsis.

spheres associated with them. Although the synaptonemal complex provides a structural framework for chromosomal rearrangements, the actual process is thought to be mediated by recombination nodules. Each nodule has a diameter of about 90 nm, which almost completely spans the distance between homologous chromosomes in a synaptonemal complex. Each recombination nodule is probably made of several enzymes that act in concert to bring together homologous segments of DNA from opposite chromatids.

Chiasmata and Chromosome Segregation

As described previously, chiasmata in late prophase I indicate that crossing-over has probably occurred. In addition to recombining genes, crossing-over helps to segregate chromosomes during the first division of meiosis. This is because chiasmata continue to join homologs on the spindle after desynapsis. In mutant or hybrid plants that have fewer chiasmata than normal, however, homologs often fail to segregate normally. The products of such an abnormal meiotic division receive the wrong number of chromosomes and usually either die or are sterile (fig. 10.12).

After bivalents align in metaphase I, chiasmata begin to dissolve. Their dissolution coincides with the detachment of the fused kinetochores of sister chromatids. The attraction between the arms of sister chromatids is also disrupted, giving them a splayed appearance as they enter anaphase I (fig. 10.8d).

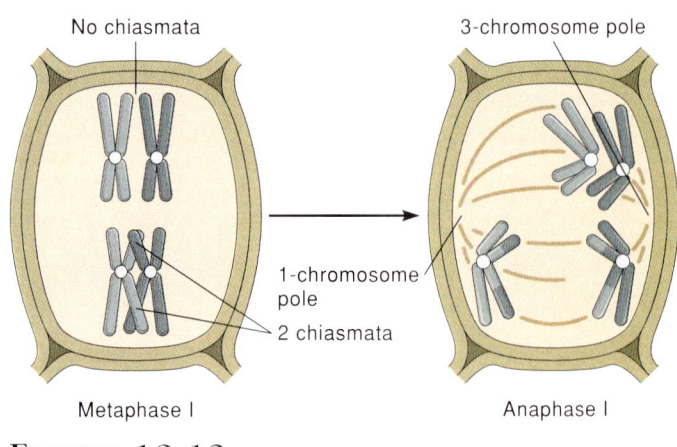

Figure 10.12

Chiasmata in prophase I are associated with proper separation and migration of homologs in anaphase I. Homologs may not separate when there are no chiasmata. Too few or too many chromosomes usually cause the new cells to die or to be sterile.

CONCEPT

Chiasmata usually indicate that crossing-over has occurred. Chiasmata ensure correct chromosome segregation because they hold chromosomes together until bivalents are properly aligned on the metaphase I plate.

GENETIC RECOMBINATION

Meiosis always produces genetic recombination, much of which comes from the exchange of chromosomal fragments during prophase I. However, genetic recombination would occur even without crossing-over. New combinations of traits would still appear in offspring because of chromosomal separation in meiosis.

Recombination by Chromosomal Separation

Homologs are not oriented along the metaphase plate until spindle fibers attach to each centromere. This means that, before spindle attachment, each homolog has an equal chance of moving to either pole. Furthermore, the direction of movement of each homolog is independent of all other bivalents. This is analogous to independent assortment in Mendelian genes, but for entire chromosomes.

How does chromosomal separation cause genetic recombination? Imagine that all the chromosomes of a sperm cell are light gray, and all the chromosomes of an egg cell are dark gray. The zygote from their fertilization would have a set of light gray and a set of dark gray chromosomes (fig. 10.13). When the mature plant from this zygote reproduces sexually, all the bivalents in meiosis would have one light gray homolog and one dark gray homolog. If,

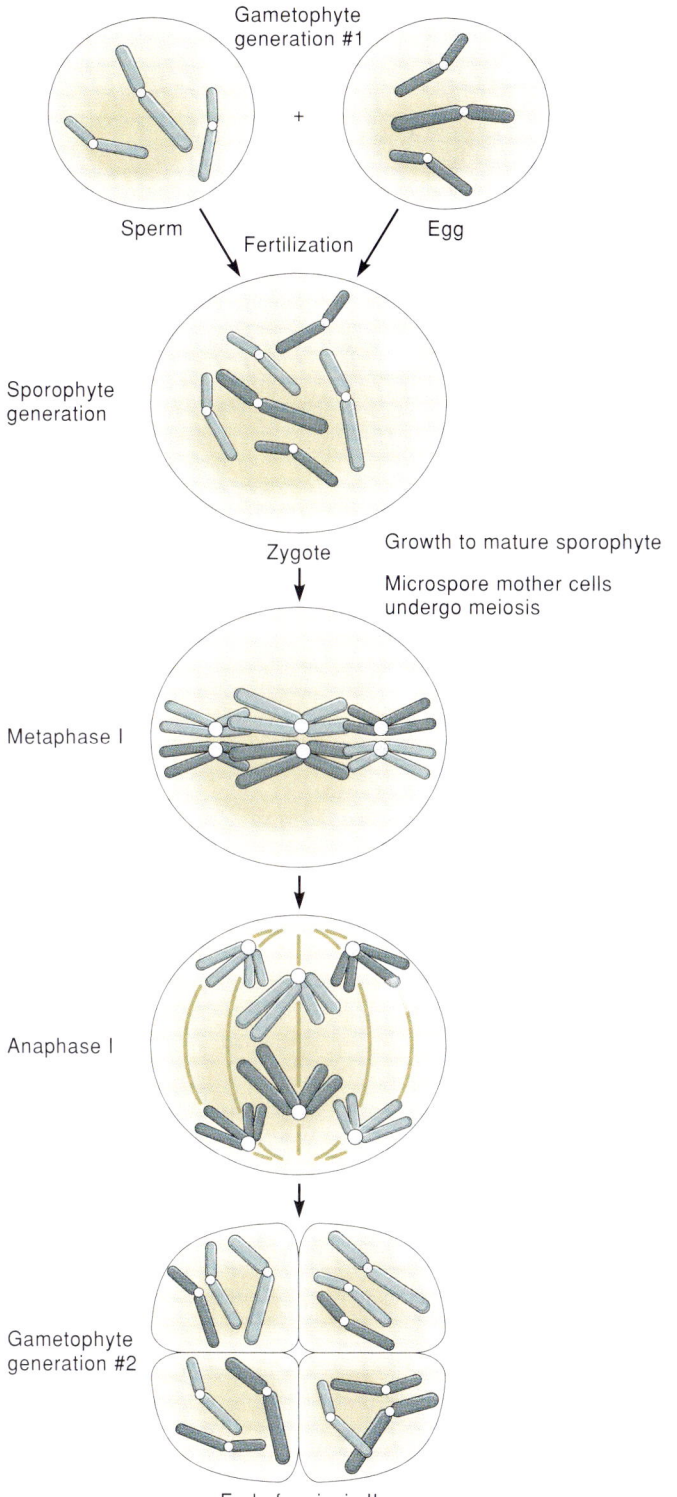

FIGURE 10.13

The pattern of separation of meiotic chromosomes is a source of genetic combination. Chromosomes from the sperm (light gray) and egg (dark gray) in the first gametophyte generation are mixed together in new combinations in the second gametophyte generation. Mixing comes from the orientations of homologous pairs in metaphase I during formation of the second gametophyte generation.

FIGURE 10.14

The desert sunflower (*Machaeranthera gracilis*) in bloom. Diploid cells of these plants have only four chromosomes, the smallest number of chromosomes known in plants.

during metaphase I, some of the dark gray chromosomes moved to the same pole as some of the light gray chromosomes, then each haploid nucleus would have light gray and dark gray chromosomes. This means that the gametophytes and the gametes produced by them would have new combinations of chromosomes, and that the offspring would therefore differ genetically from the parents.

To illustrate the potential for new genetic combinations based on chromosome separation, consider the gametes of the desert sunflower, *Machaeranthera gracilis* (fig. 10.14). The cells of the sporophyte have only four chromosomes, which is the smallest chromosome number known in plants. This means that there are only two bivalents in prophase I. If we color-code the homologs of these bivalents light gray and dark gray, we see that they can be oriented in four (i.e., 2^2) combinations in metaphase I (fig. 10.15). In anaphase I, two of the combinations will keep both dark gray chromosomes together and both light gray chromosomes together, but the other two combinations will produce nuclei with light gray and dark gray chromosomes mixed. Thus, by chance, half of the spores in the desert sunflower will have different combinations of chromosomes; that is, they will differ genetically from their parents.

The number of orientations of chromosomes in metaphase I increases dramatically for larger numbers of chromosomes. For example, there are 8 ways to orient three bivalents (i.e., $2^3 = 8$). As before, 2 of these are parental combinations (all dark gray and all light gray), so 6 are new combinations. Thus, based on random chromosomal separation alone, 75% of the spores will differ genetically from the previous gametophytic generation. However, most organisms have more than three kinds of chromosomes. For example, the garden pea has seven pairs of chromosomes,

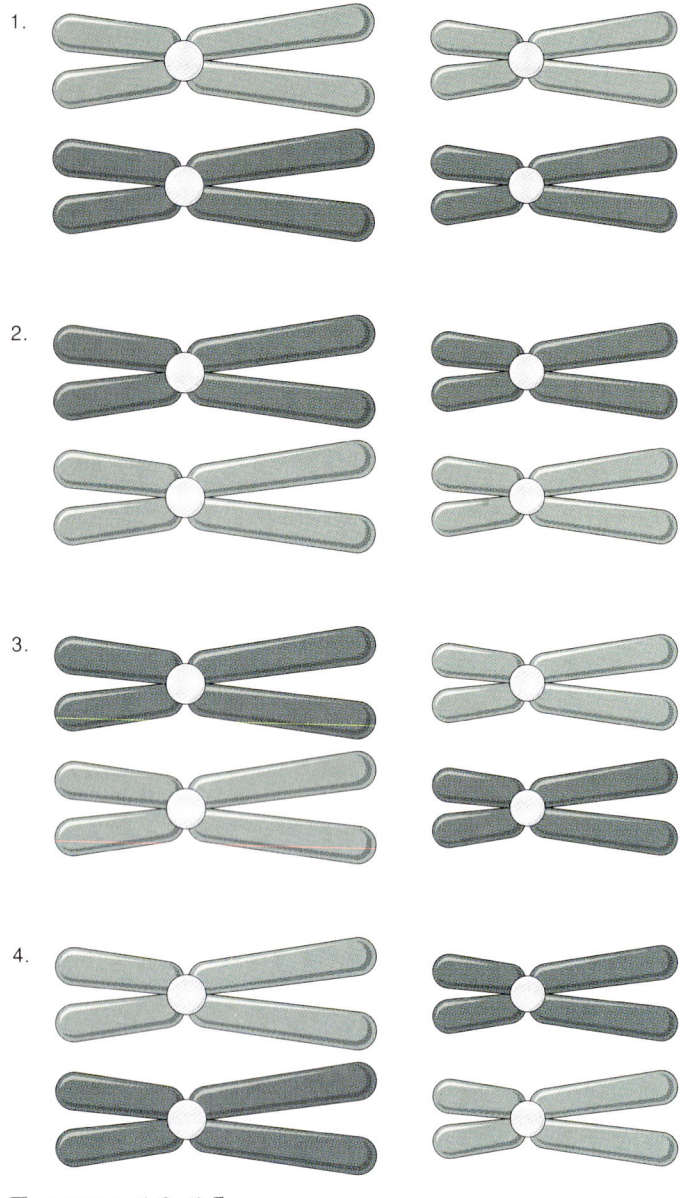

FIGURE 10.15

There are four possible orientations of two bivalents in metaphase I. Two of them (1 and 2) will segregate into parental combinations in the gametophyte generation. The other two (3 and 4) will result in new combinations of chromosomes in the gametophyte generation.

which have 128 (i.e., 2^7) possible orientations of bivalents in metaphase I. Thus, 0.78% (2/128) are parental, and 99.12% (126/128) are new combinations. Similarly, the 23 pairs of chromosomes in humans have more than 8 million possible patterns of chromosome separation (2^{23} = 8,388,608), but only two of these are parental combinations.

Remember that (1) the random separation of chromosomes during meiosis produces microspores in anthers and megaspores in ovules, and (2) the chance of reconstituting the parental chromosome combination in the microspore is independent of that in the megaspore. In the garden pea, for instance, this means that the probability of reconstituting both a parental sperm and a parental egg is 0.0078 × 0.0078 = 0.00006.

In other words, only 6 of every 100,000 fertilizations produce offspring without new combinations of chromosomes. Even without considering how two such gametes might get together, these simple calculations show that the odds of maintaining any one combination of chromosomes from parent to offspring are minuscule. The probability of genetic recombination by chromosome assortment is overwhelming.

C O N C E P T

Homologs separate independently of each other in meiosis. Spores are therefore likely to have new combinations of chromosomes, some from the maternal parent of the sporophyte, and some from the paternal parent. The number of new combinations of chromosomes is greater for higher chromosome numbers.

Doing Botany Yourself

Describe how you could use meiotic cells to determine the chromosome number of a plant. Include the materials, equipment, and skills you would need to accomplish this task. List two or three reasons why knowing the chromosome numbers of plants might be important to botanists.

Evidence for Chromosomal Exchange During Crossing-Over

The first experimental evidence that genetic recombination occurs during meiosis came from breeding experiments with corn (*Zea mays*) performed by H. S. Creighton and Barbara McClintock in the 1930s. Their evidence came from the inheritance of linked genes on a chromosome with unusual morphology. The two linked genes of interest control kernel color and texture, respectively. Kernels are either colored (C) or colorless (c) and have either standard texture (Wx) or waxy texture (wx). The chromosome that bears these genes sometimes has a darkly stained knob at one end and a piece of another chromosome at the other end. In the experimental plants, a chromosome of normal appearance carried alleles for c and Wx, and its homolog, with both the knob and the extra segment, carried C and wx.

The results of the physical exchange between homologous chromosomes was seen with a light microscope and was associated with recombination of the two gene loci. Recombined chromosomes have the knob on one homolog and the extra piece on the other (fig. 10.16). Such switching of chromosome segments explains why offspring from this cross can have colored kernels with standard texture or colorless kernels with waxy texture, even though the parents had neither of these combinations of traits.

Crossing-over is more easily observed in the bread mold *Neurospora crassa*. Spores show the results of meiotic chromosome behavior directly, without having to wait for the next diploid generation, as in plants and animals. Also, in contrast to what happens in plants, spores form in single file in small

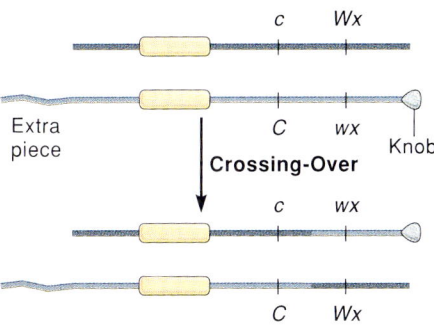

FIGURE 10.16

Evidence for crossing-over in corn (*Zea mays*). When nonsister chromatids cross over in prophase I, the chromatid bearing the allele for waxy kernel texture (*wx*) is exchanged with its homologous chromosome bearing the allele for normal kernel texture (*Wx*). Although the alleles themselves cannot be seen, the new position of the knob is direct evidence for chromosome exchange.

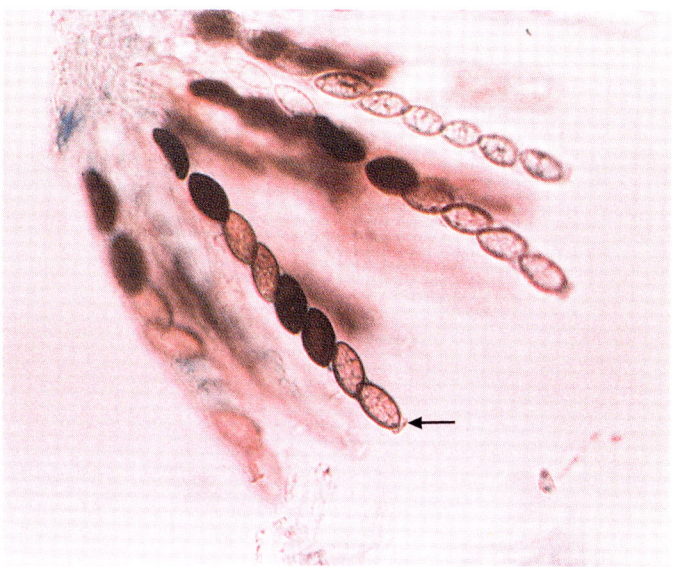

FIGURE 10.17

Light micrograph of spore sacs from the bread mold *Neurospora crassa*. Each spore sac contains eight spores in single file. When crossing-over occurs between chromosome segments that bear the gene for spore color, the sac contains an alternating pattern of spore colors (arrow).

sacs in *Neurospora*, and each spore undergoes mitosis immediately after meiosis, thereby making 8 spores in each sac (fig. 10.17). Spores can be either brown or gray, depending on one or the other allele of a single gene. After meiosis and mitosis, spore sacs usually have 4 brown spores in a row, followed by 4 gray spores in a row (fig. 10.18a). However, some sacs have the following pattern: 2 brown, 2 gray, 2 brown, 2 gray. This is evidence that crossing-over occurred between nonsister chromatids bearing the gene for spore color, as diagrammed in figure 10.18b.

Gene Conversion

Besides promoting recombination from crossing-over and chromosome separation, meiosis may also expose mistakes during DNA replication in spore mother cells. One such change was discovered in *Neurospora crassa* by Mary Mitchell in 1955. She found that one allele can apparently be changed to an alternate allele during spore production by this fungus. This change is called **gene conversion.**

Like crossing-over, gene conversion in *Neurospora* can be shown by following the inheritance of spore color (fig. 10.19). Instead of containing 4 brown and 4 gray spores, spore sacs occasionally contain 2 brown and 6 gray spores, or 6 brown and 2 gray spores. This ratio (3:1) cannot be the result of allelic interactions such as dominance, because the spores are haploid. Mitchell interpreted this pattern to mean that one allele is changed to the other, either brown to gray or gray to brown, during meiosis.

Although gene conversion probably develops by different mechanisms in different organisms, all mechanisms are believed to involve recombination and DNA repair between mismatched homologs. In one of the models for this process, gene conversion begins when a single strand in one double helix breaks and unzips from its complementary strand. The single-stranded "whisker" then invades a double strand on the other homolog, thereby displacing part of the other double helix (fig. 10.20).

When the invading DNA whisker is not perfectly complementary with its homolog, any mismatches are recognized and repaired by DNA repair enzymes. If the invader is used as the template, then the complement is changed to it, thereby undergoing gene conversion. If the complement is used as the template, then the invader is changed, which means that no gene conversion occurs. Meanwhile, the single strand that was left behind by the departing whisker is used as a template to replace the missing DNA fragment. The sequence that was displaced by the whisker on the invaded strand is removed and broken down.

The enzymes that initiate single-strand breakage of DNA and promote recombination are well-known only in bacteria, which are incapable of meiosis. Nevertheless, gene conversion probably occurs in all organisms that can repair DNA. The small amount of DNA synthesis that occurs during prophase I in eukaryotes is considered to be indirect evidence for gene conversion in these organisms.

C O N C E P T

Gene conversion is a process by which the DNA sequence of one allele is converted to that of the other allele. The process entails breakage and repair of DNA. Although it cannot be readily seen in most organisms, gene conversion is believed to occur when DNA is repaired after a single strand from one chromosome invades a complementary but nonidentical region of its homolog.

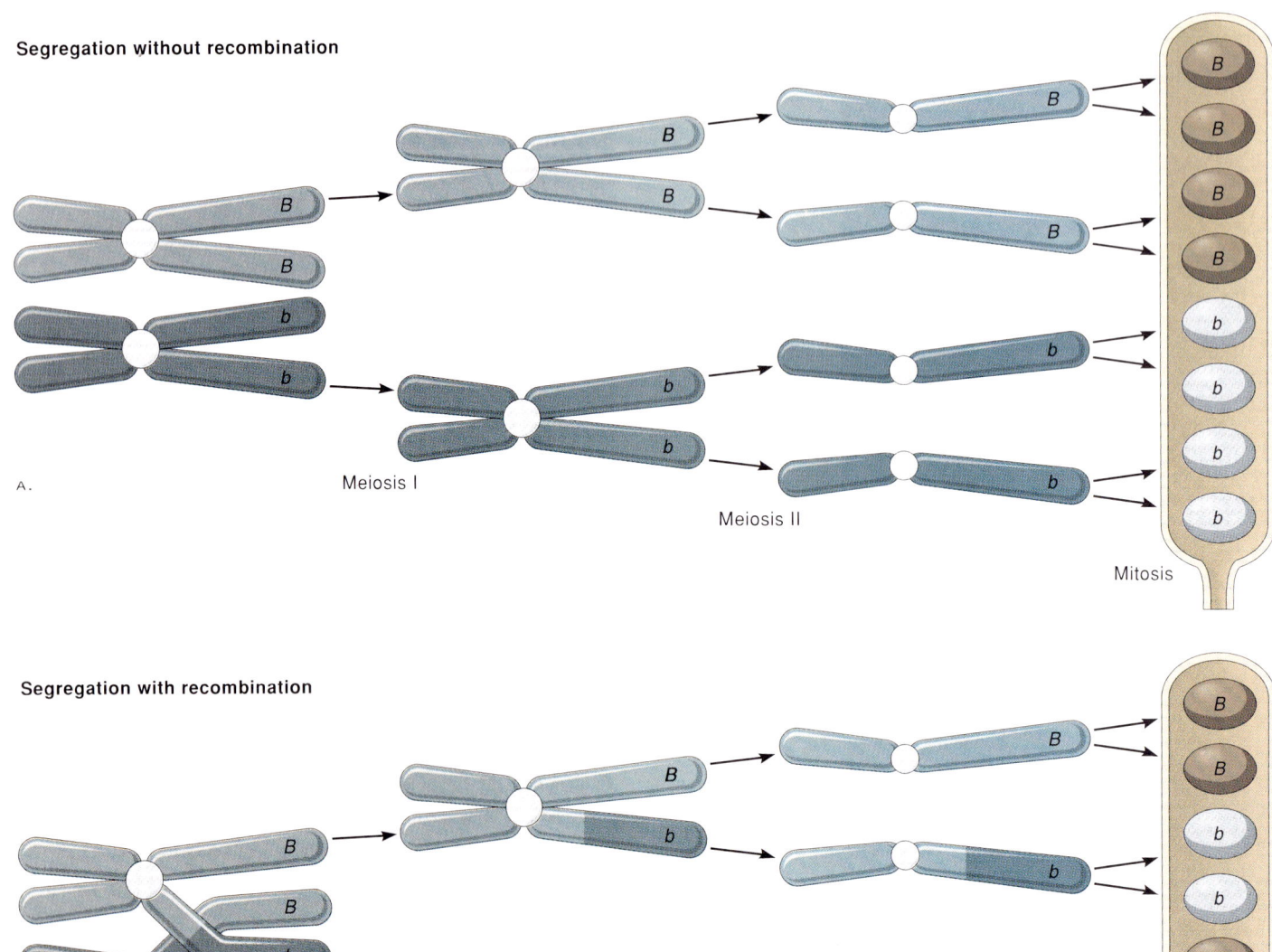

FIGURE 10.18

Evidence for crossing-over from the segregation of spore colors in bread mold. (a) Without crossing-over, brown spores are together at one end of the sac and gray spores are together at the other end. (b) After crossing-over, brown and gray spores alternate in pairs.

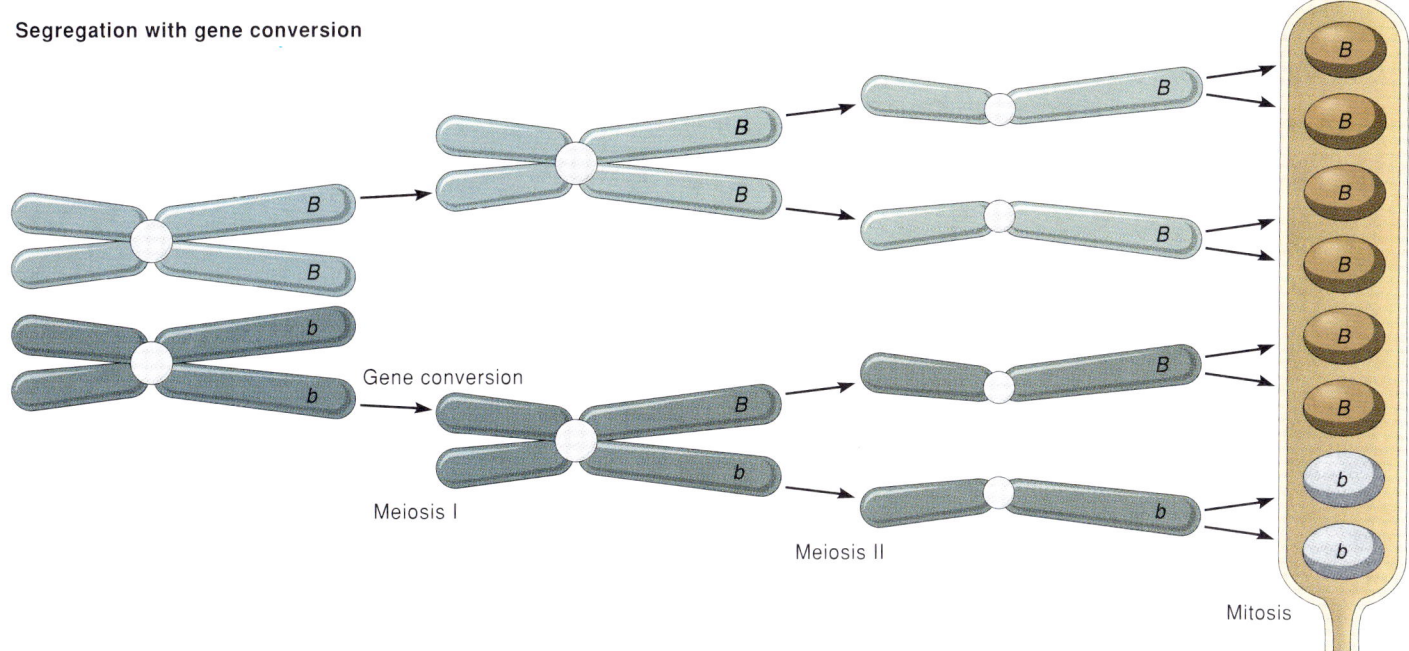

FIGURE 10.19

Evidence for gene conversion in bread mold. When one allele is changed to the other, the sac contains six spores of one color and two of the other (3:1 ratio) instead of four of each (1:1 ratio).

Unequal Crossing-Over and Gene Duplication

The discussion to this point has focused on the exchange of equal amounts of DNA by crossing-over. However, unequal amounts of DNA may be exchanged when homologs are not aligned perfectly (fig. 10.21). In such cases, crossing-over is called **unequal crossing-over,** and it probably occurs most often when a gene has already been duplicated at least once. The mechanisms for making an extra copy of a single gene are unknown, but they seem to be associated with the activity of transposable elements and with the irregular fusion of chromosome fragments during synapsis. One of the more common outcomes of such duplications is a **tandem repeat,** which is the occurrence of two or more copies of a gene in a row (fig. 10.22).

The most likely mechanism for making a third copy of two tandem genes is shown in figure 10.23. According to this model, chromosomes first misalign during synapsis so that one member of the tandem is paired and one is unpaired on each homolog. When crossing-over occurs in the paired region, genetic exchange may produce a "hybrid" gene between the tandem genes on one of the homologs, but the other homolog loses one member of the tandem at the same time. Thus, after unequal crossing-over, one homolog has three copies of the gene in a row, and the other homolog has one gene left from the original tandem.

Tandem repeats are common, which is considered to be indirect evidence for gene duplication by unequal crossing-over. For example, some strains of corn have more than 9,000 tandemly repeated genes that code for ribosomal RNA (rRNA). Tobacco has about 1,000 copies of rRNA genes, and humans have only 50 copies. These genes are the most highly duplicated in all organisms because of the high demand for ribosomes. Nevertheless, many other kinds of genes also occur in duplicate, but usually in only 5–10 copies. In addition, some copies of duplicated genes have apparently mutated independently from other copies (see box 10.2, "Multigene Families").

CONCEPT

Many genes occur in multiple copies that probably arose by genetic exchange during unequal crossing-over. Some genes, such as those for ribosomal RNA, may be present in thousands of copies. However, most duplicate genes have ten or fewer copies.

What are Botanists Doing?

Use the reference resources in your library to answer the question, "How many multigene families have been discovered in plants during the past twelve months?" As you compile your list, also note the plant names and the names or descriptions of the genes or gene products on the list.

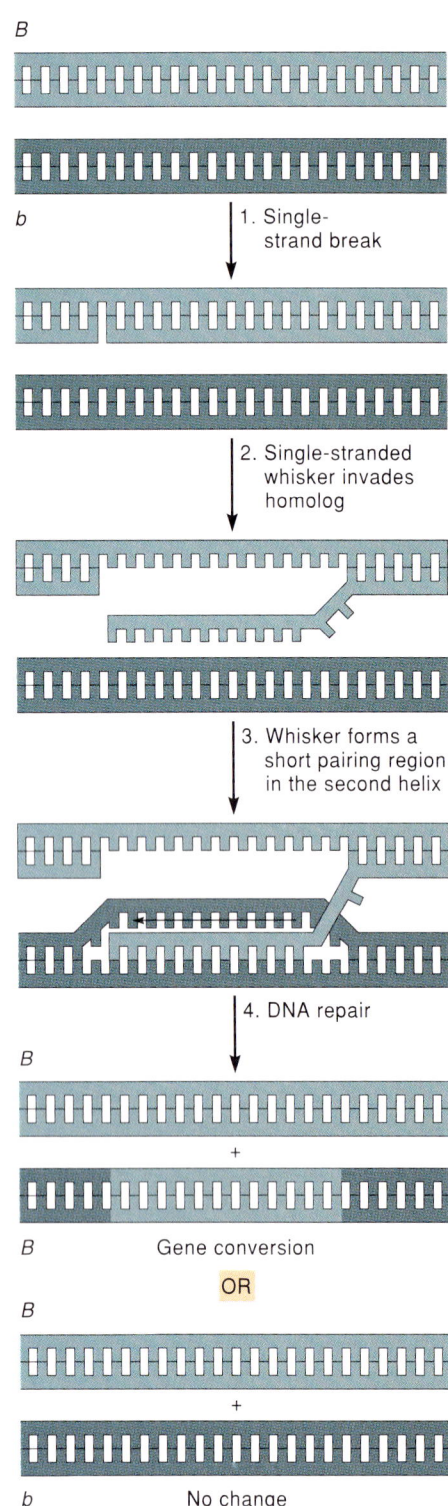

FIGURE 10.20

Model for gene conversion: 1) a single strand of one homolog breaks; 2) the whisker of single-stranded DNA invades the double strand on a nonsister chromatid; 3) the invading whisker forms a short pairing region in the second helix; 4) the mismatched region between the invading whisker and the invaded chromatid undergoes DNA repair to match both strands in the region. DNA displaced by the invasion is degraded; the gap left by the whisker is filled in by new DNA. Gene conversion occurs when the invading whisker is used as the template for DNA repair (*b* allele becomes *B* allele in second helix). There is no change when the second helix is used as the template for DNA repair.

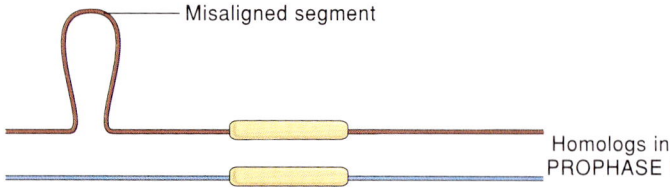

FIGURE 10.21

A chromosome may have an extra segment due to unequal crossing-over in a previous generation. During synapsis, the extra segment protrudes out into a loop because it has no matching sequence on its homolog.

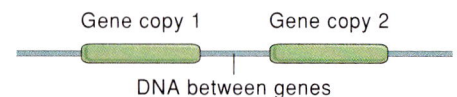

FIGURE 10.22

Model of a tandem repeat. Two or more copies in a row of the same gene are said to be in tandem.

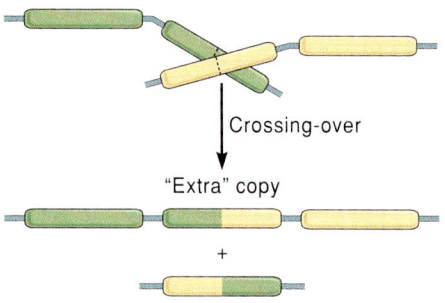

FIGURE 10.23

Model of gene duplication. Tandemly repeated genes may misalign during synapsis and exchange unequal-sized fragments. One of the homologs ends up with an "extra" copy, and the other homolog has one copy fewer than it had before.

WHY SEX?

The cycle of meiosis and fertilization depends on so many variables that it seems to flirt with failure every time sexual reproduction is attempted. For instance, most male gametes are wasted, since most pollen grains do not arrive at the stigma of a flower. This is especially true of wind-pollinated species, such as oaks, pines, walnuts, and birches. In addition, sexual reproductive structures require substantial amounts of energy. In some species, such as the small monkey flower (*Mimulus kelloggii*), more than 20% of the energy from photosynthesis is used to make flowers. Moreover, genetic recombination disrupts adaptive gene combinations. This means that although an individual plant is well-adapted to its environment, it is unlikely that its gene

BOXED READING 10.2

MULTIGENE FAMILIES

A set of duplicated genes is referred to as a **multigene family**. Although most genes are single-copy genes, multigene families probably occur in all plants. The number and diversity of genes in a multigene family represent the historical record of gene duplication in the genome. Scientists are fascinated by this phenomenon because it has the potential for revealing insights into the evolution of new genes.

Genes associated with photosynthesis are of particular interest in studying molecular evolution in plants. For example, polypeptides from two kinds of genes combine to form rubisco, the primary CO_2-fixing enzyme of photosynthesis (see Chapter 7). The genes for this enzyme have received more attention for evolutionary studies than any other genes.

One of the genes for rubisco occurs in the chloroplast and makes the large subunit of the enzyme. The other gene occurs in the nucleus and makes the small subunit. In all plants that have been examined, the gene for the small subunit, called *rbc*S, occurs in a multigene family. The tomato, for example, has five copies in its *rbc*S family, and the petunia has eight copies in its *rbc*S family. In tomatoes, three of the five copies occur in a head-to-tail sequence, close to each other on one chromosome. These are tandem repeats. A fourth copy is farther away on the same chromosome, and the fifth copy is on another chromosome entirely. By comparing the DNA sequences of each copy, the amount of similarity among the different gene copies can be determined. In so doing, evolutionary biologists have discovered that the three head-to-tail genes are almost identical to each other (99% similarity), but the other two differ by at least 10%. This means that the nontandem copies of the gene have evolved so that they differ slightly from their counterparts in tandem repeats.

Duplicated genes from unequal crossing-over should, when first formed, be identical and in the same place on one chromosome. So how can we explain the diversity of sequences and the different genomic locations of genes in the same multigene family? The best explanation for their different locations relies on the behavior of transposable elements. When mobile sequences of DNA jump from one location to another, they often take nearby DNA with them. In so doing, a transposable element can move the duplicated copy of a gene, which was originally formed by unequal crossing-over, to another location. The new copies can then accumulate random mutations, becoming different from the DNA sequence of the parent gene. In contrast, the tandem copies may remain nearly identical because gene conversion "homogenizes" them—that is, one copy acts as a template to correct any mutations that arise in another copy. Tandem genes may also be nearly identical because gene duplication was recent, and the newest copies have not yet had time to accumulate many mutations.

BOX FIGURE 10.2

Distribution of *rbc*S genes of tomato (*Lycopersicon esculentum*). Three copies are tandem repeats, one is distant from the tandem repeats on the same chromosome, and one is on a different chromosome than the other four copies.

---

combination will remain intact in the offspring. It is also unlikely that genetic recombination will produce offspring that are more fit than the parents, since the parents are already well-adapted.

If genetic recombination is often detrimental to the fitness of an individual's offspring, and if plants can be successful asexually, why is there sex at all? In trying to answer this question, biologists have given most of their attention to the evolutionary consequences of sex. The assumption is that sexual reproduction is advantageous because it is so common in virtually all kinds of organisms. Although most scientists agree that there is some underlying evolutionary benefit from sexual reproduction, they do not agree on what the benefit is. The two main hypotheses to account for the significance of sexual reproduction are the DNA repair hypothesis and the transposon hypothesis.

DNA Repair Hypothesis

Crossing-over does not occur in certain mutant strains of yeast (*Saccharomyces cerevisiae*). These strains also cannot repair double-stranded DNA. This observation is evidence that the mechanisms of crossing-over are shared with those of the DNA repair system, at least partially. Furthermore, such evidence may mean that the primary function of crossing-over is to repair DNA.

A damaged chromosome can be repaired if it has an appropriate template to guide the repair. Fusion to such a template occurs only during synapsis in prophase I, when gene conversion can change a defective chromosome back to its undamaged state. According to this reasoning, DNA repair is the main function of meiosis I.

According to the DNA repair hypothesis, damage in one haploid generation should be repaired before another haploid generation is produced. From this perspective, the diploid generation need only be transient, existing just long enough to bring homologs together, repair any damage, and segregate chromosomes back into haploid cells. The first eukaryotic organisms to reproduce sexually were haploid and had short-lived diploid generations. Among extant organisms, many algae and fungi still have this type of life cycle. The only cell in the diploid generation is the zygote, and meiosis occurs at that stage. Moreover, fertilization in these organisms is often induced by low or high temperatures, by desiccation, by nutrient depletion, or by some other kind of stress. This may be additional indirect evidence for the DNA repair hypothesis, because stressed haploid cells are more likely to suffer lethal damage to DNA than diploid cells. However, such damage can be repaired when the cells fuse and then undergo synapsis and gene conversion.

Although the DNA repair hypothesis helps to explain how sexual reproduction may have started, most of the evidence for it is circumstantial. Nevertheless, studies of mutant strains of yeast and of life cycles in single-celled eukaryotes provide evidence that is consistent with the hypothesis.

Transposon Hypothesis

Transposable elements, also called *transposons*, were discussed in Chapter 8 with regard to their ability to disrupt gene expression. Such disruption occurs when a mobile sequence of DNA moves into a gene, thereby inhibiting its function. Besides their mobility, many transposable elements are apparently duplicated during transposition; the parental copy stays in place while a duplicate copy is inserted elsewhere in the genome. Furthermore, newly made copies of transposons are believed to spread, like viruses, from one genome to another by meiosis and fertilization. That is, transposons probably spread throughout populations and species because of sexual reproduction.

The main function of transposons seems to be to make more transposons. They are therefore considered to be genomic parasites. They are efficient parasites, however, because they exploit sexual processes for their own reproduction. For this reason, they have been referred to as **selfish DNA.** Because the spread of such selfish DNA is enhanced by sexual reproduction, transposon sequences are most abundant in sexual species, especially those that rely predominantly on cross-breeding between different individuals. Indeed, in some species of plants, transposons make up about 10 percent of the genome. The abundance of transposon-derived DNA sequences explains, at least partially, why so many sexually reproducing organisms have large amounts of repetitive DNA (Chapter 9).

The transposon hypothesis asserts that the driving force for the origin of sex came from parasitic DNA sequences that could enhance their own fitness by promoting cycles of chromosome fusion and segregation. This hypothesis is supported by the behavior and widespread occurrence of transposons in sexual organisms. Furthermore, it eliminates the need to explain how cells or organisms can derive benefits from sexual reproduction. Such benefits, if any, are irrelevant to the individual plant or animal; benefits accrue instead to the genomic parasites.

Evidence that sex is an adaptation of parasitic DNA is also known in prokaryotes. In *Escherichia coli*, cell fusion and genetic recombination result from interactions between the main bacterial genome and DNA that occurs in plasmids, which are self-replicating DNA sequences that are not part of the primary bacterial genome. In addition, the DNA of bacteriophages, which are all parasitic, also causes genetic change in the host bacterium. Thus, both kinds of genomic parasites, plasmids and phages, provide forms of sexuality among prokaryotes. Thus, eukaryotic sex may have begun with a type of intracellular genetic parasitism that arose from prokaryotic ancestors.

Chapter Summary

The life cycle of plants is an alternation between diploid and haploid generations. The diploid generation produces the haploid generation by meiosis; the haploid generation produces the diploid generation by fertilization. The stages of meiosis have the same names as the stages of mitosis: prophase, metaphase, anaphase, and telophase. However, meiosis consists of two divisions. The first division is a reduction division, and the second is like mitosis. Meiosis is also fundamentally different from mitosis because, in meiosis, homologous pairs of chromosomes fuse at the beginning of the first prophase. Thus, paired chromosomes separate into haploid nuclei in the first meiotic division, and centromeres divide and separate in the second meiotic division.

During the first prophase, chromosomes exchange genetic material across the synaptonemal complex that holds them together. Chiasmata, the apparent physical indicators of chromosomal crossing-over, ensure correct chromosome segregation because they hold chromosomes together until bivalents align properly on the metaphase plate.

New genetic variation occurs during meiosis because of chromosomal separation and genetic exchange between homologs. Chromosomal separation produces new combinations of parental chromosomes in the offspring. Genetic exchange also occurs because of gene conversion by DNA repair processes. In gene conversion, a strand from one chromosome functions as a template for repairing a strand on its homolog. When crossing-over occurs between imperfectly aligned homologs, the exchange of genetic material can be unbalanced. Through such repeated unequal crossing-over, entire genes can be duplicated on one homolog and deleted from another. This process of gene duplication also explains how the number of genes for ribosomal RNA can be in the thousands in some plants. Many other kinds of genes occur in multiple copies, but the usual number of copies is ten or fewer.

Sexual reproduction benefits species and populations because it produces genetic variation, but it is not clear how individuals or their progeny benefit from sexual processes. Two hypotheses have been advanced to explain the origin and maintenance of sexual reproduction. One states that chromosome fusion and segregation evolved as a mechanism to repair DNA. An alternative hypothesis for the origin of sex is based on the evolution and spread of parasitic DNA sequences, or transposable elements.

Questions for Further Thought and Study

1. What are the major differences and similarities between meiosis and mitosis? What is the significance of each process?
2. If diploidy evolved because of its potential to repair DNA, it seems that a single-celled diploid stage would suffice to accomplish it. Yet most organisms are dominated by a complex, multicellular diploid generation. How can this discrepancy be explained?
3. Genes in a multigene family can vary from one copy to another. What are the possible reasons for this variation?
4. Two races of *Machaeranthera gracilis* exist, one with $n = 2$ chromosomes and one with $n = 3$ chromosomes. Assume that gametes from the two races are interfertile and that the hybrid from such a cross can grow into a normal-looking plant. Would you expect this hybrid to be fertile or sterile? Why?
5. Some cholla cacti are triploid with a chromosome number of 33. How might prophase I look when synapsis occurs in cells of these plants?
6. How can you explain the origin of a range of chromosome numbers—such as $2n = 18, 16, 14$—in some plant species?

Suggested Readings

Articles

Anderson, A. 1992. The evolution of sexes. *Science* 257:324–326.

Atkins, T., and J. M. Roderick. 1991. "Dropping your genes." A genetics simulation in meiosis, fertilization, & reproduction. *The American Biology Teacher* 52:164–169.

Brown, C. R. 1990. Some misconceptions in meiosis shown by students responding to an advanced level practical examination question in biology. *Journal of Biological Education* 24:182–185.

Maguire, M. P. 1992. The evolution of meiosis. *Journal of Theoretical Biology* 154:43–55.

Petrusky, B. 1990. DNA on target: Homologous recombination. *Mosaic* 21(May):44–52.

Rose, M. R. 1983. The contagion mechanism for the origin of sex. *Journal of Theoretical Biology* 101:137–146.

Simchen, G., and Y. Hugerat. 1993. What determines whether chromosomes segregate reductionally or equationally in meiosis? *BioEssays* 15:1–8.

Stahl, F. 1987. Genetic recombination. *Scientific American* (February):90–101.

Stewart, J., B. Hafner, and M. Dale. 1990. Students' alternate views of meiosis. *The American Biology Teacher* 52:228–232.

Tanksley, S. D., and E. Pichersky. 1988. Organization and evolution of sequences in the plant nuclear genome. In L. D. Gottlieb and S. K. Jain, eds., *Plant Evolutionary Biology*. New York: Chapman and Hall.

Troyer, J. R. 1989. John Henry Schaffner (1866–1939) and reduction division in plants: Legend and fact. *American Journal of Botany* 76:1229–1246.

Books

Alberts, B., D. Bray, J. Lewis, M. Raff, K. Roberts, and J. D. Watson. 1989. *Molecular Biology of the Cell*. 2d ed. New York: Garland.

Farley, J. 1982. *Gametes and Spores: Ideas about Sexual Reproduction, 1750–1914*. Baltimore: Johns Hopkins University Press.

John, B. 1990. *Meiosis*. New York: Cambridge University Press.

Marguilis, L., and D. Sagan. 1986. *Origins of Sex*. New Haven, CT: Yale University Press.

Michod, R. E., and B. R. Levin, eds. 1988. *The Evolution of Sex: An Examination of Current Ideas*. Sunderland, MA: Sinauer.

Moens, P. B., ed. 1987. *Meiosis*. New York: Academic Press.

Russell, P. J. 1990. *Genetics*. Glenview, IL: Scott, Foresman.

This corn (*Zea mays*) plant is genetically engineered for tolerance to the herbicide glyphosate.

CHAPTER 11

Molecular Genetics and Gene Technology

Chapter Outline

INTRODUCTION

HOW GENES WORK
  RNA Synthesis: Transcription
  Different Kinds of RNA

BOX 11.1
RIBOZYMES AND THE ORIGIN OF LIFE

BOX 11.2
THE DISCOVERY OF INTERRUPTED GENES

  Polypeptide Synthesis: Translation
  Making the Finished Product: Polypeptide to Protein

THE GENETIC CODE
  The Structure of Eukaryotic Genes
  Roles of Gene Interruptions
  The Exon-Shuffling Hypothesis

GENETIC ENGINEERING
  DNA Cloning
  Finding the Gene of Interest
  Industrial Uses of Genetically Engineered Bacteria
  Making Transgenic Plants: Problems and Solutions
  Monocots vs. Dicots in Genetic Engineering

BOX 11.3
GENETIC ENGINEERING WITH A GENE GUN

  Transgenic Crops
  THE MOLECULAR GENETIC BASIS OF ROUND VS.
      WRINKLED PEAS

Chapter Summary
Questions for Further Thought and Study
Suggested Readings

Chapter Overview

Imagine that the nucleus is an enormous dictionary containing the words that define life, but that the words are written in a microscopic code. As recently as the 1950s, this code seemed so complicated that scientists thought it might be beyond human comprehension. Furthermore, the code was so large for any given organism that, if its letters could be printed in the same size as this type, the dictionary would fill several hundred volumes the size of this book. Now we know that the code is made of DNA and that DNA is organized into "words" called *genes*. But how is the code translated into organisms? There is still no good answer for this question, although pieces of the answer have been discovered by finding out how genes work at the cellular level. Studies of the mechanisms of gene action constitute a relatively new subdiscipline of genetics known as *molecular genetics*, the topic of this final chapter in the unit on genetics.

The methods and discoveries of molecular genetics have greatly improved crop productivity, gene therapy, and other aspects of applied biology, which rely on our knowing how to make new genes, change existing genes, or transfer genes from one organism to another. This work is the basis for a highly sophisticated, multibillion-dollar industry called *gene technology*, which is also discussed in this chapter.

INTRODUCTION

The first trait of the garden pea that was described in 1865 by Mendel was the round versus wrinkled shape of mature seeds (fig. 11.1). The genetic basis for this trait was not described until 1990, however, by Madan K. Bhattacharyya and several colleagues at the John Innes Institute in Norwich, England. This group identified the enzyme that controls roundness, found the gene that makes it, and determined the differences between the dominant-round allele (*R*) and the recessive-wrinkled allele (*r*). Roundness is apparently controlled by one form of **starch-branching enzymes (SBEI;** I = isoform), which converts the straight chains of amylose to the branched polymers of amylopectin (see Chapter 2). This form of the enzyme is active in the early development of round seeds but absent in wrinkled seeds. Seeds with either *RR* or *Rr* genotypes have large amounts of starch with high ratios of amylopectin. Conversely, seeds that are *rr* have less starch and proportionately less amylopectin; instead they have more sucrose. Wrinkles occur because sucrose is osmotically active, which causes *rr* seeds to absorb water. When *rr* seeds dry as they mature, the loss of water makes them shrivel. In contrast, since starch is not osmotically active, *RR* and *Rr* seeds contain less water. Hence, when these seeds mature, they lose little water and remain full and round.

Bhattacharyya's study is an example of how genes are found and how they influence phenotypes. However, a complete explanation of the molecular genetics of wrinkled seeds requires an understanding of the mechanisms of gene action and how they can be exploited to find and manipulate genes of interest in the laboratory. These are the subjects of this chapter. We therefore postpone further discussion of the genetic details of wrinkled peas until you have read about how genes work.

HOW GENES WORK

Proteins are made in the cytoplasm, not in the nucleus where the DNA occurs. This observation was one of the first clues that DNA does not make proteins directly. Another clue was that protein synthesis is associated with ribosomes, which occur outside the nucleus. These clues led to the discovery that proteins are made in ribosomes. Protein synthesis is a multistep process that begins when DNA makes RNA, which is exported from the nucleus into the cytoplasm. There are three main kinds of RNA in the cytoplasm. **Messenger RNA (mRNA)** is the actual message for a protein. **Transfer RNA (tRNA)** binds to amino acids, so that they can be brought together during protein synthesis. **Ribosomal RNA (rRNA)** is one of the main ingredients of ribosomes.

The **central dogma of molecular biology** states that one gene codes for one protein. For example, the *R* allele in the garden pea codes for a starch-branching enzyme. In general terms, gene expression has the following phases:

$$\text{DNA} \rightarrow \text{mRNA} \rightarrow \text{protein}$$

The first phase, DNA to mRNA, is called **transcription** because one nucleic acid code (DNA) is transcribed into another nucleic acid code (mRNA). This means that the DNA version of *R* makes the mRNA version of *R*. The second phase, mRNA

FIGURE 11.1

The first trait that was described by Mendel in 1865 for his experiments with the garden pea was seed shape (round vs. wrinkled). The molecular genetic basis for this trait was not discovered until 1990.

to protein, is called **translation** because a nucleic acid code (mRNA) is translated into an amino acid code (protein). This means that the mRNA version of *R* is translated into the chain of amino acids that make a protein.

The immediate product of translation often requires modification before it becomes a functional protein. For example, the complete rubisco enzyme of photosynthesis consists of two kinds of polypeptides. One polypeptide comes from a nuclear gene, and the other comes from a chloroplast gene (see Chapter 10). Thus, the mRNA from each of these genes does not translate directly into a mature protein; instead, it translates into a polypeptide that becomes part of a mature protein. Therefore, an improved version of the central dogma can be stated as follows:

DNA → mRNA → polypeptide → protein

Dogma usually refers to an unquestioned doctrine, which has no place in science. If so, then why does molecular biology have a central dogma? The term *central dogma* was coined by Francis Crick more as a joke than as a serious suggestion. But the term took on a life of its own, regardless of its inappropriateness. Nevertheless, the central dogma is universally accepted because there is no evidence to refute it and no alternative proposal for how genes work (certain viruses, however, have genomes of RNA; RNA genes are reverse-transcribed to DNA, followed by transcription and translation).

RNA Synthesis: Transcription

During transcription, DNA serves as a template for making complementary RNA copies of itself. The entire gene is transcribed into molecules of RNA that are about the same length as the DNA template (fig. 11.2). The RNA differs from the gene, however, in a number of ways. One way is that RNA is the complement of a DNA code, not a duplicate of it. Also, unlike DNA, the RNA molecule is single-stranded; it contains ribose instead of deoxyribose, and it contains uracil (U) instead of thymine (T). Accordingly, a DNA sequence of ATGCCTGGA will be transcribed into an RNA sequence of UACGGACCU. The main features of RNA synthesis are similar to those of DNA replication, but RNA synthesis is up to twenty times faster than DNA synthesis. As in DNA replication, RNA synthesis proceeds from the 5′ end to the 3′ end of the molecule, so that the complementary RNA strand is assembled antiparallel to the template DNA strand. The oldest (first-made) end of the RNA is the 5′ end.

Steps in Transcription

Transcription occurs in three main steps. The initiation step of transcription begins when a special enzyme, **RNA polymerase**, attaches to the DNA sequence (fig. 11.2). The RNA polymerase attaches to the DNA at a promoter site, which is a special sequence of DNA that is the initiation signal for transcription. Only one strand of DNA is the coding strand for each gene—that is, only one strand contains the coding se-

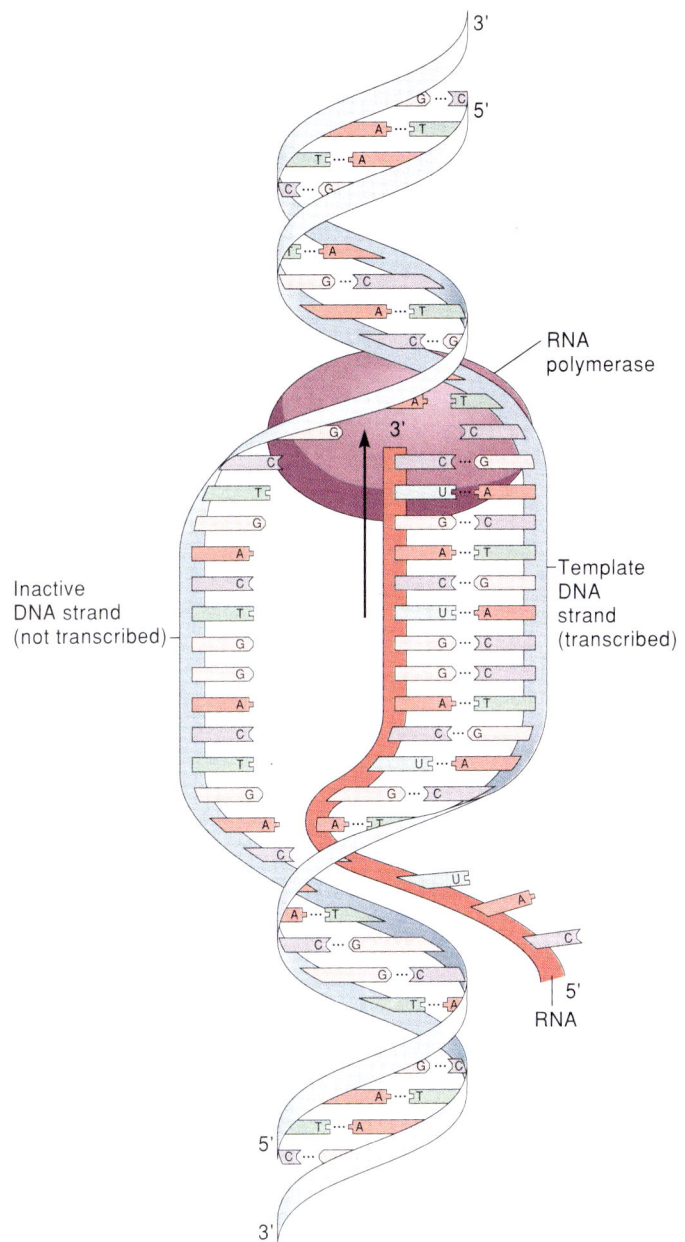

FIGURE 11.2

Diagrammatic representation of transcription. RNA polymerase assembles nucleotides in a sequence that is complementary to the DNA template. RNA is therefore identical to the nontranscribed strand of DNA, except for the substitution of uracil (U) for thymine (T) in RNA.

quence for a polypeptide. The other strand is the noncoding strand and is not transcribed. How does RNA polymerase recognize the coding strand versus the noncoding strand of double-stranded DNA? The key to recognizing the coding strand is the promoter site. Sequences vary among different promoters, but almost all of them include two specific sets of six nucleotides, TATAAT and TTGACA, to which RNA polymerase binds. The antisense strand contains their complements (ATATTA

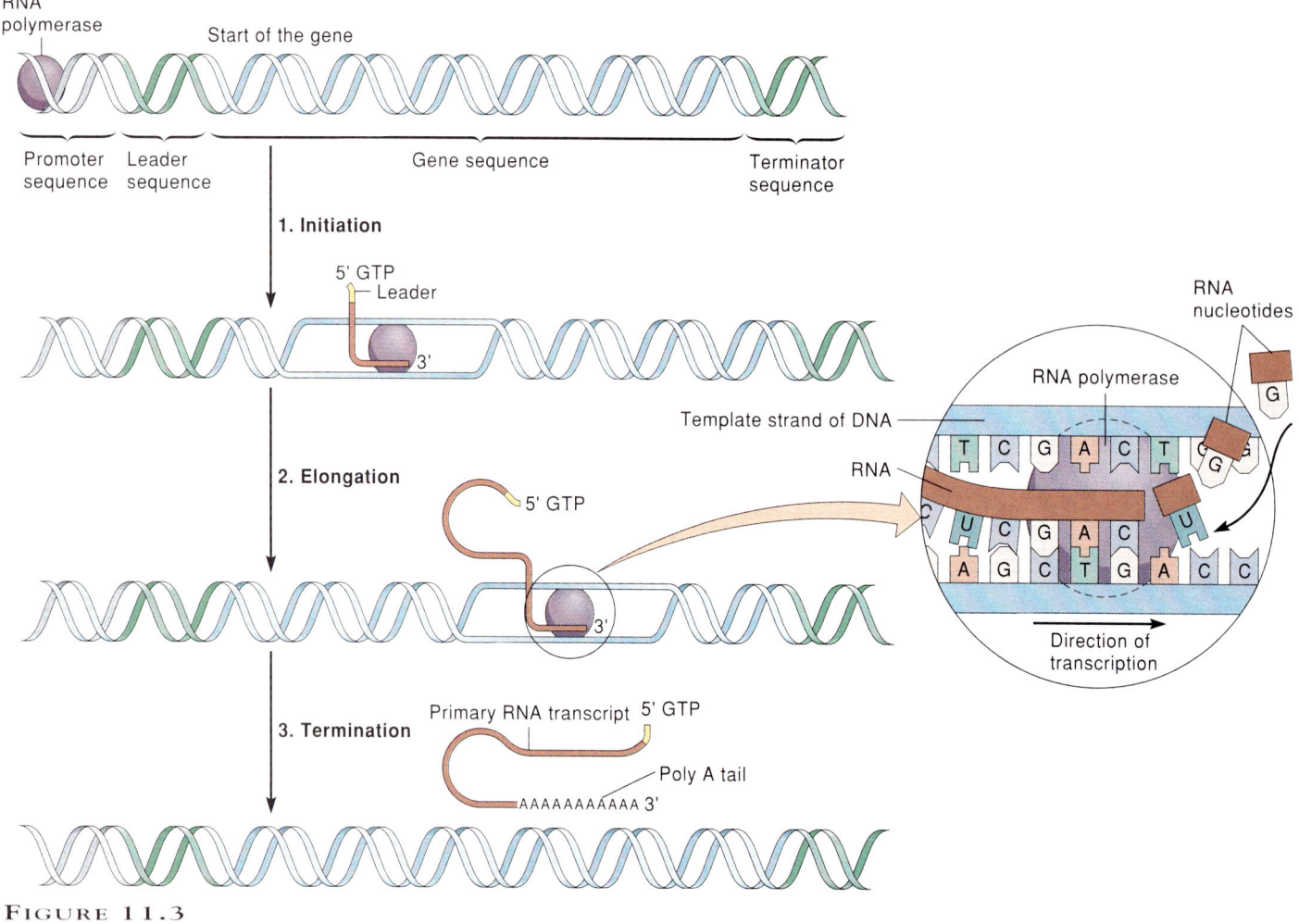

Figure 11.3

Steps in RNA transcription. 1. RNA polymerase attaches to the DNA sequence at a promoter site. 2. RNA polymerase adds nucleotides to the growing chain of RNA as the DNA unwinds ahead of it. 3. Transcription stops, and the new RNA molecule is cleaved from its template.

and AACTGT, respectively), which are not recognized by RNA polymerase. The coding strand for some genes is the noncoding strand for other genes, which means that genes may occur on either strand throughout an entire chromosome.

In the second step in transcription, the elongation step, the polymerase enzyme moves along the strand of DNA, adding nucleotides to the growing chain of RNA (fig. 11.3). The transcribed region may also include a **leader sequence** of variable length—that is, a noncoding sequence between the promoter and the beginning of the gene. After about thirty bases have been transcribed, a chemical **cap** is attached to the 5′ end of the RNA. This cap is a nucleotide triphosphate, which is analogous to ATP, called 7-methylguanosine triphosphate (**GTP**). The GTP cap protects the RNA molecule from degradation and later serves as the initiation signal for translation.

During the termination step, transcription stops and the new RNA molecule is cleaved from its DNA template. After the polymerase reaches the end of the gene, it continues transcription until it arrives at a cleavage signal. The extra amount of RNA beyond the end of the gene is called a **trailer sequence.** Like the leader sequence, the trailer sequence does not function as a coding region; it is a signal for another polymerase to cut the RNA molecule and release it from the DNA template. Immediately after cleavage, 100–200 adenylic acid molecules are added to the 3′ end of the RNA. With the addition of this **poly-A tail,** the complete transcription product is called the **primary RNA transcript.**

Speed of Transcription

The demand for certain polypeptides may be immediate and crucial for cellular metabolism. To satisfy this demand, a single gene may be transcribed by several RNA polymerases at the same time (fig. 11.4). The rate of RNA synthesis is up to 30 nucleotides per second for each RNA polymerase, which means

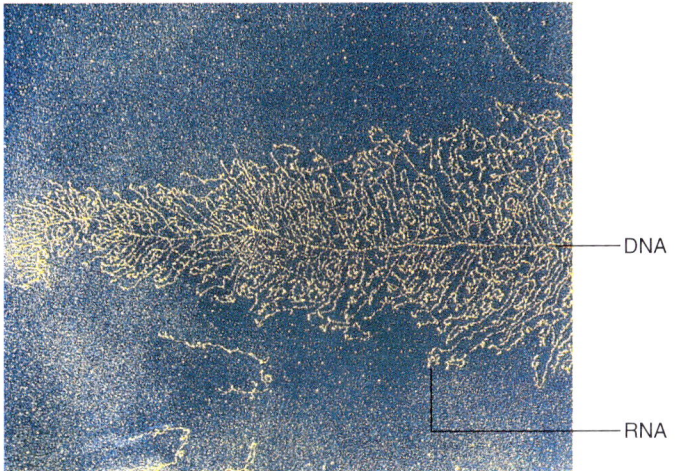

FIGURE 11.4

Electron micrograph of simultaneous transcription of a single gene by several RNA polymerases.

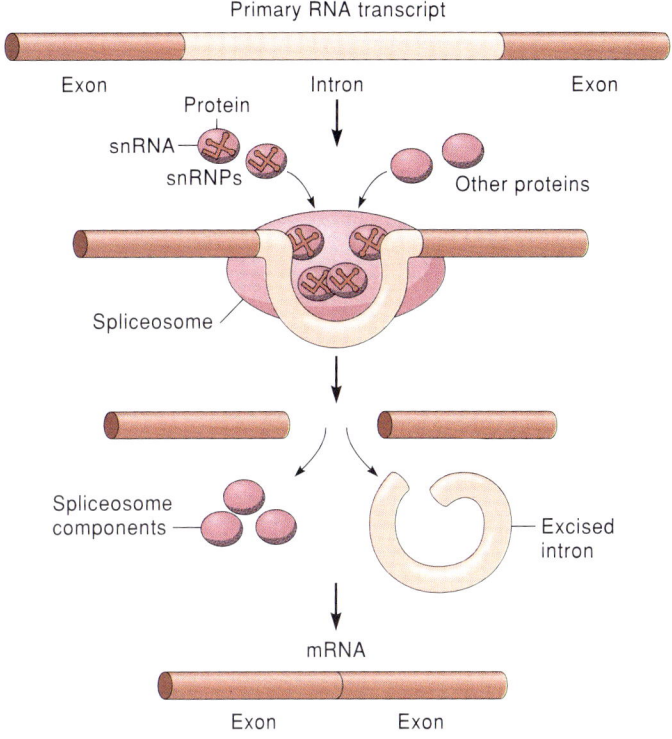

FIGURE 11.5

RNA processing. Portions of a primary RNA transcript are removed before the RNA is exported from the nucleus. Complexes of proteins and snRNA (small nuclear RNA) bind to segments to be excised (introns), cut them out, and splice the functional segments (exons) together.

that an average-size gene of about 8,000 nucleotides can be transcribed very rapidly. The first primary RNA transcript is complete in about five minutes. About 100 primary RNA transcripts can be made in less than ten minutes during one bout of RNA synthesis when there are at least 100 RNA polymerases attached to the gene. These estimates are only for transcription, and several more steps must occur before polypeptides are finally synthesized. One of the main steps after transcription is the modification of the primary RNA transcript into messenger RNA, which is discussed in the next few paragraphs.

CONCEPT

Transcription entails three main steps: initiation, elongation, and termination of RNA synthesis. The coding strand of DNA contains sequences for initiation and termination that are recognized by the enzyme RNA polymerase. In addition to transcribing the gene, RNA polymerase also transcribes a noncoding leader sequence and a noncoding trailer sequence. The primary RNA transcript is completed when it is bound to a long poly-A tail and released from the DNA template.

Different Kinds of RNA

According to the central dogma, all RNA comes from the transcription of DNA. But RNA occurs in several forms, some of which stay in the nucleus, and some of which move into the cytoplasm. In this chapter we focus on nuclear RNAs that help make polypeptides and on cytoplasmic RNAs (mRNA, tRNA, rRNA). Certain kinds of RNA molecules called **ribozymes** can also function as enzymes. These RNAs are discussed in box 11.1, "Ribozymes and the Origin of Life."

RNA in the Nucleus

Initially, each molecule of RNA is about the same size as the DNA sequence that serves as its template. A gene of 8,000 nucleotides is copied into an RNA molecule of 8,000 nucleotides, plus leader and trailer sequences. Large primary RNA transcripts from different genes occur in different sizes, from a few thousand nucleotides to more than 20,000 nucleotides. The pool of large RNA transcripts in the nucleus is called **heterogeneous nuclear RNA (hnRNA)** because of the heterogeneity of sizes among RNA molecules. Large transcripts must be trimmed by **RNA processing,** also called **RNA splicing,** before they leave the nucleus. The splicing of RNA transcripts involves small molecules of RNA that remain in the nucleus (**snRNA**), which are about 100–200 nucleotides each. These small RNA molecules condense with proteins, which are called **small nuclear ribonucleoproteins (snRNPs).** A complex of numerous snRNPs is called a **spliceosome.** Each spliceosome binds to a large primary RNA transcript, cuts out certain parts of the transcript, and splices the rest of the RNA back into a continuous strand (fig. 11.5). The parts of the primary transcript that are cut out are called **introns;** the parts that are spliced together are called **exons.** Several spliceosomes can attach to an RNA transcript

BOXED READING 11.1
RIBOZYMES AND THE ORIGIN OF LIFE

In 1981, Thomas Cech discovered that primary RNA transcripts from *Tetrahymena*, a single-celled protist, can be processed in the absence of enzymes. Processing occurs because one of the introns is self-splicing, which means that it acts like an enzyme to catalyze a chemical reaction. Unlike an enzyme, however, the intron works on itself. To acknowledge this difference, Cech named this intron a *ribozyme*. His work on ribozymes earned Cech a Nobel Prize in the "Physiology or Medicine" category in 1986.

Cech's discovery sparked a search for ribozymes in other organisms. So far, they have been found in the nuclei and organelles of plants, animals, and fungi, and in the genomes of bacteria. Additional functions have also been discovered for ribozymes. For example, they work like nucleases, which are enzymes that digest nucleic acids. Ribozymes also mimic polymerases by joining nucleotides in short chains and by joining short RNA sequences into longer ones.

What do ribozymes have to do with the origin of life? One possibility is that they may be the answer to the "chicken versus egg" paradox between nucleic acids and proteins. The paradox is that if proteins came first, how could they be encoded without DNA or RNA? Conversely, if nucleic acids came first, how could they be replicated without enzymes? This paradox rests on the dogma that only proteins can be enzymes, and only nucleic acids can contain genetic information. However, we now know that RNA can play both roles. Thus, the first precellular organisms probably contained RNA long before DNA or proteins evolved.

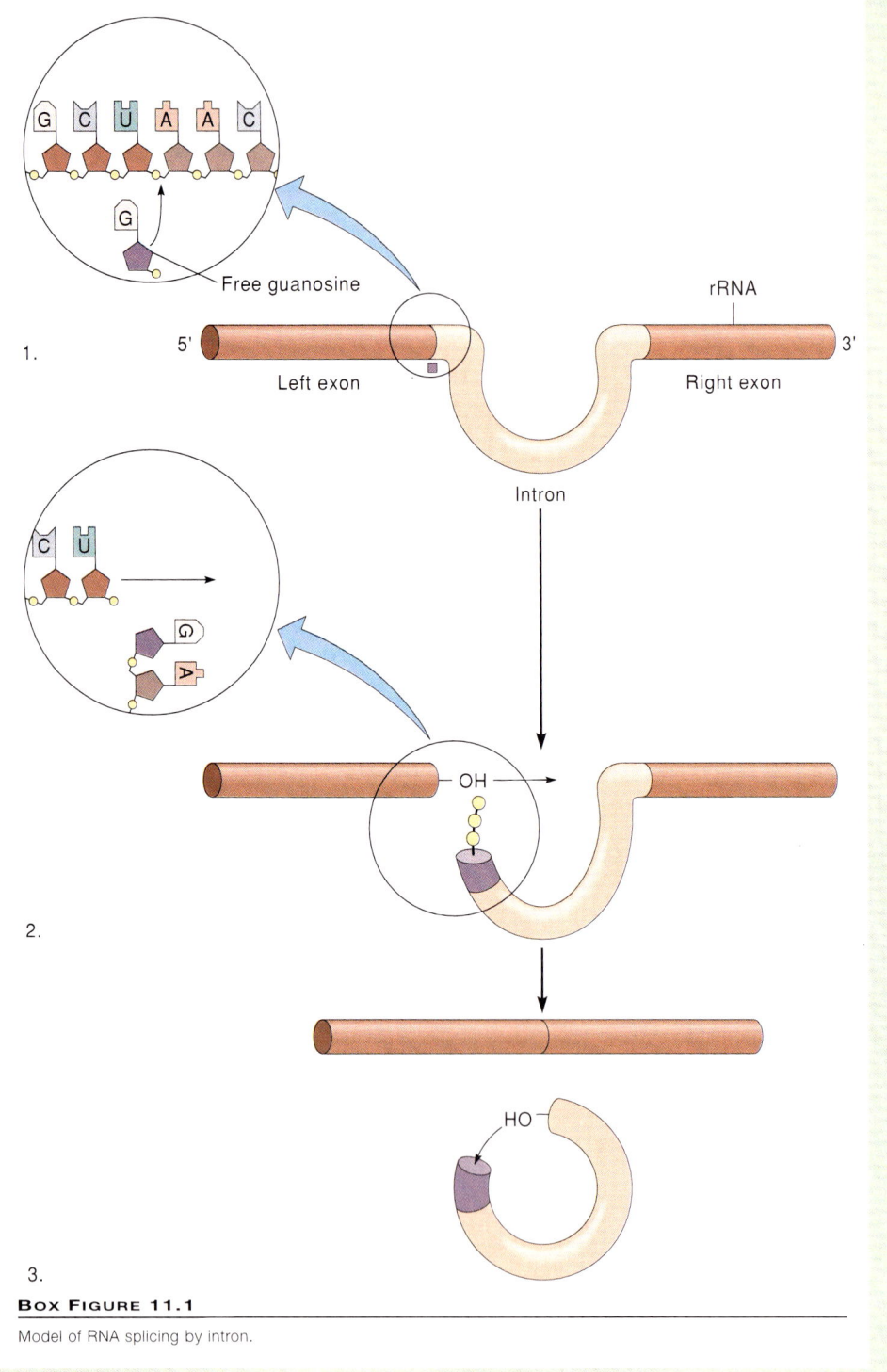

BOX FIGURE 11.1
Model of RNA splicing by intron.

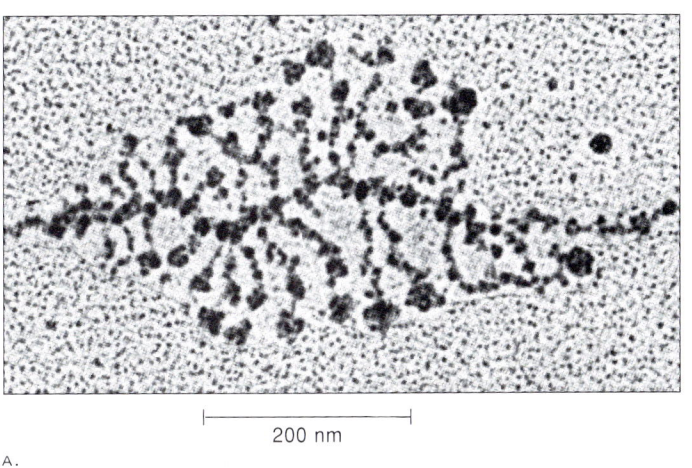
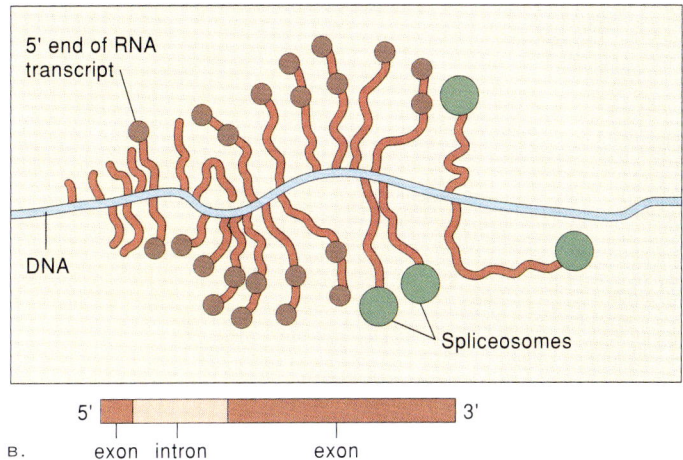

FIGURE 11.6

(a) Electron micrograph of RNA processing during transcription. (b) The larger circles represent whole spliceosomes. The smaller circles are probably snRNA-protein particles at the ends of introns. They will become spliceosomes as they get larger.
(b) Source: Alberts, Molecular Biology of the Cell, 2d ed. Copyright © Garland Publishing Company, New York, NY.

simultaneously. Spliceosomes begin processing RNA immediately after transcription, even as some parts of the RNA molecule are still being transcribed (fig. 11.6). After processing, the final form of the RNA molecule is often thousands of nucleotides smaller than the primary transcript. For example, it is common for a primary transcript of 8,000 nucleotides to be processed into an RNA molecule of about 1,200 nucleotides.

Most of the nucleotides in a primary RNA transcript do not function in polypeptide synthesis because they are cut out before the RNA leaves the nucleus. This means that primary RNA transcripts and the genes that code for them are interrupted by noncoding sequences. The functions of interruptions in genes are still unknown, but they may have roles in gene regulation or in genetic recombination. The experiments that led to the discovery of interrupted genes are described in box 11.2.

Messenger RNA

Messenger RNA (mRNA) gets its name because it contains the "message" (i.e., coding sequence) for a polypeptide. The mRNA molecule has four main parts: the GTP cap, the leader sequence, the coding sequence, and the trailer sequence with a poly-A tail (fig. 11.7). The largest part is the coding sequence, which usually ranges between a few hundred nucleotides and about 2,000 nucleotides among mRNA molecules from different genes.

Each polypeptide in a cell is coded by a separate mRNA. This means that there are thousands of different kinds of mRNA. Cells maintain at least 5–10 copies of most kinds of mRNA, but a few kinds of mRNA occur in 10,000 or more copies at a time. The number of copies of mRNA depends on the metabolic need for a particular polypeptide. However, cells usually contain only 3%–5% of their total RNA as mRNA.

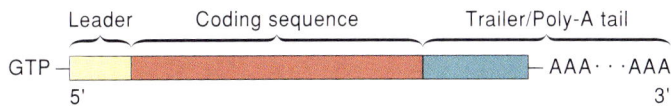

FIGURE 11.7

Model of mRNA that is exported from the nucleus. It consists of a GTP cap, a leader sequence, a coding sequence for a polypeptide, and a poly-A tail.

Transfer RNA

The *transfer* in transfer RNA (tRNA) refers to its role in bringing amino acids together during translation. Whereas there are thousands of distinct molecules of mRNA in each cell, there are only about forty different kinds of tRNA. Although the average size of a molecule of tRNA is only 80 nucleotides, tRNA makes up about 15% of total cellular RNA.

The probable three-dimensional structure of a tRNA molecule is shown by computer simulation in figure 11.8. Note that the molecule is twisted and folded back on itself, so that four double-stranded regions and three loops are formed. This is the characteristic form of most tRNA molecules.

Each tRNA molecule recognizes a certain sequence of nucleotides on an mRNA molecule and attaches to it. The recognition site on an mRNA molecule is called a **codon,** and the attachment site on the tRNA molecule is called the **anticodon.** The anticodon sequence, which occurs in one of the loops in the tRNA molecule, is complementary to the codon sequence. Each tRNA molecule also contains a specific sequence at its 3´ end, called the **amino acid acceptor site,** which recognizes and binds to one of the twenty amino acids that make up proteins.

Boxed Reading 11.2

The Discovery of Interrupted Genes

Until the 1970s, most studies of the structure and function of genes involved bacteria. Bacterial genes were found to be made of continuous coding sequences, and there seemed to be no obvious reason why genes of other organisms should be organized differently. However, by the early 1970s, hnRNAs (heterogeneous nuclear RNAs) had been discovered as the precursors of mRNAs in eukaryotes. Because hnRNAs and mRNAs have poly-A tails at their 3′ ends, geneticists assumed that mRNAs were derived by extensive degradation of a long leader sequence at the 5′ ends of hnRNAs. This hypothesis was rejected when biologists discovered that hnRNAs have 5′ caps and leader sequences that were preserved during the conversion of hnRNA to mRNA. Hindsight makes it seem obvious that nucleotides *within* the hnRNA molecule must be removed to make mRNA, thereby leaving the 5′ and 3′ ends intact. This hypothesis seemed absurd at the time, but by 1977 molecular biologists had evidence from two kinds of experiments to support it.

The first evidence for interrupted genes came from studies of a human virus called *adenovirus-2*. In these experiments, a segment of double-stranded viral DNA was denatured in a mixture that included

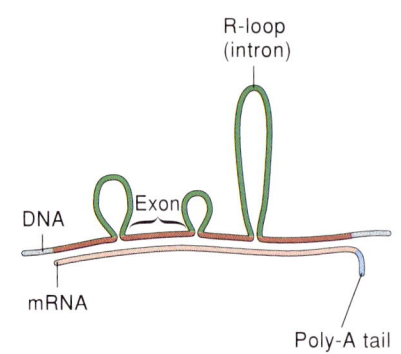

Box Figure 11.2

Evidence for introns came from DNA-mRNA binding experiments. Binding only occurs where both sequences are complementary, thereby excluding noncomplementary sequences of DNA (introns) as R-loops.

mRNA from a gene in the segment. The mixture was then cooled to a temperature that promotes annealing between RNA and DNA but not between two strands of DNA. However, "hybrids" between mRNA and its DNA template occurred only in regions where the two were complementary. Sequences of DNA within the gene that did not match the mRNA were displaced into loops, called **R-loops**, which could be seen by electron microscopy (see accompanying figure). Because they do not match the coding strand of mRNA, R-loops are noncoding sequences that interrupt the gene. Later, R-loops were also discovered in genes from plants and fungi, which means that interrupted genes are a general feature of eukaryotes.

At about the same time of the discovery of R-loops, molecular biologists discovered different but comparably surprising evidence for interrupted genes in chickens, rabbits, and mice. In these experiments, DNA was made by reverse-transcription (RNA → DNA) from an mRNA template using a special polymerase from viruses. Evidence for gene interruptions came from comparisons of this DNA with its original gene. Enzymes that cut DNA at specific sequences did not cut the reverse-transcribed DNA, but they did cut the original gene into several pieces. This observation is evidence that native genes contain sequences that are absent in the mRNA that is transcribed from them. Furthermore, since the cut sequences do not occur in mRNA, they must occur in noncoding regions—that is, in sequences that interrupt the gene. We now know that the noncoding portions of genes correspond to introns, which are removed during the processing of primary RNA transcripts into mRNA.

The discovery of introns was a major breakthrough toward understanding how genes work. In 1993 several scientists were awarded a Nobel Prize for their contribution to this discovery.

Ribosomal RNA

Ribosomes are the sites of protein synthesis. Each ribosome consists of a complex of ribosomal RNA (rRNA) and proteins, which are organized into a large subunit and a small subunit (fig. 11.9). In an average eukaryotic ribosome, the large subunit contains about 50 proteins and two kinds of single-stranded RNA: one that is about 4,700 nucleotides and one that is about 120 nucleotides. The small subunit contains more than 30 proteins and one kind of single-stranded RNA of about 1,900 nucleotides. More than half of the mass of a ribosome is RNA.

Transcription of rRNA molecules is similar to transcription of other kinds of RNA, except that rRNA transcripts are made by a different kind of RNA polymerase. Also, genes for rRNA occur in a string of tandem repeats (see Chapter 10). The number of each repeating unit varies in different plants, even within the same species. For example, rRNA genes range from as few as 3,000 copies to more than 9,000 copies in different strains of corn (*Zea mays*).

Some rRNA may be transcribed from tandem three-gene clusters on regions of chromosomes in the nucleolus (fig. 11.10). Other rRNA is transcribed from single genes in tandem but not in the nucleolus. This means that the construction of ribosomes is complicated by having to unite products from two separate gene clusters. The process is further slowed because transcribed regions are long, the primary transcripts are made of multiple subunits, and the completed rRNA must be packaged with many proteins in the cytoplasm. It may take up to an hour to construct one ribosome. Nevertheless, several million ribosomes are made during each cell cycle. Such large numbers of ribosomes can be made only because the cell contains many copies of ribosomal RNA genes and because each gene is

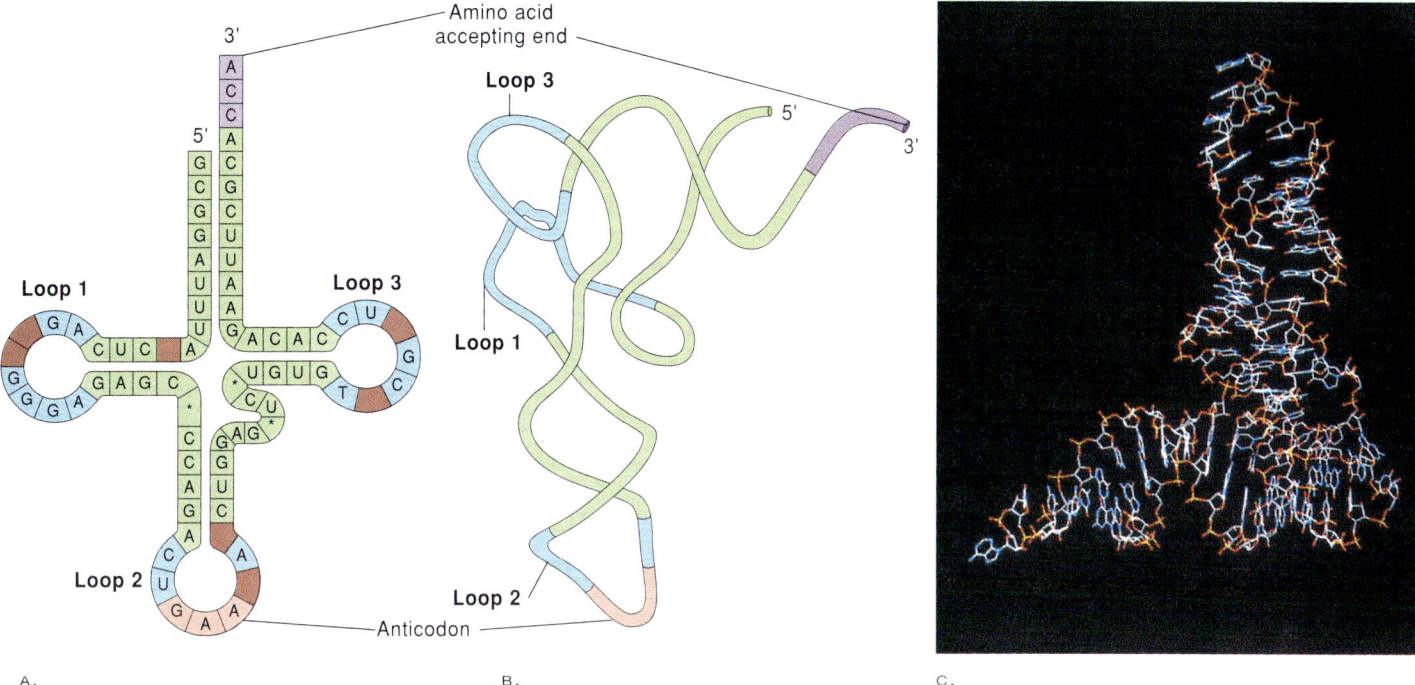

FIGURE 11.8

The structure of transfer RNA. (a) Two-dimensional representation, showing three loops and four regions of internally complementary nucleotides that form double strands. At one end of the molecule is the amino acid attachment site, and at the middle loop (Loop 2) is the mRNA recognition site (anticodon). (b) Diagram of three-dimensional representation. (c) Computer-generated three-dimensional structure of tRNA.

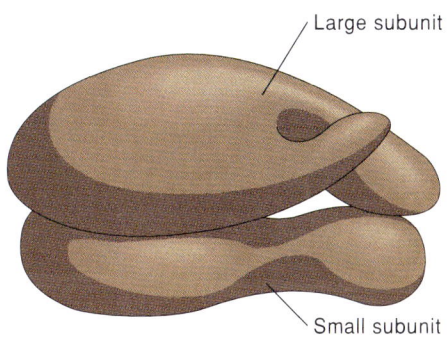

FIGURE 11.9

Structure of a eukaryotic ribosome. Each ribosome consists of a large subunit and a small subunit. The two subunits join together for protein synthesis.

transcribed simultaneously by many molecules of RNA polymerase. At least 80% of the total cellular RNA is rRNA.

CONCEPT

Three main kinds of RNA work together to make proteins. Messenger RNA carries the protein code from the gene. Transfer RNA is directed by mRNA to bring amino acids together in the proper order. Ribosomes, which contain protein and ribosomal RNA, are where mRNA and tRNA come together to make proteins in the cytoplasm.

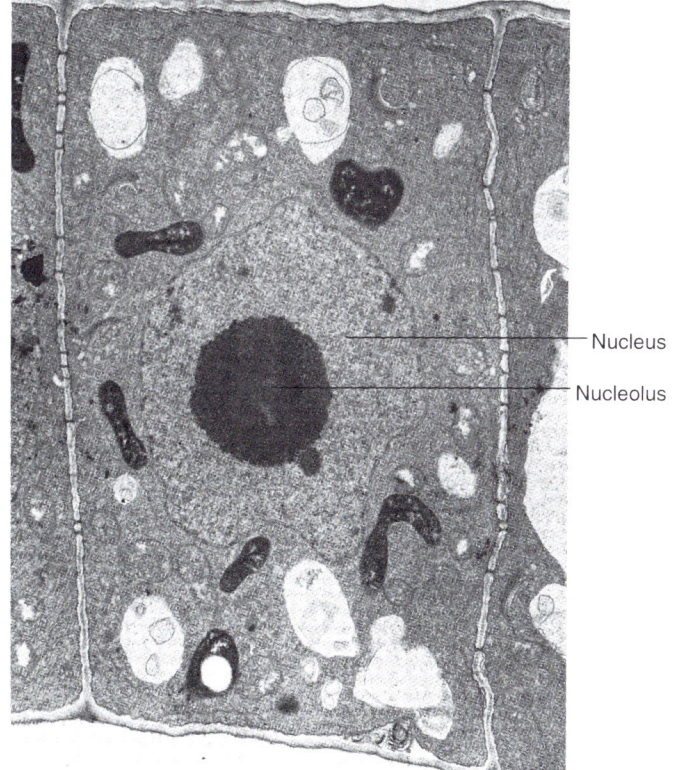

FIGURE 11.10

Electron micrograph of a nucleus with a nucleolus, ×14,000. The nucleolus is where some of the cell's rRNA is synthesized.

CHAPTER ELEVEN *Molecular Genetics and Gene Technology*

Polypeptide Synthesis: Translation

Translation is like transcription because it uses enzymes to make a polymer. Both processes occur in stages that include initiation, elongation, and termination. Both processes also require GTP. Translation differs, however, because it makes amino acid chains, not nucleotide chains.

Before Translation Starts

Molecules of tRNA must bind their appropriate amino acids before translation can start. Enzymes called **aminoacyl-tRNA synthetases** ensure that each tRNA molecule bonds covalently to a specific amino acid. There is a different synthetase for each of the twenty amino acids used in translation. Each aminoacyl-tRNA is made in a process that utilizes energy from ATP and includes an intermediate that consists of an amino acid and adenosine monophosphate (AMP) (fig. 11.11). In this way, for example, methionine is attached only to molecules of $tRNA^{Met}$, and valine is attached only to molecules of $tRNA^{Val}$.

Every aminoacyl-tRNA molecule has two main functions. One function is as a decoder of nucleotide sequences. The decoder function comes from its dual codes: the anticodon that reads a codon, and the amino acid acceptor site that reads an amino acid. The second function is to activate the carboxyl group of an amino acid so that it reacts with the amino group of another amino acid to make a peptide bond (see Chapter 2).

Steps in Translation

Translation starts when mRNA and an initiator tRNA bind to a small ribosomal subunit. Binding is aided by enzymes in the ribosomal subunit that recognize the GTP cap at the 5´ end of mRNA. The initiator tRNA is the molecule that carries the first amino acid for a polypeptide, which is usually methionine. The anticodon of the initiator tRNA is UAC, which is the complement of the **start codon** (AUG) on the mRNA. After the initiator tRNA, the mRNA, and the small subunit are in place, the large ribosomal subunit binds to the small subunit. This phase of initiation completes the assembly of a functional ribosome (fig. 11.12).

Elongation continues when a second tRNA molecule is attached next to the initiator tRNA on the large ribosomal subunit. The second tRNA is determined by the second codon on the mRNA. For example, the codon GUC will bind only to its complementary anticodon, CAG. The tRNA with the anticodon CAG carries the amino acid valine, which would therefore be the second amino acid in the growing polypeptide. The two amino acids are joined covalently when the enzyme **peptidyl transferase** moves the methionine from the initiator tRNA to the valine on the next tRNA (fig. 11.13). After methionine is bound to valine, $tRNA^{Met}$ is released from the ribosome. The ribosome then moves forward one codon along the mRNA

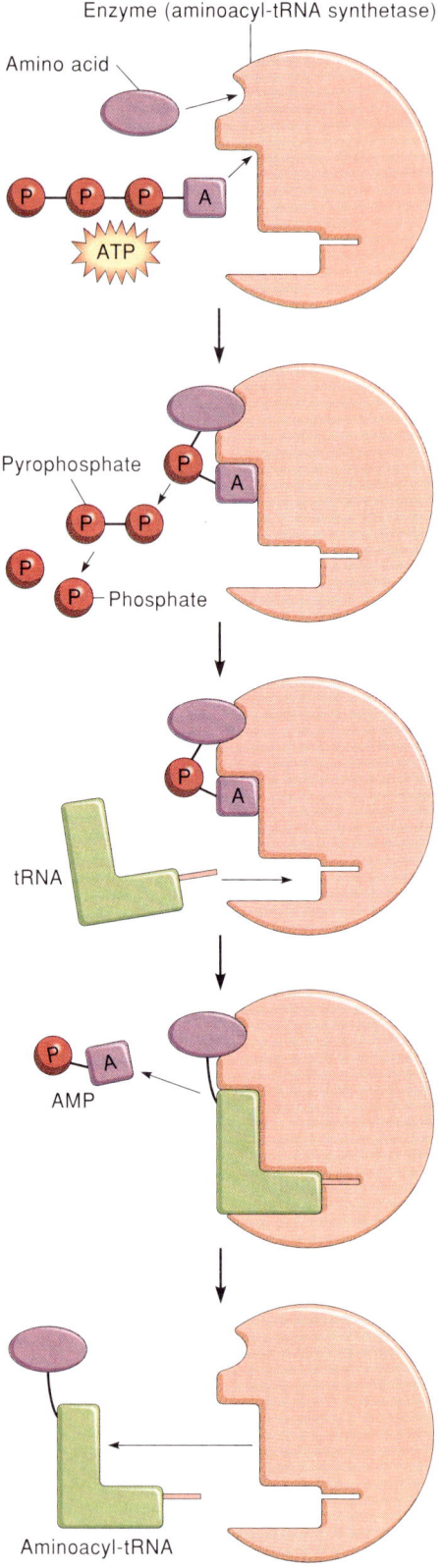

FIGURE 11.11

Synthesis of aminoacyl-tRNA. Aminoacyl-tRNA synthetase binds an amino acid and a molecule of ATP, using the energy of two phosphate bonds. The remaining AMP is replaced at the enzyme by the appropriate tRNA, which is then bound to the amino acid; AMP is released from the enzyme. The aminoacyl-tRNA complex is then released from the enzyme and can be used in protein synthesis.

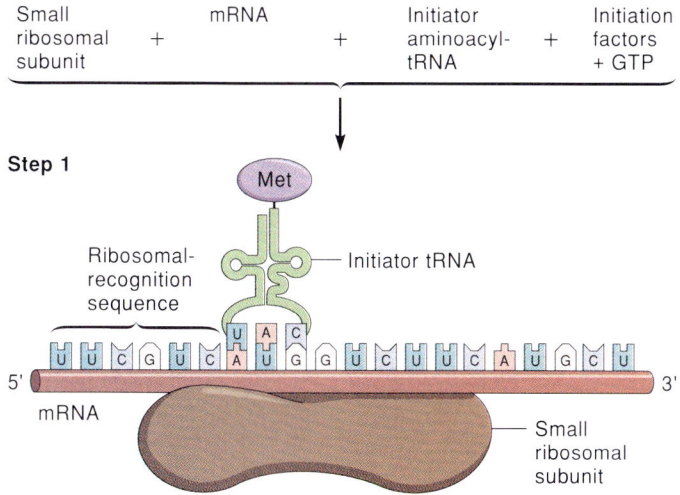

molecule. The movement of the ribosome brings the tRNA^{Val} molecule into the place vacated by the tRNA^{Met} molecule. A third tRNA then binds to the mRNA next to the tRNA^{Val}, bringing another amino acid with it. This process continues to link amino acids, one by one, until the polypeptide is complete (fig. 11.14).

Translation stops when the ribosome arrives at a **stop codon** on the mRNA molecule. The stop codon does not attach to tRNA; instead, it binds to special cytoplasmic proteins, called **release-factors.** Release-factors interrupt translation and hydrolyze the bond between the final amino acid in the new polypeptide and its tRNA.

Speed of Translation

The amount of polypeptide synthesis is the result of two levels of activity. The first level is the transcription of a gene to produce a large number of mRNA molecules. The second level is the repeated translation of each mRNA molecule. It takes less than a minute to translate mRNA into a polypeptide of 400 amino acids. Simultaneous translation by many ribosomes (*polyribosomes*) allows each mRNA molecule to make as many as ten polypeptides per minute. By using several thousand mRNA molecules at the same time, a cell can translate tens of thousands of polypeptides per minute. Continued translation at this rate produces millions of polypeptides during a single cell cycle.

Making the Finished Product: Polypeptide to Protein

Polypeptides must often be modified before they become functional proteins. For example, a polypeptide can be joined to one or more other polypeptides to make a protein having multiple subunits, as in the enzyme rubisco. Other changes occur in polypeptides that contain the amino acid cysteine. In these polypeptides, two cysteine molecules form bonds, called *disulfide bonds,* between their respective sulfur atoms. Also, membrane polypeptides bond covalently to sugars or lipids, thereby

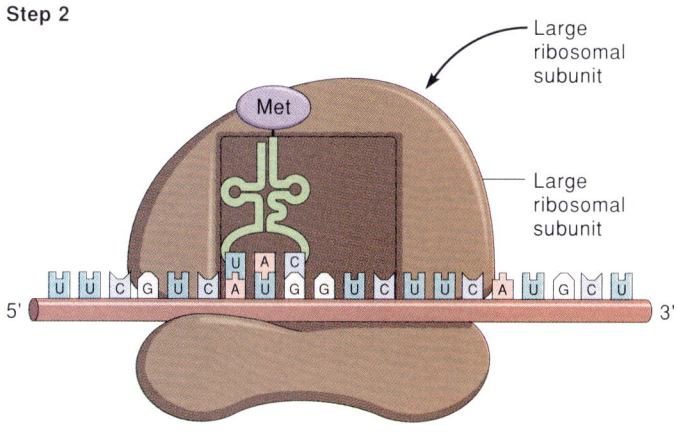

FIGURE 11.12

Initiation of protein synthesis. An initiator aminoacyl-tRNA binds to the start codon on mRNA, in a complex together with the small ribosomal subunit. Protein synthesis begins after the large ribosomal subunit is attached to this complex.

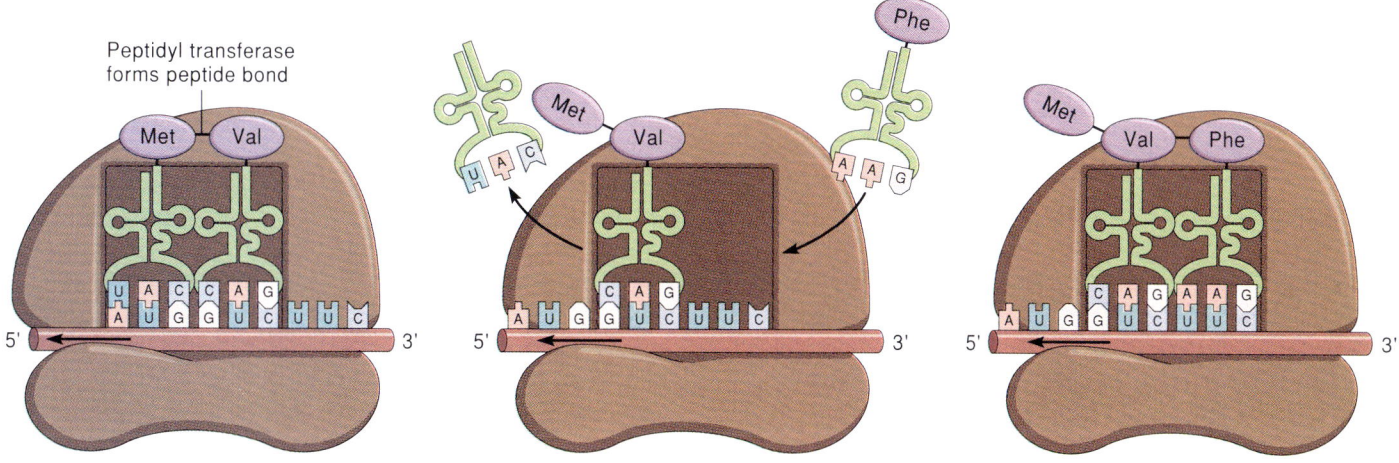

FIGURE 11.13

Steps in elongation. A second aminoacyl-tRNA binds to the ribosome next to the initiator tRNA, upon which the two amino acids are joined by a peptide bond. The initiator tRNA is then released, the mRNA advances one codon through the ribosome, and a third aminoacyl-tRNA binds to the spot on the ribosome that was vacated by the second aminoacyl-tRNA. A peptide bond forms between the second and third amino acids. These steps are repeated for each additional amino acid in the polypeptide.

CHAPTER ELEVEN *Molecular Genetics and Gene Technology*

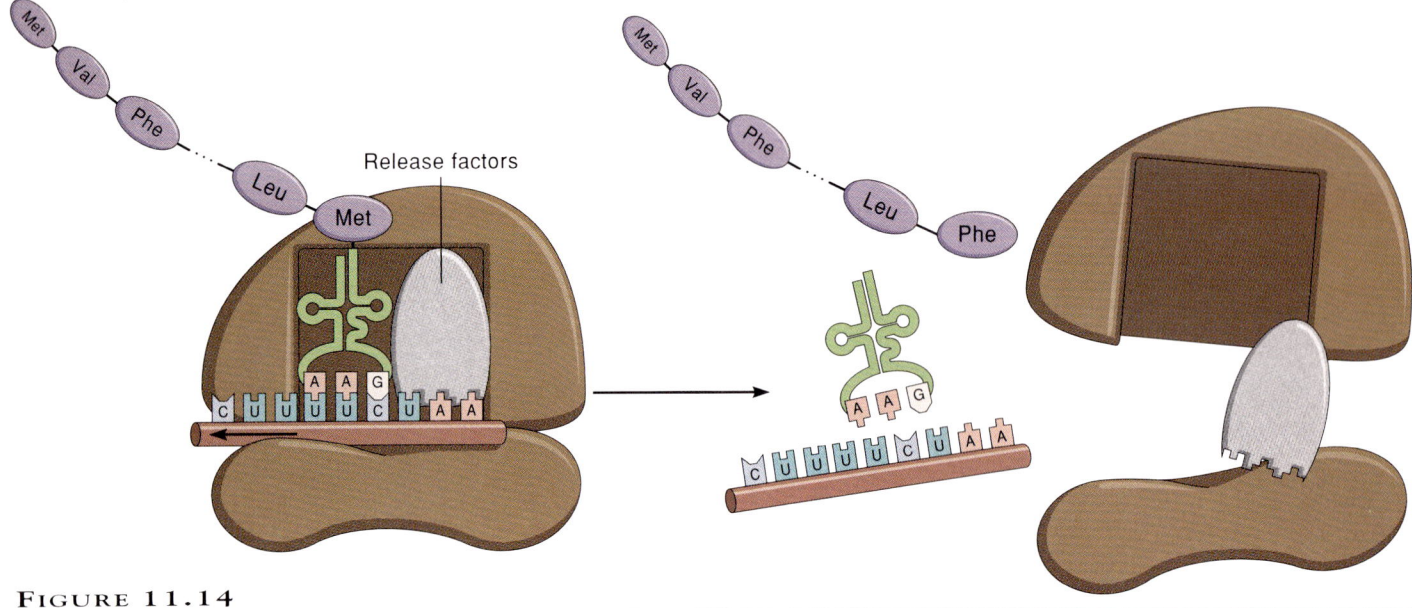

FIGURE 11.14

Translation stops when the ribosome reaches a stop codon. The stop codon binds to release-factors that hydrolyze the final aminoacyl-tRNA, thereby interrupting translation.

converting these polypeptides into glycoproteins or lipoproteins, respectively. Polypeptides that will help make cell walls or plasma membranes are often converted to such proteins in dictyosomes. Other kinds of polypeptides must bind to metal ions before becoming functional proteins. Examples include metalloenzymes such as iron- and copper-containing cytochrome oxidases and zinc-containing DNA polymerases and NADH-dehydrogenases.

C O N C E P T

Each codon in mRNA binds to a specific molecule of tRNA, which carries a certain amino acid. The mRNA sequence is translated into a polypeptide sequence when successive tRNA molecules bring amino acids together into a chain. Polypeptides often require posttranslational modification before they become functional proteins.

THE GENETIC CODE

The **genetic code** is a set of nucleotide messages that specify amino acids during translation. The smallest unit of the code is a codon. Each codon is three nucleotides long, collectively called a **triplet.** By using three nucleotides, the genetic code has 64 possible codons (4^3). (A two-nucleotide codon has only sixteen—that is 4^2—possible combinations of four nucleotides (4^2), which would not be enough for 20 amino acids.) The presence of 64 possible codons means that there are more triplets than needed for 20 amino acids and a stop signal. What do the other 43 codons do? As explained below, it took a long time to answer this question, but eventually all 64 combinations were found to be useful in the genetic code.

Although the size of a codon was easy to calculate, codon messages were not deciphered until the 1960s. The first triplet to be decoded was UUU. In 1961, Marshall Nirenberg made artificial mRNA with uracil as its only nucleotide. When he added this mRNA to the appropriate ingredients for protein synthesis, he obtained a polypeptide that contained only phenylalanine. In this way he discovered that the codon UUU designates the amino acid phenylalanine. In similar experiments, the other three uniform codons were also quickly deciphered: AAA for lysine, CCC for proline, and GGG for glycine. Codons for the other sixteen amino acids were more difficult to determine, but all codons had been deciphered by 1966, thirteen years after Watson and Crick proposed their model for the structure of DNA. The messages of all codons are listed in figure 11.15.

Note that all 64 triplets are used in the genetic code; 43 codons are not "left over" after 20 amino acids and a stop signal are coded. Instead, most amino acids are encoded by more than one codon. For example, lysine is coded by both AAA and AAG, and leucine is coded by six different triplets.

Codons that specify the same amino acid are called **synonymous codons** because they translate to the same product. The word *synonymous* may be misleading, however, because nobody knows if different codons for the same amino acid are identical metabolically. Although their products are the same, synonymous codons may be translated at different rates. Nevertheless, in all, there are 61 codons for 20 amino acids, and three codons (UAG, UAA, UGA) are stop signals. One of the coding triplets, AUG (methionine), is also a start codon.

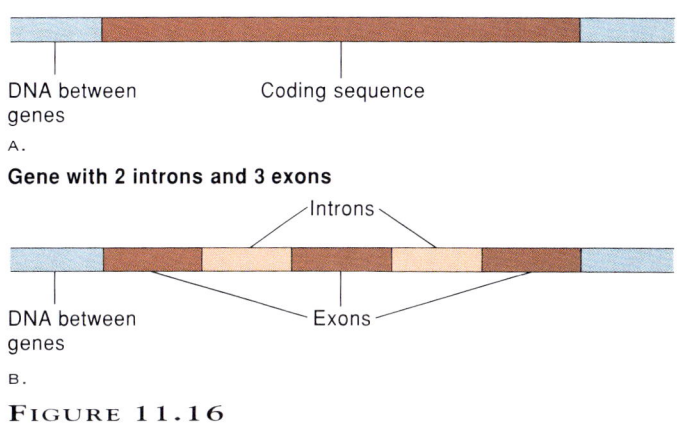

FIGURE 11.15

The genetic code. Read each codon in the 5′ to 3′ direction, starting with the first base on the chart, to find the message for each amino acid. Both the three-letter and one-letter abbreviations are given for each amino acid. Three of the 64 codons are stop signals; they do not translate into amino acids.

Knowing that the genetic code consists of triplet codons, we can see how a molecule of mRNA that is 1,200 nucleotides long makes a polypeptide having 399 amino acids. The first 1,197 nucleotides comprise 399 triplets for amino acids, and the last 3 nucleotides are a nontranslated stop codon. Also, since the start codon is usually AUG, the first amino acid in most polypeptides is methionine.

The Structure of Eukaryotic Genes

Genes that are made entirely of codons, such as genes of histone proteins, are the simplest genes (fig. 11.16a). The primary RNA transcripts of such genes do not need to be processed, because they do not contain introns. Conversely, many eukaryotic genes contain exons that are separated by introns (fig. 11.16b; look again at box 11.2, "The Discovery of Interrupted Genes").

The number of introns varies from one to several among different genes, but the number is usually stable in each gene for each organism. Conversely, the same gene may have different numbers of introns among different organisms. For example, one of the enzymes of glycolysis, triosephosphate isomerase, is coded by a gene that has eight introns in maize (*Zea mays*), six introns in the domestic chicken (*Gallus domesticus*), and five introns in the fungus *Aspergillus*. This gene has no introns in brewer's yeast (*Saccharomyces cerevisiae*) or in the gut bacterium *Escherichia coli* (fig. 11.17).

How many genes are interrupted? This question cannot be answered at present, but introns are continually being discovered in different genes. Genomes with the highest propor-

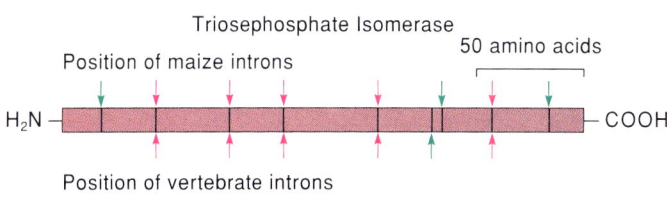

FIGURE 11.16

Diagram of (a) intron-free gene structure, and (b) intron-containing gene structure.

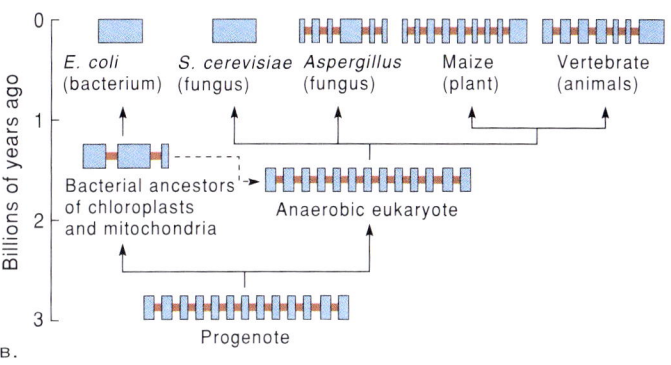

FIGURE 11.17

Evolution of interrupted genes. (a) A comparison of the enzyme triosephosphate isomerase in maize (*Zea mays*) and vertebrate animals reveals the protein to be translated from several exons that are shared between plants and animals. Five identical intron positions are indicated by red arrows; unshared intron positions are marked by green arrows. Shared intron positions are indirect evidence that these introns arose before the evolutionary divergence of plants and animals, more than 1 billion years ago. (b) Outline of the most likely path of evolution of the triosephosphate isomerase gene in different organisms. Exons are in blue and introns are in red. The common ancestor for all organisms is called a *progenote*. The dashed arrow marks the endosymbiotic invasions of prokaryotes that gave rise to chloroplasts and mitochondria (see boxed reading 3.2, p. 66).

tion of interrupted genes occur in the nuclei of complex eukaryotes—that is, in plants and animals. Fewer interrupted genes occur in genomes of chloroplasts and mitochondria, and in the nuclei of simpler eukaryotes such as yeasts. Introns are unknown in most prokaryotes.

Roles of Gene Interruptions

Although the functions of introns are unclear, scientists speculate that there are two possible roles for them. One is in regulating development. In this role, long or frequent introns slow gene expression because they take longer to process. Hence, different genes are expressed at different rates, depending on the size and number of their introns. The idea of a regulatory role for introns is supported by circumstantial evidence: Introns occur mainly in complex, multicellular organisms whose development is carefully regulated, including plants and animals. In contrast, single-celled organisms that grow as fast as possible, without complex developmental control, have either few introns or no introns at all. Bacteria that lack introns are in this group. Moreover, introns are rare in *Saccharomyces*, a single-celled fungus, but they are common in *Aspergillus*, a multicellular fungus.

Introns may also enhance genetic recombination during crossing-over in meiosis (see Chapter 10). According to this hypothesis, when unequal crossing-over occurs within a gene, homologs that break and reconnect in introns would not affect the coding sequence of the gene. In contrast, breakage and nonreciprocal exchange within an exon would probably disrupt the coding sequence and ruin the polypeptide coded by it. This means that genes with long or frequent introns have a better chance of successful genetic exchange than do genes with short or few introns.

The Exon-Shuffling Hypothesis

The genetic recombination hypothesis for the role of introns implies that each exon in a gene is somehow independent because it can move to another gene and still function. This means that structural and functional portions of a polypeptide, called **domains,** may be encoded separately by specific exons. Evidence for this hypothesis comes from comparing protein structures with gene structures. For example, the domains of triosephosphate isomerase generally correspond to specific exons (fig. 11.18).

The potential mobility of exons has led to the **exon-shuffling hypothesis,** which explains the orgin of complex new genes by the joining of independent exons into new combinations. Exon shuffling may therefore produce new proteins from already existing exons. Circumstantial evidence for this hypothesis comes from the similarity of exons in different genes. For example, exon 4 of the chloroplast gene, *psbA*, and exon 17 of the mouse band 3 protein gene both encode similar amino acid sequences.[1] Although they are in different proteins, both amino acid sequences function as membrane-spanning domains. These exons seem to be descendants of a single, ancestral exon that was reshuffled into different genes before the evolutionary divergence of plants and animals.

The idea that ancestral exons can be recombined to make new proteins has led a team of scientists, headed by Walter Gilbert of Harvard University, to suggest that existing genes consist of a much smaller number of reusable exons. By comparing amino acid sequences from all known exons, they determined which exons were most likely to come from the same ancestor. From these comparisons, they estimated that only 1,000–7,000 original exons were needed to make the hundreds of thousands of proteins that exist today. This proposal, however, is controversial.

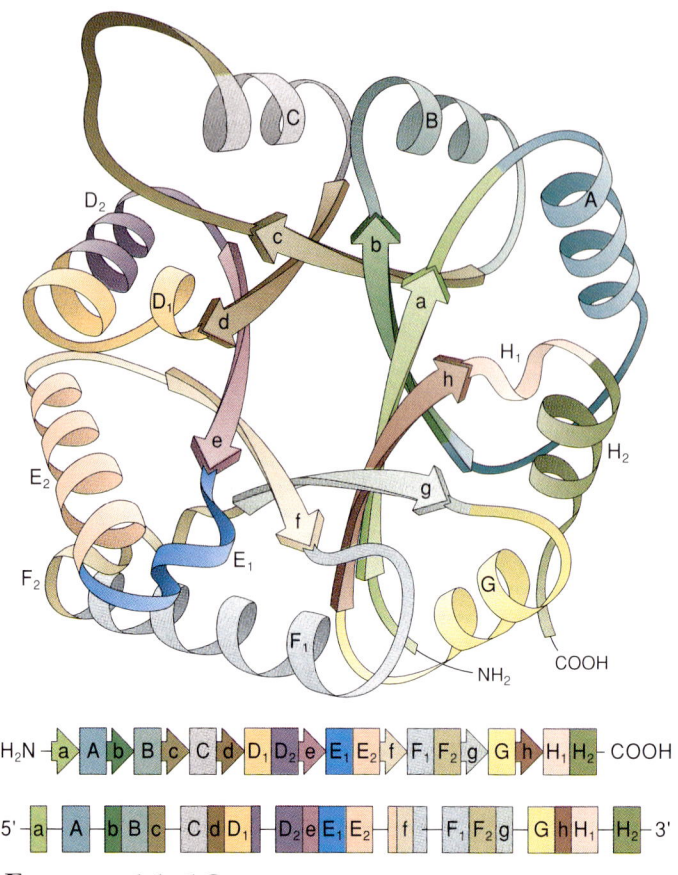

FIGURE 11.18

Computer-graphic model of triosephosphate isomerase enzyme, color-coded to show domains. Domains generally correspond to specific exons.

C O N C E P T

Genes begin with a start codon and end with a stop codon. Codons for amino acids and the start and stop signals are three nucleotides long. Many amino acids have more than one codon. In eukaryotic genes, the protein-coding sequence between the start and stop signals is often interrupted by noncoding sequences.

---

1. The name of a gene usually has a logical derivation. The gene *psbA*, for example, includes "ps" for photosystem, "b" for photosystem II, and "A" for the first enzyme of its type to be discovered (there are now more than a dozen known). These enzymes have a role in photosynthesis (see Chapter 7). Some genes, however, are nameless. For example, the gene for the third protein band on an electrophoresis gel, in an extract of mouse membrane proteins, is simply referred to as the gene for the mouse band 3 protein.

A. B. C.

FIGURE 11.19

Examples of genetically engineered plants. (a) Herbicide-resistant (right) and normal cotton. (b) Insect-resistant (left) and normal tobacco. (c) Spoilage-resistant (left) and normal tomatoes.

GENETIC ENGINEERING

Can you imagine a blue rose, a potato that makes plastic, a tobacco plant that glows in the dark, or a truly tasty tomato from the supermarket? All of these products have already been made by **genetic engineering,** which is the artificial manipulation of genes, or the transfer of genes from one organism to another (fig. 11.19). Such artificial (i.e., human-directed) exploitation of genes is also called **recombinant DNA technology.** In the above examples, the rose has genes from the petunia that control the synthesis of blue floral pigments. The potato has bacterial genes that make polymers that can be used to make biodegradable plastics. The luciferase gene, which causes fireflies to glow, has been transferred to tobacco plants as an easily identifiable marker gene for the detection of successful gene transfers; if the plants glow, then the transfer of the luciferase gene and other genes linked to it was successful (see box 11.3 entitled "Genetic Engineering with a Gene Gun, p. 253). These plants are all **transgenic,** which means that they contain genes from other organisms. Conversely, tomatoes are genetically engineered when the tomato gene for a specific carbohydrate-degrading enzyme is altered in the laboratory and reinserted into the tomato plant. The vine-ripened fruits from these plants, which have all their natural flavor, stay firm during shipping and storage.

Recombinant DNA technology relies on **vectors** (DNA carriers), such as viruses or bacterial plasmids, which are independent molecules of DNA apart from the primary bacterial genome. They are vectors because they can recombine foreign DNA into their own DNA, which is then replicated. In this way, DNA from plants or animals is **cloned** by inserting it into bacteria or viruses. As discussed below, DNA cloning is an important tool in gene technology. It is used to find and study important genes and to make genetically engineered plants.

DNA Cloning

Before recombination, both the foreign DNA and the vector DNA must be made receptive to each other. This job is done by bacterial **restriction enzymes,** which cut (i.e., restrict) DNA. Their use in DNA cloning is discussed in the next few paragraphs.

Restriction Enzymes

Different bacteria make hundreds of restriction enzymes, each of which recognizes a specific DNA sequence of 4–8 nucleotides. In nature, these enzymes protect bacteria by destroying foreign DNA. To prevent self-destruction, bacterial DNA is chemically modified to be inert to its own restriction enzymes.

An example of a restriction enzyme is **EcoRI,** which is harvested for commercial sale from *Escherichia coli*. *Eco*RI recognizes the nucleotide sequence GAATTC and then cuts the DNA between the guanine and the adenine. The six-nucleotide *Eco*RI site has identical sequences on both DNA strands, because its antiparallel complement is CTTAAG. This means that by cutting between guanine and adenine on each strand the enzyme makes a zigzag cut (fig. 11.20) which leaves short, single-stranded ends on each DNA double strand. Such

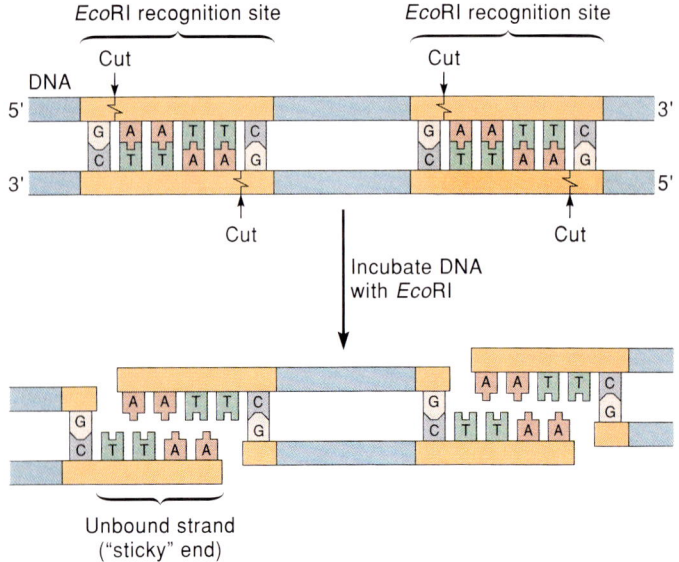

FIGURE 11.20

Like most restriction enzymes, *Eco*RI makes a zigzag cut wherever it finds a recognition site. Single-stranded portions of restriction fragments are called *sticky ends* because they can form hydrogen bonds with complementary sticky ends on other molecules of DNA.

single-stranded ends are called *sticky ends* because they can be "glued" by hydrogen bonding to complementary sticky ends of other DNA molecules.

Cloning in Plasmids

After cutting up a whole genome with a restriction enzyme, the next step in cloning DNA is to insert the DNA into a vector. One type of vector is a bacterial plasmid, which is a small, circular molecule of DNA (fig. 11.21). After the foreign DNA is inserted into isolated plasmids, the plasmids are absorbed back into bacteria and reproduced (i.e., cloned) during normal DNA replication.

Foreign DNA and plasmid DNA are receptive to each other when they are cut by the same restriction enzyme. For example, an enzyme such as *Eco*RI makes the same sticky ends on both the foreign DNA and the plasmid DNA. When the two kinds of DNA are mixed, they anneal because their sticky ends complement each other.

At first, plasmid DNA is weakly held to foreign DNA by hydrogen bonds between complementary nucleotides. The linkage between them is strengthened when sugar-phosphate bonds form between two molecules of DNA. Sugar-phosphate bonds are made by the enzyme **DNA ligase.** When this enzyme is added to the mixture, it joins (i.e., ligates) phosphate groups to deoxyribose between adjacent strands of DNA.

Cloning in Viruses

Methods for cloning DNA in bacteria also work with viruses. Like plasmid DNA, viral DNA is cut by restriction enzymes, and foreign DNA is inserted into it. Since viruses are parasites,

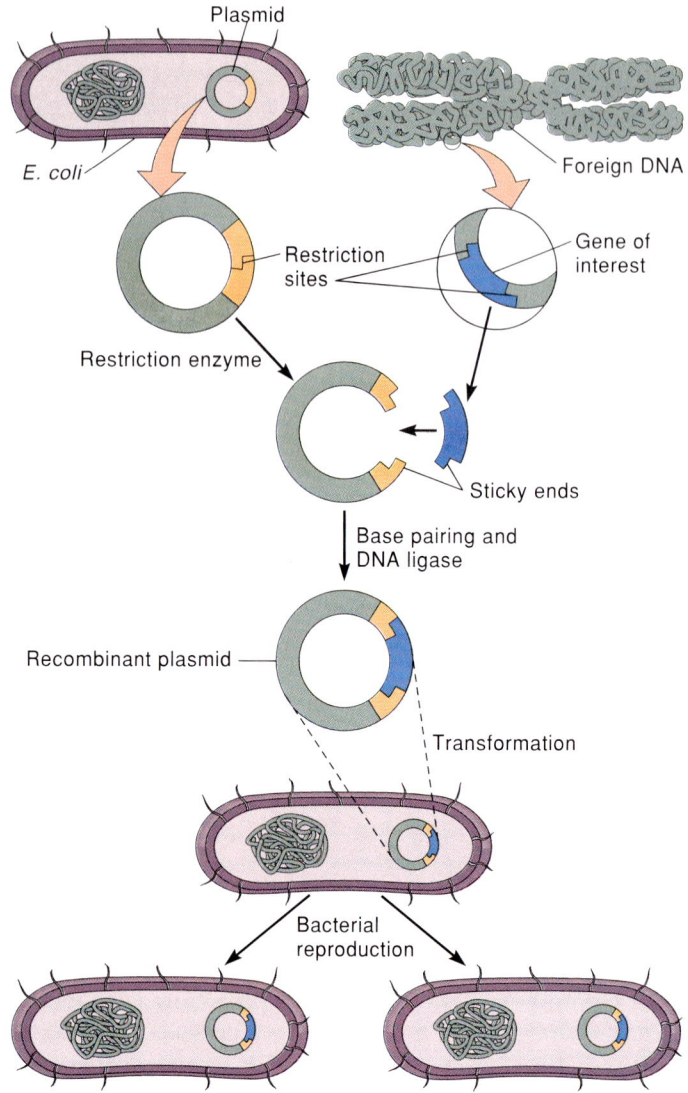

FIGURE 11.21

Gene cloning by bacteria. The gene of interest from another organism is cut out of its genome by a restriction enzyme and inserted into a bacterial plasmid that has been cut with the same restriction enzyme. Sticky ends of the foreign fragment bind to the complementary sticky ends of the plasmid, followed by the formation of new sugar-phosphate bonds through the action of DNA ligase. Finally, the altered plasmid is taken up by the bacteria, which make many copies of the foreign gene by duplicating the plasmid during reproduction.

however, they must be reintroduced into their hosts before they can replicate the recombined DNA. For cloning DNA in viruses, the most convenient hosts are bacteria such as *E. coli* that can be easily cultured in the laboratory. The most commonly used viruses for cloning DNA are bacteriophages.

Libraries of Genes

A whole genome can be cut by restriction enzymes, and all of the fragments inserted into vectors. This has been done for many plants, animals, and fungi. A fragment or gene of interest from any of these organisms can be retrieved from storage by

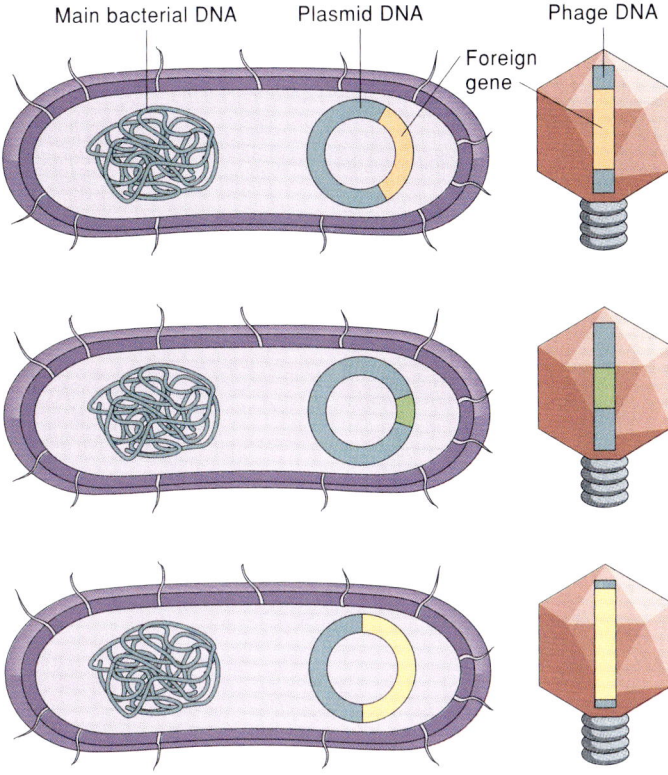

FIGURE 11.22

Genomic libraries are clones of DNA fragments, usually consisting of a single gene, that are stored in vectors. Cloning may be by plasmids or by phages. The bacterial cells on the left and the viruses on the right represent the cloning of three foreign genes by each type of vector.

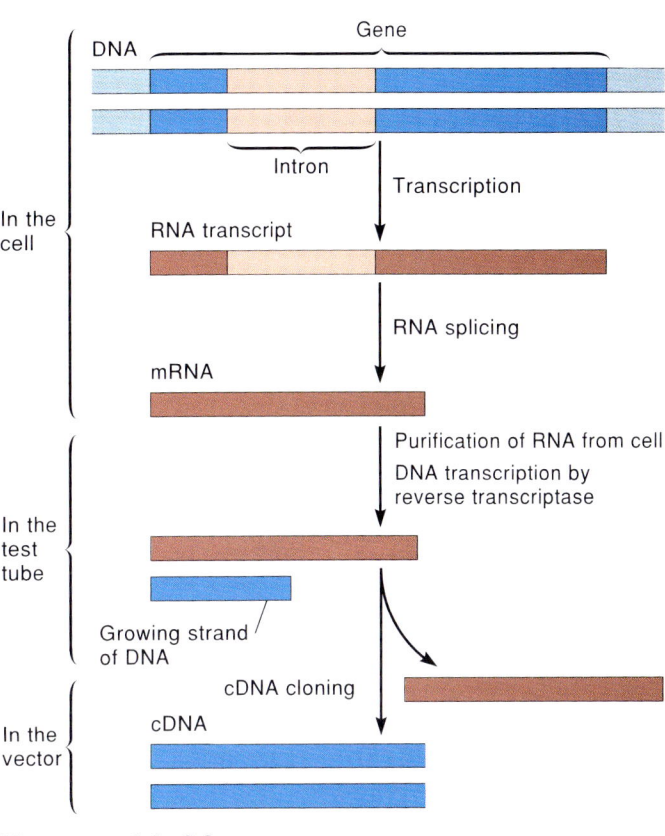

FIGURE 11.23

Steps for making complementary DNA. After transcription and editing, mRNA is used as a template for making DNA by reverse transcription. Unlike native genes, complementary DNA lacks introns because it is made from processed RNA.

culturing the appropriate vector. This storage and retrieval of genetic information is like having a library of genes, so scientists call the set of cloned fragments of a genome a **genomic library** (fig. 11.22); such libraries may have several thousand to several million entries, each a recombined plasmid or virus that contains a different restriction fragment.

Libraries are also made using complementary DNA (**cDNA**). To make a cDNA library, mRNA is reverse-transcribed to make cDNA. This reaction from RNA back to DNA is catalyzed by the enzyme **reverse transcriptase,** a viral enzyme that is available commercially (fig. 11.23). The cDNA is later modified to have the appropriate sticky ends, after which it is annealed and ligated to a vector. Libraries of cDNA are always incomplete because they contain only genes whose mRNA has been reverse-transcribed. In addition, genes are generally smaller in cDNA libraries than in their native state because, since they are made from mRNA, their introns are missing.

In addition to being clonable, some genes can make polypeptides in their host vectors. This means that intron-free genes in a genomic library and genes from a cDNA library may be translated in the vector. The products of these genes can therefore be made in large amounts for research or for industrial uses by culturing the vector. The first such product to be available commercially was human insulin, but many proteins are now being made by cloning. Unfortunately, however, intron-containing genes in genomic libraries cannot be translated in a bacterial vector, because prokaryotes lack the enzymes to process RNA.

Making and screening gene libraries is a tedious and time-consuming procedure. A plasmid or a virus can replicate only a few thousand nucleotides of foreign DNA, so many clones are needed to replicate whole genomes. For example, by chance alone, *Eco*RI will find the sequence GAATTC once out of every 4,096 nucleotides. This is 4^6, which is the number of ways four different nucleotides can be arranged in a six-base sequence. This means that the genome of *Arabidopsis thaliana*, the smallest known in plants (70 million nucleotide pairs), will be cut into more than 17,000 fragments with an average size of 4,096 nucleotides each by *Eco*RI (70,000,000/4,096 = 17,089). Thus, *Arabidopsis* would require more than 17,000 clones in its complete genomic library. By the same calculation, genomes of the garden pea (10.2 billion nucleotide pairs) would require about 2.5 million clones, and that of Easter lily (90 billion nucleotide pairs) would require about 22 million clones! However, gene libraries are incomplete because only part of an organism's genome can be efficiently taken up by vectors.

Writing to Learn Botany

Should we be concerned about releasing artificially produced genes into nature? Why or why not?

Making Artificial Chromosomes

One of the most important recent advances in genetic engineering is the construction of artificial chromosomes that can be cloned in yeast (*Saccharomyces cerevisiae*). These **yeast artificial chromosomes (YACs)** are made by putting fragments of foreign DNA into native yeast chromosomes. The YAC contains nonyeast DNA, a centromere, at least one replication origin, and two **telomeres,** which are DNA sequences that counteract chromosome condensation before mitosis (fig. 11.24). When such artificial chromosomes are introduced into yeast cells, they are replicated like native chromosomes.

Unlike bacterial or viral vectors, YACs can unite with and replicate millions of nucleotides of foreign DNA. Large fragments of foreign DNA are obtained by lightly digesting genomic DNA from the source organism, so that only some of the restriction sites are cut. The average size of restriction fragments made by *Eco*RI may therefore be in the millions of nucleotides instead of 4,096. A complete genomic library of three billion nucleotide pairs may require about a thousand YACs. Besides requiring fewer clones by this method, the construction of YACs has the advantage that intron-containing genes from the source organism can be expressed, because yeast (a eukaryote) has the appropriate enzymes for processing RNA. However, the use of YACs is so complicated and expensive that the only extensive work currently being done with them involves the human genome.

C O N C E P T

Bacteria and viruses can incorporate foreign DNA into their own. The entire genome of a plant can be cut into fragments and inserted into bacterial or viral vectors. By replicating foreign DNA with their own DNA, these vectors make clones of the genes inserted into them. The collection of bacterial or viral clones of plant DNA is like a genomic library. Genomic libraries can also be cloned in artificial chromosomes that are replicated in yeast.

Finding the Gene of Interest

Perhaps the most difficult and time-consuming step in genetic engineering is finding the gene of interest. One method is to isolate specific mRNA and then make a cDNA copy of it for cloning. For example, fungal infection stimulates soybeans to make large amounts of mRNA from the chalcone isomerase gene, a gene that is part of a biosynthetic pathway that makes

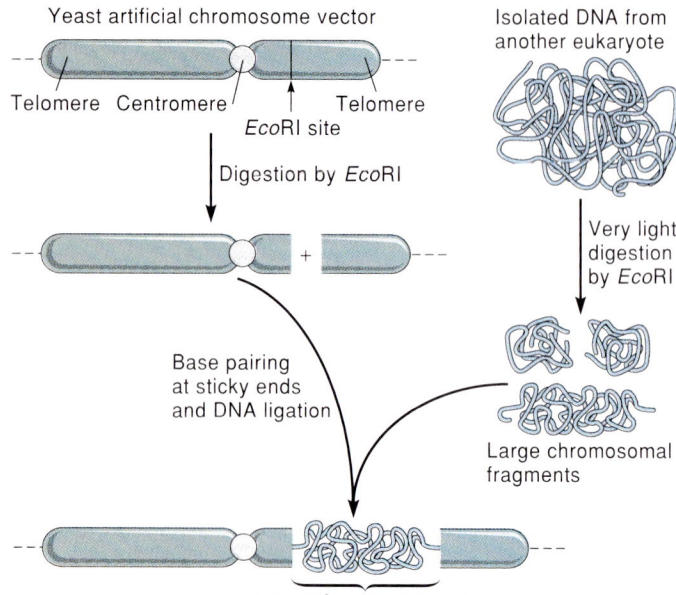

FIGURE 11.24

Yeast artificial chromosome with inserted foreign DNA. Construction of a yeast artificial chromosome (YAC). Once a YAC is made, it can be taken up by a yeast cell and replicated like a normal chromosome when the cell reproduces. YACs can replicate fragments of foreign DNA that are millions of nucleotides long.

compounds to fight the infection. After it is induced to make mRNA, the chalcone isomerase gene of the soybean can be cloned as cDNA.

Genes are found in gene libraries by using **probes** to search for them. A probe is a sequence of radioactive DNA that matches part of the gene of interest. In such cases, the amino acids are used to predict the DNA sequence that might code for the polypeptide. A short fragment containing this DNA sequence is made chemically and radiolabeled, and then used as a probe.

By using a probe, we can detect the gene of interest in bacterial cultures by **colony hybridization.** In this procedure, an agar plate containing bacterial cultures is blotted with filter paper to make a replica of the pattern of colonies. The replica is then covered in a solution containing the probe, which hybridizes (i.e., anneals) to complementary DNA in those colonies that have the target gene (fig. 11.25). Only those colonies that have the target gene will bind to the probe and become radioactive. Their location is marked by exposing the filter paper to X-ray film, which reacts with radiation from the probe-bound colonies. The radioactive colonies are then picked out of the culture and grown individually, thereby cloning the gene of interest for further study.

Gene libraries can be screened more directly by searching for specific polypeptides. For example, in the molecular genetic study of round vs. wrinkled seeds in the garden pea, the first

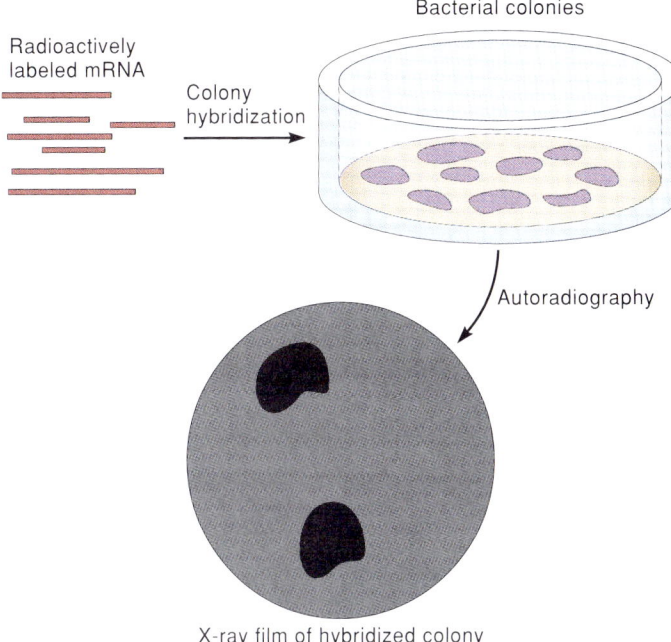

FIGURE 11.25

Colony hybridization. Only those colonies containing DNA of the target gene, which is complementary to the radioactive mRNA, will show up on X-ray film. This diagram shows that two colonies have the target gene.

step was to make an antibody against the purified starch-branching enzyme. When a solution of the antibody was infused into the vector culture of the gene library from *RR* plants, it reacted only in colonies containing the starch-branching enzyme, SBEI, thereby revealing the presence of the gene.

What are Botanists Doing?

Obtain up to five journal articles that use radioactive probes for finding a plant gene of interest. Examine the articles to find out how the probes were obtained. Based on this list, what is the most common method for obtaining probes for molecular genetic studies of plants?

Polymerase Chain Reaction

The newest technique for both finding and cloning a gene of interest is the **polymerase chain reaction (PCR)**, in which DNA polymerase assembles free nucleotides, thereby making a complementary sequence from a strand of template DNA.[2] In cells, the DNA polymerase replicates the entire genome during a cell cycle (see Chapter 9), but the goal of PCR is to replicate one gene, not the entire genome. Therefore, a specific gene is targeted by including **primers** for it in the reaction mixture. A primer is a small sequence of DNA, usually 10–30 nucleotides long, that is complementary to one end of the target gene. Once it is annealed to the gene, the DNA polymerase adds nucleotides to it along the template. Both strands of the target gene are copied by using two primers, one for the 5′ end of the coding strand and one for the 5′ end of the complementary strand.

One cycle of the PCR takes about ten minutes and requires three steps:

1. Denature the DNA at high temperature.
2. Cool the mixture to allow the primers to anneal to the target gene.
3. Incubate the mixture to allow the polymerase to make new DNA (fig. 11.26).

The amount of a target gene doubles during each cycle. Normally, the PCR runs for 30 cycles, which means that several million copies of a gene are made in a few hours. By comparison, a plasmid or a virus takes several weeks to make the same amount of a specific gene.[3]

Industrial Uses of Genetically Engineered Bacteria

As described above, bacterial vectors are used to store and retrieve genes in modern genetic research. Genetically engineered bacteria are also important for making foreign gene products on a commercial scale. The polypeptide of interest is either harvested from bacteria that are grown in large vats, or the live bacteria are used directly (fig. 11.27). The synthesis of a eukaryotic polypeptide in a vector is often enhanced by linking the gene to a highly active promoter, which helps to make the polypeptide in large amounts at relatively low cost.

Most commercial uses of genetically altered bacteria involve agriculture and medicine. In agriculture, bacteria that contain foreign genes are used directly on crops. For example, genetically engineered **ice-minus bacteria** have a foreign gene whose polypeptide inhibits the formation of ice crystals. In another example, root-colonizing bacteria have been altered to make a protein toxin that protects crops from root pests. These bacteria contain the toxin gene from another species of bacteria, *Bacillus thuringiensis*, which does not colonize roots.

---

2. Kary B. Mullis is given credit for inventing or discovering PCR, but what he actually did was to show how the mechanisms of DNA synthesis in the nucleus could be imitated in a test tube. He devised such a method in 1983, and in 1993 he was awarded a Nobel Prize for this work.

3. Techniques such as PCR and the analysis of DNA fragments from restriction enzymes are the basis for DNA typing that is now used in criminal trials, paternity suits, and the identification of abandoned babies. Gel electrophoresis of PCR products or restriction fragments looks like a bar code on a grocery store product, and each person (except identical twins) has a unique bar code. Such *DNA fingerprinting* was used recently to identify a plant for a murder case in Arizona. Seeds from a palo verde tree (*Cercidium floridum*) were found in the back of a suspect's truck. DNA from the seeds matched that of a tree near the scene of the crime.

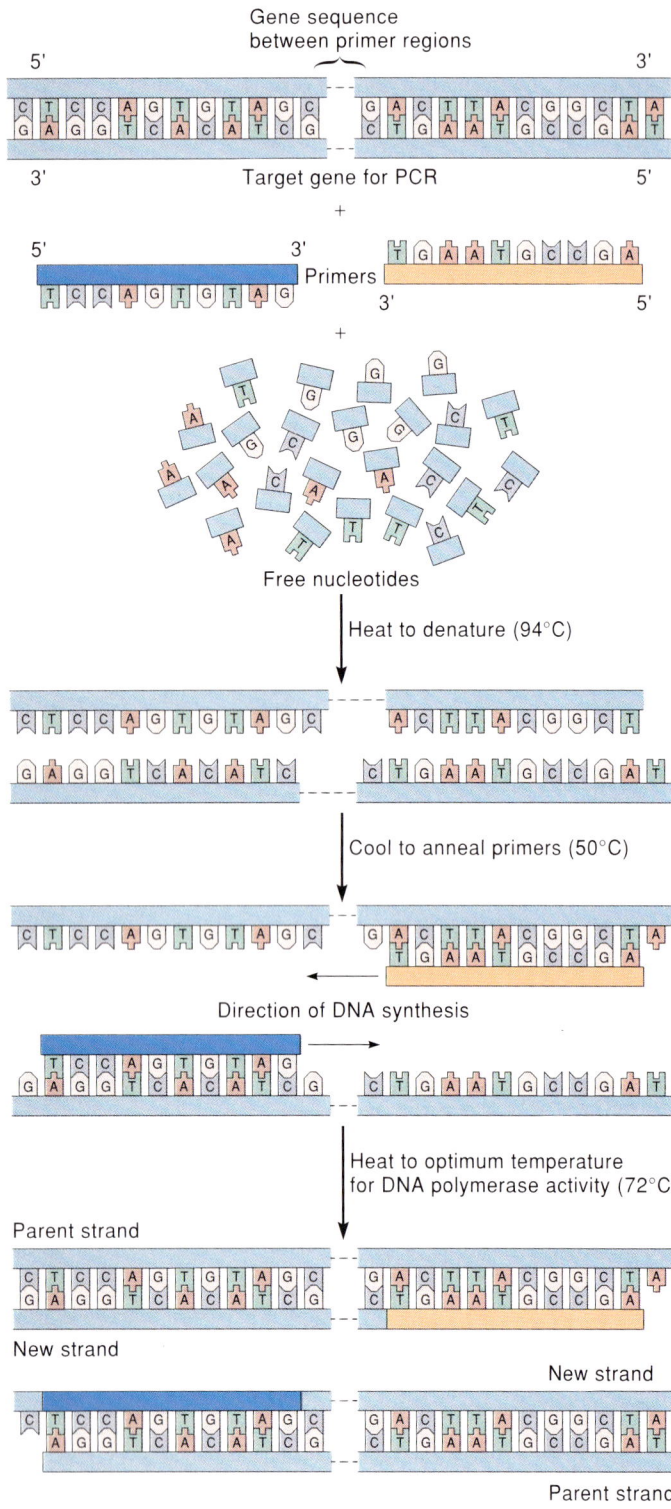

FIGURE 11.26

The polymerase chain reaction. Target DNA is amplified by DNA polymerase after primers anneal to their complementary sequences in the target. The cycle is repeated until several million copies of the target gene have been made. Temperatures in each cycle and the number of cycles vary for different target genes and primers.

FIGURE 11.27

An industrial fermenter used to grow genetically engineered bacteria. Such large-scale bacterial cultures produce large amounts of the foreign polypeptide at relatively low cost.

 *Doing Botany Yourself*

Describe how you might find a gene of interest for which there was no available probe.

In medicine, genes that code for the synthesis of several polypeptide drugs are now produced by bacteria. Unlike the bacteria used in agriculture, which contain genes from other prokaryotes, drug-producing bacteria contain genes from eukaryotes. These genes make pharmaceutically important proteins, including insulin used to treat diabetes, human growth hormone used to treat dwarfism, and interferons and interleukins used to stimulate the immune system. Before these genetically engineered products were available, all such proteins had to be obtained from animals. Human growth hormone was especially scarce because it had to come from human cadavers.

In genetic engineering, the product of interest is not always a protein. For example, genetically engineered plants are now used to make drugs (e.g., morphine, quinine, codeine) food additives (e.g., vanillin, mint oil), and fragrances (e.g., jasmine oil, rose oil). Similarly, potato plants are now used to make a type of plastic polymer. Genes for this polymer come from certain bacteria that are difficult to grow and that make only small amounts of the polymer. The polymer genes have now been inserted into *E. coli*, which is easy to grow, but polymer

FIGURE 11.28

The tumorous growths on these *Gladiolus* bulbs arose from cells that were induced to become meristematic by the bacterium *Agrobacterium tumefaciens*. Such growth is known as *crown gall disease*.

production is still low. Scientists are searching for promoters that will enhance expression of the polymer genes. If this work is successful, biodegradable plastics from bacteria may replace petroleum-based plastics, which remain in the environment too long.

Making Transgenic Plants: Problems and Solutions

The main problem in making transgenic plants is that, unlike bacterial plasmids, plant DNA cannot be recombined with foreign DNA by the direct use of restriction enzymes, nor can plant cells be transformed easily by absorbing DNA. Nevertheless, nature has its own genetic engineers that can overcome these difficulties: parasites that have special genes for inserting their DNA into plant genomes. One such parasite is the bacterium *Agrobacterium tumefaciens*, which causes plants to grow galls (tumors) (fig. 11.28). *Agrobacterium* has a plasmid, called the T_i plasmid, that combines with plant DNA (fig. 11.29).

The most common way to make a transgenic plant is to use the T_i plasmid to insert a gene of interest into a plant genome. First, the gene is recombined with the plasmid, and then the plasmid is absorbed by the bacteria. When the bacteria infect plant cells, recombined T_i plasmids combine with the plant's chromosomes. The foreign gene is then replicated in each cell cycle as if it were a normal part of the plant genome. Ideally, the newly inserted gene is also expressed in all of the cells that contain it.

When these methods are used on whole plants, only those cells infected with recombined T_i plasmids will be transgenic. To make the plant fully transgenic, however, all of its cells must be genetically altered. For many plants, this is a minor problem that is solved by genetically engineering their cells in culture and growing the cultured cells back into whole plants (fig. 11.30). The genes of this plasmid are then replicated and expressed as

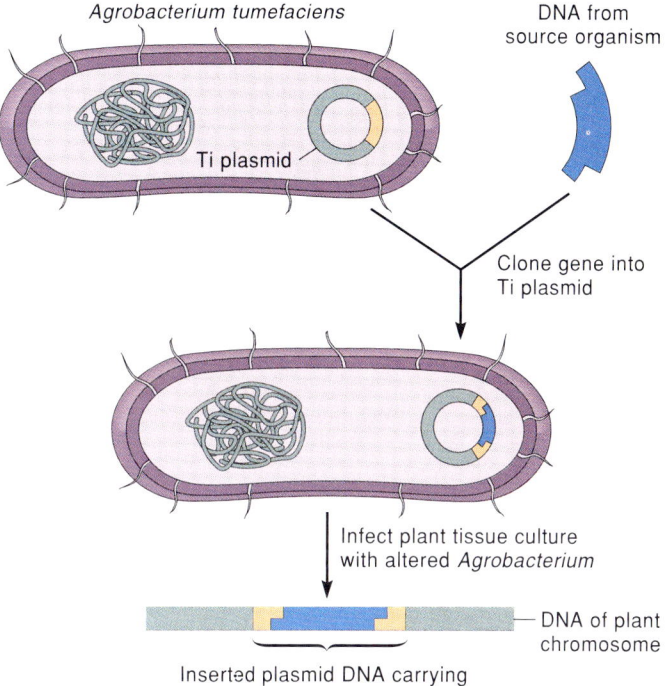

FIGURE 11.29

Inserting foreign genes by using the *Agrobacterium* T_i plasmid. First, the bacterium is transformed by the insertion of a foreign gene into its T_i plasmid, which carries genes for inserting itself into plant chromosomes. Then a plant cell that is infected with the plasmid receives the foreign gene into its genome.

if they belonged to the plant's genome. Such genes reactivate the cell cycle in cells that had stopped in the G_1 phase (see Chapter 9). These cells then divide rapidly and form plant tumors.

Antisense Technology

Although particular genes can be inactivated by inserting transposons or by mutagenic chemicals, these processes are random and inefficient, and can alter genes that we do not want to change. We can avoid this problem by using **antisense technology,** which inactivates specific genes. To appreciate this technique, recall that active genes produce proteins, usually enzymes. When one copy of a strand of DNA is transcribed, the sequence of bases in the DNA is preserved, except that the sequence is exactly complementary to the original code. The mRNA, now containing instructions from DNA in its own language, is then translated into protein. If done correctly, the message is said to be in correct **sense.** However, if an exactly complementary (i.e., antisense) RNA were present to combine with the sense RNA, the message of the sense RNA would be nullified.

An antisense message can be made from the "other" strand of DNA and introduced into a cell, or it can be transcribed by the cell after an antisense gene is introduced. When the sense RNA couples with its antisense RNA, it cannot combine with a ribosome, and is inactivated. Thus, the original gene is blocked by its own antisense RNA, without affecting other genes.

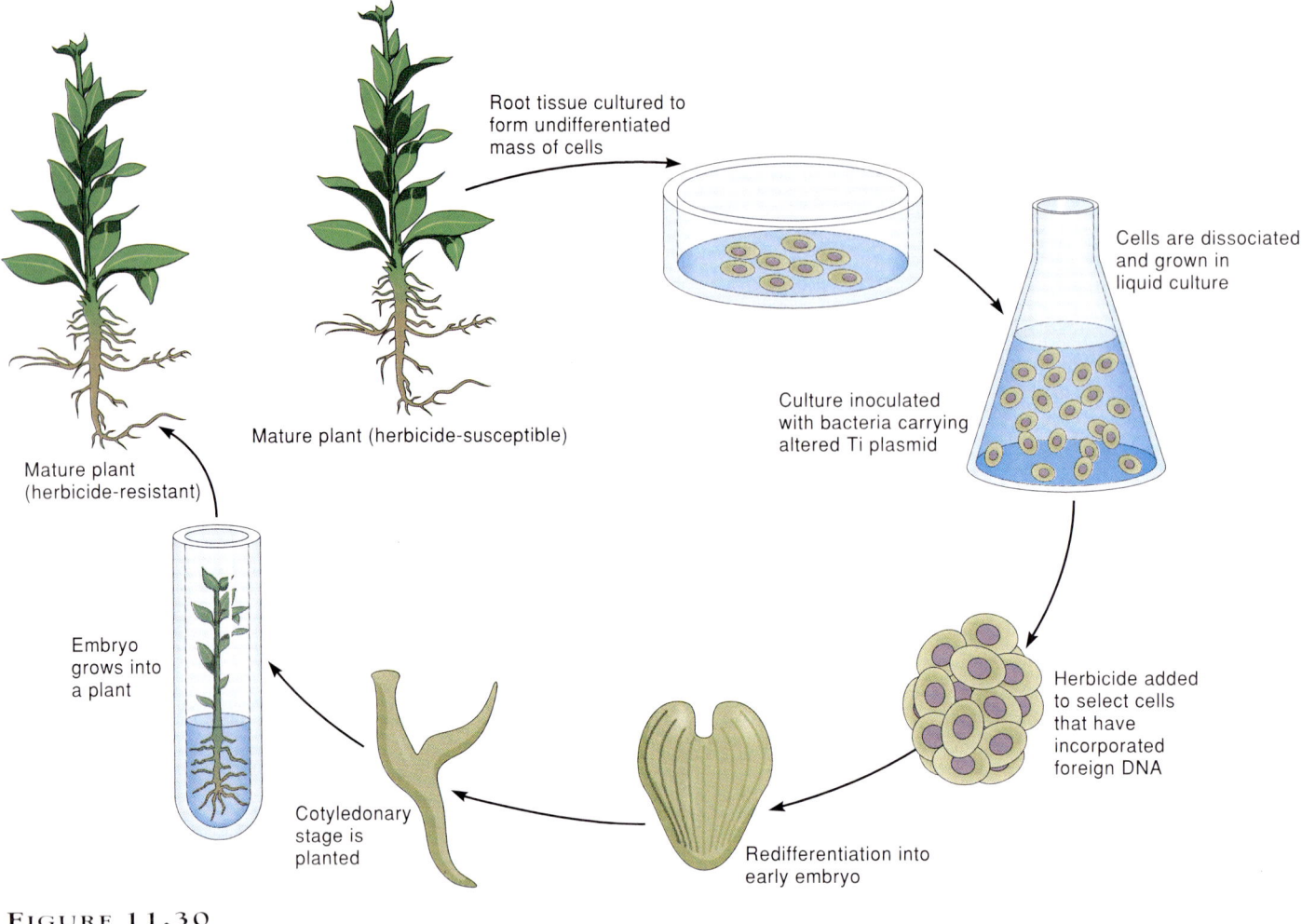

FIGURE 11.30

Steps in making transgenic plants. Differentiated cells from a plant tissue are put into cell culture, where they grow into an undifferentiated mass of cells. The culture is transferred to a liquid medium and infected with T_i plasmids carrying a gene for herbicide resistance. Cells that survive herbicide treatment are transgenic. Plants that are regenerated from these cells are also transgenic, making them resistant to the herbicide in the garden.

Antisense technology has been used to control a variety of plant processes, including fruit ripening (see "Improvement in Fruit Storage," p. 254). Because of its specificity and simplicity, antisense technology has tremendous potential for use in agriculture.

Monocots vs. Dicots in Genetic Engineering

Although gene transfer in plants is achieved by the T_i plasmid of *Agrobacterium*, this bacterium mostly infects plants in one of the two classes of flowering plants, the *dicots*, which is characterized by seeds that have two embryonic leaves (cotyledons). Dicots include tobacco, tomato, potato, and petunia. Plants in the other class, the *monocots*, have seeds with one cotyledon; they include cereal grasses, lilies, and palms.

Cereal grasses (e.g., rice, corn, and wheat) are the most important food crops worldwide, but because they are not readily susceptible to *Agrobacterium*, genetic engineering of such crops has lagged behind that of dicot crops. Cereal grasses are also difficult to grow into new plants from cell culture.

A potential solution to the problems of genetic engineering in cereal grasses was discovered in the late 1980s. This method is called *electroporation* and involves several steps. First, cells suspended in liquid culture were digested to remove their cell walls. The remaining cells, called *protoplasts*, were then shocked, either osmotically or electrically, to open temporary pores in the cell membranes. Foreign DNA, which was mixed into the culture medium, passed through these pores in some of the cells. The few cells that withstood the shock treatment continued to grow in the suspension culture. Finally, successful colonies were transferred to culture plates of nutrient agar, where some of them formed new plants.

Methods for introducing foreign genes into cereal grasses are still not very successful. Only 1 in 10,000 or 1 in 100,000 protoplasts can divide after being shocked, and most of these cells do not form plants. Nevertheless, protoplasts of corn and rice have taken up and integrated foreign DNA into their genomes, and the newly recombined genes are expressed in regenerated plants (e.g., see the photo that opens this chapter).

BOXED READING 11.3
GENETIC ENGINEERING WITH A GENE GUN

A **gene gun** is a device that shoots DNA directly into cells. The DNA is first coated onto tiny metal beads, which are placed on the end of a cylinder that is put into the barrel of a .22-caliber gun. The gun fires the cylinder into a retaining plate, but the beads keep going through a hole in the plate and into the plant tissue below. This injects DNA into cells that are hit just right by the metal beads. Some of these cells integrate the foreign DNA into their chromosomes.

The advantage of the gene gun is that it introduces DNA into differentiated tissues, such as leaves or meristems, so that there is no need for cells to be grown in an undifferentiated state. However, as with most other methods for transferring genes, blasting genes into tissues is a random process: scientists cannot predict where foreign DNA will be integrated into the plant genome. Consequently, some method of visualizing the location of a successful entry is necessary. The usual solution is to splice *reporter genes* to the DNA that is introduced; these genes "report" their presence, such as by producing colored products from chemicals supplied to the tissue. For example, the GUS reporter gene produces glucuronidase, an enzyme that produces a blue color wherever it acts, whereas the *lux* reporter gene encodes the firefly enzyme luciferase, which causes the cells to glow in the dark. Other reporter genes confer resistance to antibiotics.

The first transgenic plants made by a gene gun were grown in 1988, so this method is too new to know how useful it will be for genetically engineering crops. Estimates are that the efficiency of gene blasting may be about the same as the efficiency of making recombinant DNA by other methods.

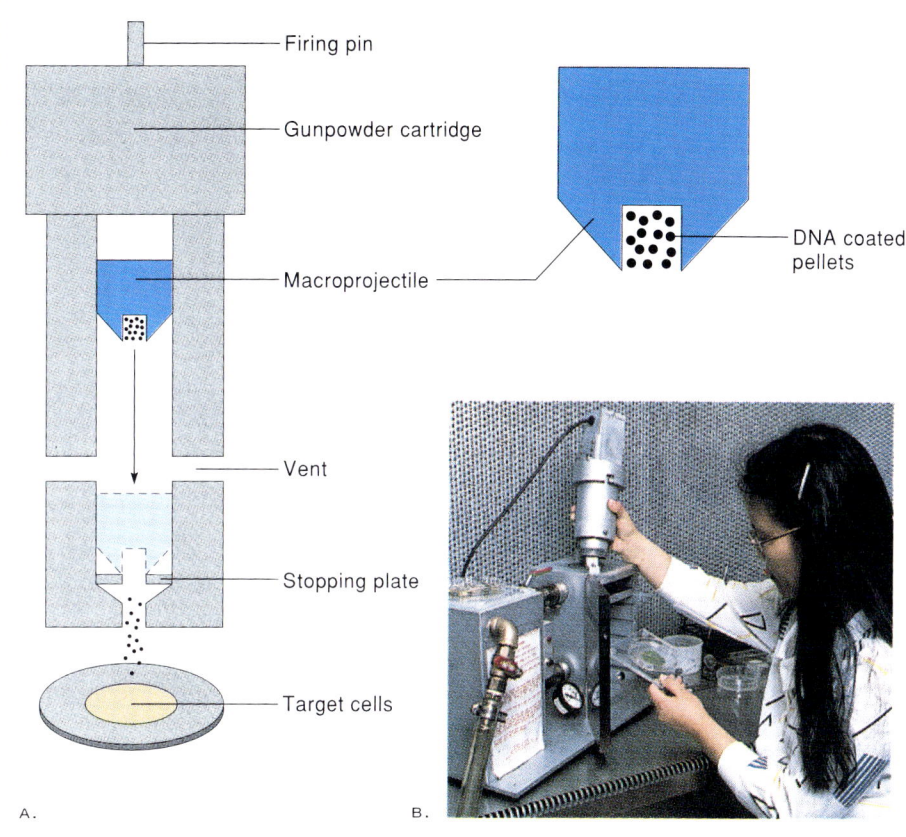

BOX FIGURE 11.3

(a) Diagram of a DNA particle gun. (b) Technician preparing to use a DNA particle gun.
(a) From "Transgenic Crops" by Charles S. Gasser and Robert T. Fraley. Copyright © 1992 by Scientific American, Inc. All rights reserved.

CONCEPT

Transgenic bacteria are made directly by inserting foreign genes into their plasmids, but transgenic plants must be grown from cell cultures that are infected by parasitic bacteria carrying foreign genes. This technique is limited to dicots, because monocots resist infection by the appropriate bacteria. This means that the world's most important food crops, which are monocots, cannot be genetically engineered as easily as dicots.

Transgenic Crops

Several goals of recombinant DNA technology in agriculture are listed in table 11.1. Of these, the development of herbicide resistance has been the most successful. Transgenic plants that resist certain insects or viruses have also been made. Most of the goals listed, however, have not yet been accomplished. Examples of those that have been achieved, at least experimentally, are discussed in the following paragraphs.

TABLE 11.1

Goals of Recombinant DNA Technology in Agriculture

1. Crop resistance to herbicides.
2. Crop resistance to insects and diseases.
3. Reduction of photorespiration in C_3 crops.
4. Atmospheric nitrogen fixation by crop plants.
5. Tolerance to high-salt soils and to flooding in crops.
6. Resistance to drought in crops.
7. Increased nutritional value of plant storage proteins.
8. Cold-tolerance in tropical and subtropical crops.
9. Longer storage life of fruits and vegetables.
10. Enhanced production of plant chemicals used in pharmaceutical and food industries.
11. Enhanced productivity in ornamental and food plants.

From D. Grierson and S. N. Covey, *Plant Molecular Biology*, 2d ed. Copyright © 1988 Blackie Academic and Professional, an imprint of Chapman & Hall. Reprinted by permission.

FIGURE 11.31

Normal tomato plants (*right*) and transgenic plants with a gene for herbicide resistance (*middle*) were treated with the herbicide glyphosate. All normal plants died. Plants in the left row were not treated with herbicide.

Herbicide Resistance

Most plants are killed by **glyphosate,** a herbicide that inhibits an enzyme called *EPSP synthetase* that is required for making aromatic amino acids. (EPSP stands for the product made by EPSP synthetase, which is 5-**E**nol**P**yruvyl-**S**hikimic acid-3-**P**hosphate.) When plants cannot make aromatic amino acids, their metabolism stops and they die.

Glyphosate-resistant petunias have been made by inserting extra copies of the EPSP synthetase genes into them. These petunias make enough enzyme to overcome inhibition by glyphosate. Plants that contain a bacterial form of EPSP synthetase also resist glyphosate (fig. 11.31).

Ideally, crops that resist glyphosate do not have to be weeded. Treating a field with the herbicide kills the weeds but does not affect the genetically engineered crop.

Disease Resistance

Tobacco mosaic virus (TMV) causes a mosaic pattern of infection on tobacco leaves (fig. 11.32). When the gene for the protein coat of TMV is transferred to tobacco, the plants become resistant to the virus. The TMV responds to the genetically engineered tobacco cells as if they were already infected, and since the virus cannot infect cells that are already infected, the plant is immune to infection.

Insect Resistance

As discussed previously in this chapter, crops can be protected against insects when they are sprayed with bacteria that contain the toxin gene from *Bacillus thuringiensis*. This gene has also been inserted into tomato, potato, and tobacco plants. Caterpillars that normally eat the leaves of these species do not eat plants that contain the toxin made by this gene (fig. 11.33).

Perhaps the most important crop for genetically engineering resistance to insects is cotton; indeed, most of the chemical pesticides used in the United States are used on cotton. Experimental cotton plants that contain the toxin gene from *B. thuringiensis* have better resistance to insects and require fewer chemical pesticides.

Drought Resistance

Genes from desert petunias have been transferred into cultivated petunias. The transgenic petunias require much less water than the normal plants. If crop plants likewise can be altered genetically to require less water, there will be less need for irrigation. Such crops will become more important as the demand for water increases in semiarid agricultural regions of the southwestern United States.

Changing Storage Proteins

Grains from cereal grasses are deficient in lysine, which is an essential amino acid in our diet. However, induced mutations in cultured rice cells have increased their lysine and protein production. Plants from these mutant cell cultures produce grains containing more lysine and protein than do normal rice plants. Ideally, the gene for the high-lysine mutation will be transferred to other cereal grasses, but the gene has not yet been isolated.

Improvement in Fruit Storage

One of the main problems with fruits is that when they are left on the plant too long, they get mushy during shipping and storage. To overcome this problem, fruits are picked and shipped while they are still immature; they are ripened later by exposure to the hormone ethylene. Unfortunately, such artificially ripened fruits often have little flavor.

FIGURE 11.32

This tobacco leaf is infected with tobacco mosaic virus (TMV). Resistance to this virus can be engineered into tobacco by using a gene from the virus itself.

FIGURE 11.33

Infrared aerial image (enhanced by false color) of potato fields. Normal potatoes were defoliated by the Colorado potato beetle, thereby allowing wet ground to be exposed (*green*). Beetles avoided transgenic potatoes that contained an insecticide-producing gene (*red*). (White areas are wheat fields.)

Genetic engineering has been used to solve this problem for tomatoes, which become mushy because a carbohydrate-digesting enzyme breaks down cell walls in the fruit. Specifically, botanists have introduced an antisense message for the enzyme system that controls over-ripening of the fruit. Consequently, these plants produce fruits that do not become mushy. Such genetically engineered tomatoes that are left to ripen on the parent plant maintain their firmness (and flavor) during storage.

The Lore of Plants

In a recent survey of consumers, only 25 percent could stomach the idea of food that is genetically engineered by mixing genes between plants and animals. In contrast, 66 percent of those polled approved of foods made by transferring genes from one plant to another. Nevertheless, the International Food Biotechnology Council lists twenty-six companies that are working on at least seventy genetically altered foods, including such cross-kingdom marvels as a tomato that makes "antifreeze protein" with a gene from an Arctic fish. In anticipating that such foods will be approved for commercial sale, some 1,500 well-known chefs across the U.S. have joined a "Pure Food Campaign" to boycott genetically engineered food.

THE MOLECULAR GENETIC BASIS OF ROUND VS. WRINKLED PEAS

Let's now return to our round and wrinkled peas. So far you have learned that the wrinkled-pea phenotype is probably caused by the absence of a certain starch-branching enzyme (p. 232), and that the gene for this enzyme was detected by colony hybridization with an antibody against the enzyme in a gene library of *RR* plants (p. 248). The question remains, What is the difference between the *R* allele and the *r* allele?

The first part of the answer is that the antibody reacts to the gene library from *RR* plants but not to the gene library from *rr* plants. This does not mean that the gene locus is missing in *rr* plants, but it does mean that the *r* allele is not translated. The next part of the answer was found when the *R* allele from the *RR* gene library was used to probe the *rr* genome.

The gene library that was used to find the *R* allele came from a digestion of the *RR* genome by *Eco*RI. This digestion gave two fragments that contained the *R* allele, which means that the allele has an *Eco*RI site (GAATTC) in it. Clones containing these *Eco*RI fragments were used as the probes.

The first probe analysis was done by a procedure called **Northern blotting.** For this procedure, all of the RNA is extracted from the plant and separated by gel electrophoresis (Chapter 8 describes gel electrophoresis of proteins). Nucleic acids, like proteins, are ionic and therefore separable in an electric field, and like proteins, they are separated on the basis of different sizes: larger fragments are slowed more by the gel, so they do not move as far. For a Northern blot analysis, after electrophoresis the RNA is transferred to filter paper (*blotting*), and the filter paper is immersed in a solution that contains the radioactive probe. The probe then binds to sequences that are complementary to it wherever they occur on the filter paper. Sequences that bind to the radioactive probe are then detected by exposing the probed filter paper to X-ray film (fig. 11.34).

The RNA extracts from *RR* and *rr* plants showed a significant difference when probed by the *Eco*RI fragment of the *R* allele. The *R* allele made large amounts of an mRNA molecule that was about 3,300 nucleotides long, but the *r* allele

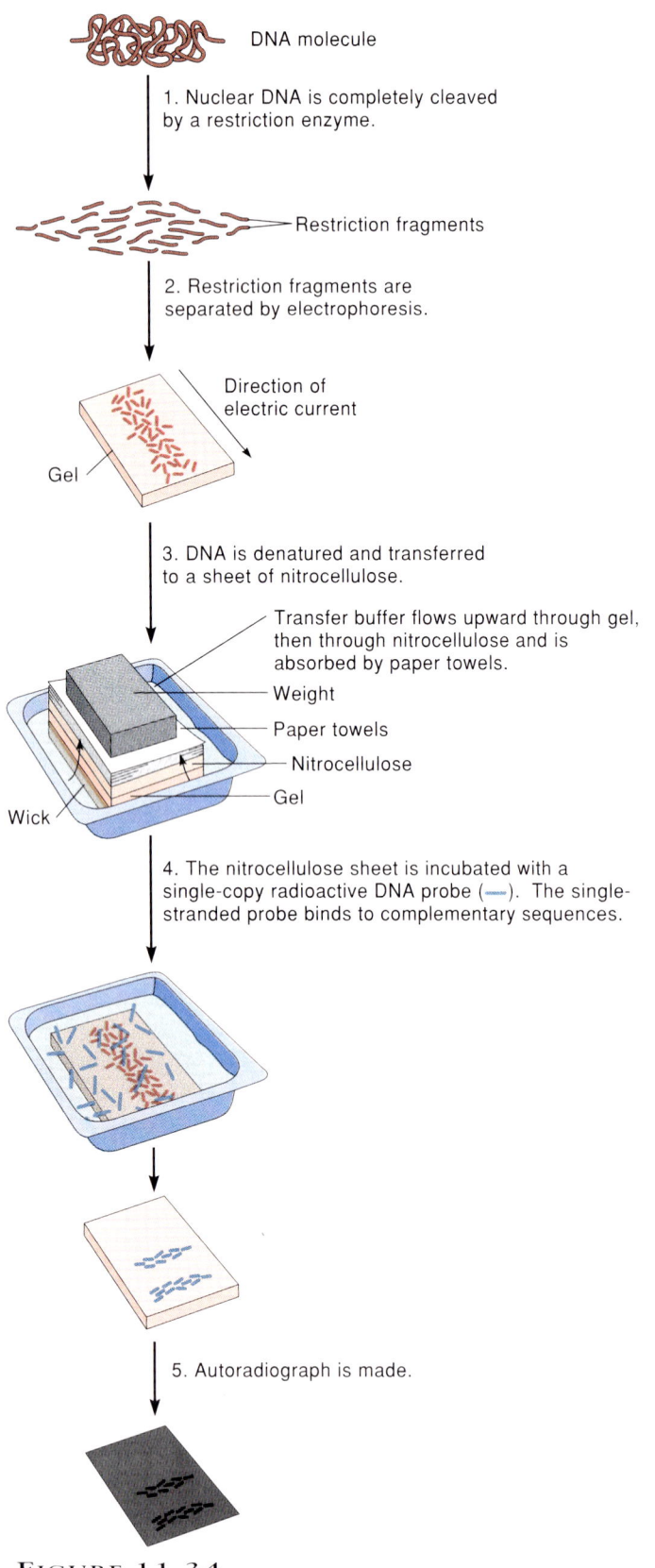

FIGURE 11.34

Detection of target DNA by Southern blotting and target mRNA by Northern blotting. Different names are used for the detection of each type of nucleic acid, although the blotting procedure is the same.

Source: Fristrom, Principles of Genetics, 2d ed. Copyright © Chiron Press, Cambridge, MA.

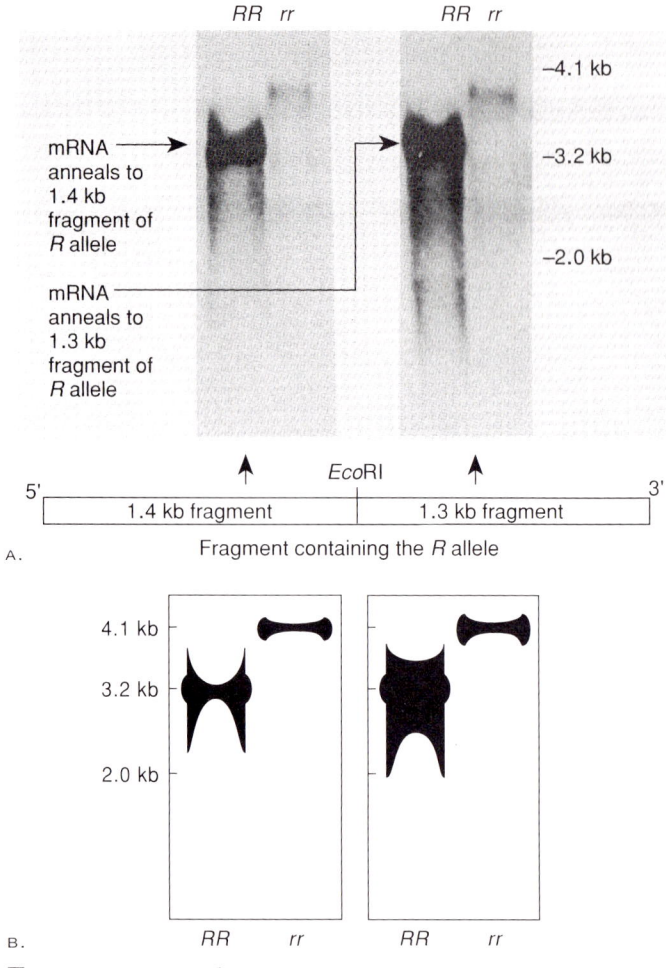

FIGURE 11.35

(a) Northern blots of mRNA extracts from garden pea embryos with either the genotype *RR* or *rr*. Blots were made using radiolabeled fragments of the *R* allele. (b) The *RR* genotype makes large amounts of mRNA that is about 3,300 nucleotides; the *rr* genotype makes small amounts of a larger mRNA molecule (about 4,100 nucleotides). This is evidence that the *r* allele contains an insertion that interrupts translation.

made much smaller amounts of an mRNA molecule that was about 4,100 nucleotides long (fig. 11.35). The first probe analysis showed that mRNA from the *r* allele is about 800 nucleotides larger than mRNA from the *R* allele. Also, the *r* allele is not transcribed as much as the *R* allele. This is evidence that the 800 extra nucleotides have been inserted into an exon and that this insert blocks translation. If the insert had been in an intron, it would have been removed by RNA processing and therefore would not have appeared in the mRNA.

In the second analysis, *RR* and *rr* genomes were digested with different restriction enzymes, and the DNA was probed directly. This type of analysis is called **Southern blotting,** which is named for its discoverer, E. M. Southern (RNA detection is called *Northern blotting* simply to distinguish it from Southern blotting). Analysis by Southern blotting showed that the *r* allele is larger than the *R* allele by about 800 nucleotide pairs, thereby confirming the results from Northern blotting.

At this point, the explanation of the molecular genetic basis for wrinkled seeds is nearly complete. The *r* allele contains an 800-nucleotide insert in a coding region, and this insert inhibits translation of mRNA into a functional starch-branching enzyme. But how did this insert get into the *r* allele in the first place? To address this question, Bhattacharyya's group used a clone containing the insert to probe the garden pea genome. They found that this probe binds to many regions of both the *RR* and the *rr* genomes, indicating that the insert is repeated many times throughout the genome of the garden pea. Furthermore, the sequence of nucleotides in the pea insert was found to be similar to sequences from snapdragon, maize, and parsley that correspond to parts of transposable elements. Therefore, the insert in the *r* allele probably came from the activity of a transposable element that is widespread in the garden pea and other plants.

Chapter Summary

Genes work by making RNA during transcription. In eukaryotes, genes are interrupted by nonsense DNA, which must be removed from RNA transcripts before they are transported out of the nucleus. The removal of nonsense sequences from RNA is called RNA processing, and is done by spliceosomes.

After messenger RNA is processed and transported out of the nucleus, it is translated into protein in the cytoplasm. Translation also requires transfer RNA and ribosomal RNA. Transfer RNA brings amino acids together. Ribosomal RNA is a component of ribosomes, which are the sites of protein synthesis.

The unit of DNA that codes for an amino acid is a codon, which is a triplet of three nucleotides. Most amino acids have more than one codon. Codons that code for the same amino acid are called synonymous codons.

Coding sequences in interrupted genes are called exons, and noncoding sequences are called introns. Genes that contain introns occur almost exclusively in eukaryotes. Introns are rare in simple eukaryotes, such as yeasts, but they are common in complex eukaryotes, such as plants and animals. Many exons code for separate domains in polypeptides.

Whole genomes can be cut into fragments by bacterial restriction enzymes and inserted into vectors, such as bacterial plasmids or viruses. Once inserted into a vector, the genome can be cloned by culturing the vector, thereby maintaining a library of the genes from that organism.

Individual genes are also cloned in test tubes by DNA polymerase in the polymerase chain reaction. Unlike cloning in vectors, which may take weeks to produce enough of the target gene for further study, the polymerase chain reaction takes only a few hours.

Genetic engineering, also called *recombinant DNA technology*, is the transfer of genes from one organism to another. Transgenic organisms are those whose genomes contain a foreign gene. Transgenic bacteria are made by inserting foreign DNA into plasmids that are incorporated back into the bacteria. Making transgenic plants requires extra steps, because plants lack plasmids. However, foreign DNA can be inserted into plant genomes by parasitic bacteria. Cultured plant cells that are infected by such a transgenic parasite can be regenerated into transgenic plants.

The molecular genetic study of round vs. wrinkled seeds in the garden pea is an example of how genes can be studied by using modern laboratory techniques. The transcription product, the translation product, and the gene itself were all manipulated in this study, which showed that the *r* allele is defective because of a DNA insert that blocks translation of the mRNA. The absence of the translation product, a starch-branching enzyme, apparently causes metabolic changes in seeds with the *rr* genotype, thereby producing the wrinkled phenotype when the seeds mature.

Questions for Further Thought and Study

1. How may introns regulate genes?
2. What are the advantages and disadvantages of a genomic library versus a cDNA library?
3. Most research in plant genetic engineering involves tobacco, tomato, or petunia. Why?
4. Starting with the gene, what are the fastest and slowest steps in protein synthesis? What makes the fast steps fast and the slow steps slow?
5. Compare the polymerase chain reaction with DNA replication.
6. What are the natural functions of restriction enzymes? How are these enzymes used in gene technology?
7. Why are transgenic monocots hard to make?
8. What is the minimum number of different mRNAs, tRNAs, rRNAs, codons, and amino acids for making all of the proteins in a cell? How many of each are known? Explain the discrepancy between the minimum and the actual numbers if they are different from each other.
9. Why do you suppose the 800-nucleotide insert in the *r* allele of the garden pea blocks translation? In other words, why does this allele not translate into a protein that is 267 amino acids longer than SBEI from the *R* allele?
10. What is the importance of reverse transcriptase in gene technology?

Suggested Readings

ARTICLES

Barton, J. H. 1991. Patenting life. *Scientific American* 264(March):40–46.

Cech, T. R. 1986. RNA as an enzyme. *Scientific American* 255(November):64–75.

Chambon, P. 1981. Split genes. *Scientific American* 244(May):60–71.

Clegg, M. T. 1990. Dating the monocot-dicot divergence. *Trends in Ecology and Evolution* 5:1–2.

Daviss, B. 1990. Super spud. *Discover* 11(5):30.

Dorit, R. L., L. Schoenbach, and W. Gilbert. 1990. How big is the universe of exons? *Science* 250:1377–1382.

Gasser, C. S., and R. T. Fraley. 1992. Transgenic crops. *Scientific American* 266(June):62–69.

Holliday, R. 1989. A different kind of inheritance. *Scientific American* 260(June):60–73.

McKinght, S. L. 1991. Molecular zippers in gene regulation. *Scientific American* 264(April):54–64.

Mullis, K. B. 1990. The unusual origin of the polymerase chain reaction. *Scientific American* 262(April):56–65.

Ogden, R., and D. A. Adams. 1989. Recombinant DNA technology: Basic techniques. *Carolina Tips* 52(4):14–16. Burlington, N.C.: Carolina Biological Supply Co.

Pace, N. R., and T. L. Marsh. 1985. RNA catalysis and the origin of life. *Origins of Life* 16:97–116.

Strange, C. 1990. Cereal progress via biotechnology. *BioScience* 40:5–9, 14.

White, R., and J.-M. LaLouel. 1988. Chromosome mapping with DNA markers. *Scientific American* 258(February):40–49.

Wills, C. 1991. Turning over an old leaf. *Discover* 12(1):78.

BOOKS

Grierson, D., and S. N. Covey. 1988. *Plant Molecular Biology*. 2d ed. New York: Chapman and Hall.

Juma, C. 1989. *The Gene Hunters: Biotechnology and the Scramble for Seeds*. Princeton, NJ: Princeton University Press.

Lewin, B. 1990. *Genes IV*. New York: Oxford University Press.

Murphy, T. M., and W. F. Thompson. 1988. *Molecular Plant Development*. Englewood Cliffs, NJ: Prentice Hall.

Watson, J. D., J. Tooze, and D. T. Kurtz. 1989. *Recombinant DNA: A Short Course*. 2d ed. New York: Scientific American Books.

Wilson, T. M. A., and J. W. Davies, eds. 1992. *Genetic Engineering with Plant Viruses*. Boca Raton, FL: CRC Press.

UNIT FOUR

The Form and Function of Plants...

Plants have a long and interesting evolutionary history. They evolved from algae 450 million years ago and then diversified in a way that forever changed our planet: they invaded land, thereby paving the way for the subsequent colonization of land by animals. Because the terrestrial habitats encountered by plants were more diverse than their ancestral aquatic environments, land plants faced many new problems and selective pressures. As a result, they evolved rapidly and became more complex and diverse than did their ancestral algae.

In this unit we present generalities, explanations, and examples to help you understand the form and function of vascular plants. Vascular plants (such as ferns, conifers, and flowering plants) have vascular tissues specialized to transport water and dissolved substances throughout the plant. (Nonvascular plants such as mosses and liverworts lack vascular tissues.) In this unit we emphasize vascular plants for two reasons:

1. Vascular plants represent several closely related lines of evolution and share a common form and function.
2. Vascular plants dominate our terrestrial ecology and our economy; they affect almost every aspect of our lives.

Since there are probably more than 300,000 species of vascular plants, you should expect some variety. We will also introduce a few of the exceptions to our generalities, because unusual plants often help us understand the norm.

The most diverse group of vascular plants is the flowering plants, which include more than a quarter of a million species. Because we know more about this group than any other, the focus in this unit will be on the flowering plants. Nevertheless, vegetative tissues develop and are organized similarly in almost all vascular plants.

Leaves of this flame bromeliad (*Neoregelia carolinae*) form a cup that gathers rain and nutrients.

Plant Growth and Development

CHAPTER 12

Chapter Outline

INTRODUCTION
A PERSPECTIVE ON PLANT FORM AND FUNCTION
THE LIVES OF PLANTS
MERISTEMS
    Functions of Apical Meristems
    Derivatives of Meristems

BOX 12.1 MAKING APPLES AND ORANGES

    How Apical Meristems Are Organized
PLANT GROWTH AND DEVELOPMENT
    Growth
    Development
SIGNALS THAT REGULATE PLANT GROWTH AND DEVELOPMENT
    Electrical Currents
    Hormones
    Positional Controls
    Biophysical Controls
    Genetic Controls
    Competence of Cells to Respond to Developmental Signals
MODULAR PLANT GROWTH
Chapter Summary
Questions for Further Thought and Study
Suggested Readings

Chapter Overview

Plant growth and development result from interactions of cellular division, cellular enlargement, and the patterned formation of tissues and organs. Roots and shoots grow from meristems, which are embryonic regions of cellular division and expansion. Although this growth enlarges a plant, it doesn't necessarily adapt a plant to its environment. Rather, adaptation results from differentiation, whereby new cells are organized into tissues and organs specialized for different functions. Because organs such as leaves, roots, and stems are produced continually, each adapts to changing environmental conditions. Growth and development are limited by a plant's genome and are integrated with environmental signals such as light and gravity.

INTRODUCTION

Roots absorb water and minerals from soil and, in doing so, also anchor a plant in place. This sedentary life-style has profound consequences for plant growth and development. Unlike advanced animals, whose development is controlled internally and usually isn't affected much by the environment, plants use environmental signals such as light and gravity to direct their growth and development. Growth and development are ultimately limited by a plant's genome, but the genome isn't a strict developmental blueprint. Rather, plants constantly adjust their growth and development to changing environmental conditions.

Living in one place makes plants unable to flee danger or pursue food. Thus, they are constantly threatened by damage and starvation. Natural selection has helped plants to cope with these threats by producing parts that are totipotent, controlled locally, and able to revert to an earlier stage of development, if necessary. This developmental flexibility adapts plants to their environment; indeed, in the worst of times, plants can become dormant or regenerate themselves. However, this adaptability has a price: plant structure must remain relatively simple (fig. 12.1). For example, seedlings consist largely of only three kinds of cells: parenchyma, collenchyma, and sclerenchyma. **Collenchyma** and **sclerenchyma** are thick-walled cells that support the plant. **Parenchyma** cells are more versatile: their functions range from protection and secretion to photosynthesis and storage. The simple design of plants also extends to tissues and organs.

- Simplicity of plant tissues. Plants consist primarily of three kinds of tissues: epidermis, vascular tissue, and ground tissue. Examine the locations of these tissues in figure 12.1. **Epidermis** covers the plant, protecting the underlying tissues and regulating the movement of gases between the plant and the atmosphere. Just interior to the epidermis is **ground tissue,** which is usually modified for photosynthesis or storage. The various parts of a plant are linked by **xylem** and **phloem,** which are **vascular tissues** that conduct water and dissolved solutes.

- Simplicity of plant organs. Epidermis, ground tissue, and vascular tissue are arranged into three kinds of vegetative organs: stems, leaves, and roots (fig. 12.1). Leaves are usually the primary photosynthetic organs and are arranged in spirals on the stem. Above the point at which a leaf attaches to a stem is an axillary bud. Although axillary buds are usually dormant, under certain circumstances they can form branches, flowers, or specialized shoots such as tendrils. Shoots, like roots, grow at their tips.

This chapter introduces the principles of development common to roots and shoots. It is meant to be an overview, so concentrate on the principles, not the specifics. In the other chapters of this unit you'll learn how these principles apply to roots and shoots.

A Perspective on Plant Form and Function

Many aspects of plant structure were discovered long ago, when people believed that all living organisms were created as a series of divine "types." Because there was no reason to expect intermediates between these divinely created types, most botany of the nineteenth century involved describing and categorizing the types of organs, tissues, and cells in various groups of plants. Had this approach continued, botany would have been little more than a description of the different plant types and lists of families and genera.

Today we explain plant structure and function not with types and divine creation, but with biochemistry, physiology, and genetics, all of which are guided by evolution. In this textbook we do not discuss types; rather, we present plant structure and development in relation to its function. Instead of reading lists of types and families, you will learn how plants are adapted to diverse environments, how their parts perform various functions, and why each function is important for a plant's survival.

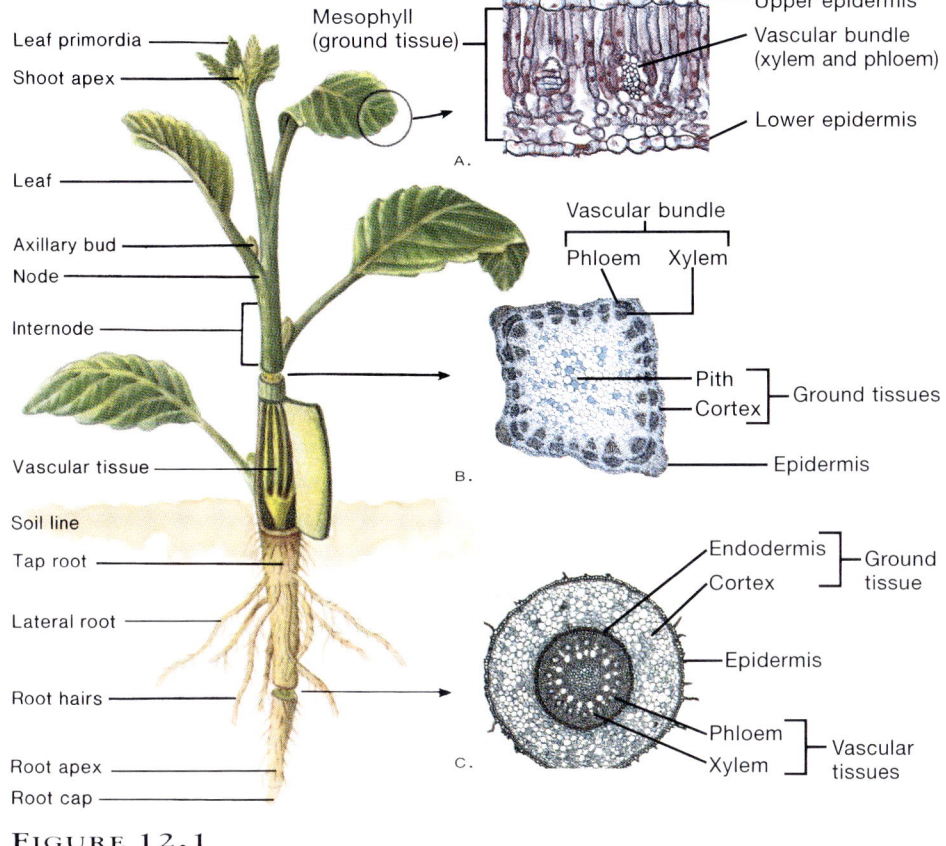

FIGURE 12.1

The tissues and organs of a herbaceous plant. Cross sections of (a) a leaf, (b) a stem, and (c) a root.

1. **Apical meristems** occur near the tips of roots and shoots, and produce *primary tissues.* These meristems account for *primary growth*, which is elongation of roots and shoots. Primary growth is important because it enables a plant to explore new environments for light, water, and nutrients.

2. **Lateral meristems** are cylindrical meristems that form in subapical regions of the roots and shoots of woody plants. Lateral meristems produce **secondary growth,** which increases the girth of the plant. Secondary growth is important because it makes a plant sturdier and thus capable of growing taller to intercept light.

3. **Intercalary meristems** occur between mature tissues. These meristems are most common in grasses, where they occur at the bases of nodes. Intercalary meristems are important because they help regenerate parts removed by grazing herbivores (or lawn mowers).

Localized growth in meristems has important implications for plant growth and development: it means that plants are a mixture of young dividing cells, maturing cells, and mature cells, all of which are derived from meristems. As a result, plants function as mature organisms while still growing. Plants are especially sensitive to damaged meristems, however, because such damage often stops growth. In shoots, plants minimize this risk by protecting their meristems with young leaves and by forming dormant reserve meristems (i.e., buds) that can take over if the apical meristem is damaged. Similarly, roots protect their apical meristems with a root cap. If the primary root is damaged, its function is often assumed by a lateral root.

We begin this chapter by discussing apical meristems, because they occur in all plants and ultimately produce the cells that form mature plants. Lateral and intercalary meristems are discussed later in this unit.

THE LIVES OF PLANTS

Multicellular plants and animals grow differently. Most animals are embryonic for only a short time, during which all parts of their body grow simultaneously. This *closed growth*, or *determinate growth*, produces an adult whose shape doesn't change much throughout the rest of its life. Conversely, most plants never attain a fixed size (i.e., become adults) because they grow indefinitely. This *open growth*, or *indeterminate growth*, is peculiar to plants; with the exception of a few organisms, such as fish, nothing like it occurs in higher animals.[1] As a result, the life spans of plants are often determined by environmental factors such as drought and disease rather than genetic limitations. For example, annuals usually complete their life cycle in one year, but many of them become perennials when protected from cold and drought. Similarly, perennials often reach astounding ages: many trees are thousands of years old. Few plants die of old age. Rather, they succumb to disease, drought, or even to their own success—some plants become so large that their transport systems can't service all of their parts.

Not all parts of a plant grow at the same time. Rather plant growth is restricted to perpetually embryonic regions called **meristems.** Plants have three types of meristems (fig. 12.2):

MERISTEMS

Functions of Apical Meristems

Roots and shoots enlarge by primary growth from apical meristems located near their tips. Apical meristems produce cells that, when they expand, lengthen the root or shoot. The cells of apical meristems usually have thin walls, prominent nuclei, and small vacuoles. Compared to other types of plant cells, meristematic cells appear rather simple and are therefore sometimes

---

1. Usually only roots and stems grow indeterminately. Leaves, fruits, and flowers grow determinately and therefore reach a mature size and shape.

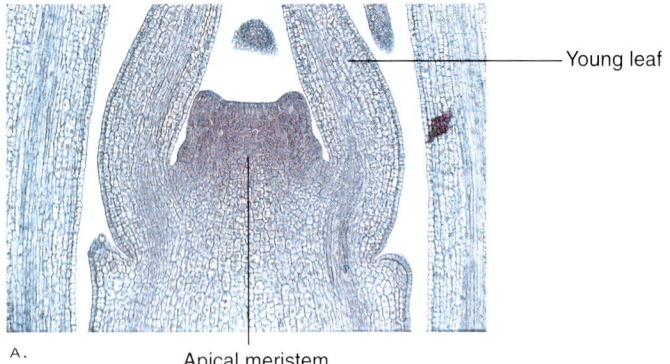

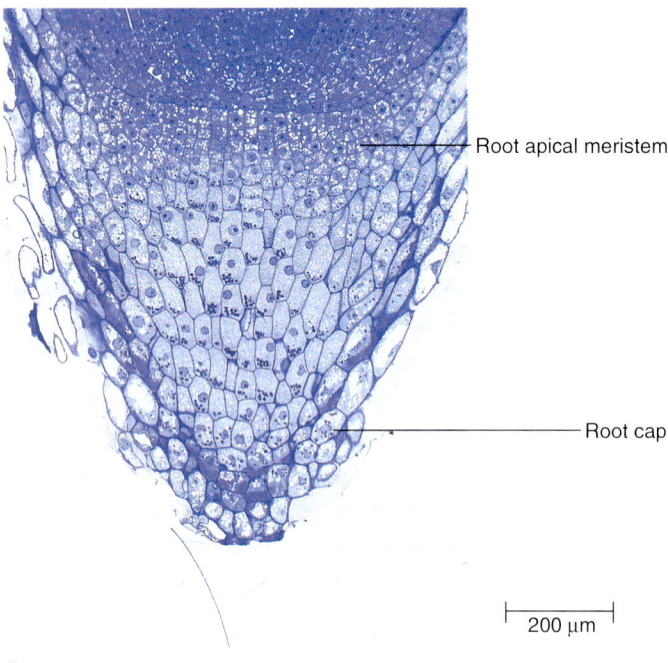

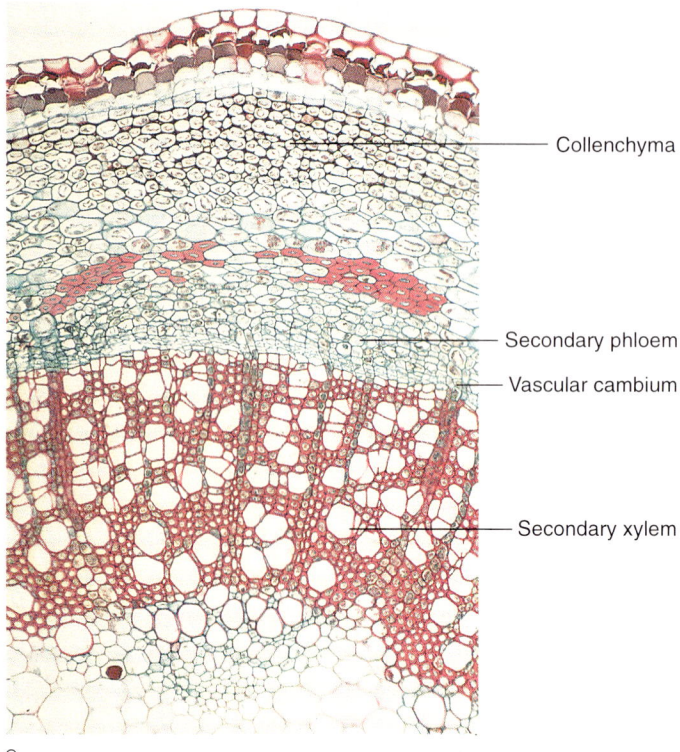

FIGURE 12.2

Types of meristems in plants. Apical meristems in shoots (a) and roots (b) produce *primary growth*, which elongates the organs. The shoot apical meristem (A, shown in longitudinal section) produces young leaves, which usually overarch the meristem. The root apical meristem (B, also shown in longitudinal section) is covered by a thimble-shaped root cap that protects the meristem as the root grows through the soil. The vascular cambium (C, shown in cross section) is a lateral meristem that produces *secondary growth*, or thickening, of the root or shoot. The vascular cambium produces wood (secondary xylem, which conducts water and dissolved minerals) to the inside, and secondary phloem (which conducts dissolved organic compounds) to the outside.

referred to as *undifferentiated*. However, this is incorrect: distinct types of meristematic cells have unique structures, even though they may be in the same plant. Moreover, cells in different zones of an apical meristem change structure as the meristem changes function. Thus, meristematic cells are unique and differentiated; none should be considered unspecialized.

Apical meristems consist of meristematic initials and their immediate derivatives. As these cells divide, they leave a cylindrical root or shoot behind them. Apical meristems have two primary functions: (1) establishing patterns and (2) producing new, genetically healthy cells.

Establishing Patterns

Apical meristems establish many patterns in plants (e.g., the arrangement of leaves). When an apical meristem is damaged, the remaining tissues continue to grow. However, only those tissues whose patterns were already set by the apical meristem develop normally. Continued patterned development is delayed until a meristem is regenerated.

Providing New, Genetically Healthy Cells

As you learned in Chapter 9, meristematic cells divide cyclically: divisions are followed by cellular enlargement, after which the cells again divide. These divisions push cells out of the apical meristem, where they become specialized *while remaining meristematic*. That is, cells continue to divide after they're pushed out of the meristem. This is important because it allows meristems to function as a reserve of mitotically young and genetically healthy cells. To understand the significance of this, recall that DNA replication can make mistakes that result in mutations. If uncorrected, these errors would accumulate to lethal levels within a few mitotic cycles. Although cells minimize these mistakes with elaborate proofreading and correction procedures, mutations nevertheless appear at rates of approximately one base-pair per billion base-pairs replicated. If plants grew only from small meristems, the millions or billions of replications necessary to produce all of its cells would produce a useless genome long before the plant could reproduce.

Plants and animals have solved the problem of accumulating mutations in different ways. In plants, apical meristems are

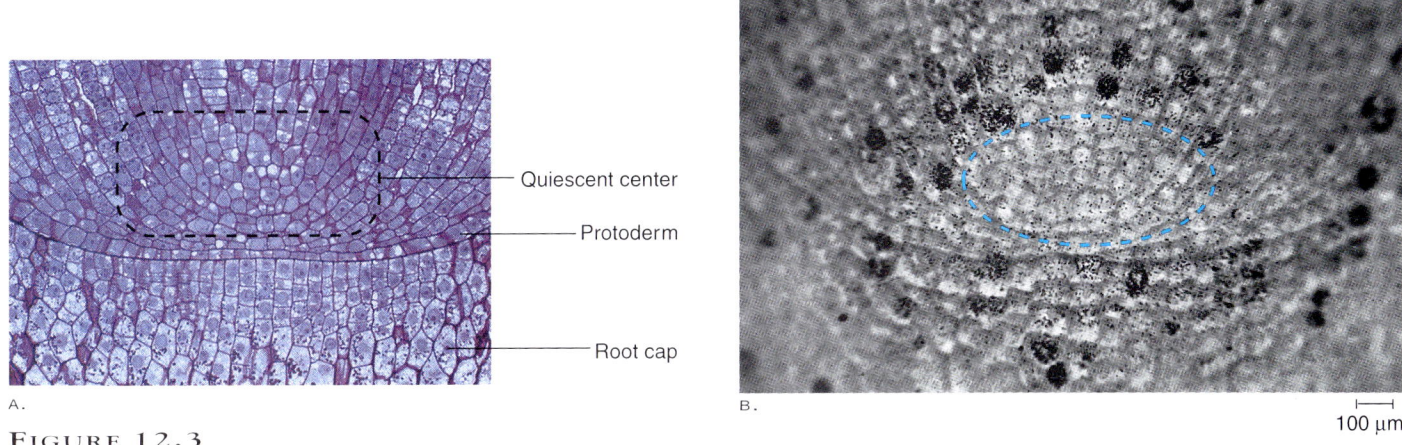

FIGURE 12.3

The root tip of corn (*Zea mays*). (a) The apical meristem and quiescent center (outlined) are located just behind the root cap. (b) The quiescent center consists of 500–1,000 apparently inactive cells. This was tested by botanists using a technique called *autoradiography*. Dividing cells exposed to radioactive thymidine incorporate the thymidine into their DNA. When sections of the labeled root are covered by photographic emulsion, producing dark grains over dividing cells, the cells of the quiescent center exposed few grains, indicating that they divide much less frequently than do the surrounding cells of the meristem.

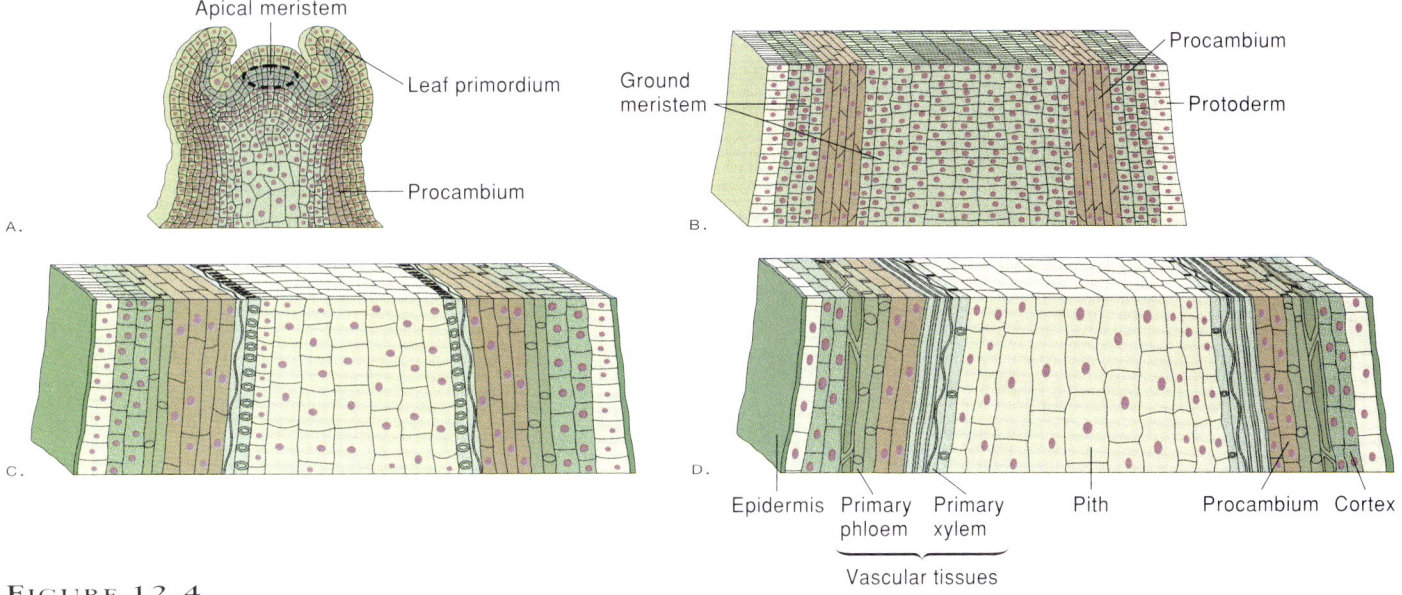

FIGURE 12.4

The three kinds of transitional meristems in plants. (a)–(d) The protoderm produces the epidermis, while the ground meristem forms the cortex and pith. The procambium forms the primary vascular tissues: primary xylem and primary phloem.

multistep meristems, in which a small group of initials divides occasionally to provide cells for the rest of the meristematic region (fig. 12.3). These derivatives, rather than cells of the meristem itself, continue to divide for several weeks or months and form the various tissues of the plant body. By not dividing rapidly, the initials ensure that the adjacent meristematic region is constantly supplied with new, genetically healthy cells (see box 12.1, "Making Apples and Oranges"). In contrast, determinate growth in animals minimizes the effects of these mutations by producing adults without any cells that go through multiple mitotic cycles. Similarly, animals set aside their sex cells as reserves early in development; because these cells do not go through many mitotic cycles, their genome usually remains error-free.

Derivatives of Meristems

Most cell divisions in tips of roots and shoots occur in derivatives of meristems rather than in the meristem's initials. These derivatives of meristems form regions of cellular division and specialization called **transitional meristems.** There are three transitional meristems in plants (fig. 12.4):

1. **Protoderm** is the transitional meristem that forms epidermis. Protodermal cells usually divide only anticlinally (i.e., perpendicular to the surface) and form a sheet of new epidermal cells that covers the growing root or shoot. Interestingly, the protoderm is irreplaceable: if damaged, no other cells can assume its functions.

BOXED READING 12.1
MAKING APPLES AND ORANGES

Humans have long exploited the apical meristem's immortality and ability to produce genetically healthy cells. For example, in 1825, a retired brewer named Richard Cox plucked an apple from a tree while walking across an abandoned farm. He was so enamored with the apple's taste that he grafted buds of the apple tree onto other "normal" apple trees to propagate his prized discovery. Subsequent growth of these grafted buds formed stems that produced apples identical to the one he'd discovered during his walk. Similar grafts have been made thousands of times. Today, Cox's orange pippin apples can be traced to the neglected tree that Richard Cox discovered more than 160 years ago. Similarly, the seedless navel orange of California can be traced to a tree that today grows in Riverside, California. This tree was sent to California as a bud-grafted tree discovered by a missionary in Bahia, Brazil in 1870.

When horticulturists realized that Cox's apples were best-sellers, they tried to hasten the cloning via tissue culture. These ex-

periments failed because the cultured tissues produced numerous new "sports" and "mutants," each having features different from that of Cox's original apple. Why don't similar changes occur in intact meristems? Recent research suggests that mutations *do* arise in meristems; however, the intact meristem recognizes and inhibits the growth of these "rogue" cells. As a result, the apical meristem remains genetically intact and continues to produce clones of the original apple. Apparently this censoring that occurs in intact meristems doesn't occur in cultured tissues, suggesting that recognizing mutant cells depends on the presence of an intact meristem.

---

2. **Procambium** is the transitional meristem that produces vascular tissues. Procambium differentiates as strands connected to the plant's mature vascular tissues.

3. **Ground meristem** produces ground tissue and consists of two other meristems: the flank meristem and the rib meristem. The *flank meristem* forms the cortex, and the *rib meristem* forms the pith. Cells of the ground meristem divide perpendicular to the stem's or root's axis and form longitudinal files (i.e., ribs) of cells. In shoots, division and subsequent elongation of these cells expand internodes (i.e., lengths of stems between leaves or branches) and separate the shoot's leaves. Dwarf and rosette plants, such as dandelion (*Taraxacum*) and *Agave*, don't have a rib meristem, and thus have short internodes and compacted leaves. The hormone gibberellin strongly influences the activity of the rib meristem; applying gibberellin to many rosettes and dwarfs induces internodal elongation (see Chapter 18).

Transitional meristems aren't the only means by which specialized cells can form. For example, masses of callus tissue grown *in vitro* via tissue culture (see fig. 12.8) lack procambium but contain isolated clumps of vascular tissue. Thus, procambium is not necessary for the differentiation of vascular cells. Rather, procambium is important because it forms vascular tissue in *patterned strands* that link the growing apex with mature vascular tissues.

The procambium and ground meristem form in response to developmental signals received from nearby organs and tissues. For example, young leaves control the differentiation of procambium: removing leaves from a shoot apex stops procambial differentiation, and grafting leaves to a mass of callus tissue induces formation of procambium and vascular tissue.

CONCEPT

Apical meristems are multistep meristems that produce orderly arrangements of new, genetically healthy cells. Most divisions occur in derivatives of meristematic initials called *transitional meristems*. Protoderm, procambium, and ground meristem are transitional meristems that produce epidermis, vascular tissues, and ground tissue, respectively.

How Apical Meristems Are Organized

Botanists have devised several models to describe the organization of apical meristems. One of the earliest models suggested that the three primary tissues of a plant (i.e., epidermal, vascular, and ground tissues) originate from distinct zones in the apex called *histogens*. This model also stated that the developmental fate of a plant cell, like that of many animal cells, is controlled by its lineage. A simple and convincing experiment discredited the histogen model: when regions of the meristem designated as histogens were surgically removed and rearranged, the root or shoot developed normally. Thus, the fate of a cell usually isn't determined by its lineage. That is, a cortical cell differentiates from a meristem not because the cell is unique, but because it receives developmental signals that change its metabolism to that of a cortical cell. When the histogen model failed to account for the development of meristems, botanists developed new models to explain how shoots and roots are organized.

Shoot Apical Meristems

Examine the shoot apical meristem shown in figure 12.5. The shoot apical meristem produces a stem and several kinds of lateral appendages, including photosynthetic leaves and modified stems such as thorns and tendrils, each of which comes from a differently organized meristem.

The first step toward our current understanding of shoot apical meristems came in 1924, with the formulation of the **tunica-corpus model.** According to this model, a shoot apex consists of two zones—a tunica and a corpus—each of which is distinguished by its planes of cellular division (fig. 12.6). The **tunica** is the outermost layer(s) of the shoot apex; these layers divide only anticlinally. That is, these divisions occur in one plane, and create sheets of cells that cover the apex of the shoot (fig. 12.6). Most angiosperms have one or two tunica layers, the outermost of which becomes protoderm. The **corpus** includes cells of the shoot tip below the tunica. These cells divide in all planes and produce bulk growth. In contrast to the histogen model, the tunica-corpus model is purely descriptive; it cannot be used to predict the developmental fate of individual cells.

The next step in understanding the organization of apical meristems was the discovery that the corpus is not homogeneous. Rather, it contains several zones with distinctive cellular features. To appreciate this, examine figure 12.6, which shows the positions of these zones and their relation to the protoderm, procambium, and ground meristem. The *central mother cells* are large, cuboidal cells in the uppermost zone of the corpus. These cells divide in all planes. Flanking the central mother cells is a cone-shaped *peripheral meristem*. These cells are small and densely cytoplasmic. The peripheral meristem, along with the protoderm and procambium, forms leaves. The leaves form in regular

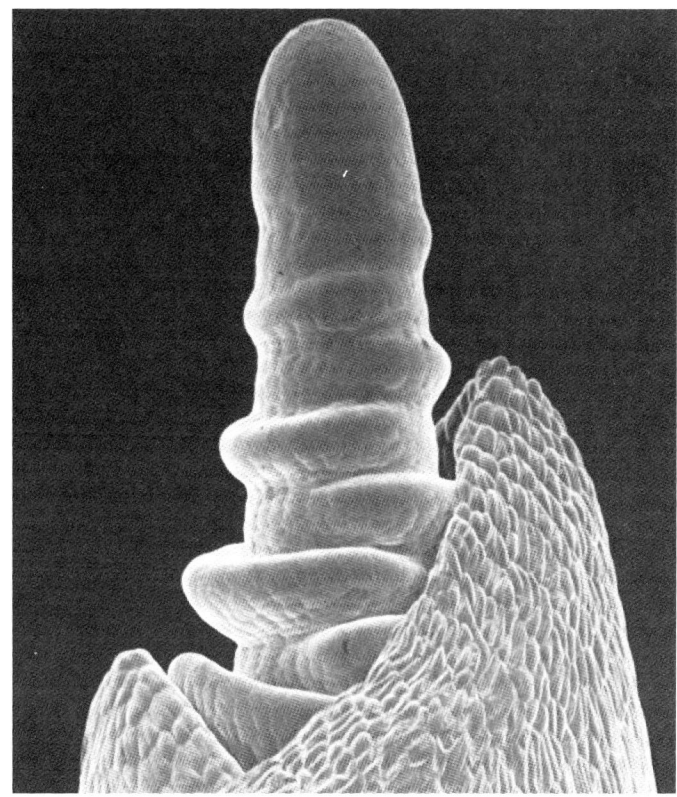

FIGURE 12.5

The shoot apex of wheat (*Triticum*). The apical meristem produces the stem and leaves, which first appear as bulges on the side of the apex. (Scanning electron micrograph, ×200.)

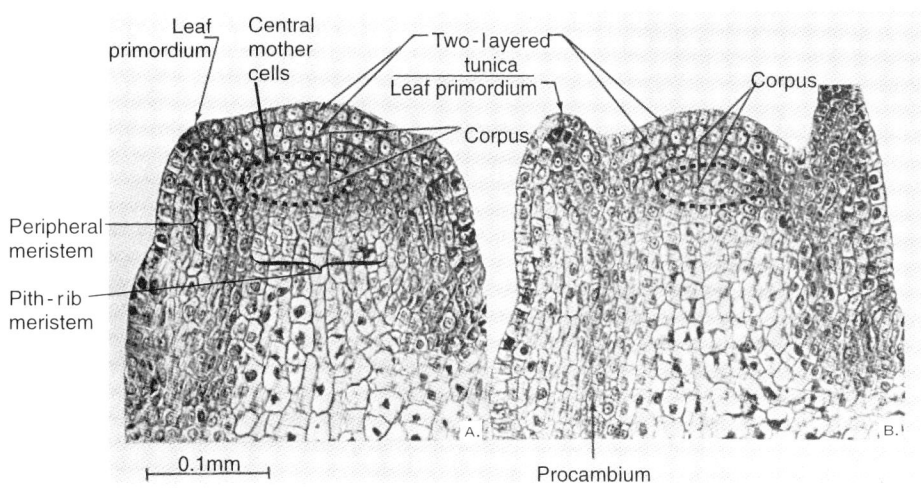

FIGURE 12.6

The shoot apex of potato. (a) The tunica-corpus organization. The two-layered tunica overlies the corpus. (b) The central mother cells form the upper zone of the corpus. Leaves are formed by the procambium and peripheral meristem, while the pith-rib meristem is responsible for internodal expansion.

patterns so rapidly that nodes (i.e., points on stems where leaves or branches attach) and internodes are indistinguishable at the apex. Internodal expansion is the function of the *pith-rib meristem*, which forms just below the central mother cells. Most cellular divisions in shoot tips occur in expanding internodes.

In plants such as corn (*Zea mays*), the entire shoot apex is mitotically active, and cells flow from the central mother cell region to the peripheral meristem. In plants such as sunflower (*Helianthus*), however, the meristems have a central region of inactivity bordered by a ring of mitotically active cells that produces cells for vegetative growth. Cells in the center of the meristem remain inactive and therefore genetically healthy as they wait for an environmental signal (e.g., changing day-length) to trigger their activation.

Root Apical Meristems

The root tip is covered by a slimy *root cap* that protects the root apical meristem and helps the root move through the soil. Behind the root cap is the root apical meristem, at the center of which is the *quiescent center*, which consists of 500–1,000 cells (fig. 12.3). As their name suggests, these cells are relatively inactive. Indeed, cells in root caps divide fifteen times more often than do cells in the quiescent center. The functional basis for these quiescent cells is that the root cap is effective but imperfect at protecting the apical meristem as the root forces its way through soil. As a result, the meristem is often damaged during growth. When this occurs, cells of the quiescent center divide and re-form the meristem and root cap. Thus, the quiescent center, like the central mother cells of plants such as sunflower, is a reservoir of genetically healthy cells. Surrounding the quiescent center are the root's transitional meristems, which produce longitudinal files of cells that differentiate into vascular tissues, cortex, and epidermis.

CONCEPT

Apical meristems are collections of several types of cells that function together to produce roots and shoots. Shoot apical meristems are protected by young leaves, while root apical meristems are protected by the root cap.

Lateral Meristems

Woody plants such as pines, oaks, and magnolias have lateral meristems that produce secondary growth, which increases the girth of the plant. There are two kinds of lateral meristems: the vascular cambium and phellogen (cork cambium). These meristems form a woody secondary body consisting of wood and bark (i.e., all tissues outside the vascular cambium). Wood (secondary xylem) and secondary phloem are derived from the vascular cambium; phellogen forms the periderm. The periderm forms a protective coat of dead, suberized cells that protect the inner tissues of the secondary plant body.

Secondary growth is important because it provides support that enables plants to grow taller, thereby increasing their chances of intercepting light for photosynthesis. You'll learn more about lateral meristems and secondary growth in Chapter 16.

PLANT GROWTH AND DEVELOPMENT

Plant growth and development typically involve quantitative increases in the *number* and *size* of cells and qualitative differences in *types* of cells.

Growth

What Is Growth?

Growth can mean several different things, such as an increase in size, volume, weight, or number of cells. For example, a piece of dry wood "grows" (i.e., swells) when placed in water, and a child who is "growing up" is becoming taller. We'll define growth as *any irreversible increase in size of an organism or its parts*. Growth is accompanied by metabolic processes that occur at the expense of metabolic energy. Thus, producing a leaf is growth, whereas the water-induced swelling of dead wood is not.

Cellular Division and Enlargement

Cellular division usually accompanies cellular enlargement. Meristematic cells repeatedly enlarge and divide, thereby maintaining an average cell-volume in the meristem. However, cellular expansion doesn't trigger division, and cell division doesn't always accompany enlargement. Therefore, cellular division and enlargement must be regulated separately. Furthermore, cellular division is not necessary for cell enlargement. This conclusion is supported by experiments involving irradiated wheat seedlings called *gamma plantlets*. You'll recall that such plants were discussed in Chapter 3 with regard to the organismic theory. Irradiation stops DNA synthesis and mitosis, but doesn't affect seed germination or growth of the seedlings. Thus, gamma plantlets develop without cellular division and grow only by cellular enlargement.

What are Botanists Doing?

Conduct a computer-aided search of scientific journals to find what research has been done in the past five years on the effects of irradiation on plants. Does any of this research relate to earlier experiments involving gamma plantlets of wheat? If so, how?

Two events affect cellular enlargement:

1. **"Loosening" the cell wall so that it can stretch.** One means by which walls are loosened is by secretion of acids into the cell walls. These acids loosen the cell wall by activating pH-dependent enzymes that break

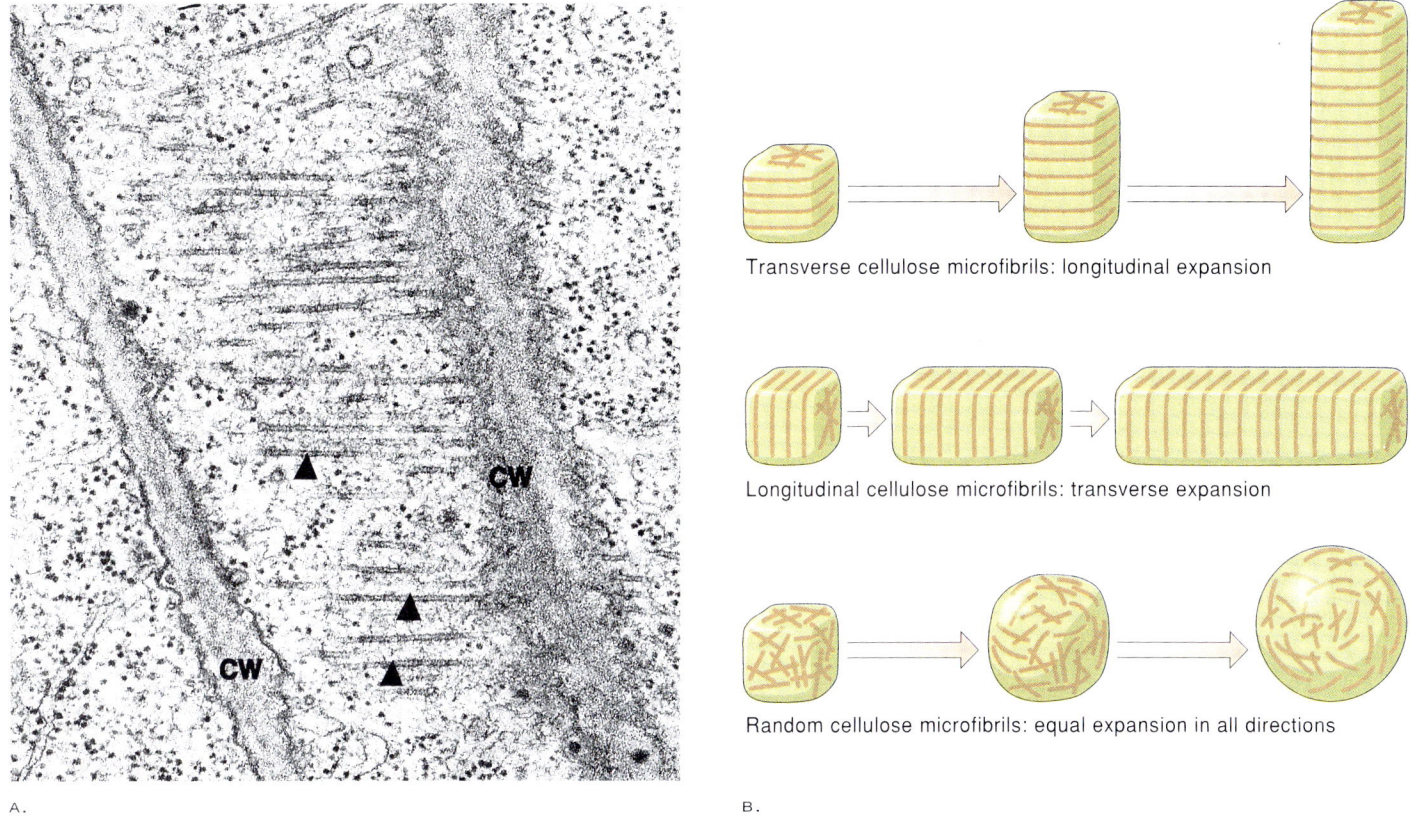

FIGURE 12.7

Microfibrils and cellular shape. (a) The microtubules (arrowheads) shown in this electron micrograph are involved in the deposition of cellulose microfibrils in the cell wall (cw). In this cell, how would the microfibrils be arranged? What are the consequences of such an arrangement? (b) How the orientation of microfibrils affects cellular shape.

bonds between cellulose molecules in the wall. Cell-wall acidification associated with cellular elongation is strongly influenced by auxin, a plant hormone. You'll learn more about auxin in Chapter 18.

2. **Positive turgor pressure.** Turgor pressure is pressure in a cell resulting from osmotically driven uptake of water, and occurs primarily in response to increasing concentrations of solutes in the vacuole. Thus, filling the vacuole with water is what "fills up" an expanding cell (e.g., cellular expansion in onion roots corresponds to a 30- to 150-fold increase in the size of the vacuole). This means of enlargement is more efficient in terms of energy use than synthesizing an equal volume of protein, enzymes, and organelles. It can also be fast; for example, petioles of a tropical water lily (*Victoria regia*) can lengthen more than 2 cm h^{-1}.

Plants can't grow without cellular enlargement. Since enlargement requires turgor, which in turn requires water, plant growth and development are intimately linked with a plant's water status. Fluctuations in water availability also influence metabolism, which controls events such as seed germination, bud growth, and the cellular elongation that causes roots and shoots to respond to environmental signals such as light and gravity.

Cells often enlarge in only one direction. The direction in which a cell elongates is determined by which cell wall is most elastic, and that in turn is determined by the orientation of the wall's cellulose microfibrils (fig. 12.7). We don't know what determines how these microfibrils are deposited in the cell wall, but we do know that microtubules are intimately involved. For example, microtubules near the plasmalemma are oriented in the same direction as microfibrils deposited in the cell wall (fig. 12.7a). Other evidence supporting the dependence of microfibril deposition on microtubules involves experiments using colchicine, a drug that disrupts microtubules but not the deposition of cellulose microfibrils. Cells treated with colchicine lack microtubules, and their cellulose microfibrils are deposited randomly in the cell wall. As a result, these cells become spherical rather than elongate (fig. 12.7b).

Cellular division and enlargement are important for growth, but do not by themselves constitute development. To understand this, consider the callus masses growing *in vitro* shown in figure 12.8a. These callus masses divide and expand rapidly but do not resemble a plant, because they lack roots, stems, and leaves. Producing these characteristic parts of a plant is referred to as *plant development* (fig. 12.8b) and requires cellular specialization; that is, it requires that cells differentiate.

CONCEPT

Growth is an irreversible increase in size of an organism or its parts, and results from increases in the number and size of cells. Cellular division and elongation are controlled separately. Cellular elongation accounts for most primary growth, and requires "loose" cell walls and positive turgor pressure.

Doing Botany Yourself

Plant one or two seeds of garden bean or some other easy-to-grow plant in each of two pots. After the seedlings emerge and leaves become well-developed, keep one pot in normal growing conditions and place the other pot in a dark cabinet. Leave this "dark" plant in the cabinet until its growth pattern is obviously different from that of the normal plant. How can you explain the difference in type of growth, if any, between the normal and the "dark" plant? How can you obtain evidence for your explanation?

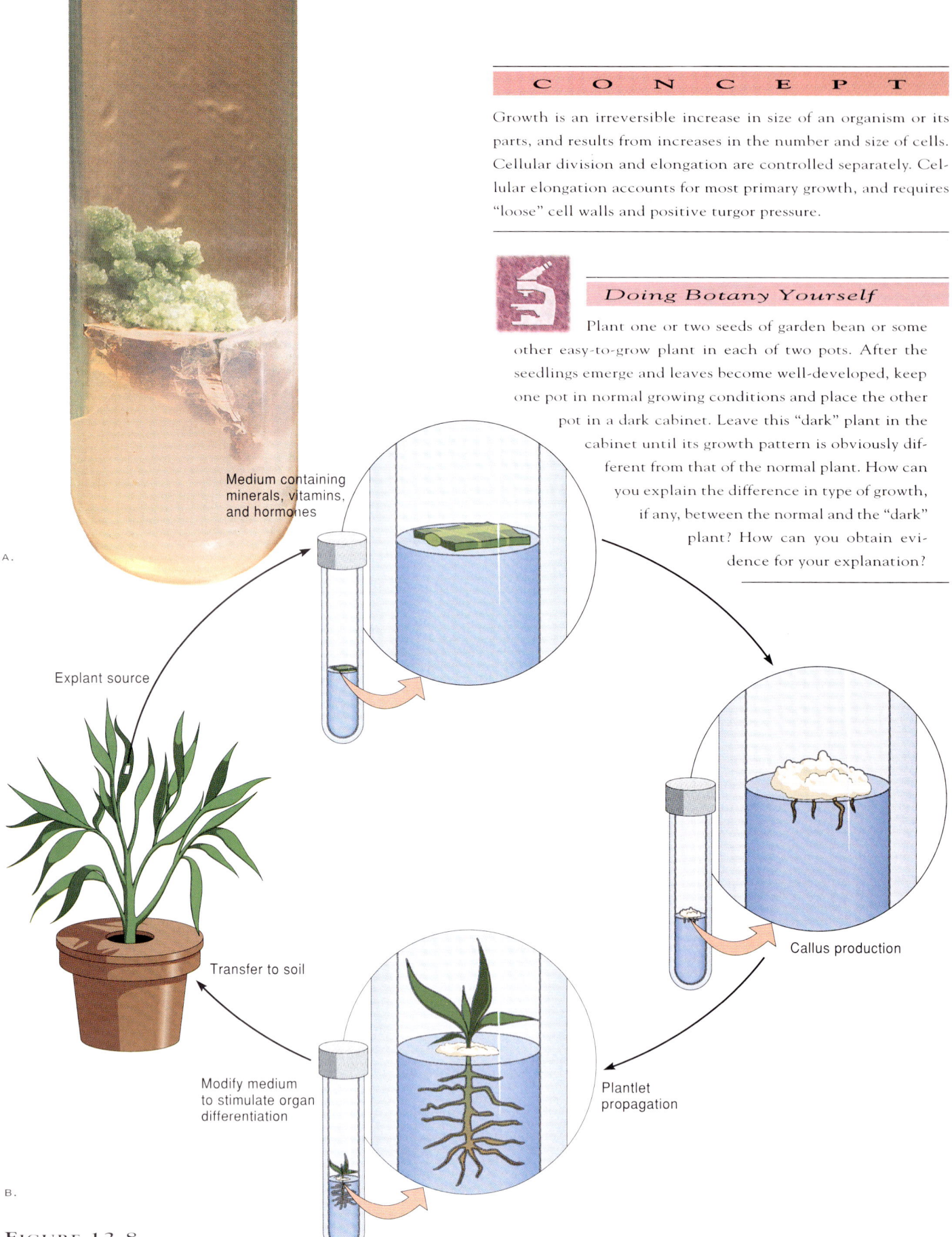

FIGURE 12.8

Plant tissue culture. (a) Large masses of callus tissue growing *in vitro*. (b) Diagram of plant development and propagation by tissue culture.

270 UNIT FOUR *The Form and Function of Plants*

Development

Roots, stems, and leaves are organized similarly—so much so that their organizing influences must be similar in different groups of plants. For example, vascular tissues consist of the same types of cells in a wide variety of plants. Furthermore, each tissue forms in a characteristic position related to its function. How do cells having similar origins (i.e., from the same meristem) become different?

Cellular Differentiation

Differentiation is an orderly process by which the structure and function of genetically identical cells become different. It occurs throughout a plant's life and is always preceded by complex biochemical changes. Differentiation produces cells, tissues, and organs specialized for different functions.

Much of what we know about differentiation in plants comes from studies using **tissue culture,** a technique for growing fragments of plants in an artificial medium. The basis for tissue culture was proposed in 1902 by the German botanist Gottleib Haberlandt, who suggested that plant cells are totipotent; that is, each living cell has the full genetic potential of the organism. Gardeners already knew that the organs of many plants were totipotent; they had cultivated potatoes and several other plants for centuries with cuttings that, when planted, grew into entire plants. However, Haberlandt's suggestion extended totipotency to individual cells. Botanists soon began testing Haberlandt's idea: evidence for totipotency would be the regeneration of an entire plant from one or a few nonzygotic cells. The earliest attempts failed; cultured cells remained alive for a while, but didn't divide and soon died. There were two possible explanations for these failures:

HYPOTHESIS 1

Plant cells are not totipotent. That is, the cells were genetically incompetent to regenerate a new plant.

HYPOTHESIS 2

The workers' failures were due not to the cells' genetic incompetence, but to other factors such as inadequate nutrients and growth conditions. Were the cultured cells being given everything they needed?

Continued research finally produced a breakthrough when, in 1958, Cornell botanist F. C. Steward regenerated a carrot plant from a tiny piece of phloem (see Chapter 9). Why was Steward successful? Like previous workers, he supplied the cultured cells with sugars (i.e., sources of reduced carbon), minerals, and vitamins, but he also added a new ingredient: coconut milk. Coconut milk contains, among other things, a substance that induces cellular division (subsequent research identified this substance as cytokinin, a plant hormone). Once the cultured cells began dividing, they were transplanted to new media, where they formed roots and shoots and developed into plants (fig. 12.8b).

Steward's work was important because it showed that plant cells are totipotent and that, given the proper environment, a zygote can be replaced by other types of cells. Today, tissue culture is an important tool for plant biotechnology and crop improvement. Tissue culture is also an important technique for studying cellular differentiation because it is relatively simple and gives a researcher a great deal of experimental control. Many plants produce a mass of callus tissue when grown in culture; the effects of compounds such as minerals, sugars, and hormones can be tested by observing what happens to the callus tissue when these compounds are added or removed from the growth medium.

Differentiation is an orderly process by which the structure and function of genetically identical cells become different. Differentiation results from differential activation of the cell's genome. Differentiation is important because it specializes cells for different functions.

Dedifferentiation

Wounds disrupt structural patterns in plants. For example, a grazing herbivore might sever the vascular tissue in a stem or leaf, or strong winds could break branches from a tree. When this occurs, cells near the wound respond to the disruption by **dedifferentiating,** meaning that they become meristematic, reprogram their genome, and prepare for differentiation. Severing a vascular bundle induces divisions of parenchyma cells near the wound, thereby forming a layer of parenchyma that links the severed ends of the vascular bundle. This new parenchyma then differentiates into vascular tissues that link the severed strands. Dedifferentiation is important because it enables plants to repair wounds and to re-form disrupted patterns.

Polarity

Many of the environmental signals that direct plant growth and development are polar (i.e., directional). For example, gravity is a unidirectional force, and light for photosynthesis often comes primarily from one direction. Since these directional signals control plant growth and development, it is not surprising that plants and their parts also exhibit a marked **polarity,** which is a directionality expressed as differences between different sides or ends of a cell, tissue, organ, or organism. For example, the root and shoot ends of a plant's axis are radically different, as are the upper and lower sides of a leaf.

Polarity is controlled by the environment and the plant's genome. The plant's genome ultimately limits growth and development; that is, genes are the ultimate arbiters of the cellular division, enlargement, and differentiation that create polarity. However, these genes are answerable to and activated by environmental signals such as light, gravity, and temperature. For example, aerial roots of *Cissus*, a relative of grapes, don't branch until their tips touch the soil; similarly, the spores of many ferns require light to germinate and grow.

Another example of environmental control of polarity occurs in the zygotes of *Fucus*, a brown alga. *Fucus* zygotes are an ideal experimental system because they are relatively easy to obtain and because they develop freely and individually in seawater. When released, *Fucus* zygotes have a random distribution of organelles (i.e., they are nonpolar). However, when illuminated by light from one direction, a cascade of events soon occurs that produces a remarkable polarity. Here's what happens (fig. 12.9):

1. Half of the zygote becomes densely cytoplasmic and electronegative relative to the other half. This electrical polarity involves an influx of Ca^{2+} at the densely cytoplasmic end and an efflux of Ca^{2+} from the other end, which drives an electrical current across the zygote.
2. When the cell divides, the daughter cell derived from the densely cytoplasmic half of the zygote (i.e, where Ca^{2+} enters) becomes a rootlike rhizoid, while the daughter cell derived from the half of the cell characterized by Ca^{2+} efflux forms a shootlike thallus.

These electrical currents and calcium asymmetries determine polarity in *Fucus* zygotes. Rhizoids can be induced to grow from the half of a zygote to which calcium is applied, and placing zygotes in an electrical current determines the site of rhizoid formation (fig. 12.9). A similar environmentally determined polarity also occurs in other plants; for example, the first cellular divisions in spores of horsetails (*Equisetum*) are always perpendicular to the direction of incoming light. Environmental signals determine the polarity of many plants and their parts.

Polarity can also be inherent and independent of environmental signals. For example, filamentous algae such as *Cladophora* and *Griffithsia* form rhizoids only at their physiological base, regardless of the filament's orientation. If the filament is broken into smaller pieces, each piece forms rhizoids only at its physiological base. This polarity extends even to individual cells of the filament, which form a "shoot" cell at their apex and rhizoids at their base. Thus, these filaments are analogous to a bar magnet: they retain their original polarity no matter how many times they are subdivided. This cellular polarity can be changed by centrifuging the cells, suggesting that it results from a differential distribution of a cell's contents.

Polarity in angiosperms is often fixed and difficult to change. To appreciate this, consider the experiment using internodal segments shown in figure 12.10. Although the physiological basal ("root") and apical ("shoot") ends of these segments are structurally indistinguishable, they respond differently during organogenesis: roots always form at the basal end, and shoots always form at the apical end of the cutting, even if the cutting is inverted. Furthermore, this polarity doesn't depend on having an entire internode, since cuttings maintain their original polarity regardless of how many times the internode is divided. These results indicate that (1) root-shoot polarity is an inherent property of a plant axis, and (2) the same part of a

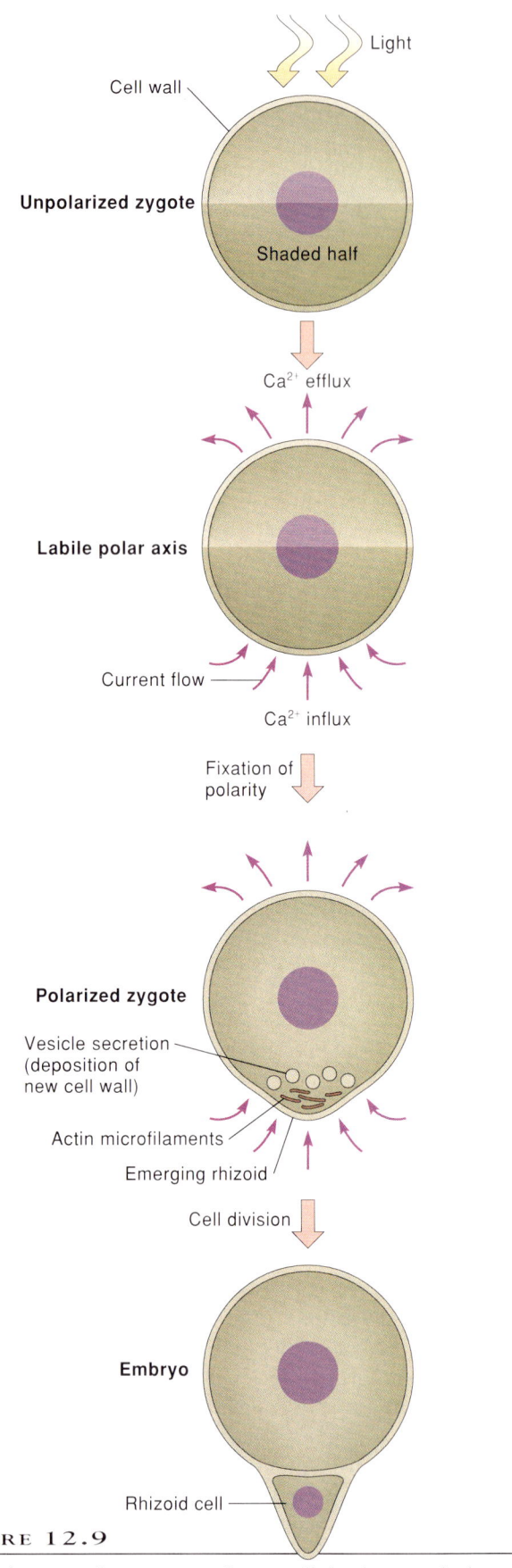

FIGURE 12.9

The development of environmentally controlled polarity in the brown alga *Fucus*. Undirectional light induces a polarity characterized by an influx and efflux of Ca^{2+} in different halves of the zygote. The rootlike rhizoid forms where there was an influx of Ca^{2+} into the zygote.

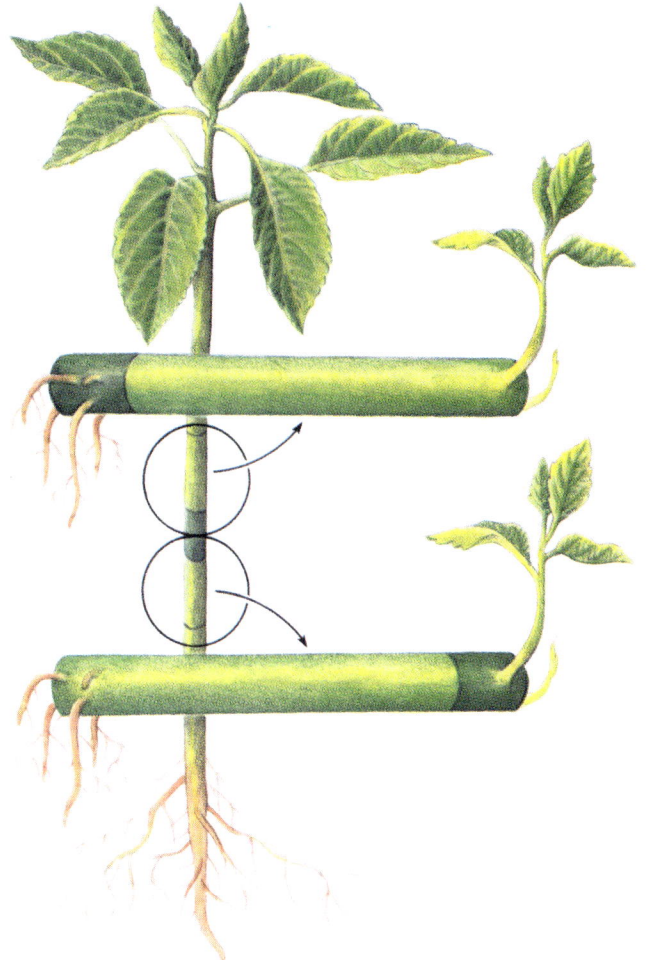

FIGURE 12.10

Polarity of plant tissues. Roots grow from the base of a cutting, while buds grow from the apex, even when the cuttings are kept horizontal and in uniform conditions. Development is determined by the location on the cutting, not on the intact plant.

plant can produce different apices, depending on its location relative to the rest of the cutting.

What could account for the inherent polarity of angiosperm tissues? Several botanists have suggested that the polarity results from metabolic gradients. For example, the hormone auxin moves polarly in stems from the apex toward the base, regardless of the stem's orientation. The resulting accumulation of auxin (or other metabolite) at the base could induce formation of roots and inhibit formation of buds, and therefore could account for the root-shoot polarity of angiosperms. However, the polarity persists from one season to the next, as well as through dormant periods when concentration gradients would be unlikely to persist. Therefore, phenomena such as polar auxin transport and gradients may *result from* rather than *cause* polarity. Today, many botanists believe that polarity results from a combination of (1) polar transport of morphological signals (e.g., auxin), and (2) formation of a developmental gradient along the plant's axis (e.g., root vs. shoot).

CONCEPT

Plants and their parts are polar, meaning that they have directional differences. In many instances, polarity is determined by environmental signals such as light and gravity, while in others it is inherent and independent of the environment.

Writing to Learn Botany

In the 1800s, German plant anatomist Anton de Bary said that "The plant forms cells, not cells the plant." What does this mean, and how is it related to plant growth and development?

The polarity of a tissue or organ reflects the polarity of its individual cells, which often begins with a choreographed rearrangement of the cell's parts. This rearrangement of organelles often precedes another polar event that influences differentiation: asymmetric cellular division.

Asymmetric Cellular Division and Differentiation

Cellular differentiation is often initiated by an asymmetric division, in which the smaller of the two daughter cells differentiates into a specialized structure. For example, asymmetric divisions of epidermal cells of leaves produce a large, vacuolate cell and a smaller, densely cytoplasmic cell (fig. 12.11). The large cell becomes an ordinary epidermal cell. The smaller cell becomes a stomatal mother cell, which then divides equally at a right angle to the first division to produce the two identical guard cells that form the stoma. Stomata regulate gas exchange in leaves.

Similar asymmetric divisions occur throughout plant growth and development. For example, the larger cell from asymmetric divisions that form pollen grains becomes the vegetative cell that forms the pollen tube. The smaller cell becomes the generative cell, which eventually divides equally to produce two identically shaped sperm cells. In asymmetric divisions of the epidermal cells of roots, the larger cell becomes an ordinary epidermal cell, while the smaller cell becomes a root hair.

It is interesting to note that structures such as stomata and root hairs, whose differentiation is preceded by asymmetric divisions, don't form in gamma plantlets, indicating that asymmetric divisions may be critical to differentiation and polarity. We know little about what controls asymmetric cellular division, but it is strongly influenced by Ca^{2+}. Asymmetric divisions are often preceded by an accumulation of Ca^{2+} in the cytoplasm and cell wall where the division will occur.

Although asymmetric cellular divisions often initiate differentiation, they don't necessarily determine the developmental fate of cells. For example, asymmetric cellular divisions in ferns produce a large and small cell. The smaller cell becomes a reproductive structure if gibberellins are present, but it

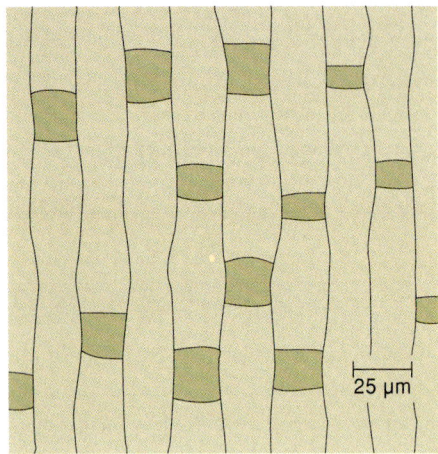

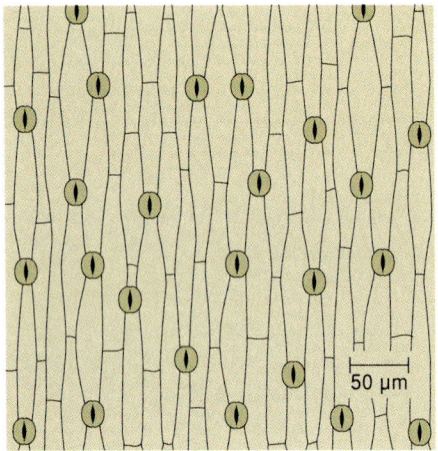

FIGURE 12.11

Stomatal differentiation in leaves of *Crinum,* a monocot. (a) Early in development, epidermal cells divide asymmetrically. The small cells (shaded) divide to form stomatal mother cells, while the larger cells become ordinary epidermal cells. (b) This type of cell lineage produces regularly spaced stomata.

becomes a hairlike rhizoid if gibberellins are absent. Therefore, although asymmetric divisions sometimes may be essential for differentiation, they may not be sufficient for complete differentiation. Additional developmental signals are necessary to direct the process.

SIGNALS THAT REGULATE PLANT GROWTH AND DEVELOPMENT

In addition to totipotency, Haberlandt suggested that cellular differentiation resulted from differential expression of the cell's genetic potential. This suggestion, like that of totipotency, has been supported by many experiments and today is a foundation of molecular biology and genetics. But if all plant cells have the same potential, and if specialization results from differential

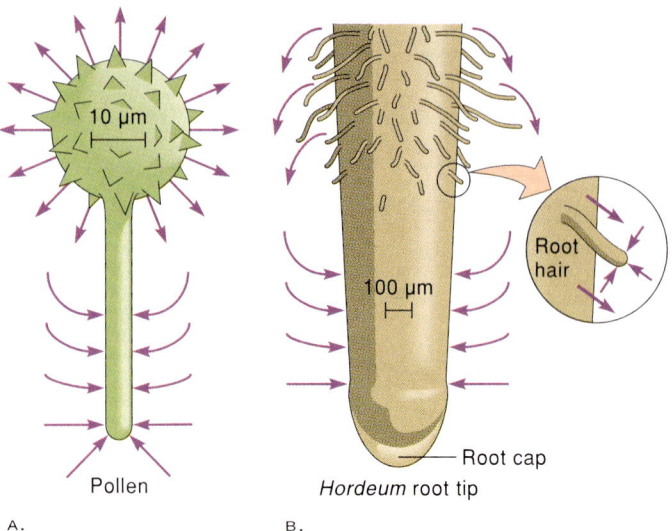

FIGURE 12.12

Growing plants generate electrical currents. (a) Currents enter tips of growing pollen grains and leave laterally behind their tips. (b) Similarly, currents enter tips of growing roots and root hairs. Abolishing these currents abolishes the polarized growth typical of the organ or cell.

expression of this potential, what controls which part of a cell's potential is expressed? Stated another way, what internal signals determine the developmental fate of a cell? The signals that control plant growth and development have intrigued botanists for centuries. Today, we know that these signals are complex and involve electrical, hormonal, positional, biophysical, and genetic controls.

Electrical Currents

Electrical currents influence several aspects of plant development, ranging from the closure of Venus's-flytraps to cellular elongation and polarity. For example, currents enter pollen tubes, roots, and root hairs at their tips and exit laterally behind their tips (fig. 12.12). When these currents are abolished, polarized growth is also abolished. Similarly, growth and development usually change when plants are grown in electrical fields, indicating that electrical currents influence development.

Hormones

Botanists have long known that one part of a plant affects other parts of the plant. For example, removing the shoot apex usually stimulates bud growth, and seeds often germinate faster when they're removed from fruit. These effects have often been attributed to **hormones,** which are organic molecules made in one part of a plant and transported to another part, where they elicit a physiological response.

There are five major groups of plant hormones: auxins, gibberellins, cytokinins, abscisic acid, and ethylene. Several aspects of growth and development have been attributed to these hormones (table 12.1). For example, auxin strongly

TABLE 12.1
Functions of Plant Hormones

Hormone	Major Functions	Where Produced or Found in Plant
Auxin (such as IAA)	Stimulates stem elongation, root growth, differentiation and branching, apical dominance, development of fruit; instrumental in phototropism and gravitropism	Endosperm and embryo of seed; meristems of apical buds and young leaves
Cytokinins (such as kinetin)	Affect root growth and differentiation; stimulate cell division and growth, germination, and flowering; delay senescence	Synthesized in roots and transported to other organs
Gibberellins (such as GA_1)	Promote seed and bud germination, stem elongation, leaf growth; stimulate flowering and development of fruit; affect root growth and differentiation	Meristems of apical buds, roots, and young leaves; embryo
Abscisic acid	Inhibits growth; closes stomata during water stress; counteracts breaking of dormancy	Leaves, stems, green fruit
Ethylene	Promotes fruit ripening; opposes or reduces some auxin effects; promotes or inhibits growth and development of roots, leaves, flowers, depending on species	Tissues of ripening fruits, nodes of stems, senescent leaves

affects the differentiation of procambium into young leaves. Here's how we think it happens:

1. Young leaves produce large amounts of auxin.
2. This auxin moves polarly down the stem.
3. Auxin moving down the stem induces differentiation of vascular tissues.

As you would predict, replacing leaves of a shoot apex with auxin stimulates differentiation of procambium. Furthermore, applying auxin to a callus mass induces differentiation of procambium at the site of application. These observations suggest that auxin formed in leaves and transported down the stem stimulates the differentiation of procambium into the leaf. Thus, young leaves use auxin production to control the differentiation of vascular tissues for service and support.

Each of the five types of hormones has many different and redundant effects in different tissues (e.g., auxin can stimulate or inhibit cellular enlargement and, like cytokinin, often stimulates cellular division). Thus, plant hormones are not specific like animal hormones,[2] and probably no phase of plant growth and development is controlled exclusively by one hormone. Rather, hormones are probably integrating agents that are necessary for but do not control a particular response. For example, cytokinins are necessary to break the dormancy of buds, but they do not control the subsequent growth of the bud.

The influences of hormones are nonspecific and are influenced by other hormones and other substances, such as

---
2. This lack of specificity has prompted many botanists to question the validity of plant hormones as regulators of plant growth and development. This and other aspects of plant hormones are discussed in Chapters 19 and 20.

FIGURE 12.13

Patterns in plants resulting from cellular differentiation and polarity. (a) The regular arrangement of stomata on leaves of corn (*Zea mays*). (b) Spiral arrangement of leaves in sedge. (c) Spines on a cactus. (d) Sunflower.

calcium. For example, asymmetries of Ca^{2+} affect events ranging from cellular elongation to responses to light and gravity. Similarly, pollen tubes have a tip-to-base gradient of Ca^{2+}, with the highest concentration of Ca^{2+} in the growing tip. Disrupting these gradients of Ca^{2+} abolishes polarity and polar growth of the pollen tube. Similarly, artificially establishing a gradient of Ca^{2+} across tips of roots or shoots induces differential growth and curvature.[3]

[3]. Calcium ions (Ca^{2+}) regulate many aspects of growth and development, including responses to gravity and light. Ca^{2+} exerts these effects by activating calmodulin, a small protein that affects several enzymes (see the discussion on p. 87).

The effects of hormones are largely a function of their targets: plant hormones affect growth and differentiation of cells that are programmed to differentiate in a certain way. This programming is usually a function of the cell's position and biophysical constraints.

Positional Controls

Cellular differentiation and polarity often create beautiful patterns in plants (fig. 12.13). Stomata in many plants are arranged in regular arrays, and leaves form in spirals so uniform that they've intrigued botanists, mathematicians, philosophers, and artists for centuries. How do these patterns form? Since plant cells don't move after they're produced, they must receive *positional signals* that inform them of their position so that they "know" where they are in the developing pattern. Cells then respond to these signals by becoming specialized (i.e., by differentiating) for their particular position in the pattern.

A cell's position also influences its rate of division and enlargement. Cells of growing carrot embryos, for instance, divide approximately twice per day. If they maintained this rate of cell division, after four weeks there would be about 2^{28} cells (about the number of cells in a mature carrot); if it continued another two weeks, the carrot would have about 2^{42} cells and weigh 2.5 tons. Obviously, carrots don't mature in only a month, and they seldom weigh more than about 150 g. The reason for these differences is that cells usually divide only when they're in an embryo or a meristem. Thus, although restricting divisions to embryos and meristems limits growth, it also ensures that other cells are present to protect and service the fragile (and vulnerable) dividing cells.

You've probably already thought of several questions about the positional signals that control differentiation. For example, what are the signals? How do they arise? And how is positional information transmitted? Unfortunately, we have few answers to these questions.

Biophysical Controls

Physical pressure generated by growing organs affects plant growth and development. For example, roots and shoots are cylinders bound by an epidermis. Cellulose microfibrils in the outer walls of epidermal cells function as reinforcing hoops and direct the growth of these cylindrical organs (fig. 12.14). Altering the arrangement of cellulose microfibrils in epidermal cells alters the directional pressure, which in turn alters growth and development. For example, one of the earliest signs of leaf initiation is altered deposition of microfibrils in protodermal cells of the shoot apex. This alteration creates a "weak spot" from which the leaf primordium bulges and grows (fig. 12.15).

Genetic Controls

Although botanists have long assumed that genetic controls underlie the specialization of cells in meristems, until recently no such developmental genes had been discovered. In 1989, however, a team of scientists led by Sarah Hake discovered a gene that causes odd growths (knots) on the leaves of maize (*Zea mays*). To Hake's surprise, the DNA sequence of this gene, called *knotted-1* or *kn1*, included a segment that was already known in the developmental genes of fruit flies. By early 1994, Hake and her colleagues had found several other maize genes containing such segments.

In fruit flies, developmental genes determine which cells in the early embryo will form legs, antennae, or other body parts. By analogy, Hake suggests that *kn1* and similar genes in maize dictate which cells in a meristem will form leaves or stems. Hake's evidence for this hypothesis is that, when the maize *kn1* gene is

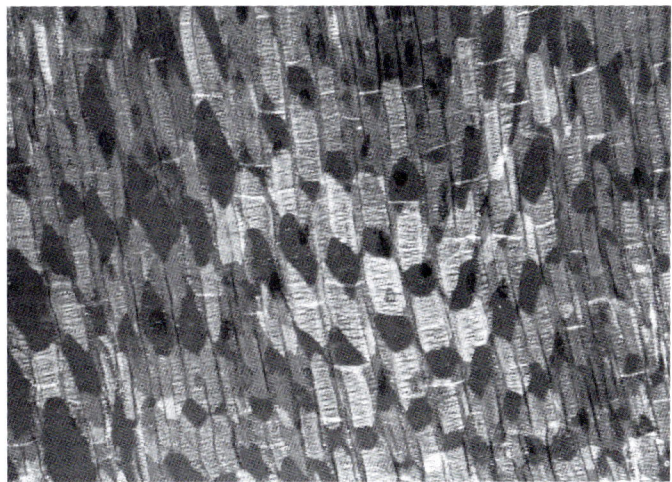

FIGURE 12.14

Hoop reinforcement in roots of *Sprekelia* as seen with polarized light. In this longitudinal section, the cell files are sectioned obliquely, so that bands of cell wall alternate with bands of cytoplasm. The cell walls are lighter than the background, indicating that most of their cellulose microfibrils are oriented transversely.

inserted into a tobacco plant, the tobacco leaf cells behave like meristems and sprout new stems and leaves. However, just how the *kn1* gene controls the meristem and how such genes are activated only in meristems remain to be explained.

Electric currents, hormones, position, and biophysical constraints are developmental signals that control growth and development of plant cells. Genetic controls of plant development resemble those of animals.

Competence of Cells to Respond to Developmental Signals

Although presumably totipotent, all plant cells do not respond similarly to developmental signals. The particular effect of any of these signals is also influenced by neighboring cells and the timing of developmental signals.

Neighboring Cells

Isolated cells and small pieces of callus (i.e., pieces less than 100–200 µm in diameter) cannot differentiate, regardless of what's in the growth medium. Competence to form roots and shoots usually develops only after a callus mass enlarges to a minimum size. Therefore, *competence to differentiate is an acquired condition and typically requires the support of neighboring cells.*

Timing of Developmental Signals

The timing of developmental signals strongly influences plant growth and development. For example, there are two meristems at the base of each leaf of a spike moss (*Selaginella*): the ventral meristem typically forms a root, and the dorsal meristem forms a shoot. Applying auxin to the dorsal meristem early in development causes it to produce a root, but similar applications late in development have little effect. Furthermore, an auxin-transformed meristem reverts to producing a shoot unless it is continually supplied with auxin. Therefore, *there is a window of time during which developmental signals can be received; these signals fix the developmental fate in a stable but not necessarily irreversible way.*

C O N C E P T

The competence of a cell to respond to developmental signals is influenced by neighboring cells and the timing of developmental signals.

The Lore of Plants

Ornamental plants are often prized for their oddities. One such oddity comes from unusual meristematic activity in cacti that causes the formation of broad crests at the tops of stems which make the plants look like some sort of spiny-headed aliens. These crests arise because of multiple apical meristems, which cause divisions to occur primarily in one direction instead of two. The origin of multiple apical meristems is one of the unsolved puzzles of plant development.

MODULAR PLANT GROWTH

Plants are modular organisms made of standardized parts: roots, shoots, and leaves. These modules are produced over and over by meristems, linked by vascular tissue, and protected by a waterproof epidermis. Once formed, each module follows a do-it-yourself strategy: each develops somewhat autonomously, which enables it to adapt to its particular environment.

Modules of plants cooperate and help plants exploit a patchy environment. For example, the twinflower (*Linnaea borealis*) is a pretty plant that grows in northern forests. As it grows across the forest floor, cooperation between modules allows roots to develop in a nutrient-rich hollow or leaves to exploit a sunfleck. Such cooperation occurs not only within but between plants. For example, Douglas fir trees and fig trees—even those of different species—graft together to form a giant supertree in which each module (i.e., tree) shares its resources with adjacent grafted trees. This cooperation accounts for why, after a patch of Douglas firs is cut down, many of the stumps remain alive for more than twenty years, sustained by root grafts to the remaining trees.

Although cooperation between plant modules helps plants survive, it has a price: exploitation by individuals that break developmental "rules." For example, mistletoes such as *Phoradendron* and *Viscum*[4] are parasitic plants that tap a host's vascular tissue, much like a developing leaf. The host plant can't recognize the incursion as that of a parasite, and so the mistletoe is accepted as another module of the plant. However, the mistletoe has a genome different from its host, and therefore it doesn't behave according to the host's developmental rules. Mistletoe leaves open their stomata very wide—much wider than stomata of their host—and therefore lose water faster than leaves of the host. As a result, the parasite receives a disproportionate share of the host's xylem sap and, with it, a disproportionate share of the nutrients carried in the sap. The vulnerability of vascular tissues to parasites such as mistletoe suggests that a plant's vascular system has few if any mechanisms for excluding parasites. However, the stability that plants gain from a modular strategy apparently outweighs the risks associated with this vulnerability.

Chapter Summary

Plants have a simple structure. They consist of three kinds of tissues: epidermal, vascular, and ground tissues. These tissues form leaves, roots, and stems, the vegetative organs of plants. Organs cooperate with each other and are precisely adapted to their particular environment.

Plants grow throughout their lives from perpetually embryonic regions called meristems. Apical meristems occur near tips of roots and shoots. Elongation of cells produced by apical meristems produces primary growth. Lateral meristems occur only in woody plants, and increase the plant's girth.

Apical meristems establish patterns and produce new, genetically healthy cells. Apical meristems are multistep meristems, meaning that a small group of meristematic initials divides and provides cells for the rest of the meristem. Most divisions occur in derivatives of meristematic initials called transitional meristems. Protoderm, procambium, and ground meristem are transitional meristems that produce epidermis, vascular tissues, and ground tissue, respectively.

Apical meristems are collections of several distinct meristems that function together to produce roots and shoots. Botanists use the tunica-corpus model to describe the organization of shoot apical meristems. The tunica is the outermost layer(s) of the meristem; these cells divide only anticlinally (i.e., perpendicular to the surface). The corpus includes cells of the shoot apex below the tunica, and is not homogeneous. Central mother cells occur in the uppermost part of the corpus, and are surrounded by a cone-shaped peripheral meristem that, with protoderm, produces leaves. Internodal expansion is a function of the pith-rib meristem, which is located just below the central mother

---

4. *Viscum* was worshipped by Druids, an order of priests assisting in religious rites in ancient Gaul, Ireland, and Britain. At certain times—probably at Midsummer Eve—Druids harvested *Viscum*. The mistletoe was cut from the parasitized tree with a golden sickle and caught in a white cloth; only then were the animals (and even humans) killed and burned. In contemporary Europe, there are many remnants of the Druidic ceremony, although they are rarely recognized for what they are. The Druidic rituals have been replaced by the newer and more sophisticated mythology of Christmas, in which the pagan mistletoe often plays a part.

cells. The shoot apical meristem is protected by leaf primordia, while the root apical meristem is covered by a protective root cap. Cells in the center of the root apical meristem are usually inactive and comprise the quiescent center. Transitional meristems adjacent to the quiescent center produce the epidermis, vascular tissues, and ground tissue (i.e., cortex) of the root.

Growth is an irreversible increase in the size of an organism or its parts. In multicellular plants, growth results primarily from cellular elongation, which requires a "loosened" cell wall and positive turgor pressure. Cells often elongate primarily in one direction, which is determined by the orientation of cellulose microfibrils in the cell wall.

Differentiation is an orderly process by which the structure and function of genetically identical cells become different, and results from the differential activation of a cell's genome. Asymmetric divisions often precede cellular differentiation, which depends more on a cell's position than its lineage. Wounds typically induce dedifferentiation, meaning that the cells resume meristematic activity, reprogram their genome, and prepare to differentiate. Dedifferentation is important because it enables plants to repair wounds and to re-form disrupted patterns.

Plants and their parts are polar, meaning that they have directional differences. Polarity in plants is often determined by environmental signals such as light and gravity, while in other instances it is inherent and independent of the environment.

Electric currents, hormones, a cell's position, biophysical conditions, and genes control plant development. The competence of a cell to respond to these signals is influenced by neighboring cells and the timing of the signal.

Questions for Further Thought and Study

1. A primary function of apical meristems is to provide new, mitotically healthy cells. How do apical meristems accomplish this?
2. Is pattern formation unique to living organisms? Give examples to support your answer.
3. What are meristems, and why are they important?
4. What is indeterminate growth, and why is it important for plant growth and development?
5. The differentiation of procambium is influenced strongly by young leaves. Why is this significant?
6. What are transitional meristems, and why are they important?
7. What is the role of asymmetric cellular division in cellular differentiation?
8. What is polarity? Discuss several examples of polarity in plants.
9. Discuss how electrical, hormonal, positional, and biophysical controls affect plant development.
10. Discuss the concept of modular growth in plants.
11. How is a callus mass like a meristem? How is it different?
12. What is dedifferentiation, and why is it important?
13. Leaves of many grasses grow from meristems located at the base of the leaf. Why don't trees grow from basal meristems at the base of their trunks?
14. The idea that the body-plan of plants is established in the embryo, as occurs in animals, contradicts the traditional view that plants have an "open" and "plastic" type of development. At first glance, plants grow by adding new structures produced by apical meristems located at opposite ends of the plant's axis. However, recent evidence suggests that the primary plant body-plan is generated in the embryo, and that meristems are merely terminal elements of the embryonic axis. How would you test this idea?

Suggested Readings

Articles

Aloni, Roni. 1987. Differentiation of vascular tissues. *Annual Review of Plant Physiology* 38:179–204.

Cosgrove, Daniel. 1986. Biophysical control of plant cell growth. *Annual Review of Plant Physiology* 37:377–405.

Feldman, Lewis J. 1984. Regulation of root development. *Annual Review of Plant Physiology* 35:223–242.

Furuya, Masaki. 1984. Cell division patterns in multicellular plants. *Annual Review of Plant Physiology* 35:349–373.

Hardwick, Richard. 1986. Construction kits for modular plants. *New Scientists* (April 10):39–42.

Jürgens, G. 1992. Genes to greens: Embryonic pattern formation in plants. *Science* 256:487–488.

Schnepf, E. 1986. Cellular polarity. *Annual Review of Plant Physiology* 37:23–47.

Books

Barlow, P. W., and D. J. Carr. 1984. *Positional Controls in Plant Development*. Cambridge: Cambridge University Press.

Lyndon, R. F. 1990. *Plant Development: The Cellular Basis*. London: Unwin Hyman.

Mauseth, James D. 1988. *Plant Anatomy*. Menlo Park, CA: Benjamin/Cummings.

Sattler, R. 1982. *Axioms and Principles of Plant Construction*. The Hague: Martinus Nijhoff/Dr. W. Junk Publishers.

Smith, R. 1992. *Plant Tissue Culture: Techniques and Experiments*. San Diego: Academic Press.

Steeves, T. A., and I. M. Sussex. 1989. *Patterns in Plant Development*, 2nd ed. Cambridge: Cambridge University Press.

Sussex, I., A. Ellingboe, M. Crouch, and R. Malmberg. 1985. *Plant Cell-Cell Interactions*. New York: Cold Spring Harbor Laboratory.

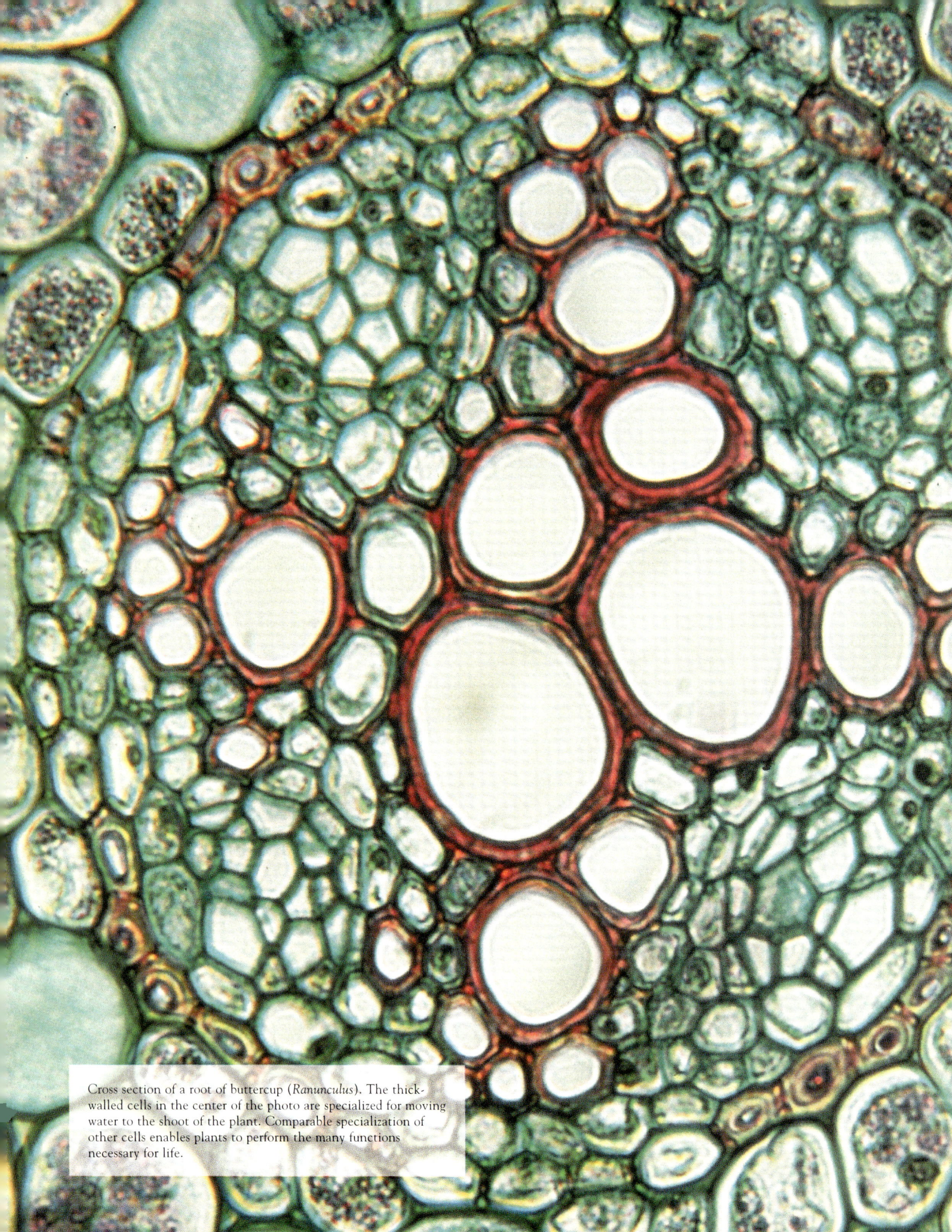

Cross section of a root of buttercup (*Ranunculus*). The thick-walled cells in the center of the photo are specialized for moving water to the shoot of the plant. Comparable specialization of other cells enables plants to perform the many functions necessary for life.

CHAPTER 13

Primary Growth: Cells and Tissues

Chapter Outline

INTRODUCTION
GROUND TISSUE
  Parenchyma
  Collenchyma
  Sclerenchyma
DERMAL TISSUE: THE EPIDERMIS
  Cuticle
  Epidermal Cells and Gas Exchange
  Trichomes

BOX 13.1
FEELING THEIR WAY

  Epidermal Cells and Cellular Recognition
  The Fate of the Epidermis
VASCULAR TISSUES: XYLEM AND PHLOEM
  Xylem
  Phloem
SECRETORY STRUCTURES
  External Secretory Structures
  Internal Secretory Structures
PLANT TISSUES AND LIFE ON LAND
Chapter Summary
Questions for Further Thought and Study
Suggested Readings

Chapter Overview

Photosynthetic cells of land plants have requirements similar to those of the green algae from which they evolved. Foremost among these requirements is an aquatic cellular environment, since unprotected cells, whether of algae or land plants, desiccate and die within minutes after being exposed to dry air. Thus, the survival of plants on land depends on their ability to establish an aquatic cellular environment in the dry land environment. Establishing this environment is a complex problem and requires specializations for gas exchange, ventilation, support, water retention, and protection. Plant tissues are the evolutionary solutions to these problems. For example, the epidermis waterproofs the plant and regulates gas exchange, vascular tissues move water and solutes throughout the plant, and secretory structures seal wounds and deter herbivores. Together, these tissues enable plants to grow in the hostile land environment.

Introduction

Chapter 12 stressed two basic features of plants: (1) their localized growth via meristems, and (2) structural and functional specializations that occur via cellular differentiation. Taken together, these processes produce the **primary body** of a plant, which is an axis consisting of a root and shoot. This axis is made of **primary tissues,** which are groups of cells having a common structure or function. In this and the following two chapters, we'll examine the structure and function of these various cells and tissues, and how they are integrated in the primary body of a plant. We'll consider secondary growth in a separate chapter because it begins only after the primary plant body has formed, produces unique cells and tissues, redirects the plant's resources for growth, and does not occur in all plants.

The primary body of plants consists of four tissues: meristems, ground tissue, dermal tissue, and vascular tissues. We've already discussed meristems (see Chapter 12), which are localized regions of cellular division that form a plant's cells. Like meristems, the ground, dermal, and vascular tissues consist of distinctive types of cells and usually have more than one function. For example, epidermal tissue absorbs nutrients and protects plants from desiccation, and vascular tissues provide support and move water and solutes throughout the plant. The functions of each of these tissues represent evolutionary adaptations to life on land. Indeed, plants are designed much like buildings: they have supporting structures, ventilating and plumbing systems, storage areas, and protective coverings. These life-support activities in plants are performed by the ground, vascular, and dermal tissues.

Ground Tissue

Ground tissue differentiates from the ground meristem and constitutes most of the primary body of a plant. The cortex and pith of stems and roots consist almost entirely of ground tissue. Ground tissue is covered by the epidermis and surrounds vascular tissues. It has several functions, including storage, basic metabolism, and support. These functions are performed by its three kinds of cells: parenchyma, collenchyma, and sclerenchyma.

Parenchyma

Parenchyma (from the Greek words *para,* meaning "beside," and *en* + *chein,* meaning "to pour in") cells are the most abundant and versatile cells in plants. They are identified by their relatively inconspicuous structure; that is, they have few distinctive structural characteristics. In practice, botanists classify as parenchyma any cell not assignable to any other structural or functional class. The shortcomings of this classification scheme become apparent when we consider some of the diverse functions of parenchyma cells:

- **Storage.** Plants store nutrients in parenchyma cells. Most nutrients in plants such as corn and potatoes are contained in starch-laden parenchyma cells (fig. 13.1). Similarly, the thick primary walls of storage parenchyma cells in the seeds of plants such as persimmon (*Diospyros virginiana*), date palm (*Phoenix dactylifera*), and coffee (*Coffea arabica*) contain hemicellulose, which is used as an energy source by germinating embryos.

- **Basic metabolism.** Parenchyma cells are the primary sites of the metabolic functions that you learned about in earlier chapters of this textbook, including photosynthesis, respiration, and protein synthesis.

Clearly, all parenchyma cells are not homogeneous; it is only our lack of understanding that prompts us to group such functionally diverse cells as a single cell type.

The more specialized cells of plants evolved from parenchyma cells, which are therefore considered to be phylogenetically primitive. As their name suggests, parenchyma cells usually form the matrix in which the other, more specialized cells are embedded. However, parenchyma cells are not merely filler, as evidenced by their many functions. Moreover, parenchyma cells can **dedifferentiate** and then **redifferentiate,** meaning that they can change activities and become more specialized. For example, wounding often stimulates parenchyma cells to divide and form masses of undifferentiated cells from which roots (fig. 13.2) develop on a cutting. The ability of parenchyma cells to dedifferentiate and redifferentiate is important because it is a primary way that plants develop and adapt to various influences such as wounding and changing environments. Indeed, parenchyma cells are the "ready reserves" from which a plant makes specialized cells to meet its changing needs.

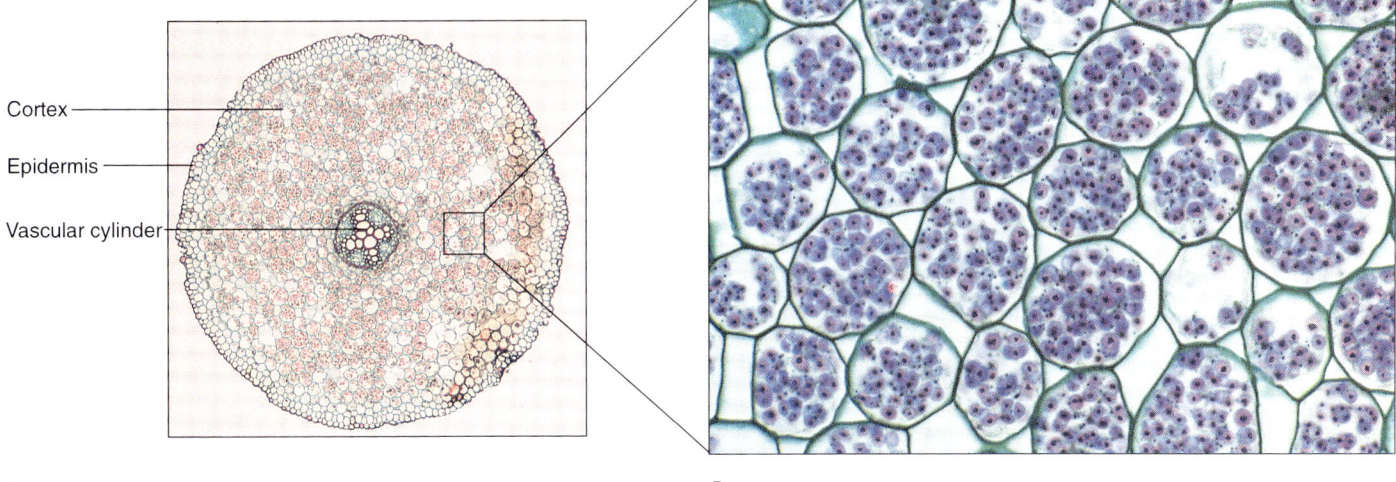

FIGURE 13.1

Transverse sections of the root of a buttercup (*Ranunculus*), ×16. (a) Overall view of mature root. The vascular cylinder includes tissues specialized for long-distance transport of water and solutes, while the epidermis forms a protective outer layer of the root. (b) Detail of cortex, ×250. Parenchyma cells in the cortex each contain many amyloplasts, which store starch.

Parenchyma cells are made by all of a plant's meristems and occur throughout the plant body. They comprise the photosynthetic tissue of a leaf, the flesh of fruit, and the storage tissue of roots and seeds. Parenchyma cells are alive at maturity and usually have only a primary cell wall. Mature parenchyma cells are more or less isodiametric (i.e., have equal diameters in all directions). This shape results from pressure exerted by adjacent cells, much as the shape of an isodiametric soap bubble results from the pressure of adjacent bubbles. Finally, parenchyma cells usually have large vacuoles. When full of water, these vacuoles make the cells turgid—this is what makes lettuce in your salad crisp (see Chapter 4).

You learned in Chapter 7 that *chlorenchyma* cells are chloroplast-containing parenchyma cells specialized for photosynthesis (fig. 13.3). Another kind of specialized parenchyma tissue is *aerenchyma* (fig. 13.4), which is characterized by prominent intercellular spaces. These spaces improve the gas-exchange capacity of the tissue and, not surprisingly, usually occur in or near metabolically active tissues. For example, the spongy mesophyll of leaves is aerenchyma that promotes the gas exchange needed for photosynthesis (fig. 13.4b, c). From a structural perspective, aerenchyma like that in *Juncus* stems (fig. 13.4a) provides maximum support with a minimum metabolic requirement.

Transfer cells are parenchyma cells specialized for short-distance transport of solutes. To complement this function, transfer cells have highly convoluted, nonlignified, secondary cell walls (fig. 13.5). This invagination of the cell wall increases the surface area of the plasmalemma as much as twentyfold, thereby increasing the cell's capacity for transport. Transfer cells occur in areas of high solute transport, such as in secretory glands of carnivorous plants, in nectar-secreting tissues of flowers, and along the conducting cells of xylem and phloem.

FIGURE 13.2

Roots at the base of a plant cutting. Wounding induces parenchyma cells near the cut surface to dedifferentiate into meristematic cells. These meristematic cells then divide and produce root apical meristems that form the roots on the cutting.

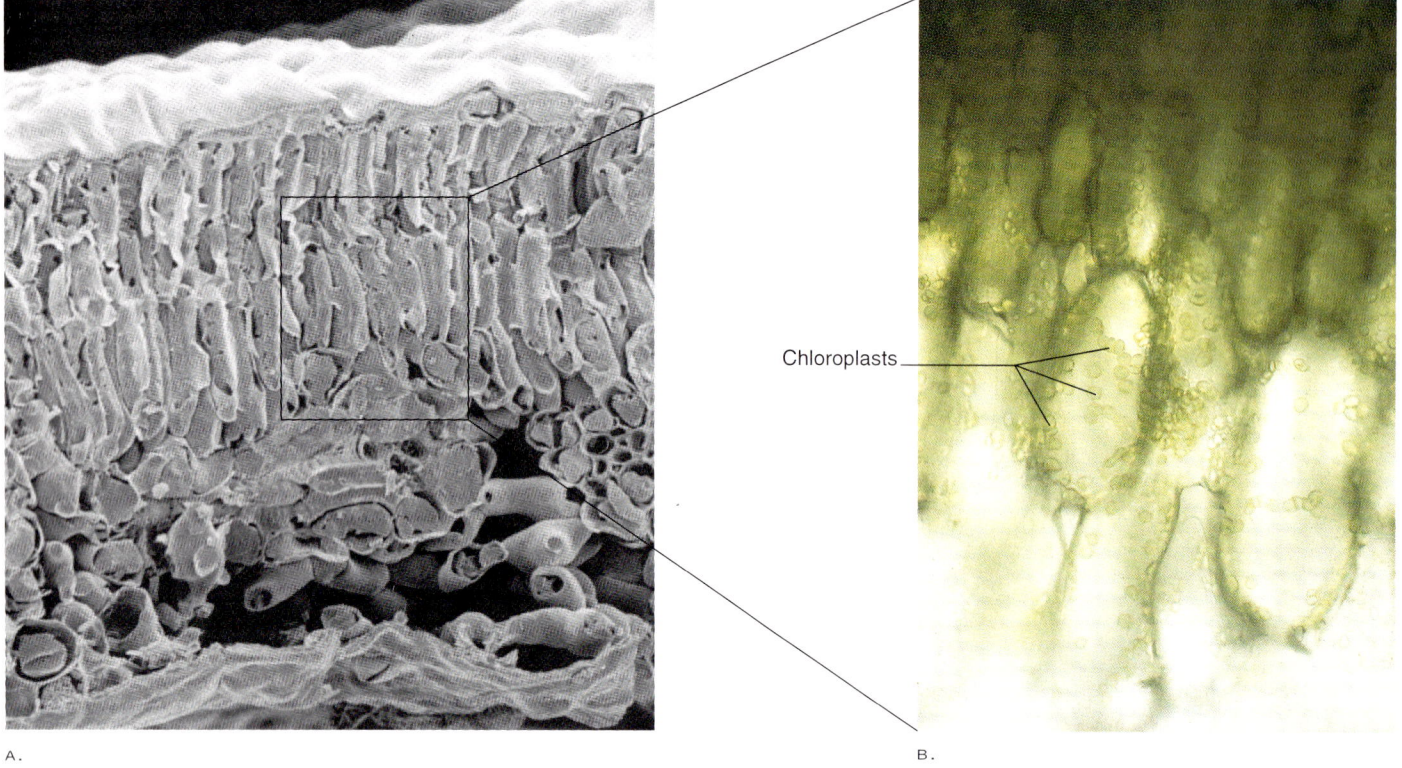

A.   B.

FIGURE 13.3

(a) Scanning electron micrograph of cross section of leaf of apple (*Malus*), ×950. Individual chloroplasts in photosynthetic cells that have been torn open by sectioning are indicated by arrows. The chloroplast-containing parenchyma cells are termed *chlorenchyma*. (b) This light micrograph shows a different view of chloroplasts in the chlorenchyma cells.

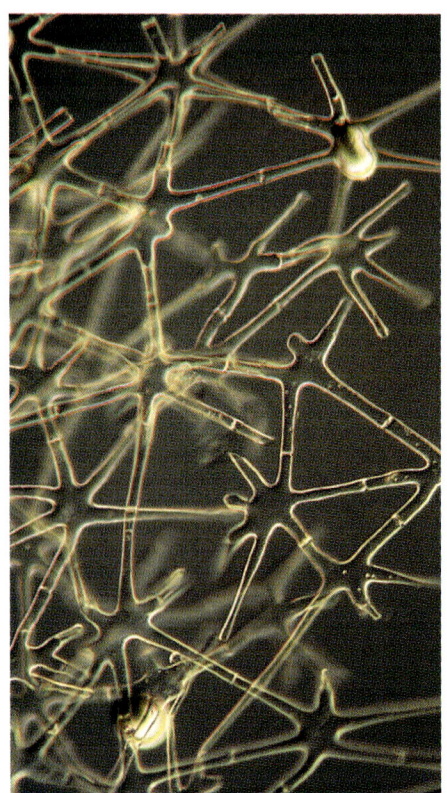

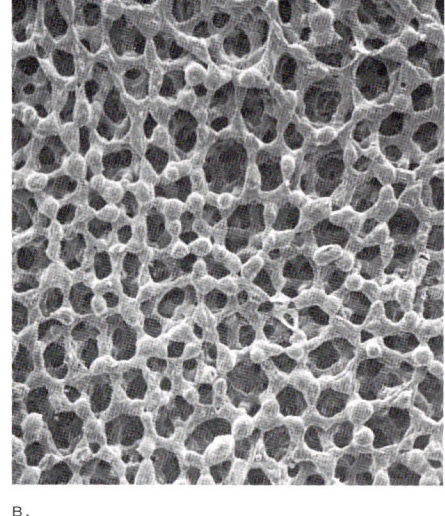

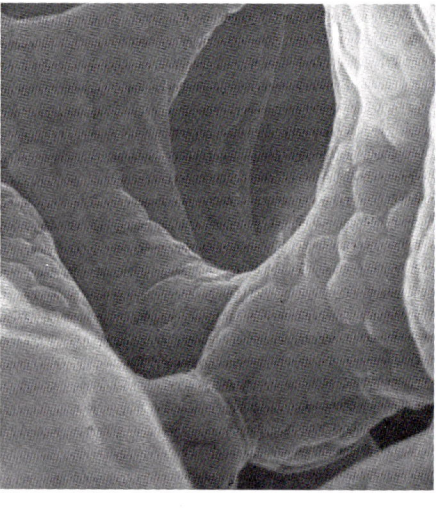

A.   B.   C.

FIGURE 13.4

Three views of aerenchyma tissue, a specialized type of parenchyma. (a) Light micrograph of aerenchyma from *Juncus*, a plant that often grows along the edges of ponds and lakes, ×400. Note the large amount of intercellular space. Aerenchyma in this plant provides support while maximizing gas exchange. (b) Aerenchyma tissue in leaves (shown here in the broad bean) is spongy mesophyll, a type of chlorenchyma. The numerous intercellular spaces increase gas exchange for photosynthesis in the chlorenchyma cells. (c) Magnified view of spongy mesophyll in a leaf of broad bean. The pattern on the cell walls results from chloroplasts in the cells being pressed outward by the vacuole. Note the labyrinth of intercellular spaces for gas exchange.

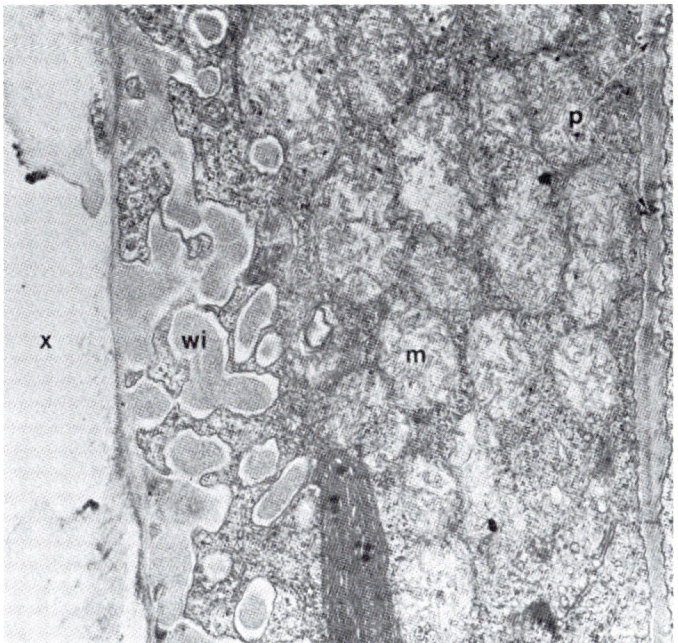

FIGURE 13.5

Part of a transfer cell in the shoot of *Galium aparine*, showing wall ingrowths (wi). These ingrowths increase the surface area available for transport. (X = xylary element; M = mitochondrion; P = plasmodesmata.)

Parenchyma cells are the primary components of ground tissue, and are the most abundant and least structurally specialized cells in plants. Parenchyma cells are the site of the basic functions of plants. Parenchyma cells can form other, more specialized tissues.

Collenchyma

Collenchyma (from the Greek word *kolla*, meaning "glue") cells are elongate (up to 2 mm long) cells having unevenly thickened primary cell walls (fig. 13.6). They support growing regions of shoots, and are therefore common in expanding leaves, petioles, and elongating stems (e.g., near the apical meristem). Collenchyma cells are exquisitely adapted for support: their nonlignified cell walls can stretch, thereby enabling the cells to elongate as the tissue they are a part of elongates. Furthermore, collenchyma cells often differentiate in strands or as a cylinder beneath the epidermis. This arrangement maximizes support, because a cylinder provides more support than does a rod located in the center of a stem or petiole. You're already familiar with strands of collenchyma—they are the resilient strings in petioles of celery (*Apium graveolens*).

Collenchyma cells differentiate from parenchyma cells and are alive at maturity. Their differentiation is strongly influenced by mechanical stress. For example, celery plants shaken for nine hours per day for a month have twice as much collenchyma as do unshaken controls. Furthermore, the walls of collenchyma cells in shaken plants are 40%–100% thicker than those of controls. This stimulation of collenchyma differentiation by mechanical stress helps a plant organ remain upright despite environmental disturbances such as wind and rain.

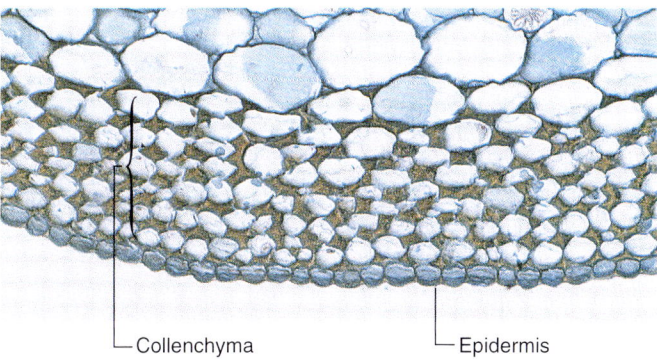

FIGURE 13.6

Light micrograph of a transverse section of collenchyma tissue from a petiole of rhubarb (*Rheum rhaponticum*). In fresh tissue like this, the unevenly thickened collenchyma cell walls have a glistening appearance. Collenchyma supports growing regions of plants.

 *Writing to Learn Botany*

Collenchyma is common in shoots, but relatively rare in roots. What could account for this?

Sclerenchyma

Sclerenchyma (from the Greek word *skleros*, meaning "hard") cells are rigid and have thick, nonstretchable secondary cell walls (figs. 13.7–10). They support and strengthen nonextending regions of plants such as mature stems, and are usually dead at maturity. Thus, the support provided by sclerenchyma cells is attributable to its cell-wall "skeleton," which is produced before the cell dies. Sclerenchyma cells occur in all mature parts of plants, including leaves, stems, roots, and bark.

There are two types of sclerenchyma cells: sclereids and fibers, both of which differentiate from parenchyma. Sclereids and fibers are distinguishable on the basis of cellular shape and grouping. Sclereids are relatively short, have variable shapes, and usually occur singly or in small groups (figs. 13.8–9), while fibers are long, slender cells typically occurring in strands (fig. 13.7).

Sclereids occur throughout plants, including in roots, leaves, stems, seed coats, and even in the hulls of peanuts (*Arachis*). They often form hard layers; for example, the tough core of an apple (*Malus*) consists mostly of sclereids, as does the shell of a walnut (*Juglans*), and they produce the gritty texture of pears (*Pyrus*; fig. 13.9). The differentiation of sclereids is influenced strongly by wounding and by the position of the

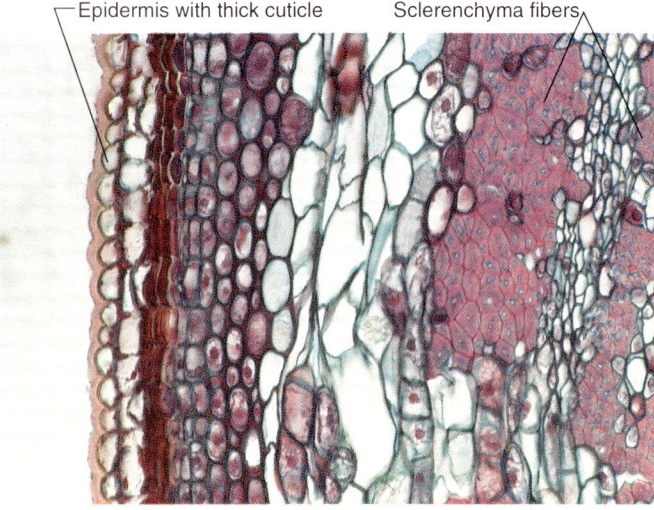

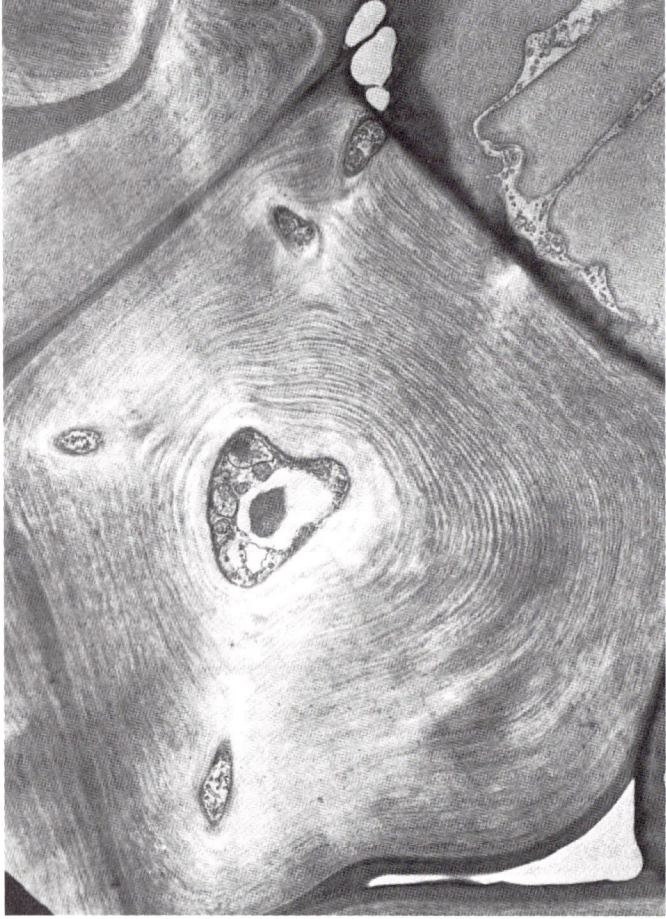

FIGURE 13.7

(a) Sclerenchyma fibers from a stem of basswood (*Tilia americana*) seen in cross-sectional view, ×250. (b) Thick, layered, secondary wall in sclerenchyma fibers, ×11,000. The layers are zones of different density due to variations in how the wall was deposited during development of the cell.

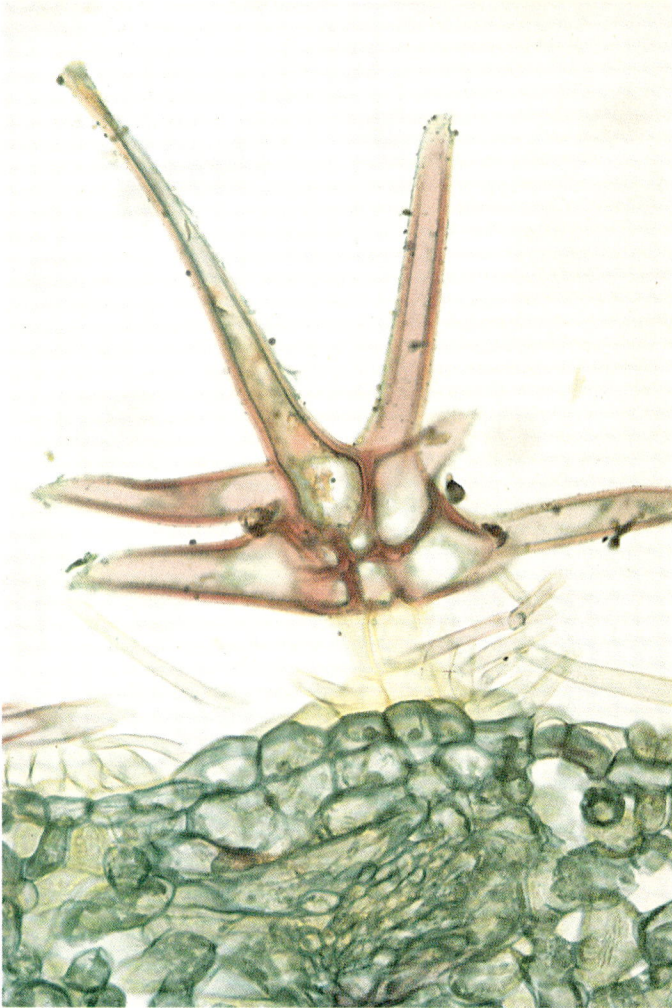

FIGURE 13.8

Branched sclereid (also called an astrosclereid) from a leaf of water lily (*Nymphaea odorata*), ×250. Branched sclereids such as this are common in petioles and leaves of many plants.

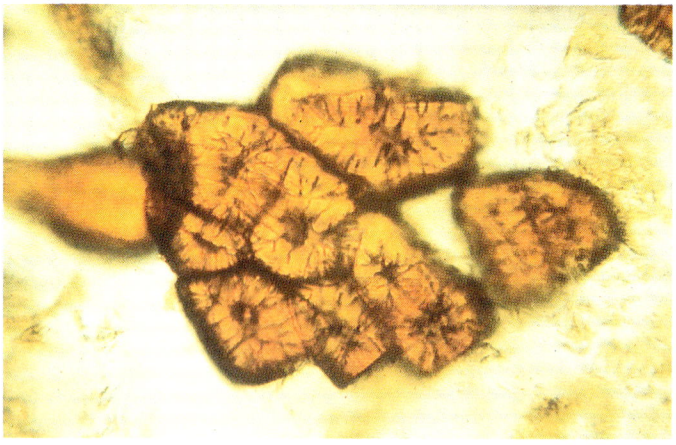

FIGURE 13.9

Sclereids in fruits of pear (*Pyrus communis*) are called brachysclereids. These sclereids give the fruit its gritty texture.

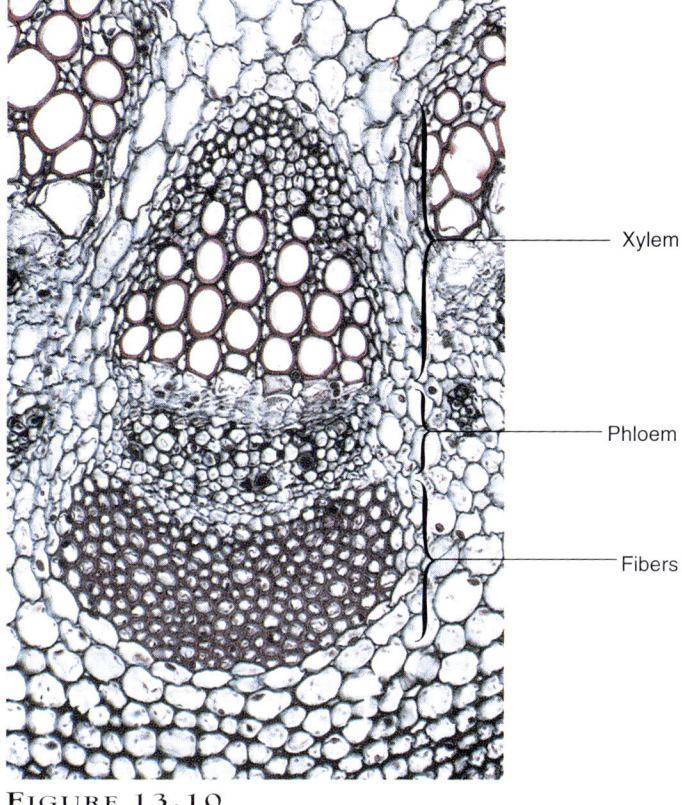

FIGURE 13.10

Cross-sectional view of fibers associated with the vascular tissue of a sunflower stem, ×485. Note that the walls of fibers are much thicker than those of adjacent cells.

Classification of Fibers

Fibers are classified several ways. One common classification scheme is based on their location. Fibers in xylem are called *xylary fibers*, while those that occur in tissues other than the xylem are called *extraxylary fibers*. Extraxylary fibers are usually longer than xylary fibers and are used in fabrics such as linen.

Commercial fibers are also classified according to their hardness. Fibers harvested from most monocots also include xylem, thereby making them lignified and stiff. Consequently, these fibers are called *hard fibers*. Conversely, fibers harvested from many dicots are called *soft fibers* because they lack lignin, or contain only small amounts of it, which makes them more flexible than hard fibers. The consequences of this difference in cell-wall structure are easily appreciated by comparing two products made of fibers: coarse rope and linen cloth. Coarse rope is made from hard, lignified fibers from sisal (*Agave sisalana*), whereas fine linen is woven from phloem fibers of flax (*Linum usitatissimum*), which are almost exclusively cellulose. In general, dicot fibers are stronger and more durable than monocot fibers.

Uses of Fibers

Humans have used fibers for more than 10,000 years. People living in the northwestern United States extracted and bound fibers into cords as early as 8,000 B.C., and a complete bag made of fibers dates to 5,000 B.C. Flax and hemp have been cultivated for fibers for more than 5,000 years (in fact, hemp was one of the largest crops grown by George Washington on his Virginia plantation). Today, humans cultivate more than forty families of plants for fibers (fig. 13.11), many of which you encounter every day. For example,

Musa textilis (Manila hemp) is used to make ropes and cords.
Agave sisalana (sisal; century plant) is used to make coarse ropes and twines.
Furcraea gigantea (Mauritius hemp) is used to make ropes, cords, and coarse fabrics.
Cannabis sativa (hemp; marijuana) is used to make twine and rope.
Linum usitatissimum (flax) is used to make linen.
Boehmeria nivea (ramie) is used to make fine Oriental textiles and, more recently, Western clothing.
Corchorus capsularis (jute) is used to make coarse fabrics, bags, burlap, and sacks.

Collenchyma and sclerenchyma cells are thick-walled cells specialized for support. Collenchyma cells support growing regions, and sclerenchyma cells support nongrowing regions. Many textiles are made of sclerenchyma fibers.

cell. For example, sclereids often occur at vein endings and near the edges of leaves.

Fibers are long, slender cells occurring in single strands or bundles often associated with vascular tissue (figs. 13.7, 13.10). They differentiate from parenchyma or from the vascular cambium, and vary in length; for example, fibers in sisal (*Agave sisalana*) are 1–8 mm long, while those of ramie (*Boehmeria nivea*) may be over half a meter long. Fibers in stems elongate as internodes grow, after which they deposit a thick secondary wall that occupies as much as 90% of the cell's volume (fig. 13.7b).

Several factors influence the differentiation of fibers. For example, fibers do not form in stems or petioles when leaves are removed, suggesting that their differentiation is promoted by a transmittable substance emanating from leaves. This substance is probably not auxin, since applying auxin does not replace the differentiating effects of leaves. Rather, gibberellins may influence fiber differentiation. Jute (*Corchorus capsularis*) and hemp (*Cannabis sativa*) plants treated with gibberellin have more fibers than do untreated controls. Furthermore, these fibers are thicker and up to four times longer than those in untreated plants. As is true for collenchyma, mechanical stress promotes the differentiation of fibers; plants grown on shakers contain more fibers than do unshaken controls.

By now you're probably wondering "What about cotton fibers?" Cotton (*Gossypium*) is the most important commercial fiber, but it is not sclerenchyma. Rather, cotton fibers are *trichomes*, which are outgrowths of epidermal cells. You'll learn about cotton fibers and trichomes later in this chapter.

FIGURE 13.11

Harvesting *Musa textilis*, commonly known as abaca, or Manila hemp. The trunk, which consists of leaf stalks, is used for its fibers; these fibers make high-grade cordage.

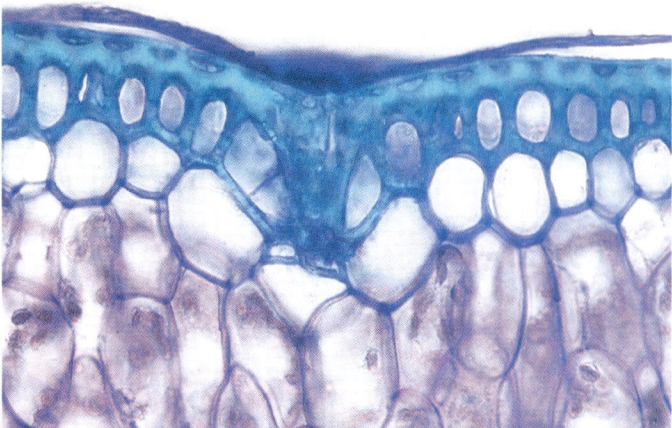

FIGURE 13.12

Epidermal cells are covered by a water-resistant cuticle that protects the underlying cells and tissues.

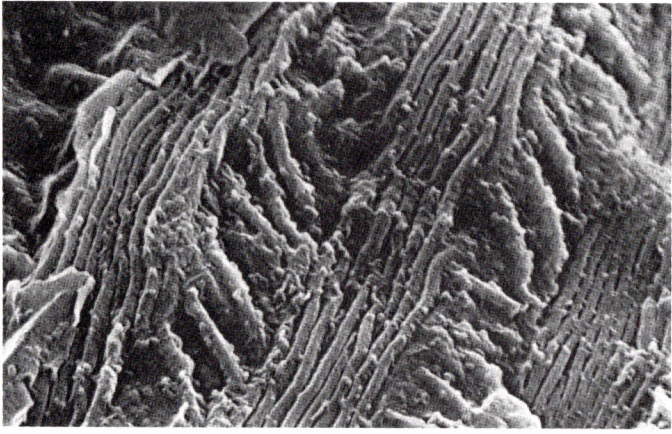

FIGURE 13.13

Scanning electron micrograph showing epicuticular wax on a leaf of corn (*Zea mays*), ×4,400. Epicuticular wax helps protect plants from desiccation and from invasion by pathogens.

DERMAL TISSUE: THE EPIDERMIS

Dermal tissues cover the plant body. The dermal tissue that covers the primary body of plants is the **epidermis.** The epidermis has several functions, including the absorption of water and minerals, secretion of cuticle, protection against herbivores, and the control of gas exchange. Each of these functions is attributable to one or more of the unique features of the epidermis, such as the presence of a cuticle, few intercellular spaces, and multifunctional outgrowths called trichomes.

Cuticle

The outer walls of epidermal cells are covered with a waterproof **cuticle** made of a fatty material called *cutin* (fig. 13.12). The cuticle protects the plant from desiccation by helping to maintain a watery, aquatic environment inside the plant. The thickness of the cuticle, and hence the ability of a plant to retain water, is strongly influenced by the environment: plants growing in dry environments typically have thick cuticles, while those growing in wet environments have thin cuticles. The cuticle also protects a plant from microbes because it resists microbial infection and degradation. This resilience also accounts for why cuticles have been recovered from fossils millions of years old.

The cuticle is often covered by **epicuticular wax,** which is deposited as smooth sheets, rods, or filaments (fig. 13.13). The wax imparts a whitish coloring to the surface of many leaves and fruits. On the undersides of leaves of wax palm, *Copernicia cerifera* (from the Latin word *cerifera*, meaning "wax bearer"), the whitish layer of epicuticular wax is often more than 5 mm thick. Wax from leaves of this plant is called *carnauba wax* and is used to make polishes, candles, and lipstick. It takes about three hundred large leaves to produce a kilogram of wax.

The cuticle and its thick underlying cell wall protect the plant from stresses such as wind, desiccation, predators, and abrasion. Some plants also use debris from the environment as a protective coating. For example, *Cerochlamys pachyphylla* is a small desert succulent whose epidermal cells secrete a sticky, gluelike substance. Windblown sand adheres to this glue and protects the plant from predators and the hot sun.

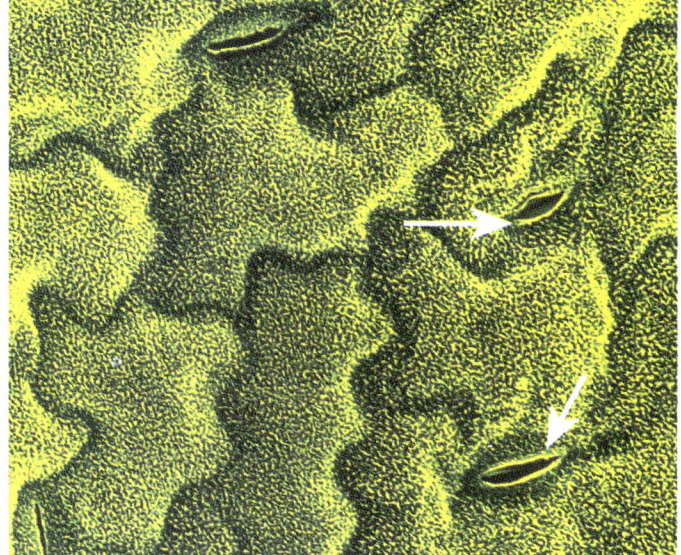

FIGURE 13.14

Scanning electron micrograph of epidermal cells on the upper surface of a leaf of *Pisum sativum* (garden pea). Epidermal cells are packed closely together; the only spaces occur between specialized cells that form stomata (arrowheads). Stomata allow for gas exchange for photosynthesis, ×100.

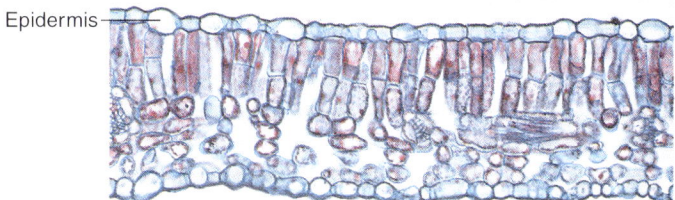

FIGURE 13.15

Cross section of a leaf, showing the epidermis. In most plants the epidermis is one cell thick.

Epidermal Cells and Gas Exchange

Most epidermal cells are flat, tilelike cells packed together like bricks in a brick wall (fig. 13.14). These cells typically lack chloroplasts and are transparent; it is the underlying chlorenchyma cells that give leaves their green color. However, the vacuoles of epidermal cells occasionally contain pigments. For example, red cabbage is colored by anthocyanins in epidermal cells, as are flowers and colored parts of variegated leaves of plants such as *Coleus*.

The epidermis differentiates from protoderm, and is usually only one cell thick (fig. 13.15). However, the epidermis of leaves of *Peperomia* and the rubber plant (*Ficus elastica*) is typically several cells thick; the additional layers of cells store water (fig. 13.16). Similarly, the roots of many epiphytic orchids have a multicellular epidermis called a *velamen* that helps the plant retain water (fig. 13.17).

The only intercellular spaces in the epidermis are **stomata** (singular: **stoma**). Each stoma (from the Greek word *stoma*, meaning "mouth") is surrounded by two *guard cells* (fig. 13.18). Stomata occur on stems, leaves, flowers, and fruits, but tend to be most

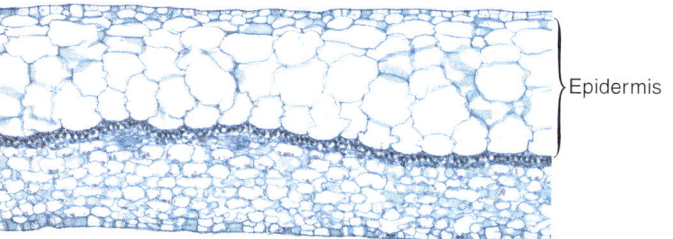

FIGURE 13.16

The epidermis of rubber plant is several cells thick and is used to store water.

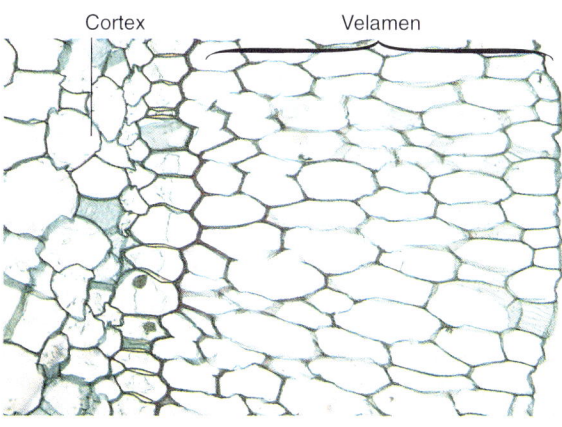

FIGURE 13.17

(a) Roots of an epiphytic orchid. (b) Cross section of the outer portion of an orchid root, showing the velamen, ×100. The velamen helps the orchid conserve water.

abundant on the undersides of leaves. Stomata usually cover less than 1% of the epidermal surface, yet they are numerous: the leaves of most plants have 10,000–80,000 stomata cm^{-2}. Stomata are often surrounded by distinctively shaped cells called *subsidiary cells* (fig. 13.19), which function as reservoirs for water and ions that

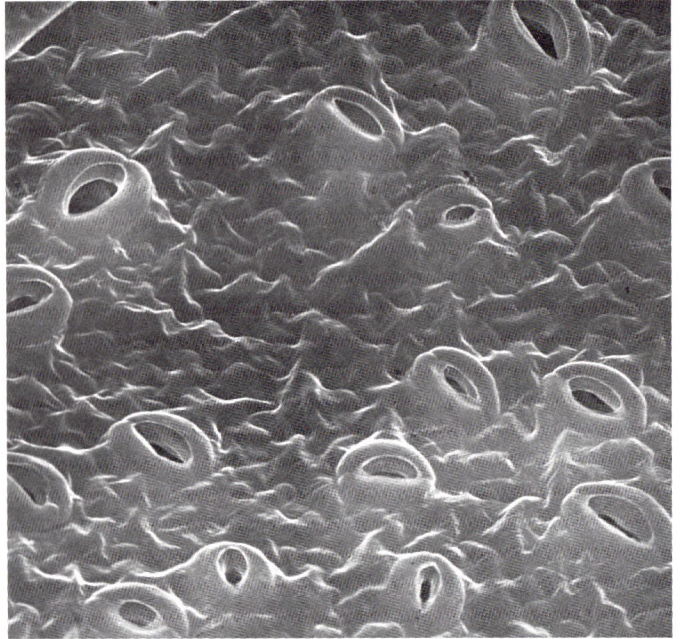
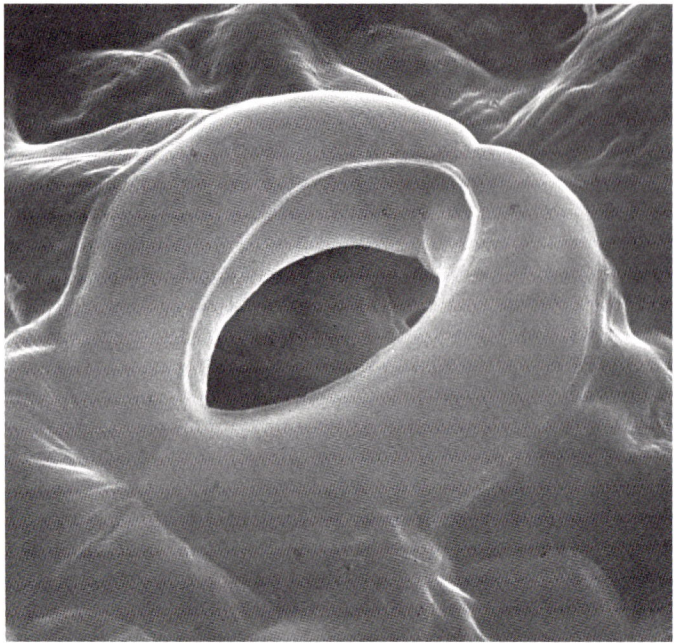

FIGURE 13.18

(a) Stomata on lower surface of a cucumber leaf, ×1,400. The pore formed by the guard cells allows for gas exchange between the atmosphere and the underlying cells. (b) Stoma in epidermis of a cucumber leaf. Two guard cells surround the pore.

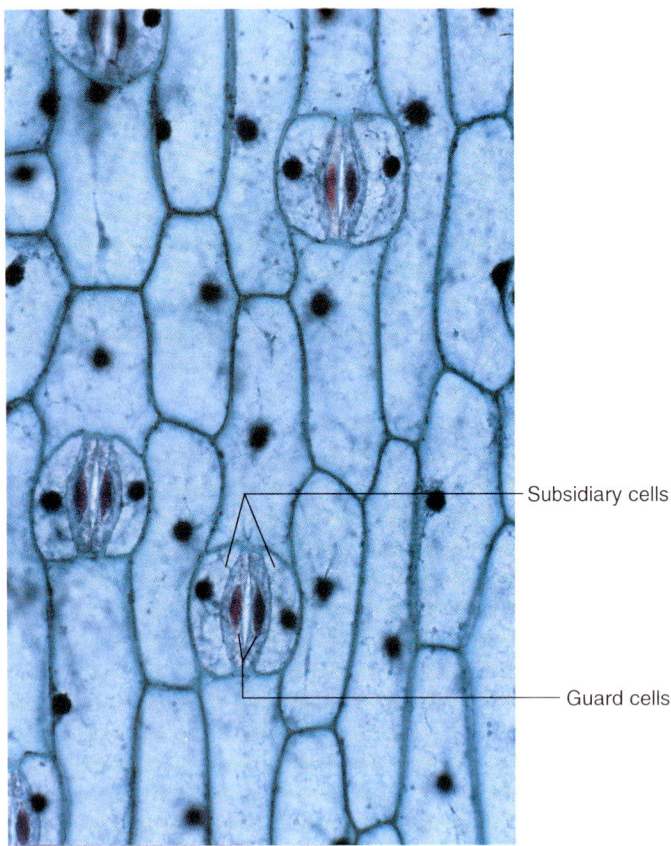

FIGURE 13.19

Epidermis of *Zebrina*. The large polygonal cells are ordinary epidermal cells. Each pair of guard cells is bordered by subsidiary cells, ×150.

enter and leave guard cells. Although substomatal cavities form after stomata differentiate, not all guard cells open to a cavity.

Guard cells in dicots are kidney-shaped, while those of monocots are shaped like dumbbells (fig. 13.20). Despite their different structures, guard cells in all plants have the same function: they regulate gas exchange by opening and closing the stomata. The opening and closing of stomata regulate the diffusion of CO_2 into the leaf for photosynthesis and also the loss of water from the leaf. Indeed, for every gram of carbon fixed via photosynthesis, the plant loses from 250 to 600 g of water, depending on the type of photosynthesis (e.g., C_3, C_4, or CAM). Such large losses of water through stomata are an inevitable cost of photosynthesis.

Stomata open and close in response to environmentally induced changes in the turgor pressure of guard cells (fig. 13.20). Light and other environmental signals cause K^+ to enter guard cells, thereby drawing water into them via osmosis. This influx of water nearly doubles the volume of the guard cells and causes them to elongate. As they elongate, they bow apart and open the stomatal pore. Similarly, stomata close in the dark when K^+ and water leave the guard cells. We'll discuss the details of the structure and function of stomata in Chapter 21. For now, remember that stomata are a critical adaptation for conserving water, and thus for maintaining life on land.

The epidermis coats the entire primary body of plants and is therefore a plant's first line of defense against herbivores. Although generally effective, the epidermis is not impenetrable; many pathogens invade plants through stomata. For example, *Pseudoperonospora humuli* and *Plasmopara viticola* are pathogens

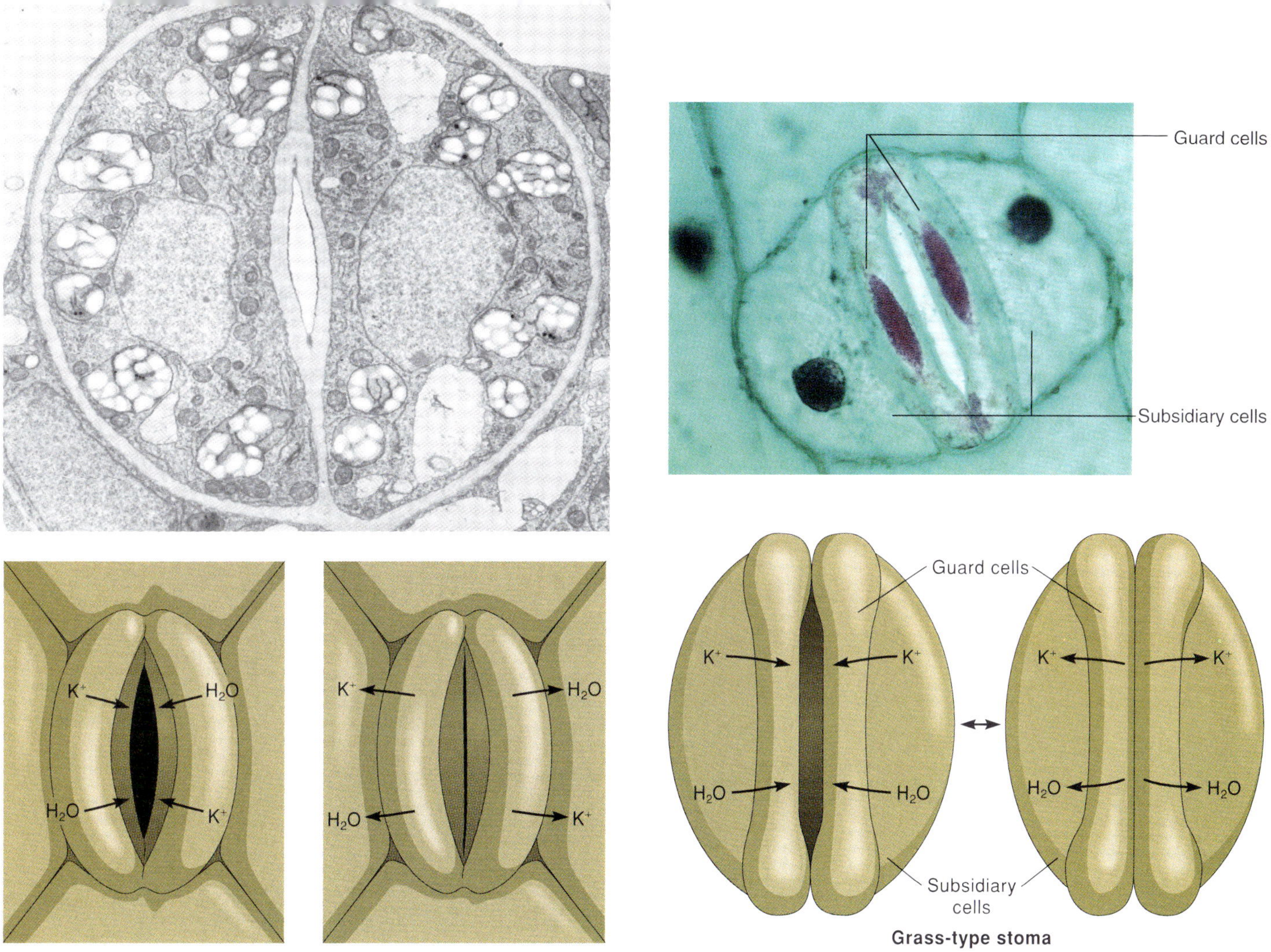

FIGURE 13.20

Stomata of (a) cow pea (*Vigna sinensis*), a dicot, ×13,000, and (b) corn (*Zea mays*), a monocot, ×400. Both kinds of stomata function similarly (see drawings). Light triggers an influx of K^+ into guard cells, which causes an influx of water (via osmosis) into the guard cells. As the cells elongate, they bow apart and form the stomatal pore. The pore closes via the reverse mechanism: K^+ exits the cells, causing water to move out of them via osmosis. This in turn shrinks the cells and closes the pore.

that cause downy mildew on hop (*Humulus lupulus*) and grape (*Vitis*). Zoospores of these pathogens settle on stomata in light (fig. 13.21). Having located an entry into the plant, the zoospores then germinate and grow into the vulnerable inner tissues of the leaf. Many fungal pathogens locate stomata by tactile stimulation (see box 13.1, "Feeling Their Way").

Guard cells are not the only specialized cells in the epidermis. The upper epidermis of the leaves of many grasses contains longitudinal rows of large, vacuolate cells called *bulliform cells*. When it is hot and dry, bulliform cells rapidly lose water from their vacuoles and shrink, causing the leaf to roll into a protected cylinder. Leaf rolling reduces the surface area that is exposed to the environmemt, thereby helping the leaves conserve water.

Trichomes

Trichomes are single-celled or multicellular outgrowths of epidermal cells. The most economically important trichomes are cotton fibers, which are up to 6 cm long and are made by

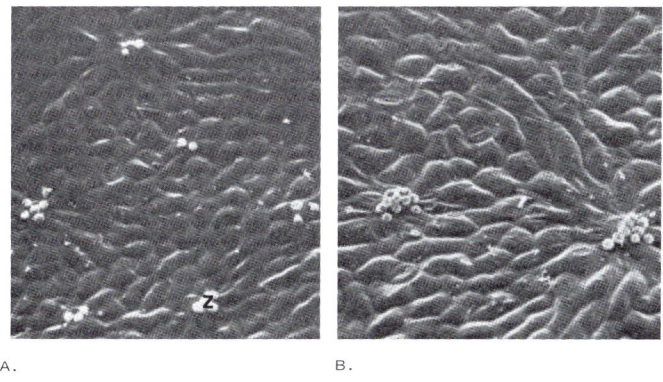

FIGURE 13.21

Invasion of downy mildew (*Plasmopara viticola*) on young grapevine leaves (*Vitis vinifera*). (a) After only one hour in light, groups of zoospores (z) have settled on many stomata. (b) In the dark, larger groups of zoospores have settled on a few stomata. The zoospores germinate and grow through the stomatal pore.

CHAPTER THIRTEEN *Primary Growth: Cells and Tissues*

BOXED READING 13.1

FEELING THEIR WAY

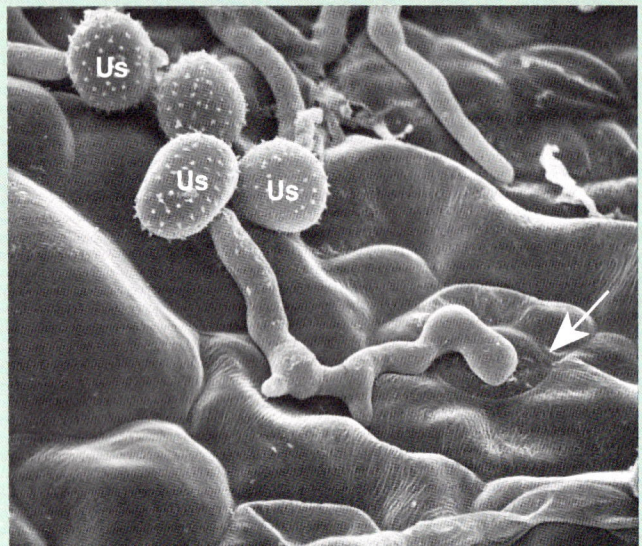
A.

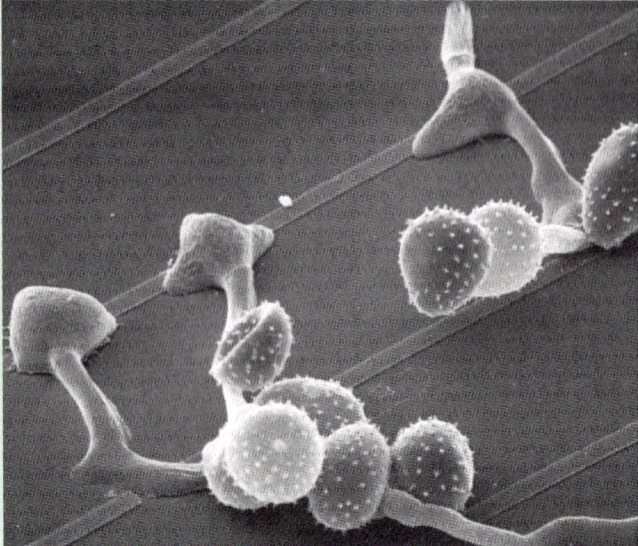
B.

BOX FIGURE 13.1

Uredospores of rust fungus (*Uromyces appendiculatus*) invade hosts such as beans (*Phaseolus vulgaris*) by distinguishing minute differences in the topology of leaves. (a) Germinated uredospore (Us) grows toward stoma (arrow), where it forms an appressorium (a primary infection structure) necessary for the later invasion of the leaf. (b) Polystyrene replicas having ridges 0.5 μm high and 4.0 μm wide induced formation of appressoria. (c) Replicas similar to those in (b) but with ridges 5.0

The epidermis of plants is an imposing barrier to most microbes and tiny insects. Its waxy cuticle forms a slick, desertlike environment that rapidly sheds most microbes and on which most others cannot grow. Furthermore, the absence of intercellular spaces in the epidermis blocks access to the underlying (and vulnerable) tissues. Many pathogens have overcome these defenses by evolving elaborate chemical systems that help them recognize and invade susceptible plant prey. Others exploit the weak link in the epidermis—its stomata. A stomatal pore provides for gas exchange between the leaf and air, but it is also a breach in the epidermis through which pathogens can invade the leaf. Many pathogens use the structural characteristics of the epidermis to help them "feel" their way to stomata that they will use to invade the plant. To appreciate this, consider the battle between the rust fungus *Uromyces appendiculatus* and its host, the common bean plant (*Phaseolus vulgaris*).

Spores of the rust that land on the leaf germinate and form threadlike structures called hyphae. When these hyphae touch a stoma, they form appressoria, which are specialized structures that start the infection of the leaf. This method of infection involves two critical events. First, the hypha must economically find a stoma before exhausting its stored food reserves; that is, it must find a stoma without expending unnecessary energy by growing aimlessly over the leaf surface. Secondly, the hypha must have some way to recognize the stoma, stop growing, and form the appressoria. How do fungi do this?

Harvey Hoch and his colleagues at Cornell University answered this question with a strikingly clever set of experiments. They first used electron-beam etching to construct replicas of leaf surfaces. They then etched polystyrene wafers and inoculated these replicas with spores of *Uromyces*. Hyphae on wafers having ridges 0.5–15 μm apart grew straight and perpendicular to the ridges as long as the ridges were less than 30 μm apart. Interestingly, straight growth decreased as distances between the ridges increased. These observations suggested to Hoch and his colleagues that hyphae use

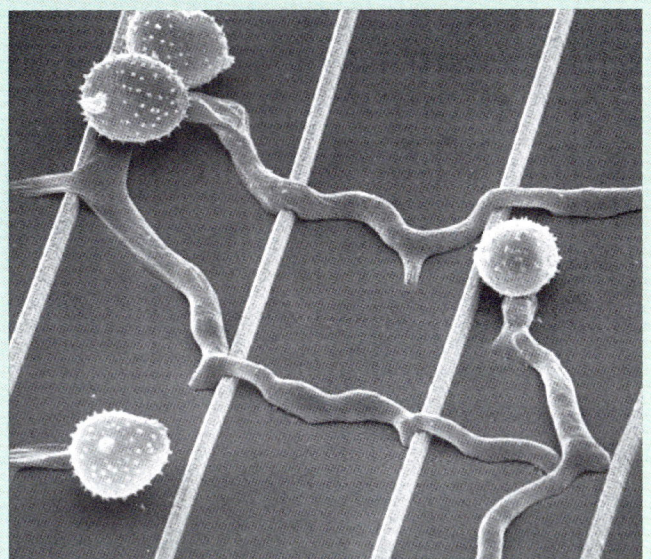
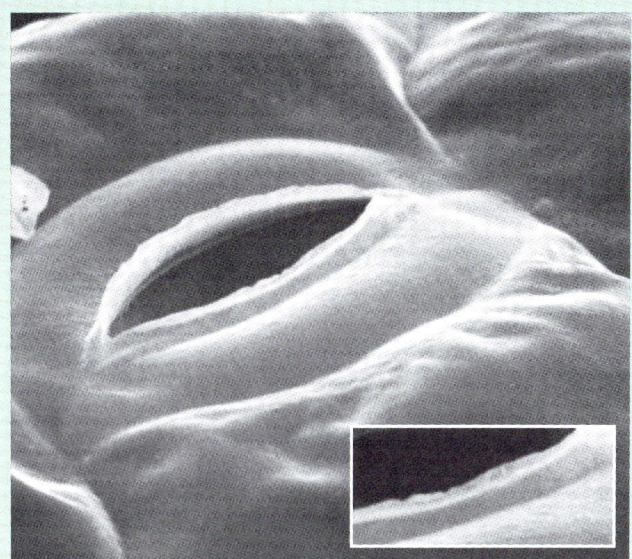

c.   d.

µm high and 4.0 µm wide did not induce formation of appressoria. Instead, the fungus germlings grew across the ridges. (d) Scanning electron micrograph of guard cells of a bean leaf. These cells have erect, ridgelike lips that function as a signal to induce formation of appressoria.

changes in surface topography to orient their growth.

With this in mind, the Cornell researchers then searched the surface of bean leaves for similar topographical cues. They did not have to look long. Indeed, epidermal cells and stomata are arranged in longitudinal rows separated by depressions about 15–30 µm apart. Therefore, hyphae exploit the patterned arrangement of epidermal cells to efficiently navigate and grow toward stomata. Hyphae growing perpendicular to epidermal cells have a better chance of finding stomata, which are also arranged in longitudinal rows.

The second stage of the infection process involves stopping growth and forming an appressorium over the stomatal pore. Experiments with etched wafers suggested that the leaf's topography also dictated the formation of these appressoria. For example, hyphae form appressoria when they encounter ridges 0.5 µm high. We do not understand how hyphae sense such minute changes in elevation, but it may involve changes in the cytoskeleton. Whatever the mechanism, it is incredibly sensitive: appressoria do not form if ridges are shorter than 0.1 µm or exceed 1 µm. Armed with this clue, Hoch and his colleagues again examined the stomata of bean leaves to see if the stomata have any structures similar to the ridges of their leaf replicas. They found that guard cells have flanges that rise 0.49 µm above the leaf's surface. Thus, the hyphae of *Uromyces* use topography as a map to navigate and infect their prey.

The findings of the Cornell researchers are more than an interesting discovery; they provide plant breeders with another means of increasing the resistance of beans to fungal infection, and therefore a means of potentially increasing yields. Previously, most breeders concentrated on altering beans' chemical defenses to fight pathogens. The Cornell workers' findings indicate that resistance can also be increased by altering the plant's structure to increase or decrease the height of the flange on guard cells. If such a change were made in bean plants, *Uromyces* hyphae would not recognize the stomatal pore, and the plants would be immune to the fungal pathogen.

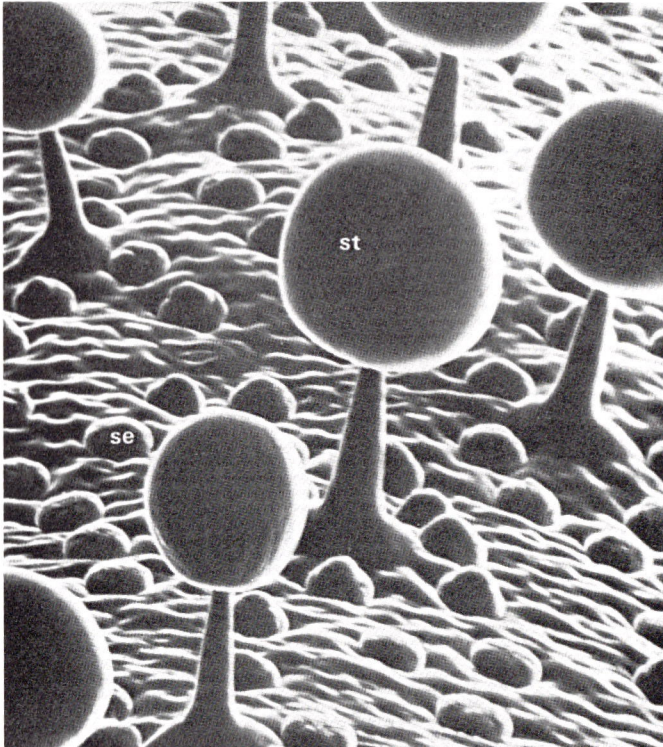

FIGURE 13.22

Scanning electron micrograph of part of a leaf of butterwort (*Pinguicula grandiflora*), an insectivorous plant. The stalked (st) and smaller sessile (se) trichomes secrete enzymes and other compounds that help trap and digest animals.

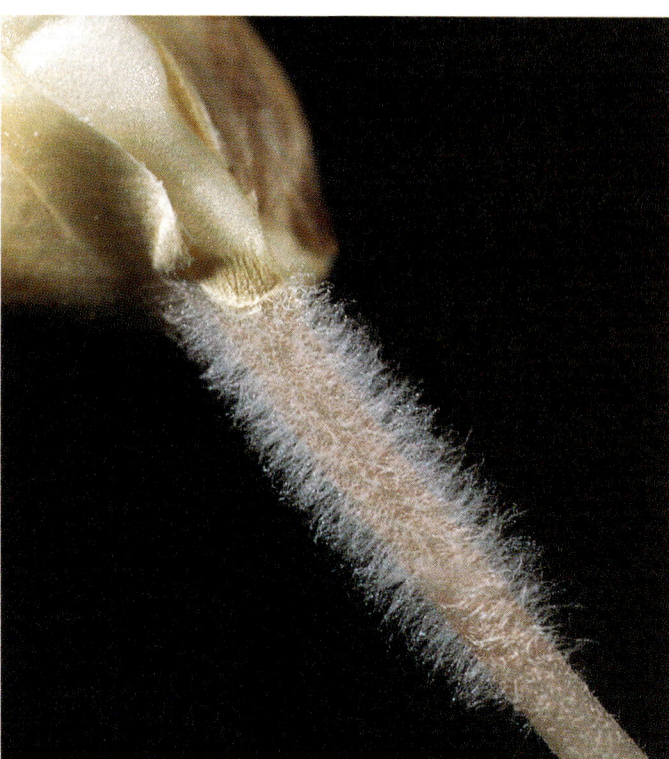

FIGURE 13.23

Corn seedling. Note the many root hairs, which increase the surface area for absorption of water and dissolved minerals. Root hairs form just behind the elongating tip of the root.

the epidermis of cotton seeds. The flexibility and soft texture of cotton is due to the absence of lignin—indeed, cotton fibers are 95% cellulose. However, trichomes on other seeds may be hard and lignified, such as those on the seed coat of *Strychnos nux-vomica*, the source of strychnine.

The contents of trichomes of some plants are used by humans. For example, menthol is a volatile oil collected from trichomes of peppermint (*Mentha piperita*), and trichomes of *Cannabis sativa* produce the tetrahydrocannabinols (THC) responsible for marijuana's soothing effects. The purified resin from *Cannabis* trichomes is called *hashish* and is a powerful drug.

Doing Botany Yourself

Trichomes of some plants reflect light, and therefore could help protect a plant against overheating and excessive water loss. Design an experiment to test this idea.

Functions of Trichomes

Nutrition and Absorption

Trichomes of carnivorous plants such as sundew (*Drosera*) and butterwort (*Pinguicula*) secrete enzymes that help the plant liquefy its trapped prey (fig. 13.22). Trichomes also absorb nutrients released as the prey are dissolved in traps.

Root hairs are outgrowths of epidermal cells specialized for absorbing water and minerals from soil (fig. 13.23). They occur near the tips of roots, where they are abundant (e.g., many plants have as many as 40,000 root hairs cm^{-2}). Root hairs increase the absorptive surface area of roots several thousandfold, thereby enabling the plant to extract water and dissolved minerals more effectively from nooks and crannies in the soil.

Protection

In many plants, trichomes deter marauding animals. For example, the resistance of cotton and soybean to leafhoppers is proportional to the density of trichomes on their leaves. Similarly, the resistance of *Passiflora adenopoda* to larvae of *Heliconius* butterflies is due to the presence of hook-shaped trichomes; larvae move only a few millimeters before being impaled by these trichomes. Similarly, aphids often inadvertently break off the heads of trichomes, thereby releasing their sticky contents. The aphids become trapped in this sticky secretion and soon starve to death.

Some trichomes protect plants from larger prey, including humans. For example, the leaves of stinging nettle (*Urtica*) have brittle, vase-shaped trichomes (fig. 13.24). When touched, the tip of the trichome breaks off, leaving behind a sharp, syringe-shaped cell that readily penetrates your skin. Pressure from the contact also injects the trichome's irritating chemicals into your skin. Ouch!

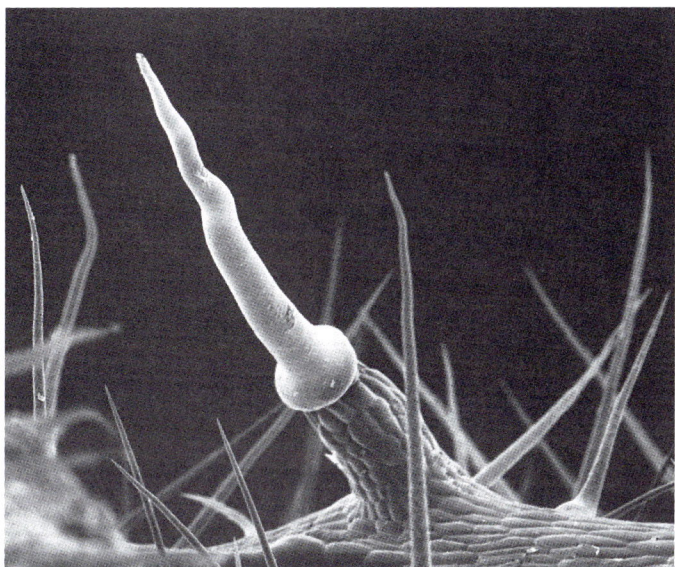

FIGURE 13.24

Stinging hairs of *Urtica dioica* (stinging nettle) are large cells filled with toxin. The tip is pointed and glassy, which makes it brittle and easily broken. The small tip allows it to catch in an animal's skin and break off, thus releasing the toxin.

The Lore of Plants

The protective function of trichomes on sunflower leaves has apparently been overcome by *Chlosyne* butterflies. Female butterflies are attracted to chemicals in the trichomes as cues for where to lay their eggs. Larvae from these eggs munch away on the leaves, apparently unaffected by the chemicals in the trichomes.

Epidermal Cells and Cellular Recognition

Epidermal cells often recognize other organisms, such as pathogenic and nonpathogenic bacteria. Recognition typically involves interactions between molecules called *lectins* on the surfaces of epidermal cells and the contacting organism; it is referred to as **cellular recognition.** These chemical identification tags on a plant's epidermis are important because they direct many aspects of plant growth and development, including germination of pollen grains, formation of nitrogen-fixing nodules in roots, and whether a pathogen can successfully invade a plant (see Chapter 4).

CONCEPT

The epidermis covers the primary body of a plant. It is covered with a waxy cuticle that decreases water loss, and is perforated by pores, called stomata, that control gas exchange. Outgrowths of epidermal cells called trichomes have many functions, including absorption and protection. The epidermis is also the site of cellular recognition, by which a plant chemically recognizes and responds to other organisms.

The Fate of the Epidermis

The epidermis is short lived in many plants. Radial expansion of stems caused by activation of the vascular cambium usually ruptures the epidermis during the first year of growth. When this occurs, the epidermis is replaced by a secondary dermal tissue called the *periderm*. You'll learn about the periderm when we discuss secondary growth in Chapter 16.

VASCULAR TISSUES: XYLEM AND PHLOEM

Vascular tissues are specialized for long-distance transport of water and dissolved solutes. They ramify throughout the plant and are easily seen as veins in leaves. Vascular tissues typically contain transfer cells, secretory cells, and fibers in addition to parenchyma and conducting cells.

Xylem and phloem are the two kinds of vascular tissues in plants. In this section, we'll concentrate on their differing structures and functions, and on the ways in which these properties adapt plants to life on land. You'll learn more about transport in the xylem and phloem in Chapter 21.

Xylem

Xylem (from the Greek word *xylos*, meaning "wood") transports water and dissolved nutrients in an unbroken stream from the roots to all parts of a plant. The water transported in xylem replaces that lost via evaporation through stomata. There are two types of xylem: primary xylem and secondary xylem.

Primary and Secondary Xylem

Primary xylem differentiates from procambium in the apical meristem and occurs throughout the primary body of a plant. Xylem in the elongating regions of plants is called *protoxylem*. During growth, protoxylem is stretched and replaced by newly differentiated protoxylem. When elongation stops, the procambium forms *metaxylem*, a relatively permanent type of primary xylem. The secondary cell walls of protoxylem and metaxylem are elaborately sculptured into hoops, bands, or springlike helices that allow elongation and prevent the cells from being crushed by adjacent cells or by internal tensions generated by water transport (fig. 13.25).

Secondary xylem differentiates from vascular cambium and is commonly called *wood*. Secondary xylem contains the same types of cells as primary xylem: the main difference is that these cells are more abundant and occur in different frequencies in secondary xylem than in primary xylem. Such differences vary in different species and account for the diverse properties of different kinds of wood. For example, balsa (*Ochroma*) wood contains many large, thin-walled parenchyma cells and relatively few thick-walled fibers. This combination of cells gives balsa wood a low specific gravity (0.1–0.16), making it lightweight and ideally suited as an insulator and component of model airplanes and lifeboats. Conversely, oak (*Quercus*)

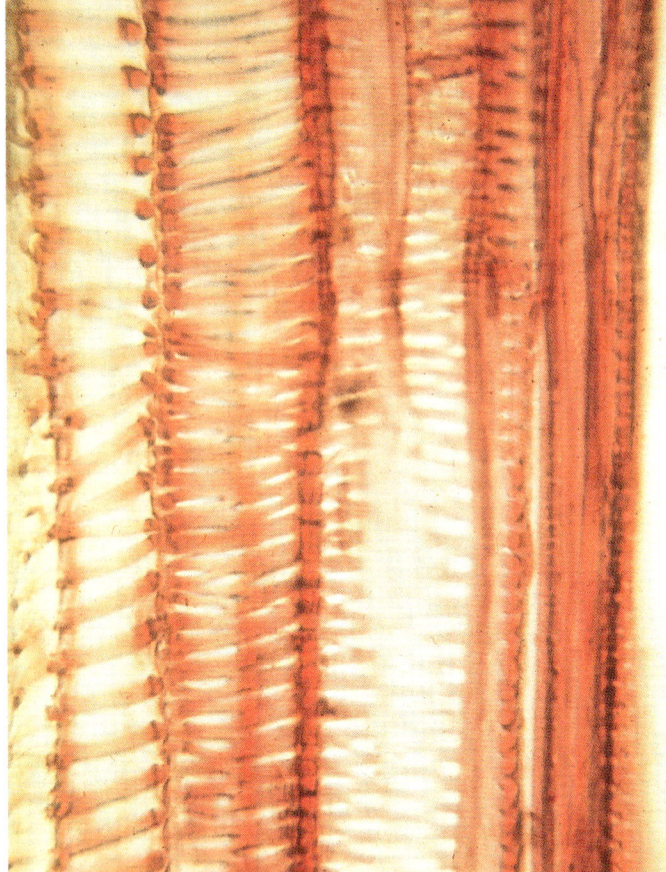

A.   B.

FIGURE 13.25

The secondary walls of some of the first-formed tracheary elements of the protoxylem are deposited in forms of (a) rings or (b) spirals. These thickenings strengthen the tracheary element while enabling the tracheary element to elongate during growth.

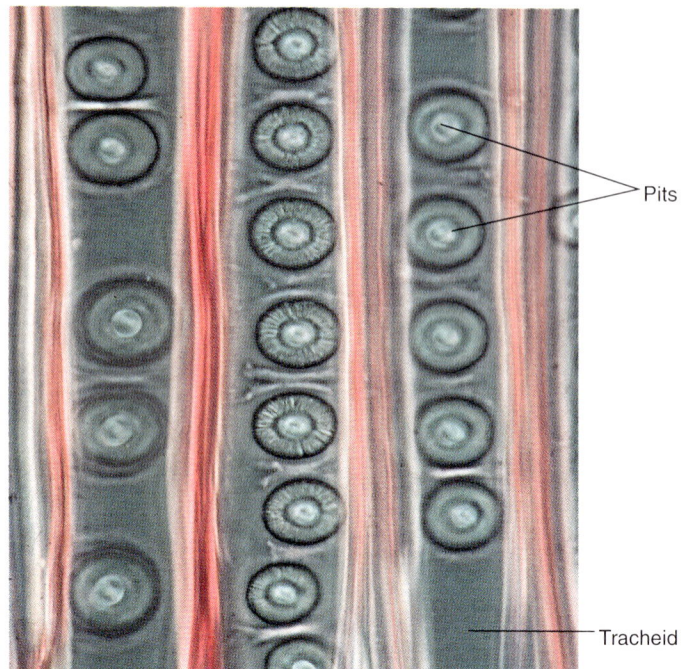

FIGURE 13.26

Light micrograph of tracheids in wood of pine (*Pinus*), ×163. Tracheids are connected by circular bordered pits, which are thin areas in adjacent cell walls. The wood of most gymnosperms (e.g., conifers) resembles that of pine; the primary conducting cells are tracheids.

wood contains many thick-walled fibers and relatively few thin-walled parenchyma cells; thus, it is much denser (its specific gravity is approximately 0.6) than balsa wood. You'll learn more about wood and its characteristics in Chapter 16 when we examine secondary growth.

Conducting Cells in Xylem

Conducting cells in xylem are called **xylary elements.** There are two kinds of xylary elements: *tracheids* and *vessel elements* (figs. 13.26, 13.27). Both are elongate, dead at maturity, and have thick, lignified secondary cell walls. Because water is typically pulled through xylary elements at a negative pressure (i.e., tension; lower than atmospheric pressure), these thick walls help prevent the cells from collapsing. The thick walls of xylary elements also help support the plant.

Tracheids are the most primitive (i.e., least specialized) xylary elements, and are the only water-conducting cells in most woody, nonflowering plants. Tracheids are long, slender cells with tapered, overlapping ends. Water moves upward from tracheid to tracheid through thin areas called *bordered pits* in their cell walls (fig. 13.26).

Vessel elements are more evolutionarily advanced than tracheids and occur in several groups of plants, including angiosperms. With only a few exceptions, all angiosperms contain vessels and tracheids (fig. 13.27). The vessel elements of many plants are barrel-shaped; that is, they are relatively short and wide. Vessel elements, which are shorter and wider than tracheids, are arranged end to end. Their transverse walls are partially or wholly dissolved, forming long, hollow vessels through which water can move as much as 3 meters without having to move through a pit (fig. 13.28). This feature of vessels, along with their relatively large diameter, enables them to transport water more rapidly than can smaller, thinner tracheids. The evolution of vessels in angiosperms and their greater water-transport capacity is a major reason why angiosperms dominate today's landscapes.

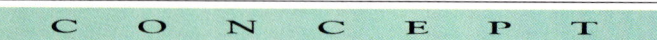

Xylem conducts water and dissolved minerals from roots to leaves. Water moves through xylary elements called *tracheids* and *vessel elements*, which are dead, hollow cells having thick secondary cell walls.

Tracheids have also evolved into thick-walled, fiberlike cells called *fiber-tracheids*. Unlike vessel elements, fiber-tracheids are specialized for support rather than water transport.

Phloem

Phloem (from the Greek word *phloios*, meaning "bark") transports dissolved organic materials (especially sucrose) throughout

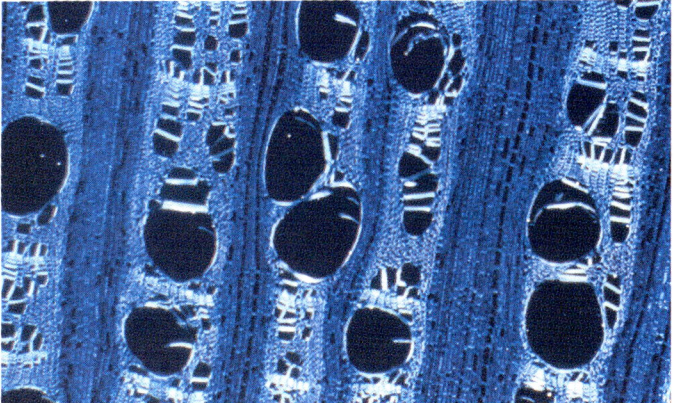

FIGURE 13.27

Light micrograph of a cross section of wood, showing tracheids, vessels, and multiserate rays. The wood of most angiosperms (e.g., oak) has tracheids and vessels to conduct water and dissolved materials from the roots to the shoot.

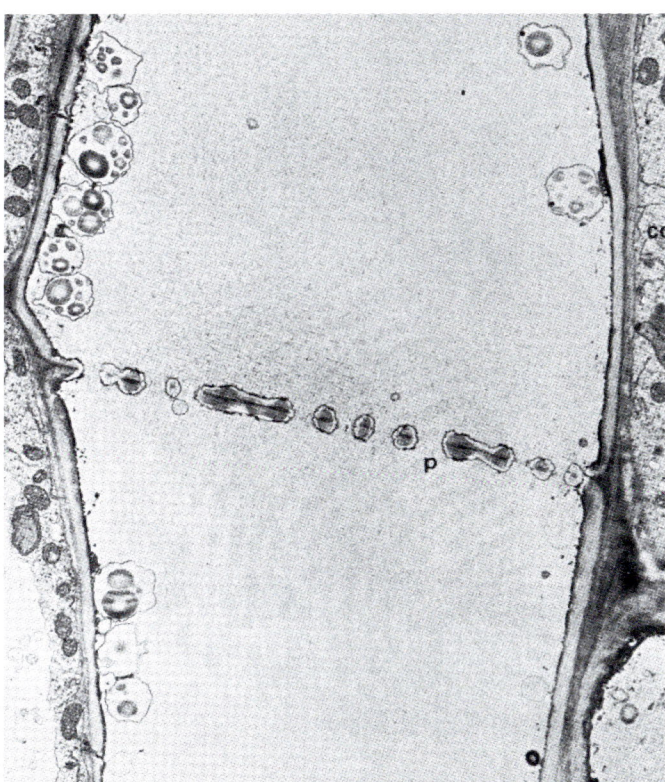

FIGURE 13.29

Sieve elements of tobacco (*Nicotiana tabacum*). Note that the pores (p) of the sieve plate are open. Sieve elements conduct water and dissolved organic matter (especially sugars) in all directions in the plant (cc = companion cell).

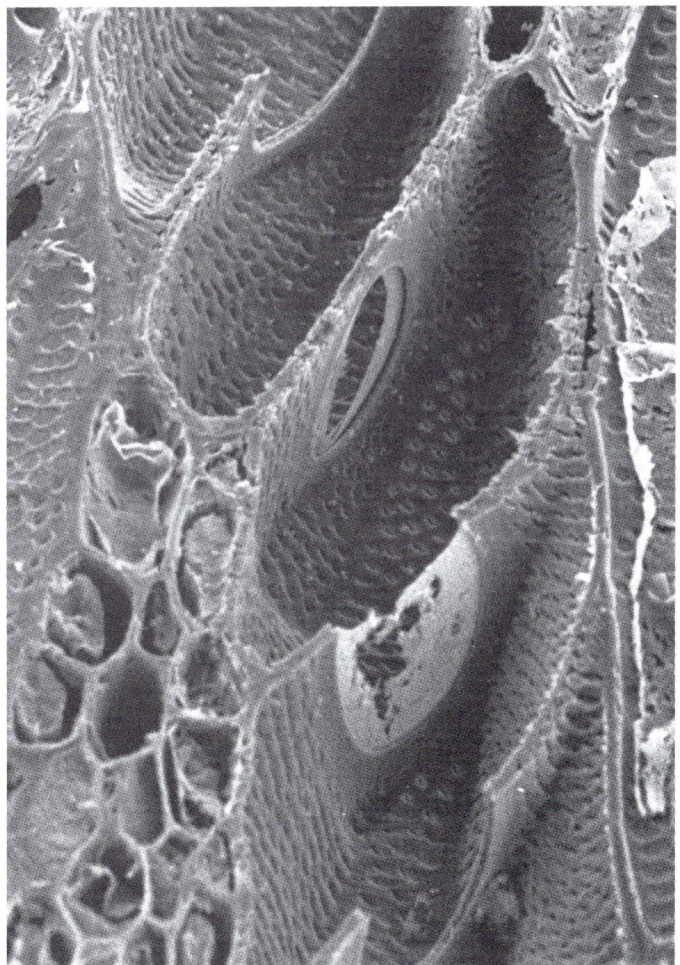

FIGURE 13.28

Vessel elements in wood of *Knishtia excelsa*, an angiosperm. The vessel elements form in long rows, are wider than they are long, and have thick, reinforced walls. The transverse walls of adjacent vessel elements are partly or wholly dissolved, forming long, hollow vessels through which water and dissolved minerals flow.

the plant. Unlike xylem, which transports water only upward, solutes in the phloem move in all directions. Furthermore, dissolved solutes move under a positive pressure in the living conducting cells of phloem, whereas they move at a negative pressure in the dead conducting cells of xylem.

Primary and Secondary Phloem

Primary phloem differentiates from procambium and extends throughout the primary body of a plant. Primary phloem in elongating regions of the plant is called *protophloem*, while that in nonelongating regions is called *metaphloem*. Cells of primary phloem have primary cell walls that are not elaborately sculptured.

Secondary phloem differentiates from the vascular cambium and constitutes the inner layer of bark. Like xylem, phloem consists of several types of cells, including transfer cells and other parenchyma cells, fibers, conducting cells, and secretory structures.

Conducting Cells of Phloem

Conducting cells of phloem are called **sieve elements.** Sieve elements lack nuclei and are alive at maturity. They also have thin areas along their cell walls called *sieve areas* that are perforated by *sieve pores* (fig. 13.29); solutes move from sieve element to sieve element through these pores. Sieve elements

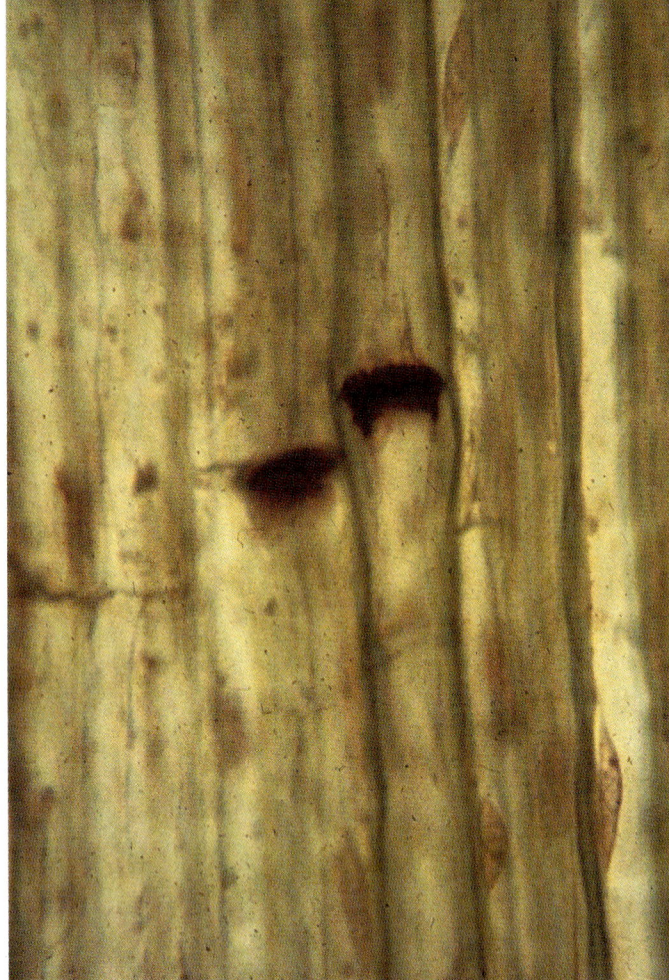

FIGURE 13.30

Sieve tube members in the phloem of squash (*Cucurbita*). The dark regions near the center of the photo are side views of sieve plates.

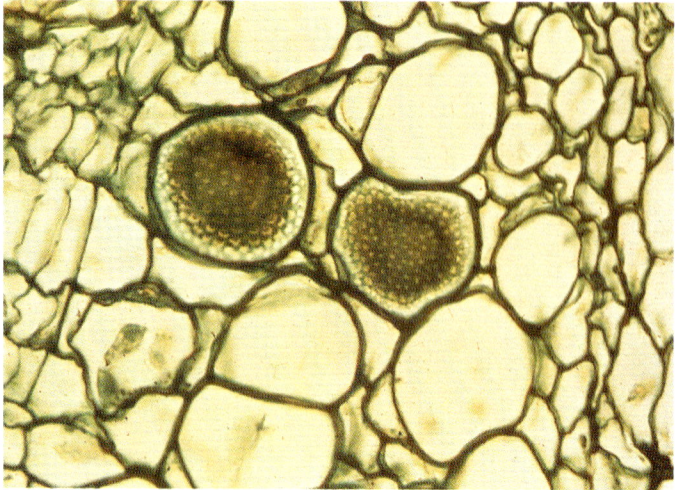

FIGURE 13.31

Face view of sieve plate of *Cucurbita*. During peak periods of transport, enough water and solutes move through a sieve plate to empty and refill a 0.5 mm-long sieve tube member every two seconds.

have thin, primary cell walls and are the most delicate cells in plants. They usually live for less than a year, although in palms they can live for more than a century.

There are two kinds of sieve elements: sieve cells and sieve tube members. **Sieve cells,** which are more primitive than sieve tube members, occur in nonflowering plants. They are long and have tapered, overlapping ends. Sieve areas occur in equal frequencies all over the surface of sieve cells, including along their longitudinal walls. Sieve cells may be associated with specialized parenchyma cells called *albuminous cells*, which help regulate the sieve cells' activities. Despite this interdependency, however, sieve cells and albuminous cells do not arise from the same mother cell.

Sieve tube members occur in all but a few of the most primitive angiosperms. They are usually shorter and wider than sieve cells, and are arranged end to end into sieve tubes (fig. 13.30). Their sieve areas have larger sieve pores than do sieve cells, and they are concentrated along the contacting end-walls of adjacent sieve tube members (fig. 13.31). These specializations allow solutes to move through sieve tubes faster than through sieve cells.

Mature sieve tube members are rather strange: they are the only cells in plants that lack nuclei but contain living cytoplasm (fig. 13.29). Furthermore, each sieve tube member is associated with at least one specialized cell called a *companion cell*, which regulates many of its activities. Companion cells help regulate the loading and unloading of carbohydrates from sieve tube members. Sieve tube members and companion cells arise from the same cell, and are linked via numerous plasmodesmata. Their interdependence is apparently absolute: the death of a sieve tube member also causes the demise of its companion cell.

The absence of nuclei is not the only unusual thing about sieve tube members. Indeed, they also contain a proteinaceous substance called *P-protein* (for *p*hloem-protein) along their longitudinal walls. Sometimes they also contain a spirally wound polymer of glucose called *callose* (fig. 13.32). P-protein and callose seal wounded sieve tubes.

CONCEPT

Phloem conducts water and organic solutes (especially sucrose) in all directions in plants. Solutes move through sieve elements called *sieve cells* and *sieve tube members*, which are living at maturity.

What are Botanists Doing?

The ultrastructure of plastids in sieve tube members has become very important for plant classification. Find one recent article on the use of variation in phloem plastids in classification and summarize the results that it presents.

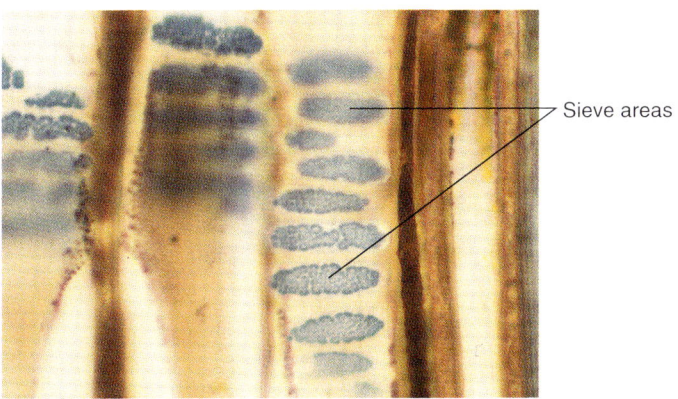

FIGURE 13.32

Sieve plates of sieve tube members of grape (*Vitis* sp.). Each sieve area consists of pores bordered by callose, which is stained blue in this preparation, ×400. Callose helps to seal wounded sieve tube members.

FIGURE 13.33

Extrafloral nectaries of passion flower (*Passiflora incarnata*). What do you suppose is the primary function of these glands? How would you test your hypothesis?

Secretory Structures

Plants secrete a variety of substances from structures called **secretory structures.** Secretory structures are seldom classified as a separate type of plant tissue because they often intergrade with other tissues. For example, some secretory structures are epidermal trichomes, while others are parts of the ground and vascular tissues. Despite their various origins in plants, secretory structures typically meet the requirement for classification as a plant tissue: they form a group of cells having a common function.

There are two types of secretory structures: external secretory structures and internal secretory structures.

External Secretory Structures

Nectaries

Nectaries are structures that secrete *nectar*, a sugary exudate that attracts insects, birds, or other animals. Most nectaries are associated with flowers and are called *floral nectaries*. Their nectar is 10%–50% sugar (especially sucrose, glucose, and fructose) and also contains amino acids. Nectar typically is pushed or diffuses through the walls of secretory cells, but may also ooze from stomata.

Nectar in floral nectaries attracts insects and other animals that pollinate the plant. Plants usually secrete small amounts of nectar, which forces foraging animals to visit several flowers before getting a full meal. Thus, a single animal can pollinate tens or hundreds of plants.

Nectaries that occur on vegetative parts of plants are called *extrafloral nectaries* (fig. 13.33). These nectaries often attract animals that defend the plant. For example, the extrafloral nectaries of plants such as trumpet creeper (*Campsis radicans*) and *Costus* attract ants that eat the nectar and, in return, defend the plant from leaf-eating insects. These tiny defenders are surprisingly effective: *Costus* plants deprived of ants are quickly devastated by fly larvae, and produce only one-third as many seeds as plants protected by ants.

Hydathodes

Hydathodes are tissues that secrete water via a process called *guttation*. Hydathodes are loose arrangements of parenchyma cells usually located at vein endings along the edges of leaves, and occur in plants ranging from ferns to flowering plants. Guttation from hydathodes occurs through modified stomata that are always open. The significance of hydathodes is unknown, but they may help relieve pressure caused by the uptake of excessive water by roots at night. You'll learn more about the uptake and transport of water in Chapter 21.

Digestive Glands of Carnivorous Plants

Carnivorous plants have trichomes that secrete enzymes and digest trapped prey (fig. 13.22). For example, trichomes on the sticky "flypaper" leaves of *Pinguicula* secrete digestive enzymes within an hour after trapping a victim; the secretion is so abundant that pools of the enzymes form on the leaf within two or three hours. These enzymes help gather nitrogen from animal protein, thereby allowing the plants to survive in nitrogen-deficient soil.

Salt Glands

Plants such as *Atriplex*, *Spartina*, and *Tamarix* that grow in salty soil have secretory structures called *salt glands*. These glands often form a glistening crust of salt that covers the leaves of the plant. In effect, salt glands are dump sites for the excess salt absorbed in water from the soil. Salt glands help plants adapt to life in saline environments.

Internal Secretory Structures

Secretory Cells

Secretory cells are large cells containing substances such as oils, tannins, resins, mucilage, and crystals. They often occur in groups and have several functions, including the storage and production of chemical deterrents to foraging animals. Secretory cells are also a source of many oils used by humans. For

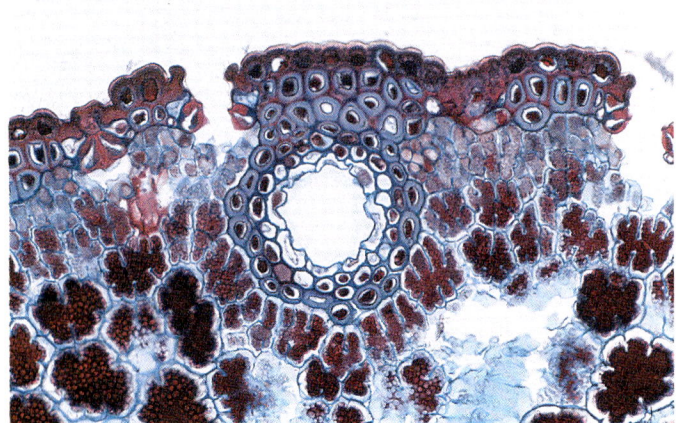

FIGURE 13.34

Resin duct from cross section of a leaf of *Pinus resinosa*. When ruptured, these ducts release resin, which deters herbivores and seals the wound.

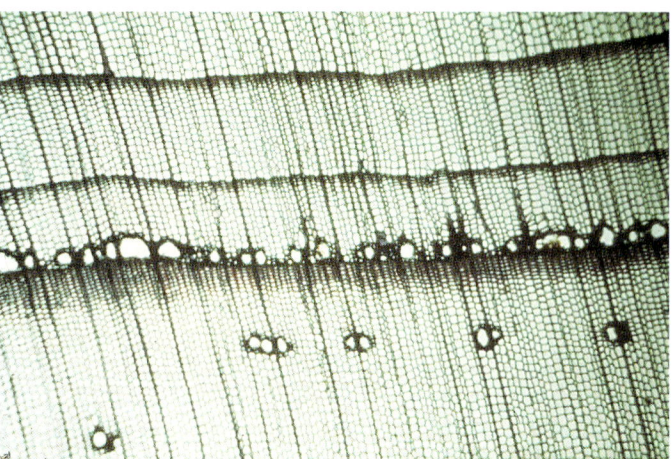

FIGURE 13.35

Resin ducts in pine (*Pinus*) wood. The band of ducts resulted from a forest fire.

example, peanut oil comes from secretory cells in cotyledons of peanut (*Arachis hypogaea*), safflower oil from seeds of safflower (*Carthamus tinctorius*), clove oil from flowers of clove (*Eugenia caryophyllata*), palm oil from fruits of palms such as *Elaeis guineensis*, and cinnamon oil from the secondary phloem of *Cinnamomum zeylanicum*.

Canals, Ducts, and Cavities

Many plants secrete oils and resins into internal canals, ducts, and cavities. For example, myrrh, frankincense, *Citrus* oils, *Eucalyptus* oil, and pine (*Pinus*) resin are extracted from internal cavities, ducts, and canals (fig. 13.34); these oils deter grazing animals, and resin rapidly seals wounds. Resin ducts form in response to wounding and can therefore be used as indicators of past trauma. For example, the layer of resin ducts shown in figure 13.35 was formed in response to a forest fire. Commercial growers wound trees to increase their harvest of resin.

As already mentioned, resins and oils help plants defend themselves. However, some animals have evolved means of not only tolerating these resins, but also using them for their own defense. Sawfly larvae, for example, eat pine and store its resin in their bodies. When disturbed by another animal, the larva rears up and secretes a drop of resin onto the intruder. This typically ensures privacy for the larva, and troubles the rude visitor.

Laticifers

Laticifers are secretory structures that contain *latex*, which is cytoplasm containing a hodgepodge of carbohydrates, organic acids, alkaloids, terpenes, oils, resins, enzymes, and rubber (fig. 13.36). Laticifers occur in several families of plants either as long, single cells called *nonarticulated laticifers* or as a series of fused cells called *articulated laticifers*. They have only a primary cell wall and can be several centimeters long. Because

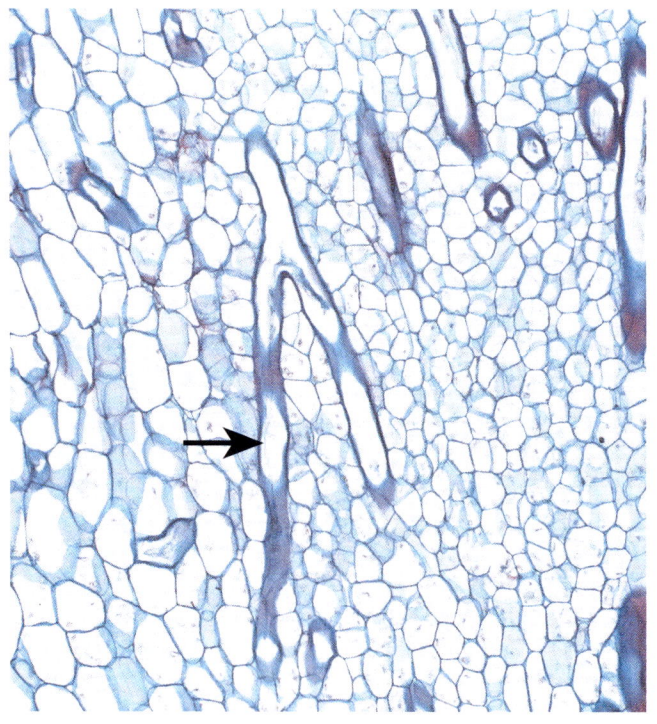

FIGURE 13.36

Longitudinal section of shoot tip of *Nerium oleander*, showing a branched, nonarticulated laticifer (arrow). Laticifers contain latex under positive pressure. When these cells are ruptured, latex therefore oozes from the wound, deterring herbivores and helping to seal the wound.

the contents of laticifers are under positive pressure, latex oozes from wounded surfaces of plants such as *Euphorbia* and dandelion (*Taraxacum*). It may be white (as in *Euphorbia*), colorless (as in some species of *Morus*), or orange-yellow (as in *Cannabis*). Latex is important because it deters grazing animals and helps seal wounds.

Humans have many uses for latex. For example, opium, morphine, and codeine are derived from latex of *Papaver somniferum,* the opium poppy (see boxed reading E.1, "Poppies, Opium, and Heroin"). Latex from the rubber tree (*Hevea brasiliensis*) is 30%–50% rubber—$(C_2H_5)_n$; Amazon Indians used this latex to make rubber balls and containers more than 450 years ago. Today, *Hevea* remains the primary source of natural rubber: yields at Central American plantations often exceed 2,000 kg of rubber per hectare per year. The latex of several other plants contains rubberlike particles: gutta percha used to make dentures and golf balls is derived from the latex of *Palaquium,* and chicle (as in Chiclets) gum comes from *Achras sapota.*

CONCEPT

Plants secrete many different kinds of substances, including resins, nectar, and digestive enzymes. These compounds are secreted internally and externally, and are important because they protect plants, attract pollinators, and seal wounds. Secretions such as rubber and oils are economically important.

PLANT TISSUES AND LIFE ON LAND

The tissues that you learned about in this chapter adapt plants for life on land. In the upcoming chapters, you will learn how plants use these tissues to build organs such as leaves, stems, and roots. However, we need to look beyond organs to understand the significance of plant tissues; that is, we need to understand tissues in the context of the entire plant and its habitat. To appreciate this perspective, consider a unicellular alga and an oak tree. Despite having drastically different sizes and being separated by hundreds of millions of years of evolution, the basic life processes of these organisms are similar: both have similar cellular organelles, trap chemical energy via photosynthesis, use respiration to free this energy for work, and have nearly identical metabolic pathways. But if these cells are so similar, why are oaks so different from algae?

The answer to this question lies not in *how* the cells function, but *where* they function (fig. 13.37). Algae live in aquatic environments where they are protected from desiccation. When removed from such environments, unprotected algal cells desiccate and die in less than a minute. Similarly, unprotected cells of an oak tree also die when placed in air. Thus, all cells, whether of a towering oak or a microscopic alga, are strictly aquatic.

The similar life processes of oak and algal cells require a delicate balance of several factors, such as light, nutrients and gases; excesses or deficiencies of any of these quickly kill each organism. Algae live in aquatic systems where these delicate balances occur naturally, such as in moist soil and surface waters. There, they conveniently exchange gases, absorb nutrients, and secrete waste products directly into the surrounding aquatic environment. Terrestrial plants such as oaks, however, face a dilemma: they extend far beyond the areas where these delicate balances naturally occur. To live outside these areas, plants must re-create an aquatic oasis in the hostile atmosphere. Plants can survive on land because they maintain an internal aquatic environment that is protected from the harsh atmosphere.

FIGURE 13.37

Photosynthetic algae in this pond live only where they can exchange gases, absorb nutrients, and secrete wastes. To survive on land, plants must re-create an aquatic environment in their leaves, where their photosynthetic, algalike cells reside. With the exception of chlorenchyma and a few reproductive cells, all of a plant's tissues function to re-create the aquatic environment necessary for life.

With the exception of chlorenchyma cells and a few reproductive cells, all of a plant's tissues function to create the aquatic environment necessary for life. Foremost among these tissues is the epidermis, which forms a waterproof, protective shell. In leaves, the epidermis encases millions of algalike photosynthetic cells. Thus, the epidermis maintains the oak tree's aquatic environment, despite the fact that the deadly, desiccating atmosphere is only fractions of a centimeter away.

The epidermis establishes and protects the leaf's aquatic environment. However, this creates other problems for the plant. As a consequence of having a protective epidermis, plants must import and export gases, nutrients, and waste products. These "services" are the functions of other plant tissues. For example, xylem supports the plant and transports water and dissolved minerals to leaves to replace that lost by evaporation. Similarly, phloem transports energy-rich sugars to nonphotosynthetic parts of the plant for storage or to enable continued growth. Finally, plants form secretory structures, produce chemicals that deter herbivores, and use other compounds to seal wounds and attract animals that pollinate and protect the plant. Although these life-support systems seem rather diverse, they share a common function: to establish, maintain, and protect the internal aquatic environment necessary for life.

Despite this commitment, however, the aquatic environment inside terrestrial plants is often threatened because plants must exchange gases directly with the atmosphere. In doing so, they lose large amounts of water through their stomata. Because photosynthesis and water conservation are diametrically opposed events, there is no perfect solution to the gas-exchange dilemma faced by plants. However, stomata are an adequate, albeit imperfect, solution to the problem. Stomata use internal and external cues to balance the photosynthesis-desiccation compromise. As a result, photosynthesis occurs only when and to the extent that the internal aquatic environment is not threatened.

Chapter Summary

The primary body of plants consists of four kinds of primary tissues: meristems, ground tissue, dermal tissue, and vascular tissues. These tissues contain distinctive types of cells and are typically multifunctional. Meristems are regions of localized cellular division and growth. They produce all of the cells of the plant body and are responsible for many instances of pattern formation in plants.

Ground tissue differentiates from the ground meristem and constitutes most of the primary body of a plant. Ground tissue consists of three kinds of cells: parenchyma, collenchyma, and sclerenchyma cells.

Parenchyma cells are the most abundant and least structurally specialized cells in plants. Parenchyma cells have many functions, including storage, photosynthesis, and basic metabolism. Parenchyma cells can dedifferentiate, meaning that they can change activities. The ability of parenchyma cells to dedifferentiate helps plants adapt to changing environmental demands.

Chlorenchyma, aerenchyma, and transfer cells are special types of parenchyma cells. Chlorenchyma cells are chloroplast-containing parenchyma cells specialized for photosynthesis. Aerenchyma tissue contains large intercellular spaces and is specialized for gas exchange. Transfer cells have elaborate ingrowths of their cell walls and are specialized for short-distance transport of solutes.

Collenchyma cells are long, living cells with thickened primary cell walls. These walls can expand and support growing regions of the plant.

Sclerenchyma cells have thick, nonstretchable secondary cell walls. They support nonextending regions of plants. There are two kinds of sclerenchyma cells: sclereids and fibers. Sclereids are relatively short cells that have a variety of shapes and occur singly or in small groups. Fibers are long, slender cells occurring as single strands or bundles in mature parts of the plant. Humans use the fibers of many plants to make textiles.

Epidermis is the dermal tissue that covers the primary body of a plant. The epidermis has several functions, including absorption, secretion, digestion, protection, cellular recognition, and the control of gas exchange. These functions are attributable to the unique features of the epidermis, including the presence of a cuticle, few intercellular spaces, and the presence of trichomes. The outer walls of epidermal cells are covered by a waterproof cuticle that protects the internal tissues from desiccation.

Stomata are direct avenues of gas exchange between the environment and the interior of the plant. Each stoma consists of two guard cells that surround a stomatal pore. This pore opens and closes as guard cells enlarge and shrink, thereby controlling gas exchange.

Trichomes are single-celled or multicellular outgrowths of epidermal cells. They have several functions, including absorbing water and dissolved nutrients, and protecting the plant from predators.

Vascular tissues are specialized for long-distance transport of water and dissolved solutes. Xylem and phloem are the two vascular tissues in plants. Both tissues occur throughout a plant and contain parenchyma cells, secretory cells, fibers, and conducting cells.

Xylem transports water and dissolved minerals from roots to leaves. Water and dissolved minerals move through xylem under negative pressure. Primary xylem differentiates from procambium. The first primary xylem that forms is called *protoxylem,* whereas that formed in more mature regions is called *metaxylem.* Secondary xylem is called *wood* and is produced by the vascular cambium, a lateral meristem.

There are two kinds of conducting cells in xylem: tracheids and vessel elements. Both are dead cells having thick, lignified, secondary cell walls. Tracheids are long, narrow cells with tapering, overlapping ends. Water moves upward from tracheid to tracheid through thin areas called pits. Tracheids occur in all vascular plants.

Vessel elements are more evolutionarily advanced than tracheids and occur mostly in angiosperms. Vessel elements are arranged end to end and are shorter and wider than tracheids. Their transverse walls are partially or wholly dissolved, thereby forming long, hollow pipes called vessels. Vessels can transport water more rapidly than can tracheids because of their width and dissolved transverse walls.

Phloem transports dissolved sugars in all directions in plants. Substances in phloem, unlike those in xylem, move under positive pressure through living cells. Primary phloem differentiates from the procambium. The primary phloem formed first is called *protophloem;* that formed later in the mature region of the plant is called *metaphloem.* Secondary phloem is the inner layer of bark and is made by the vascular cambium.

There are two kinds of conducting cells in phloem: sieve cells and sieve tube members. Both are living cells that have only a primary cell wall. Water and dissolved solutes move between adjacent sieve tube members through sieve pores of sieve plates. Sieve cells occur in all vascular plants except angiosperms. Sieve areas occur all over sieve cells and may be associated with albuminous cells. Sieve tube members occur only in angiosperms and are arranged end to end in sieve tubes. Their sieve areas are concentrated in sieve plates occurring along their transverse walls. Sieve tube members lack nuclei and are controlled by adjacent parenchyma cells called companion cells.

Secretory structures occur in ground, vascular, and dermal tissues. Secretory structures secrete substances internally into secretory cells, laticifers, or ducts, and externally from nectaries, hydathodes, salt glands, and digestive glands. Secretory structures protect plants, attract beneficial animals, and aid in the plant's nutrition.

Plant tissues have enabled vascular plants to successfully invade the land. With the exception of chlorenchyma and reproductive cells, all plant tissues serve as life-support systems, maintaining the aquatic environment necessary for life. The epidermis minimizes desiccation, xylem transports water to replace that lost by evaporation, and secretory structures seal wounds and deter grazing animals.

Questions for Further Thought and Study

1. What parenchyma cells in plants are not isodiametric? What is the functional significance of this atypical shape?
2. Justify why parenchyma can be considered a cell type or a tissue.
3. Discuss the structural and functional differences between (a) parenchyma, collenchyma, and sclerenchyma; and (b) xylem and phloem.
4. What are the functions of trichomes? Secretory structures? Epidermis?
5. How would you distinguish a vessel element from a sieve tube member? A vessel element from a tracheid?
6. List the unique properties of the epidermis. Why is each important for plant growth?
7. Discuss how the structure of xylem could be used to identify plants.
8. What is redifferentiation? Why is it important in plants?
9. Nectaries typically contain large amounts of phloem, while hydathodes contain large amounts of xylem. Discuss how these structural differences relate to the functions of these secretory structures.
10. Vessel elements evolved from tracheids in several groups of angiosperms. What do you conclude from this observation?
11. Discuss the structure and function of (a) fibers, (b) guard cells, (c) tracheids, (d) sieve cells, and (e) companion cells.
12. Of what adaptive significance is it that the differentiation sclerenchyma and collenchyma is induced to form by mechanical stress?
13. How are nectaries important to plants?
14. Discuss how plant tissues establish, maintain, or protect an aquatic environment in terrestrial plants.
15. What might be the adaptive advantage of sclereids differentiating near wounds?

Suggested Readings

Articles

Hoch, H. C., R. C. Staples, B. Whitehead, J. Comeau, and E. D. Wolf. 1987. Signaling for growth orientation and cell differentiation by surface topography in *Uromyces*. *Science* 235:1659–1662.

Walker, D. B. 1978. Plants in the hostile environment. *Natural History* 87:74–81.

Books

Behnke, H.-D., and R. D. Sjolund, eds. 1990. *Sieve Elements: Comparative Structure, Induction and Development.* New York: Springer-Verlag.

Burgess, J. 1985. *An Introduction to Plant Cell Development.* Cambridge: Cambridge University Press.

Buvat, R. 1989. *Ontogeny, Cell Differentiation, and Structure of Vascular Plants.* New York: Springer-Verlag.

Cutter, E. G. 1978. *Plant Anatomy. Part 1: Cells and Tissues.* 2d ed. Mass.: Addison-Wesley.

Fahn, A. 1979. *Secretory Tissues in Plants.* London: Academic Press.

Hall, J. L., and C. R. Hawes. 1991. *Electron Microscopy of Plant Cells.* San Diego: Academic Press.

Lydon, R. F. 1990. *Plant Development: The Cellular Basis.* London: Unwin Hyman.

Mauseth, J. D. 1988. *Plant Anatomy.* Menlo Park, CA: Benjamin/Cummings.

Sachs, T. 1991. *Pattern Formation in Plant Tissues.* New York: Cambridge.

Leaves of cottonwood (*Populus fremontii*).

CHAPTER 14

Primary Growth: Stems and Leaves

Chapter Outline

INTRODUCTION

STEMS
- Stems and Their Functions
- Control of Stem Growth
- The Structure of Stems
- Axillary Buds and Branching
- Modified Stems
- The Economic Importance of Stems

LEAVES
- How Leaves Form
- Phyllotaxis

BOX 14.1
LEONARDO THE BLOCKHEAD

- The Structure of Leaves
- Environmental Control of Leaf Variation

BOX 14.2
CLOGGING THE WATERWAYS

- Leaf Movements
- Modified Leaves
- How Leaves Defend Themselves
- Leaf Abscission
- The Economic Importance of Leaves

Chapter Summary
Questions for Further Thought and Study
Suggested Readings

Chapter Overview

The autotrophic life-style of plants requires that they expose large amounts of surface area to the environment. The most conspicuous surface area is the plant's shoot, which is an integrated system of stems and leaves. Stems support leaves—they are the "hat-racks" on which plants hang their solar collectors. The most important functions of leaves and stems are photosynthesis and support. Some plants modify stems and leaves for climbing, trapping insects, and reproduction. The structure and function of stems and leaves are sensitive to changes in light and moisture, which allow plants to adapt to different and constantly changing environments.

Introduction

Stems and leaves are the most conspicuous and diverse organs of plants. The different shapes and structures of stems and leaves do not result from the presence of different or unique tissues, but from different *arrangements* of the tissues discussed in the previous chapter. That is, what we call different organs are merely different combinations of the same building blocks. Leaves and stems are important to us because they make our food, oxygen, building materials, and many of our drugs.

We usually think of stems and leaves as different organs; after all, they usually look different, and we can easily distinguish them from each other. However, this distinction is not always as clear as you may think. Indeed, stems form leaves, and there is often no clear-cut point at which a stem ends and a leaf begins. Furthermore, the leaves of many plants look like stems, while the stems of other plants look like leaves.

Although we'll discuss leaves and stems separately in this chapter, remember that this distinction is somewhat artificial. You'll best appreciate the structure and function of stems and leaves by viewing them as parts of a single dynamic shoot system.

Stems

Stems and Their Functions

A stem is a collection of integrated tissues arranged as nodes and internodes. **Nodes** are regions where leaves attach to stems, and **internodes** are the parts of stems between nodes (fig. 14.1). Together, nodes and internodes perform several important functions:

- **Stems support leaves,** the solar collectors of plants. Turgor pressure in stems provides a hydrostatic skeleton that supports the young plant. Leaves are also supported by a stem's internal skeleton of collenchyma and sclerenchyma.

- **Stems produce carbohydrates.** Stems of plants such as *Salicornia* are green and photosynthetic. Although photosynthesis in stems is usually insignificant compared to that in leaves, in plants such as cacti it accounts for most of the plant's carbon fixation.

- **Stems store materials.** Parenchyma cells in stems store large amounts of starch and water. For example, water accounts for as much as 98% of the weight of many cactus stems.

- **Stems transport water and solutes between roots and leaves.** The vascular system of stems maintains an aquatic environment in leaves and transports sugars and other solutes between leaves and roots. Stems link leaves with the water and dissolved nutrients of the soil.

Control of Stem Growth

The growth of stems is controlled by several factors, of which leaves are perhaps the most important. Leaf primordia control the differentiation of procambium in stems. As a result, there is no procambium above the youngest primordium, and vascular tissues in developing leaves align and connect with mature vascular tissues of the stem. Light also controls stem growth: the stems of seedlings growing in darkness elongate much faster than those growing in light.

Stems elongate in subapical regions in response to auxin and gibberellins, which are made in young leaves. These hormones elongate stems by stimulating cellular elongation and division. In many plants, elongation occurs throughout the internode; in others, it proceeds as a wave originating at the base of the internode. You'll learn more about plant hormones in Chapter 18.

Grasses such as bamboo elongate at meristems intercalated between mature tissues at the bases of their nodes and leaf sheaths (fig. 14.2). These meristems are called **intercalary meristems** and remain meristematic long after tissues in the rest of the internode are fully differentiated. Intercalary meristems are important because they re-form the stem or leaf of a grass when its tip is torn off by a grazing animal or lawn mower. Because grasses are a primary food for grazing animals, and because humans and other predators depend to a greater or lesser extent on grazers for food, intercalary meristems are largely responsible for keeping our meat markets open.

Intercalary meristems can produce phenomenal rates of growth in many plants. For example, bamboo stems can elongate almost a meter per day (that's more than an average child grows between birth and its tenth birthday), and can grow to be over 35 meters tall. Intercalary meristems also have a direct impact on us—no matter how many afternoons we spend vibrating behind our lawn mowers, we can be sure that the grass blades will soon grow back, thanks to intercalary meristems.

Plants whose stems do not elongate are called **rosette plants;** these plants have short internodes with tightly packed

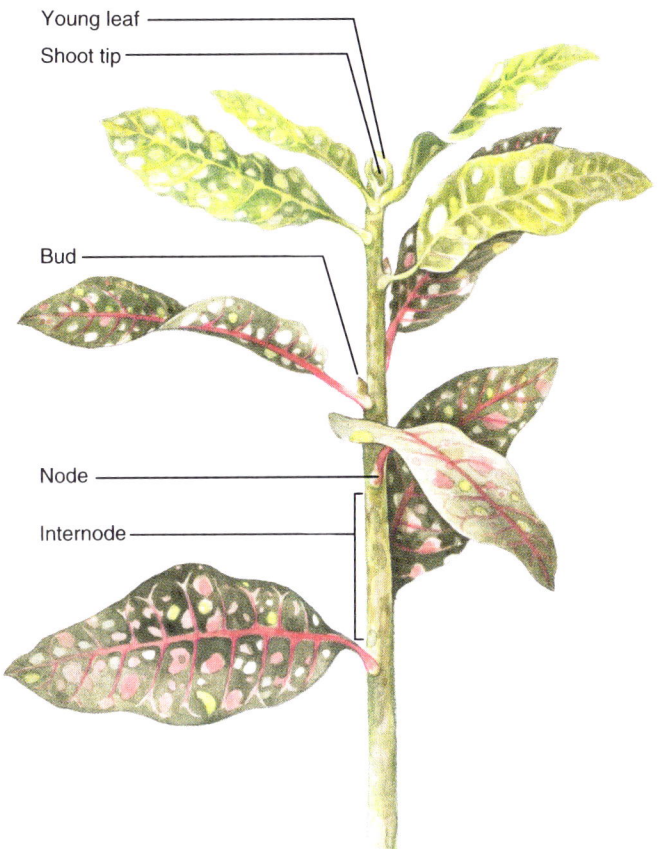

FIGURE 14.1

Stems consist of nodes and internodes. Nodes are regions where leaves attach to stems, whereas internodes are the parts of stems between nodes.

leaves. The stems of rosette plants are short and made almost entirely of overlapping leaf bases. These pseudostems can reach impressive proportions. For example, although the large, overlapping leaf bases of a banana tree form a "trunk" that can raise leaves more than 10 meters, its pyramid-shaped stem barely reaches above ground. Most rosettes fail to elongate because they do not make enough physiologically active gibberellins. Thus, treating rosettes with gibberellins typically causes internodal elongation.

CONCEPT

Stems support leaves, produce and store carbohydrates, and transport materials between roots and leaves. Stem elongation occurs in subapical regions, is influenced by hormones, and helps plants intercept light.

The Structure of Stems

Stems are made of three tissues: epidermal tissue, ground tissue, and vascular tissue (fig. 14.3).

FIGURE 14.2

Bamboo in Indonesia. These plants elongate rapidly from intercalary meristems at the bases of their nodes.

CHAPTER FOURTEEN   *Primary Growth: Stems and Leaves*

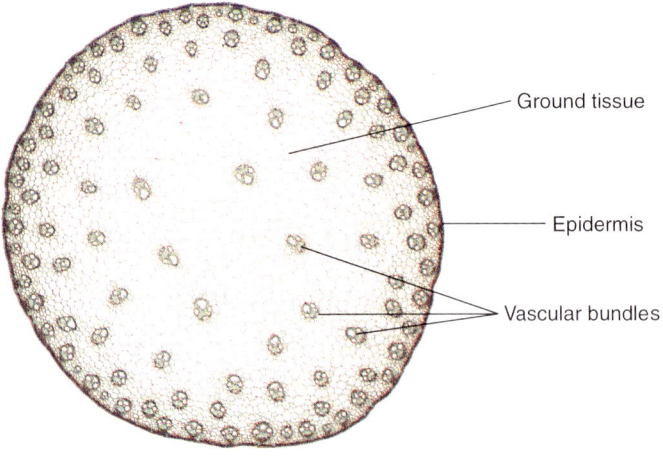

FIGURE 14.3

Cross section of a stem of corn (*Zea mays*), a monocot, ×5. Bundles of vascular tissue occur throughout the ground tissue. The stem is surrounded by an epidermis.

Epidermal Tissue

Stems are surrounded by a transparent epidermis that is usually one cell thick and often bears trichomes. The trichomes of tomato plants secrete an irritating juice that deters hungry insects, while hook-shaped trichomes often entangle insects and prevent them from feeding while they struggle to free themselves.

Vascular Tissue

As already mentioned, there is no procambium above the youngest leaf primordium, and removing a plant's leaf primordia stops vascular differentiation. Thus, substances coming from young leaves control the differentiation of procambium and vascular tissue in stems. The xylem-inducing effect of a leaf primordium can be produced by auxin, suggesting that auxin from leaf primordia controls differentiation of vascular tissue in stems. Because vascular tissue differentiates in response to leaf primordia, it is not surprising that vascular tissue in stems is arranged relative to leaves. This is most obvious at nodes, where one or more vascular bundles branch into leaves.

Xylem and phloem in stems occur in vascular bundles (fig. 14.4). Phloem forms before xylem and differentiates on the outside of the bundle. Xylem forms on the inside of the bundle.

Vascular bundles are often enclosed by or capped with sclerenchyma fibers that differentiate after the internode has finished elongating. However, a layer of cells between the xylem and phloem remains meristematic (fig. 14.4). In woody plants and some herbaceous dicots, this layer of cells later becomes part of the vascular cambium, the lateral meristem that produces secondary growth (see Chapter 16). The absence of a vascular cambium in monocots, however, is an important feature of this group of plants.

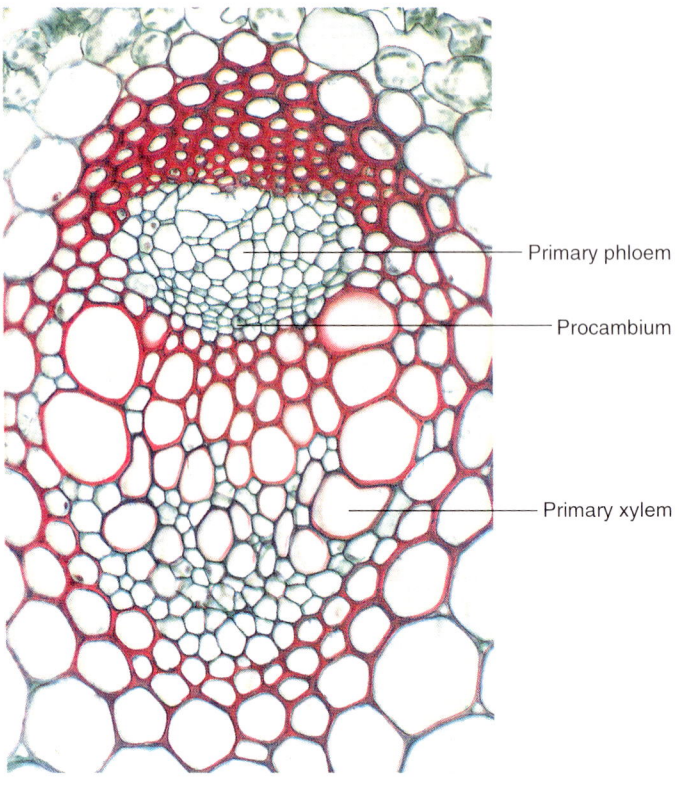

FIGURE 14.4

Vascular bundle from the stem of *Ranunculus*, ×400. The bundle consists of primary phloem, primary xylem, and procambium. The bundles are often associated with bundles of fibers.

Vascular bundles in stems are arranged differently in different groups of plants (fig. 14.5):

- **Monocots** such as corn (*Zea mays*) have vascular bundles embedded throughout the ground tissue.
- **Most dicots** such as alfalfa (*Medicago sativa*) have a single ring of vascular bundles embedded in ground tissue.
- **Many nonflowering plants and a few dicots** such as linden (*Tilia americana*) have concentric cylinders of xylem and phloem. The cylinder of phloem surrounds an inner cylinder of xylem.

These generalities are not absolute: monocots such as Job's tears (*Coix lacryma-jobi*) have a ring of vascular bundles, while dicots such as *Bougainvillea* have vascular bundles scattered throughout the stem.

Ground Tissue

Between the epidermis and ring of vascular tissue in dicots is the **cortex** (fig. 14.5a). Most cortical cells are parenchyma. Cortical cells are photosynthetic in plants such as *Pelargonium*, and often store starch.

In dicots, the parenchymatous ground tissue in the center of the stem is specialized for storage and is called **pith** (fig. 14.5a).

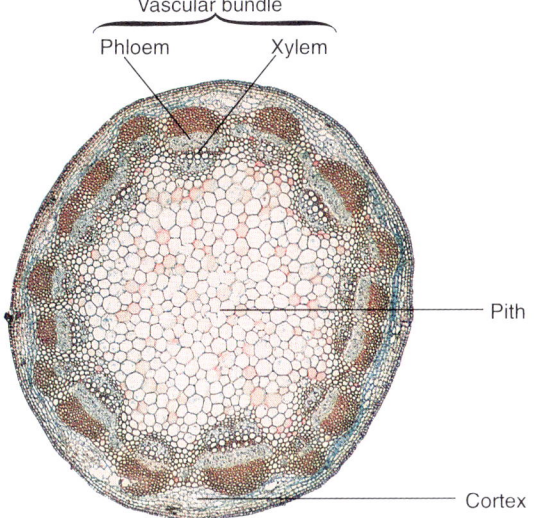

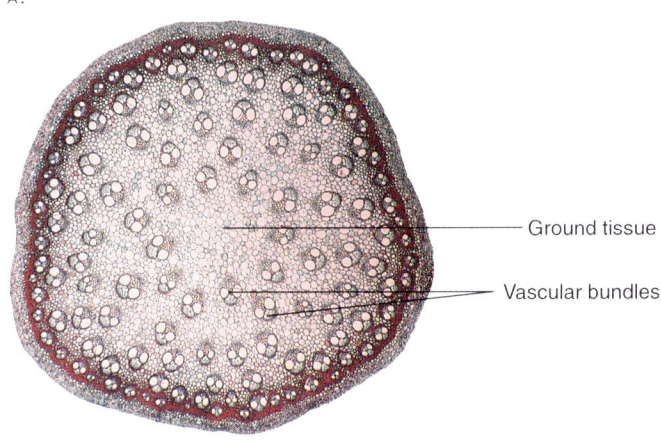

FIGURE 14.5

Cross-sectional views of the stems of (a) a dicot (sunflower) and (b) a monocot (greenbrier), ×10.

Pith cells are often lignified, arranged loosely, and the pith may contain secretory structures such as laticifers. In some plants, stresses induced by growth destroy the central part of the pith, thereby forming a hollow stem. Because monocots have vascular bundles throughout their ground tissue, their stems do not have a discernable pith (fig. 14.5b); the parenchyma cells in monocot stems are referred to simply as ground tissue.

Stems consist of epidermal, ground, and vascular tissue. The vascular tissue is embedded in ground tissue, which often produces or stores food. Stems are covered by a protective epidermis.

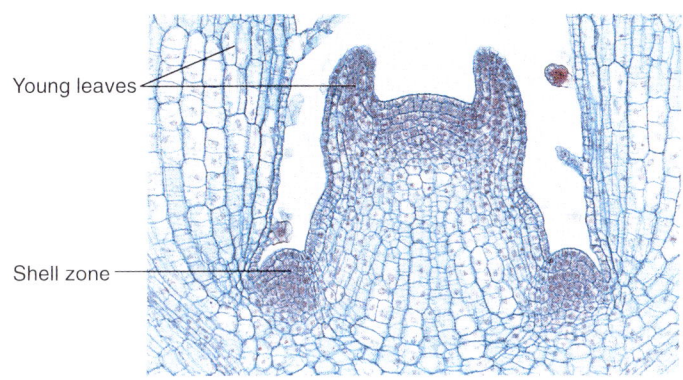

FIGURE 14.6

Axillary buds form in the axil where a leaf attaches to the stem.

Axillary Buds and Branching

Early in leaf development, a small island of meristematic cells forms in the axil where a leaf attaches to the stem. These cells rapidly form an **axillary bud** (fig. 14.6), which typically undergoes a dormant period controlled by hormones made by the shoot apex. In most plants, axillary buds near the shoot apex remain dormant, while those progressively farther away from the tip start to grow. This dominating effect of the shoot apex on growth of axillary buds is called *apical dominance* and strongly influences the symmetry of the shoot. It is strong in plants such as pine and spruce, and accounts for their tiered, Christmas-tree shape. Conversely, plants with weak apical dominance have a shrub shape. Axillary buds are important because they are a shoot's insurance policy: they are inactive (i.e., genetically resting) cells that can form a branch or flower. As such, axillary buds can replace or modify the roles of the shoot apical meristem.

Modified Stems

Plants often modify their stems for special functions. Such modified stems can have unusual and even bizarre shapes, and grow above and below ground. Refer to figures 14.7 and 14.8 as you learn about these unusual stems.

Modified Stems That Grow Above Ground

Stolons or **runners** are horizontally oriented stems that grow along the soil surface. Their function is vegetative reproduction. For example, buds at nodes of strawberry stolons produce shoots and roots that eventually form new plants (fig. 14.7a). Other plants having stolons include Boston fern (*Nephrolepis exaltata*), Bermuda grass (*Cynodon dactylon*), crabgrass (*Digitaria sanguinalis*), and spider plant (*Chlorophytum*).

Tendrils and **twining shoots** of plants such as morning glory and sweet potato coil around objects and help support the plant. Tips of tendrils of plants such as Virginia creeper (*Parthenocissus*) and grapevine (*Cissus*) have adhesive pads that stick to nearby objects. Tendrils (fig. 14.7b) and twining shoots are important because they help support plants.

FIGURE 14.7

Modified stems that grow above ground. (a) Stolons of beach strawberry (*Fragaria chilensis*). Roots that form at intervals provide new sources of water and minerals for the plant. Rooted parts of the plant can live independently if the stolon is cut. (b) Tendrils of grape (*Vitis*) help the plant cling to other objects.

Searcher shoots are stems with long internodes that move in circles through the air seeking a support. Searcher shoots may be more than 3 meters long, and may die if they do not find a substrate. Examples of plants with searcher shoots include honeysuckle (*Lonicera*) and *Wisteria*.

Cladodes or **cladophylls** are flat, leaflike stems modified for photosynthesis. Examples of plants with cladodes include butcher's broom (*Ruscus*), greenbrier (*Smilax*), asparagus (*Asparagus officinalis*), and orchids (e.g., *Epidendrum*).

Thorns are modified stems that protect plants from grazing animals. Examples of plants with thorns are grapefruit, honey locust (*Gleditsia triacanthos*), fire thorn (*Pyracantha*), hawthorn (*Crataegus*), and bougainvillea (*Bougainvillea spectabilis*).

Short and long shoots are distinguished by internodal elongation: long shoots have long internodes, while short shoots are rosettes that have short internodes. This difference in shoot length usually underlies a functional division of labor. For example, long shoots of pine (*Pinus*) produce protective scale leaves, while short shoots produce foliage leaves. Examples of plants with short and long shoots include cedar (*Cedrus*), maidenhair tree (*Ginkgo*), and barberry (*Berberis*).

Succulent stems of plants such as cacti have a low surface-to-volume ratio. Succulent stems store large amounts of water and are common in plants growing in deserts.

Modified Stems That Grow Below Ground

Bulbs are rosette stems surrounded by fleshy leaves that store nutrients. When these nutrients are removed (such as during flowering), the fleshy leaf collapses into a papery scale leaf surrounding the stem. Onion, lily, hyacinth, tulip, narcissus, and daffodil are examples of plants that produce bulbs (fig. 14.8a). Bulbs are important because they enable many plants to survive stressful periods, such as winter.

Rhizomes are underground stems that grow near the soil surface. They typically have short internodes and scale leaves, and produce roots along their lower surface. Rhizomes store food for renewing growth of the shoot after periods of stress, such as cold winters. This explains why plants such as Johnson grass (*Sorghum halepense*) are so difficult to eradicate—although their aerial shoots die back during winter, their dormant rhizomes renew growth the following spring. Other examples of plants having rhizomes include quack grass (*Agropyron repens*), *Iris*, *Canna*, *Begonia*, and ginger (*Zingiber officinalis*).

Corms are stubby, vertically oriented stems that grow underground. Corms, which have only a few thin leaves, store nutrients. Like rhizomes, corms enable many plants to survive winter. Examples of plants having corms are *Gladiolus*, *Cyclamen*, and *Crocus*.

Tubers are swollen regions of stems that store food for subsequent growth. For example, Irish potatoes (*Solanum tuberosum*) are tubers produced on stolons that burrow into the soil (fig. 14.8b). The eyes of potatoes are buds in the axils of small, scalelike leaves; the areas between a potato's eyes are internodes.

C O N C E P T

Plants use modified stems for reproduction, climbing, photosynthesis, protection, and storage. These modified stems adapt plants to different environments.

FIGURE 14.8

Modified stems that grow below ground. (a) Onions are bulbs consisting of layered, fleshy leaf bases attached to a short stem. Roots form on the underside of the stem. (b) The Irish potato is a tuber. Sprouts growing from the "eyes" of this potato will produce new plants.

The Economic Importance of Stems

We have many uses for stems. Sugar and molasses from sugarcane (*Saccharum officinalis*) go in our drinks and over our pancakes, and stem fibers are used to make several fabrics. We also eat a few stems, such as tubers of potatoes. In some parts of the world, potato tubers are the most important part of the diet. Andean tribes mash and dry tubers to form *chuño*, which is added to just about everything they eat. Indeed, members of the tribes say that "stew without *chuño* is like life without love." Although we don't expect you to take your potatoes quite that seriously, we do hope you'll remember what you're eating next time you order some fries.

Lumber, pulp for paper, charcoal, corks, insulation, life preservers, quinine, cinnamon, and even a diesel-like fuel are derived from stems. These products come mainly from wood or bark, and will be discussed in Chapter 16 when we discuss secondary growth.

What are Botanists Doing?

The potato is one of our most important food crops, so studies on how to improve it are always ongoing. Go to the library and read about some of this research. How are botanists trying to improve potatoes?

LEAVES

Leaves are the most active and conspicuous organs of plants. The most important of their functions is absorbing sunlight for photosynthesis. To do this, they expose large amounts of surface area to the environment. For example, a maple tree (*Acer*) with a trunk 1 m wide has approximately 100,000 leaves with a combined surface area exceeding 2,000 m^2—that's roughly the area of six basketball courts. Oak trees have approximately 700,000 leaves, and you'd better think twice before agreeing to rake a lawn shaded by American elm trees (*Ulmus americana*)—when mature, these trees can each produce more than 5 million leaves per season. On a global basis, leaves produce more than 200 billion tons of sugars per year. Those sugars sustain most life on this planet.

How Leaves Form

Leaves are the most diverse of all plant organs—they can be tubular, needlelike, feathery, cupped, sticky, fragrant, smooth, or waxy. Leaves range in size from the pinhead-sized leaves of watermeal (*Wolffia*) to the 20-meter fronds of tropical palms, and they range in number from the millions of leaves on American elms to those of *Welwitschia mirabilis*, a desert plant of southwest Africa that grows only two leaves during its lifetime. However, regardless of their number or size, leaves are formed by the coordinated efforts of several meristems, each of which is named for its position.

Although mature leaves often consist of millions of cells (e.g., leaves of broad bean contain about 40 million cells), the earliest stage of leaf development is a small bulge at the shoot apex called a **leaf buttress** or **leaf primordium** that consists of only 100 to 300 cells. This buttress is formed by cellular divisions one to three cell layers below the overlying protoderm. Continued cellular divisions and cellular expansion produce a radially symmetrical cone called an **apical peg.** This peg has an apical meristem and a procambial strand that, in most species, forms the leaf's midrib. The adaxial surface (i.e, the surface closest to the internode above it) of the apical peg elongates slower than the abaxial surface, thereby arching the leaf primordium over the shoot apical meristem. This differential growth is important because it positions the primordium to protect the apical meristem.

The leaf then forms an **adaxial meristem** that thickens the leaf. Soon thereafter (i.e., when the leaf primordium is about 0.2–0.5 mm high) it forms an **upper leaf zone** and a **lower leaf zone.** The upper leaf zone contains **marginal meristems** that form the flattened blade and stalklike petiole that attaches the leaf to the stem. In plants such as cocklebur (*Xanthium*), marginal growth continues for approximately three weeks, and produces a blade about six cells thick. The lower leaf zone forms the leaf base.

Continued growth of a leaf involves cellular expansion and division. For example, leaf formation in cocklebur involves almost thirty generations of cells. These divisions continue until the leaf is one-half to three-fourths grown. Cellular expansion forms most of the intercellular spaces in a leaf. Stomata differentiate soon after intercellular spaces form. Except for vascular tissue, which differentiates acropetally (i.e., from the base into the tip of the leaf), other tissues in leaves differentiate basipetally (i.e., from the tip toward the base). Until it is 30%–40% of its final size, a growing leaf depends on the rest of the plant for its nutrition.

Developmental abnormalities that alter cellular expansion and division also affect a leaf's morphology. For example, tobacco mosaic virus inhibits the marginal meristems of developing tobacco leaves. Consequently, infected leaves have deformed blades or no blades at all.

Fern leaves are called *fronds*. Unlike the leaves of other plants, fronds usually form in a curled structure called a *fiddlehead* (fig. 14.9). During development, the abaxial side of the fiddlehead elongates faster than the adaxial side, causing the fiddleheads to unroll and form a frond. Fronds of some ferns can live more than fifty years and be more than 15 meters long. We even eat some of them: for example, fiddleheads of ostrich fern (*Matteuccia struthiopteris*) are an asparaguslike delicacy enjoyed by many New Englanders.

FIGURE 14.9

Leaves of ferns are called *fronds*. Fronds unroll from base to tip. When they are young and coiled as shown in this photo, they are referred to as fiddleheads.

Phyllotaxis

Phyllotaxis (from the Greek words *phyllon*, meaning "leaf" and *taxis*, meaning "arrangement") is the arrangement of leaves on a stem. It is determined at the shoot apex and is species-specific.

- Most plants have **spiral,** or **alternate, phyllotaxis** (fig. 14.10), meaning that they have one leaf per node. Birch (*Fagus*), oak (*Quercus*), and *Agave* have a spiral phyllotaxis.

Figure 14.10

Alternate (spiral) phyllotaxis in poplar.

Figure 14.11

Coleus has decussate phyllotaxis. It has two leaves per node; these pairs of leaves at successive nodes form at right angles to each other. This view is from above the plant.

- Plants with **opposite phyllotaxis** have two leaves per node, as in maple and ash (*Fraxinus*). When pairs of leaves at successive nodes form at right angles to each other, the phyllotaxis is *decussate*, **as in** *Coleus* (fig. 14.11). If the leaves form two parallel ranks along the stem, the phyllotaxis is *distichous*, as in ginger.
- Plants with **whorled phyllotaxis** have three or more (and as many as twenty-five) leaves per node. Oleander (*Nerium oleander*), *Peperomia*, and horsetail (*Equisetum*) have a whorled phyllotaxis.

Phyllotaxis is independent of leaf shape and can be described mathematically. The leaves of plants with alternate phyllotaxis are arranged in spirals around the stem, and each leaf is vertically superimposed on other leaves of the spiral. That is, the leaves are arranged in vertical files on stems, which are called **orthostichies.** Botanists use orthostichies to describe phyllotaxis with the following equation:

$$\text{phyllotaxis} = \frac{\text{(number of turns of spiral)}}{\text{(number of leaves between successive leaves of an orthostichy)}}$$

Plants with decussate leaves, such as *Coleus*, have a phyllotaxis of (1 + 2). This means that you must circle the stem once to find two leaves that are superimposed vertically. There are two leaves on this spiral, meaning that each leaf is halfway around the stem.

Many botanists also quantify phyllotaxis based on the position of leaves at the apex. For example, examine the shoot apex of the cactus shown in figure 14.12. Each leaf is part of two opposing spirals called **parastichies.** Botanists describe a plant's phyllotaxis by counting these spirals. The cactus shown in figure 14.12 has a phyllotaxis of (13 + 21). Other common plants and their phyllotaxis include cabbage (*Brassica oleracea*: 1 + 2), flax (*Linum usitatissimum*: 3 + 5), and sunflower (*Helianthus annuus*: 34 + 55). The (2 + 3) phyllotaxis is most common in angiosperms, and almost half of all nonflowering seed plants have a (3 + 5) phyllotaxis. As you might have already noticed, each of the numbers that describe phyllotaxis is part of a mathematical progression called the *Fibonacci series*, a number series in which each number (except the first two) is the sum of the two preceding numbers: 1, 2, 3, 5, 8, 13, 21, 34, 55, etc. (see box 14.1, "Leonardo the Blockhead").

Finally, leaf formation can also be described temporally. A **plastochron** is the time required to form two successive leaves at the shoot apex. Plastochrons in most plants vary considerably, depending on the plant's growth rate and nutritional status. Plastochrons in flax (*Linum*) and spruce (*Picea*) range from 3 to 22 hours, while in clover (*Trifolium*) they range from 1.2 to 2 days.

What Controls Phyllotaxis?

Although phyllotaxis has been studied by botanists and philosophers alike, we still know relatively little about the process. Although, phyllotaxis is usually unaltered by changes in daylength, light intensity, or moisture, what causes leaves to form at precisely the "right" place? One hypothesis is that older primordia determine the sites where new primordia form by

BOXED READING 14.1

LEONARDO THE BLOCKHEAD

His real name was Leonardo, but his friends in Pisa called him "the Blockhead." History speaks of him simply as Fibonacci, a thirteenth-century mathematician who discovered one of the great mysteries of the universe. In 1202, at the age of 27, Fibonacci published *Liber Abaci* (*The Book of the Abacus*), a historic book that introduced Europeans to Arabic numbers. A small section of the book contained a puzzling problem that has fascinated mathematicians and other scientists for centuries. Fibonacci wrote:

> Someone placed a pair of rabbits in a certain place, enclosed on all sides by a wall, to find out how many pairs will be born in the course of one year, it being assumed that every month a pair of rabbits produces another pair, and that rabbits begin to bear young two months after their own birth.

Fibonacci calculated the total pairs of rabbits at the end of each month, and came up with the following sequence of numbers: 1, 2, 3, 5, 8, 13, 21, 34, 55, 89, 144, 233, 377, 610, and so on. Except for the first two, each of these numbers is the sum of the two preceding numbers. Interestingly, these numbers appear throughout the plant kingdom (e.g., see fig. 14.12). For example,

- You'd better check out the Fibonacci series before trying the "loves me, loves me not" routine—daisies usually have 21, 34, 55, or 89 petals.
- Most sunflowers have spirals of 34 and 55 seeds. Smaller sunflowers have spirals of 21 and 34 seeds, and a giant sunflower grown in the former Soviet Union reportedly had spirals containing 89 and 144 seeds.
- Pine needles usually grow in clusters of 2, 3, or 5, depending on the species. Furthermore, the modified leaves of pine cones are arranged in spirals based on Fibonacci numbers.
- There are usually 13 buds arranged between 2 vertical lines on stems of pussy willow. To count these buds, you must circle the stem 5 times.

BOX FIGURE 14.1

Part of a branch of *Prunus*. The line that passes from leaf to leaf shows a (2 + 5) phyllotaxis: you must circle the stem twice and pass five leaves before you find two leaves that are superimposed vertically.

The Fibonacci series extends far beyond botany. Look again at the Fibonacci numbers. Each number has a special relationship to the numbers surrounding it—if you divide a Fibonacci number by the next higher number, you'll see that it is precisely 0.618034 times as large as the number that follows. (This figure is valid only when the Fibonacci numbers are large enough to be precise—it works for all numbers after the 14th in the sequence.) This number occurs throughout nature with amazing precision. For example, it is the basis for Greek architecture (including the Parthenon), the Egyptian pyramids, playing cards, and the work of countless artists, including another Leonardo—Leonardo da Vinci. However, even these examples hardly touch the importance of Fibonacci's numbers. To understand this, consider the so-called golden rectangle having a length:width ratio of 0.618034. If you draw a square into one end of the rectangle, you are left with another, smaller golden rectangle. Repeating this process produces a succession of golden rectangles. If you connect the centers of the successive squares that you draw, you'll have a spiral, any part of which is 0.618034 times larger than the remainder of the spiral. This "golden spiral" underlies the design of mollusk shells, ram horns, parrot beaks, elephant tusks, lion claws, breaking waves, comet tails, spiderwebs, meteor craters, the shoreline of Cape Cod, and even the spiral galaxies of the universe.

Finally, the rabbits of Fibonacci pop up in music, a universal language that we hear and feel. For example, music is based on an eight-note octave which, on a piano keyboard, is represented by five black and eight white keys—a total of thirteen keys. Furthermore, the major third is the interval that our ears like best; in the major third from note C to note E, for example, the E vibrates at a ratio of 0.625000 to the C, only 0.006966 away from the golden mean. These notes produce good vibrations in your inner ear's cochlea, a spiral-shaped organ. So next time you catch yourself dancing or tapping your toes to the beat of your favorite song, think of Leonardo the Blockhead, and how his magic numbers describe what we like to build, look at, feel, and hear.

FIGURE 14.12

Whorls of spines (modified leaves) on cactus. Each spine is part of two opposing spirals called parastichies. This cactus has a phyllotaxis of (13 + 21).

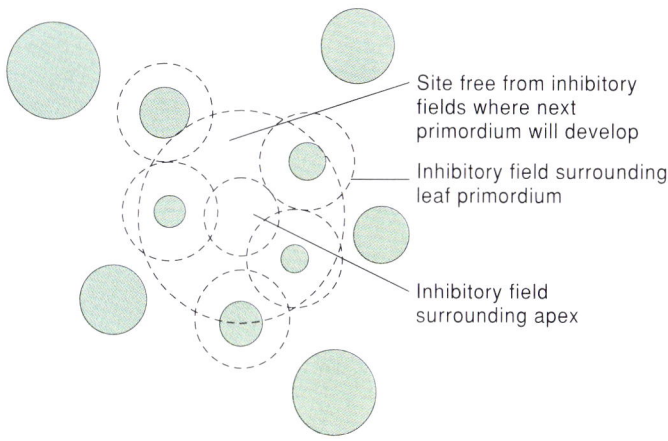

FIGURE 14.13

The position of a leaf on the mature plant is determined largely by where it forms in the apical meristem. One hypothesis to explain where leaves will form in the apex is the *inhibitor field* hypothesis, which suggests that a new leaf forms in the next available space that is free of the inhibitory influences of other leaves.

releasing compounds that inhibit the formation of other primordia. As a result, each older primordium is surrounded by an inhibitor field in which new primordia cannot form (fig. 14.13). Older leaves produce progressively smaller amounts of the inhibitor, thereby forming a moving field. New primordia form where these fields are weakest. Although we do not know what these inhibitors are or if they actually exist, plant hormones can sometimes alter phyllotaxis. For example, applying gibberellic acid to shoot tips of cocklebur changes the phyllotaxis from (2 + 3) to (3 + 5).

Another hypothesis is that phyllotaxis is controlled physically by differential reinforcement in the outermost layers of the shoot apex. According to this hypothesis, leaves form at points of least reinforcement, much as bulges form at the weakest areas of tires (see Chapter 12). Although surgical experiments to test these and other ideas have produced interesting results, we still do not understand exactly what controls phyllotaxis.

C O N C E P T

Leaves are the primary photosynthetic organs of most plants. They are formed by several meristems in specific positions. The arrangement of leaves on a stem is called *phyllotaxis* and is usually species specific.

The Structure of Leaves

External Structure

Although leaf morphology can be affected by the environment, it can often be used as a taxonomic tool to identify plants; maple leaves, for example, are easily distinguished from leaves of, say, walnut. Despite this strong genetic control of leaf morphology, no two leaves on a plant are identical.

There are four basic kinds of leaves: simple, compound, peltate, and perfoliate leaves. Refer to figures 14.14–16 as you learn about these kinds of leaves:

Simple leaves (fig. 14.14) have a flat, undivided *blade* that is supported by a stalk called a *petiole*. The petiole is typically supported by collenchyma and sclerenchyma fibers. In plants such as silk tree (*Albizia*), the petiole includes a jointlike swelling called a *pulvinus* that enables the leaf to respond to environmental stimuli such as light and gravity (see Chapter 19). The leaves of plants such as *Zinnia* that lack petioles are called *sessile* leaves. Redbud, elm, and maple have simple leaves.

Compound leaves (fig. 14.15) have blades divided into *leaflets* that form in one plane and lack axillary buds. Each compound leaf has a single bud at the base of its petiole. There are two kinds of compound leaves: pinnately compound leaves and palmately compound leaves.

Leaflets of **pinnately compound leaves** (fig. 14.15a) form in pairs along a central, stalklike *rachis*. Ash, walnut, and rose have pinnately compound leaves.

FIGURE 14.14

Simple leaves of *Thottea*. Simple leaves have a flat, undivided blade attached to the stem by a stalk called a petiole.

A.

Leaflets of **palmately compound leaves** (fig. 14.15b) attach at the same point, much as fingers are attached to your palm. Examples of palmately compound leaves include horse chestnut (*Aesculus*), marijuana (*Cannabis sativa*), and lupine (*Lupinus*). The most famous palmately compound leaves are those of the trifoliate shamrock (*Trifolium*) that St. Patrick used to explain the doctrine of the Trinity. Today, the shamrock is the national plant of Ireland and is traditionally worn on St. Patrick's Day. Similarly, the "leaves" of a four-leaf-clover are actually leaflets of a palmately compound leaf.

Peltate leaves (fig. 14.16b) have petioles that attach to the middle of the blade. Mayapple (*Podophyllum*) is an example of a plant with peltate leaves. The most extreme examples of peltate leaves are the tubular leaves of carnivorous and other plants.

Perfoliate leaves (fig. 14.16a) are sessile leaves that surround and are pierced by stems. Perfoliate leaves of yellow-wort (*Blackstonia perfoliata*) and thoroughwort (*Eupatorium perfoliatum*) form by the fusion of two opposing leaves.

B.

FIGURE 14.15

Compound leaves. (a) Pinnately compound leaves of this member of the mimosa family form in pairs along a central rachis. (b) Leaves of alfalfa (*Medicago arborea*) are palmately compound and consist of three leaflets.

FIGURE 14.16

Perfoliate leaf (a) and peltate leaf (b).

Many plants have adult and juvenile leaves that form on adult and juvenile parts of the plant. For example, juvenile leaves of some species of *Acacia* are compound, while adult leaves are simple. Similarly, elm produces progressively more teeth along the edges of its leaves as the tree ages.

Although leaf shape is characteristic for a species, leaf size can vary greatly. Clearly, the number and size of epidermal cells determine leaf area, and a small amount of evidence suggests that the ratio of epidermal cells to the total number of cells in a leaf is relatively constant after cellular divisions stop. Because cellular divisions stop first in the epidermis, the epidermis may govern leaf size.

Internal Structure

Leaves consists of epidermal, ground, and vascular tissues.

Epidermis

The epidermis is compact, transparent, and usually not photosynthetic. It also contains numerous stomata (e.g., a cabbage leaf typically has more than 11 million stomata). In horizontally oriented leaves, there are usually more stomata on the protected lower side than on the exposed upper side. Conversely, vertically oriented leaves usually have similar numbers of stomata on their adaxial and abaxial surfaces.

The frequency and distribution of stomata vary in different species and in different parts of individual leaves. For example, stomata differentiate in parallel rows in monocot leaves, while in dicots they are more scattered.

Although stomata usually occupy less than 1% of the leaf surface, they lose huge amounts of water to the atmosphere. The evaporation of water from leaves into the atmosphere is called *transpiration* and can influence patterns of rainfall. For example, about half of the moisture in rainfall in Amazon rain forests originates from transpiration. On the more local scene, your neighbor's one-acre lawn can transpire more than 100,000 liters of water per week during the summer.

Plant and Leaf Orientation	Stomatal Frequency (stomata cm^{-2})	
	Upper Epidermis	Lower Epidermis
Horizontally oriented leaves		
Apple (*Malus sylvestris*)	0	38,760
Bean (*Phaseolus vulgaris*)	4,031	24,806
Oak (*Quercus velutina*)	0	58,140
Pumpkin (*Cucurbita pepo*)	2,791	27,132
Vertically oriented leaves		
Corn (*Zea mays*)	9,800	10,800
Pine (*Pinus sylvestris*)	12,000	12,000
Onion (*Allium cepa*)	17,500	17,500

Vascular Tissues

Xylem and phloem in leaves form in strands called **veins**. Xylem forms on the upper side of a vein, and phloem forms on the lower side (fig. 14.17). Veins are often supported by fibers and are usually surrounded by a layer of parenchyma cells called the *bundle sheath*. The bundle sheath often extends to the epidermis of the leaf. These *bundle-sheath extensions* help support the veins and may conduct water to epidermal cells.

Veins are a leaf's "fingerprint" and can be used to identify plants. Most dicots and some nonflowering plants have *netted venation*, meaning that they have one or a few prominent *midveins* from which smaller *minor* veins branch into a meshed network (fig. 14.18). The leaves of most monocots have *parallel venation*, meaning that several prominent and parallel veins interconnect with smaller, inconspicuous veins (fig. 14.18). Minor veins are extensive: for example, there are about 70 cm of minor veins per cm^2 in the leaves of sugar beet (*Beta*). These minor veins end blindly in the mesophyll as *vein endings* (fig. 14.19), which are the "business end" of the leaf's vascular system. Each vein ending services a small neighborhood of cells and is where most water and solutes are exchanged with cells of the leaf. Vein endings in leaves of sugar beet each service about 30 cells, each of which is within about two cell diameters (70 μm) of the vein.

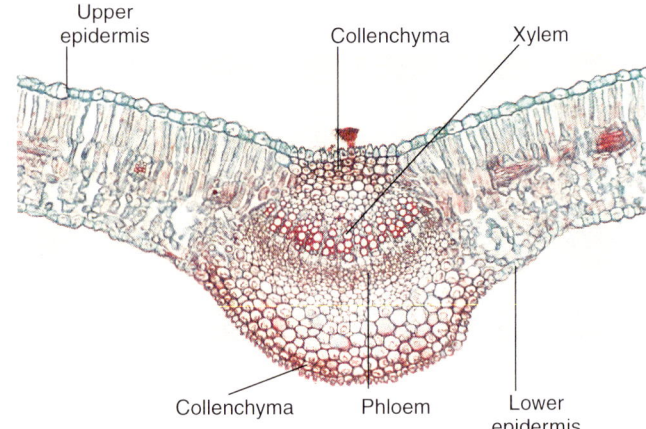

FIGURE 14.17

Cross sections of a leaf of lilac (*Syringa vulgaris*) through (a) the midvein and (b) the blade. Minor veins are visible in the blade.

FIGURE 14.18

(a) Leaves of dicots such as this pumpkin have a netted venation. (b) Leaves of monocots such as this lily have parallel venation.

Several factors influence the formation of veins, including the presence of other veins. For example, new veins form in blades of *Narcissus* as soon as cellular divisions separate existing veins by more than about eleven cells.

Although veins are prominent in most leaves, in no plant are they more obvious than in Madagascar lattice-leaf (*Aponogeton fenestrale*), an aquatic plant with large, frilly leaves. Leaves of these plants have only a few chlorenchyma cells surrounding their veins; that is, there are no chlorenchyma cells linking adjacent veins. As a result, the leaves look like a skeleton of veins, and are a natural demonstration of venation.

Ground Tissue

The ground tissue of leaves is called **mesophyll.** It contains several types of cells, including sclerenchyma (usually sclereids), storage parenchyma, and chlorenchyma. Chlorenchyma cells are photosynthetic cells which, as you learned in Chapter 7, can have differing arrangements and metabolisms. In general, the arrangement of chlorenchyma is determined genetically, and is influenced by whether the leaf is oriented horizontally or vertically when it forms.

FIGURE 14.19

Vein endings in *Diospyros kaki*. Most water and solutes are exchanged between the vein and other cells of the leaf at vein endings.

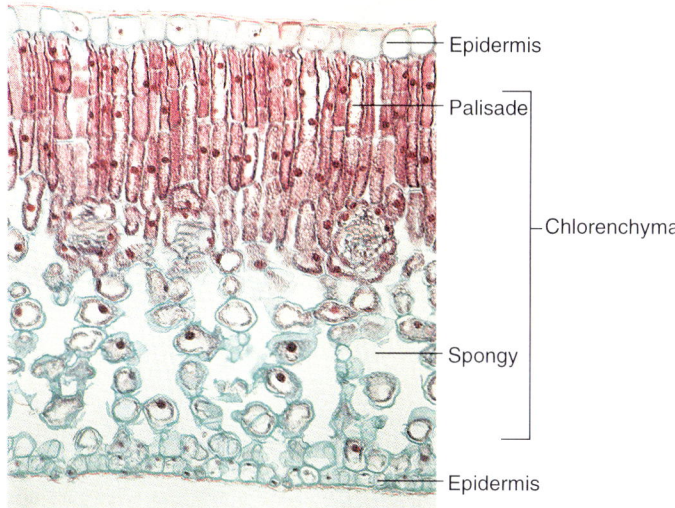

FIGURE 14.20

Cross section of a leaf of the common hedge privet (*Ligustrum*), ×50. The photosynthetic tissue consists of a densely packed palisade mesophyll and a loosely packed spongy mesophyll.

Horizontally oriented leaves. Examine the cross section of the privet (*Ligustrum*) leaf shown in figure 14.20. These leaves have two different kinds of chlorenchyma tissue. Along the upper side of the leaf are one or more layers of long, columnar chlorenchyma cells called **palisade mesophyll** cells, which are densely packed (i.e., have small intercellular spaces) and perform as much as 90% of the leaf's photosynthesis. Palisade cells contain large amounts of chlorophyll, and are thus specialized for the light-absorption and carbon-fixation portions of photosynthesis.

Along the lower side of horizontally oriented leaves are **spongy mesophyll** cells, which are a type of photosynthetic aerenchyma tissue. Spongy mesophyll cells are green, irregularly shaped cells separated by large intercellular spaces connected to stomata. Depending on the species, these air spaces account for 10%–70% of the leaf volume, thereby greatly increasing the amount of surface area for gas exchange. Spongy mesophyll is specialized for the gas-exchange portion of photosynthesis.

Vertically oriented leaves. Plants such as corn and other monocots form leaves vertically. These leaves intercept light from all directions, and their chlorenchyma is usually arranged differently than that of horizontally oriented leaves, which intercept light primarily on their upper surfaces. To appreciate this difference in leaf anatomy, examine the cross section of the corn leaf shown in figure 14.21a. These leaves lack palisade and spongy layers. Rather, most of their chlorenchyma cells appear similar. Consequently, they are called uniform mesophyll cells.

The ground tissue of many tropical grasses is arranged concentrically (fig. 14.21). Uniform chlorenchyma surrounds a photosynthetic bundle sheath containing large, active chloroplasts. This combination of uniform chlorenchyma and photosynthetic bundle-sheath cells is referred to as *Kranz* (German

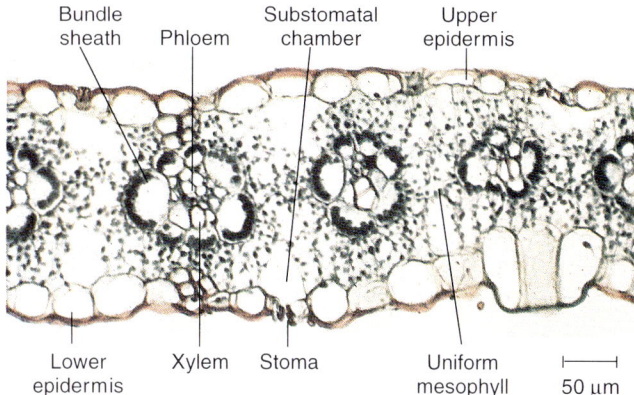

FIGURE 14.21

(a) Cross section of a leaf of corn (*Zea mays*), a C_4 plant. The vascular bundles are surrounded by photosynthetic bundle-sheath cells, which in turn are surrounded by photosynthetic uniform mesophyll. This arrangement of chlorenchyma is typical of C_4 plants and underlies a photosynthetic division of labor (see Chapter 7). (b) Veins (i.e., vascular bundles) of C_4 plants are tightly surrounded by bundle-sheath cells. Although the bundle-sheath cells appear to be on only two sides of the vein, they actually encircle the vein.

for "wreath") *anatomy*, and underlies a metabolic division of labor called C_4 photosynthesis (see Chapter 7). Pigweed (*Chenopodium*), corn (*Zea*), Bermuda grass (*Cynodon*), crabgrass (*Digitaria*), and nutsedge (*Cyperus*) are examples of C_4 plants that have vertically oriented leaves with Kranz anatomy. However, not all grasses are C_4 plants; Fescue (*Festuca*), and bluegrass (*Poa*) are C_3 plants. As you might guess, the bundle sheaths of these C_3 plants are not photosynthetic.

C O N C E P T

Leaves have many shapes and sizes, and are made of ground tissue, veins, and epidermis. Chlorenchyma cells of the ground tissue produce sugars and receive nutrients and water from vascular tissues of veins. Leaves are covered by epidermis containing stomata that regulate gas exchange.

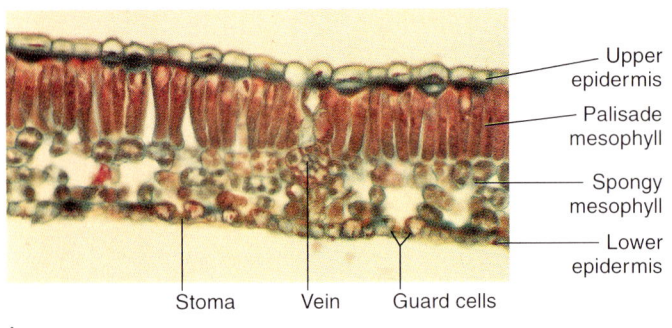
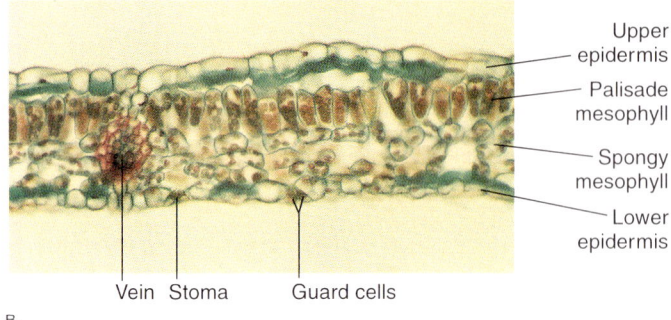

FIGURE 14.22

Cross section of (a) sun versus (b) shade leaves of maple. Sun leaves have larger mesophyll cells and more chloroplasts than do shade leaves.

Environmental Control of Leaf Variation

The morphological differences that distinguish plants growing in different habitats are most striking in leaves. Indeed, leaf development is affected by several environmental factors, the most influential of which are light and moisture.

Light

Daylength, light intensity, and the presence or absence of light strongly affect leaf development.

The Presence or Absence of Light

Leaves of most dicots require light to expand and produce chlorophyll. Light that controls leaf expansion is absorbed by phytochrome, a pigment that also influences several other aspects of plant growth and development (see Chapter 19). This light-controlled expansion of leaves is important because it ensures that a plant will not form leaves unless light is present for photosynthesis.

Daylength

Leaf development is also affected by daylength, which is the amount of light per day. During short days, plants such as *Kalanchoë* produce small, succulent leaves that are sessile and not lobed. During long days, *Kalanchoë* produces large, thin, lobed leaves with petioles. Daylength also affects a variety of other processes in plants, including flowering and seed germination (see Chapter 19).

Light Intensity

The leaves of many plants respond to differing intensities of light. Consider the plants growing in a dense rain forest. Leaves atop the plant canopy are bathed in intense light and are called **sun leaves.** These leaves have significantly different structures than do **shade leaves,** which grow in the dim light on the forest floor. Look at the sun and shade leaves shown in figure 14.22.

Sun leaves are smaller and thicker than shade leaves. For example, leaves of *Plectranthus* grown in intense light are three times thicker than leaves grown in dim light.

In intense light, sun leaves fix carbon faster than do shade leaves. For example, sun leaves fix 16–20 mg C dm^{-2} h^{-1}, while shade leaves fix only 3–5 mg C dm^{-2} h^{-1}. Similarly, sun leaves respire 3–5 times faster than shade leaves.

Sun leaves have smaller and more numerous chloroplasts than do shade leaves. Chloroplasts in sun leaves have fewer grana than do chloroplasts in shade leaves. Epidermal cells of shade leaves often contain many chloroplasts.

Sun and shade leaves can form on the same plant; leaves near the shoot apex are often sun leaves, while leaves in the dimmer light of the lower canopy are shade leaves. This flexibility in leaf development allows plants to exploit different and changing environments.

Moisture

As mentioned in the previous chapter, obtaining enough water is the biggest challenge faced by land plants. Therefore, it is not surprising that water availability strongly influences leaf development. For example, the features described earlier in this chapter typify **mesophytes,** which are plants that grow best in moist but not wet environments (i.e., with intermediate amounts of water). Most plants you are familiar with are mesophytes. The structural features of mesophyte leaves intergrade with those of plants living in more extreme environments: xerophytes and hydrophytes.

Xerophytes

The leaf in figure 14.23a is from a **xerophyte,** which is a plant that grows in habitats characterized by seasonal or persistent drought. Since these environments are usually bright deserts that support relatively little vegetation, light availability seldom limits the growth of xerophytes. Rather, their growth is limited by how efficiently they use water. The leaves of xerophytes usually have one or more of the following modifications that help conserve water.

Xerophytes have small, thick leaves with well-developed spongy and palisade layers. These leaves are often modified for storing water and typically contain relatively few intercellular spaces. Xerophytes such as sagebrush (*Artemisia*) that

BOXED READING 14.2
CLOGGING THE WATERWAYS

Most aquatic plants contain large amounts of aerenchyma, which enhances gas exchange between the atmosphere and submerged tissues. Some plants use aerenchyma for support. For example, petioles of water hyacinth (*Eichhornia crassipes*) have air-filled bladders filled with aerenchyma that buoys the plant's leaves.

Water hyacinth was introduced to North America in 1864 at a horticultural exhibit in Louisiana. One exhibitor, especially attracted to the plant's luscious leaves and elegant flowers, took some of the plants and began cultivating them near his home in Florida. There, without its natural herbivores, water hyacinth spread wildly throughout waterways, where today it clogs ponds and drainage canals. Similar problems caused by introduced water hyacinth now also plague India and the Nile and Congo Rivers of Africa.

Transpiration from dense mats of water hyacinth can drain a pond at an incredible rate; for example, ponds covered by *Eichhornia* lose water almost eight times faster than uncovered surfaces. Although its leaves can be harvested to feed cattle and generate methane, clogging of waterways by *Eichhornia* costs millions of dollars to control. Water hyacinth is not a problem in the Amazon River, its natural habitat. There, herbivores and periodic flooding prevent it from clogging the river.

BOX FIGURE 14.2
Water hyacinth was introduced into Florida in the 1860s. These plants, which can double their population in only two weeks, clog many of the waterways in Florida and other southeastern states.

BOX FIGURE 14.3
Close-up of water hyacinth.

---

grow in seasonally dry habitats produce relatively large leaves during the wet season; when drought begins, these leaves are replaced by smaller leaves. Conversely, some plants abort all of their leaves during drought. When this occurs, the survival problem is no longer transpiration, but rather how to stay alive until foliage leaves form in response to the next rainy season. Plants solve this problem by producing photosynthetic stems and storing water in their succulent stems.

Xerophytes are covered by an epidermis having thick cell walls, numerous stomata, and a thick cuticle. For example, leaves of mesquite (*Prosopis*) growing in dry areas have cuticles ten times thicker than those growing in wet soil. Their stomata are often sunken and overlaid with trichomes. The large number of stomata in leaves of xerophytes increase their photosynthetic rates during rare wet periods—a "get-it-while-you-can" strategy.

Xerophytes have large amounts of supporting tissues in their leaves. Turgor pressure supports the leaves of plants growing in most environments. Since turgor cannot always be maintained in xerophytes, their leaves usually contain large amounts of sclerenchyma for support.

Not all botanists think that these leaf modifications are adaptations for conserving water. Indeed, transpiration does not always decrease when stomata are sunken and covered with trichomes. Such observations have prompted several botanists to suggest that xerophytic traits may be caused by adaptations to intense light. Most xerophytic modifications can also be induced in well-watered plants by cold temperatures or nutrient deficiencies. For example, plants given only small amounts of nitrogen often develop more striking xerophytic features than do those deprived of water.

Hydrophytes

Look again at figure 14.23. The leaf in figure 14.23b is from a **hydrophyte,** which is a plant that grows in habitually wet environments (see box 14.2, "Clogging the Waterways" on p. 321). Aquatic plants that grow partly or completely submerged in fresh water are hydrophytes and often have unusual leaves. *Wolffia*, a hydrophyte, is the smallest flowering plant; it is common in ponds and slow-moving streams and has leaves only 1 millimeter wide. Conversely, the leaves of royal water lily (*Victoria amazonica*) are large floating discs up to 2 meters wide. Although their undersides have many protective spines, their upper surfaces are smooth and are used as floats by insects, frogs, and children. Leaves of royal water lily inspired the architecture of the Crystal Palace of London.

Obtaining enough water is never a problem for hydrophytes; however, the water surrounding submerged plants such as *Elodea* reduces the intensity of light reaching submerged leaves and severely limits gas exchange (remember that diffusion through liquid is much slower than through air). Thus, the watery environment solves one problem (i.e., desiccation), but creates two others: absorbing enough light and exchanging gases. Most leaf modifications in hydrophytes enhance light absorption and gas exchange.

Hydrophytes have large, thin leaves with poorly developed spongy and palisade layers. Epidermal cells typically contain chloroplasts and are photosynthetic, while ground tissue is modified for storage. The leaves of hydrophytes also contain large amounts of aerenchyma for gas exchange and support.

Hydrophytes are covered by a thin cuticle and have thin cell walls. Submerged leaves lack stomata, while floating leaves have stomata only on their upper surface.

Hydrophytes contain relatively little xylem and supporting tissue. Submerged leaves of hydrophytes are supported by aerenchyma and the buoyancy of the surrounding water.

As you can see, the characteristics that distinguish the leaves of hydrophytes are the opposites of those of xerophytes. Hydrophytes also have reduced root systems; nutrients are absorbed by epidermal cells, which are often modified into transfer cells.

The submerged leaves of aquatic plants often have different shapes than the floating leaves. For example, floating leaves of buttercup (*Ranunculus*) are large and flat, while submerged leaves are highly dissected and lacelike. Similarly, submerged leaves of American pondweed (*Potamogeton nodosus*) are long and narrow, thin (3–4 cells thick), have a uniform chlorenchyma, and lack stomata, cuticle, and large air spaces, whereas the floating leaves are elliptical and somewhat similar to the leaves of many terrestrial plants; that is, they have a cuticle, numerous stomata on their upper surface, large air spaces, and palisade and spongy chlorenchyma. The occurrence of morphologically distinct leaves on the same plant is called **leaf dimorphism.** Plant hormones influence the formation of dimorphic leaves. For example, pondweed produces floating leaves instead of submerged leaves when exposed to abscisic acid, a plant hormone. This effect can be overcome by simultaneously exposing the leaf to other plant hormones such as gibberellins and cytokinins.

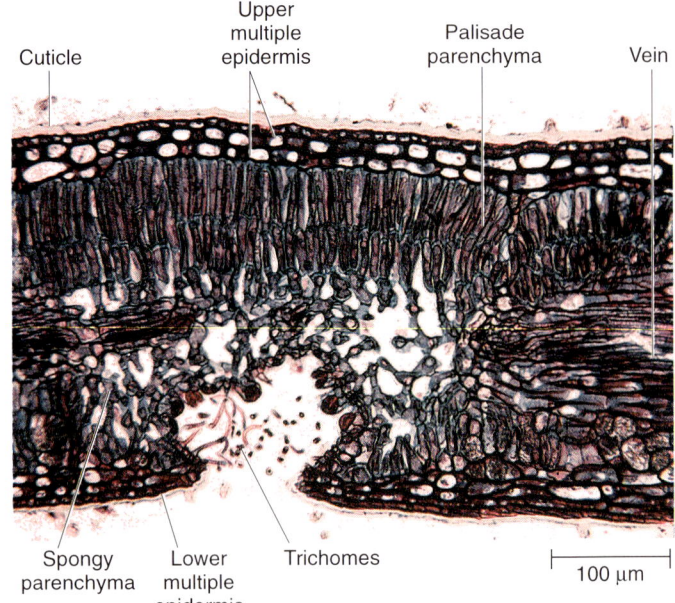

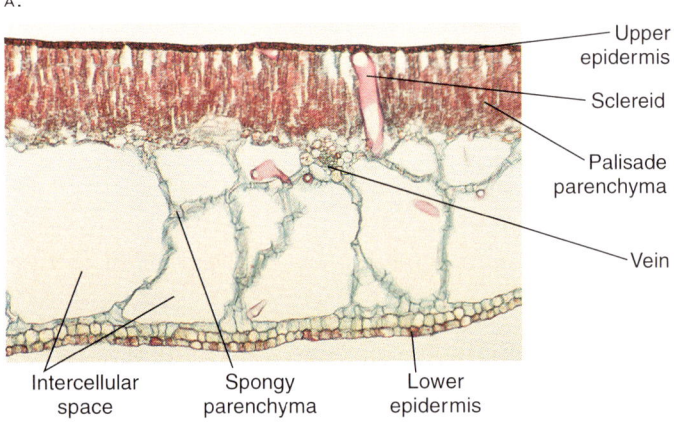

Figure 14.23

(a) Cross section of a leaf of oleander (*Nerium oleander*), a xerophyte, ×55. Note the thick cuticle and multiple epidermis. Stomata and trichomes are sunken into chambers called stomatal crypts. (b) Cross section of a leaf of water lily (*Nymphaea odorata*), a hydrophyte, ×100. Stomata form only on the upper surface of these floating leaves, and the vascular tissue (especially the xylem) is reduced. Large intercellular spaces add buoyancy to the leaves.

Light and the availability of water strongly modify leaf structure. This plasticity in leaf development adapts plants to different and changing environments.

Doing Botany Yourself

Design an experiment that would enable you to evaluate the environmentally controlled vs. genetically controlled features of leaf variation.

Leaf Movements

The leaves of many plants move. For example, leaves often orient themselves perpendicular to sunlight, thereby increasing the amount of light they absorb for photosynthesis. As a result of this light-directed movement, leaves form **mosaics** that minimize the shading of leaves by each other (fig. 14.24). Similarly, the leaves of many desert plants often orient themselves parallel to sunlight to decrease their heat load. However, movements and arrangements of leaves often go beyond even these generalizations. For example, leaves of the Ceará rubber tree (*Manihot glaziovii*) of northeastern Brazil fold like an umbrella, while those of some species of rhododendron roll into a cylinder when it's cold. In contrast, leaves of the compass plant (*Silphium laciniatum*) do not move, but use light for orientation—half of the plant's leaves point to the east, and half point to the west. A most unusual kind of leaf movement occurs in the telegraph plant (*Desmodium gyrans*), a member of the legume family. Its leaves sometimes move in circles; those on one side of the plant circle in one direction, and the rest circle in the opposite direction. At other times, the leaves seem to go berserk: they jerk and twitch wildly, move up and down, or move in circles—with rest periods between the spasms. We do not know the basis for these bizarre movements.

Modified Leaves

Like other organs, leaves are often modified for functions other than photosynthesis. Refer to figure 14.25 as you read about these modifications.

Tendrils

Tendrils of plants such as sweet pea (*Lathyrus odoratus*) and trumpet flower (*Bignonia capreolata*) are leaves modified for support (fig. 14.25a). In plants such as yellow vetchling (*Lathyrus pratensis*), the entire leaf is a tendril; photosynthesis in these plants is delegated to leaflike structures called *stipules* at the base of each leaf (see below). Conversely, only the petiole is a tendril in potato vine (*Solanum jasminoides*) and garden nasturtium (*Tropaeolum majus*). Tendrils of many plants may be up to 30 cm long, which makes them well suited for seeking support in the plant's nearby environment.

Stipules

Stipules are small, leaflike structures at the base of petioles. Stipules have a variety of functions. For example, stipules of woodruff (*Asperula*) and sweet pea are photosynthetic, while those of black locust (*Robinia pseudoacacia*) and spurge (*Euphorbia*) form protective spines. Stipules protect buds in oak and beech, and in passion flower vine (*Passiflora gracilis*) they become

FIGURE 14.24

Leaves of plants such as this Boston ivy (*Parthenocissus tricuspidata*) form mosaics that help ensure that almost all leaves are exposed to light. Leaf mosaics often form in house plants exposed to light from one direction.

tendrils that coil around objects that they touch. These tendrils are extremely sensitive: they'll coil around a platinum wire weighing only 1.23 mg—that's about 1/50 the weight of a paper clip.

Spines

The spines of plants such as ocotillo (*Fouquieria splendens*) and cacti are leaves modified for protection.

Bud Scales

Bud scales are tough, overlapping, waterproof leaves that protect buds from frost, desiccation, and pathogens. Bud scales form before the onset of unfavorable growing seasons such as winter.

Window Leaves

Window leaves (fig. 14.25b) are common in many desert plants such as fairy-elephant's-feet (*Frithia pulchra*). These leaves are shaped like tiny ice-cream cones and grow mostly underground, with only a small transparent "window" tip protruding above soil level. The covering soil shields window plants from the desert's drying winds and increases their chances of being overlooked

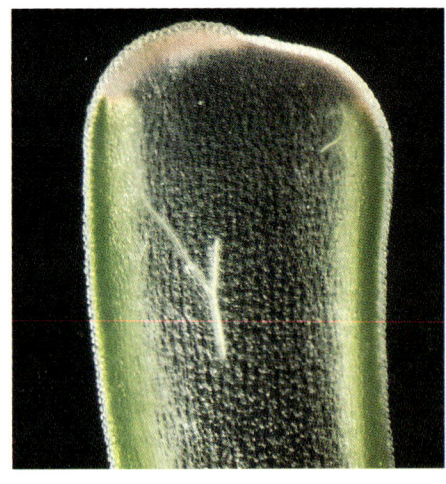

A.

B.

C.

D.

E.

FIGURE 14.25

Modified leaves. (a) Tendrils on a garden pea enable the plant to cling to other objects for support. (b) In window leaves of plants such as *Frithia pulchra*, the tip of the leaf is transparent (i.e., is a "window"), which allows the light to reach the photosynthetic tissue that is below the soil level. (c) Flowerpot leaves of *Dischidia* are homes for ants that bring soil and debris into the leaf. As the debris decays, it (and the ants' waste products) are mined by adventitious roots that grow into the hollow leaf. (d) Leaves of *Sarracenia* are modified into pitchers that trap and kill insects. The digested insects are an important source of nitrogen for the plant. (e) Leaves of *Kalanchoë* form tiny plants at their margins. When separated from the parent, they can become new individuals.

by grazing herbivores. Similarly, their windows allow light to penetrate and illuminate the chlorenchyma tissue, which is below soil level. Underground photosynthesis!

Bracts

Bracts are floral leaves that form at the base of a flower or flower stalk. They are usually small and scalelike, and protect developing flowers. Some plants have colorful bracts; for example, the colorful portions of the flowers and inflorescences of poinsettia, Indian paintbrush (*Castilleja*), bird-of-paradise (*Strelitzia*), *Bougainvillea*, and chaconia (*Warszewiczia coccinea*, the national flower of Trinidad) are bracts. In these plants, bracts replace petals and attract pollinators (the petals of these plants are small and inconspicuous). For example, the tiny flowers of dogwood (*Cornus*) are a drab, inconspicuous yellow-green and are arranged in a circle (i.e., the "eye" of the flower); the large bracts are pink or white and attract pollinating insects.

Storage Leaves

The storage leaves of plants such as onion (*Allium cepa*) and lily are fleshy, concentric leaves modified to store food. Onions have tubular leaves, the white bases of which form the bulb. The leaves of most bulbs store sugar or starch.

Flowerpot Leaves

Plants with flowerpot leaves (fig. 14.25c) are often epiphytes. They are packed tightly into a flowerpotlike structure that catches falling water and debris. Many epiphytes grow roots among the bases of their flowerpot leaves to absorb nutrients collected by the flowerpot. More ingenious plants such as *Dischidia* have hollow leaves that function as flowerpots. These leaves do not catch falling debris; rather, they are homes for ants that bring soil and debris into the leaf. Decaying debris deposited by ants is mined by roots that form at the node and grow into the hollow leaf.

Insect-Trapping Leaves

In carnivorous plants, insect-trapping leaves are modified for attracting, trapping, and digesting animals. These adaptations range from sticky "flypaper" surfaces of leaves such as those of butterwort (*Pinguicula*) to the vatlike leaves of pitcher plants such as *Sarracenia* and *Nepenthes* (fig. 14.25d). Pitcher plants typically use nectar to lure insects into the leaf chamber. About halfway down the pitcher of many pitcher plants, the epidermal surface abruptly becomes flaky wax. When the insects step onto this wax, their legs become covered with the wax, and their delicate feet are transformed into unwieldy clodhoppers. Unable to cling to the flaky wax on the side of the pitcher, the insects then slip into the pitcher's vat and die. They are then degraded and their nutrients are absorbed by the leaf.

Leaves Modified for Reproduction

Succulent plants such as *Peperomia*, *Begonia*, rock-lettuce (*Sedum*), and maternity plant (*Kalanchoë*) (fig. 14.25e) commonly have leaves that are modified for reproduction. These plants form tiny plants at the edges of their leaves. These plants become new individuals when they are shed from the parent leaves.

Cotyledons

Cotyledons are embryonic leaves. Monocots, such as corn, usually have one cotyledon, while dicots have two. However, there are some exceptions; for example, *Degeneria vitiensis*, one of the most primitive dicots, has three or four cotyledons. Cotyledons have several functions. In beans they absorb the endosperm and therefore store energy used for germination. Storage products in cotyledons are usually carbohydrates, but they may also be oils, as in peanuts (*Arachis hypogaea*). The cotyledons of the cacahuanache tree (*Licania arborea*) contain large amounts of flammable oils—enough that the seeds can be strung on sticks and used as torches. Oils extracted from *Licania* cotyledons are also used to produce candles, soaps, and grease. Although the cotyledons of some plants are green and grow above ground, they usually lack stomata and are nonphotosynthetic.

Prophylls

Prophylls are the first leaves to form on axillary buds. Monocots usually have one prophyll, whereas dicots have two, which suggests that these tiny leaves may be analogous to cotyledons. Prophylls protect axillary buds.

C O N C E P T

Plants use modified leaves for obtaining and storing nutrients, climbing, protection, and reproduction.

How Leaves Defend Themselves

Since they cannot run away or physically defend themselves from herbivores, the delicate, nutrient-rich leaves of most plants are apparently easy prey for the 300,000 known species of herbivorous insects. However, leaves are not defenseless against the prospect of being eaten. Some of a plant's defenses are obvious: their lignin makes them hard to chew, and trichomes deter many herbivores. In other plants, an attacker's first nibbles unleash a barrage of chemicals that do all sorts of awful things to the animals.

Plants that Poison Their Attackers

Plants such as white clover (*Trifolium repens*) and bird's-foot trefoil (*Lotus corniculatus*) are chemical minefields containing cyanogenic compounds, and merely touching the leaves such as stinging nettle (*Urtica*) will quickly get the attention of most animals. Even more powerful are the poisons of plants such as poison hemlock (*Conium maculatum*, an herb not to be confused with the hemlock tree, *Tsuga*, whose leaves are not poisonous). The parsleylike leaves of this plant contain coniine, a deadly alkaloid. The ancient Greeks knew about this powerful poison and used it to kill prisoners, including Socrates. Poisons of other plants are still used. For example, the Maku Indians of Colombia and Brazil use extracts from *Euphorbia cotinifolia* and *Phyllanthus brasiliensis* to kill fish. Shoots of these plants are placed on bridges over streams and beaten, so that their juices trickle into the stream and suffocate the fish. The fish are then gathered a few hundred yards downstream.

Several insects have not only circumvented leaves' poisons, but use them to defend themselves. For example, the grasshopper *Poekilocerus bufonius* eats only milkweeds that contain poisonous cardenolides. The grasshopper stores these poisons in special glands. When attacked, the grasshopper sprays the poisons on its predator, killing the predator. Eating milkweed is a critical part of the grasshopper's defense; grasshoppers fed diets lacking milkweed secrete 90% less cardenolides and are relatively defenseless against their attackers.

Many plants produce chemicals called *photosensitizers* that are toxic only when insects are in light. Photosensitizers produce highly reactive molecules that literally burn up insects. Some insects avoid the effects of photosensitizers by rolling themselves in leaves when they eat.

Plants That Change the Life Cycle of Their Attackers

Ecdysteroids and juvenile hormone are hormones that regulate insect development. Many plants produce these chemicals, and when insects eat them, their life cycle is altered, usually to the benefit of the plant. For example, swarms of locusts can denude large prairies in only a few hours, but they do not usually bother bugleweed (*Ajuga remota*). This plant laces its leaves with large amounts of ecdysonelike compounds, many of which are more potent than those made by insects. Insects that eat bugleweed develop several head capsules when they change from larvae to adults. These extra head capsules block their mouth parts, and the insects starve to death.

The discovery of substances similar to juvenile hormone in plants is an excellent example of the scientific method in action. Two researchers, one in the United States and one in Europe, were collaborating on a research project that involved growing an insect called *Pyrrhocoris apterus*. These insects developed normally when grown in Europe, but those reared in the United States failed to become adults. Puzzled, the researchers tried to discover why they could not get the same results. They used carefully designed control experiments to sequentially eliminate several factors, including temperature, light, and humidity. Finally, they discovered the culprit: the filter paper on which the insects were being grown. The active ingredient of the paper was traced to the pulp of balsam fir, a primary source of North American (but not European) paper products, including the filter paper used in laboratories. This compound, which was named *juvabione*, is also abundant in leaves; it arrests insect development and enables fir trees to eliminate many herbivores.

Plants also produce chemicals that alter the life cycles of vertebrates. For example, chemicals in pine leaves (i.e., needles) eaten by grazing cattle double the concentration of progesterone, a hormone needed to maintain normal pregnancy. As a result, pregnant cattle that eat as little as 0.7 kg of needles per day usually abort their calves.

Plants That Make Themselves Less Digestible

Many plants fight their attackers by making themselves less digestible. For example, leaves of tomato and potato attacked by chewing insects release substances that move through the plant from the site of wounding and induce the formation of enzymes that interfere with insects' digestion. As a result, tomato plants become less nutritious, and insects tend to dine elsewhere.

Plants also warn each other of impending attacks by insects. Sitka willows (*Salix sitchensis*) being attacked by tent caterpillars produce compounds that decrease the digestibility of their leaves. Uninfested plants up to 60 meters away also produce these chemicals, suggesting that willows use airborne signals to warn others that the herbivore is coming. However, the suggestion that plants emit such warning signals is still very controversial.

Finally, the leaves of many plants use animals for defense. For example, stipules of bull's-horn acacia (*Acacia cornigera*) are modified as hollow spines that resemble bull horns and serve as homes for ants. The plant's nectaries produce fatty globules, which are eaten by aggressive ants living in the spines. In return for this food, the ants defend the plant fiercely: an intruder landing on the plant is soon covered by a swarm of ants. These armies of ants are important, since *Acacia* plants without ants suffer more damage by herbivores than do plants protected by ants.

Plants That Shift Their Resources

Plants can also defend themselves by changing their production of certain chemicals. One group of leaves on a maple tree, for example, may contain large amounts of toxins, while another group may contain relatively little nitrogen. Plants change the concentrations of these chemicals regularly and in response to insect attacks. This shell-game strategy forces insects to move around constantly, thereby exposing them to *their* predators.

C O N C E P T

Leaves protect themselves by poisoning attackers, changing their life cycle, making themselves less digestible to attackers, and shifting their resources.

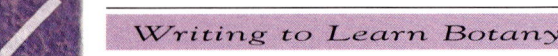

Writing to Learn Botany

Discuss the various ways in which leaves may defend themselves. How do we exploit these potential defenses?

Many of the chemical defenses of plants are elicited by attacks. For example, phytoalexins are antimicrobial compounds made in response to carbohydrates released from the cell walls of

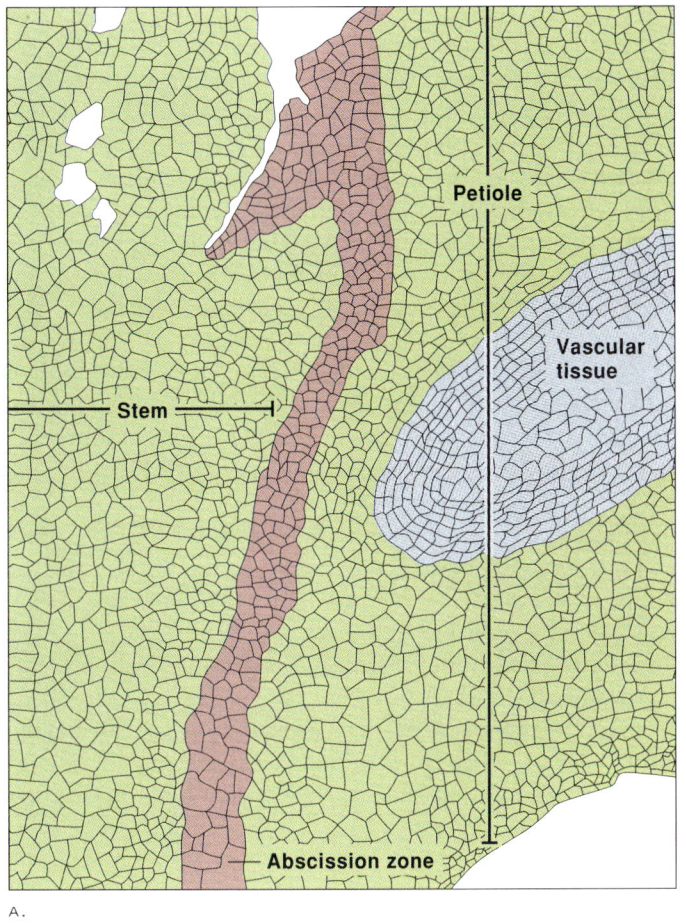
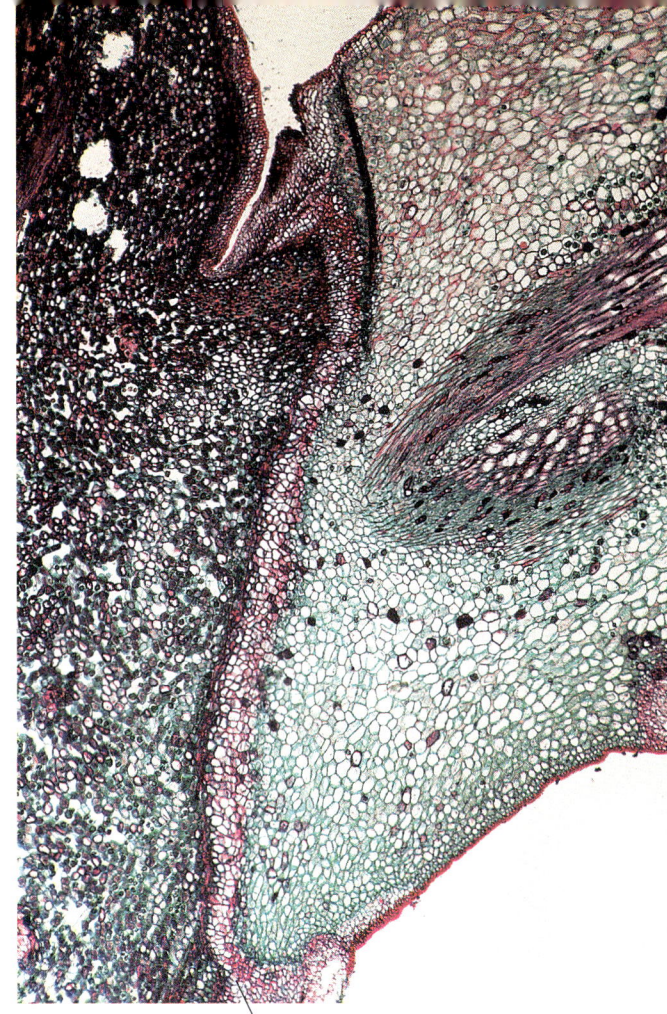

FIGURE 14.26

Leaf abscission occurs when a separation zone forms near the base of the petiole. This zone minimizes infection and loss of nutrients.

attacking bacteria and fungi. Phytoalexins have several functions, including preventing germination of spores and inhibiting the enzymes that degrade cell walls. Plants make large amounts of phytoalexins when attacked by pests. Many botanists hope to develop resistant hybrids by using naturally occurring elicitors to turn on a plant's defenses *before* an attack—something similar to vaccinating plants against their enemies.

The associations described between ants and plants are examples of **coevolution,** a reciprocal process in which characteristics of one organism evolve in response to specific characteristics of another. The interactions between the two organisms can be remarkably specific: many plants are defended by only one kind of ant, and many plants are pollinated by only one kind of insect. As a reward for their fidelity, insects get food and, occasionally, a place to live.

Leaf Abscission

Leaves typically have a limited life span, after which they are shed from a plant via a complex process called **leaf abscission.** Abscission can result from injury or seasonal changes in climate such as drought or the shortening days of autumn. Plants such as oak and pecan (*Carya*) that abscise their leaves in the fall are called *deciduous*. Leaves of *evergreens* such as pine and oleander live 3–5 years, and are shed year-round.

Leaves abscise at a predetermined region at the base of the petiole called the *abscission zone* (fig. 14.26). In most leaves, this zone contains reduced amounts of sclerenchyma, few intercellular spaces, and thin-walled cells. Injury or short days alter the production of auxin and ethylene, which are plant hormones that control abscission. As a result, a *separation layer* forms in the abscission zone. Cells along the separation layer suberize (i.e., form suberin) and, in doing so, isolate the senescing leaf from the stem. Once the separation layer forms, wind or other disturbances break the dead leaf from the stem. What's left on the stem is a suberized *leaf scar* that resists infection and desiccation. Leaf abscission is important because it prunes plants and rids them of injured or dying leaves. You'll learn more about leaf abscission in Chapter 19.

The Economic Importance of Leaves

We use leaves and their products every day. Most importantly, chlorenchyma cells produce the sugars and oxygen that we and most other organisms need to live. However, leaves also have other important, though less critical, uses.

Food, Spices, and Drinks

The leaves of plants such as cabbage (*Brassica oleracea*), lettuce (*Lactuca sativa*), spinach (*Spinacia oleracea*), chard (*Beta vulgaris*), sage (*Salvia*), thyme (*Thymus vulgaris*), celery (*Apium graveolens*), parsley (*Petroselinum crispum*), and bay laurel (*Laurus nobilis*) have long been parts of our diets and favorite items in our salads. Many of these leaves are especially nutritious; for example, spinach contains large amounts of vitamin A. Conversely, other plants are inedible in the wild; for example, wild lettuce contains large amounts of a bitter and narcotic latex.

Edible leaves of some other plants have mythical stories associated with them. Peppermint (*Mentha piperita*), for instance, derives its name from Menthe, a nymph of Greek mythology.[1] Thyme, oregano (*Origanum vulgare*), peppermint, spearmint (*Mentha spicata*), wintergreen (*Gaultheria procumbens*), basil (*Ocimum basilicum*), and sage are all derived from leaves. Similarly, tea is extracted from leaves of a relative of the garden *Camellia* (fig. 1.1), and *Agave* leaves are used to make tequila.

Dyes

Although most dyes are now made from coal tar, plants were the original sources of most colorings. The leaves of bearberry (*Arctostaphylos uva-ursi*) contain a yellow dye, while those of henna (*Lawsonia inermis*) contain a red dye that was used to stain fingernails and the cloth used to swath Egyptian mummies.

Fibers

We mentioned in Chapter 13 that fibers from plants such as flax, sisal, and Manila hemp are woven into ropes and fabrics prized by clothiers and sailors. Palm leaves are used to make Panama hats, clothing, brooms, and thatched huts in the tropics.

Fuel

The leaves of plants such as yareta (*Azorella yareta*) contain flammable resins that can be used as fuel.

Drugs

Many leaves contain poisons which, when administered in small amounts, are useful drugs. For example, digitoxin and digitalis are popular drugs extracted from foxglove (*Digitalis purpurea*) (fig. 1.1). These heart stimulants increase the relaxation of the heart without affecting its contraction, thereby increasing blood circulation. Hyoscyamine, atropine, and scopolamine are derived from deadly nightshade (*Atropa belladonna*), a poisonous plant named after Atropos, the Greek god of fate who held the shears to cut the thread of human life. *Atropa* was a savior for Macbeth, whose soldiers used it to poison the Danish army during peace talks. The *belladonna* specific epithet is Italian for "beautiful woman," a reference to Italian women's use of *Atropa belladonna* to dilate their eyes and make them brighter and more attractive. *Aloe* leaves are used to treat burns. The lobeline sulfate used in drugs for quitting smoking is derived from leaves of a relative of the garden lobelia. Coffee and tea leaves (fig. 1.1; box 1.1) contain as much as 5% caffeine, which, like drugs from nightshade, is a heart stimulant. Trichomes on *Cannabis* leaves contain narcotic tetrahydrocannabinols, and cocaine, a drug extracted from coca (*Erythroxylon*) leaves, is a dangerous yet popular drug (fig. E.5d in "Epilogue").

Tobacco leaves contain large amounts of drugs such as nicotine that cause cardiovascular problems. Thanks to billions of dollars in federal subsidies, we produce almost a billion kilograms of tobacco leaves annually. The return for this investment of tax dollars is that smoking kills 17.2% of the people in the United States (alcohol abuse, motor vehicle accidents, and illicit drugs account for only 4.9%, 2.3%, and 0.2% of U.S. deaths, respectively). But the social costs of tobacco use do not end with funerals. At last count, smoking costs Americans $53.7 billion in absenteeism, lost production, and health care—a staggering $1.79 per pack of cigarettes.

These examples underscore the indispensable roles of plants in society and modern medicine. Drugs derived from plants account for more than 25% of all prescriptions dispensed in the United States. These prescriptions do not come cheaply; in 1990, Americans spent more than $10 billion for prescription drugs derived from plants.

Other Uses of Leaves

Carnauba wax is derived from leaves of the carnauba palm, and extracts from the lancelike leaves of *Aloe* are ingredients of medicated soaps and creams. *Aloe* is sacred to many Muslims, who often hang its leaves above their doors to show that they have made a pilgrimage to Mecca. We also sculpt leaves into unusual shapes, use them as names for sports teams (e.g., the Toronto Maple Leafs hockey team), and even as parts of corporate logos. For example, the shamrock on the Boston Celtics' logo is a palmately compound leaf.

Finally, do not overlook the elegant beauty and aesthetic importance of leaves. They provide shade for us and other organisms, creating a cool oasis that we often seek during summer heat waves.

> **The Lore of Plants**
>
> The trees at about ten suburban homes can produce a ton of dead leaves each fall. Burning these leaves produces smoke that is not as nice as it smells. Indeed, burning these leaves releases 90 kg of soot, 260 kg of carbon monoxide (a poison), and various amounts of hydrocarbons that can irritate the eyes, nose, throat, and lungs. Some of these hydrocarbons are carcinogens.

---

[1]. Menthe was a nymph beloved by Pluto, alias Hades (God of the dead). She was trampled by Persephone (daughter of Zeus and the symbol of rebirth of crops in spring) and turned into mint. As was true for Menthe, the more you crush mint, the sweeter it gets.

Chapter Summary

The shoot system of a plant consists of leaves and stems. Both are made of the same tissues; the structural and functional differences between them result from different arrangements of these tissues. Stems are collections of nodes and internodes. Nodes are areas where leaves attach to stems, and internodes are the portions of stem between nodes. Stems support leaves, produce and store food, and transport water and solutes between roots and leaves.

Stem elongation occurs in subapical regions. Many grasses elongate via the activities of intercalary meristems, which are meristems intercalated between mature tissues at the bases of their nodes. Internodal elongation helps plants intercept light. Plants that produce insufficient amounts of auxins and gibberellins for elongation are called rosette plants. The internodes of these plants do not elongate, leaving the leaves packed tightly on a short stem.

Stems are made of epidermal, ground, and vascular tissues. The epidermis of stems is made of tightly packed and cutinized cells that prevent desiccation. Vascular tissues are arranged in vascular bundles: in monocots these bundles are scattered throughout the stem's ground tissue, whereas in most dicots they are arranged in a single ring. Ground tissue inside this ring is called the pith and is often modified for storage. Between the vascular bundles and the epidermis is the cortex, which stores food or is photosynthetic.

Stolons, tendrils, searcher shoots, cladodes, thorns, short and long shoots, succulent stems, bulbs, rhizomes, corms, and tubers are stems modified for reproduction, climbing, photosynthesis, protection, and storage. We use stems for food, paper, drugs, spices, and lumber.

Leaves are the most active and diverse organs on plants. Their formation depends on several meristems, and their final shape results from cellular division and expansion. Simple leaves have a flattened blade and a stalklike petiole. Blades of compound leaves are divided into leaflets. Leaflets of pinnately compound leaves form in pairs along a central rachis, while those of palmately compound leaves form at the same point. Axillary buds, which form in the axil between the petiole and stem, form flowers or branches that modify that plant's shape and ability to intercept light.

Phyllotaxis is the arrangement of leaves on stems. Plants with one leaf per node have spiral phyllotaxis, while those with two leaves per node have opposite phyllotaxis. Plants with three or more leaves per node have whorled phyllotaxis. Phyllotaxis is probably controlled by fields of inhibitors at the shoot apex or differential reinforcement of cells of the tunica layers.

Leaves are made of epidermal, ground, and vascular tissues. The epidermis is usually transparent and contains numerous stomata. Vascular tissues in leaves are arranged in veins: xylem forms on the upper side of the vein, and phloem forms on the lower side. The leaves of dicots have netted venation, in which a prominent midvein connects with a meshlike network of minor veins. Monocots have parallel venation, meaning that the major veins are oriented parallel to each other and are connected by smaller minor veins. Vein endings in the ground tissue are where water and solutes move into and out of veins.

The ground tissue in leaves is called mesophyll. Horizontally oriented leaves have densely packed, columnar palisade mesophyll cells along the upper side, and spongy mesophyll cells along the lower side. Most light absorption and carbon fixation occurs in the palisade mesophyll, and most gas exchange occurs in the spongy mesophyll. Vertically oriented leaves have a uniform mesophyll. Mesophyll cells of grasses such as corn are arranged concentrically around a photosynthetic bundle sheath. This arrangement of mesophyll underlies C_4 photosynthesis, which in hot, dry environments is more efficient than C_3 photosynthesis.

The structure and function of leaves is strongly influenced by light and moisture. Leaves require light to expand, and daylength often determines what kind of leaves a plant produces. Leaves in intense light are sun leaves, and are smaller, thicker, and capable of higher rates of photosynthesis than shade leaves, which grow in dim light.

Xerophytes grow best in dry environments and have small, thick leaves with numerous stomata, a thick cuticle, and well-developed palisade and spongy mesophyll layers. Hydrophytes grow best in wet environments and have large, thin leaves with a thin cuticle and few stomata. The environmental control of leaf formation adapts leaves to different and changing environments.

Tendrils, spines, bud scales, window leaves, bracts, storage leaves, flowerpot leaves, insect-trapping leaves, leaves modified for reproduction, stipules, prophylls, and cotyledons are leaves modified for obtaining and storing nutrients, climbing, protection, and reproduction. Leaves defend themselves by poisoning their attackers, altering the life cycle of their attackers, making themselves less digestible, and shifting their resources.

Old and injured leaves abscise from plants at a predetermined abscission zone. Leaves of evergreens live several years and are shed year-round. Deciduous plants shed their leaves seasonally, usually in response to drought or the shortening days of autumn. We use leaves for food and to make spices, drinks, dyes, fibers, and drugs.

Questions for Further Thought and Study

1. Explain why submerged leaves often resemble shade leaves, while the leaves of xerophytes often resemble sun leaves.
2. How can you distinguish a leaf from a leaflet?
3. The tiny plants that form along the edges of leaves of plants such as *Kalanchoë* (fig. 14.25) are genetically identical to the parent. Why?
4. How does the ability of a plant to form (a) tendrils, (b) thorns, (c) bulbs, and (d) rhizomes help the plant survive?
5. What is the significance of having more stomata on the lower side of a leaf than on the upper side?
6. Describe three ways that light affects leaf development. What is the significance of each?
7. How do sun leaves differ from shade leaves? What is the significance of these differences?
8. How do the leaves of xerophytes differ from those of hydrophytes? What is the significance of these differences?
9. List six kinds of modified leaves and the functional significance of each.
10. What advantages do climbing plants have over erect plants? What disadvantages?
11. What are the advantages of producing many small leaves instead of fewer large leaves? What are the disadvantages?
12. Plasmodesmata link all cells of leaves. Why, then, do leaves have veins for transporting water and solutes?
13. What are the advantages of having a palisade layer on the upper surface of a leaf?
14. Plants transpire huge amounts of water. For example, plants in Connecticut transpire approximately 50 billion gallons of water per week. What happens to all of this water?

Suggested Readings

Articles

Boardman, N. K. 1977. Comparative photosynthesis of sun and shade plants. *Annual Review of Plant Physiology* 28:355–377.

Bolz, D. M., and K. B. Sandved. 1987. A world of leaves: Familiar forms and surprising twists. *Smithsonian* 16:150–155.

Dale, J. E. 1992. How do leaves grow? *BioScience* 42:323–332.

Dussourd, D. E., and T. Eisner. 1987. Vein-cutting behavior: Insect counterploy to the latex defense of plants. *Science* 237:898–901.

Rosenthal, G. A. 1986. The chemical defenses of higher plants. *Scientific American* 254:94–99.

Books

Sandved, K. B., and G. T. Prance. 1985. *Leaves.* New York: Crown Publishers.

Williams, R. F. 1975. *The Shoot Apex and Leaf Growth.* Cambridge: Cambridge University Press.

Woods, R. K. S., ed. 1982. *Active Defense Mechanisms in Plants.* New York: Plenum Press.

Tip of a primary root of corn (*Zea mays*). Root tips produce large amounts of mucigel that help the root force its way through the soil.

CHAPTER 15

Primary Growth: Roots

Chapter Outline

INTRODUCTION
KINDS OF ROOT SYSTEMS
 Taproot System
 Fibrous Root System
 Adventitious Roots
FUNCTIONS AND STRUCTURE OF ROOTS
 Root Tip
 Subapical Region
 Mature Region
 Transition Region Between the Root and the Shoot
THE ROOT-SOIL INTERFACE
FACTORS CONTROLLING THE GROWTH AND
 DISTRIBUTION OF ROOTS
THE GROWTH OF ROOTS VS. SHOOTS
MODIFIED ROOTS
 Nutrition

Box 15.1
FINDING A HOST

THE ECONOMIC IMPORTANCE OF ROOTS
Chapter Summary
Questions for Further Thought and Study
Suggested Readings

Chapter Overview

We're all familiar with stems and leaves because they grow above ground and are conspicuous. Although we do not see much of roots, they are equally important to plant growth because they provide the chlorenchyma cells of stems and leaves with a steady supply of water and dissolved minerals. To accomplish this, roots constantly grow into new territory, where they absorb and transport water and minerals from the soil to the shoot, thereby linking photosynthetic cells with the soil's moisture and nutrients. Simultaneously, roots receive sugars and other organic compounds from the shoot. These compounds are stored, used for growth, and even released to the soil's microbes for their help during growth. As a result, much of a plant's photosynthate supports an extensive underground mining operation. The ability of roots to extract water and minerals from the soil is affected by many of the same environmental factors that affect the growth of shoots, including light and water. Some roots are modified for functions such as support, movement, propagation, aeration, and parasitism.

Introduction

Plants and animals have different strategies for growth. Animals usually expose only enough surface area to sense the environment and move about to gather food. This strategy is successful because it makes animals more efficient, decreases injury, and helps them obtain concentrated sources of nutrients. Conversely, plants absorb dilute nutrients from the environment. For example, the CO_2 essential for photosynthesis accounts for only 0.035% of air, and sunlight is often at nonsaturating intensities. Thus, exposing large amounts of surface area increases a plant's chances of absorbing light and nutrients from the environment. For the surface area of the shoot, there's usually *some* light available each day, and the breeze constantly bathes leaves in a fresh supply of CO_2. Thus, the millions of leaves on an elm tree usually have no problem producing enough sugar for growth, as long as they get enough water and nutrients from roots. Roots grow in unstirred soil, however, which makes it necessary for them to grow *to* water and nutrients.

The growth necessary to locate dispersed water and nutrients produces large roots and extensive root systems. To appreciate this, consider the simple but tedious experiment performed in the 1930s by Howard Dittmer, who grew a winter rye plant (*Secale cereale*) in a shoebox-sized container full of fertile soil. After four months he carefully unearthed the plant and measured its root and shoot systems. The shoot was 8 cm high and consisted of 80 leaves covering 5 m^2, an area about the size of a ping-pong table. However, the situation was different underground. The root system consisted of more than 13 million roots and 1.4×10^{10} root hairs having a surface area larger than two and a half tennis courts. The combined length of these roots and their root hairs spanned more than 11,000 km, a distance almost one-third that of the earth's circumference.

Dittmer's experiment emphasizes an important principle of plant growth; namely, that plants invest heavily in roots. Although roots seldom occupy more than 5% of the soil's volume, they often consume more than half of a plant's net primary production each year. Furthermore, roots account for more than 80% of plant biomass in ecosystems such as shortgrass prairies and tundra. In trees such as pecan (*Carya illinoensis*), roots are usually longer and spread wider than the shoot. As a result, roots can effectively mine dispersed water and minerals from the soil. Roots are also important to animals because they unlock the soil's store of nutrients and move elements from the soil into the food chain. Thanks to roots, animals can obtain essential elements such as calcium and sulfur by eating plants instead of dirt.

Kinds of Root Systems

Taproot System

The first root to emerge from a seed is the **radicle,** or **primary root** (fig. 15.1). In most dicots, the radicle enlarges to form a prominent **taproot** that persists throughout the life of the plant. Many progressively smaller **branch roots** grow from the taproot. This type of root system consisting of a large taproot and smaller branch roots is called a **taproot system** (fig. 15.2) and is common in conifers and many dicots. In plants such as sugar beet and carrot, fleshy taproots are the plant's food pantry: they store large reserves of food, usually as carbohydrates. Not all taproots are modified for storage, however; for example, the long taproots of poison ivy (*Rhus toxicodendron*) and mesquite (*Prosopis*) are modified for reaching water deep in the ground.

Taproots usually control the growth and development of branch roots, much as the shoot apex controls the growth of axillary buds. For example, if the radicle is damaged, a branch root enlarges and assumes the dominating role of the taproot. Because taproots of most dicots grow faster than branch roots throughout the life of the plant, many plants have long taproots. Indeed, engineers digging a mine in the southwestern United States uncovered a mesquite root 53 m down. If you'd like to see how big roots can get, visit the Roto-Rooter Monster Root Hall of Fame in Des Moines, Iowa. There you can see Moby Root, a 31-meter root pulled out of a drainage pipe of a parking garage in 1994.

FIGURE 15.1

Primary root of a corn seedling.

FIGURE 15.2

Taproot system of dandelion (*Taraxacum*). Taproot systems consist of a prominent taproot and smaller lateral roots.

FIGURE 15.3

Fibrous root system of a grass. Fibrous root systems consist of many similarly sized roots. Fibrous root systems form extensive networks in the soil.

Fibrous Root System

Most monocots have a **fibrous root system** consisting of an extensive mass of similarly sized roots (fig. 15.3). In these plants, the radicle is short-lived and is replaced by a mass of **adventitious roots** (from the Latin word *adventicius*, meaning "not belonging to"), which are roots that form on organs other than roots. Fibrous roots of a few plants are edible; for example, sweet potatoes are fleshy parts of fibrous root systems of *Ipomoea batatas*. Because the adventitious roots of monocots are so extensive and cling tenaciously to soil particles, such plants are excellent for preventing erosion.

Adventitious Roots

The adventitious roots of most monocots begin growing soon after the seed germinates. Each node in the embryo usually produces 2–6 *seminal roots* that grow from the seed. These roots are soon supplemented by an extensive system of *crown roots* that grow from nodes of the growing shoot. Within a few weeks, a corn plant's root system consists of seminal and crown roots that form an extensive network in the soil (fig. 15.3). In only four weeks the hemispherical root system of a corn seedling is more than 45 cm deep and 60 cm in diameter. By the time the plant forms fruit, the root system spans more than 1.5 m.

Recent studies of monocot root systems tell us that they are more complex than we originally thought. For example, corn plants have two types of adventitious roots, one of which is usually unbranched and has actively growing tips. These *feeder roots* are associated with a 1-mm-thick soil sheath permeated by root hairs and cemented together by secretions of the root and the soil's microbes. Other roots of the root system are long, branched, and lack an encasing soil sheath. These roots lack actively growing tips; that is, they grow determinately. Some plants have roots with characteristics of both tap and fibrous root systems. For example, clover (*Trifolium*) has a taproot and an extensive fibrous system of roots produced at nodes of its stolons.

There are several types of adventitious roots besides those of monocots. For example, adventitious roots are common along rhizomes of ferns, club mosses (*Lycopodium*), and horsetails (*Equisetum*). In many plants, adventitious roots are a primary means of vegetative reproduction; prairie grasses and forests of quaking aspen (*Populus tremuloides*) are often a single clone spread by adventitious roots. We use adventitious roots on cuttings to propagate many plants, including raspberries (*Rubus*), apples

(*Malus*), cabbage (*Brassica*), and brussels sprouts (*Brassica*). Adventitious roots form in all sorts of places on plants, including leaves, petioles, and stems. The formation of adventitious roots is controlled by hormones such as auxin, which are the active ingredients in the rooting compounds sold in stores. Although they arise at different locations, primary roots and adventitious roots have similar structures and functions.

Most dicots have a taproot system consisting of a large taproot and smaller branch roots. Taproot systems maximize support and storage. Monocots have fibrous root systems consisting of similarly sized roots that maximize absorption. Adventitious roots are roots that form on organs other than roots. They form most of the root systems of many monocots and are important for vegetative reproduction.

FUNCTIONS AND STRUCTURE OF ROOTS

We have long known that roots are critical for plant growth: Aristotle taught about the importance of roots in the fourth century B.C., and you do not have to be a botany whiz to know that most plants die when separated from their roots. Roots have four primary functions:

Anchorage. To locate water and minerals, roots permeate the soil. In doing so, they anchor the plant in one place for its entire life.

Storage. Roots store large amounts of energy reserves. In biennials (i.e., plants that complete their life cycle in two years) such as carrot and sugar beet, these reserves are concentrated in only one or a few roots. We harvest these roots after the first year of growth, before the plant uses the stored energy for vegetative growth and reproduction.

Absorption. Roots absorb large amounts of water and dissolved minerals from the soil. For example, the roots of a corn plant absorb more than 2 liters of water per day.

Conduction. Roots transport water and dissolved nutrients to and from the shoot. The roots of plants such as quillwort (*Isoetes*) and shoreweed (*Littorella*) even transport CO_2 to leaves for photosynthesis. The leaves of these plants usually have thick cuticles and lack stomata.

Roots anchor plants, store reserves, absorb water and dissolved minerals, and transport materials to and from the shoot.

Writing to Learn Botany

Explain how roots can have such large surface areas and yet occupy less than 5% of the soil's volume.

Each of these functions is linked to the unique structure of roots.

Root Tip

Root Cap

The tips of roots are covered by a thimble-shaped **root cap** (fig. 15.4a), which has its own meristem that pushes cells forward into the cap. As they move through the cap, these cells differentiate into elongate *columella cells*, so named because they are arranged in longitudinal columns (fig. 15.4b). Columella cells each contain 15–30 amyloplasts that sediment in response to gravity to the lower side of the cell (fig. 15.4b). Many botanists suspect that this sedimentation of amyloplasts is how roots perceive gravity (although other botanists now question this interpretation; see Chapter 19). Besides protecting the growing root tip and its meristem, the root cap senses light and pressure exerted by soil particles. There is no structure in shoots that corresponds to a root cap.

Within 2–3 days, continued cellular divisions in the root-cap meristem push columella cells to the periphery of the root cap, where they differentiate into *peripheral cells* (fig. 15.4b). Thus, the root cap is in constant flux: new cells constantly move through the cap and replace the thousands of peripheral cells that are shed from the cap as the root pushes its way through the soil. The root caps of corn shed as many as 10,000 peripheral cells per day.

The peripheral cells of the root cap and the epidermal cells of the root produce and secrete large amounts of **mucigel,** a slimy substance made by dictyosomes (fig. 15.5; also see photo that opens this chapter). Mucigel is a hydrated polysaccharide containing sugars, organic acids, vitamins, enzymes, and amino acids. A root weighing 1 g can secrete as much as 100 mg of mucigel per day. Although this rate of secretion may seem rather insignificant, the total amount of mucigel secreted by an actively growing group of plants can reach impressive proportions: for example, the roots of one hectare (10,000 m^2) of corn secrete more than 1,000 m^3 of mucigel during a growing season—enough to fill a typical two-story, four-bedroom house.

Mucigel has several important functions:

Protection. Mucigel protects roots from desiccation and contains compounds that diffuse into the soil and inhibit growth of other roots. For example, the mucigel of giant foxtail decreases the growth of nearby corn roots by 35%.

Lubrication. Mucigel lubricates roots as they force their way between soil particles.

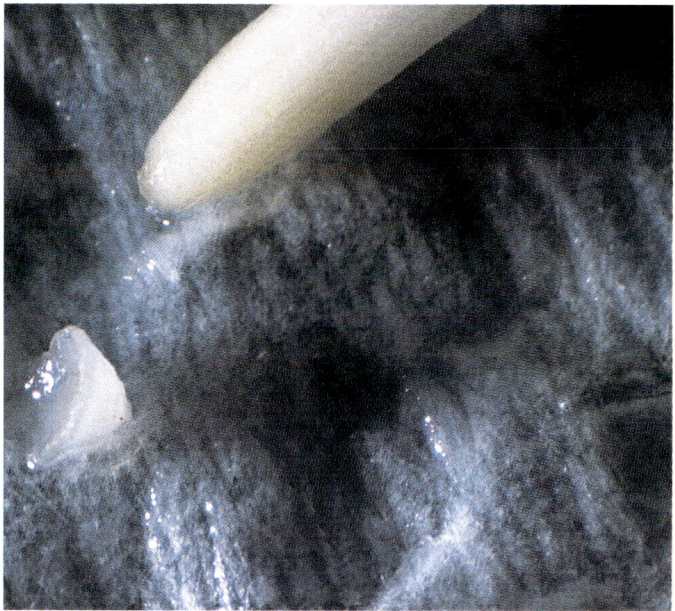
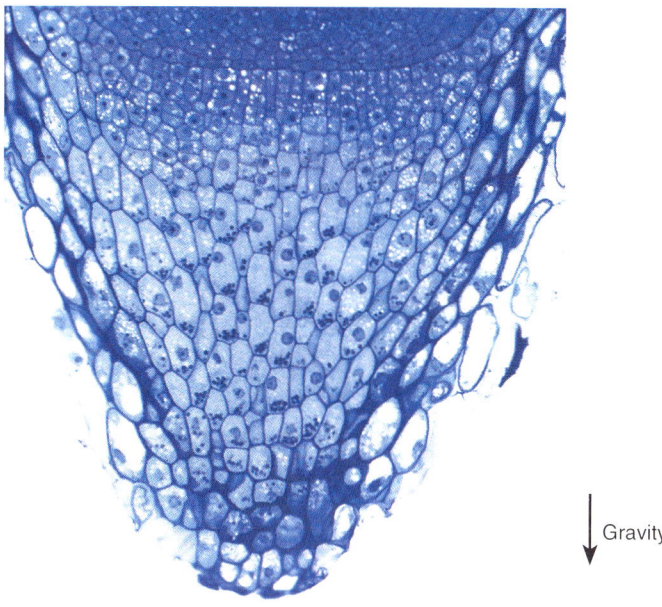

A.   B.

FIGURE 15.4

(a) Root tip with its thimble-shaped root cap removed. (b) Light micrograph of cap of primary root of corn (*Zea mays*). The cells in the center of the cap are arranged in columns and are called *columella* cells. Note the dense amyloplasts at the lower side of the columella cells; the sedimentation of amyloplasts may be involved in the perception of gravity by roots. Surrounding the columella cells are peripheral cells. These cells secrete mucigel, which lubricates the root tip and eases growth through the soil (also see photo at the beginning of this chapter).

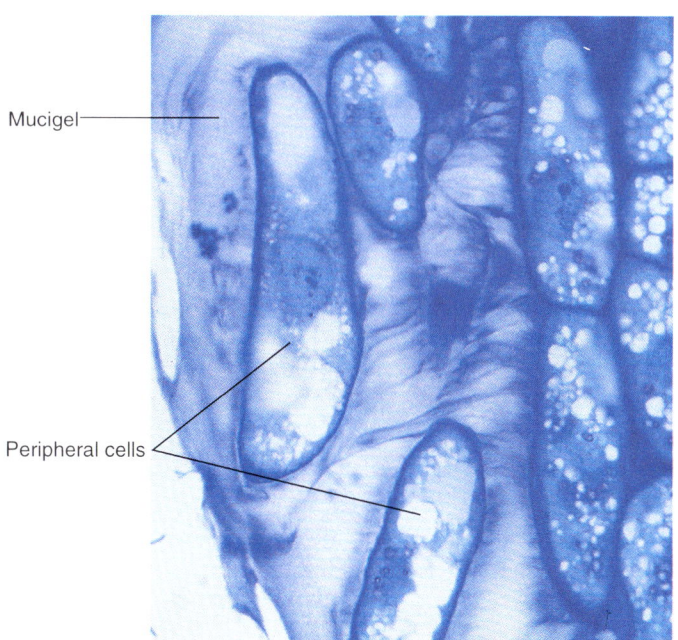

FIGURE 15.5

The outermost peripheral cells of root caps secrete mucigel, which surrounds them in this micrograph.

Water absorption. Soil particles cling to mucigel, thereby increasing the root's contact with the soil. The water-absorbing properties of mucigel help maintain the continuity between roots and soil water.

Nutrient absorption. Carboxyl groups in mucigel influence ion uptake, and organic acids in mucigel make certain ions more available to plants. Also, fatty acids, lectins, and sterols in mucigel may help establish beneficial symbioses with soil microbes. For example, nitrogen-fixing bacteria such as *Azospirillum* are attracted to and inhabit the mucigel of plants such as corn. Each gram of mucigel contains enough nutrients to feed about 2×10^8 (36 mg) hungry bacteria.

Quiescent Center

Just behind the root cap is the **quiescent center,** which consists of 500–1,000 seemingly inactive cells (fig. 15.6). These cells are typically arrested in the G_1 phase of the cell cycle and divide only about once every 15–20 days (in comparison, cells of the adjacent meristem divide more than once per day). Quiescent and meristematic cells are differentially sensitive to environmental perturbations such as radiation. For example, meristematic cells stop dividing when exposed to intense X rays, while quiescent cells are unaffected by radiation and soon begin

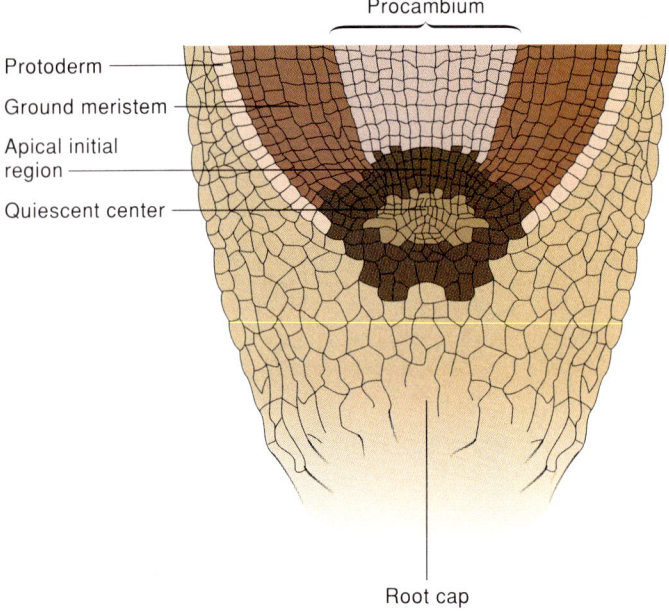

FIGURE 15.6

Diagram of median longitudinal section through a hypothetical root. Cells in the quiescent center divide much less frequently than do other cells in the root tip. The protoderm, ground meristem, procambium, and root cap are derived ultimately from the apical initials.

dividing to re-form the meristem. Thus, cells of the quiescent center are not inherently quiescent; rather, they function as a reservoir to replace damaged cells of the meristem. The quiescent center is important because it organizes the patterns of primary growth in roots.

Tips of roots are covered by a root cap, which produces mucigel that protects, lubricates, and helps absorb materials from the soil. Tips of roots are covered by a root cap, which senses environmental stimuli such as gravity. The quiescent center is located just behind the root cap and is made of seemingly inactive cells. It replaces damaged cells of the adjacent meristem.

Subapical Region

The subapical region of roots has traditionally been divided into three regions: the zones of cellular division, cellular elongation, and cellular maturation (fig. 15.7). Although such divisions are useful for teaching, these regions intergrade and are not sharply defined. Moreover, they do not always accurately describe what is happening in a particular region of the root. For example, cells of some tissues elongate in the zone of cellular division, while those of others mature in the zone of cellular elongation. Concentrate on how structure correlates with function as we examine each of the subapical regions of a root tip.

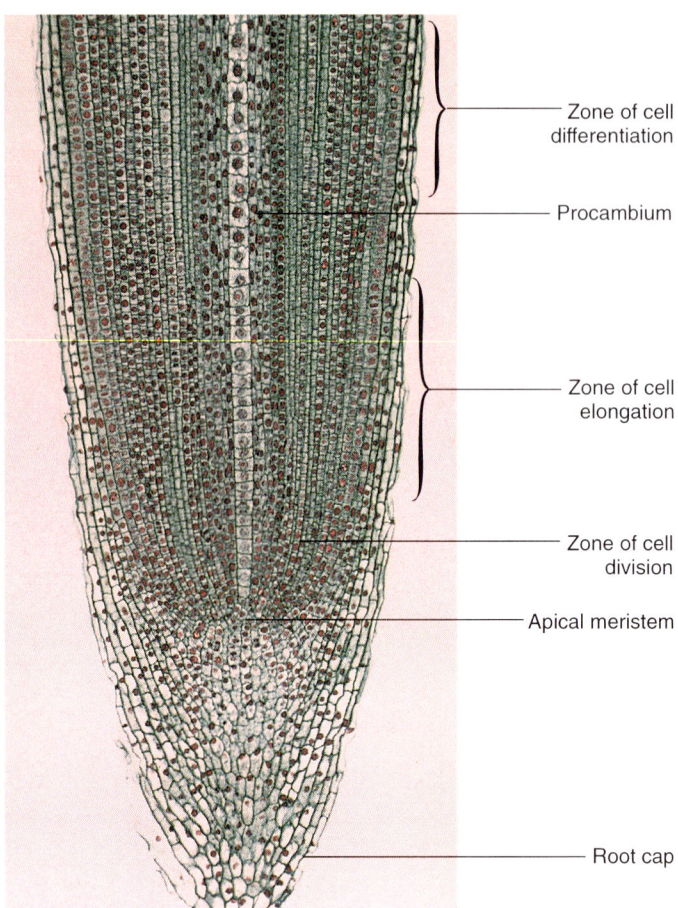

FIGURE 15.7

The subapical region of roots includes the zone of cellular division, zone of cellular elongation, and zone of cellular differentiation. Although these designations are useful for teaching, the regions intergrade and seldom describe what occurs in a particular part of a root.

Zone of Cellular Division

Surrounding the quiescent center is a dome-shaped apical meristem located 0.5–1.5 mm behind the root tip (fig. 15.6). This is the first of four unique features of roots as compared to shoots, namely, that *the apical meristem of a root is subterminal*. This meristematic region is the **zone of cellular division** and is made of small (diameter = 10–20 μm), densely cytoplasmic cells. Meristematic cells in roots divide every 12–36 hours; in some plants, the meristem produces almost 20,000 new cells each day. Divisions rarely occur past 1 cm behind the root tip.

Zone of Cellular Elongation

The **zone of cellular elongation** occurs 4–15 mm behind the root tip (fig. 15.7). Cells in this zone elongate by as much as 150-fold, primarily by filling their vacuoles with water. Thus, this zone is easily distinguished from the root cap and zone of cellular division by its long, vacuolate cells. Cellular elongation

A.

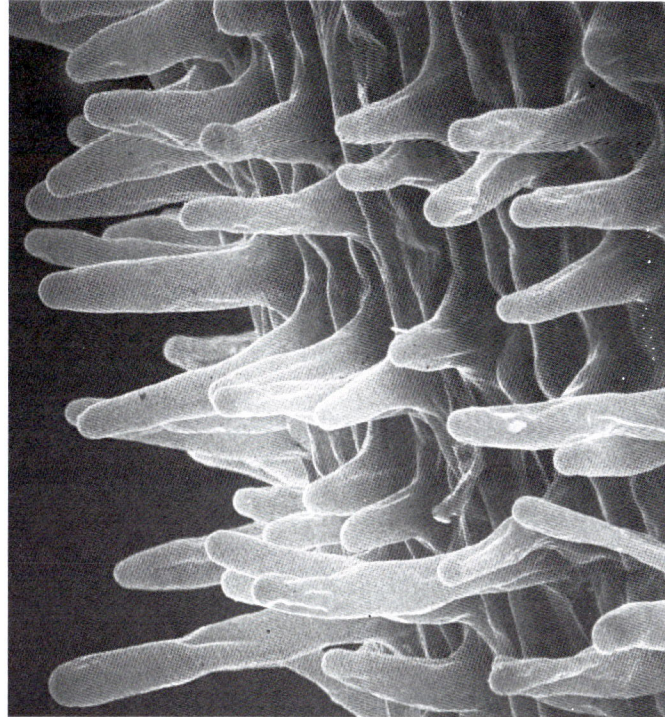
B.

FIGURE 15.8

Root hairs on a primary root of radish (*Raphanus sativus*). Root hairs, which are extensions of epidermal cells, greatly increase the absorptive surface area of the root. The photo shown in (b) is a scanning electron micrograph, ×840.

in the elongating zone shoves the root cap and apical meristem through the soil at rates as high as 4 cm per day (secondary roots grow progressively slower: 0.05–0.5 cm per day). Cells behind the elongating zone do not elongate.

Zone of Cellular Maturation

Cellular elongation typically begins the process of cellular differentiation. Differentiation is completed in the **zone of cellular maturation,** which occurs 1–5 cm behind the root tip (fig. 15.7). The maturation zone is easily distinguished by the presence of many ephemeral root hairs—as many as 40,000 cm^{-2}. Root hairs, which increase the absorptive surface area of the root several thousandfold, are usually less than a millimeter long (fig. 15.8). In most plants they form from asymmetric divisions of the protoderm (see p. 273) and usually live only a few days, with old hairs farthest from the tip constantly being replaced by new ones closer to the tip.

Root hairs form only in the maturing, nonelongating region of the root. This gives root tips a simpler structure than the tips of shoots and accounts for the second unique feature of roots: *roots have no lateral appendages at their tips*. Because root hairs are fragile extensions of epidermal cells, they usually break off when plants are transplanted. Therefore, do not wash off the adhering soil when you transplant your plants.

CONCEPT

The subapical region of roots can be divided into the zones of cellular division, elongation, and maturation. The zone of cellular division includes the root apical meristem, which is subapical. Roots elongate in the zone of cellular elongation, and cellular differentiation is completed in the zone of maturation. Root hairs form in the zone of cellular maturation and greatly increase the root's absorptive surface area. Roots have no lateral appendages at their tips.

Mature Region

Primary tissues differentiate in or distally (i.e., "behind") to the zone of cellular maturation. A cross section through this region gives us our first look at the primary structure of a root.

Epidermis

The root is surrounded by an *epidermis*, which is usually one cell thick. Epidermal cells differentiate from protoderm and usually either lack a cuticle or have a thin cuticle that does not significantly affect water absorption. The epidermis covers all of the root except the root cap and usually lacks stomata.

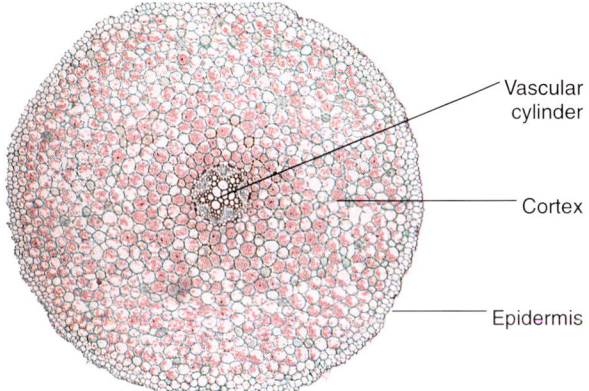

FIGURE 15.9

Cross section of a root of buttercup (*Ranunculus* sp.). At this magnification, the most obvious tissues are the epidermis, cortex, and the vascular tissues of the vascular cylinder.

Cortex

Just interior to the epidermis is the *cortex*, which is formed by the ground meristem. The cortex usually occupies the largest cross-sectional area of a root (fig. 15.9) and consists of three concentric layers: the hypodermis, storage parenchyma cells, and endodermis.

Hypodermis

In many plants, the outermost layer(s) of the cortex is a suberized, protective layer called the *hypodermis* (fig. 15.10), which is usually most prominent in roots growing in arid soil or near the soil's surface. Hypodermal cells are lined with suberin and complete their differentiation well behind the root-hair zone. Suberin in the hypodermis slows the outward movement of water and dissolved nutrients by 200-fold, thereby helping roots retain water and nutrients that they've absorbed.

Storage Parenchyma Cells

Most of the cortex consists of thin-walled *storage parenchyma cells* (figs. 15.9; 15.10). These cells often contain starch and are separated by large intercellular spaces that can occupy as much as 30% of the root's volume.

Endodermis

The innermost layer of the cortex is the *endodermis*, and differentiates 5–8 mm from the root tip. Unlike other cortical cells, endodermal cells are packed tightly together and lack intercellular spaces. Furthermore, their radial and transverse walls are impregnated with a *Casparian strip* made of lignin and suberin and arranged similarly to a rubber band around a rectangular box (fig. 15.11). This may better help you visualize the structure of the endodermis: if endodermal cells were likened to bricks in a brick wall, then the Casparian strip would be analogous to the mortar surrounding each brick. The Casparian strip differentiates around cells approximately 1 mm from the apical meristem, which is just prior to where root hairs form. It is also

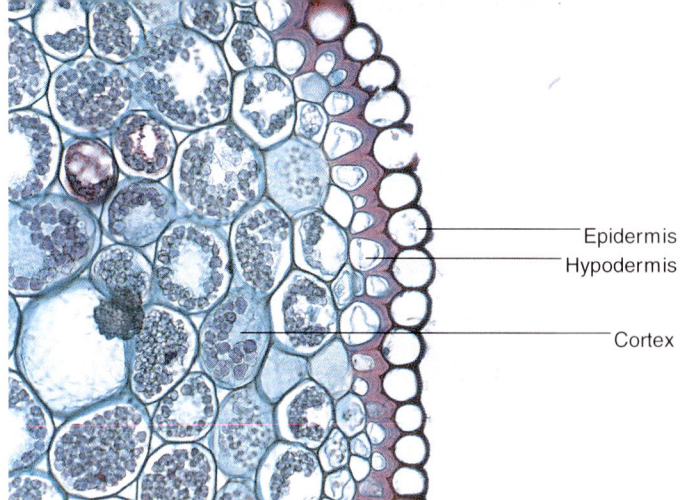

FIGURE 15.10

Cortical cells contain many amyloplasts (stained darkly in this micrograph) and are usually separated by intercellular spaces. In many plants, the outermost layer of the cortex is specialized as a hypodermis, which retards loss of water and dissolved nutrients. Surrounding the root is the epidermis.

tightly fused with the plasmalemmas of endodermal cells, thereby preventing the inward movement of water and nutrients through the cell wall and intercellular space. Thus, water and dissolved minerals must pass through the plasmalemmas of endodermal cells to reach the vascular tissues of the root (fig. 15.12). As a result, the endodermis functions somewhat as a valve that regulates the movement of nutrients into the vascular tissue via the **symplast** (i.e., membranes and living cells). This arrangement is critical for helping to eliminate leaks and for conserving ions in the vascular tissue. No such control can be exerted over transport through the **apoplast** (i.e., through cell walls and intercellular spaces).

Suberin is deposited continually in endodermal cells as a root ages. Beyond 5 cm or so from the root tip, suberin completely blocks the movement of water and solutes through the endodermis. Thus, these mature zones of the root absorb little water. Rather, they store reserves and anchor the plant in the soil. Although the precise pathway for exchange between the cortex and vascular tissues in these parts of roots is unknown, most botanists think that it occurs through *passage cells*, which are endodermal cells that do not produce excess suberin. However, this hypothesis is controversial.

C O N C E P T

The epidermis surrounds the mature region of the root. Interior to the epidermis is the cortex, which has three layers: hypodermis, storage parenchyma, and endodermis. The hypodermis protects roots, and storage parenchyma tissue stores reserves for subsequent use. The endodermis is lined with a strip of suberin called the *Casparian strip,* which diverts water and dissolved minerals into the cytoplasm of endodermal cells.

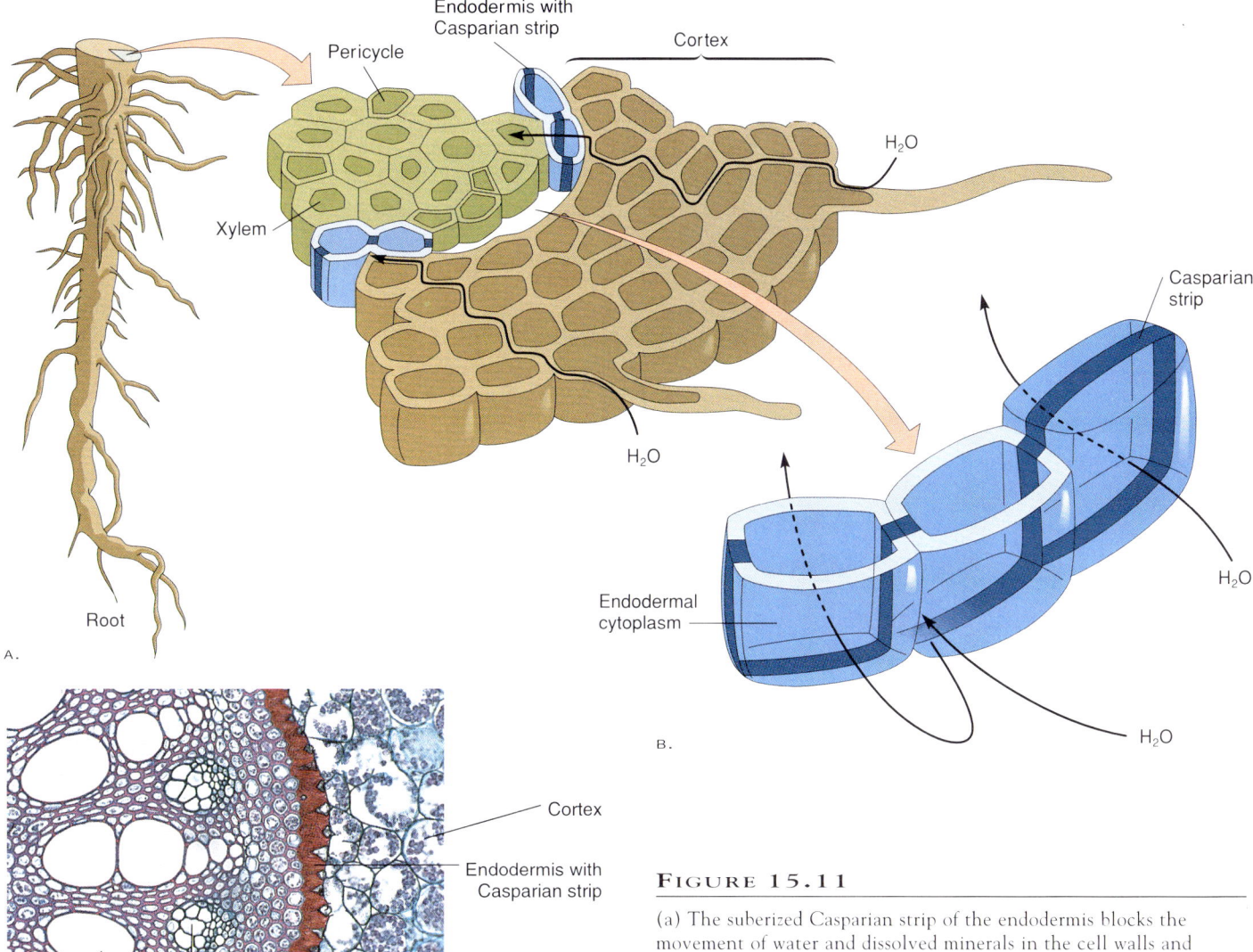

FIGURE 15.11

(a) The suberized Casparian strip of the endodermis blocks the movement of water and dissolved minerals in the cell walls and between cells. (b) As a result, water and dissolved minerals are directed into the cytoplasm of endodermal cells where subsequent use and movement is more controlled. (c) Endodermis of Smilax root. Note the darkly stained suberin of the Casparian strip.

Stele

The *stele* includes all of the tissues inside the cortex. It consists of the pericycle, vascular tissues, and sometimes a parenchymatous pith.

Pericycle

The outermost layer of the stele is the *pericycle,* a meristematic layer of thin-walled parenchyma cells one to several cells thick. The pericycle is important because it produces *branch roots* (also called *secondary* or *lateral roots*). This is the third unique feature of roots: *lateral appendages form endogenously* (fig. 15.13). Branch roots typically form 8–20 mm from the root tip.

The earliest sign of branch-root formation is cellular divisions in the pericycle (fig. 15.13). Soon thereafter, a root cap and primary tissues form, and the endodermis of the parent root becomes meristemtic and forms a sheath over the growing branch root. The branch root finally forces its way through the cortex and epidermis of the parent root, much as it will later force its way through the soil. The vascular tissue and endodermis of branch roots link with those of the parent root. Older branch roots form near the root-shoot junction; young ones form near the root tip.

Each branch root services a small area of the shoot; that is, certain parts of a shoot are targeted for deliveries from specific branch roots. This is best seen when examining plants infected with the fungus *Cephalosporium.* This pathogen typically enters the xylem of branch roots, and soon thereafter symptomatic yellow stripes form in the shoot at the delivery sites of that root.

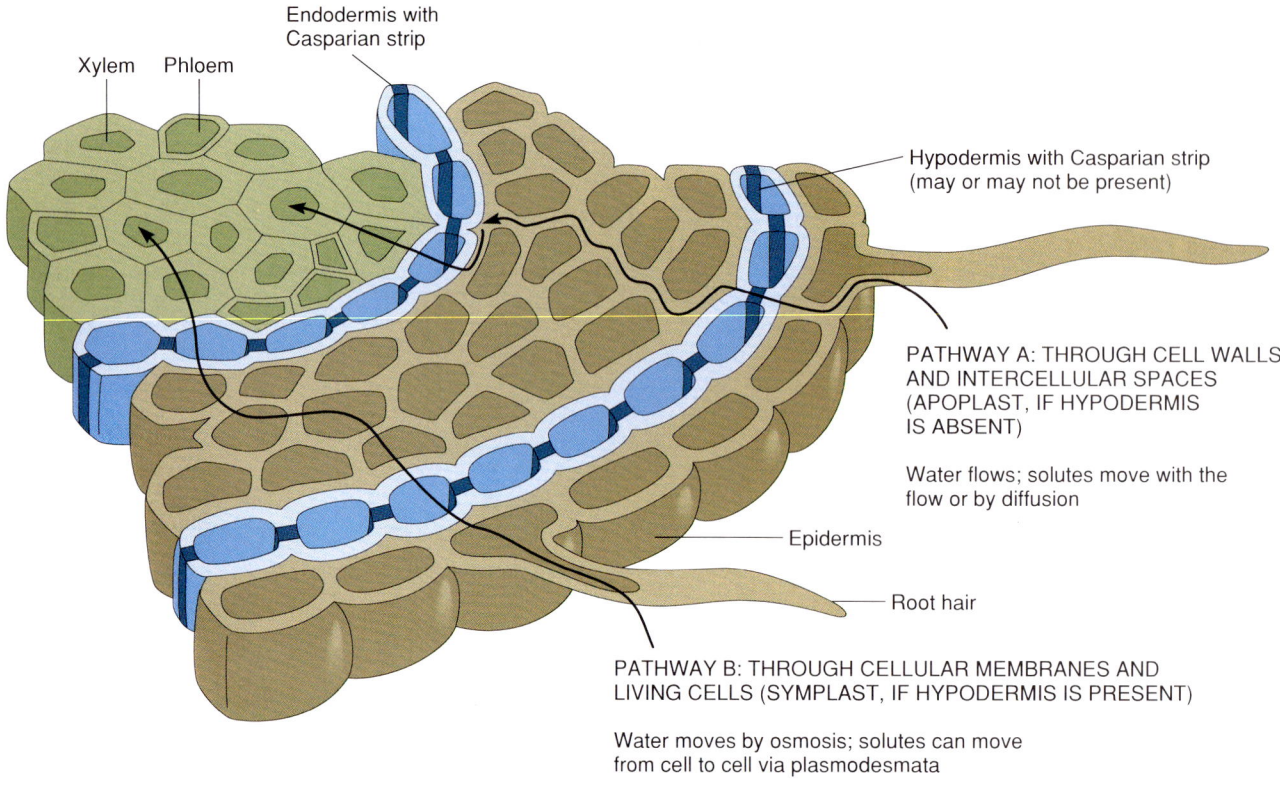

FIGURE 15.12

Diagram of cross section of a root, showing the alternative routes for uptake of water and dissolved nutrients. Note that the suberin of the Casparian strip blocks the movement of water through intercellular spaces and between cells. In roots that have a hypodermis, movement between cells and through cell walls occurs at the outer edge of the cortex, not at the endodermis (i.e., the inner edge) as in roots that lack a hypodermis.

Vascular Tissue and Pith

Inside the pericycle is the root's vascular tissue. The roots of most dicots are **protostelic,** meaning that the procambium forms a lobed, solid core of primary xylem in the center of the root. Roots with two lobes of xylem are called *diarch* (*arch* here means "first," referring to the first cells of the lobe), those with three lobes are called *triarch*, and those with many lobes are called *polyarch* (fig. 15.14). However, the number of lobes of xylem is a poor means of classifying plants because it is highly variable, even in the same plant.

The roots of monocots and a few dicots are **siphonostelic,** meaning that a ring of vascular tissue surrounds a parenchymatous pith (fig. 15.15). In these roots, primary xylem typically forms in isolated rows that radiate toward the periphery of the root. The protoxylem of roots, unlike that in shoots, differentiates centripetally, that is, from the outside toward the inside of the root. As a result, protoxylem forms next to the pericycle on the outer side of the metaxylem (fig. 15.16). This type of differentiation is called *exarch*. Thus, roots have an exarch arrangement of protoxylem, while the arrangement of protoxylem in shoots is *endarch*.

In roots, bundles of primary phloem differentiate between lobes of xylem. This is the final unique feature of roots: *xylem and phloem in roots alternate with each other*. Sieve elements typically differentiate nearer the root tip than do xylary elements. The precise position where vascular tissue differentiates depends on several factors, including the root's growth rate, type, developmental stage, and external influences such as temperature, aeration, and soil texture.

C O N C E P T

The stele includes all of the tissues inside the cortex, including the pericycle and vascular tissues. The pericycle produces branch roots, which form endogenously. Xylem and phloem form in alternating strands interior to the pericycle. The roots of most dicots have a solid core of xylem, while those of most monocots have a parenchymatous pith.

In many dicots a vascular cambium later forms between the xylem and phloem, and produces secondary growth. The pericycle also contributes to secondary growth in the roots of some plants (see Chapter 16).

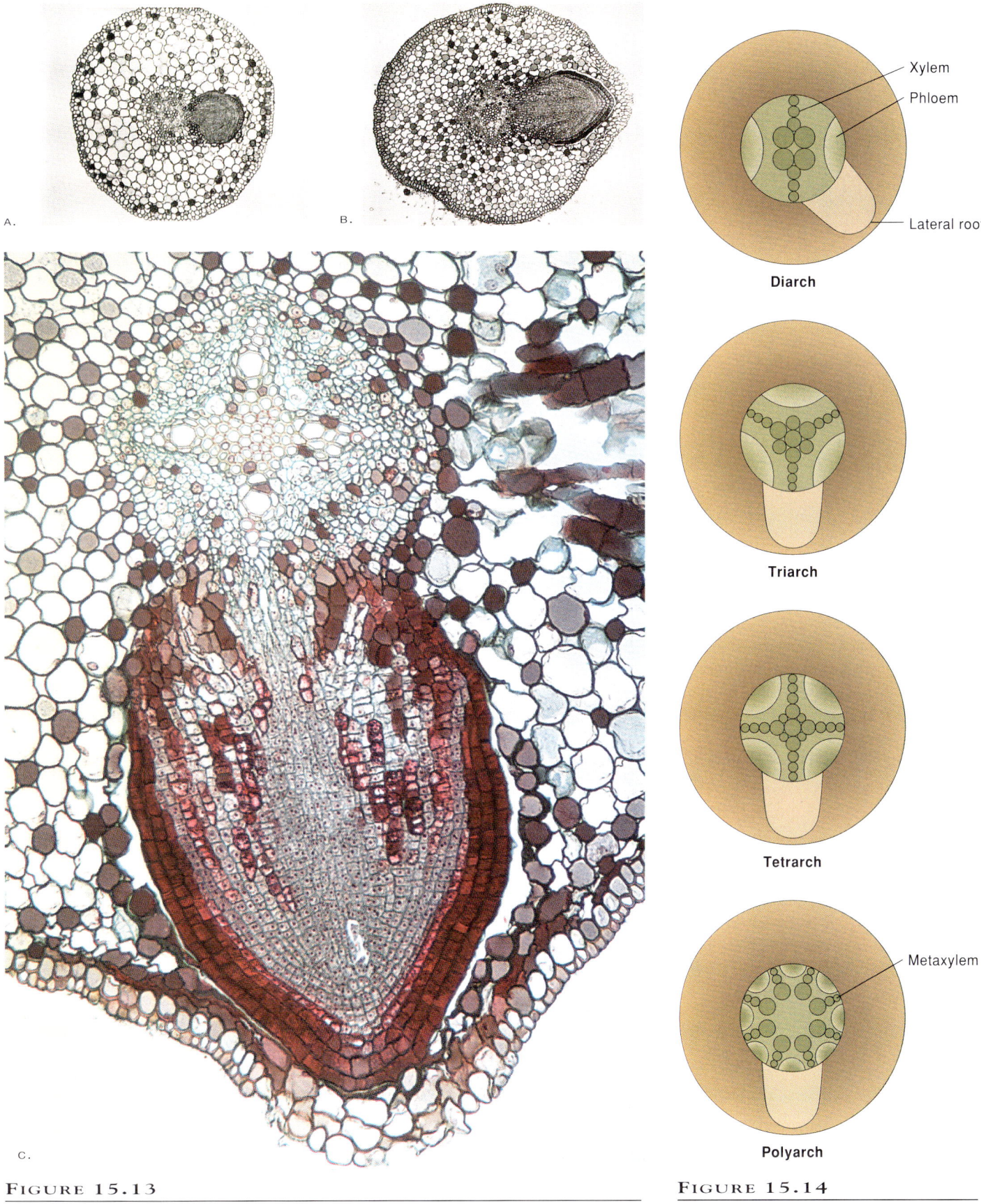

FIGURE 15.13

Formation of lateral roots. Lateral roots are formed by the pericycle, the outermost layer of the stele.

FIGURE 15.14

Different arrangements of xylem in protostelic roots.

CHAPTER FIFTEEN *Primary Growth: Roots*

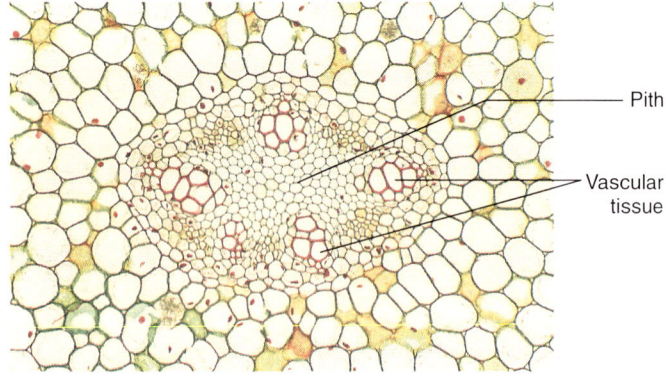

FIGURE 15.15

In siphonostelic roots such as this one of mayapple (*Podophyllum*), a ring of vascular tissue surrounds a parenchymatous pith, ×100.

Transition Region Between the Root and the Shoot

We mentioned in the previous section that vascular tissues in roots and shoots are arranged differently. For example, the vascular tissue in dicot roots is surrounded by a cortex (i.e., there is no pith), while in stems it usually encircles a pith. Moreover, strands of xylem and phloem alternate in roots, but they are arranged opposite each other in the vascular bundles of shoots.

The vascular tissues of roots and shoots join in the **transition region** located between the root and the lower internodes of the shoot. The transition region is different in different plants; for example, it may be abrupt or extend for some distance. Although terms such as inverted, twisted, and rotated are often used to describe the connection between the different vascular systems, cells in the transition region do not move. Rather, the transition between the vascular tissues of roots and shoots is established early in plant development by differentiation of the procambium.

THE ROOT-SOIL INTERFACE

The narrow zone of soil surrounding a root and subject to its influence is called the **rhizosphere.** It extends up to 5 mm from the root's surface and is a complex and ever-changing environment. Growth and metabolism of roots modify the rhizosphere in several ways.

Roots enrich the soil with organic matter. Plants transport as much as 60% of their net photosynthate to roots, which deposit more than 30% of this material in the soil as mucigel and other compounds. In plants such as wheat, the amount of carbohydrate deposited in the soil often exceeds that stored in the plant's fruits.

Roots compress the soil. Roots compress the soil as they force their way through crevices and between soil particles.

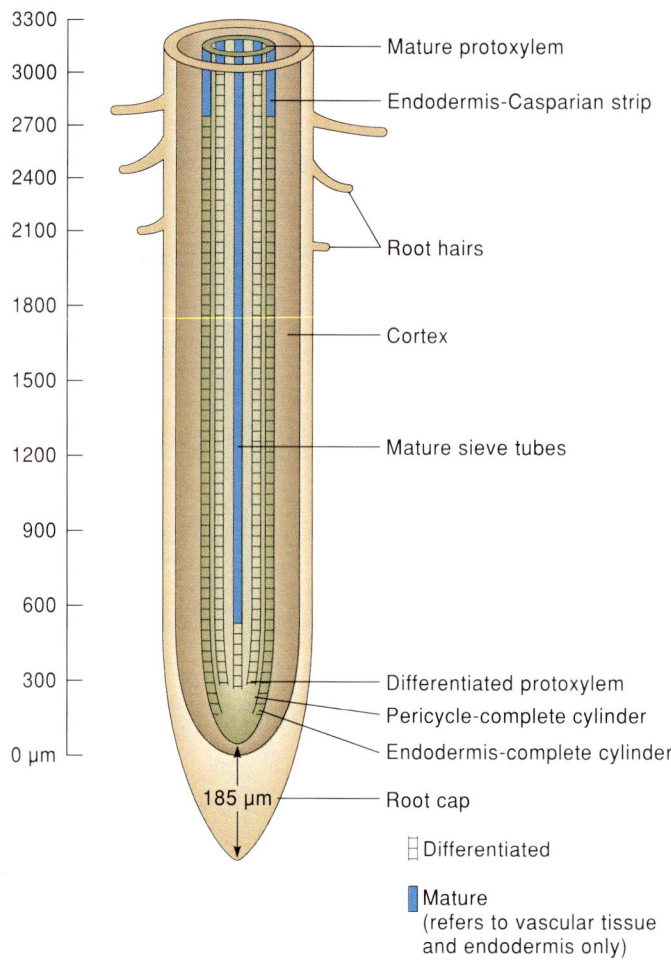

FIGURE 15.16

Diagram of root tip of mustard (*Sinapis alba*), showing where different tissues differentiate relative to the root apex. These data emphasize the artificial nature of labeling discrete zones of cellular division, elongation, and differentiation (see fig. 15.7).

Absorption of nutrients by roots alters the chemical composition of the soil. Roots secrete bicarbonate when they absorb anions such as nitrate (NO_3^-), and secrete H^+ when they absorb cations such as ammonium (NH_4^+). As a result, the pH of the rhizosphere and the surrounding soil can differ by more than 2 pH units. Hydrogen ions alter the availability of other minerals. For example, decreasing the pH increases the availability of aluminum, often to toxic amounts. The respiration of roots decreases the concentration of oxygen and increases the concentration of CO_2 in the rhizosphere.

As a result of these influences, the rhizosphere differs significantly from bulk soil. For example, the rhizosphere usually contains large amounts of energy-rich molecules. These molecules feed microbial floras often exceeding 10^{10} organisms cm^{-3} of soil—populations that are 10–100 times more dense than those in bulk soil.

CONCEPT

The narrow zone of soil surrounding a root and subject to its influence is called the rhizosphere. Roots change the rhizosphere by enriching it, compressing it, and absorbing nutrients from it. Many microbes live in the rhizosphere.

Earlier in this chapter you learned the four primary functions of roots: absorption, anchorage, conduction, and storage. Let's now consider how the structure of a root relates to these four functions (fig. 15.17).

Absorption

Most water and nutrients are absorbed by root hairs in the zone of maturation (fig. 15.8). Water enters the root via two pathways: an apoplastic route consisting of intercellular spaces and the cell wall, and a symplastic route involving movement across the plasmalemma and into the cytoplasm of cells (fig. 15.12). Water and nutrients moving through the epidermis in the apoplast are absorbed by cells in the cortex; these cells form a vast collecting system that gathers water and nutrients from the cell walls and intercellular spaces. Nutrients seeping through the apoplast toward the stele finally encounter the endodermis, which is the primary barrier to absorption. The Casparian strip in the endodermis ensures that water and nutrients enter the stele via the symplast. Most nutrients are absorbed and accumulate in the apical 3–5 cm of the root, where most growth occurs.

Anchorage

Little absorption occurs past a few centimeters beyond the root tip, because these parts of the root lack root hairs and have a heavily suberized endodermis. These nonabsorptive regions of roots anchor plants and may later produce branch roots.

Conduction and Storage

Water and dissolved minerals absorbed by roots move to the shoot in xylary elements. Similarly, roots are heterotrophic and therefore must receive sugars and other compounds from the shoot via the phloem. These nutrients are either used immediately for growth or are stored in cortical cells for future use.

FACTORS CONTROLLING THE GROWTH AND DISTRIBUTION OF ROOTS

We do not know as much about roots as we do about shoots, primarily because roots grow underground and are more difficult to study than shoots. Much of what we know comes from studies of potted plants or seeds germinated in artificial environments such as sterile dishes or moist paper towels. However, there is little evidence that these roots behave like those growing in field conditions. Understanding how roots grow in their natural environment requires much unusual work; some of it has been ingenious as well as tedious, and has involved methods ranging from underground cameras and radioactive tracers to painstaking excavations. The newest and most elaborate methods for studying roots involve laboratories called *rhizotrons*, which are underground walkways with glass walls (fig. 15.18a). As they grow, many roots press against these glass walls and can be easily observed (fig. 15.18b). Observations in rhizotrons have been supplemented with other studies to reveal that the growth and distribution of roots are controlled by several factors.

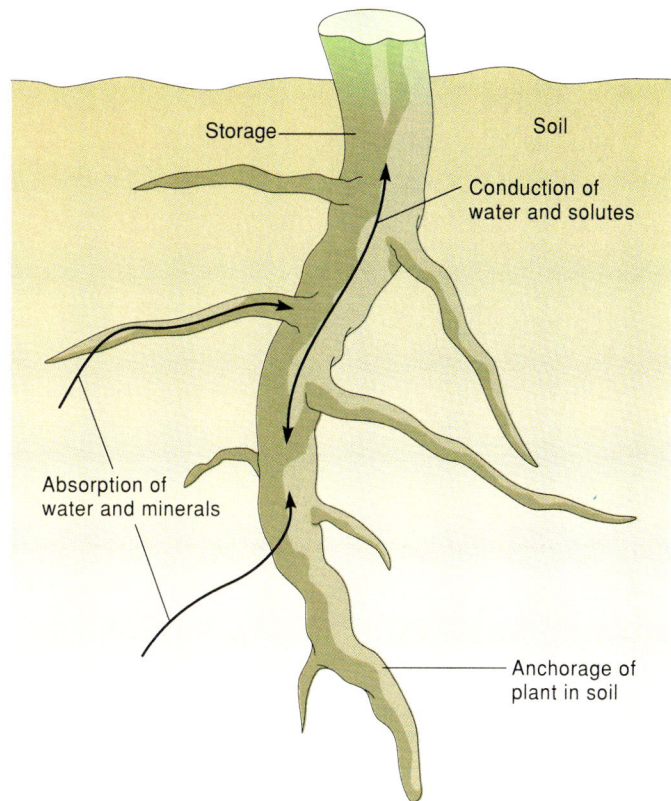

FIGURE 15.17

Diagram summarizing the major functions of roots.

Temperature

Roots usually become dormant when it gets cold. Dormancy involves sealing the root until warmer temperatures and more favorable growing conditions return.

Other Organisms

Most microbes in soil live near roots. These microbes secrete compounds that affect the growth and distribution of roots. Equally important, they increase the uptake and translocation of minerals from the soil. Plants growing in sterile soil absorb fewer minerals than do those whose rhizospheres include microbes.

Competition for water and nutrients in soil is often fierce; indeed, a single gram of fertile soil contains approximately 10^9 bacteria, 10^6 actinomycetes, 10^5 fungi, 10^3 algae, and several millimeters of roots. Although we do not understand what characteristics enable the roots of some species to outcompete those of others, different plants have evolved different strategies for competing with their neighbors:

A. B.

FIGURE 15.18

Use of rhizotrons to study root growth and development. (a) Interior of the rhizotron at Muscle Shoals, Alabama. (b) Roots of corn (*Zea mays*) growing against an observation window in a rhizotron. Note the extensive nature of the root system.

Plants produce many roots. Plants must produce extensive root systems that locate water and minerals dispersed in the soil. For example, in only six weeks a corn seedling produces more than 2,000 roots, a 100-year-old Scotch pine (*Pinus sylvestris*) more than 5 million roots, and a mature red oak (*Quercus rubra*) more than 500 million roots.

Roots permeate the soil. Roots of plants such as salt cedar (*Tamarix*) and *Retama* spread tens of meters and burrow more than 30 m deep. The taproot of a 12-m oak tree goes down more than 5 m, while its branch roots span more than 40 m. Most roots grow throughout the upper 3 m of soil where nutrients are most abundant.

Roots of different plants often grow in different zones of the soil. Plants minimize competition by growing into different areas of the soil. Consider the root systems of mesquite (*Prosopis*) and a saguaro cactus (*Carnegiea gigantea*), two plants that grow in dry environments. Mesquite produces long taproots that obtain water from deep underground. As a result, mesquite grows as a mesophyte in arid environments. Saguaro cacti survive in the same environment by producing an extensive mass of shallow roots that spread as far as 30 m. These roots maximize water absorption after infrequent rains.

How roots protect themselves from other organisms. Since they usually grow underground, roots are protected from many of the herbivores, winds, and lawnmowers that plague stems and leaves. Thus, it's not surprising that roots have fewer obvious adaptations for protection than do shoots. Most of a root's defenses against soil pathogens are chemical rather than structural. Roots often secrete phytoalexins and other obnoxious chemicals that inhibit the growth of pathogens and other organisms.

Light

Light inhibits root growth in corn, wheat, peas, and rice. Light is sensed by the root cap and inhibits growth by slowing the rates of cellular division and elongation.

Gravity

Different roots respond differently to gravity. For example, primary roots grow down; that is, they are *positively gravitropic* (fig. 15.19). Branch roots growing out of primary roots often do not respond to gravity, and grow in whatever direction they happen to diverge from the primary root. In plants such as castor bean (*Ricinus communis*), branch roots grow laterally for several centimeters, after which they become graviresponsive and grow down. As a result of this differential responsiveness to gravity, roots permeate the soil at various angles and efficiently absorb its water and dissolved nutrients.

Genetic Differences

The growth and distribution of roots are also controlled genetically. For example, plants such as locoweed (*Astragalus*) always grow deep taproots, whereas most grasses produce shallow, fibrous root systems. Similarly, the root systems of corn are denser near the soil surface than those of soybeans (*Glycine max*) regardless of the type of soil they are growing in.

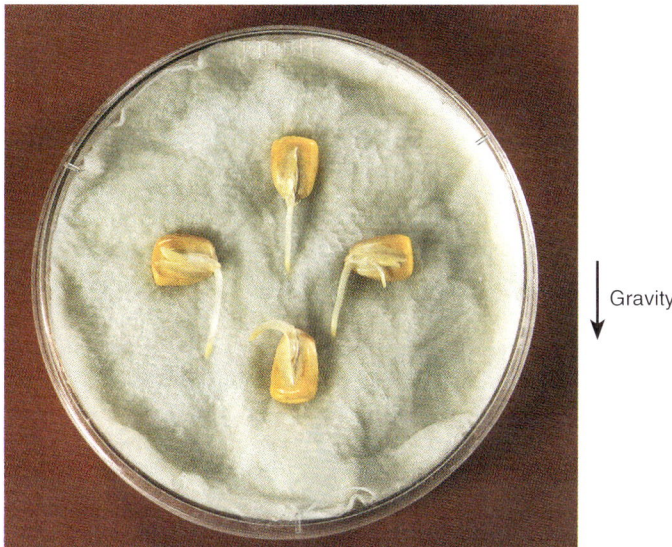

FIGURE 15.19

Primary roots of corn (*Zea mays*) are positively gravitropic; that is, they grow downward. They do so whether the grain germinates in a vertical, horizontal, or inverted position.

Stage of Plant Development

Prior to and during fruit formation, most of a plant's resources are used for shoot rather than root growth. Thus, root growth typically slows during flowering and fruit formation. This shift in the allocation of resources is most obvious in some monocots, where much of the root system dies before harvest.

Soil Properties

Soil Texture

Roots usually grow best in loosely packed soil. For example, roots of wheat grow four times faster in loose sand than in tightly packed clay. Roots that grow in tightly packed soil are usually shorter and thicker than those that grow in loosely packed soil.

Moisture and Air

Most roots grow best in moist but not wet soil. Similarly, roots usually grow deeper in moist, aerated soil than in soaked, poorly aerated soil. The slow growth of roots in poorly aerated soil may also be due to the accumulation of the plant hormone ethylene, which diffuses 10^4 times more slowly from roots in flooded soil than in well-aerated soil. As a result, ethylene accumulates in flooded soil and slows root growth. This ethylene also stimulates the formation of aerenchyma tissue that improves aeration in the root. We'll discuss other effects of ethylene in Chapter 18.

The roots of hydrophytes (i.e., plants that live in wet areas) are usually small and modified for gas exchange rather than water absorption. They contain relatively small amounts of xylem, usually lack root hairs, and are often filled with aerenchyma, which improves ventilation. Conversely, the roots of xerophytes (i.e., plants that live in arid areas) are often extensive and modified for rapid transport of water. For example, they often have a thin cortex, which shortens the distance between the soil and vascular tissue. They also contain large amounts of well-developed xylem that allows them to move water rapidly to the shoot after a short rain.

Nutrients

Roots tend to proliferate in pockets of nutrient-rich soil.

Other Factors

Root growth is also influenced by the presence of other roots, the pH of the soil, and the amount of ethylene in the soil. Ethylene levels in some soils can reach 10 ppm, which would probably slow root extension.

CONCEPT

The growth and distribution of roots is controlled by temperature, other organisms, light, gravity, genetic differences, the stage of plant development, and properties of the soil. Plants compete for the soil's nutrients by producing many roots, by permeating the soil, and by growing roots in places different from those of other plants. Most of a root's defenses are chemical rather than structural.

The Lore of Plants

Some epiphytes can only be cultivated in acidic soil (pH 4.0; that's about the pH of tomato juice). This requirement is due to the adaptation of these plants to form roots in arboreal ant nests, which are very acidic environments.

THE GROWTH OF ROOTS VS. SHOOTS

All of the preceding factors strongly influence the major management problem faced by a plant, namely, how best to allocate its resources between roots and shoots. Plants must produce shoots to intercept light for photosynthesis, but efficient photosynthesis requires an equally efficient root system for gathering water and nutrients from soil. Clearly, there is a trade-off, and efficient growth requires a mechanism that effectively allocates energy for the growth of roots vs. shoots. Botanists study this problem in a variety of ways, including by determining the ratio of the weights of roots and shoots in plants subjected to differing environmental conditions. This *root-shoot ratio* tells us how the environment affects the growth of roots and shoots. For example, the root-shoot ratio is relatively large for seedlings, and decreases gradually as a plant ages. Decreasing the amount of nutrients available to roots decreases the root-shoot ratio, as does diminishing light. That is, roots are affected more strongly by nutrients and light than are shoots. If stressed, most plants preferentially route energy and materials to their shoots.

MODIFIED ROOTS

Like stems and leaves, roots are often modified for special functions (fig. 15.20).

Storage

In plants such as beets (*Beta vulgaris*), turnips (*Brassica rapa*), radish (*Raphanus sativus*), dandelion (*Taraxacum officinale*), carrot (*Daucus carota*), and cassava (*Manihot esculenta*), roots store large amounts of starch. The roots of other plants store carbohydrates as sugars; for example, the roots of sweet potato (*Ipomoea batatas*) contain 15%–20% sucrose. Roots can also store large amounts of water; the taproots of some desert plants store more than 70 kg of water.

Propagation

The roots of cherry (*Prunus*), pear (*Pyrus*), apple (*Malus*), and teak (*Tectona grandis*) produce adventitious buds, which form aerial shoots called (rather disrespectfully) *suckers*. When separated from the parent plant, suckers become new individuals. Adventitious buds are a common means of propagating many other plants. For example, most groups of creosote bushes (*Larrea tridentata*) are clones derived from a single plant. Some of these clones are more than 12,000 years old; this means that the first seed germinated approximately 4,000 years before humans began writing.

The most massive organism in the world is believed to be a quaking aspen (*Populus tremuloides*) that has grown thousands of suckers from the same root system. This plant, which grows in the Wasatch Mountains of Utah, consists of more than 47,000 tree trunks, each with the usual complement of leaves and branches. It covers almost 43 hectares and its mass has been estimated to be about 6 million metric tons. The discoverers of this aspen have named it *Pando*, a Latin word meaning "I spread."

Aeration

Many plants grow in stagnant water and mud that contains less than 3% of the oxygen in air. Plants such as black mangrove (*Avicennia germinans*) that grow in these environments avoid suffocation by producing specialized roots called *pneumatophores* that import oxygen from the atmosphere. Pneumatophores contain as much as 80% aerenchyma and grow up into the air, where they function like snorkels through which oxygen diffuses to submerged roots.

Movement

Contractile roots are common on corm and bulb-forming plants such as lily and *Gladiolus*, as well as on nonbulbous plants such as ginseng (*Panax*) and dandelion. These roots contract by shrinking their cortical cells, which twists the xylary elements into a corkscrew, wrinkles the surface of the root, and contracts it by as much as 3 mm d^{-1}. As a result, contractile roots can shrink more than 50% in only a few weeks. Contractile roots are important because they can pull a plant into the more stable environment of deeper soil.

Nutrition
Parasitism

Plants such as witchweed (*Striga*) and broomrape (*Orobanche*) use their roots to parasitize other plants. Losses caused by these parasitic plants often devastate a crop (see box 15.1 "Finding A Host").

Mycorrhizal Roots

The epidermis and cortex of many roots are associated with beneficial fungi. Such associations between a root and a fungus occur throughout the plant kingdom and are called **mycorrhizae**. In *ectomycorrhizal roots* of plants such as orchids and maple (*Acer*), the fungus produces a mass of filaments on the surface of the root. These filaments invade the root and form an extensive netlike structure between the cortical cells. In *endomycorrhizal roots* of plants such as pine (*Pinus*) and oak (*Quercus*), the fungus forms inconspicuous filaments on the root's surface and invades cortical cells. Mycorrhizal roots often lack root hairs, suggesting that the fungi replace the absorptive functions of root hairs.

Mycorrhizae are a type of *mutualism*, meaning that both the plant and the fungus benefit from the association. The fungus absorbs nutrients from the soil, while the host plant provides the fungus with carbohydrates, amino acids, vitamins, and other organic substances. Plants infected with mycorrhizal fungi tolerate drought and other types of stress better than uninfected plants.

Mycorrhizae are a good "investment" for a plant, since growing disease-resistant fungal filaments into the soil to increase rooting density requires less energy than does producing new roots.

FIGURE 15.20

Modified roots. (a) Nodules on the roots of these soybeans (*Glycine max*) consist of cortical cells infected with nitrogen-fixing bacteria. The red color of the nodules is due to leghemoglobin, a pigment that is similar to the hemoglobin in red blood cells of vertebrates. Much of the nitrogen fixed by the bacteria is used by the host plant for growth. (b) Prop roots on corn (*Zea mays*) are adventitious roots that arise from the stem. Prop roots help support the plant. (c) Adventitious roots of Boston ivy (*Parthenocissus tricuspidata*) can penetrate tiny crevices of a supporting surface. These so-called climbing roots help a plant climb objects, thereby increasing the plant's chance of intercepting light for photosynthesis. (d) Some tropical trees produce planklike buttress roots at the base of their trunk. Buttress roots help stabilize and support the tree.

BOXED READING 15.1
FINDING A HOST

Witchweed (*Striga asiatica*) is a red-flowered parasite that uses its roots to infect cereals such as corn and sorghum. *Striga* infections can be devastating; for example, *Striga* reduces yields by as much as 70% and is now the leading cause of famine in Africa.

The key to survival for the parasite's seedlings is to find a host quickly. To solve this problem, its seeds typically germinate only in the presence of a signal released by a host. For example, a hydroquinone released by sorghum roots triggers germination of *Striga* seeds that are within a few millimeters of sorghum roots; the seeds then grow toward increasing concentrations of the hydroquinone and, in the process, toward their nearby meal ticket. Seeds farther from host roots do not germinate because the host's hydroquinone is transformed by oxygen into an inactive compound that does not stimulate germination. Strigol, another compound that is released by sorghum plants and stimulates germination of *Striga* seeds, is active at concentrations as low as 10^{-15} M.

BOX FIGURE 15.1
This red-flowered *Striga* is parasitizing a corn (*Zea mays*) plant. Worldwide, *Striga* causes millions of dollars worth of damage to crops.

Doing Botany Yourself

Design an experiment to show how mycorrhizae might be mutually beneficial both to the fungus and to its host plant.

Nodules

In the roots of legumes, cortical cells are often infected with *Rhizobium*, a nitrogen-fixing bacterium. This bacterium infects the root through root hairs and, once inside, forms infection threads that permeate the root. Localized swellings in response to these infections are called **nodules** (fig. 15.20a) and represent another type of mutualism between roots and microbes. Bacteria receive carbohydrates and other substances from the host, while the host plants receive nitrogen-containing ions from the bacteria. You'll learn more about plant nutrition in Chapter 20.

Aerial Roots

Aerial roots are adventitious roots that are formed by above-ground structures such as stems. In plants such as mangrove, they account for more than 25% of the above-ground biomass. Aerial roots have different functions in different kinds of plants:

Water retention. The multiple epidermis of roots of some epiphytic orchids, called a **velamen** (fig. 13.17), was originally thought to be a modification for absorbing water, but this conclusion was premature. Subsequent studies have shown that a velamen is almost impermeable to water. The primary role of the velamen appears to be mechanical protection and the *retention* of water in the root.

Photosynthesis. In plants such as *Philodendron*, vanilla orchid (*Vanilla planifolia*), and several aquatic plants, the roots are photosynthetic. In orchids such as *Microcoelia smithii*, aerial roots are flat and green, and look remarkably like shoots.

Support. *Prop roots* are supportive aerial roots that grow into the soil; they are common in plants such as banyan trees (*Ficus benghalensis*) that grow in mud flats, as well as in plants such as corn (fig. 15.20b). Banyan trees may produce thousands of prop roots that grow down from horizontally oriented stems and form pillarlike supports. Similarly, roots of ball moss (*Tillandsia recurvata*), Spanish moss (*Tillandsia usneoides*), and English ivy (*Hedera helix*) anchor the plant's shoot to its substrate (fig. 15.20c).

Shallow-rooted tropical trees such as *Khaya* and fig (*Ficus*) produce remarkable *buttress roots* at the base of their trunks to keep from blowing over (fig. 15.20d). These planklike roots are often more than 4 m high and are specialized for support: they contain large amounts of fibers and relatively small amounts of xylary elements.

CONCEPT

Modified roots are used for storage, propagation, aeration, movement, parasitism, nutrition, and support.

THE ECONOMIC IMPORTANCE OF ROOTS

Roots have long been among our favorite foods. For example, we've cultivated carrots (*Daucus carota*) for more than 2,000 years, and sugar beets represent a common source of the sugar that appears on our tables. Similarly, horseradish (*Rorippa armoracia*), sweet potatoes (*Ipomoea*) and turnips appear regularly on our kitchen tables. Of these roots, sweet potato is the most nutritious: it is about 5% protein and contains large amounts of calcium, iron, and other minerals. Conversely, cassava, which is used to make tapioca, contains almost no protein, yet it provides more starch per hectare than any other cultivated crop. Spices such as licorice (*Glycyrrhiza glabra*), sassafras (*Sassafras albidum*), and sarsaparilla (*Smilax*; the flavoring used to make root beer) are derived from roots.

What are Botanists Doing?

Botanists have discovered that when seedlings are infected with *Agrobacterium rhizogenes*, the plants form root hairs over the entire length of each rootlet. What does recent scientific literature contain regarding the potential economic importance of these so-called hairy roots?

Roots also provide several other economically important products. Drugs such as aconite, gentian, ipecac, ginseng, the tranquilizer reserpine, and the heart-relaxant protoveratrine are extracted from roots. Members of the coffee family provide several dyes, as do carrots: their carotene is sometimes used to color butter.

Finally, a woodland shrub called the wahoo plant (*Euonymus*) is sold in some novelty shops as a cure to "uncross" victims of witches' spells. Folklore has it that the victim will be saved from the curse by holding a piece of the plant's root overhead and screaming "wahoo" seven times.

Chapter Summary

Plants invest heavily in roots. The first root to emerge from a seed is the radicle. Most dicots produce a taproot system consisting of a prominent taproot and numerous smaller branch roots. Taproot systems maximize support and absorption of materials from deep in the soil. Monocots produce a fibrous root system made of similarly sized roots. Fibrous root systems are relatively shallow and rapidly absorb materials from near the soil surface. Root systems, especially of monocots, also include adventitious roots, which are roots that form on organs other than roots. Roots anchor plants, store food, absorb water and dissolved minerals, and transport materials to and from the shoot.

Root tips are covered by a root cap that senses light, pressure, and perhaps gravity. The root cap produces and secretes mucigel, which protects and lubricates roots, and helps them absorb water and nutrients. Just behind the root cap is the quiescent center, which is a cluster of seemingly inactive cells. The quiescent center serves as a reservoir of cells that replace damaged cells of the adjacent meristem.

The subapical region of a root can be loosely divided into the zones of cellular division, cellular elongation, and cellular maturation. The zone of cellular division includes the root apical meristem, and is characterized by high rates of mitosis. Cells produced by the meristem elongate in the zone of cellular elongation, which is located 4–15 mm behind the root tip. This elongation accounts for most of root extension. Cells complete differentiation in the zone of cellular maturation, which is located 1–5 cm behind the root tip. Root hairs and mature tissues form in the zone of cellular maturation. Roots have no apical appendages.

The primary body of roots is enclosed by an epidermis. The epidermis surrounds the cortex, which consists of three concentric layers: hypodermis, storage parenchyma, and endodermis. The hypodermis is the protective outer layer(s) of the cortex. Storage parenchyma, which comprises most of the cortex, stores starch that is used for subsequent growth. The endodermis is the inner layer of cortex, and is the primary barrier to absorption. Its tangential walls are surrounded by a Casparian strip that forces water and dissolved minerals into the cytoplasm of endodermal cells. The pericycle and vascular tissues are located inside the cortex and form the stele. Branch roots form endogenously from the pericycle. Xylem and phloem form in alternating strands interior to the pericycle. Roots of monocots have a pith; those of dicots have a solid core of xylem.

The narrow zone of soil surrounding a root and subject to its influence is called the *rhizosphere*. Roots change the rhizosphere by enriching the soil, compressing soil particles, and absorbing nutrients. The rhizosphere houses numerous microbes.

Plants compete with other organisms for the soil's water and nutrients. Most of a root's defenses are chemical rather than structural.

The growth and distribution of roots are controlled by temperature, other organisms, light, gravity, genetic differences, the stage of plant development, and properties of the soil. Modified roots are used for storage, propagation, aeration, movement, parasitism, nutrition, and support. The roots of many plants produce economically important products, such as food, spices, and drugs.

Questions for Further Thought and Study

1. Discuss how roots are structurally and functionally modified for storage, anchorage, and absorption.
2. Of what adaptive value are the compounds that inhibit the growth of other plants' roots?
3. Suppose you are given a slide of a root cross section and a slide of a stem cross section. How would you distinguish the two?
4. What's the difference between a small root and a root hair?
5. List eight types of modified roots and the functional significance of each.
6. List and discuss the functional significance of four unique features of roots.
7. What is the functional importance of mucigel?
8. List the unique features of the endodermis. How do these features relate to the absorption of water and ions by roots?
9. Pneumatophores typically have diaphragms located throughout the length of the root. What might be the functional significance of these diaphragms?
10. What are the advantages and disadvantages of a fibrous root system? A taproot system?
11. Describe seven factors that control the growth and distribution of roots.
12. What is the adaptive significance of the fact that most annuals have fibrous root systems, while many perennials have taproot systems?
13. Why are roots important to plants? Why are they important to animals?
14. Discuss the economic importance of roots.
15. What characteristics distinguish the zone of cellular division? Cellular elongation? Cellular maturation?
16. Describe how roots modify the soil.

Suggested Readings

ARTICLES

Davies, W. J., and J. Zhang. 1991. Root signals and the regulation of growth and development of plants in drying soil. *Annual Review of Plant Physiology and Plant Molecular Biology* 42:55–76.

Epstein, E. 1973. Roots. *Scientific American* 228:48–58.

Feldman, L. J. 1984. Regulation of root development. *Annual Review of Plant Physiology* 35:223–242.

Grant, M. C. 1993. The trembling giant. *Discover* 14(10):82–89.

Pennisi, E. 1992. Hairy Harvest: Bacteria turn roots into chemical factories. *Science News* 141:366–367.

Stewart, G. 1990. Witchweed: A parasitic weed of grain crops. *Outlook on Agriculture* 19:115–117.

BOOKS

Böhm, Wolfgang. 1979. *Methods of Studying Root Systems*. Berlin: Springer-Verlag.

Box, J. R., and L. C. Hammond, eds. 1990. *Rhizosphere Dynamics*. Boulder, CO: Westview Press.

Brouwer, R., O. Gasperíková, J. Kolek, and B. C. Loughman. 1981. *Structure and Function of Plant Roots*. The Hague: Martinus Nijhoff/Dr W. Junk Publishers.

Elliott, Douglas B. 1976. *Roots: An Underground Botany and Forager's Guide*. Old Greenwich, CT: The Chatham Press.

Foster, R. C., A. D. Rovira, and T. W. Cock. 1983. *Ultrastructure of the Root-Soil Interface*. St. Paul, MN: The American Phytopathological Society.

Givnish, Thomas J. 1986. *On the Economy of Plant Form and Function*. Cambridge: Cambridge University Press.

Gregory, P. J., J. V. Lake, and D. A. Rose, eds. 1987. *Root Development and Function*. Society for Experimental Biology, Seminar Series, vol. 30. Cambridge: Cambridge University Press.

Russell, R. S. 1977. *Plant Root Systems: The Function and Interaction with the Soil*. London: McGraw-Hill.

Majestic trees, such as these adult and immature sequoias, are produced by secondary growth.

Secondary Growth

CHAPTER 16

Chapter Outline

INTRODUCTION
THE VASCULAR CAMBIUM
  Where and How the Vascular Cambium Forms
  What Controls the Activity of the Vascular Cambium?
  Extent of Secondary Growth
SECONDARY XYLEM: WOOD
  Kinds of Wood
  Characteristics of Wood

Box 16.1
THE BATS OF SUMMER: BOTANY AND OUR NATIONAL PASTIME

  Reaction Wood
BARK: SECONDARY PHLOEM AND PERIDERM
  Secondary Phloem
  Periderm
UNUSUAL SECONDARY GROWTH
  Dicots

Box 16.2
CORK

  Monocots

Box 16.3
THE LORE OF TREES

USES OF SECONDARY XYLEM AND PHLOEM
  Commemorating Trees
Chapter Summary
Questions for Further Thought and Study
Suggested Readings

Chapter Overview

Woody plants are characterized by secondary growth. Secondary growth results from lateral meristems that produce large amounts of secondary phloem and secondary xylem. These secondary tissues are protected by a suberized tissue called cork and increase the plant's girth, thereby making it sturdier and able to grow taller. This increased height decreases the chances that the plant's leaves will be shaded by other plants. Humans use secondary xylem (i.e., wood) to make a variety of products, most notably paper and lumber.

Introduction

Competition among organisms for the environment's resources is fierce. In the previous chapter you learned how roots form alliances with other organisms to improve their chances of obtaining water and minerals from the soil. Competition among organisms is also keen above ground. Although winds maintain a relatively constant concentration of CO_2 throughout the atmosphere, there is only a limited amount of light available for photosynthesis. Plants compete with each other to intercept this light, and have evolved various adaptations for maximizing its absorption. For example, plants have elaborate guidance systems that direct them toward light. This light-directed growth, called phototropism, helps ensure that plants intercept light (see Chapter 19). Another means of intercepting light is simply to grow taller than nearby plants. However, primary tissues such as parenchyma and phloem cannot support a tall plant against the pull of gravity. Plants such as climbing vines have overcome this problem by clinging to other objects for support. This adaptation is relatively rare, however, and there are few herbaceous plants more than a meter or two high. Clearly, a tall plant that could support itself against gravity would have a decided advantage: it could grow taller and therefore avoid being shaded by other plants or objects.

An improved system for supporting a plant against gravity's pull evolved 370 million years ago (i.e., in the Middle Devonian) and involved the formation of two lateral meristems: the vascular cambium and phellogen (cork cambium). Unlike apical meristems that increase the length of roots and shoots, the vascular cambium and phellogen increase the girth (i.e., thickness) of stems and roots by a process called **secondary growth.** Such growth makes plants sturdier and enables them to grow taller, thereby improving the plant's chances of intercepting light. Secondary growth can also produce huge plants. For example, the canopy of a banyan tree (*Ficus benghalensis*) near Gutibayalu, India, covers more than 21,000 m^2, an area about the size of five football fields.

Although secondary growth improves a plant's chances of intercepting light, it also means that the plant must maintain and defend a larger mass of permanent tissue. Increased thickness and height also put new demands on a plant's transport systems. Thus, tissues derived from the vascular cambium and phellogen have several functions besides support, the most important of which are conduction and protection. Support and conduction are functions of derivatives of the vascular cambium, and protection is a function of derivatives of the phellogen.

Tissues produced by lateral meristems form the **secondary body** of a plant. Most of the secondary body is made by the vascular cambium.

The Vascular Cambium

The vascular cambium is a thin layer of cells arranged in an open-ended cylinder (fig. 16.1). Except for a few unusual monocots such as *Dracaena*, only dicots and nonflowering seed plants have secondary growth. There are two kinds of cells in the vascular cambium: ray initials and fusiform initials.

Ray initials are small, elongate cells oriented perpendicular to the axis of the stem (fig. 16.2). They form ray parenchyma (i.e., rays) that radially divide a stem, much like slices of a pie. Rays transport water and dissolved solutes radially.

Fusiform initials are tapered, prism-shaped cells oriented parallel to the axis of the stem (fig. 16.2). They produce secondary xylem and secondary phloem (i.e., the axial system of the stem) between the rays. The conducting cells of these tissues transport water and dissolved solutes longitudinally.

There are several differences between meristematic cells of the vascular cambium and those of apical meristems. Fusiform initials of the vascular cambium divide longitudinally toward their tips, whereas other meristematic cells usually divide along the shortest distance across the cell. Also, cells of the vascular cambium are highly vacuolate (fig. 16.3), while those of apical meristems are densely

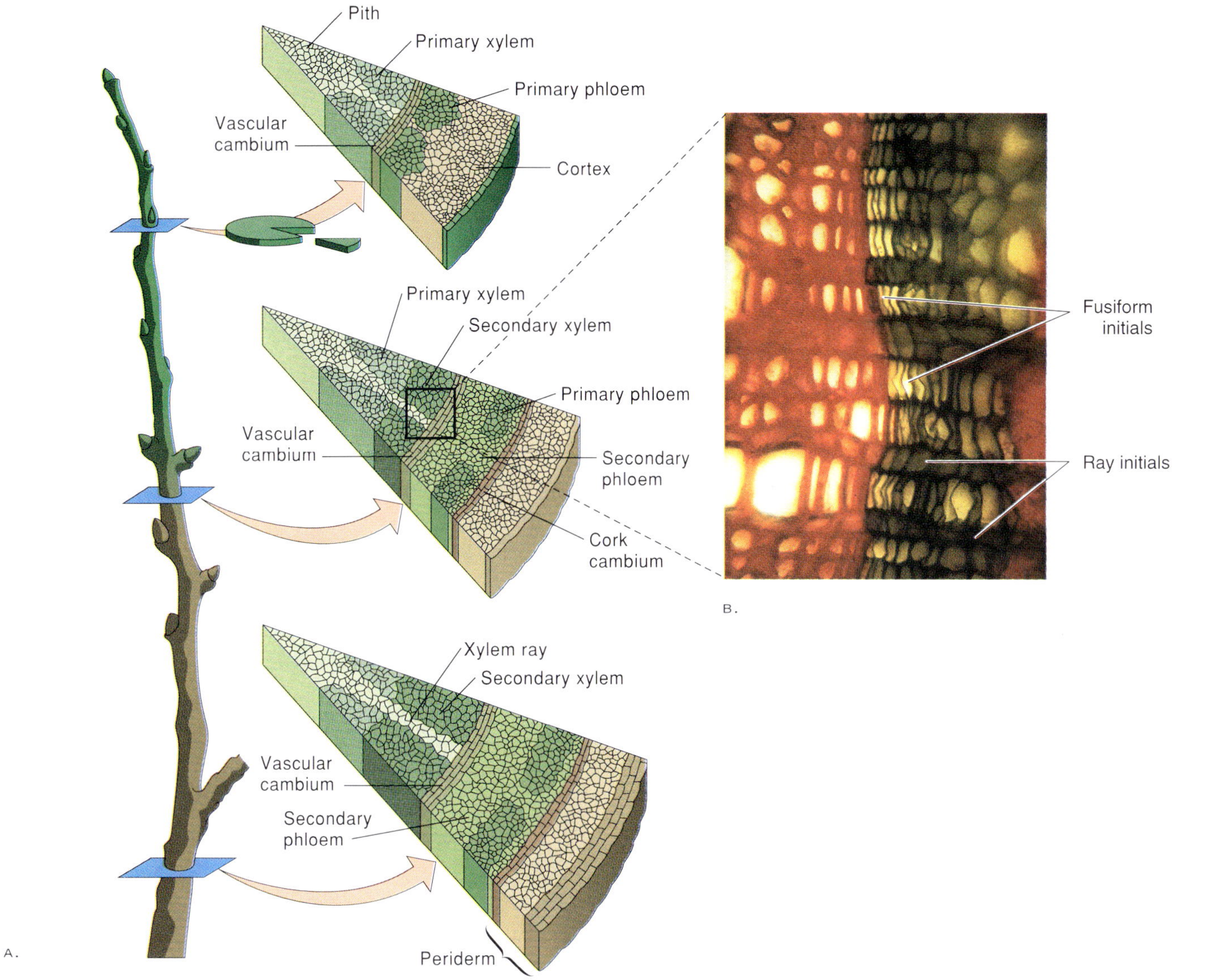

FIGURE 16.1

Secondary growth of stems. (a) The vascular cambium differentiates between the primary xylem and primary phloem, and forms an open-ended cylinder. The vascular cambium produces secondary phloem to the outside and secondary xylem (wood) to the inside. Derivatives of the secondary phloem form a protective layer called the *periderm*. (b) Cross section of the vascular cambium shows ray initials, which form rays that transport water and dissolved solutes radially. Fusiform initials form secondary xylem and secondary phloem between the rays.

cytoplasmic. The vacuolation of cambial cells presumably enables them to withstand pressure generated by the expanding secondary xylem.

Most divisions of fusiform initials are periclinal (i.e., parallel to the nearest surface) and produce radial rows of secondary xylem (wood) to the inside and secondary phloem to the outside (fig. 16.4). Derivatives of fusiform initials usually divide several times before differentiating, thus producing a cambial *zone* rather than a single layer of meristematic cells. On average, the vascular cambium produces 4–10 times more xylem than phloem—a ratio that is unaffected by environmental changes such as rain and temperature. Occasional anticlinal divisions (i.e., perpendicular to the nearest surface) expand the vascular cambium to accommodate the increasing girth of the stem.

CONCEPT

The vascular cambium consists of fusiform initials and ray initials. Ray initials produce rays that transport fluids radially. Fusiform initials produce secondary xylem and secondary phloem, the conducting cells that transport liquids longitudinally.

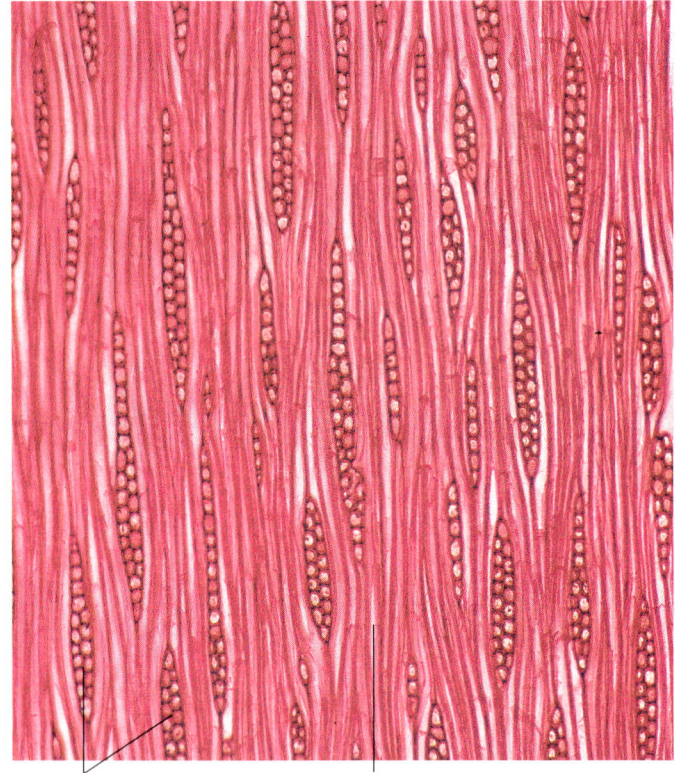

FIGURE 16.2

Longitudinal section of the vascular cambium, showing the orientation of ray initials and fusiform initials, ×100. Compare this with the cross section of the vascular cambium shown in figure 16.1.

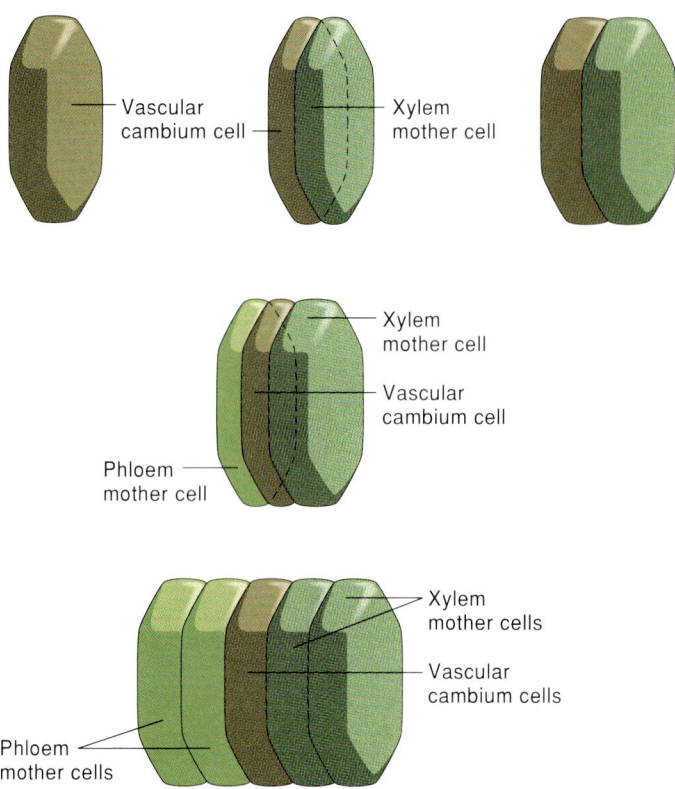

FIGURE 16.4

Differentiation of secondary xylem and secondary phloem from the vascular cambium. Derivatives of the vascular cambium—xylem mother cells and phloem mother cells—divide several times before differentiating, forming a zone of meristematic cells.

Where and How the Vascular Cambium Forms

Examine figure 16.5, which shows where the vascular cambium forms in roots and shoots. In roots, the vascular cambium differentiates from latent procambium between xylem and phloem. These regions of *fascicular cambium* differentiate simultaneously and are subsequently linked by *interfascicular cambium* that differentiates through the pericycle. Thus, the vascular cambium of roots is initially lobed and surrounds the spokes of primary xylem. The vascular cambium in roots differentiates acropetally (i.e., toward the apex) and becomes roughly circular after it begins to divide. Secondary growth in roots is usually less extensive than in shoots.

In shoots, the fascicular cambium differentiates from latent procambium between xylem and phloem of the stem's ring of vascular bundles (fig. 16.5). The interfascicular cambium then differentiates from cortical cells separating the vascular bundles, thereby forming a cylindrical cambium. As in roots, the vascular cambium in shoots differentiates acropetally (i.e., from the base toward the tip) and is continuous with the vascular cambium of branches.

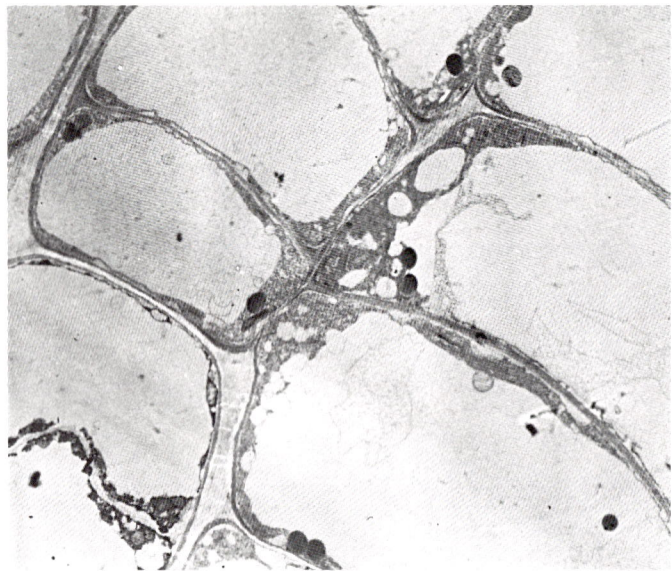

FIGURE 16.3

Unlike the cells of most meristems, the cells of the vascular cambium are highly vacuolate. The cytoplasm in each of these cells forms a thin film between the vacuole and the plasma membrane.

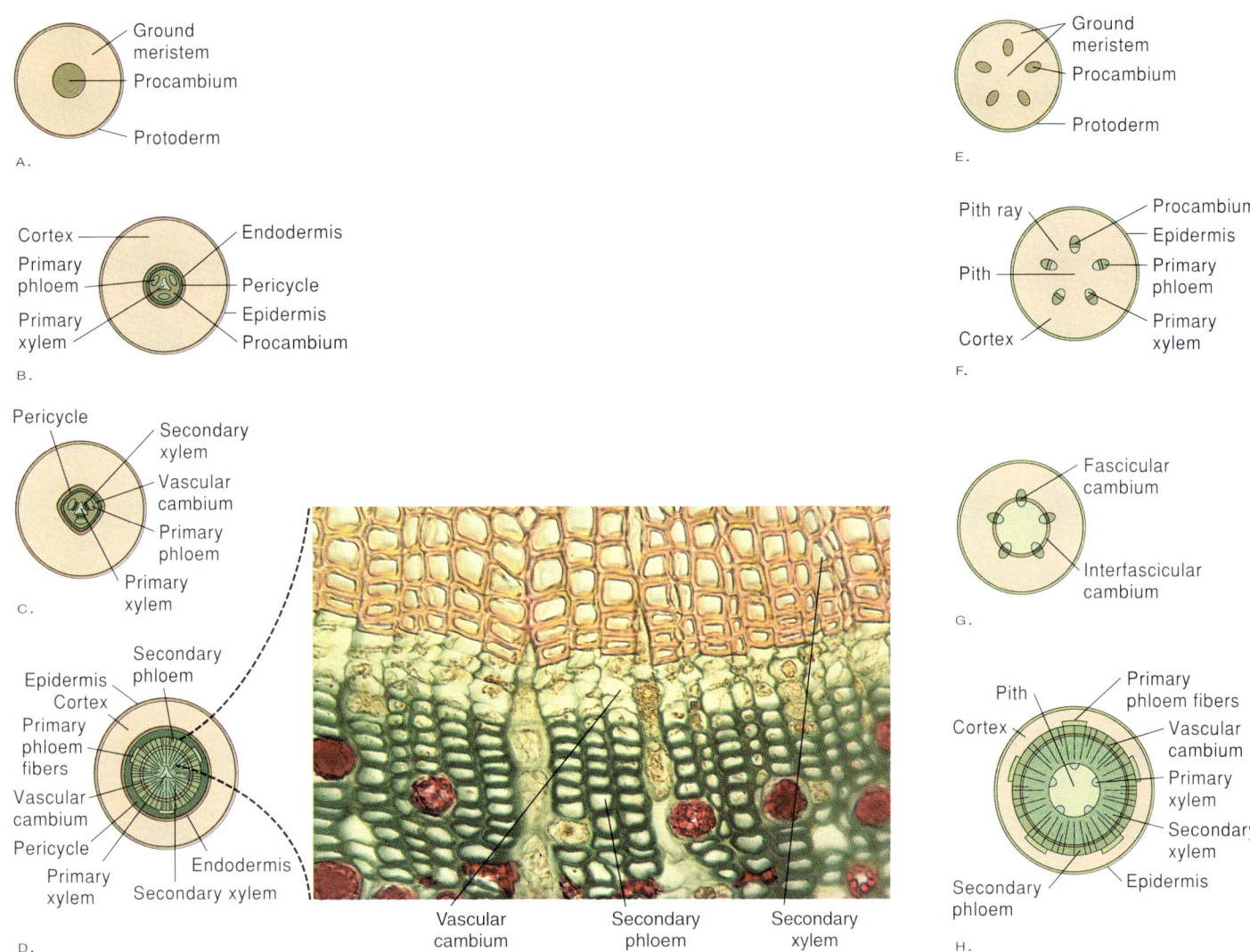

FIGURE 16.5

Differentiation of the vascular cambium in roots (a–d) and shoots (e–h). **Roots.** (a) Early stage of primary growth in roots, showing primary meristems. At this stage, the vascular cambium has not formed. (b) At the completion of primary growth, the meristematic procambium remains between the primary xylem and primary phloem. (c) In this root, the vascular cambium forms from procambium between the strands of primary xylem and primary phloem. The pericycle opposite the three poles of protoxylem also contributes to the vascular cambium. (d) The vascular cambium produces secondary phloem to the outside and secondary xylem to the inside. The radiating lines represent rays. **Shoots.** (e) Early stage of primary growth in a stem, showing primary meristems. (f) After primary growth. (g) Initiation of the vascular cambium. The fascicular cambium (between the xylem and phloem) links with the interfascicular cambium (between the different bundles) to form a cylinder. (h) The fully differentiated vascular cambium produces secondary xylem to the inside and secondary phloem to the outside, ×400.

What controls the formation and polarity of the vascular cambium? Hormones such as auxin coming from young leaves are important, as is the predisposition of cambial cells. To understand this, consider the grafting experiments shown in figure 16.6. When plugs of tissue were removed from the interfascicular region and reoriented 180° prior to differentiation of the vascular cambium, the cambium differentiated normally across the grafted tissue. However, it produced secondary tissues that were 180° off—that is, it produced secondary xylem to the *outside* and secondary phloem to the *inside*. These and other experiments show that the cortical cells that ultimately differentiate into vascular cambium are developmentally predetermined. The developmental basis for this predetermination of cells remains unknown.

What Controls the Activity of the Vascular Cambium?

During autumn and winter, the vascular cambium is inactive, and its dormant cells have thick cell walls. The moist, warmer, and longer days of spring reactivate the vascular cambium by thinning its walls and inducing cellular divisions. What controls

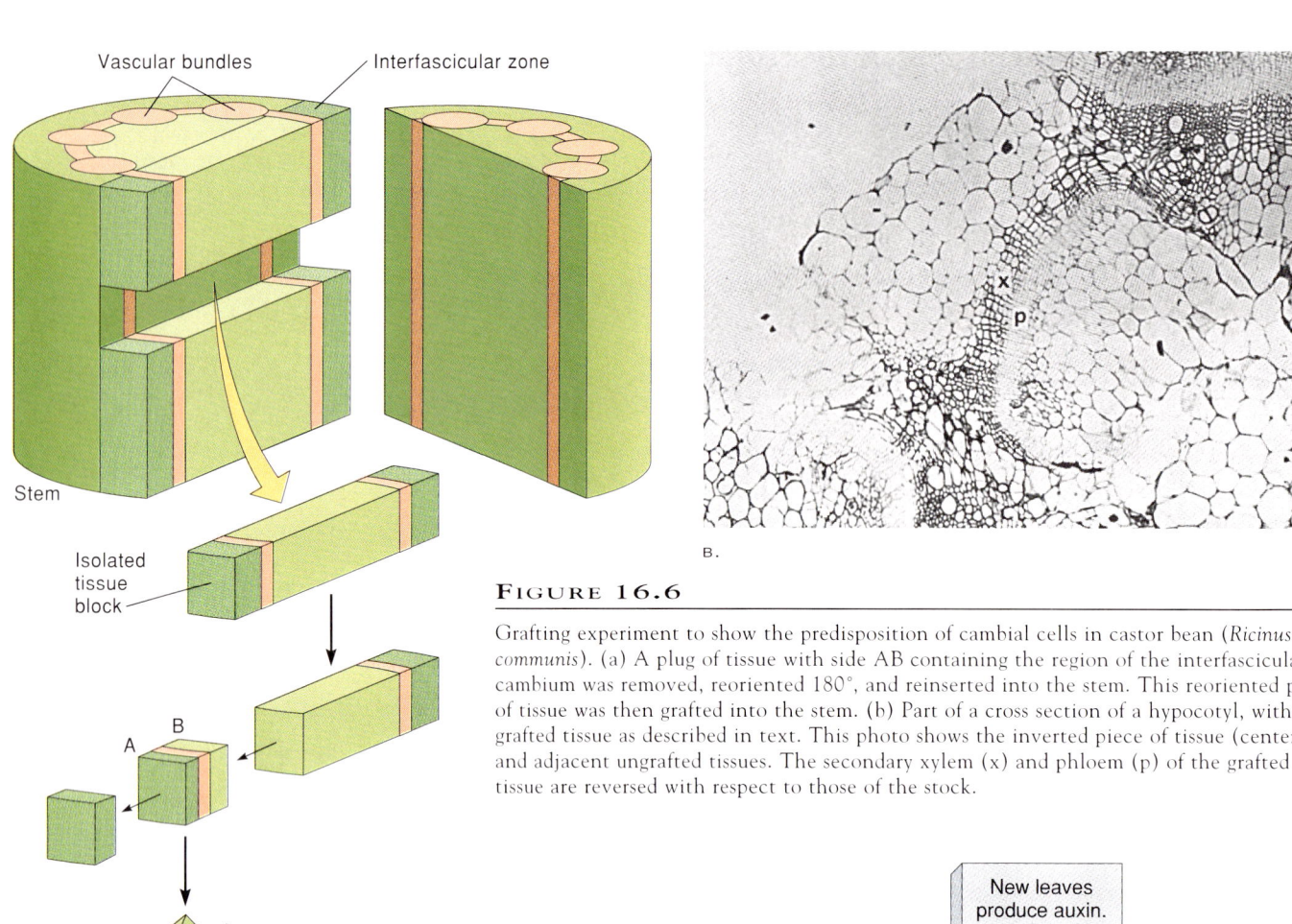

Figure 16.6

Grafting experiment to show the predisposition of cambial cells in castor bean (*Ricinus communis*). (a) A plug of tissue with side AB containing the region of the interfascicular cambium was removed, reoriented 180°, and reinserted into the stem. This reoriented plug of tissue was then grafted into the stem. (b) Part of a cross section of a hypocotyl, with grafted tissue as described in text. This photo shows the inverted piece of tissue (center) and adjacent ungrafted tissues. The secondary xylem (x) and phloem (p) of the grafted tissue are reversed with respect to those of the stock.

Figure 16.7

Hypotheses to account for how young leaves activate and influence the activity of the vascular cambium.

this reactivation? The earliest observations suggested that the activity of the cambium is controlled by auxin made by and released from buds and young leaves. Indeed, reactivation of the cambium begins in buds and twigs and moves progressively down the stem in a wave. These results prompted botanists to suggest that spring weather induces auxin synthesis in leaves and buds and that this auxin moves down the stem, activating the vascular cambium as it goes (fig. 16.7).

More recent evidence has prompted other botanists to question this model. Indeed, the inactivation of the cambium that occurs when the apex is removed cannot be replaced by auxin applied to the cut surface. Moreover, the concentration of auxin in the cambium remains high even when the cambium is dormant, suggesting that the onset and cessation of cambial activity are not due to changes in auxin concentration. Therefore, the dormancy of the vascular cambium is probably not due to a deficiency of auxin, but rather to a decreased *sensitivity* to auxin: auxin affects the cambial cells in the spring only after the cambial cells regain their ability to respond to the auxin. This increased sensitivity to auxin begins in buds and twigs and moves progressively down the stem in a wave, possibly in response to something released by buds and young leaves. In both of these models, buds and young leaves control the formation of vascular tissues for service and support.

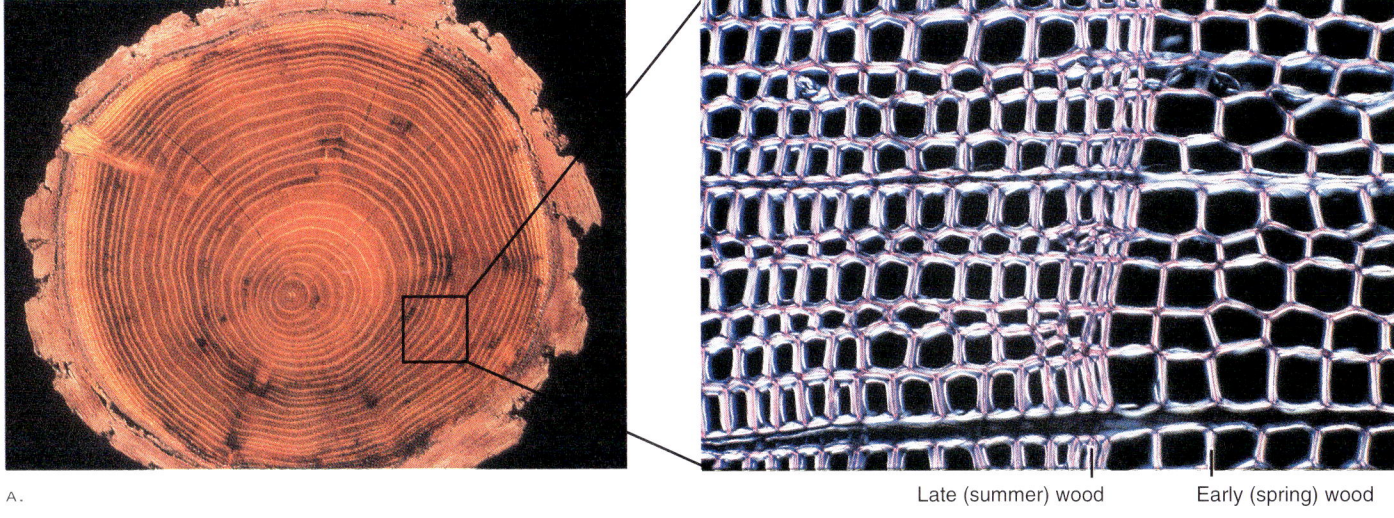

Late (summer) wood   Early (spring) wood

C.

FIGURE 16.8

(a) The seasonal activity of the vascular cambium in temperate trees produces growth rings in wood. Each ring consists of early (spring) wood made of larger, thin-walled cells, and late (summer) wood made of small, thick-walled cells. (b) Bristlecone pine (*Pinus longaeva*) in the White Mountains of eastern California. These are the oldest living trees. (c) A cross section of wood from bristlecone pine. Note the small and differing widths of the annual rings.

Wood (secondary xylem) formed during the moist days of spring (i.e., early in the growing season) is called **spring** or **early wood,** and consists of large, thin-walled cells (fig. 16.8). Later in the season, the drier days of summer gradually slow the activity of the vascular cambium and cause it to produce **summer** or **late wood** made of smaller cells with thicker walls (fig. 16.8). This environmentally induced difference between spring and summer wood is influenced strongly by auxin: auxin stimulates the formation of spring wood, and applying auxin-transport inhibitors stimulates the formation of summer wood. Differences between spring and summer wood are abrupt in trees such as pine, and are visible in most trees as **growth rings** (fig. 16.8). In temperate regions, the predictable seasons usually produce one growth ring per year, thus accounting for the term *annual ring*. However, trees such as ebony (*Diospyros ebenum*) and jacaranda (e.g., *Jacaranda arborea*) that grow in the seasonless tropics lack growth rings, whereas plants growing in arid and semiarid regions often produce more than one growth ring per year in response to sporadic rains.

Growth rings appear as concentric circles in cross section and are usually 1–10 mm wide (fig. 16.8). Since climate (especially water availability) strongly influences the formation of growth rings, a cross section of wood is a diary of the climatic history of a particular region. The science of interpreting history by studying growth rings is called *dendrochronology,* and is an important tool for meteorologists and anthropologists. For example, the White Mountains of eastern California are home of bristlecone pines (*Pinus longaeva*), whose slow growth packs almost 1,000 growth rings into only 13 cm or so of wood. By overlapping the growth rings of living and dead trees, dendrochronologists have reconstructed the area's climate for the past 10,000 years. They've used this information to date cliff-dwellings and document droughts in 840, 1067, 1379, and 1632—each 200 to 300 years apart.

Production of secondary xylem and phloem by the vascular cambium is also affected by pressure. The influence of pressure is easily demonstrated with experiments using longitudinal strips of bark partially separated from the stem. If the strip of bark is under no pressure, callus forms on the inner side of the bark, and a vascular cambium subsequently differentiates through this callus tissue. If the strip of bark is wrapped tightly back into place (i.e., placed under pressure), no callus tissue forms. Rather, the vascular cambium produces secondary xylem and phloem as usual.

Water availability, the sensitivity of cambial cells to auxin, and pressure control the activity of the vascular cambium. Wood made during the moist days of spring contains large xylary elements and is called *spring* or *early wood.* Wood made during the drier days of summer contains fewer large xylary elements and is called *summer* or *late wood.* Differences between early and late wood are visible as growth rings.

Extent of Secondary Growth

The vascular cambium is reactivated each year and is virtually immortal. Many plants are more than 1,000 years old, and a bristlecone pine named Methuselah growing in California is almost 5,000 years old, which means that the seed that produced this plant germinated more than 500 years before the pyramids were completed. The perpetual reactivation of the vascular cambium for thousands of years produces impressive trees—the largest and most conspicuous plants ever to live. The fattest tree is a chestnut (*Castanea*) named the Tree of One Hundred Horses, which lives on Sicily's Mount Etna: this tree is 58 m around. A 2,000-year-old tule tree (*Taxodium mucronatum*) in Oaxaca, Mexico, is almost 45 m in circumference, and at last measurement, a California redwood (*Sequoia sempervirens*) was more than 110 m tall. That redwood weighs approximately 1,600 tons, which is roughly equivalent to the weight of ten blue whales. (see box 16.3, "The Lore of Trees," p. 373).

Trees have attained these gigantic sizes and old ages despite their lack of an immune system and their inability to flee enemies. Trees attacked by pathogens defend themselves by **compartmentalization,** which involves walling off infections, fortifying cell walls near wounds, and making antimicrobial compounds such as phenols. This strategy is effective, but has important consequences for a tree. Phenols that kill invading microbes also damage the tree because they poison its tissues. Trees survive these sacrifices of their parts by continually producing new tissue to maintain life. However, the walled-off compartments of a tree are inaccessible, which forces it to grow a "new" tree over the sealed compartments of the old tree. This new tree is visible as growth rings and provides new tissues to perform the tree's activities. Thus, a tree's survival depends on how fast and effectively it can react to and wall off its invaders—American chestnut (*Castanea dentata*) is nearly extinct, and American elm (*Ulmus Americana*) faces possible extinction because they cannot react fast enough to pathogens. Too many walled-off compartments (i.e., too many infections) stop the movement of a tree's vital fluids and thereby kill the tree.

SECONDARY XYLEM: WOOD

Secondary xylem, or wood, is the inner derivative of the vascular cambium and comprises about 90% of a typical tree. About one-third of the United States is covered by trees, of which about 65% are farmed commercially. We have all sorts of uses for the 1.7 million cubic meters of wood produced by these forests each day. Each person in the United States uses an average of 2.25 cubic meters of wood per year. About 50% of this wood is used for lumber, 25% for pulp and paper products, 10% for plywood and veneer, and the rest for fuel and miscellaneous products such as fence posts, toothpicks, bowling pins, guitars, baskets, decks, and barrels. Each December we chop down about 50 million Christmas trees, and Americans use 250,000 tons of napkins and 2 million tons of newsprint and writing paper per year. These and other uses of wood make forestry a $36-billion-per-year industry.

Kinds of Wood

There are several kinds of wood. We classify them on the basis of the kind of plant that produces the wood (i.e., softwoods vs. hardwoods) and the location and function of the wood in the plant (i.e., sapwood vs. heartwood).

Softwoods and Hardwoods

Softwoods are nonflowering seed plants native to temperate zones, and include pine, spruce, larch, fir, and redwood. Their wood is relatively homogeneous because it is about 90% tracheids and lacks vessels (fig. 16.9). Many softwoods, such as pine and spruce, also contain vertically oriented *resin canals* filled with resin secreted by living parenchyma cells that

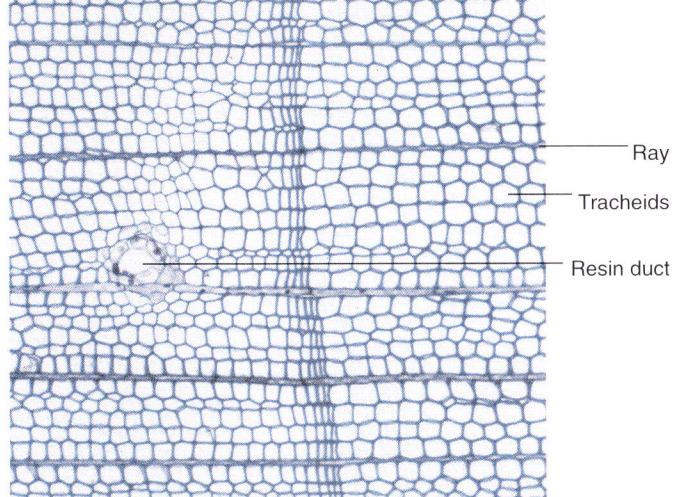

FIGURE 16.9

Cross section of softwood of eastern white pine (*Pinus strobus*). Softwood is relatively homogeneous because it lacks vessels. Note the uniseriate rays and resin ducts in the wood. The annual ring shown here results from differing diameters of tracheids produced at different times of the year.

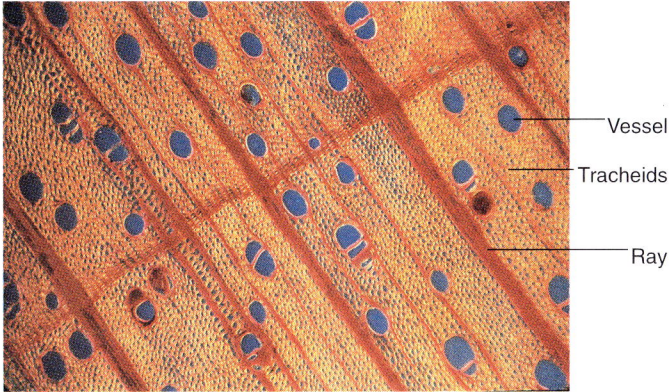

FIGURE 16.10

Hardwood of sugar maple (*Acer saccharum*), an angiosperm. Hardwoods contain tracheids as well as vessels. Because the vessels in maple are distributed throughout the wood, maple wood is referred to as diffuse-porous wood.

line the canals (fig. 16.9). Resin hardens when exposed to air and is an effective means of sealing wounds. Fossilized resin is *amber*, a substance described in some mineralogy books as a gem.

Most softwoods are relatively light (i.e., soft) and are easily penetrated by nails, which explains their widespread use in construction. The large amount of lignin in softwoods, such as white pine, makes them ideal for lumber, because lignin makes wood stable and unlikely to warp. Most commercial plantations grow pines or firs because these trees grow faster and are more profitable than dicots. Pulp and plywood come primarily from softwoods.

The uniseriate rays of softwoods are vertically oriented sheets of cells that are 1 cell wide and 1–20 cells high (fig. 16.9). Softwoods often contain resin canals that form naturally as well as in response to injury (figs. 16.9, 13.34, 13.35).

Hardwoods are dicots native to temperate and tropical regions, and include oak, maple, ash, walnut, and hickory. Unlike softwoods, which contain only tracheids and ray parenchyma cells, hardwoods contain fibers and vessels (fig. 16.10). Fibers make hardwoods stronger and denser (i.e., harder) than softwoods, which explains why woods like oak, walnut, maple, and hickory are hard to nail and seldom used for construction.

The conducting cells in hardwoods are tracheids and vessels. Vessels are often visible to the naked eye and are frequently called *pores* by woodworkers. If large vessels occur only in early wood, the wood is called *ring-porous wood* (see fig. 16.10). Because of their targetlike arrangement of vessels, ring-porous woods such as elm, chestnut, oak, and ash have easily discernible growth rings. Ring-porous wood is an evolutionary specialization and occurs only in a few species native to temperate regions. In most trees there is little contrast in the size or pattern of vessels across growth rings; that is, vessels in these trees are distributed uniformly in early and late wood (fig. 16.10). Such wood (e.g., maple and birch) is called *diffuse-porous wood*.

Rays of hardwoods are more diverse than rays of softwoods and seldom contain resin canals. These rays may be uniseriate but are usually more than twenty cells wide (i.e., are multiseriate) and hundreds of cells long.

Writing to Learn Botany

The large vessels in early ring-porous wood conduct water at 0.5–1 m h^{-1}, which is 5–20 times faster than in diffuse-porous wood. However, trees with ring-porous wood usually don't lose more water than trees with diffuse-porous wood. Explain.

Because of their many fibers, hardwoods are usually denser than softwoods. However, there are many exceptions to this generalization. For example, hardwoods such as poplar (*Populus*) and basswood (*Tilia*) are softer than softwoods such as hemlock (*Tsuga*) and yellow pine. Similarly, balsa (*Ochroma lagopus*) is a broad-leaf hardwood whose wood has a specific gravity (0.12) only half that of cork. But even balsa (Spanish for "raft") is not the lightest wood—that distinction goes to the pith-plant (*Aeschynomene aspera*), whose wood has a specific gravity of only 0.04.

CONCEPT

Softwoods are nonflowering seed plants native to temperate regions. Their wood is homogeneous and contains tracheids and uniseriate rays. Hardwoods are dicots whose wood contains multiseriate rays, tracheids, vessels specialized for transport, and fibers specialized for support.

What are Botanists Doing?

Find one example of variation in wood structure that has been found in the past 1–2 years. How is this kind of variation useful in botanical research?

Sapwood and Heartwood

We also classify wood as either sapwood or heartwood. This distinction involves differences in position, function, and appearance of a tree's wood, and is based on different parts of a woody stem being specialized for different functions.

Water and dissolved nutrients move in the outer few centimeters of secondary xylem. This wood that transports sap is called **sapwood** and is usually light, pale, and relatively weak (fig. 16.11).

The dry wood in the heart (i.e., center) of a tree is called **heartwood** and is the dump site for some of the tree's waste products (fig. 16.11). As a tree ages, metabolites such as resins, gums, oils, and tannins are gradually deposited in heartwood, where they clog the xylary elements and eventually stop up the wood. Heartwood, which can be more than a meter wide in large trees, is darker, denser, more durable, and more aromatic than sapwood. Although heartwood helps support a tree, it is not essential for growth: it can be removed without harming a tree.

CONCEPT

Sapwood transports water and dissolved nutrients, and comprises the outer few centimeters of wood. Heartwood occurs in the center of a tree and is where many waste products are stored. Heartwood does not transport water and minerals.

Characteristics of Wood

Interpreting the features of wood requires an understanding of how the wood you are examining was cut. Figure 16.12 shows the three ways that wood is cut:

Cross section. Growth rings of trunks cut in cross section (or *transverse* section) are arranged in concentric circles, much like the circles on a target. Rays radiate from the center of the section, like spokes from the center of a wheel. Cross sections of trees are seldom used because they often split after they're cut.

Radial section. A radial section is a longitudinal section that goes through the center of the stem. Boards made from radial sections are called *quartersaw cuts*. In these boards, growth rings appear as parallel lines oriented perpendicular to rays. Quartersawed boards are favorites from trees like oak, whose large rays add texture to the wood. However, only a few boards can be quartersawed from a tree trunk.

FIGURE 16.11

This cross section of the trunk of a 28-year-old coastal redwood (*Sequoia sempervirens*) shows the dark heartwood (inner 19–20 rings) and the lighter sapwood. Transport of water and dissolved minerals occurs only through the sapwood. The dark color of heartwood results from the presence of resins, gums, and other metabolites.

Tangential section. A tangential section is a longitudinal section that does not go through the center of the stem. Boards made from tangential sections are called *planesaw cuts*. Growth rings in these boards are arranged in large irregular patterns of concentric V's.

Keep these sections of wood in mind as you examine the following properties of wood, which are illustrated in figure 16.13.

Knots are bases of branches that have been covered by lateral growth of the main stem. Knots produced by dead limbs have no xylary continuity with the main stem and therefore fall out of lumber when the wood dries. Conversely, knots of living branches do not fall out because their xylem is continuous with that of the main axis. Knots usually weaken lumber and decrease its value.

Grain is the direction of axial cells relative to the longitudinal axis of a tree or a particular piece of wood. Grain may have regular patterns or, as in trees such as English elm (*Ulmus procera*), be irregular and lack patterns. In trees such as sycamore (*Platanus*), the grain changes every few years and is referred to as interlocked grain.

Texture refers to the arrangement and size of the pores in wood. Texture depends primarily on the size and distribution of vessels, tracheids, and, to a lesser extent, rays. Trees such as oak have large vessels and a coarse texture, whereas softwoods lack vessels and have a fine texture.

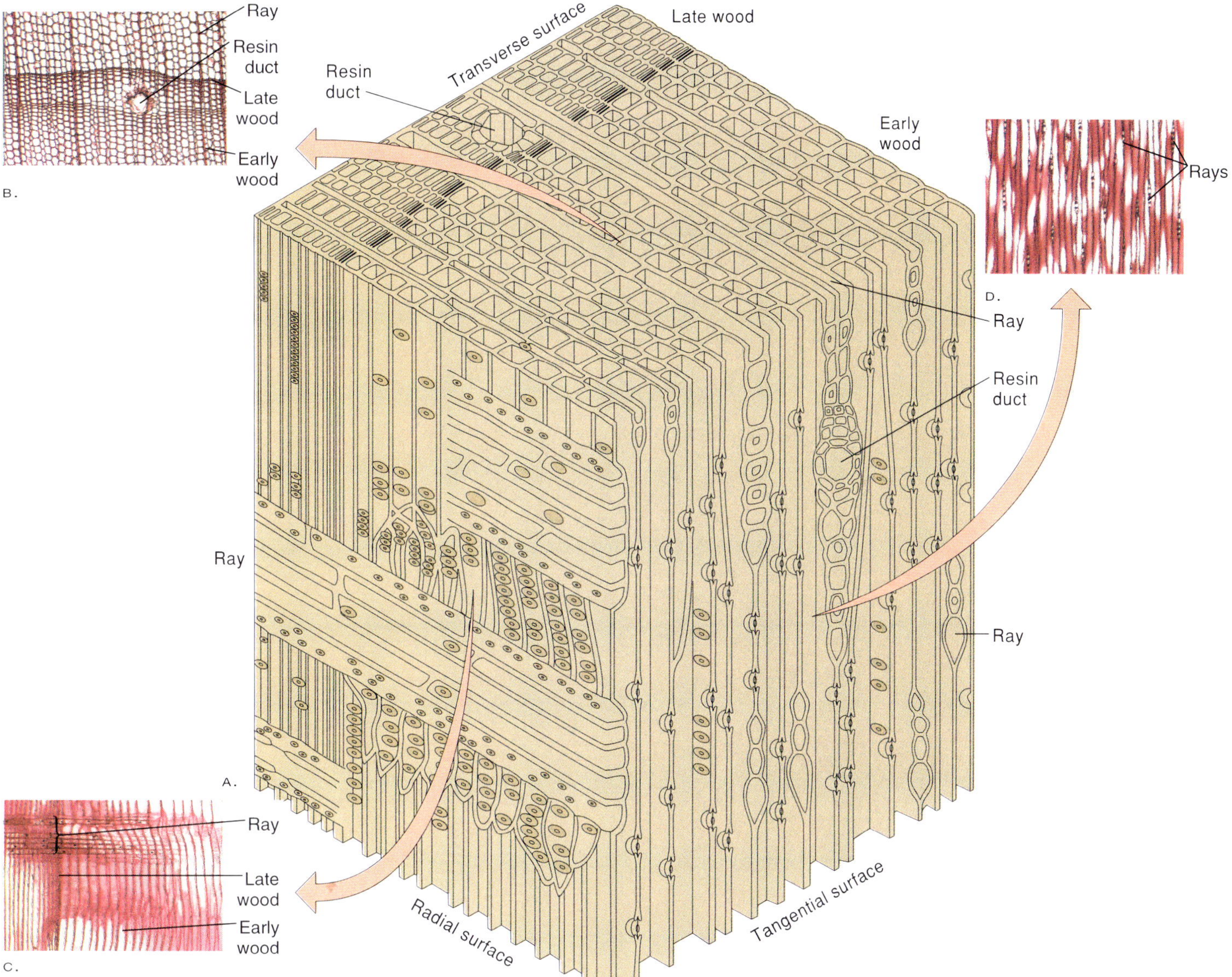

FIGURE 16.12

The different kinds of sections of wood of eastern white pine (*Pinus strobus*): transverse (i.e., cross) section, radial section (longitudinal section through the center of the stem), and a tangential section (longitudinal section not through the center of the stem) (b) ×100; (c) ×100; (d) ×100.

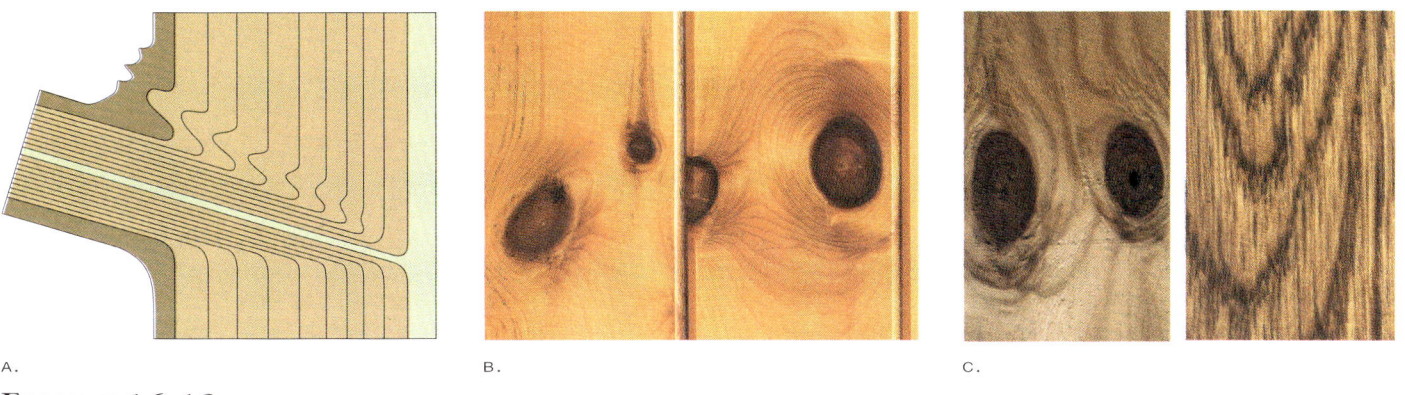

FIGURE 16.13

(a) Radial section of trunk, showing a living branch whose cambium and growth rings are continuous with the branch. Lateral growth of the main stem eventually covers the branch. (b) Tangential section of a covered branch shows a knot, which is a cross section of the branch that was covered by lateral growth of the main stem. (c) Two prominent features of finished wood are grain and figure. These features vary, depending on how the wood was cut.

CHAPTER SIXTEEN *Secondary Growth*

BOXED READING 16.1

THE BATS OF SUMMER: BOTANY AND OUR NATIONAL PASTIME

Baseball is a sport influenced heavily by plants. Pitchers use rosin to improve their grip, hitters coat their bat handles with pine tar, and an occasional cheater "corks" his bat by filling its barrel with cork to improve bat speed and drive. No part of baseball conjures up as much superstition and irrational behavior than the game's icon—the baseball bat. Players have prayed for them, coated them with manure, and even slept with them to improve their hitting. One aspiring minor leaguer in the Philadelphia Phillie's farm system even painted eyes and glasses on his bat after going 0-for-16, to help his bat see the ball better. He must have nearly blinded his bat that night when he hit a pair of home runs.

Baseball bats are made from white ash (*Fraxinus americana*) growing in the northeastern United States, especially New York and Pennsylvania. Ash is ideal for bats because it is tough, light, and will drive a ball. Ash trees harvested to make bats are each about 75 years old (40 cm in diameter) and produce about 60 bats per tree. Technicians use hand-turned lathes to produce one major league bat every 10–20 minutes. The bats on sale at your local sporting goods store receive somewhat less attention—machine lathes produce a retail bat in only 8 seconds.

Following lathing, a bat's trademark is branded on the wood's tangential surface and against the grain. Thus, a hitter who follows the adage of "keeping the trademark up" will strike a ball with the grain and where the bat is strongest. When Yogi Berra

BOX FIGURE 16.1

Wooden bats are made from wood of white ash (*Fraxinus*).

dismissed this advice by saying he went to the plate "to hit, not to read," his batmakers moved the trademark of his bats a quarter of a turn, thus saving the lives of countless bats.

Although ash bats are durable and strong, they frequently break. Each major leaguer uses six to seven dozen bats per season—approximately one bat per base-hit. Hillerich and Bradsby, the makers of Louisville Sluggers, cut down about 200,000 trees per year to meet major leaguers' demands for bats. Today, bats made of aluminum—"aerospace alloy" as their manufacturers say—have replaced wooden bats in amateur baseball. As a result, Hillerich and

Bradsby produce less than 1 million bats per year, down from 6 million several years ago. They sell these bats to major league baseball teams for $24 to $30 each.

Figure refers to patterns produced by variations in color, arrangements of tissues, growth rings, grain, and knots.

Density of wood is expressed as *specific gravity*, which is the ratio of the density of the wood to that of water. In general, fast-growing conifers and diffuse porous hardwoods are less dense (i.e., have a lower specific gravity) than slow-growing hardwoods. The specific gravity of wood is determined by the size of its cells, thickness of the cell wall, amount of lignin, and the proportions of early and late wood. Most woods have a specific gravity between 0.3 and 0.7; for example, hickory has a specific gravity of 0.65, white cedar (*Thuja*) has a specific gravity of only 0.30, and white ash (*Fraxinus americana*), the tree used to make baseball bats (see box 16.1, "The Bats of Summer: Botany and Our National Pastime") has a specific gravity of 0.55.

Thus, all of these woods float in water. Among the densest woods are lignum vitae (*Guaiacum officinale*) and black ironwood (*Krugiodendron ferreum*), which have specific gravities of 1.25 and 1.30, respectively. Blocks of these and a few other woods sink when placed in water.

Durability of wood refers to its ability to resist weathering and decay. Woods such as redwood (*Sequoia*), teak (*Tectona*), incense cedar (*Calocedrus*), and black locust (*Robinia*) contain large amounts of natural preservatives such as tannins and are extremely durable, while those such as cottonwood (*Populus*), willow (*Salix*), and fir (*Abies*) contain relatively few secondary products and decay rapidly. One of the most durable woods is lignum vitae, a hard and resinous wood that's been found in near-perfect condition in 400-year-old sunken Spanish galleons. Another is sequoia, which is rich in tannins. A well-preserved sequoia log was carbon-dated in 1964 and found to be 2,100 years old; that means that fungi and bacteria had been side-stepping it since the beginning of the Roman empire.

Water-content of wood accounts for as much as 75% of the weight of a living tree. Once harvested, tree trunks are dried (i.e., *cured* or *seasoned*) in large ovens that reduce the water-content of the wood to between 5% and 20%. Dry wood is about 65% cellulose, 20% lignin, and 15% secondary products such as resins, gums, oils, and tannins (ash, the inorganic compound in wood, seldom accounts for more than 0.5% of wood). Drying wood is important because wood with a water-content less than 20% is essentially immune to attack by most fungi.

C O N C E P T

Figure, knots, grain, texture, durability, density, and water-content are distinguishing features of wood.

Reaction Wood

Horizontal stems and stems bent by wind or other disturbances right themselves by making **reaction wood**, which is wood produced in response to stress caused by gravity. Auxin influences the formation of reaction wood. Tilting a stem causes auxin to accumulate on its lower side, where it triggers the vascular cambium to produce reaction wood.

In dicots, reaction wood forms on the upper side of the stem and is called *tension wood* (fig. 16.14a). Tension wood contains relatively few vessels and many gelatinous fibers made primarily of cellulose. These fibers resist cutting and make tension wood brittle. The gelatinous fibers of tension wood shrink and thus pull the stem upright.

Reaction wood in conifers such as pine forms on the lower side of the stem and is called *compression wood* (fig. 16.14b). Compression wood contains intercellular space and large amounts of lignin, which causes it to shrink less than normal wood on the upper side of the stem. The differential shrinkage of compression wood as compared to normal wood lifts the stem.

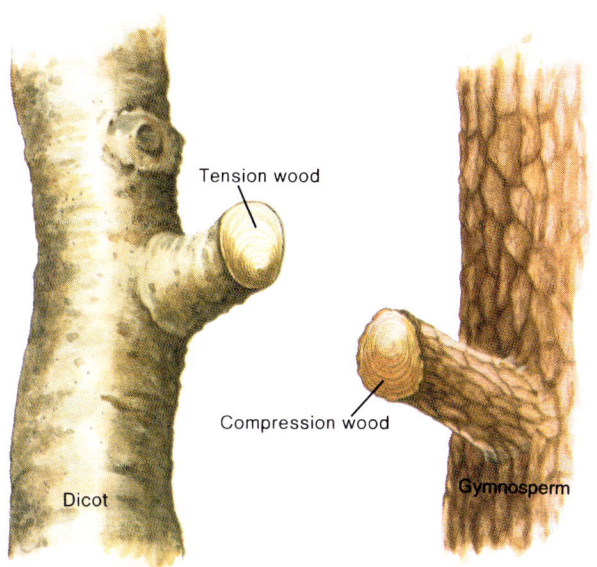

FIGURE 16.14

Formation of reaction wood. (a) In dicots, tension wood forms on the upper side of the stem. (b) In gymnosperms such as pine, compression wood forms on the lower side of the stem.

How do we know that reaction wood differentiates from the vascular cambium rather than the parenchyma in normal wood? Consider this simple experiment. Bent stems of *Eucalyptus* from which the vascular cambium is scraped from the upper side produce no tension wood and do not reorient themselves with respect to gravity. Removing the vascular cambium from the lower side of these stems has no effect on the production of tension wood or reorientation. Thus, the production of tension wood requires an intact vascular cambium. You may also wonder how we know that gravity rather than tension induces the formation of reaction wood. This was tested by bending the trunk of a sapling into a circle and determining where reaction wood formed (fig. 16.15). It formed only on the upper surface of the upper and lower side of the loop, indicating that gravity rather than tension induces the formation of reaction wood.

Reaction wood is important because it helps establish a tree's architecture, reorients stems, and maximizes the interception of light. It is useless as lumber, however, because it shrinks unpredictably when dried.

C O N C E P T

Reaction wood is abnormal wood that develops in leaning branches. In conifers it is called compression wood and forms on the lower side of a stem. In angiosperms it is called tension wood and forms on the upper side of a stem. Reaction wood is important because it helps establish a tree's architecture and reorients leaning stems.

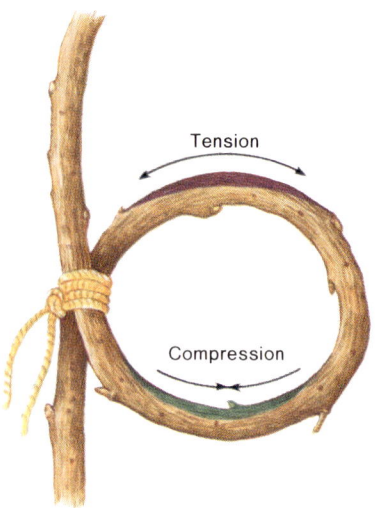

FIGURE 16.15

Testing the influence of gravity on the formation of reaction wood. When the stem of an angiosperm sapling is bent into a loop and tied into place, tension wood forms on the tops of both horizontal parts of the stem. Thus, the stimulus for the formation of tension wood must be gravity rather than the stress itself.

BARK: SECONDARY PHLOEM AND PERIDERM

All of the tissues outside of the vascular cambium constitute **bark** (fig. 16.16). Bark consists of two tissues:

Secondary phloem transports water and organic solutes between roots and leaves.

Periderm is an outer suberized layer that protects and insulates underlying tissues.

Secondary Phloem

Secondary phloem is the outer derivative of the vascular cambium (fig. 16.17). It includes sieve elements and parenchyma cells that often alternate with bands of thick-walled fibers and prevent splitting of the inner bark. Only the inner centimeter or so of secondary phloem contains functional sieve elements; sieve elements in the outer parts of secondary phloem are dead and nonfunctional, and help protect the inner tissues.

Periderm

Radial expansion resulting from secondary growth eventually ruptures the epidermis of stems and roots. The ruptured epidermis is replaced by periderm, which is a suberized layer that protects underlying tissues (fig. 16.18). Periderm consists of three tissues: phellogen (cork cambium), phellem (cork), and phelloderm (secondary cortex).

Phellogen

The **phellogen** (from the Greek words *phellos*, meaning "cork," and *genesis*, meaning "birth"), or cork cambium, is the meristem that produces the periderm (fig. 16.18). Phellogen differentiates soon after the vascular cambium forms, and originates

A.

B.

FIGURE 16.16

Varieties of bark. (a) Bark of Pacific yew (*Taxus brevifolia*). The bark of these trees contains a drug that is often effective at slowing or stopping the growth of some types of ovarian cancer. (b) Bark of aspen (*Populus tremuloides*). The markings on the bark are "graffiti" made by elk.

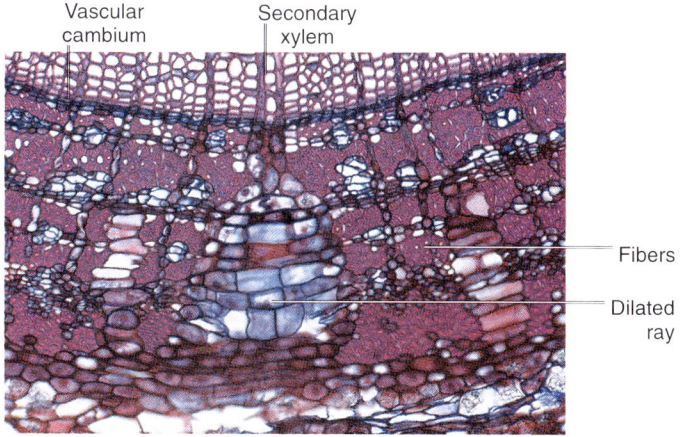

FIGURE 16.17

Cross section of secondary phloem of basswood (*Tilia*), a dicot. Secondary phloem is the outer derivative of the vascular cambium. Where in this photo are the conducting cells of the secondary phloem?

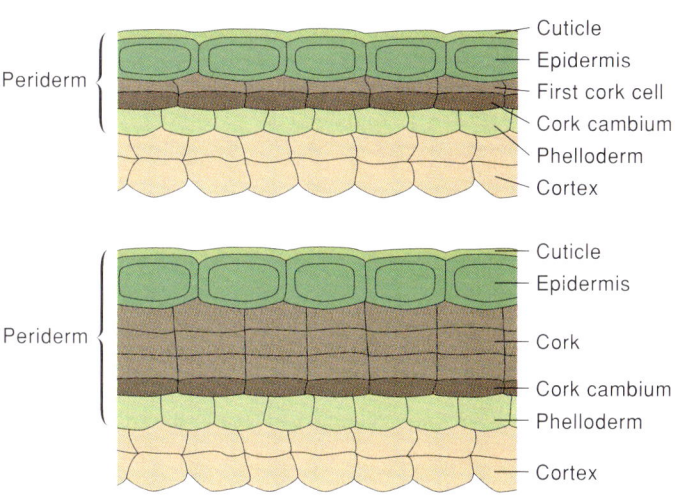

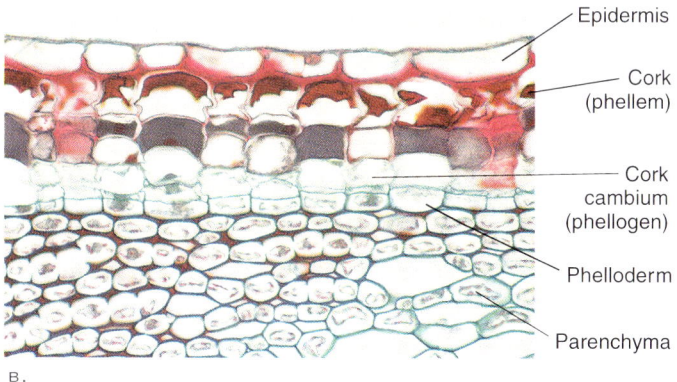

FIGURE 16.18

The periderm. (a) Diagrams showing the origin of the first cork cambium and the first layer of cork. (b) Light micrograph showing the cork cambium and newly formed cork in a stem of elderberry (*Sambucus*), × 400.

FIGURE 16.19

Ring and scale bark. (a) The thin, paperlike ring bark of paper birch (*Betula papyrifera*). (b) The scale bark of sycamore (*Platanus occidentalis*).

from cellular divisions in the cortex, secondary phloem, or epidermis. Unlike the vascular cambium, the phellogen has only one type of initial.

Secondary growth in most trees splits the outer periderm. Trees cope with this problem by continually forming new layers of phellogen deeper and deeper in the cortex. When the cortex is gone, these replacement layers of phellogen form from parenchyma in secondary phloem. In trees such as paper birch (*Betula*), the replacement phellogen is cylindrical; thus, these trees shed their bark in strips or large, hollow cylinders and are called *ring bark* trees (fig. 16.19a). Trees that continually form overlapping plates of phellem (cork) are called *scale bark* trees and include sycamore, elm, maple, and pine (fig. 16.19b). As their name implies, these trees produce scales rather than large rings of bark.

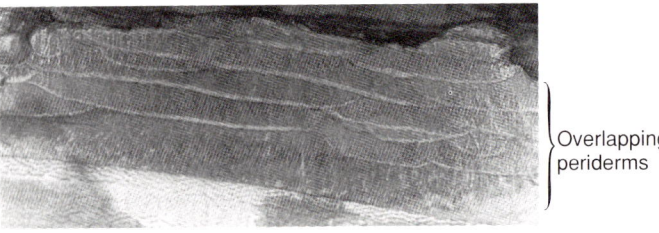

FIGURE 16.20

Overlapping periderms in bark. The lightly colored layers of this piece of outer bark are phellem. The innermost layer of phellem isolates the outer layers from the rest of the plant.

Phellem

The outer derivatives of the phellogen are radial rows of densely packed **phellem** (cork) cells (fig. 16.18), which are dead, suberized, and lack intercellular spaces. In one season, the phellogen produces as many as forty layers of cork cells that waterproof, insulate, and protect the plant. Cork also forms from phellogen on many fruits and potatoes, over wounds, and in abscission zones.

Trees typically produce several overlapping periderms (fig. 16.20). Phellem made by the innermost periderm seals the outer tissues from the water and nutrients carried in secondary xylem and phloem. As a result, all of the tissues outside of the innermost periderm layer die, and are called the *rhytidome*.

Gas exchange across the phellem occurs through **lenticels,** which are raised, localized areas of loosely packed cells (fig. 16.21). Lenticels are formed by the phellogen and range in shape from round to elliptical. Lenticels are easily visible on pears and apples, and are the dark spots and streaks on corks (see box 16.2, "Cork").

Phelloderm

The inner derivative of the phellogen is a parenchymatous **phelloderm,** or secondary cortex (fig. 16.18). Phelloderm cells are alive, are not suberized, and may be photosynthetic. Phelloderm contains many intercellular spaces for gas exchange.

C O N C E P T

All of the tissues outside of the vascular cambium constitute bark. Bark includes secondary phloem, which transports water and organic solutes, and periderm, which protects and insulates underlying tissues. Periderm is made by phellogen, or cork cambium. The outer derivative of phellogen is a suberized, protective tissue called phellem (cork), while the inner derivative is parenchymatous phelloderm. Gas exchange across the periderm occurs through lenticels.

Replacement of the epidermis by periderm during secondary growth has important consequences for plants. Epidermal cells are unique because they seldom dedifferentiate. Consequently, organs covered by an intact epidermis retain their individual integrity and do not graft, even when they

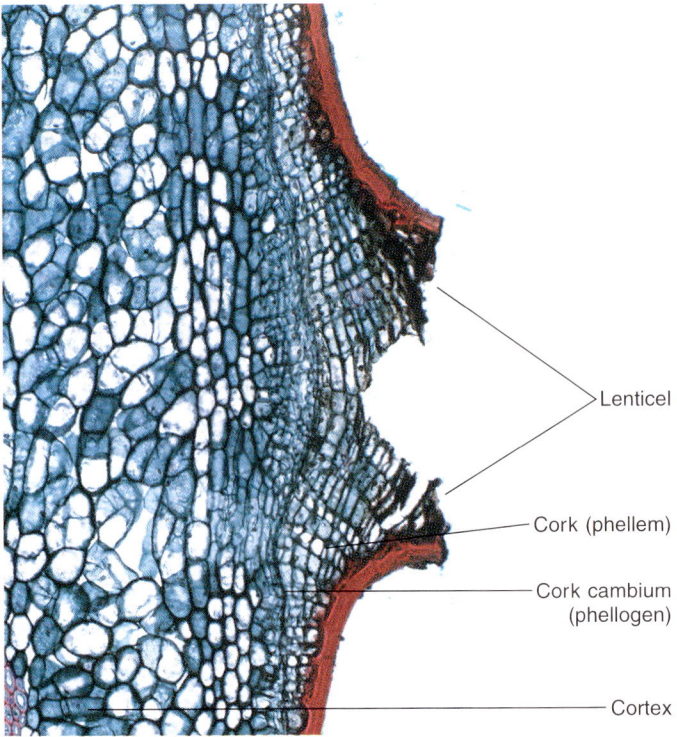

FIGURE 16.21

Cross section of part of a young stem of elderberry (*Sambucus*). Gas exchange across the phellem occurs through lenticels.

touch each other for long periods. This is why our lawns consist of thousands of individual blades of grass rather than a gigantic, fused blob of photosynthetic astroturf. Unlike epidermal cells, certain cells in the periderm can dedifferentiate and graft with members of the same or a related species. Grafts between adjacent organisms can be extensive; for example, more than half of the trees of a white-pine forest are often grafted together by secondary growth of their roots, thus creating a gigantic individual that "captures" other plants via more root grafts. These grafts establish close associations between different trees and radically change a forest. Root-grafted trees grow and develop as a single organism, and individual trees survive or die as part of a giant, interconnected system. Grafts also transmit pathogens and poisons among adjacent trees of the system; thus, applying poison to one tree often kills several adjacent trees.

UNUSUAL SECONDARY GROWTH

Dicots

Not all dicots have the typical kind of secondary growth described earlier in this chapter. The unusual secondary growth of some dicots results from one of three anomalies:

1. *Unusual behavior of a typical vascular cambium.* In plants such as carrot (*Daucus*) and some cacti, the vascular

BOXED READING 16.2

CORK

Each year we use almost 200,000 metric tons of commercial cork for things as diverse as stoppers for wine bottles, insulation for the space shuttle, and grips on symphony conductors' batons. Commercial cork is the outer bark of cork oak (*Quercus suber*), an evergreen oak grown in cork plantations in the western Mediterranean.

The first periderm, which forms from the epidermis, is commercially useless. Consequently, it is stripped and discarded when the tree is approximately 10 years old. A usable cork layer 3–10 cm thick can be harvested when a tree is 20–25 years old (i.e., when the tree is about 40 cm in diameter), and thereafter once per decade until the tree is approximately 150 years old. Since cork oaks grow crookedly, cork is harvested by hand rather than machine; workers use hatchets to peel it off in lumberlike slabs. The cork is then dried and boiled before being marketed. In contrast to that of most other trees, the cork of a cork oak breaks away at the cork cambium and can be peeled without harming the tree.

Commercial cork has several properties that make it ideal for a variety of uses. It consists of densely packed, suberized cells (~1 million cells cm^{-3}) that make it impermeable to liquids and gases. Half of a cork's volume is trapped air; thus, it is four times lighter than water. It is also compressible—cork can tolerate 400 kg cm^{-3} and expand to its original shape within one day. Cork is virtually indestructible, fire resistant, durable, resists friction, and absorbs vibration and sound.

BOX FIGURE 16.2

The bark of this cork oak (*Quercus suber*) has just been harvested for its cork. The inner bark remains, so that the tree does not die when the cork is harvested.

---

cambium produces large amounts of storage parenchyma and only a few procambial strands. In cacti, these parenchyma cells store water, while in carrots they store carbohydrates.

2. *Presence of more than one cambium.* In plants such as beets (*Beta vulgaris*), cambia are arranged in concentric circles. These *supernumerary cambia* form phloem toward the outside and xylem toward the inside, along with large amounts of storage parenchyma (fig. 16.22a). Conversely, plants such as sweet potato (*Ipomoea batatas*) have a normal cambium and several *accessory cambia* that form around xylary elements. These accessory cambia produce xylem to the inside and phloem to the outside (fig. 16.22b).

3. *Differential activity of a vascular cambium.* In plants such as Dutchman's pipe (*Aristolochia*), parts of the cambium produce parenchyma, while other parts produce vascular tissue. As a result, the stems of these plants are often ridged. In some species of *Bauhinia*, only two opposite areas of the vascular cambium are active, which produces a flattened stem.

Monocots

Secondary growth by monocots is considered unusual because most monocots form only a primary body. However, some arborescent monocots such as *Dracaena* and *Cordyline* form a vascular cambium just outside their vascular bundles (fig. 16.23a). Unlike that of dicots, the monocot cambium consists of only one irregularly shaped type of cell. The outer derivative of this cambium is a parenchymatous *secondary cortex*, while the inner derivatives are procambial strands and lignified *conjunctive tissue*. Procambial strands later produce vascular bundles that connect with the plant's primary vascular system.

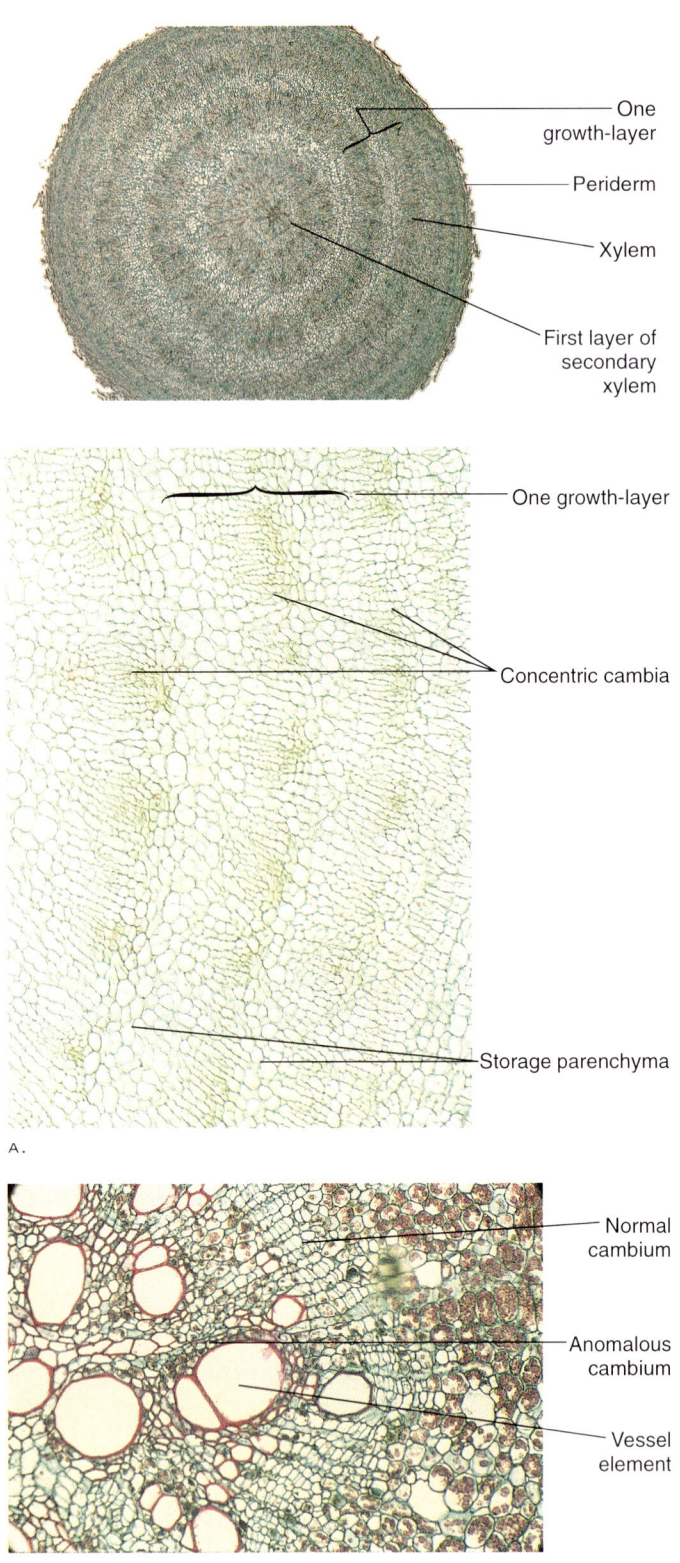

FIGURE 16.22

Unusual secondary growth. (a) In roots of sugar beet (*Beta vulgaris*), cambia form in concentric circles. Each cambium produces xylem, phloem, and storage parenchyma. (b) Roots of sweet potato (*Ipomoea batatas*) have a normal cambium plus several other anomalous cambia that form around vessels in secondary xylem. Each of these cambia produces phloem to the outside and xylem to the inside, ×100.

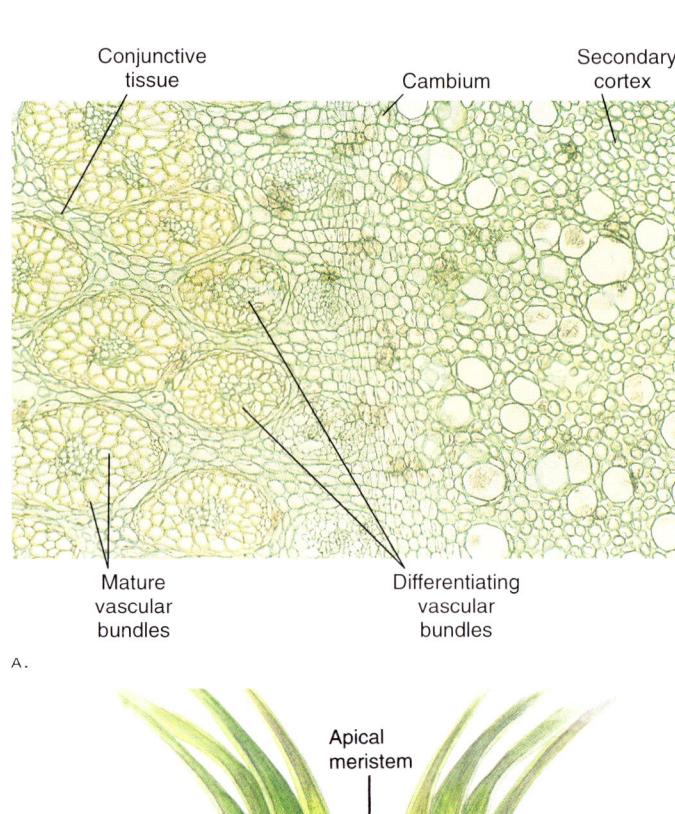

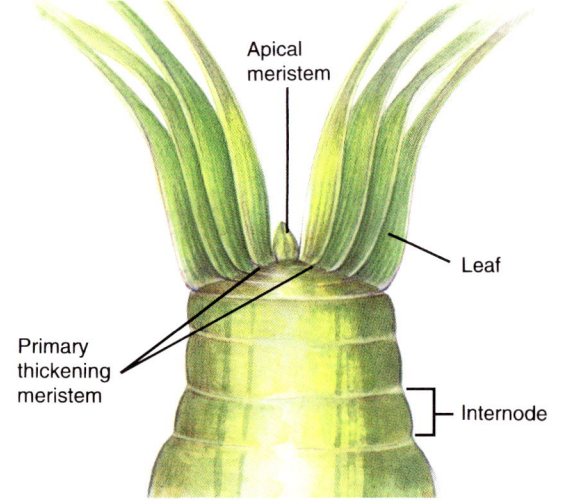

FIGURE 16.23

Vascular cambia and primary thickening in monocots. (a) *Dracaena* is a monocot with a vascular cambium that produces a secondary cortex of parenchyma to the outside and conjunctive tissue and procambial strands to the inside. The procambial strands later form vascular bundles, ×100. (b) In palms, primary thickening occurs just below the apical meristem.

Secondary growth of some monocots can be impressive: for example, some species of *Dracaena* can be more than 18 m high and 12 m around. However, not all arborescent monocots exhibit secondary growth. For example, palms are overgrown herbs that become treelike by *primary thickening*, in which intense lateral expansion occurs just behind the apical meristem (fig. 16.23b). Cells in these plants expand radially before elongating.

Periderms of monocots also differ from those of dicots. All cells of a monocot's periderm are suberized, and there are no different layers such as phellem, phellogen, or phelloderm.

BOXED READING 16.3
THE LORE OF TREES

Trees, like all plants, have a fascinating history and lore:

- There are about 1,000 species of trees in the United States and more species of trees in the Appalachian mountains than in all of Europe. However, there are more species of trees in one hectare (10,000 m²) of Malaysian rain forest than in all of the United States.

- The average tree planted today along a street in New York City lives only about seven years.

- A fully grown deciduous tree can pull more than a ton of water from the soil each day.

- Planting three or four trees around every American house would save 10%–50% on air-conditioning bills.

- The most isolated tree in the world is a Norwegian spruce growing in the wasteland of Campbell Island, Antarctica. Its nearest arboreal neighbor is 120 miles away in the Auckland Islands.

- The fastest-growing tree is *Albizia falcata,* a member of the pea family. One tree in Malaysia grew more than 10 meters in only thirteen months, and more than 30 meters in just over five years.

- Each year, the average tree takes in about 12 kg of carbon dioxide, an amount equivalent to that emitted by a car on a 7,000-km trip. The tree also releases enough oxygen to keep a family of four breathing for a year.

- Last year, people in the United States planted more than 4 billion trees. However, if you live in an urban area, you might not have noticed: in some of these areas, as many as four trees die for each one that is planted.

A.

C.

B.

D.

BOX FIGURE 16.3
(a) El Arbol del Tule, a Montezuma bald cypress (*Taxodium mucronatum*), is the oldest living organism in Mexico. (b) Red firs (*Abies magnifica*) on Mt. Shasta, California. Firs have been used to make telephone poles. (c) The world's largest Valley Oak (*Quercus lobata*) in Butte County, California. (d) Giant sequoia (*Sequoiadendron giganteum*) growing in Sequoia National Park in California. Adult trees are 10–20 times more massive than blue whales, the largest animals.

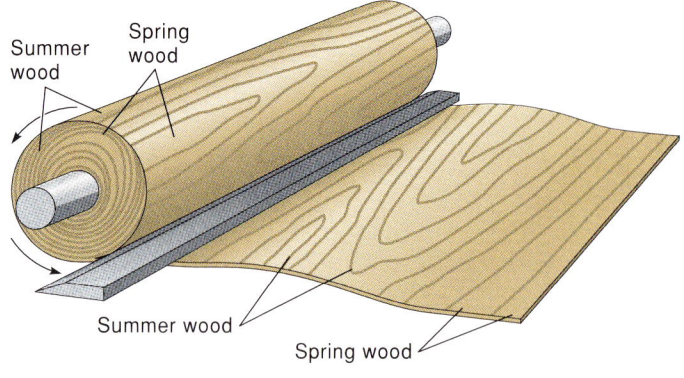

A.   B.

FIGURE 16.24

Making plywood. (a) Plywood is made by gluing together several sheets of veneer, usually at right angles. (b) Veneers are made by slowly peeling a thin sheet of wood from a log. This diagram shows how the grain in plywood relates to annual rings in the log.

USES OF SECONDARY XYLEM AND PHLOEM

We make more than 5,000 products from the secondary xylem and phloem of the 1,000 or so species of trees that grow in the United States.

Lumber

Most of the wood consumed in the United States is used for lumber and comes from about thirty-five species of trees. We use this lumber for all sorts of things, including fence posts, telephone poles, houses (white pine), furniture and paneling (black walnut), barrels and flooring (white oak), oars (white ash), skis (hickory), acoustic instruments such as violins (spruce, rosewood, maple), bowling pins, croquet balls, and cue sticks (hard maple). Since wood conducts heat poorly, wood floors are a good way to insulate a house. We also use lumber to make veneers, plywood, and particle board.

Veneers are thin (~1 mm thick) sheets of wood peeled from large logs (fig. 16.24). They are often glued over cheaper and less attractive wood to make cabinets and furniture.

Plywood is made by gluing several veneers together at right angles. These layered veneers make plywood strong, hard to split, and well suited for construction.

Particle board is made by gluing and pressing together small particles (i.e., chips) of wood into flat sheets. Particle board is used as insulation and to make boxes.

Doing Botany Yourself

Devise a method for determining the specific gravity of particle board, and explain why its specific gravity might be different from that of whole wood. How can you obtain evidence for your explanation?

Pulp and Paper

Paper is made from wood pulp, which is pulverized wood.[1] Paper mills first strip the bark from harvested trees. The trunks are then either pulverized on a grindstone or cooked in a batch of chemicals. Both of these processes degrade the wood into pulp, which is about 99% water and 1% fibers. The untreated pulp is fed into a Fourdrinier machine, which spreads and dries it into newsprint, an inexpensive type of paper containing only about 5% water. In only one minute, a paper-making machine can produce a sheet of paper covering more than 22,000 square meters, an area bigger than five football fields. An average tree harvested by a lumber company produces enough pulp to make about four-hundred copies of a forty-page newsprint tabloid.

Because newsprint is made from untreated pulp, it contains much lignin; this lignin contributes to the paper becoming yellow and brittle after a few days. To make higher quality paper, paper mills first bleach and remove the lignin from the pulp. They then add titanium dioxide to whiten the paper and, if necessary, add china clay, latex, and alginates (compounds extracted from kelps) to produce glossy paper. All of these processes increase costs and decrease yield by 20%–50%, which is why fine stationery costs more than newsprint. Although more than 90% of paper is derived from wood, special types of paper are derived from a few other plants. For example, money and cigarette paper are made from flax (*Linum*).

Most wood pulp is used to make paper, for which there is an ever-increasing demand. Each person in the United States uses an average of 681 g of paper per day for things ranging from advertisements and botany books to airplanes, greeting cards, clothes, and general paperwork. This year, almost 2 trillion pieces of paper will move through offices.

Wood pulp is also used to make several other products, including cardboard, fiberboard, nitrocellulose explosives, synthetic cattle food, fillers in ice cream and bread, linoleum, and

---

1. The writing material of the ancient world was woven reeds of papyrus, an aquatic plant that still grows in the delta of the Nile River. However, the Chinese began making paper in about 105 A.D. Before that they wrote and drew on silk.

plastic films such as cellophane. In fact, you may be wearing a derivative of wood pulp—rayon (artificial silk) and several other synthetic fibers are made from the dissolved cellulose of wood pulp.

Fuel

More than half of all wood used worldwide is used for fuel: More than 1.5 billion people get at least 90% of their energy for heating and cooking by burning wood, while another 1 billion use wood for at least half of their heating and cooking needs. In developing countries, almost 90% of wood is used for fuel, while only 10% of the wood consumed in the United States is for fuel.

Charcoal and Related Products

Charcoal is made by burning blocks of hardwood in the presence of relatively little air. Subsequent distillation of the briquettes produces other economically important products, including methanol (wood alcohol), methane, acetic acid, and acetone. Distilling blocks of softwoods produces turpentine and a sticky rosin used to make paint, ink, lacquer, soap, and polishes. Rosin is also rubbed onto the bows of violins, the shoes of boxers and ballerinas, and the hands of baseball pitchers to increase friction.

Fabrics and Rope

Fibers from the inner bark of baobab (*Adansonia digitata*) trees are woven into fabric and rope. Baobab ropes are strong—as natives of tropical Africa say, "As secure as an elephant bound with baobab rope." Baobab trees are often more than 10 m wide, and their hollowed trunks have been used for everything from morgues to city jails.

Sugar and Spice

Cinnamon is the outer bark of *Cinnamomum zeylanicum*; it is sold either as a pulverized powder or as "sticks," which are the curled outer-bark segments split from twigs. Maple syrup originates from the wood of sugar maples (see box 21.2, "Tapping Wood for Maple Syrup," in Chapter 21).

Dyes

Dyes extracted from the inner bark of alder (*Alnus*) are used to color wool and other fabrics.

Drugs

The bark of *Cinchona* trees contains quinine, a drug that kills the protozoan responsible for malaria, the world's leading killer. Although quinine has been a prized drug since the seventeenth century, it has now been replaced by the synthetic drug chloroquine. A more recent discovery is that the bark of the Pacific yew (fig. 16.16) contains a promising drug that arrests the growth of tumors. To extract only one kilogram of the drug requires that we cut down about 12,000 trees and strip from them more than 27,000 kg of bark.

Other Uses of Secondary Xylem and Phloem

Tannins from chestnut and tan oak are used to prepare animal skins for sale, and natural chewing gum is extracted from sapodillo (*Manilkara zapota*) (*Achras sapota*) trees (most chewing gums sold today are synthetic polymers). *Hevea brasiliensis* produces natural rubber, and turpentine is distilled from longleaf and slash pines. Gum arabic from *Acacia* trees is used to glaze doughnuts and as the adhesive on stamps and envelopes. Balsa wood is used to make model airplanes and lifeboats, and ambatch, which is lighter than balsa, is used to make pith helmets. The average American annually uses the equivalent of a tree more than 30 m tall and 45 cm in diameter.

CONCEPT

We have many uses for secondary xylem and bark, including lumber, pulp, fuel, charcoal and related products, fabrics and rope, spices, dyes, and drugs.

The Lore of Plants

Each year, people in the United States cut down about 3 billion trees. This harvest exceeds that of all other countries, including Brazil.

Commemorating Trees

Trees are commemorated each year on Arbor Day, a largely ignored holiday that falls on the last Friday in April. Although Arbor Day was established in 1872 by the Nebraska state legislature, it didn't go into effect until several years later, because overeager settlers had already cut down most trees in the state. The first year, repentant Nebraskans planted a million trees; within sixteen years, citizens had planted 600 million trees. The rest of the country was apparently impressed, and gradually acknowledged Arbor Day. Although Arbor Day is a feature of the federal calendar, it remains a fake holiday: no one gets the day off.

Chapter Summary

Secondary growth is growth in girth produced by two lateral meristems: the vascular cambium and phellogen (cork cambium). These meristems produce a secondary body consisting of secondary xylem (wood) and bark (secondary phloem and periderm). Secondary growth is important because it enables plants to grow taller and increases their chances of intercepting light for photosynthesis.

The vascular cambium is a meristematic cylinder made of ray initials and fusiform initials. Ray initials produce rays that transport fluids radially. Fusiform initials produce the secondary xylem and secondary phloem. The conducting cells of these tissues transport fluids longitudinally.

The vascular cambium differentiates from latent procambium between primary xylem and phloem. These regions of fascicular cambium are subsequently linked when the interfascicular cambium differentiates through the pericycle or cortex.

The activity of the vascular cambium is controlled by water availability, the sensitivity of cambial cells to auxin, and pressure. Wood made during moist days of spring contains large xylary elements, and is called *early* or *spring wood*. Wood produced during the drier days of summer contains few large vessels, and is called *late* or *summer wood*. Differences between early and late wood are visible as growth rings.

Trees fight pathogens by compartmentalization, which involves walling off infections, fortifying cell walls near wounds, and producing antimicrobial compounds such as phenols. Trees survive without these damaged compartments by continually producing new tissues to maintain their activities.

Secondary xylem, or wood, is the inner derivative of the vascular cambium. Softwoods are nonflowering plants native to temperate regions; their wood is homogeneous and contains tracheids and uniseriate rays. Hardwoods are dicots whose wood contains multiseriate rays, tracheids, vessels, and fibers.

Sapwood transports water and dissolved nutrients, and comprises the outer few centimeters of wood. Heartwood is located in the center of a tree and is where many waste products are stored. Heartwood does not transport water and nutrients.

Reaction wood develops in leaning branches. In conifers it is called *compression wood* and forms on the lower side of the stem. Reaction wood in angiosperms is called *tension wood*, and forms on the upper side of the stem. Auxin influences the formation of reaction wood. Reaction wood is important because it helps establish a tree's architecture.

All of the tissues outside of the vascular cambium constitute bark. Bark includes secondary phloem, which transports water and organic solutes, and periderm, which protects and insulates underlying tissues. Periderm is produced by phellogen, or cork cambium. The outer derivative of phellogen is a suberized, protective tissue called *phellem* (cork), while the inner derivative is parenchymatous phelloderm. Gas exchange across the periderm occurs through lenticels.

Unusual secondary growth in dicots results from the unusual behavior of a typical vascular cambium, the presence of more than one cambium, and differential activity of a vascular cambium. A few monocots have a vascular cambium that produces secondary cortex to the outside and procambial strands to the inside. These procambial strands later produce vascular bundles.

We use secondary xylem and bark to make lumber, pulp, fuel, paper, charcoal, fabrics, rope, spice, dyes, and drugs.

Questions for Further Thought and Study

1. The Pacific yew tree was once thought to be economically worthless. Now, however, we prize it for its anticancer drugs. What does this tell us about the possible consequences of our destruction of tropical rain forests?
2. Are there growth rings in the secondary phloem? Explain your answer.
3. Which is denser, early wood or late wood? Why?
4. Which would produce more uniform paper, pulp from a softwood or from a hardwood? Why?
5. What has happened to the epidermis, cortex, and primary phloem of a 200-year-old oak tree?
6. How does secondary growth in monocots differ from that of dicots?
7. Why don't liquids ooze through lenticels of bottle corks?
8. Is formation of lateral roots an example of primary or secondary growth? Why?
9. Why would extensive secondary growth in leaves or annual plants be a "poor investment"?
10. Cell walls of most plants have a specific gravity of ~1.53. How, then, can woods of different species have different specific gravities?
11. Beached blue whales are crushed by the weight of their bodies. Why aren't trees crushed by their weight?
12. What is the significance of secondary growth?
13. Discuss the contribution of each of the following to secondary growth: vascular cambium, rays, periderm, secondary phloem, heartwood, sapwood, compartmentalization of damage.
14. How would wood produced during a wet year differ from that produced during a dry year?
15. How could thick bark benefit a tree?
16. Discuss the following data relative to the energy costs and benefits of secondary growth:

Tissue	Respiratory Rate $(mm^3 O_2\ h^{-1}\ gfw^{-1})$*
Secondary phloem	88
Vascular cambium	180
Sapwood	17
Heartwood	0.3

*Cubic millimeters of oxygen consumed per hour per gram fresh weight of tissue.

17. The structural diversity of plants results from different arrangements and proportions of tissues rather than from the presence of different types of cells. Use specific examples to support this statement, and indicate why it is important for understanding how plants are designed.
18. Two major functions of a tree trunk are conduction and support. Describe the anatomical basis for each of these functions.
19. Why does removing a ring of bark from around a tree (i.e., *girdling*) kill a tree?

Suggested Readings

ARTICLES

Fritts, H. C. 1972. Tree rings and climate. *Scientific American* 226:92–101.

Hitch, Charles J. 1982. Dendrochronology and serendipity. *American Scientist* 70 (May–June):300–305.

Lauchaud, Suzanne. 1989. Participation of auxin and abscisic acid in the regulation of seasonal variations in cambial activity and xylogenesis. *Trees* 3:125–137.

Patton, Phil. 1984. Wooden bats still reign supreme at the old ball game. *Smithsonian* 15 (October):152–176.

Shigo, Alex. 1985. Compartmentalization of decay in trees. *Scientific American* 252 (April):96–103.

———. 1986. Journey to the center of a tree. *American Forests* 92 (June):18–23.

Stewart, D. 1990. Green giants. *Discover* (April):61–64.

Wilson, B. F., and R. R. Archer. 1977. Reaction wood: Induction and mechanism of action. *Annual Review of Plant Physiology* 28:23–43.

BOOKS

Adkins, Jan. 1980. *The Wood Book*. Boston: Little, Brown.

Burnie, D. 1988. *Eyewitness Books: Tree*. New York: Alfred A. Knopf.

Core, H. A., W. A. Coté, and A. C. Day. 1979. *Wood Structure and Identification*. 2d ed. Syracuse, NY: Syracuse University Press.

Hoadley, R. B. 1980. *Understanding Wood*. Newtown, CT: The Taunton Press.

Johnston, David. 1983. *The Wood Handbook for Craftsmen*. New York: Arco Publishing.

Zimmerman, M. H. 1983. *Xylem Structure and the Ascent of Sap*. Berlin: Springer-Verlag.

Flowers of the scarlet gilia (*Ipomopsis aggregata*). Flowers are the reproductive organs of flowering plants.

CHAPTER 17

Reproductive Morphology

Chapter Outline

INTRODUCTION
FLOWERS
  Stamens
  Carpels
  Petals
  Sepals
  The Nature of Flower Parts
REPRODUCTIVE MORPHOLOGY AND PLANT DIVERSITY
  Flower Variation

BOX 17.1
ORCHIDS THAT LOOK LIKE WASPS

  Inflorescences
  Breeding Systems
  Vegetative Reproduction
  Pollination Mechanisms
FRUITS
  Types of Fruit
  Fruit Development
SEEDS
  Seed Structure
  Seed Germination
  Seed Banks

BOX 17.2
THE DODO BIRD AND THE TAMBALACOQUE TREE

  Seedling Development
DISPERSAL OF FRUITS AND SEEDS
  Dispersal by Wind and Water
  Dispersal by Animals
  Self-Dispersal
Chapter Summary
Questions for Further Thought and Study
Suggested Readings

Chapter Overview

Plants have evolved an extraordinary diversity of features for reproducing sexually. The most common reproductive organ of plants is the flower. Indeed, there are more plant species with flowers than with any other kind of reproductive organ. For this reason, the focus of this chapter is on the reproductive morphology of flowering plants.

Successful reproduction in different plants depends on adaptations to various environmental factors. Much of the diversity among flowers comes from specializations for certain pollinators. Further diversity is based on adaptations of seeds for germination and of fruits and seeds for dispersal. Most of the variation in reproductive morphology presented in this chapter is adaptive. This means that features such as the colors and odors of flowers, the size of pollen, and the number of seeds in a fruit are functionally important.

The adaptive point of view is standard for general botany textbooks, probably because explanations of plant morphology are more powerful when structure can be related to function. As you read through this chapter, however, keep in mind that many of the morphologically distinctive features of different plants are not clearly distinctive in their functions. Mustard flowers, for example, have four petals and six stamens, but wintergreen flowers have five petals and ten stamens. It is not clear why, or even if, these kinds of differences are important in the reproduction of plants. Nevertheless, it does show the important and dynamic role that reproductive morphology has in our attempts to understand why there are so many kinds of plants.

Introduction

The most familiar reproductive structures of plants are flowers, fruits, and seeds. Flowering plants get more attention than any other group, largely because of their economic importance, beauty, and diversity. Flowers are distinguished from other kinds of reproductive organs by having seeds in containers, which become fruits. As a group, therefore, the flowering plants are commonly referred to as the **angiosperms,** which indicates that their seeds (*sperm*) are in vessels (*angio*) (fig. 17.1).

The role of reproductive structures is the same in all plants: to produce offspring. Moreover, the basic life cycle is the same for all plants (see Chapter 10). The sporophyte produces spore mother cells that undergo meiosis, thereby producing spores. These spores grow into gametophytes that produce gametes; male and female gametes then fuse into a zygote that becomes the next sporophyte stage. Thus, the life cycle of plants is an alternation between sporophytic and gametophytic stages.

Organs that house meiosis and fertilization range from tiny spore capsules in mosses and ferns to large cones in pines and firs. The most familiar reproductive organs of plants to you, however, are probably flowers, which are the focus of this chapter.

Flowers

Like other organs, flowers consist of different tissues and parts that collectively have one main function: in this case, to produce a new generation. A flower is essentially a short stem of several nodes with short internodes between them. However, unlike most other shoots, a floral shoot does not have an apical meristem that grows continuously, because the flower is at the tip of the shoot.

Study the flower shown in figure 17.2. In this generalized flower, the region of the floral shoot where the flower parts are attached is the **receptacle.** The various parts of flowers are considered by some botanists to be modified leaves, all of which are attached at nodes. The lowest node is a whorl of **sepals,** collectively called the **calyx.** A whorl of **petals,** together called the **corolla,** forms above the calyx. Sepals and petals are the sterile parts of flowers; the fertile parts are the **stamens,** which form just above the corolla, and the **carpels,** which form in the center of the flower (fig. 17.2).

Botanists are interested in floral variation because features of flowers are important for plant classification. Floral variation also represents the evolutionary record of flowering plants. The next few sections of this chapter present a brief overview of floral variation and the nature of different parts of flowers.

Stamens

Each stamen consists of four pollen-containing chambers that are fused into an **anther,** which is often on a stalk called a **filament** (fig. 17.3). The chambers are named **microsporangia** (singular, **microsporangium**), because that is where microspores are produced. The stamens are collectively called the **androecium** (plural, **androecia**), which means "male house." *Male* refers to the male gametophytes that develop from the microspores.

Historically, the androecium was believed to be the male part of the flower; indeed, stamens are still sometimes called the male floral parts. Reference to the "maleness" of the androecium is misleading, however, since no part of the sporophyte can have a gender. Only the gametophytes can be male or female, because only they can produce male or female gametes. Accordingly, the "male" flowers of pumpkin, date palm, and corn are more correctly called *staminate flowers*. Likewise, "male" mulberry, pistachio, and jojoba plants, which have only staminate flowers, should instead be called *staminate plants*.

Variation among androecia comes mostly from the number and arrangement of stamens, the attachment of anthers to the filament, and the way the anthers open for releasing pollen. For example, the androecium consists of one stamen in some grasses, two in ash, four in fuchsia, six in mustard, and dozens in buttercups and poppies (fig. 17.4). In addition, the stamens

FIGURE 17.1

Angiosperms are plants whose seeds are in vessels called fruits, as shown here from papaya (*Carica papaya*).

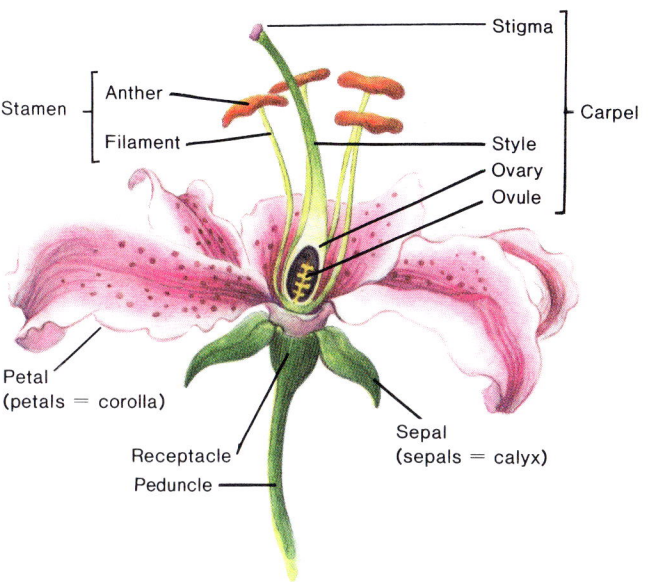

FIGURE 17.2

The parts of a flower. This is a generalized flower that has four main kinds of parts: sepals, petals, stamens, and carpel.

A.

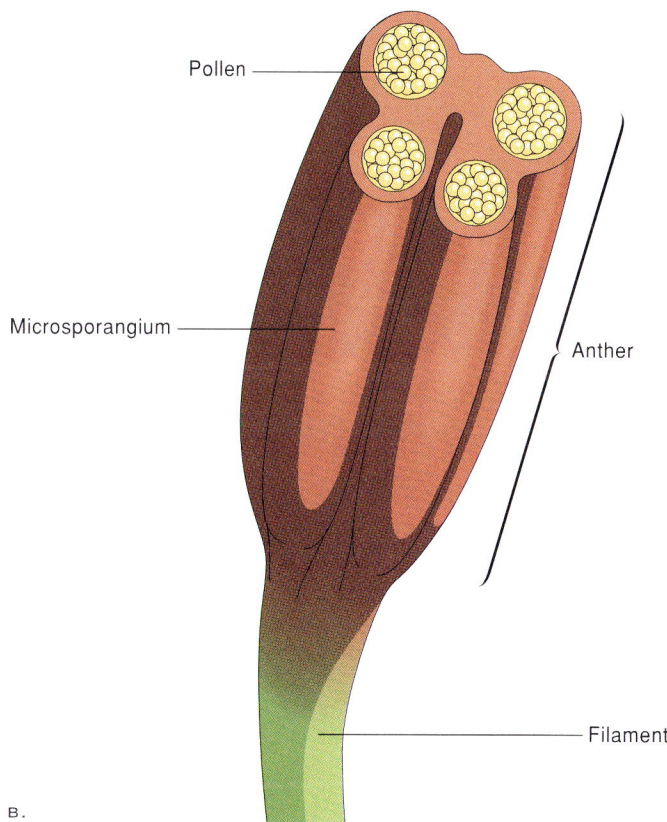

B.

FIGURE 17.3

(a) Flower of lily (*Lilium* sp.) showing six stamens each with a white filament and golden-brown anther. (b) A single stamen, showing pollen in microsporangia.

CHAPTER SEVENTEEN *Reproductive Morphology*

FIGURE 17.4

The number of stamens in an androecium varies among different kinds of flowers. This photo shows dozens of stamens in St.-John's-Wort (*Hypericum* sp.).

FIGURE 17.6

Fruit development in garden pea (*Pisum sativum*). (a) Pod containing immature seeds. (b) Ripe fruits with mature seeds.

FIGURE 17.5

Stamens are fused in some flowers. Filaments in rose mallow (*Hibiscus* sp.), for example, are fused to the long style.

of mustard, buttercup, and poppy are free from each other, but those of garden pea, cotton, and mallow are fused at their filaments (fig. 17.5).

Although anthers usually release pollen by splitting open along a slit, rhododendrons and other members of the heath family release pollen through pores at the tip of the anther.

Carpels

The main parts of an ovary are one or more **carpels,** each of which contain one to several ovules, and the **stigma,** which is the receptive area for pollen (fig. 17.2). The stigma is often borne on a stalklike **style** that extends from the ovary, thereby elevating the stigma to a level that enhances pollination (see

Chapter 10). A flower can have one or more carpels, collectively called the **gynoecium** ("female house") in reference to the female gametophytes that are produced in this structure. Like the androecium, however, the gynoecium has no gender, because it is an organ of the sporophyte. Nevertheless, carpellate flowers, as well as plants that have only carpellate flowers, are often incorrectly referred to as female.

The ovary has one or more chambers, each of which encloses ovules. In the simplest ovaries, there is one carpel with one or more ovules, as in garden pea (fig. 17.6). The ovaries of other flowers, such as those of red larkspur, consist of two or more free carpels. However, violets and many other flowers have carpels that coalesce at their edges.

Areas where ovules are attached within the carpels are the **placentae** (singular, **placenta**). In garden pea and red larkspur, the placenta occurs along the margin of the suture, which is referred to as **marginal placentation.** Ovaries derived from more than one carpel have more complex types of placentation (fig. 17.7). For example, when placentae are on the ovary wall, as in violet, the attachment of ovules is called **parietal placentation.** In contrast, placentae on the central axis of the ovary are **axial** when there is more than one chamber, as in lily, or **free-central** when there is just one chamber, as in primrose (fig. 17.7).

An ovary usually has several indicators for how many carpels it comprises. The most straightforward cases have a chamber for each carpel, as in citrus. This means that each section of an orange, for example, represents a carpel. In addition, stigmas or styles may also reflect the number of carpels. Accordingly, the lily has three seed chambers and three stigmas, but it has one style that probably evolved by the fusion of three styles from the three carpels. Likewise, some kinds of evening primrose have four lobes on the stigma, signifying four carpels. Other kinds of evening primrose, however, have only one stigma but still have

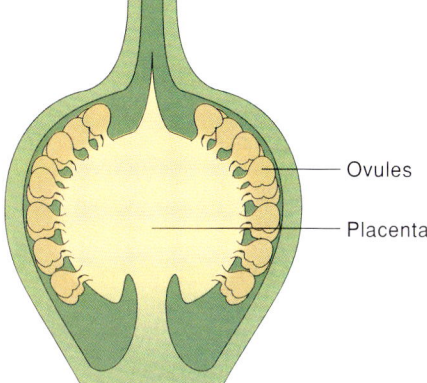

A. Free central (longitudinal section)

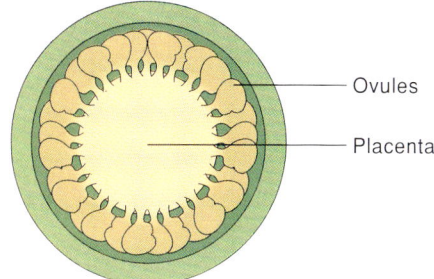

B. Free central (cross section)

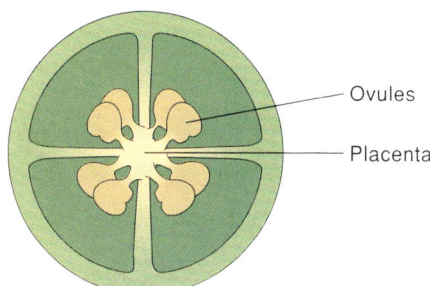

C. Axial (cross section)

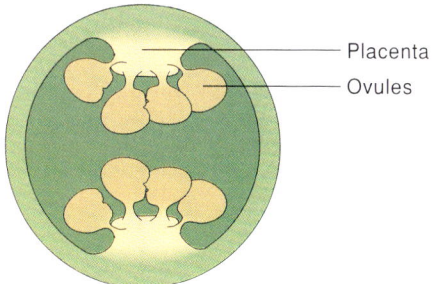

D. Parietal (cross section)

FIGURE 17.7

Ovules are attached to different places in ovaries of different kinds of plants. The area of ovular attachment is called the placenta.

four seed chambers (fig. 17.8). Conversely, in saguaro and other cacti, the ovary has one chamber and several stigmas, which suggests that the ovary evolved from several carpels. This suggestion is supported by the occurrence of several rows of ovules, each of which came from a separate placenta.

A.

B.

FIGURE 17.8

(a) Flower of evening primrose (*Oenothera hookeri*), showing stigma with four lobes. (b) Flower of evening primrose (*Oenothera biennis*), showing an unlobed stigma. Each flower has four fused carpels.

Flowers consist of two kinds of reproductive structures: the androecium, which produces the male gametophytes in pollen grains, and the gynoecium, which produces female gametophytes in ovules that are contained in ovaries.

Petals

The corolla is usually the most noticeable part of a flower. Petals are often large and showy. Differences in color and odor distinguish many kinds of flowers. In addition, petals may be free (see figs. 17.2 and 17.4) fused into a short tube with large lobes (as in oleander), or fused into a long tube that encompasses most of the corolla (as in honeysuckle; fig. 17.9). Many of the modifications of corollas are important in pollination, which is discussed later in this chapter.

A.

B.

FIGURE 17.9

(a) Flower of oleander (*Nerium oleander*), showing short tube of fused petals. (b) Flower of honeysuckle (*Lonicera* sp.), showing petals fused into a long tube.

FIGURE 17.10

Flowers of stonecrop (*Sedum* sp.) have five sepals that alternate with the five petals. Each flower also has ten stamens and five carpels.

In addition to color and odor, petal development also influences flower symmetry. When all petals develop equally, corollas are radially symmetrical (i.e., regular). However, when petals do not develop equally, corollas become bilaterally symmetrical (i.e., irregular). Mustard, lily, oleander, and poppy have regular corollas, whereas orchids, monkeyflower, and garden pea have irregular corollas.

Each flower of a plant usually has a specific number of petals. For example, wild roses have five petals. The number of petals often corresponds to the number of stamens, carpels, and sepals. However, the flowers of cacti, buttercups, and magnolias have an indefinite number of petals. Furthermore, in cacti and magnolias, petals intergrade with sepals, thereby making it impossible to distinguish all petals from all sepals.

Sepals

Most sepals are leaflike, but they may resemble petals or intergrade with them. Sepals of lily are identical in form to the petals. The calyx in four-o'clocks looks like a corolla, but flowers of these plants have no petals. Like petals, sepals may be fused into a tube, and the calyx may be regular or irregular in its symmetry.

The number of sepals, like the number of petals, often corresponds to the number of other flower parts. In most flowers, the number of petals and the number of sepals are identical. Moreover, if you look directly at the center of a flower that has the same number of sepals and petals, the sepals appear to alternate with the petals (fig. 17.10).

Sepals protect the inner parts of a flower before it opens. The calyx is especially important for keeping the unopened flower from desiccating. However, sepals may fall off the flower when it matures. Poppies, for example, appear to have no sepals because they fall off just before the flower opens.

The Nature of Flower Parts

As mentioned, each part of a flower is thought by many botanists to be a specialized leaf. This idea is best supported for sepals because they are mostly leaflike. Botanists generally agree, therefore, that sepals are specialized leaves, but there is a long and fascinating history of debate over the origins of petals, stamens, and carpels. This disagreement stems from the absence of any direct evidence, either from fossils or from living plants, for their origins. Instead, different conclusions have been drawn from indirect evidence.

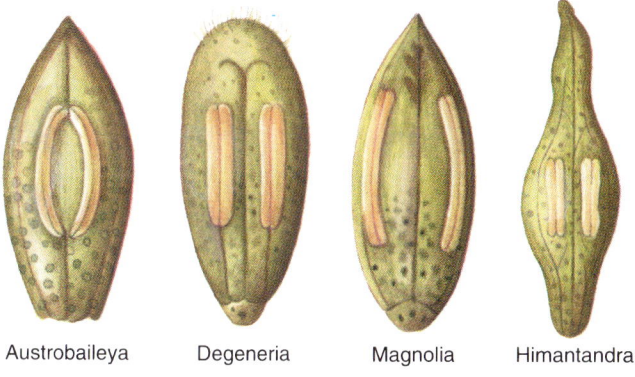

FIGURE 17.11

Stamens of *Austrobaileya* and certain other tropical trees look like leaves with microsporangia embedded in them. Such leaflike stamens are indirect evidence that stamens arose from fertile leaves.

The possibility that stamens and carpels were derived from specialized leaves was so widely accepted at one time that the terms **microsporophyll** (microspore-leaf) and **megasporophyll** (megaspore-leaf), respectively, were used to describe them. These terms are still commonly used. Support for the origin of stamens from leaves comes from the stamen morphology of certain tropical trees, whose stamens look like leaves, each with several veins and four microsporangia embedded in it (fig. 17.11). If such stamens represent primitive types, then the common stamen type evolved from them by loss of the lateral veins and by reduction of the leaf blade into a filament. However, in many flowers, the vascular bundles that lead to stamens originate in groups, not as single veins as if from a microsporophyll. This may suggest that each stamen came from a group of branching structures, which evolved into a solitary stamen by the loss of all but one member of the group. Such observations have prompted many botanists to question the leaf-origin hypothesis for stamens. Thus, the issue remains unsettled.

The argument for the origin of the carpel from a fertile leaf is the same as that for the origin of the stamen. A supportive example is the carpel of the genus *Drimys*, which is a tropical tree related to magnolia (fig. 17.12). The carpel of *Drimys* looks like it evolved from a leaf by folding in the middle and joining at the margins. The stigma in this carpel is the entire length of the margin. According to this model, carpels evolved further by sealing the margins and shrinking the stigmatic surface to a small area at the apex of the carpel. A pea pod represents the endpoint of such evolution.

Alternative views on the origin of the carpel come from fossil evidence. One view is that the carpel arose from the seed-bearing structure of an extinct group of plants called the *seed ferns*. These and other seed-bearing structures from fossil plants challenge the idea that carpels evolved from fertile leaves. Thus, like the origin of stamens, the origin of carpels remains controversial.

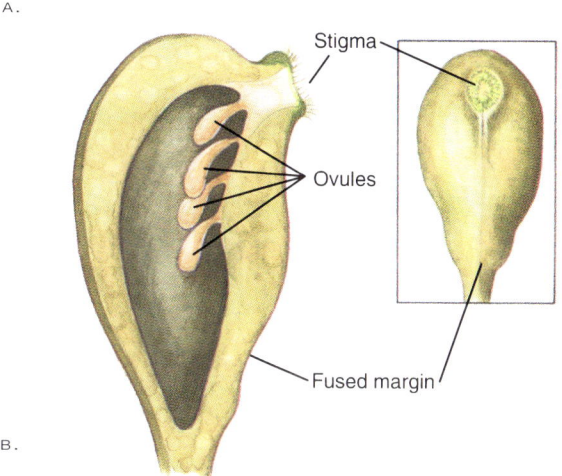

FIGURE 17.12

(a) Flowers of *Drimys winteri* have short, broad stamens and free, stalked carpels. (b) Carpels of *Drimys* resemble leaves folded at the middle and fused at the edges, with a stigma along the entire length of the fused area.

Petals are probably derived from stamens, according to evidence provided by abnormal flowers. For example, the "double flower" of cultivated roses and camellias appears to have risen by the transformation of stamens into petals (fig. 17.13). This transformation results from the broadening of the filament and anther of most or all of the stamens into petal-like structures, accompanied by a color change to that of the petals. In addition, such transformed stamens cannot

FIGURE 17.13

"Double flowers" of cultivated roses arose by transformation of stamens into petals. Such transformations may indicate that petals evolved from stamens.

FIGURE 17.14

Flowers of southern magnolia (*Magnolia grandiflora*) have many carpels, stamens, and petals, and usually three petal-like sepals.

produce pollen. Conversely, petals may have arisen from sepals. Many flowers, such as those of the saguaro cactus, lily, and magnolia, have petals that either intergrade with sepals or are identical with them. This observation supports the idea that both sepals and petals arose from specialized leaves.

C O N C E P T

Different parts of flowers may have arisen as modified leaves, or they may have arisen from the fertile structures of extinct seed plants. There is indirect evidence for both points of view.

REPRODUCTIVE MORPHOLOGY AND PLANT DIVERSITY

The diversity of flowering plants is greater than that of any other group of plants; perhaps as many as 260,000 species of flowering plants have been discovered and named. The diversity of angiosperms is reflected in their different sizes, shapes, and forms. The basis for this diversity comes from the reproductive success of flowering plants in a wide variety of habitats. Reproductive success in angiosperms is based on the evolution of the flower, which allowed this plant group to diversify into new habitats. Such diversification was made possible by the protective ovary around the seeds and the potential for efficient pollination by insects and other animals.

The following sections describe some of the variations in flowers, fruits, and pollination mechanisms that represent angiosperm diversity. Understanding this diversity and knowing some of the terms that botanists use to describe it will help you to use field guides and other books to identify plants in native habitats and gardens.

Flower Variation

A typical flower has several sepals, petals, stamens, and carpels. Much of the variation among flower types is based on variation of these basic parts. The showy, cream-colored flowers of the southern magnolia (*Magnolia grandiflora*), for example, have many carpels, stamens, and petals and usually three petal-like sepals (fig. 17.14). In contrast, the bright yellow flowers of stonecrop have five sepals, five petals, ten stamens, and five carpels (fig. 17.10). The highly reduced flowers of many grasses have three stamens, one functional carpel (and perhaps two nonfunctional ones), and no petals or sepals (fig. 17.15). Still other grass flowers have either stamens or a carpel, but not both. A flower that has all the major parts is a **complete flower.** An **incomplete flower** lacks one or more of the four whorls. However, even when sepals or petals are missing, a flower that has both an androecium and a gynoecium is called a **perfect flower.** In contrast, flowers that are only carpellate or staminate are **imperfect flowers.**

The position of the ovary also varies among different flower types. St. John's Wort, for example, has a **superior ovary**—that is, the ovary is attached above (i.e., is superior to) the other three whorls (see fig. 17.4). In flowers such as the daffodil, the other three whorls grow from the top of the ovary, which is **inferior** to them (fig. 17.16a). Some members of the rose family have intermediate flowers. For example, flowers of the rose and cherry have a superior ovary, but it is surrounded by the receptacle; the corolla and androecium branch from the receptacle above the ovary (fig. 17.16b).

FIGURE 17.15

Grass flowers are highly reduced. The flower of this grass, for example, has three stamens, one functional carpel, and no petals or sepals.

Pollen Development

As described in Chapter 10, pollen is formed by microspores that arise by meiosis in anthers. Variations on this theme occur early in development. In some plants, cell walls form after meiosis I and meiosis II, but cytokinesis in other plants is postponed until the end of meiosis II. Also early in development, sterile cells around the young microspores differentiate into a **tapetum,** which is the innermost layer of the pollen sac. The tapetum nourishes the developing pollen.

During or before the time that the microspores undergo mitosis and form two-celled pollen grains, each grain forms an outer wall, called the **exine,** and an inner wall, called the **intine.** The exine is made of a resistant polymer that protects the male gametophyte from desiccation. The intine comes from cellulosic and pectic material that is exported from the cytoplasm of the microspore.

Pollen grains released from the anthers of most plants contain two cells: a **tube cell** that will form the pollen tube, and a **generative cell** that will divide and form two sperm cells. In some plants, however, the generative cell divides to form two sperm cells before the pollen is released. Thus, depending on the plant species, some pollen is two-celled, and some is three-celled when it is transferred to the carpel. Furthermore, pollen may be shed in packets of two or more grains. Pollen is released in pairs from marsh arrow-grass (*Scheuchzeria palustris*), in tetrads from cattail (*Typha latifolia*), and in packets of sixteen grains from the silk tree (*Albizia julibrissin*). The largest packets occur in orchids, which shed all of the pollen from a flower in two or four masses that attach to the pollinator (see box 17.1, "Orchids That Look Like Wasps").

The greatest variation among pollen occurs in the morphology of the pollen wall (fig. 17.17). The pollen of grasses,

A.

B.

FIGURE 17.16

Longitudinal sections of flowers. (a) A daffodil (*Narcissus* sp.), showing the inferior ovary. (b) A rose (*Rosa* sp.), whose ovaries are surrounded by the receptacle.

for example, has a relatively smooth exine, but the pollen of morning glories is spiny. Pollen size ranges from 250 μm long in the tropical cherimoya tree (*Annona cherimola*) to a diameter of less than 20 μm in Bermuda grass (*Cynodon dactylon*). In addition, pollen grains have pores or weak spots in the pollen wall

BOXED READING 17.1

ORCHIDS THAT LOOK LIKE WASPS

Although many flowers reward their pollinators, other flowers use deception. Perhaps the most sensational cases of pollinator deception occur in orchids. For example, flowers of *Ophrys* species imitate female wasps. These orchids have a wasplike shape, often including a lower lip that is fringed with red hairs like the abdomen of the female wasp. Chemicals in the flower's fragrance are similar to those secreted by the female wasp as a sexual attractant. Furthermore, these orchids bloom before most of the female wasps have emerged in the springtime.

Male wasps are attracted to the wasplike orchid and try to copulate with the flower. A large pollen packet is dislodged from the flower during the vigorous movements of this pseudocopulation. The pollen packet sticks to the wasp and is carried to another flower, where it detaches and nestles snugly among three stigmas. One of the stigmas is sterile and modified into a small outgrowth with a sticky area. The sticky spot ensures that the pollen packet stays on the stigmas after the wasp has left.

BOX FIGURE 17.1

(a) Wasplike flower of a species of *Ophrys*. (b) A wild bee (*Eucera longicornis*) during pseudocopulation with a flower of another species of *Ophrys*.

A.

B.

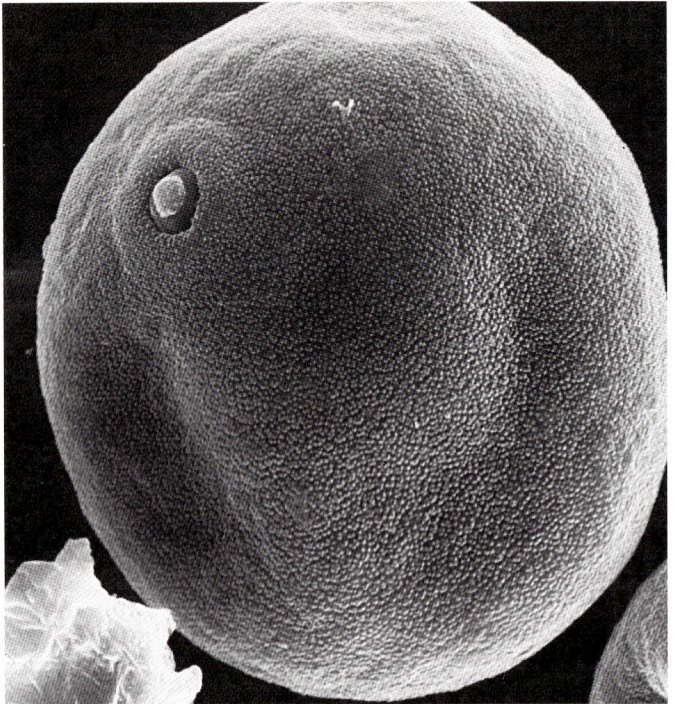

A.

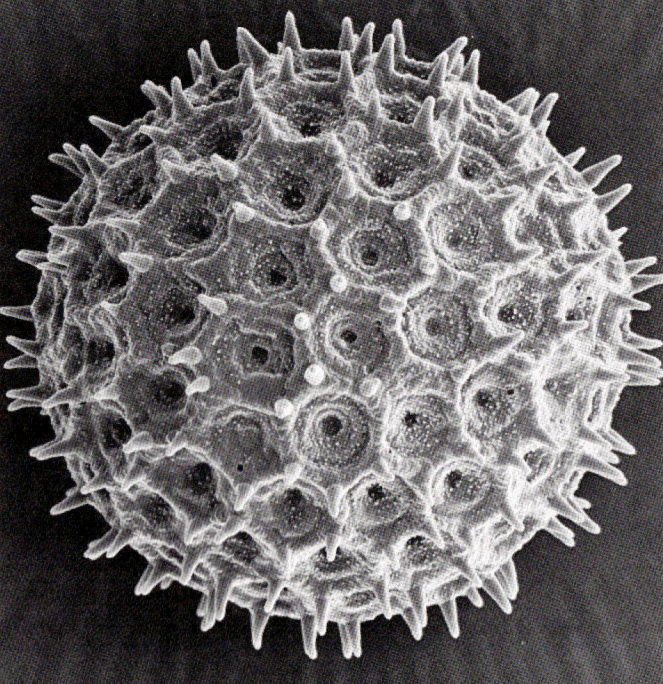

B.

FIGURE 17.17

Scanning electron micrographs of pollen from different plants, showing variation in pollen wall morphology. (a) Pollen from a grass, showing smooth exine and single pore. (b) Pollen from a morning glory, showing spiny exine, ×1,160.

388 UNIT FOUR *The Form and Function of Plants*

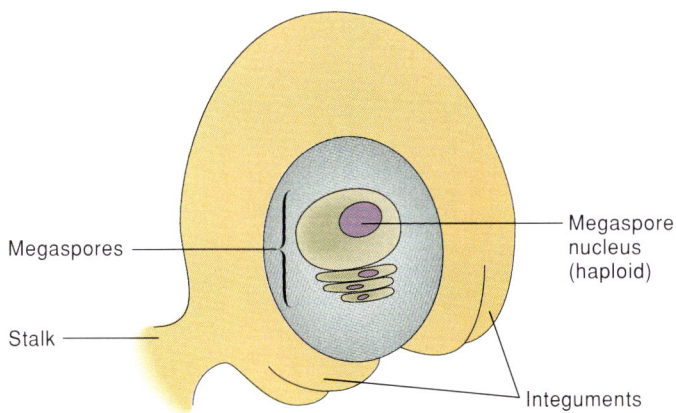

FIGURE 17.18

An ovule consists of a stalk, integuments, and megaspore-producing tissue. Meiosis has already occurred in this ovule, but the embryo sac has not yet developed.

that enable the pollen tube to emerge from the grain when it germinates after pollination. Grass pollen has one such pore, and morning glory pollen has several, but most plants have three porelike or slitlike openings (fig. 17.17).

Development of the Embryo Sac

The embryo sac develops in the ovule, a relatively complex structure that consists of a stalk and one or two **integuments** that later develop into the seed coat (fig. 17.18). The integuments surround the megaspore-producing tissue, one cell of which undergoes meiosis in preparation for development of the embryo sac.

The most common type of development of embryo sacs was briefly described in Chapter 10. It is called the **Polygonum type** of development, because it was first found in knotweed (*Polygonum* species). About 70% of angiosperms have this type of development in their embryo sacs.

Development of the *Polygonum*-type embryo sac begins at meiosis, when the megaspore mother cell produces four megaspores (fig. 17.19). Three of them disintegrate, leaving one **functional megaspore** that forms the female gametophyte. The nucleus of the functional megaspore divides by mitosis three times, thereby forming eight free nuclei. Three of the four nuclei nearest the **micropyle** (i.e., the opening in the ovule through which the pollen tube will enter) differentiate into the **egg apparatus**. The egg apparatus consists of three cells: the egg cell and two flanking cells, called **synergids.** Three of the four cells at the opposite end of the developing embryo sac, which is called the **chalazal pole,** are called the **antipodal cells.** The two remaining nuclei, one from each pole, migrate to the middle of the embryo sac and are called the **polar nuclei.**

Other plants form embryo sacs differently from the *Polygonum* type of development. Some textbooks describe the development of the embryo sac in *Lilium* as being typical of angiosperms because the female gametophyte of lily is large and has large nuclei. However, the *Lilium*-type of embryo sac differs

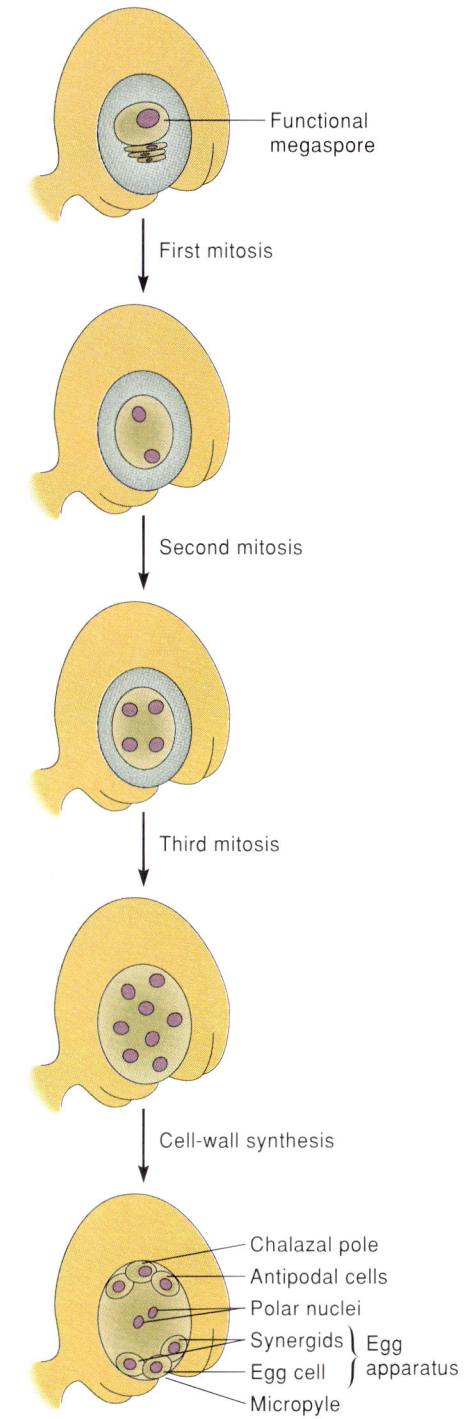

FIGURE 17.19

Development of a *Polygonum*-type embryo sac. This type of development involves a single functional megaspore that forms the embryo sac.

from that in *Polygonum* and is much rarer. The first difference in the *Lilium*-type is that meiosis is not followed by cytokinesis (fig. 17.20). Instead, all four spore nuclei remain in the cytoplasm of the megaspore mother cell and function in its development into an embryo sac. After meiosis, one nucleus migrates to the micropylar pole, and the remaining three nuclei migrate to the chalazal pole. The micropylar nucleus undergoes two

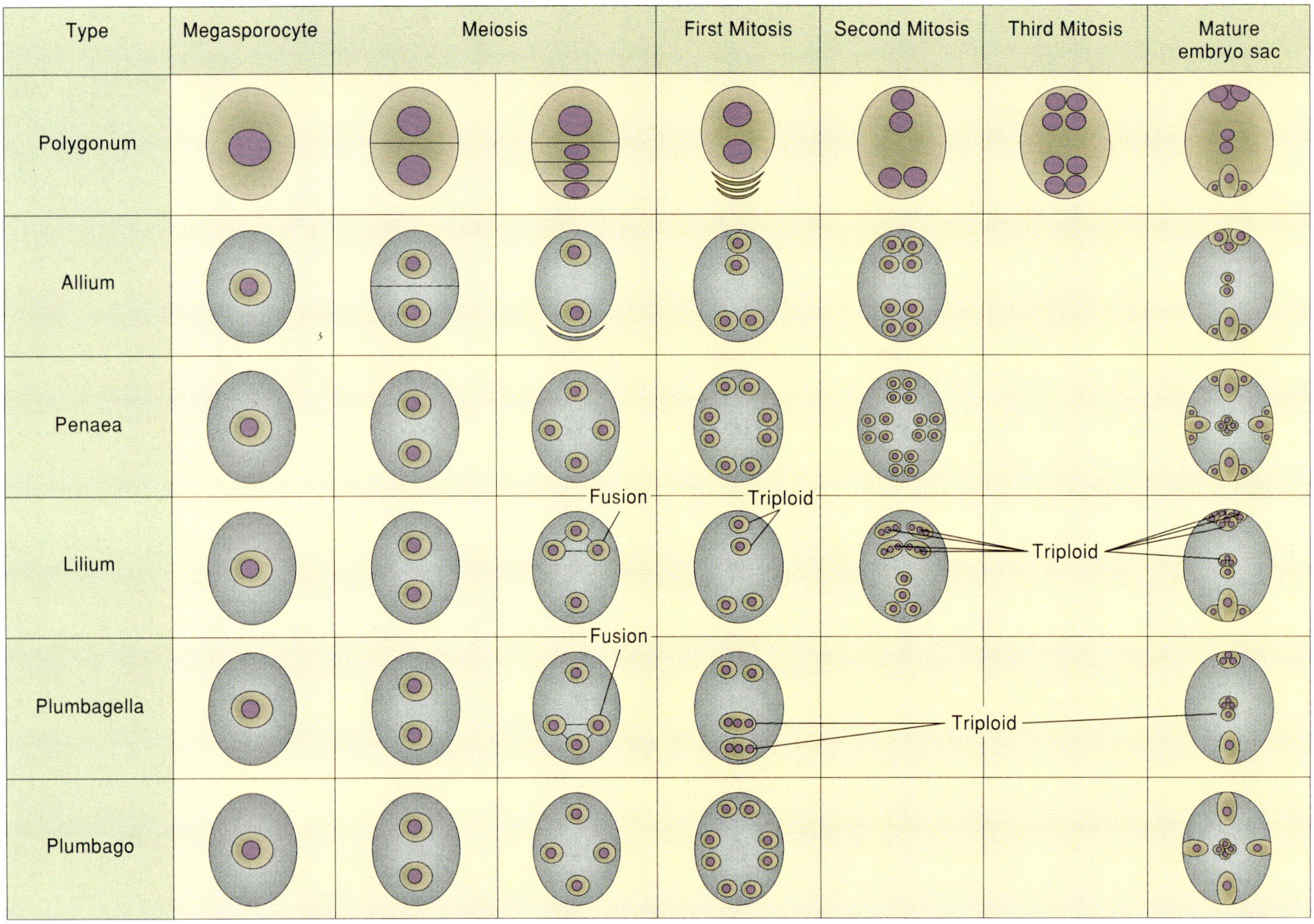

FIGURE 17.20

Variation in development of embryo sacs in angiosperms. Note that the embryo sac in *Polygonum* comes from a single megaspore nucleus, but the embryo sac in *Allium* comes from two megaspore nuclei, and the embryo sac in many other plants comes from four megaspore nuclei. Embryo sacs of *Penaea*-type plants and *Plumbago*-type apparently have four egg apparatuses, but only one egg per embryo sac normally participates in reproduction.

mitotic divisions, producing four haploid nuclei. Three of these nuclei differentiate into the egg apparatus, and one migrates to the middle of the sac and becomes a polar nucleus. Meanwhile, the three chalazal nuclei fuse into a triploid nucleus, which divides by two mitotic divisions into four triploid nuclei. Three of these nuclei become antipodal cells and the fourth becomes another polar nucleus. Thus, the *Lilium*-type embryo sac looks like the *Polygonum* type, but the *Lilium* embryo sac has triploid antipodal cells; one of the polar nuclei is also triploid.

Several other types of embryo sacs are known in angiosperms, the development of which is summarized in figure 17.20. They each have polar nuclei, antipodals, and an egg apparatus, with or without synergids. There is usually one egg, but the numbers of cells and polar nuclei vary. Nevertheless, all types of embryo sacs function similarly during fertilization. The pollen tube goes through the micropyle of the ovule to one of the synergids, if present. One sperm cell passes through the synergid and fertilizes the egg, and the other sperm cell fertilizes the polar nuclei (fig. 17.21). This is called **double fertilization.** The zygote that results from fusion of the egg and sperm is diploid (2n), but the endosperm that results from the union of sperm and polar nuclei may be diploid, triploid (3n), pentaploid (5n), or nonaploid (9n), depending on the type of development.

Flowers vary immensely in such features as numbers, sizes, shapes, and colors of flower parts. Different kinds of flowers also form different types of pollen and embryo sacs.

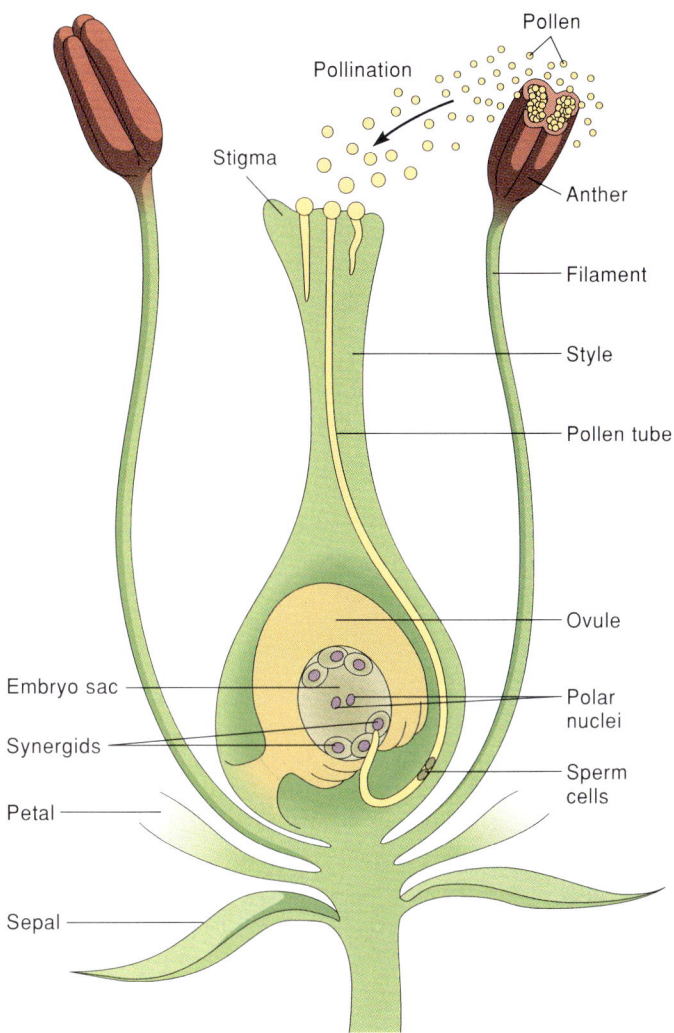

FIGURE 17.21

Pollination and fertilization. Two sperm cells move down the pollen tube. One fertilizes the egg, and the other fertilizes the polar nuclei. In this example, the embryo sac is of the *Polygonum*-type.

What are Botanists Doing?

Double fertilization is not unique to flowering plants; it also occurs in the genus *Ephedra*, a nonflowering seed plant. Recent studies of embryo sacs and fertilization in *Ephedra* help to answer the question, "Should *Ephedra* be classified as a flowering plant?" Find a journal article on this topic that was published within the past three years and see if the author(s) address this question, either directly or indirectly. How are the reproductive features of *Ephedra* used to support this view?

FIGURE 17.22

In monocots such as this lily (*Lilium longiflorum*), flower parts usually occur in multiples of three. A lily has six stamens, three petals, three petal-like sepals, and a three-chambered ovary.

Monocots and Dicots

Floral variation and other aspects of reproductive morphology provide the basis for dividing the flowering plants into two major classes: the class Dicotyledonae (*dicots*) and the class Monocotyledonae (*monocots*).[1] The dicots are named for the presence of two embryonic leaves (**cotyledons**) in their seeds; monocots have one embryonic leaf in their seeds. In monocots, floral parts occur in multiples of three: three petals, three sepals, six stamens, and a carpel that has three chambers (fig. 17.22). In dicots flower parts usually occur in multiples of four or five (see figs. 17.8 and 17.10). Although many dicots and monocots have other numbers of floral parts, many other features are unique to each class. These are listed in table 17.1, along with several examples of familiar species in each class. Note that dicots include about 80% of all angiosperm species, including many herbaceous plants and all woody, flower-bearing trees and shrubs. Monocots are primarily herbaceous, but they also include nonwoody trees such as palms and Joshua trees (see Chapter 16 regarding why such trees are not woody).

---

1. These classes are also commonly called the Magnoliopsida (dicots) and Liliopsida (monocots). Some botanists are campaigning for adoption of these names, but no general agreement has yet been reached.

CHAPTER SEVENTEEN *Reproductive Morphology*

TABLE 17.1

Main Differences between Monocots and Dicots

Characteristic	Dicots	Monocots
Flower parts	In fours or fives (usually)	In threes (usually)
Pollen	Usually having three furrows or pores	Usually having one furrow or pore
Cotyledons	Two	One
Leaf venation	Usually netlike	Usually parallel
Primary vascular bundles in stem	In a ring	Complex arrangement
True secondary growth, with vascular cambium	Commonly present	Absent
Examples	Rose, pea, sunflower, magnolia, ash	Lily, corn, palms, pineapple, banana

Inflorescences

Flowers may be solitary, or they may be grouped closely together in an **inflorescence.** The spectacular inflorescences of urn plants, lupine, snapdragon, and many other plants are popular as ornamentals. Less obviously, the flowers of hazelnut, oak, and willow also occur in inflorescences (fig. 17.23). Like a solitary flower, an inflorescence has one main stalk, or **peduncle.** It also bears numerous smaller stalks, called **pedicels,** each with a flower at its tip. The arrangement of pedicels on a peduncle characterizes different kinds of inflorescences. Some of the common types of inflorescences are diagrammed in figure 17.24. Woody plants often have staminate or carpellate flowers in spikelike tassels, whereas sunflower and other members of the family Asteraceae have reduced, compact inflorescences that may include two kinds of flowers. One kind is irregular and occurs at the periphery of the inflorescence; the other is regular and occurs in the center of the inflorescence. Thus, the "flower" of a sunflower is really a composite of many flowers packed into an inflorescence.

Breeding Systems

No discussion of reproductive morphology and plant diversity is complete without mention of how variation in flower structure can be related to its function. For example, mulberry, cottonwood, and willow have staminate flowers and carpellate flowers, each on different plants. These plants, therefore, are **dioecious;** that is, they have "two houses" (i.e., different plants) for reproduction. This means that such plants reproduce only by cross-pollination, which is the transfer of pollen from the anthers of one plant to the stigmas of another. In contrast,

A.

B.

C.

D.

FIGURE 17.23

The inflorescences of (a) urn plant (*Aechmea fasciata*) and (b) lupine (*Lupinus nootkatensis*) are spectacular because of their brightly colored flowers. The less obvious flowers of (c) hazelnut (*Corylus* sp.) and (d) willow (*Salix* sp.) are packed into tassellike inflorescences.

FIGURE 17.24

Common types of inflorescences. Each type is distinguished by how the flowers are arranged in the inflorescence.

garden pea and snapdragon have perfect flowers that are **self-compatible,** which means that reproduction can be successful following pollination within a flower or between flowers of the same plant.

Mulberry, cottonwood, and willow are plants that depend on mating between gametophytes from different kinds of plants; this is called **outcrossing.** In contrast, garden pea and snapdragon exemplify plants that are characterized by **inbreeding,** which in these plants is mating between gametophytes from the same sporophyte.

The breeding systems of plants range from complete outcrossing to complete inbreeding, but most plants use some combination of the two. For instance, flowers of the hoary plantain (*Plantago media*) are initially carpellate and later become perfect (fig. 17.25). This breeding system enhances outcrossing because most pollen is shed from a particular flower before the carpel matures; however, self-pollination is still possible as the flowers get older. In contrast, flowers of garden pea are equally receptive to self-pollination and cross-pollination throughout their development.

Some plants, such as the ground-cherry (*Physalis* species), produce perfect flowers that are **self-incompatible** (fig. 17.26). This means that, even though self-pollination can occur, the pollen does not function properly with carpels of the same plant and no seeds are formed; thus, the pollination is unsuccessful. Self-incompatibility in the ground-cherry, as it probably is in most self-incompatible plants, is controlled by two genes. Depending on the plant species, these genes may work on the interaction between the pollen and stigma, thereby preventing the pollen tube from germinating. This is called *sporophytic self-incompatibility* because it is imposed by gene action in the stigma, a sporophytic structure. Conversely, in other plants the genes

FIGURE 17.25

Inflorescence of the hoary plantain (*Plantago media*). Young flowers near the top of the inflorescence are carpellate only; the older flowers below are perfect.

FIGURE 17.26

Flowers of the ground-cherry (*Physalis subglabrata*) are self-incompatible.

TABLE 17.2
Features of Outcrossing vs. Inbreeding Plants

Outcrossing	Inbreeding
Self-incompatible	Self-compatible
Diploid sporophyte	Polyploid sporophyte
Many flowers per plant	Fewer flowers per plant
Long flower-stalks	Short flower-stalks
Large sepals	Smaller sepals
Large petals	Smaller petals
Often more than one high-contrast flower color	Usually one flower color; color of low contrast if more than one flower color
Nectaries present	Nectaries absent
Flowers scented	Flowers not scented
Nectar guides present	Nectar guides absent
Long carpel	Shorter carpel
Stamens longer or shorter than carpel	Stamens the same length as the carpel
Anthers distant from stigma	Anthers close to stigma
Many pollen grains per anther	Fewer pollen grains per anther
Style protrudes from flower	Style does not protrude from flower
Anthers do not open when stigma is receptive	Anthers open at the same time the stigma is receptive
Many ovules per flower or inflorescence	Few ovules per flower or inflorescence
Many ovules are not fertilized	All ovules fertilized
Some fruits do not mature	All fruits mature

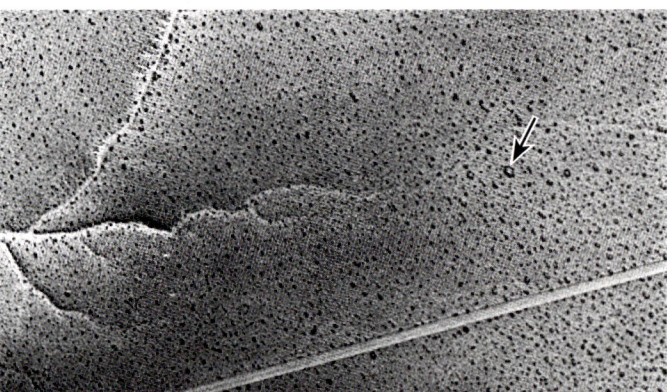

FIGURE 17.27

The circle (arrow) of creosote bush (*Larrea tridentata*) in this aerial photo is a clone that is estimated to be at least 12,000 years old. Smaller circles nearby are of younger clones.

may work in the pollen grain as the pollen tube grows through the style. Self-incompatibility that is caused by genes in the male gametophyte is called *gametophytic self-incompatibility*.

Populations or species of plants usually have some combination of outcrossing and inbreeding. However, observations of thousands of different plants reveal that their reproductive features generally fall into one of the two categories shown in table 17.2. Some of the main features of outcrossing plants include many flowers per plant, long pedicels or peduncles, scented flowers, and many pollen grains per anther. Inbreeding plants generally have few flowers per plant, short pedicels or peduncles, unscented flowers, and fewer pollen grains per anther.

C O N C E P T

Breeding systems of flowering plants include self-compatibility and self-incompatibility, which affect different combinations of inbreeding and outcrossing.

Vegetative Reproduction

Plants are characterized by indeterminate growth; that is, they have meristematic cells that can grow indefinitely (see Chapter 12). Because of this property, tissues or cells can be removed from a plant and induced to grow into a new plant that is a **clone** of the old one. This kind of reproduction is **vegetative**, since meiosis and fertilization are not involved. Cloning is a modern term that simply refers to vegetative reproduction. Most plants reproduce vegetatively to some extent, and some species reproduce almost exclusively by vegetative means. Aspens, for example, grow in large stands that consist of just one or a few clones (see discussion in Chapter 15 on *Pando*, the giant quaking aspen). Humans exploit the clonability of plants by using cuttings to grow many kinds of houseplants, garden plants, and agricultural crops, including roses, African violets, potatoes, bananas, and oranges.

In nature, the genotype of a clone may be long-lived. Clones of some grasses routinely live for a few hundred years. The oldest known clone that is still living is a creosote bush (*Larrea tridentata*) in the Mojave Desert of California (fig. 17.27); this plant is more than 12,000 years old!

Some plants seem to go through the motions of sexual reproduction without actually undergoing meiosis or fertilization. For example, the common dandelion (*Taraxacum officinale*) produces flowers, fruits, and seeds asexually more than 99% of

FIGURE 17.28

Seedless fruits of the teddy-bear cholla (*Opuntia bigelovii*) fall to the ground and grow into clones of the parent plant.

the time. In the absence of meiosis and fertilization, one of the diploid cells in the ovule develops into an embryo. Many species of cholla cacti also produce fruits without seeds. These naturally seedless fruits fall to the ground and grow into clones of the parent plant (fig. 17.28).

 **Writing to Learn Botany**

What are the possible advantages and disadvantages of vegetative reproduction?

Pollination Mechanisms

Pollination requires a *vector*, or carrier, to transfer the pollen from anther to stigma. Flowers are usually highly adapted for one type of vector, which may even be a single species of animal. Flowers and the animals that are attracted to them are often so closely coadapted that they are completely interdependent. As you read about pollination mechanisms, remember that pollination biology is a large and fascinating discipline that includes both botany and zoology. Pollination biologists are continually discovering new and complex interactions between flowers and their pollinators. Some of the major kinds of pollination mechanisms are described below.

Pollination by Wind and Water

The simplest method of cross-pollination, and the most inefficient, occurs by wind. The main features of wind-pollinated angiosperms are as follows: they produce enormous amounts of lightweight, nonsticky pollen; they lack showy floral parts or strong fragrances; they have well-exposed stamens and large stigmas; they have a single ovule in each ovary; and they have many flowers packed into each inflorescence (fig. 17.29a; see also fig. 17.23c, d).

Pollination by water is rare, simply because so few plants have flowers that are submerged under water. Such plants include seagrasses (*Zostera* species), which release pollen that is

A.

B.

FIGURE 17.29

(a) Each flower of rye grass (*Lolium* sp.) produces about 50,000 dry, dustlike pollen grains, which readily take to the air in a light breeze. (b) Staminate plants of the aquatic ribbon weed (*Vallisneria spiralis*), release their flowers as "pollen boats," which drift near the larger, surface-borne flowers of carpellate plants. Those pollen boats not eaten by fish may reach the edge of a dimple in the water that is created by surface tension around the carpellate flower. Once there, the pollen boats will slide down into the carpellate flower and be catapulted onto a receptive stigma.

carried passively by currents of water, much as wind-borne pollen is carried by wind. Other submerged aquatic plants, however, have more complicated pollination mechanisms. For example, pollen of ribbon weed (*Vallisneria spiralis*) is carried from one plant to another in "pollen boats" (fig. 17.29b).

A.

C.

B.

D.

E.

Pollination by Insects

Insects are the most common group of animals that pollinate flowers. Figure 17.30 shows several types of flowers that are adapted for different insect pollinators. Although bees pollinate more kinds of flowers than any other type of insect, flowers can also be pollinated by wasps, flies, moths, butterflies, or beetles. There is no single set of characteristics for insect-pollinated flowers, because insects are such a large and diverse group of animals. Rather, each plant may have a set of reproductive features that attracts mostly one kind of insect, as summarized in table 17.3.

Many kinds of brightly colored flowers look like targets to insects because, in ultraviolet light, their petals are darker toward the center of the flower. Such targets, or **nectar guides,** are made by UV-absorbing pigments called *flavonoids*, which are visible to insects but invisible to other kinds of animals (fig. 17.31). Insect-pollinated flowers may also secrete strong fragrances.

FIGURE 17.30

Insect-pollinated flowers. (a) Bumblebee on an aster. (b) Swallowtail butterfly on the inflorescence of an Indian paintbrush (*Castilleja* sp). (c) The large nocturnal flowers of *Hydnora africana* emit foul odors that attract carrion beetles for pollination. (d) This elephant hawkmoth has a very long tongue that reaches deep into the narrow tube of honeysuckle flower (*Lonicera* sp.). (e) Ant pollinating a flower of *Orthocarpus pusillus*. This is one of the few examples of pollination by ants.

TABLE 17.3

Floral Features Associated with Pollination by Different Kinds of Insects

Type of Insect	Flower Color	Flower Odor	Nectar Guides
Beetles	Dull	Strong and fruity	None
Carrion/Dung flies	Purple-brown or greenish	Strong and foul	None
Bee-flies	Variable	Variable	None
Bees	Variable but not solid red	Usually sweet	Present
Hawkmoths	White or pale	Strong and sweet	None
Small moths	Variable but not solid red	Usually sweet	None
Butterflies	Variable, commonly pink	Moderately strong; sweet	Present

A.

FIGURE 17.31

Flowers of the evening primrose (*Oenothera*) are uniform in color to the human eye (*left*), but insects see a different pattern in ultraviolet light (*right*).

In contrast to flowers with bright colors or nectar guides, less showy flowers can also attract nocturnal insects. Some moths, for example, are attracted to strongly scented, night-blooming flowers (fig. 17.30d), which are usually white or cream-colored. Such flowers are often tubular or trumpet-shaped, which prevents all but the long-tongued moths from reaching the nectar.

Attractiveness and sweetness of fragrance are relative terms. Some "attractive" flowers are reddish-brown and drab, and their fragrance is like that of rotting flesh (fig. 17.30c). These flowers are referred to as *carrion flowers* and are pollinated by carrion flies or beetles that are attracted to their foul odors.

Some animals get the rewards meant for pollinators by stealing the nectar without doing any work for the plant (fig. 17.32a). Some predatory animals use flowers as hunting grounds, where they prey on the pollinators attracted to the flowers (fig. 17.32b).

Pollination by Mammals

Like moth-pollinated flowers, flowers that attract bats and small rodents also open at night. Mammal-pollinated flowers are usually white and strongly scented, often with a fruity odor. Such

B.

FIGURE 17.32

Nonpollinating flower visitor. (a) The hornet in this photo has bitten a hole in the corolla of an evening primrose flower (*Oenothera* sp.) and is drinking nectar through it. (b) This bright yellow crab spider seems conspicuous on the fuchsia-colored petals of this orchid flower, but it is camouflaged in UV light to pollinators who are its prey.

flowers must be large and sturdy enough to bear the vigorous visits of these small mammals (fig. 17.33).

Pollination by Birds

Hummingbirds are the most common group of flower-visiting birds in the Americas. Honey creepers are common pollinators in Africa and Asia. The long beaks of both kinds of birds can reach to the base of long, tubular corollas to obtain nectar (fig. 17.34). Hummingbirds are mostly attracted to bright red and yellow flowers, colors that are not usually attractive

A.

B.

FIGURE 17.33

Mammal-pollinated flowers. (a) A greater short-nosed bat feeds on the pollen and nectar of a banana plant (*Musa* sp.). (b) The tiny Australian honey-possum pollinates plants such as this coral gum (*Eucalyptus* sp.) as it forages for pollen and nectar.

A.

B.

FIGURE 17.34

Bird-pollinated flowers. (a) A rufous hummingbird gets nectar from flowers of *Mimulus cardinalis*. (b) The yellow-plumed honeyeater has a brush-tipped tongue for lapping up nectar and pollen from the bell-fruited mallee (*Eucalyptus pressiana*).

to insects. Birds also have a relatively poor sense of smell, and hummingbird-pollinated flowers are generally odorless. Columbines, penstemons, and scarlet monkeyflowers are examples of flowers that attract hummingbirds.

CONCEPT

Flowers are adapted for pollination mainly by wind or by animals. Some kinds of flowers rely on pollination by water. Wind-pollinated flowers produce abundant pollen in nonshowy flowers. Animal-pollinated flowers are usually showy or strongly scented.

Doing Botany Yourself

Describe how you could determine experimentally what the pollinators are for a particular plant species. Be sure to include how you could distinguish pollinators from other visitors to the flower, if any.

FRUITS

A fruit is a seed container derived from an ovary and any tissues that surround it. As such, fruits are products of flowers and therefore only occur in flowering plants. Beyond this simple definition, angiosperms have a remarkable diversity of fruit types. Fruits are classified on the basis of the characteristics of the mature ovary tissue; for example, whether the fruit is fleshy or dry, or whether the ovary is fused to other kinds of tissues. Examples of different types of fruit and their main features are presented below.

Types of Fruit

Ovaries that have matured into a fleshy fruit often consist of three regions. These regions are the skin (**exocarp**), the fleshy part (**mesocarp**), and the interior (**endocarp**) that surrounds the seeds. Stone fruits, such as apricots, fit this description, as do tomatoes. The main difference between an apricot and a

Table 17.4

Dichotomous Key to Major Types of Fruit

Fleshy fruits	
Simple fruits (i.e., from a single ovary)	
Flesh mostly of ovary tissue	
Endocarp hard and stony; ovary superior and single-seeded (cherry, olive, coconut):	**DRUPE**
Endocarp fleshy or slimy; ovary usually many-seeded (tomato, grape, green pepper):	**BERRY**
Berry with leathery skin containing oils (orange, grapefruit, lemon):	**HESPERIDIUM**
Berry with thick rind (watermelon, pumpkin, cucumber):	**PEPO**
Flesh mostly of receptacle tissue (apple, pear, quince):	**POME**
Complex fruits (i.e., from more than one ovary)	
Fruit from many carpels on a single flower (strawberry, raspberry, blackberry):	**AGGREGATE FRUIT**
Fruit from carpels of many flowers fused together (pineapple, mulberry):	**MULTIPLE FRUIT**
Dry fruits	
Fruits that split open at maturity (usually more than one seed)	
Split occurs along two seams in the ovary	
Seeds borne on one of the halves of the split ovary (pea and bean pods, peanuts):	**LEGUME**
Seeds borne on a partition between halves of the ovary (mustard, radish):	**SILIQUE or SILICLE**
Seeds released through pores or multiple seams (poppies, irises, lilies):	**CAPSULE**
Fruits that do not split open at maturity (usually one seed)	
Pericarp hard and thick, with a cup at its base (acorn, chestnut, hickory):	**NUT**
Pericarp thin	
Ovaries often together in pairs (parsley, carrot, dill):	**SCHIZOCARP**
Ovaries occur singly	
Pericarp winged (maple, ash, elm):	**SAMARA**
Pericarp not winged	
Single seed attached to pericarp only at its base (sunflower, buttercup):	**ACHENE**
Single seed fully fused to pericarp (cereal grains):	**CARYOPSIS**

tomato is that the endocarp in an apricot is hard and stony, whereas the endocarp of a tomato is soft and slimy. The mesocarp and endocarp are fused and indistinct from each other in other types of fruit. In dry fruits, all three layers are fused into one **pericarp,** which is often a thin layer around the seed.

Fruits are classified by several main features, including whether they are fleshy or dry, whether they are derived from one or more ovaries from a single flower or an inflorescence, or whether they consist solely of ovary tissue or of ovary tissue plus the tissue of other floral parts that may be fused to the ovary. Based on these and other differences, the major fruit types can be presented in a dichotomous key (table 17.4). By using this key, you can identify the major types of fruit by choosing between successive pairs of features (dichotomies) at each of several steps. Examples of each fruit type are listed in the key and several are illustrated in figure 17.35. The key should enable you to determine the fruit types of, for example, bananas, peaches, and buckwheat.

Although the classification of fruits is informative, it is also inexact because there are so many variations in fruits that do not fit into the key to major fruit types. For example, the fleshy tissue of strawberries is an expanded receptacle. Other fruits, such as the coconut, are modified so that they partially fit the description of a certain fruit type, but not perfectly. A coconut is a drupe, and drupes normally have a fleshy mesocarp; however, the mesocarp of a coconut is a fibrous husk (fig. 17.36). Only the bony endocarp containing the seed ends up on the grocery shelf. Blackberries are aggregate fruits whose individual ovaries develop into tiny drupes. Peanuts are legumes that grow underground.

Fruit Development

Fertilization is usually a prerequisite for the development of fruits. Chemical signals, called *hormones*, are secreted by seeds as they develop. These hormones induce the ovary tissue to expand and mature into a fruit. Hormones or their synthetic analogs are applied to some crops so that fruits will form and mature in synchrony, which makes harvesting more efficient and economical. Treatment of flowers with artificial hormones can also induce the formation of seedless fruits, (e.g., seedless grapes) in the absence of fertilization. You'll learn more about the role of such plant growth regulators in fruit development in Chapter 18.

FIGURE 17.35

Examples of some of the types of fruit described in table 17.4. (a) grapes (berry), (b) apple (pome), (c) orange (hesperidium), (d) acorns (nut).

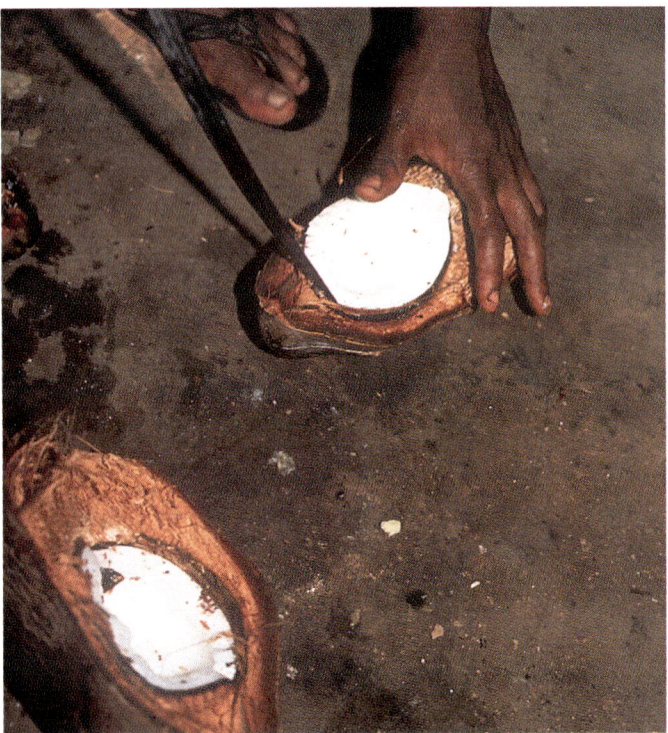

FIGURE 17.36

The fruit of the coconut (*Cocos nucifera*) is a drupe, but the mesocarp is fibrous instead of fleshy.

C O N C E P T

Ovaries develop into fruits after fertilization. Fruits are either fleshy or dry. A fleshy fruit has three distinct regions of ovary tissue, but some fleshy fruits are surrounded by accessory tissue that is not derived from the ovary. Dry fruits may remain tightly closed or they may split open at maturity.

SEEDS

Seed Structure

A distinguishing feature of a seed is the *seed coat*, which is the outer layer that develops from one or two integuments of the ovule and is often thin and papery. In addition, in some dicot seeds the embryo is surrounded by endosperm; in others the endosperm is absorbed by the cotyledons (fig. 17.37). In the seeds of monocots, the embryo is either surrounded by or off to one side of the endosperm. The single cotyledon of monocot seeds is the absorptive organ that takes in nutrients from the digested endosperm. In grasses, which have the most complex seeds of monocots, the cotyledon is so highly modified for absorption that it has a special name, the **scutellum.** In the small seeds of some plants such as orchids, the endosperm does not develop after fertilization of the polar nuclei.

The region above the attachment point of cotyledons is the **epicotyl,** and the region below the attachment point is the **hypocotyl.** An embryonic root (**radicle**) is often distinguishable at the tip of the hypocotyl. The embryos of corn and other grass seeds are partially enclosed in protective sheaths. The sheath around the embryonic shoot is called the **coleoptile,** and the sheath around the radicle is called the **coleorhiza** (fig. 17.37).

Besides varying in structure, seeds also differ in their requirements for germinating, as discussed next.

C O N C E P T

The main features of seeds are a seed coat and an embryo. Seeds may also contain endosperm or cotyledons that have already absorbed the endosperm. In angiosperms, the embryo consists of one or two cotyledons, an embryonic root, and an embryonic shoot.

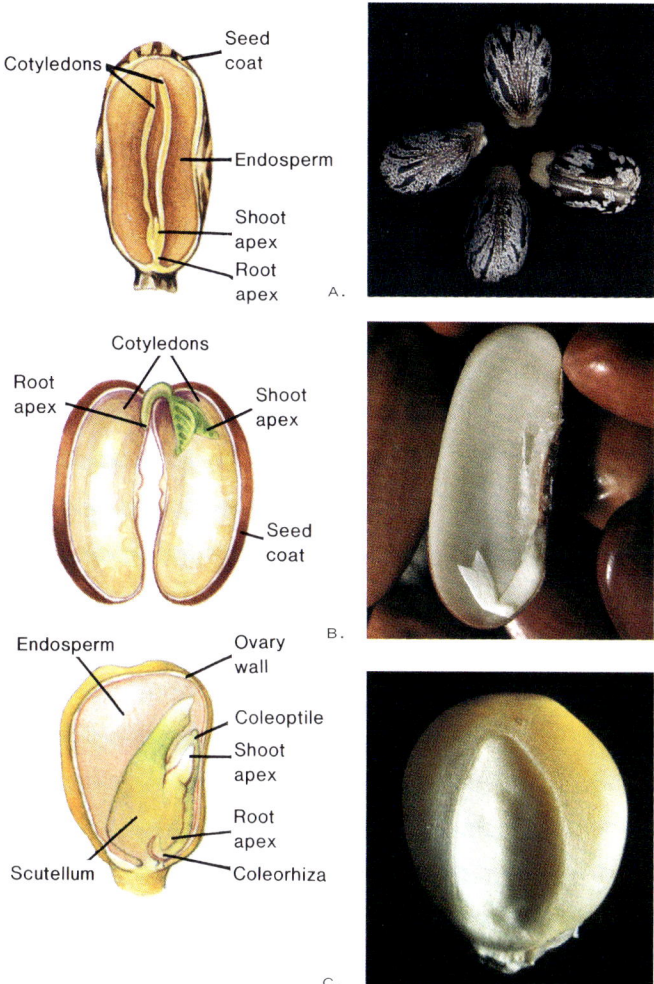

FIGURE 17.37

(a) Seeds of castor bean (*Ricinus communis*) have abundant endosperm that surrounds two thin cotyledons. (b) The two cotyledons in each seed of garden bean (*Phaseolus vulgaris*) absorb the endosperm before germination. (c) Corn (*Zea mays*) has seeds in kernels; the single cotyledon is an endosperm-absorbing structure called a scutellum.

Seed Germination

After fertilization and early development, the embryo usually stops growing. It begins to grow again when the seed germinates. This period of delayed germination is called **dormancy.** Different kinds of seeds require different internal or external stimuli to break dormancy and germinate. The most common external factors are water, temperature, and light.

Water Requirement

Perhaps the most common stimulus for breaking dormancy is water. A seed that absorbs water expands and bursts its seed coat. The seeds of many desert plants have a safeguard against premature germination after short rainfalls, which are common in deserts: they contain a germination inhibitor that must be leached out before they will germinate. Even though the seeds absorb much water, the rainfall may be insufficient to wash out the inhibitor. This adaptation ensures that seeds will germinate only when there is enough water for the successful growth of the seedling. The seeds of rice and other aquatic or semiaquatic plants germinate only in soil that is submerged under water.

Temperature Requirement

The optimum temperature for the germination of most seeds is close to room temperature (25° C), although the seeds of plants adapted to colder or warmer climates can germinate at lower or higher temperatures, respectively. Germination can occur at temperatures as low as 0° C and as high as 45° C in different species. The seeds of woody plants in temperate climates often require a wet period that is followed by several weeks of a cold period before they will germinate. These conditions activate chemicals in the seed that stimulate germination. For example, the seeds of apple trees and most pines can be artificially induced to germinate by wetting them and placing them in a refrigerator. This process is called **stratification.**

Light Requirement

Not all seeds are affected by light. However, the seeds of birch, certain grasses, and some varieties of lettuce require light for germination. These seeds will not germinate in the dark because of an inhibitor that is broken down only in the presence of light. The inhibitor breaks down when a pigment in the seed absorbs red light and sends chemical signals to other parts of the seed (you'll learn about this pigment in Chapter 18). In contrast, the seeds of other plants, such as geraniums and poppies, will germinate only in the dark. In these seeds, light stimulates the synthesis of compounds that inhibit germination.

Special Requirements

Seeds with thick or tough seed coats or pericarps that do not split open may require special conditions for germination. A thick covering may reduce the uptake of water by the seed or prevent the embryo from expanding. For example, many legumes contain seeds with hard seed coats, and the seeds of some drupes are surrounded by a tightly sealed endocarp. Many such seeds will germinate only after the seed coat is scratched or cracked, or after it is briefly soaked in concentrated acid. Such treatments are called **scarification.** In nature, seed coats or pericarps may be scarified by bacterial action, by freeze-thaw cycles, by abrasive handling from squirrels, or by passing through the digestive tracts of animals (see box 17.2, "The Dodo Bird and the Tambalacoque Tree").

Seed Banks

Many viable seeds do not germinate right after they are shed from the parent plant, nor do they germinate during the following growing season. Seeds can lie dormant for many years before conditions are suitable for their germination. Everywhere

BOXED READING 17.2

THE DODO BIRD AND THE TAMBALACOQUE TREE

On the island of Mauritius in the western Indian Ocean, the seeds of the tambalacoque tree (*Calvaria major*) apparently required passage through the digestive tract of the now-extinct dodo bird (*Raphus cucullatus*) before they could germinate. This hypothesis is based on the observation that only about a dozen tambalacoque trees occur in natural stands, all from seeds that germinated at about the time the dodo went extinct (ca. 1680). There are no natural stands of seedlings or younger trees in spite of the abundant production of tambalacoque fruits for the past three centuries.

The hypothesis of a dodo-tambalacoque interaction cannot be tested experimentally because the dodo is extinct, but much indirect evidence supports it. Tambalacoque fruits are fleshy drupes that are eaten by fruit-eating animals, such as the Mauritius parakeet and the Mauritius flying fox; these fruits were undoubtedly also eaten by dodo birds. Seeds germinate after the thick endocarp wall is abraded enough to allow the embryo to sprout through it. The endocarp resists crushing and destruction by forces that probably occurred in the gizzard of the dodo bird. When force-fed to turkeys, a small percentage of tambalacoque seeds can germinate, but turkeys are not attracted to tambalacoque fruits. The digestive systems of extant animals that do eat tambalacoque fruits do not abrade the endocarp enough for any seeds to germinate.

The dodo bird was driven to extinction by human activities. Perhaps due to the absence of this large, flightless bird, the extinction of the tambalacoque tree has also been accelerated. Although the loss of the dodo bird may have been a main factor, habitat destruction and the introduction of weedy, non-native plants by humans have also contributed to the downfall of the tambalacoque tree.

BOX FIGURE 17.2

The dodo bird.

BOX FIGURE 17.3

A three-centuries old tambalacoque tree (*Calvaria major*).

---

that seed plants grow, the soil contains viable, ungerminated seeds in natural storage—that is, a **seed bank.** Seeds in a seed bank may be dormant because of their own inhibitors, as in many desert plants. In other habitats, seed germination may be inhibited by chemicals released from nearby plants or by a lack of nutrients in the soil. Ecologists can sometimes determine what kinds of seeds are in the seed bank of a particular habitat by removing the shrubs from a small area. When a new growing season begins, the seeds of many annual plants germinate. Such experiments simulate what happens, in part, when fire sweeps through an area. In addition to eliminating the source of potential germination inhibitors, fire also releases the nutrients contained in plants. Thus, annual plants grow abundantly in burned areas during the first growing season after a fire (fig. 17.38). As perennial plants become reestablished, the newly replenished seed bank of annual plants once again goes into natural storage until the next fire.

How long can seeds remain viable? The answer is that longevity varies among different kinds of plants and with environmental conditions. Some seeds, such as those of willows and orchids, have a short life-expectancy, perhaps no more than a few weeks; others, such as those of pumpkins, squashes, and other members of the melon family (Cucurbitaceae), can germinate after several years of storage. In some parts of the Mojave Desert, the seeds of weedy plants remain dormant for at least fifteen years. In a seed-longevity experiment that began in the

FIGURE 17.38

Plants that germinated after a fire. Most of these plants grew from seeds in a seed bank and were dormant until the fire cleared the area.

FIGURE 17.39

Arctic tundra lupine (*Lupinus arcticus*). Seeds of this species germinated after being frozen in a lemming burrow for about 10,000 years.

late 1800s at Michigan State University, the longest-lived seeds were of moth mullein, mullein, and mallow. Seeds from all of these plants germinated after 101 years of storage in buried jars. The record for longevity, however, belongs to the arctic tundra lupine (*Lupinus arcticus*) (fig. 17.39). Seeds of this species were frozen in a lemming burrow that has been estimated to be about 10,000 years old. Several of the seeds from this burrow germinated within 48 hours of planting, and one of the plants produced flowers within a year after planting.

CONCEPT

Seed germination is often delayed in a period of dormancy. Germination may be induced by several factors, including changes in temperature, water availability, or light. Some seeds require scarification by animals, bacteria or abrasion during repeated freeze-thaw cycles. In nature, seeds may remain dormant for several years before environmental conditions are favorable enough to induce germination.

Seedling Development

The first part of the embryo to emerge from the seed during germination is the radicle. This first root of the seedling increases the supply of water and nutrients to the shoot before the shoot breaks through the surface of the soil. The shoot may or may not bear cotyledons when it emerges, depending on which embryonic meristems are most active. In the garden bean, for example, the fastest growth occurs in the hypocotyl. As the new shoot pushes through the soil, the hypocotyl elongates and forces the embryonic apical meristem of the shoot, along with the cotyledons, above the surface (fig. 17.40). In contrast, cotyledons of the garden pea remain below ground because the fastest growth occurs at the embryonic shoot tip. In both cases, the cotyledons supply carbohydrates from the endosperm to the growing seedling. Aboveground cotyledons also become green and photosynthetic in garden bean. Cotyledons of garden pea, on the other hand, quickly shrivel as they are used up by the seedling.

Grasses have the most complex seedling development. The seed, which remains enclosed in the ovary wall, germinates when the coleoptile that sheathes the shoot and the coleorhiza that sheathes the radicle both begin to grow. Each sheath grows several millimeters. The radicle then breaks through the tip of the coleorhiza and becomes the primary root. Similarly, when the coleoptile stops growing, the uppermost leaf pushes through it and becomes the first photosynthetic organ of the new seedling. During the germination of grass seeds, the cotyledon continues to absorb sugars from the endosperm until the new leaves make enough photosynthate for the plant to grow independently.

CONCEPT

Seedling development varies depending on the function of the endosperm or the location of the greatest meristematic activity in the embryo. Faster growth in the hypocotyl pushes the shoot and cotyledons through the soil surface; faster growth at the shoot apex keeps the cotyledons below ground.

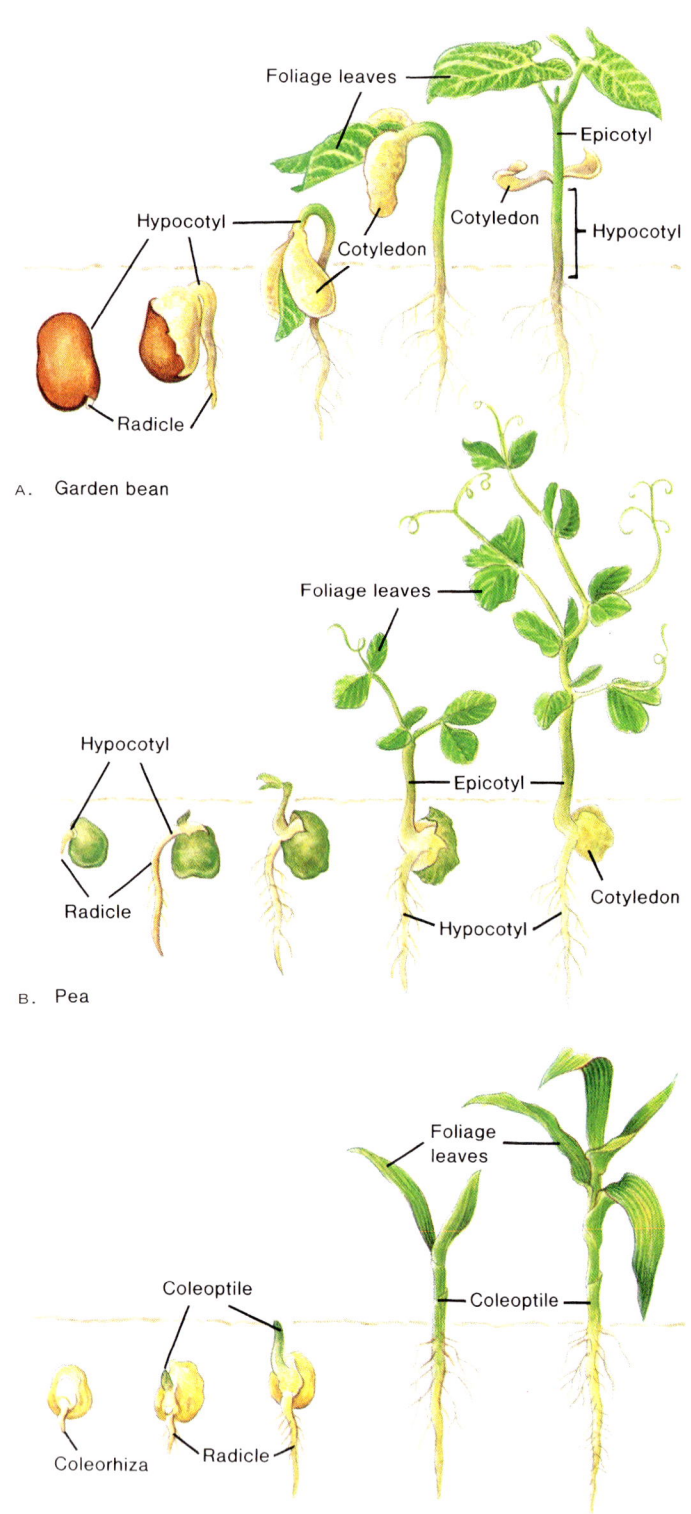

FIGURE 17.40

Seed germination. (a) Growth in the hypocotyl in garden bean forces the shoot apex and the cotyledons above the surface. (b) In garden pea, rapid growth at the shoot apex leaves the cotyledons below ground. (c) In corn and other grasses, the shoot apex grows out of the kernel through the tube of the coleoptile, and the root apex grows through the tube of the coleorhiza.

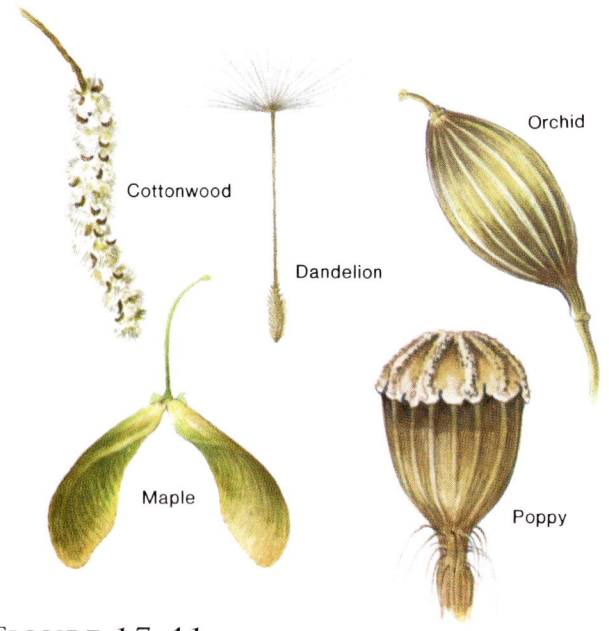

FIGURE 17.41

Wind-dispersed fruits and seeds. Small seeds of orchids and poppies can be blown around by light breezes. Seeds or fruits of other plants have plumes or wings that catch the wind.

DISPERSAL OF FRUITS AND SEEDS

Seeds must get from the parent plant to a favorable place for germination. The best place for a new plant may be where the parent plant is already growing, had it not been there. When a seed germinates next to a mature plant, however, the young seedling may not get enough sunlight, water, or nutrients for continued growth, due to competition with the mature plant. The new generation has a greater chance of surviving if the seeds are moved away from the parent plant to a place that is favorable for seedling establishment, as well as away from herbivores that have found the mature plant.

Dispersal occurs either by physical or biological carriers. Wind and water are the physical carriers. The main biological carriers of fruits and seeds are animals, but some plants can forcefully eject their seeds away from the parent plant.

Dispersal by Wind and Water

The seeds of orchids and heaths are so small that they can be blown around by light winds. The familiar plumes of dandelions, willows, milkweeds, poplars, and some buttercups are examples of further modifications for seed or fruit dispersal by wind (fig. 17.41). A maple fruit has a curved wing that causes the fruit to spin as it floats through the air. Even in a light wind, maple fruits spin away from the parent plant. In arid areas where strong winds are common, tumbleweeds break off at their main stems and whole plants are blown around, releasing seeds as they tumble.

FIGURE 17.42

Fruits of the coconut palm (*Cocos nucifera*) may float for weeks before reaching a beach where they can germinate.

A.

B.

FIGURE 17.43

(a) Sticky or prickly fruits adapted for dispersal by animals. (b) This elk cow has seeds stuck to her fur.

FIGURE 17.44

Bright red seeds of the mescal bean (*Sophora secundiflora*) attract birds.

Coconuts are perhaps the most familiar example of adaptation to dispersal by water (fig. 17.42). The buoyancy of a coconut comes from its fibrous husk, which allows the fruit to stay afloat for days or weeks before too much salt water gets into it and kills the embryo. Many other plants that grow in or near water have fruits or seeds that float. Most sedges, for example, have air pockets around their seeds, so they can float for a while before they absorb water and sink.

Dispersal by Animals

Fruits or seeds that are covered with a sticky substance (e.g., the seeds of mistletoes) or that have barbs or hooks (e.g., the fruits of bur clover and puncture vine) cling to beaks, fur, or skin (fig. 17.43). Such seeds or small fruits may be carried a few meters or many kilometers before they fall from the animal carrying them. The longest dispersal distances are probably achieved by seeds that get stuck to the muddy feet of migratory birds.

The colorful and sweet-tasting fruits of many plants attract fruit-eating animals. Seeds from such fruits usually pass unharmed through the digestive tracts of birds and mammals, or the animals may regurgitate them. A few plants, such as the mescal bean (*Sophora secundiflora*), have brightly colored seed coats that attract birds (fig. 17.44). However, birds drop these seeds after a short distance because the seed coats are too hard.

The most common seed dispersers in some habitats are ants. Seed-harvester ants obtain food from seeds, either from special oil bodies on the seeds or from the embryo. The ants remove the oil bodies and often leave the seeds unharmed. Many of these seeds are discarded by the ants after they have been carried to the nest. When the seeds germinate, dense patches of plants grow around the entrance to the nest.

A.

B.

FIGURE 17.45

Self-propelled seed dispersal. (a) Fruits of the touch-me-not (*Impatiens glandulifera*) explode open, hurling seeds as far as 2 meters. (b) The green fruits of this Mediterranean cranesbill (*Erodium botrys*) still enclose seeds, but the dried-up fruits have already shed their seeds.

Self-Dispersal

The fruits of many plants disperse their own seeds by forceful ejection. For example, the seeds of touch-me-nots (*Impatiens*), witchhazel (*Hamamelis*), dwarf mistletoe (*Arceuthobium*), and some legumes (e.g., African *Acacia*) may be flung several meters from the fruit (fig. 17.45).

CONCEPT

Successful reproduction involves transportation of seeds or fruits to favorable habitats. Seeds and fruits are dispersed either by wind, water, or animals. Some plants disperse their seeds by forceful ejection from the fruit.

The Lore of Plants

A single plant species, or two closely related ones, can be naturally separated by large distances (e.g., between North America and Asia, between South America and Africa). While some of these separations can be explained by continental drift, botanists believe that many were caused by long-distance dispersal. The mechanisms of ancient long-distance dispersal, however, have been harder to imagine; possible mechanisms have included such creative suggestions as seed transport by migratory, constipated birds or by plant migration over temporary land bridges that crisscrossed oceans.

Chapter Summary

Flowers are the reproductive organs of angiosperms. A flower is a stem tip that bears one or more of the following kinds of appendages on a receptacle: sepals, petals, stamens, and carpels. The ovary is the ovule-bearing part of the carpel. During pollination, pollen is transferred from an anther to a stigma. A pollen tube grows to the micropyle of an ovule and carries two sperms to the female gametophyte. After fertilization, ovules become seeds, and ovaries become fruits.

Both sperms from the male gametophyte fertilize cells in the female gametophyte. One sperm fertilizes the egg, which forms a zygote that develops into an embryo. The second sperm fuses with the polar nuclei, thereby forming the first cell of the endosperm. When the endosperm grows, it nourishes the embryo. The endosperm in some seeds is absorbed by the cotyledons, whereas in other seeds the endosperm is digested as the seed germinates.

Diversity in angiosperms is based largely on variation in reproductive morphology. Angiosperms are divided into two classes, the monocots and the dicots. Flowers in either class may be complete or incomplete, perfect or imperfect, regular or irregular. They also may be solitary or arranged in an inflorescence with several other flowers. Most flowers are adapted for pollination by wind or by different animals. Wind-pollinated flowers are usually incomplete and not showy. Insect-pollinated flowers and bird-pollinated flowers are often colorful. Night-blooming flowers attract nocturnal mammals or insects, so these flowers are usually aromatic and white or cream-colored.

Fruits may be either fleshy or dry. Some kinds of dry fruits remain sealed at maturity, whereas others split open at maturity. Fleshy fruits that are colorful and sweet-tasting attract animals that eat them and disperse the seeds. Dry fruits, or the seeds from them, are also adapted for dispersal. These adaptations include plumes or wings that catch the wind, air sacs or spongy tissue for flotation, barbs or spines that attach to passing animals, and edible oil bodies that attract seed-harvester ants.

The distinguishing feature of a seed is the seed coat, which develops from one or two integuments. The embryo must expand and grow through the seed coat during germination. Seed germination is often delayed through a period of dormancy, which may be broken by changes in temperature, water availability, or light. Some seeds require special treatments, such as partial digestion by bacteria or animals or abrasive handling by rodents, before they will germinate.

Many seeds remain in natural storage in seed banks until dormancy is broken or environmental conditions are favorable for germination. The longevity of seeds in storage can be several years or several decades, depending on storage conditions. Seeds remain viable longer when they are kept cold and dry. The record for seed longevity is about 10,000 years for seeds of the arctic tundra lupine.

Questions for Further Thought and Study

1. What are the components of a castor bean seed? What are the components of a garden bean? How do these two types of seeds differ in structure and function?

2. Plants that bear only ovules or only pollen are called *female* and *male*, respectively. Why do some botanists consider this technically incorrect?

3. Seeds of orchids undergo double fertilization without developing endosperm. How might the absence of endosperm be adaptive for such seeds?

4. Why are strawberries, raspberries, and mulberries not berries? Since they are not berries, what are they?

5. Many seeds store their food reserves mostly as oil instead of carbohydrate. What are some examples of plants that produce oily seeds?

6. Why is it incorrect to refer to commercial sunflower "seeds" as seeds? Since they are not seeds, what are they?

7. Apples are fleshy, sweet-tasting fruits that attract and are eaten by fruit-eating animals that disperse their seeds. How could you explain the fact that the seeds of such an attractive fruit contain cyanide-producing chemicals?

8. What advantage is gained by seeds that contain germination inhibitors?

9. Do you think there can be any functional difference between a pollen grain that has one pore versus a pollen grain that has many? Why or why not?

Suggested Readings

ARTICLES

Beattie, A. J. 1990. Seed dispersal by ants. *Scientific American* 263(2):76.

Cook, R. E. 1983. Clonal plant populations. *American Scientist* 71:244–253.

Cox, P. A. 1993. Water-pollinated plants. *Scientific American* 269(4):68–74.

Fleming, T. H. 1993. Plant-visiting bats. *American Scientist* 81:460–467.

Keddy, P. A. 1981. Why gametophytes and sporophytes are different: Form and function in a terrestrial environment. *American Naturalist* 118:452–454.

Niklas, K. J. 1985. Wind pollination—A study in controlled chaos. *American Scientist* 37:462–470.

Owadally, A. W. 1979. The dodo and the tambalacoque tree. *Science* 203:1363–1364.

Postiglione, R. A. 1993. Velcro and seed dispersal. *The American Biology Teacher* 55:44.

Robacker, D. C., J. D. B. Meeuse, and E. H. Erickson. 1988. Floral aroma. *BioScience* 38:390–396.

Stebbins, G. L. 1981. Why are there so many species of flowering plants? *BioScience* 31:573–577.

Steele, L. C., and J. E. Keeley. 1991. Chaparral and fire ecology: Role of fire in seed germination. *The American Biology Teacher* 53:432.

Stein, B. A. 1992. Sicklebill hummingbirds, ants, and flowers. *BioScience* 42:27–33.

Vasek, F. C. 1980. Creosote bush: Long-lived clones in the Mojave Desert. *American Journal of Botany* 67:246–255.

Wheelright, N. 1991. Frugivory and seed dispersal: 'La coevolucion ha muerto—iviva la coevolucion!' *Trends in Ecology and Evolution* 6:312.

Willson, M. F. 1981. The evolution of complex life cycles in plants: A review and an ecological perspective. *Annals of the Missouri Botanical Garden* 68:275–300.

BOOKS

Barth, F. G. 1991. *Insects and Flowers: The Biology of a Partnership*. Princeton, NJ: Princeton University Press.

Bell, A. D. 1990. *Plant Forms: An Illustrated Guide to Flowering Plant Morphology*. New York: Oxford University Press.

Faegri, K., and L. van der Pijl. 1979. *The Principles of Pollination Ecology*. 3d ed. New York: Pergamon Press.

Fenner, M., ed. 1992. *Seeds, the Ecology of Regeneration in Plant Communities*. Wallingford, England: CAB International.

Heywood, V. H., ed. 1985. *Flowering Plants of the World*. New York: Prentice-Hall.

Pijl, van der. L. 1982. *Principles of Dispersal in Higher Plants*. 2d ed. New York: Springer-Verlag.

Willson, M. F. 1983. *Plant Reproductive Ecology*. New York: John Wiley and Sons.

UNIT FIVE

Regulating Growth and Development...

Plants seem to keep track of time because they do certain things at specific times of the day or year. For example, morning glories open their flowers in the morning, but flowers of the Queen-of-the-night cactus open at night. Mulberries and aspens drop their leaves in the fall and grow new leaves in the spring; the timing is never the other way around. Plants also sense gravity, light, and touch. All of these activities show that plants sense their environment and respond to many different stimuli. Environmental stimuli vary, as do the responses of plants to these stimuli.

How do plants perceive a stimulus, communicate the stimulus to appropriate cells in the plant, and change their growth or development accordingly? This complex question can be addressed by learning about the roles of hormones (also called *plant growth-regulating substances*) and other chemical means of communication in plants and about the kinds of changes plants undergo in response to different stimuli. These topics are the main subject of Unit 5, Regulating Growth and Development.

As you read the chapters in this unit, keep in mind that the search for compounds that regulate plant growth and development is ongoing. New compounds that may affect plant growth and development are discovered every year. Research on the established hormones continues because there are many unanswered questions about how they work.

Leaves of Northern red oak. Leaf abscission is strongly influenced by plant hormones.

Plant Hormones

CHAPTER 18

Chapter Outline

INTRODUCTION
AUXIN
    Discovery

Box 18.1
MEASURING PLANT HORMONES

    Synthesis
    Synthetic Auxins
    Transport
    Effects of Auxin
    Auxin and Calcium
GIBBERELLINS
    Discovery
    Synthesis and Transport
    Effects of Gibberellins
CYTOKININS
    Discovery
    Synthesis and Transport
    Effects of Cytokinins
    Cytokinins and Calcium
ETHYLENE
    Discovery
    Synthesis and Transport

Box 18.2
PLANT HORMONES IN PLANT PATHOLOGY

    Effects of Ethylene
    Ethylene and Auxin
ABSCISIC ACID
    Discovery
    Synthesis and Transport
    Effects of Abscisic Acid
OLIGOSACCHARINS AND OTHER PLANT HORMONES
CONTROLLING THE AMOUNTS OF PLANT HORMONES: HORMONAL INTERACTIONS
DO PLANT HORMONES REALLY EXIST?
Chapter Summary
Questions for Further Thought and Study
Suggested Readings

Chapter Overview

Plant growth and development are strongly influenced by plant growth-regulating substances, which are organic compounds made in one part of a plant and transported to another part, where they elicit a response. These growth-regulating substances have traditionally been referred to as *hormones*. There are five major classes of plant hormones: auxin, gibberellins, cytokinins, abscisic acid, and ethylene. These hormones each elicit many responses and interact in complex ways to stimulate or inhibit growth. Unlike most animal hormones, which have specific effects, each plant hormone is made in several parts of the plant and has several effects. The effects of hormones are governed by many factors, including the presence of other hormones, nonhormonal factors such as calcium ions, and the varying sensitivity of different tissues to hormones.

Introduction

Plant growth and development result from complex, highly organized events. They involve more than forming masses of new cells or the increase in size of an organism. Rather, plants grow and develop by producing specialized cells, tissues, and organs having predictable shapes, locations, and functions. These specializations account for the form and function of a plant, which were the topics of the preceding unit. However, merely knowing something about the structure and function of these specialized structures is not enough; we are also interested in how these specializations arise. For example, what determines that a cell differentiates into a vessel element and dies, while an adjacent cell remains meristematic or becomes a sieve element in the phloem? What controls if and when a leaf is shed by a tree? Why do the buds of most trees grow only during spring instead of all year long? In short, how do plants control all of the things that they do?

One of the earliest models of plant growth and development was proposed more than a century ago by Julius von Sachs, a German botanist who speculated that each plant organ resulted from the presence of a unique substance. Despite decades of intensive research, these substances have never been found. However, we *do* know that plant growth and development are controlled by internal signals, the most important being the information contained in DNA. This information governs the basic machinery for life, including growth and development, energy transformations and reproduction. Since the signals for these processes are genetically predetermined, they provide a blueprint for many aspects of growth and development. The leaves of a corn plant, for example, always have the same basic shape, and seeds germinate in specific and predictable ways.

Plant growth is also influenced by external signals from the environment. Many genetic designs for growth and development come with several "options": different environmental conditions can activate different genes and produce different modes of growth. For example, specific genes stimulate growth during the favorable conditions of spring, whereas others induce dormancy during a harsh winter. These different modes of growth are important because they adapt plants to their changing environment. Although this adaptability is not without limits, it usually ensures growth

---

CONCEPT

Plant hormones are organic compounds made in one part of a plant and transported to another part, where they elicit a response. Plant hormones are active in small concentrations. The five major classes of plant hormones are auxin, gibberellins, cytokinins, ethylene, and abscisic acid, each of which has many effects.

AUXIN

Discovery

You're probably already familiar with Charles Darwin: his book *On the Origin of Species* formulated the theories of descent with modification and natural selection and today stands as one of the most influential publications of all time. However, Darwin would have been famous even if he had never studied evolution.

During the late 1870s, Darwin and his son Francis began studying phototropism, which is the growth of stems and leaves toward light. The Darwins wanted to answer a seemingly simple question: how do plants grow toward light? They studied coleoptiles of canary grass (*Phalaris canariensis*) and oats (*Avena sativa*), both of which grow toward light coming from one direction (fig. 18.1a). In these plants, growth toward light occurs several millimeters below the tip of the coleoptile. One of the Darwins' first experiments involved blocking the incoming light by covering the tips of coleoptiles with metal foil. To the Darwins' surprise, these coleoptiles did not grow toward light, despite the fact that the growing region where curvature occurred was illuminated. Growth toward light resumed when the metal foil was removed or replaced with transparent glass. The Darwins then repeated the experiment, this time covering the growing region of the coleoptiles rather than their tips, and observed

and survival in the various conditions in which the species evolved.

The internal and external signals that regulate plant growth are mediated, at least in part, by plant growth-regulating substances, or **hormones** (from the Greek word *hormaein*, meaning "to excite"). Plant hormones are organic compounds that are made in small amounts in one part of the plant and transported to another part, where they initiate physiological responses. However, these responses are not always an "excitation" or stimulation; the regulation of bud dormancy, for example, results from an inhibition of growth.

Botanists have identified five major classes of plant hormones: **auxin, gibberellins, cytokinins, abscisic acid,** and **ethylene.** According to our definition, these hormones have several characteristic features: they are made by the plant, are active in small quantities, are transported to other parts of the plant, and can elicit a response. Consequently, in the following sections we'll discuss each of the aspects of plant hormones listed below:

- How the hormone was discovered
- How and where in plants the hormone is made
- How the hormone is transported
- Responses elicited by the hormone

Despite the fundamental importance of hormones in plant growth and development, our knowledge of plant hormones is limited and, in many cases, controversial. Unfortunately, there are only a few generalizations that we can give you in this chapter. We'll begin with the three most important of them:

- Although a hormone may have some clearly characteristic effects, it may also have many other effects. That is, a single hormone can elicit many different responses.
- The particular effect elicited by a hormone depends on many factors, including the presence of other hormones, the amount of hormone present, and the sensitivity of the tissue to the hormone.
- Finally, it is difficult to predict a specific response, since the response changes under different conditions and in different plants. That is, the gene expression triggered by a plant hormone depends on the identity (i.e., message) of the hormone and its readability by a tissue.

You'll learn more about each of these generalizations throughout this chapter. For now, use them as a mental checklist as you learn how plant hormones were discovered and where all of the interest in them began.

---

that the coleoptile grew toward the light. They concluded that the growth of coleoptiles toward light was somehow controlled by the tip of the coleoptile.

In 1881, the Darwins published their findings in a book entitled *The Power of Movement of Plants*. They suggested that phototropism was due to an "influence" produced in the tip of a coleoptile that moved to the growing region, where it caused the coleoptile to grow toward light. Although identification of plant hormones was decades away, you can already see the importance of the Darwins' work: the phototropic "influence" that they described was made in one part of the plant and transported to another, where it elicited a response. Their work was our first step toward discovering plant hormones.

The death of Charles Darwin in 1882 abruptly ended the Darwins' studies of phototropism. However, important questions remained, the most critical of which related to the nature of the "influence" that the Darwins had described. For example, was the "influence" a chemical signal, an electrical signal, or some other kind of stimulus? In 1913, Danish plant physiologist Peter Boysen-Jensen began trying to answer this question (fig. 18.1c). When he cut off the tip of a coleoptile, he noted that the coleoptile stopped growing. This was consistent with the Darwins' observation that something from the tip of the coleoptile controlled growth. He then replaced the tip but separated it from the coleoptile with a tiny piece of agar, a gelatinous substance that keeps water in a semisolid state at room temperature. Within a few minutes, these coleoptiles resumed growth and curved toward the light (fig. 18.1c). Boysen-Jensen concluded that (1) the tips of coleoptiles do not have to be in their normal position to affect growth, and (2) the "influence" that controls phototropism can move through agar.

But what was the signal? Since it could move through agar, Boysen-Jensen reasoned that it was probably a water-soluble chemical. He tested this by replacing the agar blocks with butter. Because water is insoluble in butter, a water-soluble chemical coming from the tip could not move through the butter and

A.

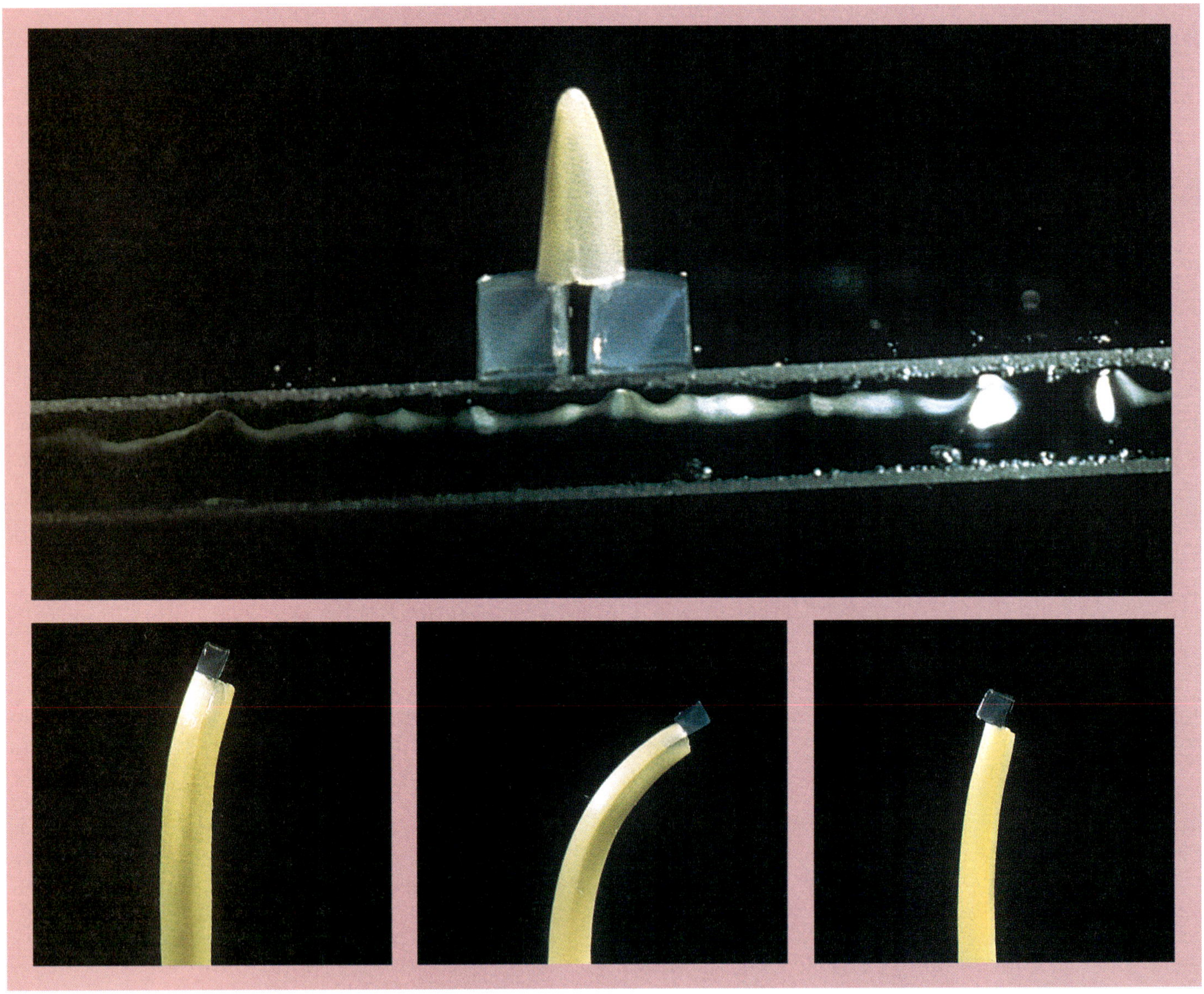

B.

FIGURE 18.1

(a) Darwin's study of phototropism. In the left photograph, the coleoptile is curving toward the light of a candle. On the coleoptile shown in the middle photograph, the top 2 mm of the plant has been covered by lightproof foil. The shoot does not bend, even though the growing zone of the coleoptile is illuminated. In the right photograph, the foil covers the growing zone but not the tip of the coleoptile. Curvature of the coleoptile toward the light shows that a signal must move from the light-sensitive tip to the growing

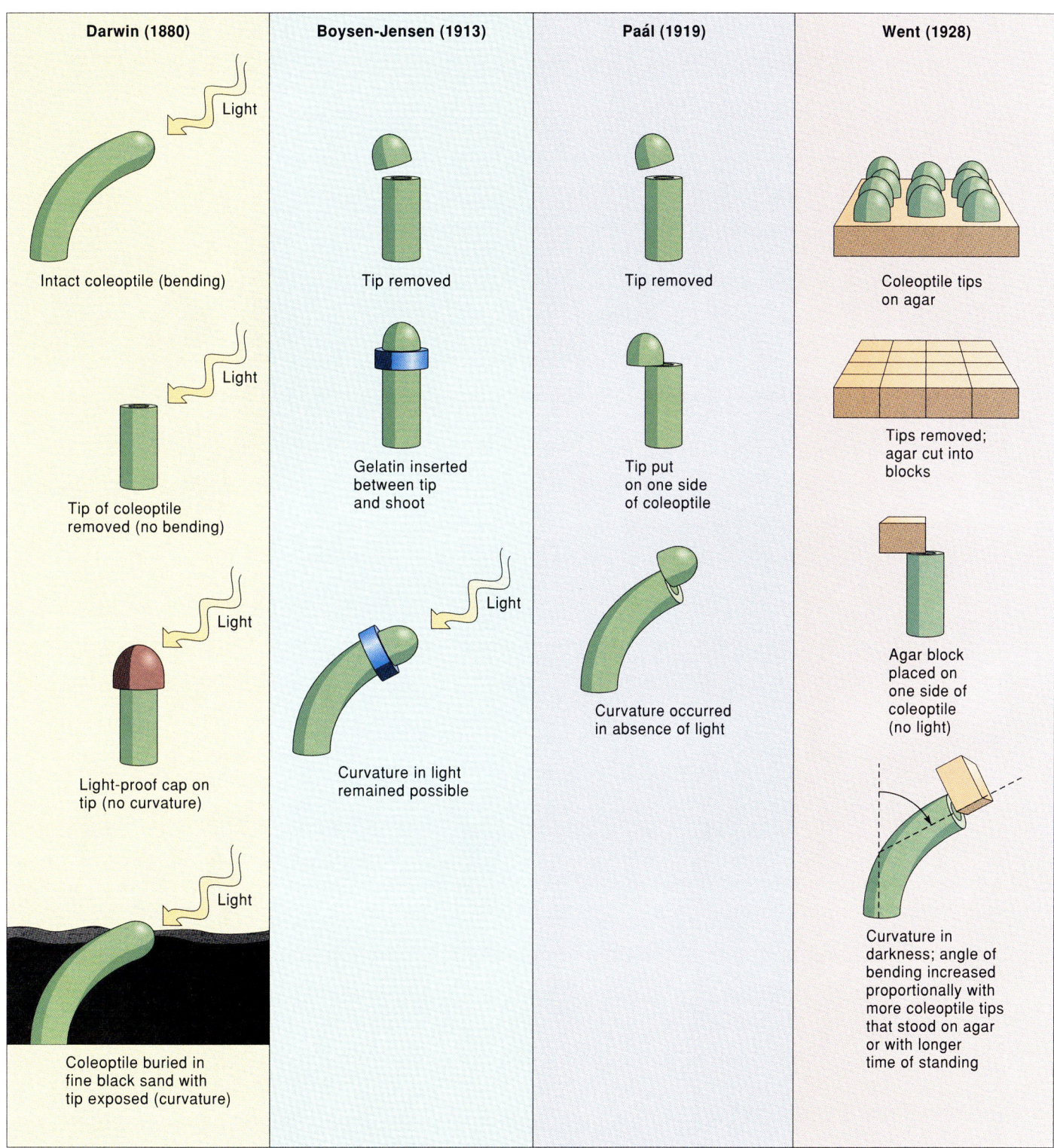

zone, which is farther down. (b) Went's study of phototropism. These experiments were the first demonstration of a hormone in plants. Went gave the name *auxin* to the hormone involved in phototropism of coleoptiles. (c) A diagrammatic summary of some of the important experiments that demonstrated the existence of auxins. The Paál and Went experiments were done in the dark.

BOXED READING 18.1
MEASURING PLANT HORMONES

Botanists have traditionally thought that hormonal responses are controlled by the amount of hormone present. Thus, if a particular response is controlled by the amount of a hormone present, then the response must change as the amount of hormone changes. Although hormone-influenced responses such as leaf abscission, apical dominance, and phototropism are relatively easy to measure, measuring how much hormone is present is a more formidable problem, because hormones are active at minute concentrations (0.01 to 1 µM). Thus, any method for measuring a plant hormone must be sensitive. It must also be specific: the effect must be attributable to a specific hormone rather than to the influence of any other hormone or substance.

Hormones were originally measured with **bioassays,** which are procedures to quantify biologically active compounds by measuring their biological effects. Bioassays exploit the extreme sensitivity and specificity of certain plant parts to particular hormones. For example, Frits Went showed that auxin could be collected from plants in agar, and that placing this agar on one side of a decapitated *Avena* coleoptile would induce curvature. Since the curvature is proportional to the amount of indoleacetic acid (IAA) in the block, the amount of IAA can be measured by measuring the amount of curvature. This procedure, the *Avena* curvature test, became the standard bioassay for IAA. Similar bioassays were later developed for other hormones. For example, the classic bioassay for gibberellins is stimulation of growth in dwarf mutants. Because they are inexpensive, relatively quick, and easy to do, bioassays are still used by many botanists to measure plant hormones. However, improved technology has resulted in more sensitive and specific techniques. Today, most hormones are measured with **gas chromatography,** a technique in which a sample is moved as a vapor through a column of liquid or particulate solids. Components of the sample are differentially adsorbed into the column, thereby separating the hormone from other parts of the sample.

A recently developed technique for measuring plant hormones is called **immunoassay.** This technique involves using animals to produce an antibody to the hormone and then using this antibody to select and bind the hormone from an extract. Thus, immunoassay uses the selectivity of antibodies to detect, identify, and measure quantities of plant hormones. Immunoassays of plant hormones are relatively inexpensive, speedy, extremely specific, and sensitive, and allow for a convenient analysis of unpurified samples.

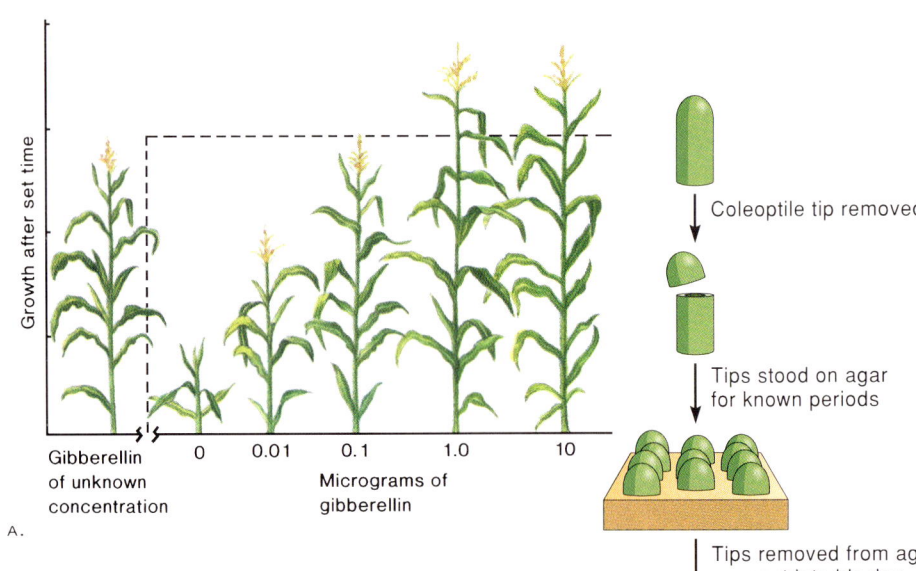

BOX FIGURE 18.1

(a) A plant's response to gibberellin can be used to measure the amount of hormone. The height of a gibberellin-treated plant is compared with that of dwarfs treated with a range of known concentrations of gibberellin. The technique of measuring the concentration of a chemical by measuring the response of living material to that chemical is called a bioassay. What was the unknown concentration of gibberellin in this bioassay? (b) The oat (*Avena*) coleoptile curvature bioassay is a popular bioassay for auxin. The amount of auxin that diffuses into an agar block can be estimated by comparing the results with a standard curve prepared from responses of coleoptiles to known concentrations of auxin. Why must this bioassay be done in the dark?

reach the growing region of the coleoptile. Sure enough, the results were identical to those in which the tip was absent: there was no growth or curvature. Boysen-Jensen concluded that the "influence" controlling growth and phototropism of the coleoptiles was a water-soluble compound. But were electrical signals also involved? He tested this by replacing the agar blocks with pieces of platinum foil, which would be expected to transmit electrical signals from the tip to the growing region of the coleoptile. Again, there was no growth or curvature, suggesting that the signal was not electrical.

In 1918, the Hungarian plant physiologist Arpad Paál continued Boysen-Jensen's efforts to identify the Darwins' "influence" responsible for phototropism. Paál studied coleoptiles grown in the dark. He cut off the tips of dark-grown coleoptiles and asymmetrically replaced them on one side of the coleoptile's cut surface (fig. 18.1c). Surprisingly, these coleoptiles curved away from the side onto which the tips were placed, despite the fact that the plants were in the dark. Furthermore, the curvature resembled that of plants growing toward light. These results suggested to Paál that the tip of the coleoptile produces a substance that moves down and stimulates growth, and, more importantly, that light must cause it to accumulate on the shaded side of the coleoptile. By 1920, Paál's work had brought botanists tantalizingly close to finally accounting for the "influence" described by Charles and Francis Darwin some forty years before.

The Darwins' observations were finally explained in 1926 by Dutch plant physiologist Frits Went. While working in his father's laboratory at the University of Utrecht in the Netherlands, Went separated the "influence" for phototropism from the plants that produced it. His experiments were elegantly simple and were influenced strongly by the work of Boysen-Jensen and Paál. He first cut off the tips of oat coleoptiles and placed them onto agar so that their cut surfaces touched the agar. After about an hour, the tips were discarded, and pieces of the agar that they had touched were placed on the cut tips of decapitated coleoptiles grown in the dark. As you can see in figures 18.1b and 18.1c, Went's results were convincing:

- Decapitated coleoptiles without agar blocks did not grow. These results confirmed the Darwins' findings that the tips of coleoptiles produced something necessary for growth.
- Agar blocks that had not contacted cut tips of coleoptiles elicited no response when they were placed on decapitated coleoptiles. Therefore, there was nothing in the agar that caused growth or curvature of the coleoptile.
- When agar blocks that had contacted cut tips were placed on the center of the decapitated coleoptiles, they grew straight up. Thus, the coleoptile tips had produced a chemical that diffused into the agar, and this chemical stimulated the growth of coleoptiles.
- When agar blocks that had contacted cut tips were placed on one side of the decapitated coleoptiles, they curved away from the agar blocks, just as Paál's seedlings did when the tip of a coleoptile was placed on one side of a decapitated coleoptile. This growth away from the agar blocks was similar to phototropic curvature, even though the plants were kept in the dark, and the tips were absent. These results indicated that the agar blocks contained a chemical that stimulated the growth of coleoptiles.

Went concluded that phototropic curvature was not due to the mere presence of the coleoptile's tip, but rather due to a chemical coming from the coleoptile's tip that stimulated growth. Went named this chemical *auxin* (from the Greek word *auxein*, meaning "to grow"). Auxin does, in fact, influence phototropism: light striking one side of a coleoptile causes auxin to migrate to the shaded side of the coleoptile, where it stimulates growth and causes growth toward light.

According to Went's explanation, auxin fit the definition of a hormone: it was made in one part of the plant (the tip of coleoptile) and transported to another part (the growing region of the coleoptile), where it caused a response (increased growth). Thus, auxin was the first plant hormone to be discovered. Since its discovery, auxins have been found throughout the plant kingdom, in fungi and in some bacteria. Animals also produce auxins, although we do not know how or if they respond to them. Interestingly, the first auxin to be identified chemically was isolated from human urine.

Synthesis

The most active naturally occurring auxin in plants is **indole-3-acetic acid,** or **IAA** (fig. 18.2). Botanists thought for a long time that auxin was made from the amino acid tryptophan; however, a study in late 1991 using a mutant incapable of making tryptophan showed that plants can make IAA without tryptophan as an intermediate. Thus, the primary pathway for IAA synthesis may not involve tryptophan. The most active areas of IAA synthesis are shoot tips, embryos, young leaves, flowers, fruits, and pollen.

There are two other naturally occurring auxins: 4-chloro-IAA and phenylacetic acid. The precise roles of these auxins in plant growth and development are unknown; however, they are generally less active than IAA.

Synthetic Auxins

Although IAA is the most active naturally occurring auxin, several synthetic compounds have auxinlike effects. Synthetic auxins like 2,4-D (2,4-dichlorophenoxyacetic acid) and NAA (naphthaleneacetic acid) have structures that resemble IAA (fig. 18.3).

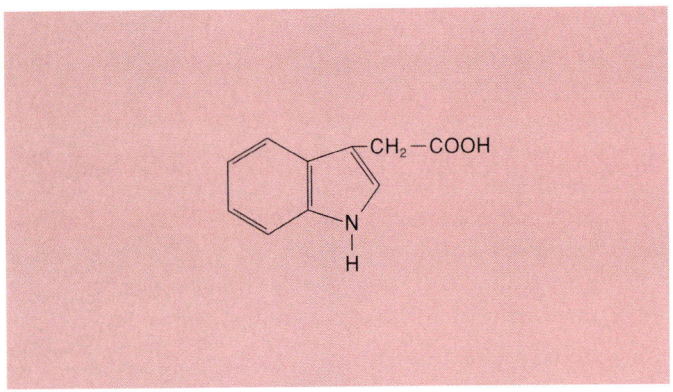

FIGURE 18.2

The structure of indole-3-acetic acid (IAA).

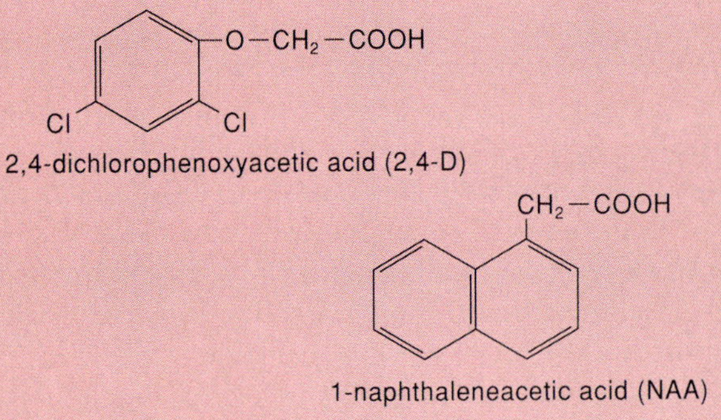

FIGURE 18.3

IAA is the most active naturally occurring auxin. Several synthetic compounds have auxinlike effects. Synthetic auxins such as 2,4-D and NAA have chemical structures similar to that of IAA.

Synthetic auxins such as 2,4-D are used extensively as herbicides because they are inexpensive, relatively nontoxic to humans, and have a selective effect: they kill broadleaf dicots, but not monocots. The exact mechanism for this selectivity is unknown, but it may involve the greater ability of broadleaf weeds to absorb and translocate the synthetic auxins. Synthetic auxins are also used to prevent preharvest dropping of fruit, to produce roots on cuttings, and to inhibit sprouting of lateral buds ("eyes") on Irish potatoes. Synthetic auxins have been used widely since the 1940s and have revolutionized weed-control in agriculture: their use has decreased production costs by reducing the amount of labor and mechanical weeding needed to grow and efficiently harvest a crop.

Unfortunately, the effects of synthetic auxins on human physiology have not all been positive. Most of the negative effects can be traced to a defoliant used in the Vietnam war known by the code name Agent Orange. Agent Orange was a 1:1 mixture of 2,4,5-T (2,4,5-trichlorophenoxyacetic acid) and 2,4-D that was sprayed throughout the jungles of Vietnam and followed by napalm bombs. These treatments destroyed hundreds of square kilometers of forest. This defoliation was protested during the war by many botanists, but the problems have proved to be more serious than the destruction of the forests. Synthesis of the 2,4,5-T used in Agent Orange also produces dioxin (2,3,7,8-tetrachloro-dibenzo-*para*-dioxin), which is one of the most toxic synthetic chemicals known. Thousands of U.S. pilots and Vietnamese citizens exposed to Agent Orange (and often contaminated with dioxin) have had increased occurrences of miscarriages, birth defects, leukemia, and other types of cancer. The Environmental Protection Agency banned the use of 2,4,5-T in the United States in 1979, but its effects continue to plague many Vietnam veterans and Vietnamese citizens.

Transport

IAA moves primarily through parenchyma cells of the cortex, pith, and vascular tissues. It moves slowly: typical rates of 1 cm hr^{-1} are slow compared to those of substances moving in xylem and phloem, but they are still about ten times faster than those predicted by diffusion.

IAA moves polarly in roots and stems (fig. 18.4). Polar transport requires energy; thus, inhibitors of ATP synthesis block the transport of IAA. In stems, IAA is transported *basipetally*, meaning that it moves toward the base. Basipetal transport in stems continues even if the stem is inverted, so that the apex is pointing down. In roots, IAA is transported *acropetally*, meaning that it moves toward the tip. The polar transport of IAA has intrigued botanists for many years and may involve auxin carrier proteins in the plasmalemma (i.e., cell membrane). IAA made in mature leaves moves nonpolarly in the phloem.

C O N C E P T

IAA is the most active naturally occurring auxin. It moves basipetally in stems and acropetally in roots. Synthetic auxins such as 2,4-D have widespread uses in modern agriculture.

Effects of Auxin

Cellular Elongation in Grass Seedlings and Herbs

Short-Term Effects

There are two requirements for cellular elongation: (1) positive turgor pressure, and (2) increased plasticity (stretchability) of the cell wall. Turgor pressure in cells results primarily from the presence of dissolved solutes and is not significantly

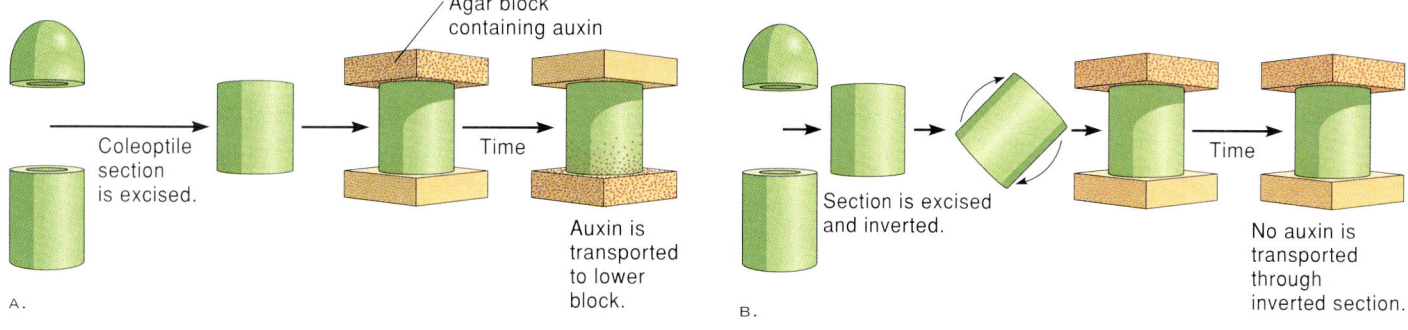

FIGURE 18.4

The polar transport of auxin (shaded). (a) When the coleoptile section remains right-side-up, auxin moves through to the receptor block below. (b) No auxin moves through an inverted coleoptile section.

affected by IAA. However, IAA does increase the plasticity of the cell wall. One explanation of IAA-induced growth is the **acid-growth hypothesis,** which is summarized as follows:

1. IAA stimulates H⁺ pumps in the plasmalemma.
2. Once activated, these pumps secrete H⁺ into the cell wall, decreasing its pH to about 5.0.
3. This acidification of the cell wall activates pH-dependent enzymes that break bonds between cellulose microfibrils.
4. When these bonds break, the wall "loosens" and turgor pressure causes the cell to expand.

Several observations are consistent with the acid-growth hypothesis for IAA-induced growth. For example, applying acid to internodal segments induces growth similar to that resulting from applying IAA. Furthermore, IAA-induced cellular elongation is inhibited by bathing the tissue in neutral buffers, which prevent acidification of the cell walls.

Although the acid-growth hypothesis is attractive, it suffers from several inconsistencies. For example, there are several instances in which IAA but not acid promotes growth. More importantly, IAA-induced growth can be separated from wall acidification; that is, IAA-induced growth and wall acidification are not causally linked. These problems have prompted some botanists to suggest that wall acidification is a result rather than a cause of acid-induced growth. Regardless of its exact mechanism, IAA-induced growth is rapid: it occurs within as little as three minutes in grass seedlings. Because it occurs so fast, many botanists thought that it probably did not involve transcription or translation of genetic information. This conclusion was premature, however, for improved techniques have shown an accumulation of IAA-induced mRNA several minutes before H⁺ secretion and wall acidification. Thus, IAA-induced growth may result from rapid gene activation, which in turn may activate H⁺ pumps in the plasmalemma. Other effects of IAA may also alter mRNA synthesis, turnover, or activation.

Long-Term Effects

According to the acid-growth hypothesis for IAA-induced growth, cellular elongation should continue as long as the cell wall is acidic. However, this does not occur: growth induced by acid stops after 1–3 hours, while growth induced by IAA continues much longer. Therefore, acid-growth may account for only the early stages of IAA-induced growth, whereas the long-term growth induced by IAA may result from another mechanism. A clue to this other mechanism was provided in the 1960s when researchers discovered that compounds that inhibit nucleic acid and protein synthesis also inhibit IAA-induced growth. This suggests that IAA may act at the gene level, possibly by activating a gene required for making a protein necessary for growth. Indeed, IAA stimulates the production of mRNA and ribosomes. Thus, long-term growth induced by IAA may result from selective activation of genes and altered patterns of protein synthesis. Among the genes turned on by IAA are those for the synthesis of cell-wall materials. These materials are made and secreted by dictyosomes, and accommodate increases in cellular size that accompany growth.

Apical Dominance

Plants form axillary buds at the base of their petioles (see Chapter 14). In many plants, the growth of axillary buds is unequal along the axis of the plant: axillary buds near the shoot apex grow little, while those progressively farther from the tip grow much more. This inhibiting influence of the shoot apex on the growth of axillary buds is called **apical dominance** and produces a cone-shaped plant.

If the shoot tip is removed, axillary buds near the tip of the stem begin growing within as little as four hours. Botanists long believed that apical dominance was due only to IAA coming from the shoot tip, since the dominance could be restored by applying IAA to the cut surface of decapitated stems. However, the mechanism for apical dominance is more complicated than this. For example, one cannot detect any radioactive IAA in axillary buds after applying radioactively labeled IAA (^{14}C-IAA) to the stem tip. Furthermore, applying IAA to buds does not inhibit growth; sometimes it even promotes bud growth. This may account for the less pronounced apical dominance typical of many plants.

The most recent hypothesis for apical dominance is based on the idea that IAA stimulates the production of ethylene, another plant hormone. According to this explanation, IAA coming from the shoot tip stimulates cells around lateral buds to make ethylene, and it is the ethylene produced in response to IAA (rather than the IAA itself) that inhibits bud growth. Cytokinins coming from roots also influence apical dominance. Indeed, cytokinins promote bud growth, even if the shoot apex is present. These observations emphasize another generalization we can make about plant hormones: namely, that *a single aspect of growth and development can be influenced by several hormones: a particular response probably results from changing ratios of hormones rather than from the presence or absence of an individual hormone.*

The Lore of Plants

The characteristic shape of most Christmas trees results from apical dominance. Except for a few nonconformists, such as pin oak, all Christmas-tree-shaped trees are conifers. Ever since President Franklin Pierce (a politician otherwise noted for putting stickum on the backs of postage stamps) first had a Christmas tree in the White House in 1856, people have been picky about their trees, preferring ones that are about 2 meters high and 1.3 meters wide at the base. Today, thanks largely to the efforts of a former car salesman named Fred Musser, Scotch pine is the most popular Christmas tree in the United States. Mr. Musser's marketing of Scotch pine (*Pinus sylvestris*) emphasized that the tree's needles, though long, stayed on the tree much longer than those of balsam fir or spruce, the previous best-sellers. The Scotch pine is also fast-growing; it takes only eight years to grow a 2-meter tree.

Abscission

Another example of the interaction of IAA and other plant hormones is **abscission**, which is the shedding of leaves or fruit by a plant. Abscission of leaves is usually preceded by **senescence**, which is a collective term for the processes contributing to the age-induced decline and ultimate death of a plant or plant part. During the final stages of senescence, leaves and fruit separate from the plant at the **abscission zone**, which is a layer of thin-walled parenchyma cells at the base of the organ that will be shed (see fig. 14.26). Abscission occurs like this:

1. Actively growing leaves and fruit produce large amounts of IAA, which is transported to the stem. This IAA, along with cytokinin and gibberellins from the roots, retards the onset of senescence and abscission.
2. Environmental stimuli such as the shorter days of fall, drought, wounding, or a nutrient deficiency cause decreased production of IAA. This begins senescence.
3. Some signal, perhaps a "senescence factor," stimulates cells in the abscission zone to produce ethylene, which stimulates cells of the abscission zone to expand, suberize, and produce cellulase and pectinase.
4. Cellulase and pectinase digest the middle lamella, which usually cements cells of the abscission zone together.
5. As a result of this wall digestion and concurrent cellular expansion, the cells separate and abscission occurs. Deciduous plants lose their leaves each fall. Conversely, evergreens usually retain their leaves for 2–3 years or more.

The inhibitory effect of IAA on abscission has been exploited by horticulturists. For example, synthetic auxins are applied to orchards of apples, oranges, and grapefruits to prevent preharvest drop. Similar applications of auxin are used to hold berries on holly, thereby making them marketable for longer periods of time. Conversely, applying large amounts of synthetic auxins stimulates the production of excess amounts of ethylene, which often hastens abscission. Heavy applications of synthetic auxins are used to coordinate the abscission of apples, olives, and other fruits, thereby allowing the crop to be harvested in a shorter period of time.

Writing to Learn Botany

What is the selective advantage to a plant of shedding its leaves? Its fruit?

Differentiation of Vascular Tissue

The vascular cambium is activated in the spring by IAA produced by young, developing leaves (fig. 16.7). Gibberellin is also probably involved. Indeed, a high auxin/gibberellin ratio promotes differentiation of xylem, while a low auxin/gibberellin ratio favors differentiation of phloem. Nonhormonal factors such as sugars produced in leaves also influence the effect of IAA on cellular differentiation. For example,

- Auxin plus small amounts (2%) of sucrose favor differentiation of xylem.
- Auxin plus moderate amounts (3%) of sucrose favor differentiation of xylem and phloem.

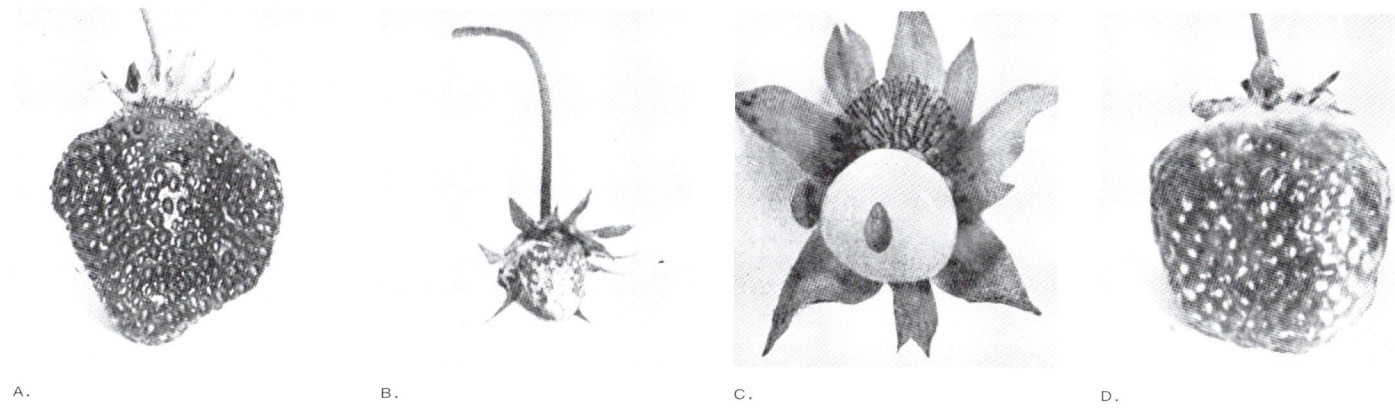

A.　　　　　　　　B.　　　　　　　　C.　　　　　　　　D.

FIGURE 18.5

Auxin stimulates development of strawberries. IAA from the seeds in each fruit of the strawberry stimulates development in the adjacent portion of the receptacle. Strawberries do not develop when fruits are removed. (a) normal; (b) all seeds removed; (c) all seeds but one removed; (d) all seeds removed and replaced by lanolin paste containing auxin.

- Auxin plus large amounts (4%) of sucrose favor differentiation of phloem.

These observations illustrate our next generalization about plant hormones: *Physiological responses elicited by hormones are strongly influenced by nonhormonal factors.*

Fruit Development

The sources of auxin that stimulate fruit development are seeds in fruit. For example, fruits of strawberry (*Fragaria ananassa*) are dispersed across a swollen receptacle: IAA from the seed in each fruit triggers development of the adjacent portion of the receptacle (fig. 18.5). Thus, strawberries do not develop when the fruits (i.e., the sources of IAA) are removed. Similarly, when fruits are removed from half of the strawberry, only the remaining half develops normally. Commercially important seedless fruits of plants such as tomato and cucumber can be induced to form if the unfertilized ovaries are treated with auxin. Such fruits are called **parthenocarpic** (from *parthenos*, meaning "virgin") fruit.

Formation of Adventitious Roots

Nurseries and amateur gardeners exploit the ability of auxin to stimulate the formation of adventitious roots to propagate plants. The procedure is simple: cut surfaces of pieces of a parent plant are dipped in a solution of synthetic auxin. The auxin stimulates the formation of adventitious roots at the cut surface (fig. 18.6), after which the so-called cutting can be transplanted and grown into a new plant.

C O N C E P T

Auxin stimulates cellular elongation, differentiation of vascular tissue, fruit development, and formation of adventitious roots.

FIGURE 18.6

Auxin promotes development of adventitious roots. To propagate plants, cut surfaces of parent plants are dipped into solutions of synthetic auxins. Auxin promotes the formation of adventitious roots at the cut surface. The stalk of the leaf on the right was placed in pure water.

Auxin and Calcium

The responses elicited by IAA depend on the presence of another nonhormonal factor: calcium ions (Ca^{2+}). For example, Ca^{2+} is required for the polar transport of IAA, and Ca^{2+}-deficient plants are usually not responsive to IAA. The most recent model accounting for the interaction of IAA and Ca^{2+} is as follows:

1. IAA stimulates release of Ca^{2+} from the vacuole and endoplasmic reticulum of a cell.

2. This release of Ca^{2+} increases the concentration of Ca^{2+} in the cytoplasm.

3. The increase in internal Ca^{2+} activates **calmodulin,** a Ca^{2+}-activated protein that regulates many processes in plants, animals, and microbes.

4. Activated calmodulin may then alter membrane transport, enzyme activity, and/or H^+ pumps, and thereby produce the physiological response. According to this model, wall acidification occurs as a result of changes in the concentration of Ca^{2+} rather than changes in IAA per se.

C O N C E P T

Auxin increases the internal concentrations of Ca^{2+}. These auxin-induced increases in internal Ca^{2+} activate calmodulin, which then may trigger the physiological response.

GIBBERELLINS

Discovery

At about the time Frits Went published the results of his auxin experiments, botanists in Japan were beginning experiments that would ultimately lead to the discovery of another plant hormone: gibberellin. Ewiti Kurosawa and his colleagues were studying rice (*Oryza sativa*) plants suffering from *bakanae*, or "foolish seedling" disease. These plants were pale and spindly, and their growth caused them to fall over and die before they could be harvested. Kurosawa soon discovered that the diseased plants were infected by a fungus named *Gibberella fujikuroi*. Knowing this allowed him to design some clever experiments. First, he grew pure cultures of the fungus in his laboratory. When he ground up the fungus and applied the extract to healthy plants, the plant quickly developed symptoms characteristic of *bakanae*. Similar symptoms also developed when healthy plants were treated with media in which the fungus had grown. These results indicated that the fungus made and secreted a compound that caused *bakanae* in infected rice plants. Kurosawa named this compound *gibberellin*.

By 1939, Japanese botanists had isolated and chemically identified gibberellin. However, their work went largely unnoticed in the United States because of U.S. botanists' preoccupation with auxin, the lack of scientific communication with Japan, and World War II. Later studies in the United States showed that gibberellins are involved in many aspects of plant growth and development.

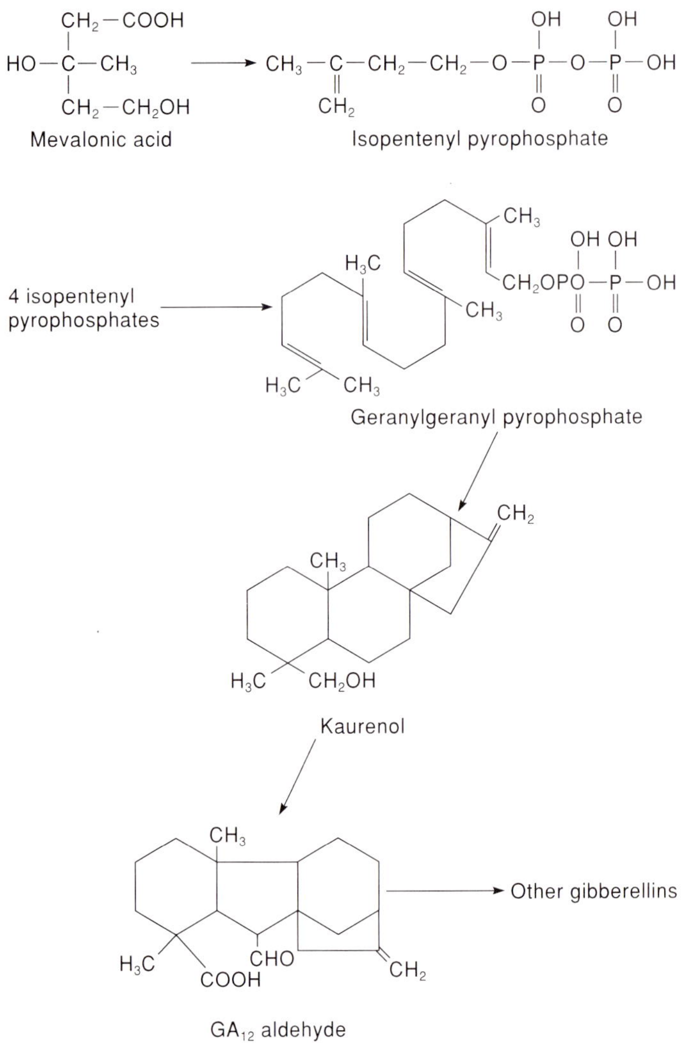

FIGURE 18.7

Gibberellin is made via the mevalonic acid pathway.

Synthesis and Transport

Gibberellin is made via the mevalonic acid pathway (fig. 18.7). Although more than eighty gibberellins have been isolated from various fungi and plants, no more than fifteen have been found in a single species. Each of the gibberellins has an interlocking ring structure and one or more carboxyl groups that impart acidic properties to the molecule (fig. 18.8). Gibberellins are abbreviated **GA** (for gibberellic acid) and assigned subscripts that distinguish them from each other. For example, GA_3 is isolated from *Gibberella fujikuroi* and is the most intensively studied gibberellin.

Several commercial compounds inhibit the synthesis of gibberellins. These inhibitors are called *growth retardants* and

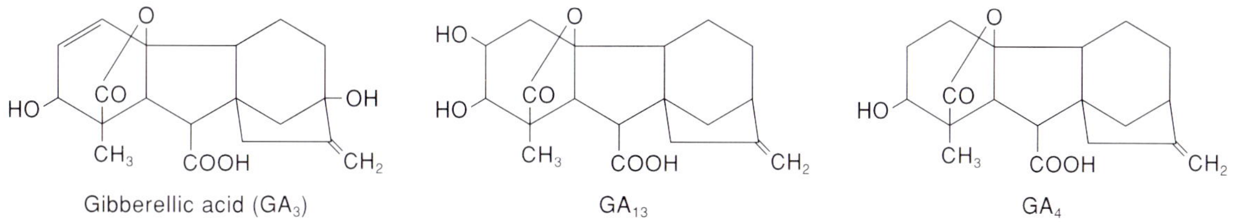

FIGURE 18.8

Three of the more than 80 gibberellins that have been isolated from various fungi and plants. All gibberellins have an interlocking ring structure and one or more carboxyl groups that impart acidic properties to the molecules.

include Phosphon D, Cycocel (CCC), Amo-1618, and Ancymidol. Growth retardants inhibit stem elongation, thereby producing stunted plants. Growth retardants are frequently sprayed on growing chrysanthemums to produce flowers with thicker, sturdier stalks. People in the turfgrass industry often use growth retardants to control the development of seeds and stem elongation.

Why are there so many different gibberellins? Most gibberellins are inactive and are probably precursors of more active gibberellins. Indeed, GA_1 is probably the only gibberellin that controls stem elongation in angiosperms.

Gibberellins occur in angiosperms, gymnosperms, mosses, ferns, algae, and fungi, but are unknown in bacteria. In angiosperms, gibberellins occur in immature seeds, apices of roots and shoots, and young leaves. Unlike the transport of IAA, the transport of gibberellin is not polar: it moves in all directions in the xylem and phloem.

CONCEPT

Although there are many different gibberellins, only GA_1 controls stem elongation. Unlike auxins, gibberellins move nonpolarly throughout plants.

Effects of Gibberellins

Extensive Growth of Intact Plants

Many dwarf mutants grow normally if given GA. The sensitivity of many dwarf plants to GA is striking: for example, dwarf pea seedlings respond to as little as one-billionth of a gram of gibberellin. Dwarfism does not result from the absence of gibberellins. In fact, dwarf plants often contain even more gibberellins than do normal plants. Rather, their stunted growth results from their inability to make enough *active* gibberellins, specifically GA_1.

Gibberellin is used commercially by Hawaii's sugarcane growers. Growing plants are sprayed with gibberellin, which stimulates shoot elongation and thereby increases the yield of sugar. However, the stimulation of shoot elongation by GA differs from that induced by IAA. For example,

- GA controls internode elongation in the mature regions of trees, shrubs, and a few grasses, whereas IAA regulates elongation in grass seedlings and herbs.

- GA induces cellular division *and* cellular elongation; IAA induces cellular elongation alone.

- GA-stimulated elongation does not involve the cell-wall acidification characteristic of IAA-induced elongation.

Thus, we identify another generalization about plant hormones: *different hormones can elicit similar effects via different mechanisms.*

Seed Germination

Gibberellins play a critical role in seed germination in cereal grasses such as barley. Here's what happens:

1. The sequence of events leading to seed germination begins with **imbibition,** which is the absorption of water by a seed. Imbibition causes the embryo to release gibberellins.

2. These gibberellins stimulate transcription of genes of hydrolytic enzymes in the seed's **aleurone layer,** a specialized layer of the endosperm 2–4 cells thick located just inside the seed coat. Gibberellin enhances the transcription of amylase mRNA.

3. These hydrolytic enzymes are secreted by dictyosomes into the seed's endosperm. One of the hydrolytic enzymes produced by the aleurone layer, amylase, catalyzes the conversion of starch to sugar, which is used as an energy source for the growing seedling.

Several relatively simple experiments provide evidence for the role of GA in the germination of barley seeds. For example, the endosperm is not broken down if the embryo (i.e., the source of GA) is removed. Similarly, the addition of GA to these embryoless seeds results in the production of amylase and the hydrolysis of endosperm starch to sugar. As is true for genetic dwarf plants, the endosperm of barley seeds is sensitive to GA: digestion begins when the endosperm is exposed to as little as nine-trillionths of a gram of GA.

As convincing as the preceding mechanism seems, it does not account for the germination of dicot seeds, nor even the germination of seeds of other cereal grains. For example, the aleurone layers of some cultivated oats and corn do not respond to added GA. In many of these seeds, the **scutellum** (i.e., cotyledon) rather than the aleurone layer is the primary source of enzymes that digest the endosperm. In the seeds of many dicots and gymnosperms, cytokinins may stimulate the breakdown of stored food reserves.

Many plants produce seeds that usually germinate only after exposure to cold or light. Gibberellins will substitute for this cold requirement in plants such as tobacco, lettuce, and oats. Gibberellins stimulate germination of these seeds by enhancing cellular elongation in the embryo, which causes the embryo to rupture the seed coat. The ability of gibberellins to stimulate germination of cereal seeds is exploited in making alcoholic beverages (see Chapter 4, box 4.1). Brewers treat barley grains with gibberellins to stimulate seed germination and speed the conversion of starch to sugar. These germinated seedlings are called *malt*. Brewing converts the sugar to alcohol by anaerobic respiration (see Chapter 6).

Juvenility

Many plants have a juvenile stage and an adult stage of growth. For example, juvenile stages of eucalyptus (*Eucalyptus globulus*) have leaves that are shaped differently than those of adult stages (fig. 18.9). Gibberellins may help determine whether a particular part of a plant is juvenile or adult. For example, the buds of adult branches usually develop only into adult branches, but treating them with GA causes them to grow into juvenile branches.

Flowering

During their first year of growth, biennial plants such as cabbage have short internodes. These plants are called **rosettes** because their tightly packed leaves are arranged like the petals of a rose. Following winter's cold temperatures, internodes of these plants expand rapidly, and the plant forms flowers during the second year of growth. The rapid expansion of internodes and formation of flowers by rosette plants in response to cold is referred to as **bolting.** Applying GA to rosette plants also induces bolting, suggesting that cold temperatures somehow stimulate the synthesis of GA during the following season (fig. 18.10).

GA-induced bolting in rosettes is commercially exploited to induce the early production of seeds by some biennials. When these plants are treated with GA during their first year of growth, they form flowers and fruit, thereby allowing growers to obtain seeds after only one season instead of the two seasons normally required by biennials. GA does not stimulate flowering in most other plants.

Fruit Formation

The most important commercial use of gibberellins involves their ability to increase the size of seedless grapes. Indeed, almost all vines of the "Thompson Seedless" grape grown in the Central and Imperial Valleys of California are sprayed with gibberellins each year. As a result of these treatments, the grapes increase in size almost threefold and are more loosely packed, making them less susceptible to fungal infections.

A.

B.

FIGURE 18.9

(a) Juvenile and (b) mature leaves in *Eucalyptus globulus* are shaped differently. Juvenile leaves are softer and opposite each other, with palisade parenchyma only on upper surfaces. Adult leaves are hard, spirally arranged and have palisade on both sides.

Gibberellins promote extensive growth of intact plants, stimulate germination in some cereal grasses, influence formation of juvenile vs. adult stages of growth, promote bolting of biennials, and increase fruit size in grapes.

FIGURE 18.10

Bolting in cabbage (*Brassica oleracea*). The plant on the left was untreated. The plant on the right was treated with gibberellin once a week for eight weeks.

CYTOKININS

Discovery

The trail leading to the discovery of cytokinins can be traced to Gottlieb Haberlandt, who reported in 1913 that an unknown compound in vascular tissue stimulated cellular division. Like the Darwins' studies of the phototropic "influence," Haberlandt's work went largely unnoticed. However, Went's discovery of auxin in 1926 stimulated botanists to search for other plant hormones, and later led to the rediscovery of Haberlandt's pioneering work. The search for the hormone controlling cellular division was a long and unusual one: it spanned several decades, involved many researchers, and included substances ranging from coconut milk to fish sperm.

In the 1940s, botanist Johannes van Overbeek made an interesting observation: plant embryos grew faster when they were supplied with coconut milk. Coconut milk is the liquid endosperm of coconut seeds and is rich in nucleic acids. This information was critical to the next set of experiments, which were done by Folke Skoog and his student Carlos Miller at the University of Wisconsin in the 1950s. Skoog and Miller were studying the influence of auxin on the growth of tobacco (*Nicotiana tabacum*) cells. They grew their cells via **tissue culture,** a technique for growing plants in artificial media in test tubes or beakers (you'll recall discussions of this technique in Chapters 9 and 12; see figs. 9.1 and 12.8). Growing plants in controlled cultures was important for Miller and Skoog because it enabled them to test the effects of individual compounds that they could add or delete from the culture medium. When auxin alone was added to the medium, tobacco cells expanded but did not divide. Auxin therefore appeared to stimulate cellular enlargement, but that alone does not explain plant growth, which in intact plants involves cellular enlargement *and* division. Apparently, the signal for cellular division was missing.

One day in the lab, Miller noticed a jar containing something you probably don't have in your kitchen: herring-sperm DNA. Miller knew of van Overbeek's research showing that the DNA-rich endosperm of coconut stimulated plant growth, and he suspected that something in the DNA could cause the increased growth. When he added the herring-sperm DNA to the culture medium, the tobacco cells began dividing. Miller was excited about these results and quickly ordered another supply of herring-sperm DNA so that he could repeat and expand his experiments. To his chagrin, cells treated with the newly ordered DNA did not divide: apparently, the growth factor present in the old DNA was not in the newly arrived DNA.

Miller reasoned that his original results were due to the age of the original DNA; that is, the signal that triggered cellular division was probably a breakdown product of the DNA rather than the DNA itself. Miller later discovered that he could also induce cellular division by supplementing the culture medium with adenine, a purine base in DNA. Similar results were later obtained by adding adeninelike compounds extracted from yeast. After painstaking work, Miller, Skoog, and their coworkers finally isolated the growth factor responsible for cellular division from a DNA preparation. They named this substance *kinetin*, and they named the class to which kinetin belongs **cytokinins,** because these substances stimulated *cytokinesis*, or cellular division.

The first naturally occurring cytokinin was isolated in 1964 from corn (*Zea mays*) and was named *zeatin* (fig. 18.11). Soon thereafter, the influence of coconut milk on cellular division was explained when it was shown to contain zeatin and zeatin riboside, another cytokinin. Since then, botanists have isolated other naturally occurring cytokinins (e.g., dihydrozeatin and isopentenyl adenine) and have made several artificial cytokinins (e.g., kinetin and 6-benzylamino purine). All of these cytokinins have structures similar to adenine: that is, they have a side chain rich in carbon and hydrogen attached to nitrogen protruding from the top of the purine ring. Cytokinins are often minor components of RNA, but we do not know if these cytokinins are related to free cytokinins in cells.

Synthesis and Transport

Contrary to what was first believed, cytokinins are not breakdown products of DNA. Rather, they are made via the mevalonate pathway, the same pathway used to make gibberellins. Cytokinins are widespread, if not universal, in plants: they have been isolated from angiosperms, gymnosperms, mosses, and ferns. In angiosperms, most cytokinins are made in roots, and also occur in seeds, fruits, and young leaves. Cytokinins move nonpolarly in xylem, phloem, and parenchyma cells.

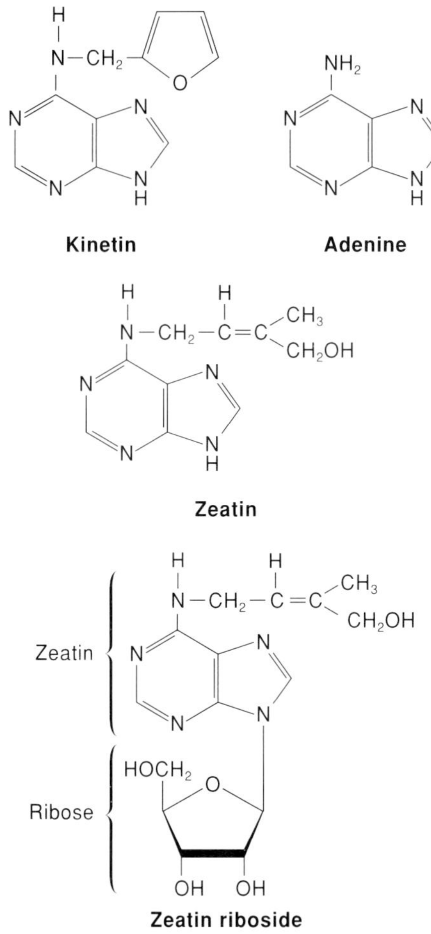

Effects of Cytokinins

Although cytokinins have relatively few uses in agriculture (e.g., turfgrass growers sometimes use them to promote root growth), they do strongly affect plant growth and development.

Cellular Division

Cytokinins stimulate cellular division by hastening the transition of cells from the G_2 phase (the growth phase following DNA replication) to the M phase (mitosis) of the cell cycle (see Chapter 9). This effect depends on the presence of auxin. For example, cytokinins alone have no effect on cultured tobacco cells; cellular division begins only when auxin is added to the culture medium.

The discovery that cytokinins promote cellular division raised another interesting question: Could cytokinins be responsible for animal cancers, which result from uncontrolled cellular division? Research has demonstrated repeatedly that cytokinins have no effect on animal cells and that there is no direct link between animal cancers and cytokinins.

Effects on Cotyledons

Cytokinins promote cellular division and expansion in cotyledons. Cellular expansion results from cytokinin-induced increases in wall plasticity that do not involve wall acidification. Cytokinins also increase the amount of sugars (especially glucose and fructose) in cells, which may account for the osmotic influx of water and the resulting expansion of cytokinin-treated cells in cotyledons.

Organogenesis

Cytokinins and IAA affect **organogenesis,** which is the formation of organs. Figure 18.12 shows the influence of changing amounts of cytokinin and IAA on the formation of shoots and roots. Cultured cells grow only in the presence of cytokinin and IAA. High cytokinin/auxin ratios favor the formation of shoots, while low ratios favor the formation of roots. Thus, a plant can be completely regenerated from single cells by varying the amounts of cytokinin and IAA. This hormonally controlled means of plant regeneration has been used to propagate plants that are resistant to pathogens, drought, and other stresses.

Senescence

Cytokinins delay the breakdown of chlorophyll in detached leaves, apparently by preventing genes that stimulate chlorophyll

FIGURE 18.11

Cytokinins. All cytokinins are structurally related to the purine adenine.

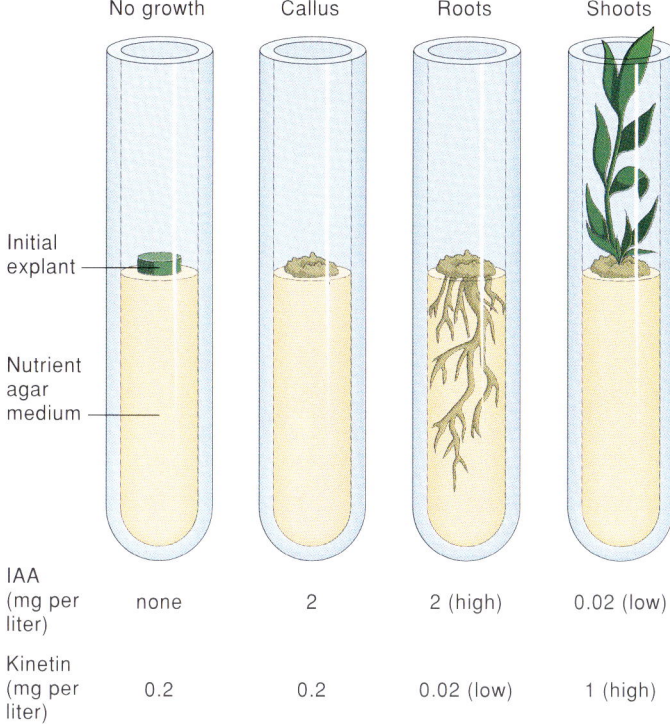

FIGURE 18.12

The responses of plant tissue culture to kinetin and auxin. The initial explant is a small piece of sterile tissue cut from the pith of a tobacco plant. High ratios of auxin/cytokinin favor the formation of roots, while low ratios of auxin/cytokinin favor the formation of shoots.

formation from being turned off. Cytokinin-treated areas of leaves remain healthy as the remaining parts of the leaf senesce. This effect of cytokinin may be due to its ability to establish a "sink" to which nutrients move. However, the cytokinin-induced delay in leaf senescence occurs only in detached leaves; cytokinins have little or no effect on senescence in attached organs. Leaf senescence is also delayed by the formation of adventitious roots. Roots, you'll recall, are rich in cytokinins; and the transport of these cytokinins to leaves could account for the delayed senescence.

Cytokinins are sometimes used commercially to maintain the greenness of excised plant parts, such as cut flowers. However, their use on edible crops such as broccoli is banned in the United States; any compound like cytokinin that resembles a nucleic acid component is automatically a suspected carcinogen.

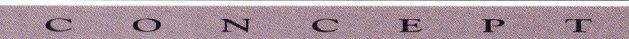

Cytokinins are adeninelike compounds that stimulate cellular division, enlargement of cotyledons, and development of lateral buds. Cytokinins delay senescence of detached leaves and, with auxin, influence the formation of shoots and roots.

Cytokinins and Calcium

Small amounts of cytokinin do not stimulate cellular division of cultured cells, which merely enlarge due to the presence of auxin. However, adding calcium switches the growth pattern from cellular enlargement to cellular division. Therefore, the presence of calcium increases a cell's sensitivity to cytokinin. Cytokinins also stimulate bud formation in mosses; substances that increase cells' permeability to Ca^{2+} mimic the effect of cytokinins, and cytokinin-induced formation of buds can be inhibited by applying inhibitors of calmodulin. Similarly, the addition of calcium enhances the cytokinin-induced delay of leaf senescence. These effects may be mediated by cytokinin-induced changes in the concentration of Ca^{2+}.

ETHYLENE

Discovery

During the 1800s, the city streets of Germany were illuminated by lamps that burned "illuminating gas." Soon after these lamps were installed, city residents made a curious observation: plants growing near the lamps had short, thick stems. Furthermore, the leaves fell off of most of these plants. Were these effects caused by the lamp's light, heat, or some other factor?

This question was answered in 1901 when Soviet plant physiologist Dimitry Neljubow identified ethylene as the combustion product of "illuminating gas" that was responsible for defoliating and inhibiting the elongation of plants growing near the lamps. Neljubow also showed that these effects could be induced by as little as 0.06 parts per million of ethylene (1 part per million = 1 milliliter in 1,000 liters). Although botanists knew in 1901 that ethylene had many effects on plant growth, it would be more than three decades before they would learn that these effects were naturally occurring aspects of growth caused by ethylene made by plants.

In 1910 an annual report to the Japanese Department of Agriculture recommended that oranges not be stored with bananas, because oranges released something that caused premature ripening of the bananas. This "something" was not identified until 1934, when R. Gane showed that ethylene is made by plants and that it causes faster ripening of many fruits, including bananas. Subsequent research showed that ethylene met the requirements of a plant hormone: it is made in one part of a plant and transported to another, where it induces a physiological response. Thus was discovered the first gaseous plant hormone: ethylene.

Synthesis and Transport

Ethylene is made from methionine, an amino acid (fig. 18.13). Its synthesis is inhibited by CO_2 and requires oxygen. When plants are placed in pure CO_2 or O_2-free air, ethylene synthesis decreases dramatically.

Boxed Reading 18.2

Plant Hormones in Plant Pathology

Cytokinins and IAA play important roles in diseases caused by several plant pathogens. For example, *Corynebacterium fascians* is a bacterium that infects plants such as peas and chrysanthemums. Symptoms usually include flattened stems and extensive branching that forms broomlike structures called *witch's brooms*. Similar branching patterns can be induced in healthy plants by applying cytokinins. Thus, pathogenic symptoms can be induced by excess amounts of cytokinins. But where do these excess cytokinins come from?

Pathogenic strains of *C. fascians* contain a plasmid (i.e., a small circle of DNA that occurs independently of the organism's main DNA—see Chapter 11) having genes that produce cytokinins. Following infection, this plasmid is incorporated into the host's DNA. Subsequent transcription and translation of this plasmid results in excessive amounts of cytokinins, which cause extensive branching. Nonpathogenic strains of *C. fascians* lack this plasmid.

A similar mechanism is used by the bacterium *Agrobacterium tumefaciens* to produce crown gall, a cancerlike disease of dicots. During infection, *A. tumefaciens* transfers to its host a fragment of a large plasmid that carries genes responsible for making cytokinins and IAA. When these genes are expressed, the amounts of IAA and cytokinin increase, thereby producing the symptoms of crown gall (for a discussion of how botanists use *Agrobacterium* in genetic engineering see Chapter 11).

Other plant pathogens do not cause their hosts to produce hormones that disrupt growth. Instead, the pathogens produce and release large amounts of IAA and cytokinin themselves. For example, the bacterium *Pseudomonas syringae* causes olive-knot disease in olive, oleander, and privet. Unlike *C. fascians* and *A. tumefaciens*, *P. syringae* does not transfer any of its DNA to the host. Instead, it exerts its pathogenic effects by secreting large amounts of IAA and cytokinins. Similarly, many rust fungi produce cytokinins that delay leaf senescence by establishing areas to which nutrients are transported. Consequently, necrotic lesions induced by the fungus are often surrounded by green islands of healthy tissue on which the fungus will subsequently feed. Witch's brooms are also common in trees infected by fungi such as *Exobasidium*. The degree of pathogenicity is roughly proportional to the amount of cytokinin produced and secreted by the fungus.

Plant abnormalities induced by excessive amounts of hormones are not limited to those produced by fungi and bacteria. For example, leaf galls form in response to parasitic insects such as mites and gall wasps. Female insects deposit their eggs in the leaf mesophyll. These eggs develop into larvae that produce cytokinins. When these cytokinins are released into the leaf, they induce cellular divisions that form a ball-like mass, or gall, that becomes the home for the developing insect. Because cytokinins also attract sugars and other nutrients, the gall provides shelter and a source of nutrition for the insect, even though the rest of the leaf may be undergoing senescence. During spring, the adult insect leaves the gall and lays eggs that infect other leaves, thereby completing its life cycle.

Box Figure 18.2
Sawfly gall on willow.

All parts of angiosperms make ethylene, but especially large amounts are released into the air by roots, the shoot apical meristem, nodes, senescing flowers, and ripening fruits (e.g., the dark flecks on a ripening banana peel are concentrated pockets of ethylene). Because most ethylene-induced effects result from ethylene in the air, the effects of ethylene can be contagious: ethylene made by one "bad" (i.e., overripe) apple *can* "spoil" (i.e., induce rapid ripening of) an entire bushel of apples.

Effects of Ethylene

Fruit Ripening

The ancient Chinese knew that fruit would ripen faster if placed in rooms containing burning incense. The factor responsible for this hastened ripening was not heat, but ethylene released as the incense burned. This stimulation of fruit ripening by ethylene is multifaceted and includes the breakdown of chlorophyll and synthesis of other pigments (e.g., apples changing from green to red during ripening), fruit softening due to the breakdown of cell walls by cellulase and pectinase, production of volatile compounds associated with the scents of fruit, and conversion of starches and acids to sugars. Ethylene stimulates each of these aspects of fruit ripening.

In fruits such as tomatoes and apples, there is a conspicuous increase in respiration immediately before fruit ripening. This increase in respiration is called a **climacteric**, and fruits that display it are referred to as **climacteric fruits**. The climacteric begins just after a huge (e.g., up to a 100-fold) increase in ethylene production. Thus, the climacteric and fruit ripening are triggered by ethylene.

Fruit growers have often (and sometimes unknowingly) taken advantage of ethylene's stimulation of fruit ripening to have fruits available for sale out-of-season. For example, many apples are picked in September and October when they are

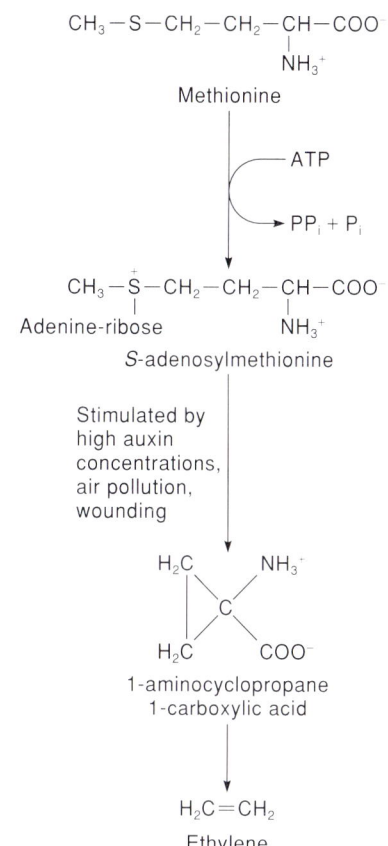

FIGURE 18.13

Biosynthesis of ethylene from methionine, which serves as a precursor of ethylene in all higher plant tissues. 1-amino-cyclopropane-1-carboxylic acid is the immediate precursor of ethylene.

FIGURE 18.14

Leaf abscission is strongly influenced by hormones, especially ethylene.

green and immature. These apples are stored in rooms containing air that has small amounts (1%–3%) of O_2, large amounts (5%–10%) of CO_2, and no ethylene. Because large amounts of CO_2 and low levels of O_2 inhibit ethylene synthesis, the fruit can be stored without ripening. When these immature fruits are needed for sale, producers expose them to normal air containing about 1 part per million of ethylene. The ethylene induces the climacteric, and ripening soon occurs. Thus, the "fresh" apples you buy in March were probably harvested in September or October of the previous year. This "ripening on demand" is also used by producers of tomatoes, lemons, and oranges. Not all fruit can be ripened by exposure to ethylene, however. Such fruits (e.g., grapes) are called **nonclimacteric fruits** and are insensitive to ethylene.

Flowering

Ethylene inhibits flowering in most species, but promotes it in a few plants, including mangos, pineapples, and some ornamental plants. This effect was known long ago by Puerto Rican pineapple growers and Filipino mango growers, who set bonfires near their crops. The fires produced ethylene, which initiated and synchronized flowering of their plants. In the 1950s, growers sprayed their crops with NAA, a synthetic auxin, which increased ethylene production and thereby stimulated flowering. Today, pineapple growers in Hawaii produce pineapple fruit all year by spraying plants with ethephon, a compound that breaks down in neutral and alkaline conditions to release ethylene.

Ethylene also promotes senescence of flowers. When pollen grains germinate, the stigmas of flowers produce large amounts of ethylene that trigger senescence of floral parts.

Abscission

The increased production of ethylene at the abcission zone triggers the breakdown of the middle lamella, and thereby initiates abscission (fig. 18.14; also see fig. 14.26). Like the influence of ethylene on flowering, this effect is used by horticulturists to increase the efficiency of harvesting fruit. Fruits such as cherries, grapes, and blueberries are sprayed with ethephon to coordinate abscission, thereby allowing growers to harvest their crops in shorter periods of time.

Sex Expression

The sex of flowers on monoecious plants (i.e., plants that have male and female flowers on the same individual) is determined by ethylene and gibberellins. For example, cucumber (*Cucumis sativus*) buds treated with ethylene become carpellate flowers,

FIGURE 18.15

Tomato (*Lycopersicon esculentum*) plants exposed to 28 consecutive days of brief, periodic shaking. Left: unstressed control; center: 30 seconds of stress once a day; right: 30 seconds of stress twice a day. Mechanical disturbances such as shaking decrease elongation.

whereas those treated with gibberellins become staminate flowers. Correspondingly, buds that ultimately become carpellate flowers produce more ethylene than do buds that become staminate flowers.

Stem Elongation

Mechanical disturbances such as shaking decrease elongation (fig. 18.15). This effect, called **thigmomorphogenesis,** is mediated by ethylene. Mechanical disturbances increase ethylene production severalfold, which causes cells to arrange their cellulose microfibrils into longitudinal hoops. This lengthwise reinforcement inhibits cellular elongation, causing cells to expand radially and form short, thick stems. This effect is the opposite of that of auxin, which causes cells to orient their microfibrils more transversely, thereby accounting for cellular elongation.

Waterlogging

When a plant's roots are submerged for long periods, water eventually fills the intercellular spaces. Since these spaces are the primary routes of gas exchange with the atmosphere, the submerged roots become waterlogged and anaerobic. As you might predict, the symptoms of waterlogged plants (e.g., leaf chlorosis, shorter and thicker shoots, and wilting) can be induced by placing roots in oxygen-free air.

Because oxygen is needed to make ethylene, its synthesis is greatly reduced in roots of waterlogged plants. The small amount of ethylene that is made in these roots is trapped, where it accumulates and eventually stimulates the activity of cellulase and pectinase. These enzymes break down the cell walls and, in doing so, form the many intercellular spaces characteristic of hydrophytes. Meanwhile, ethylene precursors in the

A.

B.

FIGURE 18.16

Severe leaf epinasty caused by ethylene. (a) A young tomato plant photographed just before dipping in a solution of 1.0 mM naphthalene acetic acid (an auxin) which, at this concentration, causes the plant to produce an excess of ethylene. (b) The same plant 24 hours later. The extreme downward bending and twisting of the leaves is called *epinasty*.

shoot are converted to ethylene, which causes parenchyma cells on the upper side of the petiole to expand and point the leaf down, a response called **epinasty** (fig. 18.16).

CONCEPT

Ethylene is a gaseous hormone that stimulates fruit ripening, promotes abscission, affects sex expression, and regulates stem elongation.

> ### What are Botanists Doing?
>
> Go to the library and read a recent article about any of the plant growth-regulating substances mentioned in this chapter. Summarize the importance of the article. How does what you learned from the article relate to what you learned from this chapter?

Ethylene and Auxin

We emphasized earlier in this chapter that IAA stimulates ethylene production, thereby linking the responses of these two hormones. But ethylene does not account for all of the effects elicited by applying IAA. Several responses of plants to IAA are unrelated to ethylene. For example, IAA's stimulation of cellular elongation and the production of lateral roots occur independently of ethylene. Conversely, leaf epinasty, decreased elongation of roots and shoots, and sex determination are reponses to ethylene rather than IAA.

ABSCISIC ACID

Discovery

The exciting discoveries of auxin, cytokinins, ethylene, and gibberellins offered new prospects for explaining plant growth and development. Most of the effects first discovered for plant hormones were stimulatory. For example, IAA stimulates cellular elongation, and cytokinins stimulate cellular division. But by the 1940s, botanists suspected that some aspects of plant growth and development resulted from inhibition rather than stimulation of growth. Near the end of that decade came a hint that this suspicion was correct: Torsten Hemberg of Sweden reported that dormant buds of ash and potato contained inhibitors that blocked the effects of IAA. When the buds germinated, the amount of these inhibitors decreased. Hemberg named these inhibitors *dormins*, after their apparent influence on the dormancy of buds.

In the early 1960s, Philip Wareing confirmed Hemberg's findings and reported that applying dormin to a bud induced dormancy. At about the same time, F. T. Addicott discovered a compound that stimulated abscission of cotton fruit. He named this substance *abscisin*. Botanists later were surprised to discover that dormin and abscisin were the same compound. This compound was named **abscisic acid,** or **ABA**—an unfortunate name, because subsequent research has shown that ethylene rather than ABA controls abscission.

Synthesis and Transport

ABA (fig. 18.17) in plants is made from carotenoids. It occurs in angiosperms, gymnosperms, and mosses, but apparently not in liverworts. Once synthesized, ABA moves throughout a plant in xylem, phloem, and parenchyma. Like gibberellins

FIGURE 18.17

The structure of abscisic acid (ABA).

and cytokinins, ABA moves nonpolarly. There are no synthetic abscisic acids.

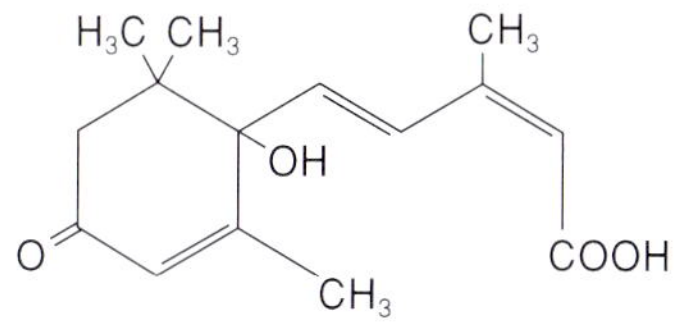

Effects of Abscisic Acid

Closure of Stomata

During drought, leaves make large amounts of ABA, which causes stomata to close. Thus, ABA functions as a messenger that enables plants to conserve water during drought. Because ABA closes stomata within 1–2 minutes, this effect probably occurs independently of protein synthesis. ABA probably exerts its effect by binding to proteins on the outer surface of the plasmalemma of guard cells. This makes the plasmalemma more positively charged, thereby stimulating transport of ions (especially K^+) from guard cells to epidermal cells. Because the loss of these ions causes water to leave guard cells via osmosis, the guard cells collapse, closing the stomatal pore.

A plant's tolerance to drought may relate to its ability to make ABA. In some species, the levels of ABA increase by as much as 10-fold within minutes after wilting occurs.

Bud Dormancy

ABA was initially thought to control bud dormancy, but recent evidence questions this conclusion. For example,

- After leaves are treated with radioactive ABA, no radioactivity can be detected in buds.
- In several plants, the induction and breaking of dormancy do not correlate with changes in endogenous amounts of ABA.
- Treatments that induce dormancy do not alter the amounts of ABA in buds.

These results suggest that bud dormancy is not controlled by ABA alone, but is probably influenced by cytokinins and IAA-induced synthesis of ethylene.

Seed Dormancy

Applying ABA delays seed germination in many species. Similarly, the amount of ABA in the seeds of many plants decreases when seeds germinate. Thus, ABA may control seed dormancy in certain species. This conclusion may not apply to all plants, however, since germination of many seeds occurs without any changes in the amount of ABA.

ABA Counteracts Stimulatory Effects of Other Hormones

ABA usually inhibits the stimulatory effects of other hormones. For example, ABA (1) inhibits amylase produced by seeds treated with gibberellin, (2) promotes chlorosis that is inhibited by cytokinins, and (3) inhibits wall elasticity and cell growth promoted by IAA. How does ABA counteract these effects? ABA is a Ca^{2+} antagonist; thus, its inhibition of the stimulatory effects of IAA and cytokinin may be due to its interference with Ca^{2+} metabolism.

Although ABA usually inhibits growth, it is not toxic to plants, as are inhibitors of energy metabolism and RNA/protein synthesis. Although, ABA often decreases gene activity, there are instances of ABA stimulating genes. For example, ABA stimulates the synthesis of mRNAs for storage proteins in developing wheat grains.

CONCEPT

Abscisic acid causes stomata to close, influences dormancy of some seeds, and generally inhibits the stimulatory effects of other hormones. These effects may involve changes in the plant's calcium metabolism.

Doing Botany Yourself

Suppose you have isolated a substance from a rare, unstudied group of plants. What experiments would you do to learn if the substance is similar to auxin? gibberellin? cytokinin? abscisic acid?

OLIGOSACCHARINS AND OTHER PLANT HORMONES

A growing number of reports indicate that plant hormones are not limited to auxin, gibberellins, cytokinins, ethylene, and abscisic acid. For example, fragments of the cell wall called **oligosaccharins** control plant growth, differentiation, reproduction, and defense against disease, and therefore function as plant hormones. However, oligosaccharins typically differ from other plant hormones because they elicit specific effects: different oligosaccharins can induce cultured cells to form undifferentiated callus, roots, shoots, or flowers in a variety of plants. Moreover, oligosaccharins that inhibit flowering and promote vegetative growth in one species have the same effects on other species. Furthermore, these effects are induced by impressively small amounts of the oligosaccharin: it takes 100 to 1,000 times less oligosaccharin than IAA or cytokinin to induce a response.

Oligosaccharins are released from cell walls by enzymes. Different oligosaccharins are released by different enzymes, and each one transmits a message that regulates a particular function. This specificity has prompted several botanists to suggest that the many effects of hormones like IAA and gibberellins may be due to the activation of enzymes that release specific oligosaccharins.

Other investigations have identified a variety of other compounds that function as hormones in various groups of plants. For example,

- Yams contain **batasins,** which induce dormancy of bulbils (vegetative reproductive structures) that form from lateral bulbs. Increased amounts of batasins induce dormancy, and cold treatments that break dormancy reduce the amounts of batasins.

- **Brassosteroids** are plant hormones in tea, bean, and rice plants that stimulate the growth of stems. These hormones have chemical structures similar to that of ecdysone, the insect molting hormone.

CONTROLLING THE AMOUNTS OF PLANT HORMONES: HORMONAL INTERACTIONS

Hormones seldom, if ever, function alone; rather, plant growth and development usually result from interactions of plant hormones (fig. 18.18). We've already discussed several of these interactions, including those responsible for organogenesis and vascular differentiation. But what controls the amounts of hormones, and thereby their ratios and resulting interactions? The amounts of hormones are controlled in two ways:

1. *Regulation of the rate of synthesis.* Several factors influence the rate of hormone production. For example, daylength can trigger the synthesis of IAA, and the synthesis of gibberellins in biennials is stimulated by cold temperatures.

2. *Regulation of the rate of breakdown or inactivation.* Inactivation of a hormone usually occurs either by oxidizing the hormone or by conjugating (i.e., combining) it with another compound. Coordinated synthesis and inactivation could control the amount of hormone present and therefore control the growth response.

DO PLANT HORMONES REALLY EXIST?

Since plant hormones were discovered after animal hormones, most early research was based on the assumption that plant and animal hormones had fundamentally similar effects. However, there are important differences in plant and animal hormones:

- There is no evidence that the fundamental actions of plant and animal hormones are the same.

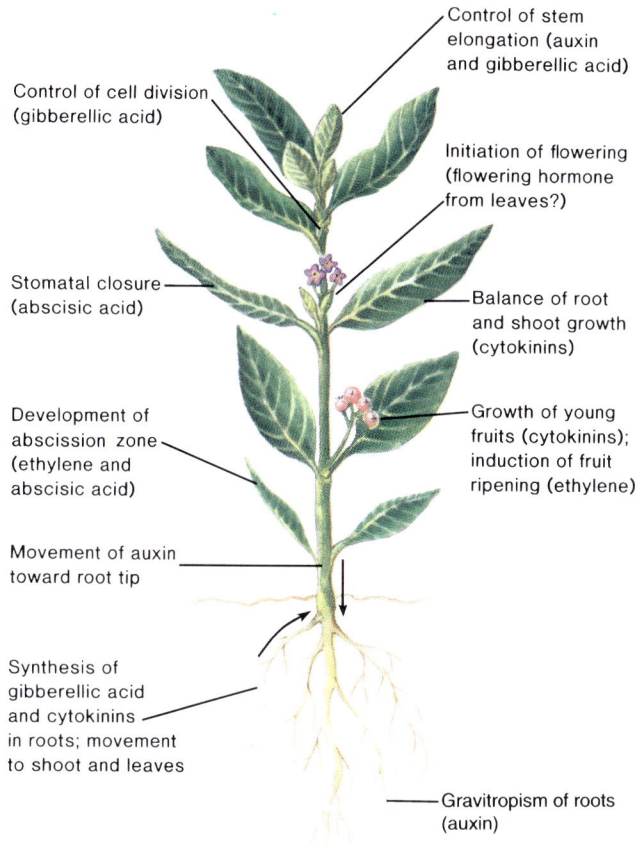

FIGURE 18.18

Numerous hormonal interactions influence plant growth and development.

- Unlike animal hormones, plant hormones are not made in tissues specialized for hormone production.
- Unlike animal hormones, plant hormones do not have definite target areas.
- Animal hormones usually have specific effects, while those of plants seldom, if ever, have specific effects.

Despite these differences, botanists have traditionally suggested that plant hormones function like animal hormones; that is, the response elicited by a plant hormone is determined by the concentration of the hormone. Certain predictable responses often *can* be obtained by applying hormones to various parts of plants, but this does not necessarily mean that such responses are naturally controlled by the hormone. As a result, some botanists now question whether the concept of hormones applies to plants as it does to animals. These botanists—who emphasize these differences by referring to plant hormones as *plant growth regulators*—claim that the traditional view of plant hormones fails to explain plant growth and development. Rather, they suggest that plant hormones are integrating agents that are *necessary* for, but do not *control*, the response. For example, cytokinins regulate the transformation of buds from a dormant to nondormant state, but do not control the subsequent growth of the bud. Thus, cytokinins are necessary for bud growth, but do not control it. According to this perspective, the response elicited by a hormone is determined not by the *amount* of hormone, but rather by the *sensitivity* of the tissue to the hormone.

Who's right? What *does* determine the response elicited by a hormone: its concentration or the tissue's sensitivity? There are several examples in which the amount of hormone determines the response. For example, inhibition of root growth is roughly proportional to the amount of IAA in the root. Conversely, other examples suggest that the sensitivity of a plant to a hormone changes with different environments and developmental stages. For example, the sensitivity of corn coleoptile segments to IAA changes more than twofold during the first three hours after excision. Also, the responses characteristic of a particular hormone often occur without an accompanying increase in the amount of the hormone. For example, applying IAA to the elongating region of stems stimulates growth, but an increase in the amount of IAA has not been detected in rapidly growing stems.

CONCEPT

Hormones are necessary for, but do not control, many aspects of plant growth and development. Plant growth and development are probably controlled by the amount of hormone present and the tissue's sensitivity to the hormone.

Chapter Summary

Plant hormones are organic compounds made in one part of a plant and transported to another part of the plant, where they elicit a response. Plant hormones are active at small concentrations. Plant hormones and the responses that they elicit have the following characteristics:

- Although a hormone may have some characteristic effects, it also has many other effects. That is, a single hormone can elicit many different responses.
- The responses elicited by a hormone depend on many factors, including the presence of other hormones, the amount of hormone present, nonhormonal factors, and the sensitivity of the tissue to the hormone.
- Hormonal responses change under different conditions and in different plants.
- A single aspect of growth and development can be influenced by several hormones.
- Responses elicited by plant hormones probably result from changing ratios of hormones rather than from the presence or absence of any one hormone.

The five major classes of plant hormones are auxin, gibberellins, cytokinins, ethylene, and abscisic acid.

Auxin stimulates cellular elongation, differentiation of vascular tissue, fruit development, formation of adventitious roots, and production of ethylene. The most active naturally occurring auxin is indoleacetic acid (IAA), which is transported polarly in roots and stems. Synthetic auxins are used extensively in modern agriculture.

Gibberellins stimulate extensive growth of intact plants, the transition from juvenile to adult growth, bolting of biennials, fruit formation, and germination of some cereal grains.

Cytokinins stimulate cellular division, expansion of cotyledons, and growth of lateral buds. Cytokinins also delay senescence of detached leaves and, in combination with IAA, may influence formation of roots and shoots.

Ethylene is a gaseous hormone that influences fruit ripening, abscission, sex expression, and the radial expansion of cells.

Abscisic acid (ABA) is an inhibitor that causes stomata to close, affects dormancy of some seeds, and, in general, counteracts the stimulatory effects of other hormones. These effects may occur because ABA is a calcium antagonist.

Oligosaccharins are fragments of cell wall that regulate plant growth and development. Unlike other plant hormones, oligosaccharins have specific effects.

Many botanists think that plant hormones are necessary for, but do not control, plant growth and development. According to this perspective, a plant's response to a hormone is not determined by the amount of hormone present, but rather by the sensitivity of the tissue to the hormone.

Questions for Further Thought and Study

1. Calcium is often referred to as a "second messenger" in plant growth and development. What does this mean?
2. Calcium influences several effects of plant hormones. Why, then, isn't calcium considered a plant hormone?
3. Abscisic acid is a calcium antagonist. Discuss how this property could be the basis for inhibition by ABA of the stimulatory effects of other hormones.
4. How are oligosaccharins different from other plant hormones?
5. Some botanists refer to ABA as a "plant tranquilizer." Is this a good analogy? Why or why not?
6. Why are immature fruits stored in 1%–3% oxygen instead of 0% oxygen?
7. How might ethylene-producing fungi in soil affect plant growth and development?
8. Provide examples for each of the following generalizations about plant hormones:

    a) A single plant hormone can produce many effects.

    b) The effects elicited by a hormone depend on many factors, including the presence of other hormones.

    c) Hormonal responses probably result from changing ratios of hormones rather than from the presence or absence of an individual hormone.

9. How is the transport of IAA different from that of other hormones?
10. What is the adaptive significance of the fact that auxin produced by young leaves influences vascular differentiation into the shoot apex?
11. What is the difference between a hormone and an enzyme?
12. Gibberellins, cytokinins, and IAA promote growth. How would you distinguish them from each other?
13. Morphactins are substances that influence plant growth and development by interfering with the transport of auxin. How would morphactins affect phototropism of coleoptiles?

Suggested Readings

ARTICLES

Albersheim, P., and A. G. Darvill. 1985. Oligosaccharins. *Scientific American* 253:58–64.

Cline, M. 1991. Apical dominance. *The Botanical Review* 57:318–358.

Evans, M. L. 1984. The action of auxin on plant cell elongation. *CRC Critical Reviews in Plant Sciences* 2:317–365.

Huddart, H., R. J. Smith, P. D. Langton, A. M. Hetherington, and T. A. Mansfield. 1986. Is abscisic acid a universally active calcium antagonist? *New Phytologist* 104:161–173.

Kende, H. 1993. Ethylene biosynthesis. *Annual Review of Plant Physiology and Plant Molecular Biology* 44:283–308.

Morris, R. O. 1986. Genes specifying auxin and cytokinin biosynthesis in phytopathogens. *Annual Review of Plant Physiology* 37:509–538.

Ryan, C. A., and E. E. Farmer. 1991. Oligosaccharide signals in plants: A current assessment. *Annual Review of Plant Physiology and Plant Molecular Biology* 42:651–674.

Sisler, E. C., and S. F. Yand. 1984. Ethylene, the gaseous hormone. *BioScience* 33:233–338.

Theologis, A. 1986. Rapid gene regulation by auxin. *Annual Review of Plant Physiology* 37:407–438.

Trewavas, A. 1991. How do plant growth substances work? *Plant, Cell and Environment* 14:1–12.

BOOKS

Addicott, F. T., ed. 1983. *Abscisic Acid*. New York: Praeger.

Galston, A. W. 1994. *Life Processes of Plants*. New York: W. H. Freeman and Co.

Nickell, L. G. 1982. *Plant Growth Regulators—Agricultural Uses*. New York: Springer-Verlag.

Salisbury, F. B., and C. W. Ross. 1992. *Plant Physiology*. 4th ed. Belmont, CA: Wadsworth.

Wareing, P. F., and I. D. J. Phillips. 1981. *Growth and Differentiation in Plants*. 3d ed. Elmsford, NY: Pergamon Press.

Flowers of a lily (*Lilium superbum*). What controls flowering in plants?

CHAPTER 19

How Plants Respond to Environmental Stimuli

Chapter Outline

INTRODUCTION
TROPISMS
  Phototropism

Box 19.1
TRACKING THE SUN

  Gravitropism
  Hydrotropism
  Thigmotropism
NASTIC MOVEMENTS
  Seismonasty
  Nyctinasty
THIGMOMORPHOGENESIS
SEASONAL RESPONSES OF PLANTS TO THE ENVIRONMENT
  Flowering

Box 19.2
IS THERE A FLOWERING HORMONE?

Box 19.3
VERNALIZATION

  Senescence
  Dormancy
CIRCADIAN RHYTHMS
  What Controls Circadian Rhythms in Plants?
  Biological Clocks and Entrainment
Chapter Summary
Questions for Further Thought and Study
Suggested Readings

Chapter Overview

Plant growth and development are strongly influenced by the environment. Some responses of plants to stimuli such as light, gravity, and touch are rapid; other responses, such as flowering, are relatively slow and are associated with changes of seasons. Regardless of their duration, most responses of plants to environmental signals are due to growth and are controlled, at least in part, by hormones. The coupling of plant growth and development with environmental signals determines the seasonality of growth, reproduction, and dormancy, ultimately fostering survival of the plant and the species in a given environment.

INTRODUCTION

Plants have evolved an elaborate set of intricate and dramatic responses to the environment. These responses allow plants not only to survive adverse conditions that would kill most animals, but also to coordinate their growth and development with appropriate environmental conditions. The degree to which plants rely on environmental cues to orchestrate development is unparalleled in the animal kingdom.

Responses of plants to environmental stimuli such as light, gravity, and touch occur in many ways and include such diverse events as flowering, the growth of stems toward light, and the trapping of food by carnivorous plants. Some of these behaviors are short-term responses. For example, a Venus's-flytrap takes less than a second to close, and curvature of a stem toward light is usually completed within a few hours. Other behaviors, such as flowering, take longer and are typically associated with changing seasons. Environmental signals trigger all of these seemingly unrelated responses: stems grow toward light, and rapid growth allows a Venus's-flytrap to feed on insects. The key element in these responses is *growth*; with few exceptions, plants respond to environmental stimuli by growing. This growth produces a variety of responses, among the most obvious of which are tropisms.

TROPISMS

Plant growth toward or away from a stimulus such as light or gravity is called a **tropism**. There are several kinds of tropisms, each of which is named for the stimulus that causes the response. For example, phototropism is the growth of a stem or coleoptile toward light, and gravitropism is growth toward or away from gravity. Tropisms result from *differential growth*, meaning that one side of the responding organ elongates faster than the other side. This causes curvature of the organ toward or away from the stimulus. Growth of an organ toward a stimulus is called a *positive tropism*. Thus, stems that grow toward light are positively phototropic. Conversely, growth of an organ away from a stimulus is called a *negative tropism*. Roots, which grow away from light, are negatively phototropic.

Phototropism

Phototropism is a tropic response to unidirectional light (light coming from one direction; see fig. 19.1). As you learned in Chapter 18, Frits Went determined that during phototropism, the rapid elongation of cells along the shaded side of coleoptiles is controlled by IAA coming from the apex. Went formulated his hypothesis jointly with Russian plant physiologist Nicolai Cholodny, and this work became known as the Cholodny-Went hypothesis. Although this hypothesis was a major step in our understanding of phototropism, the mechanism by which IAA controlled phototropism remained a mystery. Either of two hypotheses could have accounted for the increased activity of IAA on the shaded side of the coleoptile:

HYPOTHESIS 1

Light destroys IAA along the lighted side of the coleoptile.
Such a mechanism would result in more IAA along the shaded side of the coleoptile, which would account for phototropic curvature.

The evidence: In the 1950s, Winslow Briggs and his colleagues determined that the amount of IAA produced by coleoptiles grown in light is the same as that made by coleoptiles grown in the dark (fig. 19.2). Thus, light does not destroy IAA.

HYPOTHESIS 2

Light causes IAA to move to the shaded side of the coleoptile.
According to this hypothesis, the difference in IAA concentration between the lighted and shaded sides of the coleoptile would result from the movement of IAA rather than its destruction.

The evidence: Briggs and his colleagues collected more IAA from the shaded side of coleoptiles than from the lighted side, suggesting that light causes IAA to move to the shaded side of the stem. This conclusion was confirmed by inserting impermeable barriers between the split tips of coleoptiles. These barriers blocked the movement of IAA to the shaded side of the coleoptiles (fig. 19.2). As a result, the coleoptiles did not curve toward the light. More recent experiments using IAA labeled with radioactive carbon (^{14}C) have confirmed

FIGURE 19.1

Phototropic curvature of corn shoots toward light. Unilateral light causes IAA to move to the shaded side of the stem. This IAA stimulates growth on the shaded side, causing the shoot to curve toward the light.

that unidirectional light causes IAA to move to the shaded side of coleoptiles, where its increased concentration causes cells there to elongate more rapidly than cells on the lighted side of the coleoptiles. As a result, the coleoptile curves toward the light.

Although the Cholodny-Went hypothesis adequately explains phototropism in coleoptiles, it often fails to account for phototropism by stems, which contain more chlorophyll than do coleoptiles. In stems, unidirectional light triggers production of an inhibitor that slows cellular elongation on the illuminated side of the stem. Because cellular elongation does not change on the shaded side of the stem, the stem curves toward the light (also see box 19.1, "Tracking the Sun," p. 440).

How does a plant sense unidirectional light? Botanists have determined that only blue light with wavelengths less than 500 nm effectively induces phototropism (fig. 19.3). A xanthophyll called zeaxanthin is probably the photoreceptor for phototropism. This pigment may alter the rate of transport of IAA, thereby speeding its movement to the shaded side of the stem or coleoptile.

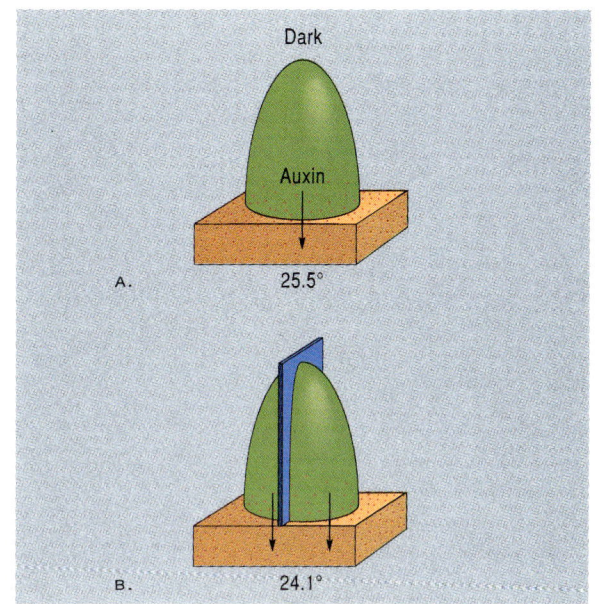
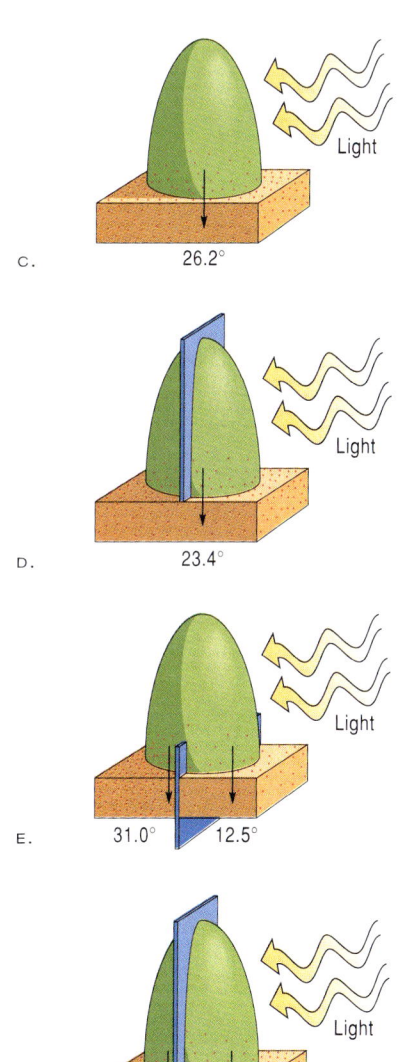

FIGURE 19.2

Briggs's experiments A and B (done in the dark) showed that splitting the tip of the coleoptile did not significantly affect the total amount of auxin that was diffusing from the tip into the agar block. The degree of curvature produced when the agar block is applied to the side of a decapitated coleoptile is directly related to amount of auxin produced, and is shown by the numbers below the agar blocks. Comparing A and B with C and D shows that auxin production is not dependent on light. If a barrier is partially inserted into the tip, auxin is displaced from the lighted to the dark side as shown in E. Experiment F shows that it was displacement that had occurred, and the results in figure E do not reflect different rates of auxin production on the lighted and dark sides.

CHAPTER NINETEEN *How Plants Respond to Environmental Stimuli*

BOXED READING 19.1

TRACKING THE SUN

One of the most unusual and adaptive movements in plants is *solar tracking*, also known as **heliotropism** (from the Greek word *helios,* meaning "sun"). Solar tracking is how the organs of plants track the sun across the sky, much as a radio telescope tracks stars or satellites. Like several other plant movements, solar tracking is caused by turgor changes in motor cells in pulvini located at the base of leaves and leaflets.

The tracking movements of leaves depend on environmental and physiological conditions. For example, the so-called compass plant (*Silphium laciniatum*) orients its leaves parallel to the sun's rays, thereby decreasing leaf temperature and minimizing desiccation. Other plants often orient their leaves perpendicular to the sun's rays, thereby increasing the amount of light intercepted by the leaf for photosynthesis. Finally, some plants orient their leaves more obliquely to the sun when it is hot than when it is cooler. This variation in solar tracking helps keep the leaf temperature near the optimal temperature for photosynthesis.

Solar tracking occurs in many plants (e.g., cotton, alfalfa, beans, and soybeans) and is not restricted to leaves. What happens on cloudy, overcast days? On these days, leaves are oriented horizontally in their "resting" position. If the sun appears from behind the clouds late in the day, leaves rapidly reorient themselves: they can move up to 60 degrees per hour, which is four times faster than the movement of the sun across the sky.

Solar tracking is controlled not only by the sun's position; leaves begin orienting themselves toward the direction of sunrise several hours *before* sunrise. How plants "remember" the position of the last sunrise is a mystery, but IAA may be involved; IAA from leaves can alter turgor of motor cells. Whatever the mechanism, plants are fast learners: only four sunrises are needed to entrain solar tracking.

BOX FIGURE 19.1

Sunflowers (*Helianthus annuus*) oriented in the same direction, toward the sun. *Helianthus* means "sunflower."

BOX FIGURE 19.2

Time-lapse photograph of a buttercup (*Ranunculus ficaria*) tracking a source of light.

CONCEPT

Phototropism is a growth response to unidirectional light. Phototropism is probably influenced by IAA and results in stems growing toward light and roots growing away from light. This important response increases the probability that stems and leaves will intercept light for photosynthesis.

Gravitropism

Charles Darwin was a phenomenally productive scientist. In addition to his landmark investigations of evolution, he and his son Francis also studied **gravitropism,** which is a growth response to gravity. One of their first experiments involved the responses of roots whose caps had been surgically removed. To the Darwins' surprise, these decapped roots continued to grow but did not respond to gravity. These clever experiments showed that the root cap is necessary for root gravitropism.

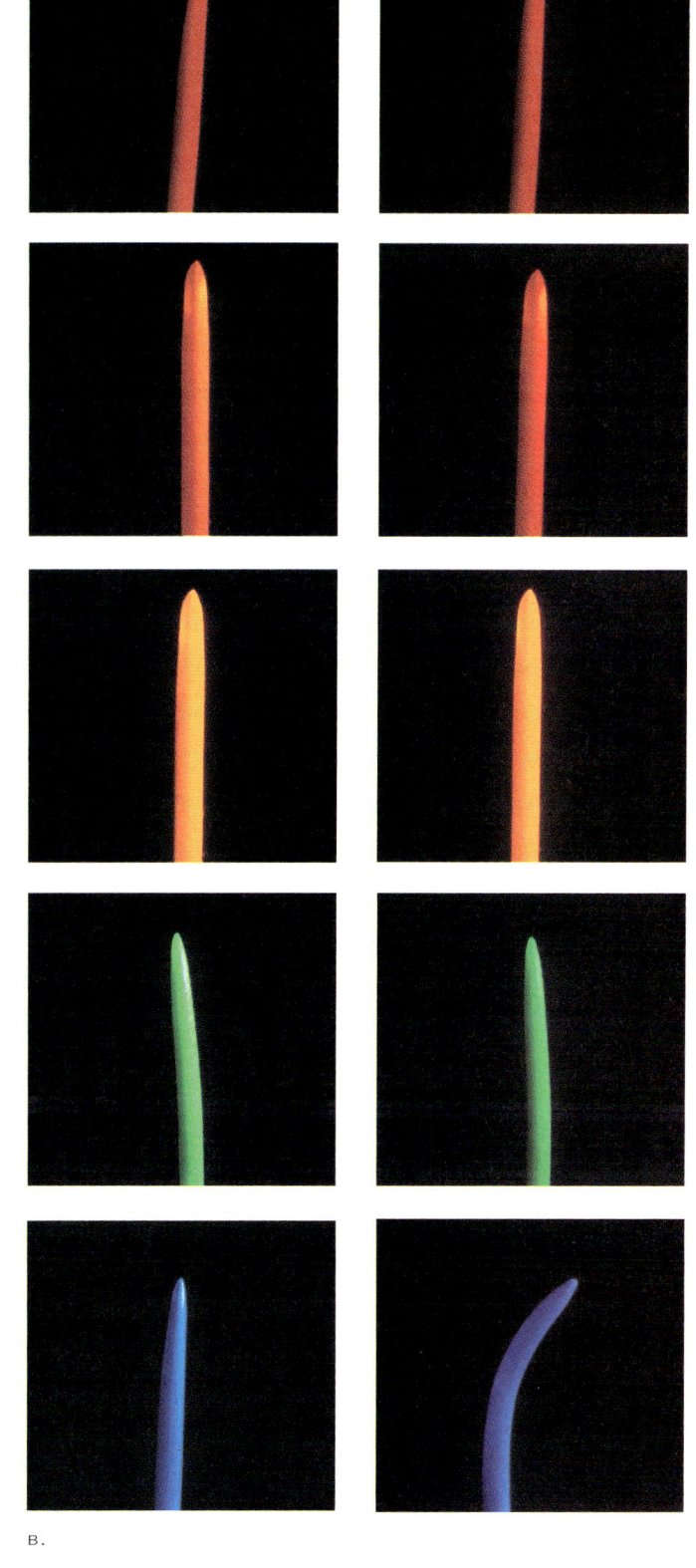

FIGURE 19.3

Sensitivity of phototropism to different wavelengths of light. Phototropism is most sensitive to blue light having wavelengths less than 500 nm.

(a) Source: Data from M. Wilkins, Plantwatching: How Plants Remember, Tell Time, Form Partnerships and More, Macmillan Publishing Company, New York, NY, 1988.

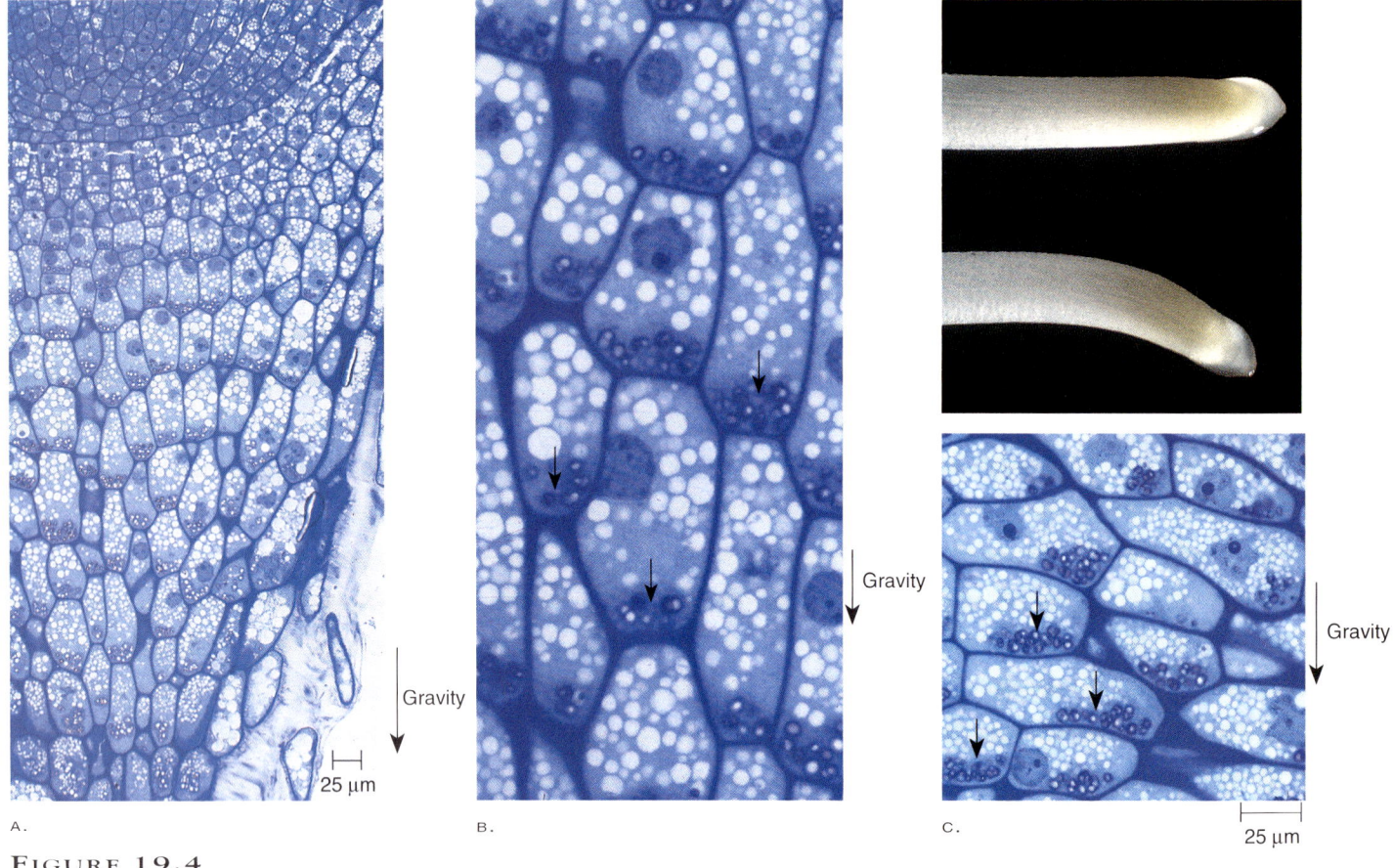

FIGURE 19.4

(a) Longitudinal section of a root cap of corn. Cells in the center of the cap contain numerous starch-laden amyloplasts located in the lower part of the cells. (b) Magnified view of (a). Arrows indicate amyloplasts that have sedimented to the bottom of the cell. (c) These amyloplasts (arrows in lower photo) quickly sediment to the lower sides of cells when roots are oriented horizontally. This sedimentation of amyloplasts has long been thought to be the basis for how roots perceive gravity (upper photo).

Other researchers, following the Darwins' lead, began studying the root cap in hopes of understanding its strong influence on gravitropism. They soon made an exciting discovery: cells in the center of the cap contain numerous starch-laden amyloplasts which, under the influence of gravity, sediment to the lower side of the cells (fig. 19.4). Could this gravity-dependent sedimentation of amyloplasts be how plants sense gravity? Several subsequent experiments suggested that this was true, which further intensified the study of amyloplasts as gravity-sensors in plant roots. Despite over fifty years of intensive research, however, no convincing explanation has emerged for how the sedimentation of amyloplasts in the root cap could induce differential growth in the elongating zone of the root, which is 3–6 mm behind the root cap.

Studies by Timothy Caspar and his coworkers at Michigan State University suggested that research on amyloplasts as gravity sensors in roots may have been misguided. Caspar took a unique approach to investigating the problem: he reasoned that a sure way to test if amyloplasts are gravity sensors in root caps would be to study gravitropism in a plant lacking amyloplasts. Although this reasoning was sound, it proved to be a formidable task, since all species that had been studied had amyloplasts in the cells of their root caps. To get around this problem, Caspar engineered his own experimental plant: a mutant of *Arabidopsis thaliana* that lacked starch and so also lacked amyloplasts (fig. 19.5). With this plant, the critical question and experiment became rather simple: Do roots of mutant seedlings respond to gravity, even though they lack amyloplasts? When Caspar oriented roots of the mutant horizontally, they curved downward, indicating that amyloplasts are not necessary for root gravitropism. These results question the long-held assumption that amyloplasts constitute the gravity-sensing apparatus in roots.

More recent research by Randy Wayne and his colleagues at Cornell University suggests that roots respond to gravity by sensing gravitational pressures exerted by the protoplast, not the sedimentation of amyloplasts. Proteins at the interface between the cell membrane and cell wall may be required for sensing gravity.

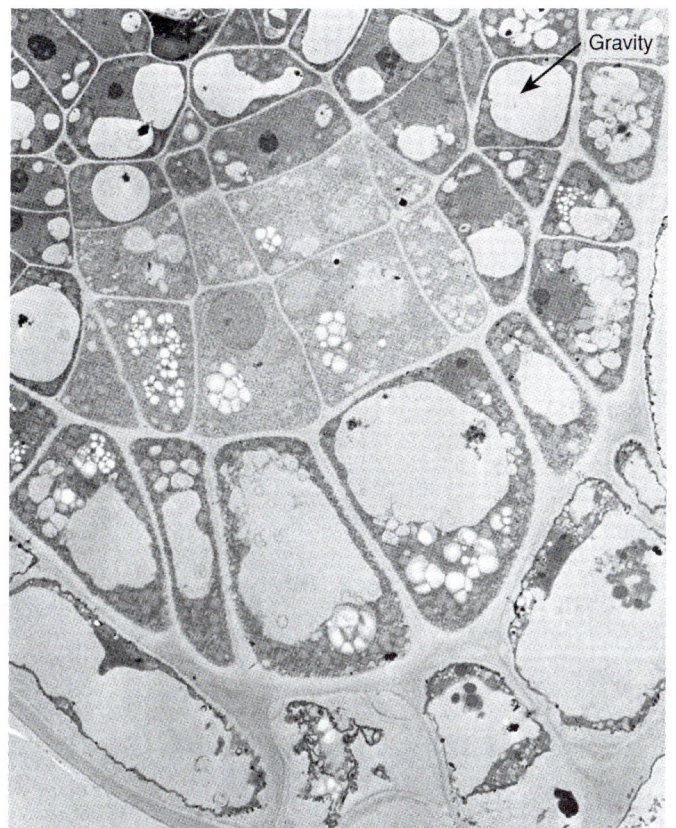 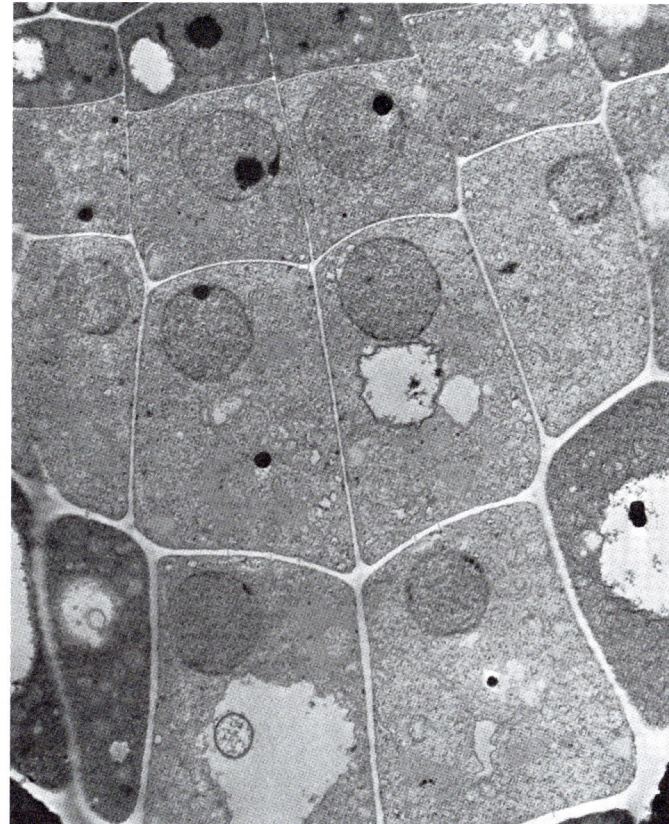

FIGURE 19.5

Electron micrographs of (a) normal (i.e., starchy), ×1,500, and (b) mutant (i.e., starchless) cells, ×3,000, from *Arabidopsis thaliana*. Roots of normal and mutant seedlings are graviresponsive, thus prompting many botanists to question the importance of root cap amyloplasts in root gravitropism.

Although we do not understand how roots *perceive* gravity, we are beginning to understand how they *respond* to gravity. When roots are oriented horizontally, growth slows along the lower side of the elongating zone, thereby causing the root to curve downward. One of the first events that ultimately causes this differential growth is the accumulation, not of a hormone, but of calcium ions (Ca^{2+}). Ca^{2+} moves to the lower side of the cap and elongating zone of horizontally oriented roots. This accumulation of Ca^{2+} along the lower side of the root triggers an accumulation of IAA along the lower side of the root tip. Since IAA inhibits cellular elongation in roots, the lower side of the root grows slower than the upper side of the root, and the root curves down (fig. 19.6). When the root reaches a vertical position, the lateral asymmetries of Ca^{2+} and IAA disappear, and straight growth resumes.

IAA and Ca^{2+} also direct the negative gravitropism of shoots. IAA accumulates along the lower side and Ca^{2+} along the upper side of horizontally oriented stems. Concurrently, auxin-induced mRNAs disappear from the cortex and epidermis of the upper (i.e., more slowly growing) side and accumulate on the lower (i.e., more rapidly growing) side of horizontally oriented hypocotyls. These mRNAs, or encoded proteins, stimulate cellular elongation along the lower side of the stem, thereby producing upward curvature (fig. 19.7).

CONCEPT

Gravitropism is a growth response to gravity. It is controlled by calcium and IAA, and results in stems growing up and roots growing down. Gravitropism is important because it increases the probability that i) roots will encounter water and minerals, and ii) that stems and leaves will intercept light for photosynthesis.

Hydrotropism

The growth of roots toward soil moisture is called **hydrotropism.** Roots whose caps have been removed are not responsive to moisture gradients, which suggests that the root cap is the site of moisture perception by roots. Interactions between light,

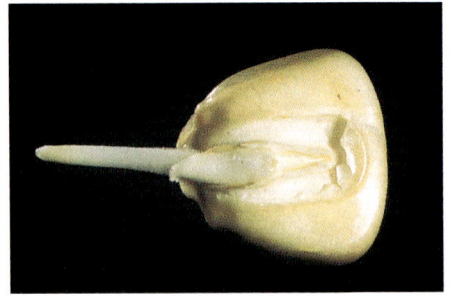
A.

B.

C.

D.

E.

FIGURE 19.6

Gravitropism by horizontally oriented roots of corn. Downward curvature begins within 30 minutes and is completed within a few hours. Curvature results from faster elongation of the upper side of the root than of the lower side.

FIGURE 19.7

Negative gravitropism by a stem of *Coleus*. This plant was placed on its side 24 hours before this picture was taken.

gravity, and soil moisture could therefore account for the occasional meandering growth of roots through soil.

Thigmotropism

Thigmotropism is a growth response of plants to touch. The most common example of thigmotropism is the coiling of tendrils or of entire stems of plants such as morning glory (*Ipomoea*) and bindweed (*Convolvulus*) (fig. 19.8). Before touching an object, tendrils and twining stems often grow in a spiral pattern called *circumnutation* that increases their chances of contacting an object to which it can cling. Contact with an object is perceived by specialized epidermal cells, which induce differential growth in the tendril. Such growth can be extremely rapid: a tendril can encircle an object within 5–10 minutes.

FIGURE 19.8

Thigmotropic coiling of a stem of bindweed.

Furthermore, thigmotropism is often long lasting: stroking a tendril of garden pea for only a couple of minutes can induce a curling response that lasts for several days. Thigmotropism is probably influenced by IAA and ethylene; these hormones induce thigmotropic-like curvature of tendrils even in the absence of touch.

Tendrils can also store the "memory" of touch. Tendrils that are touched while growing in the dark do not respond until they are illuminated. Thus, although tendrils can store the sensory information received in the dark, light is required for the growth response to proceed. This light-induced expression of thigmotropism may be due to a requirement for ATP, since ATP will substitute for light in inducing thigmotropism of dark-stimulated tendrils. Various degrees of thigmotropism are exhibited not only by tendrils, but also by leaves, stems, petioles, and roots.

CONCEPT

Thigmotropism is a growth response to touch and is probably controlled by IAA and ethylene. Thigmotropism by tendrils allows plants to climb objects, thereby increasing the plant's chances of intercepting light for photosynthesis.

NASTIC MOVEMENTS

Nastic movements also occur in response to environmental stimuli. Unlike tropisms, nastic movements are independent of the direction of the stimulus: they occur in an anatomically predetermined direction rather than toward or away from the stimulus. Nastic movements include some of the most unusual as well as spectacular responses in all of the plant kingdom.

Seismonasty

Seismonasty is a nastic movement resulting from contact or mechanical disturbances such as shaking. Seismonastic movements are based on a plant's ability to rapidly transmit a stimulus from touch-sensitive cells in one part of the plant to responding cells located elsewhere. Among the most dramatic of these responses are those exhibited by the sensitive plant (*Mimosa pudica*): touching a leaf causes the leaflets to fold and the petiole to droop (fig. 19.9). Here's how the response occurs:

1. Touching a leaf generates an electrical signal that moves along the petiole.
2. This electrical signal is translated into a chemical signal that causes cell membranes to become more permeable to K^+ and other ions. The cells that are affected are called *motor cells*, which are large, thin-walled parenchyma cells located in a jointlike structure called a *pulvinus* (fig. 19.10). In the sensitive plant, a pulvinus is located at the base of each leaflet and petiole.

A.

B.

C.

FIGURE 19.9

Seismonastic movement of leaves and leaflets of the sensitive plant (*Mimosa pudica*). In undisturbed plants (a), leaves are erect. Touching a leaf (b) causes leaflets to fold and the petiole to droop (c).

CHAPTER NINETEEN *How Plants Respond to Environmental Stimuli*

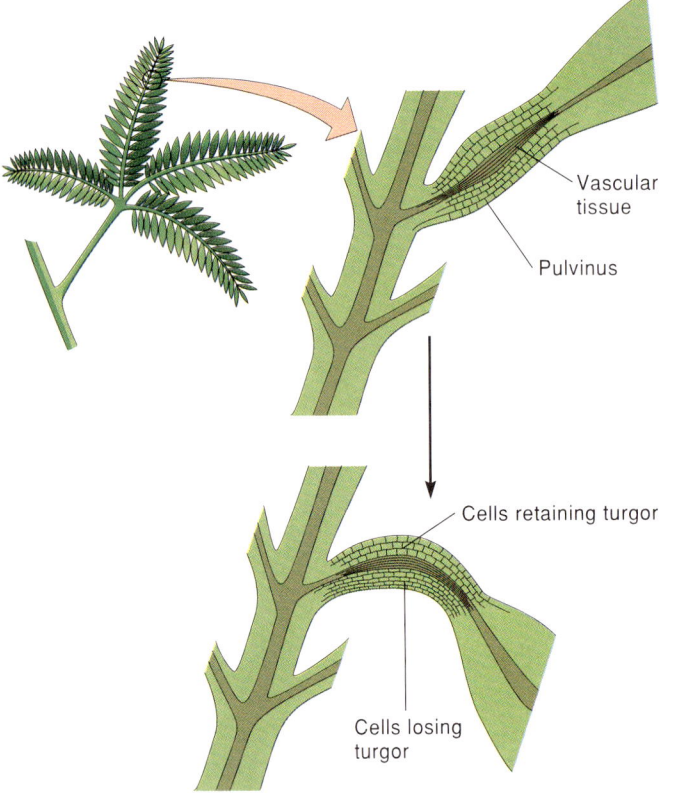

FIGURE 19.10

Seismonasty in the sensitive plant is due to turgor changes in pulvini, which are swollen structures located at the bases of leaflets and petioles. Turgor changes on one side of the pulvinus cause cells there to shrink, thereby producing movement.

FIGURE 19.11

Seismonasty enables a Venus's-flytrap to trap its prey. The unsuspecting fly shown here was attracted to the trap by compounds secreted on the leaf's surface. When this fly touched the trap's trigger hairs twice, it was snared.

3. The movement of ions out of motor cells decreases the water potential in the surrounding extracellular space, which causes water to move out of motor cells via osmosis.
4. The loss of water causes the motor cells to shrink, thereby producing the seismonastic movement.

The unfolding of leaves takes 15–30 minutes and is accomplished by reversing this process. Motor cells take up K^+ and other ions, causing water to enter the cells via osmosis. This influx of water inflates the cells to their original size, thereby unfolding the leaves to their original position.

Similar seismonastic movements occur in a few other plants. Such movements are curious and interesting, but what is their adaptive significance? Some evidence suggests that seismonastic movements often scare insects off of plants, thereby decreasing a leaf's chances of being eaten by a herbivore. Presumably, folded leaves are more difficult to see, which may also divert a herbivore's attention to some other plant. Seismonastic movements may also defend sensitive plants by exposing sharp prickles located along the leaf axis when the leaflets close.

The most spectacular and macabre of plants' responses to environmental stimuli is the seismonastic response of Venus's-flytrap (*Dionaea muscipula*), a popular novelty plant. Although Venus's-flytraps have been studied since before the time of Charles Darwin, the mechanism responsible for trap-closure remains controversial. One hypothesis for trap-closure involves acid-induced growth. That is, movements of Venus's-flytrap may involve increases in cellular size initiated by acidification of the trap's cell walls. Consistent with this hypothesis is the observation that traps will close if the walls of cells at the trap's "hinge" are acidified to pH 4.5 and below. Leaves (i.e., traps) of Venus's-flytrap have two lobes, each of which has three sensitive "trigger" hairs overlying the motor cells. When a meandering animal touches these hairs, an electrical signal moves from the trigger hairs to the motor cells and initiates transport of H^+ to the walls of epidermal cells along the outer surface of the trap. The resulting acidification of these cell walls causes the outer epidermal cells along the central portion of the leaf to expand rapidly. Because epidermal cells along the inner surface of the leaf do not expand, the flytrap snaps shut (fig. 19.11). Closure of the trap takes 0.1 to 0.5 seconds and requires a large expenditure of ATP: motor cells use almost one-third of their ATP to pump the H^+ that acidify the cell walls and close the trap.

According to a competing hypothesis, trap-closure results not from cell-wall acidification, but rather from a release of tension. Mesophyll cells of a trap are already estensible but are kept compressed in an open trap, thereby developing a tension in the tissue. This tension is somehow released by the action potential triggered by an animal that touches the trigger hairs, thereby causing the trap to close.

FIGURE 19.12

Seismonasty enables sundews to trap and digest their prey. Insects become trapped on the sticky mucilage secreted by glandular hairs. Seismonastic movement of these glandular hairs moves the insect to the center of the leaf, where it is digested.

FIGURE 19.13

Nyctinastic movement in a prayer plant (*Maranta*). (a) During the day leaves are oriented horizontally. (b) At night, the leaves become oriented in a vertical configuration resembling praying hands.

The trapping mechanism is quite sophisticated. Closure requires two stimuli, which are unlikely to result from inanimate objects such as falling leaves that touch the plant. The toothed edges of the trap mesh like a beartrap, pressing the captive animal against digestive glands along the inner surface of the trap. The prey's struggle to free itself stimulates the trap to close even tighter and stimulates glands in the trap to secrete digestive enzymes. Digestion lasts one to several days. The opening of empty traps usually requires 8–12 hours and results from the expansion of epidermal cells along the midrib of the inner surface of the leaf.

Seismonastic movements are also responsible for the feeding behavior of another carnivorous plant, the sundew (*Drosera* species, fig. 19.12; also see photo that opens Chapter 20, p. 460). The club-shaped leaves of sundew are covered by glandular hairs. Insects that land on the leaf become trapped in this nectar. The prey's contact with the leaf generates an electrical signal, which causes the surrounding tentacles to bend inward and carry the prey to the center of the leaf, where it is digested (fig. 19.12).

CONCEPT

Seismonasty is a nastic response resulting from contact or mechanical disturbance. The mechanisms for this response include reversible changes in turgor.

Nyctinasty

Nyctinasty (from Greek *nyktos*, "night," and *nastos*, "pressed together"), or "sleep movement," is a nastic response caused by daily rhythms of light and dark. One of the most common nyctinastic responses occurs in the prayer plant (*Maranta* species), an ornamental houseplant. During the day, leaves of the prayer plant are horizontal, thereby maximizing their interception of sunlight. At night, the leaves fold vertically into a shape resembling a pair of praying hands (fig. 19.13). This movement of leaves in response to light and dark results from changes in the turgor of motor cells in a pulvinus located at the base of each leaf. In the dark, K^+ ions are transported from cells of the upper side of the pulvinus to cells

CHAPTER NINETEEN *How Plants Respond to Environmental Stimuli*

along its lower side. This movement of ions causes water to move via osmosis into cells along the lower side of the pulvinus. This, in turn, causes cells along the upper side of the pulvinus to lose water and shrink as the cells along the lower side gain water and expand. Taken together, these changes in cellular volume move the leaf to a vertical position. At sunrise, the process is reversed and the leaf again assumes its horizontal position.

Similar sleep movements occur in other plants, including sorrel (*Oxalis*) and legumes such as beans. In these plants, nyctinastic movements occur at regular times each day. A clever use of these regular movements was made by Carolus Linnaeus, a famous Swedish botanist. Linnaeus filled wedge-shaped portions of a circular garden with plants having sleep movements that occurred at different times. By seeing which plants of his so-called *horologium florae* (flower clock) were "asleep," Linnaeus could determine the time of day.

CONCEPT

Nyctinasty is a nastic response resulting from daily rhythms of light and dark. The mechanism underlying nyctinasty is turgor changes in pulvini.

The Lore of Plants

Thermonasty is a nastic response caused by changes in temperature. A popular example of thermonasty is the opening and closing of tulips. Another example is the curling of the edges of *Rhododendron* leaves in response to cold. Some natives of the mountains in the Carolinas use this response as a means of estimating outdoor temperatures. Thermonasty has not been studied extensively, and its mechanism is unknown.

THIGMOMORPHOGENESIS

Plants growing in the protected environment of a greenhouse develop differently than do plants grown outside, because plants in a natural environment are often disturbed by rain, hail, wind, falling objects, and passing animals. These disturbances inhibit cellular elongation and produce shorter, stockier plants containing relatively large amounts of supportive tissue (i.e., collenchyma and sclerenchyma fibers). This response to mechanical disturbances is called **thigmomorphogenesis.** Thigmomorphogenesis is mediated by *touch-induced genes* that are activated by many stimuli, including wind and touch. Within 10–30 minutes after stimulation, mRNAs for these genes increase by as much as a hundredfold. One of these mRNAs encodes calmodulin, a calcium-binding protein that is involved in a variety of plant responses ranging from the secretion of enzymes to the actions of hormones. Ethylene is probably only one of several factors involved in the control of thigmomorphogenesis.

SEASONAL RESPONSES OF PLANTS TO THE ENVIRONMENT

The behavior of plants is strongly influenced by seasonal changes in the environment. For example, the cooler nights and shorter days of autumn bring decreased growth, beautiful coloration of leaves, and dormancy of buds in preparation for the rigors of the upcoming winter. In the spring, buds resume growth and rapidly transform a barren forest into a dynamic, photosynthetic community. Clearly, plants must sense and anticipate seasonal changes. How do they do this? To answer this question, let's consider flowering, which is the basis for sexual reproduction in flowering plants.

Flowering

One of the most striking and predictable responses of most plants to changes of season is **flowering.** Gardeners have long known that many plants flower only during certain times of the year. For example, clover flowers during summer, and asters bloom during the short days of early spring or fall.

A study of flowering provided our first clues to how plants sense and anticipate changes of the seasons. In the early 1900s, Wightman Garner and Henry Allard were studying tobacco at a research center in Beltsville, Maryland. Since tobacco plants flower during late summer in Maryland, Garner and Allard were not surprised when their tobacco plants began flowering in late August. However, one plant caught their eye: an oversized mutant that did not flower like the rest of the crop, but continued to grow vegetatively into autumn. This mutant became much larger than other tobacco plants, leading Garner and Allard to name it Maryland Mammoth. Because the oversized mutant had the potential for increasing the yield of tobacco crops, Garner and Allard moved their Mammoth plant into the greenhouse to protect it from cold, and continued to observe its growth. To their surprise, the mutant finally flowered in December. This "out-of-sync" response piqued the interest of Garner and Allard; they began studying its unusual response. First they propagated the mutant so they would have more plants to work with. Plants grown from cuttings and seeds of the parent plant also flowered in December. After confirming their earlier observations, Garner and Allard set out to answer an important question: Why didn't the mutant flower at the same time as other tobacco plants?

Photoperiodism

Garner and Allard began their studies by studying the influence of light, moisture, temperature, and nutrition on the flowering response. They discovered that the environmental signal

BOXED READING 19.2
IS THERE A FLOWERING HORMONE?

Many botanists have long suspected the existence of a hormone that controls flowering. One of the first botanists to provide evidence of such a hormone was Soviet scientist M. K. Chailakhyan, who studied short-day chrysanthemums in the 1930s. Chailakhyan separated the upper and lower halves of his chrysanthemums with a light-proof barrier so that he could expose the upper and lower halves of the plants to different photoperiods. When the upper half of the shoot was exposed to noninductive long days and the lower half to inductive short days, the entire plant flowered, even the upper portion that had been exposed to a noninductive photoperiod. Removing the leaves from the upper half of the plant did not change the results. However, flowering did not occur when leaves were stripped from the lower half of the plant (i.e., the part exposed to inductive short days). These results suggested that leaves perceive the photoperiod and that they produce a flower-inducing substance. Chailakhyan called this substance **florigen** ("flower maker").

Subsequent research has provided additional information about flowering and florigen. For example, Hamner and Bonner showed that flowering would occur if only one leaf were exposed to an inductive photoperiod. That is, the flowering stimulus can be transmitted from one leaf to the whole plant. However, the exposed leaf must be left on the plant for several hours after exposure for flowering to occur; these hours are presumably required for synthesis and transport of the stimulus from the leaf to the stem. Transport probably occurs in the phloem, since flowering does not occur if the phloem is stripped between the leaf and responding buds. Moreover, transport of the stimulus requires direct contact of living cells; flowering does not occur if the leaf and stem are separated by agar.

The most convincing evidence regarding florigen involves grafting of short-day and long-day plants. Short-day plants flower in long days if grafted to long-day plants that are flowering. Similarly, a long-day plant will flower in short days if grafted to a short-day plant that is flowering. In each instance, flowering begins nearest the graft union and then spreads progressively to the rest of the plant. These results suggest that there is a graft-transmittable substance that induces flowering and that this stimulus is similar in long-day and short-day plants.

Despite decades of intensive research, no one has yet isolated florigen. Thus, florigen remains the hypothetical flowering hormone. The inability to isolate florigen has prompted many botanists to suggest that flowering is induced, not by a single hormone, but by changes in ratios of other hormones. For example, plants such as pineapple flower when exposed to ethylene, and some long-day plants flower in short days when treated with gibberellic acid. However, short-day plants do not flower if treated with gibberellic acid. Thus, gibberellic acid is not florigen, although it may be involved in the flowering of some long-day plants.

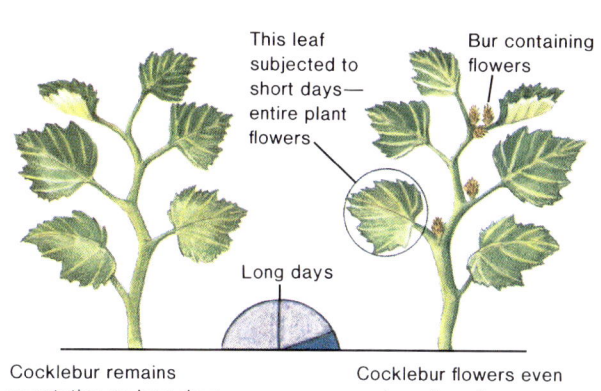

BOX FIGURE 19.3

Cocklebur (*Xanthium strumarium*) is a short-day plant (left). Hamner and Bonner showed that exposure of just one leaf to short days induces flowering in the entire plant, even on long days (right).

BOX FIGURE 19.4

Chailakhyan's experiments with short-day chrysanthemums (*Chrysanthemum* species). (a) Flowering occurred when leaves were subjected to short days. (b) Flowering did not occur when leaves were exposed to long days and upper, leafless parts were limited to short days.

for flowering was the **photoperiod,** which is the ratio of the length of day to the length of night. Most tobacco flowered only during the long days of summer. However, the Maryland Mammoth mutant was different: it flowered only during the short days of winter. These findings suggested to Garner and Allard that plants could measure daylength. They soon confirmed their hypothesis with experiments using soybeans. Garner and Allard set up several experimental plots, each planted a week or so apart. All of the plants flowered at the same time, even though the different planting times resulted in plants of different ages and sizes. These experiments convinced Garner and Allard that plants could measure seasonal changes in daylength. They called this phenomenon **photoperiodism.**

CONCEPT

Photoperiodism is a response to changes in the relative lengths of night and day. Photoperiodism is important because it allows plants to measure and anticipate seasonal changes in climate.

Writing to Learn Botany

What is the selective advantage of plants anticipating seasonal changes by measuring the photoperiod rather than rainfall, temperature, or some other weather factor?

Kinds of Flowering Responses in Plants

Subsequent studies of other flowering plants led botanists to classify plants into four groups, based on the plant's responses to photoperiod: day-neutral plants, short-day plants, long-day plants, and intermediate-day plants (fig. 19.14).

- *Day-neutral plants* flower without regard to photoperiod; that is, daylength has no effect on their flowering. Examples of day-neutral plants include roses, snapdragons, cotton, carnations, dandelions, sunflowers, tomatoes, cucumbers, and many weeds.
- *Short-day plants* flower only if light periods are shorter than some critical length. For example, ragweed plants flower only when exposed to 14 hours or less of light per day. Because of their light requirement, short-day plants usually flower in late summer or fall. Asters, strawberries, dahlias, poinsettias, potatoes, soybeans, goldenrods, and some chrysanthemums are short-day plants.
- *Long-day plants* usually flower in the spring or early summer; they flower only if light periods are longer than a critical length, which is usually 9–16 hours. For example, wheat plants flower only when light periods exceed fourteen hours. Lettuce, spinach, radish, beets, clover, corn, gladiolus, and iris are long-day plants.

Long-day length   Short-day length   **Day-neutral plant**

Shorter than critical day length   Longer than critical day length   **Short-day plant**

Longer than critical day length   Shorter than critical day length   **Long-day plant**

Longer or shorter than critical day length   Intermediate day length   **Intermediate-day plant**

FIGURE 19.14

Kinds of flowering responses to daylength. Botanists originally thought that flowering is controlled by daylength. Flowering by day-neutral plants is not affected by daylength; short-day plants require a light period that is shorter than some critical length; long-day plants flower only if light periods are longer than some critical period; intermediate-day plants do not flower if the light period is too long or too short. Subsequent studies showed that the length of the dark period is what controls flowering in most plants.

Intermediate-day plants flower only when exposed to days of intermediate length; they grow vegetatively if exposed to days that are either too long or too short. Examples of intermediate-day plants are sugarcane and purple nutsedge.

What are Botanists Doing?

Go to the library and read an article about flowering. What is the message of the article? What questions do you have about the work? Why are botanists so interested in flowering?

Do Plants Really Measure the Length of Day?

The findings of Garner and Allard stimulated many botanists to examine photoperiodism in other plants. Among these investigators were plant physiologists Karl Hamner and James Bonner. Their work is a model of how one set of observations can lead to new hypotheses and new knowledge.

Hamner and Bonner studied cocklebur (*Xanthium strumarium*), a short-day plant requiring 16 or fewer hours of light to flower. By growing plants in controlled-environment growth-chambers, Hamner and Bonner could manipulate the photoperiods. To their surprise, they discovered that the length of the light period was unimportant. Rather, plants flowered only if the *dark* period exceeded eight hours, regardless of the length of the light period. Hamner and Bonner then made another startling discovery: flowering did not occur if the dark period was interrupted by a 1-minute pulse of light, even if the regular light period remained less than 15 hours (fig. 19.15). Similar experiments in which the light period was interrupted with darkness had no effect on flowering. Other experiments with long- and short-day plants confirmed their finding: flowering requires a specific period of uninterrupted *dark* rather than uninterrupted light. Thus, short-day plants such as cocklebur are more accurately described as **long-night plants,** because they flower only if their uninterrupted dark period exceeds a critical length. Similarly, long-day plants such as clover are more accurately described as **short-night plants.**

Control and Sensitivity

The factor that determines if and when a plant flowers is not the absolute length of the photoperiod, but rather whether the photoperiod is longer or shorter than the critical length required for that species. For example, ragweed and spinach will both flower if exposed to 10 hours of dark: ragweed, a long-night plant, requires 10 hours or *more* of dark; spinach, a short-night plant, requires 10 hours or *less* of dark. Thus, spinach flowers in the short nights (i.e., long days) of summer, while ragweed blooms in the long nights (i.e., short days) of fall.

The measuring system in many plants is remarkably sensitive: henbane (*Hyoscyamus niger*), a short-night plant, will

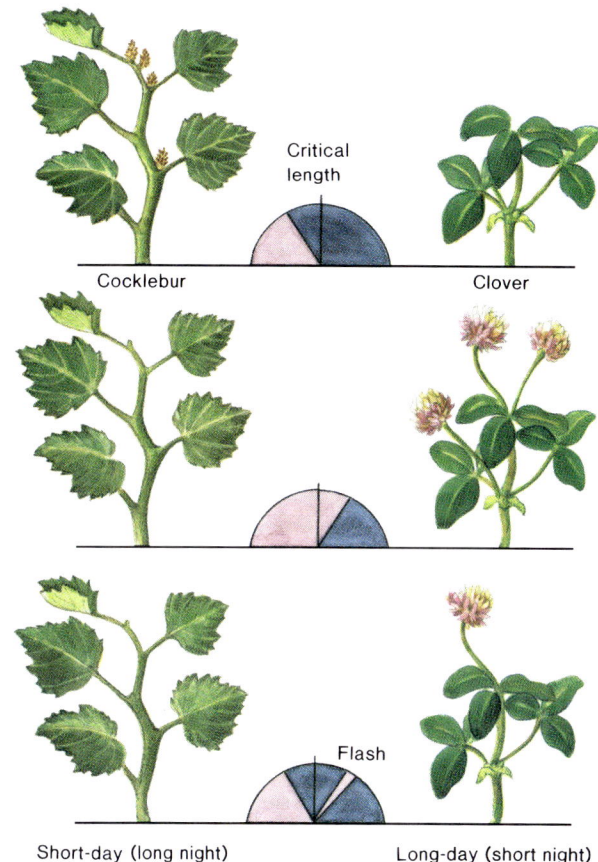

FIGURE 19.15

The influence of daylength and nightlength on flowering by short-night plants (e.g., clover) and long-night plants (e.g., cocklebur). Long-night plants require an uninterrupted dark period longer than a critical period, while short-night plants require a dark period shorter than a critical length. Interrupting the dark period of a long-night plant inhibits flowering.

flower when exposed to dark periods of 13.7 hours but will not flower when exposed to dark periods of 14.0 hours. Plants stripped of their leaves are not responsive to changes in photoperiod, suggesting that photoperiod is sensed by leaves.

How strict is the requirement for an inductive (flower-inducing) photoperiod? For some varieties of soybean (*Glycine max*), the requirement for an inductive photoperiod is absolute: these plants are long-night plants and will not flower unless exposed to long nights. This phenomenon is called *obligate photoperiodism*. Conversely, other plants such as marijuana (*Cannabis sativa*) and Christmas cactus (*Schlumbergera bridgesii*) will eventually flower without an inductive photoperiod. In these plants, the inductive photoperiod merely causes flowering to occur sooner. This response is called *facultative photoperiodism*. Finally, the photoperiodic requirement for flowering can be influenced by other features. For example, poinsettias (*Euphorbia pulcherrima*) are long-night plants at high temperatures and short-night plants at low temperatures. The mechanism underlying this temperature-controlled switch is

BOXED READING 19.3
VERNALIZATION

Most people have never seen a carrot flower, even though they may buy carrots regularly at their local market. Why not? Don't carrots ever flower?

Yes, carrots do flower, but only during their second year of growth. Like other biennials such as turnips and beets, carrots spend their first year growing vegetatively, and flower only after exposure to the cold temperatures of the ensuing winter. Without this cold treatment, carrots would not flower; they would continue to grow vegetatively for an indefinite period of time. Of course, humans harvest carrots after the first year's growth (before flowering) when food has accumulated in the root. As a result, carrots for sale in markets never have a flower stalk attached to them. Harvesting carrots in the spring of their second year of growth (when plants are flowering) would be pointless, because flowering consumes most of the food stored in the carrot.

The promotion of the springtime flowering of carrots and other plants in response to cold is called **vernalization** (from the Latin word *vernalis,* meaning "to make springlike"). Our understanding of this process is important for producing several kinds of crops, such as winter wheat, barley, and rye. For example, winter wheat is planted in autumn, and the seedlings are exposed to the cold temperatures of the following winter. These cold temperatures vernalize the wheat and cause it to flower in late winter or early spring, thereby producing an early crop. Such a crop is especially well-suited to the midwest, because it allows farmers there to harvest before summer's drought. After harvesting winter wheat, farmers plant spring wheat, which does not require a cold treatment to flower. Spring wheat, like many other cereal crops, is a long-day plant and therefore flowers and produces fruit for a summer harvest.

The vernalizing effects of winter can be simulated in the laboratory. For example, if moist seeds of winter rye are stored for a few weeks in a refrigerator, they will flower about 7 weeks after being planted in the spring. Seeds kept dry and warm throughout the winter will not flower until 14 weeks after they are planted in the spring.

BOX FIGURE 19.5
This sugar beet is 41 months old and was never exposed to low temperatures. Consequently, it has remained vegetative.

A.   B.

BOX FIGURE 19.6
(a) Winter wheat in snow. (b) Spring wheat in summer.

---

unknown. Flowering of many plants is also influenced by moisture, soil conditions, nearby plants, and temperature (see box 19.3 above, "Vernalization").

Significance of Photoperiodic Control of Flowering

The control of flowering by photoperiod influences the distribution of plants. For example, many short-night plants do not grow in the tropics because daylengths there are never long enough to induce flowering. Similarly, long-night plants such as ragweed do not grow far into northern areas because they do not have time to form seeds.

Photoperiodic initiation of flowering can occur only if a plant has passed from its juvenile stage into a phase designated as *reproductively mature*, or *ripe to flower*. For example, ragweed is a long-night plant that flowers only in the short days of autumn; why doesn't it also flower during the long nights of spring? The answer is that ragweed is not reproductively mature in the spring. The time required for reproductive maturation ranges from only a few days (as in the Japanese morning glory, *Ipomoea nil*) or weeks (as in annuals) to several years. Indeed, some species of *Agave* require decades before being ripe to flower—hence their name, *century plant*. Similarly, the giant bamboos of Asia flower only every 33 or 66 years, even when

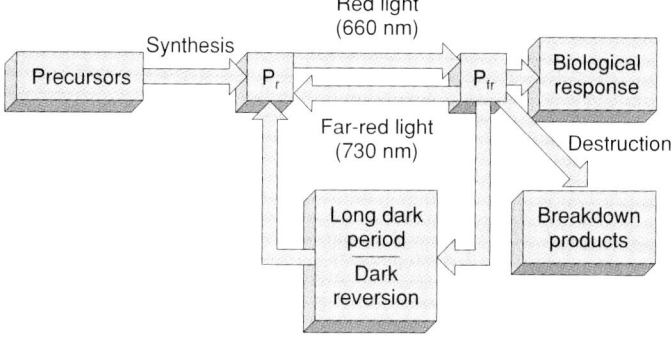

FIGURE 19.16

Phytochrome is synthesized as P_r, which absorbs red light, and is converted to P_{fr}. P_{fr} has several possible fates: it can absorb far-red light and be converted to P_r, initiate a biological response, be destroyed, or revert to P_r if maintained in darkness.

transplanted to new environments. The delay in flowering caused by reproductive maturation ensures that the plant has stored enough food reserves to form and maintain flowers.

Photoperiodism and Phytochrome

The experiments of Garner and Allard prompted botanists to search for the pigment that controlled photoperiodism. Their search was aided by two clues:

1. Flowering is inhibited most effectively by interrupting the dark period with red light.
2. An interruption of the dark period with red light can be reversed if it is followed immediately by exposure to far-red light (the shortest wavelengths of infrared light).

These clues suggested that the pigment that was sensitive to daylength existed in two forms, one that absorbed red light and another that absorbed far-red light. The pigment was soon isolated and identified as **phytochrome.** As suspected, phytochrome exists in two interconvertible forms, P_r and P_{fr}. Phytochrome is synthesized as P_r, which is the inactive form of the pigment. P_r absorbs red light and is converted to P_{fr}, the active form of phytochrome. P_{fr} promotes flowering of short-night plants and inhibits flowering of long-night plants. P_{fr} absorbs far-red light and is converted to P_r (fig. 19.16).

What is the significance of the interconvertibility of P_r and P_{fr} by red and far-red light? Sunlight contains proportionally more red light than far-red light. Thus, because P_r is converted to P_{fr} during the day, the presence of a large amount of P_{fr} could signal the plant that it is in light. When the plant is placed in darkness, P_{fr} is slowly converted to P_r. This conversion of phytochrome to P_r was originally thought to start a set of reactions that enabled the plant to measure the length of darkness relative to the length of light. Indeed, this reasoning would explain why an uninterrupted period of dark (rather than light) controls photoperiodism. However, the dark-reversion of P_r to P_{fr} requires only 3–4 hours and therefore cannot fully account for the light-dark sensing mechanism that controls flowering. Rather, flowering is probably controlled by the combined actions of phytochrome and an internal clock. To explain the coordination of phytochrome and this internal clock, consider a long-night plant in which flowering is inhibited by P_{fr}. This plant probably *measures* the longer-than-critical dark period with its internal clock, but *responds* (i.e., flowers) because the dark-reversion of P_{fr} to P_r removes an inhibition to flowering. Interrupting the dark period with red light would reconvert P_r to P_{fr}, thereby inhibiting flowering. Similarly, interrupting the dark period with a flash of red light followed by a flash of far-red light would leave most of the phytochrome in the P_r form, and flowering would occur. The internal clock involved in this process is discussed later in this chapter.

Other Responses Influenced By Photoperiod and Phytochrome

Phytochrome influences several responses of plants in addition to flowering. One of its most important roles is to control the early growth of seedlings. When a seed germinates underground or in darkness, the seedling has abnormally elongated stems, small roots and leaves, a pale color, and appears spindly. This condition is termed **etiolated.** The rapid elongation of etiolated plants helps them reach light before exhausting their stored food reserves. Once a plant is in light, etiolated growth is replaced by normal growth. The light-controlled transformation from etiolated to normal growth is controlled by red light, which converts P_r to P_{fr} and thus brings about normal growth. Etiolated plants are sensitive to red light: exposing an etiolated plant to only 1 minute of red light will initiate normal growth. If this red light is followed immediately by exposure to far-red light, P_{fr} is converted to P_r, and etiolated growth continues.

Phytochrome may also provide the plant with information regarding more subtle aspects of lighting, such as shading by plants overhead. Much of the red light of sunlight is absorbed by chlorophyll in a plant canopy. Underlying plants, such as those on a forest floor, therefore receive less red light, and less P_r is converted to P_{fr}. Because P_r promotes stem elongation (as in etiolated plants), information provided by the phytochrome system can help a plant reach sunlight more rapidly. Phytochrome may also help direct shoot phototropism. Unidirectional light could presumably create a P_r/P_{fr} gradient across the stem. Since P_r (abundant on the shaded side) promotes stem elongation, and P_{fr} (abundant on the illuminated side) inhibits stem elongation, this gradient of phytochrome could influence in shoot phototropism.

Another aspect of plant development influenced by phytochrome is seed germination. In certain types of lettuce and weeds, red light stimulates germination, and far-red light inhibits germination. Alternating exposure of seeds to red and far-red light can be repeated indefinitely, but seed germination is affected only by the last exposure (fig. 19.17). Thus, germination occurs after exposure to red/far-red/red/far-red/red light just as it does after a single exposure to red light. Treatments with far-red and red/far-red/red/far-red/red/far-red light result in relatively little germination. Seeds of many species germinate when there is enough light to stimulate the conversion of P_r to

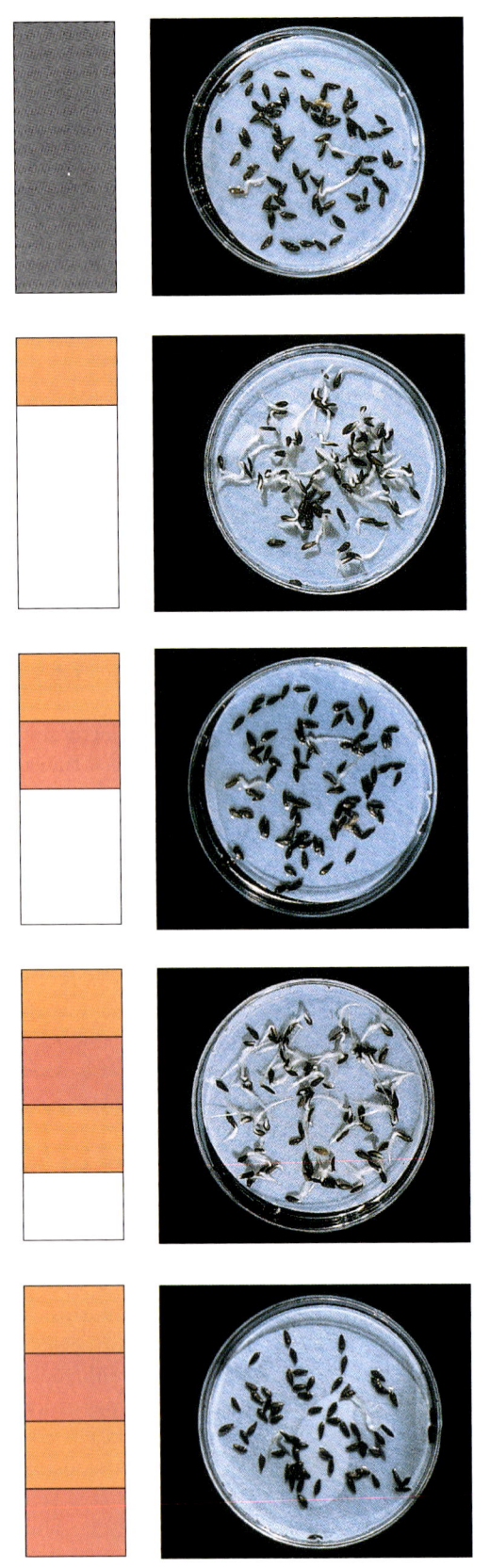

P_{fr}. Most seeds with this sensitivity contain small amounts of stored food. It is unlikely that these seeds could survive germination from deep within the soil. The absence of P_{fr} (due to no sunlight) may account for the lack of germination by seeds buried deep in soil.

Phytochrome and photoperiod also affect several other responses, including shoot gravitropism, stomatal formation, leaf abscission, chlorophyll synthesis, spore germination, sex expression, flower formation, and nyctinastic movements. In each of these responses, phytochrome is the light receptor, not the effector. How phytochrome induces these responses is unknown. It may alter membrane permeability, enzyme activity, gene expression, or the movement of hormones into and out of cells, thereby regulating cellular activities. Indeed, concentrations of most growth-promoting hormones increase immediately after exposure to red light.

There are many types of phytochrome, each of which may have a specific physiological role. We do not yet understand how P_{fr} induces the expression of responsive genes.

C O N C E P T

Phytochrome is a pigment that exists in two interconvertible forms. It absorbs red and far-red light, and influences several aspects of plant growth and development, including flowering.

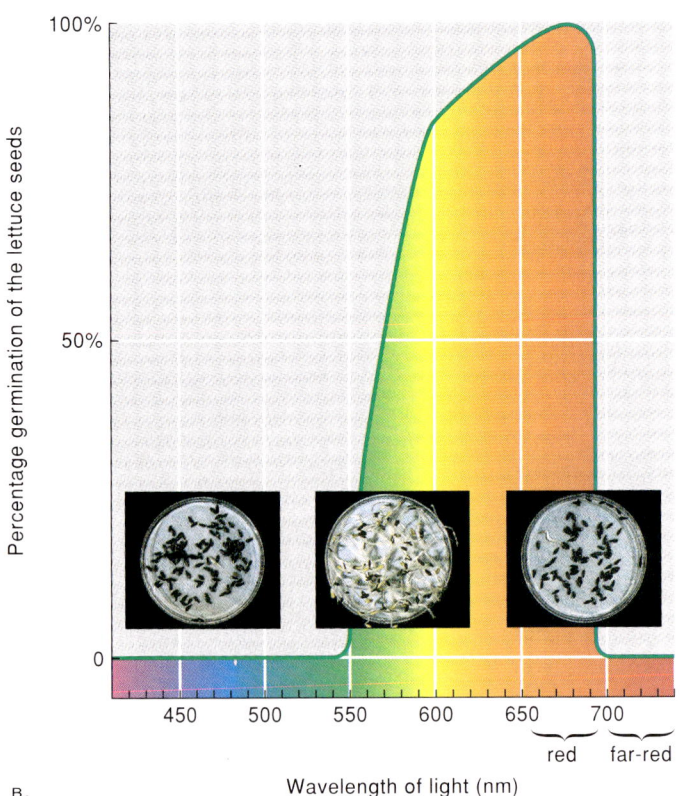

A. B.

FIGURE 19.17

Control of lettuce seed germination by red (R) and far-red (FR) light. If the last exposure is to red light, most of the seeds germinate. The seeds remain dormant if the last exposure is to far-red light. This sensitivity to red and far-red light is controlled by phytochrome.

Source: Data from M. Wilkins, Plantwatching: How Plants Remember, Tell Time, Form Partnerships and More, *Macmillan Publishing Company, New York, NY, 1988.*

Manipulating Photoperiodism

Horticulturists have long exploited our knowledge of photoperiodism to produce flowers when they are wanted for sale. For example, chrysanthemums are long-night plants that usually flower only in the fall. However, mums (i.e., flowers of chrysanthemums) can be made available year-round by using light-proof shades to create an inductive, long-night photoperiod. Similarly, supplemental light can be used to induce out-of-season flowering in winter by short-night plants such as irises.

Doing Botany Yourself

Design a set of experiments to determine if a plant is a long-night plant or a short-night plant. What would be the potential sources of error?

Senescence

Earlier in this chapter we stressed that plants usually respond to environmental stimuli by altering their growth patterns. However, some stimuli induce **senescence,** which is a collective term for the processes accompanying aging that lead to the death of a plant or plant part. In many instances, senescence is rapid; for example, flowers of plants such as wood sorrel (*Oxalis europaea*) and heron's bill (*Erodium cicutarium*) shrivel and die only a few hours after being formed. Other examples of senescence, such as the colorful "turning" of leaves in autumn (fig. 19.18) last longer and are triggered by changes in photoperiod. Whatever its duration, senescence is not merely a gradual cessation of growth; it is an energy-requiring process brought about by metabolic changes. For example, leaf senescence begins during the shortening days of late summer and involves mobilization of nutrients and breakdown of proteins. By the time a leaf is shed from a plant, it contains little more than cell walls and remnants of a nutrient-depleted protoplasm; most of its nutrients have long since been moved to the roots for storage. These nutrients are used in the spring when the plant resumes growth.

The most striking event in leaf senescence is the destruction of chlorophyll. When chlorophyll is degraded, the yellow and orange carotenoids previously masked by chlorophyll become visible. The senescing cells also produce brightly colored anthocyanins. These pigments cause leaves to change colors, often in spectacular fashion (fig. 19.18; see also "Making and Destroying Pigments" in Chapter 7).

What controls leaf senescence? Senescence is strongly influenced by plant hormones; for example, fruits of plants such as soybeans produce a "senescence factor" that is transported to leaves and induces senescence. In the laboratory, leaf senescence can be delayed by applying cytokinins, gibberellins, and/or IAA, and can be promoted by applying abscisic acid or ethylene.

Dormancy

Leaf senescence is only one aspect of a plant's preparation for cold or drought. The shortening days of autumn induce **dormancy,** which is a period of decreased metabolism. Like senescence, dormancy involves structural and chemical changes within the plant. For example, cells make sugars and amino acids that function as antifreeze (i.e., they lower the freezing point in cells) and thereby prevent or minimize damage from cold. Also, inhibitors accumulate in buds, transforming them into winter buds covered by thick, protective scales. These changes in preparation for new environmental conditions (such as winter) are called *acclimation*. As a result of acclimation, a plant can better survive a cold, dry winter, and is said to be *cold-hardy*. Cold-hardiness plays a major role in determining the distribution of a species: the poor cold-hardiness of some plants restricts their growth to tropical or warm temperate regions.

Resumption of active growth in spring is usually influenced by photoperiod and/or temperature. For example, the lengthening days of spring can release dormancy in birch and red oak, whereas fruit trees such as apple and cherry resume growth only after exposure to winter's cold. This cold requirement for breaking dormancy presents problems for apple and cherry trees planted in warm climates. Since these climates usually do not have a cold or long winter, spring growth either does not occur or is greatly delayed.

In other plants, dormancy is released by factors unrelated to photoperiod or temperature. For example, rainfall alone releases dormancy in many desert plants, while plants such as potato require a dry period before renewing growth. The mechanisms by which photoperiod or cold break dormancy are unknown, but probably involve changes in amounts of hormones or the sensitivity of tissues to hormones.

As mentioned in Chapter 18, seed dormancy in many plants is controlled by ABA. ABA begins forming in leaves when the plant senses that daylength is becoming progressively shorter. This ABA is transported to seeds, where it accumulates and prevents germination. During its dormancy, the seed slowly breaks down the ABA; when the amount of ABA falls to a permissive level, the seed can germinate. This usually occurs when the seasonal climate is best for the plant to begin growth.

CIRCADIAN RHYTHMS

Not all rhythmic responses of plants are seasonal. For example, the common four-o'clock (*Mirabilis jalapa*) opens its flowers only in late afternoon, and the yellow flowers of evening primrose (*Oenothera biennis*) open only at nightfall. Similarly, nyctinastic movements of prayer plants occur at the same time every day, whereas some dinoflagellates phosphoresce in warm ocean waters within a few minutes of midnight each night. These regular, daily rhythms are called **circadian rhythms** (from the Latin words *circa*, meaning "about," and *dies*, meaning "day"). Not all circadian rhythms are as obvious as the opening of flowers or the nyctinastic movements of leaves; subtle activities such as cellular division, stomatal opening, protein synthesis, secretion of nectar, and hormone synthesis also occur in a daily rhythm. Circadian rhythms are widespread among eukaryotes but are unknown in bacteria.

FIGURE 19.18

Leaf senescence is a complex, energy-requiring process that is often controlled by photoperiod. The spectacular colors of leaves in the fall result from destruction of chlorophyll, which reveals the presence of other pigments such as carotenoids and anthocyanins.

What Controls Circadian Rhythms in Plants?

How do plants measure a day? Do circadian rhythms originate internally, or do plants depend on some external signal to measure a day? Some of the experiments that have helped answer this question have been elegantly simple.

Are circadian rhythms exact? Usually not; different species and different individuals of the same species have similar rhythms, but these rhythms may be a few hours longer or shorter than 24 hours.

What happens if the environment changes? Circadian rhythms, such as those of prayer plants, occur even when the plants are kept under constant conditions and deprived of obvious external information about the time of day (fig. 19.19).

Other experiments have been much more complicated and somewhat bizarre, such as launching plants into outer space, growing plants in the depths of salt mines, and observing plants grown at the South Pole. The results of these experiments suggest that circadian rhythms are controlled internally rather than by external factors. Many rhythms are mediated by the translation of mRNA and involve phytochrome, which absorbs light that "resets the clock" by accelerating or delaying the onset of the rhythm. That is, phytochrome constantly adjusts the endogenous circadian rhythm to keep up with seasonal changes in daylength.

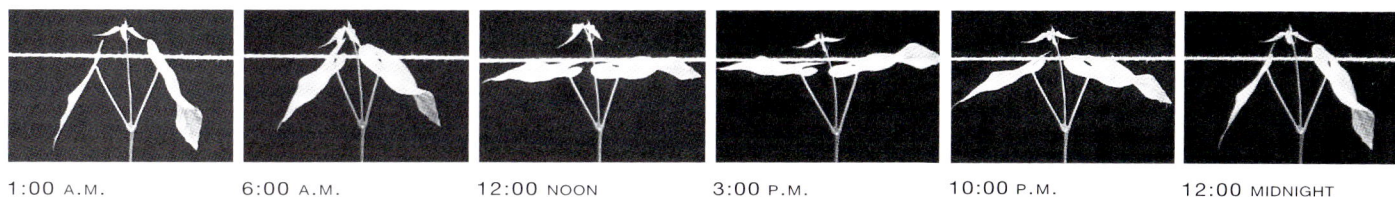

| 1:00 A.M. | 6:00 A.M. | 12:00 NOON | 3:00 P.M. | 10:00 P.M. | 12:00 MIDNIGHT |

FIGURE 19.19

Leaf movements ("sleep movements") of bean leaves. A young bean plant was photographed at hourly intervals over a 24-hour period. The plant was held at constant temperature in the dark in front of a black velvet background and a horizontal string. An electronic flash was used to take the pictures. Six photographs were selected to produce the series shown.

Biological Clocks and Entrainment

The internal timing mechanism that controls circadian rhythms is called the plant's **biological clock;** rhythmic events such as leaf movements in prayer plants can be viewed as the "hands" of this clock (also see discussion of Linnaeus' *horologium florae* on p. 448). Interestingly, biological clocks do not speed up as temperature increases. This is somewhat surprising, because most biological processes are strongly influenced by temperature. Biological clocks must have some kind of feedback mechanism to compensate for changes in temperature. We know little about this feedback mechanism.

Although the control of circadian rhythms is internal, certain environmental factors can influence plants' rhythms. For example, slight modifications of the photoperiod can change the circadian rhythm. This resynchronization of the biological clock by the environment is called **entrainment.** However, entrainment to a new environment is limited: if the new photoperiod is too different from the plant's internal clock, the plant will revert to its own internal rhythms. Plants previously entrained to a new photoperiod will also revert to their natural rhythms when placed in constant light, regardless of how long they had been maintained in the modified photoperiod.

Of what significance is a biological clock? One important function of biological clocks is that they provide a timing mechanism for photoperiodism that allows plants to synchronize their activities. For example, biological clocks ensure that all flowers of a particular species are open at a particular time of day and are available for visits by pollinators. For some plants, this timing must be precise; flowers of *Cereus* cacti are pollinated by bats and must therefore open at night when bats are active. Furthermore, because flowers of some species of *Cereus* (e.g., Queen-of-the-Night) flowers persist for only a night, different individuals must flower at the same time if they are to be cross-pollinated.

Biological clocks also allow plants to sense and anticipate changes in seasons. Phytochrome probably influences the time-keeping mechanism of biological clocks, as do changes in transport and metabolism in cells.

C O N C E P T

Circadian rhythms are rhythmic activities that occur daily. They are controlled by an internal biological clock and influence many aspects of plant growth and development.

Chapter Summary

Plants adjust their growth and development in response to environmental signals and rhythms.

Tropisms are short-term growth responses determined by the direction of an environmental stimulus. Tropisms result from differential growth and are important because they increase a plant's chances of intercepting more light for photosynthesis and encountering water and minerals in soil.

- *Phototropism*, a growth response to unidirectional light, results from IAA moving to the shaded side of a coleoptile. The resulting accumulation of IAA on the shaded side stimulates cellular elongation there and causes the coleoptile to curve toward the light. Phototropism of stems involves production of inhibitors on the illuminated side of the stem. The inhibitors slow cellular elongation, thereby causing the stem to curve toward the light.

- *Gravitropism*, a growth response to gravity, results from an accumulation of IAA along the lower side of roots and stems. In stems, this accumulation of IAA stimulates cellular elongation, which causes the stems to curve up. In roots, the accumulation of IAA inhibits cellular elongation, which causes roots to curve down.

- *Thigmotropism* is a growth response to touch. Thigmotropism is common in tendrils, which coil around objects that they touch. Thigmotropism is controlled by IAA and ethylene.

Nastic movements occur in an anatomically predetermined direction. Like tropisms, nastic movements are widespread in plant growth and include activities such as leaf movements, trapping of prey by carnivorous plants, and defense.

- *Seismonasty* is a nastic movement resulting from contact or mechanical disturbances. These stimuli initiate electrical signals, which are subsequently translated into chemical signals that alter movement of ions. In some plants (e.g., sensitive plants), this movement of ions results in reversible turgor changes that cause movement.

- *Nyctinasty* is a nastic response resulting from daily rhythms of light and dark. It is caused by movements of ions, which cause turgor changes and movement.

Thigmomorphogenesis results in the stunted growth of plants due to mechanical disturbances such as wind and rain. It involves electrical signals and probably results from changes in ethylene production.

Photoperiodism is the ability of plants to sense changes in the relative lengths of night and day. One of the responses controlled by photoperiodism is flowering. The most critical factor for flowering is the length of the uninterrupted dark period.

Plants such as corn that flower only in response to long days and short nights are called *short-night* (i.e., *long-day*) *plants*, while those that flower only in response to short days and long nights are called *long-night* (i.e., *short-day*) *plants*. Intermediate-day plants such as sugarcane flower only when exposed to days of intermediate length. Day-neutral plants such as dandelions flower without regard to photoperiod.

Phytochrome is a pigment that influences many aspects of plant growth and development, including flowering. Phytochrome absorbs red and far-red light, and exists in two interconvertible forms, P_r and P_{fr}. P_{fr} forms upon exposure to red light and is the biologically active form of the pigment.

Senescence is the process that leads to the death of a plant or plant part. Leaf senescence in the fall is triggered by shortening days, and involves mobilization of nutrients, destruction of chlorophyll, and abscission. Dormancy, a period of decreased activity, is usually controlled by photoperiod and/or temperature.

Circadian rhythms are regular rhythms of growth and activity that occur on an approximately 24-hour basis. They can be modified somewhat by the environment and are regulated by an internal biological clock.

Questions for Further Thought and Study

1. Would a short-night plant flower if its dark period were interrupted with light? Why or why not?
2. What would be the consequences of transplanting a long-night plant to the tropics? Explain your answer.
3. What would happen to IAA in a horizontally oriented stem that is illuminated from below?
4. Describe the possible significance (i.e., biological advantage) of each of the following: phototropism, thigmotropism, gravitropism, seismonasty, biological clocks, photoperiodism.
5. How does phytochrome differ from other plant pigments?
6. What is the difference between a tropism and a nastic movement?
7. Hoeing a garden or tilling a field often triggers the germination of many seeds buried in the soil. Why?
8. Why is it unnecessary to orient a seed in a particular way when you plant it?
9. Could florigen be an inhibitor rather than a promoter? Why or why not? What experiments would you do to try to isolate florigen?
10. Electrical signals are involved in several aspects of plant growth and development, including cellular differentiation, embryo development, growth of pollen tubes, attraction of symbionts, and resistance to pathogens (fig. 12.12). How are these signals similar to those in animals? How are they different?
11. In some mutants of corn and other plants, seeds geminate while still attached to the parent plant. These seedlings are referred to as being *viviparous*. Based on what you learned in this chapter, propose a hypothesis to account for viviparous seedlings.

Suggested Readings

ARTICLES

Cleland, C. F. 1978. The flowering enigma. *BioScience* 28:265.
Evans, M. L., R. Moore, and K. H. Hasenstein. 1986. How roots respond to gravity. *Scientific American* 255:112.
Lumsden, P. J. 1991. Circadian rhythms and phytochrome. *Annual Review of Plant Physiology and Plant Molecular Biology* 42:351–371.
Palmer, J. D. 1984. Biological rhythms and living clocks. *Carolina Biology Readers* no. 92. Burlington, NC: Carolina Biological Supply.
Quail, Peter H. 1991. Phytochrome: A light-activated molecular switch that regulates plant gene expression. *Annual Review of Plant Physiology and Plant Molecular Biology* 42:389–409.
Smith, H. 1984. Plants that track the sun. *Nature* 308:774.
Sussman, Michael R. 1992. Shaking *Arabidopsis thaliana*. *Science* 256:619.
Wayne, Randy, Mark P. Staves, and A. Carl Leopold. 1992. The contribution of the extracellular matrix to gravisensing in characean cells. *Journal of Cell Science* 101:611–623.

BOOKS

Galston, A. 1994. *Life Processes of Plants*. New York: W. H. Freeman and Co.
Haupt, W., and M. Feinlieb, eds. 1979. *The Physiology of Movements*. Encyclopedia of Plant Physiology. New series, vol. 7. Berlin: Springer-Verlag.
Kendrick, R. E., and B. Frankland. 1983. *Phytochrome and Plant Growth*. Studies in Biology no. 68. London: Edward Arnold.
Salisbury, F. B., and C. W. Ross. 1992. *Plant Physiology*, 4th ed. Belmont, CA: Wadsworth.
Wilkins, M. B. 1988. *Plant Watching: How Plants Live, Feel and Work*. London: Macmillan.

Unit Six

Nutrition and Transport...

Up to this point, we have focused primarily on the organic nature of plants. Chromosomes, cell walls, cell membranes, hormones, wood, and ATP are all based on carbon; that is, they are organic. However, an introductory botany book would not be complete without discussing the role of inorganic compounds in plant growth. Inorganic ions, such as calcium and potassium, are indispensable as cofactors in many biochemical reactions and as parts of organic molecules. This means that the carbon-based metabolism of photosynthesis and respiration must be supplemented with ingredients that do not enter the plant by gas exchange; instead, they must enter the plant through the roots.

Understanding the inorganic side of plant metabolism comes from learning what nutrients are essential to plants, what minerals are available in the soil, and how plants absorb and move water and dissolved minerals into and throughout the plant. These are the topics that are discussed in this unit. Our discussion of plant nutrition and mineral transport will reveal unanswered questions as well as some surprises.

A sundew (*Drosera binata*), magnified to show gland-tipped hairs. These plants obtain much of their nitrogen by trapping and digesting insects.

CHAPTER 20
Soils and Plant Nutrition

Chapter Outline

INTRODUCTION

SOILS
- How Soils Form
- Components of Soil

BOX 20.1
PUTTING THINGS BACK

BOX 20.2
LOSING THE SOIL

PLANT NUTRITION
- Essential Elements
- Obtaining Essential Elements

Chapter Summary
Questions for Further Thought and Study
Suggested Readings

Chapter Overview

Life on earth is sustained by sunlight and soil. Although soils have differing properties, each influences the distribution of plants and animals on our planet. For example, animal nutrition depends on the ability of plants to find and absorb the sixteen elements essential for growth and reproduction. Gathering these elements is an elaborate and expensive venture that depends largely on root systems that permeate and mine the soil. The mineral nutrition of most plants also depends on other organisms with whom plants establish relationships ranging from mutually beneficial symbioses to parasitisms. Some plants even get their essential nutrients by trapping and digesting animals. Nutrient absorption by plants is important because it links the living and nonliving parts of terrestrial and aquatic ecosystems.

Introduction

Throughout this book you've seen examples of how various nutrients influence plant growth and development. For example, iron is an electron carrier in respiration, and calcium plays an important role in the responses of plants to environmental stimuli. Indeed, *all* aspects of plant growth and development depend on an adequate supply of nutrients.

But what do we mean by *adequate*? To answer this question, consider a harvest of 5 metric tons by an Idaho potato farmer. About four metric tons of the crop's weight is water—that is, hydrogen and oxygen. Almost 85% of the remaining "dry weight" is carbohydrates such as cellulose, starch, and sugars. The remainder includes nitrogen (2%) and minerals such as iron (0.01%), sulfur (0.15%), and chlorine (0.2%). Thus, plants require different amounts of different nutrients. Many of these "adequate" amounts of nutrients seem trivial. For example, all of the copper in the five tons of potatoes weighs only about 2.5 grams, an amount roughly equal to that in a penny. However, no matter how small this amount of copper may seem, its role in plant growth is critical: plants *must* have copper to grow. Without that penny's worth of copper, there would be no potato plants and no harvest.

Where do nutrients come from? We have already studied an example of how plants obtain one of their nutrients from the air: carbon comes almost exclusively from atmospheric CO_2 fixed during photosynthesis. However, most nutrients are absorbed from soil by roots. Thus, roots are the interface between the living and nonliving parts of the terrestrial ecosystem: virtually all nutrients move from the nonliving environment of soil into the food chains through roots. However, soil does more than store nutrients until they can be absorbed by a plant: it also determines the availability of nutrients and thus largely determines what plants and animals live in a particular area. You can't understand plants without knowing something about soil.

Soils

How Soils Form

The earth is more than 4.5 billion years old. In its youngest stages, it was a harsh place; its "land" was little more than a mixture of igneous, sedimentary, and metamorphic rocks. Once formed, these rocks were transformed into soil by glaciers, wind, and rain, and later also by organisms and their activities. This conversion of rock into soil is called *weathering* and is a slow process. For example, in eastern North America where rainfall is relatively abundant, it can take more than two hundred years to form only 2 centimeters of topsoil. Although topsoil is usually less than a meter thick, it spans continents, bridging the rocks below with life above.

Soils have several layers, called **horizons,** each of which has several distinguishing characteristics (fig. 20.1):

The **O horizon** is the surface litter covering the soil. This horizon typically consists of fallen leaves and is only a few centimeters thick.

The **A horizon** is topsoil and usually extends 10–30 centimeters below the soil surface. In most fertile soils, the A horizon has a pH near 7 and contains 10%–15% organic matter, which gives this horizon a dark color.

The **B horizon** consists of larger soil particles than those in the A horizon, and extends 30–60 centimeters below the soil surface. This horizon usually contains relatively little organic matter and is therefore lighter in color than the overlying A horizon. In many regions, the B horizon contains large amounts of minerals and clay particles washed by rainfall from the A horizon. Mature roots commonly extend into the B horizon, where minerals accumulate. The B horizon is often called subsoil.

The **C horizon** occurs 90–120 centimeters below the soil surface and consists primarily of partially altered to unaltered rock fragments and mineral grains. This horizon usually lacks organic matter and is often referred to as the parent material, since it is the raw material from which soil forms. The C horizon extends to the underlying and often impenetrable bedrock of igneous, sedimentary, or metamorphic rock.

FIGURE 20.1

Soil horizons. The A, B, and C horizons can sometimes be seen in road cuts such as this one in Australia. The upper layers developed from bedrock. The dark upper layer is home to most of the organisms that live in the soil.

CONCEPT

Soil forms from rocks by a process called weathering. Soils are often arranged in distinctive layers called horizons, each of which has unique properties.

Soil with the horizons just described is called a *mollisol* and characterizes semiarid grasslands. Mollisols retain nutrients and are fertile. Indeed, mollisols of the breadbaskets of the Great Plains of the United States and the steppes of the Ukraine are the earth's most fertile soils. The only problem with using mollisols for growing crops is that they often occur in semiarid climates and therefore must be irrigated frequently.

Just as there is no such thing as a typical cell, there is no typical soil. Indeed, there are more than 70,000 kinds of soil in the United States alone, some of which identify places such as Redlands, California, and Black Earth, Wisconsin. Despite their diversity, however, these soils can be grouped into the following general categories:

Mollisols are nutrient-rich soils characteristic of semiarid grasslands. Most of the organic matter in a mollisol is in its A horizon.

Spodosols are light-colored soils characteristic of wet, temperate regions such as the coniferous forests of Canada. The A horizon of these soils contains small amounts of organic matter and may have a pH as low as 4. Most organic matter in a spodosol is in its B horizon.

Alfisols resemble spodosols, except that their A horizon contains more organic matter. Alfisols are relatively fertile, brown soils common in the eastern United States.

Aridosols occur in arid regions such as deserts. Since aridosols contain relatively little organic matter, they are usually light-colored.

Histosols are acidic soils typical of swamps and bogs. Histosols contain large amounts of organic matter and are usually dark brown or black.

Oxisols are acidic soils characteristic of rain forests. Despite the lush vegetation that they support, oxisols are nutrient-poor. As a result, only plants such as cassava and bananas that efficiently extract nutrients can grow well on oxisols. The lack of nutrients in oxisols makes them almost useless for modern agriculture. Indeed, the intense poverty of tropical South America and Central Africa is due largely to the inability of their oxisols to sustain the agriculture necessary for increased amounts of food and industry.

Soils are sensitive to environmental changes. This is especially true of an iron-rich oxisol called *laterite* that occurs in rainforests. When shaded by the rainforest, laterite remains damp and soil-like. However, if the forest is cut and the exposed soil dries, the iron in laterite irreversibly cements the soil into hard clumps called *ironstone*. Once formed, ironstone cannot be reconverted to oxisol. Consequently, tropical rain forests recover slowly, if at all, after being stripped of their vegetation.

CONCEPT

There are many kinds of soils, each having different properties that support different plants. These soils range from the mollisols of fertile grasslands to the nutrient-depleted oxisols of tropical rain forests. Soils are sensitive to environmental disturbances.

Components of Soil

Regardless of the different properties of their horizons, all soils contain the same five components: mineral particles, decaying organic matter (humus), air, water, and living organisms. Differing amounts of these materials define the soil's properties and, therefore, the plants that it can support.

Mineral Particles

Weathering breaks rocks into progressively smaller pieces, the smallest of which are called **soil particles.** These particles consist of **minerals,** which are naturally occurring inorganic compounds usually made of two or more elements. All soils

TABLE 20.1

Physical and Chemical Properties of Soil as Affected by Soil Texture

Soil Texture	Water Infiltration	Water-Holding Capacity	Ion Exchange	Aeration	Workability	Root Penetration
Sand	Good	Poor	Poor	Good	Good	Good
Silt	Medium	Medium	Medium	Medium	Medium	Medium
Clay	Poor	Good	Good	Poor	Poor	Poor
Loam*	Medium	Medium	Medium	Medium	Medium	Medium

*Loam soil averages out as medium in all respects because it is of variable composition, depending on the actual proportions of sand, silt, and clay particles that are present.

contain three kinds of soil particles: sand, silt, and clay. Clays are the final product of weathering, and are the smallest of these particles.

Type of Soil Particle	Diameter of Soil Particle (mm)
Sand	0.02–2
Silt	0.02–0.002
Clay	<0.002

Different kinds of soils contain different amounts of sand, silt, and clay. Clays contain more than 30% clay particles, and sandy soils contain less than 20% silt and clay. Soils having approximately equal mixtures of sand, silt, and clay—that is, the soils in which most plants grow best—are called **loams.** Soil mineral particles have many effects on plant growth, including determining the availability of nutrients and water (see table 20.1).

Doing Botany Yourself

Describe how you could estimate the percentages of sand, silt, and clay in a soil sample.

How Soil Particles Influence the Availability of Nutrients

The ability of a soil to retain nutrients depends primarily on the structure of its clay particles. Clay particles have a sheet-like structure and are called **micelles.** In their original form, micelles associate with large amounts of silica (Si^{4+}). Over time, however, the Si^{4+} of clay particles is often replaced by Al^{3+}, which in turn may be replaced by Mg^{2+} or Fe^{2+}. As a result of these substitutions, clay micelles become negatively charged, allowing them to reversibly bind cations such as Ca^{2+} and K^+ and prevent these cations from being washed from the soil by rainfall. This weak binding of cations by clay micelles is important because plants can later extract these cations by exchanging them for H^+ via a process called **cation exchange** (fig. 20.2). For example, roots can secrete H^+ that replaces and makes available the Ca^{2+} held by a clay micelle:

$$\text{micelle: } Ca^{2+} + 2H^+$$
$$\downarrow \textit{cation exchange}$$
$$H^+\text{:micelle: } H^+ + Ca^{2+}$$

Cation exchange is also enhanced by the respiratory production of CO_2 by roots. CO_2 released by respiring roots dissolves

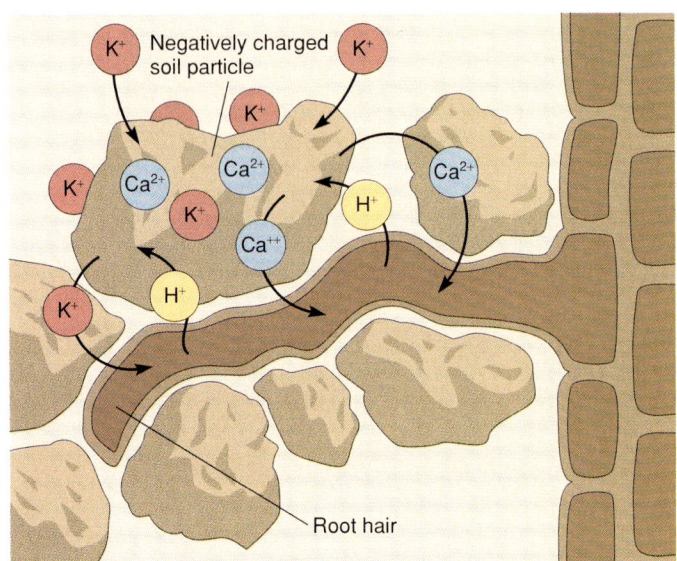

FIGURE 20.2

Cation exchange in soil. Negatively charged clay micelles reversibly bind cations such as Ca^{2+} or K^+, thereby preventing these cations from being washed from soil by rainfall. Plants can extract these cations by exchanging them for H^+ via cation exchange. The secretion of H^+ by roots facilitates this exchange.

in water to form carbonic acid (H_2CO_3), which then dissociates into HCO_3^- and H^+. These H^+ replace cations held on clay micelles, thereby making the cations available to the plant.

Because of their negative charge, clays have a high **cation-exchange capacity.** Conversely, sand particles are not charged and therefore have a low cation-exchange capacity. Because cations do not bind to sand particles, cations are rapidly leached from sandy soil after a rainfall. This explains why sandy soils are usually nutrient-deficient and poorly suited for growing crops. The high cation-exchange capacity of clays makes them well suited for plant growth. However, the cation-exchange capacity of a clay can also have negative consequences for a plant. For example, replacing clay micelles with H^+ acidifies the soil. If this process is unchecked, the pH of the soil can drop to as low as 3 or 4. This drop in the pH alters the availability of nutrients and thus can limit plant growth (see "Soil pH," p. 475).

What about anions? Since soil particles are not positively charged, rainfall can rapidly leach anions such as SO_4^{2-} and NO_3^- from the soil. Because the lack of these anions frequently

BOXED READING 20.1
PUTTING THINGS BACK

Plants extract large amounts of nutrients from soil. For example, during one growing season a wheat crop on one hectare of land removes 85 kg of nitrogen, 47 kg of potassium, and 17 kg of phosphorus from the soil. Some of these nutrients are replenished by decaying humus and by plowing under the remaining parts of the crop. However, such replenishment does not match what is lost when the crop is harvested. Consequently, these lost nutrients must be replenished with fertilizers. For example, the yield of an unfertilized soil that initially produced 100 bushels of corn per acre diminished to only 23 bushels per acre in seventy years. When this soil was fertilized, the yield increased to more than 130 bushels per acre. Although fertilization increases plant growth and crop yield, it rapidly reaches a point of diminishing returns: doubling the yield of already fertile soil often requires adding as much as five times more fertilizer.

Chemical Fertilizers
Most chemical fertilizers have a rating that consists of three numbers, such as 12-6-6. These numbers refer to the amounts of nitrogen, phosphorus, and potassium, which are the three elements most likely to be deficient in soil. Thus, a 12-6-6 fertilizer contains 12% nitrogen (usually as ammonium salts), 6% phosphorus (as phosphoric acid), and 6% potassium (as potash).

Nitrogen, which is the most expensive of these elements to produce, is incorporated into fertilizer via the *Haber-Bosch process*:

$$3H_2 + N_2 \xrightarrow{500°C,\ 300\ atm\ pressure} 2NH_3$$

BOX FIGURE 20.1
The application of fertilizer.

Nitrogen can either be added directly to the fertilizer as an ammonium salt or be converted to nitrate and then added as a nitrate salt (e.g., $NaNO_3$).

More than 40 million metric tons of nitrogen produced by the Haber-Bosch process are added to soil each year. However, this represents only about one-fifth the amount of nitrogen added to the world's soil by nitrogen-fixing bacteria. Furthermore, the Haber-Bosch process is expensive in terms of energy; producing 2.5 kg of ammonia via the Haber-Bosch process requires the energy equivalent of 1,000 kg of coal. The costs of producing nitrogen account for about half of our 14-billion-dollar fertilizer bill. Consequently, manufacturing nitrogen-containing fertilizer requires more energy than any other aspect of crop production in the United States. To compound this problem, applications of nitrogen-containing fertilizers are inefficient, because crops absorb only about half of the nitrogen that is applied. The rest is absorbed by other organisms, leached from the soil in rainfall, or reconverted to gaseous nitrogen (N_2) by denitrifying bacteria such as *Micrococcus denitrificans*.

Chemical fertilizers are concentrated, easy to apply, and allow a grower to apply specific amounts of various nutrients. However, these fertilizers do not replenish humus in the soil. To maintain humus, growers usually plow under either the unharvested plants or a subsequent cover crop of barley or rye. The latter process is called *green manuring* and provides an excellent example of another kind of fertilizer: organic fertilizer.

Organic Fertilizers
Organic fertilizers are essentially the same thing as humus. Although hardly new (planting a fish with corn seed is proverbial), the increased costs of chemical fertilizers have prompted an increasing number of gardeners and farmers to rediscover organic fertilizers, which increase both the water retention and fertility of soil. Organic fertilizers that have been used on crops include manure, dead animals and plants, fish scraps, and cottonseed meal. On a smaller scale, backyard gardeners often use compost, fish meal, lawn clippings, garbage, and a concoction called *manure tea* as organic fertilizers. We do not recommend fertilizing your houseplants with manure tea if guests are coming.

Foliar Fertilization
Despite the presence of a thick cuticle, many plants can absorb nutrients through their leaves and stems. For example, iron is sprayed on azaleas and pineapples, and copper and zinc are sprayed on citrus to prevent mineral deficiencies. This type of fertilization is called *foliar fertilization* and is restricted primarily to micronutrients.

---

limits growth of plants, anions must be replenished by adding fertilizer to the soil (see box 20.1 above, "Putting Things Back").

How Soil Particles Influence the Availability of Water
The surfaces of soil particles adsorb much of the water in soil. The amount of water held in a soil is proportional to the surface area of its particles: the larger the surface area, the greater the retention of water. Because clay micelles have a sheetlike structure, they have a large surface area; for example, the surface of clay particles in the upper few centimeters of soil in a 2-hectare cornfield equals the surface area of North America.

BOXED READING 20.2
LOSING THE SOIL

Next time you're outside, pick up a handful of soil and imagine that it is the earth's surface. Then do this:

Drop three-fourths of the soil back to the ground. That's how much of the earth is covered by water. What's left in your hand represents the land.

Drop half of the remaining soil to account for deserts, glacial poles, and mountain peaks where crops won't grow.

Drop one-tenth to account for land covered by cities, houses, roads, and parking lots.

What's left in your hand represents all the soil that supports life on earth. This soil is trickling through our fingers at an alarming rate because of erosion. Indeed, more than 6 billion tons of soil are lost in the United States each year, enough to fill 320 million dump trucks which, if parked end-to-end, would reach almost to the moon and back.

Erosion—the movement of soil by wind and water—often produces spectacular effects. For example, the breathtaking sculptures of the Grand Canyon result from millions of years of erosion. Equally spectacular, however, are other effects of erosion. For example,

BOX FIGURE 20.2

Barren fields are especially vulnerable to erosion.

dust storms in the Great Plains have deposited topsoil on ships hundreds of miles out in the Atlantic Ocean, and the Mississippi River annually dumps more than a quarter-billion tons of the Midwest's topsoil into a murky Gulf of Mexico. This loss of topsoil is costly: the eroded cropland that pollutes rivers and lakes costs Americans billions of dollars each year.

Why does all of this erosion occur? Although erosion is not caused only by humans—after all, the Grand Canyon formed long before humans began gathering at its edge with their cameras—human activities have increased erosion. Many causes underlie this human-related increase in erosion, but most are related to economics. For example, the food demands of a growing population have opened new international markets for farmers, and massive grain sales have made crop prices soar. These increased profits that resulted from these exports resulted primarily from farmers growing crops on an additional 60 million acres of land, much of which was fragile grassland. For example, the delicate grasslands of the Great Plains were stripped and used to grow wheat; when these unprotected soils dried, they were blown away by the winds.

Despite sodbuster legislation aimed at diminishing erosion, the problem persists. Today, exported crops use more than one-third of our cropland, much of which is susceptible to erosion.

Clay micelles are smaller than sand, and therefore have a larger surface area per unit soil volume than does sand. As a result, clay soils retain much more water than do sandy soils. This property of clays would seem to make them ideal for plant growth. However, this is not entirely true, for along with the huge water-retention capacity of clay come some problems. For example, the small size of clay particles results in their packing tightly together—so tightly that the clay has low amounts of oxygen, due to small air spaces. Furthermore, this packing retards the penetration of water into the soil (e.g., water penetrates clay about twenty times slower than it penetrates sand). As a result, much of the water that falls on clay soil runs off and is unavailable for plant growth. This runoff water also contributes to erosion (see box 20.2 above, "Losing the Soil"). Tightly packed clay also can impede root growth.

CONCEPT

There are three kinds of soil particles: sand, silt, and clay. Clay particles, which have a sheetlike structure, are the smallest of these particles. Because clays are negatively charged, they bind cations. Plants exchange H^+ for these cations in a process called cation exchange. Sand is not charged and does not bind cations.

Humus

Humus is the decomposing organic matter in soil. The amount of humus in soil varies: *mineral soils* contain only 1%–10% humus, while *organic soils* typically contain about 30% humus. The most extreme examples of organic soils are the histosols of swamps and bogs, which may contain more than 90% humus. These soils are usually so acidic that decomposers can hardly grow in them. As a result, humus accumulates faster than it is broken down.

The amount of humus in soil affects the soil and its plants in several ways:

Its lightweight and spongy texture increases the water-retention capacity of the soil. Water absorption by humus decreases runoff, thereby slowing erosion.
Most humus is rich in organic acids and has a negative charge. As a result, humus increases the cation-exchange capacity of the soil.
Humus swells and shrinks as it absorbs water and later dries. This periodic swelling and shrinking aerates the soil.
Humus is a reservoir of nutrients for plants. Like time-release vitamins, humus gradually releases nutrients as it is degraded by decomposers.
Most plants grow best in soil containing 10%–20% humus.

Since humus is constantly being decomposed by bacteria and fungi, it must be replenished. In natural ecosystems, humus is replenished by events such as leaf abscission and the addition of animal wastes. Many gardeners replenish humus in soil by *composting,* a method of converting organic wastes to fertilizer and a soil conditioner. The process is simple: organic wastes (e.g., lawn clippings) are gathered into a compost pile, where they are degraded by fungi and bacteria. During winter, this decomposition may take several months, while during summer it may be completed in as little as three weeks. The resulting compost costs less than commercial fertilizer, provides an efficient means of disposing of waste products, and contains nearly all of the nutrients needed by plants. Moreover, composting releases nutrients gradually as the materials decompose, thereby providing plants with a continuous balance of nutrients. On a larger scale, farmers often plow under the stubble of plants remaining after a harvest to maintain the humus in soil used to grow crops.

Humus is the decomposing organic matter in soil. Humus retains water and is a source of nutrients for plants.

Air

About 25%–50% of the volume of most soils is air. This "empty space" has a critical role: it is the conduit by which oxygen reaches the roots. When water replaces a soil's air for a long period, such as in flooded soil or overwatered houseplants, roots can become anaerobic and die.

Clay soils, with their tightly packed clay particles, contain less air than do sandy soils. Thus, aeration of a clay soil can be improved by adding silt or sand. The resulting loam often aggregates into macroscopic clumps separated by large air spaces, which have an effect similar to that of earthworms: they increase aeration and the penetration of water into the soil, thereby stimulating root growth.

Water

Plants use huge amounts of water. For example, a mature corn plant needs about 100 liters of water per month for optimal growth. This water and its dissolved minerals are obtained from the soil by roots, which are exquisitely designed for gathering water: their continuous growth exploits new territory, and their root hairs greatly increase the surface area for water absorption. Indeed, 1 cm^3 of soil under a patch of bluegrass (*Poa pratensis*) contains more than 150,000 root hairs having almost 400 cm^2 of surface area. However, even this extensive collecting system does not ensure that a plant can obtain enough water from the soil, since the soil itself also influences the amount of water available to a plant. Recall, for example, that clays hold more water than sandy soils. Another factor that affects the ability of a plant to obtain water is the type of water in the soil.

Chemically Bound Water

Some of the water in soil is bound chemically. This includes the water locked in the crystals of minerals and the water hydrated to the surface of clay micelles. Chemically bound water is unavailable to plants.

Unbound Water

The remaining water in soil is unbound water and is held by adhesion (the attraction of water molecules for solid particles) and cohesion (the attraction of water molecules for each other). Based on the tension holding unbound water in soil, we can distinguish three types of unbound water: gravitational water, capillary water, and hygroscopic water (fig. 20.3). To illustrate the differences in these types of water, consider a soil sample to which you have just added an excess amount of water. Some of this water drains from the soil due to gravity, thereby leaving the soil's air spaces filled with air. The water that drains from the soil by gravity is called **gravitational water.** In areas having large amounts of rainfall, the excess gravitational water percolates to lower soil levels, carrying with it nutrients from the upper layers of the soil. This leaching helps form many soils.

When all gravitational water has drained from the soil, the soil is saturated and is said to be at **field capacity.** Most of the remaining water is held by surface tension (i.e., capillary action) in small pores in the soil; this is called **capillary water.** Since it is held weakly in these pores, most capillary water is easily absorbed by plants. Thus, *capillary water is the main source of water and dissolved minerals for plants.* However, not all capillary water is available to plants; it is gradually lost by evaporation and absorption. When the tension at which the remaining water is held equals the ability of plants to extract it, the amount of water remaining in the soil is the **permanent wilting point.** Thus, the permanent wilting point is the lower limit of soil moisture that can support plant growth. A typical permanent

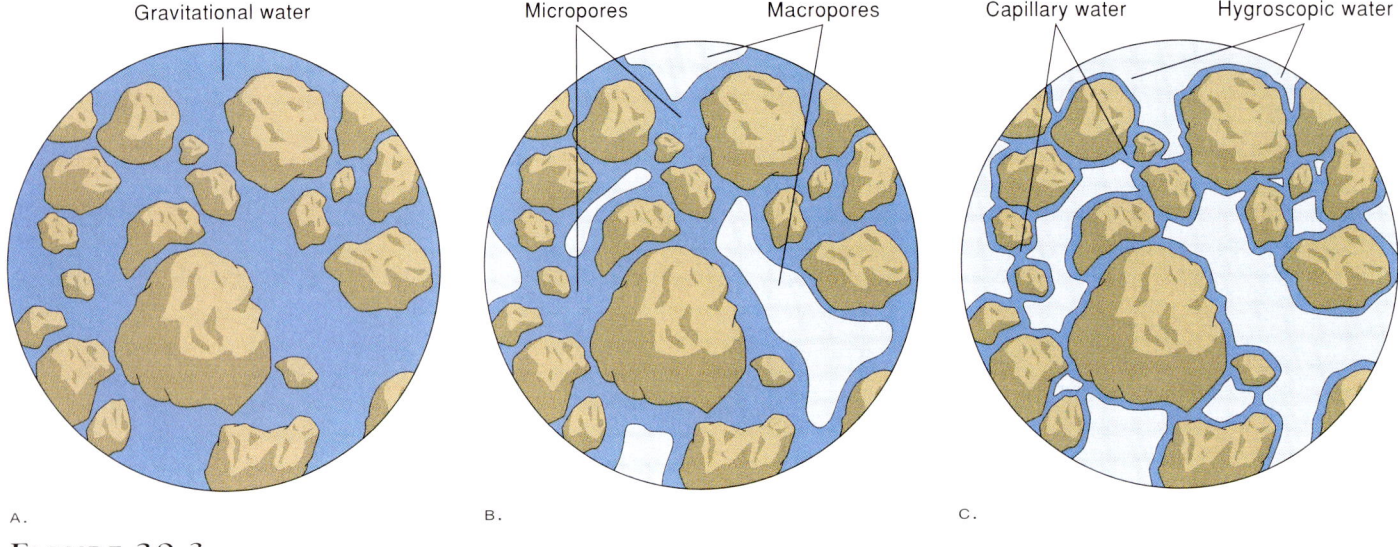

FIGURE 20.3

Soil under three conditions of hydration. (a) Water-saturated: all pores are filled with water. (b) Field capacity: macropores are air-filled and micropores are water-filled. (c) Permanent wilting point: pores are air-filled; films of hygroscopic water (which roots cannot absorb) surround soil particles, and small amounts of capillary water (which roots can absorb) occur at points of particle contact. The permanent wilting point is the smallest amount of hydration that supports some plant activity.

wilting percentage for clay may be as high as 40%, while that of sand may be as low as 4%.

When all of the capillary water is gone, the only unbound water remaining in the soil is **hygroscopic water,** which is held tightly by small soil particles. Plants cannot extract hygroscopic water from soil.

Soils contain three types of unbound water. Gravitational water is water drained by gravity from the soil, whereas capillary water is held by capillary action in pores of the soil. Hygroscopic water is held tightly by soil particles and is unavailable to plants. Almost all water absorbed by plants is capillary water.

Living Organisms

Soil is more than a collection of soil particles, humus, air, and water. It teems with life: on average, about 0.1% of the weight of soil is living organisms. Although this percentage may not impress you, consider what it means: 1 kg of fertile soil contains about 2 trillion bacteria, 400 million fungi, 50 million algae, 30 million protozoa, nematodes, other worms, and insects. Together, these organisms exert a tremendous influence on soil and the plants it supports.

Organisms mix and refine the soil. One earthworm can digest more than 10 metric tons of soil in one year. This is why Aristotle called earthworms "the intestines of the earth": they aerate and refine the soil by processing it through their gut.

Organisms add humus to the soil. Plants, especially grasses, are the primary source of organic matter in soil. For example, a four-month-old rye plant produces almost 11,000 kilometers of roots and root hairs, most of which eventually die and become humus. Because humus contains large amounts of nutrients, the biological activity of soil is usually an excellent indicator of its fertility.

Respiration by these organisms increases the amount of CO_2 in the soil. CO_2 in the soil dissolves in water to form carbonic acid, which decreases the soil pH. This in turn alters the availability of several nutrients.

Many organisms affect the availability of nutrients. For example, roots of lupine (*Lupinus albus*) growing in phosphorus-deficient soil secrete large amounts of H^+, which can decrease the pH of the soil and thus increase the availability of other nutrients to plants. Similarly, plants and microbes secrete phosphatases into the soil. These enzymes liberate phosphate from mineral particles, making more of it available to plants.

Many organisms make the soil uninhabitable for other plants. Many plant pathogens, such as nematodes, live in the soil. Furthermore, some plants produce chemicals that inhibit the growth of other plants. For example, the soft-leaved purple sage (*Salvia leucophylla*) produces terpenoids that inhibit the growth of nearby plants. As a result, colonies of *Salvia* are typically surrounded by bare zones in which no other plants grow.

Finally, soils are a repository for numerous dormant organisms, such as seeds, fungal spores, and insect eggs. Among the most influential of these dormant organisms are seeds. Indeed, a cubic meter of typical soil can contain as many as 5,000 viable seeds (see "Seed Banks" in Chapter 17). Although this number of seeds is impressive, it pales when compared to that of some fertile croplands, which contain as many as 80,000 weed seeds per cubic meter. This part of the soil ecosystem is often a problem for farmers.

Living organisms in soil refine and mix the soil, add organic matter, increase the amount of CO_2 in the soil, influence the availability of nutrients, and often make the soil uninhabitable by other organisms.

What are Botanists Doing?

The inhibition of one species of plant by chemicals from another species of plant is called **allelopathy**. Go to the library and find out why allelopathy is of concern in agriculture.

PLANT NUTRITION

In Chapter 7 you met Jan-Baptista van Helmont, a Dutch physician who measured the growth of a willow tree for several years. That plant, you will recall, gained more than 70 kg, yet reduced the weight of the soil by less than 60 grams. Van Helmont attributed the growth of his willow tree to the water he had added to the soil. This conclusion was partially correct, since most of a plant is water. However, only later did botanists discover that plant growth also depends on several other nutrients which, like water, are absorbed from the soil and are essential for growth.

Essential Elements

More than sixty elements have been found in plant tissues. These elements range from those as common as carbon and hydrogen to those as exotic as platinum, uranium, and gold. Are all of these elements essential for growth? If so, what function(s) do they perform?

Answering these questions has proven to be a formidable task. We must first define what we mean by essential. *An element is essential if*

1. it is required for normal growth and reproduction,
2. no other element can replace it and correct the deficiency, and
3. it has a direct or indirect action in plant metabolism.

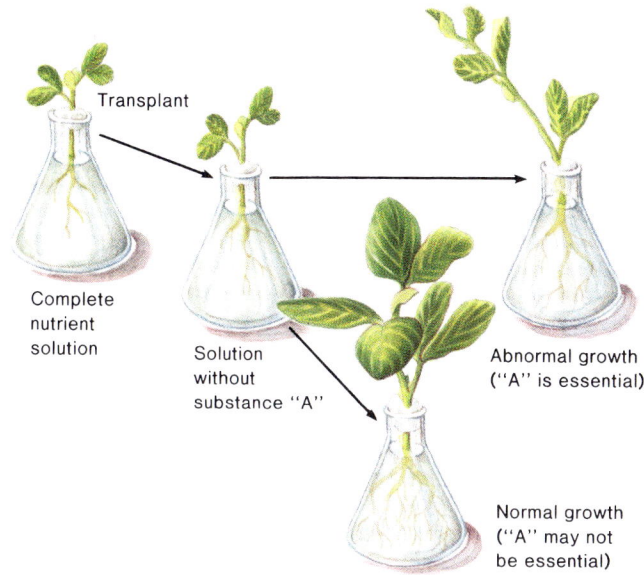

FIGURE 20.4

Discovering essential plant nutrients. Plant physiologists transplant a seedling to a solution lacking only one of the ingredients thought to be essential for growth, substance "A" in this example. If the plant grows and reproduces normally after being transplanted, the missing ingredient is assumed to be nonessential.

Thus, the mere presence of an element in a plant does not necessarily mean that it is essential. Armed with this definition, botanists began studying the ninety-two naturally occurring elements to determine how each might influence plant growth. Their approach, at least in theory, was simple: an element would be deemed essential only if a plant could not grow and reproduce without it (fig. 20.4).

Many of the early experiments to determine the essentiality of an element involved *hydroponics*, a technique for growing plants without soil in a chemically defined liquid medium (fig. 20.5). By the middle of the 1800s, these experiments had demonstrated that plants require at least nine elements: carbon, hydrogen, oxygen, phosphorus, potassium, nitrogen, sulfur, calcium, and magnesium. Since these elements are required in relatively large amounts (i.e., usually more than 0.5% of the dry weight of the plant), they became known as **macronutrients** (fig. 20.6).

Subsequent studies of the essentiality of mineral nutrients encountered several problems, perhaps the largest of which was contamination. For example, "pure" water and chemicals used to prepare nutrient solutions often contained enough impurities to satisfy plants' requirements for several elements. Furthermore, the growth containers used in hydroponics experiments were often made of glass that contained large amounts of boron. When these containers were filled with nutrient solutions, enough boron dissolved out of the glass to satisfy the plant's

FIGURE 20.5

Hydroponic farming. In this apparatus, a chemically defined solution of nutrients flows over the roots of lettuce.

requirement for boron. Such impurities, combined with insensitive methods for detecting many elements, made it impossible to determine the essentiality of many elements.

Subsequent progress in analytical chemistry has improved the purity of chemicals and increased the detectability of specific elements, thereby allowing botanists to discover seven other essential elements: iron, chlorine, copper, manganese, zinc, molybdenum, and boron. Because these nutrients are required in relatively small amounts (i.e., usually only a few parts per million), they are called **micronutrients**, or trace elements (fig. 20.6). Although required in smaller amounts than macronutrients, micronutrients are equally essential for growth. The 4 metric tons of carbon, hydrogen, and oxygen were no more important to our Idaho potato-farmer's crop than were its 2.5 grams of copper.

C O N C E P T

Plants require at least sixteen essential elements. Carbon, hydrogen, oxygen, phosphorus, potassium, nitrogen, sulfur, calcium, and magnesium are required in relatively large amounts and are called macronutrients. Iron, chlorine, copper, manganese, zinc, boron, and molybdenum are required in relatively small amounts and are called micronutrients.

There is no reason to think that the sixteen essential elements listed above are the only ones that plants require. Future improvements in analytical technology may reveal other essential elements. For now, we can only say that plants require *at least* sixteen elements.

Functions of Essential Elements

The functions of the sixteen elements essential for plant growth are summarized in table 20.2. A quick survey of this table leads to three important generalizations:

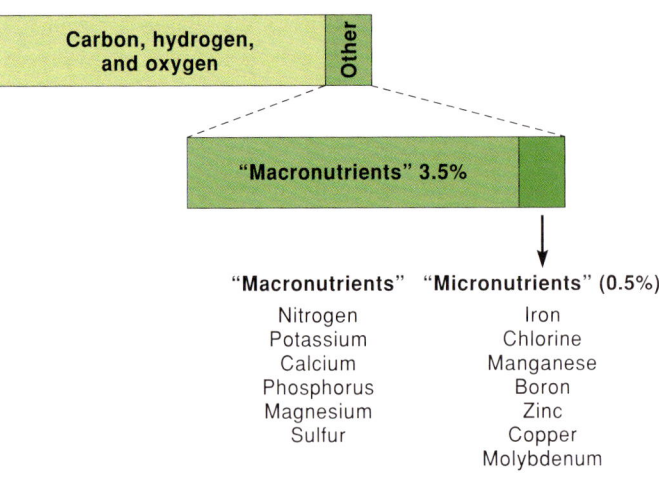

FIGURE 20.6

The proportional weights of various elements in plants. Macronutrients and micronutrients together total only 4% of the total weight of the plant, but they are absolutely essential to the plant's life and growth.

1. **Plants require different amounts of different elements.** Most plants require about 60 million times more hydrogen atoms than molybdenum atoms. These differing requirements reflect the differing uses of these elements: hydrogen is in almost all compounds in plants, while molybdenum occurs in only a few.

2. **Different elements are absorbed in different forms.** Calcium and iron are absorbed as cations, while phosphorus and sulfur are absorbed as anions.

3. **Most elements have several functions.** Potassium is involved in starch synthesis, affects protein conformation, and activates enzymes. Interestingly, the functions of most micronutrients in plants are similar to those in animals.

Although the functions of essential elements are diverse, they can be grouped into four general categories:

1. **Essential elements can be parts of structural units.** Carbon, hydrogen, and oxygen make up carbohydrates such as cellulose. Similarly, nitrogen is an integral part of proteins.

2. **Essential elements can be parts of compounds involved in metabolism.** Magnesium is part of chlorophyll, and phosphorus is part of ATP and nucleic acids.

3. **Essential elements can activate or inhibit enzymes.** Magnesium stimulates several respiratory enzymes, while calcium inhibits several enzymes. In some cases, these enzymes may be those responsible for synthesizing plant hormones.

4. **Essential elements can alter the osmotic potential of a cell.** For example, the movement of potassium into and out of guard cells is responsible for osmotic changes that cause stomata to open and close.

FIGURE 20.7

The prince's plume (*Stanleya elata*), shown here growing in Death Valley, California, requires selenium. Its selenium level is so high that this plant is toxic to browsing mammals.

CONCEPT

Essential elements are absorbed in differing amounts and forms. Once absorbed, these elements serve as parts of structural units and metabolites, activate and inhibit enzymes, and influence the osmotic potential of a cell.

The sixteen elements listed in table 20.2 are essential for all plants studied so far. Some other elements appear to be essential for only a few plants. For example,

Silicon (Si) is essential for some grasses and horsetails. Silica-containing cells comprise up to 2%–16% of the dry weight of these kinds of plants.

Sodium (Na) is essential for CAM plants, C_4 plants, and several *halophytes*, which are plants that grow in salty soils.

Nickel (Ni) may be required by soybeans, and **selenium** (Se) is required by some halophytes such as *Stanleya* (fig. 20.7).

A.

B.

C.

FIGURE 20.8

Responses of different cultivated varieties to mineral deficiencies. (a) zinc deficiency; (b) iron deficiency; (c) manganese toxicity.

These observations indicate that the essentiality of an element depends on the genotype of a plant. Similarly, different cultivars of some crop plants have differing sensitivities to mineral deficiencies (fig. 20.8). Identifying and incorporating genes for this tolerance to mineral deficiency into other crops is currently a major area of research.

CHAPTER TWENTY *Soils and Plant Nutrition* 471

TABLE 20.2

Some Functions and Deficiency Symptoms of Essential Elements

Essential Elements	Chemical Symbol	Atomic Weight	Form Available to Plants*	Percentage of Dry Tissue	Number of Atoms Relative to Molybdenum
Micronutrients					
Molybdenum	Mo	95.95	MoO_4^{2-}	0.00001	1
Copper	Cu	63.54	Cu^+, $\mathbf{Cu^{2+}}$	0.0006	100
Zinc	Zn	65.38	Zn^{2+}	0.002	300
Manganese	Mn	54.94	Mn^{2+}	0.005	1,000
Boron	B	10.83	H_3BO_3	0.002	2,000
Iron	Fe	55.85	Fe^{3+}, $\mathbf{Fe^{2+}}$	0.01	2,000
Chlorine	Cl	35.46	Cl^-	0.01	3,000
Macronutrients					
Sulfur	S	32.07	SO_4^{2-}	0.1	30,000
Phosphorus	P	30.98	$\mathbf{H_2PO_4^-}$, HPO_4^{2-}	0.2	60,000
Magnesium	Mg	24.32	Mg^{2+}	0.2	80,000
Calcium	Ca	40.08	Ca^{2+}	0.5	125,000
Potassium	K	39.10	K^+	1.0	250,000
Nitrogen	N	14.01	$\mathbf{NO_3^-}$, NH_4^+	1.5	1,000,000
Oxygen	O	16.00	O_2, H_2O	45	30,000,000
Carbon	C	12.01	CO_2	45	35,000,000
Hydrogen	H	1.01	H_2O	6	60,000,000

*Boldface indicates the most common of two forms available to plants.

Functions	Deficiency Symptoms (see figs. 20.8, 20.9, 20.12)
Part of nitrate reductase Essential for N_2 fixation	Chlorosis or twisting of young leaves
Component of plastocyanin, a plastid pigment Present in lignin of xylary elements Activates enzymes	Young leaves dark green, twisted, and wilted; tips of roots and shoots remain alive Rarely deficient
Necessary for formation of viable pollen Involved in auxin synthesis Maintenance of ribosome structure Activates enzymes	Chlorosis, smaller leaves, reduced internodes; distorted leaf margins; older leaves most affected
Photosynthetic O_2 evolution Enzyme activator Electron transfers	Interveinal chlorosis; appears first on older leaves; necrosis common; disorganization of lamellar membranes
Essential for growth of pollen tubes Regulation of enzyme function Possible role in sugar transport	Death of apical meristems; leaves twisted and pale at base; swollen discolored root tips, young tissue most affected
Required for synthesis of chlorophyll Component of cytochromes and ferredoxin Cofactor of peroxidase and some other enzymes	Interveinal chlorosis; short and slender stems; buds remain alive; affects young leaves first
Activates photosynthetic elements Functions in water balance	Wilted leaves, chlorosis, necrosis, stunted, thickened roots
Part of coenzyme A and the amino acids cysteine and methionine Can be absorbed through stomata as gaseous SO_2	Interveinal chlorosis; usually no necrosis; affects young leaves Rarely deficient
Part of nucleic acids, sugar phosphates, and ATP Component of phospholipids of membranes Coenzymes	Stunted growth, dark green pigmentation; accumulation of anthocyanin pigments, delayed maturity; affects entire plant Second to N, P is element most likely to be deficient
Part of chlorophyll Enzyme activator Protein synthesis	Chlorosis and reddening of leaves; leaf tips turn upward; older leaves most affected
Membrane integrity In middle lamella Functions as "second messenger" to coordinate plant's responses to many environmental stimuli Reversibly binds with calmodulin, which activates many enzymes	Death of root and shoot tips; young leaves and shoots most affected
Regulates osmotic pressure of guard cells, thereby controlling opening and closing of stomates Activates more than 60 enzymes Necessary for starch formation	Chlorosis and necrosis, weak stems and roots; roots more susceptible to disease; older leaves most affected
Part of nucleic acids, chlorophyll, amino acids, protein, nucleotides, and coenzymes	General chlorosis, stunted growth; purplish coloration due to accumulation of anthocyanin pigments N is element most likely deficient in soil
Major component of plant's organic compounds	Rarely limiting enough as nutrient to cause specific symptoms
Major component of plant's organic compounds	Rarely limiting enough as nutrient to cause specific symptoms
Major component of plant's organic compounds	Rarely limiting enough as nutrient to cause specific symptoms

Deficiencies of Essential Elements

Plants require large amounts of nitrogen, phosphorus, and potassium. As a result, these are the elements most likely to be deficient in soil. Remedying deficiencies of these elements often requires adding large amounts of fertilizer. Deficiencies of micronutrients are corrected more easily; for example, molybdenum-deficient soil can be replenished by annually adding less than 4 grams of molybdenum to a hectare of land, and deficiencies of zinc and iron have even been corrected by driving zinc-coated tacks or iron nails into the trunks of trees. The deficiency symptoms of essential elements are listed in table 20.2 and are shown in figures 20.8, 20.9, and 20.12.

Beneficial Elements

Several elements in plants are not essential, but will promote growth. For example, cobalt stimulates growth of legumes such as beans, clover, and alfalfa. This stimulation of growth by cobalt is due to its use, not by the plant itself, but by nitrogen-fixing bacteria that live in roots of plants. In these plants, as little as 1 µg of cobalt per liter of nutrient solution stimulates growth dramatically (fig. 20.10).

Other Elements in Plant Tissues

Many plants absorb elements for which they have no apparent use. For example:

Lead (Pb) has been until recently an antiknock component of most gasolines and is abundant along highways. Plants growing near these highways often absorb this lead, as do some grasses that grow near abandoned lead mines.

Cadmium (Cd), a contaminant in many chemical fertilizers, is common in plants growing in fertilized soils.

Gold (Au) often occurs in plants such as *Phacelia sericea* that grow near gold mines. The gold content of these plants has been used by geologists to locate deposits of gold in soil.

Strontium (Sr) is sometimes present in plants and will partially substitute for calcium. As such, radioactive strontium (^{90}Sr) is abundant in plants growing near nuclear test sites and sites of nuclear accidents such as Chernobyl.

Finally, plants called **indicator plants** grow only in soils containing large amounts of a particular element. For example, copper moss (*Merceya latifolia*) grows only in copper-enriched soils, and *Dicoma niccolifera*, a member of the sunflower family, is often abundant in soils rich in chromium and nickel. Interestingly, many indicator plants accumulate large amounts of nonessential elements. For example, *Pearsonia metallifera*, from the pea family, accumulates nickel, and selenium may account for more than 1%

FIGURE 20.9

Nutrient deficiencies produce distinctive symptoms. (a) Potassium deficiency causes leaves to curl at their edges. (b) Phosphorus deficiency leads to purple leaves in seedlings. (c) Copper-deficient plants have blue-green, curled leaves. (d) Zinc-deficient plants have small, necrotic leaves (right). (e) Leaf yellowing (chlorosis) is due to magnesium deficiency which prevents chlorophyll synthesis (recall that chlorophyll contains magnesium). (f) A deficiency of nitrogen inhibits growth.

of the dry weight of locoweed (*Astragalus*). If eaten by livestock, the large amounts of selenium in these plants affect their nervous systems, causing them to stagger about as if intoxicated. This illness, called "*blind staggers*," often kills livestock and has even caused ranchers to abandon their ranches. Similarly, bees that harvest pollen from locoweed often produce honey containing large amounts of selenium. Locoweed uses selenium to help combat the toxic levels of phosphorus characteristic of selenium-rich soils: the presence of selenium decreases the absorption of phosphorus, thereby preventing phosphorus toxicity.

Obtaining Essential Elements

Obtaining enough essential elements is an ongoing problem for most plants. Many of the tropic and nastic movements discussed in the previous chapter help solve this problem. For example, positive gravitropism in roots increases their chances of encountering dissolved minerals in soil, and hydrotropism helps ensure that roots encounter sufficient water for growth. However, plants' adaptations for absorbing nutrients are not limited to these growth responses. Rather, plant nutrition is an elaborately orchestrated process that depends on a variety of factors ranging from the architecture of plant cells to the assistance of other organisms. Plant nutrition is based not only on the presence of nutrients, but also on their *availability*.

Factors Determining the Availability of Essential Elements

The availability of an essential element is influenced by factors such as the gradual breakdown of rocks into mineral particles, the decay of humus, and the cation-exchange capacity of the soil. Four other factors that determine nutrient availability are soil pH, chelators, other elements, and the presence of other organisms.

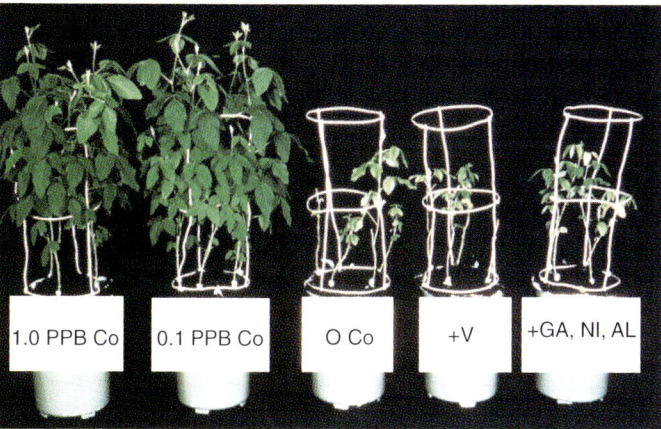

FIGURE 20.10

Cobalt is required by nitrogen-fixing bacteria in root nodules. When cobalt is available (two plants on left), bacteria fix nitrogen and stimulate growth in soybeans; when cobalt is unavailable (three plants on right), bacteria cannot fix nitrogen, so soybean growth is not stimulated.

Soil pH

Neutral and acidic soils. Most soils have a pH between 6 and 7, and thus can support almost any plant. However, cation exchange, decomposition of humus, and acidic precipitation can decrease the pH of soil to as low as 3. The pH of the soil affects the availability of many plant nutrients (fig. 20.11). For example, decreasing the pH of soil increases the availability of iron, manganese, and aluminum, often to toxic levels. Furthermore, excess amounts of these elements react with phosphate to produce insoluble precipitates. Since these precipitates are unavailable to plants, phosphorus deficiencies are common in acidic soils. Cranberry, blueberry and other members of the heath family (*Ericaceae*) are some of the plants specialized for growth in highly acidic soils.

The pH of acidic soil can be corrected by adding limestone ($CaCO_3$) or lime (CaO). This process is called *liming* and removes H^+ in the following way.

$CaCO_3$ dissociates in soil water into calcium and carbonate ions:

$$CaCO_3 \rightarrow Ca^{2+} + CO_3^{2-}$$

The calcium ions exchange for H^+ on clay micelles, thereby forming H_2CO_3 and, in the process, removing H^+:

$$\text{micelle}:2H^+ + Ca^{2+} + CO_3^{2-}$$
$$\downarrow$$
$$\text{micelle}:Ca^{2+} + H_2CO_3$$

Although liming adds calcium to soil, its primary purpose is usually to neutralize acidic soils.

Alkaline soils. Alkaline soils having a pH above 8 usually form from limestone. Alkalinity decreases the availability of nutrients. For example, the excess amounts of Ca^{2+} in alkaline soils react with phosphate to produce insoluble $Ca_3(PO_4)_2$, which is unavailable to plants. The availability of iron and manganese

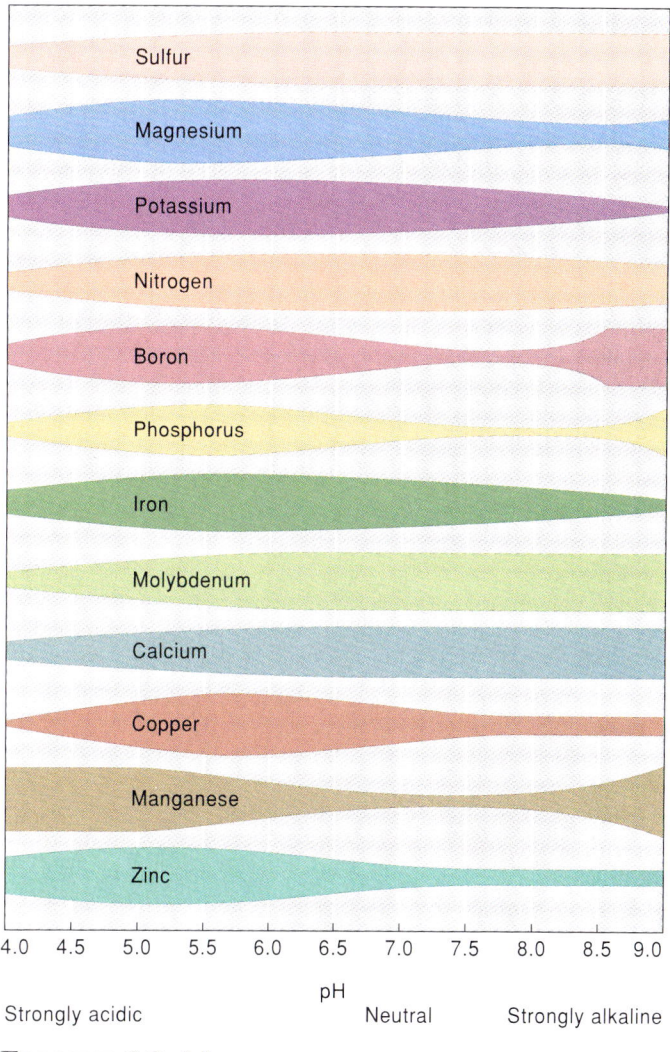

FIGURE 20.11

How the pH of soil affects the availability of nutrients. The width of the shaded areas corresponds to nutrient availability for roots.

also decreases in alkaline soil, primarily because they form insoluble precipitates. The presence of iron oxides accounts for the rust color of many alkaline soils.

Only a few plants, such as alfalfa and sweet clover, grow well in alkaline soils. The pH of alkaline soils can be neutralized by adding sulfur, which bacteria readily oxidize to sulfate. This sulfate then dissolves in water to form sulfuric acid, an acid that decreases the pH of the soil.

Chelators

A **chelator** is a compound that donates electrons to a metal cation such as Fe^{3+} and thereby increases its solubility. For example, versene is a chelator that maintains iron in soil as Fe^{2+}, which is more soluble than Fe^{3+} (fig. 20.12). Many organisms help their own cause by secreting chelators into the soil. By doing so, they increase the availability of many elements.

FIGURE 20.12

(a) Plants grown without iron chelator show symptoms of iron deficiency: veins are green but the rest of the leaf is yellow. (b) Plants grown with iron chelator; leaves are normal.

Other Elements

The requirement of a plant for a particular element often depends on the presence of other elements. For example, sodium can substitute for part of a plant's potassium requirement. Furthermore, a particular nutrient can affect different plants in different ways: the same amount of iron in soil that supports normal growth of alfalfa (*Medicago sativa*) will cause chlorosis in garden bean (*Phaseolus vulgaris*).

Other Organisms

The amounts of nutrients available to plants depend largely on the presence and activities of other organisms. For example, other organisms in soil compete with plants for nutrients, and thereby decrease the amount of nutrients available to plants. Meanwhile, decomposers such as fungi degrade humus and thereby increase the supply of nutrients available to plants.

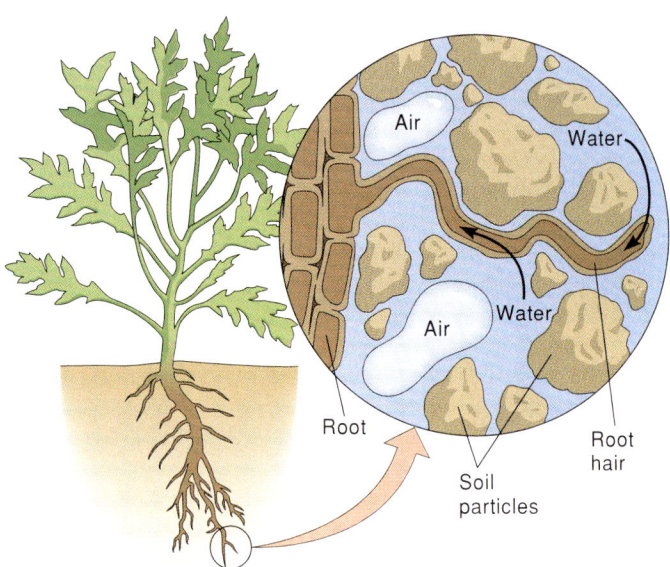

FIGURE 20.13

Root hairs absorb water and nutrients from the soil.

C O N C E P T

The availability of essential elements is influenced by mineral particles, cation exchange, decay of humus, pH, other elements, and other organisms.

The preceding discussion should convince you that nutrient availability is a complex phenomenon. Yet the mere availability of nutrients only begins the story, for to be useful to plants, these nutrients must also be located and absorbed.

Locating Essential Elements

Locating most of the essential elements is not a problem for plants. For example, CO_2 is always present in the air, and water is usually abundant enough to support at least minimal growth. As a result, deficiencies of carbon, hydrogen, and oxygen rarely occur in plants. However, nitrogen in soils almost always limits plant growth, and concentrations of some micronutrients may be undetectable. Thus, growth and development largely depend on a plant's ability to locate these nutrients. One way that plants locate essential elements is by efficiently exploring the soil. For example, a mature oak tree can have several million root tips distributed throughout the soil. Each of these root tips has thousands of root hairs, each of which mines a different area for water and dissolved minerals (fig. 20.13). As a consequence of this extensive collecting system, few available sources of nutrients go untapped by plants.

Absorbing Essential Elements

Essential elements are useless to plants until they are absorbed. Absorption of nutrients depends on (1) exposing a large surface for absorption, and (2) having a mechanism for selectively absorbing and excluding elements.

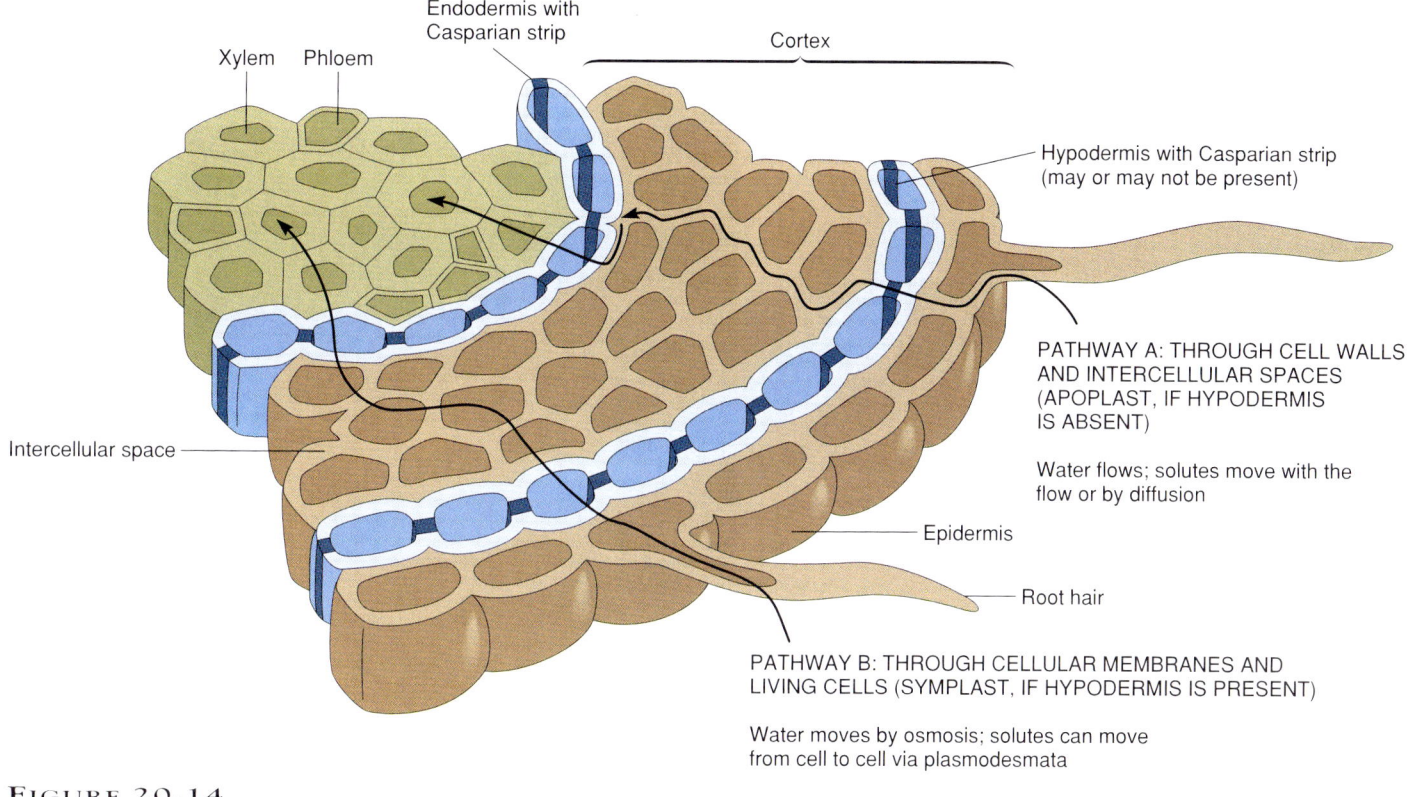

FIGURE 20.14

Apoplastic vs. symplastic pathways. Casparian strips in cell walls of the endodermis block off the apoplastic pathway all around the vascular core of the root. Water and dissolved minerals enter the vascular tissue via the symplastic pathway, ×100.

Exposing a Large Surface Area for Absorption

Roots and root hairs not only permeate the soil, but also greatly increase the surface area available for absorbing nutrients. For example, roots of a four-month-old rye plant have a surface area exceeding 600 m^2, and the 150,000 root hairs in only one cubic centimeter of soil beneath a patch of bluegrass have a surface area equal to approximately two-thirds of this page. Exposing this much surface area to the environment greatly increases the absorptive capabilities of plants.

The structural design of plants also increases the surface area available for absorbing nutrients. To understand this, re-examine the root cross section shown in figure 20.14. When water and dissolved minerals move through the apoplast, they remain outside of the cell membrane. As a result of this apoplastic movement, the cell membranes of cortical cells become absorptive surfaces equivalent to those of root hairs. Thus, the absorptive surface area of most roots is not restricted to the epidermis and its root hairs; it also includes all of the cell membranes of the cortex. Cells of the cortex tremendously increase the absorptive surface area, and thereby improve the nutrient uptake by a plant.

Absorption

The concentrations of elements in plants differ from those in soil. For example, aluminum is usually abundant in soil but rare in plant tissues. Conversely, most micronutrients are usually more concentrated in plants than in soil. Therefore, plants must *selectively* absorb elements from the soil. This selectivity depends on the presence of **ion-specific carriers** in the membranes of plant cells. For example, the membranes of cortical cells contain a K$^+$-specific carrier that requires Mg^{2+} to function. Such carriers enable a plant to absorb and maintain adequate amounts of various essential elements in their tissues, even though the elements may be more or less concentrated in the soil. Absorption of these elements requires energy; it is not clear if this energy comes directly from ATP or if it is conserved in an ATP-generated proton gradient.

C O N C E P T

All essential elements except carbon are absorbed by roots. This absorption is aided by their large surface area and the presence of ion-specific carriers in the cell membrane.

The Fate of Absorbed Nutrients

Once absorbed into the cytoplasm, a mineral can be used immediately or transported to a different part of the plant. As you learned in Chapter 15, transport from the cortex requires that the nutrient move through the endodermis and pericycle before being loaded into the xylem for transport. This transport is critical, because different parts of the plant require different amounts of essential elements. For example, nitrogen, phosphorus, and potassium are mostly used in the leaves.

Modifying the Form of Essential Elements

Most nutrients are used in the same form in which they are absorbed. For example, iron is incorporated directly into cytochromes, and phosphates are added directly to ADP to form ATP. Other elements, most notably nitrate and sulfate, are reduced before use.

Nitrate. Most nitrogen in fertilizer is in the form of ammonium salts. Ideally, this would ensure high availability of nitrogen to plants, since the NH_4^+ cation should bind to negatively charged clay micelles. However, NH_4^+ in soil is rapidly converted to NO_3^- by bacteria such as *Nitrosomonas* in a process called **nitrification** (fig. 20.15). As a result, most nitrogen in soil is in nitrate. Because nitrate is negatively charged, it does not bind to soil and is readily leached from the soil by rainfall. This is why nitrogen must be replenished by adding fertilizer.

Plants must reduce NO_3^- to ammonium before it can be incorporated into proteins. This conversion is catalyzed by a group of enzymes collectively called the **nitrate reductase system** (fig. 20.15). These enzymes produce amino acids such as glutamine, which are transported throughout the plant. This reaction is important because it provides a basic supply of nitrogen-containing compounds that can be used for protein synthesis.

Plants grown hydroponically in the absence of NO_3^- contain negligible levels of the nitrate reductase system. If given NH_4^+, plants rapidly synthesize the nitrate reductase system; that is, the nitrate reductase system is inducible. Although common in bacteria, inducible enzymes such as nitrate reductase are rare in plants.

Sulfate. Sulfur is absorbed almost exclusively as sulfate (SO_4^{2-}) and must therefore be reduced before being incorporated into proteins. The enzyme that reduces SO_4^{2-} to S^{2-} is ATP sulfurylase. Almost 90% of this reduced sulfur is incorporated into the amino acids cysteine and methionine.

CONCEPT

Many essential elements are used in the form in which they are absorbed. Other nutrients, most notably nitrate and sulfate, are reduced before use.

Writing to Learn Botany

Crop plants absorb only about half of the nitrogen supplied in fertilizer. What happens to the rest? What are the consequences of this?

Mobility of Essential Elements within Plants

Utilization of an essential element often affects its movement in plants. For example, boron, iron, and calcium are *immobile elements*: they do not move once they are used in the shoot. As a result, deficiency symptoms of immobile elements always ap-

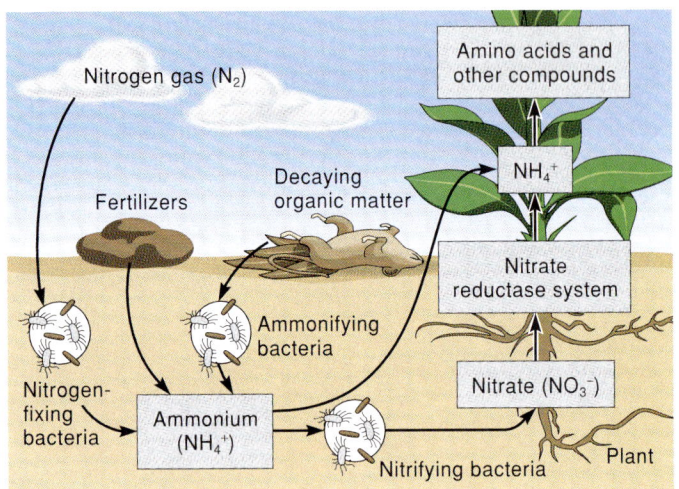

FIGURE 20.15

How plants get nitrogen from the soil.

pear first at the shoot tip and in the youngest leaves. Conversely, potassium, nitrogen, phosphorus, magnesium, and sulfur can be transported after being used in the shoot. These elements are called *mobile elements*, and their deficiency symptoms appear first in older leaves.

Associations with Other Organisms

Despite their structural and functional adaptations for obtaining nutrients, plants are not as independent as many people think they are. Indeed, their mineral nutrition often depends largely on associations with other organisms. These associations range from mutually beneficial symbioses to parasitisms, whereby plants rely on other organisms as nutrient collectors as well as prey.

Fungi

When plants invaded the land some 400 million years ago, one of the biggest problems they faced was obtaining enough nutrients. Their aquatic habitat contained many decomposers, especially fungi, that degraded large molecules, thereby increasing the availability of nutrients. As a result, aquatic organisms were literally bathed in nutrients. On land, however, the situation was drastically different: most soils contained few fungi and therefore did not contain enough nutrients to sustain plant growth.

Plants adapted to life on land by forming mutually beneficial associations called **mycorrhizae** between their roots and fungi (fig. 20.16). In effect, plants took the fungi with them onto the land, and thereby helped ensure the continued availability of nutrients for their growth. Plants benefited from the ability of the fungus to efficiently absorb and make available nutrients from the environment. The fungus also benefited, since the plant provided it with carbohydrates for growth.

Mycorrhizae occur in more than 80% of all plants. From a functional point of view, they replace root hairs; that is, mycorrhizae increase nutrient absorption. This is especially true for

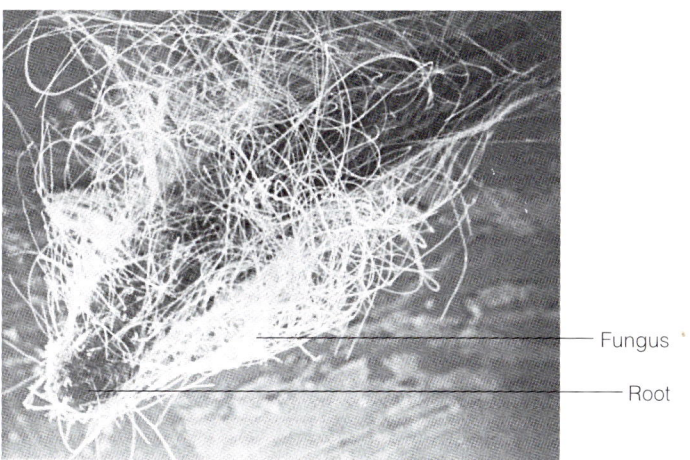

FIGURE 20.16

Scanning electron micrograph of a mycorrhizae: root of a lodgepole pine (*Pinus contorta*) associated with filaments of the fungus *Cenococcum graniforme*.

FIGURE 20.17

This comparative experiment shows the dependency of citrus (rough lemon) on mycorrhizae for growth. Plants with mycorrhizae (MYCO) grow better whether or not they are supplemented with nutrient solutions.

the absorption of phosphorus: roots with mycorrhizae absorb as much as four times more phosphorus than do roots that lack them. Two factors contributing to this efficient absorption are that (1) the fungal hyphae have a much greater surface area than roots, and (2) the fungal hyphae permeate a greater soil volume than do the roots. As a result of this increased absorption, mycorrhizae often dramatically increase plant growth; for example, they increase the growth of wheat by more than 200%, corn by 100%, and onions by more than 3,000%. Mycorrhizae also improve a plant's endurance of drought, disease, extreme temperatures, and high salinity, and are important for cycling nutrients, reclaiming strip-mined land, and establishing nursery stocks.

Mycorrhizae are especially important for plants growing in nutrient-poor soils. For example, "pioneer" plants such as alder that grow on nutrient-deficient soil invariably have mycorrhizae, as did the first plants that invaded the land. Today, many plants need and even *require* mycorrhizae for vigorous growth. For example, orchid seeds germinate only in the presence of a mycorrhizal fungus, and many citrus trees and gymnosperms are difficult to grow unless mycorrhizae are present (fig. 20.17).

CONCEPT

Mycorrhizae are mutually beneficial associations between roots and fungi. These fungi increase the absorption of nutrients and tolerance of stress.

Bacteria

Nitrogen is usually deficient in soils; this deficiency often limits the growth of plants and, as a consequence, animals. The deficiency of nitrogen in soil is somewhat paradoxical, since the atmosphere is almost 80% N_2 (N≡N). The triple bond holding the two atoms of gaseous nitrogen together makes N_2 stable and therefore unusable by most organisms. However, a few bacteria contain an enzyme complex called **nitrogenase** that can reduce atmospheric N_2 to ammonium (fig. 20.18). These bacteria live alone as well as symbiotically with plants.

Free-living nitrogen fixers. Many bacteria capable of fixing atmospheric nitrogen are free living; that is, they do not form any associations with other organisms. These prokaryotes, which include heterotrophic bacteria (e.g., *Azotobacter*), photosynthetic bacteria (e.g., *Heliobacillus*), and cyanobacteria (e.g., *Nostoc*), fix only the nitrogen that they need. On average, this amounts to about 7 kg of nitrogen per hectare per year, which is roughly equivalent to the amount of nitrogen added to soil by precipitation containing atmospheric pollutants. When these bacteria die, their nitrogen becomes available to plants.

Symbiotic nitrogen fixers. Some plants have established symbioses with nitrogen-fixing bacteria, and therefore can obtain much of their nitrogen from the atmosphere. For example, the roots of many legumes such as beans and alfalfa are often infected with *Rhizobium* bacteria (figs. 20.19; 15.20a). Unlike free living bacteria, *Rhizobium* fixes almost ten times more nitrogen than it uses. The excess nitrogen is released into the host plant, where it is used for growth. Interestingly, the presence of NO_3^- decreases infection of the host by *Rhizobium*, the growth of nodules, and the activity of nitrogenase, thereby decreasing nitrogen fixation. Consequently, applying nitrogen-containing fertilizer to legumes seldom increases yield.

Nodules containing nitrogen-fixing bacteria are not restricted to roots. For example, nodules are common on leaves of members of certain families, such as the Myrsinaceae and Rubiaceae, and removing them causes deformation, dwarfing, and chlorosis. However, nitrogen may not be the most important contribution of nodules to these plants, since applying nitrogen does not always relieve these symptoms.

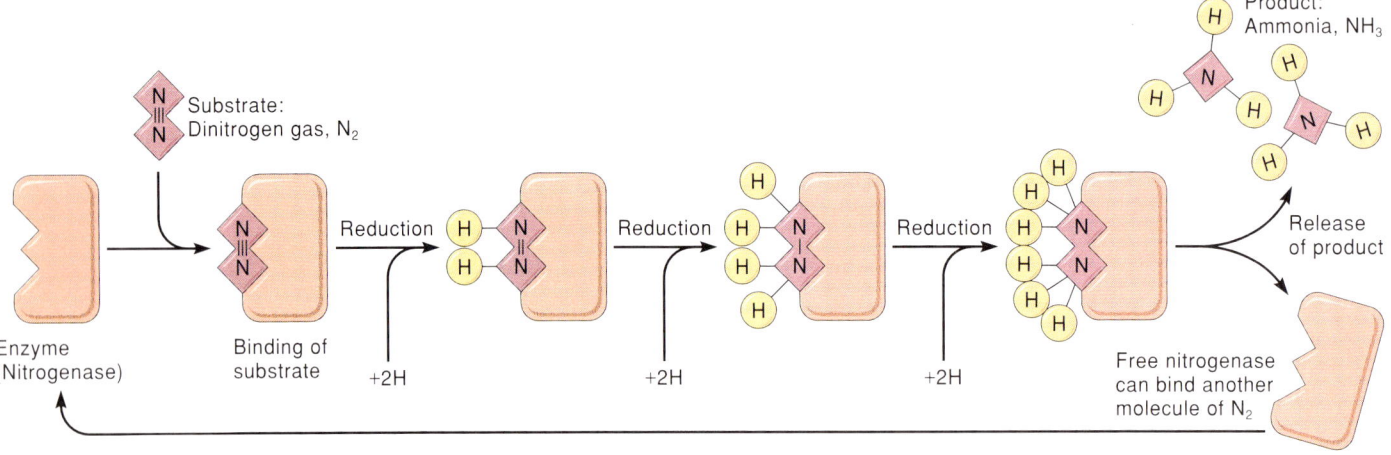

FIGURE 20.18

Nitrogenase is a bacterial enzyme that converts atmospheric nitrogen (N_2) to ammonia (NH_3).

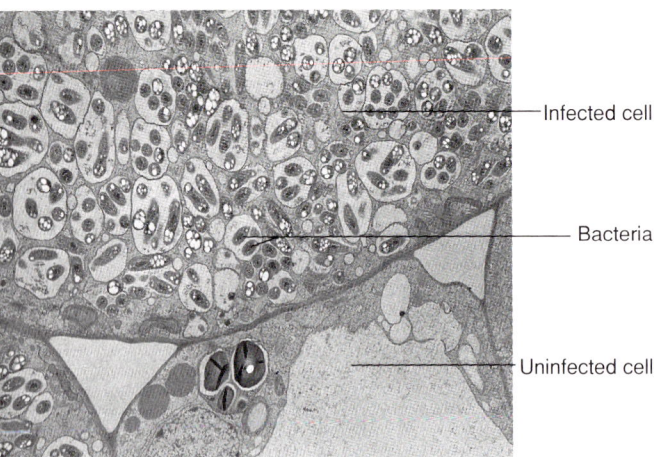

FIGURE 20.19

(a) Root nodules that were induced by *Rhizobium* bacteria.
(b) Infected cells of nodules contain nitrogen-fixing bacteria, ×11,500.

Nitrogen-fixing prokaryotes also form symbioses with plants other than legumes. For example, alders (*Alnus*) and mountain lilac (*Ceanothus*) contain *Frankia*, a nitrogen-fixing bacterium, and nitrogen-fixing bacteria form symbioses with such diverse plants as cycads (*Macrozamia*), mosses (*Sphagnum*), liverworts (*Blasia*), and tropical trees (*Parasponia*). These associations, like those with legumes, often have significant agricultural implications. For example, the cyanobacterium *Anabaena* grows symbiotically in leaf cavities of water fern (*Azolla*) (fig. 20.20). When infected *Azolla* are grown in rice paddies, some of the nitrogen fixed by *Anabaena* dissolves in the water of the paddy and becomes available to rice plants. As a result, rice crops in China and southeast Asia can be grown in nitrogen-deficient soils. Botanists are now searching for new strains of *Anabaena* to increase rice production.

CONCEPT

Many plants, especially legumes, establish mutually beneficial symbioses with nitrogen-fixing bacteria. The bacteria reduce atmospheric N_2 to ammonium, which the host uses for growth.

The Lore of Plants

Although it often seems that bacteria are the aggressors, nodule formation is an example of plants making the first step in controlling the metabolism of another organism. Roots of alfalfa and other legumes secrete flavonoids that bind to and activate a bacterial nodulation gene called nodD. The *nodD* gene product activates other *nod* genes, whose products then turn on plant genes that control the growth and function of the nodule.

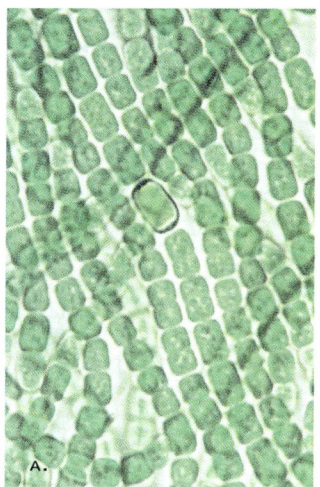

FIGURE 20.20

Anabaena and rice production. (a) Cultivation of rice is aided by nitrogen fixation by the cyanobacterium *Anabaena azollae*, ×400. (b) *Anabaena* lives and fixes nitrogen in cavities of the water fern, *Azolla*. (c) *Azolla* is grown in rice paddies of the warmer parts of Asia. Later, the paddies are drained and the *Azolla* is plowed into the soil, thereby enriching it with nitrogen for the rice crop. Nitrogen fertilizers do not have to be used in such paddies.

Not all symbioses between plants and nitrogen-fixing bacteria involve nodules. For example, you learned in Chapter 15 that roots secrete mucilage that helps roots move through soil. Mucilage also houses nitrogen-fixing bacteria such as *Spirillum*, *Azospirillum*, and *Klebsiella*. These bacteria, which are common on grasses such as corn and sorghum, fix nitrogen that can be absorbed by the host plant. In return, the bacteria receive sugars, amino acids, and proteins that the plants secrete into their mucilage.

Although each of these symbioses between nitrogen-fixing bacteria and plants evolved independently, they all have a common function: they make nitrogen from the atmosphere available to plants. This recycling of nitrogen by bacteria is an important aspect of how plants obtain nitrogen (fig. 20.15). For example, factors such as harvesting crops remove more than 21 million metric tons of nitrogen per year from the soil. Nitrogen fixation by bacteria replenishes almost two-thirds of this loss. The remaining nitrogen is resupplied to the soil in fertilizers.

Nitrogen fixation has tremendous agricultural implications. For example, much of the nitrogen fixed in nodules is released into the soil. Many farmers exploit this by practicing *crop rotation*, by which leguminous crops are alternated with nonlegumes as a means of maintaining the nitrogen in the soil. Plowing under a leguminous crop has even more dramatic effects. For example, plowing under a crop of alfalfa adds humus and more than 300 kg of nitrogen per hectare—more than forty-five times that contributed by free-living bacteria. Furthermore, nitrogen fixation eliminates the need to apply expensive fertilizers (see box 20.1, "Putting Things Back"). Molecular biologists are busy trying to incorporate nitrogen fixation into other crop plants such as corn and wheat.

Carnivorous Plants

Biologists have long been fascinated with carnivorous plants. Indeed, the botanists who discovered carnivorous plants referred to them as "the great wonder of the vegetable world" and "carnivorous vegetables." Today, the bizarre life-styles of these plants remain a favorite subject of B-movies. Although unusual, the carnivorous nature of these plants serves an important purpose: it is an adaptation to the nutrient-poor histosols in which they grow. To obtain enough nutrients, especially nitrogen, carnivorous plants have evolved a life-style that resembles that of many heterotrophs—namely, extracting nutrients from the nitrogen-rich flesh of other organisms. The modifications for this life-style include sticky, flypaperlike surfaces, springlike trapdoors, and pitfalls leading to vats filled with digestive enzymes. You were introduced to sundew (figs. 20.21; 19.12) and Venus's-flytrap (figs. 20.22; 19.11), both of which utilize seismonasty to trap and digest their prey. However, the trapping mechanisms of carnivorous plants are not limited to these kinds of traps.

Passive traps involve no movement to trap prey. For example, leaves of pitcher plants, such as the trumpet pitcher (*Sarracenia*; fig. 20.23) and cobra lily (*Darlingtonia*; fig. 20.24), are modified into tubular vats. Insects drawn into these pitfalls can initially walk across the downward-pointing hairs lining the pitcher (fig. 20.23). However, about half-way down the pitcher, the hairs are replaced by a slick wax to which insects cannot adhere, and the unsuspecting prey slips into the digestive enzymes, where it is digested within a few hours or days. Pitchers of some carnivorous plants attain impressive sizes: pitchers of *Sarracenia* can hold more than a liter, while the lavatory-shaped pitchers of *Nepenthes rajah* can be large enough to hold a rugby football. Despite the large size of such traps, their prey are usually small insects.

FIGURE 20.21

A sundew (*Drosera aliciae*) leaf in the process of surrounding its prey.

FIGURE 20.22

The Venus's-flytrap (*Dionaea muscipula*). The silhouette of a trapped fly can be seen in the closed trap.

Genlisia has unique and complex traps. The entrances of these traps contain numerous clawlike hairs that allow an animal to enter but not to leave. As a result, prey eventually die in the trap and are digested. These passive traps are designed like lobster traps, or, as Charles Darwin put it, like "the eel trap, though more complex."

Active traps move to trap prey. Sundew and Venus's-flytrap have active traps, as does butterwort (*Pinguicula*)—sticky leaves of this plant often fold over trapped prey to speed digestion in what Charles Darwin called a "temporary stomach." A butterwort such as *P. grandiflora* grows a new "trap" (i.e., leaf) about every five days, so that more than 400 cm^2 of catching surface form in a single season. In Elizabethan times, *Pinguicula* supposedly protected cattle from the mischief of elf arrows and protected humans from the spells of witches. Despite such folklore and the claims of Hollywood movies, these plants seldom trap prey larger than houseflies.

One of the most ingenious traps belongs to bladderwort (*Utricularia*; fig. 20.25), an aquatic, carnivorous plant that supplements its diet with protozoa and mosquito larvae. The traps of *Utricularia* range in diameter from 0.25–5 mm and are elastic (hence the name bladderwort). These traps involve a trapdoor mechanism set off by trigger hairs. Here's how they work:

Cells lining the trap secrete ions, which causes water to leave the trap via osmosis.

This loss of water creates a negative pressure in the trap, which establishes a balanced tension that closes the trapdoor.

When an unsuspecting prey brushes against the trap's trigger hairs, the tension is abolished, and the door swings open.

Suction created by this release of negative pressure sweeps the victim into the trap.

Release of the negative pressure also causes the trapdoor to swing back to its normal closed position, leaving the victim a prisoner.

Soon thereafter, the trap secretes enzymes that kill and digest the prey.

Carnivorous plants seduce prey with a package of handsome advertisements that include bright colors and exotic scents, much as flowers attract their pollinators. For example, the traps of Venus's-flytrap and many pitcher plants produce attractants that lure animals to their demise. Like flowers, traps also lure prey with attractive patterns invisible to our eyes, but exciting to an insect. Furthermore, the ultraviolet patterns of many pitchers appear to an insect as a guide-map to the trap's interior, and regions near the trigger hairs of a Venus's-flytrap shine bright against the trap's dark margins. Some carnivorous plants, like

FIGURE 20.23

Pitcher plants (*Sarracenia purpurea*) have leaves modified into pitchers that contain digestive enzymes (a). The downward-pointing hairs (b) help lure insects to their doom in the trap. There, the insects are digested and their nutrients are absorbed.

FIGURE 20.24

Cobra lily (*Darlingtonia californica*) lures insects by sight and smell into its pitcher trap. Light shining through the "windows" atop the dome provides a false sense of escape.

flowers, may use ultraviolet patterns to select particular prey. Sundew, for example, has a penchant for a single species of springtail insects.

Once trapped, prey are digested by enzymes such as proteases secreted by specialized glands lining the trap. Absorption of nutrients from traps is similar to that of roots in soil. Traps create their own enriched "soil" (actually a liquid fertilizer) and then absorb from it the nutrients they need. This absorption is critical to survival: plants that trap prey are more vigorous and produce more flowers and seed than do plants that do not trap prey.

CONCEPT

Carnivorous plants usually grow in nitrogen-poor soils of bogs. These plants get most of their nitrogen by trapping and digesting animals. Their traps are elaborate and range from sticky surfaces to elaborate trapdoors.

Parasitic Plants

Not all associations that improve the mineral nutrition of plants are mutually beneficial. Rather, several plants are **parasites,**

FIGURE 20.25

A bladderwort (*Utricularia minor*) trapping a mosquito larva (×6).

meaning that they harm other plants as they obtain nutrients from them (figs. 20.26; 20.27). For example, cancer root (*Orobanche*) and mistletoe (*Phoradendron*) parasitize hardwood trees (fig. 20.26), while trees such as sandalwood (*Santalum*) obtain their nutrients from nearby grasses. Indian pipe (*Monotropa*; fig. 20.27b) obtains its nutrients from mycorrhizae of trees.

CHAPTER TWENTY *Soils and Plant Nutrition*

FIGURE 20.26

Mistletoe (*Phoradendron flavescens*), a parasitic plant, growing on an oak.

A.

B.

FIGURE 20.27

Parasitic plants. (a) Orange-brown twining stems of dodder (*Cuscuta* sp.) wrap around host; dodder is nonphotosynthetic but obtains water, sugars, and nutrients through tiny, rootlike protuberances that penetrate the host. (b) Indian pipe (*Monotropa uniflora*) also does not conduct photosynthesis: it gets its nutrients indirectly from other plants, by way of a fungal bridge in the soil.

Many parasitic plants lack chlorophyll, and therefore depend entirely on their host for nutrients. However, the presence of chlorophyll does not guarantee an independent life-style. For example, mistletoe and witchweed (*Striga*) are green, yet grow only as parasites. The green portions of these parasites contain only small amounts of ribulose bisphosphate carboxylase, and therefore are not photosynthetic (see box 15.1, "Finding a Host").

The link between parasites to their hosts is called a **haustorium.** In many plants, the link is one or more xylem-to-xylem connections between the two plants (fig. 20.28). Thus,

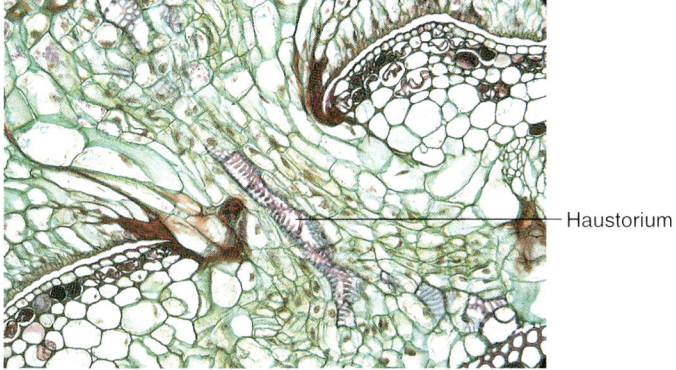

FIGURE 20.28

Haustorium of dodder (*Cuscuta campestris*) in the stem tissues of garden balsam (*Impatiens balsamina*). Note the vascular tissue in the haustorium, ×100.

the parasite depends largely on evaporation of water from its leaves as a means of pulling nutrient-containing water from the xylem of its host. To accomplish this, the stomata of many parasitic plants always remain at least partially open, even at night or during a mild drought. Evaporation of water from these stomata brings a continuous supply of nutrients from the host.

Parasitic plants obtain their nutrients directly from other plants. The parasite and host are linked by a special organ called a haustorium.

Gathering Nutrients in a Rain Forest

The transfer of nutrients from plant to plant without letting them enter the soil is not restricted to parasitic plants. Indeed, it is common in Amazon rain forests, where soils are notoriously deficient in most nutrients. The frequent deluges of rain in these forests leach large amounts of calcium, phosphorus, and magnesium from the forest canopy. Consequently, the forest's plants depend largely on their ability to gather the nutrient-rich rainwater that trickles down stems. Many plants do this by forming extensive, shallow root systems, whereas other plants intercept nutrients before they reach the soil. For example, the roots of trees such as *Eperua purpurea* grow upward in the fissures of the bark of nearby trees. These fissures are predictable pathways for flowing nutrients, which the roots absorb from the fissures without letting them enter the soil. Like their counterparts in the soil, these roots are extensively branched and occur throughout the canopy, and they may extend more than 13 meters into the air. Interestingly, the roots of *Eperua* grow up along its own stems as well as those of nearby trees, but roots of nearby trees do not grow onto *Eperua* stems. As a result, *Eperua* can exclusively recycle its own nutrients, as well as those of its neighbors.

Why do the roots of these tropical trees grow up, while most others do not? Do these roots respond to nutrients coming

FIGURE 20.29

Epiphytic bromeliad growing on a tree.

down the tree trunks, or do they fortuitously climb any vertical support that they happen to contact? To answer these questions, Robert Sanford of Stanford University sank drainpipes into the forest floor, attaching a reservoir atop each pipe. One-third of the reservoirs contained nutrient-rich cattle manure, one-third contained leaf litter, and the remaining third were empty. Within four months, climbing roots appeared on all of the "trees" containing manure, on 25% of those containing litter, and on none of the empty ones. Thus, the behavior of the roots of these plants is an adaptation to the low soil-nutrient availability of the Amazon rain forest.

Rain forests also contain many **epiphytes** (from the Greek words *epi*, meaning "upon," and *phyton*, meaning "plant"), which are plants that grow independently on other plants (fig. 20.29). Common epiphytes include bromeliads, orchids, staghorn ferns (*Platycerium bifurcatum*), and even some cacti. These plants grow slowly and must absorb their nutrients from sources other than the soil; thus, they are well adapted for aerial nutrition. Some epiphytes form their own "flower pot" to trap nutrients. Similarly, bird's-nest ferns (*Asplenium nidus*) accumulate rainfall and litter between their closely packed leaves; the ferns then get their nutrients by growing roots among these leaves. Among the most specialized of epiphytes is Spanish moss (*Tillandsia* *usneoides*), which grows throughout the southeastern United States. These flowering plants lack roots; they absorb water and nutrients through hairs that coat their stems and leaves.

Chapter Summary

Soils form from rocks by a process called weathering, a slow process that produces soil layers called horizons. Each horizon has different properties that affect plant growth and development. There are many different kinds of soils, each of which consists of mineral particles, humus, air, water, and living organisms.

- The three kinds of mineral particles are sand, silt, and clay. Clay particles have a sheetlike structure and are the smallest. Clay retains more water than does silt or sand. Also, the negative charges of clay particles retain cations in the soil. Plants can exchange H^+ for these cations via cation exchange. Because sand particles have no charge, they do not retain cations and have a low cation-exchange capacity.

- Humus is the decomposing organic matter in soil. Humus retains water and is a source of nutrients.

- Air makes up 25%–50% of soil volume. Air provides oxygen for roots and organisms living in the soil.

- Water in soil exists in two forms: chemically bound water and unbound water. Chemically bound water is unavailable to plants. Soil contains three types of unbound water. (1) Gravitational water is water drained from soil by gravity. (2) Capillary water is held by capillary action in pores in the soil. (3) Hygroscopic water, which is water held tightly by the soil particles, is unavailable to plants. Almost all water absorbed by plants is capillary water.

- Living organisms in soil refine and mix the soil, add organic matter, increase the amount of CO_2, affect the availability of nutrients, and often make the soil uninhabitable by other organisms.

Plants require at least sixteen essential elements, which are required for normal growth and reproduction, cannot be substituted for, and directly or indirectly affect metabolism. Carbon, hydrogen, oxygen, phosphorus, potassium, nitrogen, sulfur, calcium, and magnesium are required in relatively large amounts, and are called macronutrients. Iron, chlorine, copper, manganese, zinc, boron, and molybdenum are required in relatively small amounts, and are called micronutrients.

Macronutrients and micronutrients function as parts of structural units, parts of compounds involved in metabolism, activators and inhibitors of enzymes, and controlling agents of a cell's osmotic potential. Beneficial elements are not essential, but stimulate growth.

The availability of essential elements is influenced by the soil's mineral particles, cation-exchange capacity, decay of humus, pH, other elements, and organisms. Most essential elements are absorbed by roots. This absorption is facilitated by the large surface area of roots and the presence of ion-specific carriers.

Plants establish several kinds of associations with other organisms to obtain nutrients:

- The roots of almost all plants have mutually beneficial symbioses with fungi. These mutualisms, which are called mycorrhizae, increase the absorption of water and nutrients.

- Many plants, especially legumes, establish mutually beneficial symbioses with nitrogen-fixing bacteria. The bacteria reduce atmospheric N_2 to ammonium, which the host uses for growth.

- Carnivorous plants usually grow in nitrogen-poor soils of bogs. These plants get much of their nitrogen by trapping and digesting insects and some other organisms. Their traps are elaborate and range from sticky surfaces to elaborate trapdoors.

- Parasitic plants get their nutrients directly from other plants. The parasite and host are linked by a special organ called a haustorium, which links the two plants.

Rain in tropical rain forests contains many nutrients leached from the forest's dense canopy. Some trees in the rain forest have roots that grow up along the trunks of adjacent trees and absorb the nutrients from rainfall trickling down the trees.

Questions for Further Thought and Study

1. What is the difference between an essential element and a beneficial element?
2. Criteria for establishing the essentiality of an element for plants have been around for several decades. Why, then, are essential elements still being discovered?
3. How are root nodules different from mycorrhizae?
4. How does the size of soil particles affect their ability to retain water?
5. What is cation exchange, and why is it important for plant nutrition?
6. What are the consequences of acid rain on nutrient availability to plants?
7. Chlorophyll contains nitrogen, but carotenoids do not (see figs. 7.5, 7.9). Therefore, could a nitrogen-deficient plant still have color? Why or why not?

8. Traps of Venus's-flytraps triggered by inanimate objects or sugars do not secrete digestive enzymes. What is the adaptive importance of this?
9. Serpentine soils are low in calcium. What would be the consequences of replenishing calcium in this soil by adding lime? Would these consequences be different if calcium were added as $CaCl_2$ instead of lime?
10. Describe how the pH of soil affects the availability of phosphorus.
11. Describe the major types of soils and how each is suited for plant growth.
12. Explain why removing living organisms from a soil can modify the properties of a soil.
13. Explain why the histosols of bogs are nutrient poor, yet contain large amounts of organic matter.
14. Discuss the adaptations of plants for obtaining sufficient amounts of nutrients.
15. *Rhizobium* produces cytokinin. How might this relate to their ability to produce nodules in roots?
16. How does the curling of tips of root hairs relate to the establishment of root nodules? Go to the library and find the answer to this question.

Suggested Readings

Articles

Clarkson, D. T. 1985. Factors affecting mineral nutrient acquisition by plants. *Annual Review of Plant Physiology* 36: 77–115.

Gibbons, B. 1984. Do we treat our soil like dirt? *National Geographic* 166:350–388.

Hepler, P. K., and R. O. Wayne. 1985. Calcium and plant development. *Annual Review of Plant Physiology* 36:397–439.

Nap, J-P., and T. Bisseling. 1990. Developmental biology of a plant prokaryote symbiosis: The legume root nodule. *Science* 250:948–954.

Peters, G. A., and J. C. Meeks. 1989. The *Azolla-Anabaena* symbiosis: Basic biology. *Annual Review of Plant Physiology and Plant Molecular Biology* 40:193–210.

Stewart, G. R., and M. C. Press. 1990. The physiology and biochemistry of parasitic angiosperms. *Annual Review of Plant Physiology and Plant Molecular Biology* 41:127–151.

Vijn, I., et al. 1993. Nod factors and nodulation in plants. *Science* 260:1764–1765.

Books

Dixon, R. O. D., and C. T. Wheeler. 1986. *Nitrogen Fixation in Plants*. Glasgow, Scotland: Blackie & Sons, Ltd.

Galston, A. 1994. *Life Processes in Plants*. New York: W. H. Freeman and Co.

Haynes, R. J. 1986. *Mineral Nitrogen in the Plant-Soil System*. New York: Academic Press.

Lauchli, A., and R. L. Bieleski, eds. 1983. *Inorganic Plant Nutrition*. New York: Springer-Verlag.

Robb, D. A., and W. S. Pierpoint, eds. 1983. *Metals and Micronutrients: Uptake and Utilization by Plants*. New York: Academic Press.

Slack, A. 1979. *Carnivorous Plants*. London: Ebury Press.

Big trees of California (*Sequoiadendron giganteum*). How does water reach the tops of these tall trees?

CHAPTER 21

Movement of Water and Solutes

Chapter Outline

INTRODUCTION
MOVING WATER AND MINERALS IN THE XYLEM
   Leaf Architecture and Transpiration
   Structure of the Conducting Cells

BOX 21.1
THE RISKS OF HAVING VESSELS

   Water Potential: The Force Responsible for Water Movement
   How Does Water Move In Plants?

BOX 21.2
LIVING IN SALTY SOIL

   Factors Affecting Transpiration

BOX 21.3
TAPPING WOOD FOR MAPLE SYRUP

   A Model to Regulate the Photosynthesis-Transpiration Compromise
   The Adaptive Value of Transpiration
   Coping with Extremes: Too Much or Too Little Water
TRANSPORTING ORGANIC SOLUTES IN THE PHLOEM
   Structure of Conducting Cells
   How Substances Move in the Phloem
   Influence of the Environment on Phloem Transport
   What Moves in the Phloem?
EXCHANGE BETWEEN THE PHLOEM AND XYLEM
Chapter Summary
Questions for Further Thought and Study
Suggested Readings

Chapter Overview

Plants did not colonize land until about 400 million years ago. The biggest challenges that plants faced as they invaded land were obtaining and transporting water in the dry terrestrial environment. These challenges to survival eventually led to the evolution of roots and a specialized vascular system to gather and transport the water, and an epidermis and stomata to conserve it. The evolution of multicellularity and tissue specialization also produced a system to move the products of photosynthesis to distant sites of use and storage.

This chapter describes the movement of water and solutes in plants. Specifically, it describes two tissues specialized for interorgan commerce: xylem, which carries water and dissolved minerals from roots to leaves, and phloem, which transports water and organic compounds throughout the plant. The evolution of these tissues was largely responsible for the widespread invasion of land by plants.

Introduction

The survival of all plants depends on their ability to transport a variety of substances into, through, and out of their bodies. At the cellular level, diffusion is usually adequate for this movement: for example, small molecules can diffuse across a cell that is 50 μm wide in less than one second. Thus, primitive single-celled organisms such as algae typically relied on diffusion as a primary means of transport. However, the evolution of multicellularity rendered diffusion woefully inadequate for moving substances throughout a plant because the time required for diffusion is inversely proportional to the square of the distance involved. This means that a molecule diffuses across 1 μm 10^8 times faster than it diffuses across 1 cm, and the same molecule that diffuses across an algal cell in less than a second requires more than eight years to diffuse only 1 m in the watery environment of a multicellular plant.

Transport between adjacent cells of multicellular organisms improved with the advent of cytoplasmic streaming, which speeded transport to about 5 cm per hour. Although cytoplasmic streaming is faster than diffusion, it too became inadequate to meet the needs of long-distance transport in ever-larger multicellular plants.

Meanwhile, multicellularity was not the only evolutionary trend placing increased demands on the transport systems of plants. Plants colonizing the land were in a "space race" to intercept light for photosynthesis. In large part, this meant growing taller and exposing large leaves to the dry, hostile environment. However, the algaelike photosynthetic cells in leaves could function only in a highly humid environment, which was far from the soil's precious water. Absorbing this water was accomplished by roots and root hairs in ferns and other plants. However, merely absorbing water was not enough, because leaves were located at ever-increasing distances from the roots. Thus, plants required a system for transporting water from the soil to their leaves. Multicellularity also required a system for transporting the sugars made in leaves to distant sites for storage and use. Thus, multicellularity and colonization of the land were largely responsible for the evolution of xylem and phloem, the two long-distance transport systems in plants. These systems have proven to be successful, for in few instances do deficiencies in internal transport limit plant growth.

Moving Water and Minerals in the Xylem

Water is the most abundant compound in plant cells: it accounts for 85%–95% of the weight of most plants, and 5%–10% of the weight of seeds. Water is used to make organic compounds (e.g., sugars), support the plant (via turgor pressure), as a solvent in which important reactions occur, and as the medium in which solvents move. Given the critical roles played by water in plants, it seems peculiar that more than 95% of the water gathered by a plant evaporates back into the atmosphere, often within only a few hours after being absorbed. This evaporation of water from the shoot of a plant is called **transpiration**. Most transpiration from leaves is through stomata and is a result of leaf architecture.

Leaf Architecture and Transpiration

Leaves are exquisitely adapted for photosynthesis. However, the adaptations that enhance photosynthesis also enhance a plant's greatest threat: dehydration. For example, the rate of gas exchange depends, among other things, on the amount of surface area available for exchange and evaporation. The loose internal arrangement of cells in leaves produces a large internal surface area for transpiration (fig. 21.1). Indeed, the internal surface area of a leaf may be more than two hundred times greater than its external surface area; this amplifies transpiration by increasing the surface area for evaporation.

The internal surface area of a leaf is linked to the atmosphere by intercellular spaces that occupy as much as 70% of the volume of a leaf. The gates that link the internal surface area with the atmosphere are stomata. These gates are abundant: a squash leaf, for example, may have more than 80 million

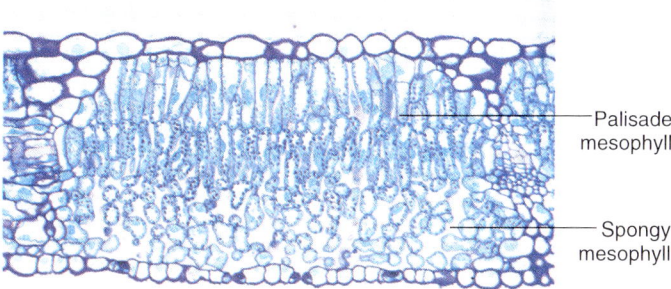

FIGURE 21.1

Cross section of a lilac leaf (*Syringa* sp.), showing palisade and spongy regions of mesophyll. The loose arrangement of mesophyll cells exposes large amounts of surface area.

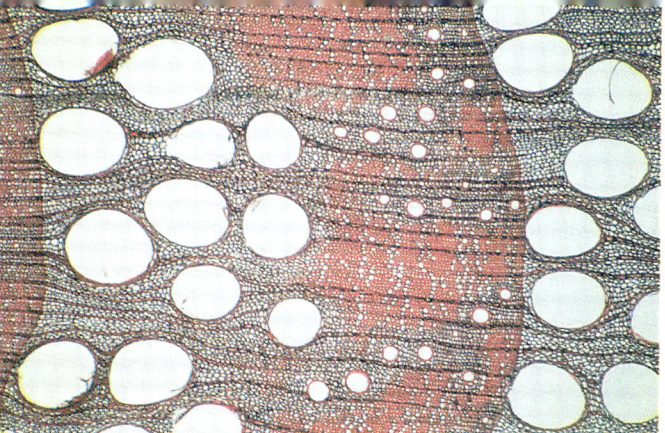

FIGURE 21.2

Cross section of wood, showing vessels and tracheids. Vessels and tracheids transport water and dissolved minerals from roots to shoots.

stomata. Leaves also have an efficient plumbing system of veins for distributing water to their internal evaporative surface: one square centimeter of leaf can have as many as 6,000 vein endings. As a result of this leaf architecture, a well-watered plant can lose tremendous amounts of water via transpiration; for example, a corn plant transpires almost 500 liters of water during a four-month growing season. If humans required the same amount of water, we would each have to drink about 40 liters of water per day (and visit the bathroom often).

Water lost via transpiration is replaced by water absorbed from the soil by roots. This water moves to the leaves as fast as 75 cm min^{-1}, which is about the speed of the tip of a second hand sweeping around a wall clock. Despite their huge requirements for water, however, plants have no active mechanism for acquiring water when they are stressed. Thus, losses of water via transpiration do not indicate an excess of water; on the contrary, water is the primary factor that limits plant growth in most areas.

Structure of the Conducting Cells

Water must move rapidly through plants to replace the water lost by transpiration. Through what tissue does this water move? To answer this question, consider the results of two simple experiments:

1. Removing the bark from a tree does not significantly alter transpiration.
2. When roots are exposed to a soluble dye, only xylary elements in the stem contain the dye.

These results suggest that water and dissolved minerals move from roots to leaves through xylary elements.

Recall from Chapter 13 that there are two kinds of xylary elements in plants: tracheids and vessel elements (fig. 21.2). Both of these cell types are well designed for conducting water: they are hollow and dead at maturity, and therefore have no cellular organelles to retard water flow. Furthermore, both have thick cell walls and can therefore withstand changes in pres-

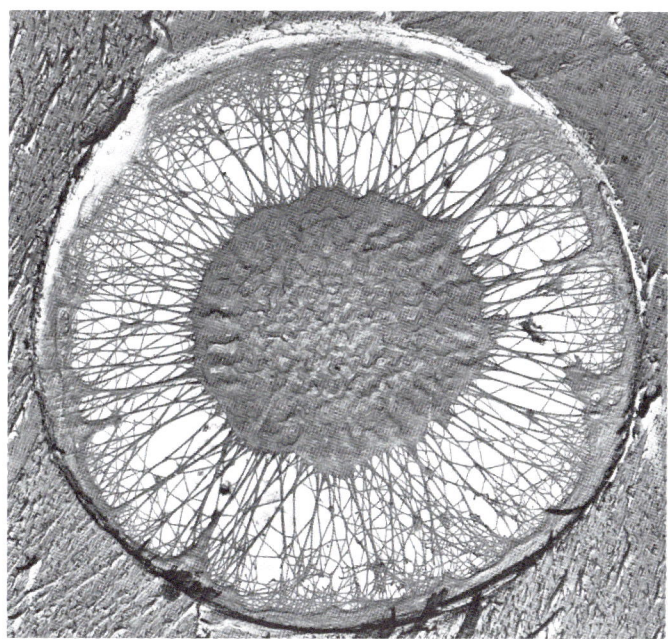

FIGURE 21.3

Bordered pits are valvelike structures that are abundant in walls where tracheids overlap one another. Pits link tracheids into long, water-conducting chains and allow water to flow through xylem, ×6,050.

sure associated with water flow caused by transpiration. However, tracheids are usually long (up to 10 mm) and thin (10–15 μm in diameter), and they overlap each other. Their walls contain many thin areas called *bordered pits* (fig. 21.3), which are valvelike structures that are especially abundant in portions of the walls where tracheids overlap. Pits link tracheids into long, water-conducting chains and allow a slow flow of water through xylem.

Vessel elements are shorter and wider than tracheids: their diameters are usually between 40 and 80 μm, and may be as large as 500 μm. Vessel elements are stacked end to end, and the end walls separating adjacent vessel elements are often

> **BOXED READING 21.1**
>
> ## THE RISKS OF HAVING VESSELS
>
> Gymnosperms have only tracheids, while angiosperms have tracheids and vessels. This difference in xylary structure provides angiosperms with tremendous potential benefits as well as risks. For example, consider a tracheid with a diameter of 10 μm and a vessel with a diameter of 80 μm. Flow through cylindrical pipes such as these is proportional to the fourth power of their radius (i.e., radius^4), so that the flow rate through these cells would be as given in the following table.
>
> Thus, water flows 4,096 times faster (i.e., 2,560,000/625) through the vessel than through the tracheid. Of what value, then, are tracheids?
>
> The increased diameter that allows such rapid water transport in the vessels of angiosperms also puts these plants at risk: the larger water-columns of vessels have a lower tensile strength than do the thinner columns of tracheids, thereby making them more likely to break when stressed by freezing or by wind-induced bending. An air bubble forms where a water column breaks, which stops the flow. Capillarity then shapes the air bubble into a sphere, which prevents it from moving through bordered pits into adjacent cells. Since the walls separating adjacent vessel elements are dissolved, an air bubble stops water flow in an entire vessel rather than in a single cell, as in tracheids. When air bubbles form in vessels, water flow is delegated to smaller vessels and tracheids of the wood. Thus, tracheids are the backup system of woody angiosperms: they ensure that water flow doesn't stop when air bubbles form in vessels.
>
> Bubbles in vessel elements are surprisingly common in many angiosperms. For example, the flow of water in virtually all of the large vessel elements of oak (*Quercus*) and ash (*Fraxinus*) is stopped by air bubbles by the end of a growing season. When this occurs, water moves in tracheids until the next spring, when the vascular cambium produces a new set of vessels.
>
> | Cell Type | Diameter | Radius | Flow Rate (radius^4) |
> |---|---|---|---|
> | Tracheid | 10 | 5 | $(5)^4 = 625$ |
> | Vessel element | 80 | 40 | $(40)^4 = 2,560,000$ |

either wholly or partially dissolved (fig. 21.4). As a result, vessel elements form cellulose pipes called *vessels* ranging in length from a centimeter to more than a meter. Water can move longer distances in vessels than in tracheids before having to traverse a pit. Moreover, their larger diameter and dissolved end walls allow water to move faster than in tracheids. This increased flow rate in vessels may help explain why angiosperms dominate today's landscapes. Angiosperms contain both tracheids and vessels, while gymnosperms contain only tracheids (see box 21.1 above, "The Risks of Having Vessels").

CONCEPT

Water and dissolved minerals move from roots to leaves in dead conducting cells of the xylem. Most of this water evaporates from leaves via a process called transpiration.

Water Potential: The Force Responsible for Water Movement

The movement of water through plants is a physical process that requires no metabolic energy. Rather, water flows passively from one place to another because of differences in potential energy. The potential energy of water in a particular system compared to pure water at atmospheric pressure and at the same temperature is termed the **water potential,** and is abbreviated by the Greek letter *psi*, Ψ. Lowering the potential energy of water lowers the water potential, and increasing the potential

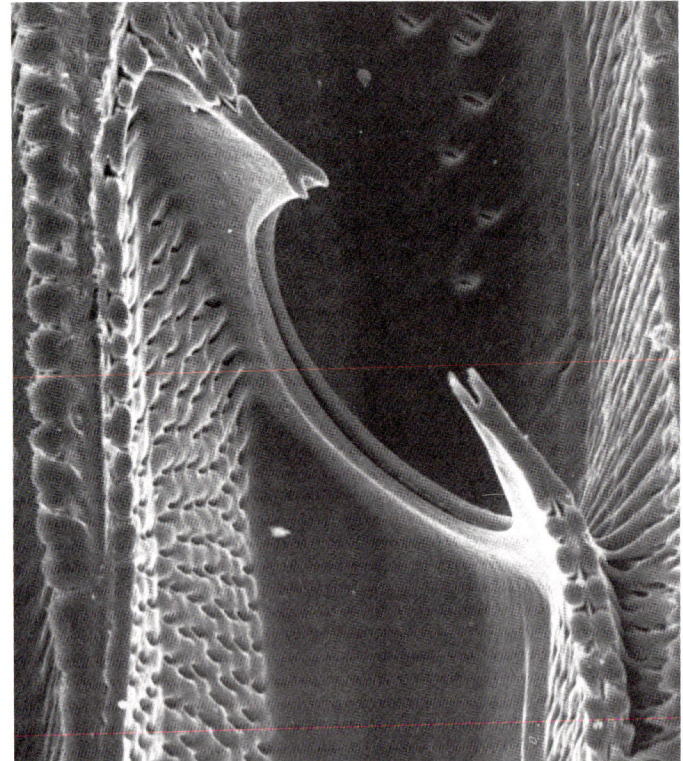

FIGURE 21.4

Scanning electron micrograph of part of a vessel. Vessel elements are stacked end to end; the walls separating adjacent vessel elements are often either wholly or partially dissolved.

energy of water increases the water potential. Differences in water potential determine the direction that water moves: *water always flows passively from areas of high water potential to areas of lower water potential.* The movement of water into, through, and out of plants is regulated by water potential.

Components of Water Potential

Water potential is a pressure (energy per unit volume) measured in units called **pascals** (Pa) or, more conveniently, **megapascals** (MPa; 10^6 Pa). These metric units relate to the more common English units for measuring pressure as follows: 1 atmosphere = 14.7 pounds per square inch = 760 mm Hg (at sea level, 45° latitude) = 1.0 bar = 0.10 MPa = 1.0×10^5 Pa. A car tire is typically inflated to about 0.2 MPa; the water pressure in home plumbing is usually 0.2 to 0.3 MPa; and the water pressure under 5 m of water is about 0.05 MPa.

The potential energy of water in an open beaker is difficult to measure, and is arbitrarily set equal to zero. The potential of this water can be modified by solutes, physical pressure, and wettable surfaces (matrices). Consequently, the water potential in a particular system (such as a cell) equals the sum of its pressure potential (Ψ_p), solute (or osmotic) potential (Ψ_π), and matric potential (Ψ_m):[1]

$$\Psi_{cell} = \Psi_p + \Psi_\pi + \Psi_m$$

Let's examine how each of these components affects the water status of plants.

Pressure Potential (Ψ_p)

The **pressure potential** is a physical pressure and may be positive or negative. For example, water in dead xylary elements of transpiring plants is often under a negative pressure (i.e., tension) of less than -2 MPa; therefore, Ψ_p of water in xylem is usually negative. Conversely, water in living cells is usually under a positive pressure, much like the air in a balloon. In these cells, Ψ_p is greater than zero. Changes in the pressure potential can alter the volume of a cell by as much as 40%.

Solute Potential (Ψ_π)

The **solute potential** is proportional to the number of dissolved solute particles and is independent of the type of particles. A particle may be a molecule of sugar, an ion of sodium, or any other dissolved chemical. These particles interact with and reduce the activity of water molecules, and thereby decrease the potential energy of the water. Thus, adding solutes always lowers the water potential, and Ψ_π is always 0. The solute potential of the leaves of most crop plants typically ranges from -1 to -2 MPa.

---

1. Many botanists also include a gravitational potential in the calculation of water potential. Gravitational potential is of negligible importance in a root or leaf, but becomes important when comparing water potentials at different heights in trees and soils. The upward movement of water in a tree must overcome a gravitational potential of approximately 0.01 MPa per meter of height.

Matric Potential (Ψ_m)

The **matric potential** results from the adhesion of water to wettable surfaces such as cell walls and the cytoplasmic matrix, and is the force required to remove water from these surfaces. This adhesion decreases the potential energy of the interacting molecules. As a result, Ψ_m is always 0.

The Ψ_m of vacuolated cells is trivial (usually between 0 and -0.01 MPa) and is therefore ignored in most calculations of water potential. However, Ψ_m is an important aspect of the water potential of densely cytoplasmic (e.g., meristematic) cells, desiccated tissues, dry seeds, and soils. For example, more than 4 MPa of pressure are required during imbibition to rupture the seed coat of a walnut; this force is generated by an uptake of water caused by the low matric potential of the seed.

Water Potential and Water Movement

To understand how Ψ_π and Ψ_p interact to account for the water potential of a cell (Ψ_{cell}), consider an algal cell with a water potential of -0.2 MPa and a solute potential (Ψ_π) of -0.5 MPa. The pressure potential (Ψ_p) of this algal cell is

$$\Psi_{cell} = \Psi_\pi + \Psi_p$$
$$-0.2 \text{ MPa} = -0.5 \text{ MPa} + \Psi_p$$
$$\Psi_p = +0.3 \text{ MPa}$$

Now suppose that this algal cell is put in a beaker of pure water ($\Psi = 0$). The water-potential gradient between the water and the cell causes water to enter the cell, inflating the cell with water and producing turgor pressure—the hydrostatic force that shapes plants and plant cells and is resisted by the rigid cell wall. Eventually, the pressure potential equals the solute potential, which prevents net uptake of water by the cell. At this point the cell is fully turgid, and its water potential equals zero. Thus, the increase in water potential from -0.2 to 0 MPa is due to a change in the Ψ_p component of water potential:

$$\Psi_{cell} = \Psi_\pi + \Psi_p$$
$$0 = -0.5 + \Psi_p$$
$$\Psi_p = +0.5 \text{ MPa}$$

If the cell is now placed on a countertop, it will begin to dry, and as it loses water, its pressure potential decreases. When $\Psi_p = 0$, the protoplast barely fills the cell, and there is no turgor. This condition is termed *incipient plasmolysis* and is the point at which $\Psi_{cell} = \Psi_\pi$ (remember, $\Psi_p = 0$ at incipient plasmolysis).

Now consider a root hair growing in soil. Suppose that the root hair has a water potential of -0.5 MPa, and the soil has a water potential of -0.2 MPa. The water-potential gradient between the soil and root hair causes water to enter the root hair. As the soil dries, however, its water potential decreases. When the water potential of the soil becomes less than that of the root hair, water moves from the root hair into the soil. Similarly, the presence of salts in soil decreases the water potential of the soil. If large amounts of solutes are present, as in salty or overfertilized soils, water moves into the soil from roots and the plants appear "burned" (see box 21.2, "Living in Salty Soil").

C O N C E P T

Water potential is the potential energy of water and indicates the tendency of water to move from one place to the next. Water potential is influenced by pressure, solutes, and wettable surfaces. Water always moves passively from areas of high water potential to areas of lower water potential.

Doing Botany Yourself

Design an experiment that would enable you to determine the concentration of a fertilizer of your choice that would cause incipient plasmolysis in a common greenhouse plant such as *Zebrina* or *Rhoeo*.

How Does Water Move In Plants?

Any hypothesis for water movement in plants must be based on water moving from areas of high water potential to areas of lower water potential. It must also account for an even more obvious requirement: water must reach the tops of tall trees such as the one pictured at the beginning of this chapter (p. 488). The forces involved in lifting water to treetops are considerable. For example, "Tallest Tree" in Redwood National Park, California, is over 110 m high; moving water to the top of this tree requires a water-potential gradient exceeding 2 MPa. Let's see how some of the hypotheses proposed for water movement in plants match up with these requirements.

HYPOTHESIS 1

Water moves up xylary elements via capillarity.
Capillarity results from the adhesion of water to the surfaces of small tubes. This adhesion pulls water up the tube, and is visible as the curved meniscus atop the water column in a glass tube. However, in tubes having the diameter of a xylary element, capillarity raises water less than 1 m. Therefore, capillarity alone cannot account for the movement of water to the tops of trees.

HYPOTHESIS 2

Water is pushed up xylary elements by atmospheric pressure.
To understand this hypothesis, imagine filling a long, hollow tube with water, closing it at one end, and placing the tube, open-end down, in a tub of water. Movement of the water column is balanced by two opposing forces: the weight of the water in the tube pulls the water column down, while atmospheric

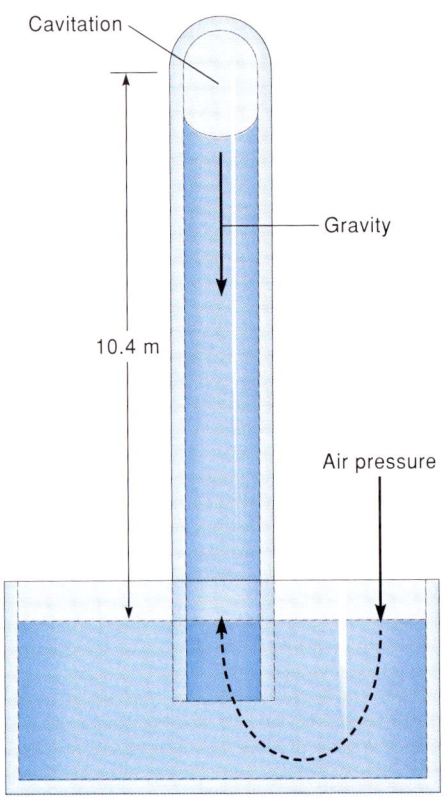

FIGURE 21.5

The effect of atmospheric pressure on a column of water. The weight of the water in the tube pulls the water column down, while atmospheric pressure pushes water up the tube. The counteracting pressures equilibrate when the water column is about 10.4 meters high.

pressure pushes water up the tube. These counteracting pressures reach equilibrium when the water column is about 10.4 m high (fig. 21.5). When the length of the tube exceeds 10.4 m, the water column *cavitates*, meaning that it forms a partial vacuum filled with water vapor in the upper, closed end of the tube. Because atmospheric pressure raises a column of water only about 10.4 m, it cannot account for the movement of water to the tops of tall trees.

HYPOTHESIS 3

Water is pumped up the xylary elements.
Water in xylem moves in xylary elements, which are dead. Furthermore, there are no "pumping cells" in xylem. Therefore, water is not actively pumped through the xylem.

HYPOTHESIS 4

Water is pushed up by root pressure.
On many mornings you have probably seen leaves like those shown in figure 21.6 that have water droplets at their edges. This loss of liquid water from the leaves of

BOXED READING 21.2
LIVING IN SALTY SOIL

Halophytes are plants that grow in salty soil. These soils occur in regions that range from hot, dry, salty deserts to moist, cool, salty marshes. One major problem faced by halophytes is obtaining enough water for growth. To do this, they must reduce their water potential below that of the soil so that water will flow into their roots. This is no small task, since the soil in which many halophytes grow often has a water potential less than −4 MPa.

Plants such as shadescale (*Atriplex*) and greasewood (*Sarcobatus*) are halophytes typical of the moderately salty soils of the deserts of the southwestern United States. However, increasingly salty soil requires that plants be able to reduce their water potential even further to obtain water. These increased demands restrict the vegetation of more saline soils to only a few plants, such as pickleweed (*Salicornia*) and iodine bush (*Allenrolfea*). These plants are adept at reducing their water potential; pickleweed, for example, can reduce its water potential to almost −8 MPa. Although these adaptations are impressive, they are not limitless: no plants grow on the salt flats of Utah.

Another problem faced by halophytes is toxic amounts of several ions, especially Na^+ and Cl^-. Unlike most other plants, halophytes not only tolerate these large amounts of ions, but often use them to their advantage. For example, halophytes such as *Spartina* and *Limonium* transport Na^+ and Cl^- to their leaves, where these ions are excreted into salt glands. Salt in the leaves reduces their water potential, which helps to pull water into them.

Understanding how halophytes tolerate such large amounts of salt is becoming increasingly important as fertilization and irrigation transform productive soils into salty, less fertile soils. Irrigation water seldom reaches the water table; rather, through evaporation and transpiration it leaves its solutes behind in the soil, where they form an encrusting layer of salt, much like the salt covering the edge of a margarita glass. In some areas, these salts are flushed from fields into rivers, where they become a problem for farmers living downstream who also use the river's water for irrigation. States such as Arizona have installed desalting facilities that purify more than 100 million gallons of water per day. Yet another approach to reclaiming saline soil is growing and harvesting halophytes; because these plants accumulate large amounts of salt, their harvest actually decreases the soil's salinity.

Many ancient civilizations arose by diverting rivers to irrigate arid lands to grow crops. The most productive use of irrigation occurred in the so-called Fertile Crescent, which is a broad valley formed by the Tigris and Euphrates rivers in what is now Iraq. At the peak of its productivity, this irrigated region supported more than a million people, and was the point from which similar civilizations spread into Afghanistan, Pakistan, and present-day Iran. Slowly but surely, however, the salts washed from rocks and soil at higher elevations became concentrated in the irrigated fields as water evaporated and was transpired by plants. As a result, all of these civilizations ultimately collapsed for the same reason: their soils became too salty to grow crops.

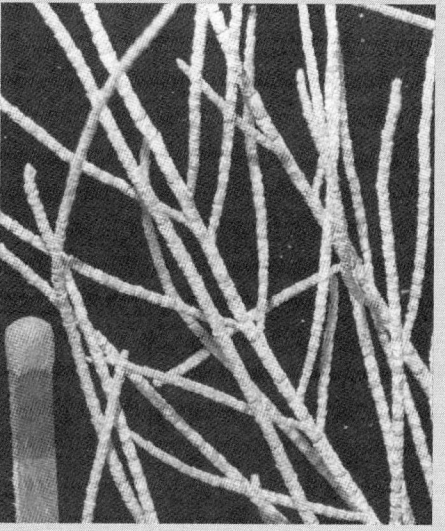

BOX FIGURE 21.1
Some halophytes excrete salt by means of specialized glands. Shown here are the salt-covered, scalelike leaves on the twigs of tamarisk (*Tamarix ramosissima*) from the Mojave Desert in California.

---

intact plants is called *guttation* and is common in herbaceous plants growing in moist soil on cool, damp mornings. Guttation is caused by root pressure that is generated as follows:

1. Minerals actively absorbed at night are pumped into the apoplast surrounding xylary elements.
2. This influx of solutes decreases the water potential of the xylary element, thereby causing water to move into it from surrounding cells.
3. Since there is only negligible transpiration at night, the pressure in the xylem increases as high as +0.2 MPa.
4. Eventually, this pressure forces liquid water out of the leaves through hydathodes.

Guttation continues as long as the plant is kept under conditions favoring rapid absorption of minerals and minimum transpiration, such as in wet soils at night. The amount of water lost by guttation from most plants is relatively small. However, a notable exception is taro (*Colocasia esculenta*), a tropical plant used by Polynesians to make *poi*. Individual leaves of taro plants can lose more than 300 ml of water per night by guttation. Although root pressure can push water several meters up a plant, it cannot push water to treetops.

FIGURE 21.6

Strawberry leaves with drops of guttation water at vein endings along the edges. Guttation is caused by root pressure.

HYPOTHESIS 5

Water is pulled up plants by evaporation.

This hypothesis was formulated more than a century ago, and today is referred to as the *transpiration-cohesion hypothesis* for water movement. It is summarized in figure 21.7 and describes the process as follows:

1. Solar-powered transpiration of water dries the cell walls of mesophyll cells.
2. This loss of water from the cell wall lowers the water potential of the cell, thereby causing it to take up water from neighboring cells that have a higher water potential because they are farther away from the air spaces.
3. Cells farther from the site of evaporation have even larger water potentials, thus causing water to move from cell to cell toward the air spaces.
4. Cells bordering tracheids replace their water with water from the xylem. This loss of water from xylary elements creates a negative pressure (i.e., $\Psi_p < 0$), thereby lifting the water column up the plant.
5. The negative pressure decreases the water potential all the way down to the tips of roots, even in the tallest trees. This tension lowers the water potential in the root xylem so much that water flows passively from the soil, across the root cortex, and into the stele. Water in the stele is then pulled up the xylem to leaves to replace water lost via transpiration.

Although this hypothesis sounds logical, it must nevertheless meet the criteria stated earlier in this section. Let's see if it does.

Can transpiration lift a column of water to the top of a tall tree? Stated another way, is there a water-potential gradient favoring movement of water from soil to plant to atmosphere? Yes. For example, the following are typical values for water potentials at various points along the transpiration path for a small tree growing in moist soil:

$\Psi_{soil} = -0.05$ MPa
$\Psi_{root} = -0.2$ MPa
$\Psi_{stem} = -0.5$ MPa
$\Psi_{leaf} = -1.5$ MPa
$\Psi_{atmosphere} = -100$ MPa (at 50% relative humidity at 22° C)

The water potential of the atmosphere is highly variable, yet it is almost always less than that of leaves. For example, air with a relative humidity of 90% has a water potential less than -13 MPa.

Is water under a negative pressure in the xylem? The transpiration-cohesion hypothesis states that water is a continuous hydraulic system that is pulled up through plants, and should therefore be under a negative pressure or tension. Tensions are greater at the tops of trees than at their bases (fig. 21.8), and the diameters of roots and tree trunks shrink when transpiration is greatest, as would be expected if the xylem were under a negative pressure. Finally, water in branches begins moving sooner than water in the trunk of the tree, as predicted if water were pulled rather than pushed up a plant. Therefore, water in xylem is under a negative pressure in transpiring plants.

Can columns of water withstand the tensions necessary to be pulled to the tops of tall trees? Yes. Water is a polar molecule, and therefore it coheres. This cohesion produces a high tensile strength—so high that it is easier to pull apart molecules in fine wires of some metals than to pull apart thin columns of water. A thin column of water in a xylary element can withstand tensions of up to -30 MPa, which far exceed the approximately -2 MPa required to raise water to the tops of trees.

Is the adhesion between water and cell walls strong enough to support a column of water? Again, yes. These forces far exceed the -2 MPa needed to support a water column. Adhesion and cell-wall hydration prevent gravity from draining the water from xylary elements.

CONCEPT

The transpiration-cohesion hypothesis best explains how water moves in xylem. The driving force for the movement of water is a water-potential gradient generated by transpiration. Transpiration from leaves lifts water up plants. Cohesion of water molecules prevents the columns of water from breaking, and adhesion of water molecules to cell walls prevents gravity from draining the water column.

Factors Affecting Transpiration

Environmental Factors

Atmospheric Humidity

Transpiration occurs as long as the water potential of the atmosphere is less (i.e., more negative) than the water potential of the leaf. Dry air increases this gradient and therefore increases transpiration. Similarly, transpiration typically slows in humid air.

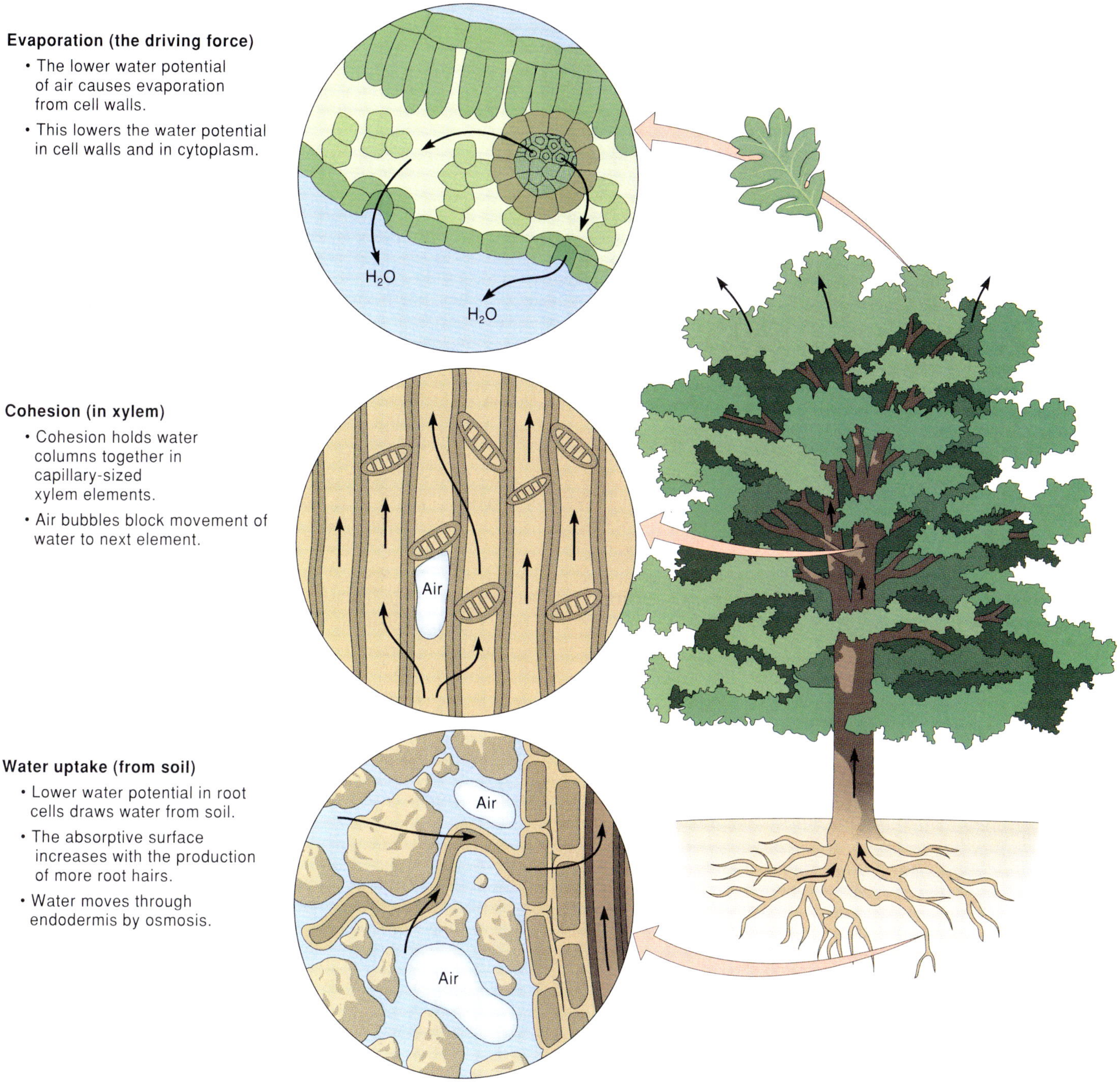

Evaporation (the driving force)
- The lower water potential of air causes evaporation from cell walls.
- This lowers the water potential in cell walls and in cytoplasm.

Cohesion (in xylem)
- Cohesion holds water columns together in capillary-sized xylem elements.
- Air bubbles block movement of water to next element.

Water uptake (from soil)
- Lower water potential in root cells draws water from soil.
- The absorptive surface increases with the production of more root hairs.
- Water moves through endodermis by osmosis.

FIGURE 21.7

A summary of the transpiration-cohesion hypothesis for the ascent of water in plants.

Internal Concentration of CO_2

The concentration of CO_2 in the atmosphere rarely deviates much from 0.03%. However, the CO_2 concentration in leaves changes considerably, especially when stomata close and photosynthesis removes CO_2 from the intercellular spaces of the leaf. Low concentrations of CO_2 in leaves cause stomata to open, whereas high concentrations cause them to close. Thus, a reduced supply of CO_2 for photosynthesis (i.e., a low internal concentration of CO_2) opens stomata and, as a result, increases transpiration.

Wind

The thin moist layer of air adjacent to a transpiring leaf is called its *boundary layer*. A thick boundary layer decreases the diffusion gradient and therefore decreases transpiration. Wind usually replaces the boundary layer with drier air, thereby increasing the water-potential gradient and increasing transpiration.

The leaves of many grasses can temporarily increase the thickness of their boundary layer and thereby temporarily reduce transpiration. For example, you learned in Chapter 13

CHAPTER TWENTY-ONE *Movement of Water and Solutes*

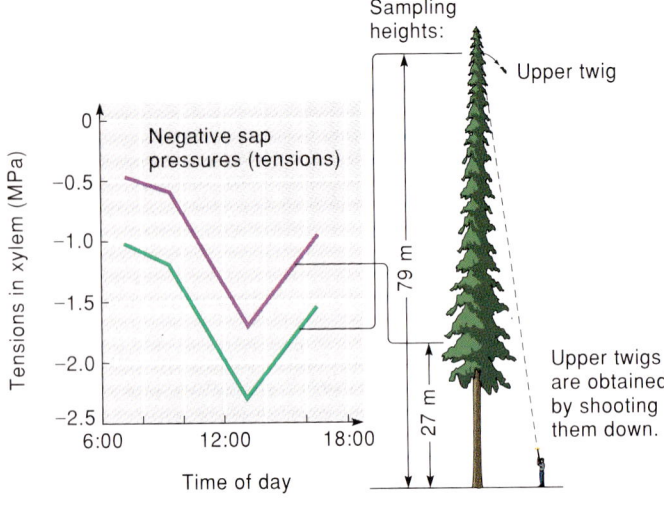

FIGURE 21.8

Throughout the day, water tensions are greater (i.e., more negative) at the tops of trees than they are toward their bases.

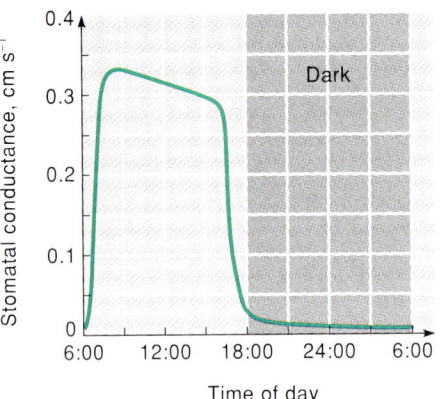

FIGURE 21.9

The diurnal curve of stomatal opening. The data are expressed as stomatal conductance (cm s^{-1}), an indication of the capacity for diffusion through stomata and an indirect measurement of stomatal opening. The stomata open rapidly in light and close at the end of the daylight period. Stomata remain closed throughout the dark period.

that the upper epidermis of the leaves of many grasses contains large, vacuolate cells called *bulliform cells*, which are sensitive to water loss. These cells shrink when they desiccate, thereby rolling the leaf into a cylinder. This shape increases the leaf's boundary layer and reduces the amount of light that reaches the leaf, thereby decreasing transpiration.

Air Temperature

In direct sunlight the temperature of a leaf may exceed that of the air by as much as 10° C. Increasing the leaf temperature increases the water-vapor pressure in the leaf, which in turn increases its water potential and leads to faster rates of transpiration. Transpiration is most rapid at 20°–30° C.

Soil

Any factor that affects water availability also affects transpiration; therefore, transpiration is affected by the water content of soil. Plants can absorb water from soil as long as their water potential is less than that of the soil.

Plants function as wicks that evaporate subsurface water from soil, which explains why soils covered by plants lose water faster than does bare soil. Indeed, almost all water lost from below 15 cm in the soil is lost via transpiration. Weeds therefore compete with crop plants not only for light and nutrients, but also decrease the availability of water in soil.

Light Intensity

Light usually causes stomata to open and therefore increases transpiration. Although stomata typically open at sunrise and close at sunset (fig. 21.9), these are not "all-or-nothing" effects; instead, stomata open gradually in the morning over a period of about one hour and gradually close throughout the afternoon. The effect of light on stomatal opening and closing is indirect: light promotes stomatal opening by stimulating photosynthesis, which decreases the internal concentration of CO_2 in the leaf. The stomatal responses of green and nongreen portions of variegated leaves support this conclusion: stomata in nongreen regions respond to light and dark more slowly than do those in green regions, presumably because light has less effect on the internal concentration of CO_2 there. The regulation of transpiration by light is important, since it prevents plants from needlessly losing water when it is too dark for photosynthesis.

CONCEPT

Transpiration is affected by humidity, temperature, light, soil, wind, and the internal concentration of CO_2 in leaves.

Transpiration is greatest in plants growing in moist soil on a sunny, dry, warm, and windy day. In these conditions, transpiration often exceeds the ability of plants to absorb water. As a result, many plants wilt at midday, even though the soil in which they are growing may be near field capacity.

Adaptations That Affect Transpiration

Physiological Adaptations

We mentioned earlier in this chapter that guard cells arbitrate the transpiration-photosynthesis compromise; that is, they control carbon fixation as well as transpiration. How efficiently do they do this?

Photosynthesis and water-use efficiency. The efficiency of plants in fixing CO_2 while losing water is termed the *water-use efficiency*, and is defined as the amount of water lost divided by the amount of CO_2 fixed:

Water-use efficiency = (g H_2O lost)/(g CO_2 fixed)

The water-use efficiency of C_3 plants (e.g., tomato) is about 600. C_4 plants (e.g., corn) usually grow in drier, hotter environments than do C_3 plants. Since C_4 plants have no apparent photorespiration, they can fix large amounts of CO_2 with

FIGURE 21.10

Opportune leaf-production. Plants in hot, dry environments lose great amounts of water through their leaves. One adaptation is to have leaves only when water is available. The ocotillo (*Fouquieria splendens*) plant shown here does this. (a) During dry periods, the thorny leafless stems appear almost dead. (b) When water is available, leaves form rapidly and become photosynthetic.

FIGURE 21.11

Leaves of eucalyptus (*Eucalyptus* sp.) hang vertically; their flat surfaces are not presented directly to the midday sun, thereby minimizing heating and water loss.

their stomata only partially open, thereby conserving water. Thus, C_4 plants need less water for growth than do C_3 plants: their water-use efficiency is about 300. CAM plants such as cacti open their stomata at night when temperatures are cooler and thus reduce transpiration. Although their limited means of CO_2 fixation results in slow growth, it also conserves water: a typical CAM plant has a water-use efficiency of about 50. This increased efficiency allows CAM plants to grow where other plants cannot.

Leaf abscission and dormancy. Another effective means of decreasing transpiration is to reduce the evaporative surface area. One way that plants accomplish this is by getting rid of their leaves when water or temperature becomes limiting, such as in the cold of winter or during prolonged drought (fig. 21.10).

Leaf position. Many plants reorient their leaves as a means of altering transpiration. For example, *Eucalyptus* trees often reposition their vertically oriented leaves, thereby avoiding the midday sun (fig. 21.11). This decreases the leaf temperature and thus reduces transpiration.

Circadian rhythms and stomatal cycling. The open-at-dawn/close-at-night cycle of stomatal opening and closing is also influenced by an internal biological clock (see Chapter 19). For example, plants grown in a photoperiod of 16 hr light:8 hr dark continue to open and close their stomata according to this photoperiod even when they are placed in the dark. This rhythmic opening and closing of stomata often occurs independently of the water status of a plant. We know little about the biological clock that controls this rhythmic opening and closing of stomata.

Abscisic acid. Mutants have played an important role in our understanding of plant physiology and plant water-relations. For example, the so-called wilty mutants of tomato helped identify the role of abscisic acid in transpiration. Stomata of these mutants are always open, which causes the plants to desiccate and appear wilty. These mutants contain only about 10% as much abscisic acid as do normal plants. Moreover, adding abscisic acid to them causes their stomata to close and the plants to regain turgor, suggesting that abscisic acid prevents desiccation. Subsequent research has supported this conclusion. For example, applying as little as 10^{-6} M abscisic acid causes the stomata of most plants to close, and desiccation stimulates the synthesis of abscisic acid in the mesophyll cells of leaves. Plants use abscisic acid as a messenger to conserve water during drought (see Chapter 18).

CONCEPT

Leaf abscission, water-use efficiency, leaf position, circadian rhythms, and the production of abscisic acid are physiological adaptations of plants that affect transpiration.

BOXED READING 21.3

TAPPING WOOD FOR MAPLE SYRUP

From early March through early April of each year, many farms in western Massachusetts display the same sign: Maple Syrup For Sale. In most parts of New England, the scene is repeated, just as it was by pilgrims and Native Americans. Although the product is delicious, most people are no longer accustomed to maple syrup and prefer the commercial syrup sold in most grocery stores, which is cheaper. Commercial syrup contains only about 2% maple syrup (the rest is corn syrup and sugar), and tastes different than pure maple syrup.

Maple syrup is harvested from the trunks of sugar maple trees, *Acer saccharum*. Unlike the production of other crops, the production of maple sugar requires widely fluctuating temperatures. From late autumn to early spring in a narrow region of North America, a brief period of daily freezes and thaws triggers the flow of sap in sugar maple trees. This freezing and thawing is important because the flow stops if temperatures are constantly below or just above freezing. During cold nights, starch made during the previous summer and stored in wood is converted to sugar. The next day's warm temperatures create a positive pressure in the xylem's sapwood. When metal tubes are driven into the sapwood, this pressure pushes the sugary sap out of the trunk at a rate of 100–400 drops per minute. Some trees produce as much as 150 liters of sap per season, and attaching vacuum pumps to the tubes can increase the seasonal yield by nearly threefold. The sugar content of maple sap ranges from 1%–10%, but is usually about 3%. Once collected, the sap is heat-distilled; about 40 liters of sap are required to make 1 liter of syrup. The characteristic taste of maple syrup is due to this heating and the presence of several amino acids in the sap.

The production of maple syrup has several other unusual characteristics. For example,

BOX FIGURE 21.2

This maple "rancher" is collecting sap from a sugar maple (*Acer saccharum*).

BOX FIGURE 21.3

Sap drips from a metal tube driven into the sapwood of a sugar maple.

- Flow usually stops in the afternoon and does not start again until the temperature rises above freezing the next morning. Maple "ranchers" harvest most sap between about 9:00 A.M. and noon.
- Because it depends on the weather, the flow of sap is often sporadic. There may be 5–10 "runs" of sap-flow during a single spring.
- The concentration of sugar in sap is low early in the season, rises to a maximum, and then gradually decreases at season's end.
- Farmers use power drills with half-inch bits to drill about three holes into each tree. Each hole is two to three inches deep. Because this tapping removes less than 10% of a maple tree's sugar, trees can be tapped repeatedly without producing significant damage. Some of the trees tapped today were probably also tapped by pilgrims in the 1600s.

Production of maple syrup has been changed little by science and modern technology. We still do not understand precisely how the freeze-thaw cycle stimulates flow of sap, and experiments aimed at increasing sap-flow have, for the most part, failed. Despite remaining a regional product surrounded by mystery, maple syrup is enjoying a renewed popularity, and is now included in foods ranging from pumpkin soup to barbecued pork ribs.

Structural Adaptations

Cuticle. The retention of water, and thus survival, would be almost impossible for plants without a cuticle. The cuticle is an effective means of conserving water: less than 5% of the water lost by a plant evaporates through the cuticle. In general, thicker cuticles provide more protection from desiccation than thin ones. Thus, desert plants typically have thick cuticles, while those of aquatic plants are thin or absent.

Trichomes. Although trichomes increase the thickness of the boundary layer overlying a leaf, the primary means by which trichomes decrease transpiration is by reflecting light and thus decreasing the temperature of a leaf.

Sunken stomata. Sunken stomata increase the boundary layer surrounding guard cells. Therefore, plants with sunken stomata typically transpire less than do plants with raised stomata.

Reduced leaf area. Many desert plants (e.g., cacti) have greatly reduced leaves, thereby decreasing their evaporative surface area. In these plants, succulent stems that store large amounts of water often replace leaves as the primary photosynthetic organs.

CONCEPT

Cuticles, trichomes, reduced leaves, and sunken stomata are structural adaptations that reduce transpiration.

Almost all of the factors that affect transpiration do so by influencing the opening or closing of stomata. For example, decreasing the internal concentration of CO_2 in a leaf opens stomata and therefore increases transpiration. Thus, understanding the movement of water through plants and into the atmosphere is largely a matter of understanding how stomata open and close.

Stomata

Stomata regulate gas exchange between the atmosphere and a plant, and are a key adaptation to life on land. That is, they moderate the compromise between photosynthesis and water loss—or, as one botanist aptly stated, "providing food while preventing thirst."

Structure and Distribution of Stomata

Leaves have anywhere from 1,000 to 100,000 stomata per square centimeter of leaf area. Each *stomatal apparatus* consists of two guard cells and adjacent epidermal cells called *subsidiary cells*, all of which surround a pore (see figs. 13.18 through 13.20). Guard cells and stomata have several distinguishing features:

- Unlike other epidermal cells, guard cells typically contain chloroplasts. The amount of starch in these chloroplasts increases at night and decreases during the day.
- Guard cells have distinctive shapes: in dicots, they are crescent-shaped, while in most grasses they are shaped like dumbbells.
- Guard cells typically lack functional plasmodesmata. Water and solutes enter guard cells from the apoplast.

Guard cells surround the only pores in an intact epidermis. When wide open, stomatal pores are usually only 3–12 μm wide and 10–40 μm long. Although these dimensions seem small, they are gigantic compared to the sizes of gas molecules that move through them. For example, a water molecule has a diameter of 0.00025 μm; even when a stoma is closed and has a diameter of only 1 μm, several thousand water molecules pass through the closed stoma at the same instant.

Opening the Stomatal Pore

Guard cells control the size of a stomatal pore by changing shape: their unusually elastic walls buckle outward when stomata open and sag inward when stomata close. What causes these changes?

HYPOTHESIS 1

Stomata close as a result of the desiccation of guard cells.
This hypothesis states that stomata close as a result of desiccation of the plant and its guard cells, thereby decreasing water loss at times of reduced availability. However, stomata often close when plants are fully turgid and well-watered—at night, for example. Therefore, desiccation is not the primary means of closing stomata.

HYPOTHESIS 2

Sugars produced by chloroplasts in guard cells reduce the solute potential of the cell, thereby drawing water into the guard cells and causing them to bow apart.
According to this hypothesis, the light-stimulated opening of guard cells results from synthesis of sugar in guard-cell chloroplasts. These sugars would lower the solute potential of the guard cells, thereby drawing water into them and making them bow apart.

This hypothesis was initially attractive because it explained the accumulation of starch in guard cells at night and the loss of starch during the day: incorporating sugars into insoluble starch at night would increase the solute potential of the guard cells and cause water to leave them, thereby explaining why stomata close at night. Although logical, this hypothesis does not explain several aspects of stomatal functioning. Specifically, chloroplasts in guard cells do not produce enough sugar to account for the decreases in solute potential necessary to open stomata. Furthermore, guard cells of plants such as onion (*Allium*) lack starch, yet function normally. Therefore, stomatal opening and closing are not caused by the production of sugars in light and their conversion to starch in the dark.

The failure of these two hypotheses to explain stomatal behavior frustrated botanists for several years. However, botanists later discovered that stomatal opening correlates with a huge increase in the amount of K^+ in guard cells, and that stomatal closing correlates with a corresponding decrease in the concentration of K^+ in guard cells. That is, the osmotic gradient responsible for drawing water into guard cells is an influx of K^+ from subsidiary cells. This discovery led to other hypotheses for stomatal function.

HYPOTHESIS 3

Light provides energy to pump K^+ into guard cells.
This hypothesis was attractive because it correlated light-stimulated opening of stomata with a mechanism for pumping K^+ into guard cells. However, botanists soon discovered that this hypothesis could not explain stomatal opening in the dark in CO_2-free air.

HYPOTHESIS 4

Guard cells pump K^+ in and H^+ out during stomatal opening.
Guard cells make large amounts of organic acids during stomatal opening. The most abundant of these is malic acid, which ionizes and releases H^+ into the cytoplasm

of the guard cell. The H⁺ is pumped out of the guard cell as K⁺ is pumped in and stored in the vacuole. The accumulation of K⁺ in guard cells may reach 0.5 M, which decreases the solute potential of the guard cell, despite a twofold dilution resulting from expansion of the guard cell. Similarly, stomata close when guard cells pump K⁺ into the apoplast: this causes water to leave the guard cells via osmosis and the pore to close. Subsidiary cells and other epidermal cells surrounding the guard cells function as reservoirs where K⁺ is stored when stomata are closed.

The active pumping of K⁺ into guard cells remains the most widely accepted mechanism for stomatal opening. Moreover, it accounts for other aspects of stomatal opening and closing. For example, flushing leaves with CO_2-free air triggers accumulation of K⁺ by guard cells, thereby causing stomata to open. Furthermore, isolated pairs of guard cells open when K⁺ is present.

Stomata open as a result of the active uptake of K⁺ by guard cells. This K⁺ decreases the solute potential of guard cells, thereby causing water to enter the guard cells and turgor pressure to increase. This increase in turgor pressure causes guard cells to bow apart and form the stomatal pore.

Early studies indicated that in some plants, the walls of guard cells adjacent to the stomatal pore were thicker and less elastic than walls along other sides of the cell. This difference in wall elasticity and thickness was presumed to be the basis for guard cells bowing apart when turgid: the cell walls adjacent to the pore stretched less than did other walls, thereby forming a pore between the cells. However, more recent studies have shown that the guard cells of most plants lack a thick, inelastic wall bordering the pore. Therefore, a differential thickness in cell walls of guard cells cannot account for stomatal opening. Rather, stomatal opening is due to radial micellation of guard cells by cellulose microfibrils arranged much like the belts in a belted tire. These microfibrils are inelastic and restrict radial expansion of guard cells while allowing increases in length. When the guard cells lengthen due to the K⁺-driven influx of water, they bow apart and form the stomatal pore (fig. 21.12).

A Model to Regulate the Photosynthesis-Transpiration Compromise

Botanists have spent decades trying to understand the interactions that affect transpiration. The most popular model to account for the compromise between photosynthesis and transpiration is shown in fig. 21.13 and involves two feedback loops: one loop monitors the needs of photosynthesis by measuring the internal concentration of CO_2:

1. Photosynthesis decreases the internal concentration of CO_2 in a leaf.

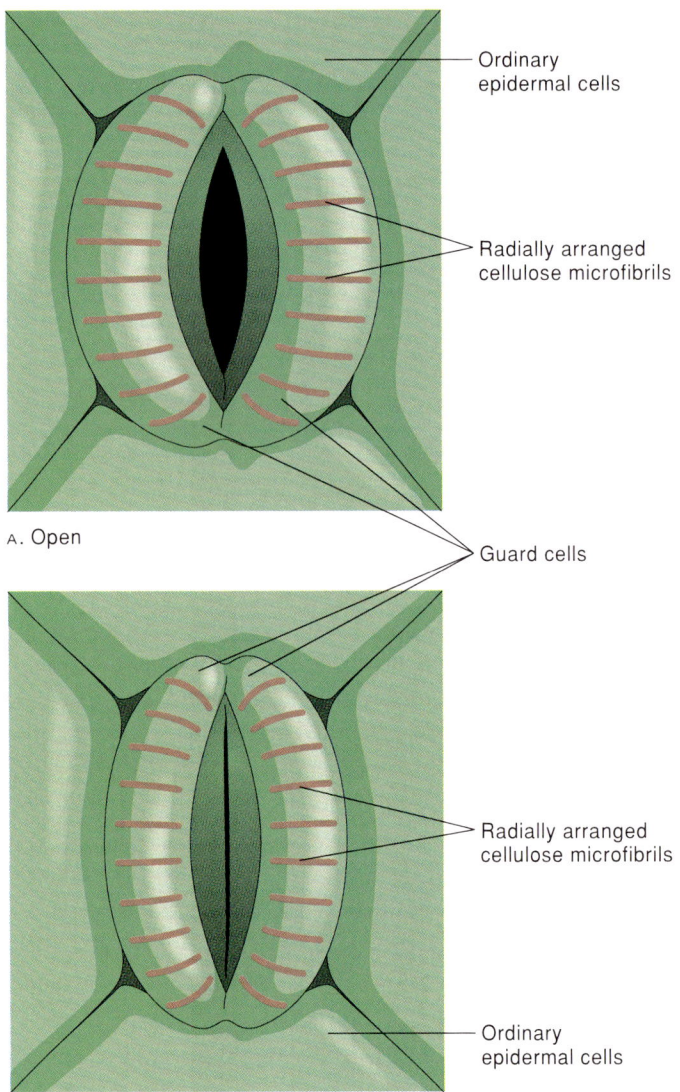

FIGURE 21.12

Stomatal opening is due to the movement of water into guard cells, which have radially arranged, inelastic microfibrils in their cell walls. (a) When water moves into guard cells, the radial reinforcements of the cells cause the cells to bow apart, thereby forming a pore. Movement of water into the guard cell occurs via osmosis triggered by the pumping of K⁺ into the cells. (b) When water leaves guard cells, the guard cells collapse and close the stomatal pore. The movement of water out of guard cells occurs via osmosis triggered by the pumping of K⁺ out of the cells.

2. This reduction in internal CO_2 causes guard cells to take up K⁺.
3. Accumulation of K⁺ in guard cells causes stomata to open.
4. CO_2 diffuses through open stomata to mesophyll cells, where it is used in photosynthesis.

The second loop protects against desiccation:

1. Desiccation stimulates synthesis of abscisic acid.
2. Abscisic acid moves to guard cells in the transpiration stream.

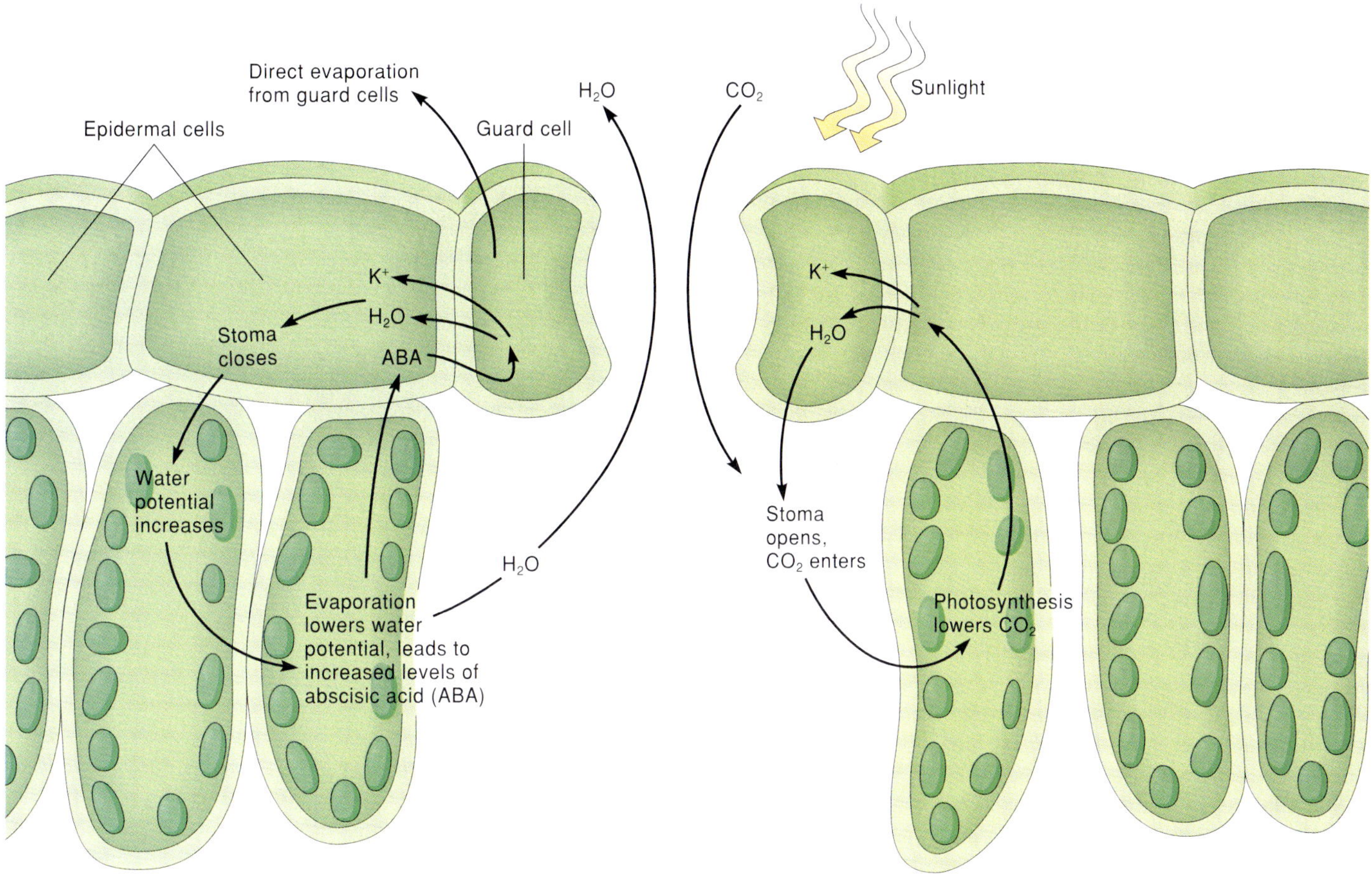

FIGURE 21.13

Model of the transpiration-photosynthesis compromise. The right side shows the feedback loop that monitors photosynthesis; the left side shows the feedback loop that protects against desiccation.

3. Abscisic acid stimulates transport of K^+ out of guard cells.
4. This loss of K^+ causes stomata to close, thereby conserving water.

These two loops cause stomata to open when light is sufficient for photosynthesis and to close when the risks of dehydration exceed the potential gains of photosynthesis.

The Adaptive Value of Transpiration

Transpiration has several beneficial effects on plants. It can keep the water column moving from roots to shoots, it can cool a leaf as much as 10–15° C below air temperature, and it can move minerals in the transpiration stream. These benefits, however, do not seem to be *essential* for plant growth. The water column can remain intact without transpiration, as it does when stomata are closed. Moreover, most minerals can move independently of transpiration, and leaves in full sunlight are seldom damaged by increases in temperature that occur when stomata close and transpiration is reduced by wilting at midday. This brings us back to our original question: Of what adaptive value is transpiration? Although in the future we may discover an essential role for transpiration, today most botanists regard transpiration as an unavoidable evil—unavoidable because of leaf structure, and evil because it desiccates and often injures leaves. Except in dry areas, the evolution of leaves capable of high rates of photosynthesis was of greater survival value than that of more water-efficient leaves.

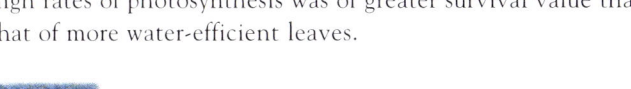

The Lore of Plants

The interior of a forest can be 10–15° C cooler than the surrounding countryside because of transpiration. Indeed, each large tree of a forest has the cooling capacity of about five air-conditioners.

Coping with Extremes: Too Much or Too Little Water

No factor is more responsible for plant diversity than the availability of water. In general, plants can be grouped into three categories based on their adaptations to the amount of water in the environment in which they live: mesophytes, hydrophytes, and xerophytes. We have already discussed many of the features of **mesophytes,** which grow in environments that have moderate amounts of water. A few plants have evolved ways of coping with extreme environments, such as those containing too much or too little water.

FIGURE 21.14

In white mangrove (*Laguncularia racemosa*), gases diffuse to submerged roots through pneumatophores.

FIGURE 21.15

Aeration in rice. Submerged leaves of rice are never wetted. Continuous air layers, which appear silvery under water, are trapped between hydrophobic leaf surfaces and the surrounding water. These layers of air are open to the atmosphere and form a low-resistance pathway for gas exchange.

Hydrophytes

Plants growing in water or excessively wet soil are called **hydrophytes.** These plants have few characteristics for conserving water; for example, they usually lack a cuticle and have reduced amounts of xylem.

The presence of too much water makes gas exchange a serious problem for hydrophytes; for example, the diffusion coefficient of O_2 is 10,000 times slower in water than in air, and the concentration of O_2 is 30 times less in water than in air. Also, the water in which hydrophytes grow often absorbs much light, thereby shading the leaves. Hydrophytes have evolved several adaptations that minimize these problems:

- Hydrophytes usually contain large amounts of aerenchyma tissue, which, you'll recall, has large amounts of air-containing intercellular spaces. In plants such as mangroves, aerenchyma-filled roots called pneumatophores project upward from the soil into the air and function as snorkels through which gases can diffuse to submerged roots (fig. 21.14).
- The underwater foliage of many hydrophytes is thin or dissected, so that no cell is far from water and its circulating currents.
- Hydrophytes use light efficiently; almost all their cells are photosynthetic, including the cells of the epidermis.

Partially submerged plants such as rice use trapped layers of air as routes of gas exchange (fig. 21.15). In these plants, layers of air continuous with the atmosphere are trapped between the hydrophobic surface of the leaf and the surrounding water. Gases diffuse from the atmosphere to submerged parts of the plant through these trapped layers of air.

Xerophytes

The problem of gathering enough water for growth is intensified by excessively dry habitats such as deserts. Plants that live in these dry environments are called **xerophytes** and have adaptations such as a reduced surface area, succulent leaves and stems, a thick cuticle, sunken stomata, and Crassulacean acid metabolism.

Plants such as mesquite *avoid* drought by growing deep roots. Succulents such as cacti *endure* drought by storing water in succulent tissues. For example, a large saguaro cactus (*Carnegiea gigantea*) may contain several tons of water in the parenchyma cells of its stem. Plants gather this water by producing hydrophilic colloids that decrease their water potential. However, not all plants that live in deserts are xerophytes. For example, desert star (*Monoptilon bellidiforme*) escapes drought by completing its life cycle during the brief rainy season in the desert. Thus, these plants are mesophytes that escape the droughts typical of xeric environments.

TRANSPORTING ORGANIC SOLUTES IN THE PHLOEM

The first serious studies of how organic solutes move in plants were done in the 1800s by Theodor Hartig, a German botanist. Hartig was a forester interested in determining how the products of photosynthesis move in trees. In 1837 Hartig discovered a new cell type in the bark of trees. He called these cells *sieve tube members* and suspected that they were the conduits for moving sugars from leaves to roots. To test his hypothesis, Hartig reasoned that if sieve tubes of bark were the cells through which nutrients moved, then removing a ring of bark from the tree trunk should cause nutrients to accumulate above this so-called girdle. This is exactly what happened (fig. 21.16). Because the wood was intact, Hartig concluded that nutrients move through the bark and not through the secondary xylem of the trunk. In

FIGURE 21.16

This girdle on a black cherry tree (*Prunus serotina*) blocked the flow of nutrients through phloem from above, stopping growth below.

the late 1850s Hartig began other experiments that eventually linked translocation with sieve tubes. He made a series of shallow cuts into the sieve tubes and, to his surprise, discovered that sap oozed from these incisions. Hartig concluded that organic solutes moved in sieve tubes. More recent studies with radioactive tracers have confirmed Hartig's conclusion: sugars and other organic substances move almost exclusively in sieve tubes of the phloem.

Structure of Conducting Cells

Hartig's experiments stimulated an enormous amount of research and eventually a storm of controversy. The reason for this research and controversy was relatively simple: structure and function are inseparable, and one cannot understand phloem transport without understanding the structure of the conducting cells.

The early studies of sieve tube structure provided valuable information; namely, that sieve tube members are arranged end to end and are associated with files of parenchymalike cells called *companion cells* (see Chapter 13). Companion cells and sieve tube members function as a single unit.

Sieve tube members are cylinders connected by sievelike areas called *sieve plates*, each of which has numerous *sieve pores* 1–5 μm in diameter. Sieve pores may occupy more than 50% of the area of a sieve plate. It is the nature of these pores that has generated so much controversy. Open sieve pores offer much less resistance to solute flow than do clogged pores. Since solutes traverse as many as 12,000 sieve plates per meter, knowing whether sieve pores are open or occluded is fundamental to understanding phloem transport.

Most of the initial studies of sieve tube structure indicated that sieve pores were clogged with a proteinaceous material called *P-protein*, or (less elegantly) *slime*. Some of these studies also showed that sieve pores were frequently clogged with *callose*, an amorphous polymer of glucose. These occlusions made it difficult to envision how solutes could move through sieve tubes. Nevertheless, many botanists accepted this structure, and began devising models for phloem transport based on clogged sieve pores. Other botanists, however, suspected that this clogging was an artifact; that is, that sieve pores are open in functional sieve tubes. If their suspicion were correct, these botanists would not only have to explain why the clogged pores were artifacts, but would also have to provide evidence that the sieve pores were, in fact, open.

Botanists explained the clogged sieve pores as follows. Sieve elements are extremely sensitive to manipulation and preparation for microscopic examination. Because sieve tubes contain large amounts of sugars, water is drawn into them by osmosis and generates a high hydrostatic pressure. When the sieve tubes are severed during sampling or when their membranes are altered by the chemical fixatives used for microscopy, this pressure is released, and their contents surge toward the site of pressure release. Surging P-protein becomes trapped in the sieve pores and gives the appearance that the pores are normally clogged, when in fact the clogging is due only to preparation of the tissue for study.

Although this explanation for clogged sieve pores seemed logical, a more formidable task remained: it had to be *shown* that the pores were open in functioning sieve tubes. Botanists attacked this challenge by using a new means of fixation: instead of fixing the severed sieve tubes with slow-acting chemicals, the tissues were fixed with rapid freezing. These fixations showed that sieve pores are not clogged by P-protein, callose, or any other material. That is, sieve pores are open in functional sieve tubes.

Sieve tubes in most plants are short-lived; they usually function only during the season in which they are formed. In these plants, sieve tubes are eventually replaced by cells derived from the vascular cambium. Plants such as basswood (*Tilia*), however, retain functional sieve elements for several years. During periods of dormancy, sieve pores become clogged with callose and become nonfunctional. When growth resumes in the spring, this callose is hydrolyzed, and the sieve tubes again transport sugars; the products of callose hydrolysis are used as substrates for the renewal of growth.

Callose and P-protein are located along the periphery of functioning sieve tube members. Moreover, callose is rapidly synthesized when sieve tubes are wounded. Callose plugs the pores of wounded sieve tubes and prevents the loss of assimilated nutrients through the wound. Similarly, P-protein rapidly clogs the pores of wounded sieve tubes and minimizes the loss of sugars.

CONCEPT

Sugars and other organic compounds move throughout plants in the phloem. These solutes move through living cells and under a positive pressure.

How Substances Move in the Phloem

Several models for phloem transport have been proposed since Hartig reported his observations in the mid-1800s. The validity of these models is judged by their ability to explain and accurately predict phloem transport. We have already established the *path* of transport: solutes move in sieve tubes that have open sieve pores. Another important fact that any model for phloem transport must explain is the *rate* of phloem transport.

Rates of Phloem Transport

Solutes move surprisingly fast in the phloem: peak rates of transport may exceed 2 m h^{-1}. The consequences of this rate of transport are impressive: indeed, as much as 20 liters of sugary sap can be collected per day from the severed stems of sugar palm. At the cellular level, solute transport is turbulent: a sieve element 0.5 mm long empties and fills every two seconds, thereby delivering about $5-10 \text{ g sugar h}^{-1} \text{ cm}^{-2}$ of phloem area to sites of sugar storage or utilization. The largest estimate for sugar transport through sieve tubes is $180 \text{ g sugar h}^{-1} \text{ cm}^{-2}$ of phloem area. Picture what this means: a gram of solid sugar passes through a square centimeter of sieve tube area every twenty seconds.

Models for Phloem Transport

The validity of a biological model depends on how well it accounts for and predicts a biological activity. Let's see how some of the proposed models for phloem transport compare with the experimental evidence.

HYPOTHESIS 1

Solutes move through the phloem via diffusion.
Diffusion is much too slow to account for phloem transport in plants. Consider a 10% sucrose solution connected to a pan of pure water by a 1-meter tube with a cross-sectional area of 1 cm^2: the diffusion of only 1 mg of sucrose to the pan of pure water would require almost three years. Therefore, mechanisms more rapid than diffusion must be involved in solute transport in phloem.

HYPOTHESIS 2

Solutes move through the phloem via cytoplasmic streaming.
Cytoplasmic streaming is too slow to account for phloem transport. Moreover, streaming apparently does not occur in mature sieve tubes. Therefore, cytoplasmic streaming alone cannot account for the movement of solutes in the phloem.

HYPOTHESIS 3

Solutes move through the phloem via pressure flow.
In 1926, German plant physiologist Ernst Münch proposed the **pressure-flow model,** which states that a turgor-pressure gradient drives the unidirectional mass flow of solutes and water through sieve tubes of the phloem. According to this model, solutes move through sieve tubes along a pressure gradient in a manner similar to the movement of water through a garden hose.

Like all good models, Münch's model is testable. Consider the experimental set-up shown in figure 21.17, in which (1) A and B are osmometers permeable only to water, (2) osmometer A contains more solutes than osmometer B, (3) both osmometers are submerged in water, and (4) the osmometers and water container have open connections that offer little resistance to the flow of water and solutes. Here's what happens:

1. Because the osmometers contain solutes, their water potential is less than that of the surrounding water. Consequently, water enters A and B by osmosis, which produces a turgor pressure.
2. Because osmometer A contains more solutes than osmometer B, the turgor pressure in A will exceed that in B. Thus, the differing solute concentrations in the two osmometers produces a pressure gradient.
3. Because the osmometers are connected, the pressure gradient is transmitted throughout the entire system. This pressure moves fluid in A through the connecting tube to B. Solutes in the flowing fluid are carried passively from A to B via a mass flow.
4. The movement of fluid from A to B builds pressure in B, which causes water to leave osmometer B, thereby creating a circulating system.

Four requirements must be satisfied for Münch's model to work:

1. There must be an osmotic gradient between the two osmometers.
2. Selectively permeable membranes must be present to establish a pressure gradient.
3. There must be an open channel between the two osmometers to allow flow.
4. The surrounding medium must have a water potential that exceeds (i.e., is less negative than) that of the most negative osmometer.

Let's now use these criteria to determine if Münch's model can explain the movement of solutes in phloem.

Does an osmotic gradient occur in the phloem? Yes; this gradient occurs between sources and sinks. *Sources* are the sites where sugars are made by photosynthesis or hydrolysis of starch. Sources are analogous to osmometer A in figure 21.17: they contain large amounts of sugar, and their solute concentration is high. Chlorenchyma cells are examples of sources. *Sinks* are sites where sugars are used or stored as insoluble starch. Sinks are analogous to osmometer B: they contain less sugar than do the sources, and

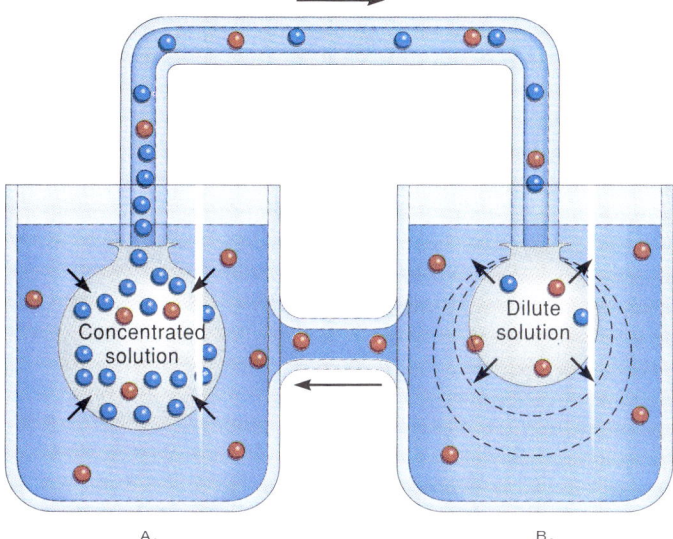

FIGURE 21.17

Model of the pressure-flow hypothesis for solute movement. Solutes move in a mass flow generated by differences in pressure in osmometers A and B.

therefore their solute concentration is relatively low. Roots, active meristems, and developing fruits are examples of sinks.

The osmotic gradients between sources and sinks are large enough to account for phloem transport. For example, the solute concentration decreases by about 0.01 M per meter in sieve tubes linking sources and sinks of white ash (*Fraxinus americana*). Furthermore, solute transport is directly proportional to the magnitude of this gradient: steeper gradients correlate with faster transport. Finally, the contents of sieve tubes are under a positive pressure, thereby explaining Hartig's observation that sap oozes from punctured sieve tubes. Taken together, these observations indicate that osmotic gradients occur in phloem and that these gradients are important for phloem transport.

Are semipermeable membranes present to form and maintain a pressure gradient? Yes. Conducting elements in the phloem must be alive and have intact membranes to function.

Does a channel exist between sources and sinks that allows for the flow of solutes? Yes. The sieve pores of sieve elements are open and form a conduit through which solutes can move.

Is the water potential of the surrounding medium more negative than that of the sink? Yes. The solute concentration in sieve tubes is 2–3 times greater than that of surrounding cells. In summary, the pressure-flow model is attractive because it explains source-to-sink movement in plants (fig. 21.18):

1. Sucrose produced at a source is loaded into a sieve tube.
2. This loading decreases the water potential, which causes water to enter the sieve tube by osmosis.
3. The influx of water into sieve tubes creates a pressure gradient that carries sucrose to a sink, where it is unloaded.
4. Removing sucrose at the sink increases the water potential there, causing water to move out of the sieve tube at the sink.

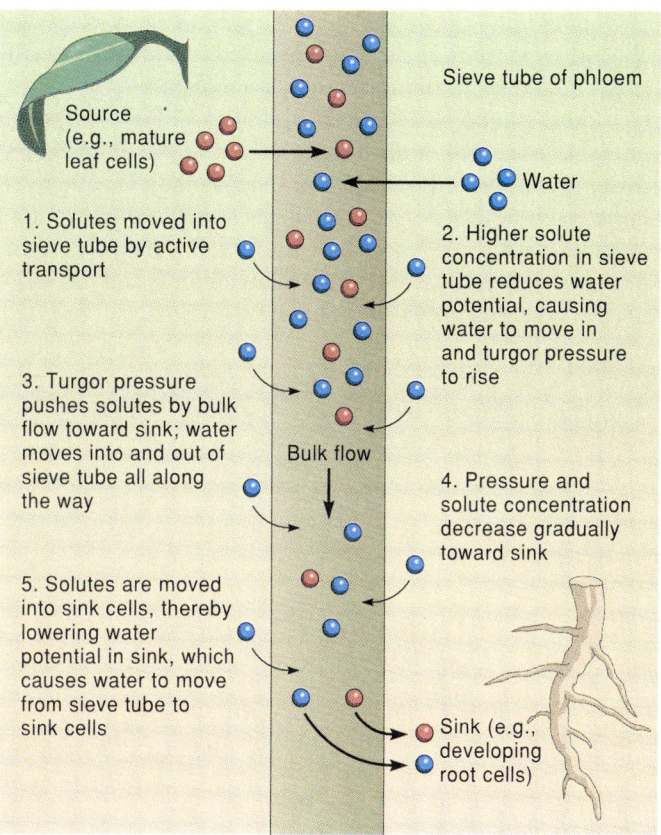

FIGURE 21.18

Proposed mechanism for pressure flow in the phloem of flowering plants. Solutes move via mass flow from sources to sinks.

5. Sucrose is either used or stored at the sink. Water exiting the sieve tube at the sink returns to the xylem and is recirculated.

The ability of the pressure-flow model to account for and accurately predict the characteristics of phloem transport makes it the most widely accepted model for phloem transport.

Transport of organic solutes in phloem is best explained by the pressure-flow model. This model states that sugars are carried by water along a gradient of turgor pressure generated by an osmotically driven influx of water at sources and an exit at sinks.

Unlike the unidirectional upward flow of water in xylem, phloem solutes move in all directions; these directions are determined by the presence of sources and sinks. However, the overall scheme of sugar transport in plants is not as simple as moving solutes from one source to one sink. Rather, sucrose may be required simultaneously by *several* sinks. The presence of numerous sinks produces two interesting questions.

Do all sinks receive equal amounts of sugars? No: larger sinks receive more sugars than other sinks. For example,

CHAPTER TWENTY-ONE *Movement of Water and Solutes*

developing fruits are strong sinks that often take priority over other sinks, thereby explaining why vegetative growth usually slows or stops until fruit formation is complete.

Do all sources contribute equally to sinks, or is a sink preferentially supplied with sugars by only a few sources? Sources contribute primarily to the nearest sink rather than equally to all sinks. For example, leaves near roots usually transport their sugars to roots, while leaves near tips of stems transport sugars to the shoot apex. Leaves midway between roots and shoot tips transport their sugars to both of these sinks.

Writing to Learn Botany

The contents of sieve tubes are under a positive pressure, while those of nearby xylary elements are under a negative pressure. How, then, could these cells have the same water potential?

Loading and Unloading the Phloem

Phloem Loading

The Münch pressure-flow model accounts for the movement of solutes once they are loaded into sieve tubes. However, sieve tubes themselves are not the sources or sinks; that is, sugars are neither produced, needed, nor stored in sieve tubes. Rather, sieve tubes *link* sources and sinks. How, then, do solutes get from sources into sieve tubes and from sieve tubes to sinks?

Consider a chlorenchyma cell (i.e., a source) in a leaf and a cortical cell (i.e., a sink) in a root. Most chlorenchyma cells are 2–4 cell layers from a vein. Thus, sugars made in chlorenchyma cells must be transported across several other chlorenchyma cells before they can be loaded into a sieve tube. The movement of solutes between adjacent chlorenchyma cells is symplastic and is enhanced by the many plasmodesmata that link these cells. That is, symplastic transport accounts for the movement of solutes to chlorenchyma cells bordering the vein. However, plasmodesmata rarely occur between mesophyll cells and companion or sieve cells. This absence of symplastic links between chlorenchyma cells and sieve tubes puzzled botanists because it suggested that sugars had to move through the cell wall (i.e., the apoplast) before being loaded into sieve tubes. Experiments with radioactive tracers confirmed that this is what occurs. For example, radioactive $^{14}CO_2$ is rapidly incorporated into sugars by chlorenchyma cells of leaves; soon thereafter, the resulting radioactive sugars can be detected in the cell walls between chlorenchyma cells and veins. Radioactive sugars cannot be detected between adjacent chlorenchyma cells, providing further evidence that solute transport between chlorenchyma cells is symplastic.

Sugars in the cell wall are loaded into sieve tubes by companion cells, which often have many plasmodesmata and include structures similar to those of transfer cells (fig. 21.19). The elaborate invaginations of the cell wall and cell membrane of transfer cells provide a large surface area for transporting sugars from the cell wall into the sieve tube. The loading of

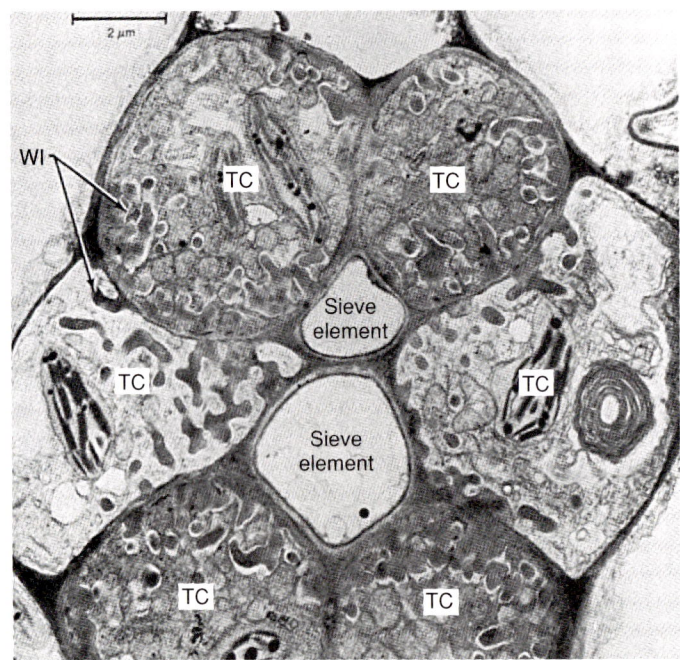

FIGURE 21.19

Cross section of a leaf vein from common groundsel (*Senecio vulgaris*), showing sieve elements and transfer cells. The wall ingrowths of transfer cells produce large surface areas that are used to enhance loading and unloading of sieve elements.

sieve tubes from the apoplast requires metabolic energy and is driven indirectly by a proton gradient (fig. 21.20) generated at the expense of ATP. Here's how it occurs:

1. ATP and an H^+-carrier in the cell membrane are used to pump protons out of the sieve tube.

2. This pumping forms a proton gradient across the membrane, with the largest concentration of H^+ being on the outside of the membrane. Electrical neutrality is maintained by K^+ entering the sieve tube, which accounts for the large amount (1–2 mg ml^{-1}) of K^+ in phloem sap.

3. Diffusion of protons back into the sieve tube is coupled to a carrier and powers the transport of sugar into the sieve tube.

How do we know that phloem loading occurs like this? The answer is experimental evidence:

- *Requirement for ATP*: ATP is common in phloem sap, and phloem loading stops when a plant is exposed to respiratory poisons such as cyanide. Furthermore, supplying sieve tubes with ATP significantly increases phloem loading. Finally, sugars are pumped into sieve tubes against a concentration gradient: the concentration of sucrose in a chlorenchyma cell is typically 10–50 mM, while that in a sieve tube of a minor vein of a leaf may be as high as 1 M. Taken together, these results indicate that energy of ATP is required for phloem loading.

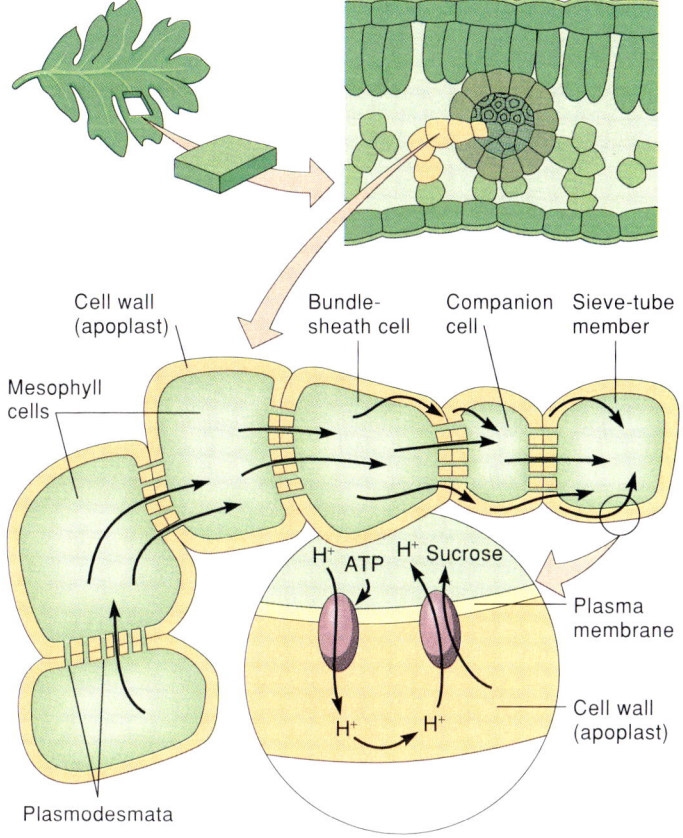

FIGURE 21.20

Model of sucrose loading into phloem. According to this model, H⁺ is pumped out of the sieve tube, using energy from ATP. Sucrose is transported into the sieve tube.

- *Proton pumping:* Phloem loading coincides with an increase in the pH of sieve tubes to about 8 and a simultaneous decrease of the apoplast pH to about 5.5. These changes are consistent with secretion of H^+ from the sieve tube into the cell wall. The loading of sieve tubes also coincides with an increase in the concentration of K^+ in the sieve tube, an observation consistent with the idea that K^+ is absorbed by sieve tubes to maintain electrical neutrality.
- *H^+ and phloem loading:* Bathing the apoplast with neutral buffers inhibits phloem loading, and acidifying the apoplast enhances phloem loading. This is consistent with a proton gradient being necessary for the loading of sugars into sieve tubes.

Sugars and other solutes are loaded selectively into sieve tubes: only those solutes that are transported will be loaded. For example, sucrose is always present in sieve tubes, while glucose is rarely present. If veins are bathed in a solution of sucrose and glucose, only sucrose will be loaded into sieve tubes. This selectivity depends on specialized membrane-carriers in the cell membrane of sieve tubes and companion cells.

Phloem Unloading

Unloading of solutes from sieve tube members can occur symplastically or apoplastically. In vegetative sinks that are growing, such as roots and young leaves, unloading is usually symplastic. In other sinks, unloading is usually apoplastic. The mechanisms underlying phloem unloading may vary in different species.

Influence of the Environment on Phloem Transport

As is true for transpiration, several environmental factors affect phloem transport. One of them is light, which promotes photosynthesis and therefore increases the production of sucrose. As a result, increased light intensity generally promotes transport to roots. Similarly, darkness stimulates translocation from roots to shoots.

Mineral deficiencies are also important in phloem transport, which is strongly affected by the nutritional status of a plant. For example, phloem transport is slow in boron-deficient plants; transport increases dramatically when these plants are supplied with boron. Potassium deficiencies also decrease phloem transport, presumably because of the dependence of phloem loading on K^+ uptake.

Finally, there is some evidence that hormones affect transport in the phloem. These effects are presumably due to altered growth and metabolism rather than to direct stimulation of phloem loading or unloading.

What are Botanists Doing?

Phloem loading and unloading are hot topics in botanical research. Using these as keywords for library searching, find out what studies have been done on these topics in the past two years.

What Moves in the Phloem?

You have already learned that sieve tubes transport sugars and some other organic solutes. But exactly which and how much of these substances move in sieve tubes?

Methods of Examining the Contents of Sieve Tubes

One seemingly easy way to determine the content of a sieve tube would be to collect phloem sap as it oozes from a wounded stem. However, there are problems with this approach, because such wounds rupture and kill other cells, and their contents mix with and contaminate the exudate from phloem. Cutting also releases the pressure in sieve tubes, which decreases their water potential and causes water to move into the cells by osmosis. Ideally, a botanist would have a microneedle that could be delicately inserted into a single sieve tube without a sudden release of pressure.

 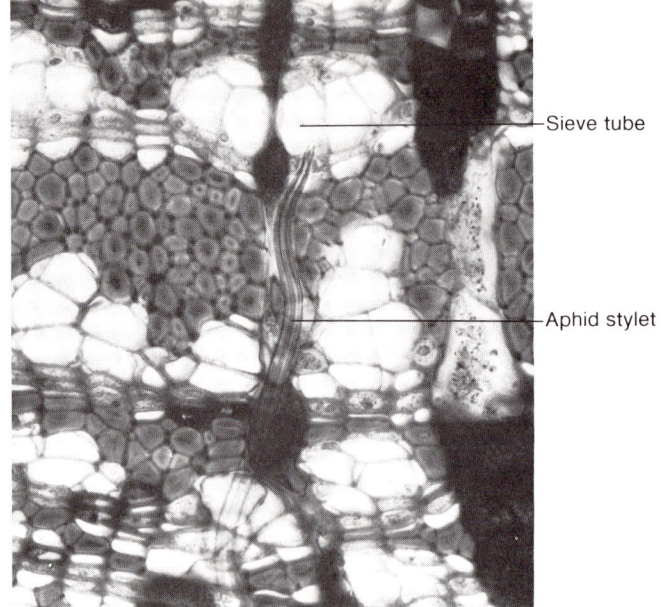

FIGURE 21.21

Aphids are phloem-feeders. (a) An aphid feeding on a plant stem; excess phloem sap (droplet of honeydew) passes through the aphid. (b) Micrograph showing an aphid stylet in a phloem cell.

Such a microneedle was discovered in 1953. Surprisingly, its discovery resulted from neither a technological breakthrough nor even the efforts of botanists. Rather, it was provided by two insect physiologists, J. S. Kennedy and T. E. Miller. These scientists were studying the feeding habits of aphids, a group of insects that obtain food from phloem. Kennedy and Miller noted that aphids are natural phloem-probes that tap into individual sieve tubes with a needlelike mouthpart called a stylet (fig. 21.21). Because the contents of sieve tubes are under positive pressure, their contents surge into the aphid, which simply sits calmly and enjoys its sugary meal. This rush of phloem sap is often overwhelming; when this occurs, the aphid secretes excess sap as a drop of sugary *honeydew*, which eventually drops onto underlying plants, sidewalks, and cars. But how can one be sure that the aphid does not alter the contents of the sap before secreting it as honeydew? This problem was solved with some clever microsurgery:

1. An aphid was allowed to insert its stylus into a sieve tube member.
2. When the feeding started, the aphid was anesthetized with a gentle stream of CO_2.
3. The stylet was then severed from the aphid's body, leaving only the open-ended stylet inserted into a sieve tube member.

Positive pressure in the sieve tube pushed sugary sap through the severed stylet for several days: collection rates averaged about 1 mm³ of phloem exudate per hour. Kennedy and Miller wrote that the technique "might also be of use to plant physiologists." They were right: botanists soon took Kennedy and Miller's suggestion, transforming the lowly aphid from a garden pest to an important research tool and technician. Thanks to a timely assist from the insect world, botanists could finally determine exactly what is in a sieve tube.

Contents of Sieve Tubes

The most abundant compound in a sieve tube is water, which is important because as much as half of the water in many fruits is delivered in sieve tubes. More than 90% of the solutes in sieve tubes are carbohydrates. In most plants, these carbohydrates move largely or entirely as sucrose. The concentration of sucrose may be as high as 30%, thereby giving phloem sap a syrupy thickness. However, not all plants transport only sucrose; a few plant families also transport other sugars, such as raffinose, stachyose, and verbascose. These sugars are similar and consist of sucrose attached to one or more D-galactose units. Like sucrose, they are all nonreducing sugars, which are less reactive and less labile to enzymatic breakdown than are reducing sugars such as glucose and fructose.

Some plants also transport sugar alcohols in their phloem. For example, apple and cherry trees transport sorbitol, and mannitol moves in the phloem of ash (*Fraxinus*). While the Biblical manna that was miraculously supplied to the Israelites may have come from various sources, commercial manna (the source of mannitol) is the dried phloem-exudate of manna ash (*Fraxinus ornus*) and related plants. Sieve tubes also contain ATP and nitrogen-containing compounds such as amino acids, especially during senescence of leaves and flowers. Sieve tubes also transport hormones, alkaloids, viruses, and inorganic ions, especially K^+.

EXCHANGE BETWEEN THE PHLOEM AND XYLEM

The contents of xylem and phloem are in aqueous equilibrium; that is, they have a similar water potential. For example, water entering loaded sieve tubes comes from the xylem, and water leaving unloaded sieve tubes returns to the xylem and is recirculated. The movement of water between xylem and phloem is occasionally accompanied by an exchange of the solutes of these tissues.

Chapter Summary

Multicellularity of plants and their colonization of the land corresponded with the evolution of systems for long-distance transport. The vascular systems responsible for this transport are xylem, which moves water and dissolved minerals from the soil to leaves, and phloem, which moves sugars and other organic compounds throughout a plant. Water movement in both of these systems occurs because of differences in water potential, which is a measure of the potential energy of the water in the system. Water potential is the sum of the energy attributable to pressure, solutes, and wettable surfaces. Water always moves from areas of high water potential to areas of lower water potential.

Water moves from roots to leaves in tracheids and vessels of xylem. These conducting cells are dead and hollow at maturity, and have thick walls that can withstand the negative pressures characteristic of xylary transport. Leaves expose large evaporative surface areas to the dry atmosphere. This evaporation of water from shoots is called transpiration.

The movement of water through the xylem under negative pressure is best explained by the transpiration-cohesion hypothesis. The driving force for water movement is the evaporation of water from the walls of leaf cells, which decreases the water potential of the leaf and thus pulls replacement water from the xylem. The water-potential gradient that lifts water through the xylem extends the length of the plant and into the soil, thereby pulling water into the plant. The movement of water in plants is also promoted by the strong cohesion of water molecules. The adhesion of water to the cell walls of tracheids and vessels helps to prevent gravity from draining the water columns.

- *Environmental factors* that affect transpiration include atmospheric humidity, internal concentration of CO_2 in the leaf, air movement, air temperature, availability of water in the soil, and light.
- *Physiological adaptations* that affect transpiration include water-use efficiency, leaf abscission, dormancy, leaf position, stomatal cycling, circadian rhythms, and the synthesis of abscisic acid.
- *Structural adaptations* that reduce transpiration include the presence of a cuticle, trichomes, sunken stomata, and decreased leaf area.

All of the factors that influence transpiration do so primarily by affecting the opening and closing of stomata. A stoma consists of two guard cells surrounded by specialized epidermal cells called subsidiary cells. Guard cells open and close a small pore; this opening and closing is controlled by the uptake and loss of K^+ from subsidiary cells. The uptake of K^+ decreases the water potential of the guard cells, which causes water to enter them and increases their turgor pressure. Radial micellations of cellulose microfibrils prevent guard cells from expanding radially. Thus, the increase in turgor pressure causes them to lengthen and bow apart, which forms the stomatal pore.

The pattern of stomatal opening and closing is a compromise between obtaining enough CO_2 for photosynthesis and conserving water to prevent desiccation. This compromise is mediated by two feedback loops. One loop meets the needs of photosynthesis by monitoring the internal concentration of CO_2; low concentrations of CO_2 in a leaf cause stomata to open. The other loop protects the plant from desiccation by producing abscisic acid during dry periods, which causes stomata to close.

Several plants are adapted to growing in the presence of too much or too little water.

- *Hydrophytes* are plants that grow in water or in excessively wet soil. These plants have few characteristics for conserving water and are specialized for gas exchange. Hydrophytes usually contain large amounts of aerenchyma tissue.
- *Xerophytes* are plants that live in unusually dry environments. These plants conserve water with thick cuticles, sunken stomata, Crassulacean acid metabolism, decreased leaf area, and succulent tissue.

Organic solutes move through the phloem in sieve tube members, which are living cells arranged into pipelike structures called sieve tubes. Solutes move under positive pressure in sieve tubes. The primary solute transported in sieve tubes is sucrose.

Phloem transport is best explained by the pressure-flow model. According to this model, companion cells load sugars into sieve tubes at sites called sources. These sugars decrease the water potential of the sieve tube, so that water enters the cell and increases its turgor pressure. Meanwhile, the turgor pressure in sieve tubes decreases at sinks, where sugars are unloaded. As a result, sugars in sieve tubes are carried along a gradient of turgor pressure generated by an osmotically driven influx of water at sources and an exit at sinks. Phloem transport is affected by temperature, light, and mineral deficiencies.

Questions for Further Thought and Study

1. Where do the ions that move during stomatal opening and closing come from?
2. Discuss how epidermal cells control photosynthesis and transpiration.
3. Define water-use efficiency. Which plants use water most efficiently: C_3, C_4, or CAM plants? What is the basis for this increased efficiency?
4. Is it a good idea to fertilize plants during a drought? Why or why not?

5. Trace the path of water from soil to plant to atmosphere. State the structure and function of each cell that the water molecule passes through.
6. Discuss how and why water potential changes during a day.
7. What is the selective advantage of producing thousands of vessels and tracheids instead of one giant conduit for water transport?
8. Suppose you collected the water droplets formed by guttation. What would you expect to find dissolved in these droplets, if anything? Why?
9. Describe what happens to the water potential of guard cells and adjacent epidermal cells when stomata open and close.
10. During periods of reduced transpiration, almost all water moves through large vessel elements and large tracheids. Smaller tracheids are used to move water only when water flow increases. Why?
11. Leaves of redwood (*Sequoia sempervirens*) can absorb water from fog that bathes them along the California coast. Explain, in terms of water potential, how this absorption could occur.
12. Adding salt to roads is a common way of minimizing ice formation during winter. What are the consequences for plants living along the side of the road?
13. Describe how leaf architecture affects transpiration.
14. Plants have been likened to an open-ended tube stuck in soil. In what ways is this analogy correct? In what ways is it incorrect?
15. Discuss the adaptations of plants for living in excessively wet and dry environments.
16. Compare and contrast the movement of substances in phloem and in xylem. Include comparisons of the conducting cells, substances transported, directions of transport, and the driving force for each.
17. Discuss how sugars made in a mesophyll cell of a leaf are transported into a sieve tube.
18. Summarize the evidence supporting the pressure-flow hypothesis for phloem transport.
19. Could a cell have a water potential greater than zero? Why or why not?
20. Stomata, when open, occupy less than 1%–2% of the area of a leaf. However, diffusion of water through stomata often exceeds 50% of the amount that evaporates from a free-standing water surface. Explain this "paradox of the pores."
21. Vessels at the base of a tree are usually longer and wider than those near the shoot tips. For example, less than 10% of the vessels that occur 11 m high in a tree are longer than 4 cm, while more than half of the vessels at the base of the trunk exceed 4 cm. What is the adaptive significance of this?

Suggested Readings

Articles

Clifford, P. E., C. E. Offler, and J. W. Patrick. 1989. How are sugars unloaded from the phloem? Some answers from a novel experimental system. *Journal of Biological Education* 23:147–151.

Evert, R. F. 1982. Sieve tube structure in relation to function. *BioScience* 32:789–95.

Kozlowski, T. T. 1984. Plant responses to flooding of soil. *BioScience* 34:162–167.

Mansfield, T. A., and W. J. Davies. 1985. Mechanisms for leaf control of gas exchange. *BioScience* 35:158–164.

Pillsbury, A. F. 1981. The salinity of rivers. *Scientific American* 245:54–65.

Van Bel, A. J. E. 1993. Strategies of phloem loading. *Annual Review of Plant Physiology and Plant Molecular Biology* 44:253–282.

Zimmermann, M. H. 1982. Piping water to the treetops. *Natural History* 91:6–13.

Books

Flowers, T. J., and A. R. Yeo. 1992. *Solute Transport in Plants*. London: Blackie Academic and Professional Publishing Co.

Kramer, P. J. 1983. *Water Relations of Plants*. New York: Academic Press.

Martin, E. S., M. E. Donkin, and R. A. Stevens. 1983. *Stomata*. Studies in Biology, no. 155. London: Edward Arnold.

Moorby, J. 1981. *Transport Systems in Plants*. New York: Longman.

Steward, F. C. 1986. *Water and Solutes in Plants*. Vol. 9 of *Plant Physiology: A Treatise*. New York: Academic Press.

Zimmerman, M. H. 1983. *Xylem Structure and the Ascent of Sap*. New York: Springer-Verlag.

Appendix A
Fundamentals of Chemistry

ATOMS AND ELEMENTS

An **element** is a substance that cannot be broken down into a simpler substance by ordinary chemical means. By this definition, there are ninety-two naturally occurring elements in the universe. Each element is represented by a symbol, usually the first one or two letters of its name. (A few of the symbols represent Latin names of elements, such as Au for *aurum*, meaning "gold," and Na for *natrium*, meaning "sodium.") About twenty-two elements are essential to life, but only six—carbon, hydrogen, oxygen, nitrogen, sulfur, and phosphorus—constitute most of what we call living matter. The rest are **trace elements**, those used by plants only in extremely small quantities.

Atomic Weight

The smallest possible amount of an element is an **atom.** Atoms are composed of even smaller particles, the most stable of which are **protons, neutrons,** and **electrons.** Atoms with different numbers of these subatomic particles make different elements. For example, an atom with one proton and one electron is elemental hydrogen. A helium atom has two protons, two neutrons, and two electrons.

Each proton or neutron has a mass of about 1.7×10^{-24} gram. For convenience, this mass is defined as one **atomic mass unit.** It is also called one **dalton,** in honor of John Dalton, who helped develop the atomic theory in the early 1800s. The mass of an electron is about 1/2000 that of a proton, so it is often disregarded when considering atomic mass. A summary of the features of several common elements is presented in Table A.1.

Structure and Activity of Atoms

Atomic structure depends on how many subatomic particles (protons, neutrons, and electrons) an atom possesses. The smallest atom is hydrogen, which has one proton, one electron, and no neutrons. The electron, which has a negative (−) electric charge, is attached to the positive (+) electric charge of the proton. (Neutrons, which occur in all elements except hydrogen, have no electric charge.) The opposing electric charges attract each other, but the particles do not touch. This is primarily because the much smaller electron moves at nearly the speed of light, flying around the proton nucleus but never slowing down enough to fall into it. The electron of a hydrogen atom orbits the proton nucleus as though it were a tiny planet orbiting its sun (fig. A.1). Although this planetary model is a useful way to envision

TABLE A.1
The Symbols, Atomic Numbers, and Atomic Masses of Some Common Elements

Element	Symbol	Atomic Number	Atomic Weight
Hydrogen	H	1	1
Boron	B	5	10.8
Carbon	C	6	12
Nitrogen	N	7	14
Oxygen	O	8	16
Sodium	Na	11	23
Magnesium	Mg	12	24.3
Phosphorus	P	15	31
Sulfur	S	16	32
Chlorine	Cl	17	35.4
Potassium	K	19	39.1
Calcium	Ca	20	40.1
Manganese	Mn	25	54.9
Iron	Fe	26	55.8
Cobalt	Co	27	58.9
Copper	Cu	29	63.5
Zinc	Zn	30	65.4
Molybdenum	Mo	42	95.9

These elements are essential for plant life; each has one or more vital roles. If any one is missing, a plant cannot survive. Atomic number corresponds to the number of protons in the atomic nucleus of each; atomic weight is the number of protons plus neutrons in each nucleus.

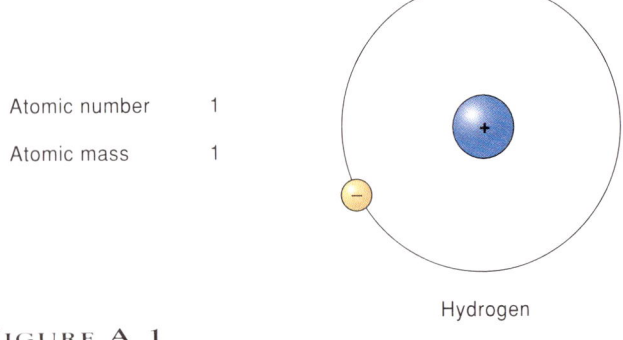

Atomic number 1
Atomic mass 1

Hydrogen

FIGURE A.1

An atom of hydrogen has one proton (+), one electron (−), and no neutrons.

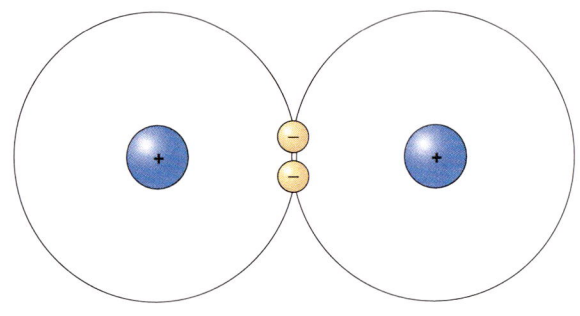

FIGURE A.2

The two atoms of hydrogen gas (H_2) are held together by a covalent bond. In covalent bonds, atoms share valence electrons.

the structure of an atom, the actual structure is unknown. Nevertheless, such models help us to understand atomic theory and to explain the physical properties of atoms.

Although the exact location of an electron cannot be pinpointed, there is a region around the nucleus where it is most likely to be. This region is called its **orbital.** For the electron of hydrogen, the shape of the orbital is approximately spherical (fig. A.1). Helium has two electrons that also occur in a spherical orbital. Atoms larger than hydrogen and helium have more than two electrons. In these atoms, additional orbitals exist further away from the nucleus to accommodate the added electrons. These outer orbitals are larger, and they occur in different shapes and sizes, in contrast to the relatively small, more-or-less spherical orbitals of hydrogen and helium.

CONCEPT

An atom is the smallest unit of an element. Each atom is characterized by its size and subatomic composition. These features are determined by the number of protons, neutrons, and electrons that an atom has.

CHEMICAL BONDING: FORCES THAT HOLD MOLECULES TOGETHER

One of the most important features of an electron orbital is that it has enough space for only two electrons. An orbital with only one electron can attract another electron to fill the available space. In the case of a hydrogen atom, which has just one electron, the orbital is filled by attracting an electron from another atom. In this way, two hydrogen atoms can share their single electrons to make a combined orbital with two electrons. The combined orbital of two hydrogen atoms makes a **molecule** of hydrogen gas, which is designated by the molecular formula H_2. A molecule of hydrogen gas is held together by the **chemical bond** between the electrons of two hydrogen atoms. This is one of several types of bonds that are based on the behavior of electrons. They are discussed in the next few paragraphs below.

Covalent Bonding

The chemical bond described above for hydrogen gas is called a **covalent bond** (fig. A.2). This is the chemical bond made between **valence electrons,** those electrons occurring in the outermost **electron shell,** or **valence shell** of the atom. A covalent bond is written as a single line that represents a pair of shared electrons. Using this notation, the structural formula for hydrogen gas is H—H.

Elements with more than one orbital, that is, all but hydrogen and helium, have more than one electron shell. Unlike the innermost shell, outer shells can contain more than one orbital, up to a maximum of four. For example, a carbon atom has six electrons, two in the innermost orbital and one in each of four outer orbitals in the valence shell. Based on this structure, carbon can share four electrons by covalent bonding. A carbon atom can, for example, bond to four hydrogen atoms. The structural formula of this molecule is shown below:

It can be written more simply as CH_4, which is the molecular formula for methane. Methane is called a compound molecule, or **compound,** because it consists of more than one kind of element. Except for the simple gases, hydrogen (H_2), oxygen (O_2), and nitrogen (N_2), virtually all biologically important molecules are compounds.

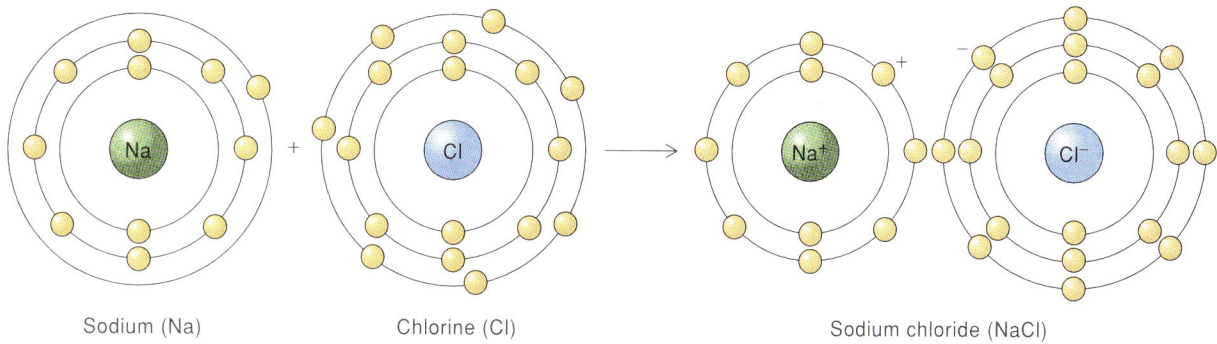

FIGURE A.3

Sodium and chlorine are bonded ionically to form sodium chloride, or table salt. In ionic bonds, atoms are held together by opposite charges that result from the loss or gain of valence electrons.

Interestingly, a carbon atom can also form a triple covalent bond with a nitrogen atom, or even with another carbon atom. When bonded to nitrogen, a carbon atom still has one free valence electron that must be shared with another atom. If the other atom is a hydrogen, the combination of hydrogen, carbon, and nitrogen makes a molecule of a poisonous gas called hydrogen cyanide. Similarly, when a carbon atom forms a triple bond with another carbon atom, each atom still requires one more covalent bond. When both bonds are made with hydrogen, the compound is a molecule of acetylene, the highly flammable gas used in blowtorches.

Ionic Bonding

In extreme cases, valence electrons are not really shared, but instead are completely removed from one atom and transferred to another. This happens between an element that can easily give up an electron, such as sodium, and one that strongly attracts an electron, such as chlorine. The result is that sodium, with 11 protons and 10 electrons, has an extra positive charge; that is, it becomes a positively charged **ion** (*cation*). Chlorine then has 17 protons and 18 electrons, which makes it a negatively charged ion (*anion*). The opposite charges of Na$^+$ and Cl$^-$ ions attract each other to form the **ionic bond** that makes sodium chloride (NaCl), or table salt (fig. A.3).

Some ions have more than one charge. Magnesium, for example, gives up two valence electrons and is therefore a doubly charged (divalent) cation, Mg^{+2}. It can form ionic bonds to two chlorine atoms to make magnesium chloride (MgCl$_2$).

Besides ionic atoms, entire molecules can become ionic by losing or gaining one or more electrons. Nitrogen and hydrogen, for example, combine into an ammonium cation (NH$_4^+$); sulfur and oxygen make a divalent sulfate anion (SO$_4^{-2}$). Ammonium sulfate, an important ingredient in nitrogen fertilizers, is a compound of these two ions with the molecular formula, (NH$_4$)$_2$SO$_4$. In this compound, the doubly charged sulfate ion is bonded to two singly charged ammonium ions.

> **CONCEPT**
>
> The two strongest types of chemical bonds are covalent bonds and ionic bonds. The forces that make these bonds come from unfilled electron orbitals. When an orbital lacks an electron, it attracts an additional electron to fill it. The orbitals of two separate atoms either share electrons (covalent bond) or transfer electrons (ionic bond).

Hydrogen Bonding and the Special Properties of Water

Nitrogen and oxygen have especially strong attractions for covalent electrons from other atoms. For this reason, they are said to be more **electronegative** than elements such as carbon. When hydrogen is covalently bonded to oxygen, the electron electronegativity of oxygen forces the hydrogen to become partially positively charged. The partial charge is strong enough to make hydrogen attractive to other electronegative atoms. This attraction is a relatively weak bond, which is neither covalent nor ionic. This type of bond is called a **hydrogen bond.**

Hydrogen bonds play several significant roles in biological processes. One of these roles is to maintain the shape of large molecules such as enzymes and other proteins that control cellular processes (see Unit Two, "Energetics"). This happens when a bond between the hydrogen of one part of the molecule and an electronegative atom in another part of the molecule holds the two parts together. Molecular shapes are important when different compounds must fit together just right to complete the chemical reactions of cellular processes.

The properties of water, which makes up over 90% of the mass of most plants, are also based on hydrogen bonding. These properties include its cohesiveness, its high heat of vaporization, and its versatility as a solvent. Each water molecule has the chemical formula H$_2$O, but the oxygen attracts the hydrogen of neighboring water molecules, and the hydrogen is attracted to other oxygen atoms. Actually, many such bonds are formed and broken and then reformed in a fraction of a second, but the overall effect is to give water a multimolecular structure. Because of this **cohesion,** plants can pull water upwards against gravity, from the roots to the tops of the tallest trees.

When molecules of a liquid absorb enough energy, the energy increases their rate of movement. Ultimately, some molecules move fast enough to leave the surface of the liquid in a gaseous state. The amount of energy (calories) required to convert 1 gram of liquid to its gaseous state is called its **heat of vaporization.** Because water has a cohesive structure, its heat of vaporization is higher than that of most other liquids. Moreover, because heat of vaporization is proportional to the amount of heat that can be removed from the liquid as it vaporizes, water functions as an **evaporative cooler.** Evaporative cooling keeps plants from becoming overheated. This is how a cotton plant, for example, can maintain its leaf temperatures below 30° C even though the air temperature around it exceeds 35° C.

Because water molecules are so much more electronegative at the oxygen ends than they are at the hydrogen end, they are said to be **polar.** When salt is added to water, Na^+ ions are attracted to the oxygen pole, and Cl^- ions are attracted to the hydrogen pole. These polar attractions are so strong that the cations and anions are separated and surrounded by water molecules, thereby causing the salt crystals to dissolve and mix into the water. Living cells contain many different kinds of salts that are dissolved in water because of its polarity. The ability of water to dissolve salts also extends to other polar compounds, even though they are not ionic. Ethyl alcohol (CH_3CH_2OH) is one such polar compound that is soluble in water. This property of water contributes to the enormous variety of compounds dissolved in cell fluids.

Hydrogen bonds are very weak but also extremely important in determining the chemical properties of water. Water is a polar solvent because of hydrogen bonding. Its polarity makes it a flexible solvent that can dissolve many substances in cell fluids.

FUNCTIONAL GROUPS AND BIOCHEMICAL DIVERSITY

The six main elements in biological molecules occur in a variety of combinations as specific subunits of molecules. These subunits are called **functional groups** because they confer chemical properties on the molecules they are part of. Ethyl alcohol is an example of a chemical with an electronegative hydroxyl group (OH) and a neutral ethyl group (CH_3CH_2). These functional groups influence the polarity, solubility in water, and reactivity of ethyl alcohol with other compounds.

The great diversity of biochemicals is based on the versatility of carbon, the element that lies at the heart of **organic chemistry** (the chemistry of carbon compounds) and **biochemistry** (the organic chemistry of biological compounds). Most functional groups either contain carbon or are bonded to carbon to make a complete molecule.

Carbon Chains and Rings

The versatility of carbon, as well as its central role in the chemistry of biological processes, comes from the ability of each carbon atom to share four covalent bonds. A carbon atom can be thought of as an intersection of covalent branches in four directions. Covalent bonding occurs in each direction, either with a different element or with another carbon. Because a carbon atom can form a covalent bond with another carbon atom, there is a seemingly infinite variety of possible compounds that have multiple carbons. Known compounds range from the smallest, with one or two carbons (methane, acetylene, ethyl alcohol), to the largest, with up to 30,000 carbons (natural rubber).

Carbon skeletons occur in straight or branched chains or even in closed rings. In each compound, every carbon is fully "satisfied" with four covalent bonds. The simplest compounds are **hydrocarbons,** those with only carbon and hydrogen. The complexity of compounds increases dramatically when these hydrocarbon functional groups are attached to other kinds of elements.

Hydroxyls, Carbonyls, and Carboxylic Acids

Oxygen, which forms two covalent bonds, has multiple roles in different functional groups. The hydroxyl group, which is the functional group of ethyl alcohol and all other alcohols, is one example. Additional roles for oxygen are derived from the double covalent bond between oxygen and carbon, called a **carbonyl group** (C=O). When a carbonyl group occurs at the end of a carbon chain, it is called an **aldehyde;** when it occurs in the interior of a carbon chain, it is called a **ketone.** Table A.2 shows the structural formulas for several kinds of functional groups, including those containing oxygen. More than one functional group can occur on the same molecule. Glyceraldehyde, for example, is a compound made of a three-carbon chain, two hydroxyl groups, and a carbonyl at the end of the chain. Compounds with both hydroxyl and carbonyl groups are sugars. Glyceraldehyde is the first sugar produced by photosynthesis in plants.

TABLE A.2

Some Functional Groups with Important Roles in Organic Compounds

Group	Name
—OH	Hydroxyl
—NH$_2$	Amino
—C(=O)—CH$_3$	Acetyl
—C—C(=O)—C—	Keto
—C(=O)—OH	Carboxyl
—O—P(=O)(OH)—OH	Phosphate

When a carbonyl group and a hydroxyl group occur on the same carbon, the combination is called a **carboxyl group.** In a carboxyl group, the collective electronegativity of the two oxygens is so strong that the hydrogen of the hydroxyl group is stripped of its covalent electron and becomes ionic. Carboxyl groups are therefore **acidic** because they increase the concentration of hydrogen ions (H$^+$) in solution (see box A.1, "Moles, Hydrogen Ions, and pH"). Compounds with carboxyl groups are called **carboxylic acids,** or **organic acids.** The most common organic acid in plants is acetic acid, which is the basic building block of fats and other lipids. It is this acid that gives vinegar its tangy flavor.

Functional Groups Containing Nitrogen

The most biologically significant functional group containing nitrogen is the **amino group** (NH$_2$). This group is important because it is one of the two functional groups that make up an **amino acid,** the other being a carboxyl group. Amino acids are used for building enzymes and other proteins that are the foundation for almost all biochemical reactions and for a variety of cellular structures.

Functional Groups Containing Sulfur or Phosphorous

Sulfur, like oxygen, has two valence electrons and forms a functional group with a hydrogen atom (SH), which is called a **sulfhydryl group.** Compounds with sulfhydryl groups are called **thiols** and are notable for their foul odors. (Butanethiol is the major ingredient of skunk spray.) Except for this property, sulfhydryls are chemically similar to hydroxyls. They are so similar that plant scientists were able to deduce how plants split water molecules (H$_2$O) during photosynthesis by analogy with the similar chemical reactions of hydrogen sulfide (H$_2$S) in bacteria.

Sulfur can also form **disulfide** bonds (S—S). These bonds form between two cysteine monomers in proteins, and are important for stabilizing a protein's three-dimensional structure.

Of all the biologically important functional groups, the most electronegative is the **phosphate group.** Each phosphate group (PO$_4^{-3}$) is made of one phosphorous atom and four oxygen atoms. It is derived from a phosphoric acid molecule (H$_3$PO$_4$) that, due to the electronegative strength of the oxygens, has released all of its hydrogen atoms as ions. One of the properties of phosphate groups is that they make short-lived, unstable covalent bonds with other phosphate groups. The ease with which these bonds can be broken means that the energy stored in them is easily released. As phosphate bonds break, their energy is used to drive cellular processes. These processes include such functions as building new molecules, moving substances from one place in the cell to another, and controlling what goes through membranes. In Unit Two ("Energetics") you will learn about a variety of cellular processes that depend on energy from phosphate bonds. The phosphate group is also important in the sugar-phosphate backbone of the nucleic acids, DNA and RNA, and in the phospholipids of membranes.

CONCEPT

The chemical properties of molecules are determined by their functional groups. These are subunits of compounds that have specific combinations of different elements. The most important molecules of biological processes are made of multiple carbons that are attached to an oxygen (hydroxyl, carbonyl, or carboxyl group), a nitrogen (amine group), a sulfur (sulfhydryl group), or a phosphorus (phosphate group). Many compounds have more than one of these functional groups.

CHEMICAL REACTIONS

Covalent bonds, functional groups, and biological polymers are all products of chemical reactions. A chemical reaction is the process of making and breaking chemical bonds, and is written as an equation that has the **reactants** on the left side

BOXED READING A.1
ISOTOPES AND RADIOACTIVITY

Each atom is characterized by how many protons, neutrons, and electrons it has. If an atom changes its number of protons, it becomes another element. However, the same element can have a different number of electrons or neutrons; for example, an element can have different atomic weights because of its variable number of neutrons. Each neutron form of an element is called an **isotope** of the element. Hydrogen has three isotopes, one without neutrons, one with one neutron, and one with two neutrons. To distinguish among them, the isotopes of hydrogen are written ^1H, ^2H, and ^3H, respectively. Similarly, carbon has the isotopes ^{12}C (6 neutrons), ^{13}C (7 neutrons), and ^{14}C (8 neutrons). Like hydrogen, all of the isotopes of carbon have the same number of protons. Notice that the isotope number and the atomic weight are about the same.

The smallest isotope of any element is generally the most abundant in nature. Of all the carbon on earth, 99% is ^{12}C and most of the remaining 1% is ^{13}C; only a trace amount is ^{14}C. The abundant smaller isotopes are also stable, but the larger isotopes of many elements are unstable. They decay by releasing high-energy subatomic particles; that is, they are **radioactive**. (This name has nothing to do with radio waves; radioactive means that an element is like radium, which was the first element discovered to be radioactive.)

Electronic instruments and photographic film can be used to detect the particles of radioactive decay. These detectors and radioactive isotopes are used to study the metabolic processes of a cell. The biosynthesis of many natural products was first discovered by following the path of radioactive precursors during metabolism. The intermediate products and chemical reactions that lead to sugars, natural rubber, steroids, and many other biological molecules were discovered with the aid of radioactive chemicals. This is still a powerful method for studying the biosynthetic pathways of plant metabolites.

Radioactive isotopes are also useful for dating fossils. This is possible because radioactive decay occurs at a constant rate for each element. For example, as a plant takes in carbon during photosynthesis, the ratio of ^{14}C to ^{12}C in its tissue is the same as the $^{14}C/^{12}C$ ratio of carbon dioxide in the atmosphere. When the organism dies, the ratio begins to change because ^{14}C decreases in abundance due to its radioactive decay. By knowing the rate of decay, we can calculate how long ago the plant died. Carbon loses half of its radioactivity in 5,600 years, the time period that is called its **half-life.** If a 10-gram sample of a fossil plant has half the amount of ^{14}C as a 10-gram sample of a living plant, that means that the fossil plant died 5,600 years ago. A 11,200-year-old fossil would have one-fourth the amount of radioactivity, and so on. Because of the limitations of electronic detectors, carbon dating is limited to about 50,000 years. Other isotopes, such as potassium-40 (half-life of 1.3 billion years), can be used to date fossils of organisms that lived hundreds of million years ago.

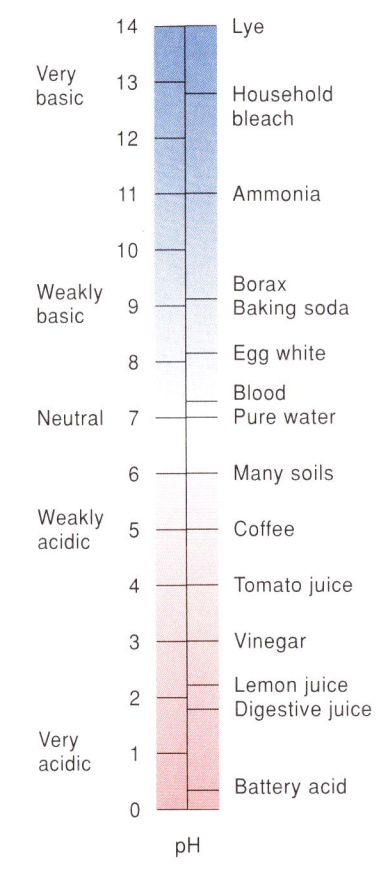

Source: Data from Essenfeld, et al., Biology, copyright © Addison-Wesley Publishing Company.

and the **products** on the right side. For example, methane and oxygen (reactants) can react to make carbon dioxide and water (products):

$$CH_4 + 2O_2 \longrightarrow CO_2 + 2H_2O$$

Notice that all atoms must be accounted for in a chemical reaction; that is, the equation must be balanced. Reactions cannot create or destroy matter, but can only rearrange it. Thus, in this reaction, each methane molecule requires two molecules of oxygen (O_2) to maintain the appropriate proportion of elements in the products.

This is a common reaction that occurs at the burners of gas stoves every day. But why does it occur? That is, what is the driving force that makes the reaction go? The simplest answer is that the stored energy (potential energy) of the covalent bonds in the products is lower than the bond energy of the reactants. Potential energy tends to run "downhill" from higher levels to lower levels. The behavior of potential energy is analogous to that of water at the top of a hill (higher potential energy), which tends to run down to the bottom of the hill (lower potential energy). The difference between the higher potential energy on the left side of the equation and the lower potential energy on the right side of the equation is accounted for by the release of heat energy. This heat energy is exactly equal to the bond energy of the reactants minus the bond energy of the products.

The physical behavior of energy is the driving force underlying all biological processes. Unit 2 ("Energetics") includes three chapters devoted to energy and its use by plants. The paragraphs below outline some of the most common chemical reactions that are driven by the energy of plant metabolism.

Oxidation and Reduction

The chemical reaction of methane and oxygen presented above is a combustion reaction. In other words, methane and oxygen burn together. In this reaction the covalent electrons of carbon, which are about evenly shared with the electrons of hydrogen in methane, are pulled further away from the carbon by the strongly electronegative oxygen to make carbon dioxide. This shift of electrons away from an element is called **oxidation**. Each hydrogen is also oxidized, since its electron goes from being equally shared with carbon to being pulled away by the oxygen in water.

When one reactant is oxidized, another reactant undergoes **reduction**. In the above reduction, oxygen is reduced; that is, it partially gains electrons from the carbon in CO_2 and from the hydrogen in H_2O. Although the example reaction involves oxygen, oxidation-reduction (*redox*) reactions do not always require it. Any partial or complete transfer of electrons is a redox reaction. For example, the ionization of sodium and chlorine fits this description (fig. A.3).

$$Na + Cl \longrightarrow Na^+ + Cl^-$$

In this reaction, sodium is oxidized (loses an electron), and chlorine is reduced (gains an electron).

The energy that cells need for growth and metabolism is derived from a redox process called **respiration**. Cellular respiration is a series of many redox reactions that begin with a carbohydrate (glucose) and oxygen and end with carbon dioxide and water. The equation below summarizes the overall process:

$$C_6H_{12}O_6 + 6O_2 \longrightarrow 6CO_2 + 6H_2O + energy$$

The many reactions that occur during respiration, the parts of cells where the reactions take place, and other details about this process are the subject of Chapter 6, "Respiration."

Cellular activities depend on the transfer of energy during chemical reactions. The driving force underlying most reactions is the loss of potential energy. This loss comes from the partial or complete transfer of electrons from one reactant (oxidation) to another (reduction). The energy lost from oxidation and reduction is often transferred to other reactions or released as heat.

Dehydration and Hydrolysis

Although oxidation and reduction occur during almost all chemical reactions, many reactions are referred to by the products they yield or by the functional groups that are involved. For example, most biological polymers are built by reactions that yield water as a by-product. These reactions are called **dehydration** reactions. The peptide bonds of proteins, the glycosidic bonds of polysaccharides, the sugar-phosphate bonds of nucleic acids, and the glycerol–fatty acid bonds of fats are all made by dehydration.

Dehydration reactions are reversible. The bond between two glucoses in starch, for example, can be broken to yield glucose monomers. This reverse reaction is called **hydrolysis** because it uses water to break the chemical bond between glucoses. All bonds formed by dehydration can be broken by hydrolysis. In cells, chemical synthesis by dehydration is enhanced by one set of enzymes, whereas chemical degradation by hydrolysis is enhanced by another set of enzymes. The metabolic needs of the cell determine which direction a reaction takes. Both kinds of reactions can occur at the same time but in different parts of the cell.

Carboxylation and Decarboxylation

Covalent bonds in carbon dioxide (O=C=O) are stable and difficult to break. The formation of carbon dioxide, with its low potential energy, is the strong driving force that removes CO_2 from organic acids:

$$R-COOH \longrightarrow R-H + CO_2$$

This **decarboxylation** reaction is common in many cellular processes (e.g., the release of carbon dioxide during respiration). In fact, several decarboxylations occur for each molecule of glucose that is oxidized during respiration. This means that organic acids play a role in the intermediate steps of this process. Organic acids and their decarboxylation are discussed in Chapter 6 ("Respiration").

Although the removal of carbon dioxide has many metabolic roles, the addition of carbon dioxide, or **carboxylation**, is restricted to photosynthesis. This reaction is the foundation for making organic molecules from carbon dioxide in the atmosphere. Energy is required to break the double covalent bonds between carbon and oxygen and to make covalent bonds between carbons. Only green plants, algae, and a few bacteria can transform light energy and use it to carboxylate organic compounds. This process is the topic of Chapter 7 ("Photosynthesis").

Phosphorylation

Reactions that make phosphate bonds are called **phosphorylation** reactions. The most common phosphorylation reaction in any cell is the addition of a phosphate group to an adenine-containing nucleotide, adenosine diphosphate (ADP). The product of this reaction is adenosine triphosphate (ATP).

Phosphorylation is accompanied by dehydration. The covalent bond between the second and third phosphate groups of ATP is easily hydrolyzed, thereby releasing energy to another molecule or chemical reaction. Because it gives up energy so easily, this phosphate bond is called a high-energy bond. The bond between the first and second phosphate groups is also a high-energy bond. High-energy phosphate bonds are designated by a "~" to distinguish them from ordinary covalent bonds. Thus, phosphorylation can be abbreviated as follows:

Because of the instability of high-energy phosphate bonds and the demand for this energy to do cell work, each ATP molecule is recycled to ADP in a fraction of a second. ATP is regenerated continuously by the energy derived from respiration and photosynthesis.

CONCEPT

The most common chemical reactions in cells are the removal of a molecule of water (dehydration) or a molecule of carbon dioxide (decarboxylation), or the addition of a molecule of water (hydrolysis). The function of molecules such as ADP and ATP depends on dehydration and hydrolysis for attaching and removing phosphate groups (phosphorylation).

Molecular Shapes and Natural Isomers

Many organic compounds can have different shapes but the same molecular formula. Glucose and fructose, for example, are both $C_6H_{12}O_6$, but they have different chemical properties. Their structural differences are easy to see when their rings are broken open by hydrolysis. The carbonyl group of fructose is a ketone, whereas the carbonyl group of glucose is an aldehyde (fig. A.4). This seemingly small difference is significant. Because of it, fructose is one of the sweetest sugars, whereas glucose has almost no sweetness at all. Glucose and fructose are **structural isomers;** that is, they have the same functional groups bonded to different carbon atoms. Galactose, on the other hand, has the same functional groups on the same carbon atoms as glucose. However, the hydroxyl group on the fourth carbon is in a different orientation on galactose than it is on glucose (fig. A.4). This means that glucose and galactose are **stereoisomers** of each other. They are seemingly identical but, like your left and right hands, are really nonidentical mirror images of each other.

By varying the orientation of hydroxyl groups on carbons 2 through 5 of glucose, there can be 16 possible stereoisomers (2^4). The natural isomers of glucose, however, all have the hydroxyl group to the right of carbon 5. They are called the D

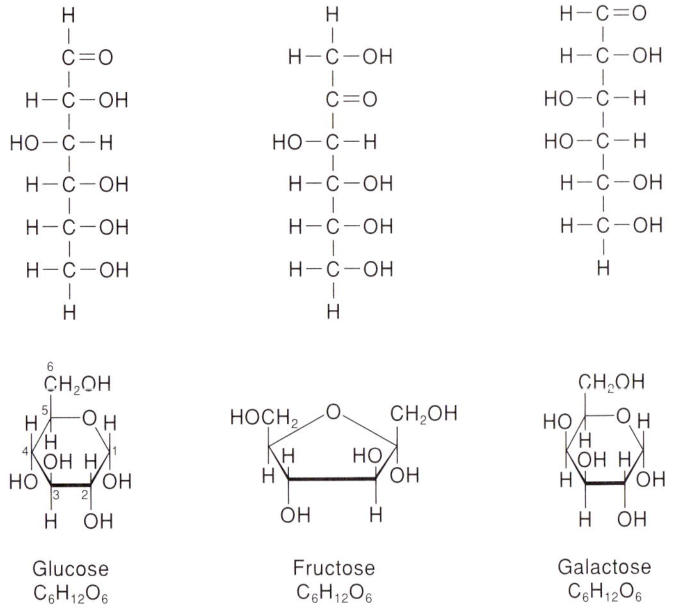

FIGURE A.4

The structures of glucose and fructose. Although these sugars are both $C_6H_{12}O_6$, they have different chemical properties.

forms, from the Latin word *dextro,* which means "right." Thus, natural glucose is D-glucose. The alternative, L-glucose (from *laevo,* meaning "left") is not made by plants. All other monosaccharides also exist only in the D form in plants. Similarly, the amine group occurs on the same side of its carbon, relative to the carboxyl group, in all amino acids. However, it is on the left side, which means that plants make only L-amino acids.

Optical Isomers

The original designation of D and L forms for sugars and amino acids was an arbitrary one, based solely on the orientation of one functional group. A more useful physical property, one which can be applied to all biological compounds, is the behavior of stereoisomers in light. When a single plane of light, called **plane-polarized light,** is aimed at a substance at one angle, it often emerges from the other side at a different angle. If the angle of the emergent plane of light is to the left of the original plane, the substance is *levorotatory.* If the emergent plane of light is to the right, the substance is *dextrorotatory.* Substances that cause plane-polarized light to rotate are called **optically active.** Optical activity occurs when a carbon atom is asymmetric; that is, it has four different functional groups attached to it. In glucose, for example, carbons 2 through 5 are asymmetric. Carbon 1 is not, because it has "two" oxygens (actually, one oxygen that is double-bonded). Carbon 6 is also inactive. Why?

BOXED READING A.2
MOLES, HYDROGEN IONS, AND pH

The molecular weight of a substance, the sum of daltons of each atom in a molecule, cannot be measured on a normal laboratory scale because it is too small. Instead, we measure the weight of a chemical in **moles**. A mole is the molecular weight in daltons expressed in grams. For example, the molecular weight of water is 18 daltons, so one mole of water is 18 grams. One mole of table salt (NaCl) is 58.4 grams, based on a molecular weight of 58.4 daltons. Moles are also convenient units for expressing the concentrations of substances in water. A solution that contains one mole of a chemical in one liter of water is a **one-molar** (1 M) solution. Therefore, a liter of water with 58.4 grams of NaCl is a 1 M NaCl solution.

One of the most important applications of molarity involves the concentration of hydrogen ions in a solution. Pure water is the standard by which all other solutions are compared, because water is an ionically neutral solution. Its neutrality is not due to the absence of ions, but to the equal concentrations of positive and negative ions. When the oxygen pulls hard enough on an electron from one of its hydrogens, two ions are formed:

$$H_2O \longleftrightarrow H^+ + OH^-$$

This dissociation of water is rare and reversible, but it happens often enough for the concentration of H^+ to be 10^{-7} M. The solution is neutral because the concentration of OH^- is also 10^{-7} M. Any substance that increases the H^+ concentration is an **acid;** any substance that decreases it is a base. For example, hydrochloric acid (HCl) or acetic acid (CH_3COOH) would increase the concentration H^+. In contrast, sodium hydroxide (NaOH) or ammonium hydroxide (NH_4OH) would increase the OH^- concentration, thus lowering the proportion of H^+.

When the concentration of one ion increases, the concentration of the other ion decreases proportionally. Thus, if enough acid is added to water to raise the H^+ concentration to 10^{-6} M, the OH^- concentration would decrease to 10^{-8} M. In the extreme, the highest possible concentration of H^+ is 10^0 M and the lowest possible concentration of OH^- is 10^{-14} M. In the other extreme, the highest concentration of OH^- is 10^0 M when the H^+ concentration is at 10^{-14} M.

By general agreement, the scale we use is based on the hydrogen ions and is expressed as a negative logarithm of the molarity of H^+. This scale, called the **pH scale,** goes from 0 ($-\log 10^0$) to 14 ($-\log 10^{-14}$). (It would be equally convenient to have a "pOH" scale, but pH was arbitrarily chosen instead.) On this scale, water has a pH of 7. An acid has a pH less than 7 (more H^+), and a base has a pH greater than 7 (less H^+). It is important to note that, because this scale is logarithmic, pH 3 is not double pH 6, but is 1,000-fold higher in H^+ concentration. Small differences in pH mean large differences in the molarity of H^+. Thus, the pH in cells is restricted to a narrow range, usually between 6 and 8.

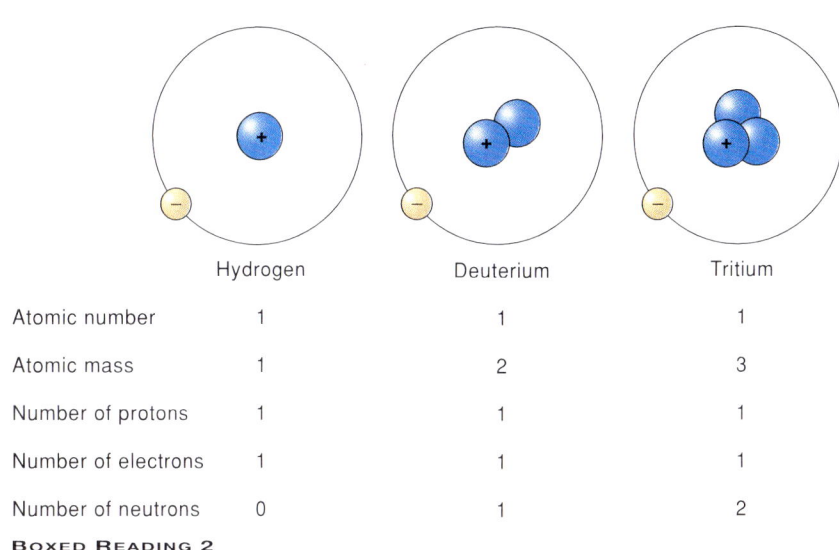

	Hydrogen	Deuterium	Tritium
Atomic number	1	1	1
Atomic mass	1	2	3
Number of protons	1	1	1
Number of electrons	1	1	1
Number of neutrons	0	1	2

BOXED READING 2

The three isotopes of hydrogen differ because of the number of neutrons. Because they all have the same number of protons (i.e., 1), they are the same chemical element (i.e., hydrogen) and have identical chemical properties.

All amino acids, except glycine ($H_2N—CH_2—COOH$), also have at least one asymmetric carbon and are therefore optically active. Nucleotides and other primary metabolites are optically active, as are most secondary metabolites. For most compounds, either the dextrorotatory (+) or the levorotatory (−) isomer occurs naturally, but not both. Chemical reactions, especially enzyme-catalyzed reactions, that depend on the fit of specific isomers to each other will not work if one compound is the "wrong" isomer. For example, when yeasts are fed a mixture of glucose isomers, they metabolize only the (+)-glucose and leave the (−)-glucose behind. In this regard, the respiration of all organisms is the same.

CONCEPT

Molecules have certain shapes because each carbon forms bonds in different directions. Compounds with the same functional groups are isomers of each other. Structural isomers have the same functional groups on different carbons. Stereoisomers have the same functional groups on the same carbons, but they are pointed in different directions. When a carbon atom is bonded to four different atoms or functional groups, it is asymmetric, and the molecule is optically active. Optical isomers rotate plane-polarized light in opposite directions.

Appendix B
The Metric System

The **metric system** is a standardized system of measurement used by scientists throughout the world. It is also the measurement system used in everyday life in most countries. Although the metric system is the only measurement system ever acknowledged by Congress, the United States remains "out of step" with the rest of the world by clinging to the antiquated English system of measurements involving pounds, inches, and so on.

Metric units commonly used in biology include

meter (m) The basic unit of length
liter (l) The basic unit of volume
gram (g) The basic unit of mass
degree Celsius (°C) The basic unit of temperature

Unlike the English system with which you are already familiar, the metric system is based on units of ten, thus simplifying interconversions. This base-ten system is similar to our monetary system, in which 10 cents equals a dime, 10 dimes equals a dollar, and so on. Units of ten in the metric system are indicated by Latin and Greek prefixes placed before the base units:

TABLE B.1
Metric System Divisions and Multiples of the Basic Unit of Length, the Meter

Prefix (Latin)	Division of Metric Unit (meter)		Multiple of Metric Unit (meter)		
deci (d)	0.1	10^{-1}	deka (da)	10	10^{1}
centi (c)	0.01	10^{-2}	hecto (h)	100	10^{2}
milli (m)	0.001	10^{-3}	kilo (k)	1000	10^{3}
micro (µ)	0.000001	10^{-6}	mega (M)	100000	10^{6}
nano (n)	0.000000001	10^{-9}	giga (G)	1000000000	10^{9}
pico (p)	0.000000000001	10^{-12}	tera (T)	1000000000000	10^{12}

Thus, we multiply by

0.01	to convert cg to g
0.1	to convert dm to m
1000	to convert kg to g
0.000001	to convert µl to l
0.1	to convert mm to cm

For example, 620 g = 0.620 kg = 620000 mg = 6200 dg = 62000 cg.

UNITS OF LENGTH

The **meter** (m) is the basic unit of length.

1 m = 39.4 inches (in.) = 1.1 yards (yd)
1 km = 1000 m = 10^{3} m = 0.62 miles
1 cm = 0.01 m = 10^{-2} m = 0.39 in. = 10 mm
1 nm = 10^{-9} m = 10^{-6} mm = 10 angstroms (Å)
470 m = 0.470 km
1 in. = 2.54 cm
1 ft = 30.5 cm
1 yd = 0.91 m
1 mi = 1.61 km

Units of area are squared (i.e., two-dimensional) units of length:

1 m^2 = 1.20 yd^2 = 1550 $in.^2$ = 1.550 × 10^{3} $in.^2$
1 cm^2 = 100 mm^2
1 hectare = 10000 square meters (m^2) = 2.47 acres

Measurements of area and volume can use the same units:

1 m^3 = 35.314 ft^3 = 1.31 yd^3
1 cm^3 (cc) = 0.000001 m^3 = 0.152 $in.^3$

UNITS OF MASS

The **gram** is the basic unit of mass.

1 g = mass of 1 cm³ of water at 4° C = 0.035 oz
1 kg = 1000 g = 10³ g = 2.2 lb
1 mg = 0.001 g = 10⁻³ g

Remember that mass is not necessarily synonymous with weight. Mass measures an object's potential to interact with gravity, whereas weight is the force exerted by gravity on an object. Thus, a weightless object in outer space has the same mass as it has on earth.

UNITS OF VOLUME

The **liter** (L) is the basic unit of volume. Units of volume are cubed (i.e., three-dimensional) units of length.

1 L = 1000 cm³ = 1000 ml
1 L = 2.1 pints = 1.06 qt = 0.26 gal = 1 dm³
1 ml = 0.035 fl oz
1 tsp = 15 ml
1 cup = 0.24 L

UNITS OF TEMPERATURE

You are probably most familiar with temperature measured with the Fahrenheit scale, which is based on water freezing at 32° F and boiling at 212° F. Celsius temperatures are synonymous with Centigrade temperatures, and these scales measure temperature in the metric system. Celsius (° C) temperatures are easier to work with than Fahrenheit temperatures since the Celsius scale is based on water freezing at 0° C and boiling at 100° C. You can interconvert ° F and ° C with the following formula:

$$5(°F) = 9(°C) + 160$$

For example,

40° C (104° F) typical for a hot summer day
75° C (167° F) hot coffee
37° C (98.6° F) human body temperature

Celsius versus Centigrade

The Celsius temperature scale was developed in 1742 by Anders Celsius, a Swedish astronomer. Interestingly, Celsius originally set 0° C as the boiling point of water and 100° C as the freezing point of water. Soon thereafter, J. P. Christine revised the scale to its present form—with 0° C as the freezing point and 100° C as the boiling point of water. The Celsius scale is identical to the Centigrade scale.

HINTS FOR USING THE METRIC SYSTEM

1. Express measurements in units requiring only a few decimal places. For example, 0.3 m is more easily manipulated and understood than 300,000,000 nm.

2. When measuring water, the metric system offers an easy and common conversion from volume measured in liters to volume measured in cubic meters to mass measured in grams: 1 ml = 1 cm³ = 1 g.

3. Familiarize yourself with manipulations within the metric system. Work within one system, and do not convert back and forth between the metric and English systems.

4. The metric system uses symbols rather than abbreviations. Therefore, do not place a period after metric symbols (e.g., 1 g, not 1 g.). Use a period after a symbol only at the end of a sentence, but always use a period with the symbol for inches (in.).

5. Do not mix units or symbols (e.g., 9.2 m, not 9 m 200 mm).

6. Metric symbols are always singular (e.g., 10 km, not 10 kms).

7. Except for degree Celsius, always leave a space between a number and a metric symbol (e.g., 20 mm, not 20mm; 10° C, not 10° C).

8. Use a zero before a decimal point when the number is less than one (e.g., 0.42 m, not .42 m).

Glossary

A

A Horizon (Topsoil) The uppermost layer of soil, usually about 10 to 20 cm. thick

ABA (see **Abscisic Acid**)

Abaxial Away from the axis

Abby Abbreviation of Abington, which is one of 19 strains of viruses found only in gypsy moth caterpillars

Abiotic Something that is non-living and never has been alive

Abscisic Acid (ABA) A plant hormone (growth regulator) associated with water stress and the inhibition of growth; also induces stomatal closing and seed dormancy in many plants

Abscission The detachment of leaves, flowers, or fruits, usually at a weak area termed the abscission zone

Absorption Spectrum The spectrum of light absorbed by a particular pigment

Accessory Pigment A pigment that captures light energy and transfers it to chlorophyll a; beta-carotene is an example of an accessory pigment

Acetyl Coenzyme A (acetyl-CoA) A two-carbon organic acid whose hydroxyl group has been replaced with coenzyme A

Acid-Growth Hypothesis The hypothesis that acidification of the cell wall leads to the breakage of restraining bonds within the wall, thereby leading to cellular elongation that is driven by turgor pressure

Acropetally Toward the apex

Actin a type of globular protein that makes up actin filaments

Actin Filament the smallest (4–7 nm in diameter) of the three types of filaments that comprise the cytoskeleton

Action Spectrum The spectrum of light that elicits a particular response

Active Transport Movement of solutes across a membrane against their concentration gradient; active transport required energy from cellular metabolism

Acylglyceride Linkage The covalent bond between the organic acid group, such as in a fatty acid, and one of the three hydroxyl groups of glycerol

Adaptive Radiation Evolution of divergent forms of a trait in several species that developed from an unspecialized or primitive common ancestor

Adaxial Toward the axis

Adenosine Triphosphate (ATP) A nucleotide consisting of adenine, ribose, and three phosphate groups; the major source of usable chemical energy in metabolism; when hydrolyzed, ATP loses a phosphate to become adenosine diphosphate (ADP) and releases usable energy

Adenovirus 2 A type of virus that causes human respiratory disease; its role in genetic research involved the discovery of introns

Adventitious Root A root that arises from a leaf or stem (i.e., not from another root)

Aerenchyma A tissue containing large amounts of intercellular spaces

Aerobic Respiration (see **Respiration**)

Agar A slimy polysaccharide, consisting mostly of a specific mixture of alpha-galactose sulfates that surround the cell walls of certain red algae; in the United States it is harvested for commerce primarily from *Gelidium robustum*

Akinete A thick-walled dormant cell derived from a vegetative cell

Albuminous Cell Certain ray and axial parenchyma cells in the phloem of gymnosperms; these cells are closely associated with sieve cells, both morphologically and physiologically

Aleurone Layer A group of protein-rich cells located at the outer edge of the endosperm of many grains

Alkaloid A nitrogen-containing base in which at least one nitrogen is part of a ring; examples include nicotine, caffeine, cocaine, and strychnine; alkaloids are often bitter and affect the physiology of vertebrates and other animals

Allele One of the alternative forms of a gene; a gene may have two or more alleles

Allopatric Occurring in different places

Allopatric Speciation Speciation induced by geographical or physical separation of the ancestral population

Allopolyploid A polyploid with multiple sets of chromosomes that originated from more than one species

Allosteric Regulation Regulation that results from a change in the shape of a protein that occurs when the protein binds a nonsubstrate molecule; in its new shape, the protein usually has different properties

Allozymes Enzymes that are coded for by different alleles of the same locus; each form is encoded by different alleles

Alpha-Amylase An enzyme that breaks down starch into smaller units by cleaving the 1,4 linkages between molecules of alpha-glucose

Alpha-Glucose The form of glucose whose structure, when drawn in flat plane, has a hydroxyl group at the first carbon that points down

Alpha-Ketoglutarate (Alpha-Ketoglutaric Acid) A five-carbon organic acid that loses a molecule of carbon dioxide and gains an acetyl-CoA group in the fourth step of the Krebs cycle, thereby being converted to succinyl-CoA; also during this conversion, one molecule of NAD^+ is reduced to NADH

Alpha-Tubulin A type of globular protein that is a main component of microtubules

Amino Acid Acceptor Site A sequence of nucleotides that recognizes and binds to a specific amino acid at the 3´ end of a molecule of transfer RNA

Aminoacyl-tRNA Synthetase A type of enzyme that catalyzes the binding of an amino acid to the amino acid acceptor site on a molecule of transfer RNA

Amylase (see **Alpha-Amylase**)

Amylopectin A highly branched polymer of up to 50,000 molecules of alpha-glucose

Amyloplast A type of plastid that stores starch

Amylose An unbranched chain of up to several thousand molecules of alpha-glucose

Anabolism Biosynthesis; the constructive part of metabolism

Anaerobic Respiration (see **Respiration**)

Anaphase The period of mitosis during which centromeres split and sister chromatids become separate chromosomes that begin to move toward opposite poles of the spindle apparatus

Anaphase I The first anaphase of meiosis; in anaphase I, homologous chromosomes move to opposite poles of the meiotic spindle apparatus, resulting in a halving of the number of chromosomes going to each daughter nucleus

Anaphase II The second anaphase of meiosis; in anaphase II, the centromeres divide, thereby allowing the separation of sister chromatids into independent chromosomes

Androecium (plural, **androecia**) Collectively, all of the stamens of a single flower

Angiosperm Any plant whose seeds are born in a fruit; an informal name for flowering plant (Division Anthophyta or Magnoliphyta)

Anther The pollen-bearing part of a stamen

Antheridiophore In some liverworts, the stalk that bears antheridia

Antheridium (plural: **Antheridia**) A unicellular or multicellular structure in which sperms are produced; may be multicellular or unicellular

Anthocyanin Any red or blue pigment that is a flavonoid; anthocyanins are the primary pigments of blue and red plant parts (e.g., flowers, fruits)

Antibody A protein whose formation is induced by an antigen and that binds to the antigen that induced it

Anticodon A sequence of three nucleotides in a molecule of transfer RNA, which is complementary to a codon sequence

Antigen A large, foreign molecule, such as a protein or polysaccharide, that induces its host to form antibodies against it

Antiparallel Refers to double-stranded DNA, in which the direction of each strand is opposite its complementary strand

Antipodal Cell Cells that form at the chalazal of the embryo sac, opposite the micropylar end

Antisense Strand In DNA, the antisense strand of a gene is the one that does not contain a coding sequence for a molecule of RNA; the antisense strand is not transcribed

Apical Dominance The influence exerted by a terminal bud in suppressing the growth of lateral buds

Apical Meristem The meristem at the tip of a root or shoot in a vascular plant

Apomictic Asexual production of seeds

Apoplastic Movement The movement of water and solutes in the free space of the tissue; the free space includes cell wells and intercellular spaces

Apothecium An open ascocarp; it is usually cup- or saucer-shaped

Archaebacteria Primitive prokaryotes with distinctive chemical and structural features

Archegoniophore In some liverworts, the stalk that bears archegonia

Archegonium A multicellular organ that produces an egg; found in bryophytes and some vascular plants

Aril A fleshy structure that may partially envelop a seed

Artificial Selection Selection by humans of specific traits in organisms being bred to produce desired characteristics

Ascocarp A reproductive structure of ascomycetes, in which asci are formed

Ascogenous Hyphae Hyphae with paired male and female nuclei; ascogenous hyphae eventually produce asci

Ascogonium (plural: **Ascogonia**) The female sexual structure of ascomycetes

Ascomycetes A large group of true fungi with septate hyphae; they produce conidiospores asexually and ascospores sexually within asci

Ascospore A spore produced within an ascus

Ascus (plural: **Asci**) A saclike cell of ascomycetes in which, following meiosis, a specific number (usually 8) of ascospores is produced

ATP Synthase A type of membrane-bound enzyme in mitochondria that phosphorylates ADP to ATP by using energy from the diffusion of protons through the enzyme

ATP (see **Adenosine Triphosphate**)

ATP Phosphohydrolase (ATPase) A type of transport protein that uses energy from the hydrolysis of ATP to actively transport ions or other solutes against their concentration gradient

Autopolyploid A polyploid with multiple sets of chromosomes that originated from more than one species

Autotroph An organism that produces its own food, usually by photosynthesis; virtually all plants are autotrophs

Auxin A plant hormone (growth regulator) that influences cellular elongation, among other things; also referred to as indole-3-acetic acid, or IAA

Axil The upper angle between a twig of leaf and the stem from which it grows

Axillary Bud Buds that occur in the axil of a leaf

Axillary Placentation Refers to the attachment of ovules along the central axis of an ovary that has more than one ovule-bearing chamber; lily is an example plant that has axillary placentation

B

B Horizon (Subsoil) The layer of soil immediately beneath the topsoil, usually about 25 to 50 cm. thick

Bacteriochlorophyll A modified chlorophyll that is the primary light-trapping pigment in green and purple photosynthetic bacteria

Bacteriophage A type of virus that parasitizes bacteria

Bar A unit of pressure; one bar is the atmospheric pressure of air at sea level and room temperature

Bark The part of the stem or trunk exterior to the vascular cambium

Basidiocarp A reproductive structure of basidiomycetes, in which basidia are formed

Basidiomycetes A large and diverse group of true fungi with septate hyphae; they produce basidiospores externally on basidia

Basidium (plural: **Basidia**) A club-shaped structure upon which, following meiosis, a specific number (usually 4 or 2) of basidiospores is produced

Basipetally Toward the base

Bedrock Solid rock beneath the layers of soil

Beta-Carotene An orange pigment that is made of eight isoprene units; it occurs in most plants as an accessory pigment to photosynthesis

Beta-Glucose The form of glucose whose structure, when drawn in flat plane, has a hydroxyl group at the first carbon that points up

Beta-Oxidation A sequence of biochemical reactions that oxidize fatty acids into a series of two-carbon compounds that are converted to acetyl-CoA

Beta-Tubulin A type of globular protein that is a main component of microtubules

Bilayer In referring to phospholipids, a bilayer is a spontaneously formed double layer of lipid, with an interior of hydrophobic hydrocarbons and an exterior of hydrophilic phosphate groups

Binomial System of Nomenclature A system of applying two-part scientific names to organisms, each name consisting of the genus (generic) name and a species (specific) epithet

Bioassay A quantitative assay of a substance using a part of or an entire organism

Biochemical Organic and inorganic chemicals that occur in living organisms and are involved in the processes of life

Biochemical Cytology Study of the biochemical properties of cell components in conjunction with techniques of microscopy to unravel the details of cell structure and function

Biochemical Reactions of Photosynthesis The temperature-dependent (i.e., "dark") reactions of photosynthesis that reduce carbon dioxide to carbohydrate; occur in the stroma of chloroplasts

Bioenergetics The energy relationships of living organisms

Biogeography The study of geographic distributions of organisms past and present, and the mechanisms that caused these distributions

Biological Clock An internal biological timing system that influences cyclic phenomena

Biological Species Concept A species consists of groups of actually or potentially interbreeding natural populations that produce viable offspring

Biomass The collective dry weight of all the organisms in a population, area, or sample

Biotic Pertaining to living organisms

Biotic Community All the populations of interactive living organisms sharing a common environment

Biotic Potential The inherent rate of natural increases, as exhibited by an individual's total number of offspring that survive long enough to reproduce

Bivalent A pair of synapsed homologous chromosomes in prophase I

Blade The broad, expanded part of a leaf

Blowout A barren area in arctic tundra caused by wind ripping out part of a vegetation mat whose edges became exposed when pulled up by grazing animals

Body Cell One of two cells produced when the generative cell of a gymnosperm male gametophyte divides; the body cell itself later divides, producing two sperm cells

Bordered Pit A pit in which the secondary wall arches over the pit membrane

Botany The scientific study of plants and plantlike organisms

Bract A structure that is usually leaflike and modified in size, shape, or color

Bracteole Diminutive form of bract

Branch Root A root that arises from another, older root; also called a branch root

Bryophyte Member of a division of nonvascular plants; the mosses, hornworts, and liverworts

Bud Scale Modified leaves that surround and protect a bud

Bulb A short underground stem covered by fleshy leaf-bases that store food

Bulliform Cells Large epidermal cells that occur in groups on the upper surface of leaves of many grasses; loss of turgor pressure in these cells causes leaves to roll up during water stress

Bundle Sheath A layer or layers of cells surrounding the vascular bundle; in C_4 plants, the bundle sheath is photosynthetic and prominent

C

C Horizon (Parent Material) The layer of soil between bedrock and the B horizon. It varies in thickness between about 10 centimeters and several meters, or it may be absent

C3 Plant Plant in which the first fixation of carbon is via the Calvin cycle; the first stable product of C_3 photosynthesis is a three-carbon compound

C4 Plant Plant in which the first fixation of carbon produces a four-carbon acid

Callose A complex carbohydrate in sieve tubes of sieve tube members; callose is especially abundant in injured sieve tubes

Calmodulin A type of protein that is activated when it binds to calcium ions (Ca^{++}); calmodulin activates enzymes in membranes; as much as 2% of the plasma membrane may be calmodulin

Calorie (Cal) 1,000 calories; the amount of heat required to raise the temperature of 1 liter of water 1° C; a slice of apple pie contains about 365 Cal.

calorie (cal) A unit of heat; one calorie is the amount of heat required to raise the temperature of 1 g of water 1° C; 1 cal = 4.12 J

Calvin Cycle Series of enzymatic reactions in which CO_2 is reduced to 3-phosphoglyceraldehyde (a three-carbon compound) and the CO_2 acceptor (ribulose, 1,5-bisphosphate) is regenerated

Calyptra The covering that partially or entirely covers the capsule of some species of mosses

Calyx Collectively, all of the sepals of a single flower

CAM (*see* **Crassulacean Acid Metabolism**)

Capsule 1) the sporangium of a bryophyte; 2) a dehiscent, dry fruit that develops from two or more carpels; 3) a slimy layer around the cells of certain bacteria

Carbohydrate An organic compound consisting of a chain of carbons with hydrogen and oxygen attached, usually in a ration of 2:1; glucose, sucrose, and starch are carbohydrates

Carotenoid Any compound in a class of yellow, orange, or red fat-soluble accessory pigments that are derived from eight isoprene units linked together; the most widespread carotenoid in plants is beta-carotene

Carpel The ovule-bearing organ of a flower; the flower of many species has more than one carpel, collectively called the gynoecium

Carpellate Flower A flower whose reproductive parts consist only of carpels; the kernel-bearing flowers on corn cobs are examples of carpellate flower; synonymous with pistillate flower

Carpellate Plant An individual plant whose flowers bear carpels but not stamens; a fruiting mulberry is an example of a plant that is exclusively carpellate (mulberries can form fertile fruits only when pollen is transferred from a staminate plant to a carpellate plant)

Carrageenan Aa slimy polysaccharide, consisting mostly of a specific mixture of alpha-galactose sulfates that surround the cell walls of certain red algae; the main commercial sources of carrageenan are species of the genus *Chondrus*

Carrion Flower A type of flower that is foul-smelling (carrion odor) and attracts flies or beetles as pollinators

Carrying Capacity The maximum number of individuals in any population of an ecosystem that can survive and reproduce

Casparian Strip The suberized layer covering the radial and transverse walls of endodermal walls

Catabolism The chemical reactions that break down complex materials

Catastrophism The concept that geologic changes result from sudden, violent, large-scale, worldwide catastrophic events

cDNA (*see* **Complementary DNA**)

Cell The structural unit of organisms; plant cells consist of a cell wall and protoplast

Cell Cycle Collectively, the repeating processes of cellular growth and division, including mitosis; the complete cell cycle occurs only in cells that divide, other cells being arrested in development at one of the phases of the cycle

Cell Fractionation The isolation of different organelles or parts of cells by centrifuging a homogenized cell extract in a concentration gradient of, for example, sucrose

Cell Membrane (*see* **Plasma Membrane**)

Cell Plate The disk-shaped structure that forms from the fusion of vesicles at the equator of the spindle apparatus during early telophase in plants and some algae; when mature, the cell plate becomes the middle lamella

Cell Theory A set of postulates describing how cells are the fundamental organizational units of living organisms

Cellulase An enzyme that breaks down cellulose into smaller units by cleaving the 1,4 linkages between molecules of beta-glucose

Central Dogma of Molecular Biology Refers to how genes work to make proteins; each protein-coding gene is transcribed into a molecule of mRNA, which is translated into a sequence of amino acids that comprise a polypeptide (i.e., a protein)

Central Placentation Refers to the attachment of ovules along the central axis of an ovary that has just one ovule-bearing chamber; primrose is an example plant that has central placentation

Centromere A constricted region of a chromosome where sister chromatids are held together

Chalazal Pole The region of the ovule where the stalk of the ovule fuses with the integument; usually the end of the embryo sac that is opposite the micropylar end

Chaparral Dense Mediterranean scrub of lower elevations of the North American Pacific Coast. The unique vegetation, which is either evergreen or deciduous in summer, is well-adapted to fires

Chemiosmosis The coupling of oxidative phosphorylation to electron transport via a proton pump

Chitin A tough, resistant, nitrogen-containing polymer of high molecular weight found in the exoskeletons of arthropods, in the cell walls of many fungi, and in a few other animals and protists

Chlorophyll The pigment responsible for trapping light energy in the primary events of photosynthesis

Chloroplast Organelle specialized for photosynthesis; chloroplasts occur in cells of aboveground parts of plants

Chromatid One of the two threads of a chromosome that has been duplicated in the S phase of the cell cycle; sister chromatids are held together at their centromere

Chromatin A DNA-protein complex that forms chromosomes

Chromatin Fiber A tightly wound coil of chromatin that is believed to consist of six nucleosomes per turn of the coil

Chromosome Theory of Heredity A set of postulates that accounts for the association of genes with chromosomes

Cisterna (plural, Cisternae) The flattened tubes and saclike regions of the endoplasmic reticulum and of dictyosomes

Citrate (citric acid) A six-carbon organic acid that is converted to isocitric acid in the second step of the Krebs cycle

Citric Acid Cycle (*see* **Krebs Cycle**)

Cladistics A method of classifying and reflecting phylogenetic relationships among organisms, based on an analysis of shared features

Cladogram A line diagram portrayal of a branching pattern of evolution, using the concepts and methods of cladistics

Cladophyll A stem or branch that resembles a leaf

Clamp Connection A looplike lateral connection between adjacent cells, occurring in the mycelium of certain basidiomycete fungi

Class A taxonomic category ranking between division and order

Classical Species Concept (*see* **Morphological Species Concept**)

Cleistothecium A closed, more or less spherical ascocarp

Climacteric Rise Point during the ripening of some fruit in which respiratory rates rise to extremely high levels

Climax Community A self-perpetuating community that becomes established at the completion of succession; its composition is strongly influenced by local climate and soils

Climax Vegetation The vegetation of a climax community

Cline Gradual differences in characteristics within a population across a geographic region

Clone An individual or group of individuals that develop vegetatively from cells or tissues of a single parent individual

cms-T cytoplasm (Texas cytoplasm) Refers to the phenotype of a variety of corn that is male-sterile due to mitochondrial (cytoplasmic) genes

CO2 Compensation Point Concentration of CO_2 at which the uptake of CO_2 equals the release of CO_2; that is, the point at which photosynthesis equals respiration

Coated Pit A bristle-like structure that occurs in clusters in certain regions of the plasma membrane; these regions form vesicles that pinch off into the cell, thereby removing excess plasma membrane; this process recycles excess plasma membrane in animal cells and it is suspected to do the same in plant cells that have coated pits

Codominance A condition that occurs when both alleles of a heterozygous gene are expressed equally

Codon A sequence of three nucleotides in a gene or molecule of mRNA that corresponds to a specific amino acid or to a stop signal at the end of a gene; of the 64 possible codons, 61 are codes for amino acids and three are stop codons

Coenocytic An organism, or part of an organism, that is multinucleate, the nuclei not being separated by membranes or crosswalls

Coenzyme An organic cofactor of enzyme-catalyzed reactions; NAD⁺ and coenzyme A are examples of coenzymes

Coenzyme Q (see Ubiquinone)

Cofactor A nonprotein substance required by enzymes for proper function

Coleoptile The protective sheath around the embryonic shoot in grass seeds

Coleorhiza The protective sheath around the embryonic root in grass seeds

Colony Hybridization A technique that uses probes to find bacterial colonies that contain a gene of interest

Columella Cells Cells in the center of the root cap; characterized by the presence of numerous amyloplasts that sediment in response to gravity

Companion Cell A small cell adjacent to a sieve tube; thought to control the function of sieve tube member

Complementary DNA (cDNA) DNA that is made by reverse-transcribing mRNA into its DNA complement; the collection of vector-cloned cDNA fragments of an organism are its cDNA library

Complete Dominance A condition that occurs when the phenotype of one allele completely masks the phenotype of another allele for a heterozygous gene

Complete Flower A flower that has all four of the main parts: calyx, corolla, androecium, and gynoecium

Compost Partially decayed organic matter used in farming and gardening to enrich the soil and increase its water-holding capacity

Compound Leaf A leaf consisting of two or more independent blades called leaflets

Compression Wood Reaction wood of conifers; compression wood forms along the lower side of leaning stems; compression wood expands and pushes the stem up against gravity

Concentration Gradient The difference in concentration of a substance over a certain distance

Conidiophore A hypha on which one or more condidia are produced

Conidium (plural: Conidia) An externally produced, asexual fungal spore

Conjugation Pilus (see Pilus)

Continuous Synthesis Refers to the uninterrupted synthesis of DNA in the 5′ to 3′ direction; continuous synthesis occurs in the same direction as a growing replication fork

Convergence The independent evolution of similar structures in organisms that are not closely related

Cork The outermost part of the periderm; the secondary tissue produced by the cork cambium

Corm An elongate, upright, underground stem

Corolla Collectively, all of the petals of a single flower

Cortex Ground tissue located between the epidermis and vascular bundles of stems and roots

Cot Curve A graph of the reassociation of DNA that has been denatured in a salt solution; "cot" comes from initial concentration of DNA (C_0) multiplied by the time required for complete reassociation (t)

Cotyledon Seed leaf; the first leaf formed in a seed; monocots have one cotyledon, and dicots have two

Coupled Reactions Reactions in which energy-requiring chemical reactions are linked to energy-releasing reactions

Coupled Cotransport System A set of active and passive transport proteins that work to actively move ions across a membrane against their gradient, then passively allow the same type of ions to diffuse back down their gradient while coupled to another type of solute that is being transported against its concentration gradient; an example of such a system is the active transport of protons against their concentration gradient by ATPase, followed by the co-transport of protons with sucrose through passive transport proteins back across the membrane

Crassulacean Acid Metabolism (CAM) A type of photosynthesis in which CO_2 is fixed at night into four-carbon acids; during the day, the stomata close and the carbon is fixed via the Calvin cycle; CAM helps plants conserve water and is often characteristic of xerophytic plants

Cristae (sing., **Crista**) The tubular or vesicle-shaped folds of the inner membrane of mitochondria; cristae contain cytochromes and other components of the electron transport chain that are involved in the synthesis of ATP

Crossing-Over The exchange of genetic material between the chromatids of homologous chromosomes during prophase I of meiosis

Cultivar A variety of plant that is selected for cultivation through hybridization and not found in nature

Cupule Refers to the seed-bearing structure of an extinct group of plants called the seed ferns

Cuticle The waxy coating on the epidermis of all aboveground parts of a plant

Cuticular Wax Wax that is embedded in a cuticle

Cutin The main waxy substance in a cuticle; it consists of hydroxylated fatty acids that are linked together in a complex array

Cyanophycin A polypeptide functioning as an energy reserve in cyanobacteria

Cyclic Photophosphorylation The light-induced flow of electrons originating from and returning to photosystem I; cyclic photophosphorylation produces ATP but no reduced NADP

Cyclosis Movement of the cytosol and the cellular components that are suspended in it; cyclosis is usually circular around a central vacuole

Cytochemistry (see Biochemical Cytology)

Cytochrome Heme-containing proteins that carry electrons in respiration and photosynthesis

Cytochrome b-c_1 Complex A cluster of cytochromes that carry electrons in the electron transport chain; the complex probably also pumps protons across the inner mitochondrial membrane

Cytochrome Oxidase Complex A cluster of cytochrome oxidases that function as the terminal electron carrier in the electron transport chain; this complex donates electrons to oxygen, which is then reduced to form water

Cytokinesis The division of cytoplasm into distinct cells; together with mitosis, cytokinesis comprises the phases of the cell cycle involved in cell division

Cytokinin Group of hormones (growth regulators) that promote growth by stimulating cellular division

Cytology The study of cell structure and function

Cytophotometry A method of studying cells by staining selected parts, such as the nucleus, and measuring how much light they absorb; the absorbance of stained chromatin in a nucleus is proportional to the amount of DNA it contains

Cytoplasmic Inheritance Refers to the inheritance of genes that occur in chloroplasts and mitochondria; it is cytoplasmic because it is not nuclear

Cytoplasmic Male Sterility (*cms*) A male-sterile condition in which sterility is controlled by mitochondrial (cytoplasmic) genes

Cytoskeleton A network of microscopy filaments that form a mechanical support system in the cell

D

Day-Neutral Plant Plant whose flowering is not affected by the length of day

Debranching Enzyme A type of enzyme that hydrolyzes the branched linkages of starch

Deciduous Plant Plant that loses all of its leaves during autumn

Decomposer An organism, such as bacterium or a fungus, that facilitates recycling of nutrients in an ecosystem through the breakdown of complex molecules to simpler ones

Denature To break bonds that maintain the three-dimensional structure of proteins or nucleic acids; also to break the hydrogen bonds that hold DNA together in a double helix

Dendrochronology The study of growth rings of trees to determine past conditions

Dentrifying Bacteria Bacteria that convert nitrates or nitrites to gaseous nitrogen

Deoxyribonucleic Acid (DNA) The nucleic acid containing four different nucleotides whose simple sugar is deoxyribose; genes are made of DNA; DNA exists as a double helix that can be unwound to replicate itself or to make RNA

Desert A biome characterized by low annual precipitation and/or porous soil, low humidity, wide daily fluctuations in temperature, high radiation, and living organisms adapted to these conditions

Desertification The conversion of non-desert biomes into deserts

Desynaapsis The unpairing and separation of homologous chromosomes upon the disintregation of the synaptomeal complex

Deuteromycetes Fungi that have no known sexual reproduction; most reproduce by conidia, and most otherwise have characteristics of ascomycetes. Deuteromycetes are also called Fungi Imperfecti

Dicot A type of angiosperm that belongs to a class whose members are characterized by having two cotyledons (seed leaves) per seed; Class Magnoliopsida

Dictyosome A stack of flattened, membranous vesicles that are often branched; dictyosomes are the sites where precursors of cell wall materials and other cellular components are assembled and prepared for secretion from the cell; dictyosomes are also called Golgi bodies

Differentiation Physical and chemical changes associated with the development and/or specialization of an organism or cell

Diffuse-Porous Wood Wood in which the vessels are distributed uniformly throughout the growth layers

Diffusion The net movement of particles, either dissolved or suspended, from a region of higher concentration to a region of lower concentration; the energy of diffusion is derived from the random motion of particles that is caused by molecular motion; diffusion tends to cause the distribution of particles to become homogenous throughout a medium

Dihybrid Cross A hybridization experiment that follows the inheritance of phenotypes that are controlled by two different genes

Dikaryotic Fungi whose hyphal cells each have two nuclei, the nuclei usually being derived from two different parents

Dioecious Having the pollen-producing and the ovule-producing organs on different individuals of the same species; mulberry is an example of a dioecious species

Diploid The condition of having two sets of chromosomes in a nucleus

Directional Selection Selection for a phenotype that is either higher or lower in frequency than the most abundant phenotype

Disaccharide A carbohydrate composed of two monosaccharides that are linked by a covalent bond; sucrose and maltose are examples of disaccharides

Discontinuous Synthesis Refers to the synthesis of DNA that occurs in the opposite direction of a growing replication fork; in discontinuous synthesis, DNA polymerase jumps ahead on one strand in the direction of fork movement (in this case, the 3´ to 5´ direction), then builds a new chain "backward" in the 5´ to 3´ direction

Disulfide Bond A type of covalent bond between the sulfur atoms of separate amino acids in the same protein; disulfide bonds strengthen the tertiary structure of proteins

Diterpene A compound that consists of four isoprene units linked together; gibberellins are examples of diterpenes

Diversifying Selection Selection for the low-frequency (extreme) phenotypes above and below the norm of the population; or selection against the high-frequency phenotype (norm)

Division A taxonomic category between kingdom and class

DNA Polymerase A type of enzyme that catalyzes the covalent bonding between nucleotides into a nucleic acid chain

DNA Ligase A type of enzyme that joins adjacent nucleotides together by catalyzing the formation of sugar-phosphate bonds in a strand of DNA

Dolipore A complex central pore occurring in the hyphal septa of many basidiomycete fungi; it is covered by a cap on both sides of the septum

Domain A structural and functional portion of a polypeptide, which may be encoded separately by a specific exon; a portion of a protein that has a globular tertiary structure

Dominant A trait that masks an alternative (recessive) trait when the gene for these traits is heterozygous

Dormancy A condition in which plant parts such as buds and seeds are temporarily arrested in their development; dormancy is typically seasonal and is broken as environmental conditions change during the year

Double Helix The spiral shape of a double strand of DNA

Double Fertilization In angiosperms, the process by which one sperm cell fertilizes the egg to a zygote and another sperm cell fertilizes the polar nuclei to form a primary endosperm nucleus

Dynein A large contractile protein that forms the connecting sidearms and spokes between microtubules in flagella

E

Ecological Race A race composed of many similar variants of the same species in several local populations distributed over a relatively large geographic area

Ecology The study of the interactions of organisms with one another and with their environment

EcoRI An example restriction enzyme that comes from the bacterium *Eschericia coli*; this restriction recognizes the DNA sequence GAATTC, then cleaves it between the guanine and the adenine

Ecosystem A major system of organisms that are interacting with one another and with their physical environment

Ecotype An individual taxon of plants adapted to a specific community within its overall distribution

Ectomycorrhizae Mycorrhizae that develop externally and do not penetrate to the interior of the cells they surround

Edaphic Factor A soil factor

Egg Apparatus A group of usually three cells in an embryo sac, one of which is the egg and two of which are synergids

Electrochemical Gradient The combination of a concentration gradient and an electrical gradient of ions across a membrane

Electrogenic Pump An active transport protein that transports (pumps) ions against their concentration gradient; the main electrogenic pumps in plants are proton pumps

Electron Microscope An instrument that uses an electron beam to magnify images of specimens

Electron Transport Chain A sequence of electron carriers that use the energy from electron flow to transport protons against a concentration gradient across the inner mitochondrial membrane

Embryo In plants, the part of the seed that will form the growing seedling after germination; includes a radicle, apical meristem, and embryonic leaf or leaves

Embryo Sac The common name for the female gametophyte of flowering plants

Endergonic A reaction that requires an input of energy before it will occur; endergonic reactions never occur spontaneously

Endocarp The innermost layer of simple fleshy fruits; the endocarp can be soft, as in tomatoes, or hard and stony, as in peaches

Endocytosis The process by which the plasma membrane invaginates and forms vesicles whose contents from outside of the cell can be brought into the cell

Endodermis Layer of cells inside the cortex and outside the pericycle of roots; radial walls of the endodermis are suberized by the casparian strip

Endomycorrhizae Mycorrhizae that develop within the interior of cells

Endoplasmic Reticulum (ER) An extensive network of sheetlike membranes distributed throughout the cytosol of eukaryotic cells; portions that are densely coated with ribosomes are the rough ER; other regions, with fewer ribosomes, are the smooth ER

Endosperm The nutritive, storage tissue that grows from the fusion of a sperm cell with polar nuclei in the embryo sac

Endosymbiotic Hypothesis An explanation for the origin of chloroplasts and mitochondria from the descendants of prokaryotes that lived symbiotically in larger prokaryotic hosts

Entrainment The process by which a periodic repetition of some signal (e.g., light, dark) produces a circadian rhythm that remains synchronized with the same cycle as the entraining (i.e., modifying) factor

Entropy The degree of orderliness in a system

Enzyme A biological catalyst, usually a protein, that can speed up a chemical reaction by lowering its energy of activation; amylase is an example of an enzyme

Epicuticular Wax The outermost layer of wax in a cuticle

Epidermis The outermost layer of cells that covers a plant

Epinasty The differential growth of petioles that causes the leaf blade to curve downward

Epiphyte An organism that is attached to another organism without parasitizing it

Epistasis The interaction of two or more genes that act together to make a phenotype; epistasis is best known for serial gene systems that control the multistep biosynthesis of a complex molecule or the sequence of steps in a metabolic pathway

EPSP Synthetase A type of enzyme that catalyzes a step in the synthesis of enolpyruvyl-shikimic acid-3-phosphate, which is a precursor to aromatic amino acids; the herbicide glyphosate works by inhibiting the activity of this enzyme

Ethylene A gaseous plant hormone (growth regulator) that promotes fruit ripening and other physiological responses

Etiolation The abnormal elongation of stems caused by insufficient light; etiolated stems usually lack chlorophyll

Eubacteria The majority of all bacteria; their cell walls contain muramic acid, certain lipids, and other features that distinguish them from archaebacteria

Euchromatin Lightly staining portion of chromatin, not easily visible by light microscopy

Evolutionary Species An ancestral-descendant sequence of populations evolving separately from others and forming a single unit

Exergonic A reaction that releases energy and occurs spontaneously

Exine The outermost layer of a spore or pollen grain; the exine consists of a resistant polymer that protects the male gametophyte from desiccation

Exocarp The outermost layer (usually the skin) of simple fleshy fruits

Exocytosis The process of expelling the contents of a vesicle from a cell by fusing the vesicle membrane with the plasma membrane and opening the inside of the vesicle to the outside of the cell

Exon A sequence of DNA within a gene that codes for an amino acid sequence

Exon-Shuffling Hypothesis An explanation for how complex new genes arise from the joining of independent exons into new combinations

Extensin A family of related glycoproteins that are structural proteins in cell walls

F

Facilitated Diffusion Passive transport through a transport protein

Fascicle A cluster of pine leaves (needles) or other needlelike leaves of gymnosperms

Fascicular Cambium The part of the vascular cambium that forms between the xylem and phloem within a vascular bundle

Fatty Acid A long, mostly hydrocarbon chain that has an organic acid group at one end; the most common fatty acids in plants are oleic acid, linoleic acid, and linolenic acid

Feedback Inhibition Control mechanism in which the increasing concentration of a molecule inhibits the further synthesis of that molecule

Fermentation A process by which energy is obtained from organic compounds without the use of oxygen as an electron acceptor

Fiber An elongated, thick-walled sclerenchyma cell; helps support or protect the plant

Fiddlehead The curled fern frond prior to unrolling and elongation; also known as a crozier

Field Capacity The water-storage capacity of soil; the amount of water in soil after gravitational percolation stops

Filament The stalk of a stamen; or, the vegetative body of filamentous algae and fungi

Filial Refers to a generation of offspring; the first set of offspring from a hybridization experiment is the first filial generation (F_1), the second set is the second filial generation (F_2), etc.

First Gap (see G_1 phase)

Fission The asexual division and formation of two similar new cells within mitosis, as seen in prokaryote reproduction

Fitness A measure of an individual's evolutionary success; number of its surviving offspring relative to the number of surviving offspring of other individual's within the population

Flagellum (plural, **Flagella**) A hairlike locomotor organelle that protrudes from the cell into the medium surrounding it; flagella enable cells to swim, but the only swimming cells of plants are the sperm cells of some plant groups; flagella also occur in algae, fungi, bacteria, and animals

Flavin Mononucleotide (FMN) The first electron acceptor in the electron transport chain; FMN takes electrons from NADH in the mitochondrial matrix, plus one proton from NADH and one proton from the matrix to become $FMNH_2$; protons from $FMNH_2$ are released into the mitochondrial intermembrane space

Flavonoid Any phenylpropanoid-derived compound that is linked to three acetate units and condensed into a multiple-ringed structure; the most common flavonoid is rutin; flavonoids also include naringin, which is a bitter substance in grapefruits

Flora The plants or organisms (other than animals) of a particular region; also: a publication devoted to the taxonomy of plants of a particular region

Florigen The hypothetical flowering hormone; florigen has never been identified or isolated

Fluid Mosaic Model A model for the structure of membranes as a fluid phospholipid bilayer through which proteins float in a continually shifting mosaic pattern

Fluorescence The release of energy at a longer wavelength than the energy that was absorbed

Food Web An interlocking flow of energy, involving producers and consumers in an ecosystem

Foot Basal part of a moss sporophyte; the foot is embedded in the gametophyte

Free Energy Energy available to do work

Free-Nuclear Embryo An early stage of embryo development in a gymnosperm, in which the zygote nucleus divides repeatedly without walls forming around the nuclei

Frond Photosynthetic leaf blade of a fern

Fumarate (Fumaric Acid) A four-carbon organic that takes on a molecule of water and becomes malic acid in the seventh step of the Krebs cycle

Functional Megaspore The megaspore that, in some types of embryo sac development, is the only one of the four meiotic products to grow into a female gametophyte; the other three spores disintegrate

Fusiform Initials Vertically elongated cells in the vascular cambium that produce cells of the axial system in the secondary xylem and secondary phloem

G

G_1 Phase During interphase, the portion of the cell cycle that occurs between the end of mitosis and the onset of DNA synthesis; G_1 refers to first gap

G_2 Phase During interphase, the portion of the cell cycle that begins at the end of the S phase and lasts until the beginning of mitosis; G_2 refers to the second gap

Gametangium (plural: **Gametangia**) A cell or structure in which gametes are produced

Gamete A haploid reproductive cell that fuses with another gamete to form a zygote; the female gamete is an egg and the male gamete is a sperm; in certain kinds of algae and fungi, however, the gametes are neither male or female

Gametophyte The phase of the plant life cycle that produces gametes

Gametophytic Self-Incompatibility A type of self-incompatibility that is imposed by gametophyic tissues or organs; an example would be incompatibility that is imposed by the pollen tube, which is a gametophytic structure

Gas Vacuole A membrane-bound bubble of gas that enables aquatic bacteria to float

Gel Electrophoresis A technique by which nucleic acids or proteins are separated in a gel that is placed in an electric field

Gemmae (sing. **Gemma**) Asexual plantlets in certain liverworts that can form new gametophytes; often form in gemmae cups

Gemmules An erroneous concept of inheritance; described as packets of heritable information produced throughout a mature organism, transported to the reproductive organs, and packaged into gametes before fertilization

Gene A sequence of DNA that codes for a molecule of mRNA, tRNA, or rRNA, or that regulates the transcription of such codes; a gene is the basic unit of heredity

Gene Conversion The change of one allele to another during crossing over

Gene Flow Introduction of genetic material into the gene pool of one population from another population

Gene Gun An instrument that shoots tiny beads coated with DNA directly into cells; some cells treated this way integrate the foreign DNA that is shot into them, thereby becoming transgenic; early models of the instrument used .22 caliber cartridges, hence the name gene gun.

Gene Pool All of the alleles within a population that are available to future generations

Generation Time In plants, the length of time it takes from seed germination to reach sexual maturity

Generative Cell The cell in the pollen grains of angiosperms that divides to form two sperm cells, or the cell in the pollen grains of gymnosperms that divides to form a sterile cell and another cell that divides to form to sperm cells

Genetic Species Concept Two species are considered distinct if their genetic makeup is sufficiently different from one another

Genetic Drift Random changes in gene frequencies within the gene pool of a population

Genetic Distance Measure of the degree of genetic difference between different populations or species

Genetic Code A system of codons (nucleotide triplets) in DNA or RNA that together code for a sequence of amino acids in a polypeptide; 61 of the possible codons are codes for amino acids, the remaining three being stop codons that are not translated

Genetic Engineering The artificial manipulation of genes, or the transfer of genes from one organism to another; synonymous with recombinant DNA technology

Genomic Library The set of fragments of an organism's genome that are cloned in a virus or bacterial plasmid

Genophore A bacterial chromosome; its DNA is not associated with the histone proteins of eukaryotic chromosomes

Genotype An organisms's genes, either individually or collectively

Genus (plural: **Genera**) A taxonomic category between family and species; it forms the first part of the binomial of the scientific name of organism

Gibberellin A type of plant hormone that affects, for example, stem elongation and seed germination

Gill One of the fleshy plates that radiate out from the stipe beneath the cap of a mushroom basidiocarp

Gliadin A storage protein in the grains of wheat

Glucose (also see **Alpha-Glucose** and **Beta-Glucose**) A common monosaccharide whose empirical formula is $C_6H_{12}O_6$

Glutelins A complex mixture of storage proteins in the grains of wheat

Glycogen A carbohydrate food reserve similar to starch in many organisms other than plants

Glycolysis The anaerobic metabolic pathway by which glucose is broken down into two molecules of pyruvic acid in the cytosol; the substrate-level phosphorylation of two molecules of ADP to ATP and the reduction of two molecules of NAD⁺ to NADH occur for the breakdown of each molecule of glucose

Glycoprotein A type of protein that has sugars attached to it; extensin in cell walls is an example of a family of glycoproteins

Glyoxylic Acid Cycle A sequence of biochemical reactions that converts acetyl-CoA into carbohydrate

Glyoxysome A type of microbody that is common is germinating oilseeds and seedlings that arise from them; glyoxysomes contain enzymes that catalyze the breakdown of fatty acids into acetyl-CoA

Glyphosate The generic name of one of the most commonly used herbicides in agriculture

Golgi Body (see **Dictyosome**)

Gram Stain A crystal violet stain that is retained by gram-positive bacteria and not retained by gram-negative bacteria, after alcohol or a similar solvent is applied

Gram-Negative (see **Gram Stain**)

Gram-Positive (see **Gram Stain**)

Grana (sing., **Granum**) Stacks of thylakoids where the photochemical (i.e., "light") reactions of photosynthesis occur

Grassland A biome characterized by the predominance of grasses

Gravitropism The curvature of roots or stems in response to gravity

Ground Meristem The fundamental tissue of the apical meristem; produces the cortex

Growth Ring A growth layer in secondary xylem or secondary phloem, as seen in cross section

GTP Cap A molecule of 7-methylguanosine triphosphate (GTP) that is attached to the 5′ end of a molecule of RNA as transcription begins; the GTP cap protects the RNA from degradation as it is being synthesized

Guard Cells Two specialized epidermal cells that form a stomatal apparatus

Gum A hemicellulose that is secreted by plants, which consists of several kinds of monosaccharides; an example is gum arabic, which is a mixture of the monosaccharides arabinose, galactose, glucose, and rhamnose

Gum Arabic A gum produced by the plant species *Acacia senegal*; this gum is a hemicellulose, which in this case is a complex branched chain consisting of arabinose, galactose, glucose, and rhamnose

Guttation The exudation of liquid water from leaves; caused by root pressure

Gynoceium (plural, **Gynoecia**) Collectively, all of the carpels of a single flower

H

Habitat The location, with its own specific set of environmental conditions, where an organism naturally occurs

Habitat, Operational The soil components and moisture, shade, associated organisms, and other habitat features that directly affect an organism

Half-Life Time required in a chemical reaction for half the original reactant material to decay or be consumed

Halon A bromine-based compound that is especially destructive of the ozone layer

Haploid The condition of having only one set chromosomes in a nucleus

Hardpan A hard soil with disrupted structure that may develop through the gradual accumulation of salt residues when inorganic fertilizers are applied annually without the addition of organic matter; it generally restricts the downward movement of water and roots

Hardwood A woody dicot

Heartwood Wood in the center of a tree trunk; usually darker due to the presence of resins, oils, and gums; does not transport water and solutes

Helicase A type of enzyme that breaks hydrogen bonds between complementary base pairs of DNA, thereby causing the double strand to split into separate single strands

Helix Anything of a spiral shape; in biology it refers to the shape of DNA molecules, which occur as double helices

Heme A complex organic ring structure, called a protoporphyrin, to which an iron atom is bound; heme occur in the cytochromes of all organisms and in the hemoglobin of animals

Hemicellulose Primarily a cell wall polysaccharide of variable composition and structure; hemicellulose that is secreted by plants is also called a gum (see **Gum** and **Gum Arabic**)

Herbarium A systematically arranged collection of dried, pressed, and mounted plant specimens

Heterochromatin A condensed, darkly staining portion of chromatin, easily visible by light microscopy

Heterocyst A relatively large, unpigmented, thick-walled, nitrogen-fixing cell that is produced within the filaments of certain cyanobacteria

Heterogenous Nuclear RNA (hnRNA) The pool of primary RNA transcripts in the nucleus, which are of various, usually large sizes

Heterosis A condition in which crossbred organisms are more fit than inbred organisms because they have more heterozygotic loci

Heterotroph An organism that obtains its food from other organisms

Heterozygote Superiority A condition in which individuals heterozygous at one or more loci have higher fitness than an individual with fewer heterozygous loci

Heterozygous A condition in which a gene has two different alleles in a diploid individual

Hill Reaction The photolysis of water and the photoreduction of an artificial electron-acceptor by chloroplasts in the absence of CO_2

Histone A type of protein that comprises the protein component of chromatin

hn RNA (see **Heterogenous Nuclear RNA**)

Homologous Refers to a pair of chromosomes that have alleles for the same genes

Homology A condition in which a common trait possessed by different species was derived from a common ancestor

Homozygous A condition in which both alleles of a gene are the same in a diploid individual

Hormone An organic molecule made in one plant that exerts an effect in another part of the plant; effective in small concentrations

Humus The organic portion of soil; derived from partially decayed plant and animal material

Hybrid Vigor (see **Heterosis**)

Hybridization Production of offspring from crossing different species or between genetically different populations

Hydrolysis Any chemical reaction that proceeds by the addition of water to break down a molecule; the breakdown of starch by amylase and the breakdown of sucrose by invertase are examples of enzyme-catalyzed hydrolyses

Hydrophilic Refers to chemicals that are freely soluble in water; sugars are examples of hydrophilic compounds

Hydrophobic Refers to chemicals that are not soluble in water but are soluble in nonpolar solvents; lipids and hydrocarbons are generally hydrophobic

Hydrophyte A plant that is adapted to submersion in water or an aquatic environment for at least part of its growing season

Hydrotropism Growth of a root toward water

Hypertonic Refers to a solution of high solute concentration in comparison with one of low solute concentration; a hypertonic solution tends to gain water across a membrane from a solution of lower solute concentration

Hypha (plural: **Hyphae**) A single tubular thread of the mycelium of a fungus or similar organism

Hypocotyl The region of an embryo that is between the radicle and the attachment point of the cotyledons

Hypodermis One or more layers of cells just beneath the epidermis that are distinct from the underlying cortical or mesophyll cells

Hypothesis A proposed solution to a scientific problem that must be tested by experimentation; a working explanation based on evidence and suggesting some principle; if disproved, a hypothesis is discarded

Hypotonic Refers to a solution of low solute concentration in comparison with one of high solute concentration; a hypotonic solution tends to lose water across a membrane to a solution of higher solute concentration

I

Ice-Minus Bacteria Genetically engineered bacteria that contain a foreign gene whose polypeptide inhibits the formation of ice crystals

Imbition The adsorption of water onto the internal surfaces of materials

Imperfect flower A flower that lacks either an androecium or a gynoecium

Imperfect Fungi (see **Deuteromycetes**)

In-Group Analysis The assumption in cladistics that the most prevalent character state is primitive

Inbreeding Mating within the same plant or between the offspring of an inbred parent

Incomplete Flower A flower that has one or more of the parts absent (calyx, corolla, androecium, or gynoecium)

Incomplete Dominance A condition that occurs when the phenotype of one allele only partly masks the phenotype of another allele for a heterozygous gene

Indeterminate Growth Growth that is not inherently limited, as with a vegetative apical meristem that produces an unrestricted number of organs indefinitely

Indole-3-Acetic Acid (IAA) A naturally occurring auxin (see **Auxin**)

Inferior Ovary An ovary located below the other flower parts on a floral axis

Inflorescence A cluster of flowers that are arranged on their axis in a specific pattern

Inorganic Compound A type of molecule that either lacks carbon or contains carbon but not hydrogen; carbon dioxide and water are examples of inorganic compounds

Integument The layer or layers of tissue that surround the megasporangium (nucellus) in an ovule; the integument becomes the seed coat

Intercalary Meristem Meristem at the base of a blade and/or sheath of many monocots

Interfasicular Cambium The part of the vascular cambium that forms between vascular bundles and connects with the fascicular cambium

Intermediate Filament The middle-sized (8–12 nm in diameter) of the three types of filaments that comprise the cytoskeleton

Internode Part of the stem between two successive nodes

Interphase Collectively, all of the phases of cell growth apart from cell division

Intine The inner layer of a spore or pollen grain; the intine consists of cellulosic and pectic material that is exported from the microspore

Introgression Back-crossing; mating of fertile hybrids with parent populations

Intron A sequence of DNA within a gene that does not code for an amino acid sequence

Invertase A type of enzyme that catalyzes the breakdown of sucrose by hydrolysis into glucose and fructose

Island Biogeography A theory explaining the relationship between defined habitat area (such as an island) available for organisms and the number and diversity of species in that area

Isocitrate (Isocitric Acid) A six-carbon organic acid that loses a molecule of carbon dioxide in the third step of the Krebs cycle, thereby being converted to alpha-ketoglutaric acid; also during this conversion, one molecule of NAD^+ is reduced to NADH

Isoprene The basic five-carbon subunit of terpenoid polymers

Isozymes Enzymes that have the same function but are encoded from different genes

J

Joule (J) The amount of energy needed to move one kilogram through one meter with an acceleration of one meter per second per second; 10^7 ergs; one watt-second; a slice of apple pie contains about 1.5×10^6 J

K

Kinetic Energy The energy of motion; a solute that moves down its concentration gradient has kinetic energy

Kinetin A purine that acts as a cytokininin

Kinetochore A disc-shaped complex of proteins that is bound on one side to a centromere and on the other side to a spindle fiber

Kingdom The highest taxonomic category

Kranz Anatomy Specialized leaf anatomy characteristic of C_4 plants; characterized by having vascular bundles surrounded by a photosynthetic bundle sheath

Krebs Cycle The metabolic pathway by which acetyl-CoA is oxidized in mitochondria to carbon dioxide; each turn of the Krebs cycle also forms one ATP by substrate-level phosphorylation, reduces one NAD^+ to HADH, and reduces one ubiquinone to ubiquinol; the Krebs cycle is also called the citric acid cycle or the tricarboxylic cycle

L

Lateral Meristem Meristem that produces secondary tissue; the vascular cambium and cork cambium are examples of lateral meristems

Leader Sequence A short non-coding sequence of DNA, immediately upstream from the beginning of a gene, that is transcribed into RNA

Leaf Buttress A lateral protrusion below the apical meristem; the initial stage in the development of a leaf primordium

Leaf Gap Region of parenchyma tissue in the primary vascular cylinder above a leaf trace

Leaf Primordium A lateral outgrowth from the apical meristem that will eventually form a leaf

Leaf Scar A scar left on a twig when a leaf falls from a stem

Leaf Trace The part of a vascular bundle that extends from the base of a leaf to its connection with a vascular bundle of a stem

Leaflet An individual blade of a compound leaf

Lectin A type of protein that binds to carbohydrates on cell surfaces; many lectins are glycoproteins; lectins occur in all parts of the cell but are mostly associated with the endoplasmic reticulum and other membranes, including the plasma membrane

Lenticel Spongy areas in the cork surfaces of stems and roots of vascular plants; allows gas to exchange to occur across the periderm

Liana A woody vine that is supported by other plants

Light Microscope An optical instrument that uses light to magnify images of specimens

Light-Compensation Point Light level at which photosynthesis equals respiration

Lignin A complex phenylpropanoid polymer that makes cell walls stronger, more waterproof, and more resistant to pests, herbivores, and disease organisms

Lilium-Type Embryo Sac Development A type of embryo sac development that entails all four spores of an ovule; in this type of development, the antipodal cells and one of the polar nuclei are triploid; the other polar nucleus and the egg apparatus are haploid (also see **Polygonum-Type Embryo Sac Development**)

Linkage The condition of having genes on the same chromosome (linked); alleles of genes that are linked tend to be inherited together

Locus (plural, **Loci**) The position of a gene on a chromosome

Long-Day Plant Plant that flowers when the length of dark is shorter than some critical value; long-day plants flower in spring and summer

Looped Domain A fold or loop in a region of packed chromatin fibers, which extends out from the main axis of the chromosome; looped domains may consist of 20,000 to 100,000 nucleotide pairs

M

Macroevolution Evolutionary changes that refer to the development of new species

Macronutrients Inorganic elements required in large amounts for plant growth (e.g., nitrogen, calcium, sulfur)

Magnification Enlargement of an object

Malate (Malic Acid) A four-carbon acid that is oxidized by the reduction of NAD^+ to NADH in the eighth step of the Krebs cycle; malic acid is also formed by the reduction of oxaloacetic acid that is derived from fixing carbon dioxide to phosphoenolpyruvic acid in C_4 and CAM photosynthesis

Marginal Placentation Refers to the attachment of ovules (placentation) along the edge (margin) of a suture; garden pea pods have marginal placentation

Matric Potential The component of water potential caused by the attraction of water molecules to a hydrophilic matrix

Mediterranean Scrub The often dense, shrubby vegetation that occurs in areas with wet winters and dry summers; it is dominated by evergreen bushes, or those that are deciduous in the summer

Megapascal (MPa) A unit of pressure; one million (10^6) pascals; 1 MPa = 10 atmospheres of pressure; a car tire is typically inflated to about 0.2 MPa, whereas the water pressure in home plumbing is 0.2–0.3 MPa

Megaspore A spore that will grow into a female gametophyte

Megaspore Mother Cell A cell that will undergo meiosis and cytokinesis to produce megaspores

Megasporophyll Refers to a leaf-like organ that bears megasporangia

Meio-Blastospore A spore that arises by budding from a haploid, meiotically produced spore

Meiosis Nuclear division in which chromosomes are doubled, then divided twice; the daughter nuclei from meiosis have half the number of chromosomes of the parent nucleus; in plants, meiosis forms spores

Meiosis I The first of two nuclear divisions that, in plants, form spores; in meiosis I, homologous chromosomes synapse, cross over, and move to opposite poles of the meiotic spindle apparatus; the separation of homologous chromosomes in meiosis results in a reduction in chromosome number by one-half in daughter nuclei

Meiosis II The second of two nuclear divisions that, in plants, form spores; in meiosis II, centromeres divide and sister chromatids become independent chromosomes that move to opposite poles of the spindle apparatus

Membrane Potential The potential electrical energy of ions across a membrane; membrane potential is measured in volts

Membrane Selectivity The control that a membrane exerts over how much and what kinds of materials pass through it

Membrane System The interconnected membranes of a cell, including the plasma membrane and the various organellar membranes

Mendelian Inheritance Refers to patterns of inheritance that were discovered by Gregor Mendel

Meristem Regions of specialized tissue whose cells undergo cell division

Mesocarp The middle layer (often fleshy) of simple fleshy fruits; the mesocarp occurs between the exocarp and the endocarp

Mesophyll Parenchyma tissue between the epidermal layers of a leaf; is usually photosynthetic

Mesophyte A plant that requires a relatively humid atmosphere and abundant soil water

Messenger RNA (mRNA) A class of RNA that carries the genetic message of genes to ribosomes, where the message is translated into the amino acid sequence of a polypeptide

Metabolism The sum of all chemical reactions occurring in a cell or organism

Metaphase The period of mitosis during which chromosomes become attached to spindle fibers, which align the chromosomes in a circular plane that is perpendicular to the microtubules of the spindle apparatus

Metaphase I The first metaphase of meiosis; in metaphase I, pairs of homologous chromosomes align along an equatorial plane that is perpendicular to the axis of the spindle apparatus

Metaphase II The second metaphase of meiosis; in metaphase II, chromosomes align along an equatorial plane that is perpendicular to the axis of the spindle apparatus

Metaphase Plate The plane of alignment of chromosomes during metaphase; the metaphase plate is perpendicular to the axis of the spindle apparatus

Metaxylem Primary xylem that differentiates after the protoxylem; reaches maturity after the part of the plant in which it is located has stopped elongating

Microbody A vesicle-like organelle that is bounded by a single membrane and is generally associated with the endoplasmic reticulum; glyoxysomes and peroxisomes are types of microbodies

Microevolution Evolutionary changes that occur within a population; may eventually lead to the formation of a new species, but not as a one-time event

Microfibril A complex of cellulose molecules that are twisted together into a strong, threadlike component of cell walls

Microhabitat The particular part of a habitat occupied by an individual

Micronutrients Inorganic elements required in small amounts for plant growth (e.g., boron, copper, zinc)

Micropyle The opening in a ovule through which the pollen tube will enter in angiosperms, or through which the pollen grains will enter in gymnosperms

Microsporangium (plural, **Microsporangia**) A microspore-containing sporangium

Microspore A spore that will grow into a male gametophyte

Microspore Mother Cell A cell that will undergo meiosis and cytokinesis to produce microspores

Microsporophyll Refers to a leaf-like organ that bears microsporangia

Microtome An instrument that is used for slicing specimens into microscopically thin sections

Microtubule The largest (18–25 nm in diameter) of three types of filaments that comprise the cytoskeleton; microtubules also move chromosomes during nuclear division and make up the internal structure of flagella

Middle Lamella The pectin-containing layer between cells that probably acts as the glue to hold cells together

Midrib The large central vein of a leaf

Mitosis The process of nuclear division in which chromosomes are first duplicated, followed by the separation of daughter chromosomes into two genetically identical nuclei; the division of nuclei; together with cytokinesis, mitosis comprises the phases of the cell cycle involved in cell division

Molecular Phylogeny A phylogeny based on molecular data

Monocot A type of angiosperm that belongs to a class whose members are characterized by having one cotyledon (seed leaf) per seed; Class Liliopsida

Monoecious Having the pollen-producing and the ovule-producing organs on the same individuals

Monokaryotic Fungi whose cells each contain a single nucleus

Monophyletic A taxon and all its descendants

Monomer The smallest subunit that is a building block of a polymer

Monosaccharide A simple sugar that cannot be broken down by hydrolysis; glucose is an example monosaccharide

Monoterpene A compound that consists of two isoprene units linked together; menthol is an example monoterpene

Morphological Species Concept Traditional concept of taxonomic species surmising that two species are considered distinct if they are sufficiently different morphologically

Morphological Plasticity Condition in which environmental factors induce different phenotypes from the same genotype

mRNA (*see* **Messenger RNA**)

Mucigel Slimy material secreted by root tips to facilitate growth of the root through soil

Multigene Family A set of duplicated genes; many genes occur in multigene families; an example is the family of genes in which each gene codes for the small subunit of ribulose-1, 5-bisphosphate carboxylase/oxygenase

Mutation A genetic change; mutations include changes in DNA sequences of genes, rearrangements of chromosomes, and the movements of transposable elements

Mycelium (plural: **Mycelia**) Collective term for the hyphae of a fungus

Mycolaminarin A carbohydrate food reserve of water molds (oomycetes)

Mycorrhizae (plural: **Mycorrhizae**) A mutualistic association between a fungus and the roots of a plant

Myxobacteria A group of complex, gram-negative soil bacteria that often form upright, multicellular, reproductive bodies

N

NADH dehydrogenase complex A complex of enzymes whose function is to transport protons from NADH across the inner mitochondrial membrane

Nastic Movement A movement that occurs in response to a stimulus, but whose direction is independent of the direction of the stimulus

Natural Selection Differential reproduction of phenotypes; genotypes and phenotypes vary among organisms and some of these phenotypes promote reproduction more than other phenotypes

Nectar A sweet exudate secreted by plants to attract insects (e.g., for pollination)

Nectary A structure in angiosperms that secretes nectar; usually (but not always) associated with flowers

Net Movement The amount of movement that goes in one direction more than another; particles diffuse in all directions, but net movement occurs away from where particles are most concentrated to where they are least concentrated

Net Productivity The energy produced in an ecosystem by photosynthesis minus the energy lost through respiration

Niche The ecological role of a species within a community

Nitrification The oxidation of ammonium ions or ammonia to nitrate, done by certain free-living bacteria in the soil

Nitrogen Fixation Incorporation of atmospheric nitrogen into nitrogenous compounds; done by certain free-living and symbiotic bacteria

Nitrogen-Fixing Bacteria Bacteria that convert gaseous nitrogen to nitrates or nitrites

Nitrogenase A complex of enzymes that convert atmospheric nitrogen gas into ammonia

Nivea Gene (*niv*) A gene in snapdragons that, when homozygous recessive, blocks the synthesis of flower pigments; plants that are homozygous recessive for this gene have white flowers

Node Point where one or more leaves attach to a stem

Nodule Tumorlike swelling on roots of certain higher plants (e.g., legumes) that houses nitrogen-fixing bacteria

Noncyclic Photophosphorylation The light-driven flow of electrons from water to $NADP^+$ in oxygen-evolving photosynthesis; requires both photosystems I and II

Nonvascular Plants Plants that lack vascular tissue (e.g., liverworts)

Northern Blotting A procedure by which molecules of RNA are separated by gel electrophoresis, transferred to a filter, and probed with DNA that is complementary to the RNA sequence of interest; the location of the target sequence is found because it becomes radioactive when the probe anneals to it (also see **Southern Blotting**)

Nuclear Envelope The double membrane that surrounds the nucleus

Nucleic Acid An organic acid that is a polymer of mostly four different nucleotides; deoxyribonucleic acid (DNA) and ribonucleic acid (RNA) are the two kinds of nucleic acids

Nucleosome The basic beadlike unit of chromatin in eukaryotes, consisting of DNA that is wound around a core of histone proteins

Nucleotide The subunit of a nucleic acid, consisting of a phosphate group, a simple sugar (either ribose or deoxyribose), and a nitrogen-containing base that is either a purine or a pyrimidine

Nyctinasty The "sleep movements" of leaves in response to changes in turgor of cells at the base of their petioles

O

Occam's Razor A principle of logic that holds that the best explanation of an event is the simplest, using the fewest assumptions of hypotheses

Operculum The lid of the sporangium in mosses

Opposite Phyllotaxis Leaves occurring in pairs at a node

Organelle A specialized part of the cell, usually bounded by a membrane; nuclei, chloroplasts, and mitochondria are membrane-bound organelles; ribosomes are membrane-free organelles

Organic Evolution Changes in the genetic composition of a population of organisms across generations

Organismal Theory A set of postulates describing how whole organisms, not cells, are the fundamental organizational units of living organisms; according to this theory, organisms develop by compartmenting the whole organism into cells, not by building the organism from cells

Osmosis The diffusion of water or other solvent through a differentially permeable membrane

Osmotic Potential The potential of solutes to cause osmotic pressure; also called solute potential

Osmotic Pressure The water potential of pure water across a membrane; osmotic pressure is an indicator of how concentrated a solution is on the other side of a membrane from pure water

Osmotically Active Solutes that can cause a change in a cell's osmotic potential; potassium (K^+) and other ions are osmotically active

Out-Breeding Mating with unrelated individuals

Out-Group Analysis The assumption in cladistics that the most prevalent character state of plants outside of a given group is primitive

Outcrossing Mating between different individual plants

Ovary The enlarged, ovule-bearing portion of a carpel or of a cluster of fused carpels; after fertilization, an ovary matures into a fruit

Ovule The structure that contains the female gametophyte in seed plants; the female gametophyte is surrounded by a nucellus (megasporangium tissue), which is covered by one or two integuments; when mature, an ovule is called a seed

Oxaloacetate (Oxaloacetic Acid) A four-carbon organic acid that is converted to citric acid by the addition of an acetyl group in the first step of the Krebs cycle; oxaloacetic acid is also the product of the carbon dioxide fixation of phosphoenolpyruvic acid in C_4 and CAM photosynthesis

Oxidation The loss of electrons from an atom or molecule that is involved in an oxidation-reduction (redox) reaction; oxidation removes energy from one substance, which is coupled with the simultaneous addition of energy to another substance by reduction (also see **Beta-Oxidation**)

Oxidative Phosphorylation Phosphorylation of ADP to ATP that uses energy from a proton pump fueled by the electron transport system

Ozone A form of oxygen (O_3) in the stratosphere that, when compared with ordinary oxygen (O_2), more effectively shields living organisms from intense ultraviolet radiation

P

Paleospecies A species defined only by fossil morphology

Palisade Mesophyll The vertical photosynthetic cells below the upper epidermis of a leaf

Parallelism In cladistics, a pattern of character evolution where the same character state arises from the primitive state more than once

Parapatric Occurring in adjoining places

Parapatric Speciation Speciation that occurs between contiguous populations, often induced by low dispersal range of the individuals

Paraphyletic Term applied to a group of organisms that does not contain all the descendants of a single ancestor

Paraphyses (sing. **Paraphysis**) Sterile filaments that grow among the reproductive cells of certain fungi and brown algae

Parenchyma The tissue type characterized by relatively simple, living cells having only primary walls

Parietal Placentation Refers to the attachment of ovules (placentation) along the wall of an ovary (i.e., parietal); violets are example plants that have parietal placentation

Parsimony In cladistics, the shortest hypothetical pathway that provides the most likely explanation of an evolutionary event

Parthenocarpy Development of fruit without fertilization

Pascal (Pa) The pressure unit (i.e., energy per unit volume) used to measure water potential; one pascal equals the force of one newton per square meter; one atmosphere of pressure equals 1.0×10^5 Pa

Passage Cell Endodermal cells of root that have a thin wall and casparian strip when other endodermal cells develop thick secondary walls

Passive Transport The unrestricted movement of a substance through a biological membrane; the energy for passive transport is the kinetic energy of movement down a concentration gradient; it is passive because it does not require energy from cellular metabolism

Pectin A gluey polysaccharide that holds cellulose fibrils together; pectins are mostly polymers of galacturonic acid monomers with alpha-1,4 linkages

Pedicel The stalk of a flower in an inflorescence

Peduncle The stalk of a flower or of an inflorescence

Pentose Phosphate Pathway A series of chemical reactions that start with glucose-6-phosphate from glycolysis and involve several five-carbon sugars (pentoses); during this pathway, NADP is reduced to NADPH, but no ATP is produced

Peptide Bond A carbon-nitrogen bond that links amino acids together in a chain

Peptidoglycan A large carbohydrate polymer found in the walls of true bacteria; it is composed of long chain molecules interconnected by short chains of peptides

Peptidyl Transferase A type of enzyme in the large ribosomal subunit that catalyzes the formation of a peptide bond between the amino acid at the end of a growing polypeptide and the next amino acid to be added to the chain

Perfect Flower A flower that has an androecium and a gynoecium

Pericarp Refers collectively to the layers of ovary tissue in a fruit; pericarp is the preferred term for fruits whose layers cannot be easily distinguished from one another

Pericycle The layer of cells surrounding the xylem and phloem of roots; produces branch roots

Periderm The protective tissue that replaces epidermis; includes cork (phellem), cork cambium (phellogen), and phelloderm

Peripheral Cells Outermost cells of the root cap that secrete mucigel; are sloughed from the root cat as the root grows through the soil

Peristome The "teeth" around the opening of the sporangium of mosses

Perithecium A flask-shaped or spherical ascocarp with a terminal opening

Permanent Wilting Point The moisture content of soil at the point when a particular plant's root system cannot absorb water, even when given water and placed in a humid chamber

Peroxisome A type of microbody that occurs primarily in leaves and contains enzymes that metabolize hydrogen peroxide and glycolic acid

Petal One of the parts of the flower that are attached immediately inside the calyx; collectively, the petals of a single flower are called the corolla; the corolla is usually the part of the flower that is conspicuously colored

Petiole The stalklike part of a leaf that connects the blade to the stem

Phagotropic Ingesting solid food particles

Phellem Cork; produced by the phellogen

Phelloderm The inner part of the periderm; forms inside of the phellogen

Phellogen Cork cambium

Phenolic Any compound that contains a fully unsaturated, six-carbon ring that is linked to an oxygen-containing side group

Phenotype An organism's observable features, either individually or collectively; the phenotype results from the interaction of the genotype of an organism with its environment

Phenylpropanoid A complex phenolic that has a three-carbon side chain; phenylpropanoids are generally derived from the amino acids phenylalanine and tyrosine; myristicin, the main flavor ingredient of nutmeg, is a phenylpropanoid

Phloem Vascular tissue that transports water and organic solutes

Phospholipid Aa lipid that has two fatty acids and a phosphate group bound to a molecule of glycerol; phospholipids are important components of membranes

Photochemical Reactions The "light" reactions of photosynthesis; these reactions occur on the grana of chloroplasts and produce ATP and reduced NADP

Photon The elementary particle of light

Photoperiodism Response to the duration and timing of day and night; the system within plants that measures seasons and coordinates seasonal events such as flowering

Photorespiration The light-dependent formation of glycolic acid in chloroplasts and its subsequent oxidation in peroxisomes

Photosynthesis The production of carbohydrates by combining CO_2 and water in the presence of light energy; occurs in chloroplasts and releases oxygen

Photosystem A complex of chlorophyll and other pigments embedded in the thylakoids of chloroplasts and involved in the photochemical (i.e., "light") reactions of photosynthesis

Phototropism Growth of a stem or root toward or away from light

Phragmoplast A set of microtubules oriented parallel to the axis of the spindle apparatus (perpendicular to the plane of cell division), which will form a cell plate; phragmoplasts occur in plants and in most green algae

Phycobilins Water-soluble accessory pigments occurring in the red algae and cyanobacteria

Phycocyanin A blue photosynthetic pigment in cyanobacteria and red algae

Phycoerythrin A red phycobilin

Phycoplast A set of microtubules oriented perpendicular to the axis of the spindle apparatus (parallel to the plane of cell division), which will form a cell plate; phycoplasts occur only in a few green algae

Phyllotaxis The arrangement of leaves on a stem

Phylogenetic Reflecting evolutionary relationships

Phylum A taxonomic category between kingdom and class in animals; it is equivalent to *division* in plants

Phytochrome A group of proteinaceous pigments involved in phenomena such as photoperiodism, the germination of seeds, and leaf formation; absorbs red and far-red light

Pigment Molecule that reflects and absorbs light at particular wavelengths

Pilus (plural: **Pili**) A minute tube between two bacterial cells, through which transfer of genetic material may occur

Pioneers The first plants to become established on new soil

Pistillate Flower (*see* **Carpellate Flower**)

Pistillate Plant (*see* **Carpellate Plant**)

Pith Parenchyma tissue in the center of a stem; located interior to the vascular bundles

Placenta (plural, **Placentae**) The area inside a carpel where the ovules are attached

Plasma Membrane The semipermeable membrane that surrounds the cytoplasm and is next to the cell wall; also called the cell membrane or the plasmalemma

Plasmid A small, circular fragment of DNA in bacteria; a plasmid can be integrated into and replicated with the rest of the bacterial genome; because of their ability to take up foreign DNA, bacterial plasmids are used as vectors for genetic engineering and research

Plasmodesma (plural, **Plasmodesmata**) A tiny, membrane-lined channel between adjacent cells

Plasmolysis Shrinkage of cytoplasm away from the cell wall due to the loss of water by osmosis

Plastid A type of organelle that is bounded by a double membrane and is associated with different pigments and storage products; chloroplasts are green, photosynthetic plastids; amyloplasts are storage plastids that contain starch

Pleiotropic Gene A gene that affects more than one phenotypic character; an example of a pleiotropic gene occurs in tobacco, in which a single gene controls the size and shapes of leaves, flowers, anthers, and fruits

Pneumatophor Upward-growing roots of some plants that grow in swamps; contain much aerenchyma and function in gas exchange

Polar Fiber A spindle fiber that does not bind to a kinetochore

Polar Nuclei Nuclei that come from opposite poles of the embryo sac and fuse with a sperm cell to form the primary endosperm nucleus

Polarity Establishment of poles of specialization at opposite ends of a cell, tissue, organ, or organism; for example, polarity leads to the differentiation of roots and shoots

Pollen Grain A male gametophyte that is surrounded by a microspore wall in seed plants

Pollen Tube The germination tube of a pollen grain, which grows from the stigma, through the style, and into the micropyle of the ovule; the pollen tube carries the sperm cells to the embryo sac

Pollination The transfer of pollen from microsporangia to the stigma in angiosperms or to directly to the ovule in gymnosperms

Pollination Droplet A sticky exudate at the mouth of the micropyle of a gymnosperm ovule; pollen grains catching in it are slowly withdrawn to the interior (pollen chamber) as the droplet recedes

Poly-A Tail A chain of adenylic acid molecules that is added to a molecule of RNA immediately after it has been transcribed and cleaved from its DNA template

Polyembryony, Cleavage The development of multiple embryos in a gymnosperm seed as a result of the differentiation of certain cells of a single embryo

Polyembryony, Simple The development of multiple embryos in a gymnosperm seed as a result of the development of two or more zygotes

Polygene A set of genes that act together, without dominance, to control a continuously variable phenotype; length, width, and oil content are examples of continuously variable phenotypes that are most like to be under polygenic control

Polygonum-Type Embryo Sac Development A type of embryo sac development from a functional megaspore that forms eight free nuclei, three of which become an egg apparatus, two of which are polar nuclei, and two of which become antipodal cells

Polymer A molecule consisting of many identical or similar monomers linked together by covalent bonds

Polymerase Chain Reaction (PCR) A procedure by which free nucleotides are assembled into a nucleic acid chain in a test tube by enabling the activity of a bacterial DNA polymerase to bind them together; the PCR is cycled 30 or more times to produce a million-fold amplification of the target DNA sequence

Polypeptide A chain of amino acids linked together by peptide bonds

Polyploid A condition in which a nucleus has more than two complete sets of chromosomes

Polysaccharide A carbohydrate polymer composed of many monosaccharides that are linked covalently into a chain; polysaccharides include starch, glycogen, and cellulose

Polysome A cluster of ribosomes on a single molecule of mRNA

Population A group of interbreeding individuals of the same species usually occupying the same territory at the same time

Population Density The number of individuals of a population within a given area

Population Genetics The application of genetic laws and principles to entire populations; assumes that evolution is the result of progressive change in the genetic composition of a population rather than individuals

Population, Local A population within a relatively small geographic area

Postulate A basic or necessary assumption; a set of postulates that address the same phenomenon can be taken together as a theory

Potential Energy The energy stored by matter because of its location or configuration; regarding a solute, the higher its concentration, the steeper is its concentration gradient and the greater is its potential energy; energy available to do work

Preprophase Band A band of microtubules that rings the cell just beneath the plasma membrane in a plane that is perpendicular to the axis of the future mitotic spindle apparatus; the preprophase band also corresponds to the orientation of the future metaphase plate and cell plate

Pressure Potential The component of water potential caused by the force created by turgor pressure against a membrane

Primary Cell Wall The usually thin cell wall that forms during cell division; it is part of all but some sperm cells in plants

Primary Consumer A consumer that feeds directly on producers

Primary Growth Growth resulting from the activity of apical meristems

Primary Pit-Field A thin area in a cell wall where clusters of plasmodesmata occur

Primary RNA Transcript A molecule of RNA that includes the GTP cap, the leader sequence, the gene sequence, the trailer sequence, and the poly-A tail

Primary Structure The sequence of amino acids in a protein

Primary Succession Succession that is initiated on bare rock or in water after a disturbance has occurred

Primary Thickening Meristem In some monocots, the meristem that increases the thickness of the shoot axis

Primer A short sequence of single-stranded DNA (e.g., 10-30 nucleotides long) that is complementary to one end of a target gene of interest; primers are annealed to their complementary sequences so that DNA polymerase can begin replicating the target gene in the polymerase chain reaction

Probe In genetic research, a sequence of radioactive DNA or RNA that is used to find the complementary sequence of a gene of interest in a culture of clones or cells

Procambrium A meristem that produces the primary vascular tissues

Producer An autotrophic organism; producers form the base of food chains in an ecosystem

Progymnosperms A group of extinct plants believed to be the ancestors of gymnosperms

Prop Roots Adventitious roots that form on a stem above the ground; help support the plant (as in corn)

Prophase The period of mitosis during which chromosomes condense, first appearing as a mass of elongated threads and later as individual chromosomes

Prophase I The first prophase of meiosis; in prophase I, homologous chromosomes condense, synapse, cross over, and desynapse; chiasmata move to the ends of chromosomes by the end of prophase I

Prophase II The second prophase of meiosis; in prophase II, chromosomes condense, the nuclear envelope disintegrates, and a spindle apparatus is assembled; in many organisms, prophase II is bypassed if telophase I is also bypassed, in which case the meiotic nuclei go directly from anaphase I to metaphase II

Protease Inhibitor Any chemical that inhibits the activity of enzymes that digest proteins (i.e., proteases); protease inhibitors can also be proteins

Protenema (pl. **Protenemata**) The early, filamentous growth of the gametophyte of bryophytes and ferns

Prothallial Cells Two of the four cells produced during the development of a gymnosperm microspore into a pollen grain. The prothallial cells are functionless.

Protoderm The outermost tissue of an apical meristem; produces the epidermis

Protoxylem The first xylem cells formed in the primary xylem

Pseudoplasmodium A phase of cellular slime molds in which the myxamoebae do not fuse but aggregate into a sluglike body that moves as a unit

Pulvinus Jointlike thickening at the base of a petiole; involved in movements of a leaf (or leaflet)

Punctuated Equilibrium A model stating that long geological time periods with little or no evolutionary change are punctuated by periods of rapid evolution

Punnett Square A gametic grid that is used to show the expected genotypic and phenotypic ratios resulting from a hybridization experiment

Purine A two-ringed nitrogen-containing base that is part of a nucleotide; the most common purines are adenine and guanine

Pyrimidine A one-ringed nitrogen-containing base that is part of a nucleotide; the most common pyrimidines are thymine, cytosine, and uracil

Q

Quaternary Structure The way that different subunits attach to one another in a multi-subunit protein

Quiescent Center The relatively inactive region in the apical meristem of a root

R

R Group A general term for the side group of a molecule, such as a methyl group, a hydroxyl group, or a monosaccharide

R-loop A sequence of DNA within a gene that is displaced into a loop-like projection when the gene is annealed to its complementary mRNA; the R-loop does not anneal with the mRNA because it is an intron whose complementary sequence has been spliced out of the mRNA molecule

Radicle The root of an embryo

Ray Initials Cells in the vascular cambium that produce the ray cells of secondary xylem and secondary phloem

Reaction Wood Wood produced in response to a stem that has lost its vertical position; reaction wood straightens the stem

Receptacle The region of the floral shoot where the parts of the flower are attached

Recessive A trait that is masked by an alternative (dominant) trait when the gene for these traits is heterozygous

Recombinant DNA Technology (*see* **Genetic Engineering**)

Recombination Nodule A cluster of enzymes in a synaptonemal complex, which are believed to act in concert to bring matching segments of homologous chromosomes together

Reduction The gain of electrons by an atom or molecule that is involved in an oxidation-reduction (redox) reaction; reduction involves the addition of energy to one substance, which is coupled with the simultaneous removal of energy from another substance by oxidation

Reduction Division A synonym for meiosis, specifically for meiosis I

Reductionism The approach of studying simpler components in order to understand the functions of complex systems

Release Factors A group of cytoplasmic proteins that bind to a stop codon on a molecule of mRNA and interrupt translation by hydrolyzing the bond between the final amino acid in a polypeptide and its transfer RNA

Repetitive DNA Sequences of DNA that occur in many copies in a genome; some sequences of repetitive DNA can occur in a million copies per nucleus

Replication Bubble A region of DNA that has been separated into single strands between opposing replication forks

Replication Fork The region where a DNA double strand is split into separate strands, creating a fork-like appearance in electron micrographs; once replication begins at a replication origin, two replication forks proceed along the double helix in opposite directions from one another

Replication Origin The point of initiation of DNA synthesis along the double helix; two replication forks form at the replication origin and move in opposite directions from one another during DNA synthesis

Replicon A block of DNA between two adjacent replication origins

Reproductive Barriers Various mechanisms that prevent reproduction between individuals, usually from different species

Resin A thick, translucent, combustible, organic fluid usually secreted into resin ducts in pines and many other seed plants

Resin Duct An elongate intercellular space lined with resin-secreting cells and containing resin

Resolving Power The minimum distance necessary to distinguish two points from each another

Respiration The process by which organic compounds are oxidized with the release of energy; respiration is aerobic if oxygen is required as the terminal electron acceptor; it is anaerobic if oxygen is not used as an electron acceptor

Restriction Enzyme A type of enzyme that recognizes a specific sequence of DNA and catalyzes the cleavage of the double helix at that site; most restriction enzymes recognize DNA sequences whose complementary sequence reads the same in the reverse direction; synonym with restriction endonuclease

Reverse Transcriptase A type of enzyme from viruses that catalyzes the synthesis of DNA from an RNA template; in genetics, reverse transcriptase is used for making cDNA of eukaryotic genes

Rhizome A fleshy, horizontal, underground stem

Ribonucleic Acid (RNA) The nucleic acid containing four different nucleotides whose simple sugar is ribose; molecules of RNA, which are made as complements of DNA segments called genes, function in protein synthesis

Ribosome An organelle that is responsible for protein synthesis; ribosomes consist of ribosomal RNA (rRNA) and proteins that are arranged into two subunits, one large and one small

Ribosomal RNA (rRNA) The type of RNA that is a component of ribosomes

Ribozyme A sequence of RNA that has enzymatic properties; first named from a self-splicing intron

Ring-Porous Wood Wood having larger vessels in early wood than in late wood, thereby producing a ring when viewed in a cross-section of wood

RNA Polymerase A type of enzyme that catalyzes the synthesis of RNA as the complement to a specific sequence of DNA

RNA Processing The trimming of larger primary RNA transcripts in the nucleus into smaller, coding sequences that are exported into the cytosol; synonymous with RNA splicing

RNA splicing (see RNA Processing)

Root Cap An organ that covers the root meristem; helps the growing root penetrate the soil

Root Hairs Epidermal cells just behind the zone of elongation in roots; increase the absorptive surface area of the root

Rosin The hard, brittle component of resin remaining after volatile parts have been removed

rRNA (see Ribosomal RNA)

Rubber A large polymer of up to 6,000 isoprene units

Runner (see Stolon)

S

S phase During interphase, the portion of the cell cycle in which DNA synthesis occurs; S refers to the synthesis of DNA

Saprobic Obtaining food directly from nonliving organic matter

Sapwood Wood found between the vascular cambium and the heartwood, transports water and solutes

Saturated Refers to fatty acids or other hydrocarbon-containing chemicals whose carbon-carbon bonds are all single bonds; palmitic acid is an example of a saturated fatty acid

Savanna A grassland with scattered trees. Many savannas are located in tropical or subtropical areas.

Scanning Electron Microscopy (SEM) Microscopy that focuses an electron beam that is reflected from a specimen, thereby showing fine details of its surface structure

Scarification The cutting, abrading, or otherwise softening of the seed coat to induce the seed to germinate

Scientific Method A way of analyzing the physical universe; observations are used to construct a hypothesis that predicts the outcome of future observations or experiments; something that cannot be verified cannot be accepted as part of a scientific hypothesis

Sclereids Sclerenchyma cells found in tissues varying from pear fruits to the hard shells of some nuts

Scutellum The cotyledon of a grass seed; the scutellum is specialized for absorbing nutrients from the endosperm as the seed germinates

Second Gap (see G_2 phase)

Secondary Cell Wall The cell wall that forms interior to the primary cell wall only after cell division is completed; it is restricted to certain cells and often contains lignin

Secondary Consumer A consumer that feeds on primary consumers

Secondary Growth Growth derived from lateral meristems (e.g., the vascular cambium and cork cambium)

Secondary Metabolism The metabolism of chemicals that occur irregularly or rarely among different plants and that usually have no known metabolic role in cells

Secondary Structure The portion of a protein's shape that is maintained by hydrogen bonds between amino acids

Secondary Succession Succession in habitats where the climax community has been disturbed or removed

Secondary Xylem Xylem formed by the vascular cambium; wood

Seed A mature ovule, consisting of a seed coat that surrounds the embryo and associated tissues

Seed Bank The ungerminated but still viable seeds that occur in natural storage in soil

Seed Coat The outer layer of a seed; the seed coat develops from the integument of the ovule

Seed Ferns An extinct group of plants that were characterized by frond-like leaves and seed-bearing structures; classified together in the Division Pteridospermophyta

Selection Pressures Those environmental factors that promote or retard reproductive success of a phenotype

Selectively Permeable Refers to a membrane that restricts the passage of some solutes through it

Self-Compatible Refers to the potential for successful reproduction between flowers of the same plant or between stamens and carpels of the same flower

Self-Incompatible Incapable of successful reproduction between flowers of the same plant or between stamens and carpels of the same flower

Self-Replication Refers to the ability of DNA to make exact copies of itself

Selfish DNA Refers to DNA that can perpetuate itself by semi-autonomous replication; transposons are considered to be selfish DNA because they can move copies of themselves to several sites in a genome

Semiconservative Replication Refers to the replication of a DNA molecule wherein half of each new double strand consists of one newly synthesized strand and one strand from the parent double helix

Sense Strand In DNA, the sense strand of a gene is the one that contains the coding sequence for a molecule of RNA and, in the case of mRNA, indirectly for a polypeptide

Sepal One of the outermost parts of a flower; collectively, the sepals of a single flower are called the calyx

Septum (plural: **Septa**) A crosswall in a fungal hypha

Serotype A protein that is a unique antigen; it induces and binds to antibodies that are specific to it alone. Serotypes are used in a classification system applied to viruses

Sessile Leaf Leaf Lacking a petiole; blades of sessile leaves attach directly to the stem

Seta The stalk that supports the capsule of a moss sporophyte

Short-Day-Plant Plant that flowers when the length of dark is longer than some critical value; short-day plants usually flower in autumn

Sieve Area Part of the wall of a sieve element containing many pores through which the protoplasts of adjacent sieve elements are connected

Sieve Cell A long sieve-element having unspecialized sieve areas and tapering end-walls that lack sieve plates; sieve cells occur in the phloem of gymnosperms and lower vascular plants

Sieve Elements Cells in the phloem that transport organic solutes; sieve cells and sieve-tube members are examples of sieve elements

Sieve Plate The part of a wall of a sieve-tube member that has one or more sieve areas

Sieve Tube A series of sieve tube members arranged end-to-end and connected by sieve plates

Simple Leaf A leaf having one blade which may be lobed or dissected

Single-Strand Binding Proteins Proteins that prevent the fusion and rewinding of DNA once the double strands are split apart for replication

Sink Where organic solutes are being transported by the phloem; where metabolites such as sugar are used or stored

Sister Chromatids A pair of chromatids in a duplicated chromosome

Sliding-Microtubule Hypothesis An explanation for how chromosomes are moved during anaphase; this hypothesis holds that opposing polar spindle fibers slide past one another, creating a force that pushes the poles of a spindle apparatus apart

Slug (see Pseudoplasmodium)

Small Nuclear Ribonucleoprotein (snRNP) A complex of small RNA molecules condensed with specific proteins in the nucleus; a snRNP is the basic unit of a spliceosome

snRNP (see Small Nuclear Ribonucleoprotein)

Softwood Coniferous gymnosperm

Solute Potential The component of water potential caused by the presence of solutes in water

Solute A substance dissolved in a solution

Solvent A liquid that dissolves solutes

Source Where organic compounds such as sugar are being made and loaded into the phloem

Southern Blotting A procedure by which fragments of DNA are separated by gel electrophoresis, transferred to a filter, and probed with DNA that is complementary to the gene of interest; the location of the target gene is found because it becomes radioactive when the probe anneals to it (also see Northern Blotting)

Speciation Evolutionary formation of a new species

Species (plural: **Species**) A species is a group of similar organisms capable of, or potentially capable of, freely interbreeding. The scientific names of species are binomials consisting of a genus (generic) name and a specific epithet.

Species Diversity The number of species and the number of individuals per species in an ecosystem

Spindle Apparatus Refers to the elliptically shaped collection of spindle fibers in a cell

Spindle Fibers Nearly parallel microtubules that form between the poles of dividing cells; some spindle fibers attach to chromosomes but fibers from opposite poles mostly interact with each other; spindle fibers are believed to move chromosomes both by pulling homologous chromosomes in opposite directions and by pushing poles apart

Spliceosome A cluster of snRNPs; a spliceosome binds to a large primary RNA transcript, cuts out certain parts of the RNA, then splices the rest of the RNA back into a continuous strand

Spongy Mesophyll Leaf tissue consisting of loosely arranged photosynthetic cells

Sporangium A structure within which the protoplasm becomes converted into an indefinite number of spores

Spore A small reproductive structure, usually consisting of a single cell, which is capable of developing independently into a much larger, mature and often multicellular body

Sporophyte The phase of the plant life cycle that produces spores

Sporophytic Self-Incompatibility A type of self-incompatibility that is imposed by sporophytic tissues or organs; an example would be incompatibility that is imposed by the stigma, which is a sporophytic structure

Spring (Early) Wood Wood produced in the spring; usually characterized by relatively large cells

Stabilizing Selection Selection for a phenotype within the norm of a population, or selection against extreme phenotypes

Stable Isotope Tracing A technique based on the typical ratio of carbon^{12} to carbon^{13} in tissue samples, which enables ecologists to determine food sources and consumption in food webs

Stalk Cell One of two cells produced when the generative cell of a gymnosperm male gametophyte divides. Immediately before fertilization the body cell divides, becoming two sperms.

Stamen The pollen-producing part of a flower, usually consisting of an anther and a filament; collectively, the stamens of a single flower are called the androecium

Staminate Flower A flower whose reproductive parts consist only of stamens; the tassels at the tops of corn plants are examples of staminate flowers

Staminate Plant An individual plant whose flowers bear stamens but not carpels; a "fruitless" mulberry is an example of a plant that is exclusively staminate (mulberries can reproduce only when pollen is transferred to a carpellate plant)

Starch-Branching Enzyme (SBEI) A type of enzyme that converts straight chains of amylose to the branched polymers of amylopectin; "I" refers to an isoform of the enzyme

Starch Phosphorylase A type of enzyme that cleaves a molecule of glucose from one end of a glucose polymer by phosphorylating the glucose that is removed from the chain

Start Codon The codon at the beginning of a polypeptide-coding gene; the start codon codes for the first amino acid in the polypeptide, which is usually methionine

Stele The central vascular cylinder of roots and stems

Sterol A compound derived from six isoprene units linked together in a multiple-ringed structure; beta-sitosterol is an example of a plant sterol; cholesterol is a widely known example of an animal sterol

Sticky End The single-stranded portion of a DNA sequence after it undergoes a zigzag by a restriction enzyme

Stigma The surface of a carpel that is receptive to pollen grains and upon which the pollen grains germinate; a photosensitive eyespot found in certain kinds of algae

Stipule A leaflike appendage that occurs on either side of the base of a leaf (or encircles the stem) in several kinds of flowering plants

Stolon A stem that grows horizontally along the ground

Stoma (pl. Stoma) The epidermal structure consisting of two guard cells and the pore between them

Stop Codon A codon that occurs at the end of a gene and signals where translation stops

Stratification The exposure of seeds to extended cold periods before they will germinate at warm temperatures

Strobilus, Compound An axis with lateral branches bearing sporophylls

Strobilus, Simple An unbranched axis bearing sporophylls

Stroma The matrix between the grana in chloroplasts; the site of the biochemical (i.e., "dark") reactions of photosynthesis

Structural Polysaccharide A polysaccharide that holds cells and organisms together; cellulose is the most abundant structural polysaccharide in plants

Style A column of carpel tissue arising from the top of an ovary and upon which is the stigma; the style raises the stigma to a receptive position for pollen grains whose pollen tubes must grow through it

Suberin Aa waxy substance that occurs in cork cells and in the cells of underground plant parts; it consists of hydroxylated fatty acids that are linked together in a complex array

Subsidiary Cella Epidermal cells that are structurally distinct from other epidermal cells and associated with guard cells

Substrate-Level Phosphorylation The transfer of a phosphate group from a substrate, such as phosphoenol pyruvic acid, to ADP, thereby making ATP

Subunit A polypeptide that combines with other polypeptides to comprise a multi-subunit protein

Succession An orderly progression of population replacements in a specific geographic area; it is initiated with pioneer species and completed with the establishment of a stable climax community

Succinate (Succinic Acid) Aa four-carbon organic acid that is oxidized by the reduction of ubiquinone to ubiquinol in the sixth step of the Krebs cycle; the product of this oxidation is fumaric acid

Succinyl-CoA An acetylated four-carbon acid that is converted to succinic acid by losing its acetyl-CoA group, thereby driving the substrate-level phosphorylation of one molecule of ADP to ATP in the fifth step of the Krebs cycle

Succulent A plant having think, fleshy leaves or stems; succulence is usually an adaptation to water or salt stress

Sucker A sprout on the roots of some plants that forms a new plant

Sucrose Synthase A type of enzyme that catalyzes the reversible breakdown of sucrose from starch by hydrolysis into free fructose and bound glucose; the glucose is bound to a carrier molecule called uridine diphosphate (UDP)

Summer (Late) Wood Wood produced in the summer; characterized by relatively small cells

Superior Ovary An ovary located above the other flower parts on a floral axis

Suspensor A group of cells at the base of the embryo of many seed plants that expands and moves the embryo into the endosperm

Suture The line along which a fruit splits when it is mature

Symbiont One of tow (or more) dissimilar organisms that live in close association with each other. The association may be beneficial to both organisms (*mutualism*) or harmful to one organism (*parasitism*)

Sympatric Having the same or overlapping geographic distribution but separated by reproductive or biotic barriers

Sympatric Speciation Formation of a new species entirely within the geographical range of its parental form

Symplast The interconnected living mass of an organism; the symplast is a continuous unit that is comprised of cells that are connected by plasmodesmata throughout the organism

Symplastic Movement of water and solutes through tissues by passing through interconnected protoplasts and their plasmodesmata

Synapsis The pairing of homologous chromosomes by their attachment along a synaptonemal complex; crossing over occurs during synapsis

Synaptonemal Complex A complex of proteins that forms a chromosome-length axis linking homologous chromosomes between the same gene loci

Synergid A type of cell that occurs next to the egg in an embryo sac; sperm cells entering the embryo sac first pass through one of the synergids

Synonymous Codon Refers to codons that code for the same amino acid

Syringomycin A toxic polypeptide that is secreted by *Pseudomonas syringae*, a species of bacteria that infects corn, beans, and many other kinds of plants

Systematics The classification of organisms into a hierarchy of categories (taxa) based on evolutionary interrelationships

T

Taiga A coniferous forest biome adjacent to arctic tundra in large areas of North America and Erasia.

Tandem Repeat The occurrence of two or more copies of a gene in a row; ribosomal RNA genes typically occur as tandem repeats

Tangential Section A longitudinal section that does not pass through the center of the structure

Tapetum A tissue of sterile cells that surrounds the microspores in a microsporangium; the tapetum acts as a nutritive tissue for the spores and pollen grains while they remain in a sporangium

Taproot A relatively large primary root that produces secondary roots

Tassel The downward-hanging inflorescence of some plants; in corn, tassel refers to an inflorescence of pollen-bearing flowers at the top of the plant

Taxol A drug obtained from the Pacific yew, and also from a fungus that grows on the yew, with potential for treating certain forms of cancer

Taxon (plural: **Taxa**) Any taxonomic category, such as species or family

Taxonomy The classification, description, and naming of organisms

Telomere Sequences of DNA at the tip of a chromosome that counteract its condensation before the onset of nuclear division

Telophase The period of mitosis during which chromosomes seem to mimic prophase in reverse; chromosomes steadily elongate and decondense back into diffuse chromatin, and each new daughter nucleus becomes surrounded by a nuclear envelope

Telophase I The first telophase of meiosis; in telophase I, chromosomes decondense, the spindle apparatus disintegrates, and a new nuclear envelope forms around each daughter nucleus; in many organisms, telophase I is bypassed and the meiotic nuclei go directly from anaphase I to metaphase II

Telophase II The second telophase of meiosis; in telophase II, chromosomes decondense, the spindle apparatus disintegrates, and a new nuclear envelope forms around each of the four new daughter nuclei

Temperate Deciduous Forest A biome dominated by deciduous hardwood trees, and located in areas with temperate climates

Tendril A modified leaf or stem in which only a slender strand of tissue constitutes the entire structure

Tensile Strength The maximum amount of lengthwise pull that a substance can bear without tearing apart

Tension Wood Reaction wood that forms along the upper side of leaning stems; straightens the stem by contracting and "pulling" the stem up

Terpenoid Any compound that is derived from five-carbon precursors called isoprene units; examples include menthol (two isoprene units), beta-carotene (eight isoprene units), and rubber (up to 6,000 isoprene units)

Tertiary Structure The portion of a protein's shape that is maintained by disulfide bonds, ionic interactions, or hydrophobic attraction between amino acids

Tetrad Scar A scar on a primitive spore at the point of attachment to three other spores, all four having developed after meiosis; germination takes place in the vicinity of the scar

Thallophytes A term once used to designate fungi and algae collectively

Thallus (plural: **Thalli**) A body not differentiated into roots, stems, or leaves, as seen in lichens and liverworts

Thigmotropism A response to contact with a solid object

Thorn A modified woody stem that terminates in a sharp point

Thylakoid A saclike membranous structure of the grana of chloroplasts; thylakoids house chlorophylls

Tissue Culture A technique for growing and manipulating pieces of tissue in a medium after their removal from the organism

Tonoplast The membrane that surrounds a vacuole; also called a vacuolar membrane

Topoisomerase A type of enzyme that relieves the kinks in DNA that would otherwise block the movement of replication forks; topoisomerases work by breaking one or both strands, thereby allowing the strands to uncoil by swiveling around one another; after uncoiling, the strands are also linked back together by topoisomerases

Totipotent Refers to the notion that every cell has the same genes and therefore the same genetic potential to make all cells other cell types

Trace Elements (see **Micronutrients**)

Tracheid Elongated, spindle-shaped cell that transports water in the xylem; also helps support the plant; occur in virtually all vascular plants

Trailer Sequence An extra amount of non-coding DNA that is transcribed into RNA beyond the end of the gene

Transcription The synthesis of a molecule of RNA as a complement to a specific sequence of DNA; transcription occurs in the nucleus

Transduction The transfer of genetic material between bacteria by bacteriophages

Transect, Line A straight line extending between two points, usually established arbitrarily for the purpose of studying vegetation immediately adjacent to it

Transfer RNA (tRNA) A type of small RNA molecule that binds to a specific amino acid and to a codon on messenger RNA; it is called transfer RNA because it is associated with the transfer of amino acids to mRNA in ribosomes; more than 40 different tRNA molecules have been found, at least one for each protein amino acid

Transformation A form of genetic transfer in bacteria by which a fragment of DNA is taken up and incorporated into the DNA of the recipient cell

Transgenic Refers to cells or organisms that contain genes that were inserted into them from other organisms by genetic engineering

Translation The synthesis of an amino acid sequence from a specific sequence of codons along a molecule of mRNA; translation occurs in ribosomes

Transmission Electron Microscopy (TEM) Microscopy that focuses an electron beam through the thin section of a specimen to study its internal structure

Transpiration Evaporation of water from leaves and stems; occurs mostly through stomata

Transport Protein Aa type of membrane protein that enables the transport of specific solutes across the membrane

Transposable Element A fragment of DNA that is able to multiply and move spontaneously among an organism's chromosomes

Tricarboxylic Cycle (see **Krebs Cycle**)

Trichogyne A receptive, slender outgrowth for spermatia or similar reproductive cells in red algae and ascomycete fungi

Trichome An epidermal outgrowth (e.g., a hair or scale)

Triplet Refers to a sequence of three nucleotides that together comprise a codon

Triterpene A compound that consists of six isoprene units linked together; sterols, such as beta-sitosterol, are triterpenes

tRNA (see **Transfer RNA**)

Tropical Rain Forest An endangered tropical biome with exceptional diversity of species

Tropism A response to an external stimulus in which the direction of the response is determined by the direction from which the stimulus comes; tropisms such as phototropism and gravitropism are produced by differential growth

True-Breeding Refers to purebred strains for a given trait, which means that the gene for that trait is homozygous

Tube Cell The cell in the pollen grains of seed plants that develops into the pollen tube

Tuber A fleshy, underground stem having an enlarged tip (e.g., potato tuber)

Tubulin (see **Alpha-Tubulin** and **Beta-Tubulin**)

Tundra A vast biome primarily above the arctic circle, and above the timberlines of mountain ranges further south, whose vegetation includes no typical trees

Tunica-Corpus The organization of the shoot apex of most angiosperms and some gymnosperms; consists of one or more peripheral layers (i.e., tunica layers) and an interior corpus

Turgid Full of water taken in by osmosis

Turgor Pressure The pressure on a cell wall that is created from within the cell by the movement of water into it

Turpentine The volatile, combustible component of resin

Type Specimen A specimen upon which the original description of species is based

U

Ubiquinol The reduced form of ubiquinone; ubiquinol donates electrons to cytochrome *b* in the electron transport chain

Ubiquinone A lipid-soluble quinone whose function is to accept electrons from electron donors like NADH and from the oxidation of fatty acids; also called coenzyme Q

Ultracentrifuge A high-speed centrifuge that is capable of spinning at more than 100,000 revolutions per minute

Unequal Crossing-Over Refers to the exchange of unequal amounts of DNA between homologous chromosomes that are not perfectly aligned

Unsaturated Refers to fatty acids or other hydrocarbon-containing chemicals whose carbon-carbon bonds include double bonds as well as single bonds; oleic acid (one double bond), linoleic acid (two double bonds), and linolenic acid (three double bonds) are examples of unsaturated fatty acids

Uridine Diphosphate (UDP) A uracil-containing nucleotide that acts as a carrier molecule for glucose and similar monosaccharides; the UDP-sugar complex is also an intermediate compound for the interconversion of one monosaccharide to another (e.g., glucose to galactose)

V

Vacuole A membrane-bound organelle that is filled mostly with water but also may contain water-soluble pigments and other substances; the vacuolar membrane is called the tonoplast

Vascular Bundle Strand of tissue containing primary phloem and primary xylem (and possibly procambrium); often enclosed by a bundle sheath

Vascular Tissue Tissue specialized for long-distance transport of water and minerals; xylem and phloem

Vector In genetics, a virus or bacterial plasmid that can take up a foreign gene and integrate it into the genome of a target organism; in plant reproduction, an animal that carries pollen (a pollinator) from a pollen sac to a stigma or to an ovule

Vegetative Cell A cell that is neither sexually reproductive nor divides to form cells that are sexually reproductive; this term particularly refers to the tube cell of angiosperm pollen grains, which is the only vegetative cell of the male gametophyte

Vein Vascular bundle that forms part of the connecting and supporting tissue of a leaf or other expanded organ

Velamen Multiple epidermis covering the aerial roots of some orchids and aroids

Venter The swollen base of an archegonium containing the egg

Vernalization The induction of flowering by cold

Vessel A tubelike structure in the xylem that consists of vessel elements placed end-to-end and connected by perforations; vessel elements conduct water and minerals; found in nearly all angiosperms and a few other vascular plants

Vessel Element One of the cells forming a vessel

Vesticular-Arbuscular (V-A) Mycorrhizae Treelike or bulblike mycorrhizae

Viroid A plant-infecting, viruslike particle with a single circular strand of RNA that is not associated with any protein

Virusoid A particle similar to a viroid but located inside the protein coat of a true virus

W

Water Potential The potential energy of water to move down its concentration gradient; water potential is expressed in units of pressure instead of units of energy, because pressure is simpler to measure

Whorl A circular group of at least three leaves or flower parts all attached to an axis at the same level

Wood Secondary xylem

X

Xanthophyll A yellow carotenoid; one xanthophyll, zeaxanthin, in the blue-light photoreceptor in shoot phototropism

Xerophyte A plant adapted for growth in arid conditions

Xylem Vascular system specialized for transporting water and dissolved minerals upward in the plant; characterized by the presence of tracheary elements

Y

Yeast Artificial Chromosome (YAC) A yeast chromosome into which large fragments of foreign DNA (millions of base pairs) have been inserted; YACs can be replicated like native chromosomes in yeast cells, thereby cloning large amounts foreign DNA as well

Z

Zeatin A natural cytokinin isolated from corn (*Zea mays*)

Zein A storage protein in the kernels of corn

Zygomycetes A large group of fungi with primarily coenocytic mycelia; they reproduce asexually by spores produced within sporangia; sexual reproduction includes the formation of zygosporangia

Zygosporangium (plural: **Zygosporangia**) A sporangium containing a thick-walled, multinucleate zygospore that develops in zygomycetes after the fusion of isogametes

Zygote The diploid cell that is formed by the fusion of two gametes

Credits

PHOTOGRAPHS

Part Openers
Part 1: © Bruce Iverson Photomicrography; **Part 2:** © Grant Heilman Photography; **Part 3:** © Sydney Karp/Photo/Nats; **Part 4:** © George Gainsburgh/N.H.P.A.; **Part 5:** © David Muench; **Part 6:** © A. Blake Gardner.

Preface
Page xxii (left): © Photo/Nats; Page xxii (right): © Ray Coleman/Photo Researchers, Inc.

Chapter One
Opener: © Doug Sherman; **1.1A:** © Muriel Orans; **1.1B:** © Noboru Komine/Photo Researchers, Inc.; **1.1C:** © Grant Heilman Photography; **1.1D:** © Verna R. Johnston/Photo Researchers, Inc.; **1.1E:** © Walter H. Hodge/Peter Arnold, Inc.; **1.1F:** © James Shaffer; **Box 1.1A:** © Fritz Prenzel/Peter Arnold, Inc.; **Box 1.1B:** © J. Villegier/Photo Researchers, Inc.; **1.2:** © Grant Heilman Photography; **1.3:** © Alan Pitcairn/Grant Heilman Photography; **Box 1.1C:** © John Neubauer; **1.4:** © Runk/Schoenberger/Grant Heilman Photography; **1.5A:** © Henry Grossman; **1.5B:** Photo © Charles R. Ervin, Ph.D., provided by James Mauseth; **1.6:** The Bettmann Archive; **1.7:** © Wm. C. Brown Communications, Inc./Michael J. Elderman, photographer; **1.9A:** UPI/Bettmann Archive; **1.9B** (all): From *Scientific American* 250(6):86, 1984. Article by Nina Fedoroff; **1.10A:** © BioPhot; **1.10B:** © Michael Goodman/Photo/Nats; **1.10C:** © John Kaprielian/Photo Researchers, Inc.; **1.10D:** © Kjell B. Sandved; **1.10E:** © Walter E. Harvey/National Audubon Society/Photo Researchers, Inc.; **1.11:** Courtesy R. Fraly, Monsanto.

Chapter Two
Opener: From David S. Goodsell, "A Look Inside The Living Cell," *American Scientist* 80(5):457–65, 1992; **2.3B:** © Biophoto Assoc./Photo Researchers, Inc.; **2.4B:** © Eric Grave/Science Source/Photo Researchers, Inc.; **2.9D:** © Nelson Max/Peter Arnold, Inc.; **2.12 A,B,C:** From D. Froelich and W. Barthlott, *Mikromorphologie der Epicuticularen Wachse und das System der Monokotylen*, © 1988; The Academy of Science & Literature, Maine. Courtesy of Dr. W. Barthlott; **Page 37** (top to bottom): © Kjell B. Sandved; © Ann Reilly/Photo/Nats; © Alan Pitcairn/Grant Heilman Photography; © Walter H. Hodge/Peter Arnold, Inc.; **2.13B:** © William E. Ferguson; **2.13C:** © Walter H. Hodge/Peter Arnold, Inc.; **2.13D:** © Runk/Schoenberger/Grant Heilman Photography; **2.14B:** © Dede Gilman/Visions From Nature; **2.14C:** © Betsy Fuchs/Photo/Nats; **2.14D:** © Alan Pitcairn/Grant Heilman Photography; **2.15B:** © Larry Lefever/Grant Heilman Photography; **2.15C:** © Grant Heilman Photography; **2.15D:** Courtesy of Hillerich & Bradsby Co., Inc.

Chapter Three
Opener: Reproduced by permission of the American Society of Plant Physiologists, from *The Plant Cell* 1(10), 1989. Courtesy of Craig Lending; **3.1 B,C:** © Runk/Schoenberger/Grant Heilman Photography; **3.6 A,B:** © Bruce Iverson Photomicrography; **3.6 C,D:** © Richard H. Gross; **3.6E:** © Manfred Kage/Peter Arnold, Inc.; **3.7:** © BioPhot; **Box 3.1 A,B:** From Misuza Baba, Norio Baba, Yoshinori Ohsumi, Koichi Kanaya and Masaka Osumi, "Three Dimensional Analysis of Morphogenesis Induced by Mating Pherome a Factor in Saccharomyces Cerevisiae," *Journal of Cell Science* 94:207–216, 1989; **3.8A:** © Biophoto Assoc./Photo Researchers, Inc.; **3.8B:** © Runk/Schoenberger/Grant Heilman Photography; **3.9A:** © Gary T. Cole, U. of Texas, Austin/Biological Photo Service; **3.10A:** © M. Schliwa/Visuals Unlimited; **3.10B:** Courtesy of Drs. R. R. Trelease & Francisco Carrapico; **3.13:** © Biophoto Assoc./Photo Researchers, Inc.; **3.14:** © E. Vigil/Biology Media/Photo Researchers, Inc.; **3.15A:** © Biophoto Assoc./Science Source/Photo Researchers, Inc.; **3.16:** Courtesy of W. A. Cote, Center for Ultrastructure Studies, College of Environmental Science & Forestry, State University of New York; **3.19A:** © E. H. Newcomb, University of Wisconsin/Biological Photo Service; **3.20:** Courtesy of Jean Whatley; **3.22A:** © E. H. Newcomb and W. P. Wergin, University of Wisconsin/Biological Photo Service; **3.23:** From C. M. Kunce, R. R. Trelease and D. C. Doman, "Ontogeny of Slyoxysomes in Maturing and Germinated Cotton Seeds—a Morphometric Analysis," *Planta* 161:156–164, 1984, Springer-Verlag. Courtesy R. R. Trelease, Arizona State University; **3.25A:** Courtesy of Jean Whatley; **3.26B:** © Biophoto Assoc./Photo Researchers, Inc.; **3.27A:** Courtesy of Knut Norstog; **3.27B:** © Robert & Linda Mitchell; **3.28A (both):** Courtesy of Gary L. Floyd.

Chapter Four

Opener: © Biophoto Assoc./Science Source/Photo Researchers, Inc.; **4.3:** © Don Fawcett/Visuals Unlimited; **4.5:** R. B. Park & D. Branton, *Brookhaven Symposia In Biology* (19), 1966. Courtesy Albert Pfeifhofer, University of California, Berkeley; **4.7:** © Runk/Schoenberger/Grant Heilman Photography; **4.12:** © Dr. Jeremy Burgess/SPL/Photo Researchers, Inc.; **4.13A:** © Dwight R. Kuhn; **4.13B:** © Alfred Owczarzak/Biological Photo Service; **4.14:** © William E. Ferguson; **Box 4.1A:** © Donald Specker/Earth Scenes; **4.19:** Courtesy of L. C. Fowke; **4.25:** © Hugh Spencer/Photo Researchers, Inc.; **4.26:** © Biophoto Assoc./Science Source/Photo Researchers, Inc.

Chapter Five

Opener: © Donald Specker/Earth Scenes; **5.1:** © Ray Coleman/Photo Researchers, Inc.; **5.2:** © David C. Bitters/Photo/Nats.

Chapter Six

Opener: © Malcolm Wilkins; **6.2:** © William E. Ferguson; **6.12A:** Courtesy of Peter Hinkle, Cornell University; **Box 6.2:** © Angelina Lax/Photo Researchers, Inc.

Chapter Seven

Opener: © Kjell B. Sandved; **7.1A:** The Bettmann Archive; **7.2A:** Northwind Picture Archives; **7.10:** © BioPhot; **7.13B:** © Dwight R. Kuhn; **7.13C:** © Dr. Jeremy Burgess/SPL/Photo Researchers, Inc.; **7.13D:** © Grant Heilman Photography; **7.20B (all):** From J.A. Bassham, 1965 in J. Bonner and J.E. Vaner (Eds.) *Plant Biochemistry*, Academic Press; **7.23:** © E. H. Newcomb and S. E. Frederick, University of Wisconsin/Biological Photo Service; **7.24A:** © BioPhot; **7.24B:** Courtesy of Raymond Chollet, University of Nebraska.

Chapter Eight

Opener: © Dwight R. Kuhn; **8.1:** The Bettmann Archive; **8.2:** Photo by Richard E. Ferguson © William E. Ferguson; **8.7:** Courtesy of David Cass; **8.9A:** © J.N.A. Lott, McMaster University/Biological Photo Service; **8.9B:** © Peter J. Bryant/Biological Photo Service; **8.12:** Photo by Susan S. Whitfield. Courtesy of C. R. Parks; **Box 8.3:** © Dr. Loren Rieseberg; **Box 8.4:** Reprinted from *Biochemical Systematics and Ecology* 15(5):546, Fig. 1., by L. Rieseberg & D. Soltis, © 1987 with permission from Pergamon Press Ltd., Headington Hill Hall, Oxford OX3 0BW, U.K.; **8.13:** © John Kaprielian/Photo Researchers, Inc.; **8.15A:** © Muriel Orans; **8.17B:** © Betty Outlaw; **Box 8.5:** © William E. Ferguson; **Box 8.6:** Courtesy U.S.D.A.; **8.18:** © Evelyne Cudel-Epperson, University of California—Riverside, 1992.

Chapter Nine

Opener: © Bruce Iverson Photomicrography; **9.4:** © Ed Reschke; **9.5:** © Bruce Iverson Photomicrography; **9.6:** © Dede Gilman/Visions From Nature; **9.7:** Courtesy of King's College, London; **9.9:** The Bettmann Archive; **9.11G:** © Science VU-NIH/Visuals Unlimited; **9.11H:** Courtesy of M.M. Schwesinger. From "Plasmids" by Dr. Richard P. Novick, *Scientific American*, Dec. 1980, p. 104.; **9.11I:** © A. L. Olins, University of Tennessee/Biological Photo Service; **9.11J:** Courtesy of Barbara Hamkalo; **9.11K:** From *Cell* 12:817, 1977 from an article by J. R. Paulsen and U. K. Laemmli © Cell Press; **9.11L:** © Science VU-NIH/Visuals Unlimited; **9.14 A,B:** From *Genetics* 79:137–150, 1975, by T. C. Hus; **9.17B:** From J. Cairns, "The Chromosome of E Coli," *Cold Spring Harbor Symposium on Quantitative Biology*, 28, © 1963 by Cold Spring Harbor Laboratory, Cold Spring, NY. Reprinted by permission of the publisher and author; **9.20:** © Ed Reschke; **9.21:** Courtesy of Susan M. Wick, University of Minnesota; **9.22 (all):** © Andrew S. Bajer, Professor of Biology, University of Oregon.

Chapter Ten

Opener: © Sheila Terry/SPL/Photo Researchers, Inc.; **10.1B:** © Robert & Linda Mitchell; **10.1C:** © Ed Reschke; **10.8 (all):** © C.A. Hasenkampf, U. of Toronto/Biological Photo Service; **10.14:** © Dr. Russell K. Monson; **10.17:** © Science Source/Photo Researchers, Inc.

Chapter Eleven

Opener: Courtesy of Discovery Research, DeKalb Plant Genetics; **11.1:** Courtesy of Madan K. Bhattachargya, PhD. From the cover of *Cell* 60(1), January 12, 1990 © Cell Press; **11.4:** © Professor Oscar Miller/SPL/Photo Researchers, Inc.; **11.6A:** Courtesy of O. L. Miller, Jr., Adapted from the *EMBO Journal* 5(13):3591–96, Fig. 1b, 1986; **11.8C:** © Tripos Assoc./Peter Arnold, Inc.; **11.10:** © David M. Phillips/Visuals Unlimited; **11.19 A,B,C:** Courtesy of Calgene, Inc. and Bill Santos Photography; **11.27:** © Science VU/Visuals Unlimited; **11.28:** Great Britain M.A.F.F., Crown copyright; **Box 11.3B:** Courtesy of John C. Sanford, Cornell University; **11.31:** Courtesy of Calgene, Inc.; **11.32:** © Runk/Schoenberger/Grant Heilman Photography; **11.33:** Photography Plus, Inc. Research conducted at Oregon State University as Research & Extension Center. Photo by Photography Plus, Inc., Umatilla, OR; **11.35A:** Courtesy Madan Bhattacharyya, from *Cell* 60(1):115–122, January 12, 1990 © Cell Press.

Chapter Twelve

Opener: © Kjell B. Sandved; **12.1A:** © BioPhot; **12.1B:** © Cabisco VU/Visuals Unlimited; **12.1C:** © BioPhot; **12.2 A,B:** © BioPhot; **12.2C:** © Bruce Iverson Photomicrography; **12.3A:** © BioPhot; **12.3B:** © F.A.L. Clowes; **Box 12.1 (both):** © Muriel Orans; **12.5:** Courtesy J. H. Troughton and L. Donaldson and Industrial Research, Ltd.; **12.6 A,B:** From Susses, "Morphogenesis in Solanum Tuberosum L: Apical Developmental Pattern of the Juvenile Shoot" *Phytomorphology* 5:253–273, 1955; **12.7A:** © E. Vigil/Biology Media/Photo Researchers, Inc.; **12.8A:** © Bruce Iverson Photomicrography; **12.13A:** Courtesy J. H. Troughton and L. Donaldson and Industrial Research, Ltd.; **12.13B:** © Kjell B. Sandved; **12.13C:** John Shaw/Bruce Coleman, Inc.; **12.13D:** © G. Biittner/Naturbild/OKAPIA/Photo Researchers, Inc.; **12.14:** Courtesy of Paul Green, Stanford University.

Chapter Thirteen

Opener: © Chuck Brown/Photo Researchers, Inc.; **13.1A:** © Ed Reschke; **13.1B, 13.2:** © Dwight R. Kuhn; **13.3A:** Courtesy Long Ashton Research Station; **13.3B:** © Randy Moore/BioPhot; **13.4A:** © R. B. Taylor/SPL/Photo Researchers, Inc.; **13.4 B,C:** Courtesy J. H. Troughton and L. Donaldson and Industrial Research, Ltd.; **13.5:** From Gunning & Pate, "Transfer Cell Plant," p. 420, *McGraw-Hill Yearbook of Science & Technology,* 1971; **13.6:** © BioPhot; **13.7A:** © Dwight R. Kuhn; **13.7B:** © Biophoto Assoc./Photo Researchers, Inc.; **13.8, 13.9, 13.10:** © BioPhot; **13.11:** © Bruno P. Zehnder/Peter Arnold, Inc.; **13.12:** © BioPhot; **13.13:** © John N. A. Lott, McMaster U./Biological Photo Service; **13.14:** © Dr. Jeremy Burgess/SPL/Photo Researchers, Inc.; **13.15, 13.16, 13.17 A,B:** © BioPhot; **13.18 A,B:** Courtesy J. H. Troughton and L. Donaldson and Industrial Research, Ltd.; **13.19:** © BioPhot; **13.20A:** Courtesy B. Galatis & K. Mitrakos, provided by James Mauseth; **13.20B:** © Dwight R. Kuhn; **13.21 A,B:** From Royle and Thomas, *Physiological Plant Pathology* 3:405–417. © 1973, Academic Press Ltd.; **Box 13.1 (all):** From Hoch, Staples, Whitehead, Comeau, Wolf, "Signaling for Growth Orientation & Cell Differentiation by Surface Topography in Uromyces," *Science* 235:1660. © 1987 by the AAAS.; **13.22:** From Yolanda Heslop-Harrison, "SEM of Fresh Leaves of Pinguicula," *Science* 667:173. © 1970 by the AAAS; **13.23:** © Runk/Schoenberger/Grant Heilman Photography; **13.24:** © Biophoto Assoc./Photo Researchers, Inc.; **13.25:** © BioPhot; **13.26:** © George Wilder/Visuals Unlimited; **13.27:** © BioPhot; **13.28:** Courtesy Drs. R. A. Meylan & B. G. Butterfield; **13.29:** From Richard Anderson & J. Cronshaw, "Sieve Plate Pores In Tobacco and Bean," *Planta* 91:173–180, 1970 © Springer-Verlag; **13.30, 13.31:** © BioPhot; **13.32:** © Bruce Iverson Photomicrography; **13.33:** © Richard F. Trump/Photo Researchers, Inc.; **13.34, 13.35, 13.36:** © BioPhot; **13.37:** © Willard Clay.

Chapter Fourteen

Opener: © Barbara Brundege Photography; **14.2:** © Bruce Berg/Visuals Unlimited; **14.3:** © Ed Reschke; **14.4:** © Bruce Iverson Photomicrography; **14.5 A,B:** © Ed Reschke; **14.6:** © BioPhot; **14.7A:** © Charles Gurche; **14.7B:** © Franz Krenn/OKAPIA 1989/Photo Researchers, Inc.; **14.8A:** © G. I. Bernard/Earth Scenes; **14.8B:** © Dwight R. Kuhn; **14.9:** © Richard D. Poe/Visuals Unlimited; **14.10A:** © William E. Ferguson; **14.11:** © Alford W. Cooper/Photo Researchers, Inc.; **14.12:** © Ray Coleman/Photo Researchers, Inc.; **14.14, 14.15 A,B:** © Kjell B. Sandved; **14.16A:** © Dede Gilman/Visions From Nature; **14.16B:** © Kjell B. Sandved; **14.17A:** © Bruce Iverson Photomicrography; **14.17B:** © Runk/Schoenberger/Grant Heilman Photography; **14.18 A,B:** © Dwight R. Kuhn; **14.19:** © William E. Ferguson; **14.20:** © Ed Reschke; **14.21A:** © BioPhot; **14.21B:** Courtesy of J. D. Mauseth, University of Texas, Austin; **14.22 A,B:** © Kingsley Stern; **14.23A:** © John D. Cunningham/Visuals Unlimited; **14.23B:** © Bruce Iverson Photomicrography; **Box 14.2, Box 14.3:** © Richard H. Gross; **14.24:** © E. Webber/Visuals Unlimited; **14.25A:** © Kingsley Stern; **14.25B:** © BioPhot; **14.25 C,D:** © Kingsley Stern; **14.25E:** © Kjell B. Sandved; **14.26B:** © Ed Reschke

Chapter Fifteen

Opener: © BioPhot; **15.1:** © Dr. Jeremy Burgess/SPL/Photo Researchers, Inc.; **15.2:** © Lynwood M. Chace/Photo Researchers, Inc.; **15.3:** © John D. Cunningham/Visuals Unlimited; **15.4 A,B, 15.5:** © BioPhot; **15.7:** © John D. Cunningham/Visuals Unlimited; **15.8A:** Courtesy John W. Kimball; **15.8B:** Courtesy J. H. Troughton and L. Donaldson and Industrial Research, Ltd.; **15.9:** © S. Elems/Visuals Unlimited; **15.10, 15.11C:** © BioPhot; **15.13 A,B:** © Omikron/Photo Researchers, Inc.; **15.13C:** © Dwight R. Kuhn; **15.15:** © Bruce Iverson Photomicrography; **15.18 A,B:** © J. M. Soileau; **15.19:** © Runk/Schoenberger/Grant Heilman Photography; **15.20 A-1:** © Biophoto Assoc./Science Source/Photo Researchers, Inc.; **15.20 A-2:** © Bruce Iverson Photomicrography; **15.20B:** © William E. Ferguson; **15.20C:** © Richard E. Ferguson/William E. Ferguson; **15.20D:** © Kingsley Stern; **Box 15.1:** Courtesy Robert E. Eplee, U.S.D.A., APHIS/S & T/WPMC.

Chapter Sixteen

Opener: © Eugene Fisher; **16.1B:** © BioPhot; **16.2:** © Bruce Iverson Photomicrography; **16.3:** © L. Murmanis/Visuals Unlimited; **16.5C:** © BioPhot; **16.6B:** A. M. Siebert, *Actabot Neer* 20:211–220, 1955; **16.8 A-1:** © Manfred Kage/Peter Arnold, Inc.; **16.8 A-2:** © BioPhot; **16.8B:** © Chris R. Simmons/Visions From Nature; **16.8C:** © Stouffer Enterprises, Inc./Earth Scenes; **16.9:** © BioPhot; **16.10:** © James Bell/Photo Researchers, Inc.; **16.11:** Courtesy National Agricultural Library, Forest Service Photographic Collection.; **Box 16.1A:** © E. R. Degginger/Earth Scenes; **Box 16.1B:** Courtesy of Hillerich & Bradsby Co., Inc.; **Box 16.1C:** © Tim Davis/Photo Researchers, Inc.; **16.12 B,C,D:** © Bruce Iverson Photomicrography; **16.13B:** © Dwight R. Kuhn; **16.13 C-1:** © John J. Smith/Photo/Nats; **16.13 C-2:** © Dwight R. Kuhn; **16.16A:** © Grant Heilman/Grant Heilman Photography; **16.16B:** © Phillip Hyde, 1994; **16.17:** © BioPhot; **16.18B:** © Bruce Iverson Photomicrography; **16.19A:** © Dwight R. Kuhn; **16.19B:** © Doug Sokell/Visuals Unlimited; **16.20:** Courtesy J. D. Mauseth/University of Texas, Austin; **16.21:** © BioPhot; **Box 16.2A:** © Kirtley-Perkins/Visuals Unlimited; **Box 16.2B:** © Porterfield-Chickering/Photo Researchers, Inc.; **16.22 A-1:** © Biophoto Assoc./Photo Researchers, Inc.; **16.22 A-2:** © Bruce Iverson Photomicrography; **16.22B, 16.23A:** © Bruce Iverson Photomicrography; **16.24A:** Courtesy of Capitol Machine; **Box 16.3A:** © F. C. F. Earney/Visuals Unlimited; **Box 16.3 B,C:** © Jeff Gnass Photography; **Box 16.3D:** © Henry Ausloos/Earth Scenes.

Chapter Seventeen

Opener: © Tom Edwards/Visuals Unlimited; **17.1:** © Scott Camazine/Photo Researchers, Inc.; **17.3A:** © Dwight R. Kuhn; **17.4:** © Professor Malcolm B. Wilkins, Botany Dept., Glasgow University; **17.5:** © John J. Smith/Photo/Nats; **17.6A:** © Richard H. Gross; **17.6B:** © Valorie Hodgson/Visuals Unlimited; **17.8A:** © Dede Gilman/Visions From

Nature; **17.8B:** © Joe McDonald/Visuals Unlimited; **17.9A:** © David M. Stone, 1981/Photo/Nats; **17.9B:** © Jane Burton/Bruce Coleman, Inc.; **17.10:** © Dede Gilman/Visions From Nature; **17.12A:** © Heather Angel/Biofotos; **17.13:** © Dede Gilman/Visions From Nature; **17.14:** © Photo/Nats; **17.15:** © David M. Stone/Photo/Nats; **17.16A:** © M. P. L. Fogden/Bruce Coleman, Inc.; **17.16B:** © William E. Ferguson; **17.17A:** © E. F. Anderson/Visuals Unlimited; **17.17B:** © R. E. Litchfield/SPL/Photo Researchers, Inc.; **Box 17.1A:** © J. A. L. Cooke/Oxford Scientific Films/Earth Scenes; **Box 17.1B:** © Nuridsany et Perennou/Photo Researchers, Inc.; **17.22:** © Runk/Schoenberger/Grant Heilman Photography; **17.23A:** © Richard Parker/Photo Researchers, Inc.; **17.23B:** © William E. Ferguson; **17.23C:** © Richard Packwood/Oxford Scientific Films; **17.23D:** © Professor Malcolm B. Wilkins, Botany Dept., Glasgow University; **17.25:** © G. I. Bernard/Oxford Scientific Films; **17.26:** © Larry Kimball/Photo/Nats; **17.27:** Courtesy Dr. Frank Uasek; **17.28:** © Doug Sokell/Visuals Unlimited; **17.29A:** © Stephen Dalton/Oxford Scientific Films; **17.29B:** © Sean Morris/Oxford Scientific Films; **17.30A:** © Photo/Nats; **17.30B:** © Pat & Tom Leeson/Photo Researchers, Inc.; **17.30C:** © M. J. Coe/Oxford Scientific Films; **17.30D:** © G. I. Bernard/Oxford Scientific Films; **17.30E:** © Sean Morris/Oxford Scientific Films; **17.31 A,B:** © Heather Angel/Biofotos; **17.32A:** © Runk/Schoenberg/Grant Heilman Photography; **17.32B:** © John Gerlach/Animals Animals; **17.33A:** © Merlin D. Tuttle/Bat Conservation International/Photo Researchers, Inc.; **17.33B:** © Gerald Thompson/Oxford Scientific Films; **17.34A:** © Sean Morris/Oxford Scientific Films; **17.34B:** © John R. Brownlie/Bruce Coleman, Inc.; **17.35A:** © Fern Stewart/Visions From Nature; **17.35 B,C:** © Muriel Orans; **17.35D:** © W. H. Hodge/Peter Arnold, Inc.; **17.36:** © Jane Thomas/Visuals Unlimited; **17.37A:** © Frank T. Awbrey/Visuals Unlimited; **17.37 B,C:** © Dwight R. Kuhn; **Box 17.2:** By permission from Webster's Third New International Dictionary © 1986 by Merriam-Webster, Inc., publisher of the Merriam-Webster dictionaries.; **Box 17.3:** © Heather Angel/Biofotos; **17.38:** © Dede Gilman/Visions From Nature; **17.39:** © Glenn Oliver/Visuals Unlimited; **17.42:** © William E. Ferguson; **17.43B:** © Harry Engles/Animals Animals; **17.44:** © Robert & Linda Mitchell; **17.45 A,B:** © Heather Angel/Biofotos.

Chapter Eighteen

Opener: © Dr. Scott Nielsen/Imagery; **18.1 (all):** © Professor Malcolm B. Wilkins, Botany Dept., Glasgow University; **18.5 A,B,C,D:** From J. P. Nitsch, "Growth & Morphogenesis of the Strawberry as Related to Auxin," *American Journal of Botany* 37:212, March 1950.; **18.6:** © Runk/Schoenberger/Grant Heilman Photography; **18.9A:** © R. F. Head/Earth Scenes; **18.9B:** © Patti Murray/Earth Scenes; **18.10:** © S. W. Wittwer, Michigan State University; **Box 18.2:** © William E. Ferguson; **18.14:** © Arthur M. Greene; **18.15:** Courtesy Cary A. Mitchell; **18.16 A,B:** © Frank B. Salisbury

Chapter Nineteen

Opener: © Dr. Scott Nielsen/Imagery; **19.1:** © Runk/Schoenberger/Grant Heilman Photography; **19.3 (all):** © Professor Malcolm B. Wilkins, Botany Dept., Glasgow University; **19.4 A,B:** © BioPhot; **19.4C (both):** © BioPhot; **19.5 A,B:** © BioPhot; **19.6 (all):** © Professor Malcolm B. Wilkins, Botany Dept., Glasgow University; **19.7:** © Kingsley Stern; **Box 19.1:** © Larry Simpson/Photo Researchers, Inc.; **Box 19.2:** © Kim Taylor/Bruce Coleman, Ltd.; **19.8:** © John D. Cunningham/Visuals Unlimited; **19.9A:** © John Kaprielian/Photo Researchers, Inc.; **19.9B:** © Richard H. Gross; **19.9C:** © John Kaprielian/Photo Researchers, Inc.; **19.11 A,B:** © Runk/Schoenberger/Grant Heilman Photography; **19.12:** © Fred Bavendam; **19.13 A,B:** © Tom McHugh/Photo Researchers, Inc.; **Box 19.5:** Albert Ulrich, Plant Physiologist, Emeritus, Dept. of Soil Science, College of Natural Resources, U. C. Berkeley, 94720; **Box 19.6A:** Larry Lefever/Grant Heilman Photography; **Box 19.6B:** © Holt Studios/Earth Scenes; **19.17 (all):** © Professor Malcolm B. Wilkins, Botany Dept., Glasgow University; **19.18 (inset-top right):** © Henry Groskinsky; **19.18 (background):** © Willard Clay; **19.18 (inset-bottom right):** © George H. H. Huey; **19.19 A–F:** © Frank Salisbury.

Chapter Twenty

Opener: © Kjell B. Sandved; **20.1:** © William E. Ferguson; **Box 20.1:** © Walt Anderson/Visuals Unlimited; **Box 20.2:** © John Colwell/Grant Heilman Photography; **20.5:** © William E. Ferguson; **20.7:** © Jack Wilburn/Earth Scenes; **20.8A:** © Holt Studios/Earth Scenes; **20.8B:** © 1991 Regents University of California Statewide IPM Project; **20.8C:** © David Newman/Visuals Unlimited; **20.9 A,B:** © 1991 Regents University of California Statewide IPM Project; **20.9C:** © Holt Studios/Earth Scenes; **20.9D:** © Science VU/Visuals Unlimited; **20.9E:** © John Colwell/Grant Heilman Photography; **20.9F:** © Kingsley R. Stern; **20.10:** © Harold J. Evans; **20.12A:** © 1991 Regents University of California Statewide IPM Project; **20.12B:** © Muriel Orans; **20.16:** Courtesy of John G. Mexal, Edwin L. Burke, C. P. P. Reid, © Wadsworth Publishing Co. 1994; **20.17:** Dr. J. Menge, Univ. of California, Riverside; **20.19A:** © Dwight R. Kuhn; **20.19B:** © E.H. Newcomb & S.R. Tardon, U. of Wisconsin/Biological Photo Service; **20.20A:** © Dwight R. Kuhn; **20.20B:** © G. I. Bernard/Earth Scenes; **20.20C:** © Kjell B. Sandved; **20.21:** © Cabisco/Visuals Unlimited; **20.22:** © Zig Leszczynski/Earth Scenes; **20.23A:** © John Gerlach/Visuals Unlimited; **20.23B:** © Dwight R. Kuhn; **20.24:** © D. Cauagnaro/Peter Arnold, Inc.; **20.25:** © Dwight R. Kuhn; **20.26A: (top):** © William E. Ferguson; **20.26B (bottom):** © Dede Gilman/Visions From Nature; **20.27 A,B:** © Barbara J. Miller/Biological Photo Service; **20.28:** © BioPhot; **20.29:** © Ed Reschke/Peter Arnold, Inc.

Chapter Twenty-One

Opener: © Richard Rowan's Collection, Inc./Photo Researchers, Inc.; **21.1, 21.2:** © BioPhot; **21.3:** Courtesy W. A. Cote; **21.4:** Courtesy of B. Meylan and B. G. Butterfield; **Box 21.1:** Courtesy Frank B. Salisbury; **21.6:** © John D. Cunningham/Visuals Unlimited; **Box 21.2:** © Steve Kaufmann/Peter Arnold, Inc.; **Box 21.3:** © Gregory K. Scott, 1985; **21.10 A,B:** © J. N. A. Lott, McMaster U./Biological Photo Service; **21.11:** © Carl W. May/Biological Photo Service; **21.14:** © Robert & Linda Mitchell; **21.15:** From I. Raskin and Hans, "Mechanism of Aeration in Rice," *Science* 228:327–329; cover, April 19, 1985. © AAAS, 1985.; **21.16:** Figure from *Botany*, by Peter M. Ray, Taylor A. Steeves and Sara A. Fultz. © 1983 by Saunders College Publishing, reprinted by permission of the publisher.; **21.19:** From Gunning and Pate, "Transfer Cells in Angiosperm Line," *Protoplasma* 68:135–156. © Springer-Verlag, 1993; **21.21 A,B:** From M. Zimmerman, "Movement of Organic Substances in Trees," *Science* 133:73–79, January 13, 1961. © AAAS, 1961.

LINE ART

Illustrious, Inc./Elizabeth Morales:

Figures 1.8; 2.1, BOX 2.1, 2.2, BOX 2.2, 2.3A, 2.4A, 2.5, 2.6, 2.7, 2.8, 2.9A–C, 2.10, 2.11, 2.13A, 2.14A, 2.15A; 3.2, BOX 3.2, 3.3, 3.4, 3.5, 3.9B, 3.11, 3.12, 3.15B, 3.17, 3.18, 3.19B, 3.21, 3.22B, 3.24, 3.25B, 3.26A, 3.28B, 3.29, 3.30; 4.1, BOX 4.1B, 4.2, 4.4, 4.6, 4.8, 4.9, 4.10, 4.11, 4.15, 4.16, 4.17, 4.18, 4.20, 4.21, 4.22, 4.23, 4.24; 5.3, 5.4, 5.5, 5.6, 5.7, 5.8, 5.9, 5.10, 5.11A–B, 5.12, 5.13, 5.14; 6.1, BOX 6.1, 6.3, 6.4, 6.5, 6.6, 6.7, 6.8, 6.9, 6.10, 6.11, 6.12B, 6.13, 6.14, 6.15, 6.16; BOX 7.1, 7.1B, 7.2B, 7.3, 7.4A–B, 7.5, 7.6, 7.7, 7.8, 7.9, 7.11, 7.12, 7.14, 7.15, 7.16, 7.17, 7.18, 7.19A–B, 7.20A, 7.21, 7.22, 7.25, 7.26A–B, 7.27; 8.6, 8.8, 8.10, 8.11, 8.14, 8.15B, 8.17A; BOX 9.1, BOX 9.2, 9.3, 9.8, 9.10, 9.11, 9.12, 9.13, 9.15, 9.16, 9.17A, 9.18, 9.19, 9.22, 9.23, 9.24, 9.25; 10.1A, 10.2, BOX 10.2, 10.3, 10.4, 10.5, 10.6, 10.7, 10.8, 10.9, 10.10, 10.11, 10.12, 10.13, 10.15, 10.16, 10.18, 10.19, 10.20, 10.21, 10.22, 10.23; BOX 11.1, 11.2, BOX 11.2, 11.3, BOX 11.3A, 11.5, 11.6B, 11.7, 11.8, 11.9, 11.11, 11.12, 11.13, 11.14, 11.15, 11.16, 11.17, 11.18, 11.20, 11.21, 11.22, 11.23, 11.24, 11.25, 11.26, 11.29, 11.30, 11.34, 11.35B; 12.4, 12.7B, 12.8B, 12.9, 12.11, 12.12, 13.20A–2, 13.20B–2; BOX 14.1, 14.13, 14.26A; 15.6, 15.11, 15.12, 15.14, 15.16, 15.17; 16.1A–C, 16.4, 16.5A–H, 16.6A, 16.7, 16.12A, 16.13A, 16.18A, 16.24B; 17.3B, 17.7, 17.18, 17.19, 17.20, 17.21; BOX 18.1B, 18.1C, 18.2, 18.3, 18.4, 18.7, 18.8, 18.11, 18.12, 18.13, 18.17; 19.2, 19.3A, 19.10, 19.16, 19.17A–1, 19.17B–1; 20.2, 20.3, 20.6, 20.11, 20.13, 20.14, 20.15, 20.18; 21.5, 21.7, 21.8, 21.9, 21.12, 21.13, 21.17, 21.18, 21.20.

Marjorie Leggitt:

Figures 3.1B–1; 7.13A; BOX 8.1, BOX 8.2, 8.3, 8.4, 8.5, 8.16; 9.1, 9.2; BOX 10.1; 12.1A, 12.10; 14.1; 16.14, 16.15, 16.23B; 17.2, 17.11, 17.12B, 17.24, 17.37A–C, 17.40, 17.41, 17.43A; BOX 18.1A, 18.18; BOX 19.3, BOX 19.4, 19.14, 19.15, 19.18B; 20.4.

TEXT/LINE ART

Chapter One

Figure 1.8: Reprinted by permission. From Gibbs and Lawson, "The Nature of Scientific Thinking" in *The American Biology Teacher*, 54(3):141. Copyright © National Association of Biology Teachers, Reston, VA.

Chapter Two

Box 2.2 (art): Reprinted with permission from *Diabetes Forecast*, Vol. 44(4). Copyright © 1991 by American Diabetes Association, Inc.

Chapter Three

Figure 3.1a: From Kingsley R. Stern, *Introductory Plant Biology*, 5th ed. Copyright © 1991 Wm. C. Brown Communications, Inc., Dubuque, Iowa. All Rights Reserved. Reprinted by permission.

Chapter Four

Figure 4.6: Courtesy of Willem F. J. Vermaas.

Chapter Five

Figures 5.5, 5.10e, 5.13: From *Biology: the Science of Life*, 3/e by Robert A. Wallace, et al. Copyright © 1991 by Harper Collins Publishers. Reprinted by permission;

Figures 5.7, 5.9: From Purves, Orians, and Heller, *Life: The Science of Biology*, 3d ed. Copyright © 1992 Sinauer Associates, Inc., Sunderland, MA. Reprinted by permission.

Chapter Six

Box figure 6.2 (art): Reprinted by permission. From Vivagg, *American Biology Teacher*, 49(2):113. Copyright © 1987 National Association of Biology Teachers, Reston, VA.

Chapter Seven

Figure 7.21: From Jensen and Salisbury, *Botany*, 2d ed. Copyright © 1981 Wadsworth Publishing Company, Belmont, CA. Reprinted by permission.

Chapter Fourteen

Figure 14.1: From Kingsley R. Stern, *Introductory Plant Biology*, 5th ed. Copyright © 1991 Wm. C. Brown Communications, Inc., Dubuque, Iowa. All Rights Reserved. Reprinted by permission;

Figure 14.13: From Jensen and Salisbury, *Botany*, 2d ed. Copyright © 1989 Wadsworth Publishing Company, Belmont, CA. Reprinted by permission;

Figures 14.26a,b: From Leland G. Johnson, *Biology*, 2d ed. Copyright © 1987 Wm. C. Brown Communications, Inc., Dubuque, Iowa. All Rights Reserved. Reprinted by permission.

Chapter Fifteen

Figures 15.11a,b: From *Biology: the Science of Life*, 3/e by Robert A. Wallace, et al. Copyright © 1991 by HarperCollins Publishers. Reprinted by permission;

Figure 15.16: From Cutter, *Plant Anatomy, Part II: Organs*. Copyright © 1971 Edward Arnold Publishers, division of Hodder & Stoughton Ltd. Reprinted with the permission of Cambridge University Press.

Chapter Sixteen

Figure 16.12a: From Raven, Evert, and Eichhorn: *Biology of Plants*, 5/ed. Worth Publishers, New York, 1992. Reprinted by permission.

Figure 16.23b: Figure from *Botany* by Peter M. Ray, copyright © 1983 by Saunders College Publishing, Reproduced by permission of the publisher.

Chapter Seventeen

Figure 17.11: Courtesy of James E. Canright, *American Journal of Botany*, 39:488, 1952.

Chapter Twenty

Figure 20.6: From Leland G. Johnson, *Biology*, 2d ed. Copyright © 1987 Wm. C. Brown Communications, Inc., Dubuque, Iowa. All Rights Reserved. Reprinted by permission.

Chapter Twenty-One

Figure 21.8: From Salisbury and Ross, *Plant Physiology*, 4th ed. Copyright © 1992 Wadsworth Publishing Company, Belmont, CA. Reprinted by permission.

Index

A

A horizon (soil), **462**, 463
Abrin, 31
Abrus precatorius, 31
Abscisic acid (ABA), **39**, 274, **413**, **431**–32
  discovery of, 431
  effects of, 275 (table), 431–32
  synthesis and transport of, 431
  transpiration and, 499
Abscission, 327, 410, **420**
  effect of hormones on, 420, 429
  transpiration and, 499
Abscission zone, **420**
Absorption, 476–77
  exposing large surface for, 477
  root function of, 336, 337, 344, 345, 476
Absorption spectrum, **139**
Acacia senegal, 25
Accessory pigments, **139**–42
Acer sp., 23, 320, 326–27, 363, 500 (box)
Acetate, 33
Acetyl-coenzyme A (acetyl-CoA), **117**
  glyoxylic acid cycle and metabolism of, 63, 64
  in lipid respiration, 126, 127
Acids, **A-5**, A-9 (box)
Acorns, 400
Actin, **54**, 55
  filaments, **54**, 55
Action spectrum, **139**
Active site (enzyme), **31**
Active transport, **81**, 82, 83
Active traps, **482**, 483
Acylglyceride linkage, 33, 34
Adaxial meristem, **312**
Addicott, F. T., 431
Adenine, 31–32, 426
Adenosine triphosphate. See ATP (adenosine triphosphate)
Adenovirus-2, 238
ADP (adenosine diphosphate), 64, 102–3
Adventitious roots, **335**–36, 348, 349
  auxin and development of, 421
Aechmea fasciata, 392
Aerenchyma cells, 283, 284, 321 (box)
Aerial roots, 349, 350–51
Aerobic respiration, **125**. See also Cellular respiration, aerobic
Agar, **26**
Agave sp., 4, 266, 287
Agriculture
  crop rotation in, 481
  fertilizers used in, 465 (box)
  genetically engineered bacteria used in, 249, 250
  genetically engineered crops, 253–55
  grafting fruit trees, 266 (box)
  hydroponic, 469, 470
  soil salinity and methods of, 495 (box)
Agrobacterium, 251, 252, 428
Air as soil component, 467
Albizia falcata, 373
Aldehyde, **A-4**
Aleurone layer, **84**
Alfalfa (*Medicago arborea*), 316
Alfisols, **463**
Algae, 301. See also names of specific algae
  agar and carrageenan from, 26
Alkaloids, 37 (table), **38**
  structures of three, 38
Allard, Henry, 448
Allele, **170**
  meiosis and separation of, 172–73, 174
  multiple, of same gene, 176–77
Allelopathy, **469**
Allium cepa, 9, 200
Allosteric regulation, **106**
Allozymes, **176**, 177
  analysis of, 178 (box)
Aloe, 328
Alpha-amylase, **27**, 31
  seed germination and, 84 (box)
Alpha-glucose, **24**, 25
Alpha-ketoglutaric acid, **117**, 118
Alpha tubulin, **54**
Alternanthera (Amaranthaceae), 158
Amaranthaceae, 158
Amaranthus, 156
Amino acid receptor site, **237**
Amino acids, 27, **A-5**
  binding of, by tRNA, 240
  nonprotein, 41
  structure of twenty most common, 28
  transfer RNA and, 237
Amino group, **A-5**
Aminoacyl-tRNA synthetases, **240**, 241
Amylases, **114**
Amylopectin, **26**, 232
Amyloplasts, 442–43
Amylose, **26**
  cellulose vs., 27
  digestion of, by alpha-amylase, 27, 31
Anabolism, **101**
Ananas comosus, 158
Anaphase I of meiosis, **215**, 217
Anaphase II of meiosis, **215**, 217, 218
Anaphase of mitosis, **190**, 205, 206
Anchorage, roots functioning as, 336, 345, 349, 350–51
Androecium, **380**
Anaerobic respiration, **125**–26
Angiosperms, **380**. See also Flowering plants
Animals
  pollination by, 397, 398
  seed dispersal by, 402 (box), 405
Anions, 85, 464, A-3
Antennae complexes, **144**
Anther, **380**, 381
Anthocyanins, **39**
Antibodies, **53**
Anticodon, **237**
Antigens, **53**
Antipodal cells, **389**
Antirrhinum majus, 177, 179, 182
Antisense technology, **251**
Apical dominance, 309, **419**–20
  Christmas trees as result of, 420
Apical meristems, **263**
  functions of, 263–65
  primary growth produced by, 264. See also Primary growth
  root, 265, 268
  shoot apical, 267–68
Apical peg, **312**
Apium graveolens, 285
Apoplast, **340**
Apple (*Malus*), 7–8, 56, 400
  chloroplasts in leaf of, 284
  Cox's orange pippin, 9, 266 (box)
Aquatic plants
  hydrophytes as, **322**
  roots of, 348
  water hyacinth, 321 (box)
Aquilegia vulgaris, 179
Arabidopsis thaliana, 200, 202, 203
  root gravitropism in, 442, 443
Arbor Day holiday, 375
Aridsols, **463**
Artificial chromosomes, 248
Arum maculatum, 196
Asclepias tuberosa, 2
Ash, white (*Fraxinus americana*), 40, 366 (box)
Aspen (*Populus tremuloides*), 348, 368
Atomic mass, **A-1**
Atoms, **A-1**
  structure and activity of, A-1–2
Atmospheric humidity, transpiration and, 496
Atmospheric pressure, water movement in plants and, 494
ATP (adenosine triphosphate), 64, **102**–3
  chemiosmosis and, 123–25
  coupled reactions and, 103
  as energy currency, 102–3

energy for active transport from hydrolysis of, 82, 83
phloem loading and requirement for, 508
production of, in aerobic respiration, 114–25
proton pumps and use or production of, 86
structure of, 102
synthases, **124**
ATPase (ATP phosphohydrolases), 64, 82, 83
Atriplex, 156
Atropa belladonna, 328
Austrobaileya, 385
Autoradiography, 150, *151*, 265
Auxin, 86, 274, **413**, 412–22
  calcium and, 422
  discovery of, 412–17
  effects of, 418–21
  ethylene production stimulated by, 431
  functions of, 275 (table)
  IAA (indole-3-acetic acid) as, 417, *418*
  influence of, on vascular cambium, 360
  as internal cellular signal, 86, 87
  proton pumping and, 87
  stem growth in response to, 306
  synthetic, 417, *418*
Axillary buds, **309**

B

B horizon (soil), **462**, *463*
Bacillus thuringiensis, 16, 254
Bacteria
  evolution of photosynthesis and, 141 (box)
  gene cloning by, 246
  ice-minus, 249
  industrial uses of genetically engineered, 249–51
  nitrogen fixing, 88, 349, 350, 479–81
  transgenic, 251–53
  treatment of Dutch elm disease with genetically altered, 7 n.2
Bacteriophages, **173**–75, *176*
Bacteriorhodopsin, 141 (box)
Bald cypress (*Taxodium mucronatum*), 362, *373*
Balsa (*Ochroma*), 295
Bamboo, stem growth in, 306, *307*
Banana plants (*Musa* sp.), 398
Banyan tree (*Ficus benghalensis*), 350, *356*
Bark, **368**–70
  periderm as, 368, 369, *370*
  secondary phloem as, 368, 369
  varieties of, 368
Barley (*Hordeum vulgare*), 84
Bars, water potential expressed in, **79**
Baseball bats, 366 (box)
Bases, A-9 (box)
Basswood (*Tilia americana*), 286, 369
Batasins, **432**
Bayberry (*Myrica pensylvanica*), 36
Bean (*Phaseolus* sp.), 60, 177, 292, 401, 457
Beer, 4, 84 (box)
Bees, 172
Beet (*Beta vulgaris*), 37, *371*, *372*
Beta-carotene, **39**, **139**, *140*
Beta-glucose, **24**, *25*
Beta-oxidation, **126**
Beta tubulin, 54
Beta vulgaris, 37, *371*, *372*
Betula papyrifera, 369
Beverages made from plants, 4, 5, 6 (box), 328
Bhattacharyya, Madan K., 232, 257
Bilayer, **74**
  selective permeability of, 75
Bindweed (*Convolvulus*), 444
Bioassays, **416**

Biochemical(s), **22**
  cytology (cytochemistry), **52**–53
  reactions of photosynthesis, **147**, 150–53
  summary of, 152, 153
Biochemistry, A-4
Bioenergetics, **93**, 94
Biological clock, **457**
Biophysical controls on plant growth and development, 277
Biotechnology. See Genetic engineering
Birch, paper (*Betula papyrifera*), 369
Birds, pollination by, 397, 398
Bird's-foot trefoil (*Lotus corniculatus*), 325
Bivalent chromosomes, **216**, 221, 222
Blackman, F. F., 147
Bladderwort (*Utricularia*), 482, 483
Blood type, determining with lectins, 88
Bolting, **424**, *425*
Bond energy, **100**
Bonner, James, 451
Bordered pits, 491
Boston ivy (*Parthenocissus tricuspidata*), 323, 349
Botany, **3**–18
  botanists and use of scientific method in, 10–14
  defined, **4**
  goals of, 16–17
  plants, significance of, and, 4–9
  unifying themes of, 14–18
Boysen-Jensen, Peter, phototropism studies by, 413, 415, 417
Bracts, 325
Branch (lateral) roots, **334**, 341, *343*
Brassica oleracea, 425
Brassosteroids, **432**
Bread mold (*Neurospora crassa*), 222, 223, 224, 225
Breakfast cereals, 13 (box)
Breeding systems, flower diversity and function as, 392–94
Briggs, Winslow, phototropism studies by, 438, 439
British thermal units (Btu), 95 n.2
Broad bean (*Vicia faba*), 70, 190
Bromeliad (*Neoregelia carolinae*), 260
Buckwheat (*Fagopyrum esculentum*), 40
Bud dormancy, 431
Bud scales, 323
Bulbs, **310**, *311*
Bulliform cells, 291, 498
Bundle sheath, 317
Buttercup (*Ranunculus* sp.), *15*, 280, 283, 308, 322, 440 (box)
Butterfly weed (*Asclepias tuberosa*), 2
Butterwort (*Pinguicula grandiflora*), 294, 299
Buttress roots, 349, 350–51

C

C horizon (soil), **462**, *463*
C_3 cycle, **152**
C_3 plants, **152**
  characteristics of, 156 (table)
  intermediates of, 158
  photosynthetic rates of, 156 (table)
C_4 plants, 154, **155**, 156
  characteristics of, 156 (table)
  efficiency of photosynthesis in, 156, 157
  intermediates of, 158
  leaf anatomy associated with, 154, *155*
  photosynthesis, 154–58
  photosynthetic rates of, 156 (table)
  summary of, 155
Cabbage (*Brassica oleracea*), 425
Cacti, 158
  crests from meristematic activity in, 278
  phyllotaxis of, 313, 315
  roots of, 346

Cadmium in plant tissue, 473
Calcium
  auxin and, 422
  cellular communication and, 87
  cytokinins and, 427
  plant growth and development and, 276
Callose, 298, 299, 505–6
Calmodulin, 87
Calorie (Cal), **95**
Calorie (cal), **94**
Calvaria major, 402 (box)
Calvin cycle, 152, *154*
Calvin, Melvin, experiments of, 150, *151*
Calyx, **380**
Camellia (*Camellia* sp.), 5, 176, 177
Camellia japonica, 176, 177
Camellia sinensis, 4, 5
CAM plants, **158**–59
  characteristics of, 156 (table)
  photosynthesis in, 158
  photosynthetic rates of, 156 (table)
Canals, 300
Candelilla (*Euphorbia antisyphilitica*), 36
Cannabis sativa, 294, 328
Capillarity, water movement by, 494
Capillary water, **467**
Carbohydrates, **22**–27
  amylose vs. cellulose, 27
  stem production of, 306
  storage polysaccharides, 26–27
  structural polysaccharides, 24–26
Carbon
  atmospheric, 153, 160
  chains and rings of, A-4
  oxidation and reduction of, 101–2
Carbon dioxide (CO_2)
  aerobic respiration and release of, 114
  as control factor in photosynthesis, 159
  photosynthesis and fixation of, 135, 153, 154
  transpiration and concentration of, 497
Carbonyl, **A-4**
Carboxyl group, **A-5**
Carboxylation, **A-7**
Carboxylic acids, **A-5**
Carica papaya, 31
Carnauba palm (*Copernicia cerifera*), 36, 288, 328
Carnivorous plants, 481–83
  active traps, 482, 483
  digestive glands of, 299
  modified leaves of, 324, 325
  passive traps, 481–82, 483
  seismonastic movements in, 446
  trichomes of, 294
Carotenoids, **39**, 64, *139*, 140, 142
  abscisic acid made from, 431
Carpels, 212, **380**, 382–83
Carrageenan, **26**
Carrion flowers, 396, 397
Carrot
  cell culture of, 188, *189*, 271
  flowering of, 452
Carson, Rachel, 17
Caspar, Timothy, 442
Casparian strip, 340, *341*, 342, 477
Castanea sp., 362
Castilleja sp., 396
Castor bean (*Ricinus communis*), 31, 346, 360, 401
Catabolism, **101**
Catalase, **63**
Cation exchange, **464**
  capacity, **464**
Cations, 85, A-3
Cavities, internal, 300
Cech, Thomas, 236 (box)

Celery (*Apium graveolens*), 285
Cell(s), 45–71
    competence of, to respond to developmental signals, 277–78
    cytology methods in studying, 48–53
    diagram of, 47
    functional organization of, 53–56
    laws of thermodynamics and, 97, 98
    membrane system of, 59–64. See also Membrane system
    microfibrils and shape of, 269
    microscopic observation of, 46, 48, 49, 50
    movements of, 65–68
    nucleus of, 58, 59
    organelles and energy conversion in, 64–65
    preparing, for microscopy, 51 (box)
    ribosomes of, 59
    in seedlings, 262
    structural polysaccharides in, 24–26
    structural proteins in, 30
    wall of, 56–58
    wall of, arrangement of cellulose in, 24, 25
Cell culture, cloning by, 188, 189
Cell cycle, 187–209
    chromosomes and mitotic spindle in, 207–8
    cytokinesis in, 204, 206–7
    defined, **188**
    interphase in, 189, 190–203
    mitosis in, 189–90, 204–6
    overview of, 188, 189–90
    periods of, 190
Cell fractionation, **52**
Cell plate, **62**, 205, **206**
Cell theory, 13, **14**, **69**–70
    alternative organismal theory, 70
    postulates and problems of, 69
Cellular
    communication, 76, 86–89
    division, cytokinins as stimulants of, 426
    effect of auxin on, 418–19
    effect of ethylene on, 430
    elongation, 269
    proton pumps and, 86
    root zone of, 338–39
    in stems, 306
Cellular enlargement, 268–69
Cellular recognition, epidermal cells and, **295**
Cellular respiration, **102**, 111, **A-7**
    aerobic, 111–25
    anaerobic, 125–26
    ATP production in, 112, 114, 115
    cyanide-resistant, 126–27
    efficiency of, 120–21 (box)
    electron transport and oxidative phosphorylation in, 112, 113, 118–25
    as energy transformation, 98, 99, 107–8
    glycolysis in, 112, 113, 115–17
    harvesting energy from glucose in, overview of, 114–15
    Krebs cycle in, 112, 113, 117–18
    of lipids, 126, 127
    photorespiration, 127
    retrieval of glucose in, 112–14
Cellulases, **27**
Cellulose, 24
    amylose vs., 27
    arrangement of, in cell walls, 25
    chemical structure of, 25
    as product of photosynthesis, 160
Celsius temperature scale, B-2
Centigrade temperature scale, B-2
Central dogma of molecular biology, **232**–33
Central mother cells, 267
Centromere, **204**, 205

Ceratocystis ulmi, 7
Cereals (food), 13 (box)
Cerochlamys pachyphylla, 288
Chaetomorpha algae, 25
Chailakhyan, M. K., 449
Chalazal pole, **389**
Charcoal, **375**
Chase, Martha, 173
Chelator, **475**, 476
Chemical bonding, A-2–4
Chemical fertilizers, 465 (box)
Chemical reactions, A-5–10
    carboxylation and decarboxylation, A-7
    coupled, 103
    dehydration and hydrolysis, A-7
    free energy and, 100–102
    oxidation and reduction, A-7
    phosphorylation, A-7–8. See also Phosphorylation
Chemical structures, 24 (box)
Chemiosmosis, **115**
    ATP synthesis and, 123–25
    in chloroplasts, 145, 146
Chemistry, fundamentals of, A-1–10
    atoms and elements, A-1–2
    chemical bonding, A-2–4
    chemical reactions, A-5–10
    functional groups and biochemical diversity, A-4–5
Cherry, 6
Chestnut (*Castanea* sp.), 362
Chiasmata, **216**, 220
Chicory (*Cichorium intybus*), 27
Chlamydomonas algae, 69, 70
Chlorella algae, 136, 150, 151
Chlorenchyma cells, 283, 284
Chlorophyll *a*, **138**, 139, 144
Chlorophyll *b*, 139
Chlorophylls, 64, 136, **138**–39
    in chloroplasts, 143–44
    as deodorizer, 8
    destruction of, in leaf senescence, 455, 456
    evolution of photosynthesis and, 141 (box)
    fate of energy in energized electrons of, 145
    reaction center, 144, 146
Chloroplasts, 64, 65, **143**–47
    complexes of pigments in, 143–44
    discovery of chlorophyll in, 136
    energy use by membranes of, 86
    light-driven reactions in, 145
    origin of, 66 (box)
    photophosphorylation in, 145, 146
Chocolate, 5
Cholesterol, blood levels of, and tropical oils, 35 (box)
Cholla, teddy-bear (*Opuntia bigelovii*), 395
Cholodny, Nicolai, 438
Chondrus, 26
Chromatid, **204**, 205
    crossing-over of genetic material in. See Crossing-over of genetic material
    sister, 206
Chromatin, **173**, 175
    fiber of, **194**, 195
    formation of, in S phase, 194–96
    looped domains of, **194**, 195, 196
Chromatography, 150, 151
Chromosomal theory of heredity, **172**
Chromosomes
    alignment of, during metaphase, 205–6
    bivalent, **216**, 221, 222
    condensation of, during prophase, 204, 205
    construction of artificial, 248
    crossing over and exchange of material from, 180, 216, 218, 219–20
    formation of daughter nuclei, during telophase, 205, 206

    homologous, 172, 174, **215**
    movement of, 207, 208
    multiple sets of (polyploidy), **218**, 219 (box)
    pairing of homologous (synaptonemal complex), 215, 216, 218–20
    segregation of, chiasmata and, 220
    separation of, in anaphase, 205, 206
    separation of, and genetic recombination, 220, 221, 222
    structure of, during interphase, 194, 195
Chrysanthemum (*Chrysanthemum* sp.), 449
Chymopapain, 31
Circadian rhythms, **455**–57
    transpiration and, 499
Cissus, 271
Citric acid, **117**, 118
Citric acid cycle, **117**, 118
Cladodes, **310**
Cladophora algae, 70, 272
Cladophylls, **310**
Claudius, 4
Cleistes rosea, 36
Climacteric, **428**
    fruits, **428**
Clone (vegetative reproduction), **394**
Cloning
    cell culture method of, 188, 189
    of DNA, 245–48
Clostridium botulinum, 32
Coated pits, **62**
Cobra plant (*Darlingtonia californica*), 483
Coca (*Erythroxylon*), 328
Cocklebur (*Xanthium strumarium*), 449, 451
Cocoa (*Theobroma cacao*), 5
Coconut (*Cocos nucifera*), 400, 405
Codominance, **176**
Codon, **237**
    in genetic code, 242, 243
    start, **240**, 241
    stop, **241**, 242
    synonymous, 242
    triplet, 242–43
Coenzyme Q, 122
Coenzymes, 103–4
Coevolution, **327**
Cofactors, 103
Coffee (*Coffea arabica*), 4, 6 (box), 282, 328
Cohesion, **A-4**
Coleoptile, **400**
    phototropism studies using, 413, 414–15, 417
Coleorhiza, **400**, 401
Coleus (*Coleus blumei*), 62, 72, 289, 313, 444
Collenchyma cells, **262**, 285
Collumella cells, 336, 337
Colony hybridization, **248**, 249
Columbine (*Aquilegia vulgaris*), 179
Communication, cellular, 76, 86–89
Companion cell, 298, 505
Compartmentalization, 362
Compensation point, **159**
Competitive inhibitors, **107**
Complementary DNA (cDNA), **247**
Complete dominance, **176**
Complete flower, **386**
Composting, 467
Compound, **A-2**
Compound leaves, **315**, 316
Compression wood, 367
Concentration gradient, **78**
Conium maculatum, 38, 325
Continuous synthesis, **199**
Convolvulus, 444
Copa iba (*Copaifera langsdorfii*), 160
Copaifera langsdorfii, 160
Copernicia cerifera, 36, 288

Coral gum (*Eucalyptus* sp.), 398
Corchorus capsularis, 46, 47
Cork, 368–70, 371 (box)
    Hooke's study of, 14, 69
Cork oak (*Quercus suber*), 56
Corms, **310**
Corn (*Zea mays*), 15, 91, 92, 190
    amylopectin starches in, 26
    cytoplasmic male sterility in, 182 (box)
    epicuticular wax on leaf of, 288
    ethanol produced from, 160
    gene introns in, 243
    genetic crossing-over in, 222, 223
    leaves of, 277
    leaves of, Kranz anatomy, 155, 319
    B. McClintock's research on, 11, 12, 181–83, 222
    root cap of, and gravitropism, 442, 444
    root tip, 265, 332, 337
    roots of, 335, 347, 349
    seeds of, 401
    stem structure, 308
    stomata of, 276, 291
    tassels, 182
    transposable elements in DNA of, 181–82
    zein protein in, 30–31
Corolla, 380
Corpus, 267
Cortex
    in roots, 340, 341
    in stems, 308, 309
Corylus sp., 392
Cot curve, **203**
Cotton (*Gossypium hirsutum*), 24, 53, 63, 254, 287
Cottonwood (*Populus fremontil*), 304
Cotyledons, 325, **391**, 400
    effects of cytokinins on, 426
Coupled cotransport system, **83**
Coupled reactions, **103**
Covalent bond, **A-2**–3
Cow pea (*Vigna sinensis*), 291
Cox, Richard, 9
Crassulacean acid metabolism (CAM), 156 (table), 158–59
Creativity in science, 13
Creighton, H. S., 222
Creosote bushes (*Larrea tridentata*), 348, 394
Crick, Francis, 191, 192, 193, 233
Criminal cases, 9, 249 n.3
Crinum, stomatal differentiation in, 274
Cristae, **65**, 67
Crop rotation, 481
Cross section (wood), 364, 365
Crossing-over of genetic material, **180**, **216**, **218**
    evidence of chromosomal exchange in, 222–23, 224
    repair of DNA and, 227–28
    unequal, and gene duplication, 225, 226
Crown gall disease, 251, 428
Crown roots, 335
Cucurbita, 298
Cultivars, 168
Cuscuta sp., 484
Cuticle, **288**
    transpiration and, 500
Cuticular wax, 36
Cutin, **36**
Cyanide, 41
Cyanide-resistant respiration, 126, **127**
Cyanogenic glycosides, **41**
Cycad (*Zamia*), 67
Cyclic photophosphorylation, **147**–48
Cyclosis, 65
Cynara scolymus, 27
Cytidine triphosphate (CTP), 105
Cytochemistry (biochemical cytology), **52**–53

Cytochrome a, 122
    b-c1 complex, **122**
    c, 122, 123
    oxidase complex, **122**
Cytochromes, 104, 105, **122**
Cytokinesis, **189**, 205, 206–7
Cytokinins, 271, 274, **413**, **426**
    calcium and, 427
    discovery of, 425–26
    effects of, 275 (table), 426–27
    synthesis and transport of, 426
Cytology, **48**–52
    biochemical, 52–53
    cell preparation in, 51 (box), 52
    electron microscopy in, 50, 51
Cytophotometry, 202–3 (box)
Cytoplasm, **53**
Cytoplasmic inheritance, **181**
Cytoplasmic male sterility (cms), **182**
Cytosine, 31–32
Cytoskeleton, **54**–56
    filaments of, 54, 55
    functions of, 55–56
Cytosol, **59**

D

Daffodil (*Narcissus* sp.), 175, 387
Dalton, John, A-1
Dandelion (*Taraxacum*), 266, 335, 394
Danielli, J. F., 74
Darlingtonia californica, 483
Darwin, Charles, 16
    gravitropism studies by, 441–42
    phototropism studies by, 412–13, 414–15
Darwin, Francis, 412, 441
Date palms (*Phoenix dactylifera*), 166, 167, 282
Davson, H., 74
Davson-Danielli model of membranes, 74–75
Daylength. See Photoperiodism
Decarboxylation, **A-7**
Dedifferentiation, **271**, 282
Defenses, plant, 38
    compartmentalization as, 362
    epidermis as, 290
    leaves as, 325–27
    in roots, 346
    thorns, 310
    trichome cells as, 294
Deforestation, 17–18
Dehydration reactions, **A-7**
DeMestral, George, 8
Denatured DNA, **203**
Denatured proteins, **29**
Density (wood), 366–67
Deoxyribonucleic acid (DNA), **31**
    amounts of, in plants and other organisms, 202
    chemical structure of, 32
    cloning of, 245–48
    complementary, 247
    denatured, **203**
    development of theory on structure and duplication of, 191–94
    double helix structure of, 31, 33, **192**
    enzymes in replication of, 198, 199, 200, 201
    exchange of. See Crossing-over of genetic material
    foreign, and construction of artificial chromosomes, 248
    as hereditary material in genes, 173–75, 176
    interphase and, 190
    molecular components of, 193
    repair of, crossing over and, 227–28
    repair of, mutations and, 181, 200 (box)
    repetitive, **203**
    replicons of, 200, 202

    selfish, 228
    self-replication of, 194, 196–202
    semi-conservative replication of, **196**, 197, 198
    transcription of, to RNA, **232**–35
    transposable elements in, 181–83, 228
Derbesia algae, 69, 70
Dermal tissues, 288–95. See also Epidermis (epidermal tissue)
    cellular recognition and, 295
    cuticle, 288
    gas exchange and, 289–91
    trichomes, 291–95
Desert sunflower (*Machaeranthera gracilis*), 221
Desmidiaceae algae, 47
Desynapsis, **215**
Development, 269, 270, 271–74. See also Growth
    adult and juvenile stages of, and gibberellin hormones, 424
    asymmetric cellular division and differentiation, 273–74
    cellular differentiation and, 271
    dedifferentiation and, 271
    meristems and patterned, 264
    polarity and, 271–73
    roots and stage of, 347
    signals regulating, 274–78. See also Hormones
De Vries, Hugo, 181
Diarch roots, 342, 343
Dicots
    genetic engineering in monocots vs., 252
    monocots vs., 391, 392 (table)
    netted venation of, 318
    roots of, 342
    stem structure of, 308, 309
    unusual secondary growth in, 370–71
Dictyosomes (Golgi body), **61**–62
Dictyostelium discoideum, 69
Diet, tropical oils in human, 35 (box)
Differential permeability of membranes, 76, 82–85
Differentiation, cellular, **271**
    asymmetrical, 273, 274
    patterns resulting from, 276
Diffuse-porous wood, 363
Diffusion, **78**
    facilitated, **82**, 83
Digestive glands, 299
Digitalis lanata, 5
Digitalis purpurea, 5, 37, 39, 328
Dihybrid cross, **171**
    Mendel's results from, 172 (table)
Dimers, **23**
Dionaea muscipula, 446, 482
Diospyros sp., 282, 318
Dioxin, 418
Diploid cells, **172**
Disaccharides, **23**
Dischidia, 324
Diseases of plants
    bacteriophages and, **173**–75, 176
    crown gall, 251, 428
    Dutch elm, 7
    epidermal entry of, 290–91
    genetically-engineered resistance to, 254, 255
    hormones and, 428 (box)
    microbe interaction with membranes and, 87–88
    rust spore invasion, 292–93 (box)
    tobacco mosaic virus (TMV), 254, 255, 312
Disulfide bonds, **A-5**
Diterpenes, 38
Ditmer, Howard, 334
DNA. See Deoxyribonucleic acid (DNA)
DNA fingerprinting, 249 n.3
DNA ligase, **246**
DNA polymerases, **199**, 200 (box)

Dodder (*Cuscuta* sp.), *484*
Dodo bird, *402* (box)
Domains, 244
Dominance
    law of, **170**
    types of, *176*
Dominant traits, **167**
Dormancy, **455**
    bud, *431*
    seed, *401*, *431*
Double fertilization, **214**, *215*, **390**, *391*
Double helix, DNA structure as, *31*, *33*, **192**
Dracaena, *371*, *372*
Drimys winteri, *385*
Drosera sp., *294*, *447*, *460*, *482*
Drought resistance, genetically-engineered, *254*
Drugs
    biotechnology and creation of, *16*
    medicinal, *4*, *5*, *328*, *368*, *375*
    morphine and codeine, *301*
    opium, *4*, *301*
    from plant leaves, *328*
    from plant roots, *351*
Drupe, *399*, *400*
Ducts, *300*
Durability (wood), *367*
Dutch elm disease, *7*
Dyes, *328*, *375*
Dynein, *66–67*, *207*

E

Easter lily (*Lilium longiflorum*), *15*, *217*
Echinochloa colonum, Kranz anatomy of, *155*
EcoRI enzyme, **245**, *246*
Egg (ovum), *214*, *215*
Egg apparatus, **389**
Einstein, Albert, *137*
Elderberry (*Sambucus*), *370*
Electrical currents, influence of, on plant growth and development, *274*
Electrochemical gradient, *83*, *86*
Electrogenic pump, **85**
Electromagnetic spectrum, *136*
Electron microscopes, *46*, *49*, *50*
    preparing cells for, *51* (box), *52*
Electron shell, **A-2**
Electron transport chain, **112**, *113*, **118**, *119*, *120* (table), *122–23*
Electronegative elements, **A-3**
Electrons, **A-1**–*2*
Electroporation, *252*
Elements, **22**, **A-1**
    atomic numbers, atomic mass, and symbols of, *A-1* (table)
    beneficial, *473*
    essential. See Essential elements
    most common, in plant, *23* (table)
    in plant tissues, *473–74*
    trace, *A-1*
Elm (*Ulmus* sp.), *6*, *7*, *362*, *364*
Embryo, *214*, *215*
Embryo sac, **212**, *214*
    development of, *389*, *390*
Embryonic leaves (cotyledons), *325*
Emerson enhancement effect, *144*
Emerson, Robert, *143*
Endergonic reactions, **101**
Endocarp, **398**
Endocytosis, **84**, *85*
Endodermis, *340*, *341*
Endoplasmic reticulum (ER), **60–61**
    connections of, *58*
Endosperm, *214*, *215*

Endosymbiotic hypothesis, **66** (box)
Energy, *91*, *93–109*
    ATP and cellular, *102–5*
    conversions, *96*
    defined, **94**
    enzymes and, *105–7*
    flow of, *107–8*
    free, *100–102*
    kinetic, *78*, *96*
    laws of thermodynamics and, *96–99*
    measurement of, *94–96*
    metabolism and, *99–100*
    plant-environment exchange of, *15*
    plant transformations of, *98*, *99*, *107–8*. See also Cellular respiration; Photosynthesis
    potential, *78*, *96*
    of sun, *95* (box)
Energy of activation, enzymes and, **105**, *106*
Engelmann, T. W., experiments of, *135*, *136*
Enthalpy, **100**
Entrainment, **457**
Entropy, **97**, *98*
Environmental stimuli, plant responses to, *15*, *262*, *437–58*
    circadian rhythms, *455–57*
    developmental polarity and, *271*, *272*
    growth as key response, *438*
    leaf variation as, *320–22*
    nastic movements, *445–48*
    phloem loading and, *509*
    seasonal, *448–55*
    thigmomorphogenesis, *448*
    tropisms, *437–45*
Enzymes, **31**, **105**. See also names of specific enzymes
    debranching, *114*
    DNA replication and role of, *198*, *199*, *200*, *201*
    energy of activation and, *105*, *106*
    isozymes and, *177*
    membranes and, *76*
    regulating metabolism with, *106–7*
    restriction, *245–46*
    retrieval of glucose using, *112*, *114*
    ribozymes functioning as, *235*, *236* (box)
    starch-branching, *232*
Epicotyl, **400**
Epicuticular wax, *34*, *36*, *288*
Epidermis (epidermal tissue), **262**, *288–95*
    in leaves, *317*
    in roots, *339*
    in stems, *308*
Epinasty, **430**
Epiphytes, *347*, *350*, **485**
Epistasis, **177**
Erg, *95* n.2
Erodium botrys, *406*
Erosion (soil), **466**
Essential elements
    absorbing, *476–77*
    associations with other organisms for obtaining, *478–85*
    deficiencies of, *471*, *473*, *474*
    defined, *469–70*
    factors determining availability of, *474–76*
    functions of, *470–71*
    functions of, and deficiency symptoms for, *472–73* (table)
    locating, *476*
    mobility of, within plants, *478*
    modifying form of, *478*
    obtaining, in rain forest, *485–86*
Ethylene, *274*, **413**, *427–31*
    auxin and, *431*
    discovery of, *427*
    effects of, *275* (table), *428–30*

    root growth and, *347*
    synthesis and transport of, *427–28*, *429*
Etiolated plants, **453**
Eucalyptus species, *39*, *398*, *424*, *499*
Euchromatin, **194**
Eulophia paivaeana, *36*
Euphorbia sp., *36*, *160*, *325*
Evaporation, water transport by, *496*
Evaporative cooler, **A-4**
Evening primrose (*Oenothera* sp.), *216*, *382*, *383*, *397*
Evolution
    common ancestry of plants and, *16*
    of photosynthesis, *141* (box)
    ribozymes and, *236* (box)
Exergonic reactions, **101**
Exine, **387**
Exocarp, **398**
Exocytosis, **61–62**, **83–84**
Exons, **235**
    potential mobility of, *244*
Exon-shuffling hypothesis, **244**
Exothermic reactions, **100**
Extrafloral nectaries, *299*
Extraxylary fibers, *287*

F

F_1 generation, **168**
Facilitated diffusion, **82**, *83*
Facultative photoperiodism, **451**
FAD (flavin adenine dinucleotide), *104*, *105*
Fagopyrum esculentum, *40*
Fascicular cambium, *358*
Fats. See Lipids
Fatty acids, *33*, *34*
Feedback inhibition, **106**
Feeder roots, *335*
Fermentation, **125**
Ferns (*Pyrrosia*), *158*, *312*
Fertilization (reproduction), *214*, *215*, **390**
    double, **214**, *215*, **390**, *391*
Fertilizers (soil), *465* (box)
Feulgen stain, *190*
Fibers, **287**
    classification of, *287*
    uses of, *9*, *46*, *287*, *288*, *328*, *375*
Fibonacci, Leonardo, *314* (box)
Fibonacci series, *313*, *314* (box)
Fibrils, *24*
    arrangement of, in cell walls, *25*
Fibrous root system, *335*
Ficus benghalensis, *350*, *356*
Ficus elastica, *289*
Fiddleheads, *312*
Figure (wood), *365*, *366*
Filament, *380*, *381*
Filial generation (F_1), **168**
Fir, *373*
Fire, seed germination requiring, *403*
First law of thermodynamics, **97**
5-methylcytosine, *32*
Flagella, *66*, *67*, *68*
Flank meristem, *266*
Flaveria (Asteraceae), *158*
Flavin, *439*
Flavin mononucleotide (FMN), **122**
Flavonoids, *39–40*, *396*
Flax (*Linum usitatissimum*), *5*, *287*
Fleming, Alexander, *13*
Fleming, Walther, *204*
Florigen, **449**
Flower pot leaves, *325*

Flowering, **448**
    cold temperatures and (vernalization), **452** *(box)*
    effect of ethylene on, 429
    effect of gibberellins on, 424, 425
    hormones affecting, 449 *(box)*
    photoperiodism and, 448–55
    types of, 450–51
Flowering plants
    diversity of, 386–98
    morphology of. See Flowers; Fruits; Seeds
    polarity in development of, 272
    pollination and fertilization of, 212–15
    spores and gametes of, 212, 213, 214
Flowers, 379, 380–86
    carpels of, 382–83
    gene epistasis and color of, 177, 179
    nature of parts of, 384–86
    nectaries of, 299
    petals of, 383–84
    sepals of, 384
    stamens of, 380–82
    variations in, 386–91
Fluid mosaic model of membrane structure, 75–76
Fluorescence, **145**
Foliar fertilization, 465
Food
    breakfast cereals, 13 *(box)*
    energy content of, 102
    genetically-engineered, 255
    leaves of plants as source of, 328
    roots as, 351
    sugar content of, 23
Food from plants, 4, 5
    genetically engineered, 16–17
Food reserves in plants, 26–27
Form and function in plants, 12, 262, 259, 262. *See also* Development; Growth; Sexual-reproduction morphology
Fouquieria splendens, 499
Four-o'clock (*Mirabilis jalapa*), 181
Foxglove, purple (*Digitalis purpurea*), 5, 37, 39, 328
Fragaria chilensis, 310, 399, 421, 496
Franklin, Rosalind, 192
Fraxinus americana, 40
Free central placenta, **382**
Free energy, **100**–102
Freeze-fracturing of cells, 52
Frithia pulchra, 324
Fronds, 312
Fructose, 23, 520
Fruits, 398–400
    climacteric and nonclimacteric, 428
    development of, 399, 421
    dichotomous key to, 399 *(table)*
    ethylene and ripening of, 427, 428–29
    genetically-engineered storage of, 254–55
    gibberellins and formation of, 424
    ovary chambers and, 382
    preventing browning of, 54
    types of, 398–99, 400
Fucus alga, 272
Fuel, plants as, 160, 328, 375
Fumeric acid, **117**, 118
Functional groups, A-4–5
Functional megaspore, **212**, 389
Fungi
    diseases caused by, 7
    Gibberella fujikuroi, 422
    mycorrhizal roots and beneficial, 348, 478, 479
Fusiform initials, **356**, 358

G

G_1 phase of interphase, **189**, 191
G_2 phase of interphase, **190**, 203
GA (gibberellic acid), 422
Gaertner, Karl Friedrich, 167
Galium aparine, 285
Gametes, **171**
Gametophyte, **212**, 213
    development of female, 214
    development of male, 214
Gametophytic self-incompatibility, 394
Gane, R., 427
Garner, Wightman, 448
Gas chromatography, **416**
Gas exchange
    in epidermal cells, 289–91
    in periderm, 370
Gasohol, 160
Gel electrophoresis, examining proteins by, 178 (box)
Gelidium robustum, 26
Gene(s), 15, 165, **170**, 175–76, 231, 232–42
    alleles of. See Allele
    crossing-over of. See Crossing-over of genetic material
    cytoplasmic inheritance of non-nuclear, 181
    discovery of transposable elements in, 11, 12
    DNA as hereditary material in, 173–75, 176. *See also Deoxyribonucleic acid (DNA)*
    engineering of. See Genetic engineering
    finding specific, 248–49
    interrupted, 237, 238 *(box)*, 243
    libraries of, 246–47
    locus of, 175
    multigene families, 227 *(box)*
    multiple isozymes of, 177
    mutations in, 181, 200 *(box)*
    nodD gene, 480
    pleiotropic, **178**–79
    polygenes, 177
    protein synthesis and, 232–42
    psbA gene, 244 n.1
    rbcL gene, 200 *(box)*
    reporter, 253
    serial systems of, 177, 179
    structure of eukaryotic, 243
    tandem repeat of, 225, 226
Gene conversion, **223**, 225, 226
Gene gun, **253**
General Sherman tree, 4, 7
Generative cell, **214**, 387
Genetic code, **242**, 243
Genetic controls on plant growth and development, 277
Genetic engineering, 16–17, **245**–55
    cell culture cloning, 188, 189
    DNA cloning, 245–48
    finding gene of interest in, 248–49
    with gene gun, 253 *(box)*
    industrial uses of bacteria made by, 249–51
    monocots vs. dicots in, 252
    transgenic crops from, 253–55
    transgenic plants from, 245, 251–52
Genetic recombination, **216**, 220–26
    by chromosomal separation, 220, 221, 222
    crossing-over in, 222–23. *See also* Crossing-over of genetic material
    gene conversion, 223, 225, 226
    unequal crossing-over and gene duplication, 225, 226
Genetics, 163
    cell cycle and. See Cell cycle
    complex inheritance patterns, 176–80
    engineering of. See Genetic engineering
    gene operation and function, 232–42
    genetic code, 242–44
    genetic recombination, 216, 220–26
    meiosis and synaptonemal complex, 215–20
    Mendel and theory of inheritance, 166–73
    non-Mendelian inheritance patterns, 180–83
    of peas, round vs. wrinkled, 232, 255–57
    root systems affected by, 346
    search for hereditary material, 173–76
    sexual reproduction in plants and, 212–15, 226–28
Genomes, size and composition of, 202–3 *(box)*
Genomic library, **247**
Genotype, **171**
Giant sequoia (*Sequoiadendron giganteum*), 4, 6, 354, 373, 488
Gibberellins, **39**, 274, 413, **422**–25
    discovery of, 422
    effects of, 423–24
    functions of, 274 *(table)*
    rib meristem affected by, 266
    seed germination and, 84 *(box)*
    stem growth in response to, 306
    synthesis and transport of, 422–23
    vascular tissue differentiation and, 420
Gibbs, Josiah, 100
Ginkgo biloba, 66
Gladiolus sp., 251
Gliadin, 31
Globe artichoke (*Cynara scolymus*), 27
Glucose, 22, 23
    alpha- and beta-, 24, 25
    breakdown of, to pyruvic acid, 115, 116, 117
    chemical structure, 24 *(box)*
    harvesting energy from, overview of, 114–15
    potential energy of, 115
    retrieval of, in respiration process, 112–14
    seed germination and, 84 *(box)*
    structure of, A-8
Glyceraldehyde-3-phosphate (G-3-P), 152
Glycine max, 349
Glycolysis, **112**, 113
    breakdown of glucose to pyruvic acid in, 115, 117
    efficiency of, 120 *(table)*
    potential energy of glucose and, 115
    ten steps of, 116
Glycoproteins, **30**, **76**
    determining human blood type with, 88
    role in sexual reproduction, 88–89
Glycosides, 37
Glyoxylic acid cycle, **63**, 64
Glyoxysomes, **63**
Glyphosate, **254**
Gold in plant tissue, **473**
Golgi body (dictyosomes), **61**–62
Gopher plants (*Euphorbia lathyris*), 160
Gossypium hirsutum, 24, 53, 63, 254, 287
Grain (wood), **364**, 365
Gram, B-2
Grana, 64, 65, **143**
Grapes (*Vitis vinifera*), 291, 310, 400
Grasses
    C_4 *photosynthesis in tropical*, 154–55
    fibrous root system of, 335
    flowers of, 386, 387
    genetic engineering in cereal, 252
    gibberellins in seed germination of, 423–24
    ground tissue of tropical, 319
    hemicelluloses in, 25
    pollen, 387, 388, 389
    scutellum of, 400, 423
    seedling development, 403
    stem growth in, 306, 307
    stomata of, 291
Gravitational water, **467**
Gravitropism, **441**–43

Gravity
 effect of, on formation of reaction wood, 367, 368
 effect of, on roots, 346, 347
 growth response to, 441–43
Griffith, Frederick, 173
Griffithsia alga, 272
Ground cherry (*Physalis subglabrata*), 393, 394
Ground meristem, **266**
Ground tissue, **262**, **282**–87
 collenchyma, 285
 in leaves, 318–19
 parenchyma, 282–85
 in roots, 340
 sclerenchyma, 285–87
 in stems, 308–9
Groundsel (*Senecio vulgaris*), 508
Growth, 268–69, 270, 412–13. See also Development
 of cell wall, 56–57
 cellular division and enlargement as, 268–79
 closed (determinate) vs. open (indeterminate), 263
 defined, 268
 effect of gibberellins on, 423
 as key plant response to environment, 438
 meristems and, 263–68
 modular, 278
 plant life spans and, 263
 primary. See Primary growth
 retardants, 422–23
 rings, 361–62
 secondary. See Secondary growth
 signals regulating, 274–78. See also Hormones
 in stems, 306–7
GTP (7-methylguanosine triphosphate), **234**
Guanine, 31–32
Guanosine triphosphate (GTP), 105
Guard cells, 289, 290, 291, 501, 502
Gum, 25
Gum arabic, 25
Guttation, 299, 495, 496
Gymnosperms, tracheids only in, 492 (box)
Gynoecium, 382

H

Haberlandt, Gottleib, 271, 274, 425
Hake, Sarah, 277
Half-life, **A-6**
Halobacterium halobium bacteria, 141 (box)
Halophytes, 495 (box)
Hamner, Karl, 451
Haploid cells, **172**
Hardwoods, 363
Hartig, Theodor, 504
Haustorium, **484**, 485
Hazelnut (*Corylus* sp.), 392
Heartwood, **364**
Heat
 energy transformations as source of, 97, 98
 as kinetic energy, 96
 produced by plants, 94, 127, 128 (box)
 temperature gradient and, 98 n.3
 of vaporization, **A-4**
Helianthus sp., 27, 176, 268, 276, 287, 440 (box)
Helicases, **198**, 199
Heliotropism, 440 (box)
Helix structure of proteins, **192**
Helmont, Jan-Baptista, 469
Hemberg, Torsten, 431
Heme, **122**, 123, 138
Hemicelluloses, **24**, 25
Hemlock (*Conium maculatum*), 38, 325
Herbicides
 creating plants resistant to, 16
 genetically-engineered resistance to, 254

 operation of, 139, 148, 150
 synthetic auxin as, 418
Heredity. See Genetics; Inheritance
Heterochromatin, **194**
Heterogeneous nuclear RNA (hnRNA), **235**, 238
Heterotrophs, 132
Heterozygous plants, **170**
Hevea brasiliensis, 301
Hexane, chemical structure of, 24 (box)
Hibiscus sp., 382
Hill reaction, **136**
Hill, Robin, 136
Histogens, 267
Histones, **194**, 195
Histosoils, 463
Hoary plantain (*Plantago media*), 393
Homologous chromosomes, **172**, **174**, **215**
 pairing (synaptonemal complex), 215, 216, 218–20
Homozygous plants, **170**
Honeysuckle (*Lonicera* sp.), 384, 396
Hooke, Robert, 14, 69
Hordeum vulgare, 84
Horizons (soil), **462**, 463
Hormones, **274**, 409, 411–35
 abscisic acid, 39, 431–32. See also Abscisic acid (ABA)
 animal hormones vs. plant, 275, 432–33
 auxin, 412–22. See also Auxin
 controlling amounts of, 432, 433
 cytokinins, 425–27. See also Cytokinins
 defined, **413**
 ethylene, 427–31. See also Ethylene
 flowering and effects of, 449 (box)
 functions of, 275 (table)
 gibberellins, 39, 422–25. See also Gibberellins
 measuring, 416 (box)
 oligosaccharins as, 432
 plant growth and development influenced by, 274–76
 plant pathology and, 428 (box)
 receptors for, 87
Horsepower, 95 n.2
Howard, Alma, 190
Hoya carnosa, 158
Humus, **466**–67
Hybrid, plant, 167, 168 (box)
Hydathodes, 299
Hydnora africana, 396
Hydrocarbons, **A-4**
 in plant latex, 160
Hydrogen bond, **A-3**–4
Hydrolysis, **A-7**
Hydrophilic chemicals, 74
Hydrophobic chemicals, 74
Hydrophytes, **322**, 347, **504**
Hydroponic farming, 469, 470
Hydrotropism, **443**–44
Hydroxyl groups, A-4, A-5 (table)
Hygroscopic water, 468
Hypericum sp., 382, 386
Hypertonic, **79**
Hypocotyl, 400
Hypodermis, root, 340
Hypotonic, **79**

I

IAA (indole-3-acetic acid), auxin and, **417**–22
Ice-minus bacteria, 249
Immune system of animals, 53
Immunoassay, **416**
Impatiens glandulifera, 406
Imperfect flower, **386**
Inbreeding, 393
 outcrossing vs., 394 (table)

Incomplete dominance, **176**
Incomplete flower, **386**
Independent assortment, law of, **170**
Indian paintbrush (*Castilleja* sp.), 396
Indian pipe (*Monotropa uniflora*), 484
Indicator plants, **473**
Inferior ovary, **386**, 387
Inflorescence, **392**, 393
Infrared radiation (IR), **137**
Ingenhousz, Jan, experiments of, 135
Inheritance, 165–85
 complex patterns of, 176–80
 history of research discoveries in, 173–76
 laws of, **170**
 Mendel's studies and theory of, 166–73
 non-Mendelian, 180–83
Inorganic compounds, **22**
Insects
 aphids, 510
 genetically-engineered resistance of plants to, 254, 255
 plant digestibility adversely affecting, 326
 plants hormones affecting life cycle of, 326
 pollination by, 388 (box), 396, 397 (table)
Insect-trapping leaves, 324, 325. See also Carnivorous plants
Insulin, **26**, 247, 250
Integument (ovule), **389**
Intercalary meristems, **263**, **306**, 307
Interfascicular cambium, 358
Intermediate filaments, 55
Internodal segments, 272, 273
Internodes, **306**, 307
Interphase, **189**, 190–203
 chromatin formation in, 194–96
 chromosome structure during, 194, 195
 DNA semiconservative replication in, 196, 198
 DNA structure and duplication in, theory of, 191–94
 enzymes in, 198–200, 201
 G_1 phase of, **189**, 191
 G_2 phase of, **190**, 203
 replicons, 200, 202
 S phase of, **189**–90, 194–96, 198–200
Interrupted genes
 evidence for, 237, 238 (box)
 evolution of, 243
 roles of, 244
Intine, **387**
Introns, **235**, 236
 evidence for, 238 (box)
 gene structure with, and without, 243
 roles of, 244
Ion pumps, 76, 85–86
Ionic bond, **A-3**
Ionizing radiation, **137**
Ions, **A-3**
 hydrogen, and pH, A-9 (box)
 movement of, across membranes, 85–86
Ion-specific carriers, 477
Ipomoea sp., 182, 183, 371, 444
Ipomopsis aggregata, 378
Iron, electron transport chain and, 122
Isocitric acid, 117, 118
Isomers, **23**, A-8
Isoprene, 38, 39
Isotopes, **A-6** (box)
Isozymes, 177

J

Jerusalem artichoke (*Helianthus tuberosum*), 27
Jojoba seed (*Simmondsia chinensis*), 36
Joule (J), **95**
Juncus, 284
Jute (*Corchorus capsularis*), 46, 47

K

Kalanchoë sp., 158, *324*
Kamen, Martin, 136
Kellogg corn flakes, 13 (box)
Ketone, **A-4**
Kilodaltons, **31**
Kilowatt hour (kWh), 95 n.2
Kinetic energy, **78**, **96**
Kinetin, **426**, *427*
Kinetochore, **205**
Knishtia excelsa, 297
Knots (wood), **364**, *365*
Kölreuter, Josef, 167
Kranz (halo or wreath) anatomy, **154**, *155*, 319
Krebs cycle, **112**, *113*, *117*, *118*, 120 (table)
Krebs, Hans, 117
Kurosawa, Ewiti, 422

L

Laplace, P. S., 135
Larrea tridentata, 348, *394*
Lateral meristems, **263**, 268
  secondary growth produced by, *263*, *264*
Lateral (branch) roots, 341, *343*
Latex, 160
  secretory cells containing, 300
Lathyrus odoratus, 180, *181*
Laticifers, **300**–301
Lavoisier, Antoine, 135
Law (term), 170 n.2
Laws of inheritance, **170**
Laws of thermodynamics, 96–99
Lead in plant tissue, **473**
Leader sequence, **234**
Leaf abscission, **327**, 410, **420**
  effect of hormones on, 420, *429*
  transpiration and, 499
Leaf buttress (leaf primordium), 312
Leaf dimorphism, **322**
Leaves, 311–28
  air pollution from burning of, 328
  of C_4 plants, *154*, 155
  deciduous and evergreen, 327
  defensive capacities of, 325–27
  economic importance of, 327–28
  environmental control of variation in, 320–22
  flowers as derived from, 384–86
  formation of, 311–12
  influence of, on vascular cambium, 360
  membrane potential in, 85
  modified, 323–25
  movements of, 323
  pH of different compartments of, 86
  phyllotaxis of, 312–15
  senescence of, 455
  structure of, 315–19
  transpiration and position of, 499
Lectins, **88**
Leeuwenhoek, Anton van, 14
Legumes
  hemicelluloses in, 25
  nitrogen fixing bacteria in, 88
Lenticels, **370**
Lettuce, phytochrome and seed germination of, 454
Light, 136–38
  absorption of, by pigments, 138, *139*, *141*
  chloroplast harvesting system of, *144*
  as control factor in photosynthesis, 159
  effect of, on roots, 346
  leaf variation as response to, 320
  microscope, 46, *49*, *50*
  photochemical reactions requiring, 147–50

plant response to. See Phototropism
  properties of, 137
  radiant energies of select wavelengths of, 137 (table)
  seed germination requirements for, 401
  transpiration and intensity of, 498
  types of, 137
Light-compensation point, **159**
Lignin, **40**
Ligustrum, 319
Lilac (*Syringa vulgaris*), 318
Lily (*Lilium* sp.), 15, 202, *213*, 217, 381, *382*, 389, *436*
Lindbergh, Charles and Anne, 9
Linkage, genetic, **180**
Linum sp., 5, 287
Lipids, 33–37
  oils, 33–34, 35 (box)
  phospholipids, *34*, *36*
  respiration of, 126, *127*
  structure of, 34
  waxes, 34, 36
Liter, **B-2**
Locus of gene, **175**
Lolium sp., 395
Long-day plants, **450**
Long-night plants, **451**
Lonicera sp., 384, *396*
Looped domain, chromatin, **194**, *195*, 196
Lords and ladies arum (*Arum maculatum*), 196
Lore of plants, 6–8, 31, 54, 88, 114, 142, 172, 196, 216, 255, 278, 295, 328, 347, 373, 406, 420, 448, 480
Lotus corniculatus, 325
Lower leaf zone, **312**
Lumber, 374
Lupine (*Lupinus* sp.), *392*, 403
Lycopene, 39
  as tomato pigment, 140, *142*
Lycopersicon esculentum. See Tomato (*Lycopersicon esculentum*)

M

Machaeranthera gracilis, 221
Macronutrients, **469**, 470
Magnifications, **46**, 48, 49. See also Appendix B
Magnolia (*Magnolia* sp.), 386
Maidenhair tree (*Ginkgo biloba*), 66
Malic acid, **117**, *118*, 154
Mallee (*Eucalyptus pressiana*), 398
Malting, 84 (box), 424
Malus. See Apple (*Malus*)
Mammoth, Maryland, 448
Manila hemp (*Musa textilis*), 287, *288*
Maple (*Acer* sp.), 23, 320, 326–27
  hardwood of, 363
  tapping for maple syrup, 500 (box)
Maranta, 447
Marginal meristems, **312**
Marginal placentation, **382**
Marijuana (*Cannabis sativa*), 294, 328
Maternity plant (*Kalanchoë daigremontiana*), 158
Mathematics of leaf arrangement, 313, 314 (box)
Matric potential, **493**
Mauritius, dodo bird and tambalacoque tree on, 402 (box)
Maxwell, James, 137
Mayapple (*Podophyllum*), 344
McClintock, Barbara, 11, *12*, 181–83
Medicago arborea, 316
Medicine, use of genetically engineered bacteria in, 250–51
Mediterranean cranesbill (*Erodium botrys*), 406
Megapascals (MPa), **79**, **493**

Megaspore(s), **212**, *214*
  functional, **212**, 389
  mother cell, 212
Megasporophyll, **385**
Meiosis, 172–73, *174*, **212**
  in evening primrose (*Oenothera* sp.), 216
  first division (meiosis I), 215–17
  prophase, 215–16
  second division (meiosis II), 218
  stages of, 215, 217
Meiosis I, **215**–17
Meiosis II, **215**, 218
Membrane potential, **85**
Membrane sensitivity, **59**
Membrane system, **59**–64
  cell compartments and, 54
  dictyosomes, 61–62
  endoplasmic reticulum, 60–61
  microbodies, 63–64
  plasma membrane, 46, 53–54, 59, 60
  structural proteins in, 30
  vacuoles, 62–63
Membranes, 73–90
  ATP production and, 115
  beer production, malting, and membrane transport, 84 (box)
  cellular communications and, 86–89
  differential permeability of, 82–85
  functions of, 76
  movement of ions across, 85–86
  movement of molecules across, 78–81
  structure of, 59, 74–76
Mendel, Gregor, 14, 166–73
  initial experiments of, 167–68
  plant hybridization prior to, 166–67
  results of, and theory of inheritance, 168–73
Mendelian inheritance, 166–73, **176**–80
Mentha piperita, 294
Menthol, 38, *39*, 294
Meristems, **188**, **263**
  apical, functions of, 263–65
  apical, organization of, 267–68
  cell cycle and, 188, *189*
  derivatives of, 265–66
  intercalary, 263, 306, *307*
  lateral, 263, 268
Meselson, Matthew, 196
Mesocarp, **398**
Mesophyll, **318**–19
Mesophytes, **320**, **503**
Mesquite (*Prosopis*), 334, 346
Messenger RNA (mRNA), 59, **232**, 237
  transcription of DNA to, **232**–35
Metabolic pathways, **99**, *100*
Metabolic sinks, 159
Metabolism, 94, **99**–100
  ATP and, 102–3
  cofactors and coenzymes in, 103–4
  enzymes and regulation of, 105–7
  FAD, nucleoside triphosphates, and cytochromes, 105
  NAD^+ and $NADP^+$, 104
  in parenchyma cells, 282
  secondary, 37
Metaphase I of meiosis, **215**, 217
Metaphase II of meiosis, **215**, 217, 218
Metaphase of mitosis, **190**, 205–6
Metaphase plate, **205**–6
Metaphloem, 297
Metaxylem, 295
Meter, **B-1**
Methionine, biosynthesis of ethylene from, 427, *429*
Metric system, B-1–2
Micelles, **464**

Microbes
  effect of, on roots, 345–46
  membrane interactions with, 87–88
Microbodies, 63–64
Microfibrils, **24**
  arrangement of, in cell walls, 25
  cellular shape and, 269
  in cell-wall growth, 56–57
Micronutrients, **470**
Micropyle, **389**
Microscopic observation
  electron microscopy, 46, 49, 50, 52
  light microscopy, 46, 49
  size scales of, 46, 48, 49
Microsporangia, **380**, *381*
Microspore mother cells, **212**
Microspores, **212**, *214*
Microsporophyll, **385**
Microtome, **51**
Microtubules, **54**, *55*
  cell-wall growth and role of, 57
  kinetochore, and chromosome movement, 207, *208*
  polar, and chromosome movement, 207
Middle lamella, **56**
Milkweed, 326
Miller, Carlos, 425–26
Miller, Stanley, 12
Mimosa, *316*
Mimosa pudica, 445
Mimulus sp., 226, 398
Minerals, **463**
  in soils, 463–66
  transporting, in xylem, 490–504
Mirabilis jalapa, 181
Mistletoe (*Phoradendron, Viscum*), 278, 483, *484*
Mitchell, Mary, 223
Mitchell, Peter, 123–24
Mitochondria, **64**, *65*, *67*
  energy use by membranes of, 86
  origin of, 66 (box)
Mitosis, **189**, 204–6
  anaphase of, **190**, 205, 206
  metaphase of, **190**, 205–6
  prophase of, **190**, 204, 205
  telophase of, **190**, 205, 206
Modules, plant growth in, 278
Molecular biology, central dogma of, **232**–33
Molecular genetics, 231–58
  basis of round vs. wrinkled peas in, 232, 255–57
  gene function and operation, 232–42. *See also* Gene(s)
  genetic code, 242–44
  genetic engineering based on, 245–55
Molecules, 21–42
  elements in, 22, 23 (table)
  polymers and monomers as large, 22–37
  secondary metabolites as, 37–41
  shapes and formulas of, A-8
Moles, A-9 (box)
Mollisols, 463
Money, paper, 5
Monkey flower (*Mimulus kelloggii*), 226
Monocots
  dicots vs., 391, 392 (table)
  genetic engineering of dicots vs., 252
  parallel venation of, 318
  roots of, 335, 342
  stem structure of, 308, 309
  unusual secondary growth in, 371, 372
Monomers, **22**, 23
Monosaccharides, **23**
Monoterpenes, 38
Monotropa uniflora, 484
Morning glories (*Ipomoea purpurea*), 182, 183, 444
Mosaics, **323**

Motor cells, 445
Mountain laurel (*Sophora secundiflora*), 405
Movement, 445–48
  of cells, 65–68
  of chromosomes, 207–8
  of leaves, 323, 445, 446, 447, 455–57
  of roots, 348
mRNA. *See* Messenger RNA (mRNA)
Mucigel, **336**
  functions of, 336–37
Multigene family, **227** (box)
Multistep meristems, **265**
Münch, Ernst, 506
Musa sp., 287, 288, 398
Mustard (*Sinapis alba*), 344
Mustard oil glycosides, 41
Mutations, **181**, 200 (box)
  caused by transposable elements, 11, 12
  meristematic handling of accumulating, 264–65, 266 (box)
Mutualism, 348
Mycorrhizae, **348**, 478, 479
Myrica pensylvanica, 36
Myristica fragrans, 37, 39, 40

N

NAD⁺ (nicotinamide adenine dinucleotide), 103
NADH dehydrogenase complex, **122**
NADP⁺ (nicotinamide adenine dinucleotide phosphate), 104
Narcissus sp., *175*, 387
Nastic movements, **445**–48
Natural selection, 262
Nectar guides, **396**, *397*
Nectaries, 299
Neljubow, Dimitry, 427
Nerium oleander, 300, 322, 384
Net movement, **78**
Netted venation, 317, *318*
Neurospora crassa, 222, 223, 224, 225
Neutrons, **A-1**
Newsprint, 8
Newton, Isaac, 8, 136
Nickel as essential element, **471**
Nicotiana longiflora, 178, 179
Nicotiana tabacum, 179, 180, 297, 328
Nightshade (*Atropa belladonna*), 328
Nirenberg, Marshall, 242
Nitrate, 478
  reductase system, **478**
Nitrification, **478**
Nitrogen-fixing bacteria, 88, 479–81
  free-living, 479
  roots associated with, 349, *350*
  symbiotic, 479, 480, 481
Nitrogen, functional groups containing, A-5
Nitrogenase, **479**, 480
Nodes, stem, **306**, 307
Nodules, **350**
  gene for controlling, 480
Nonclimacteric fruit, **429**
Noncompetitive inhibitors, **107**
Noncyclic electron flow, **148**, *149*
Non-Mendelian inheritance, 180–83
Nonprotein amino acids, 41
Northern blotting, **255**, *256*
Nuclear envelope, **58**, *59*
Nucleic acids, **31**–33. *See also* Deoxyribunucleic acid (DNA); Ribonucleic acid (RNA)
Nucleosomes, **194**, *195*
Nucleotide, **31**–32
  DNA mutations and gene sequences of, 200 (box)

Nucleus, **58**–59
  daughter, 206
  genetic role of, 173, 174
  mitosis and division of, 189, 204–7
  polar, 212, 389
  RNA in, 235–37
Nutmeg (*Myristica fragrans*), 37, 39, 40
Nutrients
  cycling of, 107, 108
  essential. See Essential elements
  returning, to soil, 465 (box)
  root absorption of, 336, 337, 340, 342, 345, 347, 348–51, 476
  soil particles and availability of, 464–65
  transport of. See Solutes transport; Water transport
Nutrition, 461–87
  essential elements for, 469–74
  obtaining essential elements for, 474–86
  soils and, 462–69
Nyctinasty, **447**–48
Nymphaea odorata, 286, 322

O

O horizon (soil), **462**
Oak (*Quercus* sp.), 6, 7, 295, 348, 373, 410
Obligate photoperiodism, 451
Ochroma, 295
Ocotillo (*Fouquieria splendens*), 499
Oenothera sp., 153, 382, 383, 397
Oils, 33–34
  tropical, 35 (box)
Oleander (*Nerium oleander*), 300, 322, 384
Oligosaccharins, **432**
One-molar solution, A-9 (box)
Onion (*Allium cepa*), 9, 200
Opium, 4
Opium poppy (*Papaver somniferum*), 4, 37, 301
Opium Wars, 4
Opposite phyllotaxis, **313**
Optical isomers, A-8–10
Optically-active substances, A-8
Opuntia bigelovii, 395
Orange, 400
  seedless navel, 9, 266 (box)
Orbital, **A-2**
Orchids (*Ophrys* sp.), 158
  epicuticular waxes in, 36
  pollination of, 387, 388 (box)
  roots of, 289
Organelles, **46**, 58–65
  energy conversion by, 64–65
Organic acids, **A-5**
Organic chemistry, **A-4**
Organic compounds, **22**
Organic fertilizers, 465 (box)
Organismal theory, **70**
Organisms
  plant association with, for nutrition, 478–85
  in soil, 468–69
Organogenesis, **426**, 427
Ornamental plants, 278
Orphrys. *See* Orchids (*Ophrys* sp.)
Orthocarpus pusillus, 396
Orthostichies, **313**
Oryza sativa (rice), 26, 481
Osmosis, **79**, 80
  inducing, 81
Osmotic potential, **79**–80
Osmotic pressure, **79**, 80
Osmotically active ions, **81**
Outcrossing, **393**
  inbreeding vs., 394 (table)

Ovary, **212**, *215*, 382–83
   superior vs. inferior, *386*, *387*
Overbeek, Johannes van, *425*
Ovules, **212**, *214*, 383
   development of embryo sac in, 389–90
Oxaloacetic acid (OAA), **117**, *118*, *154*, *155*
Oxidases, 63
Oxidation, 101–2, **A-7**
Oxidative phosphorylation, 114, **115**, 118–25
Oxisols, 463
Oxygen, 4
   photorespiration caused by increased, 153–54
   production of, in photosynthesis, 132

P

P680, **144**, *146*
P700, **144**, *146*
Paál, Árpád, phototropism studies by, *415*, *417*
Pacific yew (*Taxus brevifolia*), 18, 37, 368, 375
Palisade mesophyll cells, **319**
Palmately compound leaves, 316
Panicum, 158
Papain, 31
Papaver somniferum, 4, 37, 301
Papaya (**Carica papaya**), 31
Paper
   money, 5
   newsprint, 8
   wood as source of pulp for, 374–75
Parallel venation, 317, *318*
Parasites, **483**
   transgenic plants and bacterial, 251
Parasitic plants, 348, 350 (box), 483–85
Parastichies, **313**, *315*
Parenchyma, **262**, 282–85, 340
Parietal placentation, **382**
Parsley (*Petroselinum*), 37
Parthenocarpic fruits, **421**
Parthenocissus tricuspidata, *323*, *349*
Pascals, **493**
Passage cells, 340
Passion flowers (*Passiflora incarnata*), 299
Passive transport, **79**
Passive traps, 481–82, *483*
Pasteur, Louis, 11
Pauling, Linus, 191–92
Pea (*Pisum sativum*), *164*
   amylose starch of, 26
   artificial hybridization of, 168
   characteristics of, *169*
   epidermal cells of, *289*
   fruit development in, 382
   Mendel's studies of, 167–73
   round vs. wrinkled, 232, 255–57
   tendrils of, *324*
Pear (*Pyrus communis*), 191, *286*
Peat moss, 9
Pectins, **24**, *25*
Pedicels, **392**
Peduncle, **392**
Pelc, S. R., 190
Peltate leaves, 316, *317*
Penicillin, 13
Pentose phosphate pathway, **126**
Pentose sugars, respiration of, 126
Pepper, black (*Piper nigrum*), 5
Peppermint (*Mentha piperita*), *294*
Peptide bonds, **27**
Peptide synthesis, 27
Peptidyl transferase, **240**
Perfect flower, **386**
Perfoliate leaves, 316, *317*
Pericarp, **399**
Pericycle, **341**, *343*

Periderm, 268, **295**
   as bark, 368–70
Peripheral cells (root), *336*, *337*
Peripheral meristem, 267
Permanent wilting point, **467**, *468*
Peroxisomes, **63**
Persimmon (*Diospyros virginiana*), 282
Pesticides, 17
Petals, **380**, 383, *384*
Petiole, **315**
Petroselinum, 37
pH, A-9 (box)
   proton pumps and organelle, 86
   of soil, 475
Phaseolus sp., 60, 177, *292*, *401*, *457*
Phellem (cork), **369**, *370*
Phelloderm, 370
Phellogen (cork cambium), 268, 368, *369*
Phenolics, 37 (table), **39**–40
   structure of, *40*
Phenotype, **171**
   genes and, 175–76
Phenylpropanoids, 39
Philodendron (*Philodendron*), 94
Phloem, **262**, 296–98
   conducting cells of, 297–98, *299*
   contents exchange between xylem and, 510–11
   in leaves, 317–18
   primary and secondary, 297
   secondary, 357. See also *Secondary phloem*
   in stems, 308
   transporting organic solutes in, 504–10
Phloem loading and unloading, **508**, *509*
Phoenix dactylifera, 166, *167*, 282
Phoradendron, *278*, 484
Phosphate group, **A-5**
Phosphoenolpyruvic acid (PEP), *154*, 155
Phosphoglycerate kinase, 20
Phospholipids, **34**, *36*
   artificial membrane of, *74*, 75
Phosphorous, functional groups containing, A-5
Phosphorylation, **103**, A-7–8
   by chemiosmosis, 115
   oxidative, 114, **115**
   substrate-level, **114**, 115
Photochemical reactions of photosynthesis, **147**–50
   cyclic electron flow, 147–48
   noncyclic electron flow, 148, *149*
   solar tracking (heliotropism), 440 (box)
   summary of, 148, *149*, *150*, *152*
Photons, **137**
Photo-oxidation, carotenoid protection against, 139, *140*
Photoperiodism, 448–55
   defined, **450**
   flowering and, 448–55
   leaf variation as response to, 320
   phytochrome and, 453–54
   types of flowering responses, 450–51
Photophosphorylation, **145**, *146*
Photorespiration, **127**, **154**, *155*, 156
Photosynthates, fates of, 159–60
Photosynthesis, **96**, 131–61
   biochemical reactions of, *147*, 150–53
   C_4 photosynthesis, 154–58
   chloroplasts and, 64, 143–47
   control of, 159
   crassulacean acid metabolism (CAM), 158–59
   efficiency of, 153
   as energy transformation, 98, 99, 107–8
   evolution of, 141 (box)
   fate of, 159–60
   genes associated with, 200 (box), 227 (box)
   history of discoveries about, 132–36
   nature of light and, 136–38

   photochemical reactions of, 147–50, *152*
   photorespiration and, 153–54
   pigments and, 138–43
   roots capable of, 350
   transpiration-photosynthesis compromise, 501, *502*
Photosynthetic autotrophs, **132**, 141 (box)
Photosystem I, **144**, *146*, 149
   cyclic flow of electrons and, 147, *148*
Photosystem II, **144**, *146*, 149, 150
   noncyclic electron flow, *148*, *149*
Phototropism, 356, **437**–41
   auxin and, 417
   Cholodny-Went hypothesis on, 438, **439**
   studies of, 412–17
   varying light wavelengths and, 441
Phragmoplast, **206**
Phycocyanin, *142*
Phycoerythrin, *142*
Phycoplast, **206**
Phyllotaxis, **312**–15
   control of, 313–15
   mathematical relationships in, 314 (box)
   patterns of, 312, 313, *315*
Physalis subglabrata, *393*, *394*
Phytoalexins, **326**–27
Phytochrome, **453**–54
Picea, 8
Pigments, **64**, 138–43. See also names of specific pigments
   accessory, 139–42
   chlorophylls, 138–39
   evolution of photosynthesis and, 141 (box)
   leghemoglobin, 349
   light absorption by, 145
   making and destroying, 142–43
Pigweed (*Amaranthus*), 156
Pine (*Pinus*), 296, 348, 420
   bristlecone, 361, *362*
   mycorrhiza root of, *479*
   resin ducts in, 300, *363*
   sections of, *365*
Pineapples (*Ananas comosus*), 31, 158
Pinguicula grandiflora, 294, *299*
Pinnately compound leaves, **315**, *316*
Pinus sp. *See* Pine (*Pinus*)
Piper nigrum, 5
Pisum sativum. See Pea (*Pisum sativum*)
Pitcher plants (*Sarracenia purpurea*), *483*
Pith
   in roots, 342
   in stems, **308**, *309*
Pith-rib meristem, 268
Pittosporum undulatum, 160
Placentae, **382**, *383*
Plane-polarized light, **A-8**
Plant hybridization
   artificial, 168 (box)
   before Mendel, 166–67
Plantago media, *393*
Plants
   basic structure of, 262, *263*
   botanical themes applying to, 14–18
   carnivorous. See Carnivorous plants
   defenses of. See Defenses, plant
   energy and. See Cellular respiration; Energy; Photosynthesis
   genetics. See Genetics; Inheritance
   life span of, 263
   most common elements in, 23 (table)
   reproduction. See Reproduction; Sexual reproduction
   significance of, to all life, 4–9
Plasma membrane, **46**, 53–54. See also Membranes
Plasmids, cloning in, 246
Plasmodesmata, **57**, *58*
   formation of, 206

Plasmolysis, **81**
Plastid, **64**
Plastochron, **313**
Platanus occidentalis, 369
Pleiotropic genes, **178**–79
Plywood, 374
Pneumatophores, 348
Poaceae grasses, 32, 156, 158
Podophyllum, 344
Poison ivy (*Rhus toxicodendron*), 334
Poisons, 4, 31, 41
    in leaves, 325–26, 328
Polar nuclei, **212**, 389
Polar water molecules, **A-4**
Polarity in development, 271–73
    patterns resulting from, 276
Pollen, 210
    development of, 387–89
    gametophytes of, 213
    grain of, **212**
    interactions of glycoproteins in, 89
    sterile, 182 (box)
    variation in, 387, 388, 389
Pollen tube, **214**, 215
Pollination, **212**–15, 388, 390, 391
    mechanisms of, 395–98
    role of nectaries in, 299
Poly-A tail, **234**
Polyacetylenes, **41**
Polyarch roots, 342, 343
Polygenes, **178**
Polygenic inheritance, 177–78
Polygonum-type embryo sac, **389**
Polymerase chain reaction (PCR), **199 n.2**, **249**, 250
Polymers, **22**
    carbohydrates as, 22–27
    lipids as, 33–37
    nucleic acids as, 31–33
    proteins as, 27–31
    secondary metabolites, 37–41
Polypeptides, **27**
    synthesis of, 240–42
Polyploidy, **218**, 219 (box)
Polysaccharides, **23**
    storage, 26–27
    structural, 24–26
Polysomes, **59**
Pondweed (*Potamogeton nodosus*), 322
Populus sp., 304, 348, 368
Pores, wood, 363
Positional controls on plant growth and development, 276–77
Post, C. W., 13
Potamogeton nodosus, 322
Potassium ions (K⁺), loss of turgor and uptake of, 81
Potato (*Solanum tuberosum*)
    shoot apex of, 267
    tubers of, 310, 311
Potential energy, **78**, **96**
P-protein, 505–6
Prayer plant (*Maranta*), 447
Preprophase band, **204**
Pressure-flow model of phloem transport, **506**, 507
Pressure potential, **493**
Priestley, Joseph, experiments of, 133, 134, 135
Primary body of plant, **282**. *See also* Dermal tissues; Ground tissue; Meristems; Vascular tissues
Primary cell wall, **56**
    connections between, 57
Primary growth, 281–352
    dermal tissue, 288–95
    ground tissue, 282–87
    leaves, 311–28
    roots, 333–52
    secretory structures, 299–301
    stems, 306–11
    terrestrial adaptation, 301–2
    vascular tissue (xylem, phloem), 295–98
Primary phloem, 297
Primary pit-fields, **57**, 58
Primary (radicle) root, **334**, 335
Primary RNA transcript, **234**, 235 (box)
Primary structure of proteins, **27**, 29
Primary tissues, **282**. *See also* Dermal tissues; Ground tissue; Meristems; Vascular tissues
Primary xylem, 295
Primers, 200, **249**
Prince's plume (*Stanleya elata*), 471
Privet (*Ligustrum*), 319
Probes for genes, **248**
Procambium, **266**
Products of chemical reactions, **A-6**
Progenote, 243
Prop roots, 349, 350–51
Propagation with adventitious roots, 335–36, 348
Prophase I of meiosis, **215**–16, 217
    chiasmata in, 220
    prophase of mitosis vs., 216
Prophase II of meiosis, **215**, 218
Prophase of mitosis, **190**
    chromosome condensation in, 204, 205
    events before, 204
    prophase I of meiosis vs., 216
Prosopis, 334, 346
Protease inhibitors, **31**
Proteases, 31
Protein kinase, 87
Proteins, 27–31. *See also* Glycoproteins
    amino acids in, 27, 28
    enzymes as, 31. *See also* Enzymes
    examination of, gel electrophoresis and, 178 (box)
    helix structure of, 192
    hereditary material and, 173–75
    sexual reproduction, and role of, 88–89
    single-strand binding, **198**
    storage, 30–31
    structural, in cell walls and membranes, 30
    structure of, 27–28, 29
    structure of thylakoid, 77
    translation and synthesis of, 240–42
    transport, 82
Protoderm, **265**
Proton pumps, 83, 85–86
    auxin and, 87
    phloem loading and, 509
    two types of, 86
Protons, **A-1**
Protophloem, 297
Protoplasts, 252
Protostelic roots, **342**, 343
Protoxylem, 295
Pseudomonas syringae, 7 n.2
Pulp, wood as source of paper, 374–75
Pulvinus, 315, 445, **446**
Punnett, R. C., 171
Punnett square, **171**, 172
Purines, **31**–32. *See also* Adenine; Guanine
Pyrimidines, 31, 32. *See also* Cytosine; Thymine; Uracil
Pyrrole ring, 138
Pyrrosia, 158, 312
Pyrus communis, 191, 286
Pyruvic acid
    breakdown of glucose to, 115, 116, 117
    conversion of, to acetyl-CoA, 117, 120 (table)

Q

Quantum, **137**
Quaternary structure of proteins, **29**
Quercus sp., 6, 7, 295, 348, 373, 410
Quercus suber, 56
Quiescent center, 265, 268, **337**, 338

R

R group of amino acids, 27, 28
Radial section (wood), **364**, 365
Radicle root, **334**, 335, **400**
Radioactivity, **A-6** (box)
Radish (*Raphanus sativus*), 339
Rain forest, nutrient-gathering by plants in, 485–86
Ranunculus sp., 15, 280, 283, 308, 322, 440 (box)
Raphanus sativus, 339
Ray initials, **356**, 358
Reactants in chemical reactions, **A-5**
Reaction center, **144**, 146
Reaction wood, **367**, 368
Receptacle, **380**
Recessive traits, **167**
Reciprocal cross, 167 n.1
Recombinant DNA technology, **245**. *See also* Genetic engineering
Recombination. *See* Genetic recombination
Recombination modules, **219**–20
Redifferentiation, **282**
Reduction, **101**–2, **A-7**
Reduction-division of meiosis, 217
Reductionism, 22
Redwood tree (*Sequoia sempervirens*), 4, 7, 362, 364
Release factors, **241**
Renatured proteins, 29
Repetitive DNA, 203
Replication bubble, **199**, 201
Replication fork, **199**, 201
Replication origin, **199**
Replicons, 200, 202
Reporter genes, 253 (box)
Reproduction, 15, 163. *See also* Genetics; Inheritance
    asexual, 211
    leaves modified for, 324, 325
    sexual. *See* Sexual reproduction
    vegetative, 309, 310, 335, 394–95
Reproductive morphology. *See* Sexual-reproduction morphology
Resin, 300, 362, 363
Resolving power of microscopes, **46**
Respiration. *See* Cellular respiration; Cellular respiration, aerobic
Restriction enzymes, **245**, 246
Reverse transcriptase, **247**
Rheum rhabarbarum, 63, 285
Rhizobium bacteria, 88, 350, 479, 480
Rhizomes, **310**
Rhizosphere, **344**–35
Rhizotrons, 345, 346
Rhubarb (*Rheum rhabarbarum*), 63, 285
Rhus toxicodendron, 334
Rhytidome, 370
Rib meristem, 266
Ribbon grass (*Vallisneria spiralis*), 395
Ribonucleic acid (RNA), 31
    protein synthesis (translation) and, 240–42
    splicing (processing), 234, 237
    synthesis of (transcription), 232, 233–35
    types of, 232, 235–39
Ribosomal RNA (rRNA), 59, **232**, 238–39
Ribosomes, **59**, **238**
    protein synthesis and, 232
    structure of, 239
Ribozymes, **235**
    origins of life and, 236 (box)
Rice (*Oryza sativa*)
    aeration in, 504
    amylopectin starches in, 26
    cyanobacterium and production of, 481

Ricin D, **31**
Ricinus communis, 31, 346, 360, *401*
Ring-porous wood, 363
R-loops, **238**
RNA. *See* Ribonucleic acid (RNA)
RNA polymerase, **233**, *234*
RNA processing, **235**, *237*
RNA splicing, **235**
Root cap, 265, 268, **336**, *337*
    gravitropism and, 442–44
Root hairs, 339
Root pressure, water movement and, 494–95
Root shoot ratio, 347–48
Root tip, 336, *337*, 344
Roots, 333–52
    apical meristems of, 265, 268
    economic importance of, 351
    factors controlling growth and distribution of, 345–47
    functions and structure of, 336–44, *345*
    growth of shoots vs., 347–48
    importance of, to plant growth, 334
    modified, 348–51
    root-soil interface, 344–45
    transition between shoot and, 344
    types of, 334–36
    vascular cambium in, 359
    velamen of, 289
Rosary pea (*Abrus precatorius*), 31
Rose (*Rosa* sp.), 385, *387*
Rose mallow (*Hibiscus* sp.), *382*
Rosette plants, **306**
Rosettes, **424**
rRNA. *See* Ribosomal RNA (rRNA)
Rubber, 39
Rubber plant (*Ficus elastica*), 289
Rubber tree (*Hevea brasiliensis*), 301
Ruben, Samuel, 136
Rubisco (enzyme), 227, 233
    biochemical reactions of photosynthesis and, 150, *152*
    photorespiration and, 153–54
    subunits of, 27–28
RuBP carboxylase/oxygenase. *See* Rubisco (enzyme)
Runners, **309**, *310*
Rutin, 40
Rye grass (*Lolium* sp.), 395

S

Sachs, Julius von, 136, 412
St.-John's-Wort (*Hypericum* sp.), *382*, 386
Salicylic acid, 39, 40
Saline habitats, plants living in, 63, 81, 82, 495 (box)
Salix sp., 40, 392
Salt glands, 299
Sambucus, 370
Sapwood, **364**
Sarracenia sp., 324, 483
Saturated fatty acid, **33**
    in foods, 35 (box)
Saussure, Nicholas de, 135
Scanning electron microscopy (SEM), **50**, *52*
Scarification, **401**
Scarlet gilia (*Ipomopsis aggregata*), 378
Schleiden, Mattias Jakob, 14, 69
Schlerids, **285**, *286*
Schwann, Theodor, 14, 69
Scientific method, **10–11**
    nonscientists' use of, 14
    using, 11–14
Sclerenchyma, **285**, *286*, 287
Sclerenchyma cells, **262**
Scutellum, **400**, **423**
Searcher shoots, **310**
Seasons, plant response to changing, 448–55

Secale cereale, 334
Second law of thermodynamics, **97**–99
Secondary cell wall, **56**
Secondary growth, 355–77
    defined, **356**
    extent of, 362
    from lateral meristems, 263, 264, 268
    secondary phloem and periderm (bark), 368–70
    secondary xylem (wood), 362–67
    unusual, 370–72
    uses of secondary xylem and phloem, 374–75
    vascular cambium, 356–62
Secondary metabolism, **37**
Secondary metabolites, 37–41
    alkaloids, 38
    examples of well-known, 37 (table)
    functions of, 37–38
    latex as, 160
    minor classes of, 40–41
    phenolics, 39, 40
    terpenoids, 38, 39
Secondary phloem, **297**
    differentiation of, 357, 358
Secondary plant body, **356**
Secondary structure of proteins, 27, *29*
Secondary xylem, **295**–97, 357, *358*. *See also* Wood
Secretory cells, internal, 299–300
Secretory structures, **299**–301
Sedum sp., 384
Seed banks, 401, **402**, 403
Seed coat, 400
Seed ferns, 385
Seedling development, 403, 453
Seeds, 400–404
    banks of, 401–3
    dispersal of, 404–6
    dormancy of, 401, 431
    germination of, 84 (box), 401, 404, 423, 453, 454
    shape of (round vs. wrinkled), 232, 255–57
    storage proteins in, 31
    structure of, 400
    triacylglycerides in, 33–34
Segregation, law of, **170**
Seismonasty, 445–47
Selenium as essential element, **471**
Self-compatible flowers, 393
Self-incompatible flowers, 393, 394
Selfish DNA, **228**
Self-replication of DNA, **194**, 196–202
Semi-conservative replication of DNA, **196**, *197*, 198
Seminal roots, 335
Senebier, Jean, experiments of, 135
Senecio vulgaris, 508
Senescence, **420**, **455**, 456
    effect of cytokinins on, 426–27
Sensitive plant (*Mimosa pudica*), 445
Sepals, **380**, 384
Sequoia sempervirens, 4, 7, 362, 364
Sequoiadendron giganteum, 4, 6, 354, 373, 488
Sexual reproduction, 212–15
    cell cycle and, 188, 189–90
    effect of ethylene on sexual expression in, 429–30
    genetic recombination and, 220–26
    interphase, 189, 190–203
    meiosis, 172–73, 174, 215–18
    mitosis, 189, 204–6
    in plants vs. animals, 211
    pollination and fertilization in, 167, 212–15
    reasons for, 226–28
    role of proteins in, 88–89
    spores and gametes in flowering plants, 212, 214
    synaptonemal complex, 215, 216, 218–20
Sexual-reproduction morphology, 379–408

    dispersal of fruits and seeds, 404–6
    flowers, 380–86
    fruits, 398–400
    plant diversity and, 386–98
    seeds, 400–404
Shade leaves, 320
Shoots, 267–68
    growth of roots vs., 347–48
    transition between root and, 344
    vascular cambium in, 359
Short and long shoots, 310
Short-day plants, 450, **451**
Short-night plants, **451**
Sieve cells, **298**
Sieve elements, 297–98
Sieve tube members, **298**, *299*, 504
    contents of, 509–10
Silent Spring (Carson), 17
Silicon as essential element, **471**
Simmondsia chinensis, 36
Simple leaves, **315**, *316*
Sinapis alba, 344
Single-strand binding protein, **198**
Siphonostelic roots, **342**, 344
Sisal (*Agave sisalana*), 287
Skoog, Folke, 425, 426
Skunk cabbage (*Symplocarpus foetidus*), 127, 128 (box)
Slime mold (*Dictyostelium discoideum*), 69
Small nuclear ribonucleoproteins (snRNPs), **235**
Small RNA (snRNA), **235**
Snapdragon (*Antirrhinum majus*), 177, *179*, 182
Socrates, 4, 325
Sodium as essential element, **471**
Softwoods, **362**, 363
Soil, 462–69
    components of, 463–69
    erosion of, 466 (box)
    horizons and types of, 462–63
    pH of, and nutrient availability, 475
    properties of, effect on roots of, 347
    returning nutrients to, 465 (box)
    root interface with, 344–45
    salty. *See* Saline habitats, plants living in
    transpiration and, 498
Soil particles, **463**
    nutrient availability and, 464–65
    water availability and, 465–66
Solanum tuberosum. *See* Potato (*Solanum tuberosum*)
Solar energy, 95 (box)
Solute potential, **493**
Solutes, defined, **78**
Solutes transport, 76, 489, 504–10. *See also* Water transport
    contents of, in phloem sieve tubes, 509–10
    influence of environment on, 509
    membranes and, 78
    methods of movement in phloem, 506–9
    structure of conducting cells, 505–6
Solvent, **78**
Sophora secundiflora, 405
Southern blotting, **256**
Southern, E. M., 256
Soybeans (*Glycine max*), 349
Spanish moss (*Tillandsia usneoides*), 7, 158
Sperm, 214, 215, 390
Spices, 328, 351, 375
Spiderwort (*Tradescantia*), **69**, 190
Spinach, vitamins in, 8
Spindle apparatus, **205**
    chromosome movement and, 207–8
    fibers, **205**
Spines, 323
Spiral (alternate) phyllotaxis, **312**, *313*
Spirogyra algae, 135, *136*
Spliceosome, **235**, *237*

Spodosols, **463**
Spongomorpha algae, 69, 70
Spongy mesophyll cells, **319**
Spores, **212**
Sporophyte, **212**, *213*
Sporophytic self-incompatibility, 393
Spring (early) wood, **361**
Spruce (*Picea*), 8
    effect of acid rain on red, 17
Squash (*Cucurbita*), 298
Stahl, Franklin W., 196
Stamens, **380**–82
Staminate plants, 380
Stanleya elata, 471
Staphylococcus bacteria, 13
Starch-branching enzymes (SBEI), **232**
Starch phosphorylase, **114**
Starches, 26
    retrieval of glucose from, 114
Start codon (AUG), **240**, *241*
Stele, **341**–42
Stelis gemma, 36
Stems, **306**–11
    axillary buds and branching, 309
    control of growth in, 306–7
    economic importance of, 311
    effect of ethylene on, 430
    functions of, 306
    modified, 309–11
    secondary growth of, 357
    structure of, 307–9
Stereoisomers, **A-8**
Steroids, **38**–39
Steward, F. C., 188, 271
Sticky ends, 246
Stigma, **212**
Stinging nettle (*Urtica dioica*), 294, *295*, 325
Stipules, **323**
Stolons, **309**, *310*
Stomata, **289**, *290*, 291, **501**–2
    differentiation of, 274
    diurnal curve of opening, 498
    effect of abscisic acid on closing of, 431
    leaf orientation and frequency of, 317
    rust spore attack on, 292–93 (box)
    structure and distribution of, 501
    transpiration and, 501, 502
    turgid cells and open, 81
Stonecrop (*Sedum* sp.), 384
Stop codon, **241**, *242*
Storage function
    in leaves, 325
    parenchyma tissue and, 282, 283
    in roots, 336, 340, 345, 348
    in stems, 306
Storage polysaccharides, 26–27
Storage proteins, 30–31
    genetically engineered, 254
Stratification, **401**
Strawberry (*Fragaria chilensis*), 310, 399, 421, 496
Streptococcus pneumoniae bacterium, genetic studies of, 173, *175*
Streptomyces fradiae, 32
Striga asiatica, 350 (box)
Strobel, Gary, 7 n.2
Stroma, **64**, *65*, **143**
Strontium in plant tissue, **473**
Structural isomers, **A-8**
Structural polysaccharides, 24–26
Structural proteins, 30
Strychnine plant, 38, 294
Strychnos nux-vomica, 38, 294
Style, **212**, **382**
Subapical region of roots, 338–39
Suberin, **36**, 340

Subsidiary cells, 289, 290
Substrate-level phosphorylation, **114**, *115*
Subunits, protein, **28**–29
Succinic acid, **117**, *118*
Succinyl-CoA, **117**, *118*
Succulent stems, **310**
Sucrases, 112
Sucrose, 23
    formation of, 101
    loading of, into phloem, 508, 509
    as product of photosynthesis, 160
    retrieval of glucose from, 112, 114
    synthase, **114**
    synthesis and hydrolysis of, 25
Sugarcane, 154, *155*
Sulfate, 378
Sulfhydryl group, **A-5**
Sulfur, functional groups containing, A-5
Summer (late) wood, **361**
Sun, earth's energy from, 95 (box)
Sun leaves, **320**
Sundew (*Drosera* sp.), 294, 447, 460, 482
Sunflower (*Helianthus* sp.), 176, 268, 276, 287
    Chlosyne butterflies and epidermis of, 295
    solar tracking by, 440 (box)
Superior ovary, **386**
Supernumerary cambia, 371
Sweet pea (*Lathyrus odoratus*), 180, *181*
Sweet potatoes (*Ipomoea batatas*), 8, 371
Swift, Hewson, 190
Sycamore (*Platanus occidentalis*), 369
Symplast, **69**, **340**
Symplocarpus foetidus, 127, 128 (box)
Synapsis, **215**
Synaptonemal complex, **215**, 218–20
    chiasmata, chromosome segregation, and, 220
    model of, 219
    prophase I of meiosis and formation of, 216
    recombination nodules on, 219–20
Synergids, **389**
Synonymous codons, **242**
Syringa vulgaris, 318
System, defined, 96

T

Tamarisk (*Tamanx ramosissima*), 495
Tambalacoque tree (*Calvaria major*), 402 (box)
Tandem repeat of genes, **225**, *226*
Tangential section (wood), **364**, *365*
Tannins, 40
Tapetum, 387
Taproot, **334**
Taproot system, **334**, *335*
Taraxacum, 266, 335, 394
Tassels, corn, **182**
Taxodium mucronatum, 362, *363*
Taxol, 18
Taxus brevifolia, 18, 37, 368, 375
Taylor, J. Herbert, 196
Tea (*Camellia sinensis*), 4, 5
Technology, scientific discovery and use of, 13–14
Telomeres, **248**
Telophase I of meiosis, **215**, *217*
Telophase II of meiosis, **215**, *218*
Telophase of mitosis, **190**, **205**, *206*
Temperature
    Celsius vs. Centigrade scales of, B-2
    effect of, on roots, 345
    flowering and effect of cold, 452 (box)
    plant movement in response to, 448
    seed germination requirements for, 401
    transpiration and, 498
Tendrils
    as modified leaves, 323, 324
    as modified stems, **309**, *310*

Tension wood, 367
Terpenoids, 37 (table), **38**–39
    structure of, 39
Terrestrial plants, 281
    evolution of photosynthesis in, 141 (box)
    tissue adaptations for, 301–2
Tertiary structure of proteins, **28**, *29*
Tetrarch roots, 342, *343*
Texture (wood), **364**
Theobroma cacao, 5
Theory of inheritance, **170**
Thermodynamics, laws of, 96–99
Thermonasty, 448
Thigmomorphogenesis, **430**, 448
Thigmotropism, **444**–45
Thiols, **A-5**
Thorns, **310**
Thottea, 316
Thylakoids, **64**, *65*, **143**
    structure of, 77
Thymine, 31–32
Tilia americana, 286
Tillandsia usneoides, 7, 158
Tissue. *See* Dermal tissues; Ground tissue; Meristems; Vascular tissue
Tissue culture, 270, **271**, 425
Tobacco (*Nicotiana* sp.), 178, *179*, 180, 297, 328
Tobacco mosaic virus (TMV), 254, *255*, 312
Toluene, chemical structure, 24 (box)
Tomato (*Lycopersicon esculentum*), 38
    distribution of rbcS genes in, 227
    effect of ethylene on, 430
    lycopene pigment in, 140, 142
    transgenic, 254
Tonoplast, **62**
Topoisomerases, **199**, *201*
Totipotency, **188**, 271
Touch-me-nots (*Impatiens glandulifera*), 406
Trace elements, **A-1**
Tracheids, **296**, *297*
    water transport and, 491–92
Tradescantia, **69**, 190
Trailer sequence, **234**
Transcription, **232**–35
    speed of, 234, 235
    steps in, 233, 234
Transfer cells, **285**
Transfer RNA (tRNA), **232**, **237**
    binding of amino acids by, 240
    structure of, 239
Transgenic plants, 245, 251–52
    advantages of, as crops, 253–55
    antisense technology and, 251–52
    steps in making, 252
Transition region, **344**
Transitional meristems, **265**–66
Translation, 240–41
    amino acid binding prior to, 240
    speed of, 241
    steps in, 240, 241, 242
Transmission electron microscopy (TEM), **50**, *51*, *56*, *75*
Transpiration, **317**, **490**
    factors affecting, 496–502
    leaf architecture and, 490–91
    photosynthesis-transpiration compromise, 502, **503**
    in xerophytes, 321
Transpiration-cohesion hypothesis, **496**, *497*
Transpons. *See* Transposable elements (transpons)
Transport. *See* Solutes transport; Water transport
Transport proteins, **82**
Transposable elements (transpons), **181**–83
    sexual reproduction and, 228
Trees. *See also* Wood
    Christmas, 420

commemoration of, 375
lore of, 373 (box)
Triacylglyceride, 33
Triarch roots, 342, 343
Tricarboxylic acid cycle, **117**, *118*
Trichomes, **291**–95
functions of, 294–95
transpiration and, 500
Trifolium sp., 88, 325, 335
Triplet codons, 242
Triterpenes, 38
Triticum. See Wheat (*Triticum*)
Tropical oils, human health problems and, 35 (box)
Tropisms, **437**–45
gravitropism, 441–43
hydrotropism, 443–44
phototropism, 356, 412–17, 437–41
thigmotropism, 444–45
Tru-breeding strains, 167
Tube cell, **214**, *215*
Tubers, 310
Tule tree (*Taxodium mucronatum*), 362, 373
Tunica, 267
Tunica-corpus model, 267
Turacin, 138
Turgor pressure, **62**–**63**, 80, *81*
nastic movements and, 445–46
positive, and cellular enlargement, 269
Twining shoots, **309**, *310*
2,4-D herbicide, 418

U

Ubiquinol, **117**, *118*, *119*
Ubiquinone, **117**, *118*, *119*
Ulmus Americana, 6, 7, 362, 364
Ultracentrifuge, 52
Ultraviolet radiation (UV), **137**
Ulvaria alga, 68
Unequal crossing-over, **225**, *226*
Unsaturated fatty acid, 33
Upper leaf zone, 312
Uracil, 31
Uridine diphosphate (UDP), **114**
Uridine triphosphate (UTP), 105
Urn plant (*Aechmea fasciata*), 392
Urtica dioica, 294, 295, 325
Utricularia, 482, 483

V

Vacuoles, **62**–**63**
in vascular cambium, 356, 358
Valence electrons, **A-2**
Valence shell, **A-2**
Vallisneria spiralis, 395
Van Helmont, Jan-Baptista, experiments of, 132, *133*
Van Niel, C. B., 136
Vascular cambium, 268, **356**–62
controls on activity of, 359–62
extent of, 362
formation of, 358, 359, 360
types of, 356–57, *358*
Vascular tissues, **262**, **295**–98
auxin and differentiation of, 420–21
in leaves, 317–18
phloem, 296–98
in roots, 342
in stems, 308, 309
xylem, 295–96
Vectors
DNA, 245
pollen, 395

Vegetative cell, 387
Vegetative reproduction, **394**–95
roots, 335–36, 348
stolons and runners, 309, 310
Vein endings, 317, *318*
Veins, leaf, **317**, *318*
Velamen, 289, **350**
Velcro, 8
Venus's-flytrap (*Dionaea muscipula*), 446, 482
Vernalization, **452** (box)
Vessel elements, **296**, *297*
advantages and disadvantages of, 492 (box)
water transport and, 491–92
Vicia faba, 70, 190
Vigna sinensis, 291
Virchow, Rudolf, 189
Viruses
cloning in, 246
genetic studies of, 173, 176
Viscum, 278
Visible light, **137**
Vitamins, biotechnology and creation of, 16
Vitis vinifera, 291, 310, 400

W

Wall cress (*Arabidopsis thaliana*), 200, 202, 203
Warburg effect, 153
Warburg, Otto, 153
Wareing, Philip, 431
Water
absorption of, by vacuoles, 62–63
availability of, as control factor in photosynthesis, 159
availability of, effect of, on roots, 347
availability of, influence of soil particles on, 465–66
dispersion of fruit and seeds by, 404–5
hydrogen bonding and properties of, A-3–4
leaf variation as response to amount of, 320
as pollination mechanism, 395
seed germination requirements for, 401
as soil component, 467–68
as solvent, **78**
transport of. See Water transport
Water-conducting tissue, 57, 58
in phloem, 297, 298, 299
in roots, 336, 337, 342, 345
in stems, 306
transfer cells, 283, 285
turgor pressure in, 62–63, 269
in xylem, 296, 297
Water content of wood, 367
Water hyacinth (*Elchhornia crassipes*), 321 (box)
Water lily (*Nymphaea odorata*), 286, 322
Water potential, **78**–**79**, 80, **492**–**94**
components of, 493
water movement and, 493–94
Water transport, 76, 78–81, **489**–504. *See also* Solutes transport
factors affecting transpiration and, 496–502
leaf architecture and transpiration, 490–91
methods of water movement, 494–96
model on photosynthesis-transpiration compromise, 502–3
osmosis and, 79, *80*
solute movement and, 78
structure of conducting cells, 491–92
three plant types and, 503–4
turgor and, 80–81, 82
water potential and, 78–79, 492–94
Water-use efficiency, transpiration and, 498–99
Watson, James, 191, *192*, 193
Watts (W), 95 n.2

Wavelengths of light, **137**
absorption spectrum of chlorophyll vs., 139
Wax palm (*Copernicia cerifera*), 288
Wax plant (*Hoya carnosa*), 158
Waxes and waxlike substances, **34**–**36**, 328
Weathering, 462
Went, Frits, phototropism studies by, 415, *417*, 438–49
Wheat (*Triticum*)
gliadin protein in, 31
shoot apex meristem of, 267
Whisker (DNA), 223, *225*
White clover (*Trifolium repens*), 88, 325, 335
Whorled phyllotaxis, 313
Wilkins, Maurice, 192
Willow (*Salix* sp.), 40, 392
Wind
dispersion of fruit and seeds by, 404–5
as pollination mechanism, 395
transpiration and, 497–98
Window leaves, 323, *324*, 325
Winter rye (*Secale cereale*), 334
Witchweed (*Striga asiatica*), 350 (box)
Wood, 295, *296*, 297, **362**–67
baseball bats made of, 366 (box)
characteristics of, 364–67
differentiation of, 357, 358
growth rings in, 361–62
kinds of, 362–64
reaction, 367, *368*
spring (early) vs. summer (late), 361
in Stradivarius violin, 8
tapping for maple syrup, 500 (box)
uses of, 8, 366, 374–75
vessels and tracheids in, 491

X

Xanthium strumarium, 449, 451
Xanthophylls, **142**
Xerophytes, 320–22, 347, **504**
Xylary elements, **296**
water movement by, 494
Xylary fibers, 287
Xylem, **262**, **295**–96
conducting cells in, 296, *297*
contents exchange between phloem and, 510–11
in leaves, 317–18
primary and secondary, 295, 296
in roots, 342, 343
secondary. See Wood
in stems, 308
water transport in, 490–504
Xylose, 25

Y

Yeast artificial chromosomes (YACs), 248

Z

Z scheme, 148
Zamia, 67
Zea mays. See Corn (*Zea mays*)
Zeatin, 426
Zebrina, epidermis of, 290
Zein, 30
Zone of cellular division (root), **338**
Zone of cellular elongation (root), **338**–39
Zone of cellular maturation (root), **338**, *339*
Zygote, **188**, **214**, *215*